国外优秀物理著作
原版丛书（第一辑）

具有连续变量的量子信息形式主义概论

（英文）

An Introduction to the Formalism of Quantum Information with Continuous Variables

[西] 卡洛斯・纳瓦雷特-本洛赫（Carlos Navarrete-Benlloch） 著

HITP
哈爾濱工業大學出版社
HARBIN INSTITUTE OF TECHNOLOGY PRESS

黑版贸登字 08-2021-041 号

图书在版编目(CIP)数据

具有连续变量的量子信息形式主义概论=An Introduction to the Formalism of Quantum Information with Continuous Variables:英文/(西)卡洛斯·纳瓦雷特-本洛赫(Carlos Navarrete-Benlloch)著.—哈尔滨:哈尔滨工业大学出版社,2024.10
(国外优秀物理著作原版丛书.第一辑)
ISBN 978-7-5767-1341-1

Ⅰ.①具… Ⅱ.①卡… Ⅲ.①量子力学-形式主义(数学)-概论-英文 Ⅳ.①O413.1

中国国家版本馆 CIP 数据核字(2024)第 073592 号

JUYOU LIANXU BIANLIANG DE LIANGZI XINXI XINGSHI ZHUYI GAILUN

策划编辑 刘培杰 杜莹雪
责任编辑 刘立娟 刘家琳 李 烨 张嘉芮
封面设计 孙茵艾
出版发行 哈尔滨工业大学出版社
社 址 哈尔滨市南岗区复华四道街 10 号 邮编 150006
传 真 0451-86414749
网 址 http://hitpress. hit. edu. cn
印 刷 哈尔滨博奇印刷有限公司
开 本 787 mm×1 092 mm 1/16 印张 63.75 字数 1 195 千字
版 次 2024 年 10 月第 1 版 2024 年 10 月第 1 次印刷
书 号 ISBN 978-7-5767-1341-1
定 价 378.00 元(全 6 册)

Contents

Preface

Quantum information is an emerging field which has attracted a lot of attention in the last couple of decades. It is a broad subject which extends from the most applied questions (e.g. how to build quantum computers or secure cryptographic systems), to the most theoretical problems concerning the formalism and interpretation of quantum mechanics, its complexity, and its potential to go beyond classical physics.

This book is devoted to the introduction of the formalism behind quantum information, with special emphasis on continuous-variable systems, that is, systems (such as light) which can be described as collections of harmonic oscillators. It must not be taken as an exhaustive review of the field, but rather as an introductory text covering a selection of basic concepts, focusing on their physical meaning and mathematical treatment. The book is intended to be self-contained, in the sense that it starts from the very first principles of quantum mechanics, and tries to build up the concepts and techniques following a logical progression. Consequently, the target audience for this book is, primarily, students who have already studied a full semester of standard quantum mechanics and want to delve deeper into it, or researchers in closely related fields such as quantum optics or condensed matter, who would like to learn about the powerful formalism of quantum information. Length restrictions have precluded me from including practical examples that could clarify the definitions and statements, but these can always be found in the various references that I include in the different sections.

As for the choice of topics, I have chosen those that I found the most valuable when I first approached the field as a PhD student in quantum optics. The book starts by reviewing in the first chapter some basic notions of classical mechanics, linear algebra, and quantum mechanics that are needed throughout the text. Two chapters dealing with concepts applicable to general quantum systems follow, one related to bipartite entanglement, and the other to the formalism of general quantum operations (the most general transformations that one can induce on a system), including the important examples of channels and generalized measurements. The last chapter, which is the main chapter of the book, applies the ideas developed in the previous chapters to continuous-variable systems, that is, systems such as harmonic oscillators described by infinite-dimensional Hilbert spaces. After reviewing the physics of the quantum harmonic oscillator, a phase-space description based on the Wigner function is discussed. The relevant class of Gaussian states and operations is introduced next, emphasizing how they are very important both for theoretical and applied reasons, since they are relatively easy to treat mathematically and generate experimentally. After this, the quantum theory of measurements is particularized to continuous-variable systems, introducing typical detection strategies in scenarios such as photo-, homodyne-, heterodyne-, and on/off-detection. Finally it is shown that detection strategies can be used to generate non-Gaussian operations such as photon addition and subtraction probabilistically, which in turn can be used to enhance entanglement when applied locally, against intuition.

Thorough reviews covering further topics and techniques in the field of continuous variables, as well as some proofs that I have skipped, can be found in [1–6].

Above all, I hope interested readers will find this book an enjoyable ride through the thrilling subject of continuous-variable quantum information.

Carlos Navarrete-Benlloch
Munich, November 2015.

Bibliography

[1] Braunstein S L and van Loock P 2005 Quantum information with continuous variables *Rev. Mod. Phys.* **77** 513

[2] Weedbrook C, Pirandola S, García-Patrón R, Cerf N J, Ralph T C, Shapiro J H and Lloyd S 2012 Gaussian quantum information *Rev. Mod. Phys.* **84** 621

[3] Ferraro A, Olivares S and Paris M G A 2005 *Gaussian States in Quantum Information* (*Napoli Series on Physics and Astrophysics*) (Naples: Bibliopolis)

[4] Cerf N J, Leuchs G and Polzik E S (ed) 2007 *Quantum Information with Continuous Variables of Atoms and Light* (London: Imperial College Press)

[5] Shapiro J H 2008 Quantum optical communication *Lecture notes* http://ocw.mit.edu/courses/electrical-engineering-and-computer-science/6-453-quantum-optical-communication-fall-2008

[6] Kok P and Lovett B W 2010 *Introduction to Optical Quantum Information Processing* (Cambridge: Cambridge University Press)

Acknowledgements

I am very grateful to many people who helped me shape this project. Most of the material upon which this book is based originated from three events. First, from my close collaboration with Raúl García-Patrón, Nicolas J Cerf, and Jeffrey H Shapiro, whose guidance and clarifying explanations gave me the motivation to start writing the precursor of the original manuscript in 2010. I am especially indebted to Raúl and Nicolas for inviting me into the topics they had already been working on and for being the kindest and most patient teachers. Second, from the inspiring interaction with my PhD advisors Germán J de Valcárcel and Eugenio Roldán, in particular when writing my dissertation. And finally, from the weeks I spent in Doha with Hyunchul Nha and his group in 2012, during which many parts of the original manuscript were refined, benefiting greatly from discussions with Ho-Joon Kim and Hyunchul himself.

A special mention is well deserved by my good friend Erez Zohar, who was not only kind enough to write my biography for the monograph, but also read thoroughly the whole manuscript, helping me to improve it a great deal. Along these lines, I also appreciate the effort made by those who took some time to read through the manuscript and give me valuable feedback. These include Yue Chang, Johannes Kofler, Sebastian Pina-Otey, Ho-Joon Kim, Anaelle Hertz, Raúl García-Patrón, Eliška Greplová, Germán J de Valcárcel, Eugenio Roldán, Tao Shi, Alejandro González-Tudela, Nikolas Perakis, Marta P Estarellas, Samuel Fernández, Juan Bermejo-Vega, and Na'ama Hallakoun.

I feel blessed for having always been able to grow scientifically in inspiring environments, surrounded by nice colleagues and collaborators that I admire. I want to extend my appreciation to all of them, especially to the ones with whom I have discussed and applied many of the concepts that appear in this book: Mari Carmen Bañuls, Claude Fabre, Chiara Molinelli, Giuseppe Patera, J Ignacio Cirac, Inés de Vega, Diego Porras, Géza Giedke, Maarten Van den Nest, Peter D Drummond, Juan José García-Ripoll, Joaquín Ruiz-Rivas, Oriol Romero-Isart, Fernando Pastawski, Martin Schuetz, Jaehak Lee, Jiyong Park, Chang-Woo Lee, Se-Wan Ji, Myungshik Kim, Tao Shi, Yue Chang, Hannes Pichler, Peter Zoller, Julio T Barreiro, Alejandro González-Tudela, Eliška Greplová, Christopher A Fuchs, Christoph Marquardt, Florian Marquardt, Peter Degenfeld-Schonburg, Michael J Hartmann, Fernando Jiménez, Sebastian Pina-Otey, Mónica Benito, Carlos Sánchez Muñoz, András Molnár, Katja Kustura, Giulia Ferrini, Tom Douce, Cosimo C Rusconi, Alessandro Farace, and Julian Roos.

I have been blessed as well with a loving family and group of friends who have always supported me all the way in whatever adventure I have set sail. I am especially grateful to my parents for imprinting on me the will to make a living out of the things I am passionate about (be it music, sports, or science).

I gave the final shape to this book in the Lost Weekend cafe in Munich, whose staff provided me with the perfect warm and creative environment. Particular thanks are due to Markus, Chris, Eitel, Jaemin, Audrey, Tony, Julietta, Twana, and Lissie.

Finally, I would like to thank the editors and publishers of this synthesis series, especially Jeffrey Uhlmann and Michael B Morgan with whom I have been in direct contact, for giving me this extraordinary opportunity.

Carlos Navarrete-Benlloch
Munich, November 2015.

Author's biography

Carlos Navarrete-Benlloch by Erez Zohar

Credit: Lisa Miletic

Carlos Navarrete-Benlloch (born 1983) was born and raised in Valencia, Spain, where he attended Universitat de València for a six-year BSc+MSc degree in theoretical physics. During his studies, he became fascinated by several physical fields. One of these fields, which he later continued studying during his PhD, was quantum optics.

Carlos completed his PhD degree in 2011 under the joint supervision of Eugenio Roldán and Germán J de Valcárcel, in the quantum and nonlinear optics group of the Universitat de València, focusing on the quantum properties of multi-mode optical parametric oscillators. During his PhD studies, Carlos also maintained his interest in various other topics in quantum optics, which he had the opportunity to work on and become acquainted with during several long-term visits he made to other universities and research institutions. These included the Max-Planck Institute of Quantum Optics (MPQ) in Garching, Germany, where he worked under the supervision of J Ignacio Cirac; the Swinburne University of Technology in Melbourne, Australia, under the supervision of Peter D Drummond; and the Massachusetts Institute of Technology, USA, under the supervision of Jeffrey H Shapiro.

After receiving his PhD, Carlos joined J Ignacio Cirac's theory division at MPQ as a postdoctoral researcher, including two years as the recipient of a postdoctoral research fellowship granted by the Alexander von Humboldt foundation. At MPQ, Carlos has mainly focused on dissipative quantum optics with nonlinear optical cavities, superconducting circuits, and optomechanical systems, as well as on quantum simulation with ultracold atoms, collaborating with several scientists inside and outside MPQ. In the field of quantum information with continuous variables, Carlos has contributed to the theory of non-Gaussian states and operations, as well as the characterization of the classical capacity of Gaussian channels, in collaboration with Raúl García-Patrón, Nicolas J Cerf, Jeffrey H Shapiro, and Seth Lloyd.

Apart from quantum optics and information, Carlos has studied and been interested in several other physical fields, such as gravitation, cosmology, quantum field theory, and particle physics. He has also studied modern guitar and is a founding member and composer of the Valencia-based progressive jazz band Versus Five.

Chapter 1

Quantum-mechanical description of physical systems

The purpose of this first chapter is the introduction of the fundamental laws (axioms) of quantum mechanics as are used throughout the book. The quantum framework is far from being intuitive, but somehow feels reasonable (and even inevitable) once one understands the context in which it was created, specifically: (i) the theories that were used to describe physical systems prior to its development, (ii) the experiments which did not fit in this context, and (iii) the mathematical language that accommodates the new quantum formulation of physical phenomena. We will thus briefly review this context prior to introducing and discussing the axioms forming the quantum-mechanical framework.

1.1 Classical mechanics

In this section we go through a brief introduction to classical mechanics[1], with emphasis on analyzing the Hamiltonian formalism and how it treats observable magnitudes. We will see that a proper understanding of this formalism will make the transition to quantum mechanics more natural.

1.1.1 The Lagrangian formalism

In classical mechanics the state of a system is specified by the position of its constituent particles at all times, $\mathbf{r}_j(t) = [x_j(t), y_j(t), z_j(t)]$ with $j = 1, 2, \ldots, N$, N being the number of particles. Defining the kinetic momentum of the particles as $\mathbf{P}_j = m_j\dot{\mathbf{r}}_j$ (m_j is the mass of particle j), the evolution of the system is found from a

[1] For a more in depth text I recommend Goldstein's book [1], as well as Greiner's books [2, 3], or that by Hand and Finch [4].

doi:10.1088/978-1-6817-4405-6ch1

set of initial positions and velocities by solving the Newton equations of motion $\dot{\mathbf{P}}_j = \mathbf{F}_j$, $\mathbf{F}_j$ being the forces acting on particle j.

Most physical systems have further constraints that have to be fulfilled (for example, the distance between the particles of a rigid body cannot change with time, that is, $|\mathbf{r}_j - \mathbf{r}_l| = const$), and therefore the positions $\{\mathbf{r}_j\}_{j=1,...,N}$ are no longer independent, which makes the Newton equations difficult to solve. This calls for a new simpler theoretical framework: the so-called *analytical mechanics*. In the following we review this framework, but assuming, for simplicity, that the constraints are *holonomic*[2] and *scleronomous*[3], meaning that they can always be written in the form $f(\mathbf{r}_1, ..., \mathbf{r}_N) = 0$.

In analytical mechanics the state of the system at any time is specified by a vector $\mathbf{q}(t) = [q_1(t), q_2(t), ..., q_n(t)]$. n is the number of degrees of freedom of the system (the total number of coordinates, $3N$, minus the number of constraints), and the q_j are called the *generalized coordinates* of the system, which are compatible with the constraints and related to the usual coordinates of the particles through some smooth functions $\mathbf{q}(\{\mathbf{r}_j\}_{j=1,...,N}) \Leftrightarrow \{\mathbf{r}_j(\mathbf{q})\}_{j=1,...,N}$. The space formed by the generalized coordinates is called *coordinate space*, and $\mathbf{q}(t)$ describes a *trajectory* on it.

The basic object in analytical mechanics is the *Lagrangian*, $L[\mathbf{q}(t), \dot{\mathbf{q}}(t), t]$, which is a function of the generalized coordinates and velocities, and can even have some explicit time dependence. In general, the Lagrangian must be built based on general principles such as symmetries. However, if the forces acting on the particles of the system are conservative, that is, $\mathbf{F}_j = \nabla_j V[\{\mathbf{r}_l\}_{l=1,...,N}] = (\partial_{x_j} V, \partial_{y_j} V, \partial_{z_j} V)$ for some *potential* $V[\{\mathbf{r}_l\}_{l=1,...,N}]$, one can choose a Lagrangian with the simple form $L = T(\dot{\mathbf{q}}, \mathbf{q}) - V(\mathbf{q})$, $T(\dot{\mathbf{q}}, \mathbf{q}) = \sum_{j=1}^{N} m_j \dot{\mathbf{r}}_j^2(\mathbf{q})/2$ being the kinetic energy of the system and $V(\mathbf{q}) = V[\{\mathbf{r}_l(\mathbf{q})\}_{l=1,...,n}]$. The dynamical equations of the system are then formulated as a *variational principle* on the *action*

$$S = \int_{t_1}^{t_2} \mathrm{d}t L[\mathbf{q}(t), \dot{\mathbf{q}}(t), t], \tag{1.1}$$

by asking the trajectory of the system $\mathbf{q}(t)$ between two fixed points $\mathbf{q}(t_1)$ and $\mathbf{q}(t_2)$ to be such that the action is an extremal, $\delta S = 0$. From this principle, it is straightforward to arrive at the well known Euler–Lagrange equations

$$\frac{\partial L}{\partial q_j} - \frac{\mathrm{d}}{\mathrm{d}t}\frac{\partial L}{\partial \dot{q}_j} = 0, \tag{1.2}$$

which are a set of second-order differential equations for the generalized coordinates $\mathbf{q}$, and together with the conditions $\mathbf{q}(t_1)$ and $\mathbf{q}(t_2)$ provide the trajectory $\mathbf{q}(t)$.

[2] A constraint is *holonomic* when it can be written as $f(\mathbf{r}_1, ..., \mathbf{r}_N, t) = 0$. Non-holonomic constraints correspond, for example, to the boundary imposed by a wall that particles cannot cross, which is usually expressed in terms of inequalities [1], and requires a much more careful treatment.

[3] A constraint is called *scleronomous* when it does not depend explicitly on time. Time-dependent constraints are called *rheonomous* and correspond, for example, to a situation in which the motion of the particles is restricted to a moving surface or curve [1].

1.1.2 The Hamiltonian formalism

As we have seen, the Euler-Lagrange equations are a set of second-order differential equations which allows us to find the trajectory $\mathbf{q}(t)$ in coordinate space. We could reduce the order of the differential equations by taking the velocities $\dot{\mathbf{q}}$ as dynamical variables, arriving then at a set of $2n$ first-order differential equations. This is, however, a very naïve way of reducing the order, which leads to a non-symmetric system of equations for $\mathbf{q}$ and $\dot{\mathbf{q}}$. In this section we review Hamilton's approach to analytical mechanics, which leads to a symmetric-like first-order system of equations and will play a major role in understanding the transition from classical to quantum mechanics.

Instead of using the velocities, the Hamiltonian formalism considers the *generalized momenta*

$$p_j = \frac{\partial L}{\partial \dot{q}_j}, \tag{1.3}$$

as the dynamical variables. Note that this definition establishes a relation between these generalized momenta and the velocities $\dot{\mathbf{q}}(\mathbf{q}, \mathbf{p}) \Leftrightarrow \mathbf{p}(\mathbf{q}, \dot{\mathbf{q}})$. Note also that when the usual Cartesian coordinates of the system's particles are taken as the generalized coordinates, these momenta coincide with those of Newton's approach.

The theory is then built in terms of a new object called the *Hamiltonian*, which is defined as a Legendre transform of the Lagrangian,

$$H(\mathbf{q}, \mathbf{p}) = \mathbf{p}\dot{\mathbf{q}}(\mathbf{q}, \mathbf{p}) - L[\mathbf{q}, \dot{\mathbf{q}}(\mathbf{q}, \mathbf{p}), t], \tag{1.4}$$

and coincides with the total energy[4] for conservative systems with scleronomous constraints, that is, $H(\mathbf{q}, \mathbf{p}) = T(\mathbf{q}, \mathbf{p}) + V(\mathbf{q})$, with $T(\mathbf{q}, \mathbf{p}) = T[\mathbf{q}, \dot{\mathbf{q}}(\mathbf{q}, \mathbf{p})]$. Differentiating this expression and using the Euler–Lagrange equations (or using again the variational principle on the action), it is then straightforward to obtain the equations of motion for the generalized coordinates and momenta (the *canonical equations*),

$$\dot{q}_j = \frac{\partial H}{\partial p_j} \quad \text{and} \quad \dot{p}_j = -\frac{\partial H}{\partial q_j}, \tag{1.5}$$

which together with some initial conditions $\{\mathbf{q}(t_0), \mathbf{p}(t_0)\}$ allow us to find the trajectory $\{\mathbf{q}(t), \mathbf{p}(t)\}$ in the space formed by the generalized coordinates and momenta, which is known as *phase space*.

Another important object in the Hamiltonian formalism is the *Poisson bracket*. Given two functions of the coordinates and momenta $F(\mathbf{q}, \mathbf{p})$ and $G(\mathbf{q}, \mathbf{p})$, their Poisson bracket is defined as

$$\{F, G\} = \sum_{j=1}^{n} \frac{\partial F}{\partial q_j}\frac{\partial G}{\partial p_j} - \frac{\partial F}{\partial p_j}\frac{\partial G}{\partial q_j}. \tag{1.6}$$

The importance of this object is reflected in the fact that the evolution equation of any quantity $g(\mathbf{q}, \mathbf{p}, t)$ can be written as

[4] The general conditions under which the Hamiltonian coincides with the system's energy can be found in [1].

$$\frac{\mathrm{d}g}{\mathrm{d}t} = \{g, H\} + \frac{\partial g}{\partial t}, \tag{1.7}$$

and hence, if the quantity does not depend explicitly on time and its Poisson bracket with the Hamiltonian is zero, it is a *constant of motion.*

Of particular importance for the transition to quantum mechanics are the *canonical Poisson brackets*, that is, the Poisson brackets of the coordinates and momenta,

$$\{q_j, p_l\} = \delta_{jl}, \qquad \{q_j, q_l\} = \{p_j, p_l\} = 0, \tag{1.8}$$

which define the mathematical structure of phase space.

1.1.3 Observables and their mathematical structure

In this last section concerning classical mechanics we will discuss the mathematical structure in which observables are embedded within the Hamiltonian formalism. We will see that the mathematical objects corresponding to physical observables form a well defined mathematical structure, a real Lie algebra. Moreover, the position and momentum will be shown to be the generators of a particular Lie group, the Heisenberg group. Understanding this internal structure of *classical observables* will give us the chance to introduce the quantum description of observables in a reasonable way. Let us start by defining the concept of Lie algebra.

A *real Lie algebra* is a real vector space[5] $\mathcal{L}$ equipped with an additional operation, the *Lie product*, which takes two vectors f and g from $\mathcal{L}$, to generate another vector also in $\mathcal{L}$ denoted by[6] $\{f, g\}$. This operation must satisfy the following properties:

1. $\{f, g + h\} = \{f, g\} + \{f, h\}$ (linearity)
2. $\{f, f\} = 0 \underset{\text{with 1}}{\overset{\text{together}}{\Longrightarrow}} \{f, g\} = -\{g, f\}$ (anticommutativity)
3. $\{f, \{g, h\}\} + \{g, \{h, f\}\} + \{h, \{f, g\}\} = 0$ (Jacobi identity).

Hence, in essence a real Lie algebra is a vector space equipped with a linear, non-commutative, non-associative product. They have been a subject of study for many years, and now we know a lot about the properties of these mathematical structures. They appear in many branches of physics and geometry, in particular connected to continuous symmetry transformations, whose associated mathematical structures are actually called *Lie groups*. In particular, it is possible to show that given any Lie group with p parameters (such as, e.g., the three-parameter groups of translations or rotations in real space), any transformation onto the system in which it is acting can be generated from a set of p elements of a Lie

[5] The concept of complex vector space is defined in the next section. The definition of a *real* vector space is the same, but replacing by real numbers the complex numbers that appear in the definitions there.

[6] Note that we are using the same notation for the general definition of the Lie product and for the Poisson bracket, which is a particular case of Lie product as we will learn shortly.

algebra $\{g_1, g_2, \ldots, g_p\}$, called the *generators* of the Lie group, which satisfy some particular relations

$$\{g_j, g_k\} = \sum_{l=1}^{p} c_{jkl} g_l. \tag{1.9}$$

These relations are called the *algebra-group relations*, and the *structure constants* c_{jkl} are characteristic of the particular Lie group (for example, the generators of translations and rotations in real space are the momenta and angular momenta, respectively, and the corresponding structure constants are $c_{jkl} = 0$ for the translation group and $c_{jkl} = \epsilon_{jkl}$ for the rotation group[7]).

Coming back to the Hamiltonian formalism, we start by noting that *observables*, being *measurable* quantities, must be given by continuous, real functions in phase space. Hence they form a real vector space with respect to the usual addition of functions and multiplication of a function by a real number. Also appearing naturally in the formalism is a linear, non-commutative, non-associative operation between phase-space functions, the Poisson bracket, which applied to real functions gives another real function. It is easy to see that the Poisson bracket satisfies all the requirements of a Lie product, and hence, observables form a Lie algebra within the Hamiltonian formalism.

Moreover, the canonical Poisson brackets (1.8) show that the generalized coordinates $\mathbf{q}$ and momenta $\mathbf{p}$, together with the identity in phase space, satisfy particular algebra-group relations, namely[8] $\{q_j, p_k\} = \delta_{jk}1$ and $\{q_j, 1\} = \{p_j, 1\} = \{1, 1\} = 0$, and hence can be seen as the generators of a Lie group. This group is known as the *Heisenberg group*, and was introduced by Weyl when trying to prove the equivalence between the Schrödinger and Heisenberg pictures of quantum mechanics (which we will learn about later). It was later shown to have connections with the symplectic group, which is the basis of many physical theories. Note that we could have taken the Poisson brackets between the angular momenta associated with the possible rotations in the system of particles (which are certainly far more intuitive transformations than the one related to the Heisenberg group) as the fundamental ones. However, we have chosen the Lie algebra associated with the Heisenberg group just because it deals directly with position and momentum, allowing for a simpler connection to quantum mechanics.

---

[7] $\epsilon_{j_1 j_2 \ldots j_M}$ with all the subindices going from 1 to M is the Levi-Civita symbol in dimension M, which has $\epsilon_{12\ldots M} = 1$ and is completely antisymmetric, that is, changes its sign after permutation of any pair of indices.

[8] Ordering the generators as $\{\mathbf{q}, \mathbf{p}, 1\}$, the structure constants associated with these algebra-group relations are explicitly

$$c_{jkl} = \begin{cases} \Omega_{jk}\delta_{l,2n+1} & j, k = 1, 2, \ldots, 2n \\ 0 & j = 2n+1 \text{ or } k = 2n+1, \end{cases} \tag{1.10}$$

where $\Omega = \begin{pmatrix} 0_{n\times n} & I_{n\times n} \\ -I_{n\times n} & 0_{n\times n} \end{pmatrix}$, with $I_{n\times n}$ and $0_{n\times n}$ the $n \times n$ identity and null matrices, respectively.

Therefore, we arrive at the main conclusion of this review of classical mechanics: the mathematical framework of Hamiltonian mechanics associates physical observables with elements of a Lie algebra, with the phase-space coordinates themselves being the generators of the Heisenberg group.

Maintaining this structure for observables will help us introduce the laws of quantum mechanics in a coherent way.

1.2 The mathematical language of quantum mechanics

Just as classical mechanics is formulated in terms of the mathematical language of differential calculus and its extensions, quantum mechanics takes linear algebra (and Hilbert spaces in particular) as its fundamental grammar. In this section we will review the concept of Hilbert space, and discuss the properties of some operators which play important roles in the formalism of quantum information.

1.2.1 Finite-dimensional Hilbert spaces

In essence, a Hilbert space is a *complex vector space* in which an *inner product* is defined. Let us define first these terms as we use them in this book.

A *complex vector space* is a set $\mathcal{V}$, whose elements will be called *vectors* or *kets* and will be denoted by $\{|a\rangle, |b\rangle, |c\rangle, \ldots\}$ (a, b, and c may correspond to any suitable label), in which the following two operations are defined: the *vector addition*, which takes two vectors $|a\rangle$ and $|b\rangle$ and creates a new vector inside $\mathcal{V}$ denoted by $|a\rangle + |b\rangle$; and the *multiplication by a scalar*, which takes a complex number $\alpha \in \mathbb{C}$ (in this section Greek letters will represent complex numbers) and a vector $|a\rangle$ to generate a new vector in $\mathcal{V}$ denoted by $\alpha|a\rangle$.

The following additional properties must be satisfied:

1. The vector addition is commutative and associative, that is, $|a\rangle + |b\rangle = |b\rangle + |a\rangle$ and $(|a\rangle + |b\rangle) + |c\rangle = |a\rangle + (|b\rangle + |c\rangle)$.
2. There exists a null vector $|null\rangle$ such that $|a\rangle + |null\rangle = |a\rangle$
3. $\alpha(|a\rangle + |b\rangle) = \alpha|a\rangle + \alpha|b\rangle$
4. $(\alpha + \beta)|a\rangle = \alpha|a\rangle + \beta|a\rangle$
5. $(\alpha\beta)|a\rangle = \alpha(\beta|a\rangle)$
6. $1|a\rangle = |a\rangle$.

From these properties it can be proved that the null vector is unique, and can be built from any vector $|a\rangle$ as $0|a\rangle$; hence, in the following we denote it simply by $|null\rangle \equiv 0$. It is also readily proved that any vector $|a\rangle$ has a unique *antivector* $|-a\rangle$ such that $|a\rangle + |-a\rangle = 0$, which is given by $(-1)|a\rangle$ or simply $-|a\rangle$.

An *inner product* is an additional operation defined in the complex vector space $\mathcal{V}$, which takes two vectors $|a\rangle$ and $|b\rangle$ and associates them with a complex number. It will be denoted by $\langle a|b\rangle$ or sometimes also by $(|a\rangle, |b\rangle)$, and must satisfy the following properties:

1. $\langle a|a\rangle > 0$ if $|a\rangle \neq 0$
2. $\langle a|b\rangle = \langle b|a\rangle^*$
3. $(|a\rangle, \alpha|b\rangle) = \alpha\langle a|b\rangle$
4. $(|a\rangle, |b\rangle + |c\rangle) = \langle a|b\rangle + \langle a|c\rangle$.

The following additional properties can be proved from these ones:
- $\langle null|null\rangle = 0$
- $(\alpha|a\rangle, |b\rangle) = \alpha^*\langle a|b\rangle$
- $(|a\rangle + |b\rangle, |c\rangle) = \langle a|c\rangle + \langle b|c\rangle$
- $|\langle a|b\rangle|^2 \leqslant \langle a|a\rangle\langle b|b\rangle$ (Cauchy-Schwarz).

Note that for any vector $|a\rangle$, one can define the object $\langle a| \equiv (|a\rangle, \cdot)$, which will be called a *dual vector* or a *bra*, and which takes a vector $|b\rangle$ to generate the complex number $(|a\rangle, |b\rangle) \in \mathbb{C}$. It can be proved that the set formed by all the dual vectors corresponding to the elements in $\mathcal{V}$ is also a vector space, which will be called the *dual space* and will be denoted by $\mathcal{V}^+$. Within this picture, the inner product can be seen as an operation which takes a bra $\langle a|$ and a ket $|b\rangle$ to generate the complex number $\langle a|b\rangle$, a *bracket*. This whole *bra-c-ket* notation is due to Dirac [5].

In the following we assume that any time a bra $\langle a|$ is applied to a ket $|b\rangle$, the complex number $\langle a|b\rangle$ is formed, so that objects such as $|b\rangle\langle a|$ generate kets when applied to kets from the left, $(|b\rangle\langle a|)|c\rangle = (\langle a|c\rangle)|b\rangle$, and bras when applied to bras from the right, $\langle c|(|b\rangle\langle a|) = (\langle c|b\rangle)\langle a|$. Technically, $|b\rangle\langle a|$ is called an *outer product*.

A vector space equipped with an inner product is called a *Euclidean space* [6]. In the following we give some important definitions and properties which are needed in order to understand the concept of Hilbert space:
- The vectors $\{|a_1\rangle, |a_2\rangle, \dots, |a_m\rangle\}$ are said to be *linearly independent* if the relation $\alpha_1|a_1\rangle + \alpha_2|a_2\rangle + \cdots + \alpha_m|a_m\rangle = 0$ is satisfied only for $\alpha_1 = \alpha_2 = \cdots = \alpha_m = 0$, as otherwise one of them can be written as a linear combination of the rest.
- The *dimension* of the vector space is defined as the maximum number of linearly independent vectors that can be found in the space, and can be finite or infinite.
- If the dimension of a Euclidean space is $d < \infty$, it is always possible to build a set of d orthonormal vectors $E = \{|e_j\rangle\}_{j=1,2,\dots,d}$ satisfying $\langle e_j|e_l\rangle = \delta_{jl}$, such that any other vector $|a\rangle$ can be written as a linear superposition of them, that is, $|a\rangle = \sum_{j=1}^{d} a_j|e_j\rangle$, the a_j being some complex numbers. This set is called an *orthonormal basis* of the Euclidean space $\mathcal{V}$, and the coefficients a_j of the expansion can be found as $a_j = \langle e_j|a\rangle$. The column formed with the expansion coefficients, which is denoted by $\mathrm{col}(a_1, a_2, \dots, a_d)$, is called a *representation* of the vector $|a\rangle$ in the basis E. Note that the set $E^+ = \{\langle e_j|\}_{j=1,2,\dots,d}$ is an orthonormal basis in the dual space $\mathcal{V}^+$, so that any bra $\langle a|$ can be expanded then as $\langle a| = \sum_{j=1}^{d} a_j^*\langle e_j|$. The representation of the bra $\langle a|$ in the basis E corresponds to the row formed by its expansion coefficients, and is denoted by $(a_1^*, a_2^*, \dots, a_n^*)$. Note that if the representation of $|a\rangle$ is seen as a $d \times 1$ matrix, the representation of $\langle a|$ can be obtained as its $1 \times d$ conjugate-transpose matrix.

  Note finally that the inner product of two vectors $|a\rangle$ and $|b\rangle$ reads $\langle a|b\rangle = \sum_{j=1}^{d} a_j^* b_j$ when represented in the same basis, which is the matrix product of the representations of $\langle a|$ and $|b\rangle$.

For finite dimensions, a Euclidean space is a *Hilbert space*. However, in most applications of quantum mechanics (and certainly in continuous-variable quantum information), one has to deal with infinite-dimensional vector spaces. We will treat them after the following section.

1.2.2 Linear operators in finite dimensions

We now discuss the concept of linear operator, as well as analyze the properties of some important classes of operators. Only finite-dimensional Hilbert spaces are considered in this section, and we will generalize the discussion to infinite-dimensional Hilbert spaces in the next section.

We are interested in maps $\hat{L}$ (operators are denoted by a 'hat' throughout) which associate to any vector $|a\rangle$ of a Hilbert space $\mathcal{H}$ another vector denoted by $\hat{L}|a\rangle$ in the same Hilbert space. If the map satisfies

$$\hat{L}(\alpha|a\rangle + \beta|b\rangle) = \alpha\hat{L}|a\rangle + \beta\hat{L}|b\rangle, \tag{1.11}$$

then it is called a *linear operator*. For our purposes this is the only class of interesting operators, and hence we will simply call them *operators* in the following. Before discussing the properties of some important classes of operators, we need some definitions:

- Given an orthonormal basis $E = \{|e_j\rangle\}_{j=1,2,\ldots,d}$ in a Hilbert space $\mathcal{H}$ with dimension $d < \infty$, any operator $\hat{L}$ has a matrix representation. While bras and kets are represented by $d \times 1$ and $1 \times d$ matrices (rows and columns), respectively, an operator $\hat{L}$ is represented by a $d \times d$ matrix with *elements* $L_{jl} = (|e_j\rangle, \hat{L}|e_l\rangle) \equiv \langle e_j|\hat{L}|e_l\rangle$. An operator $\hat{L}$ can then be expanded in terms of the basis E as $\hat{L} = \sum_{j,l=1}^{d} L_{jl}|e_j\rangle\langle e_l|$. It follows that the representation of the vector $|b\rangle = \hat{L}|a\rangle$ is just the matrix multiplication of the representation of $\hat{L}$ by the representation of $|a\rangle$, that is, $b_j = \sum_{l=1}^{d} L_{jl}a_l$.
- The *addition* and *multiplication* of two operators $\hat{L}$ and $\hat{K}$, denoted by $\hat{L} + \hat{K}$ and $\hat{L}\hat{K}$, respectively, are defined by their action onto any vector $|a\rangle$: $(\hat{L} + \hat{K})|a\rangle = \hat{L}|a\rangle + \hat{K}|a\rangle$ and $\hat{L}\hat{K}|a\rangle = \hat{L}(\hat{K}|a\rangle)$. It follows that the representation of the addition and the product are, respectively, the sum and the multiplication of the corresponding matrices, that is, $(\hat{L} + \hat{K})_{jl} = L_{jl} + K_{jl}$ and $(\hat{L}\hat{K})_{jl} = \sum_{k=1}^{d} L_{jk}K_{kl}$.
- Note that while the addition is commutative, the product in general is not. This leads us to the notion of the *commutator*, defined for two operators $\hat{L}$ and $\hat{K}$ as $[\hat{L}, \hat{K}] = \hat{L}\hat{K} - \hat{K}\hat{L}$. When $[\hat{L}, \hat{K}] = 0$, we say that the operators *commute*.
- Given an operator $\hat{L}$, its *trace* is defined as the sum of the diagonal elements of its matrix representation, that is, $\mathrm{tr}\{\hat{L}\} = \sum_{j=1}^{d} L_{jj}$. It may seem that this definition is basis-dependent, as in general the elements L_{jj} are different in different bases. However, we will see later that the trace is invariant under any change of basis.

The trace has two important properties. It is *linear* and *cyclic*, that is, given two operators $\hat{L}$ and $\hat{K}$, $\text{tr}\{\hat{L}+\hat{K}\}=\text{tr}\{\hat{L}\}+\text{tr}\{\hat{K}\}$ and $\text{tr}\{\hat{L}\hat{K}\}=\text{tr}\{\hat{K}\hat{L}\}$, as is trivial to prove.

- Given an operator $\hat{L}$, we define its *determinant* as the determinant of its matrix representation, that is, $\det\{\hat{L}\}=\sum_{j_1,j_2,\ldots,j_d=1}^{d}\epsilon_{j_1j_2\ldots j_d}L_{1j_1}L_{2j_2}\ldots L_{dj_d}$. Just as the trace, we will see that it does not depend on the basis used to represent the operator.

  The determinant is a multiplicative map, that is, given two operators $\hat{L}$ and $\hat{K}$, the determinant of the product is the product of the determinants, $\det\{\hat{L}\hat{K}\}=\det\{\hat{L}\}\det\{\hat{K}\}$.
- We say that a vector $|l\rangle$ is an *eigenvector* of an operator $\hat{L}$ if there exists a $\lambda\in\mathbb{C}$ (called its associated *eigenvalue*) such that $\hat{L}|l\rangle=\lambda|l\rangle$. The set of all the eigenvalues of an operator is called its *spectrum*.

We can now move on to describe some classes of operators which play important roles in quantum mechanics.

The identity operator. The *identity operator*, denoted by $\hat{I}$, is defined as the operator which maps any vector onto itself. Its representation in any basis is then $I_{jl}=\delta_{jl}$, so that it can be expanded as

$$\hat{I}=\sum_{j=1}^{d}|e_j\rangle\langle e_j|. \tag{1.12}$$

This expression is known as the *completeness relation* of the basis E; alternatively, it is said that the set E forms a *resolution of the identity*.

Note that the expansion of a vector $|a\rangle$ and its dual $\langle a|$ in the basis E is obtained just by application of the completeness relation from the left and the right, respectively. Similarly, the expansion of an operator $\hat{L}$ is obtained by application of the completeness relation both from the right and the left at the same time.

The inverse of an operator. The *inverse* of an operator $\hat{L}$, denoted by $\hat{L}^{-1}$, is defined as that satisfying $\hat{L}^{-1}\hat{L}=\hat{L}\hat{L}^{-1}=\hat{I}$. Not every operator has an inverse. An inverse exists if and only if the operator does not have a zero eigenvalue, or, equivalently, when $\det\{\hat{L}\}\neq 0$.

An operator function. Consider a real, analytic function $f(x)$ which can be expanded in powers of x as $f(x)=\sum_{m=0}^{\infty}f_m x^m$. Given an operator $\hat{L}$, we define the *operator function* $\hat{f}(\hat{L})=\sum_{m=0}^{\infty}f_m\hat{L}^m$, where $\hat{L}^m$ means the product of $\hat{L}$ with itself m times.

The adjoint of an operator. Given an operator $\hat{L}$, we define its *adjoint*, and denote it by $\hat{L}^\dagger$, as that satisfying $(|a\rangle,\hat{L}|b\rangle)=(\hat{L}^\dagger|a\rangle,|b\rangle)$ for any two vectors $|a\rangle$ and $|b\rangle$. Note that the representation of $\hat{L}^\dagger$ corresponds to the conjugate transpose of the matrix representing $\hat{L}$, that is, $(\hat{L}^\dagger)_{jl}=L_{lj}^*$. Note also that the adjoint of a product of two operators $\hat{K}$ and $\hat{L}$ is given by $(\hat{K}\hat{L})^\dagger=\hat{L}^\dagger\hat{K}^\dagger$.

Self-adjoint operators. We say that $\hat{H}$ is *self-adjoint* when it coincides with its adjoint, that is, $\hat{H}=\hat{H}^\dagger$. A property of major importance for the construction of the laws of quantum mechanics is that the spectrum $\{h_j\}_{j=1,2,...,d}$ of a self-adjoint operator is real. Moreover, its associated eigenvectors[9] $\{|h_j\rangle\}_{j=1,2,...,d}$ form an orthonormal basis of the Hilbert space.

The representation of any operator function $\hat{f}(\hat{H})$ in the *eigenbasis* of $\hat{H}$ is then $[\hat{f}(\hat{H})]_{jl} = f(h_j)\delta_{jl}$, from which follows

$$\hat{f}(\hat{H}) = \sum_{j=1}^{d} f(h_j)|h_j\rangle\langle h_j|. \tag{1.13}$$

This result is known as the *spectral theorem.*

Unitary operators. We say that $\hat{U}$ is a *unitary operator* when $\hat{U}^\dagger = \hat{U}^{-1}$. The interest of this class of operators is that they preserve inner products, that is, for any two vectors $|a\rangle$ and $|b\rangle$ the inner product $(\hat{U}|a\rangle, \hat{U}|b\rangle)$ coincides with $\langle a|b\rangle$. Moreover, it is possible to show that given two orthonormal bases $E = \{|e_j\rangle\}_{j=1,2,...,d}$ and $E' = \{|e_j'\rangle\}_{j=1,2,...,d}$, there exists a unique unitary matrix $\hat{U}$ which connects them as $\{|e_j'\rangle = \hat{U}|e_j\rangle\}_{j=1,2,...,d}$, and then any basis of the Hilbert space is unique up to a unitary transformation.

We can now prove that both the trace and the determinant of an operator are basis-independent. Let us denote by $\mathrm{tr}\{\hat{L}\}_E$ the trace of an operator $\hat{L}$ in the basis E. The trace of this operator in the transformed basis can be written then as $\mathrm{tr}\{\hat{L}\}_{E'} = \mathrm{tr}\{\hat{U}^\dagger\hat{L}\hat{U}\}_E$, which, using the cyclic property of the trace and the unitarity of $\hat{U}$, is rewritten as $\mathrm{tr}\{\hat{U}\hat{U}^\dagger\hat{L}\}_E = \mathrm{tr}\{\hat{L}\}_E$, proving that the trace is equal in both bases. Similarly, in the case of the determinant we have $\det\{\hat{L}\}_{E'} = \det\{\hat{U}^\dagger\hat{L}\hat{U}\}_E$, which using the multiplicative property of the determinant is rewritten as $\det\{\hat{U}^\dagger\}_E \det\{\hat{L}\}_E \det\{\hat{U}\}_E = \det\{\hat{L}\}_E$, where we have used $\det\{\hat{U}^\dagger\}_E \det\{\hat{U}\}_E = 1$ as follows from $\hat{U}^\dagger\hat{U} = \hat{I}$.

Note finally that a unitary operator $\hat{U}$ can always be written as the complex exponential of a self-adjoint operator $\hat{H}$, that is, $\hat{U} = \exp(\mathrm{i}\hat{H})$.

Projection operators. In general, any self-adjoint operator $\hat{P}$ satisfying $\hat{P}^2 = \hat{P}$ is called a *projector*. We are interested only in those projectors which can be written as the outer product of a vector $|a\rangle$ with itself, that is, rank-1 projectors[10] $\hat{P}_a = |a\rangle\langle a|$. When applied to a vector $|b\rangle$, this is *projected* along the 'direction' of $|a\rangle$ as $\hat{P}_a|b\rangle = \langle a|b\rangle|a\rangle$.

Note that given an orthonormal basis E, we can use the projectors $\hat{P}_j = |e_j\rangle\langle e_j|$ to extract the components of a vector $|a\rangle$ as $\hat{P}_j|a\rangle = a_j|e_j\rangle$. Note also that the completeness and orthonormality of the basis E implies that $\sum_{j=1}^{d}\hat{P}_j = \hat{I}$ and $\hat{P}_j\hat{P}_l = \delta_{jl}\hat{P}_j$, respectively.

[9] For simplicity, we will assume that the spectrum of any operator is non-degenerate, that is, all the eigenvectors possess a distinctive eigenvalue.

[10] The term 'rank' refers to the number of non-zero eigenvalues.

Density operators. A self-adjoint operator $\hat{\rho}$ is called a *density operator* when it has unit trace and it is *positive semidefinite*, that is, $\langle a|\hat{\rho}|a\rangle \geqslant 0$ for any vector $|a\rangle$.

The interesting property of density operators is that they 'contain' probability distributions in the diagonal of its representation. To see this just note that given an orthonormal basis E, the self-adjointness and positivity of $\hat{\rho}$ ensure that all its diagonal elements $\{\rho_{jj}\}_{j=1,2,\ldots,d}$ are either positive or zero, that is, $\rho_{jj} \geqslant 0\ \forall j$, while the unit trace makes them satisfy $\sum_{j=1}^{d} \rho_{jj} = 1$. Hence, the diagonal elements of a density operator have all the properties required by a *probability distribution*.

1.2.3 Generalization to infinite dimensions

Unfortunately, not all the previous concepts and objects that we have introduced for the finite-dimensional case are trivially generalized to infinite dimensions. In this section we discuss this generalization.

The first problem that we meet when dealing with infinite-dimensional Euclidean spaces is that the existence of a basis $\{|e_j\rangle\}_{j=1,2,\ldots}$ in which any other vector can be represented as $|a\rangle = \sum_{j=1}^{\infty} a_j|e_j\rangle$ is not granted. The class of infinite-dimensional Euclidean spaces in which these infinite but countable bases exist are called *separable Hilbert spaces*, and are the ones relevant for the quantum description of physical systems.

The conditions which ensure that an infinite-dimensional Euclidean space is indeed a Hilbert space[11] can be found in, for example, reference [6]. Here we just want to stress that, quite intuitively, any infinite-dimensional Hilbert space[12] is *isomorphic* to the space called $l^2(\infty)$, which is formed by the column vectors $|a\rangle = \text{col}(a_1, a_2,\ldots)$ where the set $\{a_j \in \mathbb{C}\}_{j=1,2,\ldots}$ satisfies the restriction $\sum_{j=1}^{\infty}|a_j|^2 < \infty$, and has the operations $|a\rangle + |b\rangle = \text{col}(a_1 + b_1, a_2 + b_2,\ldots)$, $\alpha|a\rangle = \text{col}(\alpha a_1, \alpha a_2,\ldots)$, and $\langle a|b\rangle = \sum_{j=1}^{\infty} a_j^* b_j$.

Most of the previous definitions are directly generalized to Hilbert spaces by taking $d \to \infty$ (dual space, representations, operators,...). However, there is one crucial property of self-adjoint operators which does not hold in this case: their eigenvectors may not form an orthonormal basis of the Hilbert space. The remainder of this section is devoted to dealing with this problem.

[11] From now on we will assume that all the Hilbert spaces we refer to are 'separable', even if we do not write it explicitly.

[12] An example of infinite-dimensional complex Hilbert space consists in the vector space formed by the complex functions of real variable, say $|f\rangle = f(x)$ with $x \in \mathbb{R}$, with integrable square, that is

$$\int_{\mathbb{R}} \mathrm{d}x\, |f(x)|^2 < \infty, \tag{1.14}$$

together with the inner product

$$\langle g|f\rangle = \int_{\mathbb{R}} \mathrm{d}x g^*(x) f(x). \tag{1.15}$$

This Hilbert space is usually denoted by $L^2(x)$.

Just as in finite dimensions, given an infinite-dimensional Hilbert space $\mathcal{H}$, we say that one of its vectors $|d\rangle$ is an eigenvector of the self-adjoint operator $\hat{H}$ if $\hat{H}|d\rangle = \delta|d\rangle$, where $\delta \in \mathbb{R}$ is called its associated eigenvalue. Nevertheless, it can happen in infinite-dimensional spaces that some vector $|c\rangle$ not contained in $\mathcal{H}$ also satisfies the condition $\hat{H}|c\rangle = \chi|c\rangle$, in which case we call it a *generalized eigenvector*, χ being its *generalized eigenvalue*[13]. The set of all eigenvalues of the self-adjoint operator is called its *discrete* (or *point*) *spectrum* and it is a countable set, while the set of all its generalized eigenvalues is called its *continuous spectrum* and it is uncountable, that is, it forms a continuous set [6] (see also [7]).

In this monograph we only deal with two extreme cases: either the observable, say $\hat{H}$, has a pure discrete spectrum $\{h_j\}_{j=1,2,...}$; or the observable, say $\hat{X}$, has a pure continuous spectrum $\{x\}_{x\in\mathbb{R}}$. It can be shown that in the first case the eigenvectors of the observable form an orthonormal basis of the Hilbert space, so that we can build a resolution of the identity as $\hat{I} = \sum_{j=1}^{\infty}|h_j\rangle\langle h_j|$, and proceed along the lines of the previous sections.

In the second case, the set of generalized eigenvectors cannot form a basis of the Hilbert space in the strict sense, as they do not form a countable set and do not even belong to the Hilbert space. Fortunately, there are still ways to treat the generalized eigenvectors of $\hat{X}$ 'as if' they were a basis of the Hilbert space. This idea was introduced by Dirac [5], who realized that normalizing the generalized eigenvectors as[14] $\langle x|y\rangle = \delta(x - y)$, one can define the following integral operator

$$\int_{\mathbb{R}} \mathrm{d}x|x\rangle\langle x| = \hat{I}_{\mathrm{c}}, \tag{1.19}$$

which acts as the identity onto the generalized eigenvectors, that is, $\hat{I}_{\mathrm{c}}|y\rangle = |y\rangle$. It is then assumed that $\hat{I}_{\mathrm{c}}$ coincides with the identity in $\mathcal{H}$, so that any other vector $|a\rangle$ or operator $\hat{L}$ defined in the Hilbert space can be expanded as

$$|a\rangle = \int_{\mathbb{R}} \mathrm{d}x a(x)|x\rangle \qquad \text{and} \qquad \hat{L} = \int_{\mathbb{R}^2} \mathrm{d}x\mathrm{d}y L(x, y)|x\rangle\langle y|, \tag{1.20}$$

where the elements $a(x) = \langle x|a\rangle$ and $L(x, y) = \langle x|\hat{L}|y\rangle$ of these *continuous representations* form complex functions defined in $\mathbb{R}$ and $\mathbb{R}^2$, respectively. From now on, we will call *continuous basis* to the set $\{|x\rangle\}_{x\in\mathbb{R}}$.

---

[13] In $L^2(x)$ we have two simple examples of self-adjoint operators with eigenvectors not contained in $L^2(x)$: the so-called $\hat{X}$ (*position*) and $\hat{P}$ (*momentum*), which, given an arbitrary vector $|f\rangle = f(x)$, act as $\hat{X}|f\rangle = xf(x)$ and $\hat{P}|f\rangle = -\mathrm{i}\partial_x f$, respectively. This is simple to see, as the equations

$$xf_X(x) = Xf_X(x) \qquad \text{and} \qquad -\mathrm{i}\partial_x f_P(x) = Pf_P(x), \tag{1.16}$$

have

$$f_X(x) = \delta(x - X) \qquad \text{and} \qquad f_P(x) = \exp(\mathrm{i}Px), \tag{1.17}$$

as solutions, which are not square-integrable, and hence do not belong to $L^2(x)$.

[14] $\delta(x)$ is the so-called *Dirac-delta distribution* which is defined by the conditions

$$\int_{x_1}^{x_2} \mathrm{d}x\delta(x - y) = \begin{cases} 1 & \text{if } y \in [x_1, x_2] \\ 0 & \text{if } y \notin [x_1, x_2] \end{cases}. \tag{1.18}$$

Dirac introduced this continuous representation as a 'limit to the continuum' of the countable case. Even though this approach was very intuitive, it lacked mathematical rigor. Some decades after Dirac's proposal, Gel'fand showed how to generalize the concept of Hilbert space to include these generalized representations in full mathematical rigor [8]. The generalized spaces are called *rigged Hilbert spaces* (in which the algebra of Hilbert spaces joins forces with the theory of continuous probability distributions), and working on them it is possible to show that given any self-adjoint operator, one can use its eigenvectors and generalized eigenvectors to expand any vector of the Hilbert space, just as we did above. In other words, within the framework of rigged Hilbert spaces, one can prove the identity $\hat{I}_c = \hat{I}$ rigorously.

Note finally that given two vectors $|a\rangle$ and $|b\rangle$ of the Hilbert space, and a continuous basis $\{|x\rangle\}_{x\in\mathbb{R}}$, we can use their generalized representations to write their inner product as

$$\langle a|b\rangle = \int_{\mathbb{R}} \mathrm{d}x a^*(x)b(x). \tag{1.21}$$

It is also easily proved that the trace of any operator $\hat{L}$ can be evaluated from its continuous representation on $\{|x\rangle\}_{x\in\mathbb{R}}$ as

$$\mathrm{tr}\{\hat{L}\} = \int_{\mathbb{R}} \mathrm{d}x L(x, x). \tag{1.22}$$

This has important consequences for the properties of density operators, say $\hat{\rho}$ for the discussion which follows. We explained at the end of the last section that when represented on an orthonormal basis of the Hilbert space, its diagonal elements (which are real owing to its self-adjointness) can be seen as a probability distribution, because they satisfy $\sum_{j=1}^{\infty} \rho_{jj} = 1$ and $\rho_{jj} \geqslant 0\ \forall j$. Similarly, because of its unit trace and positivity, the diagonal elements of its continuous representation satisfy $\int_{\mathbb{R}} \mathrm{d}x\rho(x, x) = 1$ and $\rho(x, x) \geqslant 0\ \forall x$, and hence, the real function $\rho(x, x)$ can be seen as a *probability density function*.

1.2.4 Composite Hilbert spaces

In many moments of this monograph, we will find the need to associate a Hilbert space with a composite system, the Hilbert spaces of whose parts we know. In this section we show how to build a Hilbert space $\mathcal{H}$ starting from a set of Hilbert spaces $\{\mathcal{H}_A, \mathcal{H}_B, \mathcal{H}_C, \ldots\}$.

Let us start with only two Hilbert spaces $\mathcal{H}_A$ and $\mathcal{H}_B$ with dimensions d_A and d_B, respectively (which might be infinite); the generalization to an arbitrary number of Hilbert spaces is straightforward. Consider a vector space $\mathcal{V}$ with dimension $\dim(\mathcal{V}) = d_A \times d_B$. We define a map called the *tensor product* which associates to any pair of vectors $|a\rangle \in \mathcal{H}_A$ and $|b\rangle \in \mathcal{H}_B$ a vector in $\mathcal{V}$ which we denote by $|a\rangle \otimes |b\rangle \in \mathcal{V}$. This tensor product must satisfy the following properties:

1. $(|a\rangle + |b\rangle) \otimes |c\rangle = |a\rangle \otimes |c\rangle + |b\rangle \otimes |c\rangle$
2. $|a\rangle \otimes (|b\rangle + |c\rangle) = |a\rangle \otimes |b\rangle + |a\rangle \otimes |c\rangle$
3. $(\alpha|a\rangle) \otimes |b\rangle = |a\rangle \otimes (\alpha|b\rangle)$.

If we endorse the vector space $\mathcal{V}$ with the inner product $(|a\rangle \otimes |b\rangle, |c\rangle \otimes |d\rangle) = \langle a|c\rangle\langle b|d\rangle$, it is easy to show it becomes a Hilbert space, which in the following will be denoted by $\mathcal{H} = \mathcal{H}_A \otimes \mathcal{H}_B$. Given the bases $E_A = \{|e_j^A\rangle\}_{j=1,2,\ldots,d_A}$ and $E_B = \{|e_j^B\rangle\}_{j=1,2,\ldots,d_B}$ of the Hilbert spaces $\mathcal{H}_A$ and $\mathcal{H}_B$, respectively, a basis of the *tensor product Hilbert space* $\mathcal{H}_A \otimes \mathcal{H}_B$ can be built as $E = E_A \otimes E_B = \{|e_j^A\rangle \otimes |e_l^B\rangle\}_{l=1,2,\ldots,d_B}^{j=1,2,\ldots,d_A}$ (note that the notation after the first equality is symbolic).

We may use a more economic notation for the tensor product, namely $|a\rangle \otimes |b\rangle = |a, b\rangle$, except when the explicit tensor product symbol is needed for some special reason. With this notation the basis of the tensor product Hilbert space is written as $E = \{|e_j^A, e_l^B\rangle\}_{l=1,2,\ldots,d_B}^{j=1,2,\ldots,d_A}$.

The tensor product also maps operators acting on $\mathcal{H}_A$ and $\mathcal{H}_B$ to operators acting on $\mathcal{H}$. Given two operators $\hat{L}_A$ and $\hat{L}_B$ acting on $\mathcal{H}_A$ and $\mathcal{H}_B$, the *tensor product operator* $\hat{L} = \hat{L}_A \otimes \hat{L}_B$ is defined in $\mathcal{H}$ as that satisfying $\hat{L}|a, b\rangle = (\hat{L}_A|a\rangle) \otimes (\hat{L}_B|b\rangle)$ for any pair of vectors $|a\rangle \in \mathcal{H}_A$ and $|b\rangle \in \mathcal{H}_B$. When explicit subindices making reference to the Hilbert space on which operators act on are used, so that there is no room for confusion, we will use the shorter notations $\hat{L}_A \otimes \hat{L}_B = \hat{L}_A\hat{L}_B$, $\hat{L}_A \otimes \hat{I} = \hat{L}_A$, and $\hat{I} \otimes \hat{L}_B = \hat{L}_B$.

Note that the tensor product preserves the properties of the operators; for example, given two self-adjoint operators $\hat{H}_A$ and $\hat{H}_B$, unitary operators $\hat{U}_A$ and $\hat{U}_B$, or density operators $\hat{\rho}_A$ and $\hat{\rho}_B$, the operators $\hat{H}_A \otimes \hat{H}_B$, $\hat{U}_A \otimes \hat{U}_B$, and $\hat{\rho}_A \otimes \hat{\rho}_B$ are self-adjoint, unitary, and a density operator acting on $\mathcal{H}$, respectively. But keep in mind that this does not mean that all self-adjoint, unitary, or density operators acting on $\mathcal{H}$ can be written in a simple tensor product form $\hat{L}_A \otimes \hat{L}_B$.

1.3 The quantum-mechanical framework

In this section we review the basic postulates that describe how quantum mechanics treats physical systems. As the building blocks of the theory, these axioms cannot be *proved*. They can only be formulated following *plausibility arguments* based on the *observation* of physical phenomena and the *connection* of the theory with previous theories which are known to work in some limit. We will try to motivate (and justify to a point) these axioms as much as possible, starting with a brief historical introduction to the context in which they were created[15].

1.3.1 A brief historical introduction

By the end of the 19th century there was a great feeling of safety and confidence among the physics community: analytical mechanics (together with statistical mechanics) and Maxwell's electromagnetism (in the following *classical physics* altogether) seem to explain the whole range of physical phenomena that one could observe, and hence, in a sense, the foundations of physics were complete. There were, however, a couple of experimental observations which lacked explanation

[15] For a thorough historical overview of the birth of quantum physics see [9].

within this 'definitive' framework, which actually led to the construction of a whole new way of understanding physical phenomena: quantum mechanics.

Among this experimental evidence, the shape of the high-energy spectrum of the radiation emitted by a black body, the photoelectric effect which showed that only light exceeding some frequency can release electrons from a metal irrespective of its intensity, and the discrete set of spectral lines of hydrogen, were the principal triggers of the revolution to come in the first quarter of the 20th century. The first two led Planck and Einstein to suggest that electromagnetic energy is not continuous but divided into small packets of energy $\hbar\omega$ (ω being the angular frequency of the radiation), while Bohr succeeded in explaining the hydrogen spectrum by assuming that the electron orbiting the nucleus can occupy only a discrete set of orbits with angular momenta proportional to $\hbar$. The constant $\hbar = h/2\pi \sim 10^{-34}$ J · s, where h is now known as the Planck constant, appeared in both cases as the 'quantization unit', the value separating the quantized values that energy or angular momentum are able to take.

Even though the physicists of the time tried to understand this quantization of the physical magnitudes within the framework of classical physics, it was soon realized that a completely new theory was required. The first attempts to build such a theory (which actually worked for some particular scenarios) were based on applying ad hoc quantization rules to various mechanical variables of systems, but with a complete lack of physical interpretation for such rules [10]. However, between 1925 and 1927 the first real formulations of the necessary theory were developed: the *wave mechanics* of Schrödinger [11] and the *matrix mechanics* of Heisenberg, Born and Jordan [12–14] (see [10] for English translations), which also received independent contributions from Dirac [15]. Even though in both theories the quantization of various observable quantities appeared naturally and in correspondence with experiments, they seemed completely different, at least until Schrödinger showed the equivalence between them.

The new theory was later formalized mathematically using vector spaces by Dirac [5] (although not entirely rigorously), and a little later by von Neumann with more mathematical rigor using Hilbert spaces [16] (see [17] for an English version). They developed the laws of *quantum mechanics* basically as we know them today [18–23]. In the following sections we will introduce these rules in the form of six axioms that will set out the structure of the theory of quantum mechanics as will be used throughout this book.

1.3.2 Axiom 1. Observables and measurement outcomes

The experimental evidence for the tendency of observable physical quantities to be quantized at the microscopic level motivates the first axiom:

> **Axiom 1.** Any physical observable quantity A corresponds to a self-adjoint operator $\hat{A}$ acting on an abstract Hilbert space. After a measurement of A, the only possible outcomes are the eigenvalues of $\hat{A}$.

The quantization of physical observables is therefore directly introduced within the theory by this postulate. Note that it does not say anything about the dimension d of the Hilbert space corresponding to a given observable, and it even leaves open the possibility of observables having a continuous spectrum, rather than a discrete one. The problem of how to make the proper correspondence between observables and self-adjoint operators will be addressed in an axiom to come.

In this book we use the name 'observable' both for the physical quantity A and its associated self-adjoint operator $\hat{A}$ indistinctly. Observables having purely discrete or purely continuous spectra will be referred to as *countable* and *continuous observables*, respectively.

1.3.3 Axiom 2. The state of the system and statistics of measurements

The next axiom follows from the following question: according to the previous axiom the eigenvalues of an observable are the only values that can appear when measuring it, but what about the statistics of such a measurement? We know a class of operators in Hilbert spaces which act as probability distributions for the eigenvalues of any self-adjoint operator, density operators. This motivates the second axiom:

> **Axiom 2.** The state of the system is completely specified by a density operator $\hat{\rho}$. When measuring a countable observable A with eigenvectors $\{|a_j\rangle\}_{j=1,2,\ldots,d}$ (d might be infinite), associated with the possible outcomes $\{a_j\}_{j=1,2,\ldots,d}$ is a probability distribution $\{p_j = \rho_{jj}\}_{j=1,2,\ldots,d}$ which determines the statistics of the experiment (the *Born rule*). Similarly, when measuring a continuous observable X with eigenvectors $\{|x\rangle\}_{x\in\mathbb{R}}$, the probability density function $P(x) = \rho(x, x)$ is associated with the possible outcomes $\{x\}_{x\in\mathbb{R}}$ in the experiment.

This postulate has deep consequences that we analyze now. Contrary to classical mechanics (and intuition), even if the system is prepared in a given state, the value of an observable is in general not well defined. We can only specify with what probability a given value of the observable will come out in a measurement. Hence, this axiom proposes a change of paradigm; determinism must be abandoned: the theory is no longer able to predict with certainty the outcome of a single run of an experiment in which an observable is measured, but rather gives the statistics that will be extracted after a large number of runs.

To be fair, there is a case in which the theory allows us to predict the outcome of the measurement of an observable with certainty: when the system is prepared such that its state is an eigenvector of the observable. This seems much like when in classical mechanics the system is prepared with a given value of its observables. However, we will show that it is impossible to find a common eigenvector to *all* the available observables of a system, and hence the difference between classical and quantum mechanics is that in the latter it is impossible to prepare the system in a state which would allow us to predict with certainty the outcome of a measurement of each of its observables. Let us try to elaborate on this in a more rigorous fashion.

Let us define the *expectation value* of a given operator $\hat{B}$ as

$$\langle \hat{B} \rangle = \mathrm{tr}\{\hat{\rho}\hat{B}\}. \tag{1.23}$$

In the case of a countable observable $\hat{A}$ or a continuous observable $\hat{X}$, this expectation value can be written in their own eigenbases as

$$\langle \hat{A} \rangle = \sum_{j=1}^{d} \rho_{jj} a_j \qquad \text{and} \qquad \langle \hat{X} \rangle = \int_{-\infty}^{+\infty} \mathrm{d}x \rho(x, x) x, \tag{1.24}$$

which correspond to the mean value of the outcomes registered in a large number of measurements of the observables. We define also the *variance* $V(A)$ of the observable as the expectation value of the square of its *fluctuation operator* $\delta\hat{A} = \hat{A} - \langle \hat{A} \rangle$, that is,

$$V(A) = \mathrm{tr}\left\{\hat{\rho}\left(\delta\hat{A}\right)^2\right\} = \left\langle \hat{A}^2 \right\rangle - \left\langle \hat{A} \right\rangle^2, \tag{1.25}$$

from which we obtain the *standard deviation* or *uncertainty* as $\Delta A = \sqrt{V(A)}$, which measures how much the outcomes of the experiment deviate from the mean, and hence, somehow specifies how 'well defined' the value of the observable A is.

Note that the probability of obtaining the outcome a_j when measuring A can be written as the expectation value of the projection operator $\hat{P}_j = |a_j\rangle\langle a_j|$, that is $p_j = \langle \hat{P}_j \rangle$. Similarly, the probability density function associated with the possible outcomes $\{x\}_{x\in\mathbb{R}}$ when measuring X can be written as $P(x) = \langle \hat{P}(x) \rangle$, where $\hat{P}(x) = |x\rangle\langle x|$.

Having written all these objects (probabilities, expectation values, and variances) in terms of traces is really useful, since the trace is invariant under basis changes, and hence can be evaluated in any basis we want to work with, see section 1.2.2.

These axioms have one further counterintuitive consequence. It is possible to prove that irrespectively of the state of the system, the following relation between the variances of two non-commuting observables A and B is satisfied:

$$\Delta A \Delta B \geqslant \frac{1}{2}\left|\left\langle \left[\hat{A}, \hat{B}\right] \right\rangle\right|. \tag{1.26}$$

According to this inequality, known as the *uncertainty principle* (which was first derived by Heisenberg), in general, the only way in which the observable A can be perfectly defined ($\Delta A \to 0$) is by making observable B completely undefined ($\Delta B \to \infty$), or vice versa. Hence, in the quantum formalism one cannot, in general, prepare the system in a state in which all its observables are well defined, the complete opposite to our everyday experience.

Before moving to the third axiom, let us comment on a couple more things related to the state of the system. It is possible to show that a density operator can always be expressed as a *statistical* or *convex mixture* of projection operators, that is, $\hat{\rho} = \sum_{m=1}^{M} w_m |\varphi_m\rangle\langle\varphi_m|$, where $\{w_m\}_{m=1,2,\ldots,M}$ is a probability distribution and the vectors $\{|\varphi_m\rangle\}_{m=1}^{M}$ are normalized to one, but do not need to be orthogonal (note that in fact M does not need to be equal to d). Hence, another way of specifying the state of the system is by a set of normalized vectors together with some statistical rule for mixing them, that is, the set $\{w_m, |\varphi_m\rangle\}_{m=1,2,\ldots,M}$, known as an *ensemble decomposition* of the state $\hat{\rho}$. Such decompositions are not unique, in the sense that different ensembles can lead to the same $\hat{\rho}$. It can be proved though [24] that two ensembles $\{w_m, |\varphi_m\rangle\}_{m=1,2,\ldots,M}$ and $\{v_n, |\psi_n\rangle\}_{n=1,2,\ldots,N}$ (we take $M \leqslant N$ for definiteness) give rise to the same density operator $\hat{\rho}$ if and only if there exists a left-unitary matrix[16] U with elements $\{U_{mn}\}_{m,n=1,2,\ldots,N}$ such that [24]

$$\sqrt{w_m}|\varphi_m\rangle = \sum_{n=1}^{N} U_{mn}\sqrt{v_n}|\psi_n\rangle, \qquad m = 1, 2, \ldots, N, \tag{1.27}$$

where if $M \neq N$, $N - M$ zeros must be included in the ensemble with fewer states, so that $\mathcal{U}$ is a square matrix.

When only one vector $|\varphi\rangle$ contributes to the mixture, $\hat{\rho} = |\varphi\rangle\langle\varphi|$ is completely specified by just this single vector, and we say that the density operator is *pure*; otherwise, we say that it is *mixed*. A necessary and sufficient condition for $\hat{\rho}$ to be pure is $\hat{\rho}^2 = \hat{\rho}$. In the next chapters we will learn that the mixedness of a state always comes from the fact that some of the information of the system has been lost to some other inaccessible system with which it has interacted for a while before becoming isolated. In other words, the state of a system is pure only when it has no correlations at all with other systems.

Note, finally, that when the state of the system is in a *pure state* $|\psi\rangle$, the expectation value of an operator $\hat{B}$ takes the simple form $\langle\psi|\hat{B}|\psi\rangle$. Moreover, the pure state can be expanded in the countable and continuous bases of two observables $\hat{A}$ and $\hat{X}$ as

$$|\psi\rangle = \sum_{j=1}^{d} \psi_j |a_j\rangle \qquad \text{and} \qquad |\psi\rangle = \int_{-\infty}^{+\infty} \mathrm{d}x \psi(x)|x\rangle, \tag{1.28}$$

respectively, being $\psi_j = \langle a_j|\psi\rangle$ and $\psi(x) = \langle x|\psi\rangle$. In this case, the probability distribution for the discrete outcomes $\{a_j\}_{j=1,2,\ldots,d}$ is given by $\{p_j = |\psi_j|^2\}_{j=1,2,\ldots,d}$, while the probability density function for the continuous outcomes $\{x\}_{x\in\mathbb{R}}$ is given by $P(x) = |\psi(x)|^2$.

1.3.4 Axiom 3. Composite systems

The next axiom specifies how the theory accommodates dealing with composite systems within its mathematical framework. Of course, a composition of two

[16] U is left-unitary if $U^\dagger U = I$ but $UU^\dagger$ might not be I, where I is the identity matrix of the corresponding dimension. It is easy to prove that finite-dimensional left-unitary matrices are unitary.

systems is itself another system subject to the laws of quantum mechanics; the question is how can we construct it.

Axiom 3. Consider two systems A and B with associated Hilbert spaces $\mathcal{H}_A$ and $\mathcal{H}_B$. Then, the state of the composite system as well as its observables act onto the tensor product Hilbert space $\mathcal{H}_{AB} = \mathcal{H}_A \otimes \mathcal{H}_B$.

This axiom has the following consequence. Imagine that the systems A and B interact during some time in such a way that they can no longer be described by independent states $\hat{\rho}_A$ and $\hat{\rho}_B$ acting on $\mathcal{H}_A$ and $\mathcal{H}_B$, respectively, but by a state $\hat{\rho}_{AB}$ acting on the joint space $\mathcal{H}_{AB}$. After the interaction, system B is kept isolated from any other system, but system A is given to an observer, who is therefore able to measure observables defined in $\mathcal{H}_A$ only, and might not even know that system A is part of a larger system. The question is, is it possible to reproduce the statistics of the measurements performed on system A with some state $\hat{\rho}_A$ acting on $\mathcal{H}_A$ only? This question has a positive and *unique* answer: this state is given by the *reduced density operator* $\hat{\rho}_A = \mathrm{tr}_B\{\hat{\rho}_{AB}\}$, that is, by performing the partial trace[17] with respect to system B's subspace onto the joint state.

1.3.5 Axiom 4. Quantization rules

The introduction of the fourth axiom is motivated by the following fact. The class of self-adjoint operators forms a real vector space with respect to the addition of operators and the multiplication of an operator by a real number. Using the commutator we can also build an operation that takes two self-adjoint operators $\hat{A}$ and $\hat{B}$ to generate another self-adjoint operator $\hat{C} = \mathrm{i}[\hat{A}, \hat{B}]$, which, in addition, satisfies all the properties required by a Lie product. Hence, even if classical and quantum theories seem fundamentally different, it seems that observables are treated similarly within their corresponding mathematical frameworks: they are elements of a Lie algebra.

On the other hand, we saw that the generalized coordinates and momenta have a particular mathematical structure in the Hamiltonian formalism, they are the generators of the Heisenberg group. It seems then quite reasonable to ask for the same in the quantum theory, so that at least concerning observables both theories are equivalent. This motivates the fourth axiom:

[17] Given an orthonormal basis $\{|b_j\rangle\}_j$ of $\mathcal{H}_B$, this is defined by

$$\mathrm{tr}_B\{\hat{\rho}_{AB}\} = \sum_j \langle b_j|\hat{\rho}_{AB}|b_j\rangle, \tag{1.29}$$

which is indeed an operator acting on $\mathcal{H}_A$.

Axiom 4. Consider a physical system which is described classically within a Hamiltonian formalism by a set of generalized coordinates $\mathbf{q} = \{q_j\}_{j=1}^n$ and momenta $\mathbf{p} = \{p_j\}_{j=1}^n$ at a given time. Within the quantum formalism, the corresponding observables $\hat{\mathbf{q}} = \{\hat{q}_j\}_{j=1}^n$ and $\hat{\mathbf{p}} = \{\hat{p}_j\}_{j=1}^n$ satisfy the *canonical commutation relations*

$$\left[\hat{q}_j, \hat{p}_l\right] = i\hbar\delta_{jl} \text{ and } \left[\hat{q}_j, \hat{q}_l\right] = \left[\hat{p}_j, \hat{p}_l\right] = 0. \quad (1.30)$$

The constant $\hbar$ is included because, while the Poisson bracket $\{q_j, p_l\}$ has no units, the commutator $[\hat{q}_j, \hat{p}_l]$ has units of action. That it is exactly $\hbar$ the proper constant can be seen only once the theory is compared with experiments.

We can now discuss how to build the self-adjoint operator corresponding to a given observable. In general, meaningful observables are built from symmetry principles [23], e.g. the kinetic and angular momenta as the generators of space translations and rotations, respectively. An alternative route might be taken when the observable is well-defined classically. Suppose that in the Hamiltonian formalism the observable A is represented by the real phase-space function $A(\mathbf{q}, \mathbf{p})$. It seems quite natural to use then $A(\hat{\mathbf{q}}, \hat{\mathbf{p}})$ as the corresponding quantum operator. However, this correspondence faces a lot of troubles resulting from the fact that, while coordinates and momenta commute in classical mechanics, they do not in quantum mechanics. For example, given the classical observable $A = qp = pq$, we could be tempted to assign to it any of the quantum operators $\hat{A}_1 = \hat{q}\hat{p}$ or $\hat{A}_2 = \hat{p}\hat{q}$. These two operators are not equivalent (they do not commute) and they are not even self-adjoint, and hence cannot represent observables. One possible solution to this problem, at least for observables with a series expansion, is to always symmetrize the classical expressions with respect to coordinates and momenta, so that the resulting operator is self-adjoint. Applied to our previous example, we should take $\hat{A} = (\hat{p}\hat{q} + \hat{q}\hat{p})/2$ according to this rule. This simple procedure leads to the correct results most of the times, and when it fails (for example, if the classical observable does not have a series expansion) it was proved by Groenewold [25] that it is possible to make a faithful systematic correspondence between classical observables and self-adjoint operators by using more sophisticated correspondence rules.

Of course, when the observable corresponds to a degree of freedom which is not defined in a classical context (such as *spin*), it must be built from scratch based on experimental observations and/or first principles.

Note that the commutation relations between coordinates and momenta makes them satisfy the uncertainty relation $\Delta q \Delta p \geqslant \hbar/2$, and hence, if one of them is well defined in the system, the other must have statistics very spread around the mean.

1.3.6 Axiom 5. *Free* evolution of the system

The previous axioms have served to define the mathematical structure of the theory and its relation to physical systems. We have not said anything yet about how quantum mechanics treats the evolution of the system. As we are about to see, the formalism treats very differently the evolution due to a measurement performed by an observer, and the *free* evolution of the system when it is not subject to observation. The following axiom specifies how to deal with the latter case. Just as with the previous axiom, it feels pretty reasonable to keep the analogy with the Hamiltonian formalism, a motivation which comes also from the fact that, as stated, quantum mechanics must converge to classical mechanics in some limit. In the Hamiltonian formalism, observables evolve according to (1.7), so that making the correspondence between the classical and quantum Lie products as in the previous axiom, we enunciate the fifth axiom:

Axiom 5. The evolution of an observable $\hat{A}(\hat{\mathbf{q}}, \hat{\mathbf{p}}, \ldots; t)$ is given by

$$\mathrm{i}\hbar\frac{\mathrm{d}\hat{A}}{\mathrm{d}t} = \left[\hat{A}, \hat{H}\right] + \frac{\partial\hat{A}}{\partial t}, \tag{1.31}$$

which is known as the Heisenberg equation, and where $\hat{H}(\hat{\mathbf{q}}, \hat{\mathbf{p}},\ldots;t)$ is the self-adjoint operator corresponding to the Hamiltonian of the system. Note that the notation '$\hat{\mathbf{q}}, \hat{\mathbf{p}}, \ldots$' emphasizes the fact that the observable may depend on fundamental operators other than the generalized coordinates, e.g. purely quantum degrees of freedom such as spin.

For the case of an observable and a Hamiltonian with no explicit time-dependence (as will be assumed from now on), this evolution equation admits the explicit solution

$$\hat{A}(t) = \hat{U}^\dagger(t)\hat{A}(0)\hat{U}(t), \quad \text{being } \hat{U}(t) = \exp\left[\hat{H}t/\mathrm{i}\hbar\right], \tag{1.32}$$

a unitary operator called the *evolution operator*. For explicitly time-dependent Hamiltonians it is still possible to solve formally the Heisenberg equation as a *Dyson series*, but we will not worry about this case, as it does not appear throughout the monograph. Let us remark that this type of evolution ensures that if the canonical commutation relations (1.30) are satisfied at some time, they will be satisfied at all times.

Note that within this formalism the state $\hat{\rho}$ of the system is fixed in time, the observables are the ones which evolve. On the other hand, we have seen that, concerning observations (experiments), only expectation values of operators are relevant; and for an observable $\hat{A}$ at time t, these can be written as

$$\langle\hat{A}(t)\rangle = \mathrm{tr}\left\{\hat{\rho}\hat{A}(t)\right\} = \mathrm{tr}\left\{\hat{U}(t)\hat{\rho}\hat{U}^\dagger(t)\hat{A}(0)\right\}, \tag{1.33}$$

where in the last equality we have used the cyclic property of the trace. This expression shows that, instead of treating the observable as the evolving operator, we can define a new state at time t given by

$$\rho(t) = \hat{U}(t)\hat{\rho}(0)\hat{U}^\dagger(t), \tag{1.34}$$

while keeping fixed the operator. In differential form, this expression reads

$$i\hbar\frac{d\hat{\rho}}{dt} = \left[\hat{H}, \hat{\rho}\right], \tag{1.35}$$

which is known as the *von Neumann equation*. When the system is in a pure state $|\psi\rangle$, the following evolution equation is derived for the state vector itself

$$i\hbar\frac{d}{dt}|\psi\rangle = \hat{H}|\psi\rangle, \tag{1.36}$$

which is known as the *Schrödinger equation*, from which the state at time t is found as $|\psi(t)\rangle = \hat{U}(t)|\psi(0)\rangle$.

Therefore, we have two different but equivalent evolution formalisms or *pictures*. In one, which we will call the *Heisenberg picture*, the state of the system is fixed, while observables evolve according to the Heisenberg equation. In the other, which we will call the *Schrödinger picture*, observables are fixed, while states evolve according to the von Neumann equation.

1.3.7 Axiom 6. Post-measurement state

The previous postulate specifies how the free evolution of the system is taken into account in the quantum-mechanical formalism. It is then left to specify how the state of the system evolves after a measurement is performed on it. For reasons that we will briefly review after enunciating the axiom, this is probably the most controversial point in the quantum formalism. Indeed, while in classical physics we assume that it is possible to perform measurements onto the system without disturbing its state, this final quantum-mechanical axiom states:

> **Axiom 6.** If upon a measurement of a countable observable A the outcome a_m is obtained, then immediately after the measurement the state of the system *collapses* to $|a_m\rangle$.

Before commenting on the controversial aspects of this axiom, it is important to mention some operational and conceptual aspects that will be important later. Let us denote by $\hat{\rho}$ and $\hat{\rho}_m$ the states before and after the measurement is performed. It is convenient to define the *unnormalized post-measurement state* $\tilde{\rho}_m = \hat{P}_m\hat{\rho}\hat{P}_m$, where we recall that $\hat{P}_m = |a_m\rangle\langle a_m|$ is a projector. The normalized post-measurement state can be obtained then as $\hat{\rho}_m = p_m^{-1}\tilde{\rho}_m$, where the probability of obtaining the a_m outcome can be evaluated as the trace of the unnormalized state, $p_m = \text{tr}\{\tilde{\rho}_m\}$.

The axiom assumes that, after the measurement, the observer gains knowledge about the measurement outcome, which we will denote as a *selective* measurement. However, suppose that for some reason the user interface of the measurement device does not allow us to distinguish between a set of outcomes $\{a_{m_k}\}_{k=1,2,\ldots,K}$, which we will denote by a *partially selective* measurement. Then, after the corresponding experimental outcome is obtained, the best estimate that the observer can assign to the post-measurement state is the ensemble decomposition $\{\bar{p}_{m_k}, |a_{m_k}\rangle\}_{k=1,2,\ldots,K}$ with relative probabilities $\bar{p}_{m_k} = p_{m_k}/\sum_{k=1}^{K} p_{m_k}$, since the real outcome is unknown, but the *a priori* probabilities p_m of the possible outcomes are known. Hence, in such case we would assign the post-measurement state $\hat{\rho}_{\{m_1,m_2,\ldots,m_K\}} = \sum_{k=1}^{K} \bar{p}_{m_k}\hat{\rho}_{m_k}$ to the system. The extreme case in which the outcome of the measurement is simply not recorded, so that we cannot know which outcome occurred and the best estimate for the post-measurement state is $\hat{\rho}' = \sum_{m=1}^{d} p_m\hat{\rho}_m = \sum_{m=1}^{d} \tilde{\rho}_m$, is known as a *non-selective* measurement.

So far we have considered the post-measurement state in the case of measuring a countable observable. The continuous case is tricky, since, as mentioned in section 1.2.3, the eigenvectors of a continuous observable cannot correspond to physical states (they cannot be normalized). On the other hand, one can always argue that the detection of a single definite value out of the spectrum $\{x\}_{x\in\mathbb{R}}$ of a continuous observable $\hat{X}$ would require an infinite precision, whereas detectors always have some finite precision. Consequently, there are two natural ways of dealing with such a problem:

- Accepting that measuring continuous observables is simply not possible, and what is measured in real experiments is always some countable version of them, which only in some unphysical limit reproduce the continuous measurement precisely. An example of this consists in the process of *binning* the continuous observable, which assumes that the detector can only distinguish between pixels with width Δ_x centered at certain points $\{x_k = k\Delta_x\}_{k\in\mathbb{Z}}$ in the spectrum of the continuous observable, so what is measured is instead the countable observable

$$\hat{X}_{\text{count}} = \sum_{k=-\infty}^{\infty} x_k|x_k\rangle\langle x_k|, \quad \text{with } |x_k\rangle = \frac{1}{\sqrt{\Delta_x}}\int_{x_k-\Delta_x/2}^{x_k+\Delta_x/2} dx|x\rangle. \tag{1.37}$$

- Allowing for the possible outcomes of the measurement to still be continuous, but with an uncertainty given by the precision of the measurement device. In the case of starting with a pure state, this would simply mean that the post-measurement state is not an eigenvector of the continuous operator, but a normalizable superposition of several of them, spanning around the measured value with a width given by the measurement's precision. This intuitive approach can be formalized with the theory of generalized quantum measurements that we will see later in this book [26, 27].

In any case, it is sometimes useful for theoretical calculations to proceed as if perfectly precise measurements were possible, with the system collapsing to one eigenvector of the continuous observable. However, it is important to keep in mind that this is just an unphysical idealization, whose corrections have to be taken into account when applying it to a real situation.

We can pass now to discuss the controversial aspects of this axiom, of which a pedagogical introduction can be found in [22] (see also [28] and appendix E of [7]).

In short, the problem is that, even though it leads to predictions which fully agree with the observations, the axiom somehow creates an *inconsistency* in the theory because of the following argument. According to axiom 5, the unitary evolution of a system not subject to observation is *reversible*, that is, one can always change the sign of the relevant terms of the Hamiltonian which contributed to the evolution (at least conceptually), and come back to the original state. On the other hand, the *collapse* axiom claims that when the system is put in contact with a measurement device and an observable is measured, the state of the system collapses to some other state in an *irreversible* way[18]. However, coming back to axiom 5, the whole measurement process could be described reversibly by considering, in addition to the system's particles, the evolution of all the particles forming the measurement device (or even the human who is observing the measurement outcome if needed!), that is, the Hamiltonian for the whole 'observed system + measurement device' scenario. Hence, it seems that, when including the collapse axiom, quantum mechanics allows for two completely different descriptions of the measurement process, one reversible and one irreversible, without giving a clear rule for when to apply each. It is in this sense that the theory contains an inconsistency.

There are three[19] main positions that physicists have taken regarding how this inconsistency might be solved, which we will refer too as *objective*, *subjective*, and *apparent* collapse interpretations, and whose (highly simplified) main ideas we discuss here[20]:

- **Objective collapse.** There is a clear boundary (yet to be found) between the quantum and the classical worlds. In the classical world, to which measurement devices and observers belong, there exists some *decoherence mechanism* that prevents systems from being in a superposition of states corresponding to mutually exclusive values of their observables. When the measurement device enters in contact with the quantum system, the latter becomes a part of the classical world, and the aforementioned decoherence mechanism forces its collapse. Hence, within this interpretation the collapse is pretty much a real physical process that we still need to understand along with the quantum/classical boundary. There are several *collapse theories* available at the moment [28], some of which we expect to be able to falsify or confirm in the near future with modern quantum technologies based on, for example, opto-, electro-, or magneto-mechanics [31].

[18] Note that in the literature the terms *reversible* and *irreversible* are sometimes replaced by *linear* and *non-linear*, referring to the fact that unitary evolution comes from a linear equation (Schrödinger or von Neumann equations), while measurement-induced dynamics becomes non-linear through its dependence on the probability of the possible outcomes, which in turn depend on the state.

[19] But each with many sub-interpretations differing in subtle, or even not so subtle points. In essence, one can find a lot of truth to the saying 'Give me a room with N physicists and I'll find you $N + 1$ different interpretations of quantum mechanics'.

[20] One interpretation of quantum mechanics that is formulated with a completely different set of axioms, and hence does not fit this list is *Bohmian mechanics* [29, 30], in which particles follow deterministic trajectories, but determined from a *guiding wave* that obeys the Schrödinger equation.

- **Subjective collapse.** The state is simply a mathematical object which conveniently describes the statistics of experiments, but that otherwise has no physical significance. As such, what the quantum formalism provides is simply a set of rules for how to update our best estimate to the state according to the information that we have about the system. In this sense, the collapse is just the way that an observer subjectively updates the state of the system after gathering the information concerning the outcome. Quantum Bayesianism or *QBism* [32, 33] is possibly the most refined of such interpretations, and has gathered a lot of momentum in recent years.
- **Apparent collapse.** The measurement can be described without the need to abandon the framework of axiom 5 as a joint unitary transformation onto the system and the measurement device (even including the observer), leading to a final entangled state[21] between those in which the eigenstates of the system's observable are in one-to-one correspondence with a set of macroscopically distinct *pointer* states of the measurement device [22]. Hence, after a measurement, reality splits into many *branches* where observers experience different outcomes and which stay in a quantum superposition, and collapse appears just an illusion coming from the fact that we only see the effective dynamics projected into the corresponding branch that we are experiencing. This approach finds its best-developed expression in the so-called *many-worlds interpretation* [34], which describes the quantum-mechanical framework as unitary evolution of a pure state (*wave function*) of the whole Universe, which includes all the branches or *worlds*[22].

In any case, whether of objective, subjective, or apparent value, it is clear that the collapse axiom is of great *operational* value, that is, it is currently the easiest successful way of analyzing schemes involving measurements, and hence we will apply it when needed.

1.3.8 The von Neumann entropy

An object of fundamental relevance in the theory of quantum information is the *von Neumann entropy* of a state $\hat{\rho}$, which is defined as

$$S[\hat{\rho}] = -\mathrm{tr}\{\hat{\rho}\log\hat{\rho}\}. \tag{1.38}$$

Given the diagonal representation of the state

$$\hat{\rho} = \sum_{j=1}^{d} \lambda_j |r_j\rangle\langle r_j|, \tag{1.39}$$

[21] We will discuss the concept of entanglement in detail in the next chapter.

[22] Believers of the many-worlds interpretation, or even people who are not sure of their quantum-mechanical beliefs (better safe than sorry), are strongly encouraged to use the *quantum world splitter* when they need to choose between equally reasonable options in life: www.qol.tau.ac.il.

where its eigenvectors $\{|r_j\rangle\}_{j=1,2,\ldots,d}$ form an orthonormal basis of the Hilbert space $\mathcal{H}$ (which therefore has dimension d), the von Neumann entropy is just the Shannon entropy of the distribution[23] $\boldsymbol{\lambda} = \text{col}(\lambda_1, \lambda_2, \ldots, \lambda_d)$, that is,

$$S[\hat{\rho}] = -\sum_{j=1}^{d} \lambda_j \log \lambda_j. \tag{1.40}$$

For a pure state $|\varphi\rangle\langle\varphi|$ the entropy is zero, while it has a maximum $\log d$ for the *maximally mixed state*[24]

$$\hat{\rho}_{\text{MM}} = \frac{1}{d}\hat{I}. \tag{1.41}$$

This suggests that the von Neumann entropy can be understood as a measure of the mixedness of the state. Note that when the state of the system is $\hat{\rho}_{\text{MM}}$ and some observable $\hat{A}$ is measured, all its eigenvalues are equally likely to appear as an outcome of the measurement, that is,

$$p_j = \langle a_j|\hat{\rho}_{\text{MM}}|a_j\rangle = \frac{1}{d} \quad \forall j. \quad \text{(flat distribution)} \tag{1.42}$$

This reinforces the interpretation of mixedness as due to some kind of information loss.

The entropy is a strictly concave functional of density operators, that is, given the convex mixture of states

$$\hat{\rho} = \sum_{m=1}^{M} w_m \hat{\rho}_m, \tag{1.43}$$

where $\{w_m\}_{m=1,2,\ldots,M}$ is a probability distribution and $\{\hat{\rho}_m\}_{m=1,2,\ldots,M}$ is a set of density operators, we have [24]

$$S[\hat{\rho}] \geqslant \sum_{m=1}^{M} w_m S[\hat{\rho}_m], \tag{1.44}$$

where the equality holds only when all the states ρ_m with non-zero probability w_m are equal. This property agrees very well with the interpretation of mixedness as coming from information loss: the average of the information contained in each state $\hat{\rho}_m$ cannot be smaller than the information contained in the state $\hat{\rho}$ obtained by scrambling them according to the probability distribution $\mathbf{w}$, since the latter imposes an extra level of randomness.

It is interesting to note (and easy to prove mathematically [24]) that the entropy does not change by unitary evolution, can only increase by non-selective projective

---

[23] In the following we use the notations $\mathbf{p} = \text{col}(p_1, p_2, \ldots, p_d)$ and $\{p_j\}_{j=1,2,\ldots,d}$ for a probability distribution interchangeably. Remember that 'col' is short for 'column vector'.

[24] The case $d \to \infty$ will be discussed later when studying infinite-dimensional spaces in detail in the last chapter.

measurements, and can only decrease by selective projective measurements (indeed, when no degeneracies are present, it collapses to zero, as the state becomes pure). These properties are indeed expected from a purely informational point of view: while evolving 'freely' the system does not exchange any information with any other system; when we perform a selective measurement indeed we gain information about the system and its post-measurement state; when the measurement is non-selective, information can become 'lost' in the measurement device, but we definitely can never obtain more information about the system than we already had prior to the measurement.

Bibliography

[1] Goldstein H, Poole C and Safko J 2001 *Classical Mechanics* (Reading, MA: Addison-Wesley)
[2] Greiner W 1989 *Classical Mechanics: Point Particles and Relativity* (Berlin: Springer)
[3] Greiner W and Reinhardt J 1989 *Classical Mechanics: Systems of Particles and Hamiltonian Dynamics* (Berlin: Springer)
[4] Hand L N and Finch J D 1998 *Analytical Mechanics* (Cambridge: Cambridge University Press)
[5] Dirac P A M 1930 *The Principles of Quantum Mechanics* (Oxford: Oxford University Press)
[6] Prugovečky E 1971 *Quantum Mechanics in Hilbert Space* (New York: Academic)
[7] Galindo A and Pascual P 1990 *Quantum Mechanics* vol 1 (Berlin: Springer)
[8] Gelfand I M and Vilenkin N Y 1964 *Generalized Functions* vol 4 (New York: Academic)
[9] Whitaker A 1996 *Einstein, Bohr, and the Quantum Dilemma* (Cambridge: Cambridge University Press)
[10] van der Waerden B L 1968 *Sources of Quantum Mechanics* (New York: Dover)
[11] Schrödinger E 1926 An undulatory theory of the mechanics of atoms and molecules *Phys. Rev.* **28** 1049
[12] Heisenberg W 1925 Über quantentheoretische Umdeutung kinematischer und mechanischer Beziehungen *Z. Phys.* **33** 879
[13] Born M and Jordan P 1925 Zur Quantenmechanik *Z. Phys.* **34** 858
[14] Born M, Heisenberg W and Jordan P 1926 Zur Quantenmechanik II *Z. Phys.* **35** 557
[15] Dirac P A M 1926 The fundamental equations of quantum mechanics *Proc. R. Soc.* A **109** 642
[16] von Neumann J 1932 *Mathematische Grundlagen der Quantenmechanik* (Berlin: Springer)
[17] von Neumann J 1955 *Mathematical Foundations of Quantum Mechanics* (Princeton, NJ: Princeton University Press)
[18] Cohen-Tannoudji C, Diu B and Laloë F 1977 *Quantum Mechanics* vol 1 (New York: Wiley)
[19] Cohen-Tannoudji C, Diu B and Laloë F 1977 *Quantum Mechanics* vol 2 (New York: Wiley)
[20] Greiner W and Müller B 1989 *Quantum Mechanics: An Introduction* (Berlin: Springer)
[21] Greiner W 1989 *Quantum Mechanics: Symmetries* (Berlin: Springer)
[22] Basdevant J-L and Dalibard J 2002 *Quantum Mechanics* (Berlin: Springer)
[23] Ballentine L E 1998 *Quantum Mechanics: A Modern Development* (Singapore: World Scientific)
[24] Nielsen M A and Chuang I L 2000 *Quantum Information and Quantum Computation* (Cambridge: Cambridge University Press)
[25] Groenewold H J 1946 On the principles of elementary quantum mechanics *Physica* **12** 405
[26] Wiseman H M and Milburn G J 2009 *Quantum Measurement and Control* (Cambridge: Cambridge University Press)

[27] Jacobs K and Steck D A 2006 A straightforward introduction to continuous quantum measurement *Contemp. Phys.* **47** 279
[28] Bassi A and Ghirardi G C 2003 Dynamical reduction models *Phys. Rep.* **379** 257
[29] Holland P R 1993 *The Quantum Theory of Motion* (Cambridge: Cambridge University Press)
[30] Oriols X and Mompart J 2012 *Applied Bohmian Mechanics: From Nanoscale Systems to Cosmology* (Singapore: Pan Stanford Publishing)
[31] Romero-Isart O 2011 Quantum superposition of massive objects and collapse models *Phys. Rev.* A **84** 052121
[32] Fuchs C A, Mermin N D and Schack R 2014 An introduction to QBism with an application to the locality of quantum mechanics *Am. J. Phys.* **82** 749
[33] Fuchs C A 2012 Interview with a quantum Bayesian arXiv: 1207.2141
[34] Vaidman L 2014 Many-worlds interpretation of quantum mechanics *The Stanford Encyclopedia of Philosophy* (Stanford, CA: Stanford University) http://plato.stanford.edu/entries/qm-manyworlds

Carlos Navarrete-Benlloch

Chapter 2

Bipartite systems and entanglement

In this chapter we will introduce some general concepts of entanglement theory, of which we will learn much more in section 4.4.4 when applying them to continuous-variable systems (including the origins of the theory, which can be traced back to the seminal work of Einstein, Podolsky, and Rosen [1]). For a more exhaustive and detailed introduction to the world of entanglement, see [2–4].

2.1 Entangled states

Consider two systems A and B (named after Alice and Bob, two observers who are able to interact locally with their respective system) with associated Hilbert spaces $\mathcal{H}_A$ and $\mathcal{H}_B$ of dimensions d_A and d_B, respectively. The systems are prepared in some state $\hat{\rho}_{AB}$ acting on the joint space $\mathcal{H}_A \otimes \mathcal{H}_B$. Recall from the discussion after axiom 3 that Alice and Bob can reproduce the statistics of measurements performed on their subsystems via the reduced states $\hat{\rho}_A = \mathrm{tr}_B\{\hat{\rho}_{AB}\}$ and $\hat{\rho}_B = \mathrm{tr}_A\{\hat{\rho}_{AB}\}$, respectively.

When the state of the joint system is of the type $\hat{\rho}_{AB}^{(\mathrm{prod})} = \hat{\rho}_A \otimes \hat{\rho}_B$, that is, a tensor product of two arbitrary density operators, the actions performed by Alice on system A will not affect Bob's system, the statistics of which are given by $\hat{\rho}_B$, no matter what the actual state $\hat{\rho}_A$ is. In this case A and B are completely *uncorrelated*. For any other type of joint state, A and B will share some kind of correlation.

Correlations appear naturally also within the classical framework. Hence, a problem of paramount relevance in quantum information is understanding which types of correlations can appear at a classical level, and which are purely quantum. This is because only if the latter are present, can one expect to use the correlated systems for quantum-mechanical applications which go beyond what is classically possible, the paradigmatic example being the exponential speed up of computational algorithms.

doi:10.1088/978-1-6817-4405-6ch2

Intuitively, the state of the system will only induce classical correlations between A and B when it can be written in the form

$$\hat{\rho}_{AB}^{(\text{sep})} = \sum_{m=1}^{M} w_m \hat{\rho}_A^{(m)} \otimes \hat{\rho}_B^{(m)}, \tag{2.1}$$

where the $\hat{\rho}^{(m)}$ are density operators, $\{w_m\}_{k=1,2,\ldots,M}$ is a probability distribution, M can be infinite, and the index m can even be continuous in some range, in which case the sum turns into an integral in that range and the probability distribution into a probability density function. Indeed, a state of the type (2.1) can be prepared by a protocol involving only local actions and classical correlations: Alice and Bob can share a classical machine which randomly picks a value of m according to the distribution $\{w_m\}_{m=1,2,\ldots,M}$, and use it to trigger the preparation of the states $\hat{\rho}_A^{(m)}$ and $\hat{\rho}_B^{(m)}$, which can be done locally, and hence cannot induce further correlations. If the process is automatized so that Alice and Bob do not learn the outcome m of the random number generator, the best estimate that they can assign to the state is the mixture $\hat{\rho}_{AB}^{(\text{sep})}$. In other words, the state does not contain quantum correlations if it can be prepared using only *local operations and classical communication* (LOCC)[1].

There is yet another intuitive way of justifying that states which cannot be written in the separable form (2.1) will make A and B share quantum correlations. The idea is based on what is probably the most striking difference between classical and quantum mechanics: the *superposition principle*, that is, the possibility of states corresponding to *mutually exclusive* properties of the system to *interfere* (e.g. two different colors, $|\text{blue}\rangle + |\text{red}\rangle$). Hence, it is intuitive that correlations should have a quantum nature only when they come from some kind of superposition of joint states corresponding to mutually exclusive properties of the correlated systems (e.g. color in one system and flavor in the other, $|\text{blue}\rangle \otimes |\text{sweet}\rangle + |\text{red}\rangle \otimes |\text{sour}\rangle$), in which case the state cannot be written as a tensor product of two independent states ($|\text{blue}\rangle \otimes |\text{sour}\rangle$) or as a purely classical statistical mixture of these ($|\text{blue}\rangle\langle\text{blue}| \otimes |\text{sweet}\rangle\langle\text{sweet}| + |\text{red}\rangle\langle\text{red}| \otimes |\text{sour}\rangle\langle\text{sour}|$).

States of the type $\hat{\rho}_{AB}^{(\text{sep})}$ are called *separable*. Any *inseparable* state will induce quantum correlations between A and B. These correlations which cannot be generated by classical means are known as *entanglement*, and states which are not separable are called *entangled states*.

Given a bipartite state $\hat{\rho}_{AB}$, with corresponding reduced states $\hat{\rho}_A$ and $\hat{\rho}_B$, it is interesting to note that [5]

$$S(\hat{\rho}_{AB}) \leqslant S(\hat{\rho}_A) + S(\hat{\rho}_B), \tag{2.2}$$

with the equality holding only when the systems A and B are uncorrelated, that is, $\hat{\rho}_{AB} = \hat{\rho}_A \otimes \hat{\rho}_B$. This property is known as the *subadditivity* of the von Neumann entropy, and clearly agrees with intuitive arguments based on information loss: the information contained in the state of the whole system cannot be obtained from the

[1] We will give a more precise meaning for this class of operations in the next chapter.

information left in the reduced states, since we have lost the information present in the correlations, which could not even be judged as being either classical or quantum merely from the reduced states.

2.2 Characterizing and quantifying entanglement

In general, given a mixed state $\hat{\rho}_{AB}$ acting on $\mathcal{H}_A \otimes \mathcal{H}_B$, it is hard to find out whether it is separable or not, the difficulty coming from distinguishing between quantum and classical correlations. Indeed, the best known criterion for separability, the *Peres–Horodecki criterion* [6, 7], yields only necessary and sufficient conditions when $d_A \times d_B \leqslant 6$ (note that this includes the case of two qubits), and for a reduced class of states in infinite-dimensional Hilbert spaces (some Gaussian states, see chapter 4). This criterion states that a necessary condition for the separability of a density operator is that it remains positive after the operation of partial transposition, that is, given

$$\hat{\rho}_{AB} = \sum_{jk=1}^{d_A} \sum_{lm=1}^{d_B} \rho_{jl,km} |a_j, b_l\rangle\langle a_k, b_m|, \tag{2.3}$$

where $\{|a_j, b_l\rangle = |a_j\rangle \otimes |b_l\rangle\}_{l=1,2,\ldots,d_B}^{j=1,2,\ldots,d_A}$ is an orthonormal basis of $\mathcal{H}_A \otimes \mathcal{H}_B$,

$$\hat{\rho}_{AB}^{T_B} = \sum_{jk=1}^{d_A} \sum_{lm=1}^{d_B} \rho_{jm,kl} |a_j, b_l\rangle\langle a_k, b_m|, \tag{2.4}$$

is a positive operator. We will learn more about this criterion and some more when studying infinite-dimensional Hilbert spaces in chapter 4.

A very different problem is that of quantifying the level of correlations present in the state, and more importantly, how much of these correspond to entanglement. Even though we understand fairly well the conditions that a proper *entanglement measure* $E[\hat{\rho}_{AB}]$ must satisfy, we have not found a completely satisfactory one for general states [2, 3] (either they do not satisfy all the conditions, or/and can only be efficiently computed for restricted classes of states). It is not the intention of this book to introduce in detail all these measures and explain up to what point they are satisfactory, but it is interesting to spend a few lines thinking about this issue, as it will allow us to obtain a better picture of what entanglement means (see [2, 3] for more details).

The basic conditions that a good entanglement measure $E[\hat{\rho}_{AB}]$ should satisfy are actually quite intuitive:

1. $E[\hat{\rho}_{AB}]$ is positive definite and equal to zero for separable states.
2. Given the mixture $\hat{\rho}_{AB} = \sum_j p_j \hat{\rho}_j$, where $\mathbf{p}$ is a probability distribution and $\{\hat{\rho}_j\}_j$ are density operators acting on $\mathcal{H}_A \otimes \mathcal{H}_B$, $E[\hat{\rho}_{AB}] \leqslant \sum_j p_j E[\hat{\rho}_j]$, which is to say that the entanglement of a collection of states cannot be increased by not knowing which one of them has been prepared, since this is corresponds to classical information.
3. At least on average, the entanglement level cannot increase when Alice and Bob apply protocols involving only LOCC.

These three conditions define what is known as an *entanglement monotone*. By themselves, they are not enough to define a unique entanglement measure even for pure states. However, by adding two more conditions known as *weak additivity* and *weak continuity* [3], which find an intuitive justification in the asymptotic limit of having infinitely many copies of the state, it is possible to prove that the *entanglement entropy* is the unique entanglement measure of pure bipartite states $|\psi\rangle_{AB}$. This measure is very intuitive, as it simply evaluates how mixed the reduced density operator of one of the parties remains after tracing out the other, that is, given the reduced density operators $\hat{\rho}_A = \text{tr}_B\{|\psi\rangle_{AB}\langle\psi|\}$ or $\hat{\rho}_B = \text{tr}_A\{|\psi\rangle_{AB}\langle\psi|\}$, this entanglement measure can be evaluated as

$$E[|\psi\rangle_{AB}] = S[\hat{\rho}_A] = S[\hat{\rho}_B]. \tag{2.5}$$

The equality of the von Neumann entropies of the reduced states will be clear after the following section. Hence, the problem of quantifying entanglement is basically solved for pure bipartite states. Pure states whose corresponding reduced states are maximally mixed are known as *maximally entangled states.*

The case of pure states allows for such a simple entanglement measure because all the correlations in the state are quantum, in the sense that they come from correlations at the interference level, as the only pure separable states are of the $|\psi_A\rangle \otimes |\psi_B\rangle$ form, which shows no interference terms between the partitions. The complication with mixed states, which can always be written from some ensemble decomposition $\{w_m, |\varphi_m\rangle_{AB}\}_{m=1,2,\ldots,M}$ as explained in section 1.3.3, comes from the fact that part of the correlations can be due to the mixture of pure states according to the probability distribution $\mathbf{w}$, correlations which will therefore have a classical character. Indeed, note that for mixed states the entanglement entropy is not even an entanglement monotone, as the entanglement entropy of the product state $\hat{\rho}_A \otimes \hat{\rho}_B$ is just the entropy of the mixed states $\hat{\rho}_A$ or $\hat{\rho}_B$, which is not zero except for pure states, and furthermore, depends on which mode is traced out. However, there are many quantifiers which are entanglement monotones, of which we discuss examples in the following:

- Possibly the most natural entanglement measure for mixed states is the *distillable entanglement* [8, 9]. Suppose that we give N copies of the mixed state $\hat{\rho}_{AB}$ to Alice and Bob. The process of *distillation* refers to the conversion of these copies to copies of maximally entangled states via protocols involving only LOCC. The distillable entanglement is defined as the maximum number of maximally entangled states that can be distilled from infinitely many copies of $\hat{\rho}_{AB}$. Apart from an entanglement monotone, it can also be shown to satisfy the weak additivity and weak continuity conditions, and to be equal to the entanglement entropy for pure states. Its drawback is that it requires a maximization over all the possible distillation protocols, and it is therefore very difficult to evaluate (paraphrasing [3]: it is a problem ranging from 'difficult' to 'hopeless'.)
- The *entanglement of formation* [8, 10] can be seen as the dual to the distillable entanglement: it measures the number of maximally entangled states that are

needed to prepare infinitely many copies of the mixed state. It can be evaluated as

$$E_F[\hat{\rho}_{AB}] = \min_{\{w_k,|\psi_k\rangle\}_k} \sum_k w_k E[|\psi_k\rangle], \tag{2.6}$$

where the minimization is performed over all the possible ensemble decompositions of $\hat{\rho}_{AB}$, what makes it a very hard measure to evaluate too. Nevertheless, closed formulas have been obtained for the entanglement of formation of the general state of two qubits ($d_A \times d_B = 2 \times 2$) [11], as well as for reduced classes of higher-dimensional bipartite states with strong symmetries [12].

- There is an entanglement monotone which can be evaluated fairly efficiently, as it does not require any optimization procedure: the *logarithmic negativity* [13, 14]. It quantifies how much the state $\hat{\rho}_{AB}$ violates the Peres–Horodecki criterion via

$$E_N[\hat{\rho}_{AB}] = \log\left\|\hat{\rho}_{AB}^{T_B}\right\|_1 = \log\left[1 + \sum_j\left(|\tilde{\lambda}_j| - \tilde{\lambda}_j\right)\right], \tag{2.7}$$

where $\{\tilde{\lambda}_j\}_j$ are the eigenvalues of $\hat{\rho}_{AB}^{T_B}$, and $\|\hat{A}\|_1 = \mathrm{tr}\sqrt{\hat{A}\hat{A}^\dagger}$ denotes the so-called *trace norm*. The problem with this measure is that it does not collapse to the entanglement entropy (2.5) for pure states, and is not weakly additive in general.

2.3 Schmidt decomposition and purifications

When working with pure bipartite states, there is a very simple but powerful result known as the *Schmidt decomposition*, which says that this type of states can always be written in the form

$$|\psi\rangle = \sum_{j=1}^{d} \sqrt{\lambda_j}\,|u_j\rangle \otimes |v_j\rangle, \tag{2.8}$$

where $d = \min\{d_A, d_B\}$, $\{\lambda_j\}_{j=1,2,\ldots,d}$ is a probability distribution (the $\sqrt{\lambda_j}$ are known as *Schmidt coefficients*), and $\{|u_j\rangle\}_{j=1,2,\ldots,d}$ and $\{|v_j\rangle\}_{j=1,2,\ldots,d}$ form orthonormal sets in $\mathcal{H}_A$ and $\mathcal{H}_B$, respectively. Note that this decomposition strikes right at the heart of entanglement as a 'superposition of joint states corresponding to mutually exclusive properties of the correlated systems'.

The proof of this result is very simple, so let us sketch it here. Consider two orthonormal bases $\{|a_m\rangle\}_{m=1,2,\ldots,d_A}$ and $\{|b_n\rangle\}_{n=1,2,\ldots,d_B}$ in $\mathcal{H}_A$ and $\mathcal{H}_B$, respectively. We can expand the state as

$$|\psi\rangle = \sum_{m=1}^{d_A}\sum_{n=1}^{d_B} C_{mn}|a_m\rangle \otimes |b_n\rangle. \tag{2.9}$$

The expansion coefficients form a $d_A \times d_B$ matrix C, which according to the *singular-value decomposition theorem* [15] can always be written as $C = U\Lambda V^\dagger$. Here U and V are unitary[2] $d_A \times d_A$ and $d_B \times d_B$ matrices, respectively, and Λ is a diagonal $d_A \times d_B$ matrix with non-negative entries that we can then write as $\Lambda = \text{diag}_{d_A \times d_B}(\sqrt{\lambda_1}, \sqrt{\lambda_2}, \ldots, \sqrt{\lambda_d})$. Introducing this decomposition in (2.9), or element-wise $C_{mn} = \sum_{j=1}^{d_A}\sum_{l=1}^{d_B}\sqrt{\lambda_j}\, U_{mj}\delta_{jl}(V^\dagger)_{ln} = \sum_{j=1}^{d}\sqrt{\lambda_j}\, U_{mj} V_{nj}^*$, we obtain

$$|\psi\rangle = \sum_{j=1}^{d}\sqrt{\lambda_j}\underbrace{\sum_{m=1}^{d} U_{mj}|a_m\rangle}_{|u_j\rangle} \otimes \underbrace{\sum_{n=1}^{d} V_{nj}^*|b_n\rangle}_{|v_j\rangle} \tag{2.10}$$

which is exactly in the form (2.8). The orthonormality relations $\langle u_j|u_l\rangle = \delta_{jl} = \langle v_j|v_l\rangle$ follow directly from the unitarity of U and V, while the normalization of the *Schmidt distribution*, $\sum_{j=1}^{d}\lambda_j = 1$ follows from the normalization of the state, $\langle\psi|\psi\rangle = 1$.

With the state written in this form, it is completely trivial to evaluate the entanglement entropy: since the reduced density operators are diagonal, that is,

$$\hat{\rho}_A = \sum_{j=1}^{d}\lambda_j|u_j\rangle\langle u_j| \qquad \text{and} \qquad \hat{\rho}_B = \sum_{j=1}^{d}\lambda_j|v_j\rangle\langle v_j|, \tag{2.11}$$

their von Neumann entropies give

$$E[|\psi\rangle] = -\sum_{j=1}^{d}\lambda_j \log\lambda_j. \tag{2.12}$$

The number of non-zero Schmidt coefficients is known as the *Schmidt rank*. Separable states have then Schmidt rank equal to one. On the other hand, whenever $\lambda_j = 1/d\ \forall j$, the state will take the maximum entanglement possible $\log d$, corresponding to a maximally mixed reduced state in the Hilbert space with dimension d; under such conditions we then say that the system is in a *maximally entangled state*.

The Schmidt decomposition allows us to introduce the concept of *purification*: given a system A with associated Hilbert space $\mathcal{H}_A$ in a mixed state with diagonal representation $\hat{\rho}_A = \sum_{j=1}^{d}\lambda_j|r_j\rangle\langle r_j|$, we can always introduce another system B described by a Hilbert space $\mathcal{H}_B$ with the same dimension as $\mathcal{H}_A$, and with an orthonormal basis $\{|v_j\rangle\}_{j=1}^{d}$, and interpret the mixed state $\hat{\rho}_A$ as a reduction of the pure entangled state $|\psi\rangle_{AB} = \sum_{j=1}^{d}\sqrt{\lambda_j}|r_j\rangle \otimes |v_j\rangle$. Hence, a classical mixture of states can always be transformed into a pure entangled state in a 'doubled' Hilbert space.

[2] Note that in infinite dimensions C is a bounded matrix, since $|\psi\rangle$ must be normalized to one. Hence, a singular value decomposition still exists for it, but U and V are left-unitary.

Bibliography

[1] Einstein A, Podolsky B and Rosen N 1935 Can quantum-mechanical description of physical reality be considered complete? *Phys. Rev.* **47** 777

[2] Horodecki R, Horodecki P, Horodecki M and Horodecki K 2009 Quantum entanglement *Rev. Mod. Phys.* **81** 865

[3] Eisert J 2001 Entanglement in quantum information theory *PhD Thesis* arXiv: quant-ph/061025

[4] Eisert J and Plenio M B 2003 Introduction to the basics of entanglement theory in continuous-variable systems *Int. J. Quantum Inf.* **1** 479

[5] Nielsen M A and Chuang I L 2000 *Quantum Information and Quantum Computation* (Cambridge: Cambridge University Press)

[6] Peres A 1996 Separability criterion for density matrices *Phys. Rev. Lett.* **77** 1413

[7] Horodecki M, Horodecki P and Horodecki R 1996 Separability of mixed states: necessary and sufficient conditions *Phys. Lett.* A **223** 1

[8] Bennett C H, DiVincenzo D P, Smolin J A and Wootters W K 1996 Mixed-state entanglement and quantum error correction *Phys. Rev.* A **54** 3824

[9] Horodecki P and Horodecki R 2001 Distillation and bound entanglement *Quantum Inf. Comput.* **1** 45

[10] Wootters W K 2001 Entanglement of formation and concurrence *Quantum Inf. Comput.* **1** 27

[11] Wootters W K 1998 Entanglement of formation of an arbitrary state of two qubits *Phys. Rev. Lett.* **80** 2245

[12] Giedke G, Wolf M M, Krüger O, Werner R F and Cirac J I 2003 Entanglement of formation for symmetric Gaussian states *Phys. Rev. Lett.* **91** 107901

[13] Życzkowski K, Horodecki P, Sanpera A and Lewenstein M 1998 Volume of the set of separable states *Phys. Rev.* A **58** 883

[14] Vidal G and Werner R F 2002 Computable measure of entanglement *Phys. Rev.* A **65** 032314

[15] Bernstein D S 2005 *Matrix Mathematics: Theory Facts, and Formulas* (Princeton, NJ: Princeton University Press)

Chapter 3

Quantum operations

3.1 Basic principles of quantum operations

3.1.1 General considerations

Consider a system S with associated Hilbert space $\mathcal{H}_S$ subject to the actions of an experimentalist named Shane. *Quantum operations* refer, in essence[1], to the most general evolution that Shane can induce on the system S. As we explained in axioms 5 and 6, these surely involve unitary transformations and projective measurements[2] on the system, but are these the most general operations that Shane can apply?

The answer to this question is negative, since, as schematically shown in figure 3.1, Shane can always append an auxiliary system E with associated Hilbert space $\mathcal{H}_E$, apply unitaries and projective measurements on the joint system, and finally dismiss (trace out) the appended system E. Quantum operations then correspond to these joint unitaries and projective measurements as felt by the system S alone. In this context, it is customary to denote systems S and E by *system* and *environment*, respectively. We will refer to the combination of the system and the environment as the *joint system*.

As we will show in the last part of this section, any quantum operation can be represented by a map of the type [1, 2]

[1] In all fairness, they are not the *most* general type of evolution, since they do not capture some cases in which the system S interacts during the evolution with the systems with which it shared correlations initially, inducing a *non-completely positive map* on the state [1].

[2] We will refer to a measurement of a system observable as a *projective* measurement, to distinguish it from the *generalized* measurements that we will study in the next section, which are not described by a set of projectors.

doi:10.1088/978-1-6817-4405-6ch3

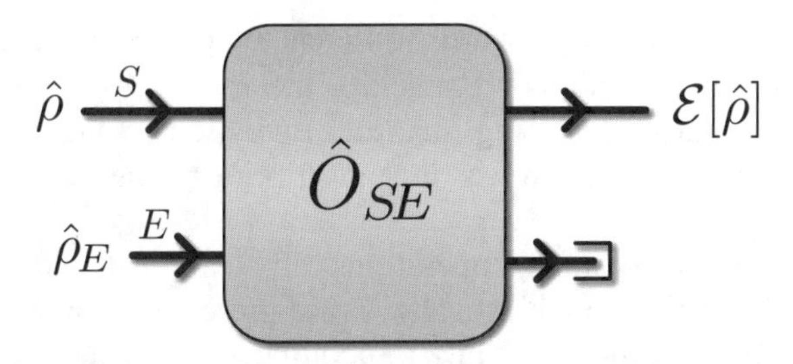

Figure 3.1. System–environment picture of quantum operations. The operator $\hat{O}_{SE}$ acting on the joint system can be a unitary or a projector, the latter being associated with a particular outcome in a measurement of a joint observable. The unitary or a non-selective measurement both lead to a trace-preserving quantum operation, while a selective measurement leads to a trace-decreasing one.

$$\mathcal{E}\left[\hat{Q}\right] = \sum_{k=1}^{K} \hat{E}_k \hat{Q} \hat{E}_k^{\dagger}, \tag{3.1}$$

where $\hat{Q}$ is any system operator, K can be selected at will by Shane, and the only restriction on the operators $\{\hat{E}_k\}_{k=1,2,\ldots,K}$ is

$$\sum_{k=1}^{K} \hat{E}_k^{\dagger} \hat{E}_k \leqslant \hat{I}, \tag{3.2}$$

where this type of operator inequalities, say $\hat{Q}_1 \leqslant \hat{Q}_2$, should be understood as $\hat{Q}_2 - \hat{Q}_1$ being positive semidefinite. If the state of the system was $\hat{\rho}$ prior to the quantum operation, it becomes

$$\hat{\rho}' = \frac{\mathcal{E}[\hat{\rho}]}{\mathrm{tr}\{\mathcal{E}[\hat{\rho}]\}}, \tag{3.3}$$

right after it. The operators $\{\hat{E}_k\}_{k=1,2,\ldots,K}$ are known as *Kraus operators*, and the expression (3.1) as an *operator-sum representation* of the quantum operation.

We can define two main types of quantum operations:

- Quantum operations which saturate (3.2) are known as *trace-preserving*, since $\mathrm{tr}\{\mathcal{E}[\hat{Q}]\} = \mathrm{tr}\{\hat{Q}\}$. As we will see in the next section, they correspond to operations that can be produced by performing a unitary transformation or a non-selective measurement on the joint system. Specifically, let us denote by $\hat{U}$ and $\hat{A} = \sum_{j=1}^{d_S \times d_E} a_j \hat{P}_j$ a unitary and an observable acting on the joint Hilbert space $\mathcal{H}_S \otimes \mathcal{H}_E$, respectively, $\{\hat{P}_j\}_{j=1,2,\ldots,d_S \times d_E}$ corresponding to the observable's eigenprojectors. Then the maps $\mathcal{E}_U[\hat{Q}] = \mathrm{tr}_E\{\hat{U}(\hat{Q} \otimes \hat{\rho}_E)\hat{U}^{\dagger}\}$ and $\mathcal{E}_A[\hat{Q}] = \mathrm{tr}_E\{\sum_{j=1}^{d_S \times d_E} \hat{P}_j(\hat{Q} \otimes \hat{\rho}_E)\hat{P}_j\}$, where $\hat{\rho}_E$ is some reference environmental state, correspond to trace-preserving quantum operations.
- Quantum operations which do not saturate (3.2) are known as *trace-decreasing*, since $\mathrm{tr}\{\mathcal{E}[\hat{Q}]\} < \mathrm{tr}\{\hat{Q}\}$. They require applying a selective measurement on the joint system, and the quantum operation is achieved only when a particular

outcome appears (in other words, different outcomes correspond to different trace-decreasing operations). Explicitly, the maps $\mathcal{E}_j[\hat{Q}] = \mathrm{tr}_E\{\hat{P}_j(\hat{Q} \otimes \hat{\rho}_E)\hat{P}_j\}$ correspond to trace-decreasing quantum operations, which occur with probability $\mathrm{tr}\{\mathcal{E}_j[\hat{\rho}]\}$ if the system was in state $\hat{\rho}$ prior to the joint measurement.

Hence, trace-preserving quantum operations can be applied deterministically, while trace-decreasing ones only probabilistically. Trace-preserving quantum operations are also known as *quantum channels*, while the combination of all the trace-decreasing operations associated to a measurement on the joint system is known as a *generalized measurement*, of which we will tell more in the next section.

3.1.2 Further properties of quantum operations

The following important properties of quantum operations can be proved [1, 2]:

- Quantum operations can be defined *axiomatically*. In particular, it is possible to show that a map $\mathcal{E}[\hat{\rho}]$ acting on the space of density operators of the system admits an operator-sum representation of the type (3.1) if and only if
  1. It is a convex-linear map, that is, $\mathcal{E}[\sum_k w_k\hat{\rho}_k] = \sum_k w_k\mathcal{E}[\hat{\rho}_k]$, where $\{w_k\}_k$ is a probability distribution and $\{\hat{\rho}_k\}_k$ are density operators.
  2. $0 \leqslant \mathrm{tr}\{\mathcal{E}[\hat{\rho}]\} \leqslant 1 \;\; \forall\hat{\rho}$.
  3. It is a completely positive (CP) map, that is, given a Hilbert space $\mathcal{H}_L$ of an arbitrary dimension, a product operator $\hat{O}_S \otimes \hat{O}_L$ and a positive-semidefinite operator $\hat{O}$ acting on the joint space $\mathcal{H}_S \otimes \mathcal{H}_L$, and the extended map defined by $\mathcal{E}_{\mathrm{ext}}[\hat{O}_S \otimes \hat{O}_L] = \mathcal{E}[\hat{O}_S] \otimes \hat{O}_L$, then $\mathcal{E}_{\mathrm{ext}}[\hat{O}]$ is also a positive-semidefinite operator for all $\hat{O}$.
- Two sets of Kraus operators $\{\hat{E}_k\}_{k=1,2,\ldots,K}$ and $\{\hat{F}_m\}_{m=1,2,\ldots,M}$ (we take $K \leqslant M$ for definiteness) lead to the same quantum operation if and only if there exists a left-unitary matrix U with elements $\{U_{mn}\}_{m,n=1,2,\ldots,M}$ for which

$$\hat{E}_k = \sum_{m=1}^{M} U_{km}\hat{F}_m, \quad k = 1, 2,\ldots,M, \tag{3.4}$$

  where if $K \neq M$, then $M - K$ zeros must be included in the set with fewer Kraus operators, so that U is a square matrix.
- Any quantum operation has an operator-sum representation with $K \leqslant d_S^2$ Kraus operators.
- Any trace-preserving quantum operation can be written as the reduced unitary evolution of the joint system with an environment having $d_E \leqslant d_S^2$. Trace-decreasing operations require an extra measurement of a joint observable, and the desired quantum operation is accomplished only when a particular outcome appears in the measurement, which happens with a probability equal to the trace of the map (we say that it requires *post-selection*). The representation of the quantum operation in terms of unitaries and measurements acting onto the joint system is known as a *Stinespring dilation*.

3.1.3 Quantum operations as reduced dynamics in an extended system

In the rest of this section we will see explicitly how, after applying unitaries and measurements to the joint system, the reduced dynamics of the system is described by a map of the type (3.1). Let us start by writing the initial state of the joint system as

$$\hat{\rho}_{SE} = \hat{\rho} \otimes |\varphi_E\rangle\langle\varphi_E|. \tag{3.5}$$

Note that we do not lose generality by assuming the environmental state to be pure, since if it is mixed, we can always purify it by adding an extra environment, and taking both environments as a new joint environment.

If Shane applies a joint unitary $\hat{U}$ the state evolves into

$$\hat{\rho}'_{SE} = \hat{U}(\hat{\rho} \otimes |\varphi_E\rangle\langle\varphi_E|)\hat{U}^\dagger. \tag{3.6}$$

Introducing now an orthonormal basis in $\mathcal{H}_E$ given by $\{|e_k\rangle\}_{k=1,2,\ldots,d_E}$, the reduced state of the system is written as

$$\hat{\rho}' = \mathrm{tr}_E\left\{\hat{\rho}'_{SE}\right\} = \sum_{k=1}^{d_E}\langle e_k|\hat{U}(\hat{\rho} \otimes |\varphi_E\rangle\langle\varphi_E|)\hat{U}^\dagger|e_k\rangle = \sum_{k=1}^{d_E}\hat{E}_k\hat{\rho}\hat{E}_k^\dagger, \tag{3.7}$$

where the operators $\hat{E}_k = \langle e_k|\hat{U}|\varphi_E\rangle$ act onto $\mathcal{H}_S$. Note that $\mathrm{tr}_S\{\hat{\rho}'\} = 1\ \forall\hat{\rho}$, and hence, the reduced unitary corresponds to a trace-preserving map such as (3.1) with a number of Kraus operators K given by the dimension of the Hilbert space of the environment.

If, on the other hand, Shane applies a non-selective measurement of a joint observable with diagonal representation

$$\hat{A} = \sum_{j=1}^{d_S\times d_E} a_j\hat{P}_j, \tag{3.8}$$

where $\{\hat{P}_j = |a_j\rangle\langle a_j|\}_{j=1,2,\ldots,d_S\times d_E}$ are the projectors onto its eigenvectors $\{|a_j\rangle \in \mathcal{H}_S \otimes \mathcal{H}_E\}_{j=1,2,\ldots,d_S\times d_E}$, the state evolves into

$$\hat{\rho}'_{SE} = \sum_{j=1}^{d_S\times d_E} \hat{P}_j\big(\hat{\rho} \otimes |\varphi_E\rangle\langle\varphi_E|\big)\hat{P}_j. \tag{3.9}$$

The reduced state of the system can then be written as

$$\hat{\rho}' = \mathrm{tr}_E\left\{\hat{\rho}'_{SE}\right\} = \sum_{j=1}^{d_S\times d_E}\sum_{k=1}^{d_E}\langle e_k|\hat{P}_j\big(\hat{\rho} \otimes |\varphi_E\rangle\langle\varphi_E|\big)\hat{P}_j|e_k\rangle = \sum_{j=1}^{d_S\times d_E}\sum_{k=1}^{d_E}\hat{E}_{jk}\hat{\rho}\hat{E}_{jk}^\dagger, \tag{3.10}$$

where $\hat{E}_{jk} = \langle e_k|\hat{P}_j|\varphi_E\rangle$. Again, it is immediate to check that $\mathrm{tr}_S\{\hat{\rho}'\} = 1\ \forall\hat{\rho}$, and hence, the reduced non-selective measurement is a trace-preserving quantum operation with a number of Kraus operators given by $d_S \times d_E^2$.

The reduced dynamics of a selective measurement of the joint observable $\hat{A}$ is a little more subtle. Assume that after the measurement Shane obtains the outcome a_j, so that, accordingly, the state $\hat{\rho}_{SE}$ collapses to

$$\hat{\rho}_{SE,j} = p_j^{-1}\hat{P}_j\left(\hat{\rho} \otimes |\varphi_E\rangle\langle\varphi_E|\right)\hat{P}_j, \tag{3.11}$$

where

$$p_j = \mathrm{tr}\left\{\hat{P}_j\hat{\rho}_{SE}\right\}, \tag{3.12}$$

is the probability for the outcome a_j to appear after the measurement. In this case it is easy to rewrite the reduced state of the system as

$$\hat{\rho}_j = \mathrm{tr}_E\left\{\hat{\rho}_{SE,j}\right\} = p_j^{-1}\sum_{k=1}^{d_E}\hat{E}_{jk}\hat{\rho}\hat{E}_{jk}^{\dagger}, \tag{3.13}$$

and the corresponding probability as

$$p_j = \mathrm{tr}_S\left\{\left[\sum_{k=1}^{d_E}\hat{E}_{jk}^{\dagger}\hat{E}_{jk}\right]\hat{\rho}\right\} \leqslant 1, \tag{3.14}$$

where $\hat{E}_{jk} = \langle e_k|\hat{P}_j|\varphi_E\rangle$. Hence, similarly to the previous cases, the reduced selective measurement is described by a map of the type (3.1) with d_E Kraus operators, but the map is trace decreasing in general.

3.2 Generalized measurements and positive operator-valued measures

As explained in the previous section, the most general type of measurement that one can perform on a system within this framework is described by a collection of trace-decreasing quantum operations $\{\mathcal{E}_j\}_{j=1,2,\ldots,J>1}$ each with an associated set of Kraus operators $\{\hat{E}_{jk}\}_{k=1,2,\ldots,K_j}$, which forms a complete set, that is,

$$\sum_{j=1}^{J}\mathrm{tr}\{\mathcal{E}_j[\hat{\rho}]\} = 1 \;\forall\hat{\rho} \in \text{density operators}. \tag{3.15}$$

Generalized measurements for which $K_j = 1 \;\forall j$, that is, all its associated quantum operations are described by a single Kraus operator,

$$\mathcal{E}_j[\hat{\rho}] = \hat{E}_j\hat{\rho}\hat{E}_j^{\dagger}, \tag{3.16}$$

with

$$\hat{E}_j^{\dagger}\hat{E}_j \leqslant \hat{I} \;\forall j \qquad \text{and} \qquad \sum_{j=1}^{J}\hat{E}_j^{\dagger}\hat{E}_j = \hat{I}, \tag{3.17}$$

are very special, because it can be shown that its simplest Stinespring dilation does not require a joint unitary, just a joint projective measurement. The set

$\{\hat{\Pi}_j = \hat{E}_j^\dagger \hat{E}_j\}_{j=1,2,\ldots,J}$ is known as a *positive operator-valued measure* (POVM), while the operators $\{\hat{E}_j\}_{j=1,2,\ldots,J}$ are called the *measurement operators.*

These generalized measurements are the closest ones to projective measurements [2]. The POVM $\{\hat{\Pi}_j\}_{j=1,2,\ldots,J}$ plays the role of the spectral decomposition of the measured observable, from which we obtain the probability of observing the outcome 'j' as

$$p_j = \mathrm{tr}\left\{\hat{\Pi}_j\hat{\rho}\right\}, \tag{3.18}$$

for a pre-measurement state $\hat{\rho}$. On the other hand, the measurement operators $\{\hat{E}_j\}_{j=1,2,\ldots,J}$ (uniquely defined from the POVM up to a left-unitary transformation) play the role of the projectors, with a post-measurement state given by

$$\hat{\rho}_j = p_j^{-1}\hat{E}_j\hat{\rho}\hat{E}_j^\dagger, \tag{3.19}$$

if the measurement is selective, or

$$\hat{\rho}' = \sum_{j=1}^{J}\hat{E}_j\hat{\rho}\hat{E}_j^\dagger, \tag{3.20}$$

if it is non-selective.

As a simple application of POVM-based measurements, consider the following problem. Suppose that someone picks one state out of the set $\{|\varphi_1\rangle, |\varphi_2\rangle\}$ and asks us to find with a single measurement which one was picked. If the states are orthogonal, this is trivial: we make a projective measurement defined by the projectors $\{\hat{P}_1 = |\varphi_1\rangle\langle\varphi_1|, \hat{P}_2 = |\varphi_2\rangle\langle\varphi_2|\}$, and check which outcome appeared. The problem is that it is simple to prove that when the states are not orthogonal, there is no strategy based on projective measurements allowing us to determine which state was given to us. However, we can design a strategy based on POVMs which will allow us to perform the needed task, although it does not work all the time.

Consider the POVM $\{\hat{\Pi}_1 = \hat{I} - |\varphi_2\rangle\langle\varphi_2|, \hat{\Pi}_2 = \hat{I} - |\varphi_1\rangle\langle\varphi_1|, \hat{\Pi}_3 = \hat{I} - \hat{\Pi}_1 - \hat{\Pi}_2\}$. Suppose that we obtain the outcome '1'; then, we know for sure that we obtained the state $|\varphi_1\rangle$, because the probability of observing '1' when the state is $|\varphi_2\rangle$ is zero, that is, $\langle\varphi_2|\hat{\Pi}_1|\varphi_2\rangle = 0$. The opposite happens when we get the outcome '2', we know for sure that $|\varphi_2\rangle$ was given to us, because $\langle\varphi_1|\hat{\Pi}_2|\varphi_1\rangle = 0$. Finally, when we get the outcome '3' we do not know which state we had, but at least we never make a misidentification of the state.

3.3 Local operations and classical communication protocols

We discuss in this section a very important class of operations performed on bipartite systems of the type discussed in the previous chapter. Suppose that Alice and Bob are in distant locations, so that one does not have access to the part of the system belonging to the other. In this scenario, it is natural to think that the most general class of operations that can be performed on the joint system are local operations (arbitrary quantum operations acting only on A or B independently) in a

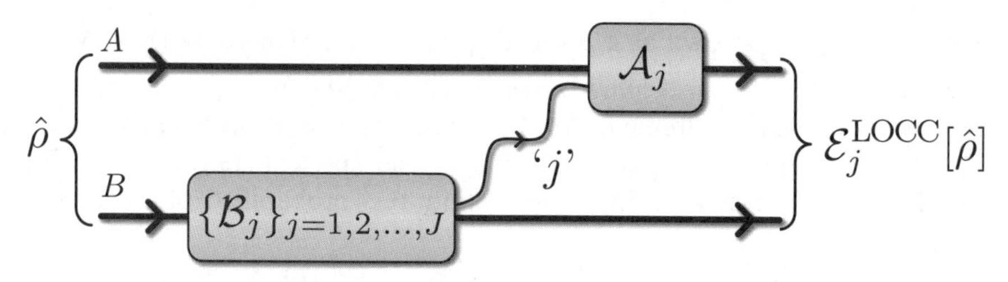

Figure 3.2. Prototypical form of a one-way-direct protocol involving only LOCC. Bob performs a generalized measurement and communicates the outcome to Alice, who applies a different trace-preserving operation depending on the information she receives from Bob.

correlated fashion (Alice and Bob can communicate by phone to decide together what to do). Quantum operations of this kind are known as *local operations and classical communication* (LOCC) *protocols*, and play a central role in many problems of quantum information (we already saw one, the characterization of entangled states).

As an example, consider the following prototypical protocol, schematically shown in figure 3.2: Bob performs a generalized measurement described by the set $\{\mathcal{B}_j\}_{j=1,2,\ldots,J}$ of trace-decreasing quantum operations, and communicates the outcome to Alice, who applies a trace-preserving quantum operation $\mathcal{A}_j$ from a pre-agreed set $\{\mathcal{A}_j\}_{j=1,2,\ldots,J}$ in one-to-one correspondence with the possible outcomes of Bob's measurements. Let us denote by $\{\hat{B}_{jk}\}_{k=1,2,\ldots,K_j}$ the Kraus operators associated with $\mathcal{B}_j$ (note that when $K_j = 1\ \forall\, j$, Bob's measurement is a POVM-based measurement), and by $\{\hat{A}_{jm}\}_{m=1,2,\ldots,M_j}$ the ones associated with $\mathcal{A}_j$. The resulting possible maps will be given by

$$\mathcal{E}_j^{\text{LOCC}}[\hat{\rho}] = \mathcal{A}_j\big[\mathcal{B}_j[\hat{\rho}]\big] = \sum_{k=1}^{K_j}\sum_{m=1}^{M_j}\big(\hat{A}_{jm}\otimes\hat{I}\big)\big(\hat{I}\otimes\hat{B}_{jk}\big)\hat{\rho}\Big(\hat{I}\otimes\hat{B}_{jk}^{\dagger}\Big)\Big(\hat{A}_{jm}^{\dagger}\otimes\hat{I}\Big). \quad (3.21)$$

We will find this type of LOCC protocols along the book (starting in the next section), which we will call *one-way-direct* LOCC protocols.

It is possible to prove that trace-preserving LOCC protocols can only decrease the entanglement of the state shared by Alice and Bob, as intuition says. What is a little more surprising is that trace-decreasing LOCC protocols can enhance the entanglement (we shall find one example of this when studying photon addition and subtraction), which is a further example of how counterintuitive quantum mechanics can be. Of course, on average a complete set of trace-decreasing LOCC maps (whose average can be seen as a trace-preserving LOCC protocol involving non-selective measurements) can only decrease the entanglement, showing that any local operation able to enhance the entanglement must be *intrinsically probabilistic*.

3.4 Majorization in quantum mechanics

In this section we introduce a relation between classical probability distributions called *majorization* [3, 4], which we will show to be connected to a couple of important questions in quantum mechanics [1], namely the freedom in the choice of ensemble decompositions of mixed states and the conversion of pure entangled states via LOCC protocols.

3.4.1 The concept of majorization

Majorization appeared as a way to order vectors or probability distributions in terms of their disorder, in an effort to understand when one can be built from another by randomizing the latter [3, 4].

Take two probability distributions $\mathbf{p} = \mathrm{col}(p_1, p_2, \ldots, p_d)$ and $\mathbf{q} = \mathrm{col}(q_1, q_2, \ldots, q_d)$, where d can be infinite (see an example with $d = 13$ in figure 3.3). We say that $\mathbf{p}$ *majorizes* $\mathbf{q}$, and denote it by $\mathbf{p} \succ \mathbf{q}$, if and only if

$$\sum_{n=1}^{m} p_n^{\downarrow} \geqslant \sum_{n=1}^{m} q_n^{\downarrow}, \ \forall\, m < d, \tag{3.22}$$

where $\mathbf{p}^{\downarrow}$ and $\mathbf{q}^{\downarrow}$ are the original vectors with their components rearranged in decreasing order.

This characterization of the majorization relation is interesting from an operational point of view, since it is easy to check numerically whether two vectors satisfy this condition or not. Nevertheless, it can be proven that $\mathbf{p} \succ \mathbf{q}$ is strictly equivalent to two other relations:

- For every concave function $h(x)$, it is satisfied $\sum_{n=1}^{d} h(p_n) \leqslant \sum_{n=1}^{d} h(q_n)$.
- $\mathbf{q}$ can be obtained from $\mathbf{p}$ as $\mathbf{q} = D\mathbf{p}$, where D is a column-stochastic matrix[3].

These relations are very interesting from an interpretational point of view. First, note that the Shannon entropy is a concave function, and hence, the first relation says that the entropy of $\mathbf{q}$ is larger than the entropy of $\mathbf{p}$; now, as we have discussed in previous chapters, larger entropy means larger lack of information, which can be somehow interpreted as more disorder. Second, it is a known result that any column-stochastic matrix can be written as a convex sum (or statistical mixture) of permutations; hence, the second relation says that $\mathbf{q}$ can be obtained from $\mathbf{p}$ by applying a random mixture of permutations on the latter. Hence, both conditions seem to justify the idea of $\mathbf{q}$ being more disordered than $\mathbf{p}$.

3.4.2 Majorization and ensemble decompositions of a state

As a first simple application of majorization to quantum mechanics, we answer the following question: under which conditions can we find an ensemble decomposition based on a probability distribution $\mathbf{w}$, say $\{w_m, |\varphi_m\rangle\}_{m=1,2,\ldots,M}$, of a density operator $\hat{\rho}$ with diagonal representation $\hat{\rho} = \sum_{n=1}^{d} \lambda_n |r_n\rangle\langle r_n|$?

[3] A square matrix is *column-stochastic* when its elements are real and positive, each of its columns sum to one, and each of its rows sum to less than one. Most of the literature on the connection between majorization and quantum information studies finite-dimensional systems, in which case it can be shown that column-stochastic matrices are also *doubly stochastic* (all columns and rows sum to one). One needs the slightly more general definition of column-stochastic to cope with infinite-dimensional spaces [4], as we will do in the next chapter.

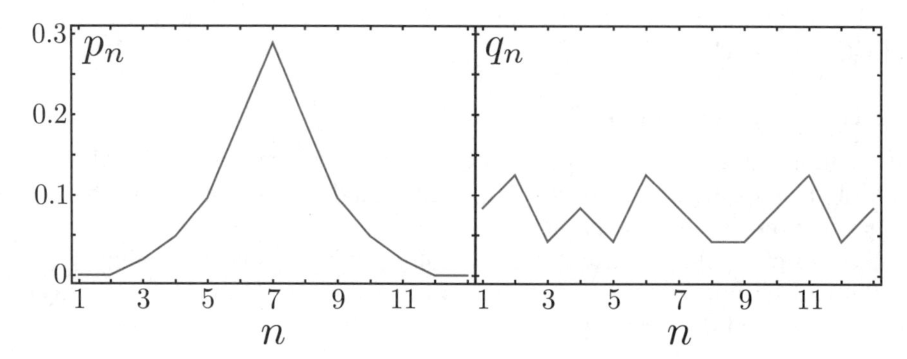

Figure 3.3. Examples of probability distributions in $d = 13$ dimensions. Clearly $\mathbf{p}$ is more ordered than $\mathbf{q}$. What is not immediate to see is whether $\mathbf{q}$ can be obtained by randomizing $\mathbf{p}$, and here is where majorization theory helps.

The answer is rather simple: only when $\boldsymbol{\lambda} \succ \mathbf{w}$ [1]. Of course, if $M \neq d$, zeros are added in the vector with smaller dimensionality to match the dimensions of $\boldsymbol{\lambda}$ and $\mathbf{w}$.

This result is indeed a simple consequence of what we learned in section 1.3.3 concerning the freedom to represent a given density operator by different ensemble decompositions. In particular, if $\hat{\rho}$ can be represented by the ensembles $\{w_m, |\varphi_m\rangle\}_{m=1,2,\ldots,M}$ and $\{\lambda_n, |r_n\rangle\}_{n=1,2,\ldots,d}$, there must exist a left-unitary matrix U with elements $\{U_{mn}\}_{m,n=1,2,\ldots,\max\{M,d\}}$ for which

$$\sqrt{w_m}|\varphi_m\rangle = \sum_{n=1}^{\max\{M,d\}} U_{mn}\sqrt{\lambda_n}|r_n\rangle, \quad m = 1, 2,\ldots, \max\{M, d\}. \tag{3.23}$$

Now, taking the inner product of this expression with itself, and using the orthonormality of the $\{|r_n\rangle\}_n$ set, we obtain

$$w_m = \sum_{n=1}^{\max\{M,d\}} |U_{mn}|^2 \lambda_n, \quad m = 1, 2,\ldots, \max\{M, d\}. \tag{3.24}$$

Finally, as U is left-unitary, the matrix with elements $\{|U_{mn}|^2\}_{m,n=1,2,\ldots,\max\{M,d\}}$ is column-stochastic.

3.4.3 Majorization and the transformation of entangled states

The next application of majorization theory to quantum mechanics appears when answering another question of paramount importance in information theory: given a bipartite state $|\psi\rangle \in \mathcal{H}_A \otimes \mathcal{H}_B$, under which conditions can it be transformed *deterministically* into another state $|\varphi\rangle \in \mathcal{H}_A \otimes \mathcal{H}_B$ if Alice and Bob are allowed to use only LOCC protocols?

By 'deterministically' we mean that given a complete set of LOCC protocols $\{\mathcal{E}_j^{\text{LOCC}}\}_j$, the transformation $|\psi\rangle \to |\varphi\rangle$ succeeds for all of them, that is, $\mathcal{E}_j^{\text{LOCC}}[|\psi\rangle\langle\psi|] \propto |\varphi\rangle\langle\varphi| \ \forall j$. If, for example, the transformation works only for

$\mathcal{E}_0^{\mathrm{LOCC}}$, but not for the rest, then Alice and Bob will be able to transform $|\psi\rangle$ into $|\varphi\rangle$ only some of the time, in particular with probability $p_0 = \mathrm{tr}\{\mathcal{E}_0^{\mathrm{LOCC}}[|\psi\rangle\langle\psi|]\}$. In other words, the transformation will fail with probability $(1 - p_0)$, that is, Alice and Bob's strategy will work only probabilistically.

To answer this question, consider the diagonal representations of the reduced states $\hat{\rho}_A^{\psi} = \mathrm{tr}_B\{|\psi\rangle\langle\psi|\} = \sum_{n=1}^{d} \lambda_n^{\psi} |r_n\rangle_{\psi}\langle r_n|$ and $\hat{\rho}_A^{\varphi} = \mathrm{tr}_B\{|\varphi\rangle\langle\varphi|\} = \sum_{n=1}^{d} \lambda_n^{\varphi} |r_n\rangle_{\varphi}\langle r_n|$. Then, Alice and Bob can transform $|\psi\rangle$ into $|\varphi\rangle$ via an LOCC strategy, if and only if $\boldsymbol{\lambda}^{\psi} \prec \boldsymbol{\lambda}^{\varphi}$ [1, 5, 6], in which case we use the symbolic notation $|\psi\rangle \prec |\varphi\rangle$.

Note that since the entanglement entropy is a concave function of the eigenvalues of the reduced density operator, this majorization relation implies that $|\psi\rangle$ cannot be transformed deterministically via an LOCC protocol into states of larger entanglement, that is, $E[|\psi\rangle] \geqslant E[|\varphi\rangle]$, as expected.

It is also possible to prove that if $|\psi\rangle$ can be transformed into $|\varphi\rangle$ deterministically via an LOCC protocol, it can always be achieved with a one-way-direct LOCC protocol of the following simple form [6]: Bob performs a measurement described by some POVM $\{\hat{\Pi}_j\}_{j=1,2,\dots,J}$ and communicates the outcome to Alice, who applies a unitary transformation from a pre-agreed set of unitaries $\{\hat{A}_j\}_{j=1,2,\dots,J}$ in one to one correspondence with the possible outcomes of Bob's measurement. Hence, if $\{\hat{B}_j\}_{j=1,2,\dots,J}$ are the measurement operators associated to Bob's POVM, the transformation is accomplished as

$$|\varphi\rangle \propto \left(\hat{A}_j \otimes \hat{I}\right)\left(\hat{I} \otimes \hat{B}_j\right)|\psi\rangle \;\forall j. \tag{3.25}$$

Bibliography

[1] Nielsen M A and Chuang I L 2000 *Quantum Information and Quantum Computation* (Cambridge: Cambridge University Press)

[2] Paris M G A 2012 The modern tools of quantum mechanics: a tutorial on quantum states, measurements, and operations *Eur. Phys. J. Spec. Top.* **203** 61

[3] Arnold B 1987 Lecture Notes in Statistics *Majorization and the Lorenz Order* vol 43 (Berlin: Springer)

[4] Kaftal V and Weiss G 2010 An infinite dimensional Schur-Horn theorem and majorization theory *J. Funct. Anal.* **259** 3115

[5] Nielsen M A 1999 Conditions for a class of entanglement transformations *Phys. Rev. Lett.* **83** 436

[6] Nielsen M A and Vidal G 2001 Majorization and the interconversion of bipartite states *Quantum Inf. Comput.* **1** 76

An Introduction to the Formalism of Quantum Information with Continuous Variables

Carlos Navarrete-Benlloch

Chapter 4

Quantum information with continuous variables

4.1 The classical harmonic oscillator

Consider the basic mechanical model of a *one-dimensional harmonic oscillator*: a particle of mass m is at rest at some equilibrium position which we take as $x = 0$; when displaced from this position by some amount a, a restoring force $F = -kx$ starts acting on the particle, trying to bring it back to $x = 0$. Newton's equation of motion for the particle is therefore $m\ddot{x} = -kx$, which together with the initial conditions $x(0) = a$ and $\dot{x}(0) = v$ gives the solution $x(t) = a\cos\omega t + (v/\omega)\sin\omega t$, $\omega = \sqrt{k/m}$ being the so-called *angular frequency*. Therefore the particle will be bouncing back and forth between positions $-\sqrt{a^2 + v^2/\omega^2}$ and $\sqrt{a^2 + v^2/\omega^2}$ with time period $2\pi/\omega$ (hence the name 'harmonic oscillator').

Let us now study the problem from a Hamiltonian point of view. For this one-dimensional problem with no constraints, we can take the position of the particle and its momentum as the generalized coordinate and momentum, that is, $q = x$ and $p = m\dot{x}$. The restoring force derives from a potential $V(x) = kx^2/2$, and hence the Hamiltonian takes the form

$$H = \frac{p^2}{2m} + \frac{m\omega^2}{2}q^2. \tag{4.1}$$

The canonical equations read

$$\dot{q} = \frac{p}{m} \quad \text{and} \quad \dot{p} = -m\omega^2 q, \tag{4.2}$$

which together with the initial conditions $q(0) = a$ and $p(0) = mv$ give the trajectory

$$\left(q, \frac{p}{m\omega}\right) = \left(a\cos\omega t + \frac{v}{\omega}\sin\omega t, \frac{v}{\omega}\cos\omega t - a\sin\omega t\right), \tag{4.3}$$

doi:10.1088/978-1-6817-4405-6ch4

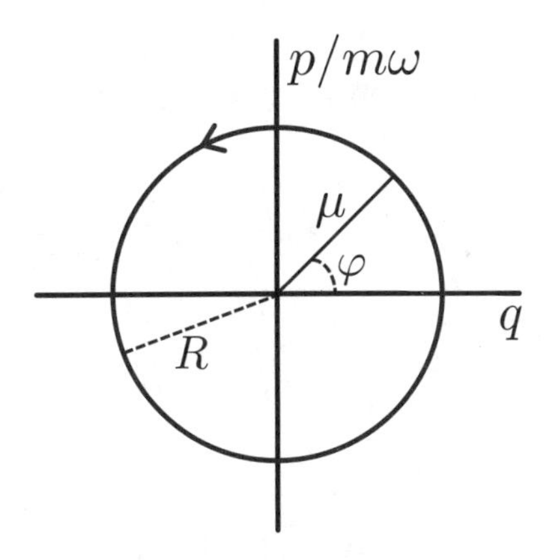

Figure 4.1. Phase-space trajectory of the classical harmonic oscillator. Both the real and complex phase-space variables, (x, p) and (μ, φ), respectively, are shown.

where we normalize the momentum by $m\omega$ for simplicity. Starting at the phase-space point $(a, v/\omega)$ the system evolves periodically drawing a circle of radius $R = \sqrt{a^2 + v^2/\omega^2}$ as shown in figure 4.1, returning to its initial point at times $t_k = 2\pi k/\omega$, with $k \in \mathbb{N}$. This circular trajectory could have been derived without even solving the equations of motion, as the conservation of the Hamiltonian $H(t) = H(0)$ leads directly to $q^2 + p^2/m^2\omega^2 = R^2$, which is exactly the circumference of figure 4.1. This is a simple manifestation of the power of the Hamiltonian formalism.

There is another useful description of the harmonic oscillator, the so-called *amplitude–phase* or *complex* representation. The amplitude and phase refer to the polar coordinates in phase space, say $\mu = \sqrt{q^2 + p^2/m^2\omega^2}$ and $\varphi = \arctan(p/m\omega q)$, as shown in figure 4.1. In terms of these variables, the trajectory reads simply $(\mu, \varphi) = (R, \varphi_0 - \omega t)$, with $\varphi_0 = \arctan(v/\omega a)$, so that the evolution is completely described by a linear time variation of the oscillator's phase. From these variables we can define the complex variable $\nu = \mu \exp(\mathrm{i}\varphi) = q + (\mathrm{i}/m\omega)p$, in terms of which the trajectory reads $\nu(t) = R \exp[\mathrm{i}(\varphi_0 - \omega t)]$, and the Hamiltonian can be written as $H = m\omega^2\nu^*\nu/2$. The complex variable ν is known as the *normal variable* of the oscillator.

4.2 The quantum harmonic oscillator

The *harmonic oscillator* is the prototype of a system described quantum mechanically by an infinite-dimensional Hilbert space. In order to see this, let us find the eigenstates of its Hamiltonian, which is given by the operator

$$\hat{H} = \frac{\hat{p}^2}{2m} + \frac{m\omega^2}{2}\hat{q}^2, \tag{4.4}$$

by virtue of the discussion after axiom 3, with the *position* $\hat{q}$ and *momentum* $\hat{p}$ satisfying the commutation relation

$$[\hat{q}, \hat{p}] = \mathrm{i}\hbar. \tag{4.5}$$

We will always work with dimensionless versions of them, the so-called X and P *quadratures* (although we may keep using the names 'position' and 'momentum' most of the time)

$$\hat{X} = \sqrt{\frac{2\omega m}{\hbar}}\hat{q} \qquad \text{and} \qquad \hat{P} = \sqrt{\frac{2}{\hbar\omega m}}\hat{p}, \tag{4.6}$$

which satisfy the commutation relation

$$\left[\hat{X}, \hat{P}\right] = 2\mathrm{i}, \tag{4.7}$$

and therefore the uncertainty relations

$$\Delta X \Delta P \geqslant 1. \tag{4.8}$$

In terms of these quadratures, the Hamiltonian reads

$$\hat{H} = \frac{\hbar\omega}{4}\left(\hat{X}^2 + \hat{P}^2\right). \tag{4.9}$$

In order to find the eigensystem of this operator, we decompose the quadratures as

$$\hat{X} = \hat{a}^\dagger + \hat{a} \qquad \text{and} \qquad \hat{P} = \mathrm{i}(\hat{a}^\dagger - \hat{a}), \tag{4.10}$$

where the operators $\hat{a}$ and $\hat{a}^\dagger$, known as the *annihilation* and *creation operators*, satisfy the commutation relation

$$[\hat{a}, \hat{a}^\dagger] = 1. \tag{4.11}$$

In terms of these operators, the Hamiltonian is rewritten as

$$\hat{H} = \hbar\omega(\hat{a}^\dagger\hat{a} + 1/2), \tag{4.12}$$

and hence the problem has been reduced to finding the eigensystem of the so-called *number operator* $\hat{N} = \hat{a}^\dagger\hat{a}$.

Let us denote by n a generic real number contained in the spectrum of $\hat{N}$, whose corresponding eigenvector we denote by $|n\rangle$, so that, $\hat{N}|n\rangle = n|n\rangle$. The eigensystem of $\hat{N}$ is readily found from the following two properties:

- $\hat{N}$ is a positive-semidefinite operator, as for any vector $|\psi\rangle$ it is satisfied $\langle\psi|\hat{N}|\psi\rangle = (\hat{a}|\psi\rangle, \hat{a}|\psi\rangle) \geqslant 0$. When applied to its eigenvectors, this property forbids the existence of negative eigenvalues, that is, $n \geqslant 0$.
- Applying the commutation relation[1] $[\hat{N}, \hat{a}] = -\hat{a}$ to $|n\rangle$, it is straightforward to show that the vector $\hat{a}|n\rangle$ is also an eigenvector of $\hat{N}$ with eigenvalue $n - 1$. Similarly, from the commutation relation $[\hat{N}, \hat{a}^\dagger] = \hat{a}^\dagger$ it is found that the vector $\hat{a}^\dagger|n\rangle$ is an eigenvector of $\hat{N}$ with eigenvalue $n + 1$.

[1] This is straightforward to find by using the property $[\hat{A}\hat{B}, \hat{C}] = \hat{A}[\hat{B}, \hat{C}] + [\hat{A}, \hat{C}]\hat{B}$, valid for any three operators $\hat{A}$, $\hat{B}$, and $\hat{C}$.

These two properties imply that the spectrum of $\hat{N}$ is the set of non-negative integers $n = 0, 1, 2, \ldots$, and that the eigenvector $|0\rangle$ corresponding to $n = 0$ must satisfy $\hat{a}|0\rangle = 0$. Otherwise, it would be possible to find negative eigenvalues, hence contradicting the positivity of $\hat{N}$. Thus, the set of eigenvectors $\{|n\rangle\}_{n=0,1,\ldots}$ is an infinite, countable set. Moreover, note that $\langle n|\hat{N}|m\rangle = n\langle n|m\rangle = m\langle n|m\rangle$ implies that the eigenvectors form an orthogonal set, that is, $\langle n|m\rangle = 0$ for $n \neq m$. In addition, we will show later by using a specific representation, see (4.27), that the eigenstate $|0\rangle$ can be normalized to one, which implies that the rest of the eigenstates can be normalized as well by virtue of (4.13). Therefore, the eigenstates of $\hat{N}$ form an orthonormal set, that is, $\langle n|m\rangle = \delta_{nm}$. Finally, according to the axioms of quantum mechanics only the vectors normalized to one are physically relevant for the description of the state of the harmonic oscillator, and hence we conclude that the vector space spanned by the eigenvectors of $\hat{N}$ is an infinite-dimensional Hilbert space, since it is isomorphic to $l^2(\infty)$, see section 1.2.3.

Summarizing, we have been able to prove that the Hilbert space associated with the one-dimensional harmonic oscillator is infinite-dimensional. In the process, we have explicitly built an orthonormal basis of this space by using the eigenvectors $\{|n\rangle\}_{n=0,1,\ldots}$ of the number operator $\hat{N}$, known as the *Fock basis*. The annihilation and creation operators, $\hat{a}$ and $\hat{a}^\dagger$, allow us to move through this basis as

$$\hat{a}|n\rangle = \sqrt{n}|n-1\rangle \qquad \text{and} \qquad \hat{a}^\dagger|n\rangle = \sqrt{n+1}|n+1\rangle, \tag{4.13}$$

the factors in the square roots being easily found from normalization requirements. The state $|0\rangle$ is known as the *vacuum* state, since it corresponds to the eigenstate of $\hat{N}$ with no *excitations*, $n = 0$.

In contrast to the number operator, which has a discrete spectrum, the quadrature operators possess a pure continuous spectrum. Let us focus on the $\hat{X}$ operator, whose eigenvectors we denote by $\{|x\rangle\}_{x\in\mathbb{R}}$ with corresponding eigenvalues $\{x\}_{x\in\mathbb{R}}$, that is,

$$\hat{X}|x\rangle = x|x\rangle. \tag{4.14}$$

In order to prove that $\hat{X}$ has a pure continuous spectrum, just note that, from the relation

$$\exp\left(\frac{\mathrm{i}}{2}y\hat{P}\right)\hat{X}\exp\left(-\frac{\mathrm{i}}{2}y\hat{P}\right) = \hat{X} + y, \tag{4.15}$$

which is easily found via the Baker–Campbell–Haussdorf lemma[2], it follows that if $|x\rangle$ is an eigenvector of $\hat{X}$ with x eigenvalue, then the vector $\exp(-\mathrm{i}y\hat{P}/2)|x\rangle$ is also an eigenvector of $\hat{X}$ with eigenvalue $x + y$. Now, as this holds for any real y, we

[2] This lemma reads

$$\mathrm{e}^{\hat{B}}\hat{A}\mathrm{e}^{-\hat{B}} = \sum_{n=0}^{\infty}\frac{1}{n!}\underbrace{\left[\hat{B},\left[\hat{B},\ldots\left[\hat{B},\right.\right.\right.}_{n}\hat{A}\underbrace{\left.\left.\left.\right]\ldots\right]\right]}_{n}, \tag{4.16}$$

and is valid for two general operators $\hat{A}$ and $\hat{B}$.

conclude that the spectrum of $\hat{X}$ is the whole real line. Moreover, as a self-adjoint operator, one can use its eigenvectors as a continuous basis of the Hilbert space of the oscillator by using the Dirac normalization $\langle x|y\rangle = \delta(x-y)$. The same results can be obtained for the $\hat{P}$ operator, whose eigenvectors we denote by $\{|p\rangle\}_{p\in\mathbb{R}}$ with corresponding eigenvalues $\{p\}_{p\in\mathbb{R}}$, that is,

$$\hat{P}|p\rangle = p|p\rangle. \tag{4.17}$$

Note that these results rely only on the canonical commutation relations, and hence are completely general, valid for any system, not only for the harmonic oscillator. Note also that not being vectors contained in the Hilbert space of the oscillator (they cannot be properly normalized), the position and momentum eigenvectors cannot correspond to physical states. Nevertheless, we will see that they can be understood as an unphysical limit of some physical states (the squeezed states).

It is not difficult to prove that there exists a Fourier transform relation between the position and momentum bases, that is,

$$|p\rangle = \int_{-\infty}^{+\infty} \frac{\mathrm{d}x}{\sqrt{4\pi}} \exp\left(\frac{\mathrm{i}}{2}px\right)|x\rangle \Longleftrightarrow |x\rangle = \int_{-\infty}^{+\infty} \frac{\mathrm{d}p}{\sqrt{4\pi}} \exp\left(-\frac{\mathrm{i}}{2}px\right)|p\rangle. \tag{4.18}$$

To this aim we now prove that

$$\langle x|p\rangle = \frac{1}{\sqrt{4\pi}}\exp(\mathrm{i}xp/2). \tag{4.19}$$

First note that the commutator $[\hat{X}, \hat{P}] = 2\mathrm{i}$ implies that

$$\langle x|\hat{P}|x'\rangle = \frac{2\mathrm{i}\delta(x-x')}{x-x'}, \tag{4.20}$$

and hence

$$\begin{aligned}\langle x|\hat{P}|\psi\rangle &= \int_{\mathbb{R}} \mathrm{d}x' \langle x|\hat{P}|x'\rangle\langle x'|\psi\rangle = \int_{\mathbb{R}} \mathrm{d}x' \frac{2\mathrm{i}\delta(x-x')}{x-x'}\langle x'|\psi\rangle \\ &= \int_{\mathbb{R}} \mathrm{d}x' \frac{2\mathrm{i}\delta(x-x')}{x-x'}\left[\langle x|\psi\rangle + (x'-x)\frac{\mathrm{d}\langle x|\psi\rangle}{\mathrm{d}x} + \sum_{n=2}^{\infty}\frac{(x'-x)^n}{n!}\frac{\mathrm{d}^n\langle x|\psi\rangle}{\mathrm{d}x^n}\right].\end{aligned} \tag{4.21}$$

The order zero of the Taylor expansion is zero because the kernel is antisymmetric around x, while the terms of order two or above give zero as well after integrating them. This means that

$$\langle x|\hat{P}|\psi\rangle = -2\mathrm{i}\frac{\mathrm{d}\langle x|\psi\rangle}{\mathrm{d}x}, \tag{4.22}$$

which applied to $|\psi\rangle = |p\rangle$ yields the differential equation

$$p\langle x|p\rangle = -2\mathrm{i}\frac{\mathrm{d}\langle x|p\rangle}{\mathrm{d}x}, \tag{4.23}$$

which has (4.19) as its solution, the factor $1/\sqrt{4\pi}$ coming from the Dirac normalization of the $|p\rangle$ vectors.

As an example of the use of these continuous representations, we now find the position representation of the number states, which we write as

$$|n\rangle = \int_{\mathbb{R}} \mathrm{d}x \psi_n(x)|x\rangle. \tag{4.24}$$

As a first step we find the projection of vacuum onto a position eigenstate, the so-called *ground state wave function* $\psi_0(x) = \langle x|0\rangle$, from

$$0 = \langle x|\hat{a}|0\rangle = \frac{1}{2}\langle x|\big(\hat{X} + \mathrm{i}\hat{P}\big)|0\rangle = \frac{1}{2}\left(x + 2\frac{\mathrm{d}}{\mathrm{d}x}\right)\psi_0(x), \tag{4.25}$$

where we have used (4.22), which is a differential equation for $\psi_0(x)$ having

$$\psi_0(x) = \frac{1}{(2\pi)^{1/4}} \exp\left(-\frac{x^2}{4}\right), \tag{4.26}$$

as its solution. The factor $(2\pi)^{-1/4}$ is found by imposing the normalization

$$\langle 0|0\rangle = \int_{\mathbb{R}} \mathrm{d}x \psi_0^2(x) = 1. \tag{4.27}$$

Now, the projection of any number state $|n\rangle$ onto a position eigenstate (the nth *excited wave function*) is found from the ground state wave function as

$$\psi_n(x) = \langle x|n\rangle = \frac{1}{\sqrt{n!}}\langle x|\hat{a}^{\dagger n}|0\rangle = \frac{1}{\sqrt{n!\,2^n}}\langle x|\big(\hat{X} - \mathrm{i}\hat{P}\big)^n|0\rangle = \frac{1}{\sqrt{n!\,2^n}}\left(x - 2\frac{\mathrm{d}}{\mathrm{d}x}\right)^n \psi_0(x), \tag{4.28}$$

which, recalling the Rodrigues formula for the Hermite polynomials

$$H_n\left(\frac{x}{\sqrt{2}}\right) = 2^{-n/2} \exp\left(\frac{x^2}{4}\right)\left(x - 2\frac{\mathrm{d}}{\mathrm{d}x}\right)^n \exp\left(-\frac{x^2}{4}\right), \tag{4.29}$$

leads to the simple expression

$$\psi_n(x) = \frac{1}{\sqrt{2^{n+1/2}\pi^{1/2}n!}} H_n\left(\frac{x}{\sqrt{2}}\right) \exp\left(-\frac{x^2}{4}\right). \tag{4.30}$$

Finally, let us stress that even though all that we are going to discuss in the following applies to a general bosonic system, that is, a system described by a collection of harmonic oscillators, we will always have in mind the electromagnetic field (light, in particular), which can be described as a set of *modes* with well defined polarization and spatio-temporal profile, each of which behaves as the mechanical harmonic oscillator that we have introduced. Consequently, we will use 'modes' and 'harmonic oscillators', as well as 'photons' and 'excitations' interchangeably. Nevertheless, it is important to keep in mind that there are many other physical systems which behave as quantum harmonic oscillators, for example, motional degrees of freedom in atoms or mesoscopic objects, polarized atomic ensembles, superconducting circuits, or even the degrees of freedom of fields other than the electromagnetic one, e.g. the ones associated to the weak and strong interactions or the Higgs boson.

4.3 The harmonic oscillator in phase space: the Wigner function

4.3.1 General considerations

As the position and momentum do not have common eigenstates and, moreover, their eigenvectors cannot correspond to physical states of the oscillator, one concludes that these observables cannot take definite values in quantum mechanics. Given the state $\hat{\rho}$, the best one can offer is the *probability density functions* which will dictate the statistics of a measurement of these observables, $P(x) = \langle x|\hat{\rho}|x\rangle$ and $P(p) = \langle p|\hat{\rho}|p\rangle$. In other words, quantum mechanically there are not well defined trajectories in phase space, the position and momentum of the oscillator are always affected by some (*quantum*) *noise*.

The following question arises naturally: is it then possible to describe quantum mechanics as a probability distribution in phase space which blurs the classical trajectories? As we are about to see, the answer is only partially positive, as quantum noise is much more subtle than common classical noise.

A logical way of building such a phase-space distribution, say $W_{\hat{\rho}}(x, p)$, is as the one having the probability density functions $P(x)$ and $P(p)$ as its marginals, that is,

$$P(x) = \int_{\mathbb{R}} \mathrm{d}p\, W_{\hat{\rho}}(x, p) \qquad \text{and} \qquad P(p) = \int_{\mathbb{R}} \mathrm{d}x\, W_{\hat{\rho}}(x, p). \tag{4.31}$$

It is possible to show that this distribution is uniquely defined by [1]

$$W_{\hat{\rho}}(x, p) = \frac{1}{4\pi} \int_{\mathbb{R}} \mathrm{d}y \exp\left(-\frac{\mathrm{i}}{2}py\right) \langle x + y/2|\hat{\rho}|x - y/2\rangle, \tag{4.32}$$

which is known as the *Wigner function*. It is immediate to check that this distribution has the proper marginals, and that it is normalized, i.e. $\int_{\mathbb{R}^2} \mathrm{d}x\mathrm{d}p\, W_{\hat{\rho}}(x, p) = 1$. The proof of its uniqueness is not that simple, however, see [1].

As we will prove in the next section, given the state $\hat{\rho}$ of the oscillator, the quantum expectation value of any operator can be found from

$$\left\langle \left(\hat{X}^m \hat{P}^n\right)^{(s)} \right\rangle = \int_{\mathbb{R}^2} \mathrm{d}x\mathrm{d}p\, W_{\hat{\rho}}(x, p) x^m p^n, \tag{4.33}$$

where $(\hat{X}^m \hat{P}^n)^{(s)}$ refers to the symmetrized version of the corresponding moment with respect to position and momentum, e.g. $(\hat{X}^2\hat{P})^{(s)} = (\hat{X}^2\hat{P} + \hat{P}\hat{X}^2 + \hat{X}\hat{P}\hat{X})/3$.

Taking into account that the prescription to find the quantum operator associated with a classical observable $O(x, p)$ consists specifically of symmetrizing it with respect to x and p, and then changing the position and momentum by the corresponding self-adjoint operators (which guarantees the self-adjointness of the remaining operator, as we saw in section 1.3.5), this result reinforces the interpretation of $W_{\hat{\rho}}(x, p)$ as a probability distribution in phase space, and hence of quantum mechanics as noise acting on the classical trajectories. However, it is readily seen that this distribution can take on negative values for many quantum-mechanical states, and hence it is not a true two-dimensional probability density function. As an example of this, let us consider the Wigner function of the number states $\hat{\rho} = |n\rangle\langle n|$. Although not straightforward [1], it is possible to show that the corresponding Wigner function is given by

$$W_{|n\rangle}(x, p) = \frac{(-1)^n}{2\pi} L_n(x^2 + p^2) \exp\left(-\frac{x^2 + p^2}{2}\right), \tag{4.34}$$

where $L_n(z)$ is the Laguerre polynomial of order n, which can be found from the Rodrigues formula

$$L_n(z) = \frac{\exp(z)}{n!} \frac{\mathrm{d}^n}{\mathrm{d}z^n}[z^n \exp(-z)]. \tag{4.35}$$

For any $n > 0$, this function has negative regions and, therefore, it cannot be simulated with any source of classical noise. For example, for odd n it is always negative at the origin of phase space, since $L_n(0) = 1 \;\; \forall\, n$. The Wigner functions of the first four Fock states are plotted in figure 4.2.

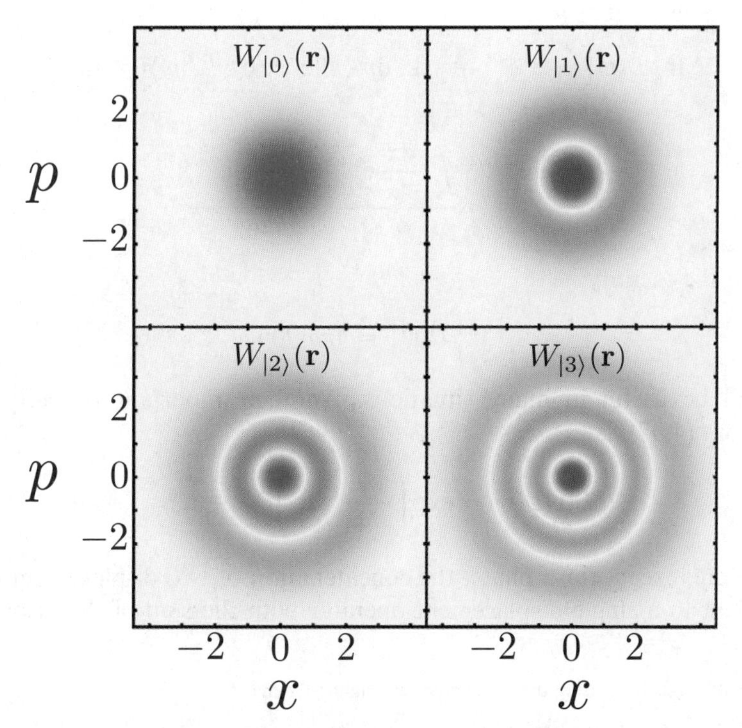

Figure 4.2. Density plot of the Wigner functions corresponding to the first four Fock states. Red and blue regions correspond to positive and negative values of the function, respectively. In both cases regions with higher contrast correspond to larger absolute values.

4.3.2 Definition based on the characteristic function

Given a state $\hat{\rho}$ of the oscillator, expression (4.32) allows us to compute the corresponding Wigner function. However, there is a much more convenient way of writing the Wigner distribution, which is based on the so-called *displacement operator*

$$\hat{D}(\mathbf{r}) = \exp\left[\frac{\mathrm{i}}{2}\hat{\mathbf{R}}^T\Omega\mathbf{r}\right], \tag{4.36}$$

where $\mathbf{r} = \mathrm{col}(x, p)$ is the coordinate vector in phase space (which is known as the *displacement* induced by the operator), $\hat{\mathbf{R}} = \mathrm{col}(\hat{X}, \hat{P})$ is the corresponding vector operator, and $\Omega = \begin{bmatrix} 0 & 1 \\ -1 & 0 \end{bmatrix}$ is known as the *symplectic form*. Note that with this matrix notation the position–momentum commutators read $[\hat{R}_j, \hat{R}_l] = 2\mathrm{i}\Omega_{jl}$. We will learn more about the physical properties of the displacement operator in the following sections, but for now, just take it as a useful mathematical object.

This operator can be written as a concatenation of two individual momentum and position translations plus some phase[3],

$$\hat{D}(\mathbf{r}) = \exp\left[-\frac{\mathrm{i}}{4}px\right]\exp\left[\frac{\mathrm{i}}{2}p\hat{X}\right]\exp\left[-\frac{\mathrm{i}}{2}x\hat{P}\right], \tag{4.38}$$

which allows us to write

$$\mathrm{tr}\left\{\hat{D}(\mathbf{r})\right\} = \mathrm{e}^{-\mathrm{i}px/4}\int_{\mathbb{R}^2} \mathrm{d}x'\mathrm{d}p'\langle x'|\mathrm{e}^{\mathrm{i}p\hat{X}/2}\mathrm{e}^{-\mathrm{i}x\hat{P}/2}|p'\rangle\langle p'|x'\rangle \tag{4.39}$$

$$= 4\pi\mathrm{e}^{-\mathrm{i}px/4}\underbrace{\int_{\mathbb{R}}\frac{\mathrm{d}x'}{4\pi}\mathrm{e}^{\mathrm{i}px'/2}}_{\delta(p)}\underbrace{\int_{\mathbb{R}}\frac{\mathrm{d}p'}{4\pi}\mathrm{e}^{\mathrm{i}xp'/2}}_{\delta(x)}, \tag{4.40}$$

arriving at the identity

$$\mathrm{tr}\left\{\hat{D}(\mathbf{r})\right\} = 4\pi\delta^{(2)}(\mathbf{r}), \tag{4.41}$$

which will be useful in many situations. Another important property, trivially proved from (4.37), is

$$\hat{D}(\mathbf{r})\hat{D}(\mathbf{r}') = \exp\left(-\frac{\mathrm{i}}{4}\mathbf{r}^T\Omega\mathbf{r}'\right)\hat{D}(\mathbf{r}+\mathbf{r}'), \tag{4.42}$$

and therefore, except for a phase, the concatenation of two displacement operators is equivalent to a single displacement operator with the sum of the displacements.

[3] This is trivially proved by using the so-called disentangling lemma

$$\exp\left(\hat{A}+\hat{B}\right) = \exp\left(-\left[\hat{A}, \hat{B}\right]/2\right)\exp\left(\hat{A}\right)\exp\left(\hat{B}\right), \tag{4.37}$$

valid for operators $\hat{A}$ and $\hat{B}$ which commute with their commutator.

Note that the phase is zero only when the condition $xp' = px'$ is met, although it plays no physical role when applied to a state of the system.

The first step in order to find the Wigner function of a given state $\hat{\rho}$ from the displacement operator is to define the *characteristic function*

$$\chi_{\hat{\rho}}(\mathbf{r}) = \mathrm{tr}\left\{\hat{D}(\mathbf{r})\hat{\rho}\right\} \Longleftrightarrow \hat{\rho} = \int_{\mathbb{R}^2} \frac{\mathrm{d}^2\mathbf{r}}{4\pi}\chi_{\hat{\rho}}(\mathbf{r})\hat{D}^{\dagger}(\mathbf{r}), \tag{4.43}$$

and then the Wigner function is obtained as its two-dimensional Fourier transform

$$W_{\hat{\rho}}(\mathbf{r}) = \int_{\mathbb{R}^2} \frac{\mathrm{d}^2\mathbf{s}}{(4\pi)^2}\chi_{\hat{\rho}}(\mathbf{s})\mathrm{e}^{\frac{\mathrm{i}}{2}\mathbf{s}^T\Omega\mathbf{r}} \Longleftrightarrow \chi_{\hat{\rho}}(\mathbf{s}) = \int_{\mathbb{R}^2} \mathrm{d}^2\mathbf{r}\, W_{\hat{\rho}}(\mathbf{r})\mathrm{e}^{-\frac{\mathrm{i}}{2}\mathbf{s}^T\Omega\mathbf{r}}. \tag{4.44}$$

It is not difficult to show that this alternative definition of the Wigner function leads to the original one given by (4.32). However, this definition simplifies many more derivations. For example, from (4.43) and (4.44), it is immediate to prove that

$$\mathrm{tr}\{\hat{\rho}\} = 1 \Longrightarrow \chi_{\hat{\rho}}(\mathbf{0}) = 1 \Longrightarrow \int_{\mathbb{R}^2} \mathrm{d}^2\mathbf{r}\, W_{\hat{\rho}}(\mathbf{r}) = 1, \tag{4.45}$$

that is, the normalization of the Wigner function. Moreover, evaluating the trace of (4.43) in the position eigenbasis, we obtain

$$\chi_{\hat{\rho}}(0, p) = \int \mathrm{d}x \langle x|\mathrm{e}^{\mathrm{i}p\hat{X}/2}\hat{\rho}|x\rangle = \int \mathrm{d}x \mathrm{e}^{\mathrm{i}px/2}\langle x|\hat{\rho}|x\rangle, \tag{4.46}$$

which, using the right-hand side of (4.44) directly implies that

$$\langle x|\hat{\rho}|x\rangle = \int_{\mathbb{R}} \mathrm{d}p\, W_{\hat{\rho}}(x, p), \tag{4.47}$$

that is, the Wigner function has the position probability density function as one of its marginals. The other marginal is the momentum probability density function, as is proved from $\chi_{\hat{\rho}}(x, 0)$ in a similar fashion.

Going one step further, using these definitions it is actually quite simple to prove (4.33). In particular, it is easy to check from (4.43) that the expectation value of the symmetrically ordered product $(\hat{X}^m\hat{P}^n)^{(s)}$ can be obtained from the characteristic function as

$$\left\langle \left(\hat{X}^m\hat{P}^n\right)^{(s)} \right\rangle = (-1)^m(2\mathrm{i})^{m+n}\frac{\partial^{m+n}}{\partial p^m\,\partial x^n}\chi_{\hat{\rho}}(\mathbf{r})\bigg|_{\mathbf{r}=\mathbf{0}}, \tag{4.48}$$

leading to (4.33) after making use of (4.43).

4.3.3 Multi-mode considerations

In general we will not deal with a single harmonic oscillator (a single mode of the light field), but with a collection of, say, N harmonic oscillators (N modes of light). Let us define again the coordinate vector in the whole phase space as $\mathbf{r} = (x_1, p_1, x_2, p_2, \ldots, x_N, p_N)$, and the corresponding vector operator

$$\hat{\mathbf{R}} = \mathrm{col}\left(\hat{X}_1, \hat{P}_1, \hat{X}_2, \hat{P}_2, \ldots, \hat{X}_N, \hat{P}_N\right), \tag{4.49}$$

in terms of which the commutation relations can be rewritten as

$$\left[\hat{R}_j, \hat{R}_l\right] = 2\mathrm{i}(\Omega_N)_{jl}, \tag{4.50}$$

where

$$\Omega_N = \bigoplus_{m=1}^{N} \Omega = \begin{bmatrix} \Omega & & & \\ & \Omega & & \\ & & \ddots & \\ & & & \Omega \end{bmatrix} \left(= -\Omega_N^T = -\Omega_N^{-1}\right), \tag{4.51}$$

is the symplectic form of N modes (in the following we will suppress the subindex N except when needed). Note that the modes are labeled here by numbers, but we can use any other suitable label; for example, when working with two modes shared by Alice and Bob it is natural to denote them by A and B, or in a system–environment scenario by S and E.

The state of the system acts now onto the tensor product of the Hilbert spaces of the modes, and so does the displacement operator, which is now defined as

$$\hat{D}(\mathbf{r}) = \exp\left[\frac{\mathrm{i}}{2}\hat{\mathbf{R}}^T\Omega\mathbf{r}\right] = \hat{D}(\mathbf{r}_1) \otimes \hat{D}(\mathbf{r}_2) \otimes \ldots \otimes \hat{D}(\mathbf{r}_N), \tag{4.52}$$

and satisfies[4] $\mathrm{tr}\{\hat{D}(\mathbf{r})\} = (4\pi)^N \delta^{(2N)}(\mathbf{r})$, as well as (4.42). The characteristic function is defined as before

$$\chi_{\hat{\rho}}(\mathbf{r}) = \mathrm{tr}\left\{\hat{D}(\mathbf{r})\hat{\rho}\right\} \iff \hat{\rho} = \int_{\mathbb{R}^{2N}} \frac{\mathrm{d}^{2N}\mathbf{r}}{(4\pi)^N}\chi_{\hat{\rho}}(\mathbf{r})\hat{D}^\dagger(\mathbf{r}), \tag{4.54}$$

and the Wigner function as its $2N$-dimensional Fourier transform

$$W_{\hat{\rho}}(\mathbf{r}) = \int_{\mathbb{R}^{2N}} \frac{\mathrm{d}^{2N}\mathbf{s}}{(4\pi)^{2N}}\chi_{\hat{\rho}}(\mathbf{s})\mathrm{e}^{\frac{\mathrm{i}}{2}\mathbf{s}^T\Omega\mathbf{r}} \iff \chi_{\hat{\rho}}(\mathbf{s}) = \int_{\mathbb{R}^{2N}} \mathrm{d}^{2N}\mathbf{r}\, W_{\hat{\rho}}(\mathbf{r})\mathrm{e}^{-\frac{\mathrm{i}}{2}\mathbf{s}^T\Omega\mathbf{r}}. \tag{4.55}$$

Symmetric multi-mode moments can be evaluated from the characteristic or Wigner functions similarly to how we did in the single-mode case,

$$\left\langle \left(\hat{X}_1^{m_1}\hat{P}_1^{n_1}\right)^{(s)} \ldots \left(\hat{X}_N^{m_N}\hat{P}_N^{n_N}\right)^{(s)} \right\rangle = \int_{\mathbb{R}^{2N}} \mathrm{d}^{2N}\mathbf{r}\, W_{\hat{\rho}}(\mathbf{r}) x_1^{m_1} p_1^{n_1} \ldots x_N^{m_N} p_N^{n_N} \tag{4.56a}$$

$$= (-1)^{m_1+\cdots+m_N}(2\mathrm{i})^{m_1+n_1+\cdots+m_N+n_N} \frac{\partial^{m_1+n_1+\cdots+m_N+n_N}}{\partial p_1^{m_1}\partial x_1^{n_1}\ldots\partial p_N^{m_N}\partial x_N^{n_N}}\chi_{\hat{\rho}}(\mathbf{r})\bigg|_{\mathbf{r}=\mathbf{0}}. \tag{4.56b}$$

[4] Note that the $2N$-dimensional Dirac delta function can be written as

$$\delta^{(2N)}(\mathbf{r}) = \int_{\mathbb{R}^{2N}} \frac{\mathrm{d}^{2N}\mathbf{s}}{(4\pi)^{2N}}\mathrm{e}^{\frac{\mathrm{i}}{2}\mathbf{s}^T\Omega\mathbf{r}}. \tag{4.53}$$

In the previous chapters, we saw that there is an operation that plays an important role when dealing with composite Hilbert spaces: the partial trace. For example, in the case of the N harmonic oscillators being in a state $\hat{\rho}$, imagine that we want to trace out the last one, obtaining the reduced state of the remaining $N-1$ oscillators, $\hat{\rho}_R = \text{tr}_N\{\hat{\rho}\}$; let us see what this means in phase space. From (4.43) we see that the characteristic function of the reduced state $\hat{\rho}_R$ is just the original one with the phase-space coordinates of the traced oscillator set to zero, that is,

$$\chi_R(\mathbf{r}_{\{N-1\}}) = \text{tr}_{\{N-1\}}\left\{\hat{D}(\mathbf{r}_{\{N-1\}})\hat{\rho}_R\right\} = \text{tr}\left\{\hat{D}(\mathbf{r}_{\{N-1\}})\hat{\rho}\right\} = \chi_{\hat{\rho}}\left(\mathbf{r}_{\{N-1\}}, \mathbf{0}\right), \tag{4.57}$$

where we use the notation $\mathbf{r}_{\{N-1\}} = \text{col}(\mathbf{r}_1, \mathbf{r}_2, \ldots, \mathbf{r}_{N-1})$. Therefore, the Wigner function associated with the reduced state $\hat{\rho}_R$ can be found by integrating out the phase-space variables of the corresponding oscillator, that is,

$$\begin{aligned} W_R(\mathbf{r}_{\{N-1\}}) &= \int_{\mathbb{R}^{2(N-1)}} \frac{\mathrm{d}^{2(N-1)}\mathbf{s}_{\{N-1\}}}{(4\pi)^{2(N-1)}} \chi_R(\mathbf{s}_{\{N-1\}}) e^{\frac{\mathrm{i}}{2}\mathbf{s}_{\{N-1\}}^T \Omega_{N-1} \mathbf{r}_{\{N-1\}}} \\ &= \int_{\mathbb{R}^2} \mathrm{d}^2\mathbf{r}_N \, W_{\hat{\rho}}(\mathbf{r}_{\{N-1\}}, \mathbf{r}_N). \end{aligned} \tag{4.58}$$

4.4 Gaussian continuous-variable systems

4.4.1 Gaussian states

General definition

A particularly important class of quantum-mechanical states of the harmonic oscillator are the so-called *Gaussian states*, that is, states which have a Gaussian Wigner function. As we will see, this is the type of states which are most naturally generated in the laboratory, although we will also show how to design experimental schemes whose purpose is the generation of non-Gaussian states.

The Wigner function of an arbitrary single-mode Gaussian state has the form[5]

$$W(\mathbf{r}) = \frac{1}{2\pi\sqrt{\det V}} \exp\left[-\frac{1}{2}(\mathbf{r} - \bar{\mathbf{r}})^T V^{-1}(\mathbf{r} - \bar{\mathbf{r}})\right], \tag{4.60}$$

where we have defined the *mean vector*

$$\bar{\mathbf{r}} = \langle \hat{\mathbf{R}} \rangle = \text{col}\left(\langle \hat{X} \rangle, \langle \hat{P} \rangle\right), \tag{4.61}$$

[5] When dealing with Gaussian states, the following integral is quite useful:

$$\int_{\mathbb{R}^{2N}} \mathrm{d}^{2N}\mathbf{r} \exp\left(-\frac{1}{2}\mathbf{r}^T A \mathbf{r} + \mathbf{x}^T \mathbf{r}\right) = \frac{(2\pi)^N}{\sqrt{\det A}} \exp\left(\frac{1}{2}\mathbf{x}^T A^{-1} \mathbf{x}\right), \tag{4.59}$$

where $\mathbf{r} \in \mathbb{R}^{2N}$ and A is a non-singular $2N \times 2N$ matrix.

and the *covariance matrix*

$$V = \begin{bmatrix} \langle \delta\hat{X}^2 \rangle & \frac{1}{2}\langle \{ \delta\hat{X}, \delta\hat{P} \} \rangle \\ \frac{1}{2}\langle \{ \delta\hat{X}, \delta\hat{P} \} \rangle & \langle \delta\hat{P}^2 \rangle \end{bmatrix}, \tag{4.62}$$

whose elements are given by

$$V_{jl} = \frac{1}{2}\langle \{ \delta\hat{R}_j, \delta\hat{R}_l \} \rangle. \tag{4.63}$$

In this expression we have used the fluctuation operator $\delta\hat{A} = \hat{A} - \langle \hat{A} \rangle$, and denoted the anticommutator by curly brackets, $\{\hat{A}, \hat{B}\} = \hat{A}\hat{B} + \hat{B}\hat{A}$. Note that number states do not belong to this class of states (save the vacuum, as we show below), but we will show that many other interesting states do. Note also that Gaussian states are completely defined by their first and second moments, which is why we will denote by $\hat{\rho}_{\mathrm{G}}(\bar{\mathbf{r}}, V)$ a given Gaussian state.

Even though the covariance matrix appears here simply as a mathematical object with which we can write the Wigner function in a compact way, we will see that it has a huge physical significance. Indeed, note that Gaussian states have a positive Wigner function, and hence, the intuition of quantum mechanics corresponding to noise on phase space applies to them. As will be clear in the next sections, the covariance matrix contains precisely the way in which this noise is distributed in phase space, that is, it contains all the information about *quantum fluctuations*.

If the the world was classical, any covariance matrix would be allowed, as long as it was real, symmetric, and positive semidefinite. A quantum-mechanical harmonic oscillator has the added constraint $\det\{V\} \geqslant 1$, which comes from the uncertainty principle between position and momentum. Indeed, the following proof is quite reminiscent of the standard proof of the uncertainty principle:

$$\begin{aligned} \det\{V\} &= \langle \delta\hat{X}^2 \rangle \langle \delta\hat{P}^2 \rangle - \frac{1}{4}\langle \{ \delta\hat{X}, \delta\hat{P} \} \rangle^2 \geqslant \left| \langle \delta\hat{X}\delta\hat{P} \rangle \right|^2 - \frac{1}{4}\langle \{ \delta\hat{X}, \delta\hat{P} \} \rangle^2 \\ &= \frac{1}{4}\left| \underbrace{\langle [\delta\hat{X}, \delta\hat{P}] \rangle}_{\text{imaginary}} + \underbrace{\langle \{ \delta\hat{X}, \delta\hat{P} \} \rangle}_{\text{real}} \right|^2 - \frac{1}{4}\langle \{ \delta\hat{X}, \delta\hat{P} \} \rangle^2 = 1, \end{aligned} \tag{4.64}$$

where the inequality comes from the Cauchy–Schwarz inequality. A Gaussian state corresponding to any real, symmetric, positive-semidefinite covariance matrix satisfying this condition is physically achievable.

In the case of Gaussian states of N modes, their Wigner function takes the following form in the whole phase space:

$$W(\mathbf{r}) = \frac{1}{(2\pi)^N \sqrt{\det V}} \exp\left[-\frac{1}{2}(\mathbf{r} - \bar{\mathbf{r}})^T V^{-1} (\mathbf{r} - \bar{\mathbf{r}}) \right], \tag{4.65}$$

where now $\bar{\mathbf{r}}$ is a vector with $2N$ components $\bar{r}_j = \langle \hat{R}_j \rangle$, and V is a $2N \times 2N$ matrix with elements $V_{jl} = \langle \{\delta\hat{R}_j, \delta\hat{R}_l\} \rangle / 2$. It is interesting to note that the mean photon number can be written as

$$\sum_{j=1}^{N} \langle \hat{N}_j \rangle = \frac{1}{4}\mathrm{tr}\{V\} + \frac{1}{4}\bar{\mathbf{r}}^2 - \frac{N}{2}. \tag{4.66}$$

In this case, the necessary and sufficient condition for a real, symmetric $2N \times 2N$ matrix V to correspond to a physical quantum state of N oscillators is [2]

$$V + \mathrm{i}\Omega \geqslant 0, \tag{4.67}$$

which is again linked to the uncertainty principle, and implies the positivity of V. We will prove in section 4.4.3 that any Gaussian state for which the eigenvalues of $V + \mathrm{i}\Omega$ are zero (equivalently, which saturates the uncertainty principle) is pure.

In many situations it is interesting to understand the state of the N oscillators as a bipartite state of M oscillators plus another M' oscillators ($M + M' = N$), in which case we talk about an $M \times M'$ continuous-variable system. Consider a Gaussian state $\hat{\rho}_{\mathrm{G}}(\bar{\mathbf{r}}, V)$ of the N oscillators, whose mean vector and covariance matrix we write as

$$\bar{\mathbf{r}} = \mathrm{col}(\bar{\mathbf{x}}, \bar{\mathbf{x}}') \qquad \text{and} \qquad V = \begin{bmatrix} W & C \\ C^T & W' \end{bmatrix}, \tag{4.68}$$

where $\bar{\mathbf{x}} \in \mathbb{R}^{2M}$, $\bar{\mathbf{x}}' \in \mathbb{R}^{2M'}$, W and W' are real, symmetric matrices of dimensions $2M \times 2M$ and $2M' \times 2M'$, respectively, and C is a real matrix of dimensions $2M \times 2M'$. Then, it is easy to prove that the state of the first M modes after tracing out the remaining M' modes, is the Gaussian state

$$\hat{\rho}_{\mathrm{G}}(\bar{\mathbf{x}}, W) = \mathrm{tr}_{M+1,M+2,\ldots,M+M'}\{\hat{\rho}_{\mathrm{G}}(\bar{\mathbf{r}}, V)\}, \tag{4.69}$$

that is, tracing out a mode in a Gaussian state is equivalent to removing its corresponding entries in the mean vector, as well as its rows and columns in the covariance matrix. In order to prove this we make use of the characteristic function, which for the general Gaussian Wigner function (4.65) takes the form

$$\chi(\mathbf{s}) = \exp\left[-\frac{1}{8}\mathbf{s}^T\Omega V \Omega^T \mathbf{s} + \frac{\mathrm{i}}{4}\bar{\mathbf{r}}^T\Omega\mathbf{s}\right]. \tag{4.70}$$

Then, by substituting (4.68) in this expression, and remembering that tracing out any mode is equivalent to setting to zero the corresponding phase-space variables in the characteristic function, we can write the characteristic function of the reduced state of the first M modes as

$$\chi_R(\mathbf{s}_{\{M\}}) = \exp\left[-\frac{1}{8}\mathbf{s}_{\{M\}}^T\Omega_M W \Omega_M \mathbf{s}_{\{M\}} + \frac{\mathrm{i}}{4}\bar{\mathbf{x}}^T\Omega_M\mathbf{s}_{\{M\}}\right], \tag{4.71}$$

which proves (4.69).

Finally, note that when the states of the partitions are uncorrelated, that is,

$$\hat{\rho}_G(\bar{\mathbf{r}}, V) = \hat{\rho}_G(\bar{\mathbf{x}}, W) \otimes \hat{\rho}_G(\bar{\mathbf{x}}', W'), \tag{4.72}$$

the covariance matrix of the Wigner function (4.65) can be written as a direct sum

$$V = W \oplus W'. \tag{4.73}$$

Examples of Gaussian states

The vacuum state. As commented above, number states are not Gaussian, with one exception: the *vacuum state* $|0\rangle$. To see this, just note that its Wigner function (4.34) can be written as

$$W_{|0\rangle}(x, p) = \frac{1}{2\pi} \exp\left(-\frac{x^2 + p^2}{2}\right), \tag{4.74}$$

that is, a Gaussian distribution such as (4.60) with

$$\bar{\mathbf{r}} = \mathbf{0} \qquad \text{and} \qquad V = \begin{bmatrix} 1 & 0 \\ 0 & 1 \end{bmatrix} \equiv I_{2\times2}. \tag{4.75}$$

Hence, in Gaussian notation $|0\rangle\langle 0| = \hat{\rho}_G(\mathbf{0}, I_{2\times2})$.

It is interesting to note that the vacuum state is a *minimum-uncertainty state*, that is, it carries the minimum amount of noise allowed by the uncertainty relations, $\Delta X \Delta P = 1$. Moreover, when the oscillator is in the vacuum state, its quantum fluctuations are equally distributed among position and momentum, $\Delta X = \Delta P = 1$. Indeed, this minimal quantum noise (commonly denoted by *shot noise* in quantum optics) is distributed homogeneously along all directions of phase space, as can be appreciated in figure 4.2.

Thermal states. As explained in sections 1.3.8 and 2.1, the mixedness of the state of a system can be related to the amount of information which has been lost to another system inaccessible to us, that is, to the correlations shared with this second system. Given a system whose associated Hilbert space has dimension d, and is spanned by some orthonormal basis $\{|j\rangle\}_{j=1,2,\ldots,d}$, we have already seen that its maximally mixed state is

$$\hat{\rho}_{MM} = \frac{1}{d}\sum_{j=1}^{d} |j\rangle\langle j| = \frac{1}{d}\hat{I}. \tag{4.76}$$

As it is proportional to the identity, this state is invariant under changes of basis and, hence, the eigenvalues of any observable of the system are equally likely. This is in concordance with what one expects intuitively from a state which has leaked the maximum amount of information to another system.

For infinite-dimensional Hilbert spaces ($d \to \infty$) this state is not physical since it has infinite excitations, that is, $\mathrm{tr}\{\hat{\rho}\hat{N}\} \to \infty$. Hence, finding the maximally mixed state in infinite dimensions makes sense only if one adds an 'energy' constraint such as $\text{tr}\{\hat{\rho}\hat{N}\} = \bar{n}$, where $\bar{n}$ is positive real. It is possible to show that the state which maximizes the von Neumann entropy subject to this constraint is

$$\hat{\rho}_{\text{th}}(\bar{n}) = \sum_{n=0}^{\infty} \frac{\bar{n}^n}{(1+\bar{n})^{1+n}} |n\rangle\langle n|, \tag{4.77}$$

which is still diagonal in the Fock basis, but does not have a flat distribution for the photon number. Interestingly, given the free Hamiltonian of the oscillator $\hat{H} = \hbar\omega(\hat{a}^\dagger\hat{a} + 1/2)$, the state can alternatively be written as

$$\hat{\rho}_{\text{th}}(\bar{n}) = \frac{\exp\left(-\beta\hat{H}\right)}{\text{tr}\left\{\exp\left(-\beta\hat{H}\right)\right\}}, \tag{4.78}$$

which, with the identification $\bar{n} = 1/(e^{\beta\hbar\omega} - 1)$, corresponds to the state expected for bosons at thermal equilibrium at temperature $T = 1/k_B\beta$, where k_B is the Boltzmann constant. This is why this state is known as the *thermal state*, whose von Neumann entropy reads

$$S\left[\hat{\rho}_{\text{th}}(\bar{n})\right] = (\bar{n}+1)\log(\bar{n}+1) - \bar{n}\log\bar{n} \equiv S_{\text{th}}(\bar{n}), \tag{4.79}$$

as is easily proven given that the state is diagonal in the Fock basis.

It is not difficult to see that this state is Gaussian (later on we will actually prove it by simple means), and that it is defined by a zero mean vector, and a covariance matrix

$$V_{\text{th}}(\bar{n}) = (2\bar{n}+1)\mathcal{I}_{2\times 2}, \tag{4.80}$$

that is, $\hat{\rho}_{\text{th}}(\bar{n}) = \hat{\rho}_G[\mathbf{0}, V_{\text{th}}(\bar{n})]$. The corresponding Wigner function can be seen in figure 4.3, where it can be appreciated that a thermal state is similar to a vacuum state, but with more noise. Consequently, the vacuum state can be seen as a thermal state with zero mean photon number.

In the next section we will learn that any N-mode Gaussian state can be decomposed into N uncorrelated thermal states (Williamson's theorem), and hence thermal states can be seen as the most fundamental Gaussian states.

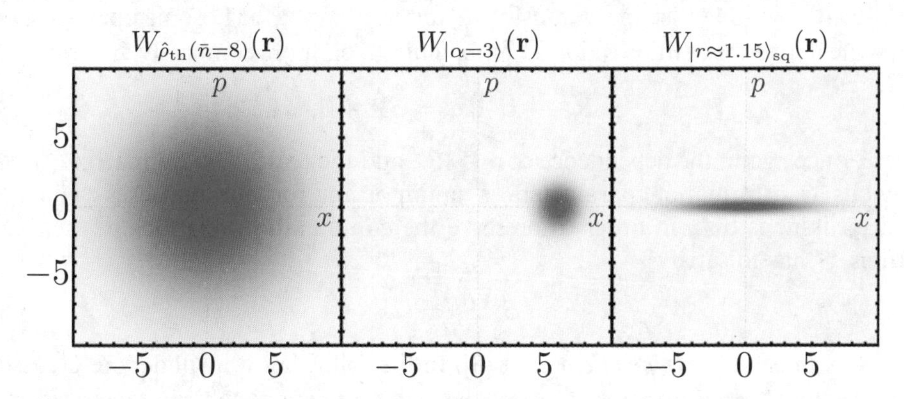

Figure 4.3. Density plot of the Wigner functions corresponding to a thermal state $\hat{\rho}_{\text{th}}(\bar{n})$, a coherent state $|\alpha\rangle$, and a squeezed state $|r\rangle_{\text{sq}}$.

4.4.2 Gaussian unitaries

General definition

Consider a unitary transformation $\hat{U} = \exp(-\mathrm{i}\hat{H})$ acting on the state of the oscillator. We say that such a unitary is Gaussian when it maps Gaussian states into Gaussian states.

Let us define the vector operator $\hat{\mathbf{a}} = \mathrm{col}(\hat{a}_1, \hat{a}_2, \ldots, \hat{a}_N)$. It is quite intuitive that any Gaussian unitary will come from a Hamiltonian having only linear or bilinear terms, that is,

$$\hat{H} = \hat{\mathbf{a}}^{\dagger}\boldsymbol{\alpha} + \hat{\mathbf{a}}^{\dagger}\mathcal{F}\hat{\mathbf{a}} + \hat{\mathbf{a}}^{\dagger}\mathcal{G}\hat{\mathbf{a}}^{\dagger T} + \mathrm{H.\ c.}, \tag{4.81}$$

for some vector $\boldsymbol{\alpha} \in \mathbb{C}^N$, and some symmetric, complex $N \times N$ matrices $\mathcal{F}$ and $\mathcal{G}$. Let us try to understand why this is so by analyzing the physical meaning of each term. The first term corresponds to the injection of excitations in the modes 'one by one'. The second term comprises all the energy shifts $\hat{a}_j^{\dagger}\hat{a}_j$, as well as the creation of a photon in one mode via the annihilation of a photon of another mode, $\hat{a}_j^{\dagger}\hat{a}_l$, that is, a 'photon-by-photon' exchange. The last term takes into account the possibility of generating two photons simultaneously, $\hat{a}_j^{\dagger}\hat{a}_l^{\dagger}$. In other words, all the possible one-photon and two-photon processes are taken into account in this Hamiltonian. It is intuitive that if we want to transform Gaussian states into Gaussian states, the Hamiltonian cannot contain multi-photon processes beyond these, because otherwise one would create correlations which go beyond the ones captured by first and second moments.

In the Heisenberg picture the Gaussian unitary induces then a linear transformation (known as *Bogoliubov transformation*) of the type

$$\hat{\mathbf{a}} \longrightarrow \hat{U}^{\dagger}\hat{\mathbf{a}}\hat{U} = \mathcal{A}\hat{\mathbf{a}} + \hat{\mathbf{a}}^{\dagger}\mathcal{B} + \boldsymbol{\alpha}, \tag{4.82}$$

where the form of the complex $N \times N$ matrices $\mathcal{A}$ and $\mathcal{B}$ in terms of $(\boldsymbol{\alpha}, \mathcal{F}, \mathcal{G})$ is unimportant for our purposes. The only restriction on these is that they have to satisfy $\mathcal{A}\mathcal{B} = \mathcal{B}^T\mathcal{A}^T$ and $\mathcal{A}\mathcal{A}^{\dagger} = \mathcal{B}^{\dagger}\mathcal{B} + \mathcal{I}_{N\times N}$, in order to preserve the commutation relations of the creation and annihilation operators ($\mathcal{I}_{N\times N}$ is the identity matrix of dimension N).

Instead of writing this linear transformation in terms of the bosonic operators, one can write it in terms of the position and momentum, or, more compactly, in terms of $\hat{\mathbf{R}}$:

$$\hat{\mathbf{R}} \longrightarrow \hat{U}^{\dagger}\hat{\mathbf{R}}\hat{U} = \mathcal{S}\hat{\mathbf{R}} + \mathbf{d}, \tag{4.83}$$

where, once again, the dependence of $\mathbf{d} \in \mathbb{R}^{2N}$ and the real $2N \times 2N$ matrix $\mathcal{S}$ in the previous transformation parameters is unimportant for our purposes. The only relevant issue is that, in order to preserve the commutation relations of the quadratures, $\mathcal{S}$ has to satisfy

$$\mathcal{S}\Omega\mathcal{S}^T = \Omega, \tag{4.84}$$

that is, $\mathcal{S}$ must be a *symplectic matrix* [3]. In the following we will denote Gaussian unitaries by $\hat{U}_{\mathrm{G}}(\mathbf{d}, \mathcal{S})$ to stress the fact that they are completely characterized by $\mathbf{d}$ and $\mathcal{S}$.

The transformation induced onto the system by the Gaussian unitary $\hat{U}_{\mathrm{G}}(\mathbf{d}, \mathcal{S})$ is easily described in the Schrödinger picture as well if the states are represented by the Wigner function. To see this, let us first note that the displacement operator is transformed by the action of this unitary as

$$\hat{U}_{\mathrm{G}}^{\dagger}\hat{D}(\mathbf{r})\hat{U}_{\mathrm{G}} = \exp\left[\frac{\mathrm{i}}{2}\left(\mathcal{S}\hat{\mathbf{R}} + \mathbf{d}\right)^T \Omega\mathbf{r}\right]$$

$$\underset{\mathcal{S}^T\Omega=\Omega\mathcal{S}^{-1}}{\equiv} \exp\left[\frac{\mathrm{i}}{2}\left(\hat{\mathbf{R}}^T\Omega\mathcal{S}^{-1} + \mathbf{d}^T\Omega\right)\mathbf{r}\right] = \hat{D}\left(\mathcal{S}^{-1}\mathbf{r}\right)\exp\left[\frac{\mathrm{i}}{2}\mathbf{d}^T\Omega\mathbf{r}\right], \quad (4.85)$$

where the identity $\mathcal{S}^T\Omega = \Omega\mathcal{S}^{-1}$ follows from (4.84) and (4.51). Therefore, given the initial state $\hat{\rho}$ of the system, the corresponding characteristic function is transformed as

$$\chi_{\hat{U}_{\mathrm{G}}\hat{\rho}\hat{U}_{\mathrm{G}}^{\dagger}}(\mathbf{r}) = \mathrm{tr}\left\{\hat{D}(\mathbf{r})\hat{U}_{\mathrm{G}}\hat{\rho}\hat{U}_{\mathrm{G}}^{\dagger}\right\} = \mathrm{e}^{\frac{\mathrm{i}}{2}\mathbf{d}^T\Omega\mathbf{r}}\mathrm{tr}\left\{\hat{D}\left(\mathcal{S}^{-1}\mathbf{r}\right)\hat{\rho}\right\} = \mathrm{e}^{-\frac{\mathrm{i}}{2}\mathbf{r}^T\Omega\mathbf{d}}\chi_{\hat{\rho}}(\mathcal{S}^{-1}\mathbf{r}), \quad (4.86)$$

and the Wigner function as

$$W_{\hat{U}_{\mathrm{G}}\hat{\rho}\hat{U}_{\mathrm{G}}^{\dagger}}(\mathbf{r}) = \int_{\mathbb{R}^{2N}} \frac{\mathrm{d}^{2N}\mathbf{s}}{(4\pi)^{2N}}\chi_{\hat{\rho}}\left(\mathcal{S}^{-1}\mathbf{s}\right)\mathrm{e}^{\frac{\mathrm{i}}{2}\mathbf{s}^T\Omega(\mathbf{r}-\mathbf{d})} \underset{\mathbf{z}=\mathcal{S}^{-1}\mathbf{s}}{\equiv} \int_{\mathbb{R}^{2N}} \frac{\mathrm{d}^{2N}\mathbf{z}}{(4\pi)^{2N}}\chi_{\hat{\rho}}(\mathbf{z})\mathrm{e}^{\frac{\mathrm{i}}{2}\mathbf{z}^T\mathcal{S}^T\Omega(\mathbf{r}-\mathbf{d})}$$

$$\underset{\mathcal{S}^T\Omega=\Omega\mathcal{S}^{-1}}{\equiv} W_{\hat{\rho}}[\mathcal{S}^{-1}(\mathbf{r} - \mathbf{d})]. \quad (4.87)$$

In the case of Gaussian states, the situation is even simpler: one only needs to find the effect of the transformation onto the first and second moments of the state. Using (4.83), it is straightforward to show that the transformation induced by the Gaussian unitary $\hat{U}_{\mathrm{G}}(\mathbf{d}, \mathcal{S})$ on the mean vector $\bar{\mathbf{r}}$ and the covariance matrix V of any state is

$$\bar{\mathbf{r}} \longrightarrow \mathcal{S}\bar{\mathbf{r}} + \mathbf{d} \qquad \text{and} \qquad V \longrightarrow \mathcal{S}V\mathcal{S}^T. \quad (4.88)$$

Taking the $M \times M'$ partition of the multi-mode system as in the previous section, note that when the unitary transformation acts independently on each partition, that is,

$$\hat{U}_{\mathrm{G}}(\mathbf{d}, \mathcal{S}) = \hat{U}_{\mathrm{G}}(\mathbf{l}, \mathcal{L}) \otimes \hat{U}_{\mathrm{G}}(\mathbf{l}', \mathcal{L}'), \quad (4.89)$$

where $\mathbf{l} \in \mathbb{R}^{2M}$, $\mathbf{l}' \in \mathbb{R}^{2M'}$, and $\mathcal{L}$ and $\mathcal{L}'$ are symplectic matrices of dimensions $2M \times 2M$ and $2M' \times 2M'$, respectively, we can write

$$\mathbf{d} = (\mathbf{l}, \mathbf{l}') \qquad \text{and} \qquad \mathcal{S} = \mathcal{L} \oplus \mathcal{L}'. \quad (4.90)$$

Finally, its important to remark that a Gaussian unitary transformation is called *passive* when it conserves the mean photon number $\sum_{j=1}^{N}\langle\hat{N}_j\rangle$ of Gaussian states, and *active* if it changes it. Now, given the transformation (4.88), and keeping in mind that the mean photon number is proportional to the square of the mean vector, $\bar{\mathbf{r}}^2$, and the trace of the covariance matrix, $\text{tr}\{V\}$, see (4.66), a Gaussian unitary will be passive if and only if

$$\mathbf{d} = 0 \qquad \text{and} \qquad \mathcal{S}^T\mathcal{S} = \mathcal{I}_{2N\times 2N}, \tag{4.91}$$

the second condition meaning that its associated symplectic transformation must be *orthogonal*, that is, $\mathcal{S}^T = \mathcal{S}^{-1}$.

Examples of Gaussian unitaries and more Gaussian states

The displacement operator and coherent states. Consider the unitary operator

$$\hat{D}(\alpha) = \exp(\alpha\hat{a}^\dagger - \alpha^*\hat{a}), \tag{4.92}$$

which, using the formula (4.37) can be written in the equivalent way

$$\hat{D}(\alpha) = \exp(-|\alpha|^2/2)\exp(\alpha\hat{a}^\dagger)\exp(-\alpha^*\hat{a}), \tag{4.93}$$

which we will refer to as its *normal form.*

Using the Baker–Campbell–Haussdorf lemma (4.16), it is fairly simple to prove that this operator transforms the annihilation operator as

$$\hat{a} \to \hat{D}^\dagger(\alpha)\hat{a}\hat{D}(\alpha) = \hat{a} + \alpha, \tag{4.94}$$

or, in terms of the quadratures

$$\hat{X} \to \hat{D}^\dagger(\alpha)\hat{X}\hat{D}(\alpha) = \hat{X} + x_\alpha, \tag{4.95a}$$

$$\hat{P} \to \hat{D}^\dagger(\alpha)\hat{P}\hat{D}(\alpha) = \hat{P} + p_\alpha, \tag{4.95b}$$

where $x_\alpha = \alpha^* + \alpha$ and $p_\alpha = \mathrm{i}(\alpha^* - \alpha)$. This unitary operator is then called the *displacement operator* because it allows us to perform translations in phase space. Indeed, it is exactly the operator that we defined in the previous section, see (4.36), what is easily shown by rewriting (4.92) in terms of the position and momentum operators.

As a Gaussian unitary we then write $\hat{D}(\alpha) = \hat{U}_\mathrm{G}(\mathbf{d}_\alpha, \mathcal{I}_{2\times 2})$ with

$$\mathbf{d}_\alpha = \text{col}(x_\alpha, p_\alpha). \tag{4.96}$$

The states obtained by displacing the vacuum state are known as *coherent states.* Using the normal form of the displacement operator, it is easy to obtain

$$|\alpha\rangle = \hat{D}(\alpha)|0\rangle = \sum_{n=0}^{\infty}\frac{\exp(-|\alpha|^2/2)\alpha^{n/2}}{\sqrt{n!}}|n\rangle. \tag{4.97}$$

These are Gaussian states with the same covariance matrix as the vacuum, but with a non-zero mean vector, that is,

$$\bar{\mathbf{r}} = \mathbf{d}_\alpha \qquad \text{and} \qquad V = \mathcal{I}_{2\times 2}, \tag{4.98}$$

or, in Gaussian notation, $|\alpha\rangle\langle\alpha| = \hat{\rho}_\mathrm{G}(\mathbf{d}_\alpha, \mathcal{I}_{2\times 2})$. Hence, these states have the same noise properties as the vacuum; however, they describe a bright mode whose mean is shifted away from the origin of phase space, as shown in figure 4.3. In this sense, they can be interpreted as the 'most classical' states, since they carry the minimal amount of noise that quantum mechanics allows (the shot noise), equally distributed along all directions of phase space, with α playing the role of the (normalized) classical normal variable of the oscillator or the average field that would be measured in an experiment. In the case of light, they are a fair approximation to the state describing the beam coming out from a (phase-locked) laser.

From a mathematical point of view, they are the eigenstates of the annihilation operator, that is, $\hat{a}|\alpha\rangle = \alpha|\alpha\rangle$. Later on we will learn that, even though the annihilation operator is not self-adjoint, we can build a POVM-based measurement whose possible outcomes are the eigenvalues α (*heterodyne detection).*

The squeezing operator and squeezed states. Consider now the *squeezing operator*

$$\hat{S}(r) = \exp\left(\frac{r}{2}\hat{a}^{\dagger 2} - \frac{r}{2}\hat{a}^2\right), \tag{4.99}$$

where $r \in [0, \infty[$. This operator is implemented experimentally for an optical mode of frequency ω_0 by pumping a non-linear crystal with a strong laser beam of twice that frequency; pairs of photons at frequency ω_0 are generated via the so-called *spontaneous parametric down-conversion process* (see [4] and references therein).

Using the Baker–Campbell–Haussdorf lemma (4.16), it is again simple to prove that this operator transforms the annihilation operator as

$$\hat{a} \to \hat{S}^{\dagger}(r)\hat{a}\hat{S}(r) = \hat{a}\cosh r + \hat{a}^{\dagger}\sinh r, \tag{4.100}$$

or, in terms of the quadratures

$$\hat{X} \to \hat{S}^{\dagger}(r)\hat{X}\hat{S}(r) = \mathrm{e}^{r}\hat{X}, \tag{4.101a}$$

$$\hat{P} \to \hat{S}^{\dagger}(r)\hat{P}\hat{S}(r) = \mathrm{e}^{-r}\hat{P}, \tag{4.101b}$$

so that it is characterized as a Gaussian unitary by $\hat{S}(r) = \hat{U}_{\mathrm{G}}[\mathbf{0}, Q(r)]$ with

$$Q(r) = \begin{bmatrix} \mathrm{e}^{r} & 0 \\ 0 & \mathrm{e}^{-r} \end{bmatrix}. \tag{4.102}$$

The application of the squeezing operator to the vacuum state leads to a so-called *squeezed vacuum state*. In the Fock basis, this state is characterized by containing only even number states, which arises from the fact that the squeezing operator generates pairs of excitations. Its explicit representation in this basis is [5]

$$|r\rangle_{\mathrm{sq}} = \hat{S}(r)|0\rangle = \sum_{n=0}^{\infty}\frac{1}{2^n n!}\sqrt{\frac{(2n)!}{\cosh r}}\tanh^n r\;|2n\rangle. \tag{4.103}$$

This Gaussian state has a zero mean, and its covariance matrix is

$$V_{\mathrm{sq}}(r) = Q(r)Q^{T}(r) = \begin{bmatrix} \mathrm{e}^{2r} & 0 \\ 0 & \mathrm{e}^{-2r} \end{bmatrix}, \tag{4.104}$$

that is, $|r\rangle_{\rm sq}\langle r| = \hat{\rho}_{\rm G}[\mathbf{0},\ V_{\rm sq}(r)]$. This state is then a minimum-uncertainty state (since it saturates the uncertainty relation, $\Delta X \Delta P = 1$), which shows a reduced variance of the momentum with respect to the vacuum state, as shown in figure 4.3. Note that in the limit $r \to \infty$ the variance of the momentum goes to zero, while the variance of the position goes to infinity, and hence in the limit of infinite squeezing the state (4.103) is an eigenstate of the momentum operator. Note, however, that this limit is unphysical, as the mean photon number $\langle \hat{N} \rangle = \sinh^2 r$ diverges, and hence, an infinite amount of energy is required for the generation of a position or momentum eigenstate.

The phase-shift operator. The free evolution of an oscillator (corresponding to the free propagation of an optical mode through a linear medium) induces the unitary transformation

$$\hat{R}(\theta) = \exp(-\mathrm{i}\theta \hat{a}^\dagger \hat{a}), \tag{4.105}$$

known as the *phase-shift operator*, which transforms the annihilation operator as

$$\hat{a} \to \hat{R}^\dagger(\theta)\hat{a}\hat{R}(\theta) = \exp(\mathrm{i}\theta)\hat{a}, \tag{4.106}$$

or in terms of the position and momentum

$$\hat{X} \to \hat{R}^\dagger(\theta)\hat{X}\hat{R}(\theta) = \hat{X}\cos\theta + \hat{P}\sin\theta, \tag{4.107a}$$

$$\hat{P} \to \hat{R}^\dagger(\theta)\hat{P}\hat{R}(\theta) = \hat{P}\cos\theta - \hat{X}\sin\theta. \tag{4.107b}$$

Hence, as a Gaussian unitary this transformation is characterized by $\hat{R}(\theta) = \hat{U}_{\rm G}[\mathbf{0}, \mathcal{R}(\theta)]$, where

$$\mathcal{R}(\theta) = \begin{bmatrix} \cos\theta & \sin\theta \\ -\sin\theta & \cos\theta \end{bmatrix}, \tag{4.108}$$

which shows that a phase shift is equivalent to a *proper rotation* in phase space.

Note that number states are invariant under this transformation since they are eigenstates of $\hat{R}(\theta)$, and hence, thermal states are invariant under rotations in phase space. This is not the case for coherent or squeezed states, as is clear from figure 4.3.

The two-mode squeezing operator and two-mode squeezed states. All the unitaries considered so far act on a single mode, and hence, they cannot be used to induce entanglement between several modes. In this example we consider the *two-mode squeezing operator*

$$\hat{S}_{12}(r) = \exp\left(r\hat{a}_1^\dagger\hat{a}_2^\dagger - r\hat{a}_1\hat{a}_2\right), \tag{4.109}$$

which can be implemented experimentally via a non-linear crystal similarly to the squeezing operator (4.99), but now in a regime in which the down-converted photons are distinguishable either in frequency, and/or polarization, and/or spatial mode [4].

Under the action of this operator, the annihilation operators transform as

$$\hat{a}_1 \to \hat{S}_{12}^\dagger(r)\hat{a}_1\hat{S}_{12}(r) = \hat{a}_1\cosh r + \hat{a}_2^\dagger \sinh r, \tag{4.110a}$$

$$\hat{a}_2 \rightarrow \hat{S}_{12}^{\dagger}(r)\hat{a}_2\hat{S}_{12}(r) = \hat{a}_2 \cosh r + \hat{a}_1^{\dagger} \sinh r, \tag{4.110b}$$

or, in terms of the quadratures

$$\hat{X}_1 \rightarrow \hat{S}_{12}^{\dagger}(r)\hat{X}_1\hat{S}_{12}(r) = \hat{X}_1 \cosh r + \hat{X}_2 \sinh r, \tag{4.111a}$$

$$\hat{P}_1 \rightarrow \hat{S}_{12}^{\dagger}(r)\hat{P}_1\hat{S}_{12}(r) = \hat{P}_1 \cosh r - \hat{P}_2 \sinh r, \tag{4.111b}$$

$$\hat{X}_2 \rightarrow \hat{S}_{12}^{\dagger}(r)\hat{X}_2\hat{S}_{12}(r) = \hat{X}_2 \cosh r + \hat{X}_1 \sinh r, \tag{4.111c}$$

$$\hat{P}_2 \rightarrow \hat{S}_{12}^{\dagger}(r)\hat{P}_2\hat{S}_{12}(r) = \hat{P}_2 \cosh r - \hat{P}_1 \sinh r. \tag{4.111d}$$

Hence, as a Gaussian unitary, this transformation is characterized by $\hat{S}_{12}(r) = \hat{U}_G[\mathbf{0}, \mathcal{Q}_{12}(r)]$ with

$$\mathcal{Q}_{12}(r) = \begin{bmatrix} \mathcal{I}_{2\times 2} \cosh r & \mathcal{Z} \sinh r \\ \mathcal{Z} \sinh r & \mathcal{I}_{2\times 2} \cosh r \end{bmatrix}, \tag{4.112}$$

where $\mathcal{Z} = \text{diag}(1, -1)$.

Applying the two-mode squeezing operator to a vacuum state, one obtains a so-called *two-mode squeezed vacuum state*. In the number state basis, this state is characterized by a perfectly correlated statistics of the number of quanta in the modes, which, once again, is a result of the fact that the two-mode squeezing operator generates pairs of excitations. Its explicit representation in this basis is [5]

$$|r\rangle_{2\text{sq}} = \hat{S}_{12}(r)|0, 0\rangle = \frac{1}{\cosh r} \sum_{n=0}^{\infty} \tanh^n r |n, n\rangle, \tag{4.113}$$

where we have used the notation $|n\rangle \otimes |m\rangle \equiv |n, m\rangle$. This Gaussian state has a zero mean, and its covariance matrix is

$$\mathcal{V}_{2\text{sq}}(r) = \mathcal{Q}_{12}(r)\mathcal{Q}_{12}^T(r) = \begin{bmatrix} \mathcal{I}_{2\times 2} \cosh 2r & \mathcal{Z} \sinh 2r \\ \mathcal{Z} \sinh 2r & \mathcal{I}_{2\times 2} \cosh 2r \end{bmatrix}, \tag{4.114}$$

that is, $|r\rangle_{2\text{sq}}\langle r| = \hat{\rho}[\mathbf{0}, \mathcal{V}_{2\text{sq}}(r)]$.

Note that by taking the partial trace with respect to any of its two modes, the two-mode squeezed vacuum state becomes a thermal state with mean photon number $\bar{n} = \sinh^2 r$, that is $\text{tr}_2\{|r\rangle_{2\text{sq}}\langle r|\} = \text{tr}_1\{|r\rangle_{2\text{sq}}\langle r|\} = \hat{\rho}_{\text{th}}(\sinh^2 r)$, and hence the two-mode squeezed vacuum state can be seen as the purification of a thermal state, which in addition shows that it is the maximally entangled state in infinite dimensions for a fixed energy. We will come back to the entanglement properties of the two-mode squeezed vacuum state in section 4.4.4.

The beam-splitter operator. We are going to analyze only one more type of two-mode unitary transformations, the one induced by the so-called *beam-splitter operator*

$$\hat{B}_{12}(\beta) = \exp\left(\beta \hat{a}_1 \hat{a}_2^\dagger - \beta \hat{a}_1^\dagger \hat{a}_2\right), \tag{4.115a}$$

which can be implemented experimentally on light by, for example, mixing two optical beams in a beam splitter of *transmissivity* $T = \cos^2\beta$.

Under the action of this operator, the annihilation operators transform as

$$\hat{a}_1 \to \hat{B}^\dagger(\beta)\hat{a}_1\hat{B}(\beta) = \hat{a}_1 \cos\beta + \hat{a}_2 \sin\beta \tag{4.116}$$

$$\hat{a}_2 \to \hat{B}^\dagger(\beta)\hat{a}_2\hat{B}(\beta) = \hat{a}_2 \cos\beta - \hat{a}_1 \sin\beta. \tag{4.117}$$

or, in terms of the quadratures

$$\hat{X}_1 \to \hat{B}_{12}^\dagger(\beta)\hat{X}_1\hat{B}_{12}(\beta) = \hat{X}_1 \cos\beta - \hat{X}_2 \sin\beta, \tag{4.118}$$

$$\hat{P}_1 \to \hat{B}_{12}^\dagger(\beta)\hat{P}_1\hat{B}_{12}(\beta) = \hat{P}_1 \cos\beta - \hat{P}_2 \sin\beta, \tag{4.119}$$

$$\hat{X}_2 \to \hat{B}_{12}^\dagger(\beta)\hat{X}_2\hat{B}_{12}(\beta) = \hat{X}_2 \cos\beta + \hat{X}_1 \sin\beta, \tag{4.120}$$

$$\hat{P}_2 \to \hat{B}_{12}^\dagger(\beta)\hat{P}_2\hat{B}_{12}(\beta) = \hat{P}_2 \cos\beta + \hat{P}_1 \sin\beta. \tag{4.121}$$

Hence, as a Gaussian unitary, this transformation is characterized by $\hat{B}_{12}(\beta) = \hat{U}_G[\mathbf{0}, \mathcal{B}_{12}(\beta)]$ with

$$\mathcal{B}_{12}(\beta) = \begin{bmatrix} \mathcal{I}_{2\times2}\cos\beta & -\mathcal{I}_{2\times2}\sin\beta \\ \mathcal{I}_{2\times2}\sin\beta & \mathcal{I}_{2\times2}\cosh\beta \end{bmatrix}. \tag{4.122}$$

Interestingly, note that when the states of both modes are coherent, they remain coherent after the action of the beam-splitter transformation, as $\mathcal{B}_{12}(\beta)\mathcal{B}_{12}^T(\beta) = \mathcal{I}_{4\times4}$. As an example, consider the state $|\alpha\rangle \otimes |0\rangle$ (one mode in an arbitrary coherent state, and the other in vacuum), which has the Gaussian representation $\hat{\rho}_G(\mathbf{d}, \mathcal{I}_{4\times4})$ with

$$\mathbf{d} = 2\text{col}(\text{Re}\{\alpha\}, \text{Im}\{\alpha\}, 0, 0); \tag{4.123}$$

after the action of the beam-splitter operator, it becomes $\hat{\rho}'_G(\mathbf{d}', \mathcal{I}_{4\times4})$ with

$$\mathbf{d}' = 2\,\text{col}(\text{Re}\{\alpha\}\cos\beta, \text{Im}\{\alpha\}\cos\beta, \text{Re}\{\alpha\}\sin\beta, \text{Im}\{\alpha\}\sin\beta), \tag{4.124}$$

which is the tensor product of two coherent states, in particular, $|\alpha\cos\beta\rangle \otimes |\alpha\sin\beta\rangle$. This is exactly what one expects when a laser field is sent through a beam splitter: part of the laser is transmitted, and part is reflected.

4.4.3 General Gaussian unitaries and states

In this section we will use symplectic analysis (or better, 'symplectic tricks'), to find interesting facts about general Gaussian unitary transformations and Gaussian states.

Bloch–Messiah reduction: connection between single- and two-mode squeezed states, and the canonical form of general single-mode Gaussian states
It is well known in symplectic analysis [3, 6] that any $2N \times 2N$ symplectic matrix $\mathcal{S}$ can be decomposed as

$$\mathcal{S} = \mathcal{K}\left[\bigoplus_{j=1}^{N} \mathcal{Q}(r_j)\right]\mathcal{L}, \tag{4.125}$$

where $\mathcal{K}$ and $\mathcal{L}$ are orthogonal, symplectic matrices (this is known as the *Euler decomposition* of a symplectic matrix, or as its *Bloch–Messiah reduction*). Physically, this means that a general N-mode unitary transformation can be seen as the concatenation of three operations: an N-port interferometer mixing all the modes[6], N single-mode squeezers acting independently on each mode, and a second N-port interferometer.

As an important example involving two modes, note that the two-mode squeezing transformation can be written as

$$\mathcal{Q}_{12}(r) = \mathcal{B}_{12}\left(-\frac{\pi}{4}\right)[\mathcal{Q}(-r) \oplus \mathcal{Q}(r)]\mathcal{B}_{12}\left(\frac{\pi}{4}\right), \tag{4.126}$$

which, in the Hilbert space, means that two-mode squeezed vacuum states can be obtained by mixing a position-squeezed state with a momentum-squeezed state in a 50/50 beam splitter, that is,

$$|r\rangle_{2\text{sq}} = \hat{B}_{12}\left(-\tfrac{\pi}{4}\right)\left[|-r\rangle_{\text{sq}} \otimes |r\rangle_{\text{sq}}\right]. \tag{4.127}$$

Note that the first beam splitter disappears because the two-mode vacuum state $|0\rangle \otimes |0\rangle$ is invariant under passive transformations.

As a second example, note that, for a single mode, the only passive transformations are the rotations in phase space, which means that an arbitrary single-mode Gaussian unitary can be written as the concatenation of a phase shift, a squeezing operation, a second phase shift, and a final displacement, that is,

$$\hat{U}_{\text{G}}(\alpha, \theta, r, \phi) = \hat{D}(\alpha)\hat{R}(\theta)\hat{S}(r)\hat{R}(\phi). \tag{4.128}$$

Now, it is quite intuitive (and we formalize it in the next section) that any Gaussian state $\hat{\rho}_{\text{G}}$ having von Neumann entropy S_0 can be obtained by applying a unitary transformation on the thermal state $\hat{\rho}_{\text{th}}(\bar{n}_0)$ with the same entropy, $S_{\text{th}}(\bar{n}_0) = S_0$, that is,

$$\hat{\rho}_{\text{G}}(\alpha, \theta, r, \bar{n}_0) = \hat{D}(\alpha)\hat{R}(\theta)\hat{S}(r)\hat{\rho}_{\text{th}}(\bar{n}_0)\hat{S}^{\dagger}(r)\hat{R}^{\dagger}(\theta)\hat{D}^{\dagger}(\alpha), \tag{4.129}$$

[6] In optics, an interferometer is just a collection of beam splitters which mix optical beams entering through their input ports. They correspond to the most general passive Gaussian unitary, and are described by a concatenation of single-mode phase shifts and two-mode beam splitters.

which implies that the covariance matrix of any single-mode Gaussian state can always be decomposed as

$$V(\theta, r, \bar{n}_0) = (2\bar{n}_0 + 1)\mathcal{R}(\theta)\mathcal{Q}(2r)\mathcal{R}^T(\theta)$$
$$= (2\bar{n}_0 + 1)\begin{bmatrix} \cosh 2r + \cos 2\theta \sinh 2r & -\sin 2\theta \sinh 2r \\ -\sin 2\theta \sinh 2r & \cosh 2r - \cos 2\theta \sinh 2r \end{bmatrix}. \quad (4.130)$$

Note that the first phase shift has disappeared because thermal states are invariant under such transformations.

Williamson's theorem: symplectic eigenvalues and thermal decomposition
A second interesting result of symplectic analysis is *Williamson's theorem* [7], which states that any positive $2N \times 2N$ symmetric matrix V can be brought to its diagonal form $V^{\oplus}$ by a symplectic transformation $\mathcal{W}$, that is,

$$V = \mathcal{W}V^{\oplus}\mathcal{W}^T, \quad \text{with} \quad V^{\oplus} = \bigoplus_{j=1}^{N} \nu_j \mathcal{I}_{2\times 2}. \quad (4.131)$$

This theorem has enormous applicability in the world of Gaussian states. Note that, physically, $V^{\oplus}$ can be seen as the covariance matrix of N independent modes in a thermal state with mean photon numbers $\{\bar{n}_j = (\nu_j - 1)/2\}_{j=1,2,\ldots,N}$, while the symplectic transformation $\mathcal{W}$ corresponds to a Gaussian unitary transformation. Williamson's theorem is then completely equivalent to stating that any N-mode Gaussian state $\hat{\rho}_G(\bar{\mathbf{r}}, V)$ can be obtained as

$$\hat{\rho}_G(\bar{\mathbf{r}}, V) = \hat{U}_G(\bar{\mathbf{r}}, \mathcal{W})\left\{\bigotimes_{j=1}^{N} \hat{\rho}_{\text{th}}\left[(\nu_j - 1)/2\right]\right\}\hat{U}_G^{\dagger}(\bar{\mathbf{r}}, \mathcal{W}). \quad (4.132)$$

The set $\{\nu_j\}_{j=1,2,\ldots,N}$ is called the *symplectic spectrum* of V, and each ν_j is a *symplectic eigenvalue*. It is possible to show that the symplectic spectrum of V can be computed as the absolute values of the eigenvalues of the self-adjoint matrix $\mathrm{i}\Omega V$. Note that this expression, together with Euler's decomposition (4.128) of general Gaussian unitaries proves that any single-mode Gaussian state can be written as we did in (4.129).

This decomposition is very important, since it allows us to write many properties of Gaussian states and covariance matrices in an easy manner. For example, the condition (4.67) which ensures that V is the covariance matrix of a physical Gaussian state is simply rewritten as

$$V > 0 \quad \text{and} \quad \nu_j \geqslant 1 \ \forall j, \quad (4.133)$$

which further shows that any state which saturates the physicality conditions ($\nu_j = 1 \ \forall j$) is pure according to (4.132). As a second important example, note that as unitary transformations do not change the von Neumann entropy, the entropy of $\hat{\rho}_G(\bar{\mathbf{r}}, V)$ can be directly computed as the sum of the entropies of the corresponding thermal states, that is,

$$S\left[\hat{\rho}_G(\bar{\mathbf{r}}, V)\right] = \sum_{j=1}^{N} S_{\text{th}}\left[(\nu_j - 1)/2\right] = \sum_{j=1}^{N} g(\nu_j), \quad (4.134)$$

where we have defined the function

$$g(x) = \left(\frac{x+1}{2}\right)\log\left(\frac{x+1}{2}\right) - \left(\frac{x-1}{2}\right)\log\left(\frac{x-1}{2}\right), \tag{4.135}$$

which is positive and monotonically increasing for $x \geqslant 1$.

It is particularly simple to evaluate the symplectic eigenvalues in the case of one or two modes. In the single-mode case, the trick is to realize that the determinant of the covariance matrix V is invariant under symplectic transformations, and hence the sole symplectic eigenvalue reads in this case

$$\nu = \sqrt{\det V}. \tag{4.136}$$

For two modes, let us write the covariance matrix in the block form

$$V = \begin{bmatrix} A & C \\ C^T & B \end{bmatrix}, \tag{4.137}$$

where $A = A^T$, $B = B^T$, and C are 2×2 real matrices. In this case, there is an extra symplectic invariant [8] denoted by $\Delta(V) = \det A + \det B + 2\det C$, and hence the symplectic eigenvalues of a general two-mode Gaussian state can be obtained from

$$\det V = \nu_+^2 \nu_-^2 \qquad \text{and} \qquad \Delta(V) = \nu_+^2 + \nu_-^2, \tag{4.138}$$

leading to

$$\nu_\pm^2 = \frac{\Delta(V) \pm \sqrt{\Delta^2(V) - 4\det V}}{2}. \tag{4.139}$$

In terms of the two-mode symplectic invariants, the second condition in (4.133) is rewritten as

$$\det V \geqslant 1 \qquad \text{and} \qquad \Delta(V) \leqslant 1 + \det V. \tag{4.140}$$

It is particularly relevant the case in which the covariance matrix of the two-mode Gaussian state is in the so-called *standard form*

$$V = \begin{bmatrix} a & 0 & c_1 & 0 \\ 0 & a & 0 & c_2 \\ c_1 & 0 & b & 0 \\ 0 & c_2 & 0 & b \end{bmatrix}. \tag{4.141}$$

Indeed, it is possible to show [9] that the covariance matrix of any bipartite Gaussian state can be brought to this standard form via a local Gaussian unitary transformation $\hat{U}_G = \hat{U}_G(\mathbf{0}, S_1) \otimes \hat{U}_G(\mathbf{0}, S_2)$. The symplectic eigenvalues take a particularly simple form for states which are invariant under correlated or anticorrelated phase shifts of the type $\hat{R}^{(\pm)}(\theta) = \hat{R}(\theta) \otimes \hat{R}(\pm\theta)$, corresponding to states with $c_1 = \pm c_2 = c > 0$, respectively, whose covariance matrices we will denote by

$$V^{(\pm)} = \begin{bmatrix} a & 0 & c & 0 \\ 0 & a & 0 & \pm c \\ c & 0 & b & 0 \\ 0 & \pm c & 0 & b \end{bmatrix}, \tag{4.142}$$

and we will assume $a \geqslant b$ for definiteness and without loss of generality. In the first case, the symplectic eigenvalues read

$$\nu_{\pm}^{(+)} = \frac{a + b \pm \sqrt{(a-b)^2 + 4c^2}}{2}, \tag{4.143}$$

while in the second case they are

$$\nu_{\pm}^{(-)} = \frac{\sqrt{(a+b)^2 - 4c^2} \pm (a-b)}{2}. \tag{4.144}$$

Together with the trivial conditions $a \geqslant 1$ and $b \geqslant 1$ obtained from demanding the reduced states for modes 1 and 2 to be physical, the physicality conditions $\nu_{-}^{(\pm)} \geqslant 1$ imply then $(a \mp 1)(b-1) \geqslant c^2$ for states invariant under $\hat{R}^{(\pm)}$. Note that $V^{(\pm)}$ have both the same doubly degenerate eigenvalues $\lambda_{\pm} = (a + b \pm \sqrt{(a-b)^2 + 4c^2})/2$, and hence the condition for their positivity is $ab \geqslant c^2$, which is always granted from the conditions obtained before.

4.4.4 Gaussian bipartite states and Gaussian entanglement

In chapter 2 we introduced the concept of entanglement as correlations between two systems A and B which go beyond the ones allowed classically. In this section we particularize those ideas to continuous-variable states, with emphasis on Gaussian states. In the following we consider only two modes, that is, a 1×1 continuous-variable system, although we shall briefly discuss general $N \times M$ systems as well.

Einstein–Podolsky–Rosen and the birth of entanglement

Even though Einstein is considered as one of the founding fathers of quantum mechanics, he always felt uncomfortable with its probabilistic character. As a result of this criticism, in 1935, with Podolsky and Rosen, he introduced an argument which was supposed to tumble down the foundations of quantum mechanics, showing, in particular, how the theory was both *incomplete* and *inconsistent with causality* [10]. Looking from our current perspective, it is quite ironic how the very same ideas they introduced, far from destroying the theory, are now the ones that fuel many of the most promising applications and deep results in quantum physics.

In this section we will review some of the EPR (for Einstein, Podolsky, and Rosen) arguments in an oversimplified manner, just to obtain a feeling for their ideas and the way they introduced, almost without noticing, the concept of entanglement.

For the sake of argument, let us consider the two-mode squeezed vacuum state (4.113) in the unphysical limit of infinite squeezing, which can be written (up to normalization) as

$$|r \to \infty\rangle_{2\text{sq}} = \sum_{n=0}^{\infty} |n, n\rangle \equiv |\text{EPR}\rangle, \tag{4.145}$$

where we denote the state by $|\text{EPR}\rangle$ because it coincides with the state that EPR used in their seminal paper, as we will now discuss. Let us write the state in terms of the momentum and position eigenstates as follows. First, let us simply apply the identity operator $\int_{\mathbb{R}^2} \mathrm{d}x_1 \mathrm{d}x_2 |x_1, x_2\rangle\langle x_1, x_2|$, where $|x_1, x_2\rangle = |x_1\rangle \otimes |x_2\rangle$, to the state, obtaining

$$|\text{EPR}\rangle = \int_{\mathbb{R}^2} \mathrm{d}x_1 \mathrm{d}x_2 \left(\sum_{n=0}^{\infty} \langle x_1|n\rangle\langle x_2|n\rangle \right) |x_1, x_2\rangle. \tag{4.146}$$

Now, using the reality of the Fock wave functions (4.30) to write $\langle n|x\rangle^* = \langle n|x\rangle$, the completeness relation of the Fock basis, and the Dirac normalization of the position eigenstates, we turn the former expression into

$$|\text{EPR}\rangle = \int_{\mathbb{R}} \mathrm{d}x |x, x\rangle = \int_{\mathbb{R}} \mathrm{d}p |p, -p\rangle, \tag{4.147}$$

where the last equality is straightforwardly proved by using the Fourier transform relation between the position and momentum eigenbasis. It can be checked that this is precisely the state introduced by EPR in [10].

EPR argue then as follows. Modes 1 and 2 are given, respectively, to *Alice* and *Bob*, two observers placed at distant locations, so that they are not able to interact. Imagine that Alice measures the position and obtains the result[7] x_0; according to quantum mechanics the state of the oscillators collapses to $|x_0, x_0\rangle$, and hence, any subsequent measurement of the position performed by Bob will reveal that his mode has a well defined position x_0. However, Alice could have measured the momentum instead, obtaining for example the result p_0; in this case, quantum mechanics says that the state would have collapsed to $|p_0, -p_0\rangle$, after which Bob would have concluded that his mode had a definite momentum $-p_0$. Now, and this is the center of the argument, *assuming that nothing Alice may do can alter the physical state of Bob's mode* (the modes are separated, even *space-like* or *causally* during the life of Alice and Bob if we like!), one must conclude that Bob's mode must have had well defined values of both its position and momentum from the beginning, hence *violating the quantum-mechanical uncertainty relation* $\Delta X_2 \Delta P_2 \geqslant 1$, and showing that quantum mechanics is inconsistent.

Even though it seems a completely reasonable statement (in particular in 1935, just a decade after the true birth of quantum mechanics), the center of their argument can also be seen as its flaw. The reason is that the state of the system is

[7] Again, this is an idealized situation used just for the sake of argument, since having a continuous spectrum, a measurement of $\hat{X}$ cannot give a definite number x_0 as discussed in section 1.3.7.

not an *element of reality* (in EPR's words), it is just a mathematically convenient object which describes the statistics that would be obtained if a physical observable were measured. Consequently, causality does not apply to it: *the actions of Alice can indeed alter Bob's state*, even if these are causally disconnected. Of course, a completely different matter is whether Alice and Bob can use this *spooky action at a distance* (in Einstein's words) to transmit information superluminally. Even though there is no rigorous proof for the negative answer to this question, such a violation of causality has never been observed or even predicted beyond doubt, and hence, most physicists believe that despite non-local effects at the level of states, quantum mechanics cannot violate causality in any way.

The EPR work is the very best example of how one can make advances in a theory by trying to disprove it, since in order to do so one needs to understand it at the deepest level. Even if their motivation was based on 'incorrect' prejudices, they were the first ones to realize that in quantum mechanics it is possible to create correlations which go beyond those admitted in the classical world. States with such property were coined *entangled states* by Schrödinger, who paradoxically was also supportive of the EPR ideas, and a strong believer of the incompleteness of quantum mechanics. Even though there is still a great deal of valuable effort directed towards developing a coherent interpretation of quantum mechanics, over the last few decades physicists have stopped looking at entangled states as the puzzle EPR suggested they were, and started searching for possible applications of them to various problems. Bell was the first who realized the potential of such states in the 1960s, proving that they could be used to rule out the incompleteness of quantum mechanics [11, 12] (exactly the opposite of what EPR created them for!), or, in other words, to prove that the probabilistic character of quantum mechanics does not come from some missing information we fail to account for (at least not without abandoning causality), but from a probabilistic character of nature itself[8]. Throughout the last decades, entangled states have been shown to be a resource for remarkable applications that would be impossible to develop was the world ruled simply by classical physics.

Let us now move to the characterization of entangled states in the continuous-variable regime.

Entanglement criteria for continuous-variable systems

As explained in chapter 2, the Peres–Horodecki criterion, that is, the positivity of the partial transpose of the state, is a necessary condition for a state to be separable. It turns out that it is also a sufficient criterion for 1×1 Gaussian states, and we proceed now to explain how to evaluate it for this special class of states.

It is straightforward to prove from (4.32) that for continuous variables, transposition is equivalent to changing the sign of the momenta [19]. Consequently, the partial transposition operation corresponds to a change of sign in the momentum of

[8] Strictly speaking, he proved that no *local hidden-variable theory* is consistent with the predictions of quantum mechanics. These predictions started being tested soon after Bell's discovery [13–15], but we had to wait until very recently [16–18] to see experiments ruling in favor of quantum mechanics beyond any reasonable doubt.

the corresponding modes. In the case of a 1×1 Gaussian state $\hat{\rho}_G(\bar{\mathbf{r}}, V)$, this means that partial transposition of the second mode turns the state into $\hat{\rho}_G(\tilde{\mathbf{r}}, \tilde{V})$ with

$$\tilde{\mathbf{r}} = (\mathcal{I}_{2\times 2} \oplus \mathcal{Z})\bar{\mathbf{r}} \qquad \text{and} \qquad \tilde{V} = (\mathcal{I}_{2\times 2} \oplus \mathcal{Z})V(\mathcal{I}_{2\times 2} \oplus \mathcal{Z}). \tag{4.148}$$

The Peres–Horodecki criterion is then reduced to checking whether $\tilde{V}$ is a physical covariance matrix. It is not difficult to prove that $\tilde{V}$ is positive definite and, hence, the criterion is equivalent to checking the condition $\tilde{V} + \mathrm{i}\Omega \geqslant 0$, or, equivalently, $\tilde{\nu}_\pm \geqslant 1$ in terms of the symplectic eigenvalues of $\tilde{V}$ (note that being symmetric and real, $\tilde{V}$ satisfies Williamson's theorem as well).

As a first interesting example, we consider the states invariant under correlated or anticorrelated phase shifts $\hat{R}^{(\pm)}$, whose covariance matrices can be written in the standard form (4.142). In this case, it is simple to check that partial transposition simply maps one covariance matrix to the other, that is, $\tilde{V}^{(\pm)} = V^{(\mp)}$. Hence, the separability conditions for $V^{(\pm)}$, become the physicality conditions for $V^{(\mp)}$; in other words, $V^{(\pm)}$ corresponds to a separable state if and only if $(a \pm 1)(b - 1) \geqslant c^2$. This implies that states of the type $V^{(+)}$ are always separable, since they must satisfy the physicality condition $c^2 \leqslant (a - 1)(b - 1)$, meaning that $(a + 1)(b - 1)$ cannot ever be smaller than c^2.

Let us consider now a more concrete example consisting of a two-mode squeezed thermal state $\hat{\rho}_{\text{th2sq}} = \hat{S}_{12}(r)[\hat{\rho}_{\text{th}}(\bar{n}) \otimes \hat{\rho}_{\text{th}}(\bar{n})]\hat{S}_{12}^\dagger(r)$, whose covariance matrix is

$$V_{\text{th2sq}}(r, \bar{n}) = Q_{12}(r)[V_{\text{th}}(\bar{n}) \oplus V_{\text{th}}(\bar{n})]Q_{12}^T(r) = (2\bar{n} + 1)V_{2\text{sq}}(r). \tag{4.149}$$

In this case we obtain

$$\tilde{V}_{\text{th2sq}} = (2\bar{n} + 1)\begin{bmatrix} \mathcal{I}_{2\times 2}\cosh 2r & \mathcal{I}_{2\times 2}\sinh 2r \\ \mathcal{I}_{2\times 2}\sinh 2r & \mathcal{I}_{2\times 2}\cosh 2r \end{bmatrix}, \tag{4.150}$$

which has

$$\det\left\{\tilde{V}_{\text{th2sq}}\right\} = (2\bar{n} + 1)^4 \qquad \text{and} \qquad \Delta\left(\tilde{V}_{\text{th2sq}}\right) = 2(2\bar{n} + 1)^2\cosh(4r), \tag{4.151}$$

and therefore symplectic eigenvalues (4.139)

$$\tilde{\nu}_\pm = (2\bar{n} + 1)^2 \mathrm{e}^{\pm 4r}. \tag{4.152}$$

For 'zero temperature', $\bar{n} = 0$, we see that $\tilde{\nu}_- < 1$ for any $r > 0$, showing that the two-mode squeezed vacuum state $|r\rangle_{2\text{sq}}$ is an entangled state. But for a given squeezing parameter r, we see that there exists a critical thermal occupation $\bar{n}_{\text{crit}} = [\exp(2r) - 1]/2$, above which thermal noise is able to destroy the quantum correlations, as $\tilde{\nu}_-$ becomes larger than 1.

From an experimental point of view, the Peres–Horodecki criterion is quite demanding, as it requires the full reconstruction of the state, or at least the full covariance matrix in the case of Gaussian states. However, there is a simpler separability criterion which requires only the analysis of the variance of a suitable pair of joint quadratures (which can be checked experimentally via two homodyne measurements, as we will see later). According to this criterion, which was introduced simultaneously by Duan, Giedke, Cirac, and Zoller [9] and by Simon [19], a sufficient condition for a 1×1 continuous-variable state to be entangled is

$$W^{\phi_1,\phi_2}(\kappa) = V\left(\frac{\kappa\hat{X}_1^{\phi_1} - \kappa^{-1}\hat{X}_2^{\phi_2}}{\sqrt{2}}\right) + V\left(\frac{\kappa\hat{P}_1^{\phi_1} + \kappa^{-1}\hat{P}_2^{\phi_2}}{\sqrt{2}}\right) < \kappa^2 + \kappa^{-2}, \quad (4.153)$$

for some combination of the real parameters ϕ_1, ϕ_2, and $\kappa \neq 0$. In this expression we have defined

$$\hat{X}_j^{\varphi} = \hat{R}_j^{\dagger}(\varphi)\hat{X}_j\hat{R}_j(\varphi) = \mathrm{e}^{\mathrm{i}\varphi}\hat{a}_j^{\dagger} + \mathrm{e}^{-\mathrm{i}\varphi}\hat{a}_j = \hat{X}_j\cos\varphi + \hat{P}_j\sin\varphi, \quad (4.154a)$$

$$\hat{P}_j^{\varphi} = \hat{R}_j^{\dagger}(\varphi)\hat{P}_j\hat{R}_j(\varphi) = \mathrm{i}\left(\mathrm{e}^{\mathrm{i}\varphi}\hat{a}_j^{\dagger} - \mathrm{e}^{-\mathrm{i}\varphi}\hat{a}_j\right) = \hat{P}_j\cos\varphi - \hat{X}_j\sin\varphi, \quad (4.154b)$$

where $\hat{R}_j(\varphi) = \exp(-\mathrm{i}\varphi\hat{a}_j^{\dagger}\hat{a}_j)$ induces a phase shift in the corresponding mode, so that the operators above are just the position and momentum quadratures in a rotated phase-space frame. The criterion becomes also a necessary condition for Gaussian states.

Note that for covariance matrices written in the standard form (4.141), the so-called *witness* $W^{\phi_1,\phi_2}(\kappa)$ reduces to

$$W^{\phi_1,\phi_2}(\kappa) = \kappa^2 a + \kappa^{-2} b + (c_2 - c_1)\cos\left(\phi_1 + \phi_2\right), \quad (4.155)$$

again showing that states of the type $c_1 = c_2$, that is, states invariant under correlated phase shifts, cannot be entangled, since a and b are larger than 1.

Let us consider the example of the two-mode squeezed vacuum state $|r\rangle_{2\mathrm{sq}}$. As its covariance matrix (4.114) is already in standard form, we obtain

$$W^{\phi_1,\phi_2}(\kappa) = (\kappa^2 + \kappa^{-2})\cosh 2r - 2\cos\left(\phi_1 + \phi_2\right)\sinh 2r. \quad (4.156)$$

We see that for $\phi_1 = -\phi_2 \equiv \phi$ and $\kappa = 1$ the witness reads

$$W^{\phi,-\phi}(1) = 2\exp(-2r), \quad (4.157)$$

which is clearly below 2 for every $r > 0$, hence showing once again that $|r\rangle_{2\mathrm{sq}}$ is indeed an entangled state. Note, in particular, that for $\phi = 0$ the witness is a sum of the variances of the combined quadratures $\hat{X}_- = (\hat{X}_1 - \hat{X}_2)/\sqrt{2}$ and $\hat{P}_+ = (\hat{P}_1 + \hat{P}_2)/\sqrt{2}$. When the modes are in a vacuum or a coherent state, we have $V(X_-) = V(P_+) = 1$, while in the two-mode squeezed vacuum state the variances are squeezed below this coherent level, $V(X_-) = V(P_+) = \exp(-2r)$. Hence, this state is characterized by the presence of quantum correlations between the modes' positions and anticorrelations between their momenta, as clearly appreciated from its form (4.147) in the $r \to \infty$ limit, and from figure 4.4.

Let us finally stress that a necessary and sufficient criterion for separability has been found for the Gaussian states of a general $N \times M$ bipartite continuous-variable system. This criterion [20] states that the Gaussian state $\hat{\rho}_{\mathrm{G}}(\bar{\mathbf{r}}, V)$ is separable if and

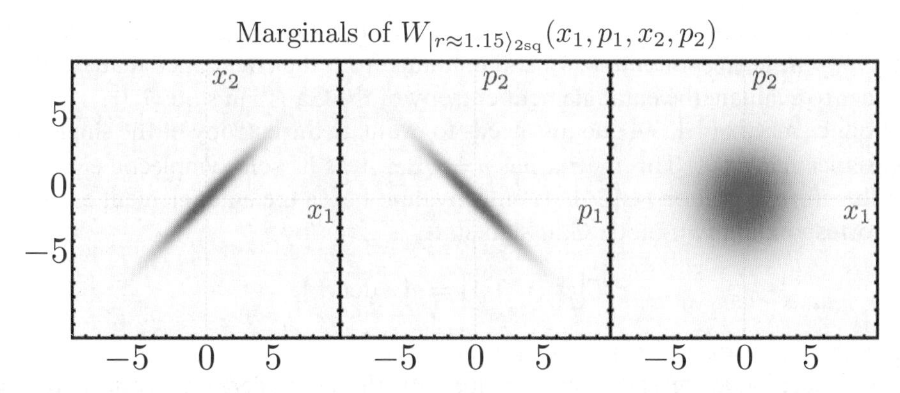

Figure 4.4. Density plot of some characteristic marginals of the Wigner function corresponding to a two-mode squeezed state $|r \approx 1.15\rangle_{2sq}$. In particular, we show $\int_{\mathbb{R}^2} dp_1 dp_2\, W_{|r\rangle_{2sq}}(\mathbf{r})$, $\int_{\mathbb{R}^2} dx_1 dx_2\, W_{|r\rangle_{2sq}}(\mathbf{r})$, and $\int_{\mathbb{R}^2} dp_1 dx_2\, W_{|r\rangle_{2sq}}(\mathbf{r})$, in the left, central, and right panels, respectively. The position correlations and momentum anticorrelations between the modes are apparent, while no special correlation is found between the position of one and the momentum of the other.

only if there exists a pair of matrices V_A and V_B with dimensions $2N \times 2N$ and $2M \times 2M$, respectively, for which

$$V \geqslant V_A \oplus V_B. \tag{4.158}$$

Of course, this criterion is quite difficult to handle in practice, but fortunately an equivalent, operationally friendly criterion was introduced by Giedke, Kraus, Lewenstein, and Cirac, based on the concept of *non-linear maps*. However, we will not go through this criterion which can be consulted in [21], or in the original reference [22].

Quantification of continuous-variable entanglement

Let us move now to the quantification of the entanglement present in a continuous-variable state. As we commented in chapter 2, this problem has been fully solved only for pure states, for which the entanglement entropy is the unique measure of quantum correlations. For mixed states, however, we have not found a completely satisfactory measure (the distillable entanglement and the entanglement of formation are reasonable measures, but cannot be computed for most states, while the logarithmic negativity is easy to compute but does not satisfy all the conditions required of a proper entanglement measure), not even for the reduced class of continuous-variable Gaussian states.

In the following we explain how to compute the entanglement entropy and the logarithmic negativity for 1×1 Gaussian states $\hat{\rho}_G(\bar{\mathbf{r}}, V)$, whose covariance matrix may be written in the same block form as before

$$V = \begin{bmatrix} A & C \\ C^T & B \end{bmatrix}, \tag{4.159}$$

where $A = A^T$, $B = B^T$, and C are 2×2 real matrices (the generalization to $N \times M$ Gaussian states is straightforward).

As commented in (4.69), tracing out one mode of a Gaussian state is equivalent to removing the corresponding rows and columns from the covariance matrix. Hence, in order to evaluate the entanglement entropy of the Gaussian state $\hat{\rho}_G(\bar{\mathbf{r}}, V)$ having covariance matrix (4.159), one just needs to evaluate the entropy of the single-mode covariance matrix A. This matrix has $\nu = \sqrt{\det A}$ as its sole symplectic eigenvalue, and therefore, based on (4.134), its entropy, and hence the entanglement entropy of the corresponding two-mode Gaussian state, is given by

$$E\left[\hat{\rho}_G(\bar{\mathbf{r}}, V)\right] = g\left(\sqrt{\det A}\right), \tag{4.160}$$

where the function $g(x)$ was defined in (4.135).

As a pure bipartite state, we can quantify the entanglement of the two-mode squeezed vacuum state $|r\rangle_{2\text{sq}}$ via its entanglement entropy, which in this case reads

$$E\left[|r\rangle_{2\text{sq}}\right] = g(\cosh 2r), \tag{4.161}$$

and is nothing but the entropy of the reduced thermal state. Starting at zero for $r = 0$, this is a monotonically increasing function of r, as expected.

For mixed states the entanglement entropy is not even an entanglement monotone, and hence, it cannot be considered a proper entanglement measure for such states. One then has to consider other measures, and here we focus on the logarithmic negativity $E_N[\hat{\rho}]$. It is possible to show [23] that for an arbitrary Gaussian state $\hat{\rho}_G(\bar{\mathbf{r}}, V)$, this entanglement measure can be computed as

$$E_N\left[\hat{\rho}_G(\bar{\mathbf{r}}, V)\right] = \sum_j F(\tilde{\nu}_j), \tag{4.162}$$

where

$$F(x) = \begin{cases} -\log x & x < 1 \\ 0 & x \geqslant 1 \end{cases}, \tag{4.163}$$

and $\{\tilde{\nu}_j\}_j$ is the symplectic spectrum of the covariance matrix $\tilde{V}$ corresponding to the partial transposition of $\hat{\rho}_G(\bar{\mathbf{r}}, V)$, which is defined in (4.148) for a 1×1 system.

In the case of the two-mode squeezed thermal state (4.149), we obtain $E_N[\hat{\rho}_{\text{th2sq}}] = 2r - \log_e(2\bar{n} + 1)$ below the critical thermal occupation number ($\bar{n} < \bar{n}_{\text{crit}}$).

4.4.5 Gaussian channels

General definition

In section 3.1.1 we introduced general *quantum channels* as trace-preserving quantum operations. Applied to continuous-variable systems, we say that such operations are *Gaussian channels* when they map Gaussian states into Gaussian states. It is in this context where the name ‘channel’ is best suited, since Gaussian channels indeed model some of the most important *communication channels* such as fibers, wires, and amplifiers.

As we saw, a way of characterizing an arbitrary trace-preserving quantum operation $\mathcal{E}$ is by giving a complete set of Kraus operators $\{\hat{E}_k\}_{k=1,2,\ldots,K}$ which transform a state $\hat{\rho}$ into the state

$$\hat{\rho} \to \mathcal{E}[\hat{\rho}] = \sum_{k=1}^{K} \hat{E}_k \hat{\rho} \hat{E}_k^{\dagger}. \tag{4.164}$$

Gaussian channels, on the other hand, can be characterized by their action on the first and second moments of the state. In particular, similarly to Gaussian unitaries, and as will be clear from the following discussion, Gaussian channels acting on N modes are characterized by a vector $\mathbf{d} \in \mathbb{R}^{2N}$, plus two real $2N \times 2N$ matrices $\mathcal{K}$ and $\mathcal{N}$, which transform the mean vector $\bar{\mathbf{r}}$ and covariance matrix V of the state as

$$\bar{\mathbf{r}} \to \mathcal{K}\bar{\mathbf{r}} + \mathbf{d} \qquad \text{and} \qquad V \to \mathcal{K}V\mathcal{K}^T + \mathcal{N}. \tag{4.165}$$

The matrices $\mathcal{K}$ and $\mathcal{N}$ must satisfy certain conditions in order to correspond to a true trace-preserving quantum operation. First, as the covariance matrix is symmetric, so must be the matrix $\mathcal{N}$. Second, in order to be a completely-positive map, they have to satisfy the following restriction [24]

$$\mathcal{N} + \mathrm{i}\Omega - \mathrm{i}\mathcal{K}\Omega\mathcal{K}^T \geqslant 0. \tag{4.166}$$

Note that Gaussian unitaries correspond to a Gaussian channel for which $\mathcal{N}$ is zero, and $\mathcal{K}$ is symplectic. It is possible to show that for single-mode channels ($N = 1$) this last condition can be rewritten as

$$\mathcal{N} \geqslant 0 \qquad \text{and} \qquad \det \mathcal{N} \geqslant (\det \mathcal{K} - 1)^2. \tag{4.167}$$

Roughly speaking, $\mathcal{K}$ plays the role of the amplification or attenuation of the channel (plus a possible rotation), while $\mathcal{N}$ includes any source of quantum or classical noise added by the channel. We shall return to their physical interpretation in the next section via a particular example.

Indeed, it is quite simple to understand why Gaussian channels correspond to a transformation of the type (4.165). To this aim, we just need to remember that any trace-preserving quantum operation can be seen as the reduced dynamics induced by a unitary transformation acting on the system and some environment in a pure state (Stinespring dilation). It is obvious that in order for the channel to be Gaussian, both the state of the environment $|\psi_E\rangle$ and the joint unitary transformation $\hat{U}_{\mathrm{G}}(\mathbf{s}, \mathcal{S})$ must be Gaussian. Moreover, as every pure Gaussian state is connected to the vacuum state via some Gaussian unitary transformation which can be included in the joint unitary $\hat{U}_{\mathrm{G}}(\mathbf{s}, \mathcal{S})$, we can take the initial state of the environment as the multi-mode vacuum state, that is,

$$|\psi_E\rangle = \bigotimes_{j=1}^{N_E} |0\rangle \equiv |\mathrm{vac}\rangle, \tag{4.168}$$

where we have assumed that the environment consists in N_E modes. The state $\hat{\rho}$ of the system is then transformed into

$$\hat{\rho}' = \mathrm{tr}_E\left\{\hat{U}_{\mathrm{G}}(\mathbf{s}, \mathcal{S})[\hat{\rho} \otimes |\mathrm{vac}\rangle\langle\mathrm{vac}|]\hat{U}_{\mathrm{G}}^{\dagger}(\mathbf{s}, \mathcal{S})\right\}. \tag{4.169}$$

Let us write the Gaussian parameters associated to the joint unitary as

$$\mathbf{s} = \mathrm{col}(\mathbf{d}, \mathbf{d}_E) \qquad \text{and} \qquad \mathcal{S} = \begin{bmatrix} \mathcal{S}_S & \mathcal{S}_{SE} \\ \mathcal{S}_{ES} & \mathcal{S}_E \end{bmatrix}, \tag{4.170}$$

where $\mathbf{d} \in \mathbb{R}^{2N}$, and the real matrices $\mathcal{S}_S$, $\mathcal{S}_E$, $\mathcal{S}_{SE}$, and $\mathcal{S}_{ES}$, have dimensions $2N \times 2N$, $2N_E \times 2N_E$, $2N \times 2N_E$, and $2N_E \times 2N$, respectively. Let us write also the mean and the covariance matrix of the initial separable joint state of the system plus the environment as

$$\bar{\mathbf{r}}_{SE} = \mathrm{col}(\bar{\mathbf{r}}, \mathbf{0}) \qquad \text{and} \qquad V_{SE} = V \oplus \mathcal{I}_{2N_E \times 2N_E}. \tag{4.171}$$

After the unitary, these are transformed into

$$\bar{\mathbf{r}}'_{SE} = \mathcal{S}\,\bar{\mathbf{r}}_{SE} = \mathrm{col}(\mathcal{S}_S\bar{\mathbf{r}} + \mathbf{d}, \mathcal{S}_{ES}\bar{\mathbf{r}} + \mathbf{d}_E), \tag{4.172a}$$

$$V'_{SE} = \mathcal{S}V_{SE}\mathcal{S}^T = \begin{bmatrix} \mathcal{S}_S V \mathcal{S}_S^T + \mathcal{S}_{SE}\mathcal{S}_{SE}^T & \mathcal{S}_S V \mathcal{S}_S + \mathcal{S}_{SE}\mathcal{S}_E^T \\ \mathcal{S}_{ES} V \mathcal{S}_S^T + \mathcal{S}_E\mathcal{S}_{SE}^T & \mathcal{S}_{ES} V \mathcal{S}_{ES}^T + \mathcal{S}_E\mathcal{S}_E^T \end{bmatrix}, \tag{4.172b}$$

so that by tracing out the environment, the transformation onto the mean vector and the covariance matrix of the system is

$$\bar{\mathbf{r}}' = \mathcal{S}_S\bar{\mathbf{r}} + \mathbf{d} \qquad \text{and} \qquad V' = \mathcal{S}_S V \mathcal{S}_S^T + \mathcal{S}_{SE}\mathcal{S}_{SE}^T, \tag{4.173}$$

which is exactly the type of transformation introduced in (4.165), where we now make the identifications

$$\mathcal{K} = \mathcal{S}_S \qquad \text{and} \qquad \mathcal{N} = \mathcal{S}_{SE}\mathcal{S}_{SE}^T. \tag{4.174}$$

Interestingly, it can be proved that the Stinespring dilation of any Gaussian channel can be generated by choosing an environment with less than twice the number of modes of the system, that is, $N_E \leqslant 2N$ [25, 26].

Note that the Gaussianity is a property of the channel, not of the state of the system, that is, one can consider the action of the Gaussian channel on non-Gaussian states. Indeed, the transformation of a general state $\hat{\rho}$ after passing through the channel receives a very simple description in terms of characteristic functions. To see this, using the *block inversion formula* [27], let us write the inverse of the symplectic matrix $\mathcal{S}$ as

$$\mathcal{S}^{-1} = \begin{bmatrix} \mathcal{T}_S & \mathcal{T}_{SE} \\ \mathcal{T}_{ES} & \mathcal{T}_E \end{bmatrix} = \begin{bmatrix} (\mathcal{S}/\mathcal{S}_E)^{-1} & -\mathcal{S}_S^{-1}\mathcal{S}_{SE}(\mathcal{S}/\mathcal{S}_E)^{-1} \\ -(\mathcal{S}/\mathcal{S}_E)^{-1}\mathcal{S}_{ES}\mathcal{S}_S^{-1} & (\mathcal{S}/\mathcal{S}_S)^{-1} \end{bmatrix}, \tag{4.175}$$

where

$$\mathcal{S}/\mathcal{S}_S = \mathcal{S}_E - \mathcal{S}_{ES}\mathcal{S}_S^{-1}\mathcal{S}_{SE} \qquad \text{and} \qquad \mathcal{S}/\mathcal{S}_E = \mathcal{S}_S - \mathcal{S}_{SE}\mathcal{S}_E^{-1}\mathcal{S}_{ES}, \tag{4.176}$$

are the so-called *Schur complements* of $\mathcal{S}_S$ and $\mathcal{S}_E$, respectively. Let us also denote by $\mathbf{r}$ and $\mathbf{r}_E$ the phase-space coordinates of the system and environment modes, respectively, so that the initial characteristic function can be written as

$$\chi_0(\mathbf{r}, \mathbf{r}_E) = \chi_{\hat{\rho}}(\mathbf{r})\chi_{\text{vac}}(\mathbf{r}_E). \tag{4.177}$$

Then, recalling the transformation of the characteristic function under Gaussian unitaries (4.86), and after tracing out the environmental modes (that is, setting to zero their phase-space variables), we obtain the output characteristic function

$$\chi_{\mathcal{E}[\hat{\rho}]}(\mathbf{r}) = \mathrm{e}^{-\frac{\mathrm{i}}{2}\mathbf{r}^T\Omega\mathbf{d}}\chi_{\hat{\rho}}(\mathcal{T}_S\mathbf{r})\chi_{\text{vac}}(\mathcal{T}_{ES}\mathbf{r}). \tag{4.178}$$

In the following, we will denote by $\mathcal{E}(\mathcal{K}, \mathcal{N})$ any Gaussian channel, obviating the displacement $\mathbf{d}$ which, at least for Gaussian states, can actually be generated after the channel via a unitary displacement transformation (4.92), since this does not change the covariance matrix in any way.

An example: phase-insensitive Gaussian channels
There is a particularly simple class of single-mode Gaussian channels which plays an important role in communication technologies: *phase-insensitive Gaussian channels*, which are defined by

$$\mathcal{K} = \sqrt{\tau}I_{2\times 2} \qquad \text{and} \qquad \mathcal{N} = \mu I_{2\times 2}, \tag{4.179}$$

where $\tau \geqslant 0$ and $\mu \geqslant 0$ satisfy

$$\mu \geqslant |\tau - 1|, \tag{4.180}$$

by virtue of the positivity condition (4.167). Note that these channels are called 'phase insensitive' because they are invariant under rotations in phase space. We will denote them by $C(\tau, \mu)$.

After crossing the channel, the covariance matrix V of any state is transformed into

$$V' = \mathcal{K}V\mathcal{K}^T + \mathcal{N} = \begin{bmatrix} \tau V_{11} + \mu & \tau V_{12} \\ \tau V_{21} & \tau V_{22} + \mu \end{bmatrix}, \tag{4.181}$$

and hence

$$\operatorname{tr}\{V'\} = \tau \operatorname{tr}\{V\} + 2\mu, \tag{4.182}$$

$$\det\{V'\} = \tau^2 \det\{V\} + \mu(\tau \operatorname{tr}\{V\} + \mu). \tag{4.183}$$

Taking into account that the mean photon number is proportional to the trace of the covariance matrix, see (4.66), while the von Neumann entropy is a monotonically increasing function of its determinant, see (4.134) and (4.136), we conclude that τ acts as an attenuation (for $0 \leqslant \tau < 1$) or amplification (for $\tau > 1$) factor, while μ adds noise (mixedness) to the state. Note that this implies that quantum mechanics does not allow for the attenuation or amplification of a signal without introducing noise

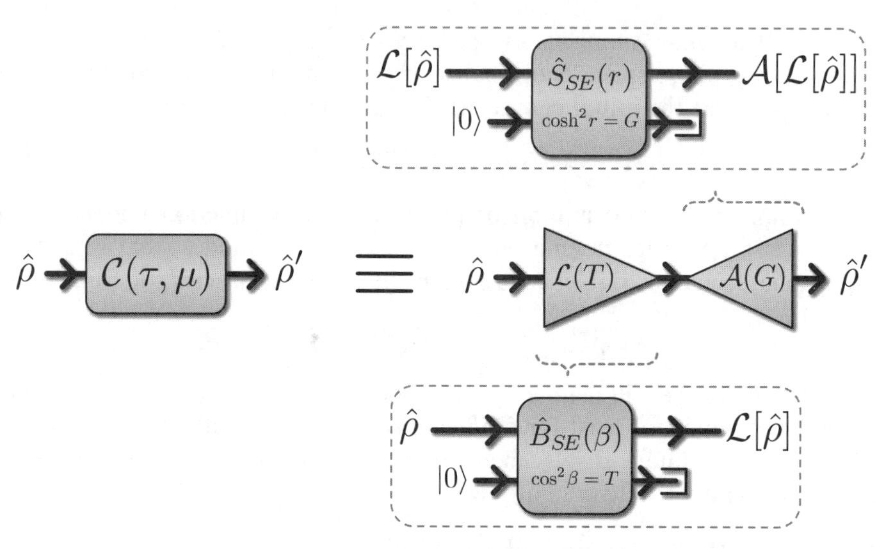

Figure 4.5. Decomposition of a general phase-insensitive Gaussian channel as a concatenation of a pure-loss channel and a quantum-limited amplifier, and Stinespring dilations of the latter.

(at least deterministically, that is, via trace-preserving operations), which comes from the fact that the uncertainty principle must be satisfied at all times.

There are two interesting limiting cases of this class of channels:

- For $0 \leqslant \tau < 1$ and $\mu = 1 - \tau$ we obtain the so-called *pure-loss channels*, which are a good approximation of the fibers used in current optical communication technologies. It is fairly simple to check that the simplest Stinespring dilation of such channels consists in mixing the input mode with a single environmental mode in a beam splitter (4.115*a*) with mixing angle $\cos^2 \beta = \tau$ (see figure 4.5). In this context, it is usual to call τ the *attenuation factor* or *transmissivity*, T, and the channel is usually denoted by $\mathcal{L}(T)$.
- For $\tau > 1$ and $\mu = \tau - 1$, we obtain the so-called *quantum-limited amplifiers*, which are the less noisy (deterministic) amplifiers that quantum mechanics allows. Again, it is simple to check that the simplest Stinespring dilation of these channels consists in mixing the input mode with a single environmental mode in a two-mode squeezer (4.109) with squeezing parameter satisfying $\cosh^2 r = \tau$ (see figure 4.5). The parameter τ receives here the name *amplification factor* or *gain*, G, and the channel is usually denoted by $\mathcal{A}(G)$.

Any phase-insensitive Gaussian channel can be written as the concatenation of a pure-loss channel and a quantum-limited amplifier, that is, $\mathcal{C}(\tau, \mu) = \mathcal{L}(T) \circ \mathcal{A}(G)$, as depicted in figure 4.5, where

$$\tau = TG \qquad \text{and} \qquad \mu = G(1 - T) + (G - 1). \tag{4.184}$$

This is trivially proven by directly applying both channel transformations subsequently onto a general covariance matrix, and then comparing the result with (4.181).

4.5 Measuring continuous-variable systems

4.5.1 Description of measurements in phase space

In section 3.2 we learned that the most general measurement that one can perform in a quantum system can always be described by a complete set of trace-decreasing operations $\{\mathcal{E}_j\}_{j=1,2,\ldots,J>1}$, each corresponding to one of the possible measurement outcomes. In this context we defined POVM-based measurements as those generalized measurements whose quantum operations can each be described by a single Kraus operator, which were shown to be the simplest generalizations of projective measurements. In this section we will learn a convenient way of describing this type of generalized measurement for continuous-variable systems.

Let us start with some useful definitions. Given the initial state of a system $\hat{\rho}$, and the POVM $\{\hat{\Pi}_j = \hat{E}_j^\dagger \hat{E}_j\}_{j=1,2,\ldots,J}$, we will denote by

$$\tilde{\rho}_j = \mathcal{E}_j[\hat{\rho}] = \hat{E}_j \hat{\rho} \hat{E}_j^\dagger, \tag{4.185}$$

the unnormalized state obtained after the outcome j appears. Such an outcome appears with probability $p_j = \mathrm{tr}\{\tilde{\rho}_j\}$, and the normalized state of the system reads $\hat{\rho}_j = p_j^{-1}\tilde{\rho}_j$. Similarly, and assuming that the system is described as a collection of N modes, we define the corresponding unnormalized characteristic and Wigner functions as

$$\tilde{\chi}_j(\mathbf{r}) = \mathrm{tr}\left\{\hat{D}(\mathbf{r})\tilde{\rho}_j\right\} \qquad \text{and} \qquad \tilde{W}_j(\mathbf{r}) = \int_{\mathbb{R}^{2N}} \frac{\mathrm{d}^{2N}\mathbf{s}}{(4\pi)^{2N}} \tilde{\chi}_j(\mathbf{s}) \mathrm{e}^{\frac{\mathrm{i}}{2}\mathbf{s}^T\Omega\mathbf{r}}, \tag{4.186}$$

from which the probability of the corresponding outcome can be obtained as

$$p_j = \tilde{\chi}_j(\mathbf{0}) = \int_{\mathbb{R}^{2N}} \mathrm{d}^{2N}\mathbf{r}\, \tilde{W}_j(\mathbf{r}), \tag{4.187}$$

and the normalized functions as

$$\chi_j(\mathbf{r}) = \mathrm{tr}\left\{\hat{D}(\mathbf{r})\hat{\rho}_j\right\} = p_j^{-1}\tilde{\chi}_j(\mathbf{r}), \tag{4.188a}$$

$$W_j(\mathbf{r}) = \int_{\mathbb{R}^{2N}} \frac{\mathrm{d}^{2N}\mathbf{s}}{(4\pi)^{2N}} \chi_j(\mathbf{s}) \mathrm{e}^{\frac{\mathrm{i}}{2}\mathbf{s}^T\Omega\mathbf{r}} = p_j^{-1}\tilde{W}_j(\mathbf{r}). \tag{4.188b}$$

There are many situations in which the measurement is not applied to the whole system, but only to one of the modes that conform to it. We talk in those cases about *partial measurements*. Moreover, as we will see in the next sections, the measurement performed on a light beam is usually *destructive*, that is, the mode disappears after the measurement is performed, so that one has to trace it out of the system. Assuming that the system has $N + 1$ modes, and that the measurement is applied on the last mode, this means that the (unnormalized) state of the N modes remaining after the measurement will be[9]

[9] Note that the cyclic property applies also to the partial trace when the joint operator acts as the identity on the non-traced subspaces. In particular, for a bipartite Hilbert space $\mathcal{H}_A \otimes \mathcal{H}_B$, it is simple to prove (for example by expanding the expression explicitly in a basis) that $\mathrm{tr}_B\{\hat{\rho}(\hat{I}_A \otimes \hat{B})\} = \mathrm{tr}_B\{(\hat{I}_A \otimes \hat{B})\hat{\rho}\}$, where $\hat{\rho}$ is any operator acting on the joint Hilbert space and $\hat{I}_A$ is the identity operator acting on $\mathcal{H}_A$.

$$\tilde{\rho}_j = \mathrm{tr}_{N+1}\left\{\left(\hat{I}_N \otimes \hat{E}_j\right)\hat{\rho}\left(\hat{I}_N \otimes \hat{E}_j^\dagger\right)\right\} = \mathrm{tr}_{N+1}\left\{\left(\hat{I}_N \otimes \hat{\Pi}_j\right)\hat{\rho}\right\}, \tag{4.189}$$

where $\hat{\rho}$ is the initial state of the $N + 1$ modes. The first thing to note is that, obviously, the first N modes only 'feel' the measurement if they share some correlations with the measured mode. On the other hand, an interesting feature of such partial, destructive measurements is that one only needs the POVM $\{\hat{\Pi}_j\}_{j=1,2,\ldots,J}$ to evaluate the final state of the non-measured modes; this is in contrast to non-destructive measurements, which require knowledge of the measurement operators $\{\hat{E}_j\}_{j=1,2,\ldots,J}$ in order to compute the post-measurement state.

Partial measurements are easily described in terms of characteristic and Wigner functions. In the case of the characteristic function the derivation is simple by using its relations with the density operator (4.54):

$$\begin{aligned}\tilde{\chi}_j(\mathbf{r}_{\{N\}}) &= \mathrm{tr}_{\{N\}}\left\{\hat{D}(\mathbf{r}_{\{N\}})\tilde{\rho}_j\right\} = \mathrm{tr}\left\{\left[\hat{D}(\mathbf{r}_{\{N\}}) \otimes \hat{\Pi}_j\right]\hat{\rho}\right\} \\ &= \int_{\mathbb{R}^{2(N+1)}} \frac{\mathrm{d}^{2(N+1)}\mathbf{s}}{(4\pi)^{N+1}}\chi_{\hat{\rho}}(\mathbf{s})\underbrace{\mathrm{tr}_{\{N\}}\left\{\hat{D}(\mathbf{r}_{\{N\}} - \mathbf{s}_{\{N\}})\right\}}_{(4\pi)^N\delta^{(2N)}(\mathbf{r}_{\{N\}}-\mathbf{s}_{\{N\}})}\underbrace{\mathrm{tr}_{N+1}\left\{\hat{\Pi}_j\hat{D}^\dagger(\mathbf{s}_{N+1})\right\}}_{\chi_{\hat{\Pi}_j}(-\mathbf{s}_{N+1})} \\ &= \int_{\mathbb{R}^2} \frac{\mathrm{d}^2\mathbf{r}_{N+1}}{4\pi}\chi_{\hat{\rho}}(\mathbf{r})\chi_{\hat{\Pi}_j}(-\mathbf{r}_{N+1}).\end{aligned} \tag{4.190}$$

Using (4.186) and (4.55), we derive now the transformation rule for the Wigner function:

$$\begin{aligned}\tilde{W}_j(\mathbf{r}_{\{N\}}) &= \int_{\mathbb{R}^{2N}} \frac{\mathrm{d}^{2N}\mathbf{s}_{\{N\}}}{(4\pi)^{2N}}\int_{\mathbb{R}^2}\frac{\mathrm{d}^2\mathbf{s}_{N+1}}{4\pi}\chi_{\hat{\rho}}(\mathbf{s}_{\{N\}}, \mathbf{s}_{N+1})\chi_{\hat{\Pi}_j}(-\mathbf{s}_{N+1})e^{\frac{i}{2}\mathbf{s}_{\{N\}}^T\Omega_N\mathbf{r}_{\{N\}}} \\ &= \int_{\mathbb{R}^2}\mathrm{d}^2\mathbf{r}_{N+1}\int_{\mathbb{R}^{2(N+1)}}\frac{\mathrm{d}^{2(N+1)}\mathbf{s}}{(4\pi)^{2N+1}}\chi_{\hat{\rho}}(\mathbf{s})W_{\hat{\Pi}_j}(\mathbf{r}_{N+1})e^{\frac{i}{2}\mathbf{s}^T\Omega\mathbf{r}} \\ &= 4\pi\int_{\mathbb{R}^2}\mathrm{d}^2\mathbf{r}_{N+1}W_{\hat{\rho}}(\mathbf{r})W_{\hat{\Pi}_j}(\mathbf{r}_{N+1}).\end{aligned} \tag{4.191}$$

Hence, both for the characteristic and the Wigner functions, the transformation is obtained by multiplying the initial function of the $N + 1$ modes by the function associated with the POVM element $\hat{\Pi}_j$ (with the suitable sign in the argument), and integrating out the measured mode.

Let us now go on to describe some particular types of measurements which find many applications in the world of continuous variables, using light modes as the guiding context.

4.5.2 Photodetection: measuring the photon number

The most fundamental measurement technique for light is photodetection. As we shall see with a couple of examples (homodyne detection and on/off detection), virtually any other scheme used for measuring different properties of light makes use of photodetection.

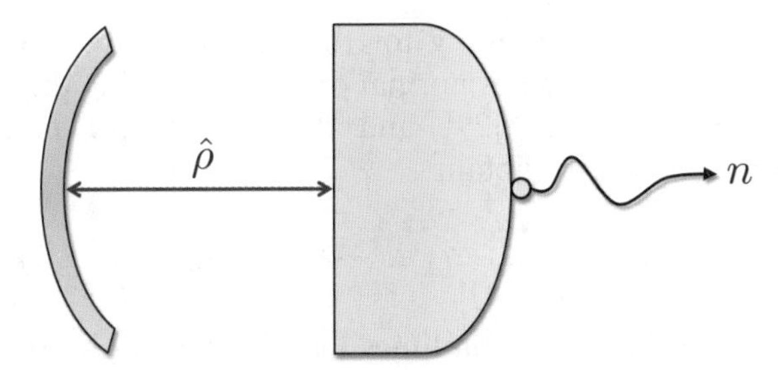

Figure 4.6. Schematic representation of Mollow's idea of photodetection, ideally implementing a measurement of the number operator associated with a single mode.

Ideal photodetection

This technique is based on the photoelectric effect or variations of it[10]. The idea is that when the light beam that we want to detect impinges a metallic surface, it is able to release some of the bound electrons of the metal, which are then collected by an anode. The same happens if light impinges on a semiconductor surface, though in this case instead of becoming free, valence electrons are promoted to the conduction band. The most widely used metallic photodetectors are known as *photo-multiplier tubes*, while those based on semiconducting films are the so-called *avalanche photodiodes*. In both cases, each photon is able to create a single electron, whose associated current would be equally difficult to measure by electronic means. For this reason, each photoelectron is accelerated towards a series of metallic plates at increasing positive voltages, releasing then more electrons via scattering which contribute to generating a measurable electric pulse, the *photopulse*.

It is then customarily said that counting photopulses is equivalent to counting photons, and hence, photodetection is equivalent to a measurement of the number of photons of the light field. This is a highly idealized situation, valid only in some limits which we will try to understand now.

Consider the following model for a perfectly efficient detection scheme. A single-mode field initially in some state $\hat{\rho}$ is kept in continuous interaction with a photodetector during a time interval T. The intuitive picture of such a scenario is shown in figure 4.6: a cavity formed by the photodetector itself and an extra perfectly reflecting mirror contains a single mode. By developing a microscopic model of the detector and its interaction with the light mode, Mollow was able to show that the probability of generating n photoelectrons (equivalently, the probability of observing n photopulses) during the time interval T is given by [28]

$$p_n = \mathrm{tr}\left\{\hat{\rho}\colon \frac{(1 - \mathrm{e}^{-\kappa T})^n \hat{a}^{\dagger n}\hat{a}^n}{n!} \exp[-(1 - \mathrm{e}^{-\kappa T})\hat{a}^\dagger\hat{a}]\colon\right\}, \tag{4.192}$$

[10] In the last few years superconducting photodetectors have received a lot of attention. However, we will stick to the photoelectric ones both for simplicity, and because they are the ones currently found in most laboratories around the world.

where κ is some parameter accounting for the light–detector interaction, and the double-dots refer to the operation of normal ordering[11]. Using the operator identity $:\exp[-(1-e^{-\lambda})\hat{a}^\dagger\hat{a}]: = \exp(-\lambda\hat{a}^\dagger\hat{a})$ [29], and the help of the number state basis $\{|n\rangle\}_{n\in\mathbb{N}}$, it is straightforward to obtain

$$p_n = \sum_{m=n}^{\infty}\langle m|\hat{\rho}|m\rangle\frac{m!}{n!(m-n)!}(1-e^{-\kappa T})^m(e^{-\kappa T})^{m-n}\xrightarrow[T\gg\kappa^{-1}]{}\langle n|\hat{\rho}|n\rangle, \quad (4.193)$$

and hence, for large enough detection times the number of observed pulses follows the statistics of the number of photons. In other words, this ideal photodetection scheme is equivalent to measuring the number operator $\hat{N} = \hat{a}^\dagger\hat{a}$ as already commented, that is, a projective measurement with projectors $\{\hat{P}_n = |n\rangle\langle n|\}_{n=0,1,2,\ldots}$.

Photodetection in the presence of finite efficiency and dark counts
However, in real photodetectors the condition $T \gg \kappa^{-1}$ is hardly met. One usually defines the *quantum efficiency* $\eta = 1 - e^{-\kappa T}$, which in current photodetectors varies from one wavelength to another, and then, according to (4.193) photodetection is equivalent to a generalized measurement with POVM elements

$$\hat{\Pi}_n = \sum_{m=n}^{\infty}\binom{m}{n}\eta^n(1-\eta)^{m-n}|n\rangle\langle n|, \quad n = 0, 1, 2, \ldots. \quad (4.194)$$

This POVM-based measurement admits a very simple Stinespring dilation, which is quite convenient to gain some intuition about optical measurement schemes: before arriving at a photodetector with unit quantum efficiency, the optical mode is mixed with an ancillary vacuum mode in a beam splitter of transmissivity $\cos^2\beta = \eta$. In order to prove that this scheme leads to the same POVM as the one associated with a detector with finite efficiency, let us compute the probability of observing n photo-pulses in the detector. Using the identity [30]

$$\hat{B}(\beta) = e^{\beta\hat{a}\hat{a}_E^\dagger - \beta\hat{a}^\dagger\hat{a}_E} = e^{\hat{a}\hat{a}_E^\dagger\tan\beta}(\cos\beta)^{\hat{a}^\dagger\hat{a}-\hat{a}_E^\dagger\hat{a}_E}e^{-\hat{a}^\dagger\hat{a}_E\tan\beta}, \quad (4.195)$$

and taking into account that

$$e^{\zeta\hat{a}\hat{a}_E^\dagger}|k,0\rangle = \sum_{j=0}^{k}\frac{\zeta^j}{j!}\hat{a}^j\hat{a}_E^{\dagger j}|k,0\rangle = \sum_{j=0}^{k}\zeta^j\sqrt{\binom{k}{j}}|k-j,j\rangle, \quad (4.196)$$

the state of the system after the beam splitter can be written as

$$\begin{aligned}\hat{\rho}_{SE} &= \hat{B}(\beta)(\hat{\rho}\otimes|0\rangle\langle 0|)\hat{B}^\dagger(\beta)\\ &= \sum_{l,m=0}^{\infty}\rho_{lm}\cos^{l+m}\beta\sum_{j=0}^{l}\sum_{k=0}^{m}\sqrt{\binom{l}{j}\binom{m}{k}}\tan^{j+k}\beta|l-j,j\rangle\langle m-k,k|. \quad (4.197)\end{aligned}$$

The reduced state of the detected mode then reads,

$$\hat{\rho}' = \mathrm{tr}_E\{\hat{\rho}_{SE}\} = \sum_{l,m=0}^{\infty}\rho_{lm}\cos^{l+m}\beta\sum_{k=0}^{\min\{l,m\}}\sqrt{\binom{l}{k}\binom{m}{k}}\tan^{2k}\beta|l-k\rangle\langle m-k|, \quad (4.198)$$

[11] Given any multiplication of creation and annihilation operators, this operation amounts to bringing all the creation (annihilation) operators to the left (right) as if they commute. Hence, for example, $:\hat{a}\hat{a}^\dagger: = \hat{a}^\dagger\hat{a}$.

and since the detector is considered ideal, the probability of observing n photopulses is equal to

$$\begin{aligned} p_n = \langle n|\hat{\rho}'|n\rangle &= \sum_{l,m=0}^{\infty} \rho_{lm} \cos^{l+m}\beta \sum_{k=0}^{\min\{l,m\}} \sqrt{\binom{l}{k}\binom{m}{k}} \tan^{2k}\beta \underbrace{\delta_{l-k,n}\delta_{m-k,n}}_{\delta_{lm}\delta_{k,l-n}} \\ &= \sum_{m=n}^{\infty} \rho_{mm} \binom{m}{m-n} \cos^{2m}\beta \tan^{2(m-n)}\beta, \end{aligned} \quad (4.199)$$

which coincides with (4.193) once the identification $\cos^2\beta = \eta$ is performed.

Apart from the finite quantum efficiency, which accounts for the missed photons which do not generate photoelectrons in the detector, there is another source of imperfection that can be understood as the dual to the latter: electrons which are pulled out from the detector without interacting with any photon of the detected mode. One refers to the corresponding photopulses as *dark counts*, and they can be modeled within the previous Stinespring dilation in a very simple way: by assuming that the ancilla mode is not in a vacuum but rather in some other state, say $\hat{\rho}_E$, usually taken as a thermal state. In this scenario, the POVM elements become

$$\hat{\Pi}_n = \mathrm{tr}_E\left\{\hat{B}(\beta)\left(\hat{I} \otimes \hat{\rho}_E\right)\hat{B}^\dagger(\beta)\left(|n\rangle\langle n| \otimes \hat{I}\right)\right\}. \quad (4.200)$$

In the following all these imperfections will be ignored, so that we will assume that photodetection is equivalent to a measurement of the number of photons of the field impinging the detector. However, it is important to understand how to deal with these experimental limitations before proposing any interesting theoretical protocol, in case one has the need to perform a more realistic analysis.

4.5.3 Homodyne and heterodyne detection: measuring the quadratures and the annihilation operator

Basic principles behind homodyning

Even though the output of the photodetectors can take only integer values (the number of recorded photopulses), they can be arranged to approximately measure the quadratures of light, which we recall are continuous observables. This arrangement is called *homodyne detection.* The basic scheme is shown in figure 4.7. The mode we want to measure is mixed in a beam splitter with another mode, called the *local oscillator*, which is in a coherent state $|\alpha_{\mathrm{LO}}\rangle$. When the beam splitter is 50/50 the homodyne scheme is said to be *balanced*, and the annihilation operators associated with the modes transformed through the beam splitter are given by

$$\hat{a}_\pm = \frac{1}{\sqrt{2}}(\hat{a} \pm \hat{a}_{\mathrm{LO}}), \quad (4.201)$$

where $\hat{a}_{\mathrm{LO}}$ is the annihilation operator of the local oscillator mode. These modes are measured with independent photodetectors, and then the corresponding signals are subtracted. Based on the idealized photodetection picture of the previous section,

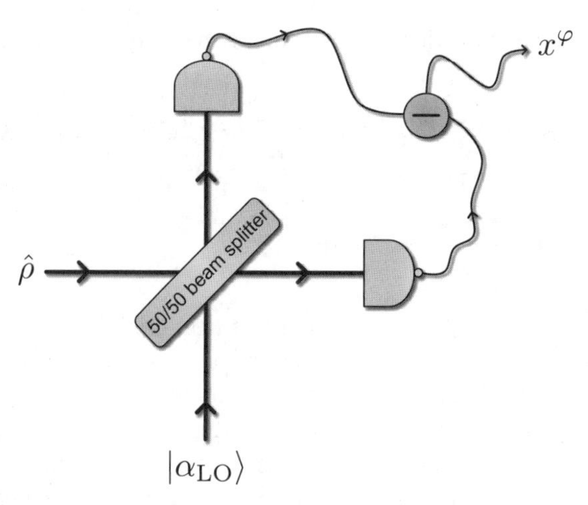

Figure 4.7. Homodyne detection scheme with ideal photodetectors. When the local oscillator is in a strong coherent state, this set up gives access to the quadratures of light.

this scheme is analogous to a measurement of the photon-number difference

$$\hat{N}_{\text{D}} = \hat{a}_+^\dagger \hat{a}_+ - \hat{a}_-^\dagger \hat{a}_- = \hat{a}_{\text{LO}}^\dagger \hat{a} + \hat{a}_{\text{LO}} \hat{a}^\dagger. \tag{4.202}$$

Taking into account that the local oscillator is in a coherent state with an amplitude $\alpha_{\text{LO}} = |\alpha_{\text{LO}}| \exp(\mathrm{i}\varphi)$, and that it is not correlated with the mode we want to measure, it is not difficult to show that the first moments of this operator can be written as

$$\langle \hat{N}_{\text{D}} \rangle = |\alpha_{\text{LO}}| \langle \hat{X}^\varphi \rangle, \tag{4.203a}$$

$$\left\langle \hat{N}_{\text{D}}^2 \right\rangle = |\alpha_{\text{LO}}|^2 \left[\left\langle \hat{X}^{\varphi 2} \right\rangle + \frac{\langle \hat{a}^\dagger \hat{a} \rangle}{|\alpha_{\text{LO}}|^2} \right], \tag{4.203b}$$

where

$$\hat{X}^\varphi = \mathrm{e}^{-\mathrm{i}\varphi} \hat{a} + \mathrm{e}^{\mathrm{i}\varphi} \hat{a}^\dagger, \tag{4.204}$$

is a generalized quadrature which coincides with the position and momentum for $\varphi = 0$ and $\pi/2$, respectively. Hence, in the *strong local oscillator limit* $|\alpha_{\text{LO}}|^2 \gg \langle \hat{a}^\dagger \hat{a} \rangle / \langle \hat{X}^{\varphi 2} \rangle$, the output signal of the homodyne scheme has the mean of a quadrature $\hat{X}^\varphi$ of the analyzed mode (the one selected by the phase of the local oscillator), as well as its same variance. Moreover, it is simple but tedious to check that all the moments of $\hat{N}_{\text{D}}$ coincide with those of $\hat{X}^\varphi$ in the strong local oscillator limit, and therefore, balanced homodyne detection can be seen as a measurement of the corresponding quadrature.

It might be difficult to accept that a measurement of $\hat{N}_{\text{D}}$, which has a discrete spectrum, can be equivalent to a measurement of $\hat{X}^\varphi$, which has a continuous spectrum. The reconciliation between these two pictures comes from the strong local oscillator condition, which essentially means that the local oscillator is very intense and, therefore, there are so many photons impinging the detectors that the photo-pulses are generated at a rate much faster than the response time of the photo-detectors, and the output signal is, basically, experienced by the observer as a continuous photocurrent.

Partial homodyne detection

Just as we explained in section 4.5.1, sometimes it is interesting to apply a measurement onto one mode out of a collection of modes (say the last mode of a system with $N+1$ modes); this is what we defined as a partial measurement. Partial homodyne detection receives a very simple treatment in terms of Wigner functions. For example, in the ideal case explained above (strong local oscillator limit), homodyne detection of the position quadrature is described by the continuous set of projectors $\{\hat{P}(x) = |x\rangle\langle x|\}_{x\in\mathbb{R}}$. Assume that the measurement pops out the outcome[12] x_0, so that, according to (4.191) and (4.190), the unnormalized characteristic and Wigner functions of the remaining N modes collapse to

$$\tilde{\chi}_{x_0}(\mathbf{r}_{\{N\}}) = \int_{\mathbb{R}^2} \frac{d^2\mathbf{r}_{N+1}}{4\pi} \chi_{\hat{\rho}}(\mathbf{r}) \chi_{|x_0\rangle\langle x_0|}(-\mathbf{r}_{N+1}), \tag{4.205a}$$

$$\tilde{W}_{x_0}(\mathbf{r}_{\{N\}}) = 4\pi \int_{\mathbb{R}^2} d^2\mathbf{r}_{N+1} W_{\hat{\rho}}(\mathbf{r}) W_{|x_0\rangle\langle x_0|}(\mathbf{r}_{N+1}), \tag{4.205b}$$

where $\hat{\rho}$ is the initial state of the $N+1$ modes. The characteristic function of the projector $|x_0\rangle\langle x_0|$ is easily found as

$$\begin{aligned} \chi_{|x_0\rangle\langle x_0|}(\mathbf{r}) &= \mathrm{tr}\left\{\hat{D}(\mathbf{r})|x_0\rangle\langle x_0|\right\} = e^{-ipx/4}\langle x_0|e^{ip\hat{X}/2}e^{-ix\hat{P}/2}|x_0\rangle \\ &= e^{-ipx/4}e^{ipx_0/2} \underbrace{\int_{\mathbb{R}} dp_0 e^{-ixp_0/2}}_{4\pi\delta(x)} \underbrace{\langle x_0|p_0\rangle\langle p_0|x_0\rangle}_{1/4\pi} = \delta(x)e^{ipx_0/2}, \end{aligned} \tag{4.206}$$

where we have used (4.19), while Fourier transforming this expression we obtain the corresponding Wigner function

$$W_{|x_0\rangle\langle x_0|}(x, p) = \int_{\mathbb{R}^2} \frac{dx'dp'}{(4\pi)^2} \chi_{|x_0\rangle\langle x_0|}(x', p') e^{\frac{i}{2}x'p - \frac{i}{2}xp'} = \frac{\delta(x - x_0)}{4\pi}. \tag{4.207}$$

These expressions lead us to the following characteristic and Wigner functions of the remaining modes:

$$\tilde{\chi}_{x_0}(\mathbf{r}_{\{N\}}) = \int_{\mathbb{R}^2} \frac{dp}{4\pi} \chi_{\hat{\rho}}(\mathbf{r}_{\{N\}}, 0, p) e^{-ipx_0/2}, \tag{4.208}$$

$$\tilde{W}_{x_0}(\mathbf{r}_{\{N\}}) = \int_{\mathbb{R}} dp W_{\hat{\rho}}(\mathbf{r}_{\{N\}}, x_0, p). \tag{4.209}$$

[12] As explained in detail after introducing axiom 6, for the measurement of a continuous observable this is an idealized situation taken here for simplicity.

Note that, even though the characteristic (4.206) and Wigner (4.207) functions of the projector are not normalizable (which makes sense, since the position eigenstate is also not), the characteristic and Wigner functions of the remaining modes can be normalized, so that the probability density function associated with the possible outcomes x_0 is given by

$$P(x_0) = \tilde{\chi}_{x_0}(\mathbf{0}) = \int_{\mathbb{R}^{2N}} \mathrm{d}^{2N}\mathbf{r}_{\{N\}} \tilde{W}_{x_0}(\mathbf{r}_{\{N\}}). \tag{4.210}$$

Note that the projector $|x_0\rangle\langle x_0|$ is Gaussian, and therefore, partial homodyne measurements map Gaussian states into Gaussian states.

Heterodyne detection from homodyne detection

Along with photodetection (or on/off detection, which we will study in the next section), homodyne detection is one of the fundamental measurement schemes from which many other more complicated ones are generated. Among these, a most relevant example is *heterodyne detection*, described by a POVM[13] $\{\hat{\Pi}(\alpha) = |\alpha\rangle\langle\alpha|/\pi\}_{\alpha\in\mathbb{C}}$.

In loose terms, one could say that heterodyne detection corresponds to a measurement of the annihilation operator $\hat{a}$, since indeed it generates the suggestive probability distribution $P(\alpha) = \mathrm{tr}\{\hat{\Pi}(\alpha)\hat{\rho}\} = \langle\alpha|\hat{\rho}|\alpha\rangle$. However, note that the post-measurement state does not correspond in general to an eigenstate of $\hat{a}$.

Heterodyne detection can be implemented from homodyne detection as follows (see figure 4.8): the mode we want to measure is mixed on a 50/50 beam splitter with another mode in vacuum, and the two output modes are homodyned one in position $\hat{x}$ and the other in momentum $\hat{p}$. It is possible to show then that the probability of recording an outcome $\{x, p\}$ is precisely described by the POVM introduced above with $\alpha = (x + \mathrm{i}p)/2$.

4.5.4 On/off detection and de-Gaussification by vacuum removal

Partial on/off detection

Despite the incredible advances in photodetector technologies, practical photon counters are still out of reach in most laboratories. In a very simplified manner, the problem arises from distinguishing when the photopulse, which is a macroscopic current created by random collisions between the free electrons and the ones bound to the metallic plates, was created from the amplification of n or $n + 1$ photoelectrons. Modern superconducting photodetectors, whose operating principle is not based on the photoelectric effect plus an amplification stage, allow for such photon-number resolution, but they are still far too expensive or come with other problems depending on what they are needed for in the particular experiment.

A less demanding detection strategy is the so-called on/off detection, in which the detector gives a signal whenever one or more photons reach it, but this signal is identical no matter how many photons triggered it. In the next section we will see

[13] It is immediate to show that the POVM is complete, that is, $\int_{\mathbb{C}} \mathrm{d}^2\alpha\hat{\Pi}(\alpha) = \hat{I}$, by expanding the coherent states in the Fock basis and integrating in polar complex coordinates $\alpha = r\exp(\mathrm{i}\theta)$, i.e. $\int_{\mathbb{C}} \mathrm{d}^2\alpha = \int_0^\infty r\mathrm{d}r \int_0^{2\pi} \mathrm{d}\theta$.

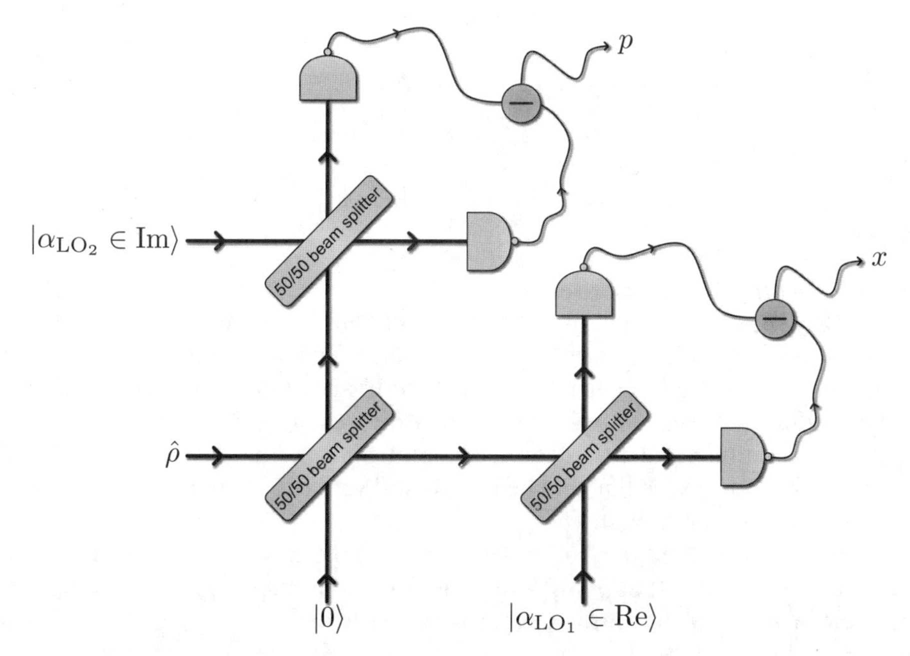

Figure 4.8. Heterodyne detection with a combination of two homodyne detection schemes.

that on/off detection allows for the implementation of interesting operations such as photon addition and subtraction onto a light field, which are the most fundamental non-Gaussian and non-unitary operations that one can think of.

On/off detection is then a projective measurement with two possible outcomes 'off' (no-click) and 'on' (click!), with the corresponding projectors

$$\hat{P}_{\text{off}} = |0\rangle\langle 0| \qquad \text{and} \qquad \hat{P}_{\text{on}} = \hat{I} - |0\rangle\langle 0| = \sum_{n=1}^{\infty} |n\rangle\langle n|. \tag{4.211}$$

These projectors have very simple characteristic functions

$$\chi_{\hat{P}_{\text{off}}}(\mathbf{r}) = \chi_{|0\rangle\langle 0|}(\mathbf{r}) \qquad \text{and} \qquad \chi_{\hat{P}_{\text{on}}}(\mathbf{r}) = 4\pi\delta^{(2)}(\mathbf{r}) - \chi_{|0\rangle\langle 0|}(\mathbf{r}), \tag{4.212}$$

and very simple Wigner functions as well,

$$W_{\hat{P}_{\text{off}}}(\mathbf{r}) = W_{|0\rangle\langle 0|}(\mathbf{r}) \qquad \text{and} \qquad W_{\hat{P}_{\text{on}}}(\mathbf{r}) = \frac{1}{4\pi} - W_{|0\rangle\langle 0|}(\mathbf{r}). \tag{4.213}$$

Applied as a partial measurement (as usual, the measurement is applied onto the last mode of a system with $N + 1$ modes), on/off detection makes the system evolve from some initial state $\hat{\rho}$ of the $N + 1$ oscillators, to a reduced state of the first N oscillators with (unnormalized) characteristic and Wigner functions

$$\tilde{\chi}_{\text{off}}(\mathbf{r}_{\{N\}}) = \int_{\mathbb{R}^2} \frac{\mathrm{d}^2\mathbf{r}_{N+1}}{4\pi} \chi_{\hat{\rho}}(\mathbf{r}) \chi_{|0\rangle\langle 0|}(-\mathbf{r}_{N+1}), \tag{4.214a}$$

$$\tilde{W}_{\text{off}}(\mathbf{r}_{\{N\}}) = 4\pi \int_{\mathbb{R}^2} \mathrm{d}^2\mathbf{r}_{N+1} W_{\hat{\rho}}(\mathbf{r}) W_{|0\rangle\langle 0|}(\mathbf{r}_{N+1}) \tag{4.214b}$$

when the outcome is 'off' or

$$\tilde{\chi}_{\text{on}}(\mathbf{r}_{\{N\}}) = \chi_{\hat{\rho}_{\{N\}}}(\mathbf{r}_{\{N\}}) - \tilde{\chi}_{\text{off}}(\mathbf{r}_{\{N\}}), \quad (4.215a)$$

$$\tilde{W}_{\text{on}}(\mathbf{r}_{\{N\}}) = W_{\hat{\rho}_{\{N\}}}(\mathbf{r}) - \tilde{W}_{\text{off}}(\mathbf{r}_{\{N\}}), \quad (4.215b)$$

when the outcome is 'on', where $\hat{\rho}_{\{N\}} = \text{tr}_{N+1}\{\hat{\rho}\}$ is the reduced initial state of the non-measured modes.

De-Gaussification by vacuum removal

Note that $W_{\hat{\rho}_{\text{off}}}(\mathbf{r})$ is Gaussian, and therefore, the 'off' event projects the state of the non-measured modes into another Gaussian state. This is not the case for the 'on' event, whose associated Wigner function $W_{\hat{\rho}_{\text{on}}}(\mathbf{r})$ is not Gaussian, and therefore, it can be used as a *de-Gaussifying* operation. Moreover, it is a very convenient non-Gaussian operation from the theoretical point of view, because it is a sum of a constant term and a Gaussian, and hence it is still very easy to treat by extending the Gaussian formalism minimally.

In order to understand this better, let us analyze the case in which the initial state is a general Gaussian state $\hat{\rho}_{\text{G}}(\mathbf{d}, V)$ of the form (4.68) with $M = N$ and $M' = 1$, whose mean vector and covariance matrix we write as

$$\bar{\mathbf{r}} = (\bar{\mathbf{r}}_{\{N\}}, \bar{\mathbf{r}}_{N+1}) \qquad \text{and} \qquad V = \begin{bmatrix} V_{\{N\}} & C \\ C^T & V_{N+1} \end{bmatrix}, \quad (4.216)$$

where $\bar{\mathbf{r}}_{\{N\}} \in \mathbb{R}^{2N}$, $\bar{\mathbf{r}}_{N+1} \in \mathbb{R}^2$, $V_{\{N\}}$ and V_{N+1} are real, symmetric matrices of dimensions $2N \times 2N$ and 2×2, respectively, while C is a real $2N \times 2$ matrix. Using the Gaussian integral (4.59), it is straightforward to prove that the probability of the 'off' event is

$$p_{\text{off}} = \tilde{\chi}_{\text{off}}(\mathbf{0}) = \frac{2}{\sqrt{\det(V_{N+1} + I_{2\times2})}} \exp\left[-\frac{1}{8}\bar{\mathbf{r}}_{N+1}^T(V_{N+1} + I_{2\times2})\bar{\mathbf{r}}_{N+1}\right], \quad (4.217)$$

while the corresponding output state is the Gaussian $\hat{\rho}_{\text{off}} = \hat{\rho}_{\text{G}}(\bar{\mathbf{r}}_{\text{off}}, V_{\text{off}})$ with

$$\bar{\mathbf{r}}_{\text{off}} = \bar{\mathbf{r}}_{\{N\}} - C(V_{N+1} + I_{2\times2})^{-1}\bar{\mathbf{r}}_{N+1}, \quad (4.218a)$$

$$V_{\text{off}} = V_{\{N\}} + C(V_{N+1} + I_{2\times2})^{-1}C^T. \quad (4.218b)$$

The probability of the 'on' event is then $p_{\text{on}} = 1 - p_{\text{off}}$, which, based on (4.215), has an associated post-measurement Wigner function

$$W_{\text{on}}(\mathbf{r}_{\{N\}}) = \left(1 - p_{\text{off}}\right)^{-1}\left[W_{\hat{\rho}_{\text{G}}(\mathbf{d}_{\{N\}}, V_{\{N\}})}(\mathbf{r}_{\{N\}}) - p_{\text{off}}\, W_{\hat{\rho}_{\text{G}}(\mathbf{d}_{\text{off}}, V_{\text{off}})}(\mathbf{r}_{\{N\}})\right]. \quad (4.219)$$

Note that even though this Wigner function is not Gaussian, it is a simple combination of two Gaussians (specifically a 'negative mixture') and, hence, as already explained, this approach to de-Gaussification is very convenient since all the tools of Gaussian states and operations can be used.

Let us consider a simple example: we have two modes in the two-mode squeezed vacuum state (4.113), and we perform an on/off detection onto the second mode.

Given the photon-number correlation between the modes, it is obvious that whenever the outcome is 'off', the first mode is projected into the vacuum state, $\hat{\rho}_{\text{off}} = |0\rangle\langle 0|$. More interestingly, if the outcome is 'on' the state of the first mode will collapse to the mixture

$$\hat{\rho}_{\text{on}} \propto \sum_{n=1}^{\infty} \tanh^{2n} r|n\rangle\langle n|, \tag{4.220}$$

which is a thermal state with the vacuum component removed.

According to (4.217), and using the mean vector and covariance matrix of the two-mode squeezed vacuum state (4.114), the probability for the 'off' event is $p_{\text{off}} = 1/\cosh^2 r$, and it is simple to check that (4.219) leads to $\bar{\mathbf{r}}_{\text{off}} = \mathbf{0}$ and $V_{\text{off}} = \mathcal{I}_{2\times 2}$, corresponding to the vacuum state. The probability of the 'on' event reads then $p_{\text{on}} = \tanh^2 r$, and the corresponding Wigner function is

$$W_{\text{on}}(\mathbf{r}_1) = \sinh^{-2} r\left[\cosh^2 r\, W_{\hat{\rho}_{\text{th}}(\sinh^2 r)}(\mathbf{r}_1) - W_{|0\rangle\langle 0|}(\mathbf{r}_1)\right], \tag{4.221}$$

that is, a 'negative mixture' of a thermal and a vacuum state. It is simple to check that, at the origin of phase space, $\mathbf{r}_1 = \mathbf{0}$, the weight of the vacuum state is always larger than that of the thermal state, and hence this Wigner function always has a negative central region, surrounded by a positive one, which shows the non-Gaussian character of the state. In particular, we find that $W_{\text{on}}(\mathbf{0}) = -1/4\pi \cosh(2r)$. Moreover, this central negative region has more or less the same size irrespective of the squeezing value. Indeed, it is very simple to show that the radius of the central negative region is given by

$$\sqrt{\frac{2\log_e(1 + \tanh^2 r)}{\tanh^2 r}}, \tag{4.222}$$

which is a monotonically increasing function of the squeezing, but is bounded by 1 from below and by $\log_e^{1/2} 4 \approx 1.18$ from above, so that it varies very little with r. In contrast, the positive region becomes larger as the squeezing increases, which arises from the thermal component of the state.

Finally, note that for a small squeezing parameter, the state tends to the $|1\rangle$ Fock state, as easily seen from (4.220). This is not an 'accident'; in fact, as we show in the next section, when a mode interacts very weakly via the two-mode squeezing or beam-splitter operations with a second mode in a vacuum, the 'on' detection of this second mode signals, respectively, the approximate application of the $\hat{a}^\dagger$ or $\hat{a}$ operators on the principal mode.

4.6 Non-Gaussian scenarios: photon addition, subtraction, and majorization properties of two-mode squeezed states

4.6.1 Photon addition and subtraction

In spite of their non-unitary character, photon addition and subtraction, which correspond to the application of $\hat{a}$ and $\hat{a}^\dagger$ as evolution onto a continuous-variable state, can be performed on a system conditioned to particular outcomes in a measurement,

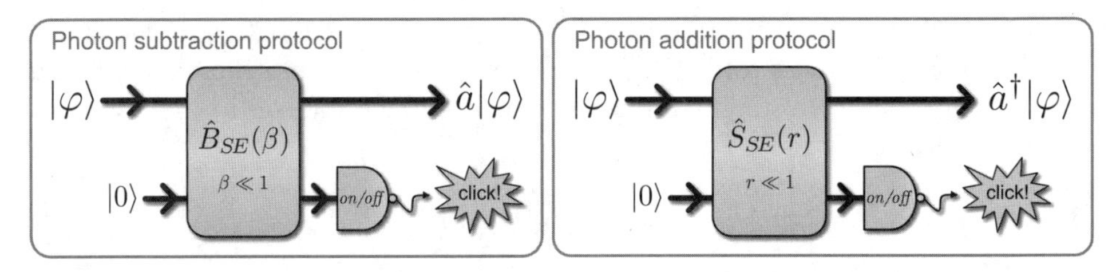

Figure 4.9. Protocols for the probabilistic implementation of photon subtraction and addition based on a click in an on/off detection.

that is, probabilistically. In this section we are going to explain how this can be achieved by using schemes based on a click in an on/off detection. Being able to apply these non-Gaussian, non-unitary operations onto a system opens the way to an exciting world in which many fundamental tests of quantum mechanics can be performed.

Consider a single mode S in some pure state $|\varphi\rangle$ (everything that follows is trivially generalized to general mixed states, or even to multi-mode states). As shown in figure 4.9, the subtraction protocol starts by mixing this mode via a weak beam-splitter transformation $\hat{B}_{SE}(\beta \ll 1)$ with another oscillator E in a vacuum state. We can make a series expansion of the beam-splitter operator (4.115*a*) up to first order in β, so that the joint state $|\varphi\rangle \otimes |0\rangle$ evolves into

$$|\psi\rangle_{SE} = |\varphi\rangle \otimes |0\rangle + \beta(\hat{a}|\varphi\rangle \otimes |1\rangle) + O(\beta^2), \tag{4.223}$$

after the beam splitter. The last step of the protocol consists in applying an on/off measurement to the E mode, so that whenever the detector *clicks*, the state of the S oscillator collapses to the unnormalized state

$$\tilde{\rho} = \text{tr}_E\left\{\left(\hat{I} \otimes \hat{P}_{\text{on}}\right)|\psi\rangle_{SE}\langle\psi|\right\} = \beta^2\hat{a}|\varphi\rangle\langle\varphi|\hat{a}^\dagger + O(\beta^4), \tag{4.224}$$

which converges to the subtracted state $\hat{a}|\varphi\rangle$ as β goes to zero. Note, however, that the probability $p_{\text{on}} = \text{tr}\{\tilde{\rho}\}$ of the 'on' event vanishes in the $\beta \to 0$ limit, and therefore, one has to find a balance between having a non-zero *success probability* and a good approximation of the subtracted state. We will return to this point shortly.

The scheme that implements photon addition is very similar, except for the fact that now the beam-splitter transformation is replaced by a two-mode squeezing transformation $\hat{S}_{SE}(r \ll 1)$, see (4.109). For a small squeezing parameter, the operator can be expanded again to first order in the squeezing parameter r, leading to the joint state after the two-mode squeezer

$$|\psi\rangle_{SE} = |\varphi\rangle \otimes |0\rangle + r(\hat{a}^\dagger|\varphi\rangle \otimes |1\rangle) + O(r^2). \tag{4.225}$$

Then, after a click in the on/off detector the unnormalized state of the S oscillator collapses to

$$\tilde{\rho} = \text{tr}_E\left\{\left(\hat{I} \otimes \hat{P}_{\text{on}}\right)|\psi\rangle_{SE}\langle\psi|\right\} = r^2\hat{a}^\dagger|\varphi\rangle\langle\varphi|\hat{a} + O(r^4), \tag{4.226}$$

which again converges to the added state $\hat{a}^\dagger|\varphi\rangle$ as r goes to zero.

Note that in the language of quantum operations, based on expressions (4.224) and (4.226), one would be tempted to define quantum operations $\mathcal{E}_{\text{sub}}$ and $\mathcal{E}_{\text{add}}$ associated with photon subtraction and addition, respectively, with corresponding Kraus operators $\hat{E}_{\text{sub}} = \beta\hat{a}$ and $\hat{E}_{\text{add}} = r\hat{a}^{\dagger}$. Note, however, that these Kraus operators do not lead to trace-decreasing quantum operations in general; they do so only provided that β and r are small enough. Indeed, note that denoting by $\bar{N} = \text{trace}\{\hat{a}^{\dagger}\hat{a}\hat{\rho}\}$ the mean photon number of the state, we obtain the success probabilities

$$p_{\text{sub}} = \text{tr}\left\{\hat{E}_{\text{sub}}^{\dagger}\hat{E}_{\text{sub}}\hat{\rho}\right\} = \beta^2\bar{N} \tag{4.227a}$$

$$p_{\text{add}} = \text{tr}\left\{\hat{E}_{\text{add}}^{\dagger}\hat{E}_{\text{add}}\hat{\rho}\right\} = r^2(\bar{N}+1). \tag{4.227b}$$

This shows that the operations are trace-decreasing, and hence physical, only when $\beta^2 < \bar{N}^{-1}$ and $r^2 < (\bar{N}+1)^{-1}$. Nevertheless, a more careful analysis of when the higher orders in the states (4.224) and (4.226) can be neglected would reveal that the conditions $\beta^2 \ll \bar{N}^{-1}$ and $r^2 \ll (\bar{N}+1)^{-1}$, and not just $\beta \ll 1$ and $r \ll 1$, are needed, so that the quantum operations $\mathcal{E}_{\text{sub}}$ and $\mathcal{E}_{\text{add}}$ are indeed trace-decreasing whenever they are accurately implemented. Note that this also means that the success probability has to be small, as we already pointed out above.

This simplified picture captures all the qualitative features that one needs to know about how the addition and subtraction operations can be applied probabilistically, that is, as trace-decreasing operations. Nevertheless, using the tools that we developed above, in particular the phase-space description of on/off detectors, one can treat them carefully from a quantitative point of view when benchmarking particular setups.

4.6.2 Increasing entanglement by local addition or subtraction

Throughout this book, we have introduced the idea that the entanglement of a state cannot be increased by acting locally on its entangled parts. However, this is only true deterministically, that is, as counterintuitive as it might seem, local strategies based on measurements can indeed enhance entanglement for some of the measurement outcomes (but, of course, on average the entanglement cannot increase). As an example of this, now we study how photon addition and subtraction can indeed achieve this task. In particular, we will study how the entanglement of the two-mode squeezed vacuum state increases when photon addition or subtraction are applied to either of the modes [31], which we denote by A and B, for Alice and Bob. It will be convenient to define the parameter $\lambda = \tanh r$.

Before addressing the problem, it will be useful to prove the following properties of the two-mode squeezed vacuum state under the action of annihilation and creation operators:

$$\hat{a}_B^k\,|r\rangle_{2\text{sq}} = \tanh^k(r)\hat{a}_A^{\dagger k}\,|r\rangle_{2\text{sq}}, \tag{4.228a}$$

$$\hat{a}_A^{\dagger k}\,|r\rangle_{2\text{sq}} = \cosh^k(r)\hat{S}_{AB}(r)\hat{a}_A^{\dagger k}|0\rangle. \tag{4.228b}$$

The first identity tells us that subtracting excitations in one mode of the pair is equivalent to adding excitations in the other when they are in a two-mode squeezed vacuum state. The second means that adding the excitations before or after the two-mode squeezing transformation is all the same. Let us now prove these relations.

The first one is easily proved for $k = 1$ by simply operating on the two-mode squeezed vacuum state (4.113) expanded in the Fock basis,

$$\hat{a}_B\,|r\rangle_{2\mathrm{sq}} = \frac{1}{\cosh r}\sum_{n=1}^{\infty}\sqrt{n}\,\tanh^n r|n,\,n-1\rangle$$
$$= \frac{1}{\cosh r}\sum_{n=0}^{\infty}\sqrt{n+1}\,\tanh^{n+1} r|n+1,\,n\rangle = \tanh(r)\hat{a}_A^\dagger\,|r\rangle_{2\mathrm{sq}}. \quad (4.229)$$

Since $\hat{a}_B$ and $\hat{a}_A^\dagger$ commute, iterating this identity we prove (4.228*a*) for any k. The second identity (4.228*b*) is also easily proven as follows:

$$\hat{S}_{AB}(r)\hat{a}_A^{\dagger k}|0,\,0\rangle = \left[\hat{S}_{AB}^\dagger(-r)\hat{a}_A^\dagger\hat{S}_{AB}(-r)\right]^k\hat{S}_{AB}(r)|0,\,0\rangle \underset{(4.110)}{=} \left(\hat{a}_A^\dagger\cosh r - \hat{a}_B\sinh r\right)^k|r\rangle_{2\mathrm{sq}}$$
$$\underset{\substack{\text{binomial}\\\text{expansion}}}{=} \sum_{j=0}^{k}(-1)^{k-j}\binom{k}{j}\sinh^{k-j} r\cosh^j r\hat{a}_A^{\dagger j}\hat{a}_B^{k-j}\,|r\rangle_{2\mathrm{sq}}$$
$$\underset{(4.228a)}{=} \frac{1}{\cosh^k r}\sum_{j=0}^{k}(-1)^{k-j}\binom{k}{j}\sinh^{2k-2j} r\cosh^{2j} r\hat{a}_A^{\dagger k}\,|r\rangle_{2\mathrm{sq}}$$
$$\underset{\substack{\text{binomial}\\\text{expansion}}}{=} \frac{1}{\cosh^k r}\hat{a}_A^{\dagger k}\,|r\rangle_{2\mathrm{sq}}, \quad (4.230)$$

where we have used the binomial expansion $(\hat{A}+\hat{B})^k = \sum_{j=0}^{k}\binom{k}{j}\hat{A}^j\hat{B}^{k-j}$, valid for any two commuting operators $\hat{A}$ and $\hat{B}$.

Equipped with these identities, we can now consider how the entanglement of the two-mode squeezed vacuum state is affected by photon addition and subtraction. To this aim, we start by defining the two-mode squeezed Fock state with k excitations in the first mode, which using the identities above is equivalent to a two-mode squeezed vacuum state with k-additions or subtractions in the first or second modes, respectively; explicitly:

$$\left|\Psi_\lambda^{(k)}\right\rangle = \hat{S}_{AB}(r)|k,\,0\rangle = \frac{1}{\sqrt{k!}\,\cosh^k r}\hat{a}_A^{\dagger k}\,|r\rangle_{2\mathrm{sq}} = \frac{1}{\sqrt{k!}\,\sinh^k r}\hat{a}_B^k\,|r\rangle_{2\mathrm{sq}}. \quad (4.231)$$

In order to proceed, we write the state in the Fock basis using (4.113) and the second form of the state presented above. It is straightforward to obtain

$$\left|\Psi_\lambda^{(k)}\right\rangle = \sum_{n=0}^{\infty}\sqrt{p_n^{(k)}(\lambda)}\,|n+k,\,n\rangle, \quad (4.232)$$

with

$$p_n^{(k)}(\lambda) = (1 - \lambda^2)^{k+1}\lambda^{2n}\binom{n+k}{n}. \tag{4.233}$$

This is a very convenient representation of the state, since it is already in the Schmidt form. Therefore, its entanglement entropy can be evaluated as

$$E\left[\left|\Psi_\lambda^{(k)}\right\rangle\right] = -\sum_{n=0}^{\infty} p_n^{(k)} \log p_n^{(k)}. \tag{4.234}$$

In order to prove that this is a monotonically increasing function of k, we proceed as follows. Using the Pascal identity

$$\binom{n+k+1}{k+1} = \binom{n+k}{k+1} + \binom{n+k}{k}, \tag{4.235}$$

we can write

$$p_n^{(k+1)}(\lambda) = \lambda^2 p_{n-1}^{(k+1)}(\lambda) + (1 - \lambda^2)p_n^{(k)}(\lambda), \tag{4.236}$$

where we set $p_n^{(k)} = 0$ for $n < 0$ by definiteness. Next we use the strict concavity[14] of the function $h(x) = -x\log x$ to write

$$\sum_{n=0}^{\infty} h\left[p_n^{(k+1)}(\lambda)\right] > \lambda^2 \sum_{n=0}^{\infty} h\left[p_{n-1}^{(k+1)}(\lambda)\right] + (1 - \lambda^2)\sum_{n=0}^{\infty} h\left[p_n^{(k)}(\lambda)\right], \tag{4.237}$$

for $0 < \lambda < 1$. The final step consists in noting that $p_{n-1}^{(k+1)}$ is equivalent to $p_n^{(k+1)}$ up to a shift to the right in the Fock basis, which does not change the entropy, i.e. $\sum_{n=0}^{\infty} h[p_{n-1}^{(k+1)}] = \sum_{n=0}^{\infty} h[p_n^{(k+1)}]$, so that the previous expression is simply equivalent to

$$E\left[\left|\Psi_\lambda^{(k+1)}\right\rangle\right] > E\left[\left|\Psi_\lambda^{(k)}\right\rangle\right] \tag{4.238}$$

for $0 < \lambda < 1$. This shows that the entanglement of the two-mode squeezed vacuum state is enhanced by photon addition or subtraction, at least when acting on one mode only. Results for the simultaneous action on both modes can be consulted in the original reference [31].

4.6.3 Majorization properties of two-mode squeezed Fock states

As we discussed in section 3.4, we can use majorization theory to find relations between pure bipartite states which are stronger than simply 'which one is more entangled'. In particular, we saw that given two bipartite states $|\psi\rangle_{AB}$ and $|\varphi\rangle_{AB}$, $|\psi\rangle_{AB} \prec |\varphi\rangle_{AB}$ implies not only that $|\psi\rangle_{AB}$ is more entangled than $|\varphi\rangle_{AB}$, but also that it can be transformed into the latter deterministically via LOCC protocols. As an example of this in

[14] A real function is $f(x)$ is strictly concave when $f(\sum_{n=1}^{N} p_n x_N) > \sum_{n=1}^{N} p_n f(x_n)$ for any non-trivial probability distribution $\{p_n\}_{n=1,2,\ldots,N}$ and set of non-equal points $\{x_n\}_{n=1,2,\ldots,N}$.

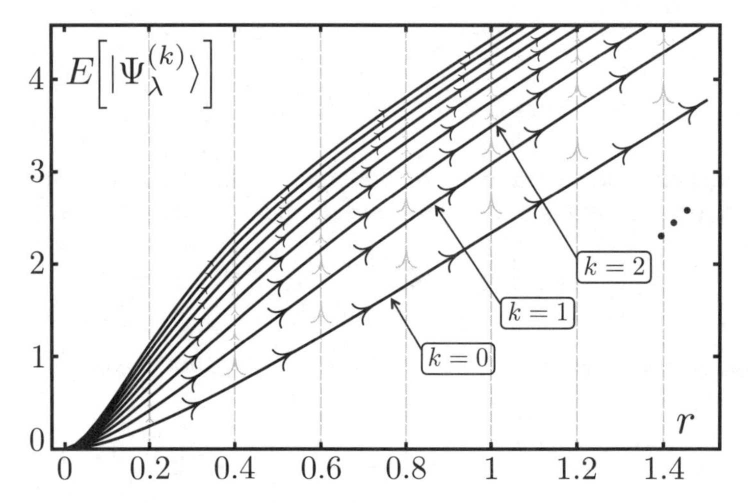

Figure 4.10. Entanglement entropy of the two-mode squeezed Fock states as a function of the squeezing parameter. The majorization relations are denoted in the curves, which imply stronger partial order relations between the states than the value of their entanglement (in particular, states with larger entanglement can be transformed into states with a lower one via deterministic LOCC protocols).

continuous-variable systems, we will now study the different majorization relations that exist between the two-mode squeezed Fock states defined above $|\Psi_\lambda^{(k)}\rangle$.

We are going to discuss two types of majorization relations [32], which are summarized in figure 4.10:

$$|\Psi_\lambda^{(k)}\rangle \prec |\Psi_\lambda^{(k'<k)}\rangle \qquad \text{and} \qquad |\Psi_\lambda^{(k)}\rangle \prec |\Psi_{\lambda'<\lambda}^{(k)}\rangle. \tag{4.239}$$

The first one shows that for a fixed squeezing λ, a two-mode squeezed k-Fock state can be transformed via an LOCC protocol into states with smaller k. Similarly, the second one means that a two-mode squeezed Fock state can be transformed into states with smaller squeezing parameter.

This result can be proven in two ways: either by finding explicit LOCC protocols performing the corresponding transformation between the states, or the column-stochastic matrices connecting the corresponding Schmidt distributions. Regarding the column-stochastic matrices, the following results can be found [32]:

$$\mathbf{p}^{(k)}(\lambda) = D^{(k-k')}(\lambda)\mathbf{p}^{(k')}(\lambda), \tag{4.240a}$$

$$\mathbf{p}^{(k)}(\lambda) = R^{(k)}(\lambda, \lambda')\mathbf{p}^{(k)}(\lambda'), \tag{4.240b}$$

with matrices D and R having matrix elements

$$D_{nm}^{(k-k')}(\lambda) = (1-\lambda^2)^{k-k'}\binom{m+k-k'-1}{k-k'-1}\lambda^{2(n-m)}H(n-m), \tag{4.241a}$$

$$R_{nm}^{(k)}(\lambda, \lambda') = \binom{m+k}{m}^{-1}\left(\frac{1-\lambda^2}{1-\lambda'^2}\right)\left[L_{n-m}^{(k,m)}\lambda^2 - L_{n-m-1}^{(k,m+1)}\lambda'^2\right]\lambda^{2(n-m-1)}H(n-m), \tag{4.241b}$$

where $H(x)$ is the Heaviside step function defined as $H(x) = 0$ for $x < 0$ and $H(x) = 1$ for $x \geqslant 0$, and $L_m^{(k,n)} = n\binom{n+k}{k}\binom{m+k}{k}\lambda'^{-2n}B(\lambda'^2; n, 1+k)$, $B(z; a, b) = \int_0^z \mathrm{d}x x^{a-1}(1-x)^{b-1}$ being the incomplete beta function. Note that both matrices are lower-triangular, that is, their elements above the diagonal are zero. This is a typical form of column-stochastic matrices in infinite dimension, which makes them a bit more simple and systematic to find recursively.

As for the LOCC protocols, let us start with the one corresponding to the transformation $|\Psi_\lambda^{(k+\Delta k)}\rangle \xrightarrow[\text{LOCC}]{} |\Psi_\lambda^{(k)}\rangle$, where Δk is a positive integer. Inspired by the column-stochastic matrix (4.240a), we can build the following protocol. Bob starts by performing a POVM-based measurement described by the measurement operators

$$\hat{B}_m = \sum_{l=m}^{\infty}\sqrt{\frac{(1-\lambda^2)^{\Delta k}\binom{m+\Delta k-1}{\Delta k-1}\lambda^{2m}p_{l-m}^{(k)}}{p_l^{(k+\Delta k)}}}|l-m\rangle\langle l|. \tag{4.242}$$

It is easy to verify the condition $\sum_{m=0}^{\infty}\hat{B}_m^\dagger\hat{B}_m = \hat{I}$ by using the (4.240a). Suppose that Bob obtains the outcome m, after which the state collapses to

$$\left(\hat{I}\otimes\hat{B}_m\right)|\Psi_\lambda^{(k+\Delta k)}\rangle \propto \sum_{l=m}^{\infty}\sqrt{p_{l-m}^{(k)}}|l+k+\Delta k, l+m\rangle = \sum_{n=0}^{\infty}\sqrt{p_n^{(k)}}|n+k+m+\Delta k, n\rangle. \tag{4.243}$$

Bob then communicates the outcome m of his measurement to Alice, who performs the shift operation

$$\hat{A}_m = \sum_{l=0}^{\infty}|l\rangle\langle l+m+\Delta k|, \tag{4.244}$$

yielding the desired state $|\Psi_\lambda^{(k)}\rangle$ for all m. Note that the shift operator is trace preserving in the subspace spanned by $\{|j+m+\Delta k\rangle\}_{j=0,1,..}$, which is the support of $(\hat{I}\otimes\hat{B}_m)|\Psi_\lambda^{(k+\Delta k)}\rangle$ on Alice's space.

Let us move on now to the LOCC protocol associated to the transformation $|\Psi_\lambda^{(k)}\rangle \xrightarrow[\text{LOCC}]{} |\Psi_{\lambda'<\lambda}^{(k)}\rangle$. Given the complicated form of the column-stochastic matrix $R^{(k)}(\lambda, \lambda')$, it seems hard in this case to get inspiration from the relation (4.240b). However, in [32] a successful LOCC strategy was found, which is the one we will discuss now. The protocol starts with Bob mixing his mode B with an ancillary mode C on a beam splitter of transmissivity T. The initial state is written in the Fock basis as

$$|\psi\rangle_{ABC} = |\Psi_\lambda^{(k)}\rangle\otimes|0\rangle = \mathcal{N}(k,\lambda)\sum_{n=0}^{\infty}\lambda^n\binom{n+k}{k}^{1/2}|n+k, n, 0\rangle, \tag{4.245}$$

with $\mathcal{N}(k, \lambda) = (1 - \lambda^2)^{(k+1)/2}$, which becomes

$$|\psi'\rangle_{ABC} = \mathcal{N}(k, \lambda) \sum_{n,m=0}^{\infty} \sqrt{(T\lambda^2)^n \left(\frac{1-T}{T}\right)^m \binom{n+k}{k} \binom{n}{m}} |n+k, n-m, m\rangle, \quad (4.246)$$

after passing through the beam splitter. The second step consists on Bob measuring the number of photons reflected by the beam splitter, that is, on mode C. With probability

$$\mathscr{P}(l) = {}_{ABC}\langle\psi'|\left(\hat{I} \otimes \hat{I} \otimes |l\rangle\langle l|\right) |\psi'\rangle_{ABC} = (1-T)^l \lambda^{2l} \binom{k+l}{l} \frac{\mathcal{N}^2(k, \lambda)}{\mathcal{N}^2\left(k+l, \sqrt{T}\lambda\right)}, \quad (4.247)$$

Bob will obtain l photons as an outcome, and the state of modes A and B will collapse in that case to

$$\begin{aligned} \sqrt{\mathscr{P}(l)}\ |\psi''\rangle_{AB} &= {}_C\langle l|\psi'\rangle_{ABC} \\ &= \mathcal{N}(k, \lambda)\left(\frac{1-T}{T}\right)^{l/2} \sum_{n=l}^{\infty} (T\lambda^2)^{n/2} \binom{n+k}{k}^{1/2} \binom{n}{l}^{1/2} |n+k, n-l\rangle. \quad (4.248) \end{aligned}$$

Making the variable change $n - l \to n$ in the sum, and using the relation

$$\binom{n+l+k}{k}\binom{n+l}{l} = \binom{n+k+l}{n}\binom{k+l}{l}, \quad (4.249)$$

this state can be rewritten as

$$\begin{aligned} \sqrt{\mathscr{P}(l)}\ |\psi''\rangle_{AB} &= \mathcal{N}(k, \lambda)(1-T)^{l/2}\lambda^l \binom{k+l}{l}^{1/2} \sum_{n=0}^{\infty} (T\lambda^2)^{n/2} \binom{n+k+l}{n}^{1/2} |n+k+l, n\rangle \\ &= \sqrt{\mathscr{P}(l)} \left|\Psi_{\sqrt{T}\lambda}^{(k+l)}\right\rangle. \quad (4.250) \end{aligned}$$

Therefore, by properly choosing the transmissivity of the beam splitter so that $\lambda' = \sqrt{T}\lambda$, the final state is $|\Psi_{\lambda'}^{(k+l)}\rangle$. The last step of the protocol consists then on the application of the LOCC-based transformation $|\Psi_{\lambda'}^{(k+l)}\rangle \to |\Psi_{\lambda'}^{(k)}\rangle$ that we introduced above. Note that the proposed protocol requires two steps,

$$|\Psi_{\lambda}^{(k)}\rangle \xrightarrow[\text{LOCC}]{\text{probabilistic}} |\Psi_{\lambda'}^{(k+l)}\rangle \qquad \text{and then} \qquad |\Psi_{\lambda'}^{(k+l)}\rangle \xrightarrow[\text{LOCC}]{\text{deterministic}} |\Psi_{\lambda'}^{(k)}\rangle, \quad (4.251)$$

meaning that it is not the minimal one, since according to what we saw in section 3.4.3, there always exists a one-way-direct LOCC protocol. This is an example of the clear dichotomy that exists in the world of quantum information (and mathematics in general): it is very useful to know that something exists, but that does not always mean that we know how to find it.

Bibliography

[1] Schleich W P 2001 *Quantum Optics in Phase Space* (New York: Wiley)

[2] Simon R, Mukunda N and Dutta B 1994 Quantum-noise matrix for multimode systems: $U(n)$ invariance, squeezing, and normal forms *Phys. Rev.* A **49** 1567

[3] Arvind Dutta B, Mukunda N and Simon R 1995 The real symplectic groups in quantum mechanics and optics *Pramana J. Phys.* **45** 471

[4] Navarrete-Benlloch C 2011 Contributions to the quantum optics of multimode optical parametric oscillators *PhD Thesis* arXiv:1504.05917

[5] Gerry C C and Knight P L 2005 *Introductory Quantum Optics* (Cambridge: Cambridge University Press)

[6] Braunstein S L 2005 Squeezing as an irreducible resource *Phys. Rev.* A **71** 055801

[7] Williamson J 1936 On the algebraic problem concerning the normal forms of linear dynamical systems *Am. J. Math.* **58** 141

[8] Serafini A, Illuminati F and De Siena S 2004 Symplectic invariants, entropic measures and correlations of Gaussian states *J. Phys.* B **37** L21

[9] Duan L-M, Giedke G, Cirac J I and Zoller P 2000 Inseparability criterion for continuous variable systems *Phys. Rev. Lett.* **84** 2722

[10] Einstein A, Podolsky B and Rosen N 1935 Can quantum-mechanical description of physical reality be considered complete? *Phys. Rev.* **47** 777

[11] Bell J S 1964 On the Einstein Podolsky Rosen paradox *Physics* **1** 195

[12] Bell J S 2004 *Speakable and Unspeakable in Quantum Mechanics: Collected Papers on Quantum Philosophy* (Cambridge: Cambridge University Press)

[13] Freedman S J and Clauser J F 1972 Experimental test of local hidden-variable theories *Phys. Rev. Lett.* **28** 938

[14] Aspect A, Dalibard J and Roger G 1982 Experimental test of Bells inequalities using time-varying analyzers *Phys. Rev. Lett.* **49** 1804

[15] Brunner N, Cavalcanti D, Pironio S, Scarani V and Wehner S 2014 Bell nonlocality *Rev. Mod. Phys.* **86** 419

[16] Hensen B *et al* 2015 Experimental loophole-free violation of a Bell inequality using entangled electron spins separated by 1.3 km *Nature* **526** 682

[17] Giustina M *et al* 2015 Significant-loophole-free test of Bell's theorem *Phys. Rev. Lett.* **115** 250401

[18] Shalm L K *et al* 2015 Strong loophole-free test of local realism *Phys. Rev. Lett.* **115** 250402

[19] Simon R 2000 Peres-Horodecki separability criterion for continuous variable systems *Phys. Rev. Lett.* **84** 2726

[20] Werner R F and Wolf M M 2001 Bound entangled Gaussian states *Phys. Rev. Lett.* **86** 3658

[21] Ferraro A, Olivares S and Paris M G A 2005 *Gaussian States in Quantum Information* (*Napoli Series on Physics and Astrophysics*) (Naples: Bibliopolis)

[22] Giedke G, Kraus B, Lewenstein M and Cirac J I 2001 Entanglement criteria for all bipartite Gaussian states *Phys. Rev. Lett.* **87** 167904

[23] Vidal G and Werner R F 2002 Computable measure of entanglement *Phys. Rev.* A **65** 032314

[24] Holevo A S and Werner R F 2001 Evaluating capacities of bosonic Gaussian channels *Phys. Rev.* A **63** 032312

[25] Caruso F, Eisert J, Giovannetti V and Holevo A S 2008 Multi-mode bosonic Gaussian channels *New J. Phys.* **10** 083030

[26] Caruso F, Eisert J, Giovannetti V and Holevo A S 2011 Optimal unitary dilation for bosonic Gaussian channels *Phys. Rev.* A **84** 022306

[27] Bernstein D S 2005 *Matrix Mathematics: Theory Facts, and Formulas* (Princeton, NJ: Princeton University Press)

[28] Mollow B R 1968 Quantum theory of field attenuation *Phys. Rev.* **168** 1896
[29] Louisell W H 1973 *Quantum Statistical Properties of Radiation* (New York: Wiley)
[30] Puri R R 2001 *Mathematical Methods of Quantum Optics* (Berlin: Springer)
[31] Navarrete-Benlloch C, García-Patrón R, Shapiro J H and Cerf N J 2012 Enhancing quantum entanglement by photon addition and subtraction *Phys. Rev.* A **86** 012328
[32] García-Patrón R, Navarrete-Benlloch C, Lloyd S, Shapiro J H and Cerf N J 2012 Majorization theory approach to the Gaussian channel minimum entropy conjecture *Phys. Rev. Lett.* **108** 110505

◎

编辑手记

世界著名物理学家冯·诺伊曼(Von Neumann)指出:

> 今天的优秀理论物理学家可能还具有其学科工作知识的大部分,但是我怀疑现在活着的数学家只具有其学科知识的四分之一以上.

本书是一部英文版关于量子信息方面的学术著作,是我们工作室极为庞大的国外优秀学术著作引进计划中的一部.

本书的中文书名或可译为《具有连续变量的量子信息形式主义概论》,作者为卡洛斯·纳瓦雷特-本洛赫(Carlos Navarrete-Benlloch),西班牙人,德国加兴马普量子光学研究所(MPQ)的博士后研究员.他曾获得Alexander von Humboldt基金会授予的博士后研究奖学金.他主要进行具有非线性光学腔、超导电路和光机械系统的耗散量子光学,以及超冷原子的量子模拟的研究.除了量子光学和量子信息,卡洛斯还对其他几个物理领域的研究感兴趣,包括引力、宇宙学、量子场论和粒子物理学.

正如作者在前言中所述:

> 量子信息是一个新兴的领域,在过去的几十年中引起了很多关注.它是一个广泛的主题,可以从最实用的问题(例如,如何构建量子计算机或安全的加密系统)延伸到最具理论性的问题(例如,量子力学的形式主义和解释、复杂性以及超越经典物理学的潜能).
>
> 本书介绍了量子信息背后的形式主义,特别强调了连续变量系统,即可以描述为谐振子集合的系统(例如,光).本书不应被视为对该领域的详尽的回顾,而应作为涵盖了一系列基本概念的介绍性文

本,重点关注其物理意义和数学处理方法.本书是独立的,从某种意义上说,它开始于量子力学的第一个原则,并尝试按照逻辑顺序去构建概念和方法.本书的目标读者主要是已经学习了一整个学期的标准量子力学课程并希望进行深入研究的学生,或者是与量子光学或凝聚物理学等领域相关的研究人员,他们希望了解量子信息的强大的形式主义.由于本书篇幅的限制,使我不能在书中包含可以阐明定义和陈述的实际案例,但这些案例通常可以在书中不同部分中列出的相关的参考文献中找到.

至于主题的选择,我选择了当我作为一名量子光学的博士生第一次涉足该领域时发现的最有价值的主题.本书的第一章通过回顾经典力学、线性代数和量子力学中的一些基本概念开始,这些概念在整本书中都会用到.接下来的两章处理了适用于一般量子系统的概念,一个关系到两体纠缠,另一个关系到一般量子运算的形式主义(可以在系统中产生的最一般的转换),包括通道和广义测量的最重要的例子.最后一章是本书的重要章节,将前面章节中的概念应用到连续变量系统中,比如通过无限维Hilbert空间描述的谐振子系统.在回顾了量子谐振子的物理性质之后,我们描述了基于Wigner函数的相空间,并介绍了相关的一类Gauss态及其运算,强调了它们在理论和应用上的重要性,它们在数学上处理起来相对容易,且可以通过实验生成.在这之后,书中提到的测量的量子理论专门针对连续变量系统,其介绍了光电检测、零差检测、外差检测、开(关)检测等场景中的典型检测策略,最后表明检测策略可用于概率性地生成非Gauss运算,例如光子加减法,而这反过来又可用于在局部应用时增强纠缠,而不是直觉.

涵盖连续变量领域的扩展主题和技术的全面评论,以及我跳过的一些证明,可以在文献①②③④⑤⑥中找到.

---

① Braunstein S L and van Loock P. 2005 Quantum information with continuous variables, Rev. Mod. Phys. 77,513.

② Weedbrook C, Pirandola S, Gariía-Patrón R, Cerf N J. Ralph T C, Shapiro J H and Lloyd S. 2012 Gaussian quantum information, Rev. Mod. Phys. 84,621.

③ Ferraro A, Olivares S and Paris M G A. 2005 Gaussian States in Quantum Information (Napoli Series on Physics and Astrophysics) (Naples: Bibliopolis).

④ Cerf N J, Leuchs G and Polzik E S (ed). 2007 Quantum Information with Continuous Variables of Atoms and Light (London: Imperial College Press).

⑤ Shapiro J H. 2008 Quantum optical communication Lecture notes http://ocw. mit. edu/courses/electrical-engineering-and-computer-science/6-453-quantum-optical-communication-fall-2008.

⑥ Kok P and Lovett B W. 2010 Introduction to Optical Quantum Information Processing (Cambridge: Cambridge University Press).

综上,通过连续变量的量子信息这个主题,我希望读者可以发现阅读本书是一个令人愉悦的过程.

本书的中文目录为:

关于连续变量系统的量子信息处理与非定域性①,中国科学技术大学近代物理系的逯怀新、郁司夏、杨洁、陈增兵、张永德五位教授于 2003 年发表过一篇长文,综述了连续变量系统的纠缠、量子非定域性及其动力学、量子计算和量子通讯等问题.关于连续变量系统的纠缠,他们从原始的 Einstein-Podolsky-Rosen 佯谬开始,讨

① 摘自《量子力学新进展(第三辑)》,曾谨言,龙桂鲁,裴寿镛主编,清华大学出版社,2003.

论了连续变量纠缠态的产生、两体连续变量量子态的可分离判据和纠缠度量、Bures 保真度的计算. 对于连续变量系统的量子非定域性,分别介绍了相空间形式和宇旋算符形式的连续变量系统的 Bell 不等式,并研究了杂化纠缠态的量子非定域性的动力学演化. 最后,他们对连续变量的 Deutsch-Josza 算法、连续变量的纠错、连续变量的隐形传态(teleportation)和连续变量的纠缠交换(swapping)分别做了介绍.

1. 引　言

量子力学自 1900 年建立至今,理所当然地成了现代物理的两大支柱之一. 它在人类认识物质世界的思维过程中引进了崭新的革命性的框架,成为人类拓展认识疆界的利器. 量子力学已经在认识各个物质层次(微观粒子、凝聚态物质、星体乃至整个宇宙)的物理规律方面扮演了核心角色. 与此同时,量子力学也是人类改造世界、创造物质文明的利器,没有量子力学,现代物质文明的物质成就是无法想象的. 原子能的应用、超导和超流的认识及应用、半导体技术的大规模发展等,无一不是量子力学的产物.

最近,由于量子信息论①②③的飞快发展,使人们认识到,量子力学还隐藏着很多新的奥秘和惊奇. 量子信息论充分地利用了量子力学中的基本原理(如量子态叠加原理)和基本概念(如量子纠缠)来实现信息的处理. 虽然目前量子信息论仍处于实验和理论物理学家的原创性研究阶段,但它为量子论的实际应用带来了全新的更为广阔的前景. 同时,通过与信息论的交叉,量子信息论也为量子论提供了一个全新的视点与生长点,对它的深入研究必将拓展和深化量子论本身. 与以前应用量子力学完全不同的是,在量子信息论中人们利用的是量子态本身,其基本任务是量子态的存储、操纵、传输与读取. 我们可以谨慎地预言,量子信息论的发展很可能会导致一个新的量子技术时代.

最初的一些量子信息处理方案都是针对离散变量(如自旋和极化)的量子体系(即量子比特)提出的. 近几年,连续变量(如位置和动量)的

---

① D. Bouwmeester, A. Ekert and A. Zeilinger (eds). The Physics of Quantum In formation. Berlin: Springer-Verlag, 2000.

② C. H. Bennett and D. P. DiVincenzo. Nature (London) 2000,404:247-255.

③ M. A. Nielsen and I. L. Chuang. Quantum Computation and Quantum Information. Cambridge Univ. Press, 2000.

量子信息处理方案引起了广泛的关注.连续量子变量体系的隐形传态①、纠缠交换②、量子克隆③、量子计算④、量子纠错⑤、纠缠纯化⑥等被相继提出;我们还提出了新的用“杂化”(连续变量与离散变量)纠缠实现量子信息处理的方案⑦.经过数十年的发展,量子光学⑧⑨已经是一门非常成熟而又充满活力的学科.它为检验量子力学的一些基本问题提供了必不可少的精密手段.连续变量的量子信息处理的一个突出特点是它可以在量子光学实验中利用线性光学元件(如相移和分束器)操纵压缩态来实现;线性光学元件易于实现较高效率和精度的量子操作.因此,量子光学为各种连续变量量子信息处理方案提供了可行的手段.但在 Bell 不等式的实验检验中⑩,人们大多使用那些具有离散量子变量的量子系统.运用非简并光学参数放大过程,可以产生双模压缩真空态,从而实现连续量子变量(如位置和动量)的 Einstein-Podolsky-Rosen 佯谬.在此基础上,连续量子变量系统的量子纠缠和非局域性及它们之间的关系就成了极大的理论兴趣之所在.

下面将就连续变量系统的若干问题展开讨论.

2. 连续变量量子纠缠

本部分将讨论连续变量量子纠缠的几个问题.因为(连续变量)纠缠态首先是由 Einstein,Podolsky 和 Rosen(EPR)在考虑量子力学的完备性的时候引进的,因此我们的讨论从著名的 EPR 佯谬(或 EPR 论证)开始.随后,我们将依次介绍连续变量纠缠态的产生、两体连续变量量子态的可

---

① L. Vaidman. Phys. Rev. 1994, A 49: 1473; S. L. Braunstein and H. J. Kimble. Phys. Rev. Lett. 1998, 80: 869; A. Furusawa et al., Science. 1998, 282: 706; G. J. Milburn and S. L. Braunstein, Phys. Rev. 1999, A 60: 937.

② R. E. S. Polkinghorne and T. C. Ralph. Phys. Rev. Lett., 1999, 83: 2095.

③ N. J. Cerf, A. Ipe, and X. Rottenberg. Phys. Rev. Lett. 2000, 85: 1754.

④ S. Lloyd and S. L. Braunstein. Phys. Rev. Lett. 1999, 82: 1784.

⑤ S. L. Braunstein. Phys. Rev. Lett. 1998, 80: 4084; S. Lloyd and J. J. -E. Slotine, Phys. Rev. Lett. 1998, 80: 4088; S. L. Braunstein. Nature (London), 1998, 394: 47.

⑥ L. -M. Duan, G. Giedke, J. I. Cirac, and P. Zoller. Phys. Rev. Lett. 2000, 84: 4002; Phys. Rev. 2000, A 62: 032304.

⑦ Z. -B. Chen, G. Hou and Y. D. Zhang. Phys. Rev. 2002, A 65: 032318.

⑧ D. F. Walls and G. J. Milburn. Quantum Optics. Berlin/Heidelberg: Springer-Verlag, 1994.

⑨ M. O. Scully and M. S. Zubairy. Quantum Optics. Cambridge: Cambridge Univ. Press, 1997.

⑩ A. Aspect. Nature (London), 1999,398:189.

分离判据和纠缠度量、Bures 保真度的计算.

2.1 原始的 Einstein-Podolsky-Rosen 佯谬

由 Einstein,Podolsky,Rosen 提出的 EPR 态①,是最早的连续变量纠缠态. 这里先简要介绍一下该佯谬. EPR 认为,一个物理理论完备的必要条件是每一个物理实在元素必须在该理论中有一个对应. 而物理实在元素中的实在是指:如果系统没有任何扰动,那么我们可以肯定地预测物理量的值,于是该物理量存在一个相应的物理实在元素.

量子力学中,对于两个非对易的算符所表示的物理量,若精确得知其中之一的值,则不可能精确得知另一个的值. 这说明要么波函数所表述的实在性是不完备的,要么这两个物理量不能同时具备实在性. 在证明其观点时,EPR 用了如下的波函数

$$|\,\mathrm{EPR}\rangle = \int \frac{\mathrm{d}p}{2\pi} |\, p, -p\rangle = \int \mathrm{d}q \,|\, q, q\rangle \tag{1}$$

式(1) 中波函数描述了一个两粒子 a,b 系统的纠缠态,满足

$$(\hat{q}_{\mathrm{a}} - \hat{q}_{\mathrm{b}})\,|\,\mathrm{EPR}\rangle = 0, (\hat{p}_{\mathrm{a}} + \hat{p}_{\mathrm{b}})\,|\,\mathrm{EPR}\rangle = 0 \tag{2}$$

由这一性质,我们可以通过测量一个粒子的坐标(或者动量)而肯定地知道另一个粒子的坐标(或者动量). 同时,如果这两个粒子是类空分离的,那么根据相对论,对一个粒子的测量应该不影响另一个粒子的状态(Einstein 定域性). 比如,我们可以同时测量 $\hat{q}_{\mathrm{a}}$ 和 $\hat{p}_{\mathrm{b}}$(因为这两个算符对易),并分别得到 q 和 p. 于是我们可以确切地知道粒子 a 的坐标和动量似乎同时具有确定值 q 和 $-p$;但是根据 Heisenberg 不确定性关系,这是不可能的. 这就是著名的 EPR 佯谬.

EPR 佯谬显示了量子力学和定域实在论观念之间的深刻矛盾,因此从该佯谬提出至今,一直备受关注. EPR 的文章把两个极为奇妙的现象——量子纠缠和非局域性——引进到量子力学中来,并导致了旷日持久的争论和大量研究工作的涌现②③. 从 EPR 开始,量子纠缠、非局域性及它们之间的关系就一直是量子论中的基本问题,最近量子信息论的兴起,这些问题更成了物理学的热点之一. 特别是 Bell 不等式④⑤及其推

---

① A. Einstein, B. Podolsky and N. Rosen. Phys. Rev., 1935,47:777.

② N. Bohr. Phys. Rev. 1935, 48:696.

③ E. Schrödinger. Proc Cambridge Phil. Soc. 1935, 31: 555; Naturwissenschaften. 1935, 23: 807.

④ J. S. Bell. Physics., 1964, 1:195.

⑤ J. S. Bell. Speakable and Unspeakable in Quantum Mechanics. Cambridge: Cambridge Univ. Press, 1987.

广 ——Clauser-Horne-Shimony-Holt(CHSH) 不等式[1] —— 的提出,使得原来只能停留在哲学层面上的 Einstein-Bohr 之争变成一个可以从实验上加以定量检验的问题,从而激发了一大批构思巧妙的实验工作. 但在 Bell 不等式的实验检验中,人们大多使用那些具有离散量子变量的量子系统,如自旋单态. 直到最近,Bell 不等式向连续变量系统的推广才得以实现[2][3][4][5][6][7].

2.2 连续变量纠缠态的产生

由于纠缠在量子信息处理中是一种基本的资源,因此纠缠态的产生是一个关键问题. 关于连续变量纠缠态的产生[8][9][10][11][12][13][14],我们这里介绍两种办法.

光学 EPR 态 —— 在量子光学中,一种产生连续变量纠缠的方法是非简并光学参数放大过程(NOPA),它是强压缩极限下的光学类比 EPR 态. NOPA 过程是双模的非线性相互作用. 设相互作用强度用参数χ 来表述,则相互作用的 Hamilton 量为

$$H = \mathrm{i}\chi(\hat{a}^{+}\hat{b}^{+} - \hat{a}\hat{b}) \tag{3}$$

设初态系统包含两个真空模式,经过 NOPA 过程后所得的态为

$$|\,\mathrm{NOPA}\rangle = \mathrm{e}^{r(\hat{a}^{+}\hat{b}^{+}-\hat{a}\hat{b})}\,|\,0,0\rangle \tag{4}$$

其中 $r=\chi t$. $|\,\mathrm{NOPA}\rangle$ 中的连续变量纠缠是连续变量量子信息处理的基本

① J. F. Clauser, M. A. Horne, A. Shimony and R. A. Holt. Phys. Rev. Lett. 1969, 23: 880.

② K. Banaszek and K. Wódkiewicz. Phys. Rev. 1998, A 58: 4345.

③ K. Banaszek and K. Wódkiewicz. Phys. Rev. Lett. 1999, 82: 2009.

④ K. Banaszek and K. Wódkiewicz. Acta Phys. Slovaca. 1999, 49: 491.

⑤ H. Jeong, J. Lee ,and M. S. Kim. Phys. Rev. 2000, A 61: 052101.

⑥ A. Kuzmich, I. A. Walmsley, and L. Mandel. Phys. Rev. Lett. 2000, 85: 1349; Phys. Rev. 2001, A 64: 063804.

⑦ Z. -B. Chen, J. W. Pan, G. Hou, and Y. D. Zhang. Phys. Rev. Lett. 2002, 88: 040406.

⑧ M. D. Reid and P. D. Drummond. Phys. Rev. Lett. 1988, 60: 2731; M. D. Reid, Phys. Rev. 1989, A 40: 913.

⑨ Z. Y. Ou, S. F. Pereira, H. J. Kimble, K. C. Peng. Phys. Rev. Lett. 1992, 68: 3663.

⑩ Z. Y. Ou, S. F. Pereira, and H. J. Kimble. Appl. Phys. B: Photophys. Laser Chem. 1992, 55: 265.

⑪ Ch. Silberhorn, P. K. Lam, O. Wei B, F. König, N. Korolkova, and G. Leuchs. Phys. Rev. Lett. 2001, 86: 4267.

⑫ A. S. Parkins, H. J. Kimble. quant-ph/9907049.

⑬ S. Scheel and D. -G. Welsch. Phys. Rev. 2001, A 64:063811.

⑭ M. S. Kim, W. Son, V. Bužek and P. L. Knight. Phys. Rev. 2002, A 65: 032323.

资源.利用等式

$$e^{r(\hat{a}^+\hat{b}^+-\hat{a}\hat{b})}=e^{\hat{a}^+\hat{b}^+\tanh r}\left(\frac{1}{\cosh r}\right)^{\hat{a}^+\hat{a}+\hat{b}^+\hat{b}+1}e^{-\hat{a}\hat{b}\tanh r} \tag{5}$$

得

$$|\,\text{NOPA}\rangle=\frac{1}{\cosh r}\sum_{n=0}^{\infty}(\tanh r)^n|\,n,n\rangle \tag{6}$$

将上式插入完备条件

$$|\,\text{NOPA}\rangle=\frac{1}{\cosh r}\sum_{n=0}^{\infty}(\tanh r)^n\int dq\int dq'|\,q,q'\rangle\langle q,q'|\,n,n\rangle \tag{7}$$

其中

$$\langle q|\,n\rangle=(2^n n!\ \sqrt{\pi})^{-1/2}H_n(q)\exp\left(-\frac{q^2}{2}\right)$$

H_n 为 Hermite 多项式.利用求和公式

$$\sum_{n=0}^{\infty}\lambda^n\langle q|\,n\rangle\langle n|\,q'\rangle=\frac{1}{\sqrt{\pi(1-\lambda^2)}}\exp\left(-\frac{q^2+q'^2-2\lambda qq'}{2(1-\lambda^2)}\right) \tag{8}$$

得

$$|\,\text{NOPA}\rangle=\frac{1}{\sqrt{\pi}}\int dq\int dq'\exp\left(-\frac{q^2+q'^2-2qq'\tanh r}{2(1-\tanh^2 r)}\right)|\,q,q'\rangle \tag{9}$$

强压缩极限下($r\to\infty$),中间的exp函数变成$\delta(q-q')$.式(9)变成式(1)

$$|\,\text{EPR}\rangle\sim\lim_{r\to\infty}|\,\text{NOPA}\rangle\sim|\,0,0\rangle+|\,1,1\rangle+|\,2,2\rangle+\cdots \tag{10}$$

因此,在该极限下,$|\,\text{NOPA}\rangle$ 态转化为原始的归一化的 $|\,\text{EPR}\rangle$ 态.

用分束器产生连续变量纠缠——分束器(图1)是一种线性光学器件,它对输入态的作用由分束器算符给出.

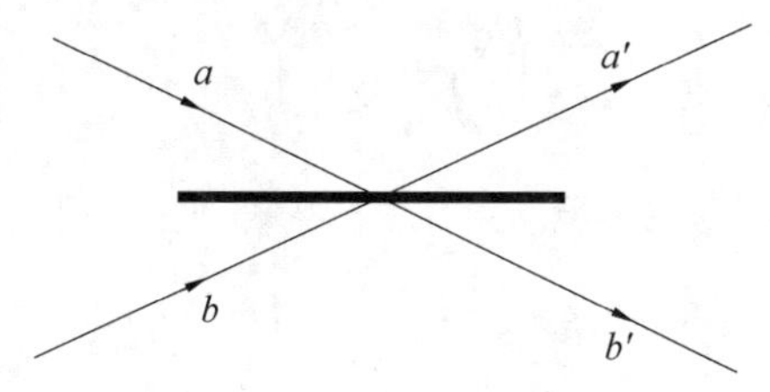

图1　分束器的示意图

$$\hat{B}=\exp\left[\frac{\theta}{2}(\hat{a}^+\hat{b}e^{i\phi}-\hat{a}\hat{b}^+e^{-i\phi})\right] \tag{11}$$

其中,振幅反射系数和透射系数分别为

$$t=\cos\frac{\theta}{2},r=\sin\frac{\theta}{2} \tag{12}$$

这里 ϕ 是反射光和透射光之间的相位差.

考虑如下输入态的情况

(1) 若入态是 Fock 态 $|n_1,n_2\rangle$,则出态为

$$\hat{B}|n_1,n_2\rangle = \sum_{N_1N_2}|N_1,N_2\rangle\langle N_1,N_2|\hat{B}|n_1,n_2\rangle$$

$$= e^{-i\phi(n_1-N_1)}\sum_{k=0}^{n_1}\sum_{l=0}^{n_2}(-1)^{n_1-k}r^{n_1+n_2-k-l}t^{k+l}\cdot$$

$$\frac{\sqrt{n_1!\,n_2!\,N_1!\,N_2!}}{k!\,(n_1-k)!\,l!\,(n_2-l)!}\delta_{N_1,n_2+k-l}\delta_{N_2,n_1-k+l}|N_1,N_2\rangle \quad (13)$$

当总的输入的光子数 $N=n_1+n_2$ 时,输出态为 $N+1$ 维纠缠态. 下面给出两种特殊的输入态.

① 当 $|n_1,n_2\rangle = |0,N\rangle$ 时,出态

$$\hat{B}|0,N\rangle = \sum_{k=0}^{N}\binom{N}{k}^{1/2}r^k t^{N-k}e^{ik\phi}|k,N-k\rangle \quad (14)$$

对于50/50分束器,且 $N=1$,出态为 $\frac{1}{\sqrt{2}}(|0,1\rangle+|1,0\rangle)$.

② 当 $|n_1,n_2\rangle = |n,n\rangle$ 时,对于50/50分束器,出态

$$\hat{B}|n,n\rangle = \sum_{m=0}^{n}e^{-i(n-2m)\phi}\left(\frac{1}{2}\right)^n\sum_{k=0}^{n}(-1)^{n-k}\binom{n}{k}\binom{n}{2m-k}\cdot$$

$$\frac{\sqrt{(2m)!\,(2n-2m)!}}{n!}|2m,2n-2m\rangle \quad (15)$$

(2) 入态是压缩态,$\Psi_{in} = \hat{S}_1\hat{S}_2|0,0\rangle$,$\hat{S}_i$ 为压缩算符

$$\hat{S}_i = e^{-(\xi_i\hat{a}_i^{+2}+\xi_i^*\hat{a}_i^2)/2} = e^{-q_i\hat{a}_i^{+2}/2}(1-|q_i|^2)^{(2\hat{n}_i+1)/4}e^{q_i^*\hat{a}_i^2/2} \quad (16)$$

其中 $q_i = \tanh|\xi_i|e^{-i\phi_i}$,$\phi_i = \arg\xi_i$. 分束器对输入模式的变换满足

$$\begin{pmatrix}\hat{b}_1\\ \hat{b}_2\end{pmatrix} = \boldsymbol{T}\begin{pmatrix}\hat{a}_1\\ \hat{a}_2\end{pmatrix} \quad (17)$$

出态的密度算符为

$$\rho_{out} = \rho_{in} = \left[\boldsymbol{T}^+\begin{pmatrix}\hat{a}_1\\ \hat{a}_2\end{pmatrix},\boldsymbol{T}^T\begin{pmatrix}\hat{a}_1^+\\ \hat{a}_2^+\end{pmatrix}\right] \quad (18)$$

得到出态

$$|\Psi_{out}\rangle = [(1-|q_1|^2)(1-|q_2|^2)]^{1/4}\exp\left[\frac{1}{2}q_1(\boldsymbol{T}_{11}\hat{a}_1^+ + \boldsymbol{T}_{21}\hat{a}_2^+)^2 - \right.$$

$$\left.\frac{1}{2}q_2(\boldsymbol{T}_{12}\hat{a}_1^+ + \boldsymbol{T}_{22}\hat{a}_2^+)^2\right]|0,0\rangle \quad (19)$$

其中 $\boldsymbol{T}$ 矩阵的矩阵元用透射反射系数表示为

$$\boldsymbol{T}_{11} = \boldsymbol{T},\boldsymbol{T}_{12} = \boldsymbol{R},\boldsymbol{T}_{21} = -\boldsymbol{R}^*,\boldsymbol{T}_{22} = \boldsymbol{T}^*,\boldsymbol{T} = |\boldsymbol{T}|e^{i\phi_T},\boldsymbol{R} = |\boldsymbol{R}|e^{i\phi_R}$$

对于50/50分束器,且

$$|\xi_1| = |\xi_2|, (\phi_1 - \phi_2) + 2(\phi_T - \phi_R) = \pm\pi$$

则出态为双模压缩真空态(TMSV)

$$|\mathrm{TMSV}\rangle = \sqrt{1 - |q|^2} e^{-q\hat{a}_1^\dagger \hat{a}_2^\dagger} |0,0\rangle \tag{20}$$

它等效于式(6).

值得指出的是,利用分束器阵列,可以产生多体的连续变量纠缠①.

2.3 两体连续变量量子态的可分离判据

判断一个量子态是否可分离是利用该态做量子信息处理的前提. 关于连续变量态的可分离判据,目前已有一些进展②③④⑤⑥⑦. 段路明等人给出了 Gauss 型连续变量态可分离的充分必要判据. 由于他们的判据具有比较明显的物理意义,同时 Gauss 型态在量子光学中有广泛的应用,实验上比较易于检验,因此这里给出段路明等人的判据.

一个双模量子态 ρ,当且仅当

$$\rho = \sum_i p_i \rho_{i1} \otimes \rho_{i2} \tag{21}$$

时,其可分离;其中,ρ_{i1},ρ_{i2} 分别表示模 1,模 2 的归一化的态,$p_i \geqslant 0$,$\sum_i p_i = 1$.

原始 EPR 态为连续变量的量大纠缠态,定义 EPR 算符 $\hat{x}_1 - \hat{x}_2$ 及 $\hat{p}_1 + \hat{p}_2$,它们有共同本征态,即原始 EPR 态. 对于一般的连续变量态,定义类 EPR 算符

$$\begin{cases} \hat{u} = |a| \hat{x}_1 + \dfrac{1}{a}\hat{x}_2 \\ \hat{v} = |a| \hat{p}_1 - \dfrac{1}{a}\hat{p}_2 \end{cases} \tag{22}$$

定理 1 可分离的充分判据:对任何可分离态,类 EPR 算符满足

$$\langle(\Delta\hat{u})^2\rangle_\rho + \langle(\Delta\hat{v})^2\rangle_\rho \geqslant a^2 + \frac{1}{a^2} \tag{23}$$

---

① P. van Loock and S. L. Braunstein. Phys. Rev. Lett. 2000, 84: 3482; Phys. Rev. 1999, A 61: 010302(R).

② L. M. Duan, G. Giedke, J. I. Cirac, and P. Zoller. Phys. Rev. Lett. 2000, 84: 2722.

③ R. Simon. Phys. Rev. Lett. 2000, 84: 2726.

④ Wang Xiang-Bin et al.. Phys. Rev. Lett. 2001, 87:137903.

⑤ G. Giedke et al.. Phys. Rev. Lett. 2001, 87: 167904.

⑥ G. Giedke et al.. Phys. Rev. 2001, A 64: 052303.

⑦ B. -G. Englert et al.. Phys. Rev. 2002, A 65: 054303.

证明

$$
\begin{aligned}
&\langle(\Delta\hat{u})^2\rangle_\rho + \langle(\Delta\hat{v})^2\rangle_\rho \\
={}& \sum_i p_i(\langle\hat{u}^2\rangle_i + \langle\hat{v}^2\rangle_i) - \langle\hat{u}\rangle_\rho^2 - \langle\hat{v}\rangle_\rho^2 \\
={}& \sum_i p_i\left(a^2\langle\hat{x}_1^2\rangle_i + \frac{1}{a^2}\langle\hat{x}_2^2\rangle_i + a^2\langle\hat{p}_1^2\rangle_i + \frac{1}{a^2}\langle\hat{p}_2^2\rangle_i\right) + \\
& 2\frac{a}{|a|}\left(\sum_i p_i\langle\hat{x}_1\rangle_i\langle\hat{x}_2\rangle_i - \sum_i p_i\langle\hat{p}_1\rangle_i\langle\hat{p}_2\rangle_i\right) - \langle\hat{u}_\rho\rangle^2\langle\hat{v}_\rho\rangle^2 \\
={}& \sum_i p_i\left[a^2\langle(\Delta\hat{x}_1)^2\rangle_i + \frac{1}{a^2}\langle(\Delta\hat{x}_2)^2\rangle_i + \right. \\
& \left. a^2\langle(\Delta\hat{p}_1)^2\rangle_i + \frac{1}{a^2}\langle(\Delta\hat{p}_2)^2\rangle_i\right] + \\
& \sum_i p_i\langle\hat{u}\rangle_i^2 - \left(\sum_i p_i\langle\hat{u}\rangle_i\right)^2 + \\
& \sum_i p_i\langle\hat{v}\rangle_i^2 - \left(\sum_i p_i\langle\hat{v}\rangle_i\right)^2
\end{aligned}
\tag{24}
$$

在方程(24)中,$\langle\cdots\rangle_i$ 表示在态 $\rho_{i1}\otimes\rho_{i2}$ 中求平均值. 由不确定关系

$$
\langle(\Delta\hat{x}_j)^2\rangle_i + \langle(\Delta\hat{p}_j)^2\rangle_i \geqslant |[\hat{x}_j,\hat{p}_j]| = 1 \quad (j=1,2)
$$

再由 Cauchy-Schwarz 不等式

$$
\left(\sum_i p_i\right)\left(\sum_i p_i\langle\hat{u}\rangle_i^2\right) \geqslant \left(\sum_i p_i|\langle\hat{u}\rangle_i|\right)^2
$$

即可得证.

一个 Gauss 型连续变量态的 Wigner 特征函数表示为

$$
\begin{aligned}
\chi^{(w)}(\lambda_1,\lambda_2) &= \mathrm{tr}[\rho\exp(\lambda_1\hat{a}_1 - \lambda_1^*\hat{a}_1^+ + \lambda_2\hat{a}_2 - \lambda_2^*\hat{a}_2^+)] \\
&= \mathrm{tr}\{\rho\exp[i\sqrt{2}(\lambda_1^I\hat{x}_1 + \lambda_1^R\hat{p}_1 + \lambda_2^I\hat{x}_2 + \lambda_2^R\hat{p}_2)]\}
\end{aligned}
\tag{25}
$$

其中,记

$$
\lambda_j = \lambda_j^R + i\lambda_j^I, \hat{a}_j = \frac{1}{\sqrt{2}}(\hat{x}_j + i\hat{p}_j)
$$

则

$$
\chi^{(w)}(\lambda_1,\lambda_2) = \exp\left[-\frac{1}{2}(\lambda_1^I,\lambda_1^R,\lambda_2^I,\lambda_2^R)M(\lambda_1^I,\lambda_1^R,\lambda_2^I,\lambda_2^R)^{\mathrm{T}}\right] \tag{26}
$$

$$
\boldsymbol{M} = \begin{bmatrix} \boldsymbol{G}_1 & \boldsymbol{C} \\ \boldsymbol{C}_{pt}^{\mathrm{T}} & \boldsymbol{G}_2 \end{bmatrix} \tag{27}
$$

$\boldsymbol{G}_1,\boldsymbol{G}_2,\boldsymbol{C}$ 是 2×2 的实矩阵,pt 表示部分转置.

为了研究可分离的性质,先将此 Gauss 型态通过局域线性幺正 Bogoliubov 操作(LLUBO)$U=U_1\otimes U_2$ 化简为标准型. 因为局域幺正变换不会改变态的可分离性质,所以总可以将 $\boldsymbol{M}$ 变成对称形式.

下面证明两个关于标准型的引理.

引理1　对于 Gauss 型态,$\boldsymbol{M}$ 的标准型 Ⅰ 是

$$\boldsymbol{M}_s^{\mathrm{I}}=\begin{bmatrix} n & & c & \\ & n & & c' \\ c & & m & \\ & c' & & m \end{bmatrix}\quad (n,m\geqslant 1) \tag{28}$$

证明　一个 LLUBO 作用在态 ρ_G 上,将 Wigner 特征函数中的 $\boldsymbol{M}$ 矩阵变换为

$$\begin{bmatrix} \boldsymbol{V}_1 & \\ & \boldsymbol{V}_2 \end{bmatrix}\boldsymbol{M}\begin{bmatrix} \boldsymbol{V}_1^{\mathrm{T}} & \\ & \boldsymbol{V}_2^{\mathrm{T}} \end{bmatrix} \tag{29}$$

这里,$\boldsymbol{V}_1,\boldsymbol{V}_2$ 是实矩阵,$\det \boldsymbol{V}_1=\det \boldsymbol{V}_2=1$. 由于 $\boldsymbol{G}_1,\boldsymbol{G}_2$ 是对称矩阵,可以选择正交矩阵 $\boldsymbol{V}_1,\boldsymbol{V}_2$,将 $\boldsymbol{G}_1,\boldsymbol{G}_2$ 对角化. 且局域压缩操作可以将对角后的 $\boldsymbol{G}_1,\boldsymbol{G}_2$ 变为 $\boldsymbol{G}_1'=n\boldsymbol{I},\boldsymbol{G}_2'=m\boldsymbol{I}$. 再选择 $\boldsymbol{V}_1,\boldsymbol{V}_2$ 将 $\boldsymbol{C}$ 对角化.

引理2　$\boldsymbol{M}$ 的标准型 Ⅱ 是

$$\boldsymbol{M}_s^{\mathrm{II}}=\begin{bmatrix} n_1 & & c_1 & \\ & n_2 & & c_2 \\ c_1 & & m_1 & \\ & c_2 & & m_2 \end{bmatrix} \tag{30}$$

其中,$n_i,m_i,c_i(i=1,2)$ 满足

$$\frac{n_1-1}{m_1-1}=\frac{n_2-1}{m_2-1}$$

$$|c_1|-|c_2|=\sqrt{(n_1-1)(m_1-1)}-\sqrt{(n_2-1)(m_2-1)} \tag{31}$$

证明　对标准型 Ⅰ 作局域压缩变化

$$\boldsymbol{M}'=\begin{bmatrix} nr_1 & & \sqrt{r_1r_2}c & \\ & \dfrac{n}{r_1} & & \dfrac{c'}{\sqrt{r_1r_2}} \\ \sqrt{r_1r_2}c & & mr_2 & \\ & \dfrac{c'}{\sqrt{r_1r_2}} & & \dfrac{m}{r_2} \end{bmatrix} \tag{32}$$

r_1,r_2 为压缩参数,$r_1,r_2>0$. 不失一般性,设 $|c|\geqslant|c'|,n\geqslant m$.

要使 $\boldsymbol{M}'$ 成为标准型 Ⅱ,则 r_1,r_2 须满足

$$\frac{n/r_1-1}{nr_1-1}=\frac{m/r_2-1}{mr_2-1} \tag{33}$$

$$\sqrt{r_1r_2}\,|c|-\frac{|c'|}{\sqrt{r_1r_2}}=\sqrt{(nr_1-1)(mr_2-1)}-\sqrt{\left(\frac{n}{r_1}-1\right)\left(\frac{m}{r_2}-1\right)} \tag{34}$$

由式(33),当 $r_1=1$ 时,$r_2=1$;当 $r_1\to\infty$ 时,$r_2\to m$. 设

$$f(r_1)=\sqrt{r_1r_2}\,|c|-\frac{|c'|}{\sqrt{r_1r_2}}-$$

$$\left[\sqrt{(nr_1-1)(mr_2-1)}-\sqrt{\left(\frac{n}{r_1}-1\right)\left(\frac{m}{r_2}-1\right)}\,\right]$$

$$f(r_1=1)=|c|-|c'|\geqslant 0$$

取

$$\hat{u}_0=\sqrt{m-\frac{1}{m}}\hat{x}_1-\frac{c}{|c|}\sqrt{n}\hat{x}_2,\hat{v}_0=\frac{\sqrt{n}}{m}\hat{p}_2$$

由不确定关系

$$\langle(\Delta\hat{u}_0)^2\rangle+\langle(\Delta\hat{v}_0)^2\rangle\geqslant|[\hat{u}_0,\hat{v}_0]|$$

得到

$$|c|\leqslant\sqrt{n\left(m-\frac{1}{m}\right)}$$

又因为

$$f(r_1\to\infty)\to\sqrt{r_1m}\left[|c|-\sqrt{n\left(m-\frac{1}{m}\right)}\,\right]\leqslant 0$$

由 f 的连续性,存在 r^*,使得 $f(r_1-r^*)=0$,方程(33)(34) 给出引理的证明.

根据上面的几个定理和引理,得到关于 Gauss 型态的可分离的充分必要条件.

定理 2 Gauss 型态 ρ_G 可分离,当且仅当该态用 Wigner 函数表示,$\boldsymbol{M}$ 为标准型 Ⅱ 时,类 EPR 算符满足式(23),其中,类 EPR 算符为

$$\begin{cases}\hat{u}=a_0\hat{x}_1-\dfrac{c_1}{|c_1|}\dfrac{1}{a_0}\hat{x}_2\\[2ex]\hat{v}=a_0\hat{p}_1-\dfrac{c_2}{|c_2|}\dfrac{1}{a_0}\hat{p}_2\end{cases}\tag{35}$$

这里

$$a_0^2=\sqrt{\frac{m_1-1}{n_1-1}}=\sqrt{\frac{m_2-1}{n_2-1}}$$

证明 将 Gauss 型态写成标准型 Ⅱ,将系数代入式(23),得到不等式

$$a_0^2\frac{n_1+n_2}{2}+\frac{m_1+m_2}{2a_0^2}-|c_1|-|c_2|\geqslant a_0^2+\frac{1}{a_0^2}\tag{36}$$

联立式(33)(34) 得

$$\begin{cases} |c_1| \leqslant \sqrt{(n_1-1)(m_1-1)} \\ |c_2| \leqslant \sqrt{(n_2-1)(m_2-1)} \end{cases} \tag{37}$$

所以($\boldsymbol{M}_s^{\text{II}}$ − Ⅰ)半正定,Gauss 型态 ρ_G^{II} 的正则特征函数表示为

$$\chi_{\text{II}}^{(n)}(\lambda_1,\lambda_2)=\chi_{\text{II}}^{(w)}(\lambda_1,\lambda_2)\exp\left[\frac{1}{2}(|\lambda_1|^2+|\lambda_2|^2)\right] \tag{38}$$

所以

$$\rho_G^{\text{II}}=\int d^2\alpha d^2\beta P(\alpha,\beta)|\alpha,\beta\rangle\langle\alpha,\beta| \tag{39}$$

其中,$P(\alpha,\beta)$ 是 $\chi_{\text{II}}^{(n)}(\lambda_1,\lambda_2)$ 的 Fourier 变换,所以是正 Gauss 函数,ρ_G^{II} 可分离,又因为 ρ 与 ρ_G^{II} 只相差一个 LLUBO,所以 ρ 可分离.

2.4 两体连续变量量子态的纠缠度量

量子纠缠作为一种重要资源已被广泛地应用于量子信息领域. 最初的量子信息处理方案都是针对离散变量(如自旋和极化)的量子体系提出的,因而相关的量子态纠缠度的计算也都是离散变量体系的. 近几年,连续变量的量子体系的量子信息处理方案被相继提出,引起了广泛的关注. 显然,量子纠缠在量子信息论中的这一广泛应用使得对连续变量体系量子纠缠的定量描述显得尤为重要,如何更方便地给出纠缠度的计算已成为迫在眉睫的任务. Parker 等人①在这方面迈出了很重要的一步,他们利用积分本征值方程的方法给出了具有连续变量的两体纠缠纯态纠缠度的计算公式. 这一方法为定量研究连续变量体系的量子纠缠提供了有用的工具. 然而,由于这一方法的基本思想是基于 Schmidt 分解而建立积分本征值方程,因此可以通过解这一积分本征值方程从而求得该纠缠态的纠缠度. 可是,在多数情况下 Schmidt 基是不好找的,更何况是这一积分本征值方程在一般情况下只能给出数值解而不能给出解析解. 所以,这一方法在操作上有一定的难度. 基于这一事实,在 Parker 等人方法的基础上,我们在 Fock 空间中导出了求解两体 Gauss 纠缠纯态纠缠度的解析计算公式②. 这一方法在操作上极为方便,利用我们给出的公式,在不知道约化密度算符本征值的情况下,可以非常容易地直接求出两体 Gauss 纠缠态的纠缠度. 为使读者对连续变量的两体纠缠纯态纠缠度的计算有一个全面的了解,便于比较,下面我们将对这两种方法作一系统的介绍. 先介绍 Parker 等人的方法.

---

① S. Parker, S. Bose, and M. B. Plenio. Phys. Rev. 2000, A 61: 032305.

② H. X. Lu, Z. -B. Chen, J. W. Pan and Y. D. Zhang, LANL eprint quant-ph/0204098.

在坐标 - 动量空间中,具有连续变量的两体纠缠纯态可表示为

$$|\psi\rangle_{12} = \int\psi(x,y)\,|x\rangle_1\,|y\rangle_2\mathrm{d}x\mathrm{d}y \tag{40}$$

这里$|x\rangle$是连续变量体系的本征态. 由式(40)可求得约化密度算符为

$$\rho_1 = \int_2 \langle x|\psi\rangle_{1212}\langle\psi|x\rangle_2\mathrm{d}x = \int\rho_1(x,y)\,|x\rangle_{11}\langle y|\,\mathrm{d}x\mathrm{d}y \tag{41}$$

其中

$$\rho_1(x,y) = \int\psi(x,z)\psi^*(y,z)\mathrm{d}z \tag{42}$$

显然,如果能找到一组 Schmidt 基使其约化密度算符在此基下是对角的,那么可得该体系的纠缠度. 为此,可通过求解核算符$\rho(x,y)$所满足的积分本征值方程来求得ρ_1的本征值从而求得该体系的纠缠度,即求解下列积分本征值方程

$$\int\rho_1(x,y)\phi_i(y)\mathrm{d}y = \lambda_i\phi_i(x) \tag{43}$$

其中λ_i是相应于$\phi_i(x)$的本征值,$\phi_i(x)$是厄米核算符$\rho(x,y)$的一组完备且正交的本征函数族,即

$$\int\phi_i(x)\phi_j^*(x)\mathrm{d}x = \delta_{ij} \tag{44}$$

于是,寻求Schmidt基的问题已转化为寻求厄米核算符$\rho(x,y)$的基. 而数学上可以证明①,对于L_2空间的厄米核算符,总能找到一组正交的基使其核算符成为对角形式,即

$$\rho_1(x,y) = \sum_{i=1}^{\infty}\lambda_i\phi_i(x)\phi_i^*(y) \tag{45}$$

对于满足式(45)的厄米核算符,原则上可通过求解积分本征值方程求得λ_i,从而可求得该体系的纠缠度

$$E(\rho_{12}) \equiv S(\rho_1) = S(\rho_2) = -\sum_i\lambda_i\log_2\lambda_i \tag{46}$$

对于给定的态,可以从上面给出的公式写出积分本征值方程. 然后求解该方程的本征值,最后可得纠缠度. 例如,考虑两体 Gauss 纠缠态

$$|B_{\alpha\beta}(x_1,x_2)\rangle_{12} = \int\exp\left[\frac{1}{\sigma^2}\left(-x^2\alpha^2 - \frac{y^2}{\beta^2} + 2\mathrm{i}x_1x\right)\right]\cdot |x\rangle_1\,|x+y+x_2\rangle_2\mathrm{d}x\mathrm{d}y \tag{47}$$

由式(40) ~ (43)及式(47),可得下面两个方程

$$\int\exp\left[\frac{1}{\sigma^2}\left(-\left(\alpha^2+\frac{1}{2\beta^2}\right)(x^2+x'^2) + 2\mathrm{i}x_1(x-x') + \frac{xx'}{\beta^2}\right)\right]\phi_i^{(1)}(x')\mathrm{d}x'$$

① F. G. Tricmi. Integral Equation, New York: Wiley, 1967.

$$= \lambda_i^{(1)} \phi_i^{(1)}(x) \tag{48}$$

$$\int \exp\left[\frac{1}{\sigma^2}\left(\left(\frac{1}{2(\alpha^2\beta^2+1)}-1\right)\frac{y^2+y'^2}{\beta^2}+\frac{yy'}{\beta^2(\beta^2\alpha^2+1)}\right)\right]\phi_i^{(2)}(y')\mathrm{d}y'$$

$$= \lambda_i^{(2)} \phi_i^{(2)}(x) \tag{49}$$

作适当的变数变换，可将上面两式合为一式. 例如，对式(48) 作变换

$$\phi_i^{(1)}(x') \to \phi_i^{(1)}(x')\exp(-2\mathrm{i}x_1x')$$

$$x \to \sqrt{2}x\beta\sigma, x' \to \sqrt{2}x'\beta\sigma \tag{50}$$

对于式(47)，作变换

$$x \to x\sqrt{2(\alpha^2\beta^2+1)}\beta\sigma, x' \to x'\sqrt{2(\alpha^2\beta^2+1)}\beta\sigma \tag{51}$$

则式(48)(49) 可合写为

$$\int_{-\infty}^{\infty} \exp[-(1+2\alpha^2\beta^2)(x^2+x'^2)+2xx']\phi(x')\mathrm{d}x' = \lambda\phi(x) \tag{52}$$

显然，式(52) 的求解只能用数值解法，为此将式(52) 写成分离形式

$$\delta\sum_{p=-n}^{n}\rho_{pq}\phi_p = \lambda\phi_q \tag{53}$$

其中

$$\rho_{pq} = \exp[(-(1+2\alpha^2\beta^2)(p^2+q^2)+2pq)\delta^2] \tag{54}$$

于是从式(53) 中解出 λ，可由式(46) 求得纠缠度，我们从中可以看出这一方法在操作上确有一定的难度. 为此，我们在 Fock 空间中导出了计算任意连续变量两体 Gauss 纠缠纯态纠缠度的公式及可分离判据. 利用这一公式，由于无须知道约化密度算符的本征值就可直接求出该体系的纠缠度，因此使用上非常方便，下面给出这一方法的详细推导.

为方便地导出任意连续变量两体 Gauss 纠缠纯态纠缠度的计算公式，我们利用场论中常用的一个公式①

$$\Omega \coloneqq \exp\left[(a_1^+, a_2^+)\begin{pmatrix}\partial_{z_1^*}\\ \partial_{z_2^*}\end{pmatrix} + (\partial_{z_1}, \partial_{z_2})\begin{pmatrix}a_1\\ a_2\end{pmatrix}\right] :\langle Z|\Omega|Z\rangle|_{z=0} \tag{55}$$

这里，Ω 是任意的玻色算符，$a_i^+(i=1,2)$ 是通常的玻色产生算符，$|Z\rangle = |Z_1\rangle \otimes |Z_2\rangle$ 是相干态，而 $z=0$ 表示 $z_{1(2)} = z_{1(2)}^* = 0$，“::”表示正规乘积符号. 利用式(55)，可将任意两体纠缠的密度算符表示为

$$\rho_{12} \coloneqq \exp\left[(a_1^+, a_2^+)\begin{pmatrix}\partial_{z_1^*}\\ \partial_{z_2^*}\end{pmatrix} + (\partial_{z_1}, \partial_{z_2})\begin{pmatrix}a_1\\ a_2\end{pmatrix}\right] :\langle Z|\rho_{12}|Z\rangle|_{z=0} \tag{56}$$

利用相干态的超完备关系

① C. Itzykson and J. B. Zuber. Quantum Fileld Theory. New York: McGraw-Hill, 1980.

$$\int \frac{d^2 Z_2}{\pi} | Z_2\rangle\langle Z_2 | = 1$$

对式(56) 部分求迹可得

$$\begin{aligned}
\rho_1 &= \mathrm{tr}_2 \rho_{12} \\
&= \mathrm{tr}_2 \int \frac{d^2 Z_2'}{\pi} | Z_2'\rangle\langle Z_2'| :\exp\left[(a_1^+, a_2^+)\begin{pmatrix} \partial_{z_1^*} \\ \partial_{z_2^*} \end{pmatrix} + (\partial_{z_1}, \partial_{z_2})\begin{pmatrix} a_1 \\ a_2 \end{pmatrix} \right] :\langle Z | \rho_{12} | Z\rangle |_{z=0} \\
&= \mathrm{tr}_2 \int \frac{d^2 Z_2'}{\pi} :\exp(a_1^+ \partial_{z_1^*} + \partial_{z_1} a_1) :\exp(z_2'^* \partial_{z_2^*} + \partial_{z_2} z_2')\langle Z | \rho_{12} | Z\rangle |_{z=0}
\end{aligned} \tag{57}$$

式(57) 可以看作任意两体连续变量纠缠纯态部分求迹的一般公式. 因此,对于给定的任意两体连续变量纠缠纯态,可以由此式方便地求得该体系的约化密度算符.

下面我们对任意两体 Gauss 纠缠纯态作详细推导. 对于任意的两体 Gauss 纠缠纯态,其一般形式为

$$\begin{aligned}
\rho_{12} = A_0 :\exp\Bigg\{ \frac{1}{2}\Bigg[& (a_1^+, a_1) \boldsymbol{M}_1 \begin{pmatrix} a_1^+ \\ a_1 \end{pmatrix} + (a_2^+, a_2) \boldsymbol{M}_2 \begin{pmatrix} a_2^+ \\ a_2 \end{pmatrix} + \\
& 2(a_1^+, a_1) \boldsymbol{M}_{12} \begin{pmatrix} a_2^+ \\ a_2 \end{pmatrix} \Bigg] \Bigg\} :
\end{aligned} \tag{58}$$

其中

$$\boldsymbol{M}_{12} = \begin{pmatrix} e & f \\ f^* & e^* \end{pmatrix} \tag{59}$$

而 e 和 f 是任意的两个复数. 由于线性项 a^+ 和 a 对纠缠度无贡献,故在式(58) 中没有这些项. 由式(58),可计算密度算符在相干态中的矩阵元为

$$\begin{aligned}
\langle Z | \rho_{12} | Z\rangle = A_0 \exp\Bigg\{ \frac{1}{2}\Bigg[& (z_1^*, z_1) \boldsymbol{M}_1 \begin{pmatrix} z_1^* \\ z_1 \end{pmatrix} + (z_2^*, z_2) \boldsymbol{M}_2 \begin{pmatrix} z_2^* \\ z_2 \end{pmatrix} + \\
& 2(z_1^*, z_1) \boldsymbol{M}_{12} \begin{pmatrix} z_2^* \\ z_2 \end{pmatrix} \Bigg] \Bigg\}
\end{aligned} \tag{60}$$

利用 Gauss 积分公式①

$$\begin{aligned}
& \int \frac{d^2 z}{\pi} \exp\left\{ -\frac{1}{2}(z^*, z) \boldsymbol{Q} \begin{pmatrix} z^* \\ z \end{pmatrix} + (u, v)\begin{pmatrix} z^* \\ z \end{pmatrix} \right\} \\
&= [-\det \boldsymbol{Q}]^{-1/2} \exp\left\{ \frac{1}{2}(u, v) \boldsymbol{Q}^{-1} \begin{pmatrix} u \\ v \end{pmatrix} \right\}
\end{aligned} \tag{61}$$

① J. W. Pan, Q. X. Dong, and Y. D. Zhang et al. Phys. Rev. 1997, E 56: 2553.

(式中 $\boldsymbol{Q}=\widetilde{\boldsymbol{Q}}$ 是非奇异的矩阵,而 u 和 v 是任意的复数)及下列公式

$$\exp(z_2'^{*}\partial_{z_2^*}+\partial_{z_2}z_2')\Psi(z_1^*,z_1;z_2^*,z_2)\big|_{z_2=z_2^*=0}=\Psi(z_1^*,z_1;z_2'^{*},z_2') \tag{62}$$

将式(60)代入式(57)中,然后积分得

$$\rho_1=\frac{A_0}{\sqrt{-\det \boldsymbol{M}_2}}:\exp\left\{\frac{1}{2}\left[(a_1^+,a_1)(\boldsymbol{M}_1-\boldsymbol{M}_{12}\boldsymbol{M}_2^{-1}\widetilde{\boldsymbol{M}}_{12})\begin{pmatrix}a_1^+\\a_1\end{pmatrix}\right]\right\}: \tag{63}$$

为将式(63)写成一般的指数二次型形式,我们记

$$\boldsymbol{M}=\begin{pmatrix}a & d\\ b & c\end{pmatrix} \tag{64}$$

其中复数 a,b,c,d 由已知矩阵 $\boldsymbol{M}_1,\boldsymbol{M}_2,\boldsymbol{M}_{12}$ 决定. 按照线性量子变换理论①②,矩阵 $\boldsymbol{M}$ 的映射为

$$D(\boldsymbol{M})=\begin{pmatrix}c^{-1}-1 & c^{-1}d\\ c^{-1}b & 1-c^{-1}\end{pmatrix}\boldsymbol{\Sigma}_B^{-1}$$

如果记

$$D(\boldsymbol{M})=(\boldsymbol{M}_1-\boldsymbol{M}_{12}\boldsymbol{M}_2^{-1}\widetilde{\boldsymbol{M}}_{12})\boldsymbol{\Sigma}_B^{-1}=\begin{pmatrix}c^{-1}-1 & c^{-1}d\\ c^{-1}b & 1-c^{-1}\end{pmatrix}\boldsymbol{\Sigma}_B^{-1} \tag{65}$$

那么利用线性量子变换理论可直接写出式(63)的一般的指数二次型形式为

$$\rho_1=A\exp\left[\frac{1}{2}(a_1^+,a_1)N\boldsymbol{\Sigma}_B\begin{pmatrix}a_1^+\\a_1\end{pmatrix}\right] \tag{66}$$

这里 $\boldsymbol{N}=\ln\boldsymbol{M}$ 为负厄米矩阵,而 $\boldsymbol{\Sigma}_B=\begin{pmatrix}0 & 1\\ -1 & 0\end{pmatrix}$,常数

$$A=A_0\sqrt{\frac{c}{-\det \boldsymbol{M}_2}} \tag{67}$$

将式(66)代入纠缠度公式

$$E=-\operatorname{tr}\rho_1\ln\rho_1=-\operatorname{tr}\rho_2\ln\rho_2 \tag{68}$$

中,计算可得

$$E=-A\left[Z(\beta)\ln A-\frac{\mathrm{d}}{\mathrm{d}\beta}Z(\beta)\right]_{\beta=-1} \tag{69}$$

其中

① Y. D. Zhang and Z. Tang. J. Math. Phys. 1993, 34: 5639.

② Y. D. Zhang and Z. Tang. Nuovo Cimento. 1994, B 109: 387.

$$Z(\beta) = \mathrm{tr}\,\exp\left\{-\frac{\beta}{2}(a_1^+, a_1)\boldsymbol{N}\boldsymbol{\Sigma}_B\begin{pmatrix}a_1^+\\a_1\end{pmatrix}\right\} \tag{70}$$

利用

$$Z(\beta) = \mathrm{tr}\,\exp\left\{-\frac{\beta}{2}(a_1^+, a_1)\boldsymbol{N}\boldsymbol{\Sigma}_B\begin{pmatrix}a_1^+\\a_1\end{pmatrix}\right\} = |\det(\mathrm{e}^{\beta N} - 1)|^{-\frac{1}{2}} \tag{71}$$

直接计算可得

$$\frac{\mathrm{d}}{\mathrm{d}\beta}Z(\beta)\bigg|_{\beta=-1} = -\frac{1}{2}|\det(\mathrm{e}^{-N} - 1)|^{-\frac{1}{2}}\mathrm{tr}\frac{\boldsymbol{N}}{1 - \mathrm{e}^{N}} \tag{72}$$

将上述公式代入式(69)中,最后可得计算纠缠度的解析表达式为

$$E = -A|\det(\mathrm{e}^{-N} - 1)|^{-\frac{1}{2}}\left(\ln A + \frac{1}{2}\mathrm{tr}\frac{\boldsymbol{N}}{1 - \mathrm{e}^{N}}\right) \tag{73}$$

从式(73)我们可以清楚地看出,对于任意给定的形如式(58)的两体纠缠Gauss纯态,可以从中求出负厄米矩阵$\boldsymbol{N}$的本征值直接代入式(73)便可求得该体系的纠缠度,而不必求约化密度算符的本征值. 这一方法显然大大缩短了计算过程,从而为研究体系的纠缠度提供了极大的方便. 同时,我们也要指出,从式(73)还可以直接得到一个重要的结论:即对于形如式(58)的两体Gauss纠缠纯态,其可分离的充分必要条件为

$$\ln A = \frac{1}{2}\mathrm{tr}\frac{\boldsymbol{N}}{\mathrm{e}^{N} - 1} \tag{74}$$

下面通过一个具体的例子进一步说明这一方法的优越性. 考虑一由分束器产生的两体Gauss纠缠纯态

$$|\psi\rangle_{12} = \hat{B}(\theta,\phi)\hat{S}_1(\zeta_1)\hat{S}_2(\zeta_2)|00\rangle \tag{75}$$

其中$\hat{B}(\theta,\phi)$和$\hat{S}(\zeta)$分别代表分束器算符和压缩算符,具体表示为

$$\hat{B}(\theta,\phi) = \exp[\theta(a_1^+a_2\mathrm{e}^{\mathrm{i}\varphi} - a_1a_2^+\mathrm{e}^{-\mathrm{i}\varphi})] \tag{76}$$

$$\hat{S}(\zeta) = \exp\left[\frac{1}{2}(\zeta^*a^2 - \zeta a^{+2})\right] \tag{77}$$

由线性量子变换理论,可以写出式(75)密度算符的正规乘积形式为

$$\begin{aligned}\rho_{12} &= |\psi\rangle_{12\,12}\langle\psi| \\ &= A_0 :\exp\left\{\frac{1}{2}\left[(a_1^+, a_1)M_1\begin{pmatrix}a_1^+\\a_1\end{pmatrix} + (a_2^+, a_2)M_2\begin{pmatrix}a_2^+\\a_2\end{pmatrix} + \right.\right. \\ &\quad \left.\left. 2(a_1^+, a_1)M_{12}\begin{pmatrix}a_2^+\\a_2\end{pmatrix}\right]\right\}:\end{aligned} \tag{78}$$

式中各矩阵表示为

$$\boldsymbol{M}_1 = -\begin{pmatrix}\alpha & 1\\1 & \alpha^*\end{pmatrix}, \boldsymbol{M}_2 = -\begin{pmatrix}\beta & 1\\1 & \beta^*\end{pmatrix}, \boldsymbol{M}_{12} = -\begin{pmatrix}\delta & 0\\0 & \delta^*\end{pmatrix}$$

其中

$$\alpha=\frac{\zeta_1}{|\zeta_1|}\tanh|\zeta_1|\cos^2\theta+e^{2i\varphi}\frac{\zeta_2}{|\zeta_2|}\tanh|\zeta_2|\sin^2\theta$$

$$\beta=e^{-2i\varphi}\frac{\zeta_1}{|\zeta_1|}\tanh|\zeta_1|\sin^2\theta+\frac{\zeta_2}{|\zeta_2|}\tanh|\zeta_2|\cos^2\theta$$

$$\delta=\frac{1}{2}\sin(2\theta)\left(\frac{\zeta_2}{|\zeta_2|}\tanh|\zeta_2|e^{2i\varphi}+\frac{\zeta_1}{|\zeta_1|}\tanh|\zeta_1|e^{-i\varphi}\right)$$

利用式(65)和(67)可计算出系数 A 与矩阵 $\boldsymbol{M}$ 为

$$A=\frac{1}{|\delta|\cosh|\zeta_1|\cosh|\zeta_2|},\boldsymbol{M}=\begin{pmatrix}a & d\\ b & c\end{pmatrix}$$

$$a=\frac{1}{c}(1+bd),c=\frac{1-|\beta|^2}{|\delta|^2}$$

$$d=\frac{1}{|\delta|^2}[\delta^2\beta^*+\alpha(1-|\beta|^2)],b=-d^* \tag{79}$$

最后得式(75)纠缠度的解析表达式为

$$E=\frac{1}{|\delta|\cosh|\zeta_1|\cosh|\zeta_2|}\cdot$$

$$\frac{\sqrt{\lambda}}{|\lambda-1|}\left[\ln(|\delta|\cosh|\zeta_1|\cosh|\zeta_2|)-\frac{1+\lambda}{2(1-\lambda)}\ln\lambda\right] \tag{80}$$

λ 是下列方程中的任意一个根

$$\lambda^2-(a+c)\lambda+1=0 \tag{81}$$

现在考虑几种特例. 首先考虑情况

$$\begin{cases}\theta=\dfrac{\pi}{4},\zeta_i=s_ie^{i\varphi} \quad (i=1,2)\\ \varphi=\dfrac{l\pi}{2} \quad (l=0,1,2,\cdots)\end{cases} \tag{82}$$

在这种情况下

$$E=\cosh^2|s|\ln(\cosh^2|s|)-\sinh^2|s|\ln(\sinh^2|s|) \tag{83}$$

式中

$$s=\frac{1}{2}(s_1e^{i\varphi}+s_2e^{-i\varphi})$$

考虑更特殊的情况,即令 $\varphi=0,\zeta_2=-\zeta_1=r$ 及 $\theta=\frac{\pi}{4}$,这时可求得 $\lambda_1=\tanh^2 r$, 因此式(80)成为

$$E=\cosh^2 r\ln(\cosh^2 r)-\sinh^2 r\ln(\sinh^2 r) \tag{84}$$

这即为众所周知的双模压缩真空态的纠缠度. 通过以上例子的计算可以看出该方法在使用上的确具有极大的方便性.

对于非 Gauss 纠缠态,其纠缠度虽然没有上面导出的 Gauss 纠缠态所具有的一般解析表达式,但上面我们给出的约化密度算符的部分求迹公式(57) 是具有普适性的. 因此,对于给定的任意两体非 Gauss 纠缠态,可以利用公式(57) 求出约化密度算符,进而求出本征值以求得该体系的纠缠度. 为此,我们考虑下面的非 Gauss 纠缠态

$$|\psi\rangle=\hat{B}|n_1,n_2\rangle=\mathrm{e}^{\theta(a_1^\dagger a_2-a_2^\dagger a_1)}|n_1,n_2\rangle \tag{85}$$

其密度算符可表示为

$$\begin{aligned}\rho_{12}&=\hat{B}|n_1,n_2\rangle\langle n_2,n_1|\hat{B}^+\\&=\frac{1}{n_1!\ n_2!}\frac{\mathrm{d}^{n_1}}{\mathrm{d}\alpha^{n_1}}\frac{\mathrm{d}^{n_2}}{\mathrm{d}\beta^{n_2}}:\exp\left\{\frac{1}{2}\left[(a_1^+,a_1)\boldsymbol{M}_1\begin{pmatrix}a_1^+\\a_1\end{pmatrix}+\right.\right.\\&\left.\left.(a_2^+,a_2)\boldsymbol{M}_2(a_2^+a_2)+2(a_1^+,a_1)\boldsymbol{M}_{12}\begin{pmatrix}a_2^+\\a_2\end{pmatrix}\right]\right\}:\Bigg|_{\alpha=\beta=0}\end{aligned} \tag{86}$$

其中各矩阵可表示为

$$\begin{cases}\boldsymbol{M}_1=(\alpha\cos^2\theta+\beta\sin^2\theta-1)\boldsymbol{\sigma}_1\\\boldsymbol{M}_2=(\beta\cos^2\theta+\alpha\sin^2\theta-1)\boldsymbol{\sigma}_1\\\boldsymbol{M}_3=\sin\theta\cos\theta(\beta-\alpha)\boldsymbol{\sigma}_1\end{cases} \tag{87}$$

式中的 $\boldsymbol{\sigma}_1$ 是 Pauli 矩阵. 利用式(57),计算得

$$\begin{aligned}\rho_1=&\frac{1}{n_1!\ n_2!}\frac{\mathrm{d}^{n_1}}{\mathrm{d}\alpha^{n_1}}\frac{\mathrm{d}^{n_2}}{\mathrm{d}\beta^{n_2}}\frac{1}{\beta\cos^2\theta+\alpha\sin^2\theta-1}\cdot\\&\left(\frac{\alpha\beta-(\alpha\cos^2\theta+\beta\sin^2\theta)}{\beta\cos^2\theta+\alpha\sin^2\theta-1}\right)^{a_1^\dagger a_1}\Bigg|_{\alpha=\beta=0}\end{aligned} \tag{88}$$

显然,由式(88) 容易求得本征值为

$$\begin{aligned}\lambda_{N_1}=&\frac{1}{n_1!\ n_2!}\frac{\mathrm{d}^{n_1}}{\mathrm{d}\alpha^{n_1}}\frac{\mathrm{d}^{n_2}}{\mathrm{d}\beta^{n_2}}\frac{1}{\beta\cos^2\theta+\alpha\sin^2\theta}\cdot\\&\left[1+\frac{\alpha\beta-1}{\beta\cos^2\theta+\alpha\sin^2\theta}\right]^{N_1}\Bigg|_{\alpha=\beta=-1}\end{aligned} \tag{89}$$

得纠缠度为

$$E=-\sum_{N_1}\lambda_{N_1}\ln\lambda_{N_1} \tag{90}$$

顺便指出,对于非 Gauss 型的两体纠缠纯态,其纠缠度的计算由于无统一的一般形式,为方便读者,下面我们再提供一种计算这类纠缠态纠缠度的方法. 对于量子光学中经常遇到的纠缠体系,诸如粒子数之和或粒子数之差为守恒量的两体纠缠纯态,我们发现,其纠缠度的一般计算公式可以用一般指数二次型的矩阵元公式给出. 下面给出简单的介绍. 对于形为

$$\Omega=\exp\left[\frac{1}{2}\Lambda(\ln\boldsymbol{M})\Sigma_B\widetilde{\Lambda}\right] \tag{91}$$

的指数二次型,给出了其矩阵元的一般计算公式

$$\langle m|\Omega|m'\rangle=\left[\det \boldsymbol{A}\det\begin{pmatrix}\widetilde{\boldsymbol{A}}^{-1} & -\widetilde{\boldsymbol{B}}\boldsymbol{A}^{-1}\\ \boldsymbol{A}^{-1}\boldsymbol{D} & \boldsymbol{A}^{-1}\end{pmatrix}\right]^{-\frac{1}{2}}\cdot \left[\prod_{i=1}^{n}\frac{1}{\sqrt{m_i!}}\left(\frac{\mathrm{d}}{\mathrm{d}r_i}\right)^{m_i}\right]\left[\prod_{j=1}^{n}\frac{1}{\sqrt{m_j'!}}\left(\frac{\mathrm{d}}{\mathrm{d}s_j}\right)^{m_j'}\right]\cdot \exp\left[\frac{1}{2}(\tilde{r},\tilde{s})\begin{pmatrix}\boldsymbol{A}^{-1}\boldsymbol{D} & \boldsymbol{A}^{-1}\\ \widetilde{\boldsymbol{A}}^{-1} & -\widetilde{\boldsymbol{B}}\boldsymbol{A}^{-1}\end{pmatrix}^{-1}\begin{pmatrix}r\\ s\end{pmatrix}\right]\Bigg|_{r=s=0} \tag{92}$$

其中

$$\Lambda=(a_1^+,a_2^+,\cdots,a_n^+,a_1,a_2,\cdots,a_n)\equiv(a^+,a^-)$$

而辛群 $\mathrm{Sp}(2n,C)$ 中 M 及粒子数态 $|m\rangle$ 表示为

$$\boldsymbol{M}=\begin{pmatrix}\boldsymbol{A} & \boldsymbol{D}\\ \widetilde{\boldsymbol{B}} & \widetilde{\boldsymbol{C}}\end{pmatrix},\ |m\rangle=\prod_{j=1}^{n}\frac{(a_j^+)^{m_j}}{\sqrt{m_j!}}|0\rangle \tag{93}$$

$\boldsymbol{A},\boldsymbol{D},\widetilde{\boldsymbol{B}},\widetilde{\boldsymbol{C}}$ 是 $n\times n$ 阶复矩阵. 式(92) 可以看作计算上面提到的一类纠缠体系的纠缠度的一般表达式,即对于如下形式的两体纠缠态

$$|\psi\rangle_{AB}=\hat{B}|n_1n_2\rangle \tag{94}$$

其中 $|n_1n_2\rangle$ 为通常的粒子数态,而 $\hat{B}$ 是满足一定条件的幺正算符,即

$$[\hat{B},\hat{N}_1+\hat{N}_2]=0\quad 或\quad [\hat{B},\hat{N}_1-\hat{N}_2]=0 \tag{95}$$

这里的 $\hat{N}_1,\hat{N}_2$ 是粒子数算符. 显然,当 $\hat{B}=\mathrm{e}^{\theta(a_1^\dagger a_2-a_2^\dagger a_1)}$ 为 Bogoliubov 算符①,或当 $\hat{B}=\mathrm{e}^{r(a_1^\dagger a_2^\dagger-a_1a_2)}$ 为压缩算符时,式(95) 均满足. 对于这样的纠缠态,部分求迹后有

$$\rho_A=\mathrm{tr}_B\rho_{AB}=\mathrm{tr}_B\hat{B}|n_1,n_2\rangle\langle n_1,n_2|\hat{B}^+ \tag{96}$$

插入完备基

$$\sum_{N_1N_2}|N_1,N_2\rangle\langle N_1N_2|=I$$

得

$$\rho_A=\sum_{N_1N_2N_1'N_2'}\mathrm{tr}_B|N_1,N_2\rangle\langle N_1',N_2'|\cdot\langle N_1,N_2|\hat{B}|n_1n_2\rangle\langle n_1n_2|\hat{B}^+|N_1',N_2'\rangle$$

由条件式(95) 可知,当 $N_2=N_2'$时,必有 $N_1=N_1'$. 因此有

$$\rho_A=\sum_{N_1N_2}|B_{n_1n_2}^{N_1N_2}|^2|N_1\rangle\langle N_2| \tag{97}$$

最后得纠缠度为

$$E=-\mathrm{tr}\,\rho_A\ln\rho_A=-\sum_{N_1N_2}|B_{n_1n_2}^{N_1N_2}|^2\ln|B_{n_1n_2}^{N_1N_2}|^2 \tag{98}$$

① N. N. Bogoliubov. Nuovo Cimento. 1958, 6: 794.

其中

$$B_{n_1 n_2}^{N_1 N_2} = \langle N_1, N_2 \mid \hat{B} \mid n_1 n_2 \rangle \tag{99}$$

很显然,对于这一类纠缠态纠缠度的计算实际上是对指数二次型算符矩阵元的计算. 因此式(92) 可以看作计算这一类纠缠态纠缠度的一般公式. 例如,当算符 $\hat{B}$ 取 Bogoliubov 算符时,仔细计算可得

$$B_{n_1 n_2}^{N_1 N_2} = \sum_{k=0}^{n_1} \sum_{l=0}^{n_2} (-1)^{n_1-k} (\cos\theta)^{k+l} (\sin\theta)^{n_1+n_2-k-l} \cdot \frac{\sqrt{N_1!\ N_2!\ n_1!\ n_2!}}{k!\ (n_1-k)!\ l!\ (n_2-l)!} \delta_{N_1, n_2-l+k} \delta_{N_2, n_1-k+l} \tag{100}$$

我们必须指出,本文导出的求迹公式(57) 对求任意两体纠缠态的纠缠度是具有普适性的,因为由此式可以导出任意两体 Gauss 纠缠纯态纠缠度的一般解析表达式,而对于非 Gauss 纠缠纯态可从该式得到约化密度算符的本征值从而求得纠缠度.

最近,我们证明了①,任意两体 Gauss 纠缠纯态都可以通过局域幺正变换转化成 $|\text{NOPA}\rangle$,而 $|\text{NOPA}\rangle$ 的纠缠度是已知的. 这一结果从另外一个角度解决了任意两体 Gauss 纠缠纯态的纠缠度量问题. 该结果也意味着两体 Gauss 纠缠纯态的纠缠形式只有一种,即 $|\text{NOPA}\rangle$ 态.

2.5 Bures 保真度的计算

保真度是描述量子态接近程度的物理量,是量子信息论中一个极为重要的概念. 人们借助于这一基本的物理量,可以在量子信息处理过程中做一些重要的定量计算,因此,保真度的计算无疑在这一过程中有着重要的地位. 近年来,关于保真度的计算,人们对如何给出其更方便的方法已做了大量的研究②③④⑤⑥⑦. 本节将主要介绍 Bures 保真度的计算,因为 Bures 保真度不仅能给出两混态接近程度及量子距离的测量,而且当量子态为纯态时,它能约化为通常的 Hilbert-Schmidt 保真度⑧,所以对 Bures 保真度计算的研究具有更广泛的意义.

---

① S. Yu et al.. to be published.

② J. Twamley. J. Phys. A: Math. Gen. 1996, 29: 3723.

③ H. Scutaru. J. Phys. A: Math. Gen. 1998, 31: 3659.

④ X. B. Wang, C. H. Oh and L. C. Kwek. Phys. Rev. 1998, A 58: 4186.

⑤ L. C. Kwek, C. H. Oh, X. B. Wang, and Y. Yeo. Phys. Rev. 2000, A 62: 052313.

⑥ G. S. Paraoanu and H. Scutaru. Phys. Rev. 1998, A 58: 869.

⑦ X. B. Wang, L. C. Kwek, and C. H. Oh. J. Phys. A: Math. Gen. 2000, 33: 4925.

⑧ M. B. Ruskai, Rev. Math. Phys. 1994, 6(5A): 1147.

Bures 保真度的定义

$$F=(\mathrm{tr}\sqrt{\hat{\rho}_1^{\frac{1}{2}}\hat{\rho}_2\hat{\rho}_1^{\frac{1}{2}}})^2 \tag{101}$$

其中,ρ_1,ρ_2是两量子混态的密度算符. 对于由n-模二次型Hamilton量描述的热平衡态,其保真度的定义式为

$$F=Z(\beta_1)Z(\beta_B)(\mathrm{tr}\sqrt{\mathrm{e}^{-\frac{\beta_1}{2}\hat{H}_1}\mathrm{e}^{-\beta_2\hat{H}_2}\mathrm{e}^{-\frac{\beta_1}{2}\hat{H}_1}})^2 \tag{102}$$

其中

$$\hat{H}_i=\hat{H}(N_i)=\frac{1}{2}\Lambda N_i\widetilde{\Lambda}=\frac{1}{2}\Lambda N_i'\Sigma\widetilde{\Lambda}\quad(i=1,2) \tag{103}$$

$\boldsymbol{N}$ 为 $2n\times 2n$ 阶厄米矩阵,$\boldsymbol{N}'=\boldsymbol{N}\boldsymbol{\Sigma}^{-1}$(其中 $\boldsymbol{\Sigma}=\begin{pmatrix}0 & I\\ -I & 0\end{pmatrix}$)为 $2n\times 2n$ 阶负厄米矩阵,而 Λ 表示为

$$\Lambda=(a_1^+,a_2^+,\cdots,a_n^+,a_1,a_2,\cdots,a_n)\equiv(a^+,\widetilde{a})$$

玻色算符a_i^+,a_i满足对易关系$[a_i,a_j^+]=\delta_{ij}$,$\widetilde{\Lambda}$是Λ的转置,$Z(\beta_i)=\dfrac{1}{\mathrm{tr}\,\mathrm{e}^{-\beta_i\hat{H}_i}}$是归一化因子.

对于由式(102) 定义的保真度,给出了计算的一般表达式

$$F=\frac{|\det(\mathrm{e}^{-\beta_1N_1\Sigma^{-1}}-I)\det(\mathrm{e}^{-\beta_2N_2\Sigma^{-1}}-\boldsymbol{I})|^{\frac{1}{2}}}{|\det(\sqrt{\mathrm{e}^{-\frac{\beta_1}{2}N_1\Sigma^{-1}}\mathrm{e}^{-\beta_2N_2\Sigma^{-1}}\mathrm{e}^{-\frac{\beta_1}{2}N_1\Sigma^{-1}}}-\boldsymbol{I})|} \tag{104}$$

这一公式的确为人们计算多模热平衡态的 Bures 保真度提供了极大的方便,但我们这里必须指出,在推导式(104) 的过程中所使用的方法过于烦琐,给人们在理解上带来了不必要的困难. 基于这一事实,本文将简化这一公式的推导过程. 为此,我们将利用线性量子变换理论这一工具,以非常简洁的证明方法给出计算保真度的公式(104). 通过下面的推导,我们所使用的方法不仅简单,更重要的是再一次显示了量子变换理论的优越性. 下面给出其证明过程.

为方便起见,我们令

$$U_1=\mathrm{e}^{-\frac{\beta_1}{2}\hat{H}_1},U_2=\mathrm{e}^{-\beta_2\hat{H}_2} \tag{105}$$

则由量子变换理论知

$$U_1\Lambda U_1^{-1}=\Lambda M_1=\Lambda\mathrm{e}^{-\frac{\beta_1}{2}N_1'},U_2\Lambda U_2^{-1}=\Lambda M_2=\Lambda\mathrm{e}^{-\beta_2N_2'} \tag{106}$$

所以有

$$U_1U_2U_1\Lambda(U_1U_2U_1)^{-1}=\Lambda(M_1M_2M_1)=\Lambda M \tag{107}$$

由此得

$$U_1U_2U_1=\exp\left[\frac{1}{2}\Lambda\ln M\boldsymbol{\Sigma}\widetilde{\Lambda}\right]=\exp\left[\frac{1}{2}\Lambda\ln(M_1M_2M_1)\boldsymbol{\Sigma}\widetilde{\Lambda}\right] \tag{108}$$

对于形如式(103)的二次型,给出的公式

$$\mathrm{tr}\,\mathrm{e}^{-\beta_i \hat{H}_i} = \left| \det(\mathrm{e}^{\beta_i N_i'} - \boldsymbol{I}) \right|^{-\frac{1}{2}} \tag{109}$$

我们可得

$$\begin{aligned}\mathrm{tr}\sqrt{U_1 U_2 U_1} &= \left| \det\left(\frac{1}{\sqrt{\mathrm{e}^{-\frac{\beta_1 N_1'}{2}} \mathrm{e}^{-\beta_2 N_2'} \mathrm{e}^{-\frac{\beta_1 N_1'}{2}}}} - \boldsymbol{I} \right) \right|^{-\frac{1}{2}} \\ &= \left| \det\left(\frac{1}{\sqrt{\mathrm{e}^{-\frac{\beta_1 N_1}{2}\Sigma^{-1}} \mathrm{e}^{-\beta_2 N_2 \Sigma^{-1}} \mathrm{e}^{-\frac{\beta_1 N_1}{2}\Sigma^{-1}}}} - \boldsymbol{I} \right) \right|^{-\frac{1}{2}}\end{aligned} \tag{110}$$

最后得保真度的解析表达式

$$\begin{aligned}F &= Z(\beta_1) Z(\beta_2) \left(\mathrm{tr}\sqrt{\mathrm{e}^{-\frac{\beta_1}{2}\hat{H}_1} \mathrm{e}^{-\beta_2 \hat{H}_2} \mathrm{e}^{-\frac{\beta_1}{2}\hat{H}_1}} \right)^2 \\ &= \frac{\left| \det(\mathrm{e}^{-\beta_1 N_1 \Sigma^{-1}} - I) \det(\mathrm{e}^{-\beta_2 N_2 \Sigma^{-1}} - \boldsymbol{I}) \right|^{\frac{1}{2}}}{\left| \det\left(\sqrt{\mathrm{e}^{-\frac{\beta_1}{2} N_1 \Sigma^{-1}} \mathrm{e}^{-\beta_2 N_2 \Sigma^{-1}} \mathrm{e}^{-\frac{\beta_1}{2} N_1 \Sigma^{-1}}} - \boldsymbol{I}\right) \right|}\end{aligned} \tag{111}$$

即式(104).

从上面的证明过程可以清楚地看出,我们所采用的证明方法简单得多,这也充分体现了量子变换理论的优越性及在众多领域中应用的广泛性.随着量子信息学这一学科的不断发展,我们深信量子变换理论在这一新兴学科中的应用也会越来越多.

下面给出几个例子.首先,考虑双模压缩热态,其密度算符由下式给出

$$\rho_i = Z(\beta_i) \hat{S}_i \hat{T}_i \hat{S}_i^+ \tag{112}$$

其中

$$S_i = \exp[(\zeta_i^* a_1^+ a_2^+ - \zeta_i a_1 a_2)]$$

$$T_i = \exp\left[\frac{-\beta_i}{2} \sum_{j=1}^{2} (a_j^+ a_j + a_j a_j^+) \right]$$

由量子变换理论,不难算得

$$N_i = M(\hat{S}_i) \begin{pmatrix} 0 & -\beta_i I \\ -\beta_i I & 0 \end{pmatrix} \widetilde{M}(\hat{S}_i)$$

其中

$$M(\hat{S}_i) = \begin{pmatrix} \cosh r_i I & -\mathrm{e}^{i\theta} \sinh r\sigma \\ -\mathrm{e}^{-i\theta} \sinh r\sigma & \cosh r_i I \end{pmatrix}, \boldsymbol{I} = \begin{pmatrix} 1 & 0 \\ 0 & 1 \end{pmatrix}$$

$$\boldsymbol{\sigma} = \begin{pmatrix} 0 & 1 \\ 1 & 0 \end{pmatrix}, r_i = |\zeta_i|, \theta_i = \frac{\zeta_i}{r_i}$$

将上面的结果代入式(111)中得

$$F = \left(\frac{2\sinh \frac{\beta_1}{2} \sinh \frac{\beta_2}{2}}{\cosh \frac{\beta_3}{2} - 1} \right)^2 \tag{113}$$

式(113) 中的 β_3 由下式决定

$$\cosh \beta_3 = \cosh(\beta_1 + \beta_2) \left[\cosh^2(r_1 + r_2) \sin^2 \frac{\Delta\theta}{2} + \cosh^2(r_2 - r_1) \cos^2 \frac{\Delta\theta}{2} \right] - \cosh(\beta_2 - \beta_1) \left[\sinh^2(r_1 + r_2) \sin^2 \frac{\Delta\theta}{2} + \sinh^2(r_2 - r_1) \cos^2 \frac{\Delta\theta}{2} \right]$$

其中,$\Delta\theta = \theta_1 - \theta_2$.

其次,考虑两维 Jump 谐振子系统[①②],在 $t = 0$ 时,体系的 Hamilton 量为

$$\hat{H}_1 = \frac{1}{2}\tilde{p}p + \frac{1}{2}\tilde{x}x \tag{114}$$

相应的密度算符为

$$\rho_1(t = 0) = Z(\beta) e^{-\beta \hat{H}_1}, Z(\beta) = \left(2\sinh \frac{\beta}{2} \right)^2$$

在 $t > 0$ 时,体系的密度算符为

$$\rho_2(t > 0) = e^{-i\hat{H}_2 t} \rho_1 e^{i\hat{H}_2 t} \tag{115}$$

$$\hat{H}_2 = \frac{1}{2}\tilde{p}p + \frac{1}{2}\tilde{x}x + \lambda x_1 x_2 \tag{116}$$

这时,其保真度的计算公式为

$$F = \left(Z(\beta) \operatorname{tr} \sqrt{ e^{-\frac{1}{2}\beta \hat{H}_1} e^{-i\hat{H}_2 t} e^{-\beta \hat{H}_1} e^{i\hat{H}_2 t} e^{-\frac{1}{2}\beta \hat{H}_1} } \right)^2 \tag{117}$$

此时,同样可用式(111) 计算出式(117) 的结果. 利用线性量子变换理论及下列变换关系

$$(\tilde{q}, \tilde{p}) = (a^+, \tilde{a}) \frac{1}{\sqrt{2}} \begin{pmatrix} \boldsymbol{I} & i\boldsymbol{I} \\ \boldsymbol{I} & -i\boldsymbol{I} \end{pmatrix} = (a^+, \tilde{a}) \boldsymbol{K} \tag{118}$$

可得

$$\boldsymbol{N}_1 = \boldsymbol{K} \begin{pmatrix} \boldsymbol{I} & 0 \\ 0 & \boldsymbol{I} \end{pmatrix} \tilde{\boldsymbol{K}}, \boldsymbol{N}_2 = \boldsymbol{K} \begin{pmatrix} 1 & 0 & 0 & 0 \\ 0 & 1 & 0 & 0 \\ 0 & 0 & 1 & \lambda \\ 0 & 0 & \lambda & 1 \end{pmatrix} \tilde{\boldsymbol{K}}$$

于是,得式(117) 的保真度为

① J. Janszky and Y. Y. Yushin. Opt. Commun. 1986, 59: 151.

② C. J. Ballhausen, Chem. Phys. Lett. 1993, 201: 269.

$$F = \frac{4\sinh^4 \frac{\beta}{2}}{\left(\cosh \frac{r}{2} - 1\right)^2} \tag{119}$$

式中 r 由下式决定

$$\begin{aligned}\cosh r = \frac{1}{16(1-\lambda^2)}[& -2\lambda^2 + \lambda^2(1+\lambda)\cos(2t\sqrt{1-\lambda}) + \\ & \lambda^2(1-\lambda)\cos(2t\sqrt{1+\lambda}) + \\ & \cosh 2\beta(16 - 14\lambda^2 - \lambda^2(1+\lambda)\cos(2t\sqrt{1-\lambda}) - \\ & \lambda^2(1-\lambda)\cos(2t\sqrt{1+\lambda}))]\end{aligned}$$

上面我们用量子变换理论导出了Bures保真度的解析表达式,其导出过程非常简洁,本方法的优越性一目了然. 同时,我们也清楚地看到量子变换理论在量子信息领域中的确有着十分广泛的应用前景.

3. 连续变量系统的量子非定域性

关于量子态的空间非定域性问题,已有大量的理论和实验工作①. 对于连续变量情形下的空间非定域性,这里我们仍将从 Bell 非定域性这一角度出发,继续对连续变量情形下量子态的非定域性作一些理论上的探讨. 为了内容上的连续性,我们先对 $|\mathrm{NOPA}\rangle$ 态(由非简并光学参量放大器产生的态)的 Bell 不等式的破坏程度作一简短回顾. 然后引入超算符的主方程求解,从而对杂化纠缠态②与热库相互作用时量子态的非定域性动力学进行深入的分析.

Banaszek 和 Wódkiewic 利用宇称算符的联合测量,证实了 $|\mathrm{NOPA}\rangle$ 态的Bell不等式破坏,而当压缩参量 $r\to\infty$ 时,$|\mathrm{NOPA}\rangle$ 态即转化为原始的 $|\mathrm{EPR}\rangle$ 态,从而证实了以位置 - 动量为参量的 $|\mathrm{EPR}\rangle$ 态也具有非定域性. 在此基础上,我们提出了用宇旋算符的方法,进一步最大限度地暴露了 $|\mathrm{NOPA}\rangle$ 态的非定域性. 我们的这一方法不仅可以给出原始 $|\mathrm{EPR}\rangle$ 态的 Bell 不等式的最大破坏,而且给非定域性动力学的研究提供了一条方便的途径. 下面我们先对 Banaszek 等人及我们的工作分别作一简单的回顾.

---

① 陈增兵,逯怀新,吴盛俊,张永德. 量子纠缠与空间非定域性;曾谨言,裴寿镛,龙桂鲁主编. 量子力学新进展(第二辑). 北京:北京大学出版社,2001.

② C. Monroe, D. M. Meekhof, B. E. King, and D. J. Wineland, Science. 1996, 272: 1131.

3.1 相空间形式

考虑如下的双模压缩真空态

$$|\text{NOPA}\rangle = e^{r(a_1^\dagger a_2^\dagger - a_1 a_2)}|00\rangle = \frac{1}{\cosh r}\sum_{n=0}^{\infty}(\tanh r)^n |nn\rangle \quad (120)$$

即式(4) 和式(6),其中 $r > 0$ 为压缩参量,有

$$|nn\rangle = |n_1\rangle \otimes |n_2\rangle = \frac{1}{n!}(a_1^+)^n(a_2^+)^n|00\rangle \quad (121)$$

为验证式(120) 的非定域性,Banaszek 等人引进如下形式的相空间位移宇称算符

$$\hat{\Pi}(\alpha,\beta) = \hat{D}_1(\alpha)(-1)^{\hat{n}_1}\hat{D}_1^+(\alpha) \otimes \hat{D}_2(\beta)(-1)^{\hat{n}_2}\hat{D}_2^+(\beta) \quad (122)$$

其中 $\hat{D}(\alpha)$ 为通常的相空间位移算符,$\hat{n}$ 为粒子数算符. 定义相应的 Bell 算符为

$$B_{\text{CHSH}} = \hat{\Pi}(\alpha,\beta) + \hat{\Pi}(\alpha',\beta) + \hat{\Pi}(\alpha,\beta') - \hat{\Pi}(\alpha',\beta') \quad (123)$$

则定域实在论预言如下的 Bell 不等式

$$|\langle \text{NOPA}|B_{\text{CHSH}}|\text{NOPA}\rangle| \leqslant 2 \quad (124)$$

经过详细地计算可得关联函数为

$$\begin{aligned}\Pi(\alpha,\beta) &= \langle \text{NOPA}|\hat{\Pi}\hat{\Pi}(\alpha,\beta)|\text{NOPA}\rangle \\ &= \exp[-2\cosh 2r(|\alpha|^2 + |\beta|^2) + \\ &\quad 2\sinh 2r(\alpha\beta + \alpha^*\beta^*)]\end{aligned}$$

如果选取 $\alpha = 0, \alpha' = \sqrt{J}, \beta = 0, \beta' = -\sqrt{J}$,可得

$$\begin{aligned}B &= \langle \text{NOPA}|B_{\text{CHSH}}|\text{NOPA}\rangle \\ &= \Pi(0,0) + \Pi(\sqrt{J},0) + \Pi(0,-\sqrt{J}) - \Pi(\sqrt{J},-\sqrt{J}) \\ &= 1 + 2\exp(-2J\cosh 2r) - \exp(-4Je^{2r})\end{aligned} \quad (125)$$

在上式中,取

$$Je^{2r} = \frac{1}{3}\ln 2 \quad (126)$$

时,可得式(125) 中的最大值为 $B = 1 + 3 \times 2^{-4/3} \approx 2.19$. 即得到了 $|\text{NOPA}\rangle$ 态对 Bell 不等式(124) 的破坏.

3.2 宇旋算符形式

为了给出上面 $|\text{NOPA}\rangle$ 态的 Bell 不等式的最大破坏,我们利用如下形式的宇旋算符

$$\begin{cases} s_z = \sum_{n=0}^{\infty} [|2n+1\rangle\langle 2n+1| - |2n\rangle\langle 2n|] \\ s_- = \sum_{n=0}^{\infty} |2n\rangle\langle 2n+1| = (s_+)^+ \end{cases} \tag{127}$$

其中 $|n\rangle$ 是通常的 Fock 态. 容易看出 $s_z=(-1)^N$ 是通常的宇称算符,而 s_+ 和 s_- 则是“宇称反转”算符. 利用 Fock 态的性质可以证明

$$[s_z, s_\pm] = \pm 2s_\pm$$
$$[s_+, s_-] = s_z$$

式中的对易关系与通常的$\frac{1}{2}$自旋的体系具有相同的对易关系. 因此,准自旋算符

$$\boldsymbol{s} = (s_x, s_y, s_z), s_x \pm \mathrm{i}s_y = 2s_\pm \tag{128}$$

是 Pauli 自旋算符的光学对应,是一种作用于光子的宇称空间的自旋算符,因此我们可以称之为光子宇旋.

取一任意的单位矢量

$$\boldsymbol{a} = (\sin\theta_a \sin\varphi_a, \sin\theta_a \sin\varphi_a, \cos\theta_a)$$

容易写出

$$\boldsymbol{a}\cdot\boldsymbol{s} = s_z\cos\theta_a + \sin\theta_a(\mathrm{e}^{\mathrm{i}\varphi_a}s_- + \mathrm{e}^{-\mathrm{i}\varphi_a}s_+)$$

利用宇旋算符的对易关系可证

$$(\boldsymbol{a}\cdot\boldsymbol{s})^2 = \boldsymbol{I}$$

由此可以看出,厄米算符 $\boldsymbol{a}\cdot\boldsymbol{s}$ 的本征值必为 ± 1,其相应的测量结果也必为 ± 1. 以上的一系列分析显示,连续变量体系与离散变量体系(如通常的$\frac{1}{2}$自旋体系)之间存在一种很好的类比. 因此,适用于离散变量体系的所有 Bell 不等式均可以在连续变量体系中找到其对应.

特别地,我们可以对两模光场定义如下 Bell 算符

$$\boldsymbol{B}_{\mathrm{CHSH}} = (\boldsymbol{a}\cdot\boldsymbol{s}_1)(\boldsymbol{b}\cdot\boldsymbol{s}_2) + (\boldsymbol{a}'\cdot\boldsymbol{s}_1)(\boldsymbol{b}\cdot\boldsymbol{s}_2) + (\boldsymbol{a}\cdot\boldsymbol{s}_1)(\boldsymbol{b}'\cdot\boldsymbol{s}_2) - (\boldsymbol{a}'\cdot\boldsymbol{s}_1)(\boldsymbol{b}'\cdot\boldsymbol{s}_2) \tag{129}$$

其中,$\boldsymbol{a}$,$\boldsymbol{a}'$,$\boldsymbol{b}$ 和 $\boldsymbol{b}'$ 均为单位矢量,$\boldsymbol{s}_1$ 和 $\boldsymbol{s}_1$ 如式(128)中所定义. 根据定域实在论,我们有如下的 Bell 不等式

$$|\langle \boldsymbol{B}_{\mathrm{CHSH}}\rangle| \leqslant 2 \tag{130}$$

其中,$\langle \boldsymbol{B}_{\mathrm{CHSH}}\rangle$ 是 $\boldsymbol{B}_{\mathrm{CHSH}}$ 在一个给定的连续变量量子态中的期望值.

但量子力学给出不同的预言. 可以证明上面定义的 Bell 算符与自旋$\frac{1}{2}$系统一样,即平均值上限为 $2\sqrt{2}$. 这是因为我们有

$$\boldsymbol{B}_{\mathrm{CHSH}}^2 = 4\boldsymbol{I} + 4[\boldsymbol{a}\times\boldsymbol{a}'\cdot\boldsymbol{s}_1]\otimes[\boldsymbol{b}\times\boldsymbol{b}'\cdot\boldsymbol{s}_2]$$

从而

$$\langle \boldsymbol{B}_{\text{CHSH}}^2 \rangle \leqslant 4 + 4 = 8$$

所以有

$$|\langle \boldsymbol{B}_{\text{CHSH}} \rangle| \leqslant 2\sqrt{2}$$

由上面的定义,对$|\text{NOPA}\rangle$态容易算出关联函数为

$$\begin{aligned} E(\theta_a, \theta_b) &= \langle \text{NOPA} | S_{\theta_a}^{(1)} \otimes S_{\theta_b}^{(2)} | \text{NOPA} \rangle \\ &= \cos\theta_a \cos\theta_b + K(r) \sin\theta_a \sin\theta_b \end{aligned}$$

式中

$$S_{\theta_a}^{(j)} = s_{jz} \cos\theta_a + s_{jx} \sin\theta_a \quad (j = 1, 2)$$

而$K = \tanh(2r) \leqslant 1$为非定域度. 如果取$\theta_a = 0, \theta_{a'} = \frac{\pi}{2}, \theta_b = -\theta_{b'}$,可得

$$\boldsymbol{B} = \langle \text{NOPA} | \boldsymbol{B}_{\text{CHSH}} | \text{NOPA} \rangle = 2(\cos\theta_b + K \sin\theta_b) \qquad (131)$$

显然,当$\theta_b = \tan^{-1} K$时,上式有最大值,即

$$B_{\max} = 2\sqrt{1 + K^2} \qquad (132)$$

因此$|\text{NOPA}\rangle$态总是破坏Bell不等式(130). 当$r \to \infty$时,有$K \to 1$,这时上式成为$B_{\max} = 2\sqrt{2}$,而相应的$|\text{NOPA}\rangle$态转化为原始的归一化的$|\text{EPR}\rangle$态. 由此我们得出,归一化的$|\text{EPR}\rangle$态最大地破坏Bell不等式(130).

值得指出的是,式(127)可以看成一个无穷维Hilbert空间到2维Hilbert空间的映射,这一看法有很有趣的应用①和推广②. 利用宇旋算符,文献③中讨论了连续变量的Greenberger-Horne-Zeilinger定理④.

3.3 量子非定域性的动力学演化

众所周知,当研究的量子系统与环境相互作用时,它的许多量子特性,如相干、纠缠及非定域性等都会有不同程度上的丢失. 研究这些量子现象对量子信息处理起着重要的作用. 显然,非定域性动力学的研究就显得更为重要. 文献⑤对双模压缩真空态、纠缠相干态与辐射场相互作用情

---

① L. Mišta, Jr. et al., Phys. Rev. 2002, A 65: 062315.

② C. Brukner, M. S. Kim, J. -W. Pan, A. Zeilinger, quant-ph0208116/.

③ Z. -B. Chen and Y. D. Zhang, Phys. Rev. 2002, A 65: 044102.

④ D. M. Greenberger, M. A. Horne, and A. Zeilinger, in Bell's Theorem, Quantum Theory, and Conceptions of the Universe, edited by M. Kafatos (Kluwer Academic, Dordrecht, 1989); D. M. Greenberger, M. A. Horne, A. Shimony, and A. Zeilinger, Am. J. Phys. 1990, 58: 1131.

⑤ H. Jeong, J. Lee and M. S. Kim, Phys. Rev. 2000, A 61: 052101; D. Wilson, H. Jeong and M. S. Kim, J. Mod. Opt. 2002, 49: 851.

况下的非定域性动力学进行了研究,给出了量子态在这一非幺正演化过程中的重要结果. 这里,我们将对杂化纠缠态与热库相互作用时量子态的非定域性及纠缠随时间的演化特性作一系统研究. 由于这时所研究的量子系统作为一个子系统所遵从的运动方程即为我们熟悉的主方程①②,而研究以上诸量子现象首先要知道的就是约化密度算符 $\rho(t)$ 的解析表达式,因此,如何方便地求解主方程也是这一研究过程中重要的一步. 所以,在下面的分析中,我们先介绍最近我们提出的超算符求解主方程的方法③.

考虑一单原子与单模腔场相耦合的量子体系与热库相互作用,在旋转波近似下,这一耦合体系所遵从的主方程由下式给出④

$$\dot{\rho}(t)=\frac{\gamma}{2}(2\sigma_-\rho\sigma_+-\sigma_+\sigma_-\rho-\rho\sigma_+\sigma_-)+\frac{k}{2}(2a\rho a^+-a^+a\rho-\rho a^+a) \tag{133}$$

其中 γ 为原子的线宽度,而 k 为腔模的损失速率. 为了方便地求解这一主方程,我们引入下列超算符

$$K_-\rho=a^+\rho a,K_+\rho=a\rho a^+,K_0\rho=-\frac{1}{2}(a^+a\rho+\rho a^+a+\rho) \tag{134}$$

$$J_-\rho=\sigma_-\rho\sigma_+,J_+\rho=\sigma_+\rho\sigma_-,J_0\rho=\frac{1}{2}(\sigma_+\sigma_-\rho+\rho\sigma_+\sigma_--\rho) \tag{135}$$

容易证明超算符 $K_{\pm,0}$ 和 $J_{\pm,0}$ 分别满足 $su(1,1)$ 及 $su(2)$ 李代数对易关系,即

$$[K_-,K_+]\rho=2K_0\rho,[K_0,K_\pm]\rho=\pm K_\pm\rho \tag{136}$$

$$[J_-,J_+]\rho=-2J_0\rho,[J_0,J_\pm]\rho=\pm J_\pm\rho \tag{137}$$

由于超算符 $K_{\pm,0}$ 和 $J_{\pm,0}$ 互相对易,所以式(133)的形式解为

$$\rho(t)=\mathrm{e}^{\frac{1}{2}(k-r)}\exp[r(J_-+J_0)t]\exp[k(K_++K_0)t]\rho(0) \tag{138}$$

对于给定的任意初态 $\rho(0)$,仅从式(138)是无法知道量子态 $\rho(t)$ 的全部信息的. 因此,必须对式(138)进行分解以使得能从任意初态 $\rho(0)$ 方便地得到 $\rho(t)$ 的解析表达式. 为此,我们利用 $su(1,1)$ 及 $su(2)$ 李代数

---

① W. H. Louisell. Quantum Statistical Properties of Radiation. New York: Wiley, 1973.

② J. Preskill. Quantum information and Quantum Computation. California Institute of Technology, 1998.

③ H. X. Lu, J. Yang, Y. D. Zhang, and Z.-B. Chen, Phys. Rev. 2003, A6F 024101.

④ W. Ren and H. J. Carmichael. Phys. Rev. 1995, A 51: 752.

通常的生成元组成的指数算符的分解定理①②，对于具有 $su(1,1)$ 及 $su(2)$ 李代数超算符生成元组成的指数算符，得到了这一超算符形式的指数算符的分解定理——即正规排列形式和反正规排列形式. 我们得式(138) 的正规排列形式为

$$\rho(t) = e^{\frac{1}{2}(\kappa-\gamma)t}\exp[\ln(x_0)J_0]\exp[x_- J_-]\cdot \exp[\ln(y_0)K_0]\exp[y_+ K_+]\rho(0) \tag{139}$$

式中各系数的表达式为

$$y_0 = e^{\kappa t}, y_+ = 1 - e^{-\kappa t}, x_0 = e^{-\gamma t}, x_- = 1 - e^{-\gamma t} \tag{140}$$

这样，利用(139) 就可以从任意的初态容易地得到 $\rho(t)$ 的显示表达式，因而可方便地研究该量子态的各种量子现象.

下面我们讨论初态为杂化纠缠态的情况，即考虑原子与腔场纠缠态

$$|\psi\rangle = \frac{1}{\sqrt{2}}(|e\rangle|\alpha\rangle_e + |g\rangle|\alpha\rangle_o) \tag{141}$$

式中，$|e\rangle$，$|g\rangle$ 分别代表原子的激发态和基态，而 $|\alpha\rangle_o$，$|\alpha\rangle_e$ 分别代表腔场的奇、偶相干态，其表达式为

$$|\alpha\rangle_e = \frac{1}{\sqrt{N_+}}(|\alpha\rangle + |-\alpha\rangle)$$

$$|\alpha\rangle_o = \frac{1}{\sqrt{N_-}}(|\alpha\rangle - |-\alpha\rangle) \tag{142}$$

其中的系数 $N_\pm = 2 \pm 2e^{-2|\alpha|^2}$. 取式(141) 作为式(139) 中的初态 $\rho(0)$，经过详细地计算，可算得下列结果

$$e^{y_+K_+}|\alpha\rangle_{e(o)e(o)}\langle\alpha| = \cosh(y_+|\alpha|^2)|\alpha\rangle_{e(o)e(o)}\langle\alpha| + \frac{N_{-(+)}}{N_{+(-)}}\sinh(y_+|\alpha|^2)|\alpha\rangle_{o(e)o(e)}\langle\alpha| \tag{143}$$

$$e^{y_+K_+}|\alpha\rangle_{e(o)o(e)}\langle\alpha| = \cosh(y_+|\alpha|^2)|\alpha\rangle_{e(o)o(e)}\langle\alpha| + \sinh(y_+|\alpha|^2)|\alpha\rangle_{o(e)e(o)}\langle\alpha| \tag{144}$$

$$e^{\ln(y_0)K_0}|\alpha\rangle_{e(o)e(o)}\langle\alpha| = e^{-\frac{1}{2}\kappa t}e^{-y_+|\alpha|^2}\frac{N_{+(-)}(t)}{N_{+(-)}}\cdot |\alpha(t)\rangle_{e(o)e(o)}\langle\alpha(t)| \tag{145}$$

$$e^{\ln(y_0)K_0}|\alpha\rangle_{e(o)o(e)}\langle\alpha| = e^{-\frac{1}{2}\kappa t}e^{-y_+|\alpha|^2}\sqrt{\frac{N_+(t)N_-(t)}{N_+N_-}}\cdot |\alpha(t)\rangle_{e(o)o(e)}\langle\alpha(t)| \tag{146}$$

---

① K. Wòdkiewicz and J. H. Eberly, J. Opt. Soc. Am. 1985, B 2: 458.

② X. Ma and W. Rhodes. Phys. Rev. 1990, A 41: 4625.

$$\begin{cases} e^{x_- J} -| e\rangle\langle e| = | e\rangle\langle e| + x_- | g\rangle\langle g|, e^{x_- J_-} | e\rangle\langle g| = | e\rangle\langle g| \\ e^{x_- J} -| g\rangle\langle g| = | g\rangle\langle g|, e^{x_- J_-} | g\rangle\langle e| = | g\rangle\langle e| \end{cases} \tag{147}$$

$$\begin{cases} e^{\ln(x_0)J_0} | e\rangle\langle e| = e^{-\frac{1}{2}\gamma t} | e\rangle\langle e|, e^{\ln(x_0)J_0} | g\rangle\langle g| = e^{\frac{1}{2}\gamma t} | g\rangle\langle g| \\ e^{\ln(x_0)J_0} | e\rangle\langle g| = | e\rangle\langle g|, e^{\ln(x_0)J_0} | g\rangle\langle e| = | g\rangle\langle e| \end{cases} \tag{148}$$

上面各式中的 $N_{\pm}(t)$ 及 $\alpha(t)$ 表示为 $N_{\pm}(t) = 2 \pm 2e^{-2|\alpha|^2}e^{-kt}$,$\alpha(t) = \alpha e^{-\frac{1}{2}\kappa t}$,将上面的结果代入式(139) 中得

$$\begin{aligned} \rho(t) = & \frac{1}{2N_+}e^{-y_+|\alpha|^2}[N_+(t)\cosh(y_+|\alpha|^2)|\alpha(t)\rangle_{ee}\langle\alpha(t)| + \\ & N_-(t)\sinh(y_+|\alpha|^2)|\alpha(t)\rangle_{oo}\langle\alpha(t)|]\cdot \\ & [e^{-\gamma t}|e\rangle\langle e| + x_-|g\rangle\langle g|] + \\ & \frac{1}{2N_-}e^{-y_+|\alpha|^2}[N_-(t)\cosh y_+|\alpha|^2)|\alpha(t)\rangle_{oo}\langle\alpha(t)| + \\ & N_+(t)\sinh(y_+|\alpha|^2)|\alpha(t)\rangle_{ee}\langle\alpha(t)|]|g\rangle\langle g| + \\ & \frac{1}{2}\sqrt{\frac{N_+(t)N_-(t)}{N_+N_-}}e^{-\frac{1}{2}\gamma t}e^{-y_+|\alpha|^2}\cdot \\ & [\cosh(y_+|\alpha|^2)|\alpha(t)\rangle_{eo}\langle\alpha(t)| + \\ & \sinh(y_+|\alpha|^2)|\alpha(t)\rangle_{oe}\langle\alpha(t)|]|e\rangle\langle g| + \\ & \frac{1}{2}\sqrt{\frac{N_+(t)N_-(t)}{N_+N_-}}e^{-\frac{1}{2}\gamma t}e^{-y_+|\alpha|^2}\cdot \\ & [\cosh(y_+|\alpha|^2)|\alpha(t)\rangle_{oe}\langle\alpha(t)| + \\ & \sinh(y_+|\alpha|^2)|\alpha(t)\rangle_{eo}\langle\alpha(t)|]|g\rangle\langle e| \end{aligned} \tag{149}$$

现在我们来研究式(149) 中量子态的非定域性,为此,对于腔场我们用式(127) 给出的宇旋算符. 于是 Bell 算符可定义为

$$\begin{aligned} \hat{\boldsymbol{B}} = & (\boldsymbol{a}\cdot\boldsymbol{\sigma})(\boldsymbol{b}\cdot\boldsymbol{s}) + (\boldsymbol{a}'\cdot\boldsymbol{\sigma})(\boldsymbol{b}\cdot\boldsymbol{s}) + \\ & (\boldsymbol{a}\cdot\boldsymbol{\sigma})(\boldsymbol{b}'\cdot\boldsymbol{s}) - (\boldsymbol{a}'\cdot\boldsymbol{\sigma})(\boldsymbol{b}'\cdot\boldsymbol{s}) \end{aligned} \tag{150}$$

其中的 $\boldsymbol{a},\boldsymbol{a}',\boldsymbol{b}$ 和 $\boldsymbol{b}'$ 均为单位矢量,例如,取

$$\boldsymbol{a} = (\sin\theta_a, 0, \sin\theta_b) \tag{151}$$

并且对于测量原子的 Pauli 算符 $\boldsymbol{\sigma}$ 及腔场的宇旋算符 $\boldsymbol{s}$ 均有

$$(\boldsymbol{a}\cdot\boldsymbol{\sigma})^2 = (\boldsymbol{a}'\cdot\boldsymbol{\sigma})^2 = \boldsymbol{I}_{2\times2}$$

$$(\boldsymbol{b}\cdot\boldsymbol{s})^2 = (\boldsymbol{b}'\cdot\boldsymbol{s})^2 = \boldsymbol{I}$$

利用式(149) ~ (151) 可得 Bell 函数为

$$\langle\hat{B}\rangle = \mathrm{tr}\,\hat{B}\rho(t) = E(\theta_a,\theta_b) + E(\theta_a,\theta_b') +$$

$$E(\theta'_a,\theta_b) - E(\theta'_a,\theta'_b) \tag{152}$$

相应的 Bell 不等式为

$$|\langle \hat{B}\rangle| \leqslant 2$$

与前面一样,其中的 $E(\theta_a,\theta_b)$ 为关联函数,经过详细地计算可得

$$E(\theta_a,\theta_b)=f(t)\cos\theta_a\cos\theta_b + g(t)\sin\theta_a\sin\theta_b \tag{153}$$

其中 $f(t)$,$g(t)$ 分别表示

$$f(t)=\frac{N_+(t)}{N_+}\mathrm{e}^{-\gamma t} - \frac{\sinh(2y_+\alpha^2)}{\sinh(2\alpha^2)} \tag{154}$$

$$g(t)=\mathrm{e}^{-\frac{1}{2}\gamma t}\sqrt{\frac{1-\mathrm{e}^{-4\alpha^2\mathrm{e}^{-\kappa t}}}{1-\mathrm{e}^{-4\alpha^2}}}\cdot \left[\frac{1}{2}\sinh(2\alpha^2\mathrm{e}^{-\kappa t})\right]^{\frac{1}{2}}\sum_{n=0}^{\infty}\frac{(\alpha\mathrm{e}^{\frac{1}{2}\kappa t})^{4n+1}}{\sqrt{(2n)!\ (2n+1)!}} \tag{155}$$

同前面的处理一样,如果选取 $\theta_a=0,\theta'_a=\dfrac{\pi}{2},\theta_b=-\theta'_b$,那么可由式(152)中得 Bell 函数为

$$\langle\hat{B}\rangle = 2\sqrt{f(t)^2+g(t)^2}\cos(\theta_b-\varphi) \tag{156}$$

其中

$$\tan\varphi=\frac{g(t)}{f(t)} \tag{157}$$

如果我们进一步取 $\theta_b=\varphi$,那么可得式(156)的最大值为

$$\langle\hat{B}\rangle_{\max} = 2\sqrt{f(t)^2+g(t)^2} \tag{158}$$

式(158)可以看作描述杂化纠缠态式(141)非定域性动力学的一般方程.例如,如果我们取 $t=0$,那么可由式(154)和式(155)得

$$f(0)=1,g(0)=\left[\frac{1}{2}\sinh(2\alpha^2)\right]^{-\frac{1}{2}}\sum_{n=0}^{\infty}\frac{(\alpha)^{4n+1}}{\sqrt{(2n)!\ (2n+1)!}}<1 \tag{159}$$

因而得相应的 Bell 函数是

$$\langle\hat{B}\rangle_{\max}=2\sqrt{1+g(0)^2} \tag{160}$$

而对于 $t\to\infty$,可得结果

$$f(t\to\infty)=-1,g(t\to\infty)=0 \tag{161}$$

$$\langle\hat{B}\rangle_{\max}=2 \tag{162}$$

这正是我们所预期的结果.因为这时混态已转化为直积态,即

$$\rho(t\to\infty)=|0\rangle\langle 0|\otimes|g\rangle\langle g| \tag{163}$$

此时没有 Bell 不等式的破坏.

图2给出了式(158)的Bell函数$\langle\hat{B}\rangle_{\max}$随时间$t$的变化曲线,从该曲线中可以看出,对于式(149)中的混态,正如文献①中所指出的那样,当它失去其非定域性时仍保持一些纠缠.这一现象说明了我们所选取的Bell算符在整个演化过程中不能很好地揭示该量子态的非定域性,同时也再一次说明了Bell算符的选取对揭示量子态非定域性的重要性.

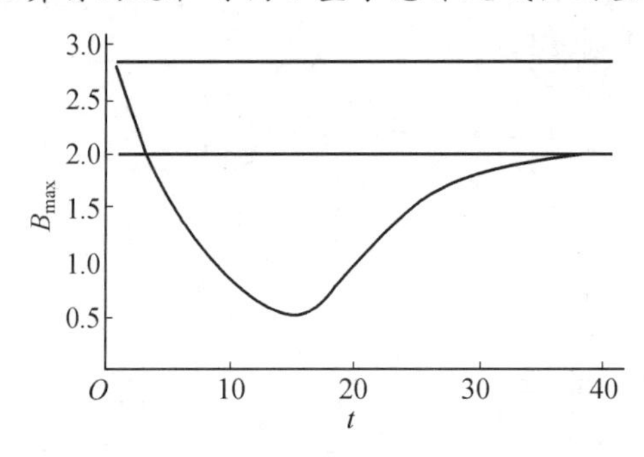

图2　Bell函数$\langle B\rangle_{\max}$随重新标度的时间t的变化曲线

(取$|\alpha|=4,\gamma=\kappa=0.2$)

为了将此问题看得更清楚一些,我们对腔场分别取如下形式的宇旋算符

$$\begin{cases}s'_z=|\alpha\rangle_{ee}\langle\alpha|-|\alpha\rangle_{oo}\langle\alpha|\\ s'_+=|\alpha\rangle_{eo}\langle\alpha|=(s'_-)^+\end{cases}\tag{164}$$

$$\begin{cases}s''_z=|\alpha(t)\rangle_{ee}\langle\alpha(t)|-|\alpha(t)\rangle_{oo}\langle\alpha(t)|\\ s''_+=|\alpha(t)\rangle_{eo}\langle\alpha(t)|=(s''_-)^+\end{cases}\tag{165}$$

按照上面同样的做法可以求得

$$\langle B'\rangle_{\max}=2\sqrt{f(t)^2+g(t)'^2}\tag{166}$$

$$\langle B''\rangle_{\max}=2\sqrt{f(t)^2+g(t)''^2}\tag{167}$$

$$g(t)'=\mathrm{e}^{-\frac{1}{2}\gamma t}\sqrt{\frac{1-\mathrm{e}^{-4\alpha^2\mathrm{e}^{-\kappa t}}}{1-\mathrm{e}^{-4\alpha^2}}}\cdot\sqrt{\frac{\sinh(2\alpha^2\mathrm{e}^{-\kappa t/2})}{\sinh(2\alpha^2)\sinh(2\alpha^2\mathrm{e}^{-\kappa t})}}\tag{168}$$

$$g(t)''=\mathrm{e}^{-\frac{1}{2}\gamma t}\sqrt{\frac{1-\mathrm{e}^{-4\alpha^2\mathrm{e}^{-\kappa t}}}{1-\mathrm{e}^{-4\alpha^2}}}\tag{169}$$

在图3(图中Bell算符的定义分别基于s(_),s'(◆)和s''(★))中将这三种Bell算符进行比较,可以看出当选取式(141)中的Bell算符时其量子态的非定域性更持久些.这些结果为实验提供了重要的参考依据.

① S. Popescu. Phys. Rev. Lett. 1994, 72: 797; R Horodecki, P. Horodecki, and M. Horodecki., Phys. Lett. 1995, A 200: 340.

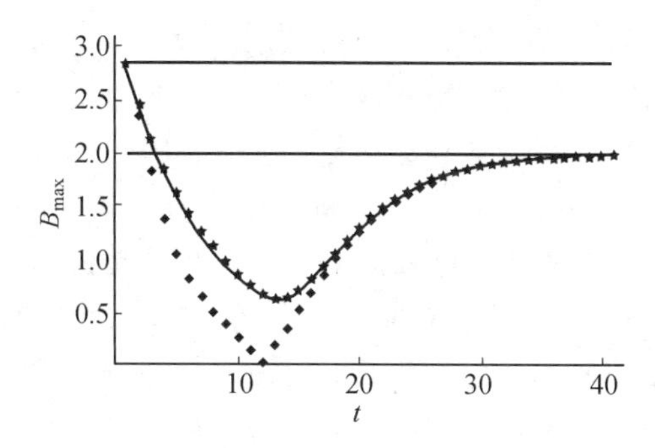

图 3　Bell 函数$\langle B\rangle_{max}$随重新标度的时间 t 的变化曲线

(取$|\alpha| = 3, \gamma = \kappa = 0.2$)

上面我们利用宇旋算符的概念对杂化纠缠态与热库相互作用时的非定域性动力学进行了全面地分析. 通过引入超算符的概念,提出了求解一类主方程(由 $su(2)$ 或 $su(1,1)$ 李代数的超生成元组成的主方程)的一般方法,从而为研究纯态在热库中的演化提供了一条方便的途径.

4. 连续变量的量子计算和量子通讯

量子计算和量子信息是用物理原理 - 量子力学的原理更有效地解决一些数学问题和进行信息处理过程①. 所以,量子计算和量子信息研究的是如何用实实在在的量子系统的物理过程完成一定的计算和信息处理任务.

通常所考虑的是离散的量子系统,即具有分立能级的系统,最简单的是二能级系统,构成了所谓的量子位(qubit). 与经典位相比,单个的量子位是一个二能级系统所有可能的量子状态,这是一个球面,而经典位只是两极. 多出来的叠加态使得并行运算变得可能,即在一次运算过程中,对所有的输入,一次性地计算输出的结果,是在一个叠加态中. 由于量子测量只能得到其中一个结果,而且具有不可预测的随机性,所以我们并不能读出所有的结果,但是可以得到它的某些整体性质,而这不依赖于过程中量子测量的结果.

这里所讨论的是建立在具有连续变量的量子系统之上的量子计算和量子信息. 这是另一个极端,系统具有无穷多个能级. 一个物理系统的连续变量指的是具有连续谱分布的力学量,如位置、动量、能量、电磁场的振

---

① M. A. Neilsen and I. Chung. Quantum Computation and Quantum Information. England: Cambridge University Press, 2000.

幅等. 连续变量存在着通常意义下的正则共轭对,即这两个连续变量的对易子为常数. 由此得到的我们熟知的一些性质,如它们的本征态互为连续的 Fourier 变换,即为一对互补力学量. 好处是利用正则变量的不确定性关系对量子状态是否可分做出判断和对纠缠的定量刻画,如对于两个系统的 Gauss 型的量子态的可分性完全由正则变量之间的不确定性关系所决定,并且纯态的纠缠度完全由正则变量的涨落来决定,以及一对相互对易的正则变量组合的本征态就是最大纠缠态,即一个 EPR 态.

带来的不利之处有两方面:一个是如何解决对连续变量测量的问题,例如位置投影算符$|x\rangle\langle x|$的平方不再是很好定义的量;另一个是在量子计算中,通常要用到对两个系统中的一个取部分迹,而对于连续变量,取迹的运算要十分小心,例如对正则对易关系取迹就会得到类似$0=\infty$这样矛盾的结果. 解决这些矛盾的关键是注意到在实验中,我们对一个连续变量的初态的制备和测量从来不可能达到精确到一个数学上的实数,而总是有一定的误差.

实际上,噪声和有限的精度使得连续变量的量子计算变成离散的,这样如果一个离散变量的量子计算不能完成的任务,用连续变量也无法完成①. 但是连续变量的量子计算会比利用离散变量的量子计算更加有效地完成计算任务. 下面我们选取 Deutsch-Jozsa 算法, 量子隐形传态(quantum teleportation),纠缠交换(entanglement swapping) 和量子纠错来介绍如何用连续变量系统实现量子计算和量子通讯.

4.1 连续变量的 Deutsch - Josza 算法

我们先来看一下 Deutsch - Josza 算法②,这是利用量子叠加态的性质所完成的量子计算中的第一个任务,它所解决的是如下一个数学问题:在一个含有n个比特的离散系统B^n上定义一个二进制的函数$f:B^n\mapsto B$,其中$B=\{0,1\}$. 已知这个函数具有如下两种性质之一,或者是常数型的,即对任意$x,x'\in B^n$我们有$f(x)=f(x')$;或者是平衡型的,即集合$\{x|f(x)=0,x\in B^n\}$与集合$\{x|f(x)=1,x\in B^n\}$具有相同个数的元素. 我们关心的是如何用最少的步骤找出函数$f(x)$的这个整体性质.

经典算法是一个一个计算所有的函数值,如果发现得到的函数值有变化,那么就可以说是平衡型,但有可能检查完了一半的函数值发现仍没有变化,那么就得再往下算一次,才能肯定是哪种类型. 所以在最坏的情

---

① S. Lloyd and S. L. Braunstein. Phys. Rev. Lett. 1999, 82: 1784.

② D. Deutsch and R. Jozsa. Proc. R. Soc. London. 1992, A 439: 553.

况下需要算 $2^{n-1}+1$ 次才能找出来. 而 Deutsch - Josza 算法利用量子态的叠加性质所提供的并行运算,只用一次运算就可找出. 下面我们详细地介绍一下连续变量的 Deutsch - Josza 算法.

连续系统中的数学问题:假设我们有一个定义在实数轴上的二进制函数 $f:R\mapsto B$,并且我们知道这个函数是常数型或者是平衡型,如何用量少的方法找出这个函数的整体性质. 这里,平衡型意味着集合 $\{x\mid f(x)=0,x\in R\}$ 与集合 $\{x\mid f(x)=1,x\in R\}$ 具有相同的测度.

建立一个物理模型:一个一维粒子处于 x 轴上,位置算符记作 $\hat{x}$. 它的 Hilbert 空间由位置的本征态 $|x\rangle$ 张成,正交归一关系为 $\langle x|y\rangle=\delta(x-y)$. 由于一个算法就是一个物理过程,因此我们可以做的是制备初态、幺正变换(包括与这个函数 $f(x)$ 有关的幺正变换)、测量.

在这样一个有连续变量系统上的 Deutsch - Josza 算法如下①. 第一步是制备初态,让这个粒子处于位置的一个本征态上. 实际上,对于一个连续变量来说,由于正则变量之间的不确定性关系,我们没法精确地让一个粒子处于位置的本征态上,而可以做到的是一个 Gauss 波包

$$|\bar{x}_0\rangle=\frac{1}{\sqrt[4]{2\pi\varepsilon}}\int_{-\infty}^{+\infty}\mathrm{d}x\mathrm{e}^{-\frac{(x-x_0)^2}{4\varepsilon}}|x\rangle \tag{170}$$

波包的中心在 x_0 处,宽度 ε 可以视为一个小量,而且可以控制它,让它任意地小. 显然这个态是归一的,而不像位置的本征态归一到 δ 函数,投影 $|\bar{x}_0\rangle\langle\bar{x}_0|$ 可以很好地定义,从而得到有限的几率. 对初态我们要进行 Fourier 变换,产生所有位置状态的一个相干叠加. 这是一个线性变换,对于位置本征态 $|x\rangle$,我们通常定义 Fourier 变换如下

$$\boldsymbol{F}|x\rangle=\frac{1}{\sqrt{2\pi}}\int_{-\infty}^{+\infty}\mathrm{d}y\mathrm{e}^{\mathrm{i}xy}|y\rangle \tag{171}$$

这实际上是动量的本征态. 其逆变换 $\boldsymbol{F}^{-1}=\boldsymbol{F}^{+}$ 为

$$\boldsymbol{F}^{-1}|x\rangle=\frac{1}{\sqrt{2\pi}}\int_{-\infty}^{+\infty}\mathrm{d}y\mathrm{e}^{\mathrm{i}xy}|y\rangle \tag{172}$$

到这一步系统的状态为

$$\begin{aligned}\boldsymbol{F}|\bar{x}_0\rangle&=\frac{1}{\sqrt[4]{2\pi\varepsilon}}\int_{-\infty}^{+\infty}\mathrm{d}x\mathrm{e}^{-\frac{(x-x_0)^2}{4\varepsilon}}\frac{1}{\sqrt{2\pi}}\int_{-\infty}^{+\infty}\mathrm{d}y\mathrm{e}^{\mathrm{i}xy}|y\rangle\\&=\sqrt[4]{2\varepsilon\pi^{-1}}\int_{-\infty}^{+\infty}\mathrm{d}y\mathrm{e}^{-\varepsilon y^2+\mathrm{i}x_0y}|y\rangle\end{aligned} \tag{173}$$

第二步是同时对所有位置的 x 计算函数值 $f(x)$,因为函数的取值只

① A. K. Pati and L. Braunstein. quant-ph/0207108.

有0或1,这可以由幺正变换$(-1)^{f(\hat{x})}$来实现.计算完之后系统的状态为

$$(-1)^{f(\hat{x})}\boldsymbol{F}|\bar{x}_0\rangle=\sqrt[4]{2\varepsilon\pi^{-1}}\int_{-\infty}^{+\infty}\mathrm{d}y\mathrm{e}^{-\varepsilon y^2+\mathrm{i}x_0y}(-1)^{f(y)}|y\rangle \tag{174}$$

第三步是判断这个粒子是否仍处于初态$\boldsymbol{F}|\bar{x}_0\rangle$,即作投影$\boldsymbol{F}|\bar{x}_0\rangle\langle\bar{x}_0|\boldsymbol{F}^\dagger$.根据测量结果就可以判断$f(x)$的性质,因为我们有

$$\begin{aligned}&\langle\bar{x}_0|\boldsymbol{F}^\dagger(-1)^{f(\hat{x})}\boldsymbol{F}|\bar{x}_0\rangle\\&=\sqrt{2\varepsilon\pi^{-1}}\int_{-\infty}^{+\infty}\mathrm{d}wy\mathrm{e}^{-\varepsilon w^2-\mathrm{i}x_0w}\mathrm{e}^{-\varepsilon y^2+\mathrm{i}x_0y}\langle w|(-1)^{f(y)}|y\rangle\\&=\sqrt{2\varepsilon\pi^{-1}}\int_{-\infty}^{+\infty}\mathrm{d}y\mathrm{e}^{-2\varepsilon y^2}(-1)^{f(y)}\end{aligned} \tag{175}$$

如果$f(x)$是常数,那么我们以几率1得到粒子仍处于初态.如果$f(x)$是平衡的,那么当波包的宽度ε趋于0时,我们得到粒子处于初态的几率是0.所以一旦发现粒子不处于初态,就可以断定$f(x)$是平衡型的,而发现粒子处于初态不变则可断定$f(x)$是常数型的.因此我们有一个确定的结果

$$\lim_{\varepsilon\to0}\langle\bar{x}_0|\boldsymbol{F}^\dagger(-1)^{f(\hat{x})}\boldsymbol{F}|\bar{x}_0\rangle=\begin{cases}1 & (f(x)\text{ 是常数型})\\0 & (f(x)\text{ 是平衡型})\end{cases} \tag{176}$$

4.2 连续变量的纠错

信息处理系统要克服的一个重大的问题是噪声.只要有可能我们就完全把噪声排除在外,如果不可能,就尽量使信息处理系统不受噪声的影响.解决办法之一是纠错码,即加入一些冗余的信息.对于经典系统,纠错码简单地说就是重复.这在日常生活中也会遇到,在一个很吵闹的环境中交谈,如果一句话重复几遍,那么对方就能理解得好一些.

噪声的问题在量子信息中更加突出,这里面临着三个问题①:其一是量子状态的不可克隆,这意味着我们不能将量子信息重复;其二是如同量子位是一个连续统,发生在一个量子位上的错误也是一个连续统;其三是量子测量可能会破坏信息.最后一个问题实际上是量子信息中的关键问题,虽然量子叠加态提供了量子并联计算的可能性,但是量子测量只会读出一个结果,而且是随机的.这样在任何的信息处理过程中只要有量子测量,就要使得不论测量得到什么结果,都有相应的方法提取我们感兴趣的信息.

量子计算是用物理系统解决数学问题.我们的物理系统不是孤立的,而是和环境相互作用的,这就带来噪声.从原则上来讲,这些噪声也是一

① S. L. Braunstein. Phys. Rev. Lett. 1998, 80: 4084.

些物理过程,也服从量子力学的规律,只是对其中的一些细节不知道罢了. 因为量子力学中发生的过程是从一个量子状态到另一个量子状态的线性过程,而量子状态是正定的(本征值大于或等于0),所以一个一般的量子力学过程可以用一个正定算符映射(completely positive map) 来描述. 这里完全性是指,如果把这个系统作为一个大系统的子系统,那么对于这个大系统的所有状态而言,这个映射仍然是状态到状态的映射,也就是正的. 也就是说,任何的噪声都可以用一个完全正映射来反映.

并不是所有的正映射都能描述合理的物理过程,如转置映射,只有完全正映射能描述合理的物理过程. 正定而非完全正定的映射会将一些纠缠态变成非正定的,反过来这提供了判定状态是否可分的一个方法. 而所有保迹的完全正定映射 $\varepsilon(\rho)$ 都有如下的幺正表示

$$\varepsilon(\rho) = \mathrm{tr}_{\text{环境}}(U\rho\rho_{\text{环境}}U^{\dagger}) \tag{177}$$

这反映了如下的物理过程:在一个包括我们感兴趣的系统和一个环境的大系统中,从一个可分的状态出发,经过一个幺正演化,一段时间后对环境求迹,得到原来系统的状态. 我们看到这个映射是保迹的且 $\mathrm{tr}\,\varepsilon(\rho)=1$.

选取环境的一个完备基 $|i\rangle$,不失一般性假设开始环境处于 $|0\rangle$,将以上对环境的求迹展开得到完全正映射的算符和表示(operator sum representation)

$$\varepsilon(\rho) = \sum_i A_i\rho A_i^{\dagger}, A_i = \langle i|U|0\rangle_{\text{环境}} \tag{178}$$

完全正映射的算符和表示在如下意义下是唯一的,即它的任何一个算符和表示 $\{B_i\}$ 都可以表示成 $B_i = \sum_j u_{ij}A_j$,其中 $u = \{u_{ij}\}$ 构成一个幺正矩阵. 保迹性要求 $\sum_i A_i^{\dagger}A_i = 1$. 反之也成立,从完全正映射的算符和形式也可以构造一个幺正表示①. 即保迹的完全正映射的这两种形式是等价的. 如果完全正映射不保迹,那么可以有算符和表示,但没有幺正表示. 不是完全正的正映射可以表示成两个完全正映射的差②.

噪声是用完全正映射来表示,一般是不保迹的,不是一个可逆的过程. 然而在特定情况下,却存在一个子空间,对这个子空间的任何态来讲,代表噪声的正定映射是可逆的! 这里可逆是指存在另一个保迹的完全正映射 R 使得 $R\circ\varepsilon(\rho) \propto \rho$. 我们有如下定理.

定理 完全正定映射 $\varepsilon = \{A_i\}$ 在一个子空间中可逆的充要条件是对

---

① B. Schumacher Phys. Rev. 1996, A 54: 2614.

② S. Yu. Phys. Rev. 2000, A 62: 024302.

于任何 i,j 有

$$PA_i^\dagger A_j P = \alpha_{ij} P \tag{179}$$

其中,P 是向这个子空间的投影算符,而 $\{\alpha_{ij}\}$ 构成一个复系数的正定厄米矩阵 $\boldsymbol{\alpha}$. 如果 R 是 ε 的一个逆,我们有 $R\circ\varepsilon(\rho)=\mathrm{tr}\,\boldsymbol{\alpha}\rho$.

证明如下. 首先证明必要条件,如果 $\varepsilon=\{A_i \mid i\in I\}$ 在子空间 P 上存在保迹的逆映射 $R=\{B_i \mid i\in J\}$,那么我们有 $R\circ\varepsilon(\rho)=cP\rho P$,其中 c 为任意一个大于零的实数. 因此我们有复合映射 $R\circ\varepsilon$ 的两个算符和表示 $\{B_jA_i \mid i\in I, j\in J\}$ 和 $\{\sqrt{c}P\}$,从完全正映射算符和表示的唯一性和 $B_jA_iP=c_{ji}P$,其中 $\sum_{ij} c_{ij}c_{ij}^*=C$. 所以 $PA_j^\dagger A_iP=\alpha_{ji}P$,其中 $\alpha_{ji}=\sum_k c_{kj}^* c_{ki}$ 并且用到 $\sum_j B_j^\dagger B_j=1$,显然有 $c=\mathrm{tr}\,\boldsymbol{\alpha}$.

其次证明充分条件. 只用从条件式(178)出发构造一个保迹的完全正映射 $R=\{B_i \mid i\in J\}$. 因为 $\boldsymbol{\alpha}$ 厄米存在幺正 u 使得 $\boldsymbol{\alpha}$ 对角化. 令 $F_j=\sum_i u_{ji}A_i$,于是 $PF_i^\dagger F_jP=d_i\delta_{ij}P$,其中 $\{d_k \mid k\in J'\}$ 为 $\boldsymbol{\alpha}$ 大于0的本征值. 将 F_iP 分解 $F_iP=\sqrt{d_i}\,U_iP$,则定义完全正映射 $R'=\{B_j \mid j\in J'\}$ 的算符和表示为 $B_j=U_j^\dagger P_j=PU_j^\dagger$,这里投影 $P_j=U_jPU_j^\dagger$ 相互正交. 可以验证

$$\begin{aligned}R'\circ\varepsilon(\rho) &= \sum_{i\in I, j\in J'} B_jF_i\rho F_i^\dagger B_j^\dagger = \sum_{i\in I, j\in J'} PU_j^\dagger F_i\rho F_i^\dagger U_jP \\ &= \sum_{i\in I, j\in J'} \frac{1}{d_j} PF_j^\dagger F_iP\rho PF_i^\dagger F_jP \\ &= \sum_{j\in J'} d_jP\rho P=\mathrm{tr}\,\boldsymbol{\alpha}\rho\end{aligned} \tag{180}$$

此外,我们有 $\sum_j B_j^\dagger B_j=\sum_j P_j$. 令 $R=\{B, B_j \mid j\in J'\}$,其中 $B_0=1-\sum_j P_j$. 由于单位算符显然在任何错误中有 $B\rho B=0$,因此 $R\circ\varepsilon=R'\circ\varepsilon$,并且 R 是保迹的.

从以上构造逆映射的过程我们也可以看出实际的操作过程,假设我们有一个可逆子空间,那么我们将信息编在这个空间中,也就是将物理的状态映射到这个子空间中,当发生噪声 ε 时都可以将原来的状态还原如下:首先作一个测量,是向 P_j 投影,根据测量结果作一个幺正变换 U_j,就可以得到原来的状态. 我们注意到不论测量结果如何,都可以得到正确的状态是因为幺正变换 U_j 与状态无关.

量子纠错码的任务就是对于特定的噪声找出满足上述条件的子空间 —— 编码空间. 当然,错误越普遍,编码空间就越小. 下面我们讨论由正映射 $\varepsilon=\{E_{1i}, E_{2i}, E_{3i}, \cdots\}$ 描述的噪声,其中

$$E_{1i}=A_i\otimes I\otimes I\otimes\cdots,\quad E_{2i}=I\otimes A_i\otimes I\otimes\cdots,\quad E_{3i}=I\otimes I\otimes A_i\otimes\cdots$$

依此类推.

这意味着假设我们有多个系统,错误只会影响其中之一,但是哪一个就不知道了,而且对这个系统而言是最一般的错误. 对于这一类噪声,最少要用5个系统才可以找到编码空间①. 下面就连续变量来具体构造这样一个编码空间.

假设我们有5个连续变量的系统, 它们的正则变量为 $x_m, p_m(m = 1,\cdots,5)$, 满足通常的正则对易关系. 系统的 Hilbert 空间为 $H^{\otimes 5}$, 由粒子位置的本征态张成. 我们定义一个子空间由以下的基张成

$$|x_{码}\rangle = \frac{1}{(2\pi)^{3/2}}\int d(wyz)e^{iwy+ixz}|z\rangle_1|y+x\rangle_2 \cdot |w+x\rangle_3|w-z\rangle_4|y-z\rangle_5 \tag{181}$$

这里所有的基都是位置的本征态. 可以证明对于不同的 m,n,我们有

$$\langle x_{码}|e^{-iX\hat{p}_m}e^{-iP\hat{x}_m}e^{-iY\hat{p}_n}e^{-iQ\hat{x}_n}|x'_{码}\rangle = \delta(x-x')\delta(X)\delta(P)\delta(Y)\delta(Q) \tag{182}$$

而

$$\langle x_{码}|e^{-iX\hat{p}_m}e^{-iP\hat{x}_m}|x'_{码}\rangle = \delta(x-x')\delta(X)\delta(P)C^2 \tag{183}$$

这里 C^2 是一个形式上的无穷大. 如果考虑到产生编码态时是用 Gauss 波包,最后令波包的宽度趋于0,我们可以令 $C^2 = 1$.

任意一个算符 $\hat{A}$ 都可以作如下展开

$$\hat{A} = \int d(XP)A(X,P)e^{-iX\hat{p}}e^{-iP\hat{x}},A(X,P) = \mathrm{tr}(\hat{A}e^{iX\hat{p}}e^{iP\hat{x}}) \tag{184}$$

显然有 $\mathrm{tr}\,\hat{A} = A(0,0)$. 对于发生在不同空间的错误 $m \neq n$,我们有

$$\begin{aligned}
&\langle x_{码}|E^{\dagger}_{mi}E_{nj}|x'_{码}\rangle \\
&= \int d(XYPQ)A^*_i(X,P)A_j(Y,Q)\langle x_{码}|e^{-iX\hat{p}_m}e^{-iP\hat{x}_m}e^{-iY\hat{p}_n}e^{-iQ\hat{x}_n}|x'_{码}\rangle \\
&= \delta(x-x')\int d(XYPQ)A^*_i(X,P)A_j(Y,Q)\delta(X)\delta(P)\delta(Y)\delta(Q) \\
&= \delta(x-x')\,\mathrm{tr}\,A^{\dagger}_i\mathrm{tr}\,A_j
\end{aligned} \tag{185}$$

对于同一空间发生的错误,我们有

$$\begin{aligned}
\langle x_{码}|E^{\dagger}_{mi}E_{mj}|x'_{码}\rangle &= \int d(XP)A_{ij}(X,P)\langle x_{码}|e^{-iX\hat{p}_m}e^{-iP\hat{x}_m}|x'_{码}\rangle \\
&= \delta(x-x')\int d(XY)A_{ij}(X,P)\delta(X)\delta(P) \\
&= \delta(x-x')\,\mathrm{tr}\,(A^{\dagger}_iA_j)
\end{aligned} \tag{186}$$

① H. F. Chau. Phys. Rev. 1997, A 56: R1.

至此已验证$|x_{码}\rangle$确实张成一个可逆子空间. 那么所有1维的状态都可以用这个5维空间的子空间表示出来,而这个空间的状态不怕任何发生在这5个系统上的任何一个系统上的错误,因为都可以纠正过来.

从前面的分析可以看出,用正则变量来描述量子信息处理的过程,我们可以同时描述不同能级系统上的量子计算,包括二能级系统和连续变量系统. 特别是在纠错码的构造 ,利用这种正则对应,可以把已有的离散系统的编码翻译成连续变量的情况. 在量子通讯中正则变量的作用可以看得更加明显. 下面我们以量子隐形传态和纠缠交换为例来说明.

4.3 连续变量的隐形传态

任意一个未知的量子态是不能复制的,这样使得量子态及其上携带的信息的移动性就会很差. 假如处于特定状态的物理系统不便于移动,那么其上的量子态就不能从一个地方转移到另一个地方. 而量子隐形传态①正可以解决这个问题,通过这个过程,我们可以将任意一个未知的量子状态从一个地方转移到另一个地方,从一个系统到另一个系统,而不需要移动量子态的载体. 在实验上用离散②和连续变量③的量子隐形传态都已分别实现.

一般来讲,实现量子隐形传态的过程涉及两地三个系统,有如下三个步骤:第一步,在A地有两个系统,其中之一处于我们要转移的未知态,而另一个系统与在B地的第三个系统处于一个最大纠缠态,建立起所谓的量子通道. 第二步,我们对处于A地的两个系统作一个量子测量,测量的力学量要求它的所有本征态都是这两个系统之间的最大纠缠态,对二能级系统,这就是一个Bell基的测量,测量结果通过适合的经典通道传到B地. 第三步,B地的实验者根据A地传过来的测量结果,对当地的量子系统作一个相应的幺正变换,就会得到那个未知态. 原来具有未知状态的系统在测量后处于一个纠缠态,这并没有违反不可克隆定理.

在以上的三个过程中,我们发现正则变量起着至关重要的作用,在连续变量的情况下更加明显. 下面我们考虑三个由连续变量描述的系统,它

---

① C. H. Bennett, G. Brassard, C. Crepeau, R. Jozsa, A. Peres, and W. K. Wootters. Phys. Rev. Lett. 1993, 70: 1895.

② D. Boumeester, J. Pan, K. Mattle, M. Eibl, H. Weinfurter, and A. Zeilinger. Nature (London), 1997, 309: 575.

③ A. Furasawa, J. L. Sorensen, S. L. Braunstein, C. A. Fuchs, H. J. Kimble, and E. S. Polzik. Scince, 1998, 282: 706.

们的基本正则变量位置 $\hat{x}_i$ 和动量 $\hat{p}_i(i=1,2,3)$ 有连续的谱分布,并且满足正则对易关系 $[\hat{x}_i,\hat{p}_j]=i\delta_{ij}$,这里已经使得位置和动量无量纲化.

我们假设系统 1 和系统 2 处于这两个系统的一个最大纠缠态——EPR 态,位置差 $\hat{x}_1-\hat{x}_2$ 与动量和 $\hat{p}_1+\hat{p}_2$ 的共同本征态为

$$|x_{12};p_{12}\rangle_{12}=\mathrm{e}^{-\mathrm{i}\hat{p}_1\hat{x}_2}|x_{12}\rangle_1\otimes|p_{12}\rangle_2 \tag{187}$$

这里 $|x_{12}\rangle_1$ 是系统 1 位置 $\hat{x}_1$ 的本征态,本征值为 x_{12},而 $|p_{12}\rangle$ 是系统 2 动量 $\hat{p}_2$ 的本征态,本征值为 p_{12}. 这两个系统被分别送到 A 地和 B 地,建立起一个量子通道,从而我们将把处于 B 地的系统 3 的一个未知量子态 $|\psi\rangle_3$ 转移到 A 地的系统 1 上. 为实现这一目的,我们只用在 B 地测量系统 2 和系统 3 的一对相互对易的正则变量即可,这里我们仍选它们的位置差 $\hat{x}_2-\hat{x}_3$ 与动量和 $\hat{p}_2+\hat{p}_3$. 测量后的系统 3 和系统 2 处于这两个正则变量共同的本征态

$$|x_{23};p_{23}\rangle_{23}=\mathrm{e}^{-\mathrm{i}\hat{p}_2\hat{x}_3}|x_{23}\rangle_2\otimes|p_{23}\rangle_3 \tag{188}$$

这里 $|x_{23}\rangle_2$ 是本征值为 x_{23} 系统 2 位置 $\hat{x}_2$ 的本征态,$|p_{23}\rangle$ 是系统 3 动量 $\hat{p}_3$ 本征值为 p_{23} 的本征态. 而系统 1 在测量之后处于如下状态

$$\begin{aligned}
&(\langle x_{23};p_{23}|_{23})|x_{12};p_{12}\rangle_{12}\otimes|\psi\rangle_3\\
&=(\langle x_{23}|_2\otimes\langle p_{23}|_3\mathrm{e}^{-\mathrm{i}\hat{p}_2\hat{x}_3})\mathrm{e}^{-\mathrm{i}\hat{p}_1\hat{x}_2}|x_{12}\rangle_1\otimes\\
&\quad|p_{12}\rangle_2\otimes\int\mathrm{d}x|x\rangle\langle x|\psi\rangle_3\\
&=\int\mathrm{d}x\langle x_{23}+x|p_{12}\rangle_2\langle p_{23}|x\rangle_3\langle x|\psi\rangle_3|x+x_{12}+x_{23}\rangle_1\\
&=\frac{1}{2\pi}\mathrm{e}^{\mathrm{i}p_{12}x_{23}}\mathrm{e}^{-\mathrm{i}\hat{p}_1(x_{12}+x_{23})}\int\mathrm{d}x\mathrm{e}^{\mathrm{i}x(p_{12}-p_{23})}\langle x|\psi\rangle|x\rangle_1\\
&=\frac{1}{2\pi}\mathrm{e}^{\mathrm{i}p_{12}x_{23}}\mathrm{e}^{-\mathrm{i}\hat{p}_1(x_{12}+x_{23})}\mathrm{e}^{\mathrm{i}\hat{x}_1(p_{12}-p_{23})}\int\mathrm{d}x|x\rangle\langle x|\psi\rangle_1\\
&=\frac{1}{2\pi}\mathrm{e}^{\mathrm{i}p_{12}x_{23}}\mathrm{e}^{-\mathrm{i}\hat{p}_1(x_{12}+x_{23})}\mathrm{e}^{\mathrm{i}\hat{x}_1(p_{12}-p_{23})}|\psi\rangle_1\\
&=\frac{1}{2\pi}\mathrm{e}^{\mathrm{i}p_{12}x_{23}}\mathscr{O}_c|\psi\rangle_1
\end{aligned} \tag{189}$$

我们看到,归一化后,系统 1 的状态为 $\mathscr{O}_c|\psi\rangle$,与未知状态差一个幺正变换 $\mathscr{O}_c$. 这个幺正变换只与测量系统 2 和系统 3 的结果和最开始的纠缠态有关,与未知的状态没有关系,所以对系统 1 根据系统 2 和系统 3 的测量结果实施幺正变换 $\mathscr{O}_c^\dagger$ 系统 1 就处于未知的状态. 而且它的物理意义也很明显,$x_{12}+x_{23}$ 意味着系统 2 和系统 3 的位置差,$p_{12}-p_{23}$ 是系统 2 和系统 3 的动量差,幺正变换 $\mathscr{O}_c^\dagger$ 正好是补偿了这些差异.

总之,我们通过以上的分析可以看到,正则变量在量子隐形传态过程中起着三重作用. 首先,量子通道的建立是产生两个系统之间的一个最大

纠缠态,而这是这两个系统一对相互对易的正则变量共同的本征态;其次,第二步的量子测量的力学量正好是相关两个系统的一对相互对易的正则变量;最后,用来补偿的幺正变换是由这个系统的正则变量生成的平移.

表面上看起来,对离散系统不存在正则变量,但实际上如果考虑Weyl对易关系,即把变量放到指数上去,然后看这些变换之间的对易关系,我们就会发现,离散系统的粒子数和位相算符之间的Weyl对易关系也是一个常数.在这种意义下,粒子数和位相算符也是一对正则变量.那么,对系统之间一对相互对易的正则变量就可以是粒子数差与位相和,或者是粒子数和与位相差,在二能级的情况下,它们的本征态正好是Bell基.所以,以上结论对离散变量的QT也成立.并且,QT过程中的每一步物理意义也清楚了,特别是最后一步用来还原初态的幺正算符,正好补偿了粒子数差和位相差①.

以上讨论的是理想情况下的量子隐形传态,由于这里不涉及无穷大的几率,所以当考虑到最大纠缠态不是EPR态而是有一定的宽度的波包时,所得到的结果将会趋近于理想的情况,当波包的宽度趋向于0时.有限精度的测量也一样,会影响到最后态的保真度,而当波包的宽度趋向于0时,这种影响也会消失.

4.4 连续变量的纠缠交换

不仅态可以远程传送,纠缠也可以远程传送,这就是交换②,也可称之为纠缠的量子隐形传态.纠缠态反映的是两个系统之间的量子关联,人们很容易认为如果两个系统没有直接的相互作用就不能产生纠缠态.然而用交换的方法,可以不通过直接相互作用就建立起两个系统的纠缠态,不论它们相距多远.这也是在实验中得到验证的一个现象③,下面就连续变量来详细看一看交换的具体过程.

假设我们有4个连续变量的量子系统,它们的基本正则变量分别为位置 $\hat{x}_i$ 和动量 $\hat{p}_i(i=1,2,3,4)$,它们满足通常的正则对易关系.假设在两个地方A和B分别是有两个系统,在A地的系统是1和2,在B地的系统是3和4.在异地的两个系统1和3,系统2和4分别处于最大纠缠态,这里也假设是EPR态,它们位置和与动量差有共同的本征态,这样整个系统的状

① S. Yu and C. Sun. Phys. Rev. 2000, A 61: 022310.

② M. Zukowski, A. Zeilinger, M. A. Horne, and A. K. Ekert. Phys. Rev. Lett. 1993, 71: 4287.

③ J. W. Pan, D. Bouwmeester, H. Weinfurter, and A. Zeilinger. Phys. Rev. Lett. 1998, 80: 3891.

态为

$$|x_{13},p_{13}\rangle_{13}|x_{24},p_{24}\rangle_{24}=\mathrm{e}^{\mathrm{i}\hat{p}_1\hat{x}_3}\mathrm{e}^{\mathrm{i}\hat{p}_2\hat{x}_4}|x_{13}\rangle_1|x_{24}\rangle_2|p_{13}\rangle_3|p_{24}\rangle_4 \tag{190}$$

像以前一样,这里$|x_{13}\rangle_1|x_{24}\rangle_2$分别为系统1和2的位置的本征态,而$|p_{13}\rangle_3|p_{24}\rangle_4$分别为系统3和4动量的本征态. 虽然同在A地的1和2两个系统没有任何的相互作用,但是我们可以通过对在B地的系统3和4的量子测量来建立起它们之间的纠缠. 这个力学量我们也选取系统3和4的位置差和动量和,这样在测量后系统3和4处于态$|x_{34},p_{34}\rangle_{34}$,而系统1和2则处于如下的状态

$$\begin{aligned}
&(\langle x_{34},p_{34}|_{34})|x_{13},p_{13}\rangle_{13}|x_{24},p_{24}\rangle_{24}\\
&=\langle x_{34}|_3\langle p_{34}|_4\mathrm{e}^{\mathrm{i}\hat{p}_3\hat{x}_4}\mathrm{e}^{-\mathrm{i}\hat{p}_1\hat{x}_3}\mathrm{e}^{-\mathrm{i}\hat{p}_2\hat{x}_4}\cdot\\
&\quad|x_{13}\rangle_1|x_{24}\rangle_2|p_{13}\rangle_3\int\mathrm{d}x|x\rangle\langle x|p_{24}\rangle_4\\
&=\int\mathrm{d}x\langle x_{34}+x|p_{13}\rangle_3\langle p_{34}|x\rangle_4\cdot\\
&\quad\langle x|p_{24}\rangle_4|x_{13}+x_{34}+x\rangle_1\mathrm{e}^{-\mathrm{i}\hat{p}_2x}|x_{24}\rangle_2\\
&=\frac{1}{2\pi}\mathrm{e}^{\mathrm{i}x_{34}p_{13}}\mathrm{e}^{\mathrm{i}\hat{p}_1(x_{13}+x_{34})}\mathrm{e}^{-\mathrm{i}\hat{x}_1(p_{13}-p_{34})}\mathrm{e}^{-\mathrm{i}\hat{p}_2\hat{x}_1}|p_{24}\rangle_1|x_{24}\rangle_2\\
&=\frac{1}{2\pi}\mathrm{e}^{-\mathrm{i}x_{13}p_{13}}\mathrm{e}^{\mathrm{i}(p_{14}+p_{34})(x_{13}+x_{34})}\mathrm{e}^{-\mathrm{i}\hat{p}_2\hat{x}_1}|p_{21}\rangle_1|x_{21}\rangle_2\\
&\equiv\frac{1}{2\pi}\mathrm{e}^{-\mathrm{i}x_{13}p_{13}}\mathrm{e}^{\mathrm{i}(p_{14}+p_{34})(x_{13}+x_{34})}|x_{21},p_{21}\rangle_{12}
\end{aligned}\tag{191}$$

我们看到这正是系统1和2之间的一个最大纠缠态,即系统1和2位置差$\hat{x}_2-\hat{x}_1$与动量和$\hat{p}_1+\hat{p}_2$的本征态,相应的本征值为

$$x_{21}=x_{24}-x_{13}-x_{34}\text{ 和 }p_{21}=p_{24}-p_{13}+p_{34}$$

正如我们所期待的那样. 简言之,实现如下的变换

$$|x_{13},p_{13}\rangle_{13}|x_{24},p_{24}\rangle_{24}\mapsto|x_{21},p_{21}\rangle_{12}|x_{34},p_{34}\rangle_{34} \tag{192}$$

位置和动量之间满足$x_{21}+x_{34}=x_{24}-x_{13}$和$p_{12}-p_{34}=p_{13}-p_{24}$,就像系统2和3对换一样.

5 结束语

我们就连续变量系统的纠缠、量子非定域性及其动力学、量子计算和量子通讯等问题进行了一些讨论. 关于连续变量系统的纠缠,我们介绍了原始的EPR佯谬(或EPR论证)、连续变量纠缠态的产生、两体连续变量量子态的可分离判据和纠缠度量、Bures保真度的计算. 我们发现,线性量

子变换理论对研究连续变量系统的很多问题是一种方便而有效的方法.随后,我们讨论了连续变量系统的量子非定域性,分别介绍了相空间形式和宇旋算符形式的连续变量系统的Bell不等式,并研究了杂化纠缠态的量子非定域性的动力学演化.最后,在连续变量的量子计算和量子通讯方面,我们主要就连续变量的Deutsch - Josza算法、连续变量的纠错、连续变量的隐形传态和连续变理的纠缠交换分别做了介绍.

应该指出,连续变量量子信息处理是整个量子信息论的一个热点,其内涵极为丰富,对相关问题的认识也在不断地深化中.因此,有很多相关的重要问题并未涉及.这些问题包括连续变量的纠缠纯化①②、连续变量的量子克隆③④⑤⑥⑦、连续变量的密集编码⑧⑨、连续变量的量子密码⑩⑪⑫⑬⑭以及连续变量的量子计算的某些方面⑮,等等.随着研究工作的深入,我们相信在连续变量系统方面一定还会有更多的新发现.

密码学曾是数论中一个古老的分支,近年来随着全球经济的飞速发展又重新焕发了生机,其中"基于连续变量的真实量子签名"⑯就是一个亮点.数学家们研究了仲裁量子签名,由于仲裁的直接参与,这种签名算法在实际应用中将受到一些限制.从而有人研究了真实量子签名算法,这种签名算法在算法的执行过程中不需要仲裁参与,只有签字者和验证者之间产生了纠纷后,仲裁才作为公证系统参与裁决.

---

① J. Eisert, S. Scheel, and M. B. Plenio. Phys. Rev. Lett. 2002, 89:137903.

② J. Fiurášek. Phys. Rev. Lett. 2002, 89: 137904.

③ N. J. Cerf, A. Ipe, and X. Rottenberg. Phys. Rev. Lett. 2000, 85: 1754.

④ G. M. D'Ariano, F. De Martini, and M. F. Sacchi, Phys. Rev. Lett. 2001, 86: 914.

⑤ S. L. Braunstein et al.. Phys. Rev. Lett. 2001, 86: 4938.

⑥ J. Fiurášek. Phys. Rev. Lett. 2001, 86: 4942.

⑦ P. van Loock and S. L. Braunstein. Phys. Rev. Lett. 2001, 87: 247901.

⑧ S. L. Braunstein and H. J. Kimble. Phys. Rev. 2000, A 61: 042302.

⑨ Xiaoying Li, Qing Pan, Jietai Jing, Jing Zhang, Changde Xie, and Kunchi Peng, Phys. Rev. Lett. 2002, 88: 047904;彭堃墀,利用明亮EPR光束实现量子密集编码和量子离物传态;曾谨言,裴寿镛,龙桂鲁主编.量子力学新进展(第二辑).北京:北京大学出版社,2001.

⑩ M. Hillery. Phys. Rev. 2000, A 61: 022309.

⑪ M. D. Reid. Phys. Rev. 2000, A 62: 062308.

⑫ N. J. Cerf, M. Lévy, and G. Van Assche. Phys. Rev. 2001, A 63: 052311.

⑬ F. Grosshans and P. Grangier. Phys. Rev. Lett. 2002, 88: 057902.

⑭ Ch. Silberhorn, N. Korolkova, and G. Leuchs. Phys. Rev. Lett. 2002, 88, 167902.

⑮ S. D. Bartlett and B. C. Sanders, quant-ph/0110039.

⑯ 摘自《量子密码学》,曾贵华著,科学出版社,2006.

1. 算法结构描述

基于严格单向函数 G 的性质,曾贵华教授设计了一个基于量子公钥体制的真实量子签名算法. 算法包括初始阶段、签名阶段和验证阶段. 物理上,算法以连续变量量子比特的属性作为算法的物理实现基础.

1.1 初始阶段

在本算法中,初始阶段主要考虑如何获取用于签字的私钥 K_s 和用于验证签字的公钥 K_p. 为了构造密钥对 $\{K_s, K_p\}$,首先构造一个变换 G

$$G: \{\boldsymbol{x}, \boldsymbol{L}, P_{ij}\} \rightarrow \{\boldsymbol{U}, \|\boldsymbol{P}\|^{1/2}\} \tag{1}$$

其中,$\boldsymbol{x} \in \mathbf{R}^{\otimes k}$ 为 k 维实空间的矢量,即 $\boldsymbol{x} = \{x_0, x_2, \cdots, x_{k-1}\}$,$\boldsymbol{L}: \boldsymbol{x} \rightarrow y(\boldsymbol{x})$ 为一线性变换,$y(\boldsymbol{x}) = \{x_0, y_1(\boldsymbol{x}), \cdots, y_{2k-1}(\boldsymbol{x})\}$ 且 $y(\boldsymbol{x})$ 中任意 k 个矢量线性独立. 变换 G 具有下面的性质.

定理 1 满足式(1) 的变换 G 是非线性变换,且为单向映射. 其中对线性变换 $\boldsymbol{L}: \boldsymbol{x} \rightarrow y(\boldsymbol{x})$ 的约束是 $y(\boldsymbol{x})$ 中任意 k 个矢量是线性独立的.

证明 易知 $\boldsymbol{U}$ 是 $\boldsymbol{P} \times \boldsymbol{x}$ 的函数,因此 G 是非线性变换. 下面证明 G 为单向映射. 若给定 $\boldsymbol{x}$ 和 $\boldsymbol{L}$,则 $y(\boldsymbol{x})$ 可求出,从而 $\boldsymbol{P}$ 可求出,还可求出 $\boldsymbol{U}$. 相反地,若给定 U_{ij},由于 U_{ij} 是 x_i,P_{ij} 和 $\mathscr{P}$ 的函数,即

$$U_{ij} = g(x_i, P_{ij}, \mathscr{P}) \tag{2}$$

x_i 和 P_{ij} 不可唯一地解出. 证毕.

由定理 1 可知,实现变换 G 是容易的,而实现逆变换 G^{-1} 是不可能的. 根据这个特点,可构造出如下的公钥和私钥对

$$K_p = \boldsymbol{U}(\boldsymbol{P}) \tag{3}$$

$$K_s = \{\boldsymbol{L}, x_i \mid i = 1, 2, \cdots, k\} \tag{4}$$

式(3) 中 $\boldsymbol{U}$ 为幺正变换,消息拥有者 Alice 利用她的私钥 K_s 对消息进行编码,从而实现对消息的签字,然后利用公钥可实现对编码后的量子比特解码,在此基础上可实现对签字消息的验证.

1.2 签名算法

待签消息为非纠缠态 $|m\rangle$,量子比特 $|m\rangle$ 的状态可以是已知的也可以是未知的. 根据私钥 K_s 构造一个满足定理 1 中条件的线性映射 $\boldsymbol{L}$. 随机选取一个任意 k 维矢量 $\boldsymbol{x}$,计算出对应的 $2k$ 维矢量 $y(\boldsymbol{x})$. 根据矢量 $y(\boldsymbol{x})$ 的元素 $\{y_1(\boldsymbol{x}), \cdots, y_{2k-1}(\boldsymbol{x})\}$ 制备量子比特

$$|\Omega(\boldsymbol{x})\rangle = \{|y_1(\boldsymbol{x})\rangle, \cdots, |y_{2k-1}(\boldsymbol{x})\rangle\}$$

并利用这些量子比特对待签消息$|m\rangle$编码,得到待签消息$|m\rangle$的编码态$|\tilde{S}\rangle$

$$K_s|\psi\rangle \rightarrow |\tilde{S}\rangle = \int_{\mathbf{R}^k} |m\rangle |y_1(\boldsymbol{x})\rangle_1, \cdots, |y_{2k-1}(\boldsymbol{x})\rangle_{2k-1} \mathrm{d}^k \boldsymbol{x} \tag{5}$$

上述过程通过量子编码将$2k$个独立的量子比特编码成为一个具有$2k$粒子的纠缠比特.量子编码过程的实现与量子纠错码中的编码方式相似,物理上可采用相似的方法处理.图1是式(5)对应的量子编码的示意图,图中量子编码器针对连续量子变量,物理上可采用线性光学器件(如光分束器等)实现.因此上述的编码过程利用当前的技术即可实现.

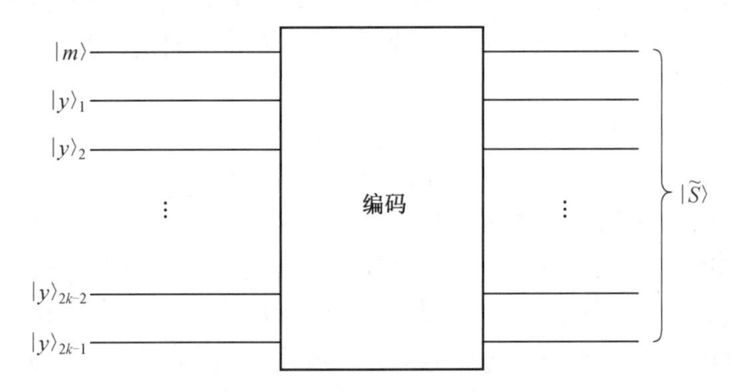

图1　消息$|m\rangle$的编码过程

利用与私钥关联的两个量子态$|y_{r_k+1}\rangle_{r_2}$和$|y_{r_k+1}\rangle_{r_k+1}$,签字者制备一个两粒子纠缠态

$$|\tilde{\Omega}\rangle = \int_{\mathbf{R}} |y_{r_k+1}\rangle_{r_2} |y_{r_k+1}\rangle_{r_k+1} \mathrm{d}\boldsymbol{x} \tag{6}$$

并产生签字量子比特

$$|S\rangle = |\tilde{\Omega}\rangle \otimes |\tilde{S}\rangle \tag{7}$$

将签字$|S\rangle$和待签消息$|m\rangle$一并发送给验证者,签字者完成对消息的签名.式(5)给出的签字$|S\rangle$是一个未知量子态,根据量子不可克隆定理,除签字者外任何人不可能得到签名$|S\rangle$的具体状态.

1.3　验证算法

验证者得到消息$|m\rangle$和签名$|S\rangle$后,使用公钥K_p,即幺正变换$\boldsymbol{U}$对$|\tilde{S}\rangle$态进行解码.参照式(17),得到

$$K_p|S\rangle \rightarrow \boldsymbol{U}|\tilde{S}\rangle = |P\rangle \otimes (J\|\boldsymbol{P}\|^{1/2}|\Theta\rangle_{2,k+1}|\Theta\rangle_{3,k+2}\cdots|\Theta\rangle_{2,2k-1}) \tag{8}$$

其中

$$|\Theta\rangle_{i,j} = \int_{\mathbf{R}} |x_i\rangle |x_j\rangle \mathrm{d}\boldsymbol{x}$$

式(8)表明可解码出消息,考虑到可能存在伪造,记解码得出的消息态为$|m'\rangle$. 若无伪造和欺骗,应有下面的结论

$$|m\rangle = |m'\rangle \tag{9}$$

式(9)表明,如何比较解码得到的消息态$|m'\rangle$和原始消息态$|m\rangle$成为验证签字合法性的关键,可采用图2所示的方法进行比较.

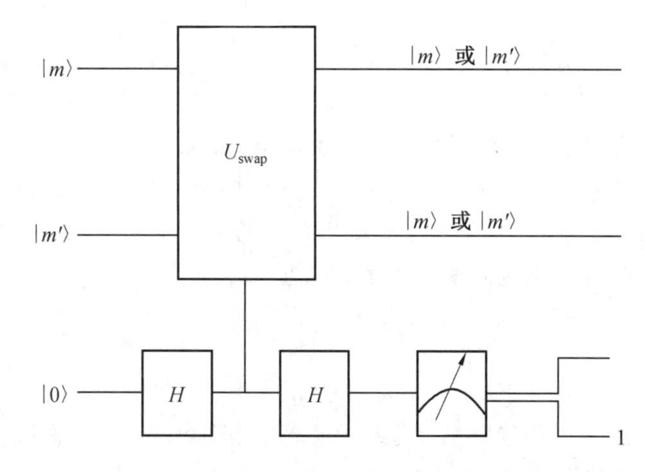

图2 两个任意未知量子比特的比较

$$\begin{aligned}
&(\boldsymbol{H}\otimes\boldsymbol{I})\boldsymbol{U}_{\mathrm{swap}}(\boldsymbol{H}\otimes\boldsymbol{I})|0\rangle|\psi\rangle|\psi'\rangle\\
&=\frac{1}{\sqrt{2}}(\boldsymbol{H}\otimes\boldsymbol{I})\boldsymbol{U}_{\mathrm{swap}}(|0\rangle|\psi\rangle|\psi'\rangle+|1\rangle|\psi\rangle|\psi'\rangle)\\
&=\frac{1}{\sqrt{2}}(\boldsymbol{H}\otimes\boldsymbol{I})\boldsymbol{U}_{\mathrm{swap}}(|0\rangle|\psi\rangle|\psi'\rangle+|1\rangle|\psi'\rangle|\psi\rangle)\\
&=\frac{1}{2}(|0\rangle(|\psi\rangle|\psi'\rangle+|\psi'\rangle|\psi\rangle)+\\
&\quad|1\rangle(|\psi\rangle|\psi'\rangle-|\psi'\rangle|\psi\rangle))
\end{aligned} \tag{10}$$

推导中使用了下面的关系式

$$\boldsymbol{U}_{\mathrm{swap}}|0\rangle|\psi\rangle|\psi'\rangle \to |0\rangle|\psi\rangle|\psi'\rangle$$
$$\boldsymbol{U}_{\mathrm{swap}}|1\rangle|\psi\rangle|\psi'\rangle \to |1\rangle|\psi'\rangle|\psi\rangle$$

由式(10)可知,当图2中测量结果为1时,必有$|\psi\rangle \neq |\psi'\rangle$;当测量结果为0时,式(9)可能成立,不过判定结果会有一定的错误η

$$\eta = \frac{1+\varepsilon^2}{2} \tag{11}$$

式中

$$\varepsilon = \|\langle\psi|\psi'\rangle\|$$

为了获得更精确的判定结果,可以图 2 中的量子线路为单元构建一个具有 k 级的量子线路,如图 3 所示.

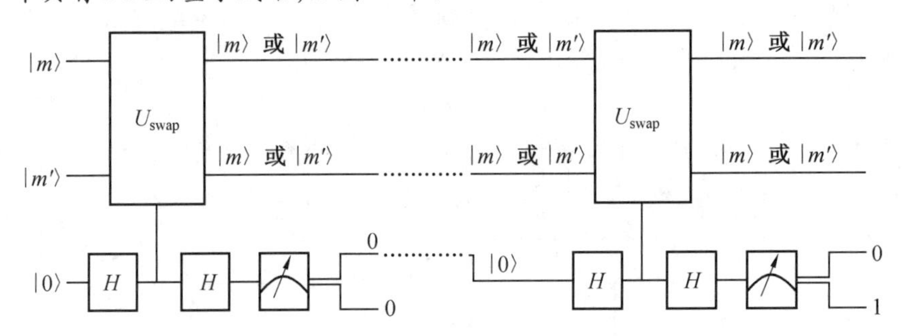

图 3　量子签名的精确验证量子线路图

图中,在前一个单元中若测量结果为 1,停止运算,因为这种情况下必有式(9) 成立. 否则,将前一级的输出作为后一级的输入. 如此连续下去直至第 k 级. 经过上述处理后的错误率为

$$\tilde{\eta} = \left(\frac{1+\varepsilon^2}{2}\right)^k \tag{12}$$

显然,错误概率明显下降. 当 k 较大时,$\tilde{\eta} \to 0$.

然后,验证者比较解码后得到的两粒子态 $|\Theta\rangle_{r_2,r_k+1}$ 和收到的两粒子态 $|\tilde{\Omega}\rangle$, 比较的方法与上述比较 $|m\rangle$ 和 $|m'\rangle$ 的方法一样. 如果 $|\Theta\rangle_{r_2,r_k+1} = |\tilde{\Omega}\rangle$, 意味着解码后得到的两粒子纠缠态和收到的两粒子纠缠态相同,验证者接受签字 $|S\rangle$;否则,验证者拒绝. 上面的操作主要用于防止如下的攻击策略:攻击者制备一个假消息态 $|M\rangle$ 和 $k-1$ 个两粒子纠缠态,将这些态做直积得到一个直积态,然后将 $\boldsymbol{U}^{-1}$ 作用在这个直积态上得到一个态 $|X\rangle$,最后将 $|X\rangle$ 和 $|M\rangle$ 发送给验证者.

2. 安全性分析

与数字签名一样,量子签名的安全性主要集中在两个方面:任何攻击者都不能获取密钥,以及签名者不能否定其所产生的签名. 对于后者是显然的. 下面主要证明算法中签名的不可伪造性.

定理2　给定消息量子比特 $|m\rangle$ 以及经过式(7) 产生的签字 $|S\rangle$. 令 $|S'\rangle = |\tilde{S}'\rangle \otimes |\tilde{\Omega}'\rangle$,其中 $|\tilde{S}'\rangle$ 和 $|\tilde{\Omega}'\rangle$ 分别为任意 $2k$ 粒子纠缠比特和两粒子纠缠态. 当且仅当 $|S'\rangle = |S\rangle$ 时, $|S'\rangle$ 为 $|m\rangle$ 在密钥 K_s 下产生的签字.

证明　首先考虑密钥确定的情况,也就是任何签字依赖于同一个密钥. 假定给定的消息量子比特 $|m\rangle$ 存在另一个签字 $|S'\rangle$,且 $|S'\rangle \neq |S\rangle$.

根据式(7)得到

$$|\tilde{S}'\rangle\otimes|\tilde{\Omega}'\rangle\neq|\tilde{S}\rangle\otimes|\tilde{\Omega}\rangle \tag{13}$$

将 $\boldsymbol{U}$ 作用于式(13)得到

$$|\tilde{m}'\rangle\otimes|\Gamma'\rangle\otimes|\tilde{\Omega}'\rangle\neq|m\rangle\otimes|\Gamma\rangle\otimes|\tilde{\Omega}\rangle \tag{14}$$

式中

$$|\Gamma'\rangle = J'\|\boldsymbol{P}'\|^{1/2}\prod_{i=2}^{k}|\Theta'\rangle_{r_i,r_{i+k-1}}$$

$$|\Gamma\rangle = J\|\boldsymbol{m}\|^{1/2}\prod_{i=2}^{k}|\Theta\rangle_{r_i,r_{i+k-1}}$$

对于固定的密钥,式(14)给出

$$|m\rangle\neq|m'\rangle \tag{15}$$

显然式(15)与假设矛盾,因此 $|S'\rangle=|S\rangle$.

此外,假设存在两个密钥 K_s^1 和 K_s^2,且 $K_s^1\neq K_s^2$,而它们可在消息 $|P\rangle$ 的基础上产生相同的签字,即 $|S\rangle_{K_s^1}=|S'\rangle_{K_s^2}$. 因为量子态 $|\Omega(\boldsymbol{x})\rangle$ 与密钥一一对应,由密钥 K_s^1 和 K_s^2 分别制备出量子态 $|\Omega(\boldsymbol{x})^1\rangle$ 和 $|\Omega(\boldsymbol{x})^2\rangle$,可给出

$$|S\rangle_{K_s^1}=\int|m\rangle\Omega(x)^1\mathrm{d}x \tag{16}$$

$$|S'\rangle_{K_s^2}=\int|m\rangle\Omega(x)^2\mathrm{d}x \tag{17}$$

根据假设 $|S\rangle_{K_s^1}=|S'\rangle_{K_s^2}$ 和上面两式得到

$$\left[\int|m\rangle\Omega(x)^1-\int|m\rangle\Omega(x)^2\right]\mathrm{d}x=0 \tag{18}$$

式(18)给出 $|\Omega(x)^1\rangle=|\Omega(x)^2\rangle$,从而 $K_s^1=K_s^2$,与假设矛盾.

综合上述两种情况,在同一个密钥产生的 $2k$ 粒子纠缠态才是消息 $|m\rangle$ 的签字. 证毕.

定理表明,要产生消息 $|m\rangle$ 的签字,必须获得密钥 K_s. 获得密钥的方法有两种:从签字者产生的签字 $|S\rangle$ 和信道中传送的消息 $|m\rangle$ 中获取密钥信息,以及从公钥中获取密钥信息. 下面证明攻击者获取密钥 K_s 的可能性. 为方便起见,记 $\boldsymbol{E},\boldsymbol{K}_s,\boldsymbol{K}_p,\boldsymbol{S},\boldsymbol{m}$ 分别为攻击的任意操作 ε、密钥 K_s,公钥 K_p,签字 $|S\rangle$ 和消息 $|m\rangle$ 对应的随机变量.

定理3 令 $I(\boldsymbol{K}_s,\boldsymbol{E}|\boldsymbol{K}_p,\boldsymbol{S},\boldsymbol{m})$ 为给定 $\boldsymbol{K}_p,\boldsymbol{S},\boldsymbol{m}$ 在操作 $\boldsymbol{E}$ 的作用下获取密钥的条件交互信息量. 对任意正数 $\sigma>0,\xi>0$ 和 $L_{\max}>0$ 有

$$\boldsymbol{I}(\boldsymbol{K}_s,\boldsymbol{E}|\boldsymbol{K}_p,\boldsymbol{S},\boldsymbol{m})\leqslant\frac{\sigma}{\ln 2}+L_{\max}\xi \tag{19}$$

证明 Shannon 信息论中,条件交互信息量定义为

$$\boldsymbol{I}(\boldsymbol{X},\boldsymbol{Y}|\boldsymbol{Z})=\sum p(z)[H_z(\boldsymbol{X})-H_z(\boldsymbol{X}|\boldsymbol{Y})] \tag{20}$$

式中,$H_z(\boldsymbol{X}|\ \boldsymbol{Y})$ 定义为

$$H_z(\boldsymbol{X}|\ \boldsymbol{Y}) = -\sum_{x,y} p(x,y|\ z)\log_2 p(x|\ y) \tag{21}$$

由定理2可知$\boldsymbol{S}(|\ m\rangle) = |\ S\rangle$,于是$p(\boldsymbol{K}_p,\boldsymbol{S},\boldsymbol{m}) = p(\boldsymbol{K}_p,\boldsymbol{S})$. 由公钥$K_p$和签字$|\ S\rangle$ 的独立性,以及$\boldsymbol{K}_p(\varepsilon) = K_p$ 和$p(\boldsymbol{K}_p) = 1$,得到$p(\boldsymbol{K}_p,\boldsymbol{S}) = p(\boldsymbol{S})$. 给定消息$|\ \psi\rangle$ 的条件下,由于密钥与签字的一一对应关系,容易得到

$$p(\boldsymbol{K}_p,\boldsymbol{S},\boldsymbol{m}) = p(\boldsymbol{K}_s) \tag{22}$$

为书写方便,记$\Xi = \boldsymbol{K}_p,\boldsymbol{S},\boldsymbol{m}$. 由式(20) 得到

$$I(\boldsymbol{K}_s,\boldsymbol{\varepsilon}|\ \Xi) = \sum_{K_s} p(K_s)[H_{K_s}(\boldsymbol{K}_s) - H_{K_s}(\boldsymbol{K}_s|\ E)] \tag{23}$$

因为

$$p(K_s) = \sum_{(K_s,\boldsymbol{\varepsilon})|\ K_s(\boldsymbol{\varepsilon}) = K_s} P(K_s,\boldsymbol{\varepsilon})$$

式(23) 可改写为

$$\begin{aligned} I(\boldsymbol{K}_s,\boldsymbol{E}|\ \Xi) &= \sum_{K_s,\boldsymbol{\varepsilon}} p(K_s,\boldsymbol{\varepsilon})(H(\boldsymbol{K}_s) + \log_2 p(K_s|\ \boldsymbol{\varepsilon})) \\ &= \sum_{K_s,\boldsymbol{\varepsilon}} p(K_s,\boldsymbol{\varepsilon})H(\boldsymbol{K}_s) + \sum_{K_s,\boldsymbol{\varepsilon}} p(K_s,\boldsymbol{\varepsilon})\log_2 p(K_s|\ \boldsymbol{\varepsilon}) \end{aligned} \tag{24}$$

用Λ表示攻击在任何攻击策略下能够从公开参数中获取信息的事件为真,则

$$\sum_{K_s,\boldsymbol{\varepsilon}} p(K_s,\boldsymbol{\varepsilon})H(\boldsymbol{K}_s) = \sum_{(K_s,\boldsymbol{\varepsilon})|\ \Lambda} p(K_s,\boldsymbol{\varepsilon})H(\boldsymbol{K}_s) + \sum_{(K_s,\boldsymbol{\varepsilon})|\ \overline{\Lambda}} p(K_s,\boldsymbol{\varepsilon})H(\boldsymbol{K}_s) \tag{25}$$

式中,$\overline{\Lambda}$ 表示事件Λ为假的事件. 同样,式(24) 中右边第二项也可按照事件$\overline{\Lambda}$和Λ拆分. 因为$\boldsymbol{\varepsilon} \in \overline{\Lambda}$ 意味着$H(\boldsymbol{K}_s) = 0$ 和$p(K_s|\ \boldsymbol{\varepsilon}) = 1$,所以

$$I(\boldsymbol{K}_s,\boldsymbol{E}|\ \Xi) = \sum_{(K_s,\boldsymbol{\varepsilon})|\ \Lambda} p(K_s,\boldsymbol{\varepsilon})\{H(\boldsymbol{K}_s) + \log_2 p(K_s|\ \boldsymbol{\varepsilon}))\} \tag{26}$$

可以得到

$$|\ p(K_s|\ \boldsymbol{\varepsilon}) - 2^{-H(K_s)} \leqslant 2^{-H(K_s)}\sigma \tag{27}$$

于是

$$\begin{aligned} I(\boldsymbol{K}_s,\boldsymbol{E}|\ \Xi) &= \sum_{(K_s,\boldsymbol{\varepsilon})|\ \&} P(K_s,\boldsymbol{\varepsilon})\{H(\boldsymbol{K}_s) + \log_2 p(K_s|\ \boldsymbol{\varepsilon})\} + \\ &\quad \sum_{(K_s,\boldsymbol{\varepsilon})|\ \Lambda\cap\overline{\&}} P(K_s,\boldsymbol{\varepsilon})\{H(\boldsymbol{K}_s) + \log_2 p(K_s|\ \boldsymbol{\varepsilon})\} \\ &\leqslant \sum_{(K_s,\boldsymbol{\varepsilon})|\ \overline{\&}} p(K_s,\boldsymbol{\varepsilon})\log_2(1 + \lambda_{K_s,\boldsymbol{\varepsilon}}) + \\ &\quad \sum_{(K_s,\boldsymbol{\varepsilon})|\ \Lambda\cap\overline{\&}} P(K_s,\boldsymbol{\varepsilon})H(\boldsymbol{K}_s) \end{aligned} \tag{28}$$

式中,$\& = \Lambda \cap \mathscr{N}_{\sigma}', \lambda_{K_s,\varepsilon} \leqslant \sigma$. 令 $L_{\max}$ 为 $H(\boldsymbol{K}_s)$ 的最大值,定义随机变量的概率 $\Pr(\boldsymbol{X}) = \Pr(\boldsymbol{X} = x)$,则 $\Pr(\Lambda \cap \overline{\&}) \leqslant \xi$. 利用关系式

$$\log_2(1+x) \leqslant |x| / \ln 2 \quad (x > 1)$$

容易得出式(19). 证毕.

定理3表明攻击者采取任何攻击策略$\boldsymbol{\varepsilon}$(包括经典策略和量子策略),从公钥以及其他公开的参数中获得的关于密钥的信息趋于0. 下面再考虑从签字者产生的签字$|S\rangle$和信道中传送的消息$|m\rangle$中获取密钥信息的可能性.

定理4 给定消息$|m\rangle$以及消息的签字$|S\rangle$,量子态$|m\rangle$和$|S\rangle$是不可区分的.

证明 利用式(5),容易计算消息和签字的内积

$$\begin{aligned}\langle m | S\rangle &= \int \langle m(x_0') | m(x_0)\rangle | \Omega(x)\rangle \mathrm{d}x \\ &= \int \delta(x_0' - x_0) | \Omega(x)\rangle \mathrm{d}x \\ &= | \Omega(x_0')\rangle \neq | \Omega(x)\rangle \end{aligned} \tag{29}$$

式(29)表明$\langle m | S\rangle$不为0,因此消息$|m\rangle$以及消息的签字$|S\rangle$不正交,因而它们是不可区分的. 另外,从式(24)可以看出,$|m\rangle$和$|S\rangle$不会泄漏密钥的信息.

事实上,经过编码后得到的$|S\rangle$相当于$|m\rangle$的纠错码,已知$|S\rangle$是不可克隆的,精确获得$|S\rangle$的状态需要无限计算资源. 因此,攻击者不能从$|m\rangle$和$|S\rangle$获取有效信息. 证毕.

本书中的方法还可应用于量子密码通信,这是近四十年才发展起来的一种新的通信技术. 它利用量子特性来得到或提高通信的保密性. 例如,可以利用连续变量进行量子密码通信①.

单光子量子密码通信的主要弱点是信号太弱,容易衰减及受干扰,从而限制了它的码率及传送距离. 况且,目前还没有简单、可靠的真正单光子源,绝大多数都是利用激光源经过衰减来得到单光子. 激光源的光子数服从泊松分布,除了含有单光子脉冲还含有多光子脉冲. 这些多光子脉冲被窃听后不会被察觉,从而影响安全性. 为了减少多光子脉冲的影响,通常使用的激光脉冲平均只含有约0.1光子/脉冲. 使用具有一定强度的激

---

① 摘自《量子密码通信》,马瑞霖编著,科学出版社,2006.

光进行量子密码通信一直是人们追求的目标，利用连续变量进行量子密码通信是其中的一个主要研究方向. 这方面已有很大的进展，但还未成熟，至今尚未见到有关实际通信的报道.

1. 连续变量量子密码通信的基本原理

我们曾利用复振幅算符 $a(0)=\hat{X}_1+\mathrm{i}\hat{X}_2$，它的两个分量 $\hat{X}_1$ 及 $\hat{X}_2$ 都是厄米算符，分别与无量纲坐标 $\hat{X}$ 及动量 $\hat{P}$ 对应，它们的对易关系为 $[\hat{X}_1\cdot\hat{X}_2]=\mathrm{i}/2$. 它们的测量误差满足 Heisenberg 不确定性原理，其标准差之积 $\Delta X_1\cdot\Delta X_2\geqslant 1/4$，当不确定性达到最小时 $\Delta X_1\cdot\Delta X_2\geqslant 1/4$. 许多文献将复振幅算符定义为

$$\hat{a}(0)=\frac{\hat{X}_1+\mathrm{i}\hat{X}_2}{2}$$

相应地有

$$[\hat{X}_1\cdot\hat{X}_2]=2\mathrm{i},\Delta X_1\cdot\Delta X_2=1$$

这样做的好处是使方差 $(\Delta X_1)^2$ 与 $(\Delta X_2)^2$ 有互为倒数的简单关系. 以后除非特别声明我们也这样做. 为了简化符号，我们将 $\hat{a}^{\dagger}(0)$ 表示为 a，将 $\hat{a}^{\dagger}(0)$ 表示为 $a^{\dagger}$，将 $\hat{X}_1$ 表示为 x，将 $\hat{X}_2$ 表示为 p，于是有

$$a=\frac{x+\mathrm{i}p}{2},a^{\dagger}=\frac{x-\mathrm{i}p}{2} \tag{1}$$

$$x=a+a^{\dagger},p=\frac{1}{\mathrm{i}}(a-a^{\dagger}) \tag{2}$$

$$[x,p]=2\mathrm{i} \tag{3}$$

$$(\Delta x)^2\cdot(\Delta p)^2\geqslant 1 \tag{4}$$

当不确定性达到最小时，x 及 p 的方差满足

$$(\Delta x)^2\cdot(\Delta p)^2=1$$

设通信协议如下：① 发送方 A 将相干态 $|x_{\mathrm{A}}+\mathrm{i}p_{\mathrm{A}}\rangle$ 发往 B，x_{A} 及 p_{A} 均是以零为中心作 Gauss 分布的随机变量，它们是由 A 发出的 x 及 p，分别对应于无量纲位置及动量，并且满足式(2). x_{A} 分布的方差记为 $V(x_{\mathrm{A}})$，p_{A} 分布的方差记为 $V(p_{\mathrm{A}})$，$V(x_{\mathrm{A}})$ 及 $V(p_{\mathrm{A}})$ 代表 A 发出的信号功率. 为了使窃听者无法区分，规定 $V(x_{\mathrm{A}})=V(p_{\mathrm{A}})$. ② 接收方 B 收到信号后用“零差检测”随机地测量 x 或 p 分量. ③ 测量后，B 通过公开信道告知 A，他测量了哪一个分量(x 或 p). A 得知后只保留 B 测量的那些 x 或 p，舍弃未经测量的数据. 于是 A 及 B 共同拥有一串紧密相关的数据，类似于前述的单光子通信的筛后数据. ④A 与 B 利用公开信道交换部分数据，判断通信是否安全、有效. ⑤ 若通信有效，继续通过公开信道进行纠错及密性放大.

现在针对上述通信协议,讨论连续变量密码通信的原理.假定窃听者E对A发出的量子态全部拦截,逐个地测量,并且如B那样只测x或p,那么E只能随意猜想没有测量的另一半,构成量子态$|x+ip\rangle$发给B.于是A及B将在上述的步骤④中发现,约有一半数据是不相关的,从而认定窃听者存在,舍弃通信.

E的另一策略是全部拦截逐个测量,但不是只测x或p而是两者都测量(例如,用50∶50的分光器,将输入分为两半,一半测x,另一半测p).根据Heisenberg不确定性原理,无论E的测量手段如何完善,也不能将x及p二者都准确地确定,充其量只能达到式(4)的结果.亦即,如果x的测量方差为$(\Delta x)^2$,那么p的测量方差是$(\Delta p)^2=1/(\Delta x)^2$.这些测量误差将作为新的噪声源一并发往B,使B收到的信噪比显著下降,从而暴露了窃听的存在.

由结果

$$I_{AB}=\frac{1}{2}\log_2\left(1+\frac{P_A}{N_B}\right) \tag{5}$$

其中,I_{AB}代表A与B之间的最大的互信息,单位是比特/符号.P_A是A发出的作Gauss分布的随机变量的功率,信道无衰减,故P_A也是B收到的信号功率.N_B是B通道的噪声功率,噪声也作Gauss分布.P_A/N_B代表B的信噪比.也可以利用式(5)求出A与E的互信息I_{AE},这时式中的B的信噪比要用E的信噪比代替.假定B的噪声功率达到最小,$N_B=N_0$,N_0是作Gauss分布的真空噪声的功率.以N_0为单位,A发出的信号功率可表示为V_AN_0,V_A是A发出的随机变量x(或p)作Gauss分布所对应的方差

$$V_A=V(x_A)=V(p_A)$$

在信道无衰减及未被拦截时,B的信噪比为V_A.

如果信道有衰减,传输率为η,B的信噪比减低为ηV_A.信噪比的降低,可以等效为B的噪声增加,即B的信噪比

$$S_B=\eta V_A=\frac{V_A}{1+d},d=\frac{1-\eta}{\eta} \tag{6}$$

当信道有衰减($\eta<1$)时,窃听的最佳策略是只截听A发出的部分信号$(1-\eta)V_AN_0$,而让其余部分(ηV_AN_0)无干扰地到达B,这样B及A都无法察觉.E的信噪比$S_E=(1-\eta)V_A$,利用式(5)可得

$$I_{AB}=\frac{1}{2}\log_2\left(1+\frac{V_A}{1+d}\right),d=\frac{1-\eta}{\eta} \tag{7}$$

$$I_{AE}=\frac{1}{2}\log_2\left(1+\frac{V_A}{1+\frac{1}{d}}\right),\frac{1}{d}=\frac{\eta}{1-\eta} \tag{8}$$

当 $I_{AB} > I_{AE}$,通信将是安全的.也就是说,经过后续的纠错及密性放大,A与B将拥有彼此一致的保密数据,其大小为 $I_{AB} - I_{AE}$,需扣除纠错及密性放大的耗费.从式(7)及式(8)可得

$$\Delta I = I_{AB} - I_{AE} = \frac{1}{2}\log_2 \frac{1 + d + V_A}{1 + d + V_A d} = \frac{1}{2}\log_2 \frac{V + d}{1 + Vd} \tag{9}$$

式中,$V = V_A + 1$.已知 $d > 0$,故得 $\Delta I > 0$ 的必要条件是 $d < 1$. $d = (1 - \eta)/\eta$,条件是 $d < 1$,相当于 $\eta > 1/2$,亦即传输率大于3 dB.

以后我们会看到,3 dB极限是可以打破的.问题在于,上述推导以 V_A 为参考基准,如果改为以B接收到的信号为基准,问题就可以解决了.回顾前述的单光子密码通信,如果信道有衰减,使某脉冲变为空脉冲,那么B会告知A,不把该脉冲计算在内.这实际上是以B收到的信号为参考,安全判据改为 $I_{BA} > I_{BE}$,从而避免了3 dB的极限.在信道衰减及窃听均存在的情形下,分析下述的模型可以得到适用性相当普遍的结果.

设A发出的场分量(quadrature)为 x_A 及 p_A,相应地,B收到的是 x_B 及 p_B,E收到的是 x_E 及 p_E,它们满足下面的关系式

$$x_B = g_{B_x} x_A + x_{BN} \tag{10}$$

$$p_B = g_{B_p} p_A + p_{BN} \tag{11}$$

$$x_E = g_{E_x} x_A + x_{EN} \tag{12}$$

$$p_E = g_{E_p} p_A + p_{EN} \tag{13}$$

式(10)中,x_B 包括两部分,一部分与 x_A 成比例,比例系数为 g_{Bx},另一部分为 x_{BN},与A发出的信号 x_A 无关,代表噪声.同样地,议论适用于式(11)(12)及式(13).A发出的量子态为 $|x_A + ip_A\rangle$.B获得的样本为 $|x_B + ip_B\rangle$,E获得的样本为 $|x_E + ip_E\rangle$.B及E的样本都是衍生自A的原样.量子力学的基本原理告诉我们,量子态是不能复制的,因此B的样本一定不同于E的样本,x_B 及 p_E 属于不同样本,彼此对易,$[x_B, p_E] = 0$,故有

$$x_B p_E - p_E x_B = 0 = g_{Bx} g_{Ep} [x_A, p_A] + g_{Bx} [x_A, p_{EN}] + g_{Ep} [x_{BN}, p_A] + [x_{BN}, p_{EN}]$$

注意到,$[x_{BN}, p_A]$,$[x_A, p_{EN}]$ 均为零,可得

$$[x_{BN}, p_{EN}] = - g_{Bx} g_{Ep} [x_A, p_A] \tag{14}$$

根据Heisenberg不确定性原理(参考式(12)),并应用式(13),$[x, p] = 2i$,我们得到

$$\Delta x_{BN} \cdot \Delta p_{EN} \geqslant | g_{Bx} g_{Ep} |$$

$$(\Delta x_{BN})^2 \cdot (\Delta p_{EN})^2 \geqslant | g_{Bx} |^2 | g_{Ep} |^2 \tag{15}$$

$(\Delta x_{BN})^2$ 是B测量时 x_{BN} 项引起的方差,是B新增加的噪声.$(\Delta p_{EN})^2$ 是E测量时 p_{EN} 项引起的方差,是E新增加的噪声.将噪声等效到A发出的信

号,并假定各种条件都很完善,不确定性达到最小,我们得到

$$d_{xB} \cdot d_{pE} = 1 \tag{16}$$

$$d_{xB} = \left(\frac{\Delta x_{BN}}{|g_{Bx}|}\right)^2 = \left(\frac{\Delta x_B}{|g_{Bx}|}\right)^2 - (\Delta x_A)^2$$

$$d_{pE} = \left(\frac{\Delta p_{EN}}{|g_{Ep}|}\right)^2 = \left(\frac{\Delta p_E}{|g_{Ep}|}\right)^2 - (\Delta p_A)^2$$

d_{xB} 是 x 分量在 B 的样本新增加的噪声方差. d_{pE} 是 p 分量在 E 的样本新增加的噪声方差. 所谓新增加是相对于 A 的原样噪声方差$(\Delta x_A)^2$ 而言.

相同的推导,可得

$$d_{pB} \cdot d_{xE} = 1 \tag{17}$$

$$d_{pB} = \left(\frac{\Delta p_{BN}}{|g_{Bp}|}\right)^2 = \left(\frac{\Delta p_B}{|g_{Bp}|}\right)^2 - (\Delta p_A)^2$$

$$d_{xE} = \left(\frac{\Delta x_{EN}}{|g_{Ex}|}\right)^2 = \left(\frac{\Delta x_E}{|g_{Ex}|}\right)^2 - (\Delta x_A)^2$$

d_{pB} 是 p 分量在 B 的样本新增加的噪声,d_{xE} 是 x 分量在 E 的样本新增加的噪声. 各个 d 均为无量纲量,当用噪声功率表示时相应的单位是 N_0(真空噪声功率).

回到上述的通信协议,A 发出相干态$|x_A + \mathrm{i}p_A\rangle$,$x_A$ 与 p_A 是对称的. 分布方差

$$V(x_A) = V(p_A) = V_A$$

相应的信号功率 $V_A N_0$. 最小测量方差

$$(\Delta x_A)^2 = (\Delta p_A)^2 = 1$$

对应的噪声功率是 N_0,在这种情况下,窃听的最佳策略也应是对称的. 通道衰减一般是对称的. 因此,将有

$$d_{xB} = d_{pB} = d_B \tag{18}$$

$$d_{xE} = d_{pE} = d_E \tag{19}$$

$$d_B \cdot d_E = 1 \tag{20}$$

也就是说,不论是 x 分量或 p 分量,B 的噪声量反比 E 的噪声增量. 将这个结果用于式(5) 可求得互信息

$$I_{AB} = \frac{1}{2}\log_2\left(1 + \frac{V_A}{1 + d_B}\right) = \frac{1}{2}\log_2 \frac{V + d_B}{1 + d_B} \quad (V = V_A + 1) \tag{21}$$

$$I_{AE} = \frac{1}{2}\log_2\left(1 + \frac{V_A}{1 + (1/d_B)}\right) = \frac{1}{2}\log_2 \frac{1 + Vd_B}{1 + d_B} \tag{22}$$

$$\Delta I = I_{AB} - I_{AE} = \frac{1}{2}\log_2 \frac{V + d_B}{1 + Vd_B} \tag{23}$$

安全判据是 $\Delta I > 0$,即 $d_B < 1$. 式(9) 与式(23) 形式相同,其实式(9) 是式(23) 的一个特例.

2. 利用压缩态的连续变量密码通信

上面针对相干态讨论了连续变量密码通信的原理,所采用的方法也适用于压缩态,只需要略作调整. 历史上,提出用压缩态连续变量进行量子密码通信比用相干态还要早. 一些研究曾认为,相干态连续变量密码通信不够安全,只有用压缩态才能得到不为零的安全码率,下面的分析表明,相干光密码通信的安全性可与压缩光相媲美.

对上一节所用的通信协议稍作调整,可以得到下述的使用压缩态连续变量进行密码通信的协议:

(1) 发送方 A 将压缩态 $|(x_A + ip_A), s\rangle$ 发往 B,x_A 及 p_A 均是以零为中心作 Gauss 分布的随机变量,它们分别对应于无量纲位置及动量,并且满足式(2). 在每次发送的量子态中随机选择 x_A 或 p_A 进行压缩,压缩因子是 s, $s < 1$. 如果选择 x_A 进行压缩,那么 x_A 的测量方差是 s,而 p_A 的测量方差为 $1/s$. 设 x_A 及 p_A 的分布方差为 $V(x_A)$ 及 $V(p_A)$,亦即信号功率为 $V(x_A)N_0$ 及 $V(p_A)N_0$,噪声功率为 sN_0 及 N_0/s,N_0 是真空噪声功率. 为了不让窃听者 E 知道 A 压缩的是 x_A 抑或 p_A,当压缩 x_A 时,应保持

$$V_s(x_A)N_0 + sN_0 = V_{1/s}(p_A)N_0 + \frac{1}{s}N_0 = VN_0 \tag{24}$$

当压缩 p_A 时,应保持

$$V_s(p_A)N_0 + sN_0 = V_{1/s}(x_A)N_0 + \frac{1}{s}N_0 = VN_0 \tag{25}$$

式中

$$V_s(p_A) = V_s(x_A) = V_s$$

代表 A 发出的被压缩的分量的分布方差.

$$V_{1/s}(x_A) = V_{1/s}(p_A) = V_{1/s}$$

代表被扩展的分量的分布方差.

(2) 接收方收到信号后用“零差探测” 随机测量 x 或 p 分量.

(3) 测量后,B 通过公开信道告知 A,他测量了哪一个分量(x 或 p). A 得知后只保留 B 测量的那个分量,舍弃约占一半的未经测量的无用数据. 于是 A 与 B 共同拥有一套紧密相关的数据.

(4) A 与 B 利用公开信道交换部分数据,判断通信是否安全、有效.

(5) 若通信有效,则继续通过公开信道进行纠错及密性放大. 最后获

得可靠的保密数据.

现在针对上述的压缩态通信协议计算互信息I_{AB}及I_{AE},B随机地选择测量x_A或p_A,B不知道A究竟是压缩了x抑或是p,B测量的与A压缩的互相符合的机会只占一半,因此我们用彼此相同以及不同二者平均值来计算I_{AB}

$$I_{AB} = \frac{1}{2}[I_{ABR} + I_{ABW}] \tag{26}$$

式中,I_{ABR}表示B测量的与A压缩的彼此一致时的I_{AB},I_{ABW}表示B测量的不是被A压缩的而是被扩展的. 注意,在上述的通信协议中,x分量与p分量从统计角度看仍然是对称的,因此式(20)仍然适用,$d_B \cdot d_E = 1$. 利用此结果及式(24)和式(25),可得

$$I_{ABR} = \frac{1}{2}\log_2\left(1 + \frac{V - s}{s + d_B}\right) \tag{27}$$

$$I_{ABW} = \frac{1}{2}\log_2\left(1 + \frac{V - 1/s}{1/s + d_B}\right) \tag{28}$$

d_B是样本B的附加噪声(等效到A发出的信号功率),包括通道损耗及窃听等影响. 于是有

$$\begin{aligned} I_{AB} &= \frac{1}{4}\log_2\left(\frac{V + d_B}{s + d_B} \cdot \frac{V + d_B}{1/s + d_B}\right) \\ &\frac{1}{4}\log_2 \frac{(V + d_B)^2}{d_B} - \frac{1}{4}\log_2\left(d_B + \frac{1}{d_B} + s + \frac{1}{s}\right) \end{aligned} \tag{29}$$

同理可求得

$$I_{AE} = \frac{1}{4}\log_2 \frac{(V + d_E)^2}{d_E} - \frac{1}{4}\log_2\left(d_E + \frac{1}{d_E} + s + \frac{1}{s}\right) \tag{30}$$

注意到$d_E = 1/d_B$,可得

$$\begin{aligned} \Delta I = I_{AB} - I_{AE} &= \frac{1}{4}\log_2\left(\frac{(V + d_B)^2}{d_B} \cdot \frac{d_E}{(V + d_E)^2}\right) \\ &= \frac{1}{2}\log_2 \frac{V + d_B}{1 + Vd_B} \end{aligned} \tag{31}$$

有趣的是式(31)与式(23)形式上完全相同,只是V的定义式略作调整. 从式(31)可知,安全判据是$\Delta I > 0$,即$d_B < 1$,与前述的相干光密码通信的要求完全相同. 就是说在上述的通信协议的框架下,使用相干光与使用压缩光就安全判据而言是完全相同的,都要求B的附加噪声(单位为真空噪声N_0)$d_B < 1$. 这是一个很有价值的结果.

产生压缩光比产生相干光困难,在远距离的传输中保持其压缩性更

难.因此,相对于压缩光而言,利用相干光的连续变量密码通信,实用前景较佳.

不少研究提出利用互相纠缠的压缩光束进行连续变量密码通信.就近期而言,其意义主要是在理论上,我们不在此论述.

在北京大学赵凯华教授的讲义中有一段专门介绍 Shor 算法①的.赵教授认为:

量子计算最重要的优点体现在量子并行运算上,特别突出的是经典计算机只能进行指数算法的问题,量子计算机有可能用多项式算法来完成.

举例来说,我们进行 4 位数乘 4 位数的乘法计算,最多只需二三十步(即步数是4的多项式)就行了,小学生也能在不长的时间里算完.但是反过来,4 位数与 4 位数的乘积是 8 位数,几乎上亿的数量级,给你一个上亿的数字 N 让你做因子分解,可就难了.一般除了逐个用小于 $\sqrt{N}$ 的素数试着去除它,别无其他妙法.这样的素数有 10^4,即上万个(4 在指数上),一个一个去试,得花相当长的一段时间.这个问题用现代计算机去算当然不成问题,那么给你一个 60 位的大数做因子分解又怎么样?现在世界上最快的计算机每秒作 10^{11} 次运算,每天 86 400 s,每年约 3×10^7 s,不停地计算,可作 3×10^{18} 次运算.要作 $\sqrt{10^{60}}=10^{30}$ 次运算,约需 3×10^{11} 年,宇宙年龄的 20 倍!然而量子计算机可望在一段很短的时间,譬如 10^{-8} s 里解决问题.

人们普遍相信,大数的因子分解不存在经典的多项式算法(或者说,有效算法).这一点在密码学中有着重要的应用.1977 年 Rivest, Shamir 和 Adelman 三人发明的 RSA 公钥系统,就是利用两个大素数的乘积难以分解来加密的.

1994 年 Shor 等人提出了一种大数因子分解的量子多项式算法,引起了轰动.Shor 算法的核心是利用数论中的一些定理,将大数因子分解转化为求某个函数的周期.现将 Shor 算法的梗概作一介绍.设待因数分解的大数为 N,它的平方用二进制来表示有 L 位,即 $N^2<2^L<2N^2$.选用的周期性函数为余函数

$$f(x)=a^x(\bmod N) \tag{1}$$

① 摘自《量子物理》(第二版),赵凯华,罗蔚茵,高等教育出版社,2008.

这里 $a(a < N)$ 是任选的一个与 N 互素的整数, x 取从 0 到 2^L 的整数值, mod N 表示取前面的数被 N 除的余数. 显然 $f(x)$ 所取的值是小于 N 的正整数, 它是一个周期性的函数. 举例来说, 令 $N = 14$, 取 $a = 3$, 则

$$f(0)=1, f(1)=3, f(2)=9, f(3)=13, f(4)=11,$$
$$f(5)=5, f(6)=1, f(7)=3, f(8)=9, f(9)=13,$$
$$f(10)=11, f(11)=5, f(12)=1, f(13)=3$$

它的周期 $T=6$. 一般说来, 对于大数 N, 选定一个 a, 若能求得式(1) 中余函数的周期 T, 设 T 为偶数(若求得的周期 T 为奇数, 另选一个 a 重来), 则令 $A=a^{T/2}+1, B=a^{T/2}-1$, 求(A,N) 和(B,N) 的最大公约数 C 和 D, 它们就是 N 的素因子, 即 $N=C\times D$. 例如, 当 $N=14$ 并选 $a=3$ 时, 求得 $T=6$, 于是 $A=28, B=26, C=7, D=2, N=7\times 2$.

虽然用量子计算机计算时, 一次测量中我们只能得到非常有限的信息, 此后该量子态就坍缩了. 在 Shor 算法中, 我们只需有关 $f(x)$ 周期的信息, 这可通过如下的 Fourier 变换来提取.

取两组各有 L 量子比特的存储器, 通过幺正变换实现交缠态, 其中 $f(x)$ 是由式(1) 定义的余函数, $n=2^L$. 对存储器 x 作离散 Fourier 变换

$$|x\rangle\rangle = \frac{1}{\sqrt{2^L}}\sum_{k=0}^{2L-1} e^{2\pi i kx/2L} |k\rangle\rangle \tag{2}$$

于是, 两存储器里的交缠态化为

$$|\Psi\rangle = \frac{1}{2^L}\sum_{x=0}^{2L-1}\sum_{k=0}^{2L-1} e^{2\pi i kx/2L} |k\rangle\rangle \otimes |f(x)\rangle\rangle \tag{3}$$

这时第一个存储器(x 存储器) 变为 k 存储器. 由于 $f(x)$ 的周期性, 上式中许多项可以合并, 而且大部分项相消或近似相消. 只有 k 取下列各值时系数(概率幅) 明显不为 0

$$k = \left[m\frac{2^L}{T}\right] \quad (m=0,1,\cdots,T-1) \tag{4}$$

式中, T 是 $f(x)$ 的周期, 方括号表示取向上靠拢的整数. 因此, 除 $k=0$ 外

$$\frac{2^L}{k} \approx \frac{T}{m} \quad (m=1,\cdots,T-1) \tag{5}$$

以 $N=14, T=6$ 的例子来说, $N^2=196$, 需要取 $L=8, 2^L=256, 2^L/T=42.667$, 系数(概率幅) 明显不为 0 的 k 值有

$$k=0,[42.667]=43,[85.333]=86,128,$$
$$[170.667]=171,[213.333]=214$$

这些是对 k 存储器进行测量时, 实际上可能测到的 k 的本征值. 要想求周期 T, 就反过来计算 $2^L/k$, 除 $k=0$ 外, 有

$$\frac{2^L}{k}=\frac{256}{43}=5.953\approx 6,\frac{256}{86}=2.977\approx\frac{6}{2},\frac{256}{128}=2\approx\frac{6}{3}$$

$$\frac{256}{171}=1.497\approx\frac{6}{4},\frac{256}{214}=1.196\approx\frac{6}{5}$$

从若干个这样的数值不难推算出 $T=6$.

在量子计算机中Shor算法的每一个步骤都是可以通过多项式算法来完成的. 所以,在量子计算机中 Shor 算法是有效的算法. 如果量子计算机能够实现,那么世界上许多保密系统将受到严重威胁.

近年来,量子计算机理论上的研究取得了重大的进展. 在实验上,根据目前正在开发的情况看,它现在有三种类型:核磁共振量子计算机、硅基半导体量子计算机和离子阱量子计算机. 量子计算机的构建和运行面临的主要困难是克服退相干和量子纠错码的问题. 量子并行计算的基础是量子相干性. 系统和外界环境的耦合会导致量子相干性急剧衰退. 相干性的衰退和其他技术原因将导致运算出错,多种量子纠错码的研究正在进行. 总之,实现量子计算已不存在不可跨越的障碍,但离真正在实际中应用还有相当长一段距离. 2000 年 8 月 15 日,IBM公司宣布造成由5个原子作处理器和存储器的实验性量子计算机. 这标志着量子计算机在走向实用化的道路上迈了一大步.

本书涉及的内容是相当前沿的. 比如,关于量子形式主义. *Nature* 杂志在 2021 年 12 月 23 日出版的一期中刊登了一篇题为"基于实数的量子理论可通过实验证伪"的文章. 文章中指出:

虽然复数在数学中必不可少,但在描述物理实验时却并不需要,因为物理实验是用概率来表示的,因此使用实数. 然而,物理学的目标是通过理论来解释而非描述实验.

虽然大多数物理理论都基于实数,但量子理论是第一个用复 Hilbert 空间的算符来表述的. 这让无数物理学家感到困惑,包括量子理论的创始人,对他们来说,量子理论的真实版本,用真实的算符来表示,似乎更为自然.

事实上,此前的研究已经表明,只要各部分共享任意真实量子态,这种"真实量子理论" 就可重现任何多体实验的结果.

研究组探讨了在量子形式主义中是否真的需要复数. 他们通过证明量子理论的实 Hilbert 空间公式和复 Hilbert 空间公式在包含独立状态和度量的网络场景中可做出不同的预测,证明了这一点.

这使人们能够设计出一个类似 Bell 测试的实验，实验成功则推翻真正的量子理论，就像标准 Bell 测试推翻局部物理学一样.

量子力学在中国是个“显学”. 特别是在民间. 一旦某种事物被冠之以量子之名，必应者云集. 类似古代文人吟诗作对，东坡居士词云：“几时归去，作个闲人. 对一张琴、一壶酒、一溪云.” 江湖上有大把的人雅好之，时常吟咏，今番闲谈明礼，逸兴不浅，江湖上必有相知者. 碌碌如我，唯“叹隙中驹、石中火、梦中身” 而已.

刘培杰

2024 年 10 月 8 日

于哈工大

国外优秀物理著作
原版丛书（第一辑）

拓扑绝缘体
Topological Insulators

（英文）

●［美］帕纳约蒂斯·科泰特斯（Panagiotis Kotetes）著

黑版贸登字 08-2021-050 号

图书在版编目(CIP)数据

拓扑绝缘体=Topological Insulators:英文/（美）帕纳约蒂斯·科泰特斯(Panagiotis Kotetes)著.—哈尔滨:哈尔滨工业大学出版社,2024.10
（国外优秀物理著作原版丛书.第一辑）
ISBN 978-7-5767-1341-1

Ⅰ.①拓… Ⅱ.①帕… Ⅲ.①拓扑-绝缘体-英文 Ⅳ.①TM21

中国国家版本馆 CIP 数据核字(2024)第 073581 号

TUOPU JUEYUANTI

策划编辑 刘培杰 杜莹雪
责任编辑 刘立娟 刘家琳 李 烨 张嘉芮
封面设计 孙茵艾
出版发行 哈尔滨工业大学出版社
社 址 哈尔滨市南岗区复华四道街 10 号 邮编 150006
传 真 0451-86414749
网 址 http://hitpress. hit. edu. cn
印 刷 哈尔滨博奇印刷有限公司
开 本 787 mm×1 092 mm 1/16 印张 63.75 字数 1 195 千字
版 次 2024 年 10 月第 1 版 2024 年 10 月第 1 次印刷
书 号 ISBN 978-7-5767-1341-1
定 价 378.00 元(全 6 册)

For my parents Αγγελική & Θοδωρή, my sister Μαρία,
and my amazing nephew Γιάννη

Contents

Preface

This book aims at introducing the reader to the field of topological matter, which has recently witnessed rapid and important developments, also reflected in the 2016 Nobel prize in physics awarded to David J Thouless, F Duncan M Haldane and J Michael Kosterlitz. While the first encounter of topological phenomena can be traced back to the detection of the integer quantum Hall effect by Klaus von Klitzing in the early 1980 s, the systematic progress of this field happened only after the prediction and subsequent experimental discovery of time-reversal invariant topological insulators. This advancement ignited developments in other topological systems, including intrinsic and artificial superconductors, semimetals, superfluids, and others. Currently, the study of topological phenomena is driving a fairly large portion of the research activities of the physics community.

My personal engagement with this research field began originally during my PhD time in Athens under the guidance of Georgios Varelogiannis. However, the trigger to prepare an early version of this book came a little later, when I was a postdoctoral researcher in the group of Gerd Schön in Karlsruhe. There, Gerd gave me the opportunity to be a guest lecturer of the Condensed Matter Physics course that he taught during the winter semester of 2013–2014. As a support to these lectures, I prepared a set of notes that constituted the seed for the present book, which is the first part of a two-volume series, with a book entitled *Topological Superconductors* to appear in the near future by the same publishers. The present book was mainly written during my three-year long postdoctoral appointment at the Niels Bohr Institute, and I could not have carried out this task without the great support and understanding of my advisers Karsten Flensberg and Brian Andersen.

This book is designed for graduate students or researchers who wish to become familiar with the field of topological matter. Despite the long list of prerequisite concepts, I tried to make this book concise and self-contained. Given the limited space, I included references to suitable textbooks or reviews to help the non-expert reader cover the unavoidable gaps in my presentation. One will naturally ask what is novel about this book, and how does it compare with the existing literature. Certainly, there exists a plethora of excellent materials that I definitely recommend, e.g., the books by: J K Asbóth, L Oroszlány and A Pályi; B A Bernevig and T Hughes; E Fradkin; S-Q Shen; T Stanescu; G Volovik; the Les Houches Summer School lecture notes edited by C Chamon, M Goerbig, R Moessner and L Cugliandolo; the TU Delft online open course on topology coordinated by A Akhmerov *et al*; great reviews that I cite throughout this manuscript, and a number of other works that I unfortunately cannot mention here due to lack of space. The obvious answer to this question is that a research topic can be never covered by a single book and alternative presentations always come in handy, if not required. Specifically, in my take on introducing the topic I put more weight on the notion of symmetries, attempted a tighter interconnection of the different chapters and the Hands-on sections, and tried to be thorough on the technical aspects and calculations. Moreover, I have put together a number of topics that are rarely presented in a combined manner, e.g., unitary symmetries, thermal and thermoelectric transport, a fresh approach to the theory of orbital magnetisation, the polar Kerr effect, and finally the tenfold classification. In fact, in my attempt to make the presentation more pedagogical, I tried to arrange the materials following a bottom-up approach, avoiding a historically-linear structure.

Completing this book concludes a long journey of mine, where I learned a lot more than I would ever imagine. I hope that I can transfer some of this knowledge to the readers of this book. Lastly, I am grateful in advance for any feedback and criticism that can teach me even more and help me further improve the presentation of this work.

Sincerely yours,
Παναγιώτης Κοτετές
Beijing, February 2019

Acknowledgements

This moment that I am bringing the writing of this book to a close, I have to admit that thanking the people who supported my efforts is by far the most enjoyable and rewarding task of this long process. I will start by thanking Gerd Schön, who gave me the opportunity to give lectures on this topic in one of his courses some years ago, and motivated me to prepare a set of notes, which constituted the basis of this book. Next in line to be thanked are Karsten Flensberg and Brian Andersen, who showed great patience and support when I had to devote a fairly large part of my time to prepare this book, while putting on ice some of our common research projects. I would also like to acknowledge my PhD adviser Georgios Varelogiannis who, while not immediately involved in this book, was the one who inspired me to work in this field. Needless to say, the paramount support that I have received from my mentors goes beyond the bounds of their contributions to this book, and has been constantly present throughout my academic efforts and career so far. Apart from the above, I would also like to gratefully thank Mario Cuoco, Alexandros Kehagias, Peter Krogstrup, Boris Narozhny, Jens Paaske, Eleftherios Papantonopoulos and Alexander Shnirman, who greatly contributed to my scientific development and education in relation to the concepts of symmetry and topology in physics.

At this point, I would like to focus on the people who contributed to the proofreading and refinement of this book. The main hero here is William (Bill) Coish. Bill, during the time he spent in Copenhagen as part of his sabbatical leave in 2017, helped me to improve this manuscript substantially, and most importantly, with his advice and encouragement he contributed significantly to the advancement of my academic career. Another important figure behind this book is Ervand Kandelaki, who served as an incredibly supportive office mate during our time in Copenhagen. Besides the help of these two, the manuscript enjoyed great improvements thanks to the efforts and comments of Brian Andersen, Philip Brydon, Michele Burrello, Morten Christensen, Ajit Coimbatore Balram, Karsten Flensberg, Andreas Heimes, Stefanos Kourtis, Tommy Li, Daniel Mendler, Maria Teresa Mercaldo, Jens Rix, Mark Rudner, Mike Schecter and Daniel Steffensen.

Now I would like to proceed with acknowledging the people at Morgan & Claypool and the Institute of Physics who worked on the production of this book. First, I would like to thank Nicki Dennis, who was the person that discovered these notes a few years ago. Nicki is to be gratefully thanked for having the idea and giving me the opportunity to make a book out of these notes. On the book production side, I would like to thank Chris Benson. Nicki and Chris deserve my biggest thanks, as well as my deepest apologies for my delay regarding the publication of this work. Moreover, I would like to acknowledge all the typesetters who worked on this project, including Karen Donnison who was responsible for the permissions of materials that I borrowed from other authors and Brent Beckley with whom I was happy to collaborate on finalising my book-cover idea.

Last but not least, I would like to thank my family, and a certain number of friends who supported me during the time I was preparing this book. Among these friends, first I would like to thank David, with whom I shared a flat around the same time when the book project took off, and he was very supportive when things got quite stressful. Finally, I would like to gratefully thank Alex, Anton, Apollo, Astrid, August, Basti, Chris, Cecy, Clara, Elias, Ermina, Eva, Fani, Fathi, Finn, Garazi, George, Giuseppe, Harry, Julia, Julie, Katrin, Konstantinos, Lejla, Maria, Maria, Mary, Minas, Moreno, Nikos, Olga, Paris, Paul, Raffael, Rihan, Ruochan, Sarah, Spyros, Tilen, Ufuk, Vaidotas, Xiaoyu and Yunpeng.

Author biography

Panagiotis Kotetes

Dr Panagiotis Kotetes recently embarked on his five-year faculty appointment at the Institute of Theoretical Physics of the Chinese Academy of Sciences in Beijing. During 2015–2018 he was a postdoctoral researcher at the Niels Bohr Institute of the University of Copenhagen, where this book was mainly written. His first postdoctoral appointment was at the Karlsruhe Institute of Technology, where he worked for five years. Panagiotis carried out his Diploma, Masters and PhD studies at the National Technical University of Athens in Greece. His research interests and activity cover the topics of topological systems, unconventional superconductivity, exotic magnetism, and quantum computing.

Symbols to topological insulators

I. Abbreviations

BdG	Bogoliubov - de Gennes
BHZ	Bernevig-Hughes-Zhang
BZ	Brillouin zone
CA	Chiral anomaly
CC	Charge conjugation
CCS	Charge-conjugation symmetry
CS	Chern-Simons
HH	Heavy hole
IR	Irreducible representation
LDOS	Local density of states
LH	Light hole
MCS	Maxwell-Chern-Simons
QAHE	Quantum anomalous Hall effect
QHE	Quantum Hall effect
RWA	Rotating wave approximation
SOC	Spin-orbit coupling
SOH	Split-off hole
SSH	Su-Schrieffer-Heeger
TKI	Topological Kondo insulator
TQD	Triple quantum dot
TR	Time reversal
TRI	Time-reversal invariant
TRS	Time-reversal symmetry
lhs	Left-hand side
occ	Occupied
rhs	Right-hand side

II. Basic mathematical notation

$\mathbb{N}$	Set of natural numbers
$\mathbb{R}$	Set of real numbers
$\mathbb{Z}$	Set of integer numbers
i	Imaginary unit defined as $\mathrm{i}^2 = -1$
$\mathbb{S}^d$	d-dimensional sphere
$\mathbb{T}^d$	d-dimensional torus
$\pi_{\mathrm{Base}}[\mathrm{Target}]$	Homotopy mapping from the Base space to the Target space
$\boldsymbol{n}$	Vector defined in a d-dimensional space with components n_i with $i = 1, 2, \ldots, d$
$\lvert\boldsymbol{n}\rvert$	Modulus of $\boldsymbol{n}$
$\check{\boldsymbol{n}}$	Unit vector $\boldsymbol{n}/\lvert\boldsymbol{n}\rvert$
$\bar{\boldsymbol{n}}$	Rank-2 tensor defined in a d-dimensional space with components n_{ij} with $i, j = 1, 2, \ldots, d$
$\mathcal{C}$	Curve or path

S	Area
$\mathcal{V}$	Volume
Ω_{d+1}	Solid angle subtended by a d-dimensional sphere
∂V	Boundary of space V
$f'(x)$	Derivative $df(x)/dx$
$\dot{\theta}$	Time derivative of θ
∇	Nabla operator defined in three-dimensional coordinate space $(\partial_x, \partial_y, \partial_z)$
$\nabla_{\boldsymbol{k}}$	Nabla operator defined in three-dimensional wavevector space $(\partial_{k_x}, \partial_{k_y}, \partial_{k_z})$
$\int d^d x$	Integral over a d-dimensional volume
$\int d\boldsymbol{r}$	Integral over a volume without specifying the dimensionality of space
$\int_S d\boldsymbol{S} \cdot \boldsymbol{f}$	Surface integral of a vector function $\boldsymbol{f}$
$\int_C d\boldsymbol{r} \cdot \boldsymbol{f}$	Line integral of a vector function $\boldsymbol{f}$
P_C	Path-ordering operator for a line integral
$\mathcal{P}\int$	Principal value of an integral
$n!$	Factorial of n
$\mathrm{Im}(z)$	Imaginary part of the complex number z
$\mathrm{Re}(z)$	Real part of the complex number z
$\mathrm{sgn}(z)$	Sign function of $z \in \mathbb{R}$
$\Gamma(z)$	Gamma function of $z \in \mathbb{R}$
$\delta(z)$	Dirac delta function/distribution of $z \in \mathbb{R}$
$\delta_{a,b}$	Kronecker delta
$\varepsilon_{a_1 a_2 a_3 \ldots a_d}$	Levi-Civita anti-symmetric symbol defined in d spatial dimensions
$\Theta(z)$	Heaviside step function of $z \in \mathbb{R}$

III. Matrix and operator notation

$\hat{A}$	Matrix with label A; Quantum-mechanical operator of the physical observable A
$\hat{A}_a$	Operator effecting the antiunitary symmetry transformation A_a
$\hat{A}_u$	Operator effecting the unitary symmetry transformation A_u
$\hat{A}^{\mathrm{T}}$	Transpose of matrix $\hat{A}$
$\hat{A}^*$	Complex conjugate of $\hat{A}$
$\hat{A}^\dagger$	Hermitian conjugate of $\hat{A}$
$\hat{A}^{-1}$	Inverse of $\hat{A}$
$\mathrm{diag}\{A_1, A_2, \ldots, A_N\}$	Diagonal matrix with entries A_s where $s = 1, 2, \ldots, N$
$\det(\hat{A})$	Determinant of matrix $\hat{A}$
$\mathrm{Pf}(\hat{A})$	Pfaffian of an anti-symmetric matrix $\hat{A}$
$[\hat{A}, \hat{B}] = \hat{A}\hat{B} - \hat{B}\hat{A}$	Commutator of two operators or matrices $\hat{A}$ and $\hat{B}$
$\{\hat{A}, \hat{B}\} = \hat{A}\hat{B} + \hat{B}\hat{A}$	Anti-commutator of two operators or matrices $\hat{A}$ and $\hat{B}$
$\otimes$	Kronecker product symbol
$\hat{\boldsymbol{\Gamma}}$	Vector notation of five anti-commuting matrices generating the so(5) Clifford algebra
$\hat{\gamma}^\mu$	

	The matrices appearing in Dirac's equation with the values of μ determined by the spacetime dimensionality
$\hat{\gamma}^5$	Chirality matrix which satisfies $\{\hat{\gamma}^5, \hat{\gamma}^\mu\} = \hat{0}$ and is definable in even-dimensional spacetimes
$\hat{\boldsymbol{\lambda}}$	Vector notation of the eight SU(3) Gell-Mann matrices
$\hat{\boldsymbol{\Lambda}}$ and $\hat{\boldsymbol{\lambda}}^{\pm}$	Vector notation of matrices which are particular linear combinations of the SU(3) Gell-Mann matrices
$\hat{\boldsymbol{\zeta}}, \hat{\boldsymbol{\eta}}, \hat{\boldsymbol{\sigma}}, \hat{\boldsymbol{\tau}}$	Pauli matrices
$\hat{1}$	Identity matrix or operator
$\hat{1}_\Gamma$	Identity matrix in the space spanned by the $\hat{\boldsymbol{\Gamma}}$ matrices
$\hat{1}_\zeta, \hat{1}_\eta, \hat{1}_\sigma, \hat{1}_\tau$	Unit matrices in the spaces defined by the respective Pauli matrices
$\hat{0}$	Null matrix or operator

IV. Symmetry operations

C_n	Counterclockwise rotation by $2\pi/n$ about the out-of-plane axis of a planar system
I	Inversion
K	Complex conjugation
$\mathcal{T}$	Time reversal
σ_{xz}	Reflection in (x, y, z) space with xz as the mirror plane, i.e., $y \mapsto -y$
σ_v	Reflection with a vertical mirror plane

V. Symmetry groups and representations

E	Identity element of a symmetry group
E'	The element of a double-covering symmetry group satisfying $(E')^2 = E$
$\mathcal{G}$	Symmetry group
g	Element of a symmetry group; Gerade irreducible representation, i.e., even under inversion operation
$h_{\mathcal{G}}$	Order of a symmetry group $\mathcal{G}$
u	Ungerade irreducible representation, i.e., odd under inversion operation
$\chi(g)$	Character of the symmetry-group element g
C_{2v}, C_3, C_{3v}, C_{4v}, C_{6v}, O_h, T_d	Types of point groups
SO(N), SU(N), U(M)	Rotation or gauge groups with $N = 2, 3, \ldots$ and $M = 1, 2, 3, \ldots$
$A_{1,2}$	One-dimensional irreducible representations of particular point groups
A_{1g}, A_{1u}, B_{1g}, B_{2g}	One-dimensional gerade or ungerade irreducible representations of particular point groups
$\Gamma_{6,7,8}$	Irreducible representations of the double-covering T_d point group
E	Two-dimensional irreducible representation of particular point groups
$T_{1,2}$	Three-dimensional irreducible representations of particular point groups
T_{1g}, T_{1u}	Three-dimensional gerade or ungerade irreducible representations of particular point groups

Γ, X, Y, M	The four high-symmetry points of a tetragonal Brillouin zone
Γ, M, Z, A	Four of the high-symmetry points of a cubic Brillouin zone
$\boldsymbol{K}$ and $\boldsymbol{K}'$	The two Brillouin-zone wavevectors where graphene's Dirac cones are located at

VI. Symmetry classes

A, AIII, AI, BDI, D, DIII, AII, CII, C, CI	The ten symmetry classes

VII. Fundamental constants

e	Unit of charge. The electron's charge is $-e < 0$
h and $\hbar$	Planck's constant
k_B	Boltzmann's constant
α_s	Fine-structure constant
Φ_0	Magnetic-flux quantum h/e

VIII. Physical quantities – Latin symbols

$\boldsymbol{A}(\boldsymbol{r}, t)$	Electromagnetic vector potential at position $\boldsymbol{r}$ and time instant t
$\boldsymbol{A}^a(\boldsymbol{R})$	Abelian Berry vector potential of the a-th eigenstate defined in the parameter space spanned by $\boldsymbol{R}$
$\boldsymbol{A}^{a,S}(\boldsymbol{R})$ and $\boldsymbol{A}^{a,N}(\boldsymbol{R})$	Abelian Berry vector potential of the a-th eigenstate per south and north patch of a $\mathbb{S}^2$
$\hat{\boldsymbol{A}}(\boldsymbol{R})$	Non-Abelian Berry vector potential (matrix) defined in the parameter space spanned by $\boldsymbol{R}$
$A_s(\boldsymbol{R})$ and $\boldsymbol{a}_s(\boldsymbol{R})$	U(1) and SU(2) parts of the non-Abelian 2×2 matrix Berry vector potential's s-th component
$\hat{A}(E)$	Energy-resolved matrix spectral function
$\hat{A}(\boldsymbol{r}, E)$	Spatially- and energy-resolved matrix spectral function
$\hat{A}^{(0)}_{\text{col}}(\boldsymbol{k}, E)$	Wavevector- and energy-resolved matrix spectral function of a disordered system
$\mathcal{A}(x, t)$	Anomalous contribution to the action due to chiral anomaly in two spacetime dimensions
$\boldsymbol{a}_{1,2,3}$	Direct-lattice vectors connecting nearest-neighbour sites of a hexagonal lattice
a, b, c	Lattice constants of a crystal in the x, y, z directions
$\boldsymbol{B}(\boldsymbol{r}, t)$	Magnetic field (induction) at position $\boldsymbol{r}$ and time instant t
$B_{\text{inc},\pm}$, $B_{\text{ref},\pm}$ and $B_{\text{trs},\pm}$	Incident, reflected and transmitted magnetic field in the circular-polarisation basis
$\hat{B}$	Sewing matrix residing on the presence of a charge-conjugation symmetry
$\boldsymbol{b}$	Burgers vector describing a dislocation defect
C_n	n-th Chern number
$C_n(\hat{F})$	

	n-th Chern character with $\hat{F}$ the corresponding non-Abelian Berry field strength tensor
CS_d	$\mathbb{Z}_2$ Chern-Simons invariant defined in d spatial dimensions
$\hat{c}_a(\boldsymbol{k})$	Annihilation operator of a spinful ($a = \uparrow, \downarrow$) wide-band (itinerant) electron with wavevector $\boldsymbol{k}$
c	Speed of light; Coefficients of wavefuctions
$\boldsymbol{D}(\boldsymbol{r}, t)$	Dielectric displacement at position $\boldsymbol{r}$ and time instant t
$\mathrm{D}\tilde{\psi}\mathrm{D}\psi$	Path integral measure for two spinless Grassmann fields
D	Dimensionality of the sphere $\mathbb{S}^D$ which encloses a topological defect
d	Number of spatial dimensions; Quantum-well width
d_c	Critical quantum-well width for realizing a topological phase transition
$\hat{d}_{12}$	Zero-energy fermionic operator constructed by two Majorana operators $\hat{\gamma}_{1,2}$
$\boldsymbol{E}(\boldsymbol{r}, t)$	Electric field at position $\boldsymbol{r}$ and time instant t
$\boldsymbol{E}_{\mathrm{conf}}(\boldsymbol{r})$	Confinement electric field at position $\boldsymbol{r}$
$\boldsymbol{E}_{\mathrm{inc}}$, $\boldsymbol{E}_{\mathrm{ref}}$ and $\boldsymbol{E}_{\mathrm{trs}}$	Incident, reflected and transmitted electric field
$E_{\pm}$	Right and left circularly-polarised electric field
$E_{\mathrm{inc},\pm}$, $E_{\mathrm{ref},\pm}$ and $E_{\mathrm{trs},\pm}$	Incident, reflected and transmitted electric field in the circular-polarisation basis
E	Energy
E_F	Fermi energy
E_c	Conduction-band energy offset
$\boldsymbol{F}$	Force
$\hat{F}_{ij}(\boldsymbol{k})$	Non-Abelian Berry field strength tensor (matrix in the occupied-states space) for a given wavevector $\boldsymbol{k}$
$f(\varepsilon)$	Fermi–Dirac distribution function
$f(\hat{\boldsymbol{p}}, \boldsymbol{r})$	Form factor of the superconducting order parameter
$\hat{f}_{n,a}$	Annihilation operator of a spinful ($a = \uparrow, \downarrow$) narrow-band (localised) electron at the lattice site $\boldsymbol{R}_n$
$\hat{f}_a(\boldsymbol{k})$	Annihilation operator of a spinful ($a = \uparrow, \downarrow$) narrow-band (localised) electron with wavevector $\boldsymbol{k}$
G_H or G_{xy}	Hall conductance
$\hat{G}$	Green function (operator or matrix)
$\hat{G}^{(0)}$	Bare Green function (operator or matrix)
$\hat{G}_{\mathrm{mf}}$	Mean-field Green function (operator or matrix)
$\hat{G}^{\mathrm{R}}$	Retarded Green function (operator or matrix)
$\hat{G}^{\mathrm{R},(0)}_{\mathrm{col}}$	Retarded Green function (operator or matrix) of a disordered system
$\boldsymbol{g}$	Vector parametrising the SU(2) part of a 2×2 matrix
g	Strength of a four-fermion-operator interaction
$\boldsymbol{H}(\boldsymbol{r}, t)$	Magnetic field at position $\boldsymbol{r}$ and time instant t
H	Hamiltonian
$\hat{H}_0$	Bare or unperturbed Hamiltonian operator
$\hat{H}_{\boldsymbol{A}}$	Hamiltonian operator coupling of the electron to the electromagnetic vector potential
$\hat{H}^{\mathrm{p}}_{\boldsymbol{A}}$	Paramagnetic Hamiltonian operator coupling of the electron to the electromagnetic vector potential

$\hat{H}_{\boldsymbol{A}}^{\mathrm{d}}$	Diamagnetic Hamiltonian operator coupling of the electron to the electromagnetic vector potential
$\hat{H}_{\mathrm{ext}}$	Hamiltonian operator describing the coupling of electrons to external perturbations
$\hat{h}(\boldsymbol{k})$	Upper block of a block off-diagonalised chiral-symmetric Hamiltonian (matrix) for a given wavevector $\boldsymbol{k}$
$\boldsymbol{J}$	Total angular momentum
$\boldsymbol{J}(t)$	Particle current at time instant t
$\boldsymbol{J}(\boldsymbol{r}, t)$	Particle current density at time instant t
$\boldsymbol{J}_c(t)$	Electric current at time instant t
$\boldsymbol{J}_c(\boldsymbol{r}, t)$	Electric current density at time instant t
$\boldsymbol{J}_c^{\mathrm{p}}(\boldsymbol{r}, t)$	Paramagnetic electric current density at time instant t
$\boldsymbol{J}_c^{\mathrm{d}}(\boldsymbol{r}, t)$	Diamagnetic electric current density at time instant t
$\boldsymbol{J}_c(\boldsymbol{q}, t)$	Electric current density in Fourier space at time instant t
$\boldsymbol{J}_{c,b}$	Bound electric current
$\boldsymbol{J}_{c,f}$	Free electric current
$\boldsymbol{J}_E$	Energy current
$\boldsymbol{J}_Q$	Heat current
$\hat{J}_{\pm}$	Raising and lowering total angular momentum operators
$\hat{J}_{c,\pm}$	Electric current operator in the circular-polarisation basis
j	Quantum number of the total angular momentum of a particle
$\boldsymbol{k}$	Wavevector
$\boldsymbol{k}_I$	Inversion-symmetric wavevector
$\boldsymbol{k}_{I,C_4}$	Inversion- and fourfold-rotation-symmetric wavevector
$\boldsymbol{k}_{\perp}$	Wavevector confined to the xy plane, i.e., (k_x, k_y)
k_c	Cutoff wavevector
k_F	Fermi wavevector
k_n	Matsubara frequencies for fermions with n an integer
$k_{x,y,z}$	Wavenumbers
k_0	Real frequency obtained as the continuous limit of the fermionic Matsubara frequencies
$\boldsymbol{L}$	Orbital angular momentum
$L(\boldsymbol{r}, t)$	Lagrangian density
$L_{\mathrm{CA}}(\boldsymbol{r}, t)$	Chiral-anomaly contribution to the Lagrangian density
$\hat{L}_a^{(0)}(\boldsymbol{k}, \omega)$	Operator appearing in the calculation of the dynamical Hall conductivity
$L_{x,y,z}$	System's length in the x, y, z directions
ℓ	Quantum number of the orbital angular momentum of a particle; Domain wall width
ℓ_B	Magnetic length
$\boldsymbol{M}(\boldsymbol{r}, t)$	Magnetisation at position $\boldsymbol{r}$ and time instant t
$\boldsymbol{M}_{\mathrm{orb}}(\boldsymbol{r}, t)$	Orbital magnetisation at position $\boldsymbol{r}$ and time instant t
$\boldsymbol{M}(\boldsymbol{p})$	Momentum dependent magnetisation
$\mathcal{M}$	$\mathbb{Z}_2$ invariant residing on the presence of a charge-conjugation symmetry
$\mathcal{M}(k)$	$\mathbb{Z}_2$ invariant residing on the presence of a charge-conjugation symmetry for a given wavevector k

$\hat{\boldsymbol{m}}_{\text{orb}}$	Orbital magnetic moment operator
$\hat{\boldsymbol{m}}_a(\boldsymbol{k})$	Orbital magnetic moment of the a-th eigenstate for a given wavevector $\boldsymbol{k}$
$\hat{\mathrm{m}}(\boldsymbol{k})$	Kane-Mele anti-symmetric matrix employed to define a $\mathbb{Z}_2$ invariant
$\tilde{m}$	Conjugate thermodynamic field of m, which is defined via a Legendre transform
m	Mass; Magnetisation energy scale
m_c and $m_c{}'$	Critical magnetisation energy scales for realizing a topological phase transition
m_j	Azimuthal quantum number of the total angular momentum of a total-angular-momentum-j particle
m_ℓ	Azimuthal quantum number of the orbital angular momentum of a particle
m_s	Azimuthal quantum number of the spin angular momentum of a spin-s particle
N	Particle number; Variable denoting one of the dimensions of a square matrix
$\mathcal{N}$	Number of lattice sites
N	Nernst signal
N_{occ}	Number of occupied Landau levels
$N_{x,y,z}$	Number of unit cells of a crystal in the x, y, z directions
$N_{2,3}$	$\mathbb{Z}_2$ topological invariants
$N_\pm$	Refractive index (complex) in the circular-polarisation basis
$n_\pm$	Real part of the refractive index $N_\pm$
n	Real part of the refractive index for an isotropic material
$\lvert n(\boldsymbol{R})\rangle$	Adiabatic state vector defined for the adiabatically- evolving parameters $\boldsymbol{R}$
n_f	Occupation number of f electrons
$\boldsymbol{P}$	Electric polarisation
$\boldsymbol{P}(\boldsymbol{r})$	Electric polarisation density
$\boldsymbol{P}_n$	Electric polarisation at the lattice site $\boldsymbol{R}_n$
P_d	Polarisation field in d spatial dimensions
P_2 or P_θ	$\mathbb{Z}_2$ polarisation
P_3	Magnetoelectric polarisation
$\hat{P}_a(\boldsymbol{k})$	Projector operator onto the a-th eigenstate for a wavevector $\boldsymbol{k}$
$\boldsymbol{p}$	Momentum
p_F	Fermi momentum
$\boldsymbol{Q}_{1,2,3}$	Magnetic/Nesting wavevector
Q	Electric charge; Wavevector of a one-dimensional charge-density wave system
$\hat{Q}(\boldsymbol{k})$	Projection operator onto the occupied bands of a chiral-symmetric Hamiltonian for a wavevector $\boldsymbol{k}$
Q_n	n-th Chern-Simons form
$\boldsymbol{q}$	Wavevector of a bosonic particle
$q_\pm$	Wavevectors of a right and left circularly-polarised electromagnetic wave
$\hat{q}(\boldsymbol{k})$ or $q(\boldsymbol{k})$	Upper block of the block off-diagonalised projection operator (matrix) $\hat{Q}(\boldsymbol{k})$
$\boldsymbol{R}$	Vector spanning a parameter space
$\boldsymbol{R}_n$	Position vector of the n-th direct-lattice site
$R_\theta^{\check{\boldsymbol{n}}}$	Rotation by an angle θ about an axis directed along $\check{\boldsymbol{n}}$

$\boldsymbol{r}$	Particle's position vector; Coordinate
$r_\pm$	Reflection coefficient for right and left circularly-polarised electromagnetic waves
$\boldsymbol{S}$	Spin angular momentum
S	Action
S	Seebeck coefficient
$S_a(\boldsymbol{k})$	Electronic entropy of the a-th eigenstate for a given wavevector $\boldsymbol{k}$
S_A^{p}	Action in the presence of the paramagnetic coupling to the vector potential
S_{CA}	Chiral-anomaly action
S_{CS}	Chern-Simons action
S_{MCS}	Maxwell-Chern-Simons action
S_{mf}	Mean-field action
Sp	Trace over the quantum-mechanical degrees of freedom combined with statistical averaging
s	Quantum number of the spin angular momentum of a spin-s particle; Variable parametrising a dislocation
T	Temperature; Period
T_{drive}	Driving period
T_{kick}	Kicking period
T_{pump}	Charge-pumping period
Tr	Trace over internal degrees of freedom
$\hat{T}_\tau$	Time-ordering operator in imaginary time τ
$\hat{T}_{T_{\mathrm{kick}}}$	Time-ordering operator for a kicking period
$\hat{T}(\boldsymbol{r}_1, \boldsymbol{r}_2, E)$	Spatially- and energy-resolved T-matrix function
$\hat{T}(\boldsymbol{k}_1, \boldsymbol{k}_2, E)$	Wavevector- and energy-resolved T-matrix function
$\boldsymbol{t}_{1,2,3}$	Lattice vectors of the Bi_2Se_3 unit cell
t	Time
$t_a^{\check{\boldsymbol{n}}}$	Translation by a parallel to $\check{\boldsymbol{n}}$
t_{dd}	Matrix element for interdot electron hopping
t_f and $\tilde{t}_f$	Bare and renormalised hopping strength of f electrons
t_{nm}	Matrix elements for electron hopping between the lattice sites $\boldsymbol{R}_{n,m}$
$t_{1,2}$	Nearest and next-nearest neighbour hopping on a hexagonal lattice; Time instants
$t_\pm$	Transmission coefficient for right and left circularly-polarised electromagnetic waves
$\hat{U}$	Unitary transformation; Impurity potential operator
$\hat{U}(\boldsymbol{k}, t)$	Quantum-mechanical time-evolution operator for a given wavevector $\boldsymbol{k}$
U_f	On-site Hubbard interaction strength
$\hat{U}_\Theta$	Unitary part of the operator $\hat{\Theta}$ effecting a generalised time reversal
$\hat{U}_\Xi$	Unitary part of the operator $\hat{\Xi}$ effecting charge conjugation
$\boldsymbol{u}_{\boldsymbol{k}}$	Bloch wavefunction of a particle with nonzero spin and wavevector $\boldsymbol{k}$
$u_{\boldsymbol{k}}$	Bloch wavefunction of a spin-zero (scalar) particle and wavevector $\boldsymbol{k}$
$\boldsymbol{V}$ and $\tilde{\boldsymbol{V}}$	Bare and renormalised hybridisation vector function
$\hat{V}_{\mathrm{int}}$	Four-fermion-operator interaction Hamiltonian

$V(\boldsymbol{r}, t)$	Electromagnetic scalar potential at position $\boldsymbol{r}$ and time instant t
$V_{\text{conf}}(\boldsymbol{r})$	Electrostatic confinement potential at position $\boldsymbol{r}$
$\boldsymbol{v}(t)$	Velocity of a point particle
v_F	Fermi velocity
$\hat{W}$	Sewing matrix residing on a time-reversal symmetry
W_C	Wilson loop
W_d	$\mathbb{Z}_2$ topological invariant defined in d spatial dimensions based on the related Chern-Simons invariant
w_d	Winding number defined in d spatial dimensions
$x_s(\Phi)$	s-th guiding center for a given magnetic flux Φ
Y_n	Quantities for electric, thermal and thermoelectric transport

IX. Physical quantities – Greek symbols

$\bar{\boldsymbol{\alpha}}$	Thermoelectric conductivity tensor
$\bar{\boldsymbol{\alpha}}^{\text{P}}$	Parathermal component of the thermoelectric conductivity tensor
α	Thermoelectric conductivity/coefficient for an isotropic system; Strength of Rashba spin–orbit coupling
α'	Strength of Rashba spin–orbit coupling
β, β'	Inverse masses
$\hat{\gamma}_E$	Annihilation operator of a Majorana quasiparticle for a given energy E
$\hat{\gamma}_E^\dagger$	Creation operator of a Majorana quasiparticle for a given energy E satisfying $\hat{\gamma}_E^\dagger = \hat{\gamma}_{-E}$
γ	Strength of spin–orbit coupling leading to hexagonal warping
$\gamma_{1,2}$	Coefficients appearing in the Luttinger Hamiltonian
$\gamma_a(t)$	Berry phase of the a-th eigenstate at time instant t
ΔA	Change of quantity A during a pumping process
ΔQ	Transferred charge during a pumping process
$\hat{\Delta}(\hat{\boldsymbol{p}}, \boldsymbol{r})$	Superconducting spin-matrix order parameter
$\Delta(q)$	Order parameter of a charge-density wave phase
$\Delta\kappa$	Difference between the imaginary parts of the refractive index for right and left circularly-polarised light
Δn	Difference between the real parts of the refractive index for right and left circularly-polarised light
$\boldsymbol{\delta}_{1,2,3}$	Direct-lattice vectors connecting next-nearest neighbour sites of a hexagonal lattice
δA	Variation of a quantity A
$\delta_{\boldsymbol{k}_l}$	Signs of the Pfaffian of the sewing matrix computed at inversion-symmetric wavevectors
$\bar{\boldsymbol{\varepsilon}}$	Dielectric (relative) tensor
ε	Onsite energy scale
$\varepsilon(\omega)$	Longitudinal dielectric constant for a frequency ω
$\varepsilon_c(\boldsymbol{k})$	Energy dispersion of the conduction band for a given wavevector $\boldsymbol{k}$
$\varepsilon_H(\omega)$	Hall dielectric constant for a frequency ω
$\varepsilon_\pm$	

	Dielectric constant for right and left circularly-polarised electromagnetic waves
ε_0	Vacuum permittivity
Z	Partition function
Z_0	Vacuum impedance
$Z_a(\boldsymbol{k})$	Renormalization factor of the a-th eigenstate for a given wavevector $\boldsymbol{k}$
ζ	Wavenumber or decay length of a localised zero-energy solution; Coefficient of spin–orbit coupling
Θ	Generalised time reversal
θ	Phase or angle; Chern-Simons action coefficient; Variable for dimensional extension or pumping processes
$\theta(\boldsymbol{r}, t)$	Phason Goldstone mode/field defined at position $\boldsymbol{r}$ and time instant t
θ_K	Kerr angle
$\theta_\pm$	Polar angles of the complex reflection coefficients $r_\pm$
ϑ	Phase or angle
$\bar{\boldsymbol{\kappa}}$	Thermal conductivity tensor
$\bar{\boldsymbol{\kappa}}^{\mathrm{p}}$	Parathermal component of the thermal conductivity tensor
$\bar{\boldsymbol{\kappa}}_e$	Electron thermal conductivity tensor
κ	Thermal conductivity for an isotropic system; Imaginary part of the refractive index for an isotropic material
κ_e	Electron thermal conductivity for an isotropic system
$\kappa_\pm$	Imaginary part of the refractive index $N_\pm$
λ	Strength of Dresselhaus spin–orbit coupling
μ	Chemical potential
μ_0	Vacuum permeability
ν_{B_z}, ν_{flux} and ν_Φ	Number of magnetic flux quanta in different situations
$\nu_{d=2,3}$	$\mathbb{Z}_2$ Fu-Kane invariant in $d = 2, 3$
ν_0	Strong $\mathbb{Z}_2$ topological invariant of a three-dimensional TRI topological insulator
$\nu_{1,2,3}$	Weak $\mathbb{Z}_2$ topological invariants of a three-dimensional TRI topological insulator
Ξ	Charge conjugation
ξ	Decay length of a localised zero-energy solution
$\xi_a(\boldsymbol{k})$	Eigenvalue of the inversion operator when acting on one of the two Kramers partners of the a-th Bloch band for a given $\boldsymbol{k}$
$\bar{\boldsymbol{\Pi}}^{\mathrm{cc}}$	Electric charge polarisation tensor
$\bar{\boldsymbol{\Pi}}^{\mathrm{Ec}}$	Energy-charge polarisation tensor
$\bar{\boldsymbol{\Pi}}^{\mathrm{EE}}$	Energy-energy polarisation tensor
$\Pi_{\theta A}$, $\Pi_{\theta V}$, $\Pi_{\theta\theta}$	Polarisation tensors of coupled electromagnetic and phason fields
$\hat{\Pi}$	Chiral-symmetry operator
$\hat{\boldsymbol{\pi}}$	Gauge invariant momentum operator
ρ	Average particle density
$\rho_{c,b}$	Bound electric charge
$\rho_{c,f}$	Free electric charge
$\rho_c(\boldsymbol{r}, t)$	Electric charge density at time instant t
$\rho_c(\boldsymbol{q})$	Electric charge density in Fourier space
$\rho_E(\boldsymbol{r}, t)$	Energy density at time instant t

$\hat{\Sigma}$	Self-energy (operator or matrix)
$\bar{\boldsymbol{\sigma}}$	Electric charge conductivity tensor
$\bar{\boldsymbol{\sigma}}^{\mathrm{p}}$	Paramagnetic component of the electric charge conductivity tensor
σ	Longitudinal electric charge conductivity; Label of spin, chirality and other eigenstates
σ_H	Hall electric charge conductivity
$\sigma_{++}, \sigma_{--}, \sigma_{+-}$ and σ_{-+}	Electric charge conductivity tensor elements in the circular-polarisation basis
τ	Imaginary time
τ_{col}	Collision time
$\boldsymbol{\Phi}$	Multicomponent wavefunction describing spinful electrons of a triple-quantum dot; Gravitational vector field
Φ	Magnetic flux; Gravitational scalar field
$\lvert\boldsymbol{\phi}>$	State vector of a particle with nonzero spin
$\boldsymbol{\phi}$	Wavefunction of a particle with nonzero spin
$\phi_{\uparrow,\downarrow}$	Upper and lower spinor components of a spin-1/2 particle
$\lvert\phi>$	State vector of a spin-zero (scalar) particle
ϕ	Wavefunction of a spin-zero (scalar) particle; Phase appearing in Haldane's model
ϕ_{nm}	Line integral of the electromagnetic vector potential for a path connecting the lattice sites $\boldsymbol{R}_{n,m}$
φ	Phase or angle
$\chi_a(\boldsymbol{k})$	Phase factor relating Kramers partners of the a-th eigenstate for a wavevector $\boldsymbol{k}$
$\chi(q)$	Charge-charge correlation function (susceptibility) for a wavevector q
χ_m	Susceptibility in response to the field m
$\hat{\boldsymbol{\Psi}}(\boldsymbol{r})$	Bogoliubov - de Gennes annihilation operator at position $\boldsymbol{r}$
$\hat{\psi}(\boldsymbol{r})$ / $\hat{\psi}^{\dagger}(\boldsymbol{r})$	Annihilation / Creation operator of a spinless electron at position $\boldsymbol{r}$
$\hat{\psi}(\boldsymbol{k})$ / $\hat{\psi}^{\dagger}(\boldsymbol{k})$	Annihilation / Creation operator of a spinless electron with a continuous wavevector $\boldsymbol{k}$
$\hat{\psi}_{\boldsymbol{k}}$ / $\hat{\psi}_{\boldsymbol{k}}^{\dagger}$	Annihilation / Creation operator of a spinless electron with a discrete wavevector $\boldsymbol{k}$
$\hat{\psi}_n$ / $\hat{\psi}_n^{\dagger}$	Annihilation / Creation operator of a spinless electron at the lattice site $\boldsymbol{R}_n$
ψ and $\tilde{\psi}$	Spinless Grassmann fields
$\boldsymbol{\psi}$ and $\tilde{\boldsymbol{\psi}}$	Spinful Grassmann fields
Ω	Thermodynamic potential
$\Omega^{(0)}$	Thermodynamic potential for a non-interacting system
Ω^{orb}	Thermodynamic potential of an electronic system obtained by considering only the orbital coupling to a magnetic field
$\Omega^a_{R_i,R_j}(\boldsymbol{R})$	Berry curvature of the a-th eigenstate defined in the parameter space $\boldsymbol{R}$
$\Omega_z(\boldsymbol{k})$	Berry curvature in $\boldsymbol{k} = (k_x, k_y)$ space
ω	Real-time frequency
ω_c	Cyclotron frequency
ω_p	Plasma frequency
ω_s	Bosonic Matsubara frequency with s an integer
ω_0	Driving frequency

X. Relativistic notation

$x^{\mu} = (t, \boldsymbol{r})$	
$\eta_{\mu\nu}$	Spacetime metric; Minkowski spacetime metric: $\eta_{\mu\nu} = \mathrm{diag}\{1, -1, -1, -1\}$
$\varepsilon_{\alpha_1,\alpha_2,\ldots,\alpha_\nu}$	Levi-Civita tensor in a ν-dimensional spacetime
A^{μ}	Electromagnetic d-potential $(V, \boldsymbol{A})$ defined in a d-dimensional spacetime
$F^{\mu\nu}$	Electromagnetic field strength tensor
J_c^{μ}	Electric d-current $(\rho_c, \boldsymbol{J}_c)$ defined in a d-dimensional spacetime
$J_{c,\mathrm{CS}}^{\mu}$	Chern-Simons electric d-current defined in a d-dimensional spacetime
$J_{c,b}^{\mu}$	Bound electric d-current defined in a d-dimensional spacetime
$J_{c,f}^{\mu}$	Free electric d-current defined in a d-dimensional spacetime
J_5^{μ}	Chiral d-current defined in a d-dimensional spacetime
$\hat{v}^{\mu}$	Matrix vertex functions

Topological Insulators

Panagiotis Kotetes

Chapter 1

Symmetries and effective Hamiltonians

In this chapter we introduce a number of symmetry concepts that play a key role in predicting and analysing topological phenomena. First we provide an introduction to unitary and antiunitary symmetry transformations and afterwards unfold a programme of how to employ them for retrieving symmetry-invariant effective Hamiltonians. We specifically focus on models for III–V semiconductors, which are particularly relevant for the topological systems to be studied later. One should note that the invariance of a system under a set of symmetries not only serves as an indispensable tool for performing technical analysis, but is essentially decisive for the arising topological properties.

1.1 Crash course on symmetry transformations

By and large, the behaviour of material systems and other abstract objects under the action of a group of symmetry transformations, allows us to categorise them into distinct classes. The members of a given class share common features and characteristics tied to the ensuing symmetry. An abstract object is said to exhibit a particular symmetry when it remains invariant under the corresponding symmetry transformation or operation, cf textbooks such as [1–5]. For example, an ideal planar disc is symmetric or invariant under arbitrary rotations about its out-of-plane axis. This, in turn, implies that the coordinate vector of a given point on the disc transforms in a specific fashion under rotations. Apart from spatial rotations, symmetry tranformations additionally include operations such as reflections, inversion and translations, while the last two are also extended to the time dimension. More importantly, one can also define symmetries not associated with the physical spacetime but rather with a type of internal or parameter space related, for instance, to isospin, flavour, colour, valley or other degrees of freedom.

When symmetry transformations are defined in the Hilbert space of a given quantum-mechanical system, they are usually considered to yield an O-symmetry when the respective operator effecting the symmetry $\hat{O}$, leaves the Hamiltonian ($\hat{H}$)

doi:10.1088/978-1-68174-517-6ch1

invariant, i.e. $\hat{O}^{\dagger}\hat{H}\hat{O} = \hat{H} \Rightarrow [\hat{O}, \hat{H}] = \hat{0}$ [1]. As a result, the presence of symmetry leads to energy degeneracies for two (or more) Hamiltonian eigenstates with quantum numbers $a \neq b$, i.e. $E_a = E_b$. While this corresponds to the conventional and most often used definition, one can extend the notion of symmetry to include also those which instead yield a sign change of the Hamiltonian, i.e. $\hat{O}^{\dagger}\hat{H}\hat{O} = -\hat{H} \Rightarrow \{\hat{O}, \hat{H}\} = \hat{0}$ [6–8]. Symmetries of this kind are crucial for predicting and describing the topological properties of a system. This will become clear in the upcoming chapters and especially in chapter 8. While these symmetries do not generally imply degeneracies, they still impose constraints on the energy spectrum, e.g. $E_a = -E_b$. Finally, the operators effecting symmetry transformations can be split into two categories, the unitary (O_u) and antiunitary (O_a), which can be distinguished by their different action on a complex c-number z, i.e. $\hat{O}_u^{\dagger} z \hat{O}_u = z\hat{1}$ and $\hat{O}_a^{\dagger} z \hat{O}_a = z^*\hat{1}$, with $\hat{1}$ the identity operator.

1.1.1 Unitary symmetry transformations

Here we introduce the unitary symmetries by considering the case of a spinless particle described by a wavefunction ϕ, that is a solution of the time-independent single-particle Schrödinger equation $\hat{H}\phi = E\phi$ [1], with E denoting the particle's energy. Within the quantum-mechanical description, a symmetry transformation acts on a wavefunction which depends on the particle's coordinate $\boldsymbol{r}$ in the following manner

$$\phi' := O\phi \quad \text{and} \quad \phi'(\boldsymbol{r}) = \hat{O}\phi(\boldsymbol{r}), \tag{1.1}$$

where $\hat{O}$ corresponds to the symmetry-transformation operator, defined in the given $\boldsymbol{r}$ basis. For instance, if we wish to translate the system by a along the direction defined by the unit vector $\check{\boldsymbol{n}}$, i.e. $\boldsymbol{r}' = \boldsymbol{r} + a\check{\boldsymbol{n}}$, we must act on $\phi(\boldsymbol{r})$ with the translation operator $\hat{t}_a^{\check{\boldsymbol{n}}} = e^{-\mathrm{i}a\check{\boldsymbol{n}}\cdot\hat{\boldsymbol{p}}/\hbar}$ [1], where $\hat{\boldsymbol{p}}$ defines the momentum operator which, in the position basis, is represented as $\hat{\boldsymbol{p}} = \hbar\nabla/\mathrm{i}$. It is straightforward to see that for an infinitesimal translation a along $\check{\boldsymbol{n}}$, we have

$$\begin{aligned}\phi'(\boldsymbol{r}) &= e^{-\mathrm{i}a\check{\boldsymbol{n}}\cdot\hat{\boldsymbol{p}}/\hbar}\phi(\boldsymbol{r}) \approx (1 - \mathrm{i}a\check{\boldsymbol{n}} \cdot \hat{\boldsymbol{p}}/\hbar)\phi(\boldsymbol{r}) \\ &= (1 - a\check{\boldsymbol{n}} \cdot \nabla)\phi(\boldsymbol{r}) \approx \phi(\boldsymbol{r} - a\check{\boldsymbol{n}}) \equiv \phi\left(t_a^{-\check{\boldsymbol{n}}}\boldsymbol{r}\right).\end{aligned} \tag{1.2}$$

Note that, in the above, the transformation acts in the argument in the inverse sense compared to the wavefunction. Such a transformation property was anticipated, since the existence of a symmetry transformation implies that we can equivalently write for the scalar wavefunction

$$\phi'(\boldsymbol{r}') = \phi(\boldsymbol{r}) \quad \Rightarrow \quad \phi'(\boldsymbol{r}) = \phi(\boldsymbol{r} - a\check{\boldsymbol{n}}). \tag{1.3}$$

In a similar fashion, a rotation by an angle θ about an axis with direction $\check{\boldsymbol{n}}$ would be effected using the rotation operator $\hat{R}_\theta^{\check{\boldsymbol{n}}} = e^{-\mathrm{i}\theta\check{\boldsymbol{n}}\cdot\hat{\boldsymbol{L}}/\hbar}$ [1], where $\hat{\boldsymbol{L}} = \hat{\boldsymbol{r}} \times \hat{\boldsymbol{p}}$ defines the orbital angular momentum operator. If the wavefunction $\phi(\boldsymbol{r})$ is instead a spinor $\boldsymbol{\phi}(\boldsymbol{r})$, then a rotation will be generated by the total angular momentum $\hat{\boldsymbol{J}} = \hat{\boldsymbol{L}} + \hat{\boldsymbol{S}}$, with $\hat{\boldsymbol{S}}$ denoting the spin angular momentum operator. A spinor corresponding to spin s, has $2s + 1$ components labelled by the quantum number $m_s = s, s-1, \ldots, -s$:

$$\boldsymbol{\phi}(\boldsymbol{r}) = \left(\phi_s(\boldsymbol{r}), \phi_{s-1}(\boldsymbol{r}), \ldots, \phi_{-s}(\boldsymbol{r})\right)^{\mathrm{T}}, \tag{1.4}$$

with T denoting matrix transposition. For a spin-1/2 particle, $m_s = \pm\frac{1}{2}$, and we have

$$\boldsymbol{\phi}(\boldsymbol{r}) = \left(\phi_{+\frac{1}{2}}(\boldsymbol{r}), \phi_{-\frac{1}{2}}(\boldsymbol{r})\right)^{\mathrm{T}} := \left(\phi_\uparrow(\boldsymbol{r}), \phi_\downarrow(\boldsymbol{r})\right)^{\mathrm{T}} \tag{1.5}$$

and the spin operator assumes the form $\hat{\boldsymbol{S}} = \hbar\hat{\boldsymbol{\sigma}}/2$, where $\hat{\boldsymbol{\sigma}}$ define the Pauli matrices [1]

$$\hat{\sigma}_x = \begin{pmatrix} 0 & 1 \\ 1 & 0 \end{pmatrix}, \quad \hat{\sigma}_y = \begin{pmatrix} 0 & -\mathrm{i} \\ \mathrm{i} & 0 \end{pmatrix} \quad \text{and} \quad \hat{\sigma}_z = \begin{pmatrix} 1 & 0 \\ 0 & -1 \end{pmatrix}. \tag{1.6}$$

The corresponding rotation operator $e^{-\mathrm{i}\theta\check{\boldsymbol{n}}\cdot\hat{\boldsymbol{S}}/\hbar}$, reads

$$e^{-\mathrm{i}\theta\check{\boldsymbol{n}}\cdot\hat{\boldsymbol{\sigma}}/2} = \cos(\theta/2)\hat{1}_\sigma - \mathrm{i}\sin(\theta/2)\check{\boldsymbol{n}}\cdot\hat{\boldsymbol{\sigma}}, \tag{1.7}$$

with $\hat{1}_\sigma$ denoting the identity matrix in spin space. The factor of 1/2 implies that the above operator exhibits a 4π, instead of 2π, periodicity with respect to θ. Contrary to the wavefuctions and spin rotation operator, the spin operators $\hat{\boldsymbol{S}}$ are connected to physical observables and thus exhibit a 2π periodicity, as shown in section 1.1.2. For more details see section 1.2.2.

In addition to the above continuous symmetry transformations, one can also introduce discrete ones, such as spatial inversion and reflections. In three spatial dimensions ($d = 3$), inversion (I) has the following action: $I\{\boldsymbol{r}, \boldsymbol{p}, \boldsymbol{L}\} = \{-\boldsymbol{r}, -\boldsymbol{p}, \boldsymbol{L}\}$. The angular momentum does not change sign, since it constitutes a pseudovector. Remarkably, the situation changes for a strictly two-dimensional system, where I can flip the sign of the angular momentum components. The reason is that in $d = 2$, I describes a transformation which is equivalent to a proper rotation, similar to that generated by $\hat{\boldsymbol{J}}$. This should be contrasted with $d = 3$, in which I effects an improper rotation. The distinction between proper and improper rotations relies on the sign of the determinant of the rotation matrix, being positive and negative, in each case respectively. As a matter of fact, for systems defined in odd spatial dimensions, inversion also effects parity. The latter is defined as a discrete improper rotation in all dimensions. For a system confined in the xy plane, the two-dimensional version of parity coincides with reflections σ_v, with 'v' denoting the presence of a vertical mirror symmetry plane. For example, the reflection or mirror operation σ_{xz}, effects the transformation $(x, y) \mapsto (x, -y)$. In general, an improper rotation defined in even spatial dimensions is equivalent to the combination of a proper rotation and an inversion, when the system is embedded in a space with one extra spatial dimension. For an example see Hands-on section 1.3.

1.1.2 Action of symmetry transformations on operators

Having retrieved the transformation properties of the wavefunctions under a symmetry operation, we can readily determine the transformation of any operator,

$\hat{A}$, associated with a physical observable. This is achieved by requiring that the matrix elements of the operator remain unchanged. We define:

$$\int d\boldsymbol{r}\ \phi^*(\boldsymbol{r})\hat{A}'\psi(\boldsymbol{r}) := \int d\boldsymbol{r}\ [\phi'(\boldsymbol{r})]^*\hat{A}\psi'(\boldsymbol{r}) = \int d\boldsymbol{r}\ \phi^*(\boldsymbol{r})\hat{O}^\dagger\hat{A}\hat{O}\psi(\boldsymbol{r}), \tag{1.8}$$

for arbitrary $\phi(\boldsymbol{r})$ and $\psi(\boldsymbol{r})$ which implies

$$\hat{A}' := O\hat{A} \equiv \hat{O}^\dagger\hat{A}\hat{O}. \tag{1.9}$$

Interestingly, for the particular description of our coordinate-space-defined wavefunctions, we can also view the position vector $\boldsymbol{r}$, as an operator $\hat{\boldsymbol{r}}$. Therefore, equation (1.9) allows us to verify that for a translation $\hat{t}_a^{\check{\boldsymbol{n}}}$, one obtains

$$\hat{\boldsymbol{r}}' = \left(\hat{t}_a^{\check{\boldsymbol{n}}}\right)^\dagger\hat{\boldsymbol{r}}\hat{t}_a^{\check{\boldsymbol{n}}} = e^{\mathrm{i}a\check{\boldsymbol{n}}\cdot\hat{\boldsymbol{p}}/\hbar}\hat{\boldsymbol{r}}\ e^{-\mathrm{i}a\check{\boldsymbol{n}}\cdot\hat{\boldsymbol{p}}/\hbar} = \hat{\boldsymbol{r}} + a\check{\boldsymbol{n}}. \tag{1.10}$$

This result was expected, as according to the definition of $\hat{A}'$ given in equation 1.8, we essentially consider that we transform the Hamiltonian system rather than the coordinate system, i.e. we employ the so-called active view of symmetry transformations [1, 3]. We could alternatively employ the passive point of view of symmetry transformations via the definition $\int d\boldsymbol{r}\ \phi^*(\boldsymbol{r})\hat{A}\psi(\boldsymbol{r}) := \int d\boldsymbol{r}\ [\phi'(\boldsymbol{r})]^*\hat{A}'\psi'(\boldsymbol{r})$ that instead yields $\hat{A}' = \hat{O}\hat{A}\hat{O}^\dagger$. Throughout this book we always employ the *active* view. Furthermore, note that for more general operators, which are not defined in a Hilbert space, we have $\hat{A}' = \hat{O}^{-1}\hat{A}\hat{O}$ [3, 5].

As an example, we investigate the transformation of the spin operators $(\hat{S}_x, \hat{S}_y)$, under a rotation by an angle θ about the z axis. For simplicity, we consider the case of a spin-1/2 particle, where we can equivalently study the transformation of the $\hat{\sigma}_{x,y}$ Pauli matrices. Equation (1.9) yields:

$$R_\theta^{\check{z}}\begin{pmatrix}\hat{\sigma}_x\\ \hat{\sigma}_y\end{pmatrix} = e^{\mathrm{i}\theta\check{z}\cdot\hat{\sigma}/2}\begin{pmatrix}\hat{\sigma}_x\\ \hat{\sigma}_y\end{pmatrix}e^{-\mathrm{i}\theta\check{z}\cdot\hat{\sigma}/2} = \begin{pmatrix}\cos\theta & -\sin\theta\\ \sin\theta & \cos\theta\end{pmatrix}\begin{pmatrix}\hat{\sigma}_x\\ \hat{\sigma}_y\end{pmatrix}. \tag{1.11}$$

From this we observe that the spin operator $\hat{\boldsymbol{S}}$, exhibits a 2π periodicity under spatial rotations and transforms in a similar way to $\hat{\boldsymbol{r}}$.

1.1.3 Antiunitary symmetry transformations: time reversal

The most familiar example of an antiunitary symmetry transformation is time reversal (TR) $\mathcal{T}$, effected by taking $t \to -t$. The time-reversed partner of a spinless particle's wavefunction $\phi(\boldsymbol{r}, t)$, is $\mathcal{T}\phi(\boldsymbol{r}, t) = \phi^*(\boldsymbol{r}, -t)$. The time-reversed Hamiltonian simply reads $\mathcal{T}\hat{H}(t) = \hat{H}^*(-t)$, since we are dealing with an antiunitary symmetry transformation operator. The explicit action of complex conjugation can be replaced by introducing the respective operation K, thus $\mathcal{T}\hat{H}(t) = K\hat{H}(-t)$. Note that $\hat{K}^2 = \hat{1}$ and its action on the position, momentum and orbital angular momentum operators is [1]

$$K\hat{\boldsymbol{r}} = \hat{\boldsymbol{r}}, \qquad K\hat{\boldsymbol{p}} = -\hat{\boldsymbol{p}} \qquad \text{and} \qquad K\hat{\boldsymbol{L}} = -\hat{\boldsymbol{L}}. \tag{1.12}$$

From this we conclude that for spinless particles governed by time-independent Hamiltonians, $\mathcal{T} \equiv K$. Consequently, $\mathcal{T}$-symmetric time-independent Hamiltonians for spinless particles are necessarily real, while the action of $\mathcal{T}$ on the above operators also reads

$$\mathcal{T}\hat{\boldsymbol{r}} = \hat{\boldsymbol{r}}, \qquad \mathcal{T}\hat{\boldsymbol{p}} = -\hat{\boldsymbol{p}} \qquad \text{and} \qquad \mathcal{T}\hat{\boldsymbol{L}} = -\hat{\boldsymbol{L}}. \tag{1.13}$$

In the case of spinful particles we have to retrieve the action of $\mathcal{T}$ on the spin operator $\hat{\boldsymbol{S}}$. Since the latter represents a type of angular momentum, it transforms in the same manner as $\hat{\boldsymbol{L}}$. Thus $\mathcal{T}\hat{\boldsymbol{S}} = -\hat{\boldsymbol{S}}$. Given the freedom to choose a representation in which $\hat{S}_{x,z}$ are real and $\hat{S}_y$ imaginary (see [1]), we can employ K and rewrite the $\hat{\mathcal{T}}$ operator as

$$\hat{\mathcal{T}} = e^{\mathrm{i}\pi\hat{S}_y/\hbar}\hat{K}. \tag{1.14}$$

Given this representation, $\mathcal{T}$ is generated by the successive operations of K and a π rotation in spin space about the y axis, since $K(\hat{S}_x, \hat{S}_y, \hat{S}_z) = (\hat{S}_x, -\hat{S}_y, \hat{S}_z)$.

For spin-1/2 fermions (e.g. electrons) we have $\hat{S}_y = \hbar\hat{\sigma}_y/2$, which implies $\hat{\mathcal{T}} = \mathrm{i}\hat{\sigma}_y\hat{K}$ and $\mathcal{T}\hat{H} = \hat{\sigma}_y\hat{K}^\dagger\hat{H}\hat{K}\hat{\sigma}_y$. Notice that $\hat{\mathcal{T}}^2 = -\hat{1}$, which is a characteristic of systems with half-integer spin. In contrast, for systems with integer spin, we obtain $\hat{\mathcal{T}}^2 = \hat{1}$. In the first case, the negative sign yields the celebrated Kramers degeneracy, i.e. every Hamiltonian eigenenergy is doubly degenerate and the two eigenstates are connected by TR. As an example consider the case of electrons in the absence of a magnetic field, in which the spin-up and -down states are degenerate. However, if $\hat{\mathcal{T}}^2 = \hat{1}$, $\mathcal{T}$ behaves in a similar way to K. In this case, the presence of time-reversal symmetry (TRS), with an operator squaring to identity, imposes a reality constraint on the Hamiltonian; i.e. by an appropriate choice of basis, the Hamiltonian can be represented as a real matrix and characterised by real eigenvectors [1].

1.1.4 Symmetry groups

In the previous paragraphs we provided some examples of symmetry transformations. However, a real material or abstract system can be simultaneously invariant under a set of symmetries. One finds that the various symmetry transformations may be interrelated, thus construing symmetry groups. A collection of elements $A, B, C, \ldots$ form a group $\mathcal{G}$, when the following conditions are met [4]:

1. The product of any two group elements yields another group element, i.e. $AB = C$, with $A, B, C \in \mathcal{G}$.
2. The elements satisfy the associative law: $(AB)C = A(BC)$.
3. There exists an identity element, E, commuting with any other group element A, i.e. $EA = AE$.
4. Every element has an inverse: $AA^{-1} = A^{-1}A = E$.

In order to make the term symmetry group more transparent, let us give the following example. Assume the time-independent Schrödinger equation $\hat{H}\phi = E\phi$

for a localised spin-1/2 particle in a magnetic field oriented along the z spin axis, for which, the Zeeman coupling yields $\hat{H} = -E_{\text{Zeeman}}\hat{\sigma}_z$. Due to the properties of the Pauli matrix, $\hat{\sigma}_z^2 = \hat{1}$, and one finds that $\hat{H}^2 = E_{\text{Zeeman}}^2\hat{1}$. Thus, the eigenvalues of the Hamiltonian are given by $E_\pm = \pm E_{\text{Zeeman}}$, implying that the respective eigenstates $|\phi_\pm\rangle$ satisfy $\hat{H}|\phi_\pm\rangle = \pm E_{\text{Zeeman}}|\phi_\pm\rangle$. Note that every Hamiltonian squaring to a constant will have the same eigenvalue structure and will lead to a similar relation for its eigenvectors. The details of the physical system under consideration solely fix the energy scale and the exact expression of the eigenvectors. These statements can be also extended to operators effecting symmetry transformations. For example, inversion satisfies $\hat{I}^2 = \hat{1}$ and thus it is characterised by the same general type of eigenvalues and eigenvectors as the Hamiltonian above.

Instead of eigenvalues and eigenvectors, a more appropriate nomenclature applying for symmetry transformations is characters and irreducible representations. In fact, the characters and irreducible representations characterise a symmetry group and not a single symmetry transformation. Notably, a group structure has made its appearance already in the discussion above. For the case of inversion, I and I^2 form a group. I^2 constitutes the identity element of the group, commmonly denoted with E. Since the two symmetry tranformations commute, i.e. $EI = IE$, the group is *Abelian*. In general, a symmetry group consists of a finite or an infinite number of elements, which defines the *order* of the group, $h_{\mathcal{G}}$. The elements of a group do not necessarily commute and the group is in this case called *non-Abelian*.

For the group $\{E, I\}$ there exist two irreducible representations A_1 and A_2, coinciding with the representations of the group elements when acting on $|\phi_+\rangle$ and $|\phi_-\rangle$, respectively. In the A_1 irreducible representation: $\{E, I\} = \{1, 1\}$, while in the A_2: $\{E, I\} = \{1, -1\}$. Note that these irreducible representations are one-dimensional, since the $|\phi_\pm\rangle$ transform into themselves under the action of the symmetry-group elements. However, non-Abelian groups support multi-dimensional irreducible representations, in which, each symmetry group element is represented by a matrix. The trace of the representation matrix defines the so-called character $\chi_{\text{IR}}(g)$ of the symmetry group element g, in the given irreducible representation IR. For one-dimensional irreducible representations the character coincides with the representation itself. Consequently, here we find $\chi_{A_1} = \{1, 1\}$ and $\chi_{A_2} = \{1, -1\}$.

While the scope of this work is not an in-depth group theory study (for more details refer to [1–5]), we present in the following section an example of how to analyse the C_{3v} symmetry point group, arising for instance in a triangularly-arranged triple-quantum-dot device.

C_{3v} point-group symmetry example: triangular triple-quantum-dot device

In the following paragraphs we provide an example regarding symmetry groups and, in particular, how to derive their irreducible representations. The model system for this purpose is a lateral triple-quantum-dot (TQD) device where the three dots are assumed identical and arranged in a triangular fashion [9, 10] as in

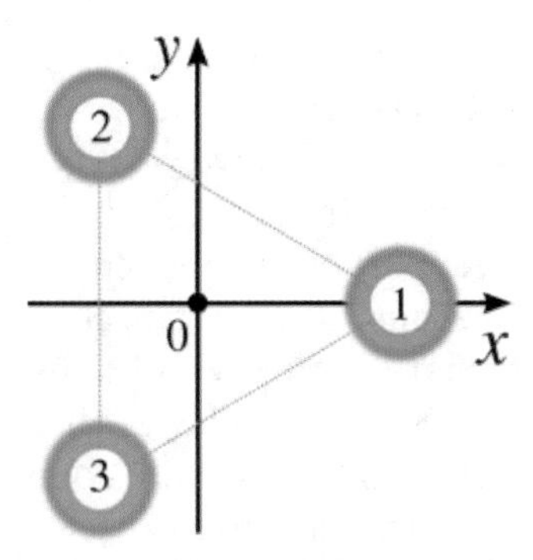

Figure 1.1. Three identical quantum dots arranged in a triangular fashion. The emergent triangle is equilateral, thus leading to a C_{3v} point group symmetry. This point-group consists of the identity element E, the C_3 counterclockwise rotation ($\{1, 2, 3\}\mapsto\{2, 3, 1\}$), the $C_3^2 \equiv C_3^{-1}$ clockwise rotation ($\{1, 2, 3\}\mapsto\{3, 1, 2\}$), the σ_{xz} mirror operation ($\{1, 2, 3\}\mapsto\{1, 3, 2\}$), the $C_3\sigma_{xz}$ mirror operation ($\{1, 2, 3\}\mapsto\{3, 2, 1\}$) and the $C_3^2\sigma_{xz}$ mirror operation ($\{1, 2, 3\}\mapsto\{2, 1, 3\}$).

figure 1.1. In the Hands-on section 1.3, we additionally demonstrate how to obtain effective Hamiltonians for the TQD system using group theoretical approaches.

Quantum dots constitute 'zero'-dimensional, electrostatically-confined (see section 1.2.4) or self-assembled, islands of electrons in which the finite-size effects lead to a discrete-only energy spectrum, since their diameter is of the order of a few to a few-hundred nanometers. Due to the complete confinement-induced energy level quantisation, they have been termed as *artificial atoms* [11]. Notably, while the number of electrons in a quantum dot can be large, the energy for adding an extra electron relative to a reference number can be finely tuned via electrostatic gating, thus allowing one to controllably create a few-excess-electron subsystem on the dot.

In the following, we consider a TQD with a single excess electron relative to the reference. Thus, a single-particle Hamiltonian is sufficient for examining the properties of the system. Nonetheless, even without specifying any further details for the TQD, we can already phenomenologically write down the most general Hamiltonian that may describe this system, solely based on the notion of its symmetries. We unfold the programme for reaching this goal below and in Hands-on section 1.3.

The three identical dots are deposited on a substrate and assumed to be perfectly aligned so to give rise to an equilateral triangle. The given structure is characterised by invariance under a set of $2\pi/3$ rotations about the out-of-plane z axis. A counterclockwise rotation of the TQD is termed a C_3 rotation, with the order n, of a rotation $2\pi/n$, being $n = 3$. One also finds the inverse rotation C_3^{-1} which rotates the TQD in a clockwise fashion. Furthermore, one finds $C_3^3 = E$ and $C_3^2 = C_3^{-1}$. Thus, the set $\{E, C_3, C_3^2\}$ forms a symmetry group, termed C_3. Nevertheless, the TQD is also dictated by the set $\{\sigma_{xz}, C_3\sigma_{xz}, C_3^2\sigma_{xz}\}$ of mirror symmetries.

These symmetry transformations exhaust all the possible unitary operations that leave the TQD invariant and form the so-called C_{3v} group. Note that C_3 is a subgroup of C_{3v}. Both groups constitute *point groups*, since they consist of elements that, when they act on the system, keep at least one point fixed. After having identified the symmetry-group elements we have to: (i) construct the multiplication table, (ii) retrieve the conjugacy classes, and (iii) obtain the irreducible representations that provides the character table. Below we appose a step-by-step presentation of this programme; for more details see [2–5].

1. *Multiplication table.* The particular table consists of all the products of two elements of the group and it also reflects its Abelian or non-Abelian structure. In the particular example we obtain table 1.1.
2. *Conjugacy classes.* The conjugacy classes divide the symmetry-group elements into sets of transformations with common features and the same character χ. For the given example one expects to find three classes. The first solely consists of the identity element $\{E\}$ and is present in every group. The second is built by the two rotations $\{C_3, C_3^2\}$, and the last is spanned by the mirror operations $\{\sigma_{xz}, C_3\sigma_{xz}, C_3^2\sigma_{xz}\}$. Mathematically, two group elements $g_{1,2}$ are conjugate if they satisfy $g_1 = g_3 g_2 g_3^{-1}$, where g_3 is an arbitrary element of the group [4]. Based on this definition, we indeed recover the three conjugacy classes mentioned above. Identifying the conjugacy classes is of utter importance, since the number of conjugacy classes coincides with the number of irreducible representations.
3. *Irreducible representations and character table.* After having identified the conjugacy classes of the symmetry group we can proceed with identifying the irreducible representations. This is greatly facilitated by the fact that the number of irreducible representations equals the number of conjugacy classes and if l_j denotes the dimensionality of the jth irreducible representation, then $\sum_j l_j^2 = h_{\mathcal{G}}$. In our case the order of the symmetry group is $h_{\mathcal{G}} = 6$, and we have three irreducible representations, therefore $l_1^2 + l_2^2 + l_3^2 = 6$. One of these irreducible representations is the identity one (A_1), which is one-dimensional. In the A_1 representation all the elements are equal to one, by definition. The above relation yields the additional one-dimensional (A_2) and two-dimensional (E) irreducible representations. Note that the characters of the irreducible representations satisfy the orthogonality condition [2–5]

$$\sum_g \chi^*_{\mathrm{IR}_i}(g)\chi_{\mathrm{IR}_j}(g) = h_{\mathcal{G}}\delta_{i,j}, \tag{1.15}$$

Table 1.1. Multiplication table of the C_{3v} point group.

$\cdot$	E	C_3^2	C_3	σ_{xz}	$C_3\sigma_{xz}$	$C_3^2\sigma_{xz}$
E	E	C_3^2	C_3	σ_{xz}	$C_3\sigma_{xz}$	$C_3^2\sigma_{xz}$
C_3	C_3	E	C_3^2	$C_3\sigma_{xz}$	$C_3^2\sigma_{xz}$	σ_{xz}
C_3^2	C_3^2	C_3	E	$C_3^2\sigma_{xz}$	σ_{xz}	$C_3\sigma_{xz}$
σ_{xz}	σ_{xz}	$C_3\sigma_{xz}$	$C_3^2\sigma_{xz}$	E	C_3^2	C_3
$C_3\sigma_{xz}$	$C_3\sigma_{xz}$	$C_3^2\sigma_{xz}$	σ_{xz}	C_3	E	C_3^2
$C_3^2\sigma_{xz}$	$C_3^2\sigma_{xz}$	σ_{xz}	$C_3\sigma_{xz}$	C_3^2	C_3	E

with χ the respective characters. The sum is over all the elements $g \in \mathcal{G}$. The above orthogonality condition can directly provide the A_2 one-dimensional irreducible representation. The orthogonality of $A_{1,2}$ yields the character relation: $1 + 2\chi_{A_2}(C_3) + 3\chi_{A_2}(\sigma_{xz}) = 0$. The latter immediately provides the A_2 irreducible representation in which $\chi_{A_2}(E) = 1$, $\chi_{A_2}(C_3) = 1$ and $\chi_{A_2}(\sigma_{xz}) = -1$. To identify the two-dimensional irreducible representation, we do not need to further stick to the particular TQD system, since the former constitutes a property of the group and not of the particular physical system under consideration. Thus, we can alternatively consider any two abstract objects, of our convenience, that transform into each other under the action of C_{3v} according to the two-dimensional irreducible representation. For example, we can simply study how the above rotations and mirror symmetries act on the position vector in two dimensions (x, y). The result in this case is easy to find, since a counterclockwise rotation of angle θ about the z axis reads

$$R_\theta^{\hat{z}}\begin{pmatrix} x \\ y \end{pmatrix} = \begin{pmatrix} \cos\theta & -\sin\theta \\ \sin\theta & \cos\theta \end{pmatrix}\begin{pmatrix} x \\ y \end{pmatrix}. \tag{1.16}$$

We therefore find the two matrices

$$\hat{C}_3 = \begin{pmatrix} -1/2 & -\sqrt{3}/2 \\ \sqrt{3}/2 & -1/2 \end{pmatrix} \quad \text{and} \quad \hat{\sigma}_{xz} = \begin{pmatrix} 1 & 0 \\ 0 & -1 \end{pmatrix}. \tag{1.17}$$

The above matrix representations of the operators generating C_3 and σ_{xz} yield $\chi_E(E) = 2$, $\chi_E(C_3) = -1$ and $\chi_E(\sigma_{xz}) = 0$. Note, however, that we have not yet verified if the above representation is indeed irreducible. In fact, instead it can be a *reducible* one, decomposable according to one of the following patterns: (i) $A_1 + A_1$, (ii) $A_2 + A_2$, or (iii) $A_1 + A_2$. To exclude this possibility we verify via equation (1.15) that the given representation is indeed orthogonal to A_1 and A_2 and thus inequivalent to both. In tables 1.2 and 1.5 we present the character table for C_{3v}.

1.1.5 Translations, Bloch's theorem and space groups

A large part of this book focuses on crystalline systems which, apart from point-group symmetries, additionally exhibit invariance under translations. The combination of the latter two kinds of symmetries yields the so-called space groups. In $d = 2$ there exist 17 space groups, also called wallpaper groups. However, in $d = 3$ we find 230 space

Table 1.2. Character table of the C_{3v} point group.

Irr. Rep.	E	$2C_3$	$3\sigma_v$
A_1	1	1	1
A_2	1	1	−1
E	2	−1	0

groups, see [4]. While studying space groups is not on the agenda of this work, we will examine some properties of translations and introduce Bloch's theorem [12].

Translations, represented here as $t_a^{\check{n}}$, constitute an Abelian group since two translation operations commute. Assume a finite-sized three-dimensional crystal with dimensions $L_x = N_x a$, $L_y = N_y b$ and $L_z = N_z c$, where a, b, c denote the lattice constants and $N_{x,y,z}$ the respective number of unit cells along the $\check{x}$, $\check{y}$ and $\check{z}$ directions. By employing periodic boundary conditions one finds that $(t_a^{\check{x}})^{N_x} = E$. Similar conditions hold for the other two operations. The particular constraint on the translation operations yields their one-dimensional irreducible representations, which coincide with the respective characters. For $t_a^{\check{x}}$, the irreducible representations are essentially retrieved using the roots of $t^{N_x} = 1 \Rightarrow t_n = e^{i2\pi n/N_x}$ with $n = 0, 1, 2, \ldots, N_x - 1$. For convenience, one introduces the wavenumber $k_x := 2\pi n/L_x$ $(n = 0, 1, 2, \ldots, N_x - 1)$ and thus the irreducible representations become $t_{k_x} = e^{ik_x a}$. When $L_x \to \infty$ we may treat k_x as a continuous variable. A similar procedure for the remaining two operators allows us to introduce the wavevector (or quasimomentum) $\boldsymbol{k} := 2\pi(\frac{n}{L_x}, \frac{m}{L_y}, \frac{l}{L_z})$ with $n = 0, 1, 2, \ldots, N_x - 1$, $m = 0, 1, 2, \ldots, N_y - 1$ and $l = 0, 1, 2, \ldots, N_z - 1$.

With the above set of wavevectors we can introduce the so-called 1st Brillouin zone (BZ), which in $d = 1$ is defined as the interval $k_x \in (-\pi/a, \pi/a]$. The emergence of a BZ is a result of the periodicity of the Hamiltonian and connects to the Bloch theorem [12]. According to the latter, the eigenvector $\boldsymbol{\phi}(\boldsymbol{r})$ of a periodic Hamiltonian satisfying $\hat{H}(\boldsymbol{r} + na\check{x} + mb\check{y} + lc\check{z}) = \hat{H}(\boldsymbol{r})$, with $n, l, m, \in \mathbb{Z}$, can be decomposed into two parts: $\boldsymbol{\phi}(\boldsymbol{r}) = \sum_k \boldsymbol{u}_k(\boldsymbol{r}) e^{i\boldsymbol{k}\cdot\boldsymbol{r}}$. Here $\boldsymbol{u}_k(\boldsymbol{r})$ has the same periodicity as the Hamiltonian and $\boldsymbol{k}$ is defined in the 1st BZ. In the case of a crystal, the electrons feel a periodic ionic potential, $V(\boldsymbol{r} + na\check{x} + mb\check{y} + lc\check{z}) = V(\boldsymbol{r})$, and the corresponding eigenvectors $\boldsymbol{u}_k(\boldsymbol{r})$ satisfy the time-independent Schrödinger equation

$$\left[\frac{(\hat{\boldsymbol{p}} + \hbar \boldsymbol{k})^2}{2m} + V(\boldsymbol{r})\right] \boldsymbol{u}_k(\boldsymbol{r}) = E_k \boldsymbol{u}_k(\boldsymbol{r}). \tag{1.18}$$

The energy bandstructure $E_{k,n}$ of a crystal is determined by retrieving the eigenstates $\boldsymbol{u}_{k,n}(\boldsymbol{r})$ of the Bloch Hamiltonian, identified as the operator on the left-hand side (lhs) of the above equation.

Before concluding this paragraph, let us retrieve the transformation properties of the single-particle spinor wavefunction $\boldsymbol{u}_k(\boldsymbol{r})$ and the Bloch Hamiltonian under K, which is useful for analysing the topological properties of the crystalline materials to be discussed in the upcoming chapters. The complete electronic wavefunction is $\boldsymbol{\phi}(\boldsymbol{r}) = \sum_k \boldsymbol{u}_k(\boldsymbol{r}) e^{i\boldsymbol{k}\cdot\boldsymbol{r}}$ and under the action of K one obtains

$$K\boldsymbol{\phi}(\boldsymbol{r}) = \sum_k \boldsymbol{u}_k^*(\boldsymbol{r}) e^{-i\boldsymbol{k}\cdot\boldsymbol{r}} \equiv \sum_k \boldsymbol{u}_{-k}^*(\boldsymbol{r}) e^{i\boldsymbol{k}\cdot\boldsymbol{r}}. \tag{1.19}$$

One observes that the action of K on the complete wavefunction can be completely absorbed in taking the complex conjugate of $\boldsymbol{u}_k(\boldsymbol{r})$ and inverting its wavevector index. The latter implies the following action of K on a Bloch Hamiltonian

$$K\hat{H}_k(\boldsymbol{r}) = \hat{H}^*_{-k}(\boldsymbol{r}). \quad (1.20)$$

Note that the wavevector satisfies $K\boldsymbol{k} = \boldsymbol{k}$. In contrast, $K\hbar\boldsymbol{k} = -\hbar\boldsymbol{k}$, which follows from $K\hat{\boldsymbol{p}} = -\hat{\boldsymbol{p}}$. Due to this difference, equation (1.19) leads to the inversion of the wavevector appearing in the right-hand side (rhs) of equation (1.20), and guarantees the consistency between the $\boldsymbol{r}$- and $\boldsymbol{k}$-space descriptions.

1.2 Effective Hamiltonians for bulk III–V semiconductors

In the following paragraphs we demonstrate how to employ the symmetry artillery developed above, to phenomenologically construct effective Hamiltonians for III–V semiconductors. These materials are particularly interesting, since they can feature strong spin–orbit coupling (SOC) [1, 4], which is a key ingredient for engineering topological systems [13, 14]. Typically, a semiconductor belonging to this family consists of two chemical elements, having 3 and 5 electrons in their outer atomic shell, which mainly consists of s and p atomic orbitals. Characteristic examples are: GaAs, InAs and InP. The electrons in these semiconductors are not free but instead feel a crystalline potential that results in an energy bandstructure, as in equation (1.18). These semiconductors usually crystallise in the so-called zincblende or wurtzite structures, invariant correspondingly under the T_d and C_{6v} point groups, both lacking a center of inversion [15, 16]. As we discuss in section 1.2.3, bulk inversion asymmetry results in a non-negligible SOC, which can play a pivotal role for crafting topologically non-trivial materials.

Despite the complex spaghetti-like energy bandstructure obtained for these systems, most of the time we are interested in the energetically low lying properties and thus restrict ourselves to the vicinity of particular $\boldsymbol{k}$-space points, about which we obtain a Taylor expansion of the energy dispersion. Usually, these $\boldsymbol{k}$-points constitute local bandstructure extrema and are governed by the same or reduced crystalline symmetry of the system, which sets constraints to the allowed terms in this expansion. This perturbative approach is termed '$\boldsymbol{k} \cdot \boldsymbol{p}$ expansion'. Such an effective Hamiltonian for the given $\boldsymbol{k}$-region can be extracted using a first-principles calculation. However, another approach is to solely rely on the symmetry properties of the system and phenomenologically derive an effective low energy Hamiltonian, sufficient for making qualitative predictions. Importantly, the latter approach can be useful for understanding the universal properties emerging in a broader class of systems, all sharing the same symmetry characteristics and being described by the same type of low energy Hamiltonian.

1.2.1 Effective Hamiltonian about the Γ-point: plain vanilla model

In the following paragraphs we construct an effective model for a semiconductor in $d = 3$, valid near the vicinity of the Γ-point ($\boldsymbol{k} = \boldsymbol{0}$) of the bandstructure. The semiconductor is here assumed to be invariant under inversion and arbitrary rotations. The discussion of crystalline effects, and also bulk and structural inversion asymmetry, is postponed for sections 1.2.2, 1.2.3 and 1.2.4. Our model Hamiltonian builds upon s and p atomic orbitals characterised by the $\ell = 0$ and $\ell = 1$ quantum numbers of the orbital angular momentum. However, the electron spin is characterised by a spin quantum number $s = 1/2$. If we further consider the case of a non-negligible SOC, the orbital and spin angular momenta couple. Under the artificially-enhanced

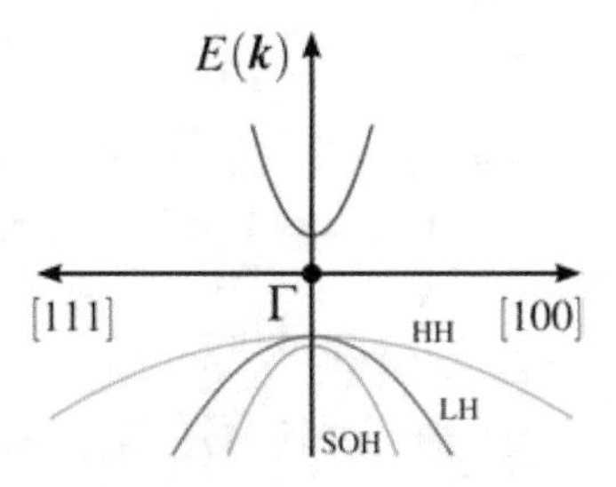

Figure 1.2. Typical bandstructure for a III–V cubic semiconductor near the Γ-point, located at $\boldsymbol{k} = (0, 0, 0)$, that enjoys the full symmetry of the respective point group. The s orbitals give rise to the conduction band that appears for positive energies. In contrast, the p orbitals yield the three valence hole bands. The latter consist of the heavy (HH), light (LH) and split-off (SOH) hole bands. The HH and LH belong to the same irreducible representation of the point group, which implies their degeneracy at the Γ-point. In addition, all the bands are two-fold degenerate by virtue of time-reversal symmetry. Note, that, there exist materials in this family which exhibit an inverted bandstructure, i.e. the conduction and the top-hole (HH and LH) bands exchange positions, see figure 7.1. As we discuss in chapter 7, the phenomenon of band inversion plays an important role in obtaining topologically-non-trivial semiconductors.

symmetry considered here, it is convenient to introduce the total angular momentum, $\hat{\boldsymbol{J}} = \hat{\boldsymbol{L}} + \hat{\boldsymbol{S}}$. By employing the total angular momentum, we can label the states using the corresponding quantum number j. For the s orbitals we obtain $j = 1/2$ and for the p orbitals $j = 3/2$, $1/2$. The former and latter yield correspondingly the so-called conduction and valence bands. One observes that the valence band consists of two different kinds of states. The $j = 3/2$ leads to the heavy (HH) and light (LH) holes, while the remaining $j = 1/2$ corresponds to the so-called split-off hole (SOH) band, as shown in figure 1.2. Since the HH and LH belong to a different irreducible representation compared to the SOH, the energy values of the bandstructure at the Γ-point of the former and the latter will generally differ unless an accidental degeneracy occurs. Below, we obtain effective Hamiltonians describing the conduction, HH and LH, and SOH bands, in the case where the respective energy splittings at the Γ-point are sufficiently large. Otherwise all these bands are described by an 8×8 Kane model, see [15].

Conduction and split-off hole bands

The effective Hamiltonian has the same general form for these bands, since they are dictated by the same total angular momentum quantum number, i.e. $j = 1/2$. By virtue of inversion symmetry, the terms appearing in the effective Hamiltonian of the conduction and SOH bands have to be even in $\boldsymbol{k}$. In addition, TRS implies that for every Hamiltonian term we should have the same order of powers in $\boldsymbol{k}$ and $\hat{\boldsymbol{J}}$. By additionally taking into account that both $\boldsymbol{k}$ and $\hat{\boldsymbol{J}}$ transform as vectors under rotations, the only couplings allowed have the form $(\boldsymbol{k} \cdot \hat{\boldsymbol{J}})^{2n}$ with $n \in \mathbb{N}$. However, for $j = 1/2$ we obtain $(\boldsymbol{k} \cdot \hat{\boldsymbol{J}})^{2n} \equiv (\hbar^2 k^2/4)^n$, since $\hat{\boldsymbol{J}} = \hbar\hat{\boldsymbol{\tau}}/2$. Here $\hat{\boldsymbol{\tau}}$ correspond to Pauli matrices satisfying $(\boldsymbol{a} \cdot \hat{\boldsymbol{\tau}})^2 = \boldsymbol{a}^2$. Thus, we conclude that the effective Hamiltonians for the conduction and SOH bands, at lowest order in $\boldsymbol{k}$, take the simple form of a quadratic energy dispersion, i.e. $\sim\boldsymbol{k}^2$, as in figure 1.2.

Heavy- and light-hole bands: Luttinger Hamiltonian

Similar symmetry arguments hold for the HH and LH bands, and the allowed Hamiltonian terms have once again the form $(\boldsymbol{k} \cdot \hat{\boldsymbol{J}})^{2n}$ with $n \in \mathbb{N}$. However, for $j = 3/2$, the representation of $\hat{\boldsymbol{J}}$ is such that the simplifications encountered in the previous paragraph are not met here. Thus, one retrieves the so-called Luttinger Hamiltonian [17] for a paramagnetic semiconductor, a specific form of which reads

$$\hat{H}(\boldsymbol{k}) = \frac{1}{2m}\left[\left(\gamma_1 + \frac{5}{2}\gamma_2\right)(\hbar \boldsymbol{k})^2 - 2\gamma_2(\boldsymbol{k} \cdot \hat{\boldsymbol{J}})^2\right]. \tag{1.21}$$

In more detail, the total angular momentum components can be calculated using the formulas [1]:

$$\begin{aligned} &\hat{J}_\pm |j, m_j\rangle = \hbar\sqrt{j(j+1) - m_j(m_j \pm 1)}\,|j, m_j \pm 1\rangle \quad \text{and} \\ &\hat{J}_z |j, m_j\rangle = \hbar m_j |j, m_j\rangle, \end{aligned} \tag{1.22}$$

with $\hat{J}_\pm = \hat{J}_x \pm \mathrm{i}\hat{J}_y$. By considering the basis $\{|3/2, 3/2\rangle, |3/2, 1/2\rangle, |3/2, -1/2\rangle, |3/2, -3/2\rangle\}$, the above lead to the matrix representations

$$\hat{J}_x = \hbar\begin{pmatrix} 0 & \frac{\sqrt{3}}{2} & 0 & 0 \\ \frac{\sqrt{3}}{2} & 0 & 1 & 0 \\ 0 & 1 & 0 & \frac{\sqrt{3}}{2} \\ 0 & 0 & \frac{\sqrt{3}}{2} & 0 \end{pmatrix}, \quad \hat{J}_y = \hbar\begin{pmatrix} 0 & -\frac{\sqrt{3}}{2}\mathrm{i} & 0 & 0 \\ +\frac{\sqrt{3}}{2}\mathrm{i} & 0 & -\mathrm{i} & 0 \\ 0 & +\mathrm{i} & 0 & -\frac{\sqrt{3}}{2}\mathrm{i} \\ 0 & 0 & +\frac{\sqrt{3}}{2}\mathrm{i} & 0 \end{pmatrix} \quad \text{and} \tag{1.23}$$

$$\hat{J}_z = \hbar\begin{pmatrix} +\frac{3}{2} & 0 & 0 & 0 \\ 0 & +\frac{1}{2} & 0 & 0 \\ 0 & 0 & -\frac{1}{2} & 0 \\ 0 & 0 & 0 & -\frac{3}{2} \end{pmatrix}. \tag{1.24}$$

By employing Kronecker products of the Pauli matrices, $\hat{\boldsymbol{\tau}}$ and $\hat{\boldsymbol{\sigma}}$, accompanied by the corresponding unit matrices, $\hat{1}_\tau$ and $\hat{1}_\sigma$, we may rewrite the above total angular momentum operators as follows

$$\begin{aligned} &\hat{J}_x/\hbar = \frac{\sqrt{3}}{2}\hat{\sigma}_x + \frac{1}{2}\left(\hat{\tau}_x\hat{\sigma}_x + \hat{\tau}_y\hat{\sigma}_y\right), \quad \hat{J}_y/\hbar = \frac{\sqrt{3}}{2}\hat{\sigma}_y + \frac{1}{2}\left(\hat{\tau}_y\hat{\sigma}_x - \hat{\tau}_x\hat{\sigma}_y\right) \quad \text{and} \\ &\hat{J}_z/\hbar = \frac{1}{2}\hat{\sigma}_z + \hat{\tau}_z. \end{aligned} \tag{1.25}$$

Note that throughout this book we adopt this simplified notation, where Kronecker products and unit matrices are omitted, e.g. $\frac{1}{2}\hat{1}_\tau \otimes \hat{\sigma}_z + \hat{\tau}_z \otimes \hat{1}_\sigma \mapsto \frac{1}{2}\hat{\sigma}_z + \hat{\tau}_z$.

Since the Luttinger Hamiltonian is made of products of the different components of the total angular momentum, i.e. $\hat{J}_a\hat{J}_b$ with $a, b = x, y, z$, it is convenient to introduce the $\hat{\Gamma}_{1,2,3,4,5}$ matrices, as employed in reference [18]

$$\hat{\Gamma}_1 = \hat{\tau}_z\hat{\sigma}_x, \quad \hat{\Gamma}_2 = \hat{\tau}_z\hat{\sigma}_y, \quad \hat{\Gamma}_3 = \hat{\tau}_z\hat{\sigma}_z, \quad \hat{\Gamma}_4 = \hat{\tau}_x \quad \text{and} \quad \hat{\Gamma}_5 = \hat{\tau}_y. \tag{1.26}$$

The $\hat{\Gamma}$ matrices can be viewed as a vector $\hat{\boldsymbol{\Gamma}}$ in a $d = 5$ space, and constitute the natural extension of the Pauli matrices $\hat{\boldsymbol{\sigma}}$, as they satisfy $\{\hat{\Gamma}_\alpha, \hat{\Gamma}_\beta\} = 2\delta_{\alpha,\beta}\hat{1}_\Gamma$. Furthermore, they also connect to the $\hat{\gamma}$ matrices appearing in the Dirac equation. With the use of $\hat{\boldsymbol{\Gamma}}$, the Hamiltonian is compactly written as

$$\hat{H}(\boldsymbol{k}) = \varepsilon(\boldsymbol{k})\hat{1}_\Gamma - V\boldsymbol{g}(\boldsymbol{k}) \cdot \hat{\boldsymbol{\Gamma}}, \tag{1.27}$$

with $\varepsilon(\boldsymbol{k}) = \frac{\gamma_1}{2m}(\hbar \boldsymbol{k})^2$ and $V = \frac{\gamma_2}{2m}\sqrt{3}\,\hbar^2$, while the vector $\boldsymbol{g}(\boldsymbol{k})$ is defined as

$$g_1(\boldsymbol{k}) = 2k_xk_z, \quad g_2(\boldsymbol{k}) = 2k_yk_z, \quad g_3(\boldsymbol{k}) = \frac{2k_z^2 - k_x^2 - k_y^2}{\sqrt{3}}, \tag{1.28}$$

$$g_4(\boldsymbol{k}) = k_x^2 - k_y^2 \quad \text{and} \quad g_5(\boldsymbol{k}) = 2k_xk_y. \tag{1.29}$$

Interestingly, the $g_a(\boldsymbol{k})$ correspond to the $\ell = 2$ spherical harmonics that transform according to the respective irreducible representation of SO(3). The eigenenergies of the above Hamiltonian read

$$E_\pm(\boldsymbol{k}) = \varepsilon(\boldsymbol{k}) \pm V|\boldsymbol{g}(\boldsymbol{k})| = \frac{\gamma_1 \pm 2\gamma_2}{2m}(\hbar \boldsymbol{k})^2, \tag{1.30}$$

and exhibit a Kramers degeneracy due to TRS. The labels + and – correspond to the LH and HH bands, and one observes that the energy dispersions reflect the bandstructure depicted in figure 1.2.

1.2.2 Cubic crystalline effects and double covering groups

While in the above we neglected the crystalline effects for simplicity, in this paragraph we restore them and understand their impact on the form of the effective Hamiltonian derived. Here we continue to keep the inversion-symmetry intact and discuss the consequences of its violation in the upcoming subsections. The relevant cubic point group preserving I is the O_h, which consists of 48 elements as shown by its character table presented in table 1.3.

One finds that the wavevector $\boldsymbol{k}$ transforms according to the T_{1u} irreducible representation, while the total angular momentum $\hat{\boldsymbol{J}}$ belongs to the T_{1g}. Note that the index *g*/*u* (gerade/ungerade) reflects that the respective quantity is even/odd under I. Due to this difference, no linear terms in $\boldsymbol{k}$ are allowed. Specifically, the product of two operators transforming according to the T_{1u} and T_{1g} irreducible representations yield operators that transform according to one of the following: A_{1u}, E_u, T_{1u}

Table 1.3. Character table of the O_h point group. See reference [2].

O_h	E	$8C_3$	$6C_2$	$6C_4$	$3C_2$	I	$6S_4$	$8S_6$	$3\sigma_h$	$6\sigma_d$
A_{1g}	1	1	1	1	1	1	1	1	1	1
A_{2g}	1	1	−1	−1	1	1	−1	1	1	−1
E_g	2	−1	0	0	2	2	0	−1	2	0
T_{1g}	3	0	−1	1	−1	3	1	0	−1	−1
T_{2g}	3	0	1	−1	−1	3	−1	0	−1	1
A_{1u}	1	1	1	1	1	−1	−1	−1	−1	−1
A_{2u}	1	1	−1	−1	1	−1	1	−1	−1	1
E_u	2	−1	0	0	2	−2	0	1	−2	0
T_{1u}	3	0	−1	1	−1	−3	−1	0	1	1
T_{2u}	3	0	1	−1	−1	−3	1	0	1	−1

and T_{2u} irreducible representations. The latter is expressed as $T_{1u} \times T_{1g} = A_{1u} + E_u + T_{1u} + T_{2u}$. Thus, only one combination of these three-dimensional irreducible representations can lead to a scalar or equivalently a one-dimensional irreducible representation. This occurs via the inner product $\boldsymbol{k} \cdot \hat{\boldsymbol{J}}$, that belongs to the A_{1u}. However, only terms transforming according to the identity irreducible representation (A_{1g}) can appear in the Hamiltonian. We observe that $A_{1u} \times A_{1u} = A_{1g}$ and consequently even powers of $\boldsymbol{k} \cdot \hat{\boldsymbol{J}}$ are legitimate candidates. This is in agreement with the Luttinger Hamiltonian retrieved earlier and arises due to the fact that the O_h point group supports three-dimensional irreducible representations, similar to the continuous rotational groups.

At this point a remark is in place. In section 1.1 we argued that the rotation operator for a half-integer spin is 4π-periodic with respect to the rotation angle. From a mathematical point of view, this stems from the fact that the elements of spin rotations in this case belong to the so-called SU(2) group, being the double covering of the SO(3) group of rotations in Euclidean $d = 3$ space [1]. As we remarked in section 1.1, the wavefuctions can be labelled by the irreducible representations of the total angular momentum, while the observables by the irreducible representations of the orbital angular momentum. In other words, the spinors transform according to the SU(2) group and the observables according to the SO(3) group. In a similar fashion, when crystalline symmetry is introduced, the observables transform according to the irreducible representations of the point group and the spinors according to the ones of the double covering point group. To retrieve the irreducible representations of the latter, one has to extend the domain of the rotation angles from $[0, 2\pi)$ to $[0, 4\pi)$. This becomes possible by introducing an additional point-group element E', that corresponds to a rotation of 2π and therefore satisfies $(E')^2 = E$. In this manner, if the ensuing point group consists of h_G elements labelled by g_i, the introduction of E' will double them, since the elements $E'g_i$ will be added. Thereafter, the procedure for determining the new irreducible representations and character table follows section 1.1.4. In the Hands-on section 1.3 we give instructions for obtaining the double covering group of C_{3v}.

Table 1.4. Character table of the T_d point group. See reference [2].

T_d	E	$8C_3$	$3C_2$	$6S_4$	$6\sigma_d$	Linear	Higher order
A_1	1	1	1	1	1	–	$k_x^2 + k_y^2 + k_z^2,\ k_x k_y k_z$
A_2	1	1	1	−1	−1	–	–
E	2	−1	2	0	0	–	$\left(2k_z^2 - k_x^2 - k_y^2,\ k_x^2 - k_y^2\right)$
T_1	3	0	−1	1	−1	$\hat{\boldsymbol{J}}$	$\left(k_x\left(k_y^2 - k_z^2\right),\ k_y\left(k_z^2 - k_x^2\right),\ k_z\left(k_x^2 - k_y^2\right)\right)$
T_2	3	0	−1	−1	1	$\boldsymbol{k}$	$(k_x k_y,\ k_x k_z,\ k_y k_z)$

1.2.3 Bulk inversion asymmetry

In the previous paragraphs we considered an I-symmetric semiconductor. However, this symmetry is violated in reality. Breaking inversion reduces the point-group symmetry from O_h, to its T_d subgroup, having a character table presented in table 1.4. In the former table we also added examples of linear, quadratic and cubic terms transforming according to the irreducible representations of T_d. We immediately observe that even when inversion is broken, no bilinear coupling of the form $\boldsymbol{k} \cdot \hat{\boldsymbol{J}}$ is allowed, since $T_1 \times T_2 = A_2 + E + T_1 + T_2$ does not include A_1. However, a coupling cubic in $\boldsymbol{k}$ and linear in $\hat{\boldsymbol{J}}$ is now permitted. The latter is the so-called Dresselhaus SOC [15] and has the following form

$$\hat{H}_D(\boldsymbol{k}) = \lambda\left[\left(k_y^2 - k_z^2\right)k_x\hat{J}_x + \left(k_z^2 - k_x^2\right)k_y\hat{J}_y + \left(k_x^2 - k_y^2\right)k_z\hat{J}_z\right], \tag{1.31}$$

with λ denoting a variable parametrising its strength. Due to the linear coupling to the total angular momentum, such a Hamiltonian can be equivalently viewed as a $\boldsymbol{k}$-dependent magnetic field. Note that such a term appears for both the conduction and valence bands, while it constitutes the lowest order SOC term in $\boldsymbol{k}$, appearing for the conduction band. Finally, we have to remark that a coupling linear in $\boldsymbol{k}$ and cubic in $\hat{\boldsymbol{J}}$ is also possible for the HH and LH bands.

1.2.4 Confinement and structural inversion asymmetry

Apart from bulk semiconductors, confined systems such as quantum wells, nanowires and dots are also of exceptionally high importance and relevance for designing topological systems [13, 14]. A confinement potential $V_{\rm conf}(\boldsymbol{r})$ leads to an electric field $\boldsymbol{E}_{\rm conf}(\boldsymbol{r}) = -\nabla V_{\rm conf}(\boldsymbol{r})$ that violates inversion and translational symmetries. If we assume for simplicity that $V_{\rm conf}(\boldsymbol{r}) = V_{\rm conf}(z)$ implying $\boldsymbol{E}_{\rm conf}(\boldsymbol{r}) = E(z)\check{z}$, then translational invariance is only broken along the $\check{z}$ direction and solely the k_z wavevector does not constitute a good quantum number any more. In contrast, the perpendicular wavevector $\boldsymbol{k}_\perp = (k_x, k_y)$ and energy are conserved. In this case energy level quantisation takes place, and the k_z quantum number is replaced by a new quantum number associated with confinement.

As an example, let us consider a three-dimensional electron gas with kinetic energy $\hat{\boldsymbol{p}}^2/(2m)$ in the presence of a confinement potential $V_{\rm conf}(z) = 0\ \forall\, z \in [0, L_z]$ and $V_{\rm conf}(z) = +\infty\ \forall\, z \in (-\infty, 0) \cup (L_z, +\infty)$. The problem is identical to a particle in a box with the difference that the particle can still move freely in the xy plane. The eigenstates and eigenergies of the resulting quantum well read

$$\phi_{k_\perp,n}(\boldsymbol{r}) = \sqrt{\frac{2}{L_z}}\sin\left(\frac{n\pi z}{L_z}\right)e^{\mathrm{i}(k_x x+k_y y)} \quad \text{and} \quad E_n(\boldsymbol{k}_\perp) = \frac{\hbar^2(n\pi)^2}{2mL_z^2} + \frac{(\hbar \boldsymbol{k}_\perp)^2}{2m}, \tag{1.32}$$

with $n = 1, 2, \ldots, +\infty$. A quantum nanowire (dot) is obtained when further confining one (two) extra dimension(s).

While the above appears to be the whole story, this is fortunately not true. More interesting phenomena occur due to confinement and the resulting *structural inversion asymmetry*. The latter can be taken into account by recalling that the non-relativistic Schrödinger equation is obtained as a limit of Dirac's equation. As a result of this limiting procedure an additional term appears which is proportional to $(\hat{\boldsymbol{p}} \times \hat{\boldsymbol{\sigma}}) \cdot \boldsymbol{E}$ [1], with $\hat{\boldsymbol{\sigma}}$ denoting the spin Pauli matrices. When inversion symmetry is present, $\boldsymbol{E} = \boldsymbol{0}$, and thus this term drops out. For the quantum well case, this term is present and leads to the so-called Rashba SOC given by the Hamiltonian

$$\hat{H}_R(\boldsymbol{k}) = \alpha\hbar\left(k_x\hat{\sigma}_y - k_y\hat{\sigma}_x\right). \tag{1.33}$$

Similar effects arise in crystalline systems, the only difference being that the ionic potential generally yields more complicated SOC terms. In the following paragraphs we focus, for simplicity, on the conduction band near the Γ-point of a III–V semiconductor with bulk inversion asymmetry for different directions of the confinement-imposed electric field.

Confinement within the (001) plane: C_{2v} point-group symmetry

In this paragraph we consider the Γ-point conduction band of a zincblende III–V semiconductor with T_d point-group symmetry, whose Hamiltonian reads

$$\hat{H}_c(\boldsymbol{k}) = \frac{(\hbar \boldsymbol{k})^2}{2m} + E_c + \lambda\left[\left(k_y^2 - k_z^2\right)k_x\hat{\tau}_x + \left(k_z^2 - k_x^2\right)k_y\hat{\tau}_y + \left(k_x^2 - k_y^2\right)k_z\hat{\tau}_z\right]. \tag{1.34}$$

The $\hat{\boldsymbol{\tau}}$ Pauli matrices above act on the total angular momentum Kramers doublet, which corresponds to the Γ_6 irreducible representation of the double covering T_d group[1] [15]. We further assume that the system is confined, by applying an electric field along the $\hat{z}$ direction. Using Miller indices[2] [12], the xy confinement plane is denoted by (001). According to [19], the confinement reduces the point-group symmetry of the system to C_{2v}, and invalidates k_z as a good quantum number. As a

[1] In contrast, the SOH (HH and LH) band(s) transform according to the Γ_7 (Γ_8) irreducible representation of the double covering T_d point group.

[2] The Miller indices (hkl), define a plane in reciprocal space which is normal to the reciprocal ($\boldsymbol{k}$) space vector $h\boldsymbol{b}_1 + k\boldsymbol{b}_2 + l\boldsymbol{b}_3$, with $\boldsymbol{b}_{1,2,3}$ denoting the reciprocal lattice basis vectors. Note that for a cubic system, the reciprocal lattice vectors coincide with the direct ($\boldsymbol{r}$-space) lattice basis vectors, generally denoted with $\boldsymbol{a}_{1,2,3}$. With the Miller indices [hkl], we denote the direction $h\boldsymbol{a}_1 + k\boldsymbol{a}_2 + l\boldsymbol{a}_3$, in the direct lattice.

result of confinement, the Hamiltonian above becomes a matrix in the confinement eigenstate basis, with matrix elements

$$\begin{aligned}\hat{H}_{c,nm}(\boldsymbol{k}_\perp) &= \left[\frac{(\hbar \boldsymbol{k}_\perp)^2}{2m} + E_{c,n}\right]\delta_{n,m} \\ &+ \lambda\left\{\left[k_y^2\delta_{n,m} - \left\langle k_z^2\right\rangle_{nm}\right]k_x\hat{\tau}_x + \left[\left\langle k_z^2\right\rangle_{nm} - k_x^2\delta_{n,m}\right]k_y\hat{\tau}_y + \left(k_x^2 - k_y^2\right)\langle k_z\rangle_{nm}\hat{\tau}_z\right\}.\end{aligned} \tag{1.35}$$

The matrix elements $\langle k_z\rangle_{nm}$ and $\langle k_z^2\rangle_{nm}$ are understood as appropriate matrix elements in the confinement eigenstate basis that replace k_z and k_z^2, respectively. In the extreme limit that crystalline effects along the z axis become negligible, the former are matrix elements of the rescaled momentum operator $\hat{p}_z/\hbar$. Since here we attempt to acquire a phenomenological understanding, it only suffices that these matrix elements are non zero. When confinement is strong, we may restrict to the energetically lowest confinement eigenstate, as given by the $E_{c,n}$ hierarchy. In the single channel approximation, we obtain $\langle k_z\rangle_{nm} \mapsto \langle k_z\rangle$ and $\langle k_z^2\rangle_{nm} \mapsto \langle k_z^2\rangle$ and we may drop the n, m indices in the Hamiltonian of equation (1.35). For a single channel $\langle k_z\rangle = 0$, since the confinement channel wavefunction is real. Thus, by restricting to up to the quadratic order in the wavevector, we obtain the effective Hamiltonian

$$\hat{H}_{c,\mathrm{lowest}}(\boldsymbol{k}_\perp) = \frac{(\hbar \boldsymbol{k}_\perp)^2}{2m} + E_{c,\mathrm{lowest}} - \lambda\langle k_z^2\rangle\left(k_x\hat{\tau}_x - k_y\hat{\tau}_y\right). \tag{1.36}$$

Note that this result contains only the projected bulk Hamiltonian and the Dresselhaus effect. The effects of structural inversion asymmetry have not been included yet and after adding them we obtain the final effective (001) Hamiltonian up to the quadratic order

$$\hat{H}^{(001)}_{c,\mathrm{lowest}}(\boldsymbol{k}_\perp) = \frac{(\hbar \boldsymbol{k}_\perp)^2}{2m} + E_{c,\mathrm{lowest}} - \lambda\langle k_z^2\rangle\left(k_x\hat{\tau}_x - k_y\hat{\tau}_y\right) + \left(|\alpha_{xy}|k_x\hat{\tau}_y - |\alpha_{yx}|k_y\hat{\tau}_x\right). \tag{1.37}$$

Note that the C_{2v} point group allows for an anisotropic Rashba SOC term.

Finally, note that in the case of an inversion-symmetric bulk semiconductor, $\lambda = 0$, and in the presence of the confinement potential, the point-group symmetry would be reduced to C_{4v}, which would only lead to a Rashba term with $|\alpha_{xy}| = |\alpha_{yx}| \equiv \alpha$.

Confinement within the (111) plane: C_{3v} point-group symmetry

If the bulk semiconductor could be cut or confined in the (111) plane, a C_{3v} point-group symmetry emerges [19], whose properties were studied in section 1.1.4. In table 1.5 we present a list of quantities that transform according to the irreducible representations of this symmetry group.

As in the previous paragraph, one can appropriately project the Hamiltonian of equation (1.34) in order to obtain the Dresselhaus SOC for the quantum well. However, here we employ the predictive power given by knowing the ensuing point-group symmetry, in combination with TRS, so to arrive at the effective Hamiltonian which incorporates both structural and bulk inversion asymmetry types of SOC.

Table 1.5. Quantities transforming according to the irreducible representations of C_{3v}.

Irr. Rep.	E	$2C_3$	$3\sigma_v$	Linear	Higher order
A_1	1	1	1	–	$k_x^2 + k_y^2,\ k_y\left(k_y^2 - 3k_x^2\right)$
A_2	1	1	−1	$\hat{J}_z$	$k_x\left(k_x^2 - 3k_y^2\right)$
E	2	−1	0	$(k_x, k_y), (\hat{J}_y, -\hat{J}_x)$	$\left(2k_xk_y, k_x^2 - k_y^2\right), (k_xk_z, k_yk_z)$

Since there exist only linear terms in $\hat{\boldsymbol{J}} = \hbar\hat{\boldsymbol{\tau}}/2$ for the given spinor representation, we find the following Hamiltonian

$$\hat{H}^{(111)}_{c,\text{lowest}}(\boldsymbol{k}_\perp) = \frac{(\hbar \boldsymbol{k}_\perp)^2}{2m} + E_{c,\text{lowest}} + \alpha\hbar\left(k_x\hat{\tau}_y - k_y\hat{\tau}_x\right) + \gamma k_x\left(k_x^2 - 3k_y^2\right)\hat{\tau}_z. \tag{1.38}$$

As one observes, for the given symmetry group we obtain a Rashba-like term, as also a cubic in $\boldsymbol{k}$ contribution. By projecting the original Dresselhaus Hamiltonian onto the energetically-lowest confinement eigenstate, one finds that it contributes to both terms.

Finally, note that a Hamiltonian identical to the one presented above, appears also for the topologically protected states on the (111) surface of a bulk topological insulator [20, 21], such as Bi_2Te_3 and Bi_2Se_3. Aspects of these so-called warped topological insulators are discussed in section 7.2.3.

1.3 Hands-on: symmetry analysis of a triple quantum dot

The scope of this section is to help the interested reader become more familiar with the symmetry classification and the construction of effective Hamiltonians. We suggest a number of tasks in relation to the triple quantum dot system introduced in section 1.1.4. For comparison, we provide the answers to the posed questions, and guidance for the required intermediate steps.

To carry out the proposed tasks, it is recommended to introduce the following multicomponent wavefunction:

$$\boldsymbol{\Phi}^{\mathrm{T}} = \left(\phi_{1,\uparrow}, \phi_{1,\downarrow}, \phi_{2,\uparrow}, \phi_{2,\downarrow}, \phi_{3,\uparrow}, \phi_{3,\downarrow}\right), \tag{1.39}$$

in order to describe the spinful electrons defined for the three quantum dots labelled by 1, 2, 3. A general 6×6 Hamiltonian acting on this wavefunction is expressed as a Kronecker product $\hat{\lambda}_a \otimes \hat{\sigma}_b$ ($a = 1, \ldots, 8$ and $b = x, y, z$) of the following SU(3) Gell-Mann [22] $\hat{\lambda}$ matrices

$$\begin{gathered}
\hat{\lambda}_1 = \begin{pmatrix} 0 & 1 & 0 \\ 1 & 0 & 0 \\ 0 & 0 & 0 \end{pmatrix},\ \hat{\lambda}_2 = \begin{pmatrix} 0 & -\mathrm{i} & 0 \\ \mathrm{i} & 0 & 0 \\ 0 & 0 & 0 \end{pmatrix},\ \hat{\lambda}_3 = \begin{pmatrix} 1 & 0 & 0 \\ 0 & -1 & 0 \\ 0 & 0 & 0 \end{pmatrix},\ \hat{\lambda}_4 = \begin{pmatrix} 0 & 0 & 1 \\ 0 & 0 & 0 \\ 1 & 0 & 0 \end{pmatrix} \\
\hat{\lambda}_5 = \begin{pmatrix} 0 & 0 & -\mathrm{i} \\ 0 & 0 & 0 \\ \mathrm{i} & 0 & 0 \end{pmatrix},\ \hat{\lambda}_6 = \begin{pmatrix} 0 & 0 & 0 \\ 0 & 0 & 1 \\ 0 & 1 & 0 \end{pmatrix},\ \hat{\lambda}_7 = \begin{pmatrix} 0 & 0 & 0 \\ 0 & 0 & -\mathrm{i} \\ 0 & \mathrm{i} & 0 \end{pmatrix},\ \hat{\lambda}_8 = \frac{1}{\sqrt{3}}\begin{pmatrix} 1 & 0 & 0 \\ 0 & 1 & 0 \\ 0 & 0 & -2 \end{pmatrix}
\end{gathered} \tag{1.40}$$

Table 1.6. Classification of $\hat{\boldsymbol{\lambda}}$ and $\hat{\boldsymbol{\sigma}}$ matrices under $C_{3v} \times \mathcal{T}$.

Irr. Rep.	E	$2C_3$	$3\sigma_v$	$\mathcal{T} = +1$	$\mathcal{T} = -1$
A_1	1	1	1	$\hat{\Lambda}_z, \hat{1}_\sigma, \hat{\lambda}_z^+$	–
A_2	1	1	−1	–	$\hat{\sigma}_z, \hat{\lambda}_z^-$
E	2	−1	0	$(\hat{\Lambda}_x, \hat{\Lambda}_y), (\hat{\lambda}_x^+, \hat{\lambda}_y^+)$	$(\hat{\sigma}_y, -\hat{\sigma}_x), (\hat{\lambda}_y^-, -\hat{\lambda}_x^-)$

acting on the 1, 2, 3 quantum-dot index and the usual $\hat{\boldsymbol{\sigma}}$ Pauli matrices acting on the spin index ↑, ↓. We additionally have the respective unit matrices $\hat{1}_\lambda$ and $\hat{1}_\sigma$. For simplicity we omit the Kronecker-product symbol ⊗, and the unit matrices $\hat{1}_\lambda$ and $\hat{1}_\sigma$. By adopting the active view of symmetry transformations for the C_{3v} point group, proceed with carrying out the tasks.

Task 1: Show that the operators generating the C_3 rotation and the σ_{xz} mirror operation read

$$\hat{C}_3 = \begin{pmatrix} 0 & 0 & 1 \\ 1 & 0 & 0 \\ 0 & 1 & 0 \end{pmatrix} e^{-i\pi\hat{\sigma}_z/3} \quad \text{and} \quad \hat{\sigma}_{xz} = \begin{pmatrix} 1 & 0 & 0 \\ 0 & 0 & 1 \\ 0 & 1 & 0 \end{pmatrix} i\hat{\sigma}_y. \tag{1.41}$$

Explain the imaginary 'i' appearing in the expression for $\hat{\sigma}_{xz}$, by embedding the system in $d = 3$ and expressing σ_{xz} as a combination of inversion and a proper rotation. See also section 1.1.

Task 2: Show that the $\hat{\lambda}_{1,\dots,8}$ and $\hat{\sigma}_{x,y,z}$ matrices are classified according to the irreducible representations of the $C_{3v} \times \mathcal{T}$ symmetry group, as presented in table 1.6 where we have introduced

$$\hat{\boldsymbol{\Lambda}} \equiv \left(\hat{\Lambda}_x, \hat{\Lambda}_y, \hat{\Lambda}_z\right) := \left(\frac{\hat{\lambda}_8 + \sqrt{3}\hat{\lambda}_3}{2}, \frac{\sqrt{3}\hat{\lambda}_8 - \hat{\lambda}_3}{2}, \hat{1}_\lambda\right), \tag{1.42}$$

$$\hat{\boldsymbol{\lambda}}^+ \equiv \left(\hat{\lambda}_x^+, \hat{\lambda}_y^+, \hat{\lambda}_z^+\right) := \left(\frac{\hat{\lambda}_1 + \hat{\lambda}_4 - 2\hat{\lambda}_6}{\sqrt{3}}, \hat{\lambda}_1 - \hat{\lambda}_4, \hat{\lambda}_1 + \hat{\lambda}_4 + \hat{\lambda}_6\right), \tag{1.43}$$

$$\hat{\boldsymbol{\lambda}}^- \equiv \left(\hat{\lambda}_x^-, \hat{\lambda}_y^-, \hat{\lambda}_z^-\right) := \left(\frac{\hat{\lambda}_2 - \hat{\lambda}_5 - 2\hat{\lambda}_7}{\sqrt{3}}, \hat{\lambda}_2 + \hat{\lambda}_5, \hat{\lambda}_2 - \hat{\lambda}_5 + \hat{\lambda}_7\right). \tag{1.44}$$

Task 3: Verify that the most general $C_{3v} \times \mathcal{T}$-symmetric single-particle Hamiltonian has the form

$$\hat{H} = \varepsilon - t_{dd}\hat{\lambda}_z^+ + \alpha\hbar\left(\hat{\lambda}_x^-\hat{\sigma}_x + \hat{\lambda}_y^-\hat{\sigma}_y\right) - \beta\hat{\lambda}_z^-\hat{\sigma}_z. \tag{1.45}$$

Notice the analogy to equation (1.38).

Table 1.7. Character table of the double covering C_{3v} point group.

Irr. Rep.	E	$2C_3$	$3\sigma_v$	E'	$2E'C_3$	$3E'\sigma_v$
A_1	1	1	1	1	1	1
A_2	1	1	−1	1	1	−1
E	2	−1	0	2	−1	0
A_1'	1	−1	i	−1	1	−i
A_2'	1	−1	−i	−1	1	i
E′	2	1	0	−2	−1	0

Task 4: Retrieve the character table for the irreducible representations of the double covering C_{3v} point group, shown in table 1.7. To carry out this task, follow sections 1.1.4 and 1.2.2. Note that the addition of the group element E', doubles the number of elements. In fact, in the present case it additionally doubles the number of irreducible representations. The latter doubling is not generic though, see [4].

References

[1] Messiah A 1999 *Quantum Mechanics* Vol 1 and 2 (New York: Dover Publications)
[2] Tinkham M 2003 *Group Theory and Quantum Mechanics* (New York: Dover Publications)
[3] Lax M J 2012 *Symmetry Principles in Solid State and Molecular Physics* (New York: Dover Publications)
[4] Dresselhaus M S, Dresselhaus G and Jorio A 2008 *Group Theory: Application to the Physics of Condensed Matter* (Berlin, Heidelberg: Springer)
[5] Vergados J D 2017 *Group and Representation theory* (Hackensack, NJ: World Scientific)
[6] Altland A and Zirnbauer M R 1997 *Phys. Rev. B* **55** 1142
[7] Kitaev A 2009 *AIP Conf. Proc.* **1134** 22
[8] Ryu S, Schnyder A P, Furusaki A and Ludwig A W W 2010 *New J. Phys.* **12** 065010
[9] Hsieh C-Y, Shim Y-P, Korkusinski M and Hawrylak P 2012 *Rep. Prog. Phys.* **75** 114501
[10] Kotetes P, Jin P-Q, Marthaler M and Schön G 2014 *Phys. Rev. Lett.* **113** 236801
[11] Ashoori R C 1996 *Nature* **379** 413
[12] Grosso G and Parravicini G P 2014 *Solid State Physics* 2nd edn (Amsterdam: Elsevier)
[13] Hasan M Z and Kane C L 2010 *Rev. Mod. Phys.* **82** 3045
[14] Qi X-L and Zhang S-C 2011 *Rev. Mod. Phys.* **83** 1057
[15] Winkler R 2003 *Spin-Orbit Coupling Effects in Two-Dimensional Electron and Hole Systems* (Berlin: Springer)
[16] Gmitra M and Fabian J 2016 *Phys. Rev. B* **94** 165202
[17] Luttinger J M 1956 *Phys. Rev.* **102** 1030
[18] Murakami S, Nagaosa N and Zhang S-C 2004 *Phys. Rev. B* **69** 235206
[19] Yu P Y and Cardona M 2010 *Fundamentals of Semiconductors: Physics and Materials Properties* (Berlin: Springer)
[20] Chen Y L *et al* 2009 *Science* **325** 178
[21] Fu L 2009 *Phys. Rev. Lett.* **103** 266801
[22] Gell-Mann M 1962 *Phys. Rev.* **125** 1067

Chapter 2

Electron-coupling to external fields and transport theory

The central topic of this chapter is the coupling of a generic electronic system to electromagnetic fields and temperature gradients. The behaviour and especially the linear response of an electronic system in the presence of such external perturbations can provide valuable information regarding its topological properties. This is possible by detecting particular current-carrying boundary modes, being present only when the system resides in a topologically non-trivial phase. These boundary modes are robust against certain types of weak disorder, by virtue of the so-called topological protection provided by the bulk. Remarkably, they can mediate dissipationless transport, which open perspectives for technological applications. A prerequisite for exploring these exciting phenomena is to develop tools for investigating topological transport. Here, we adopt a quantum-field-theoretical approach and focus on the Matsubara technique, which is particularly suitable for studying the linear response of systems in thermal equilibrium. We address systems with continuum and lattice descriptions. As an example, we compute the charge conductivity tensor of an electron gas. Finally, in the Hands-on sections, we study the magnetoconductivity of a triple quantum dot, while we briefly present the Boltzmann-equation approach to transport phenomena.

2.1 Electromagnetic potentials, fields and currents

Putting the symmetries examined in the previous chapter aside, electronic systems are also characterised by invariance under another type of transformation, the so-called *gauge transformation*. The gauge under discussion concerns the particular gauge choice for the electromagnetic scalar and vector potentials, $V(\mathbf{r}, t)$ and $\mathbf{A}(\mathbf{r}, t)$. For an arbitrary scalar function $\varphi(\mathbf{r}, t)$, the transformation

$$V'(\mathbf{r}, t) = V(\mathbf{r}, t) - \partial_t\varphi(\mathbf{r}, t) \quad \text{and} \quad \mathbf{A}'(\mathbf{r}, t) = \mathbf{A}(\mathbf{r}, t) + \nabla\varphi(\mathbf{r}, t) \tag{2.1}$$

doi:10.1088/978-1-68174-517-6ch2

leads to a new gauge that constitutes an equivalent description of the system, since Maxwell's equations (see [1] and section 6.1.3) remain unaffected. The latter follows from the fact that the electric and magnetic fields

$$\boldsymbol{E}(\boldsymbol{r}, t) = -\nabla V(\boldsymbol{r}, t) - \partial_t \boldsymbol{A}(\boldsymbol{r}, t) \quad \text{and} \quad \boldsymbol{B}(\boldsymbol{r}, t) = \nabla \times \boldsymbol{A}(\boldsymbol{r}, t) \tag{2.2}$$

remain unchanged, thus guaranteeing that the physical observables are gauge-invariant quantities.

Classically, the scalar and vector potentials couple in a linear fashion to the charge $-e$ ($e > 0$), and current $-e\boldsymbol{v}(t)$, of an electron. The velocity $\boldsymbol{v}(t)$, may also implicitly depend on the vector potential in the general case. In the following paragraphs, we present different types of systems and respective methods on how to retrieve the electric charge and current densities, within a quantum-mechanical framework.

2.2 Minimal coupling and electric charge conservation law

For the case of a single-electron Hamiltonian $\hat{H}(\hat{\boldsymbol{p}}, \boldsymbol{r})$, defined in coordinate space, one introduces the electromagnetic potentials via the minimal coupling

$$\hat{H}(\hat{\boldsymbol{p}}, \boldsymbol{r}) \mapsto \hat{H}[\hat{\boldsymbol{p}} + e\boldsymbol{A}(\boldsymbol{r}, t), \boldsymbol{r}] - eV(\boldsymbol{r}, t). \tag{2.3}$$

According to Noether's theorem [2, 3], the electric current density operator in the Schrödinger picture $\hat{\boldsymbol{J}}_c(\boldsymbol{r}, t)$, is defined via the variation of the Hamiltonian

$$\delta\hat{H}[\hat{\boldsymbol{p}} + e\boldsymbol{A}(\boldsymbol{r}, t), \boldsymbol{r}] := -\delta\boldsymbol{A}(\boldsymbol{r}, t) \cdot \hat{\boldsymbol{J}}_c(\boldsymbol{r}, t), \tag{2.4}$$

induced by a variation $\delta\boldsymbol{A}(\boldsymbol{r}, t)$ of the vector potential [3–7].

Let us now continue with a concrete example and, in particular, consider the spinless free-electron Hamiltonian in second quantisation, that in the presence of $\boldsymbol{A}$ reads

$$\hat{H} = \int d\boldsymbol{r}\, \hat{\psi}^\dagger(\boldsymbol{r}) \frac{[\hat{\boldsymbol{p}} + e\boldsymbol{A}(\boldsymbol{r}, t)]^2}{2m} \hat{\psi}(\boldsymbol{r}). \tag{2.5}$$

Variation with respect to the vector potential yields the electric current density operator

$$\hat{\boldsymbol{J}}_c(\boldsymbol{r}, t) = -\frac{e}{2m}\left\{\hat{\psi}^\dagger(\boldsymbol{r})\hat{\boldsymbol{p}}\hat{\psi}(\boldsymbol{r}) - [\hat{\boldsymbol{p}}\hat{\psi}^\dagger(\boldsymbol{r})]\hat{\psi}(\boldsymbol{r})\right\} - \hat{\psi}^\dagger(\boldsymbol{r})\frac{e^2}{m}\boldsymbol{A}(\boldsymbol{r}, t)\hat{\psi}(\boldsymbol{r}). \tag{2.6}$$

As we observe, the current consists of two terms. The first (second) defines the paramagnetic (diamagnetic) component $\hat{\boldsymbol{J}}_c^{\mathrm{p}}(\boldsymbol{r}, t)$ ($\hat{\boldsymbol{J}}_c^{\mathrm{d}}(\boldsymbol{r}, t)$). Remarkably, the diamagnetic current is proportional to the vector potential itself. Note that since the current operator above is defined in the Schrödinger picture, its time dependence stems solely from the explicit time dependence of the vector potential which appears in the current's expression. The electric current operator in the formalism of second quantisation is obtained by integrating the respective density over the system's volume, i.e. $\hat{\boldsymbol{J}}_c(t) = \int d\boldsymbol{r} \hat{\boldsymbol{J}}_c(\boldsymbol{r}, t)$. In addition, the charge density operator defined in the Schrödinger picture reads

$$\hat{\rho}_c(\boldsymbol{r}) = -e\hat{\psi}^\dagger(\boldsymbol{r})\hat{\psi}(\boldsymbol{r}). \tag{2.7}$$

At this point, we are in a position to introduce the term U(1) gauge transformations. Its meaning becomes transparent when effecting a gauge transformation on a minimally-coupled Hamiltonian in second quantisation, as in equation (2.3). Via a time-independent (for convenience) transformation, one finds

$$\begin{aligned}\hat{H} &= \int d\boldsymbol{r}\, \hat{\psi}^\dagger(\boldsymbol{r})\hat{H}[\hat{\boldsymbol{p}} + e\boldsymbol{A}(\boldsymbol{r}, t) + e\nabla\varphi(\boldsymbol{r}), \boldsymbol{r}]\hat{\psi}(\boldsymbol{r}) \\ &\equiv \int d\boldsymbol{r}\, \hat{\psi}^\dagger(\boldsymbol{r})e^{-ie\varphi(\boldsymbol{r})/\hbar}\hat{H}[\hat{\boldsymbol{p}} + e\boldsymbol{A}(\boldsymbol{r}, t), \boldsymbol{r}]e^{ie\varphi(\boldsymbol{r})/\hbar}\hat{\psi}(\boldsymbol{r}).\end{aligned} \tag{2.8}$$

Therefore, the gauge transformation results in a phase factor on the matter field operator $\hat{\psi}(\boldsymbol{r})$. Such a phase factor corresponds to a U(1) transformation and stems from the non-zero electric charge of the electrons. The U(1) gauge transformation can be generalised for multicomponent spinors. The existence of global or local gauge invariance, defined by a homogeneous or a spatially dependent φ, gives rise to conserved global or local currents, as Noether's theorem [2] predicts. In solid-state physics there exists only the local U(1) electromagnetic invariance. Nonetheless, invariance under more complicated local gauge groups can be found in high-energy physics, e.g. the color SU(3) group of quantum chromodynamics and others [3].

One may transfer the above expressions for the charge and current density operators to momentum space, by introducing the discrete and continuum plane wave bases:

$$\hat{\psi}(\boldsymbol{r}) = \frac{1}{\sqrt{\mathcal{V}}}\sum_{\boldsymbol{k}} \hat{\psi}_{\boldsymbol{k}} e^{i\boldsymbol{k}\cdot\boldsymbol{r}} \quad \text{and} \quad \hat{\psi}(\boldsymbol{r}) = \frac{1}{\sqrt{(2\pi)^d}}\int d\boldsymbol{k}\, \hat{\psi}(\boldsymbol{k})e^{i\boldsymbol{k}\cdot\boldsymbol{r}}, \tag{2.9}$$

as well as the discrete and the continuous Fourier transforms[1] of a function $g(\boldsymbol{r})$

$$\begin{aligned} &g(\boldsymbol{r}) = \frac{1}{\mathcal{V}}\sum\nolimits_{\boldsymbol{q}} g_{\boldsymbol{q}} e^{i\boldsymbol{q}\cdot\boldsymbol{r}}, \quad g_{\boldsymbol{q}} = \int d\boldsymbol{r}\, g(\boldsymbol{r})e^{-i\boldsymbol{q}\cdot\boldsymbol{r}} \quad \text{and} \\ &g(\boldsymbol{r}) = \int \frac{d\boldsymbol{q}}{(2\pi)^d}\, g(\boldsymbol{q})e^{i\boldsymbol{q}\cdot\boldsymbol{r}}, \quad g(\boldsymbol{q}) = \int d\boldsymbol{r}\, g(\boldsymbol{r})e^{-i\boldsymbol{q}\cdot\boldsymbol{r}},\end{aligned} \tag{2.10}$$

with $\mathcal{V}$ denoting the system's volume and d the number of spatial dimensions. The latter Fourier transform is relevant in the limit $\mathcal{V} \to \infty$, in which, $\boldsymbol{q}$ becomes a continuous variable and

$$\frac{1}{\mathcal{V}}\sum_{\boldsymbol{q}} \to \int \frac{d\boldsymbol{q}}{(2\pi)^d}. \tag{2.11}$$

[1] For a lattice system we have to replace the volume element $\mathcal{V}$, appearing in equations (2.9) and (2.10), by the number of sites $\mathcal{N}$.

One obtains a correspondence between the discrete and continuum descriptions via $\left\{g_q, \mathcal{V}, \sum_q\right\} \mapsto \left\{g(\boldsymbol{q}), (2\pi)^d, \int d\boldsymbol{q}\right\}$. We now focus on the charge and current densities in the continuum case

$$\hat{\rho}_c(\boldsymbol{q}) = -e \int d\boldsymbol{k}\, \hat{\psi}^\dagger(\boldsymbol{k})\hat{\psi}(\boldsymbol{k} + \boldsymbol{q}), \tag{2.12}$$

$$\begin{aligned}\hat{\boldsymbol{J}}_c(\boldsymbol{q}, t) = &-\frac{e\hbar}{m} \int d\boldsymbol{k}\, \hat{\psi}^\dagger(\boldsymbol{k}) \\ &\left[\left(\boldsymbol{k} + \frac{\boldsymbol{q}}{2}\right)\hat{\psi}(\boldsymbol{k} + \boldsymbol{q}) + \frac{e}{\hbar} \int \frac{d\boldsymbol{q}'}{(2\pi)^d} \boldsymbol{A}(-\boldsymbol{q}', t)\hat{\psi}(\boldsymbol{k} + \boldsymbol{q}' + \boldsymbol{q})\right].\end{aligned} \tag{2.13}$$

While the minimal-coupling strategy is exact, it is mainly useful for Hamiltonians that are well accessible for manipulations in coordinate space, e.g. when translational invariance is present and momentum is conserved. Thus, it appears at first sight that the minimal-coupling approach fails to provide understanding in the case of crystalline systems.

Luckily, this is not the case if we are interested in wavelengths much longer than the lattice constants of the crystal. In this case, we turn to the local conservation of the electric charge, expressed via the classical law

$$\partial_t \rho_c(\boldsymbol{r}, t) + \nabla \cdot \boldsymbol{J}_c(\boldsymbol{r}, t) = 0. \tag{2.14}$$

For a quantum-mechanical description, this has to be extended to operators defined in the Heisenberg picture. This simply yields

$$\frac{d}{dt}\hat{\rho}_c(\boldsymbol{r}, t) + \nabla \cdot \hat{\boldsymbol{J}}_c(\boldsymbol{r}, t) = \hat{0} \quad \Rightarrow \quad \nabla \cdot \hat{\boldsymbol{J}}_c(\boldsymbol{r}, t) = \frac{\mathrm{i}}{\hbar}\left[\hat{\rho}_c(\boldsymbol{r}, t), \hat{H}(\hat{\boldsymbol{p}}, \boldsymbol{r})\right]. \tag{2.15}$$

For adequately small wavevector transfers $\boldsymbol{q}$, relative to the characteristic lengthscales of the system, we Fourier transform the above quantities and obtain the current operator in second quantisation:

$$\boldsymbol{q} \cdot \hat{\boldsymbol{J}}_c^{(0)}(\boldsymbol{q}) = -\frac{e}{\hbar} \int d\boldsymbol{k}\, \hat{\psi}^\dagger(\boldsymbol{k}) \left[\hat{H}(\boldsymbol{k} + \boldsymbol{q}) - \hat{H}(\boldsymbol{k})\right]^{(0)} \hat{\psi}(\boldsymbol{k} + \boldsymbol{q}). \tag{2.16}$$

The above is defined in the Schrödinger picture and at zeroth-order with respect to the vector potential. This expression is sufficient for retrieving the paramagnetic component of the linear response of a system coupled to an applied electromagnetic field. In fact, in many situations we are solely interested in the dynamical response, in which $\boldsymbol{q} \to \boldsymbol{0}$, and we find

$$\hat{\boldsymbol{J}}_c^{(0)}(\boldsymbol{q}) = -\frac{e}{\hbar} \int d\boldsymbol{k}\, \hat{\psi}^\dagger(\boldsymbol{k}) \frac{\partial \hat{H}^{(0)}(\boldsymbol{k})}{\partial \boldsymbol{k}} \hat{\psi}(\boldsymbol{k} + \boldsymbol{q}). \tag{2.17}$$

As an example, consider a quantum well obtained by confining a III–V semiconductor in the (001) plane, as discussed in section 1.2.4. For simplicity we assume bulk $\mathcal{I}$-symmetry. Thus, the Hamiltonian describing the Γ-point conduction band electrons, restricted to the lowest confinement channel, only consists of the quadratic kinetic energy term and the Rashba SOC of equation 1.33. Under these conditions, equation (2.17) yields the corresponding current operator in second quantisation

$$\hat{J}_{c,\perp}(\boldsymbol{q}_\perp) = -e\int d\boldsymbol{k}_\perp\, \hat{\psi}^\dagger(\boldsymbol{k}_\perp)\left(\frac{\hbar \boldsymbol{k}_\perp}{m} + \alpha\hat{\boldsymbol{\sigma}}\times\check{\boldsymbol{z}}\right)\hat{\psi}\left(\boldsymbol{k}_\perp + \boldsymbol{q}_\perp\right). \tag{2.18}$$

Despite the fact that the formula of equation (2.17) appears to be very useful in the case of free and Bloch electrons, it certainly has its limitations. In the following paragraphs we elaborate on other systems as well as alternative, though exact, methods for obtaining the current operator.

2.3 Charge current in lattice systems

In the present paragraph we address the case of lattice systems. In analogy to equation (2.8) we have

$$\begin{aligned}\hat{H} &= \int d\boldsymbol{r}\, \hat{\psi}^\dagger(\boldsymbol{r})\hat{H}[\hat{\boldsymbol{p}} + e\boldsymbol{A}(\boldsymbol{r}, t), \boldsymbol{r}]\hat{\psi}(\boldsymbol{r}) \\ &\equiv \int d\boldsymbol{r}\, \hat{\psi}^\dagger(\boldsymbol{r})e^{-\frac{ie}{\hbar}\int_{r_0}^{r} d\boldsymbol{r}'\cdot\boldsymbol{A}(\boldsymbol{r}', t)}\hat{H}(\hat{\boldsymbol{p}}, \boldsymbol{r})e^{\frac{ie}{\hbar}\int_{r_0}^{r} d\boldsymbol{r}'\cdot\boldsymbol{A}(\boldsymbol{r}', t)}\hat{\psi}(\boldsymbol{r}),\end{aligned} \tag{2.19}$$

with $\boldsymbol{r}_0$ the coordinate vector of an arbitrarily chosen reference point in space. By assuming that the coordinate vector is defined on a lattice, the above form leads to the so-called Peierls substitution [7, 8]

$$\hat{\psi}(\boldsymbol{r}) \mapsto e^{\frac{ie}{\hbar}\int_{r_0}^{r} d\boldsymbol{r}'\cdot\boldsymbol{A}(\boldsymbol{r}', t)}\hat{\psi}(\boldsymbol{r}) \quad\Rightarrow\quad \hat{\psi}_n \mapsto e^{\frac{ie}{\hbar}\int_{r_0}^{\boldsymbol{R}_n} d\boldsymbol{r}'\cdot\boldsymbol{A}(\boldsymbol{r}', t)}\hat{\psi}_n, \tag{2.20}$$

with n, m labelling the direct lattice vectors $\boldsymbol{R}_{n,m}$. The above substitution implies that the matrix elements (t_{nm}) of a Hamiltonian describing electron-hopping on the lattice, must be replaced by

$$\hat{\psi}_n^\dagger t_{nm}\hat{\psi}_m \mapsto \hat{\psi}_n^\dagger t_{nm}e^{-\frac{ie}{\hbar}\int_{\boldsymbol{R}_m}^{\boldsymbol{R}_n} d\boldsymbol{r}\cdot\boldsymbol{A}(\boldsymbol{r}, t)}\hat{\psi}_m. \tag{2.21}$$

At this point, it is useful to extend the charge conservation law of equation (2.15) to lattice systems. For this purpose one makes use of the definition of the current in terms of the particle's velocity. For the single-particle case, and in the Heisenberg picture, the charge current operator reads

$$-e\hat{\boldsymbol{v}}(t) = -e\frac{d\hat{\boldsymbol{r}}}{dt} \equiv \frac{\mathrm{i}}{\hbar}[\hat{H}, -e\hat{\boldsymbol{r}}(t)]. \tag{2.22}$$

In the above we encountered the electric polarisation, also called the electric dipole moment operator $\hat{\boldsymbol{P}} = -e\hat{\boldsymbol{r}}$. Note that the above expression is applicable to both continuum, lattice or other models, since it is basis independent. In addition, it can yield the current at any order with respect to the vector potential. The order obtained in terms of the vector potential is directly related to the respective order retained in the Hamiltonian (see section 2.9). In second quantisation, the corresponding density version of the polarisation operator, $\hat{\boldsymbol{P}}(\boldsymbol{r})$, reads

$$\hat{\boldsymbol{P}}(\boldsymbol{r}) = -e\hat{\psi}^{\dagger}(\boldsymbol{r})\hat{\boldsymbol{r}}\hat{\psi}(\boldsymbol{r}). \tag{2.23}$$

For lattice systems, one directly obtains the polarisation operator at a given lattice site, given by

$$\hat{\boldsymbol{P}}_n = -e\boldsymbol{R}_n\hat{\psi}_n^{\dagger}\hat{\psi}_n. \tag{2.24}$$

2.4 Linear response and current–current correlation functions

While we have retrieved the coupling of a generic electronic system to an external electromagnetic field, it is important to note that the resulting quantum-mechanical problem is rather cumbersome or practically impossible to solve in the majority of cases, and we therefore have to restrict ourselves to a perturbative approach. Thus, we are primarily interested in the charge and current response of the system to the externally applied fields. In particular, when the external fields are time-dependent we usually focus on the linear response, see [9], which encodes information regarding the properties of the isolated electronic system. In contrast, higher orders include additional effects, specific to the external fields.

In most cases we are interested in the current response of an electronic system when an electric field $\boldsymbol{E}(\boldsymbol{r}, t)$ is applied. For such situations one employs a generalised Ohm's law [9]

$$\boldsymbol{J}_c(\boldsymbol{r}, t) = \langle\hat{\boldsymbol{J}}_c(\boldsymbol{r}, t)\rangle := \int d\boldsymbol{r}' \int dt'\, \bar{\boldsymbol{\sigma}}(\boldsymbol{r}, t; \boldsymbol{r}', t')\boldsymbol{E}(\boldsymbol{r}', t'), \tag{2.25}$$

with $\bar{\boldsymbol{\sigma}}$ corresponding to the charge conductivity tensor. To calculate $\bar{\boldsymbol{\sigma}}$, it is convenient to choose the Coulomb gauge, i.e. $\nabla \cdot \boldsymbol{A}(\boldsymbol{r}, t) = 0$ and $V(\boldsymbol{r}, t) = 0$, which implies that $\boldsymbol{E}(\boldsymbol{r}, t) = -\partial_t\boldsymbol{A}(\boldsymbol{r}, t) \Rightarrow \boldsymbol{E}(\boldsymbol{q}, \omega) = \mathrm{i}\omega\boldsymbol{A}(\boldsymbol{q}, \omega)$. In the above, we employed the Fourier transform

$$g(t) = \int \frac{d\omega}{2\pi}\, g(\omega)e^{-\mathrm{i}\omega t} \quad \text{and} \quad g(\omega) = \int dt\, g(t)e^{\mathrm{i}\omega t}. \tag{2.26}$$

The external vector potential is assumed to be switched on at the time instant t_0. We obtain the expectation value of the current operator via the evolution of the quantum-mechanical system in the presence of the external perturbation. For the present discussion we consider $T = 0$ for simplicity. To calculate the required expectation value, we employ the so-called interaction picture [4–6, 10, 11], in which, operators evolve with the unperturbed Hamiltonian $\hat{H}_0$ and the state vectors with the perturbation part of the Hamiltonian termed $\hat{H}_A$. In fact, within linear response, we have [4]

$$
\begin{aligned}
\langle \hat{J}_c(\boldsymbol{r}, t)\rangle &\approx \left\langle \left[1 + \frac{\mathrm{i}}{\hbar}\int_{-\infty}^{t} dt'\, \hat{H}_A^{\mathrm{p}}(t')\right]\hat{J}_c^{\mathrm{p}}(\boldsymbol{r}, t)\left[1 - \frac{\mathrm{i}}{\hbar}\int_{-\infty}^{t} dt'\, \hat{H}_A^{\mathrm{p}}(t')\right]\right\rangle_0 \\
&\quad + \left\langle \hat{J}_c^{\mathrm{d}}(\boldsymbol{r}, t)\right\rangle_0 \\
&\approx \left\langle \hat{J}_c^{(0)}(\boldsymbol{r}, t)\right\rangle_0 - \frac{\mathrm{i}}{\hbar}\int_{-\infty}^{t} dt' \left\langle \left[\hat{J}_c^{\mathrm{p}}(\boldsymbol{r}, t), \hat{H}_A^{\mathrm{p}}(t')\right]\right\rangle_0 + \left\langle \hat{J}_c^{\mathrm{d}}(\boldsymbol{r}, t)\right\rangle_0
\end{aligned}
\tag{2.27}
$$

with $\langle \hat{J}_c^{(0)}\rangle_0$ denoting the expectation value of the current in the absence of the external field, which is zero if the ground state preserves TRS. Moreover, we have split the current and perturbation terms into paramagnetic and diamagnetic contributions, labelled 'p' and 'd'. By assuming TRS in the absence of the field, $\langle \hat{J}_c^{(0)}(\boldsymbol{r}, t)\rangle_0 = 0$. The paramagnetic contribution can be rewritten as

$$
\left\langle \hat{J}_{c,i}^{\mathrm{p}}(\boldsymbol{r}, t)\right\rangle \approx \frac{\mathrm{i}}{\hbar}\int d\boldsymbol{r}' \int_{-\infty}^{+\infty} dt'\, \Theta(t - t')\left\langle \left[\hat{J}_{c,i}^{\mathrm{p}}(\boldsymbol{r}, t), \hat{J}_{c,j}^{\mathrm{p}}(\boldsymbol{r}', t')\right]\right\rangle_0 A_j(\boldsymbol{r}', t'). \tag{2.28}
$$

The appearance of the Heaviside step function $\Theta(t - t')$, follows from causality. The current response to the applied field should be *retarded*, i.e. only appearing after the perturbation is switched on.

For a static and translationally-invariant Hamiltonian $\hat{H}_0$, the expectation value above depends only on the differences $\boldsymbol{r} - \boldsymbol{r}'$ and $t - t'$. Therefore, equation (2.25) reads $\boldsymbol{J}_c(\boldsymbol{q}, \omega) = \bar{\sigma}(\boldsymbol{q}, \omega)\boldsymbol{E}(\boldsymbol{q}, \omega)$,[2] in Fourier space. The paramagnetic conductivity tensor elements can be written in Fourier space as

$$
\sigma_{ij}^{\mathrm{p}}(\boldsymbol{q}, \omega) = \frac{\mathrm{i}}{\omega}\Pi_{ij}^{\mathrm{cc}}(\boldsymbol{q}, \omega) \quad \text{with} \quad \Pi_{ij}^{\mathrm{cc}}(\boldsymbol{q}, \omega) = \iint d\boldsymbol{r}dt\, e^{-\mathrm{i}(\boldsymbol{q}\cdot\boldsymbol{r} - \omega t)}\Pi_{ij}^{\mathrm{cc}}(\boldsymbol{r}, t) \tag{2.29}
$$

with $i, j = x, y, z$ and we introduced the charge current–current correlation function, also referred to as the polarisation tensor, defined as

$$
\Pi_{ij}^{\mathrm{cc}}(\boldsymbol{r} - \boldsymbol{r}', t - t') := -\frac{\mathrm{i}}{\hbar}\Theta(t - t')\left\langle \left[\hat{J}_{c,i}^{\mathrm{p}}(\boldsymbol{r}, t), \hat{J}_{c,j}^{\mathrm{p}}(\boldsymbol{r}', t')\right]\right\rangle_0 . \tag{2.30}
$$

Notably, being a retarded response function, the conductivity tensor is analytic in the upper half of the complex ω-plane. As a result, it satisfies the Kramers–Kronig relations [6]. For an arbitrary function $\chi(\omega)$ fulfilling the above constraint, the Sokhotski–Plemelj theorem [12, 13] yields

$$
\oint_{\mathrm{Im}(\omega')\geqslant 0} d\omega' \frac{\chi(\omega')}{\omega' - \omega + \mathrm{i}0^+} = 0 \quad \Rightarrow \quad \chi(\omega) = \frac{1}{\mathrm{i}\pi}\mathcal{P}\int_{-\infty}^{+\infty} d\omega' \frac{\chi(\omega')}{\omega' - \omega}, \tag{2.31}
$$

[2] The corresponding relative dielectric tensor is obtained via $\bar{\varepsilon}(\boldsymbol{q}, \omega) = \bar{1} - \bar{\sigma}(\boldsymbol{q}, \omega)/(\mathrm{i}\omega\varepsilon_0)$, with ε_0 the vacuum permittivity. For more details see also section 6.1.3.

thus implying that the real and imaginary parts of the paramagnetic component of the conductivity tensor are related. In the above, $\mathcal{P}\int_{-\infty}^{+\infty}$ denotes the principal value of the integral, i.e. the value obtained by excluding the contribution of the pole $\omega' = \omega$.

It has become obvious that at the level of linear response the conductivity tensor yields information solely regarding the isolated electronic system, which is assumed to be in thermodynamic equilibrium. Since this is the case, one expects that we should be able to retrieve the polarisation tensor via the thermodynamic potential of the system. In fact, there is a close connection between retarded response functions and their counterparts defined via the so-called finite temperature Matsubara technique, that is presented below.

2.5 Matsubara technique and thermal Green functions

The Matsubara approach is particularly suitable for studying systems in thermal equilibrium. While it is out of our scope to descibe this method in detail, we will try to put forward the key concepts. The reader can turn to the textbooks in references [4–6, 10, 11] for an in-depth presentation.

All the thermodynamic properties of a system in equilibrium can be retrieved via the *thermodynamic potential* Ω, defined for a given statistical ensemble [5]. Most of the time in this book we work in the grand-canonical ensemble, where the temperature and chemical potential (T, μ) are fixed. As a consequence, statistical mechanics implies that their conjugate fields, i.e. the energy and number of particles in the system (E, N), can fluctuate. The thermodynamic potential reads

$$\Omega = -k_B T \ln Z = -k_B T \ln\left\{ \mathrm{Sp}\left[e^{-(\hat{H}-\mu\hat{N})/(k_B T)} \right] \right\}, \tag{2.32}$$

with the quantity inside the logarithm defining the partition function in the grand-canonical ensemble Z. Moreover, 'Sp' denotes a trace over all the internal quantum-mechanical degrees of freedom, combined with statistical ensemble averaging [10]. $\hat{N}$ defines the particle number operator and k_B is the Boltzmann constant. For the remainder we set $k_B = 1$. For a finite-sized system the potential above is *extensive*, i.e. proportional to the volume of the system $\mathcal{V}$.

The turning point for the development of the Matsubara method was the observation that the term $e^{-(\hat{H}-\mu\hat{N})/T}$, appearing in equation (2.32), is very similar to the evolution operator encountered in quantum mechanics $e^{-\mathrm{i}\hat{H}t/\hbar}$. Therefore, by introducing the imaginary time $\tau := \mathrm{i}t/\hbar$, one can link the two formalisms, thus extending the well-established and successful techniques employed in quantum-field theory [3], to study systems in thermal equilibrium.

Within the Matsubara technique the fermionic (bosonic) fields are described by an imaginary time, and in the respective Fourier frequency domain, by the Matsubara frequencies $k_n = (2n + 1)\pi T$ $(\omega_n = 2\pi n T)$, with $n \in \mathbb{Z}$. Remarkably, the latter frequencies are discrete, in stark contrast to the continuous frequency ω encountered in the real time description. The discretisation is an immediate result of the fact that

$\tau \in [0, 1/T]$. The latter additionally implies an (anti)periodicity property of the (fermionic) bosonic fields in imaginary time, see [4–6, 10, 11].

A fundamental block of the formalism is the Matsubara or thermal single-particle Green function, which is defined in frequency space[3]

$$\hat{G}(k_n, k_n + \omega_s) := \frac{1}{(\mathrm{i}k_n + \mu - \hat{H}_0)\delta_{\omega_s,0} - \hat{\Sigma}(\mathrm{i}k_n, \mathrm{i}\omega_s) - \hat{H}_{\mathrm{ext}}(\mathrm{i}\omega_s)}, \tag{2.33}$$

where $\hat{H}_0$ denotes the single-particle Hamiltonian of a non-interacting fermionic system, $\hat{\Sigma}$ defines the matrix *self-energy* incorporating the effects of interactions, cf section 2.7, and $\hat{H}_{\mathrm{ext}}$ the coupling to external fields. We demonstrate how to retrieve the latter in section 2.6. Note that the bosonic Matsubara frequencies ω_s, allow for time-dependent perturbations. The Green function can be employed for calculating the thermodynamic potential via [6]

$$\Omega = -T \ln\left[\det(-\hat{G}^{-1}/T)\right] = \Omega^{(0)} - T \ln\left\{\det\left[\hat{1} - \hat{G}^{(0)}(\hat{\Sigma} + \hat{H}_{\mathrm{ext}})\right]\right\}, \tag{2.34}$$

where we introduced the bare Green function $\hat{G}^{(0)}(k_n) = 1/(\mathrm{i}k_n + \mu - \hat{H}_0)$ and the respective thermodynamic potential $\Omega^{(0)}$. Remarkably, the above can serve as a starting point for a perturbation theory in terms of the self-energy, or/and the Hamiltonian describing the coupling to external fields, cf equation (2.52). Even more importantly, the single-particle Green function plays a crucial role for calculating correlation functions appearing in linear response theory.

Besides equation (2.33), the thermal Green function can be also defined in imaginary time via

$$\hat{G}_{ab}(\tau, \tau') := -\langle \hat{T}_\tau \hat{\psi}_a(\tau) \hat{\psi}_b^\dagger(\tau') \rangle. \tag{2.35}$$

The expectation value is evaluated at thermal equilibrium and involves statistical ensemble averaging and quantum-mechanical tracing [5]. The fermionic operators are defined in the imaginary time Heisenberg picture. We also introduced time-ordering in imaginary time, effected by the operator

$$\hat{T}_\tau A(\tau)B(\tau') = \Theta(\tau - \tau')A(\tau)B(\tau') \pm \Theta(\tau' - \tau)B(\tau')A(\tau), \tag{2.36}$$

with + (−) when A and B are bosonic (fermionic) operators. Interestingly, the above imaginary time Green function can be connected to its retarded analog in real time, $\hat{G}^{\mathrm{R}}$, which for fermions reads

[3] Despite the discrete character of k_n, for notational convenience we write $\hat{G}(k_n)$ instead of $\hat{G}_{k_n}$. This should lead to no confusion, since for real frequencies we consider the Green function $\hat{G}(\omega)$, with ω being the continuous frequency variable.

$$\hat{G}^{\mathrm{R}}_{ab}(t, t') := -\frac{\mathrm{i}}{\hbar}\Theta(t - t')\Big\langle\big\{\hat{\psi}_a(t), \hat{\psi}_b^\dagger(t')\big\}\Big\rangle. \tag{2.37}$$

This connection is achieved via the so-called *analytic continuation*, which in frequency space reads

$$\hat{G}^{\mathrm{R}}(\omega) = \hat{G}(\mathrm{i}k_n \mapsto \hbar\omega + \mathrm{i}0^+). \tag{2.38}$$

The above connection reveals the power of the Matsubara technique, since it allows studying the real-time evolution of quantum-mechanical systems, solely through an equilibrium picture. However, the Matsubara method is not adequate for investigating non-equilibrium properties, which is instead made possible by the Keldysh technique, see [4, 6].

One of the quantities that is straightforward to retrieve via performing analytic continuation to the thermal Green function, is the energy-resolved matrix *spectral function*:

$$\hat{A}(E) := -2\,\mathrm{Im}\Big[\hat{G}^{\mathrm{R}}(E)\Big] \equiv -2\,\mathrm{Im}\,[\hat{G}(\mathrm{i}k_n \mapsto E + \mathrm{i}0^+)]. \tag{2.39}$$

The meaning of the above quantity can be made transparent by considering the case of a system of non-interacting single-band fermions, with energy dispersion $E(\boldsymbol{k})$. The two functions now read:

$$G(\boldsymbol{k}, \mathrm{i}k_n) = \frac{1}{\mathrm{i}k_n + \mu - E(\boldsymbol{k})} \quad \text{and} \quad A(\boldsymbol{k}, E) = 2\pi\delta[E + \mu - E(\boldsymbol{k})], \tag{2.40}$$

thus yielding the energy of the quasiparticles and the weight of the pole. The spectral function satisfies the sum rule

$$\int \frac{dE}{2\pi} A(\boldsymbol{k}, E) = 1, \tag{2.41}$$

reflecting its role as a probability distribution (only in the case of fermions), while it also yields the density of states for the state with energy $E(\boldsymbol{k})$ [5].

2.6 Matsubara formulation of linear response

Having laid the foundations of the Matsubara formalism, we proceed with demonstrating how it can be employed to calculate retarded correlation functions, such as the polarisation tensor of equation (2.30). Following the spirit of the previous section, the retarded polarisation tensor can be obtained via the analytic continuation $\bar{\mathbf{\Pi}}^{\mathrm{cc}}(\boldsymbol{q}, \omega) \equiv \bar{\mathbf{\Pi}}^{\mathrm{cc}}(\boldsymbol{q}, \mathrm{i}\omega_s \to \hbar\omega + \mathrm{i}0^+)$, with ω_s the bosonic frequency of the vector potential. Here we introduced the so-called thermal polarisation tensor, $\bar{\mathbf{\Pi}}^{\mathrm{cc}}(\boldsymbol{q}, \mathrm{i}\omega_s)$, calculated via the Matsubara method. The latter is retrieved by solely considering the linear (paramagnetic) coupling of the electrons to the vector potential. In particular, the above polarisation tensor in the Matsubara formalism is defined as the Fourier transform of its imaginary time counterpart

$$\Pi^{cc}_{ij}(\mathbf{r}-\mathbf{r}',\tau-\tau') := -\left\langle \hat{\mathcal{T}}_\tau \hat{J}^{p}_{c,i}(\mathbf{r},\tau)\hat{J}^{p}_{c,j}(\mathbf{r}',\tau')\right\rangle_0 \text{ and}$$
$$\Pi^{cc}_{ij}(\mathbf{r}-\mathbf{r}',\mathrm{i}\omega_s) := \int_0^{1/T} d\tau\, e^{\mathrm{i}\omega_s(\tau-\tau')}\Pi^{cc}_{ij}(\mathbf{r}-\mathbf{r}',\tau-\tau'). \tag{2.42}$$

In order to evaluate the polarisation tensor $\bar{\mathbf{\Pi}}^{cc}(\mathbf{q},\tau-\tau')$, one employs Wick's theorem [4–6, 10, 11], according to which, every equilibrium expectation value of a time-ordered product of operators defined in imaginary time, can be written as the sum of products of single-particle Green functions in imaginary time. For the case of interest we have:

$$\begin{aligned}-\left\langle \hat{\mathcal{T}}_\tau \hat{J}^{p}_{c,i}(\mathbf{r},\tau)\hat{J}^{p}_{c,j}(\mathbf{r}',\tau')\right\rangle_0 &= -\left\langle \hat{\mathcal{T}}_\tau \hat{\psi}^\dagger(\mathbf{r},\tau)\hat{\mathcal{J}}^{p}_{c,i}\,\hat{\psi}(\mathbf{r},\tau)\,\hat{\psi}^\dagger(\mathbf{r}',\tau')\hat{\mathcal{J}}^{p}_{c,j}\,\hat{\psi}(\mathbf{r}',\tau')\right\rangle_0 \\ &= \mathrm{Tr}\left[\hat{G}^{(0)}(\mathbf{r}'-\mathbf{r},\tau'-\tau)\hat{\mathcal{J}}^{p}_{c,i}\hat{G}^{(0)}(\mathbf{r}-\mathbf{r}',\tau-\tau')\hat{\mathcal{J}}^{p}_{c,j}\right].\end{aligned} \tag{2.43}$$

We introduced appropriate single-particle paramagnetic current operators $\hat{\mathcal{J}}^{p}_{c,i}$, defined in the Schrödinger picture. In addition, 'Tr' denotes a trace over all the internal degrees of freedom. Therefore, we find

$$\begin{aligned}\Pi^{cc}_{ij}(\mathbf{q},\mathrm{i}\omega_s) &= \left(\frac{e}{\hbar}\right)^2 \int\frac{d\mathbf{k}}{(2\pi)^d} T\sum\nolimits_{k_n} \\ &\times \mathrm{Tr}\left[\hat{G}^{(0)}(\mathbf{k},\mathrm{i}k_n)\frac{\partial\hat{H}_0(\mathbf{k})}{\partial k_i}\hat{G}^{(0)}(\mathbf{k}+\mathbf{q},\mathrm{i}k_n+\mathrm{i}\omega_s)\frac{\partial\hat{H}_0(\mathbf{k})}{\partial k_j}\right],\end{aligned} \tag{2.44}$$

which, in the case of crystalline systems, is valid for long wavelengths (see section 2.2).

We move on with demonstrating how to alternatively study the linear response theory by employing a Matsubara version of the Hamiltonian, where the real time, t, is replaced with the imaginary time τ. This builds upon the mapping between retarded and Matsubara Green functions, achieved via the analytic continuation. Thus, the total Hamiltonian is rewritten as: $\hat{H}(\tau)=\hat{H}_0+\hat{H}_A(\tau)$, with $\hat{H}_A(\tau)$ denoting the imaginary time Hamiltonian describing the coupling of the electronic system to a time-dependent vector potential. We wish to employ a perturbative approach relying on equation (2.34). Therefore, we have to identify the matrix elements of the Hamiltonian $\hat{H}_{\text{ext}}$. For this purpose, we start from the coupling to the vector potential in the Schrödinger picture

$$\hat{H}_A(\tau) = \int\frac{d\mathbf{q}}{(2\pi)^d}\int d\mathbf{k}\,\hat{\psi}^\dagger(\mathbf{k}+\mathbf{q})\hat{\mathcal{H}}_A(\mathbf{q},\tau)\hat{\psi}(\mathbf{k}). \tag{2.45}$$

In order to proceed, one can employ the path integral formalism and define the Matsubara (or Euclidean) action $S_{\psi-A}$, defined as

$$\begin{aligned}S_{\psi-A} &:= \int_0^{1/T} d\tau\, H_A(\tau) \\ &= T\sum\nolimits_{\omega_s}\int\frac{d\mathbf{q}}{(2\pi)^d}\sum\nolimits_{k_n}\int d\mathbf{k}\,\bar{\psi}(\mathbf{k}+\mathbf{q},k_n+\omega_s)\hat{\mathcal{H}}_A(\mathbf{q},\mathrm{i}\omega_s)\psi(\mathbf{k},k_n),\end{aligned} \tag{2.46}$$

where the field operators are now replaced by the so-called Grassmann fields, ψ and $\tilde{\psi}$. A similar replacement took place for the Hamiltonian, i.e. $\hat{H}_A(\tau) \mapsto H_A(\tau)$. Here we do not aim at further explaining the motivation and grounds for introducing the above construction, and prompt the reader to refer to [6] for a detailed presentation. Based on the above, the desired matrix elements for the coupling Hamiltonian $\hat{H}_{\text{ext}}$ of equation (2.34), read:

$$\langle \boldsymbol{k} + \boldsymbol{q}, \mathrm{i}k_n + \mathrm{i}\omega_s | \hat{H}_{\text{ext}} | \boldsymbol{k}, \mathrm{i}k_n \rangle := \frac{T}{(2\pi)^d} \hat{H}_A(\boldsymbol{q}, \mathrm{i}\omega_s). \tag{2.47}$$

Note that we could have alternatively obtained the above by combining equations (2.33) and (2.35).

To this end, one employs equation (2.34) and obtains an effective action for the vector potential:

$$\begin{aligned} S_A := -\ln\left\{\det\left[\hat{1} - \hat{G}^{(0)}\hat{H}_{\text{ext}}\right]\right\} &\equiv -\operatorname{Tr}\left\{\ln\left[\hat{1} - \hat{G}^{(0)}\hat{H}_{\text{ext}}\right]\right\} \\ &\equiv \sum_{n=1}^{\infty} \frac{\operatorname{Tr}(\hat{G}^{(0)}\hat{H}_{\text{ext}})^n}{n}, \end{aligned} \tag{2.48}$$

where the trace is over all the electronic degrees of freedom. For simplicity we set $\hat{\Sigma} = \hat{0}$. The Matsubara effective action S_A^{p}, retrieved for the electronic system solely under the influence of the paramagnetic coupling to the vector potential, can be alternatively written in terms of the expectation value of the paramagnetic electric current operator, i.e.:

$$S_A^{\mathrm{p}} \equiv -T \sum_{\omega_s} \int \frac{d\boldsymbol{q}}{(2\pi)^d} \boldsymbol{J}_c^{\mathrm{p}}(\boldsymbol{q}, \mathrm{i}\omega_s) \cdot \boldsymbol{A}(-\boldsymbol{q}, -\mathrm{i}\omega_s). \tag{2.49}$$

This relation allows us to link the polarisation tensor to the effective action obtained for the vector potential [6], since:

$$\Pi_{ij}^{cc}(\boldsymbol{q}, \mathrm{i}\omega_s) = -\frac{\partial J_{c,i}^{\mathrm{p}}(\boldsymbol{q}, \mathrm{i}\omega_s)}{\partial A_j(\boldsymbol{q}, \mathrm{i}\omega_s)} = \frac{(2\pi)^d}{T} \frac{\partial^2 S_A^{\mathrm{p}}}{\partial A_j(\boldsymbol{q}, \mathrm{i}\omega_s)\partial A_i(-\boldsymbol{q}, -\mathrm{i}\omega_s)}\Bigg|_{\boldsymbol{A}=\boldsymbol{0}}. \tag{2.50}$$

Based on the properties of the mixed derivative above, we obtain a set of constraints

$$\Pi_{ij}^{cc}(\boldsymbol{q}, \mathrm{i}\omega_s) = \Pi_{ji}^{cc}(-\boldsymbol{q}, -\mathrm{i}\omega_s) \tag{2.51}$$

that correspond to the Onsager reciprocity relations [4, 14]. They describe the relation between the conductivity tensor elements for two reciprocal situations in which: (i) a field E_i is applied and a current $J_{c,j}^{\mathrm{p}}$ is measured and (ii) a field E_j is applied and a current $J_{c,i}^{\mathrm{p}}$ is measured.

By virtue of the second-order functional derivative appearing in equation (2.50), one infers that for studying linear response, it is sufficient to restrict to up to the quadratic order in the vector potential. Therefore, for a translationally invariant system, the second-order effective action assumes the form:

$$S_A^{(2)} = T\sum_{k_n}\int\frac{d\boldsymbol{k}}{(2\pi)^d}\,\mathrm{Tr}\left[\hat{G}^{(0)}(\boldsymbol{k},\mathrm{i}k_n)\hat{H}_A^{\mathrm{d}}(\boldsymbol{0},0)\right]$$
$$+\frac{1}{2}T\sum_{\omega_s}\int\frac{d\boldsymbol{q}}{(2\pi)^d}T\sum_{k_n}\int\frac{d\boldsymbol{k}}{(2\pi)^d} \tag{2.52}$$
$$\mathrm{Tr}\left[\hat{G}^{(0)}(\boldsymbol{k},\mathrm{i}k_n)\hat{H}_A^{\mathrm{p}}(-\boldsymbol{q},-\mathrm{i}\omega_s)\hat{G}^{(0)}(\boldsymbol{k}+\boldsymbol{q},\mathrm{i}k_n+\mathrm{i}\omega_s)\hat{H}_A^{\mathrm{p}}(\boldsymbol{q},\mathrm{i}\omega_s)\right],$$

where we decomposed the Hamiltonian $\hat{H}_A$, into paramagnetic ('p') and diamagnetic ('d') components $\hat{H}_A^{\mathrm{p}}$ and $\hat{H}_A^{\mathrm{d}}$. Note, that, for the diamagnetic coupling we solely retained the quadratic term with respect to the vector potential.

Let us make a final remark. Usually, when the electromagnetic system is coupled to a radiation field and the dynamical response is sought after, the diamagnetic contribution is customarily neglected, since it comes about in the Hamiltonian at least at second or higher order with respect to the vector potential. Nonetheless, there exist cases where the real-time effective action of the electromagnetic field in the hydrodynamic limit is pursued, cf section 3.5. In the hydrodynamic limit, the analytic continuation simply becomes: $\mathrm{i}\omega_s \mapsto \hbar\omega$. In such a situation, one must take into account that the diamagnetic contribution is required to ensure the electric charge conservation and retain the U(1) gauge invariance.

2.7 Charge conductivity of an electron gas

In order to demonstrate how to apply the aforementioned method, let us consider a simple system, and in particular the Γ-point conduction band electrons of a III–V semiconductor in the absence of SOC, with energy dispersion $\varepsilon_c(\boldsymbol{k}) = (\hbar\boldsymbol{k})^2/(2m)$. In this case we have the bare Green function per spin

$$G^{(0)}(\boldsymbol{k},\mathrm{i}k_n) = \frac{1}{\mathrm{i}k_n+\mu-\varepsilon_c(\boldsymbol{k})} \tag{2.53}$$

and the single-particle Matsubara version of the Hamiltonian is described by

$$\hat{H}_A^{\mathrm{p}}(\boldsymbol{q},\mathrm{i}\omega_s) = \frac{e\hbar}{m}\left(\boldsymbol{k}+\frac{\boldsymbol{q}}{2}\right)\cdot\boldsymbol{A}(\boldsymbol{q},\mathrm{i}\omega_s), \tag{2.54}$$

$$\hat{H}_A^{\mathrm{d}}(\boldsymbol{q},\mathrm{i}\omega_s) = \frac{1}{2}T\sum_{\omega_s'}\int\frac{d\boldsymbol{q}'}{(2\pi)^d}\frac{e^2}{m}\boldsymbol{A}(\boldsymbol{q}-\boldsymbol{q}',\mathrm{i}\omega_s-\mathrm{i}\omega_s')\cdot\boldsymbol{A}(\boldsymbol{q}',\mathrm{i}\omega_s'). \tag{2.55}$$

In order to proceed we make use of the following formula regarding the Matsubara summation of the product of two fermionic Green functions with simple poles [4–6]

$$T\sum_{k_n}\frac{1}{\mathrm{i}k_n+\mu-E_a(\boldsymbol{k})}\frac{1}{\mathrm{i}\omega_s+\mathrm{i}k_n+\mu-E_b(\boldsymbol{k}+\boldsymbol{q})}=\frac{f[E_a(\boldsymbol{k})]-f[E_b(\boldsymbol{k}+\boldsymbol{q})]}{\mathrm{i}\omega_s+E_a(\boldsymbol{k})-E_b(\boldsymbol{k}+\boldsymbol{q})}, \quad (2.56)$$

where we introduced the Fermi–Dirac distribution function in the grand-canonical ensemble

$$f(E)=\frac{1}{1+e^{(E-\mu)/T}}. \quad (2.57)$$

By employing the above, one obtains $\Pi_{ij}^{cc}\sim\delta_{i,j}$. After including both spin contributions, we have

$$\Pi_{ii}^{cc}(\boldsymbol{q},\mathrm{i}\omega_s)=2\left(\frac{e\hbar}{m}\right)^2\int\frac{d\boldsymbol{k}}{(2\pi)^d}\left(k_i+\frac{q_i}{2}\right)^2\frac{f[\varepsilon_c(\boldsymbol{k})]-f[\varepsilon_c(\boldsymbol{k}+\boldsymbol{q})]}{\mathrm{i}\omega_s-\hbar(\boldsymbol{k}+\boldsymbol{q}/2)\cdot\hbar\boldsymbol{q}/m} \quad (2.58)$$

and the paramagnetic component of the conductivity in real frequency reads

$$\sigma_{ii}^{\mathrm{P}}(\boldsymbol{q},\omega)=\mathrm{i}2\left(\frac{e\hbar}{m}\right)^2\int\frac{d\boldsymbol{k}}{(2\pi)^d}\left(k_i+\frac{q_i}{2}\right)^2\frac{f[\varepsilon_c(\boldsymbol{k})]-f[\varepsilon_c(\boldsymbol{k}+\boldsymbol{q})]}{\omega[\hbar\omega-\hbar(\boldsymbol{k}+\boldsymbol{q}/2)\cdot\hbar\boldsymbol{q}/m+\mathrm{i}0^+]}. \quad (2.59)$$

With the help of the Sokhotski–Plemelj theorem [12, 13], we find the Dirac identity: $1/(\omega+\mathrm{i}0^+)=\mathcal{P}-\mathrm{i}\pi\delta(\omega)$. Therefore, the real and imaginary parts of the conductivity read

$$\begin{aligned}\mathrm{Re}[\sigma_{ii}^{\mathrm{P}}(\boldsymbol{q},\omega)]=2\left(\frac{e\hbar}{m}\right)^2\int\frac{d\boldsymbol{k}}{(2\pi)^d}\left(k_i+\frac{q_i}{2}\right)^2\frac{f[\varepsilon_c(\boldsymbol{k})]-f[\varepsilon_c(\boldsymbol{k})+\hbar\omega]}{\hbar\omega}\\ \times\pi\delta[\omega-(\boldsymbol{k}+\boldsymbol{q}/2)\cdot\hbar\boldsymbol{q}/m],\end{aligned} \quad (2.60)$$

$$\mathrm{Im}[\sigma_{ii}^{\mathrm{P}}(\boldsymbol{q},\omega)]=2\left(\frac{e\hbar}{m}\right)^2\mathcal{P}\int\frac{d\boldsymbol{k}}{(2\pi)^d}\left(k_i+\frac{q_i}{2}\right)^2\frac{f[\varepsilon_c(\boldsymbol{k})]-f[\varepsilon_c(\boldsymbol{k}+\boldsymbol{q})]}{\hbar\omega[\omega-(\boldsymbol{k}+\boldsymbol{q}/2)\cdot\hbar\boldsymbol{q}/m]}. \quad (2.61)$$

To obtain the principal value of the second integral, the pole $\omega=(\boldsymbol{k}+\boldsymbol{q}/2)\cdot\hbar\boldsymbol{q}/m$ is not taken into account while integrating. In the dynamical-reponse limit we first take $\boldsymbol{q}\to\boldsymbol{0}$, and we obtain that $\mathrm{Im}[\sigma_{ii}^{\mathrm{P}}(\boldsymbol{q},\omega)]\to 0$, while the real part reads

$$\mathrm{Re}[\sigma_{ii}^{\mathrm{P}}(\boldsymbol{0},\omega)]=-2\left(\frac{e\hbar}{m}\right)^2\int\frac{d\boldsymbol{k}}{(2\pi)^d}k_i^2f'[\varepsilon_c(\boldsymbol{k})]\pi\delta(\omega), \quad (2.62)$$

where $f'(z)$ stands for the derivative with respect to the argument z. At $T=0$ we have $-f'(z)=\delta(z-\mu)$. Notably, the above conductivity is infinite for a static electric field. However, this is a mere artifact of the model that neglects the scattering between the electrons and the ion cores or possible impurities present, that will render the conductivity finite. One can compare the above microscopic results, with the so-called Drude model [4, 6, 15] for an electron in a static electric field, that reads:

$$m\left(\dot{\boldsymbol{v}} + \frac{\boldsymbol{v}}{\tau_{\text{col}}}\right) = -e\boldsymbol{E} \quad \Rightarrow \quad \boldsymbol{J}_c(\omega) = -e\rho\boldsymbol{v}(\omega) = \frac{\rho e^2\tau_{\text{col}}}{m}\frac{1 + \mathrm{i}\omega\tau_{\text{col}}}{1 + (\omega\tau_{\text{col}})^2}\boldsymbol{E}, \tag{2.63}$$

with ρ denoting the average particle density. Within this model, one includes a phenomenological friction term with a characteristic 'collision' time τ_{col}. Note that for $\omega\tau_{\text{col}} \gg 1$ we find that the conductivity consists only of an imaginary part which reads $\text{Im}[\sigma_{ii}(\boldsymbol{0}, \omega)] = \rho e^2/(m\omega)$. Interestingly, the above result is identical to the conductivity obtained from the diamagnetic contribution found in the microscopic theory presented earlier. For $d = 3$, ρ is given by

$$\rho = 2\int\frac{d\boldsymbol{k}}{(2\pi)^d}T\sum_{k_n}G^{(0)}(\boldsymbol{k}, \mathrm{i}k_n) = \frac{1}{\pi^2}\int dk\; k^2 f[\varepsilon_c(k)]. \tag{2.64}$$

For $T = 0$ one obtains $k_F^3 = 3\pi^2\rho$, with $k_F = \sqrt{2m\mu}/\hbar$.

In this limit one can recover the classical picture from the microscopics, since the frequency of the external field is much larger than the collision frequency, $1/\tau_{\text{col}}$, of the electrons with the ions. Thus, in this regime the electrons respond to the external field before colliding with other obstacles. In the opposite limit, where $\omega\tau_{\text{col}} \ll 1$, the conductivity is finite and tends to a constant value given by $\rho e^2\tau_{\text{col}}/m$. In this case we cannot neglect the collisions with the ions or impurities, which dominate, and lead to Ohmic losses and Joule heating. It is in this limit that our microscopic model fails, and the presence of the δ-function spike appears at first sight unphysical. Nonetheless, its presence is compatible with the causality constraints on the conductivity. By employing the Kramers–Kronig relations for $\text{Re}[\sigma_{ii}(\boldsymbol{0}, \omega)] = \alpha\pi\delta(\omega)$, we find $\text{Im}[\sigma_{ii}(\boldsymbol{0}, \omega)] = \alpha/\omega$. Thus, causality is ensured for the model under consideration only if

$$-2\frac{\hbar^2}{m}\int\frac{d\boldsymbol{k}}{(2\pi)^3}\, k_i^2 f'[\varepsilon_c(\boldsymbol{k})] = \rho. \tag{2.65}$$

The above equality can be indeed verified to hold, by employing the general relation for the Green functions

$$\hat{G}\hat{G}^{-1} = \hat{1} \quad \Rightarrow \quad \hat{G}(\partial\hat{G}^{-1})\hat{G} = -\partial\hat{G}, \tag{2.66}$$

in conjuction with equation (2.56) for $\boldsymbol{q} = \boldsymbol{0}$, $\omega_s = 0$ and $a = b$, that yield

$$\begin{aligned} -2\frac{\hbar^2}{m}\int\frac{d\boldsymbol{k}}{(2\pi)^3}\, k_i^2 f'[\varepsilon_c(\boldsymbol{k})] &= -\frac{1}{3\pi^2}\frac{\hbar^2}{m}\int dk\; k^2\, k^2\, f'[\varepsilon_c(k)] \\ &= -\frac{2m}{3\pi^2}T\sum_{k_n}\int dk\; k^2 G^{(0)}(k, \mathrm{i}k_n)\frac{\partial[G^{(0)}(k, \mathrm{i}k_n)]^{-1}}{\partial m}G^{(0)}(k, \mathrm{i}k_n) \equiv \frac{2m}{3}\frac{\partial\rho}{\partial m}. \end{aligned} \tag{2.67}$$

For $T = 0$ one finds $\partial\rho/\partial m = 3\rho/(2m)$, thus confirming the required equality.

Remarkably, the particular equality of coefficients between the paramagnetic and diamagnetic terms is essentially ensured by the U(1) gauge invariance of our theory. This can be understood by evaluating equation (2.52) for $\omega_s = 0$ and $\boldsymbol{q} \to \boldsymbol{0}$. In this limit one would generally obtain a term proportional to A^2. However, if the latter term were present it would lead to the violation of gauge invariance and at the same time generate a non-zero mass for the photon [3, 6, 7]. Nonetheless, this only occurs in superconducting systems, where gauge invariance is spontaneously broken [4–6, 10]. For a non-superconducting system, as the one examined here, the paramagnetic and diamagnetic contributions must cancel, thus imposing the equality of coefficients shown earlier [6].

Before moving on to the next section, let us mention that for retrieving the Drude model from a microscopic model, one can in the simplest case include the effect of collisions phenomenologically, by introducing a finite lifetime for the electrons [4, 6], which enters in the full Green function as an imaginary self-energy term, i.e.

$$G_{\mathrm{col}}^{(0)}(\boldsymbol{k}, \mathrm{i}k_n) = \frac{1}{\mathrm{i}k_n + \mu - \varepsilon_c(\boldsymbol{k}) - \Sigma_{\mathrm{col}}(\boldsymbol{k}, \mathrm{i}k_n)}. \tag{2.68}$$

Through the presence of self-energy, the introduction of collisions can (i) yield a finite lifetime for the quasiparticles, (ii) reduce their spectral weight and (iii) modify the bandstructure. To obtain a qualitative agreement with the Drude model it is sufficient to assume that the effects of collisions with the ions are reflected in the following retarded Green and spectral functions:

$$\begin{aligned} G_{\mathrm{col}}^{R,(0)}(\boldsymbol{k}, E) &= \frac{1}{E + \mu - \varepsilon_c(\boldsymbol{k}) + \mathrm{i}\dfrac{\hbar}{2\tau_{\mathrm{col}}}} \quad \text{with} \\ A_{\mathrm{col}}^{(0)}(\boldsymbol{k}, E) &= \frac{\hbar}{\tau_{\mathrm{col}}} \frac{1}{[E + \mu - \varepsilon_c(\boldsymbol{k})]^2 + \left(\dfrac{\hbar}{2\tau_{\mathrm{col}}}\right)^2}. \end{aligned} \tag{2.69}$$

However, a further discussion on this topic is out of the scope of this section and the reader is urged to turn to references [4, 6]. Note also that in section 2.10 we discuss an approach to studying linear response based on the Boltzmann transport equation.

2.8 Thermoelectric and thermal transport

Applying an electric field to a material not only leads to electric charge transport, but also generates a heat current. In turn, a heat current can lead to temperature gradients in the system. By reciprocity, gradients of the temperature can also lead to electric currents. The combined effects that can take place in presence of these electric and thermal perturbations are contained in the equations

$$J_c = \bar{\boldsymbol{\sigma}} \cdot \boldsymbol{E} + \bar{\boldsymbol{\alpha}} \cdot (-\nabla T), \tag{2.70}$$

$$J_Q = T\bar{\boldsymbol{\alpha}} \cdot \boldsymbol{E} + \bar{\boldsymbol{\kappa}} \cdot (-\nabla T), \tag{2.71}$$

where we additionally introduced the thermoelectric and thermal conductivity tensors, $\bar{\boldsymbol{\alpha}}$ and $\bar{\boldsymbol{\kappa}}$, respectively. These conductivity tensors are retrieved via suitable

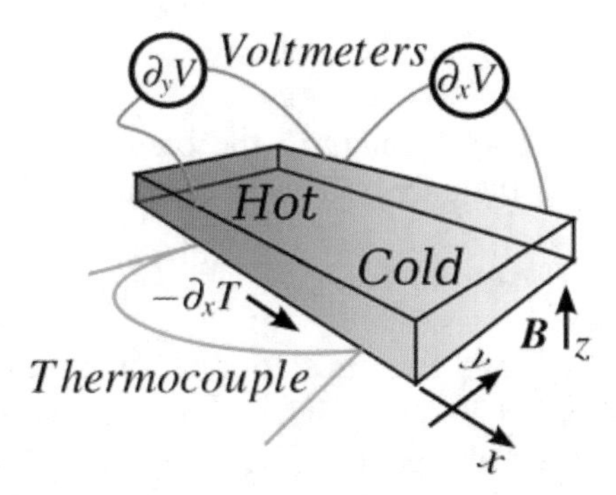

Figure 2.1. Typical experimental setup for studying thermoelectric and thermal transport. A temperature gradient is applied, which generates electric (J_c) and heat (J_Q) currents. Under conditions of vanishing electric currents, i.e. $\boldsymbol{J}_c = \boldsymbol{0}$, longitudinal and transverse voltage gradients appear, $-\nabla V = \boldsymbol{E} \neq \boldsymbol{0}$. One can then experimentally infer the Seebeck (S) coefficient and Nernst (N) signal, as also the electron thermal conductivity tensor elements ($\bar{\boldsymbol{\kappa}}_e$). Investigating Hall transport requires either the application of an external out-of-plane magnetic field, e.g. $\boldsymbol{B} = B_z\check{z}$, or the presence of a non-zero intrinsic magnetisation in the material.

current–current correlations functions, similar to the case of the electric charge conductivity [4, 16–19].

Most of the time in these experiments, cf figure 2.1, one applies a temperature gradient along one direction (e.g. x axis) and measures the induced electric fields or voltage drops under vanishing charge currents. For such situations, one additionally introduces the *electron thermal conductivity* tensor $\bar{\boldsymbol{\kappa}}_e$, defined via the relation $\boldsymbol{J}_Q = \bar{\boldsymbol{\kappa}}_e(-\nabla T)$. In the general case $\bar{\boldsymbol{\kappa}}_e \neq \bar{\boldsymbol{\kappa}}$, since both $\bar{\boldsymbol{\alpha}}$ and $\bar{\boldsymbol{\kappa}}$ contribute to $\bar{\boldsymbol{\kappa}}_e$ (see [15] and the Hands-on section 2.10). Moreover, by means of the two coupled equations, $\bar{\boldsymbol{\sigma}} \cdot \boldsymbol{E} + \bar{\boldsymbol{\alpha}} \cdot (-\partial_x T, 0)^{\mathrm{T}} = \boldsymbol{0}$, we also obtain the so-called: Seebeck (S) coefficient and Nernst (N) signal

$$\mathrm{S} := \frac{E_x}{\partial_x T} = \frac{\alpha_{xx}\sigma_{yy} - \sigma_{xy}\alpha_{yx}}{\sigma_{xx}\sigma_{yy} - \sigma_{xy}\sigma_{yx}} \quad \text{and} \quad \mathrm{N} := \frac{E_y}{-\partial_x T} = \frac{\sigma_{yx}\alpha_{xx} - \alpha_{yx}\sigma_{xx}}{\sigma_{xx}\sigma_{yy} - \sigma_{xy}\sigma_{yx}}. \quad (2.72)$$

Notably, for systems with inversion and rotational symmetries but broken TRS, one finds $\sigma_{xx} = \sigma_{yy}$, $\alpha_{xx} = \alpha_{yy}$, $\sigma_{xy} = -\sigma_{yx}$ and $\alpha_{xy} = -\alpha_{yx}$. For more details, see section 2.9. Below, we demonstrate how to determine the heat current operator that is required for inferring the thermoelectric response.

2.8.1 Energy conservation and heat current

To calculate the thermoelectric and thermal conductivity tensors, one is first required to identify the corresponding quantum-mechanical heat current operator. The heat current is customarily defined as $\boldsymbol{J}_Q = \boldsymbol{J}_E - \mu\boldsymbol{J}$,[4] thus it is the fraction of the energy current $\boldsymbol{J}_E$, obtained after subtracting the contribution stemming from the

[4] Note that the chemical potential can be assumed constant only for weak external perturbations and thus solely within the linear response regime. If non-linear effects become important, it can become spatially inhomogeneous, and its feedback to equation (2.71) has to be taken into account. See section 2.8.3.

particle-density current $\boldsymbol{J} = -\boldsymbol{J}_c/e$. Therefore it is essential to develop a strategy for identifying the energy current.

Here we follow an approach similar to the ones employed in section 2.2 and reference [4], and introduce the energy density operator

$$\hat{\rho}_E(\boldsymbol{r}, t) = \hat{\psi}^\dagger(\boldsymbol{r})\hat{H}(\hat{\boldsymbol{p}}, \boldsymbol{r}, t)\hat{\psi}(\boldsymbol{r}) \quad \text{and} \quad \hat{\rho}_E(\boldsymbol{q}, t) = \int d\boldsymbol{k}\, \hat{\psi}^\dagger(\boldsymbol{k})\hat{H}(\boldsymbol{k}, t)\hat{\psi}(\boldsymbol{k} + \boldsymbol{q}). \tag{2.73}$$

The second operator is obtained for translationally-invariant or crystalline systems described by a Hamiltonian $\hat{H}(\boldsymbol{k}, t)$. The above operators are defined in the Schrödinger picture and have an explicit time dependence due to the external fields. By transferring to the Heisenberg picture, we can employ the energy conservation law of equation (2.74) and obtain the Schrödinger-picture energy current, which is shown in equation (2.75)

$$\frac{d}{dt}\hat{\rho}_E(\boldsymbol{r}, t) + \nabla \cdot \hat{\boldsymbol{J}}_E(\boldsymbol{r}, t) = 0 \Rightarrow \tag{2.74}$$

$$\begin{aligned} \boldsymbol{q} \cdot \hat{\boldsymbol{J}}_E(\boldsymbol{q}, t) = \int d\boldsymbol{k}\, \hat{\psi}^\dagger(\boldsymbol{k}) \Bigg\{ & \frac{1}{\hbar}\hat{H}(\boldsymbol{k}, t)[\hat{H}(\boldsymbol{k} + \boldsymbol{q}, t) - \hat{H}(\boldsymbol{k}, t)] \\ & + \mathrm{i}\frac{\partial \hat{H}(\boldsymbol{k}, t)}{\partial t} \Bigg\} \hat{\psi}(\boldsymbol{k} + \boldsymbol{q}). \end{aligned} \tag{2.75}$$

2.8.2 Luttinger's gravitational field approach

While we previously introduced the energy and heat currents via a conservation law, we can alternatively introduce auxiliary potentials, similar to the electromagnetic ones, that can yield the energy current via the functional derivative of the effective action with respect to the auxiliary fields. Such a method was first introduced by Luttinger [16] in order to investigate thermoelectric phenomena. He introduced a scalar 'gravitational' field that coupled to the energy density. Similarly to the scalar and vector potentials of electromagnetism, V and $\boldsymbol{A}$, one can introduce the scalar and vector gravitational potentials, that here we denote with Φ and $\boldsymbol{\Phi}$. Here Φ corresponds to Luttinger's gravitational potential that yields the following Hamiltonian coupling:

$$\hat{H}_\Phi(t) = \int d\boldsymbol{r}\, \frac{\{\hat{\rho}_E(\boldsymbol{r}, t), \Phi(\boldsymbol{r}, t)\}}{2} = \frac{1}{2}\int d\boldsymbol{r}\, \hat{\psi}^\dagger(\boldsymbol{r})\left\{\hat{H}(\hat{\boldsymbol{p}}, \boldsymbol{r}, t), \Phi(\boldsymbol{r}, t)\right\}\hat{\psi}(\boldsymbol{r}), \tag{2.76}$$

where the anticommutator guarantees the Hermitian character of the above term. Note that the above Hamiltonian can also generally depend on the gravitational potentials, due to the diathermal contributions to the current. In addition, the potential $\boldsymbol{\Phi}$ couples to the energy current density. In Fourier space, and in the limit $\boldsymbol{q} \to \boldsymbol{0}$, we find at first order with respect to the gravitational potentials:

$$\hat{H}_{\Phi,\boldsymbol{\Phi}}(t) = \int \frac{d\boldsymbol{q}}{(2\pi)^d} \int d\boldsymbol{k}\, \hat{\psi}^\dagger(\boldsymbol{k}+\boldsymbol{q})\hat{H}^{(0)}(\boldsymbol{k}) \left[\Phi(\boldsymbol{q}, t) - \frac{1}{\hbar}\frac{\partial \hat{H}^{(0)}(\boldsymbol{k})}{\partial \boldsymbol{k}} \cdot \boldsymbol{\Phi}(\boldsymbol{q}, t) \right] \hat{\psi}(\boldsymbol{k}), \quad (2.77)$$

since at zeroth order we have

$$\begin{aligned} \hat{\rho}_E^{(0)}(\boldsymbol{q}) &= \int d\boldsymbol{k}\, \hat{\psi}^\dagger(\boldsymbol{k})\hat{H}^{(0)}(\boldsymbol{k})\hat{\psi}(\boldsymbol{k}+\boldsymbol{q}) \quad \text{and} \\ \hat{\boldsymbol{J}}_E^{(0)}(\boldsymbol{q}) &= \frac{1}{\hbar}\int d\boldsymbol{k}\, \hat{\psi}^\dagger(\boldsymbol{k})\hat{H}^{(0)}(\boldsymbol{k})\frac{\partial \hat{H}^{(0)}(\boldsymbol{k})}{\partial \boldsymbol{k}}\hat{\psi}(\boldsymbol{k}+\boldsymbol{q}). \end{aligned} \quad (2.78)$$

For most cases of interest, it is sufficient to retain the linear response. Therefore we are interested in inferring the modified free energy of the system in the presence of the gravitational potentials, in order to retrieve the respective polarisation tensors. As we observed in the case of the charge conductivity tensor, the diamagnetic terms are equally important for the response. This implies that the so-called diathermal currents must be included when analysing the thermoelectric transport. Note also, that, the electromagnetic and gravitational potentials have to be treated on the same footing. Therefore, we further need to add to the Hamiltonian the contribution

$$\hat{H}_{V,A}(t) = -e\int d\boldsymbol{k}\, \hat{\psi}^\dagger(\boldsymbol{k}+\boldsymbol{q}) \left[V(\boldsymbol{q}, t) - \frac{1}{\hbar}\frac{\partial \hat{H}^{(0)}(\boldsymbol{k})}{\partial \boldsymbol{k}} \cdot \boldsymbol{A}(\boldsymbol{q}, t) \right] \hat{\psi}(\boldsymbol{k}) \quad (2.79)$$

and employ the charge and energy conservation laws of equations (2.15) and (2.74) so to obtain the current diamagnetic and diathermal contributions to the charge and energy current operators.

To proceed, we first note that the transport phenomena considered here are induced by a constant electric (gravitational) field, that can be generated by ∇V and/or $\partial_t \boldsymbol{A}$ ($\nabla\Phi$ and/or $\partial_t \boldsymbol{\Phi}$). Therefore, one observes that considering non-zero time-derivatives for the vector fields will generally lead to unnecessarily-involved expressions for the diathermal contribution to the energy currents, due to the explicit time dependence of the fields and thus the Hamiltonian (see equation (2.75)). To circumvent these complications, we obtain the linear-order contributions of the external perturbations to the current by instead considering constant gradients of the scalar potentials assumed here, i.e. $V(\boldsymbol{r}) = \boldsymbol{r}\cdot\nabla V \equiv -\boldsymbol{r}\cdot\boldsymbol{E}$ and $\Phi(\boldsymbol{r}) = \boldsymbol{r}\cdot\nabla\Phi$. By only taking into account $V(\boldsymbol{r})$ and $\Phi(\boldsymbol{r})$, the conservation laws of equations (2.15) and (2.74) lead to the currents [16–19]

$$\hat{\boldsymbol{J}}_c^{V,\Phi}(\boldsymbol{r}) = [1+\Phi(\boldsymbol{r})]\hat{\boldsymbol{J}}_c(\boldsymbol{r}) \quad \text{and} \quad \hat{\boldsymbol{J}}_E^{V,\Phi}(\boldsymbol{r}) = [1+2\Phi(\boldsymbol{r})]\hat{\boldsymbol{J}}_E(\boldsymbol{r}) + V(\boldsymbol{r})\hat{\boldsymbol{J}}_c(\boldsymbol{r}). \quad (2.80)$$

Above, the terms that are proportional to the external potentials constitute the diathermal contributions. Remarkably they turn out to be very important when the system under discussion is subject to an external magnetic field or it has a non-zero orbital magnetisation [17–20]. In fact, the latter situation is relevant for the Chern insulators that are studied in section 6.3.3, and other topological systems discussed in [21, 22].

2.8.3 Nature of the gravitational field

So far, we have introduced the gravitational fields and exposed their role as contributors to the diathermal currents, but we have not yet commented on their nature. For instance, how can we obtain a non-zero but constant $\nabla\Phi$ or $\partial_t\Phi$, so to induce an electric current? To answer this question we assume an equilibrium situation in which Φ as also V are present and homogeneous. In this case, it is straightforward to obtain the thermodynamic potential of the system, and we find that the two potentials modify the temperature and the chemical potential [19], as follows:

$$\tilde{T}^{-1} = T^{-1}(1+\Phi) \quad \text{and} \quad \tilde{\mu} = (\mu + eV)/(1+\Phi) \Rightarrow \widetilde{\left(\frac{\mu}{T}\right)} = \frac{\mu + eV}{T}. \tag{2.81}$$

Consequently, when Φ and V become slowly varying in space, they give rise to smooth spatial modifications of the chemical potential and temperature, yielding

$$-\frac{\nabla T}{T} = \nabla\Phi \qquad \text{and} \qquad -\frac{T}{e}\nabla\left(\frac{\mu}{T}\right) = \boldsymbol{E}. \tag{2.82}$$

The equations above, also called Einstein relations, illustrate the equivalence between the customarily termed 'statistical' fields ($\nabla(\mu/T)$ and $\nabla T/T$) and the 'mechanical' fields (∇V and $\nabla\Phi$).

2.9 Hands-on: magnetoconductivity tensor of a triangular triple quantum dot

In this section we aim at further familiarising the reader with obtaining the coupling of an electronic system to static and time-dependent electromagnetic fields. Here we focus on lattice systems, and for concreteness and connection to previous paragraphs we study the linear response of a triangular triple-quantum-dot device coupled to a radiation field [23]. The planar system additionally feels the presence of an out-of-plane static external magnetic field. For simplicity, we consider the case of spinless electrons and thus the microscopic Hamiltonian in the absence of the external fields boils down to (for details see section 1.3)

$$\hat{H}_{\mathrm{TQD}} = \varepsilon - t_{dd}\hat{\lambda}_z^{+}. \tag{2.83}$$

According to the above, every dot is dictated by an onsite energy ε, while an electron can hop between successive dots with a matrix element t_{dd}. The external magnetic field couples to the spinless electrons solely via the orbital term, since the Zeeman coupling is quenched. Below we focus on the linear response when a time-dependent electric field is additionally taken into account.

Task 1: Consider a particular gauge for the vector potential $\boldsymbol{A}(\boldsymbol{r})$. One can choose either the symmetric or the Landau gauge, with $\boldsymbol{A}(\boldsymbol{r}) = \frac{B_z}{2}(-y, x, 0) \equiv B_z\check{z} \times \boldsymbol{r}/2$ and $\boldsymbol{A}(\boldsymbol{r}) = B_z(-y, 0, 0)$, respectively. For convenience, pick the symmetric gauge and calculate the line integrals of the vector potential $\phi_{nm} = \frac{e}{\hbar}\int_{R_m}^{R_n} d\boldsymbol{r}\cdot \boldsymbol{A}(\boldsymbol{r})$, linking two dots. In the present case, assume the coordinate vector $\boldsymbol{R}_1 = a(1, 0)$ for quantum dot 1 of figure 1.1. Calculate these line integrals and show that $\phi_{21} = 2\pi\nu_{\text{flux}}/3$, where ν_{flux} defines the total number of flux quanta piercing the triangular area formed by the dots. The flux quantum here is h/e. Via the results of section 2.3, one should end up with the Hamiltonian

$$\hat{H}_{\text{TQD}}(B_z) = \varepsilon - t_{dd}\left[\cos\left(\frac{2\pi\nu_{\text{flux}}}{3}\right)\hat{\lambda}_z^+ - \sin\left(\frac{2\pi\nu_{\text{flux}}}{3}\right)\hat{\lambda}_z^-\right]. \tag{2.84}$$

One observes the recovery of TRS for particular values of the field and a periodicity with the applied flux. Therefore, one can restrict to a single flux-period.

Task 2: Show, using equation (2.24), that the in-plane single-particle electric polarisation operator acting on the orbital part of the wavefunction of equation 1.39, has the form

$$\hat{P}_{x,y}^{(0)} = -\frac{\sqrt{3}}{2}ea\hat{\Lambda}_{x,y}. \tag{2.85}$$

Note that the proportionality relation between $\hat{P}_{x,y}$ and $\hat{\Lambda}_{x,y}$ could have been predicted using table 1.6. Since the polarisation operator is proportional to the position vector, it transforms according to the E irreducible representation of the C_{3v} point group. Moreover, it is even under TR. Out of the two possible candidates, $\hat{\Lambda}_{x,y}$ and $\hat{\lambda}_{x,y}^+$, only the former is acceptable since it is diagonal in the chosen position basis.

Task 3: Using equation (2.22), show that the in-plane single-particle current operator for the Hamiltonian in equation (2.84) reads

$$\hat{\boldsymbol{J}}_c^{(0)} = \frac{3t_{dd}ea}{2\hbar}\left[\sin\left(\frac{2\pi\nu_{\text{flux}}}{3}\right)\hat{\lambda}^+ + \cos\left(\frac{2\pi\nu_{\text{flux}}}{3}\right)\hat{\lambda}^-\right]\times\check{z}. \tag{2.86}$$

Similarly to the dipole operator, the current operator also transforms according to the E irreducible representation, but is in contrast odd under TR. Verify that a two-component object transforming according to E, *for the given symmetry group*, can arise either: (i) as the product of a one-dimensional and a two-dimensional irreducible representation or (ii) as the following product $(a_x b_y + a_y b_x, a_y b_y - a_x b_x)$ obtained by two objects (a_x, a_y) and (b_x, b_y) that transform according to E.

Task 4: Retrieve the orbital coupling to the external field of equation (2.84), via alternatively employing the polarisation operator and equation (2.22) up to all orders. For instance, using the definition of the current we find that the Hamiltonian, corrected at first order in the external field, reads

$$\hat{H}^{(1)}_{\mathrm{TQD}}(B_z) = \hat{H}^{(0)}_{\mathrm{TQD}} - \frac{1}{2}\left[\left\{\hat{J}^{(0)}_{c,x}, \hat{A}_x\right\} + \left\{\hat{J}^{(0)}_{c,y}, \hat{A}_y\right\}\right], \tag{2.87}$$

where $\hat{A}$ is the vector potential in the quantum-dot index basis.

Task 5: Verify that in the presence of a static external magnetic field the remnant symmetry of the TQD is given by the C_3 point group. Consider the state vectors $|\nu\rangle$, with $\nu = 0, \pm 1$:

$$|0\rangle = \frac{1}{\sqrt{3}}\begin{pmatrix}1\\1\\1\end{pmatrix}, \quad |+1\rangle = \frac{1}{\sqrt{3}}\begin{pmatrix}1\\ \zeta^2\\ \zeta\end{pmatrix}, \quad |-1\rangle = \frac{1}{\sqrt{3}}\begin{pmatrix}1\\ \zeta\\ \zeta^2\end{pmatrix}. \tag{2.88}$$

Show that they transform according to the irreducible representations of the C_3 point group, i.e. they satisfy $\hat{C}_3|\nu\rangle = \zeta^\nu|\nu\rangle$, with $\zeta = e^{2\pi i/3}$ and $\hat{C}_3$ given by the spinless version of the one presented in equation 1.41. In addition, employ the latter for diagonalising the Hamiltonian of equation (2.84) and verify that the respective energy eigenvalues have the form

$$E_\nu(B_z) = \varepsilon - 2t_{dd}\cos\left[\frac{2\pi(\nu - \nu_{\mathrm{flux}})}{3}\right]. \tag{2.89}$$

Task 6: By employing the $|\nu\rangle$ basis, and following equations (2.43) and (2.44), show that the polarisation tensor has the form

$$\Pi^{cc}_{ij}(i\omega_s) = \sum_{\nu,\nu'}\langle\nu \mid \hat{J}^{(0)}_{c,i} \mid \nu'\rangle\langle\nu' \mid \hat{J}^{(0)}_{c,j} \mid \nu\rangle\frac{f(E_\nu) - f(E_{\nu'})}{i\omega_s + E_\nu - E_{\nu'}}. \tag{2.90}$$

Calculate the conductivity originating from the above polarisation tensor by considering the circular polarisation basis. In the latter, the current operator is defined as $\hat{J}^{(0)}_{c,\pm} = \hat{J}^{(0)}_{c,x} \pm i\hat{J}^{(0)}_{c,y}$. Show that the conductivity tensor components in the latter basis read

$$\sigma_{--} = \sigma_{xx} - \sigma_{yy} + \frac{\sigma_{xy} + \sigma_{yx}}{i} \quad \text{and} \quad \sigma_{++} = \sigma_{xx} - \sigma_{yy} - \frac{\sigma_{xy} + \sigma_{yx}}{i}, \tag{2.91}$$

$$\sigma_{+-} = \sigma_{xx} + \sigma_{yy} + \frac{\sigma_{xy} - \sigma_{yx}}{i} \quad \text{and} \quad \sigma_{-+} = \sigma_{xx} + \sigma_{yy} - \frac{\sigma_{xy} - \sigma_{yx}}{i}. \tag{2.92}$$

Based on your results for the polarization tensor in the new basis, show that $\sigma_{++} = \sigma_{--} = 0$ and thus: $\sigma_{xx} = \sigma_{yy}$ and $\sigma_{xy} = -\sigma_{yx}$. These results were anticipated based on the symmetry properties of the conductivity tensor. In more detail, in the absence of the external fields one can show that the following linear combinations of the tensor elements:

$$\sigma_{xx} + \sigma_{yy}, \quad \sigma_{xy} - \sigma_{yx}, \quad \text{and} \quad (\sigma_{xx} - \sigma_{yy}, \sigma_{xy} + \sigma_{yx}) \tag{2.93}$$

transform according to the $\{A_1, A_2, E\}$ irreducible representations of the C_{3v} point group. As a result, for $B_z = 0$, only the combination $\sigma_{xx} + \sigma_{yy}$ is invariant under the point-group symmetry and thus non-zero. Setting the remaining elements equal to zero yields: $\sigma_{xx} = \sigma_{yy}$ and $\sigma_{xy} = 0$. However, B_z transforms according to the A_2 irreducible representation of the C_{3v} point group and thus can lead to a non-zero value for the term $\sigma_{xy} - \sigma_{yx}$. We set once again the components transforming according to E equal to zero, and find $\sigma_{xx} = \sigma_{yy}$ and $\sigma_{xy} = -\sigma_{yx}$. Given these conditions and via the Onsager relations in equation (2.51), one can predict the behaviour of the non-zero conductivity tensor elements under $\omega \to -\omega$.

2.10 Hands-on: Boltzmann transport equation

Despite the fact that the quantum-field-theoretical methods described in this chapter provide a direct way to tackle transport problems, they often become technically too cumbersome to handle. Nevertheless, there exist alternative simpler approaches that can capture the salient features of the transport properties.

Among the alternatives methods, one finds the Boltzmann transport equation, which relies on the modification of the Fermi–Dirac distribution in the presence of the external electric fields and temperature gradients [6, 15]. While the Boltzmann equation can be derived from microscopic principles [6], most of the time it can be treated within a phenomenological framework [15]. A key concept in building the equation under discussion is to assume that the system can be still described locally by a distribution function, that however now depends on $\boldsymbol{r}$, $\boldsymbol{k}$ and t. Moreover, in the additional presence of interactions or disorder the electrons acquire a finite lifetime, which implies that the distribution function decays in time. Therefore, for a system of single-band electrons one, within this semiclassical approach, writes

$$f[\boldsymbol{r} + \boldsymbol{v}(\boldsymbol{k})dt,\, \hbar\boldsymbol{k} + \boldsymbol{F}dt,\, t + dt] - f(\boldsymbol{r}, \boldsymbol{k}, t) = \left[\frac{\partial f}{\partial t}\right]_{\text{col}} dt, \tag{2.94}$$

with $\boldsymbol{v}(\boldsymbol{k})$ denoting the velocity of the electron in the presence of an external force $\boldsymbol{F}$. We made use of the relation $\hbar\dot{\boldsymbol{k}} = \boldsymbol{F}$, that holds for Bloch electrons [15]. In addition, the rhs of the equation, the so-called collision term, incorporates the collisions with ions or scattering by other electrons. Via a Taylor expansion of the lhs of equation (2.94), we end up with the desired Boltzmann transport equation

$$\frac{\partial f}{\partial \boldsymbol{r}} \cdot \boldsymbol{v}(\boldsymbol{k}) + \frac{\partial f}{\partial \boldsymbol{k}} \cdot \frac{\boldsymbol{F}}{\hbar} + \frac{\partial f}{\partial t} = \left[\frac{\partial f}{\partial t}\right]_{\text{col}}. \tag{2.95}$$

In most cases, one employs the so-called relaxation time (τ_{col}) approximation, where the collision term is replaced by $[\partial f/\partial t]_{\text{col}} \mapsto -(f - f^{(0)})/\tau_{\text{col}}$, where $f^{(0)}$ defines the distribution in the absence of the external perturbations. In the rest, we follow closely the strategy of reference [15], and provide guidance for deriving the electric, thermoelectric and thermal conductivity tensors via the approach above.

Task 1: Recover the Drude conductivity for an applied electric field $\boldsymbol{E}(t) = \boldsymbol{E}e^{-i\omega t}$. Retain a linear correction of the distribution function with respect to the external field, i.e. $f(\boldsymbol{E}) = f^{(0)} + e^{-i\omega t}f^{(1)}(\boldsymbol{E})$. Show that the conductivity for the system of section 2.7 assumes the form:

$$\sigma_{ij}(\omega) = \sigma\frac{1 + i\omega\tau_{\rm col}}{1 + (\omega\tau_{\rm col})^2}\delta_{i,j}, \tag{2.96}$$

where we introduced the static longitudinal electric charge conductivity, $\sigma = e^2 Y_0$, with Y_n defined in equation (2.98).

Task 2: Assume a constant temperature gradient ∇T. Retain a linear correction of the distribution function with respect to the external perturbation, i.e. $f(\nabla T) = f^{(0)} + f^{(1)}(\nabla T)$. Show that $\alpha_{ij} = \alpha\delta_{i,j}$ and $\kappa_{ij} = \kappa\delta_{i,j}$ with the conductivities [15]

$$\alpha = -eY_1/T, \quad \kappa = Y_2/T \quad \text{and} \quad \kappa_e = \kappa - \frac{T\alpha^2}{\sigma} = \frac{1}{T}\left(Y_2 - \frac{Y_1^2}{Y_0}\right). \tag{2.97}$$

In the above we introduced the quantities [15]:

$$Y_n = -2\tau_{\rm col}\int\frac{d\boldsymbol{k}}{(2\pi)^3}\left[\frac{1}{\hbar}\frac{\partial\varepsilon_c(\boldsymbol{k})}{\partial k_x}\right]^2 f'[\varepsilon_c(\boldsymbol{k})][\varepsilon_c(\boldsymbol{k}) - \mu]^n. \tag{2.98}$$

References

[1] Jackson J D 1999 *Classical Electrodynamics* 3rd edn (New York: Wiley)

[2] Noether E 1918 *Invariante Variationsprobleme* (Math-phys Klasse) 235 p

[3] Peskin M E and Schroeder D V 1995 *An Introduction to Quantum Field Theory* (New York: Perseus Books)

[4] Mahan G D 2000 *Many-Particle Physics* 3rd edn (Berlin also Heidelberg, London, New York *et al*: Springer)

[5] Bruus H and Flensberg K 2004 *Many-Body Quantum Theory in Condensed Matter Physics* (Oxford also New York *et al*: Oxford University Press)

[6] Altland A and Simons B D 2010 *Condensed Matter Field Theory* 2nd edn (Cambridge: Cambridge University Press)

[7] Fradkin E 2013 *Field Theories of Condensed Matter Physics* 2nd edn (Cambridge: Cambridge University Press)

[8] Peierls R 1933 *Z. Phys.* **80** 763

[9] Kubo R 1957 *J. Phys. Soc. Jpn.* **12** 570

[10] Abrikosov A A, Gorkov L P and Dzyaloshinski I E 1963 *Methods of Quantum Field Theory in Statistical Physics* (New York: Dover)

[11] Negele J W and Orland H 1998 *Quantum Many-Particle Systems* (Boulder, CO: Westview Press)

[12] Sokhotskii Y W 1873 *On Definite Integrals and Functions used in Series Expansions* (St. Petersburg)

[13] Plemelj J 1964 *Problems in the Sense of Riemann and Klein* (New York: Interscience)

[14] Onsager L 1931 *Phys. Rev.* **37** 405

[15] Grosso G and Parravicini G P 2014 *Solid State Physics* 2nd edn (Amsterdam (also New York *et al*): Elsevier)

[16] Luttinger J M 1964 *Phys. Rev.* **135** A1505

[17] Jonson M and Girvin S M 1984 *Phys. Rev.* B **29** 1939

[18] Oji H and Streda P 1985 *Phys. Rev.* B **31** 7291

[19] Cooper N R, Halperin B I and Ruzin I M 1997 *Phys. Rev.* B **55** 2344

[20] Xiao D, Chang M-C and Niu Q 2010 *Rev. Mod. Phys.* **82** 1959

[21] Wang Z, Qi X-L and Zhang S-C 2011 *Phys. Rev.* B **84** 014527

[22] Ryu S, Moore J E and Ludwig A W W 2012 *Phys. Rev.* B **85** 045104

[23] Kotetes P, Jin P-Q, Marthaler M and Schön G 2014 *Phys. Rev. Lett.* **113** 236801

Chapter 3

Jackiw–Rebbi model and Goldstone–Wilczek formula

In this chapter we have the first encounter with topological phenomena, associated with the emergence of Dirac physics in two spacetime dimensions. The first part focusses on the Jackiw–Rebbi model, which addresses the appearance of topologically-protected zero-energy states possessing fractional quantum numbers. These become trapped in mass solitons of the Dirac particle. This model constitutes a cornerstone of topological physics and finds several applications in condensed matter systems, including solitons in polyacetylene and sliding charge density waves in low-dimensional conductors. We study aspects of topological transport for the Jackiw–Rebbi model and its connection to the Dirac-equation-related chiral anomaly. In the second part, we provide an alternative understanding of these transport phenomena and fractionalisation via the Goldstone–Wilczek formula, which we derive in the Hands-on section using a microscopic model describing sliding charge density wave systems.

3.1 Helical electrons in nanowires: emergent Jackiw–Rebbi model

In this section we build upon the semiconductor physics studied in chapter 1 and investigate a simple nanowire model that, under specific conditions, exhibits non-trivial topological properties. We consider that the nanowire electrons originate from the Γ-point conduction band of a III–V semiconductor with bulk I-symmetry. The electrons move solely along the [100] direction, i.e. the x axis here, under the additional influence of a Rashba SOC. The setup is depicted in figure 3.1(a). The nanowire can be viewed as originating from the successive confinement of the bulk system, first to the (001) plane, and subsequently along the [010] direction. By assuming strong confinement, we rely on the results of section 1.2.4 and obtain the Hamiltonian for the lowest confinement channel

doi:10.1088/978-1-68174-517-6ch3

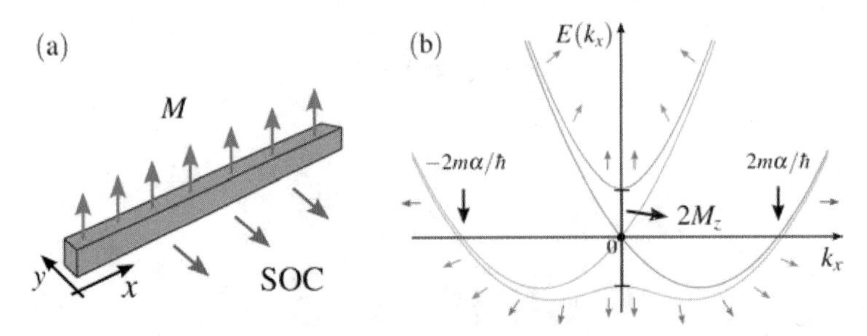

Figure 3.1. (a) Semiconducting nanowire grown along the [100] direction (x axis). The electrons experience a Rashba SOC and an external homogeneous magnetisation field, forcing the electron spin to align along the y and z axis, respectively. (b) Energy dispersions of the nanowire electrons in (a) described by the Hamiltonian of equation (3.1) with and without the influence of a magnetisation field $\boldsymbol{M} = M_z\hat{\boldsymbol{z}}$. The Rashba SOC shifts the two otherwise spin degenerate parabolas, depicted by green and red, and leads to an inner ($k_x = 0$) and an outer ($k_x = \pm 2m\alpha/\hbar$) helical electron branch. The addition of $\boldsymbol{M}$ lifts the degeneracy at $k_x = 0$ and gaps out the inner branch. The term helical reflects the winding of the spin along k_x, depicted in the figure by the coloured arrows. The colour coding reflects the degree of the admixture of the red and green parabolas.

$$\hat{H}_c(k_x) = \frac{(\hbar k_x)^2}{2m} + E_c + \alpha\hbar k_x\hat{\sigma}_y. \tag{3.1}$$

Since the conduction band originates predominantly from s atomic orbitals, cf section 1.2, the above $\hat{\boldsymbol{\sigma}}$ Pauli matrices relate to the spin angular momentum. Here we work in the canonical statistical ensemble with a fixed particle number, instead of a chemical potential, cf sections 2.5 and 2.7.

In the upcoming paragraphs we assume $E_c = 0$. In this case, the Fermi 'surface' consists of the $\hbar k_x = \pm 2m\alpha$ and $k_x = 0$ points. The former (latter) give(s) rise to the so-called *helical* outer (inner) energy branch, as depicted in figure 3.1(b). For the present example, the spin of a helical electron points parallel or antiparallel to the y axis, depending on the sign of k_x. According to the Nielsen–Ninomiya no-go theorem [1], helical (branches of) electrons have to come in pairs, otherwise inconsistent phenomena take place, associated with the so-called 'anomalies' [2, 3].

For massless Dirac particles such a problem is circumvented, since the respective equation describes both electrons and positrons, thus two helical branches. Nevertheless, an odd number of helical electron branches can still appear in nature, but as quasiparticle excitations, due to the presence of interactions or other external means. For instance, in figure 3.1(b) we present a situation in which the inner branch becomes gapped out and only the outer survives. Interestingly, an odd number of helical branches exhibits the so-called *chiral anomaly*, which we discuss in section 3.4.1.

For the remainder, we assume that the inner branch becomes gapped out by a *homogeneous* magnetisation field $\boldsymbol{M}$ with orientation along the z axis, as in figure 3.1(a). This field can be induced by proximity to a ferromagnetic insulator or via the Zeeman coupling to an external magnetic field. However, this field fails to gap out the outer branch, thus, leading to a single helical branch crossing the Fermi level, as shown in figure 3.1(b). Nonetheless, we assume that the outer branch also becomes gapped out, but apparently due to a different mechanism[1], that we do not specify here. In addition, we assume that the gapped-out outer branch is pushed to energies much higher than M. In this manner, one can restrict to the vicinity of the Γ-point and the effective Hamiltonian reads

$$\hat{H}(k_x) = \alpha\hbar k_x\hat{\sigma}_y + M\hat{\sigma}_z. \tag{3.2}$$

The above connects to the bulk version of the Jackiw–Rebbi model [4] in two spacetime dimensions (1 + 1d), and also describes the topologically-protected edge modes of a time-reversal invariant (TRI) 2 + 1d topological insulator, that we introduce in chapter 7.

3.2 Zero-energy solutions in the Jackiw–Rebbi model

Motivated by the linearised version of the model presented in the above paragraph, we now consider the following Hamiltonian defined in coordinate space

$$\hat{H}(\hat{p}_x, x) = \alpha\hat{p}_x\hat{\sigma}_y + M(x)\hat{\sigma}_z, \tag{3.3}$$

where the inner helical branch feels an inhomogeneous magnetisation field. The magnetisation profile is chosen to satisfy the following boundary conditions (see figure 3.2(a)):

$$M(x) < 0 \quad \forall x \in (-\infty, x_1) \cup (x_2, +\infty) \quad \text{and} \quad M(x) > 0 \quad \forall x \in (x_1, x_2). \tag{3.4}$$

Given the above, the model under consideration connects to the Jackiw–Rebbi model proposed in 1976 [4], in the context of high-energy physics. The central theme of that work was the 1 + 1d Dirac equation in the presence of solitons of the particle's mass. Later it was revisited by Su, Schrieffer and Heeger (SSH model) to study solitons in polyacetylene [5].

Essentially, the particular profile dictates that the magnetisation is characterized by two domain walls at the points x_1, x_2. Each domain wall corresponds to a solitonic configuration of the magnetisation field, where we interpolate from a solution with $\text{sgn}(M) = +1$ to a solution with $\text{sgn}(M) = -1$. Below we show that the particular Hamiltonian supports zero-energy solutions localised at the magnetisation domain walls, irrespective of the remaining spatial-profile details of the magnetisation field.

The emergence of the zero-energy solutions can be linked to the presence of a particular type of symmetry that the Hamiltonian possesses. This is the so-called chiral symmetry, effected by a unitary operator $\hat{\Pi}$, that anticommutes with the Hamiltonian $\hat{H}$, i.e.

$$\left\{\hat{H}, \hat{\Pi}\right\} = \hat{0}. \tag{3.5}$$

For the Hamiltonian of equation (3.3), $\hat{\Pi} = \hat{\sigma}_x$. As already announced in section 1.1, the presence of such a symmetry is expected to impose constraints on the energy spectrum. In fact, chiral symmetry implies that for every solution $\boldsymbol{\phi}_n(x)$ of the

[1] For example, a magnetisation field with periodic spatial dependence (spin-density wave) of the form, $\cos(4m\alpha x/\hbar)$, is capable of gapping out the Fermi points of the outer branch located at $\hbar k_x = \pm 2m\alpha$, since it yields the required momentum transfer of $4m\alpha$. A similar situation takes place in section 3.3.2, where a charge-density wave develops instead.

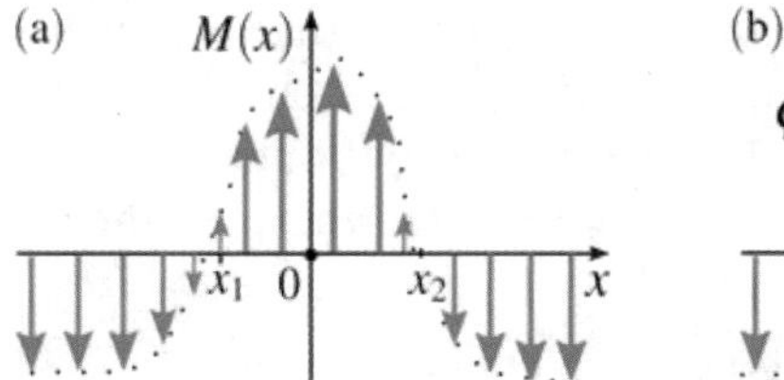

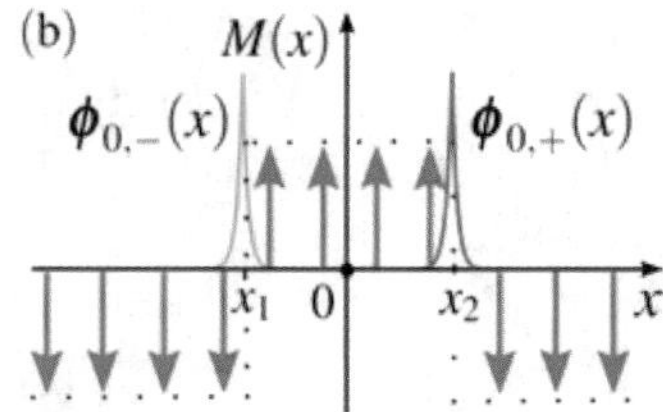

Figure 3.2. Magnetisation profiles which lead to zero-energy solutions confined at the boundary points, $x_{1,2}$, separating regions with an opposite sign for the magnetisation. In (a) ((b)) we depict an example of such a general (stepwise) profile. In addition, (b) shows the two zero-energy solutions, which are found to be localized near the boundaries separating the areas for different magnetisation. They possess specific chirality $\sigma = \pm 1$, as well as spin-projection. In general, the two solutions have an exponentially decaying overlap $\propto e^{-\text{const}\cdot(x_2 - x_1)}$ and therefore correspond to exactly zero-energy solutions only for $x_2 - x_1 \to \infty$.

Hamiltonian with eigenenergy E_n, there exists another solution with eigenenergy $-E_n$, which is given by $\hat{\Pi}\boldsymbol{\phi}_n(x)$. This is straightforward to demonstrate, since

$$\hat{\Pi}\hat{H}(\hat{p}_x, x)\boldsymbol{\phi}_n(x) = E_n\hat{\Pi}\boldsymbol{\phi}_n(x) \Rightarrow \hat{H}(\hat{p}_x, x)[\hat{\Pi}\boldsymbol{\phi}_n(x)] = -E_n[\hat{\Pi}\boldsymbol{\phi}_n(x)]. \tag{3.6}$$

In addition, if there exists a non-degenerate zero-energy solution, then $E_n \leftrightarrow -E_n$ and the same holds for the eigenvectors $\boldsymbol{\phi}_n(x) \leftrightarrow \hat{\Pi}\boldsymbol{\phi}_n(x)$, implying that the zero-energy solutions constitute eigenstates of the chiral symmetry operator $\hat{\Pi}$, and thus of $\hat{\sigma}_x$ for the present model.

Let us now retrieve the bound-state spectrum for the above Hamiltonian, given the stepwise magnetisation profile depicted in figure 3.2(b), expressed as:

$$M(x) = -M + 2M\Theta(x - x_1) - 2M\Theta(x - x_2), \tag{3.7}$$

where we assume that $x_2 - x_1 \to \infty$ and $M > 0$. For each left, middle and right region, denoted by $s = l, m, r$, the magnetisation is constant $M_l = M_r = -M$ and $M_m = +M$. Therefore, we Fourier transform, $\hat{p}_x \to \hbar k_x$, and find the solutions for each one of the translationally-invariant Schrödinger equations:

$$\hat{H}_s(k_x) = \alpha\hbar k_x\hat{\sigma}_y + M_s\hat{\sigma}_z, \qquad s = l, m, r. \tag{3.8}$$

Note, however, that k_x is not a good quantum number anymore, since translational invariance is broken. Nonetheless, the energy is still a good quantum number. This implies that instead of $E(k_x)$, we have $k_{x,s}(E)$, for $s = l, m, r$. Using the arising relativistic energy spectrum, which has the form

$$E = \pm\sqrt{(\alpha\hbar k_{x,s})^2 + M_s^2}, \tag{3.9}$$

we find that for a specific energy E, the corresponding $k_{x,s}$ are given by

$$\alpha\hbar k_{x,s} = \pm\sqrt{E^2 - M_s^2}. \tag{3.10}$$

Since for the particular selection $M_s^2 = M^2$ for $s = l, m, r$, we simply have for all regions

$$\alpha\hbar k_{x,s} = \pm\sqrt{E^2 - M^2}. \tag{3.11}$$

Based on the above relation, we find that for $E > M$ the wavevector is real, thus, yielding dispersive solutions. In contrast, for $E < M$ one obtains an imaginary wavevector leading to solutions exponentially decaying or increasing.

Since here we are primarily interested in zero-energy solutions, it is preferrable to directly set $E = 0$ in the Schrödinger equation of the Hamiltonian in equation (3.8), that leads to:

$$\begin{aligned}&\left(\alpha\hbar k_x\hat{\sigma}_y + M_s\hat{\sigma}_z\right)\boldsymbol{\phi}_{0,s}(k_x) = \mathbf{0} \Rightarrow (\alpha\hbar k_x + \mathrm{i}M_s\hat{\sigma}_x)\,\boldsymbol{\phi}_{0,s}(k_x) = \mathbf{0} \Rightarrow\\ &(\alpha\hbar k_x + \mathrm{i}\sigma M_s)\,\boldsymbol{\phi}_{0,s,\sigma}(k_x) = \mathbf{0} \Rightarrow \alpha\hbar k_{x,s,\sigma} = -\mathrm{i}\sigma M_s,\end{aligned} \tag{3.12}$$

where the quantum number $\sigma = \pm 1$, labels the eigenvectors of the $\hat{\sigma}_x$ matrix. We directly confirm that the zero-energy solutions of the particular model constitute eigenvectors of the chiral symmetry operator. Now that we have found the expression for the zero-energy solutions for each segment, we employ the wavefunction matching technique to find the zero-energy solution defined in the entire coordinate space. The wavefunctions for the three segments read:

$$\boldsymbol{\phi}_{0,s}(x) = c_{s,+}e^{+\frac{M_s}{\alpha\hbar}x}\frac{1}{\sqrt{2}}\begin{pmatrix}1\\1\end{pmatrix} + c_{s,-}e^{-\frac{M_s}{\alpha\hbar}x}\frac{1}{\sqrt{2}}\begin{pmatrix}1\\-1\end{pmatrix}. \tag{3.13}$$

They additionaly have to satisfy the boundary conditions $\boldsymbol{\phi}_{0,l}(x \to -\infty) = \boldsymbol{\phi}_{0,r}(x \to +\infty) = 0$. The above yield the following expressions

$$\begin{aligned}\boldsymbol{\phi}_{0,l}(x) &= c_{l,-}e^{+\frac{M}{\alpha\hbar}x}\frac{1}{\sqrt{2}}\begin{pmatrix}1\\-1\end{pmatrix},\\ \boldsymbol{\phi}_{0,m}(x) &= c_{m,-}e^{-\frac{M}{\alpha\hbar}x}\frac{1}{\sqrt{2}}\begin{pmatrix}1\\-1\end{pmatrix} + c_{m,+}e^{+\frac{M}{\alpha\hbar}x}\frac{1}{\sqrt{2}}\begin{pmatrix}1\\1\end{pmatrix},\\ \boldsymbol{\phi}_{0,r}(x) &= c_{r,+}e^{-\frac{M}{\alpha\hbar}x}\frac{1}{\sqrt{2}}\begin{pmatrix}1\\1\end{pmatrix}.\end{aligned} \tag{3.14}$$

Since the Hamiltonian of equation (3.3) is linear in derivatives, we determine the remaining coefficients using the wavefunction's continuity at $x_{1,2}$:

$$\boldsymbol{\phi}_{0,l}(x_1) = \boldsymbol{\phi}_{0,m}(x_1) \quad \text{and} \quad \boldsymbol{\phi}_{0,m}(x_2) = \boldsymbol{\phi}_{0,r}(x_2) \tag{3.15}$$

supplemented by the normalisation condition of the wavefunction. Furthermore, we assume that $x_2 - x_1 \to \infty$, so that in the vicinity of $x = x_1$ and $x = x_2$, we make the respective approximations

$$\boldsymbol{\phi}_{0,m}(x) \approx c_{m,-}e^{-\frac{M}{\alpha\hbar}x}\frac{1}{\sqrt{2}}\begin{pmatrix}1\\-1\end{pmatrix} \quad \text{and} \quad \boldsymbol{\phi}_{0,m}(x) \approx c_{m,+}e^{+\frac{M}{\alpha\hbar}x}\frac{1}{\sqrt{2}}\begin{pmatrix}1\\1\end{pmatrix}. \tag{3.16}$$

Under these conditions we obtain:

$$c_{l,-} = c_- e^{-Mx_1/(\alpha\hbar)},\ c_{m,-} = c_- e^{Mx_1/(\alpha\hbar)},\ c_{m,+} = c_+ e^{-Mx_2/(\alpha\hbar)},\ c_{r,+} = c_+ e^{Mx_2/(\alpha\hbar)}, \tag{3.17}$$

with the $c_{\pm}$ to be determined by the normalisation of the wavefunctions. After these steps, we end up with the following two localised solutions at the points x_1 and x_2:

$$\boldsymbol{\phi}_{0,-}(x) = c_{-}e^{-\frac{M}{\alpha\hbar}|x-x_1|}\frac{1}{\sqrt{2}}\begin{pmatrix}1\\-1\end{pmatrix} \quad \text{and} \quad \boldsymbol{\phi}_{0,+}(x) = c_{+}e^{-\frac{M}{\alpha\hbar}|x-x_2|}\frac{1}{\sqrt{2}}\begin{pmatrix}1\\1\end{pmatrix}. \tag{3.18}$$

The spatial profiles of these solutions are depicted in figure 3.2(b). One observes that such solutions are always present as long as $M \neq 0$. In fact, it is the bulk energy gap in conjuction with the chiral symmetry that renders these solutions topologically protected and robust against perturbations that are 'weak', i.e. with an energy scale sufficiently smaller than the bulk gap. Notably, the bulk gap also sets the scale for the localisation length of these bound states.

While in the above we employed a particular magnetisation profile to demonstrate the appearance of zero-energy solutions, it is straightforward to show that the bound states appear independent of the exact form of the solitonic profile. It is only required that the domain walls are present. For instance, we assume a general magnetisation profile as in figure 3.2(a). If we focus on the zero-energy solutions of the Hamiltonian, we follow reference [4], and obtain

$$\begin{gathered}\left[\alpha\hat{p}_x\hat{\sigma}_y + M(x)\hat{\sigma}_z\right]\boldsymbol{\phi}_0(x) = \mathbf{0} \quad \Rightarrow \frac{d\boldsymbol{\phi}_0(x)}{dx} = \frac{M(x)}{\alpha\hbar}\hat{\sigma}_x\,\boldsymbol{\phi}_0(x) \quad \Rightarrow \\ \boldsymbol{\phi}_{0,\sigma}(x) = \boldsymbol{\phi}_{0,\sigma}(x_0)e^{\sigma\int_{x_0}^{x} dx' M(x')/(\alpha\hbar)}.\end{gathered} \tag{3.19}$$

One observes that for a given σ, a zero-energy solution can appear only if normalisability becomes ensured. For the latter requirement to be fulfilled, $M(x)$ is solely required to undergo a sign change. This minimal requirement is another manifestation of the topological origin of these solutions. Their appearance does not depend on the details of the magnetisation profile, but only on the presence of domain walls for the Dirac model. As a matter of fact, these zero-energy solutions are localised at regions or points where $M(x) = 0$, i.e. where the sign change takes place. Such points can be also viewed as points where the gap of the bulk energy spectrum closes. This feature is quite generic; bulk energy gap closings are associated with a topological change in the system and the emergence of topologically-protected localised states. The emergence of boundary modes can be alternatively predicted based on the index theorems by Atiyah–Singer [6] (see also [7]), relying on the spectral properties of the Dirac operator.

One final, but very important comment on these localised solutions, relates to the quantum numbers that they carry. As first pointed out in the seminal work by Jackiw and Rebbi [4], the solitons trap bound states characterised by fractional quantum numbers of the original electrons. In the above, each localised solution is spin-filtered, i.e. associated with a particular spin-projection $\sigma = \pm 1$. This is quite remarkable, since both spin directions are locally accessible to the original electron. However, here we find that these two possible values of the spin-degree of freedom are found in two well separated regions of space. Therefore, the SU(2) spin degree of freedom at $E = 0$, becomes non-local. Thus, in order to access it, one has to couple simultaneously to both localised solutions. This is a recurrent theme in topological systems; the topologically-protected boundary modes carry half of the electron's degrees of freedom, thus exhibiting the phenomena of non-locality and fractionalisation.

3.3 The Jackiw–Rebbi model in condensed matter physics

Below we discuss two well-established condensed matter systems in which the Jackiw–Rebbi model finds application. These two systems include solitons in polyacetylene and one-dimensional conductors exhibiting charge density wave order.

3.3.1 Polyacetylene and the Su–Schrieffer–Heeger model

The first encounter of Jackiw–Rebbi type of phenomena in condensed matter physics was in the context of *trans*-polyacetylene. Its structure is shown in figure 3.3(a). One observes that the lattice is dimerised and can be viewed as consisting of two sublattices. *Trans*-polyacetylene possesses two energetically degenerate structural configurations, termed the A- and B-phases. At the point where the polyacetylene chain changes configuration from A $\leftrightarrow$ B, as in figure 3.3(b), a domain wall is formed and a single quasiparticle with fractional electric charge is trapped at the wall. However, due to the spin degeneracy, the charge carried by a soliton ends up being an integer and direct charge-fractionalisation fingerprints are not observed.

The formation of domain walls can be understood via the bulk model proposed by Su, Schrieffer and Heeger (SSH) [5]. According to the SSH model, the electron motion in a chain is characterised by two hopping coefficients, due to the sublattice structure. The single-particle Hamiltonian reads

$$\begin{aligned}\hat{H}_{\rm SSH}(k_x) &= (u + ve^{ik_xa})\frac{\hat{\tau}_x + i\hat{\tau}_y}{2} + (u + ve^{-ik_xa})\frac{\hat{\tau}_x - i\hat{\tau}_y}{2} \\ &= [u + v\cos(k_xa)]\hat{\tau}_x - v\sin(k_xa)\hat{\tau}_y,\end{aligned} \tag{3.20}$$

with the $\hat{\tau}$ Pauli matrices acting on the sublattice space. Note that a unitary transformation $\hat{U} = (\hat{\tau}_x + \hat{\tau}_z)/\sqrt{2}$ transforms the present Hamiltonian into a lattice version of equation (3.2) with $\hat{\boldsymbol{\tau}} \leftrightarrow \hat{\boldsymbol{\sigma}}$. In fact, in the case of $u + v \simeq 0$, one may linearise about $k_x = 0$, and essentially obtain the model studied in section 3.2, for a single region with constant magnetisation. The presence of a structural soliton appears when a sign change of the effective 'mass' parameter $u + v$ takes place. Thus, a structural soliton is associated with a mass soliton for the SSH model.

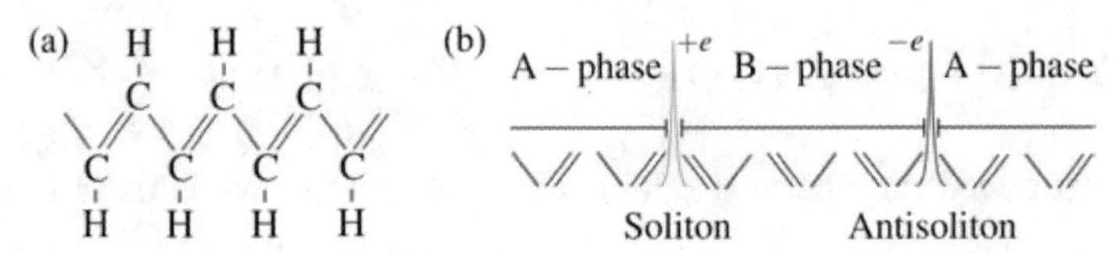

Figure 3.3. (a) Structure of a *trans*-polyacetylene chain. The chain is dimerised due to the coexistence of single (red) and double (blue) C–C bonds. (b) Soliton (A–B) and antisoliton (B–A). Similar to section 3.2, each defect traps a single zero-energy bound state. Each one of the corresponding quasiparticles possesses fractional electric charge $\pm e/2$. However, spin degeneracy doubles the number of bound states and thus the trapped electric charge, i.e. one instead measures $\pm e$ at the defects.

3.3.2 One-dimensional conductors and sliding charge density waves

Another class of systems in which 1 + 1d Dirac physics becomes manifest, includes one-dimensional conductors, which exhibit the spontaneous modulation of the charge density, i.e. a so-called charge density wave [8]. In the presence of the charge density wave order, the conductor is described in second quantisation by the Hamiltonian

$$\hat{H} = \int dx\, \hat{\psi}^{\dagger}(x)\left[\frac{\hat{p}_x^2 - p_F^2}{2m} + 2\Delta\cos(Qx - \theta)\right]\hat{\psi}(x), \tag{3.21}$$

where here we neglected spin for simplicity. A non-zero Δ induces a modulation in the electronic density, $\hat{\psi}^{\dagger}(x)\hat{\psi}(x)$, with wavevector Q and offset defined by the phase θ. Such a modulation appears due to interactions and functions as a means for minimising the system's energy. In fact, equation (3.21) was obtained by decoupling a suitable four-fermion interaction, along the lines of mean-field theory as described in section 3.5. Within the weak-coupling limit, the value of the wavevector Q which minimises the energy is equal to $2k_F$, since in this case one finds an energy gap at the two Fermi points $k_x = \pm k_F$. Such a system was first put forward by Frölich [9] as a candidate for realising superconductivity. Nevertheless, a sliding charge density wave does not constitute a superconductor, but an ideal conductor instead. Indeed, in the ideal case an external electric field forces the static charge density wave to slide, thus carrying an electric current in a collective and dissipationless manner [9].

In this section we focus on the leading instability in which $Q = 2k_F$ (see the Hands-on section 3.5). Since we are interested in the emergence of zero-energy solutions, and thus in the low energy regime, we can focus on the two Fermi points $k_x = \pm k_F$ comprising the Fermi 'surface'. In fact, we may retrieve a solvable approximate model by linearising about the Fermi energy. This is achieved via introducing the right (+) and left (−) mover fields

$$\hat{\psi}(x) \approx \hat{\psi}_+(x)e^{ik_Fx} + \hat{\psi}_-(x)e^{-ik_Fx}, \tag{3.22}$$

with $\hat{\psi}_\pm(x)$ varying slowly, possessing a Fourier decomposition consisting of wavevectors much smaller than k_F. The above approximation for the field operator leads to the Hamiltonian

$$\begin{aligned}\hat{H} &\approx \int dx\, \left(\hat{\psi}_+^{\dagger}(x)\; \hat{\psi}_-^{\dagger}(x)\right)\begin{pmatrix} v_F\delta\hat{p}_x & \Delta e^{-i\theta} \\ \Delta e^{i\theta} & -v_F\delta\hat{p}_x \end{pmatrix}\begin{pmatrix} \hat{\psi}_+(x) \\ \hat{\psi}_-(x) \end{pmatrix} \\ &\equiv \int dx\, \hat{\boldsymbol{\psi}}^{\dagger}(x)(v_F\delta\hat{p}_x\hat{\tau}_z + \Delta\cos\theta\hat{\tau}_x + \Delta\sin\theta\hat{\tau}_y)\hat{\boldsymbol{\psi}}(x),\end{aligned} \tag{3.23}$$

where we introduced the operator $\hat{\boldsymbol{\psi}}^{\dagger}(x) = (\hat{\psi}_+^{\dagger}(x), \hat{\psi}_-^{\dagger}(x))$, while the $\hat{\boldsymbol{\tau}}$ Pauli matrices are defined in the movers' space. In addition, $\delta\hat{p}_x$ is the operator of the momentum measured relative to the Fermi momenta and v_F corresponds to the Fermi velocity, $v_F = \hbar k_F/m$. At first sight, the above model appears not to be directly related to the Jackiw–Rebbi model. However, when the phase transition occurs, the system spontaneously and arbitrarily chooses a value for the phase θ. For the clean and infinite system assumed here, the value chosen is not of any particular interest, as any choice is equivalent. It merely reflects the spontaneous symmetry breaking character of the transition. Before the onset of the charge order, the single-particle Hamiltonian of the low energy model in equation (3.23) exhibits a U(1) symmetry, generated by $\hat{\tau}_z$. The latter corresponds to the emergent translational invariance of the low energy model.

Notably, the spontaneous violation of a continuous symmetry leads to a Goldstone mode, associated with the phase θ, which shows spatiotemporal variations that dynamically restore the broken symmetry [3, 10]. In fact, according to the Mermin–Wagner theorem [11], quantum (or thermal) fluctuations, do not allow the development of long range order and the spontaneous violation of a continuous symmetry below three spatial dimensions. Thus, it appears that such one-dimensional phases should be elusive. However, the materials in which such phases have been experimentally recorded [8] circumvent the above theorem, since they consist of a two-dimensional network of weakly coupled one-dimensional systems, similar to the one discussed here.

We therefore conclude from these considerations that, at the level of the mean-field approximation and in the simultaneous absence of disorder or boundaries, we may freely choose θ or equivalently gauge it away. Under these conditions, we obtain a one-to-one correspondence to the bulk Jackiw–Rebbi model of equation (3.2). Nonetheless, there is an important difference; the field θ is dynamical. As a result, its fluctuations give rise to the dissipationless electric currents predicted by Frölich, i.e. the sliding charge density waves.

3.4 Goldstone–Wilczek formula and dissipationless current

As presaged from the conclusions of the previous paragraph, one expects to obtain dissipationless electric currents in charge density wave systems when the phase field θ acquires a spatiotemporal variation. Nonetheless, this behaviour is also anticipated for the similar extension of the Jackiw–Rebbi model. As we discuss later, such an electric response is ultimately related to the phenomenon of chiral anomaly in 1 + 1d, which constitutes a distinctive feature of Dirac physics in even spacetime dimensions, see [3] and section 5.5.

To study the above phenomena, we consider the following extension to the Jackiw–Rebbi model

$$\hat{H} = \int dx\, \hat{\psi}^{\dagger}(x)\Big\{ \alpha[\hat{p}_x + eA_x(x,t)]\hat{\sigma}_y + M\cos[\theta(x,t)]\hat{\sigma}_z + M\sin[\theta(x,t)]\hat{\sigma}_x - eV(x,t)\Big\}\hat{\psi}(x), \tag{3.24}$$

where we have minimally coupled the electromagnetic field to the electrons (see section 2.2) and assumed a spatiotemporally varying θ field. In the Hands-on section 3.5 we provide the steps for investigating the aforementioned dissipationless transport in charge density waves, where θ constitutes a Goldstone mode. Nonetheless, here we present only the final expression for a θ that is a background field without intrinsic dynamics. Note that for the system described by equation

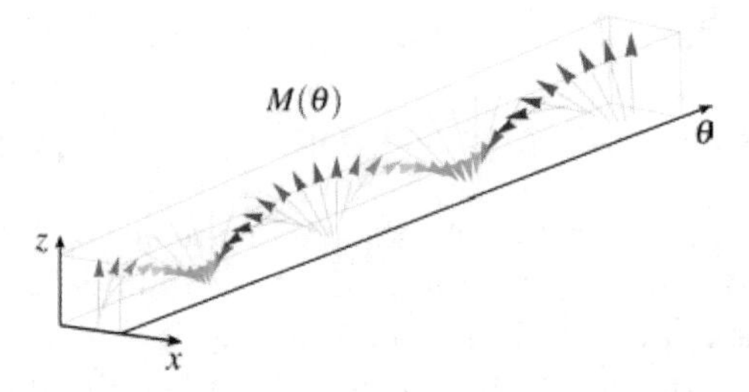

Figure 3.4. A magnetisation field rotating in the xz spin plane according to a spiral profile. The winding is generated by a spatiotemporally varying field $\theta(x, t)$, as in equation (3.24). Spatial (temporal) variations lead to an electric charge density (current) described by equation (3.26).

(3.24), the θ field can be related to a spatiotemporally rotating magnetisation field as depicted in figure 3.4.

We continue with assuming that $\theta(x, t)$ is small and fluctuates slowly about $\theta = 0$ (alternatively for $\theta = \pi$). Thus, we have $\cos\theta \approx 1$ and $\sin\theta \approx \theta$, valid for wavevectors $q_x \ll k_F$ and frequencies $\omega \ll |M|/\hbar$. We additionally consider weak and similarly slowly varying electromagnetic potentials. In the latter regime we can employ a hydrodynamic approach and consider an expansion with respect to small wavevectors and frequencies. Under these conditions, one obtains an action[2] describing the coupled phase field and electromagnetic potentials

$$S = -\text{sgn}(\alpha)\frac{e}{2\pi}\iint dxdt\ \varepsilon_{\mu\nu}A^{\mu}\partial^{\nu}\theta \quad \text{with} \quad \mu, \nu = 0, 1 \tag{3.25}$$

while the Einstein convention of repeated indices summation is implied. Based on the definition of the 2-current for a spacetime metric $\eta_{\mu\nu} = \text{diag}\{1, -1\}$, we have $J_c^{\mu} = (\rho_c, J_{c,x})$ and obtain

$$J_c^{\mu} := -\frac{\delta S}{\delta A_{\mu}} = \text{sgn}(\alpha)\frac{e}{2\pi}\varepsilon^{\mu\nu}\partial_{\nu}\theta, \tag{3.26}$$

that constitutes the so-called Goldstone–Wilczek formula [12]. The 2-current is proportional to the universal value $e/(2\pi)$, with only its sign (sgn) depending on the signs of α and θ. As we show below, this universal character can be linked to the structure of the Dirac equation itself.

3.4.1 Connection to Dirac physics and chiral anomaly

Let us now bring the Hamiltonian of the previous section in a Lorentz covariant form [2, 3], routinely used when studying Dirac's equation. We start from the time-dependent Schrödinger equation

$$\left(\alpha\hat{p}_x\hat{\sigma}_y + M\cos\theta\hat{\sigma}_z + M\sin\theta\hat{\sigma}_x\right)\boldsymbol{\phi}(x, t) = i\hbar\partial_t\boldsymbol{\phi}(x, t) \quad \Rightarrow \tag{3.27}$$

[2] The action S is defined as $S = \iint drdtL(\boldsymbol{r}, t)$, with $L(\boldsymbol{r}, t)$ defining the Lagrangian density. See also section 2.6.

$$\left(-\alpha\hbar\partial_x\hat{\sigma}_x + M\cos\theta + \mathrm{i}M\sin\theta\hat{\sigma}_y\right)\boldsymbol{\phi}(x, t) = \mathrm{i}\hbar\partial_t\hat{\sigma}_z\,\boldsymbol{\phi}(x, t) \tag{3.28}$$

and introduce the $\hat{\gamma}$ matrices

$$\hat{\gamma}^0 = \hat{\sigma}_z, \quad \hat{\gamma}^1 = -\mathrm{i}\hat{\sigma}_x \quad \text{and} \quad \hat{\gamma}_5 = \hat{\sigma}_y. \tag{3.29}$$

After setting $\hbar = 1$ and $\alpha = 1$, for convenience, we obtain the 1 + 1d Dirac equation

$$\left(\mathrm{i}\hat{\gamma}^\mu\partial_\mu - Me^{\mathrm{i}\hat{\gamma}_5\theta}\right)\boldsymbol{\phi}(x, t) = \mathbf{0}. \tag{3.30}$$

By employing the path integral formalism [3, 13], and also after coupling the U(1) electromagnetic gauge field A_μ, one obtains the Lagrangian density

$$L(x, t) = \tilde{\boldsymbol{\psi}}(x, t)\left\{\mathrm{i}\hat{\gamma}^\mu\left[\partial_\mu - \mathrm{i}eA_\mu(x, t)\right] - Me^{\mathrm{i}\hat{\gamma}_5\theta(x,t)}\right\}\boldsymbol{\psi}(x, t), \tag{3.31}$$

with $\boldsymbol{\psi}$ and $\tilde{\boldsymbol{\psi}}$ appropriate Grassmann fields [2, 3, 10]. At this point we perform the chiral gauge transformation $\boldsymbol{\psi}(x, t) \mapsto e^{-\mathrm{i}\hat{\gamma}_5\theta(x,t)/2}\boldsymbol{\psi}(x, t)$ and obtain

$$L'(x, t) = \tilde{\boldsymbol{\psi}}(x, t)\left\{\mathrm{i}\hat{\gamma}^\mu\left[\partial_\mu - \mathrm{i}eA_\mu(x, t) - \frac{\mathrm{i}}{2}\hat{\gamma}^5\partial_\mu\theta(x, t)\right] - M\right\}\boldsymbol{\psi}(x, t). \tag{3.32}$$

However, one has to also pay attention to the path integral measure, $\mathrm{D}\tilde{\boldsymbol{\psi}}\mathrm{D}\boldsymbol{\psi}$, that becomes modified [13] under this transformation $\mathrm{D}\tilde{\boldsymbol{\psi}}\mathrm{D}\boldsymbol{\psi} \mapsto \mathrm{D}\tilde{\boldsymbol{\psi}}\mathrm{D}\boldsymbol{\psi}e^{-\mathrm{i}\hat{\gamma}_5\theta(x,t)}$. This modification contributes with an additional term to the Lagrangian density, $\mathscr{A}(x, t)\theta(x, t)$, which is customarily termed *anomalous*. It can be shown, see [3], that the anomalous term is proportional to the electric field, i.e.

$$\mathscr{A}(x, t) \sim \mathrm{Tr}\left(\hat{\gamma}_5[\hat{\gamma}_\mu, \hat{\gamma}_\nu]F^{\mu\nu}\right) \sim F^{10} = E_x(x, t), \tag{3.33}$$

with $F^{\mu\nu} := \partial^\mu A^\nu - \partial^\nu A^\mu$ defining the strength tensor of the electromagnetic field. This term is the manifestation of the so-called chiral anomaly and is a direct consequence of the unbounded relativistic spectrum. The anomaly is related to the fact that while the chiral current $J_5^\mu \sim \hat{\gamma}^5\hat{\gamma}^\mu$ is conserved at the classical level, it does not continue to do so at the quantum-mechanical level. Using equation (3.25), we have

$$\partial_\mu J_c^\mu = 0 \quad \text{while} \quad \partial_\mu J_5^\mu = \partial_\mu\frac{\partial S}{\partial(\partial_\mu\theta/2)} = -\frac{e}{\pi}E_x. \tag{3.34}$$

Remarkably, the anomalous term appears even in the massless limit $M \to 0$. Via an integration by parts, one observes that the anomalous term, θE_x, yields the Goldstone–Wilczek action of equation (3.25). A term proportional to the electric field, appearing either in the thermodynamic potential or the action of the system, implies a non-zero electric polarisation density $\boldsymbol{P}(x, t)$. In the present situation, we find $P_x = e\theta/(2\pi)$. In sections 4.3.1 and 4.3.3, we provide extensions of the above result regarding the induced polarisation for arbitrary θ.

3.4.2 Fractional electric charge at solitons and electric charge pumping

Based on the Goldstone–Wilczek formula we find the electric charge and current densities (assuming $\alpha > 0$)

$$\rho_c(x, t) = -\frac{\delta S}{\delta V} = -\frac{e}{2\pi}\partial_x\theta \quad \text{and} \quad J_{c,x}(x, t) = \frac{\delta S}{\delta A_x} = \frac{e}{2\pi}\partial_t\theta. \tag{3.35}$$

Therefore, the transferred charge ΔQ for a spatially-varying θ, or, the change of the electric-polarisation density $\Delta P_x(x)$ for a time-dependent θ, are determined by (see also [14])

$$\Delta Q = -\frac{e}{2\pi}\int dx\, \partial_x\theta(x) = -\frac{e}{2\pi}[\theta(x) - \theta(x_0)] \equiv -\frac{e}{2\pi}\Delta\theta, \tag{3.36}$$

$$\Delta P_x(x) = +\frac{e}{2\pi}\int dt\, \partial_t\theta(t) = +\frac{e}{2\pi}[\theta(t) - \theta(t_0)] \equiv +\frac{e}{2\pi}\Delta\theta. \tag{3.37}$$

This implies that for a domain wall, across which the phase changes from $0 \mapsto \pi$, a half-unit of electric charge becomes accumulated[3]. Thus, one observes that the fractional charges trapped at solitons encountered in section 3.3, also arise here by employing an appropriate phase twist. Similarly, for a sinusoidally modulated phase effected for half a time period, half a unit of electric charge per unit length is transferred.

So far, our analysis focused on phase fields without any intrinsic dynamics. In the following Hands-on section, we derive the Goldstone–Wilczek formula for a sliding charge density wave conductor, in which the phase θ has its own dynamics as it constitutes a Goldstone mode. In addition, we discuss the differences brought about due to the Goldstone mode dynamics.

3.5 Hands-on: derivation of the Goldstone–Wilczek formula for a sliding charge density wave conductor

In this section we provide the steps for retrieving the Goldstone–Wilczek formula of equation (3.25). We consider the case in which $\theta(x, t)$ is the offset phase of the charge density wave order parameter. As a byproduct, we discuss aspects of mean-field theory and derive the dynamics of the θ field, as it constitutes a Goldstone mode.

Consider an effective four-fermion-operator interaction of the form

$$\hat{V}_{\text{int}} = -\frac{g}{2\pi}\iiint dq_x dk_x dk'_x\, \hat{\psi}^\dagger(k_x)\hat{\psi}(k_x + q_x)\hat{\psi}^\dagger(k'_x + q_x)\hat{\psi}(k'_x), \tag{3.38}$$

with $g > 0$, that can drive a transition to a charge density wave phase with wavevector q_x. In order to proceed, we apply mean-field theory[4] and assume that

[3] One has to bear in mind that the Goldstone–Wilczek formula was obtained perturbatively for small θ. Nonetheless, as we show in section 4.3.1, its validity can be extended to arbitrary values of θ and variations $\Delta\theta$.

[4] For simplicity we assume that the above interaction is already brought to the form to be decoupled in the charge density wave channel, which allows us to neglect the exhange term. For a more general treatment see [10, 15].

the q_x Fourier component of the charge density operator can be approximately replaced by its expectation value plus a small deviation about it. Therefore:

$$\hat{\psi}^\dagger(k_x)\hat{\psi}(k_x+q_x) \approx \langle\hat{\psi}^\dagger(k_x)\hat{\psi}(k_x+q_x)\rangle \\ +\left[\hat{\psi}^\dagger(k_x)\hat{\psi}(k_x+q_x) - \langle\hat{\psi}^\dagger(k_x)\hat{\psi}(k_x+q_x)\rangle\right]. \tag{3.39}$$

We plug the above expression in the interaction term and introduce the complex charge density wave order parameter $\Delta(q_x)$

$$\Delta(q_x) := -g\int_{-\infty}^{+\infty}\frac{dk_x}{2\pi}\langle\hat{\psi}^\dagger(k_x)\hat{\psi}(k_x+q_x)\rangle \tag{3.40}$$

for which $\Delta^*(q_x) = \Delta(-q_x)$. The mean-field-decoupled interaction now becomes

$$\hat{V}_{\text{int}} \approx \iint dq_x dk_x\left[\Delta(-q_x)\hat{\psi}^\dagger(k_x)\hat{\psi}(k_x+q_x) + \Delta(q_x)\hat{\psi}^\dagger(k_x+q_x)\hat{\psi}(k_x)\right] \\ +2\pi\int dq_x\frac{|\Delta(q_x)|^2}{g}. \tag{3.41}$$

Note that the last term in the above expression is not accompanied by electronic operators, and corresponds to the energy cost required for the order parameter to develop. This implies that a phase transition can occur only below a critical temperature or above a critical interaction strength g. When the transition takes place, the energy cost for the appearance of the order parameter is superseded by the energy lowering, which occurs in the electronic sector.

One can infer the phase transition properties through the defining equation of $\Delta(q_x)$, i.e. equation (3.40), also called the self-consistency equation. In order to evaluate the respective expectation value, one can employ the Matsubara method that was presented in section 2.5, and write

$$\langle\hat{\psi}^\dagger(k_x)\hat{\psi}(k_x+q_x)\rangle = T\sum_{k_n}G(k_x+q_x, k_x, \text{i}k_n). \tag{3.42}$$

The Green function above is determined by the mean-field Hamiltonian. If we restrict to a single $q_x = Q$ and consider $\Delta(q_x) = \Delta e^{-\text{i}\theta}\delta(q_x - Q)$ so to connect with section 3.3.2, the Hamiltonian reads

$$\hat{H} = \int dk_x \\ \left[\frac{\hbar^2\left(k_x^2 - k_F^2\right)}{2m}\hat{\psi}^\dagger(k_x)\hat{\psi}(k_x) + \Delta e^{\text{i}\theta}\hat{\psi}^\dagger(k_x)\hat{\psi}(k_x+Q) + \Delta e^{-\text{i}\theta}\hat{\psi}^\dagger(k_x+Q)\hat{\psi}(k_x)\right]. \tag{3.43}$$

Throughout the discussion, we assume that the system is described by the canonical ensemble. Since the order parameter is assumed to be static, the involved Green functions satisfy the coupled Dyson equations [10, 15]

$$
\begin{aligned}
G(k_x + q_x, k_x, ik_n) &= G^{(0)}(k_x + q_x, k_x, ik_n) + \iint dk_x' dk_x''\, G^{(0)}(k_x + q_x, k_x', ik_n) \\
&\qquad \Delta(k_x' - k_x'')G(k_x'', k_x, ik_n), \\
G(k_x, k_x, ik_n) &= G^{(0)}(k_x, k_x, ik_n) \\
&\quad + \iint dk_x' dk_x''\, G^{(0)}(k_x, k_x', ik_n)\Delta(k_x' - k_x'')G(k_x'', k_x, ik_n).
\end{aligned}
\tag{3.44}
$$

The bare Matsubara Green function is defined as

$$
\begin{aligned}
G^{(0)}(k_x + q_x, k_x, ik_n) &= \frac{1}{ik_n - \varepsilon(k_x)}\delta(q_x) := G^{(0)}(k_x, ik_n)\delta(q_x) \\
\text{and} \quad \varepsilon(k_x) &:= \frac{\hbar^2\left(k_x^2 - k_F^2\right)}{2m}.
\end{aligned}
\tag{3.45}
$$

Task 1: Assume that only the component $\Delta(Q)$ is non-zero, with $Q \neq 0$. Retrieve the Green function at first order, $G^{(1)}(k_x + Q, k_x, ik_n)$, and show that the self-consistency equation yields the so-called Stoner criterion for the phase transition [15], $1 = g\chi(Q)$, with the static charge-charge correlation function or susceptibility

$$
\begin{aligned}
\chi(Q) &= -\int \frac{dk_x}{2\pi} T \sum_{k_n} G^{(0)}(k_x + Q, ik_n)G^{(0)}(k_x, ik_n) \\
&= -\int \frac{dk_x}{2\pi} \frac{f[\varepsilon(k_x + Q)] - f[\varepsilon(k_x)]}{\varepsilon(k_x + Q) - \varepsilon(k_x)},
\end{aligned}
\tag{3.46}
$$

with the Fermi–Dirac distribution defined in the canonical ensemble, i.e. as in equation (2.57) but with $\mu = 0$. Show that for the given energy dispersion, the above exhibits a logarithmic singularity at $Q = 2k_F$ for $T = 0$. Thus, even an interaction of an infinitesimally small strength g, can stabilise a charge density wave order with $Q = 2k_F$.

Task 2: Start from equation (3.23) and consider that the phase exhibits spatiotemporal fluctuations, i.e. $\theta(x, t)$, while at the same time the system is additionally coupled to the scalar $V(x, t)$ and vector $A_x(x, t)$ potentials. Assume that all these fields are slowly varying in both space and time with frequencies $\omega \ll |\Delta|/\hbar$ and wavevectors $q \ll k_F$. In addition, linearise the order parameter with respect to the phase field, θ, i.e. $\cos\theta \approx 1$ and $\sin\theta \approx \theta$.

Retrieve the real-time action for the coupled bosonic-electronic fields by adopting a hydrodynamic approach that is perturbative in both θ, V, A_x and their spatiotemporal derivatives. Follow the Matsubara formalism described in section 2.6 and suitably employ equations (2.48) and (2.52), by making the correspondence $\theta(x, t) \mapsto \theta(q_x, i\omega_s)$, with the bosonic frequency ω_s. The real-time result is retrieved via the analytic continuation $i\omega_s \mapsto \hbar\omega + i0^+$. In the hydrodynamic limit one drops the $i0^+$ term.

Show that the mean-field Matsubara action, up to second order with respect to the fields, reads

$$S^{(2)}_{\mathrm{mf}}[\theta, V, A_x] = \frac{T}{2\pi}\frac{\Delta^2}{g}\sum_{\omega_s}\int dq_x|\theta(q_x, \mathrm{i}\omega_s)|^2 + \frac{1}{2}\sum_{\omega_s}\int dq_x\sum_{k_n}\int d\delta k_x$$
$$\times \mathrm{Tr}_\tau\Big[\hat{G}^{(0)}_{\mathrm{mf}}(\delta k_x, \mathrm{i}k_n)\hat{H}_{\mathrm{ext}}(-q_x, -\mathrm{i}\omega_s)\hat{G}^{(0)}_{\mathrm{mf}}(\delta k_x + q_x, \mathrm{i}k_n + \mathrm{i}\omega_s)\hat{H}_{\mathrm{ext}}(q_x, \mathrm{i}\omega_s)\Big], \tag{3.47}$$

with δk_x constituting the wavenumber corresponding to $\delta\hat{p}_x$ of equation (3.23). Here, δk_x and q_x are defined in the interval $[-k_c, k_c]$, with k_c a cut-off wavevector. The latter controls the window of validity for the right/left-movers approximation considered in section 3.3.2. In addition, we introduced the mean-field Green function, at zeroth-order in θ, V, A:

$$\hat{G}^{(0)}_{\mathrm{mf}}(\delta k_x, \mathrm{i}k_n) = \frac{\mathrm{i}k_n + \boldsymbol{g}(\delta k_x)\cdot\hat{\boldsymbol{\tau}}}{(\mathrm{i}k_n)^2 - |\boldsymbol{g}(\delta k_x)|^2} = \sum_{a=\pm}\frac{\hat{P}_a(\delta k_x)}{\mathrm{i}k_n - E_a(\delta k_x)} \tag{3.48}$$

with $\boldsymbol{g}(\delta k_x) = (\Delta, 0, v_F\hbar\delta k_x)$. We introduced the operators $\hat{P}_a(\delta k_x)$:

$$\hat{P}_a(\delta k_x) = \frac{1 + a\check{\boldsymbol{g}}(\delta k_x)\cdot\hat{\boldsymbol{\tau}}}{2} \tag{3.49}$$

which define projectors for the two bands with $E_\pm(\delta k_x) = \pm E(\delta k_x)$ where

$$E(\delta k_x) := |\boldsymbol{g}(\delta k_x)| = \sqrt{(v_F\hbar\delta k_x)^2 + \Delta^2}\,. \tag{3.50}$$

Moreover, we introduced the coupling Hamiltonian

$$\hat{H}_{\mathrm{ext}}(q_x, \mathrm{i}\omega_s) = \frac{T}{2\pi}\Big[-eV(q_x, \mathrm{i}\omega_s)\hat{1}_\tau + ev_F A_x(q_x, \mathrm{i}\omega_s)\hat{\tau}_z + \theta(q_x, \mathrm{i}\omega_s)\Delta\hat{\tau}_y\Big]. \tag{3.51}$$

Task 3: Expand equation (3.47) up to second order in θ and $(q_x, \mathrm{i}\omega_s)$, as also up to linear order in the electromagnetic potentials. Show that the thermodynamic potential assumes the form

$$S_{\mathrm{mf}}[\theta, V, A_x] \approx \frac{1}{2}T\sum_{\omega_s}\int\frac{dq_x}{2\pi}\,\theta(q_x, \mathrm{i}\omega_s)\left[\frac{2\Delta^2}{g} + \Pi_{\theta\theta}(q_x, \mathrm{i}\omega_s)\right]\theta(-q_x, -\mathrm{i}\omega_s)$$
$$+ T\sum_{\omega_s}\int\frac{dq_x}{2\pi}\,\theta(q_x, \mathrm{i}\omega_s)$$
$$\Big[\Pi_{\theta V}(q_x, \mathrm{i}\omega_s)V(-q_x, -\mathrm{i}\omega_s) + \Pi_{\theta A}(q_x, \mathrm{i}\omega_s)A_x(-q_x, -\mathrm{i}\omega_s)\Big] \tag{3.52}$$

with the respective ‘polarisation’ tensor elements defined as

$$\Pi_{\theta\theta}(q_x, \mathrm{i}\omega_s) := \Delta^2 T \sum_{k_n} \int \frac{d\delta k_x}{2\pi} \mathrm{Tr}_\tau\Big[\hat{G}^{(0)}_{\mathrm{mf}}(\delta k_x, \mathrm{i}k_n)\hat{\tau}_y \hat{G}^{(0)}_{\mathrm{mf}}(\delta k_x + q_x, \mathrm{i}k_n + \mathrm{i}\omega_s)\hat{\tau}_y\Big]$$
$$= -\frac{2\Delta^2}{g} + \frac{\omega_s^2}{4\pi v_F \hbar} + \frac{v_F \hbar q_x^2}{4\pi}, \tag{3.53}$$

$$\Pi_{\theta V}(q_x, \mathrm{i}\omega_s) := -e\Delta T \sum_{k_n} \int \frac{d\delta k_x}{2\pi} \mathrm{Tr}_\tau\Big[\hat{G}^{(0)}_{\mathrm{mf}}(\delta k_x, \mathrm{i}k_n)\hat{\tau}_y \hat{G}^{(0)}_{\mathrm{mf}}(\delta k_x + q_x, \mathrm{i}k_n + \mathrm{i}\omega_s)\Big]$$
$$= e\mathrm{i}q_x \int \frac{d\delta k_x}{4\pi} \frac{\Delta}{E(\delta k_x)} \left[\check{g}(\delta k_x) \times \frac{d\check{g}(\delta k_x)}{d\delta k_x}\right]_y = -\frac{e}{2\pi}(\mathrm{i}q_x), \tag{3.54}$$

$$\Pi_{\theta A}(q_x, \mathrm{i}\omega_s) := ev_F \Delta T \sum_{k_n} \int \frac{d\delta k_x}{2\pi} \mathrm{Tr}_\tau\Big[\hat{G}^{(0)}_{\mathrm{mf}}(\delta k_x, \mathrm{i}k_n)\hat{\tau}_y \hat{G}^{(0)}_{\mathrm{mf}}(\delta k_x + q_x, \mathrm{i}k_n + \mathrm{i}\omega_s)\hat{\tau}_z\Big]$$
$$\equiv -\frac{\mathrm{i}\omega_s}{\hbar q_x} \Pi_{\theta V}(q_x, \mathrm{i}\omega_s) = -\frac{e}{2\pi}\frac{\omega_s}{\hbar}. \tag{3.55}$$

where we assumed $v_F > 0$ and made use of the integrals

$$g \int \frac{d\delta k_x}{2\pi} \frac{1}{2E(\delta k_x)} = \int d\delta k_x \frac{v_F \hbar \Delta^2}{2[E(\delta k_x)]^3} = \int d\delta k_x \frac{3 v_F \hbar \Delta^4}{4[E(\delta k_x)]^5} = 1. \tag{3.56}$$

To calculate the first, we employed the self-consistency equation, while for the latter two, we took the limit $k_c \to \infty$. See also section 4.3.1 and equation (4.36). Therefore, we conclude that the effective action reads

$$\begin{aligned} S_{\mathrm{mf}}[\theta, V, A_x] = {} & \frac{1}{2} T \sum_{\omega_s} \int \frac{dq_x}{2\pi} \theta(q_x, \mathrm{i}\omega_s) \left[\frac{\hbar}{4\pi v_F}\left(\frac{\omega_s}{\hbar}\right)^2 + \frac{v_F \hbar q_x^2}{4\pi}\right] \theta(-q_x, -\mathrm{i}\omega_s) \\ & - \frac{e}{2\pi} T \sum_{\omega_s} \int \frac{dq_x}{2\pi} \theta(q_x, \mathrm{i}\omega_s) E_x(-q_x, -\mathrm{i}\omega_s), \end{aligned} \tag{3.57}$$

which after transferring to the real frequency domain, via the analytical continuation $\mathrm{i}\omega_s \mapsto \hbar\omega$, yields the equation of motion for the phason Goldstone mode [16–18]

$$\left(\omega^2 - v_F^2 q_x^2\right)\theta(q_x, \omega) = -\frac{2ev_F}{\hbar} E_x(q_x, \omega). \tag{3.58}$$

The above corresponds to the Goldstone–Wilczek formula when the phason field has intrinsic dynamics, with an energy dispersion $\omega(q_x) = v_F q_x$. For a given external electric field one can obtain the conductivity. By assuming a time-dependent homogeneous electric field $E_x(t)$, the equations of motion yield

$$\theta(\omega) = -\frac{2ev_F}{\hbar\omega^2}E_x(\omega), \tag{3.59}$$

which after being combined with equation (3.35) leads to the charge current

$$J_{c,x}(\omega) = -\frac{e}{2\pi}\mathrm{i}\omega\theta(\omega) = \frac{\mathrm{i}}{\omega}\frac{e^2}{\hbar}\frac{v_F}{\pi}E_x(\omega) \equiv \sigma(\omega)E_x(\omega). \tag{3.60}$$

The above expression is reminiscent of the one encountered in the case of a free electron gas in section 2.7, since one finds a pole at $\omega = 0$ for the imaginary part of the conductivity. Via the Kramers–Kronig relations discussed in section 2.4, one obtains an infinite static real conductivity for $\omega = 0$, similar to the one in equation (2.62). The infinite conductivity stems here from the collective response of the charge density wave condensate, which practically slides, and carries electric current under the influence of the external field. Nonetheless, the presence of the lattice or disorder can pin the phase field, which can only become unpinned above the so-called depinning electric field. In this case, the conductivity it not infinite anymore, since the $\delta(\omega)$ function is replaced by a Lorentzian centered at a finite frequency $\omega = \omega_0$ [8].

References

[1] Nielsen H B and Ninomiya M 1981 *Phys. Lett. B* **105** 219
[2] Peskin M E and Schroeder D V 1995 *An Introduction to Quantum Field Theory* (New York: Perseus Books)
[3] Weinberg S 1996 *The Quantum Theory of Fields: Volume 2. Modern Applications* (New York: Cambridge University Press)
[4] Jackiw R and Rebbi C 1976 *Phys. Rev. D* **13** 3398
[5] Su W P, Schrieffer J R and Heeger A J 1979 *Phys. Rev. Lett.* **42** 1698
[6] Atiyah M F and Singer I M 1968 *Ann. Math.* **87** 484
[7] Volovik G E 2003 *The Universe in a Helium Droplet* (Oxford: Clarendon Press)
[8] Grüner G 1988 *Rev. Mod. Phys.* **60** 1129
[9] Fröhlich H 1954 *Proc. Royal Soc. A.* **223** 296
[10] Altland A and Simons B D 2010 *Condensed Matter Field Theory* 2nd edn (Cambridge: Cambridge University Press)
[11] Mermin N D and Wagner H 1966 *Phys. Rev. Lett.* **17** 1133
[12] Goldstone J and Wilczek F 1981 *Phys. Rev. Lett.* **47** 986
[13] Fujikawa K 1980 *Phys. Rev. D* **21** 2848
[14] Qi X-L, Hughes T L and Zhang S-C 2008 *Phys. Rev. B* **78** 195424
[15] Bruus H and Flensberg K 2004 *Many-Body Quantum Theory in Condensed Matter Physics* (Oxford: Oxford University Press)
[16] Su Z-B and Sakita B 1986 *Phys. Rev. Lett.* **56** 780
[17] Ishikawa M and Takayama H 1988 *Prog. Theor. Phys.* **79** 359
[18] Yakovenko V M and Goan H S 1998 *Phys. Rev. B* **58** 10648

IOP Concise Physics

Topological Insulators

Panagiotis Kotetes

Chapter 4

Topological insulators in 1+1 dimensions

This chapter discusses concrete continuum and lattice models for topological insulators in 1+1d spacetime, constructed as the non-relativistic extensions of the Jackiw–Rebbi model investigated earlier. We study the characteristics of the zero-energy bound states emerging when termination edges are present. The topological features of these edge states are encoded in a winding number which constitutes the bulk topological invariant for this class of systems. We further generalise the winding-number construction in order to classify systems of this type defined in higher even-dimensional spacetimes. To understand the topological properties of such insulators better, we additionally focus on the adiabatic electric current generated by a slowly-varying time-dependent parameter. We show that the adiabatic current can be expressed in terms of the so-called Berry potential and curvature, which are linked to the winding number. Motivated by this finding, we introduce the concept of the Berry phase within a general framework. Finally, the Hands-on sections discuss further aspects related to the winding number, adiabatic transport and electric polarisation.

4.1 Prototypical topological-insulator model in 1+1 dimensions

In the following paragraphs we proceed with examining a non-relativistic extension of the Jackiw–Rebbi model of equation (3.2) that also exhibits zero-energy bound states in $d = 1$. As a matter of fact, this model constitutes the first and simplest topological-insulator model to be examined in this book.

4.1.1 Hamiltonian and zero-energy edge states

The model under discussion is obtained by replacing the constant magnetisation of equation (3.2), M, with a momentum dependent magnetisation $M(\hat{p}_x) = m - \beta\hat{p}_x^2$. The latter yields the Hamiltonian

$$\hat{H}(\hat{p}_x) = \alpha\hat{p}_x\hat{\sigma}_y + \left(m - \beta\hat{p}_x^2\right)\hat{\sigma}_z, \tag{4.1}$$

with m and $\beta > 0$ constants with dimensions of energy and inverse mass, respectively. Since the Hamiltonian of equation (4.1) depends solely on momentum, we can readily Fourier transform it and obtain

$$\hat{H}(k_x) = \alpha\hbar k_x\hat{\sigma}_y + [m - \beta(\hbar k_x)^2]\hat{\sigma}_z. \tag{4.2}$$

One observes that the above Hamiltonian maps to the bulk Jackiw–Rebbi model for small $\beta(\hbar k_x)^2$, and is also characterised by a chiral symmetry with $\hat{\Pi} = \hat{\sigma}_x$. Thus, at first sight the quadratic term appears unnecessary for investigating the properties of the bulk system that are associated with $k_x = 0$. While this may be true for some of the thermodynamic properties, it certainly does not hold for the topological ones. In fact, the $\beta(\hbar k_x)^2$ term is of principal importance when it comes to studying the topological features of the system, since the latter are determined by the characteristics of the Hamiltonian in the entire k_x-space. The inclusion of the quadratic term is sufficient to provide information regarding the high-energy sector of the system. Essentially, its presence reflects the fact that the outer helical branch, discussed in section 3.1, is gapped out and pushed to infinite energy. In this manner, the inclusion of the quadratic term allows us to solely focus on the inner helical electron branch and extend the wavevector to the entire k_x-space, i.e. $k_x \in (-\infty, +\infty)$. Quite often, such a quadratic term is introduced as an effective mechanism for *regularising* models of the Jackiw–Rebbi type, which are linear in momenta.

The similarities between the regularised Hamiltonian above and the bulk Jackiw–Rebbi model, suggest that zero-energy boundary solutions could also appear in the present model. Nonetheless, a question which naturally arises is what kind of boundaries are required for this purpose. By virtue of the quadratic term in momentum, we here have the possibility to obtain zero-energy bound state solutions by simply confining the electrons in a box. For concreteness we consider that the electrons are restricted to move inside the region $x \in [x_1, x_2]$, due to the presence of an infinitely-high positive confining potential $V(x) = +\infty$ for $x \in (-\infty, x_1] \cup [x_2, +\infty)$. Since the confining potential is infinite, the electronic wavefunction has to satisfy a hard-wall condition, i.e. it should be exactly zero at these points: $\boldsymbol{\phi}(x_1) = \boldsymbol{\phi}(x_2) = 0$.[1]

The introduction of infinitely-high boundaries at x_1 and x_2 invalidates the wavevector k_x as a good quantum number. Only the energy can be employed to label the eigensolutions. Since we are looking for zero-energy solutions we restrict to the equation

$$\left\{\alpha\hbar k_x\hat{\sigma}_y + [m - \beta(\hbar k_x)^2]\hat{\sigma}_z\right\}\boldsymbol{\phi}_0 = \mathbf{0} \quad \Rightarrow \quad \hbar k_x = -\mathrm{i}\sigma\frac{\alpha}{2\beta} \pm \sqrt{\frac{m}{\beta} - \left(\frac{\alpha}{2\beta}\right)^2}. \tag{4.3}$$

Similarly to the Jackiw–Rebbi model, the wavevectors above depend also here on the chirality $\sigma = \pm 1$. However, we observe an important difference. In the Jackiw–

[1] One can still obtain zero-energy edge states by considering domain walls similar to the ones employed in the Jackiw–Rebbi model via setting $m(x) = -m + 2m\Theta(x - x_1) - 2m\Theta(x - x_2)$. In this case the quadratic term in momentum becomes irrelevant, as at the domain walls $m \to 0$ and we can expand about $\hat{p}_x = 0$, thus, recovering the Jackiw–Rebbi model.

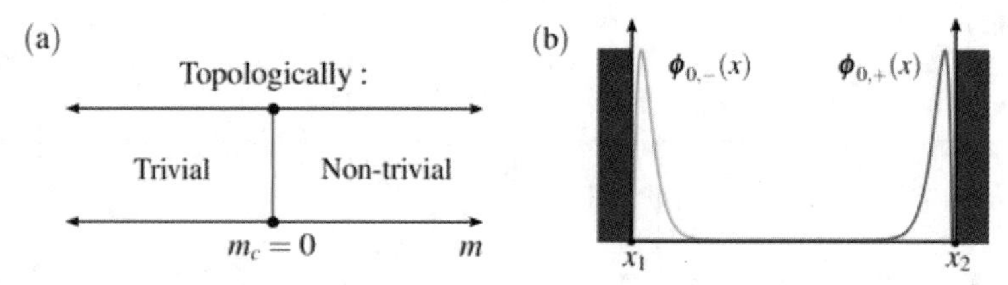

Figure 4.1. Topological phase diagram in the parameter space spanned by the values of m. For $m > m_c$ the system becomes topologically non-trivial with zero-energy bound state solutions, as in (b). For $m < m_c$ the system lies in the topologically trivial phase. For $m = m_c = 0$ the bulk spectrum exhibits a gap closing at $k_x = 0$, which triggers the topological transition from the trivial to the non-trivial phase and vice versa. (b) Here we show the zero-energy bound state solutions for a specific set of parameter values. These solutions are present only for $m > 0$ and satisfy the hard-wall conditions $\phi_{0,-}(x_1) = 0$ and $\phi_{0,+}(x_2) = 0$. In addition, they constitute eigenstates (indexed by $\pm$) of the chiral symmetry operator $\hat{\Pi} = \hat{\sigma}_x$.

Rebbi model the wavevectors corresponding to zero energy were strictly imaginary. In contrast, here the quadratic term generally allows for complex wavevectors. Nevertheless, we find in both models that the zero-energy states corresponding to real wavevectors are no longer localised at the boundaries. Instead, they leak into the bulk and become dispersive modes. The parameter values for which such a delocalisation is observed additionally mark the occurence of a topological phase transition in the bulk system. In the present case there is only one such parameter, m, and its critical value m_c is retrieved by setting $\mathrm{Im}(k_x) = 0$. Via equation (4.3) we find $m_c = 0$.

The values of m span a 1d *parameter* space, and thus, we infer that m_c constitutes the boundary point separating the topologically trivial phase ($m < m_c$) from the topologically non-trivial phase ($m > m_c$) (see figure 4.1(a)). One directly observes that, for $m = m_c = 0$, the bulk spectrum exhibits a gap closing at $k_x = 0$. For the infinite (bulk) system this gap closing leads to a zero-energy mode which is the analog of the zero-energy bound state solution for the finite system. This is essentially a consequence of the so-called bulk-boundary correspondence, which can be alternatively understood via the construction of an appropriate topological invariant quantity [1, 2] that should be non-zero only for $m > m_c$. The gap closing mechanism discussed here is often referred to as 'band inversion'[2], since the positive and negative energy bands exchange at the transition. See figure 7.1(b) for an example of band inversion in HgTe/CdTe quantum wells.

We now proceed with retrieving the zero-energy bound state solutions in the topologically non-trivial phase, $m > 0$. For concreteness, we consider $\alpha > 0$ and introduce: $\xi := 2\beta\hbar/\alpha$ and $\zeta := \sqrt{m\xi^2/(\beta\hbar^2) - 1}/\xi$. With their help, we rewrite the wavevector solutions as $k_x \equiv -i\sigma/\xi \pm \zeta$. The wavevectors corresponding to zero energy have a specific decaying behaviour which depends on the chirality $\sigma = \pm 1$. This property implies that a solution of a given chirality can only live on a given boundary side. Since for $\sigma = +1$ ($\sigma = -1$) the wavevector has a negative (positive) imaginary part, the respective zero-energy chiral solution is located at the right (left) boundary, i.e. $x = x_2$ ($x = x_1$). By considering that $x_2 - x_1 \to +\infty$, we can

[2] For more details regarding the 'band inversion', see section 7.1.1.

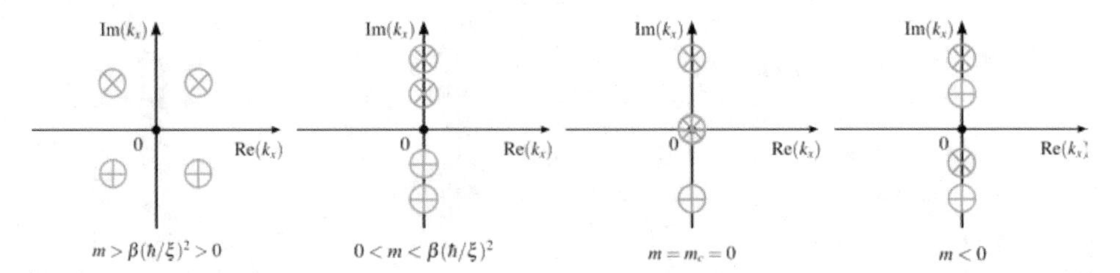

Figure 4.2. Wavevector solutions of equation (4.3) in the complex k_x-plane. The solutions in the upper (lower) half-plane lead to zero-energy bound states localized at the left (right) boundary. The orange and purple colours encode the type of chirality ($\mp$). The system lies in the topologically non-trivial phase when two solutions of the same chirality exist in each upper/lower half-plane. The transition from the topologically non-trivial ($m > 0$) to the trivial ($m < 0$) phase is accompanied by the disappearance of the zero-energy bound states and occurs when $k_x = 0$ and $m_c = 0$. For the latter values the bulk gap closes. This is a manifestation of the so-called bulk-boundary correspondence.

independently obtain the solutions for each boundary, which are given by the expressions:

$$\boldsymbol{\phi}_{0,-}(x) = \left[c_{-,+}e^{\mathrm{i}\zeta(x-x_1)} + c_{-,-}e^{-\mathrm{i}\zeta(x-x_1)}\right]e^{-(x-x_1)/\xi}\frac{1}{\sqrt{2}}\begin{pmatrix}1\\-1\end{pmatrix}, \tag{4.4}$$

$$\boldsymbol{\phi}_{0,+}(x) = \left[c_{+,+}e^{\mathrm{i}\zeta(x-x_2)} + c_{+,-}e^{-\mathrm{i}\zeta(x-x_2)}\right]e^{+(x-x_2)/\xi}\frac{1}{\sqrt{2}}\begin{pmatrix}1\\1\end{pmatrix}. \tag{4.5}$$

By applying the hard-wall boundary condition, we have $\boldsymbol{\phi}_{0,-}(x_1) = 0$ and $\boldsymbol{\phi}_{0,+}(x_2) = 0$ which is satisfied only if $c_{+,-} = -c_{+,+}$ and $c_{-,-} = -c_{-,+}$.

In figure 4.1(b) we depict the zero-energy bound states for a particular set of parameter values. Generally, ζ can be either real or imaginary, which leads to the cases presented below and in figure 4.2:

- $m > \beta(\hbar/\xi)^2 > 0$. Here we find: $c_{\pm,+}e^{\mathrm{i}\zeta(x-x_{2,1})} + c_{\pm,-}e^{-\mathrm{i}\zeta(x-x_{2,1})} \propto \sin[|\zeta|(x - x_{2,1})]$.
- $0 < m < \beta(\hbar/\xi)^2$. Here we find: $c_{\pm,+}e^{\mathrm{i}\zeta(x-x_{2,1})} + c_{\pm,-}e^{-\mathrm{i}\zeta(x-x_{2,1})} \propto \sinh[|\zeta|(x - x_{2,1})]$.
- $m = m_c = 0$. Here: $k_x = -\mathrm{i}(\sigma \pm 1)/\xi$. This implies that for $m = 0$ one solution per chirality extends through the whole bulk since it corresponds to $k_x = 0$ and, thus, is no longer 'bound'.
- $m < 0$. We observe that for each chirality the two related solutions have a decay factor that places them at opposite edges. This implies that the hard-wall conditions can only be satisfied if the corresponding eigenvectors are null vectors. In this case, the system transits to the topologically trivial phase with no zero-energy bound states.

4.1.2 Topological invariant

So far, we have studied the zero-energy bound states by directly finding the respective eigenvectors, without relying on topology. Nonetheless, the topological properties of the system have been underlying a number of characteristics of the zero-energy boundary solutions. An example is their robustness against arbitrary[3] smooth

[3] With the only constraint that the magnetization profile still exhibits the domain walls.

modifications of the magnetisation profile in section 3.2, which we called topological protection. However, it is desirable to identify a tool for diagnosing the bulk topological properties and predicting the appearance of the associated boundary zero-energy bound states, without having to resort to the brute-force diagonalisation of the finite-sized Hamiltonian of the system.

4.1.3 Homotopy mapping and winding number

To identify the desired quantity, termed also the topological invariant [1, 2], it is helpful to rewrite the Hamiltonian of equation (4.2) in the following form

$$\hat{H}(k_x) = \alpha\hbar k_x\hat{\sigma}_y + [m - \beta(\hbar k_x)^2]\hat{\sigma}_z \equiv \boldsymbol{g}(k_x)\cdot\hat{\boldsymbol{\sigma}} = E(k_x)\check{\boldsymbol{g}}(k_x)\cdot\hat{\boldsymbol{\sigma}}, \tag{4.6}$$

with

$$\boldsymbol{g}(k_x) = (0, \alpha\hbar k_x, m - \beta(\hbar k_x)^2) \equiv E(k_x)(0, -\sin[\varphi(k_x)], \cos[\varphi(k_x)]) \tag{4.7}$$

where we introduced $E(k_x) := |\boldsymbol{g}(k_x)|$ and $\varphi(k_x) \in [0, 2\pi)$. At this point, we investigate the boundary values of $\check{\boldsymbol{g}}(k_x)$. We have $\check{\boldsymbol{g}}(|k_x| \to +\infty) = (0, 0, -1)$. The latter implies that $k_x = \pm\infty$ behave exactly as if they were the same k_x-point, and so we can effectively identify them[4]. This leads to the compactification[5] of k_x-space, also termed *base manifold*, from $\mathbb{R}$ to a $\mathbb{S}^1$ sphere (circle). Moreover, $\check{g}_z(k_x = 0) = \text{sgn}(m)$. We recall that, according to the previous analysis, it is solely the sign of m that determines whether the system lies in the topologically trivial phase. If $m < 0$ the system is in the trivial phase, in which the $\check{\boldsymbol{g}}$-vector generally points towards the $-\check{\boldsymbol{z}}$ direction. In this case, we may deform our Hamiltonian to yield a constant $\check{\boldsymbol{g}}(k_x) = (0, 0, -1)$, leading to the configuration of figure 4.3(a). In contrast, for $m > 0$ the system resides in the topologically non-trivial phase, for which the $\check{\boldsymbol{g}}$-vector performs a twist, as shown in figure 4.3(b). Remarkably, the twist cannot be undone by any smooth Hamiltonian deformation that does not lead to a gap closing, i.e. $|\boldsymbol{g}(k_x)| = 0$. Essentially, it is exactly this robustness of the $\check{\boldsymbol{g}}$-vector's k_x-space texture that reflects the topologically non-trivial properties of the system.

Since the unit $\check{\boldsymbol{g}}$-vector is defined in the *yz* spin plane, it takes values on a circle $\mathbb{S}^1$, which we will refer to as the *target space*. Therefore, one finds that if we cover the base k_x-manifold once, i.e. from $-\infty$ to $+\infty$, the $\check{\boldsymbol{g}}$-vector performs a complete twist in its target space when $m > 0$. Consequently, the $\check{\boldsymbol{g}}$-vector effects a *homotopy mapping* between the two spaces. The number of twists it performs is an integer that constitutes the index of the mapping and defines the topological invariant of the system. This becomes more transparent from figure 4.3(c), where we illustrate the mapping $\mathbb{S}^1 \mapsto \mathbb{S}^1$. In general, two $\mathbb{S}^d$ spheres are always linked via an integer. As a

[4] Note that there exist other situations in which $\boldsymbol{k}$-space is already compact. This is always the case for periodic systems where the wavevector is defined in a Brillouin zone (see section 1.1.5), which in d spatial dimensions defines a torus $\mathbb{T}^d$.

[5] The concept of compactification appeared almost one century ago in high-energy physics, in terms of the Kaluza–Klein theory [3, 4] which aimed at the unification of gravity and electromagnetism. Although unsuccessful, it was a breakthrough, since it opened new routes in including additional dimensions beyond the usual 3 + 1d Minkowski spacetime.

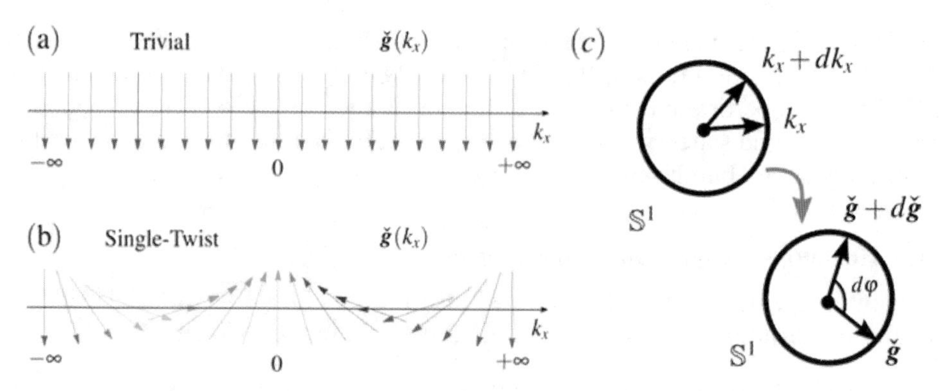

Figure 4.3. (a) Trivial configuration of the topologically-deformed $\check{g}(k_x)$-vector. (b) Topologically non-trivial configuration of the unit $\check{g}(k_x)$-vector with a single twist. In both cases, $\check{g}(k_x)$ has the same orientation for $k_x \to \pm\infty$, thus allowing the effective compactification of the one-dimensional k_x-space, from $\mathbb{R}$ to $\mathbb{S}^1$. (c) The two $\mathbb{S}^1$ spheres are homotopically mapped with the winding number w_1. The latter counts how many times an angle of 2π has been covered in $\check{g}$-space, when an angle of 2π has been covered in the compactified k_x-space.

matter of fact, the topological invariant in the present $d = 1$ case, coincides with the angle $\varphi(k_x)$ that is covered when we cover an angle of 2π in the $\mathbb{S}^1$ compactified k_x-space.

The findings above lead us to introduce the topological invariant, also called the winding number [5]

$$w_1 = \frac{1}{2\pi}\int d\varphi \equiv \frac{1}{2\pi}\int dk_x \frac{d\varphi}{dk_x} = \frac{1}{2\pi}\int dk_x \left[\check{g}(k_x) \times \frac{\partial \check{g}(k_x)}{\partial k_x}\right]_x. \tag{4.8}$$

For the case considered here, it is straighforward to verify that $w_1 = -1$.

Notably, if we calculate the above topological invariant for the bulk Hamiltonian of the Jackiw–Rebbi model dicussed in equation (3.2), we observe that $w_1 = -\text{sgn}(M)/2$, i.e. it yields a fractional number instead of an integer. This contradiction can be attributed to the fact that Hamiltonians which are linear in momenta, as for instance $\hat{H}(\hat{p}_x) = \alpha\hat{p}_x\hat{\sigma}_y + M\hat{\sigma}_z$, are pathological as far as this invariant is concerned [2]. The reason is that in these cases $\check{g}(k_x)$ does not satisfy the appropriate boundary conditions. In fact, $\check{g}(k_x)$ points along different directions for $k_x \to \pm\infty$. Consequently, the k_x-space cannot be compactified into a circle and therefore we cannot define a topologically invariant quantity. This is essentially why the Jackiw–Rebbi model cannot exhibit edge modes without the consideration of an inhomogeneous $M(x)$. Specifically, for the stepwise double domain wall magnetisation profile of section 3.2, we find that each bulk phase is characterised by such a fractional topological invariant, so that the changes of the topological invariant yield an integer number. This is exactly the way in which the integer nature of the topological invariant is restored. Consequently, we have to introduce the domain walls in order to obtain a well-defined integer number of zero-energy bound states (see figure 4.4), whose number is given by the difference of the fractional invariants. This so-called index theorem was first put forward by Atiyah and Singer in several works (e.g. [6]).

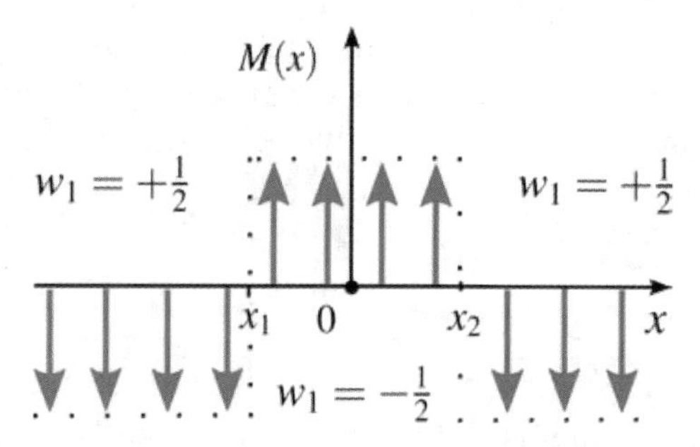

Figure 4.4. The calculated winding number w_1 for each one of the three bulk segments within the Jackiw–Rebbi model of section 3.2. The sign of w_1 depends on the sign of the magnetisation. Moreover, w_1 is fractionalised due to the linear momentum dependence of the Hamiltonian [2]. Nevertheless, the difference of w_1 across a domain wall is an integer, and yields the number of zero-energy bound state solutions appearing at the interface. The sign of this difference provides the chirality of the zero-energy bound state.

4.1.4 Topological invariance

It is straightforward to show that the winding number is a topological invariant, i.e., that it is robust under smooth deformations of the Hamiltonian which preserve chiral symmetry. 'Smooth' implies that the changes which we subject the Hamiltonian to, are such that they do not close the bulk gap, since this would lead to an indefinite $\check{g}(k_x)$-vector. To prove the above, we start from the expression for the topological invariant

$$w_1' = \frac{1}{2\pi}\int dk_x \left[\check{g}'(k_x) \times \frac{\partial \check{g}'(k_x)}{\partial k_x}\right]_x, \tag{4.9}$$

where we have smoothly deformed our Hamiltonian by infinitesimally modifying the $\check{g}(k_x)$-vector in the following manner: $\check{g}'(k_x) = \check{g}(k_x) + \delta\check{g}(k_x)$, with $\check{g}(k_x) \cdot \delta\check{g}(k_x) = 0$. The latter constraint implies that we add an orthogonal contribution to the previous Hamiltonian, which allows $\check{g}'(k_x)$ to remain a unit vector, i.e., $\check{g}'(k_x) \cdot \check{g}'(k_x) = 1 \to \check{g}(k_x) \cdot \delta\check{g}(k_x) = 0$. Making this replacement, we have

$$w_1' = w_1 + \frac{1}{\pi}\int dk_x \left[\delta\check{g}(k_x) \times \frac{\partial \check{g}(k_x)}{\partial k_x}\right]_x + \text{h.o.t.} = w_1, \tag{4.10}$$

where we used the antisymmetry of the cross product and the following relations

$$\begin{aligned} &\check{g}(k_x) \cdot \check{g}(k_x) = 1 \to \check{g}(k_x) \cdot \frac{\partial \check{g}(k_x)}{\partial k_x} = 0 \Rightarrow \\ &\qquad \frac{\partial \check{g}(k_x)}{\partial k_x} || \delta\check{g}(k_x) \Rightarrow \frac{\partial \check{g}(k_x)}{\partial k_x} \times \delta\check{g}(k_x) = \mathbf{0}. \end{aligned} \tag{4.11}$$

4.1.5 Generalised winding number

In the preceding paragraphs we introduced the winding number solely based on geometric arguments and the special form of the Hamiltonian. Nonetheless, this is not always possible and thus a general construction of a topological invariant is required. For this purpose, one should rely on the universal properties of Hamiltonians with chiral symmetry. The presence of the latter guarantees that the Hamiltonian can be always brought to a block off-diagonal form [5], i.e.:

$$\hat{H}(\boldsymbol{k}) \mapsto \underline{\hat{H}}(\boldsymbol{k}) = \begin{pmatrix} \hat{0} & \hat{h}(\boldsymbol{k}) \\ \hat{h}^{\dagger}(\boldsymbol{k}) & \hat{0} \end{pmatrix}. \tag{4.12}$$

Notably, $\det \hat{H}(\boldsymbol{k}) = |\det \hat{h}(\boldsymbol{k})|^2$. For the Hamiltonian of equations (4.6) and (4.7) this is achievable via the unitary transformation $\hat{U} = (\hat{\sigma}_z + \hat{\sigma}_x)/\sqrt{2}$, leading to

$$\underline{\hat{H}}(k_x) = \hat{U}^{\dagger}\hat{H}(k_x)\hat{U} = E(k_x)\begin{pmatrix} 0 & e^{-\mathrm{i}\varphi(k_x)} \\ e^{\mathrm{i}\varphi(k_x)} & 0 \end{pmatrix}. \tag{4.13}$$

The second property that one has to take into account is that the topological invariant is associated with the occupied energy dispersions, since we are dealing with a gapped bulk spectrum. Here, we have only one occupied dispersion and its projector reads, $\hat{P}(k_x) = |-, k_x\rangle\langle -, k_x|$. Along the lines of reference [5], we introduce the projection operator $\hat{Q}(k_x)$, on to the occupied (occ) dispersions, which in the block off-diagonal basis reads

$$\hat{Q}(k_x) := 2\sum_{a}^{\mathrm{occ}}\hat{P}_a(k_x) - \hat{1} = -\begin{pmatrix} 0 & e^{-\mathrm{i}\varphi(k_x)} \\ e^{\mathrm{i}\varphi(k_x)} & 0 \end{pmatrix} \equiv \begin{pmatrix} 0 & q(k_x) \\ q^{*}(k_x) & 0 \end{pmatrix} \tag{4.14}$$

with the properties $\hat{Q}(k_x) = \hat{Q}^{\dagger}(k_x)$ and $\hat{Q}^2(k_x) = \hat{1}$. Interestingly, $\hat{Q}(k_x)$ is here proportional to the Hamiltonian and evidently block off-diagonal. This allowed us to rewrite it in terms of the complex term[6] $q(k_x)$ which belongs to the U(1) group of unitary matrices and satisfies $|q(k_x)|^2 = 1$. The latter performs the required homotopy mapping, denoted by $\pi_1[\mathrm{U}(1)] = \mathbb{Z}$, from the base manifold (k_x) to the target space $e^{-\mathrm{i}\varphi(k_x)}$. The winding number can be written in terms of the projection operator as

$$w_1 = \frac{\mathrm{i}}{2\pi}\int dk_x\, q^{-1}(k_x)\frac{dq(k_x)}{dk_x}. \tag{4.15}$$

For a generic $2N \times 2N$ chiral symmetric Hamiltonian $\hat{H}(\boldsymbol{k})$, the operator $q(k_x)$ is represented with a complex $N \times N$ matrix $\hat{q}(\boldsymbol{k})$. Thus, it belongs to the so-called U(N) matrices and performs the homotopy mapping $\pi_{2n+1}[U(N)] \simeq \mathbb{Z}$ for $n \in \mathbb{N}$. The mapping is possible only for a spacetime with an odd number of spatial dimensions ($d = 2n + 1$) and is given by the general expression [5]

[6] In the general case, the quantity $q(k_x)$ is an operator $\hat{q}(k_x)$, represented by a complex unitary matrix belonging to the group U(N) of square complex matrices of dimensions $N \times N$.

$$w_{2n+1} = \frac{(-1)^n n!}{(2n+1)!}\left(\frac{\mathrm{i}}{2\pi}\right)^{n+1} \int d\boldsymbol{k}\, \varepsilon_{ij\ldots l}\, \mathrm{Tr} \left[\hat{q}^{-1}(\boldsymbol{k})\frac{\partial \hat{q}(\boldsymbol{k})}{\partial k_i}\hat{q}^{-1}(\boldsymbol{k})\frac{\partial \hat{q}(\boldsymbol{k})}{\partial k_j}\ldots\hat{q}^{-1}(\boldsymbol{k})\frac{\partial \hat{q}(\boldsymbol{k})}{\partial k_l}\right]. \tag{4.16}$$

4.2 Lattice topological-insulator model and higher winding numbers

In this paragraph we proceed with considering the lattice variant for the continuum topological-insulator model of equation (4.2). In the simplest case we have

$$\hat{H}(k_x) = \frac{\alpha\hbar}{a}\sin(k_x a)\hat{\sigma}_y + \left\{m - 2\beta\left(\frac{\hbar}{a}\right)^2\left[1 - \cos(k_x a)\right]\right\}\hat{\sigma}_z, \tag{4.17}$$

where we introduced the lattice constant a. One observes that if we perform a Taylor expansion of the above Hamiltonian about $k_x = 0$, we retrieve the above mentioned continuum model. The system transits to the topologically non-trivial phase supporting zero-energy bound states for $m > m_c = 0$. However, in contrast to the continuum model, it additionally supports a topological phase transition upon increasing m. This occurs when the bulk gap closes at $k_x = \pi/a$, which happens for the critical value $m_c' = 4\beta(\hbar/a)^2$. Strictly speaking, such a second gap closing point exists also in the continuum limit, since $k_x = +\infty$ corresponds to $k_x = \pi/a$ for $a \to 0$, but appears for $m_c' = +\infty$. Therefore, it is essentially inacessible. After calculating the corresponding winding number, with the integration now being over the first Brillouin zone (BZ), we find that the system becomes topologically trivial for $m > m_c'$. These results can be alternatively confirmed via the linearisation of the Hamiltonian about the two gap closing points $k_x = 0, \pi/a$. We therefore obtain the following two Jackiw–Rebbi models

$$\hat{H}_0(\delta k_x) \approx \alpha\hbar\delta k_x\hat{\sigma}_y + m\hat{\sigma}_z \quad \text{and} \quad \hat{H}_{\pi/a}(\delta k_x) \approx -\alpha\hbar\delta k_x\hat{\sigma}_y + \left(m - m_c'\right)\hat{\sigma}_z, \tag{4.18}$$

with δk_x measured here relative to the expansion points. Each one of the gap closing points contributes with half a unit to the winding number (see also section 4.1.3), with the sign depending on the value of m. Straightforward calculations yield

$$w_1 = \mathrm{sgn}(\alpha)\frac{\mathrm{sgn}(m - m_c') - \mathrm{sgn}(m)}{2}. \tag{4.19}$$

Note that for the continuum model the first contribution always has a fixed sign $\mathrm{sgn}(m - m_c') < 0$, since the gap does not close at infinity.

While for the given lattice version of the continuum model of equation (4.2) we found that the additional gap closings only reduce the topological regime, alternative lattice extensions could lead to a winding number $|w_1| > 1$. This is possible by considering higher lattice harmonics, as in the Hamiltonian below:

$$\hat{H}(k_x) = \frac{\alpha\hbar}{a}\sin(nk_xa)\hat{\sigma}_y + \left\{m - 2\beta\left(\frac{\hbar}{a}\right)^2\left[1 - \cos(nk_xa)\right]\right\}\hat{\sigma}_z \text{ with } n \in \mathbb{Z}. \quad (4.20)$$

When the system lies in the topologically non-trivial phase, $\check{g}(k_x)$ performs n full twists in target space for a single sweep in the base manifold, thus leading to $|w_1| = |n|$. In the most general case, each Hamiltonian term consists of a series of harmonics that lead to an interplay of topological phases with different values for w_1 and a corresponding number of zero-energy bound states for a given edge. Notably, in spite of the fact that a number of $|w_1|$ boundary states are located at the same edge, they do not hybridise as long as chiral symmetry is intact.

4.3 Adiabatic transport: Thouless pump and Berry curvature

Let us now revisit the dissipationless transport examined in terms of the Goldstone–Wilczek formula in section 3.4 from a different perspective, that will allow us to introduce the important concepts of the adiabatic current and the Berry curvature. In section 3.4 we found that an electric charge current is generated when the Hamiltonian is a function of a time-dependent θ parameter. In fact, we examined the case of slow variations with frequencies $\hbar\omega \ll |M|$. In the given limit, and by assuming zero temperature, the current solely consists of its adiabatic contribution, stemming from the electrons of the occupied band with bulk energy dispersion $E\_(k_x) = -E(k_x)$ where $E(k_x) = \sqrt{(\alpha\hbar k_x)^2 + M^2}$. The adiabaticity implies that the time variation of θ is sufficiently slow, so that the system's eigenstates at each time instant have the same form as the time-independent ones calculated in the presence of θ. We now proceed with calculating the adiabatic current for both the continuum and lattice models introduced in the previous paragraphs of this chapter.

4.3.1 Continuum model

In the spirit of section 3.4 and equation (3.24), we consider the parameter-dependent single-particle Hamiltonian

$$\hat{H}(k_x, \theta) = \alpha\hbar k_x\hat{\sigma}_y + \left[m\cos\theta - \beta(\hbar k_x)^2\right]\hat{\sigma}_z + m\sin\theta\hat{\sigma}_x. \quad (4.21)$$

To facilitate the discussion and connect to previous results, we employ the unitary transformation of section 4.1.5, $\hat{U} = (\hat{\sigma}_z + \hat{\sigma}_x)/\sqrt{2}$, and obtain the Hamiltonian

$$\begin{aligned}\underline{\hat{H}}(k_x, \theta) &= \left[m\cos\theta - \beta(\hbar k_x)^2\right]\hat{\sigma}_x - \alpha\hbar k_x\hat{\sigma}_y + m\sin\theta\hat{\sigma}_z \\ &\equiv E(k_x, \theta)\check{g}(k_x, \theta)\cdot\hat{\boldsymbol{\sigma}},\end{aligned} \quad (4.22)$$

where we introduced the unit vector

$$\check{g}(k_x, \theta) := \left(\sin[\vartheta(k_x, \theta)]\cos[\varphi(k_x, \theta)], \sin[\vartheta(k_x, \theta)]\sin[\varphi(k_x, \theta)], \cos[\vartheta(k_x, \theta)]\right) \quad (4.23)$$

with the parametrising angles $\varphi(k_x, \theta) \in [0, 2\pi)$ and $\vartheta(k_x, \theta) \in [0, \pi]$. We additionally defined the energy scale:

$$E(k_x, \theta) := \sqrt{[m\cos\theta - \beta(\hbar k_x)^2]^2 + (\alpha\hbar k_x)^2 + (m\sin\theta)^2}. \tag{4.24}$$

The eigenvectors of the parametric Hamiltonian in equation (4.22), corresponding to energy $E_\pm(k_x, \theta) = \pm E(k_x, \theta)$, assume the form

$$|\phi_-(k_x, \theta)\rangle = \begin{pmatrix} \sin\left[\frac{\vartheta(k_x, \theta)}{2}\right] e^{-i\varphi(k_x,\theta)} \\ -\cos\left[\frac{\vartheta(k_x, \theta)}{2}\right] \end{pmatrix} \quad \text{and}$$
$$|\phi_+(k_x, \theta)\rangle = \begin{pmatrix} \cos\left[\frac{\vartheta(k_x, \theta)}{2}\right] \\ \sin\left[\frac{\vartheta(k_x, \theta)}{2}\right] e^{i\varphi(k_x,\theta)} \end{pmatrix}, \tag{4.25}$$

where we have normalised them to unity, i.e., $\langle\phi_\pm(k_x, \theta)|\phi_\pm(k_x, \theta)\rangle = 1$. One observes that for $\vartheta(k_x, \theta) = \pi/2$ we retain the 1+1d topological insulator model of equation (4.1), but in the rotated basis. In addition, the structure of the $\breve{g}$-vector in equation (4.23) implies that the angle $\varphi(k_x, \theta)$ becomes ill-defined for $\vartheta(k_x, \theta) = 0$ or $\vartheta(k_x, \theta) = \pi$. Equivalently, for one of the latter values the angle $\varphi(k_x, \theta)$ appears in the above eigenvectors but not in the Hamiltonian. For instance, this occurs for $|\phi_-(k_x, \theta)\rangle$ when $\vartheta(k_x, \theta) = \pi$. Nevertheless, one can pick a different gauge in which $\varphi(k_x, \theta)$ is well-defined for $\vartheta(k_x, \theta) = \pi$. However, in the latter gauge $\varphi(k_x, \theta)$ will be unavoidably ill-defined for $\vartheta(k_x, \theta) = 0$. Therefore, one cannot introduce a well-defined gauge for all the eigenvectors. Using the Stokes theorem, we show in section 4.3.2 that this property is crucial for obtaining a non-zero adiabatic current.

To calculate the adiabatic current, we start from the polarisation operator $\hat{P}_x = -e\hat{x}$. In the adiabatic limit its expectation value is solely determined by the occupied states, i.e., essentially by the family of the negative energy eigenvectors $|\phi_-(k_x, \theta)\rangle$, given by equation (4.25). Via $\hat{x} = i\hbar\partial_{p_x} \equiv i\partial_{k_x}$, the expectation value of the polarisation operator reads

$$P_x(\theta) = -ie\int_{-\infty}^{+\infty} \frac{dk_x}{2\pi} \langle\phi_-(k_x, \theta)|\partial_{k_x}\phi_-(k_x, \theta)\rangle. \tag{4.26}$$

By making use of equation (2.22), the respective adiabatic current assumes the form

$$J_{c,x}(t) := \frac{dP_x(\theta)}{dt} = -ie\frac{d\theta}{dt}\frac{\partial}{\partial\theta}\int_{-\infty}^{+\infty} \frac{dk_x}{2\pi} \langle\phi_-(k_x, \theta)|\partial_{k_x}\phi_-(k_x, \theta)\rangle. \tag{4.27}$$

Let us now consider that θ varies in time in an interval $\Delta t = t_f - t_i$, and takes the initial and final values θ_i and θ_f, respectively. To calculate the change of the electric polarisation ΔP_x, that developed during this time interval, we integrate the above expression and obtain

$$\Delta P_x = -\,\mathrm{ie}\int_{\theta_i}^{\theta_f} d\theta \int_{-\infty}^{+\infty}\frac{dk_x}{2\pi} \times\left[\left\langle \partial_\theta \phi_-(k_x,\theta)\middle|\partial_{k_x}\phi_-(k_x,\theta)\right\rangle + \left\langle \phi_-(k_x,\theta)\middle|\partial^2_{\theta,k_x}\phi_-(k_x,\theta)\right\rangle\right]. \tag{4.28}$$

We employ the normalisation and completeness relations of the two parametric eigenvectors, i.e.

$$\langle\phi_a(k_x,\theta)|\phi_b(k_x,\theta)\rangle = \delta_{a,b} \quad\text{and}\quad \sum_a |\phi_a(k_x,\theta)\rangle\langle\phi_a(k_x,\theta)| = \hat{1} \tag{4.29}$$

with $a, b = \pm 1$, and rewrite the second term of equation (4.28). We obtain (cf Thouless's seminal work in [7])

$$\Delta P_x = e\int_{\theta_i}^{\theta_f} d\theta \int_{-\infty}^{+\infty}\frac{dk_x}{2\pi}\,\Omega^-_{k_x,\theta}(k_x,\theta), \tag{4.30}$$

where we introduced the so-called Berry curvature [7–9] of the family of eigenvectors $|\phi_-(k_x,\theta)\rangle$, generally defined as

$$\Omega^a_{k_x,\theta}(k_x,\theta) = \mathrm{i}\left[\left\langle \partial_{k_x}\phi_a(k_x,\theta)\middle|\partial_\theta\phi_a(k_x,\theta)\right\rangle - \left\langle \partial_\theta\phi_a(k_x,\theta)\middle|\partial_{k_x}\phi_a(k_x,\theta)\right\rangle\right]. \tag{4.31}$$

The latter is antisymmetric under the exchange $k_x \leftrightarrow \theta$. Therefore, via the correspondence $(k_x,\theta)\to(x,y)$, the Berry curvature maps to the z component of the usual magnetic field $B_z = \partial_x A_y - \partial_y A_x$. This connection becomes more transparent by additionally introducing the so-called Berry vector potential [8, 9]

$$\begin{aligned} A^a_{k_x}(k_x,\theta) &= \mathrm{i}\left\langle \phi_a(k_x,\theta)\middle|\partial_{k_x}\phi_a(k_x,\theta)\right\rangle \quad\text{and}\\ A^a_{\theta}(k_x,\theta) &= \mathrm{i}\left\langle \phi_a(k_x,\theta)\middle|\partial_{\theta}\phi_a(k_x,\theta)\right\rangle, \end{aligned} \tag{4.32}$$

where $a = \pm 1$, that allows us to write the Berry curvature as:

$$\Omega^a_{k_x,\theta}(k_x,\theta) = \partial_{k_x}A^a_\theta(k_x,\theta) - \partial_\theta A^a_{k_x}(k_x,\theta). \tag{4.33}$$

One notes that the Berry vector potential behaves as a U(1) gauge potential, similar to the usual electromagnetic gauge potential $\boldsymbol{A}(\boldsymbol{r})$, since if we perform a local U(1) transformation $|\phi_a(k_x,\theta)\rangle \to e^{-\mathrm{i}\varphi(k_x,\theta)}|\phi_a(k_x,\theta)\rangle$, the Berry vector potential becomes modified in the following fashion

$$A^a_{k_x}(k_x,\theta)\to A^a_{k_x}(k_x,\theta) + \partial_{k_x}\varphi(k_x,\theta),\quad A^a_\theta(k_x,\theta)\to A^a_\theta(k_x,\theta) + \partial_\theta\varphi(k_x,\theta). \tag{4.34}$$

In stark contrast, the Berry curvature remains invariant under gauge transformations, similar to the electromagnetic fields, cf section 2.1.

Quite remarkably, the adiabatic approach reveals that we can describe the electric charge transport via emergent magnetic-like phenomena associated with the internal space (k_x,θ) [2]. In fact, the adiabatic current is given by an appropriate integration of the Berry curvature in the internal space, which in the usual electromagnetic picture would correspond to the magnetic flux $\int_S \boldsymbol{B}\cdot d\boldsymbol{S}$ piercing an area S. If the integration involves a closed surface, then this integral yields the Gauss law for magnetism and, thus, provides the enclosed amount of magnetic charge. However, in Maxwell's electromagnetism, magnetic monopoles do not exist (see [10]). Nonetheless, this is not generally the case for the emergent magnetism that we encounter here. In fact, magnetic monopoles, also customarily called Berry magnetic monopoles, are indeed present here. To understand the nature of these magnetic monopoles further, let us calculate the Berry curvature for the specific example.

Since the Berry curvature of the family of eigenvectors $|\phi_-(k_x, \theta)\rangle$ is a gauge invariant quantity, we can calculate it using the gauge choice of equation (4.25). A straightforward calculation yields that the Berry curvature has the form

$$\Omega^-_{k_x,\theta}(k_x, \theta) = -\frac{\alpha\hbar m\,[m + \cos\theta\beta(\hbar k_x)^2]}{2[E(k_x, \theta)]^3}. \tag{4.35}$$

Similar to the case of the electric Gauss law, in which electric point charges appear as singularities in the electric field, here, Berry magnetic point charges appear as the singularities of the Berry curvature. This implies that the Berry magnetic monopoles are sourced by points where energy gap closings occur and bands touch. Therefore, one obtains the number of Berry monopoles existing in the parametric bandstructure, or more precisely in the (k_x, θ, m) space, by suitably integrating the Berry curvature over a surface in (k_x, θ, m) space which encloses the singularities. One observes that the expression retrieved above for the Berry curvature becomes singular only for $k_x = m = 0$, resulting in a single monopole.

Let us now focus on the vicinity of the monopole, and linearise the Berry curvature with respect to k_x or equivalently set $\beta = 0$. We integrate the linearised Berry curvature $\Omega^-_{k_x,\theta}(k_x, \theta, \beta = 0)$ by employing a cut-off k_c. The related outcome reads

$$\int_{\theta_i}^{\theta_f} \frac{d\theta}{2\pi} \int_{-k_c}^{+k_c} dk_x\, \Omega^-_{k_x,\theta}(k_x, \theta, \beta = 0) = -\frac{\Delta\theta}{2\pi}\frac{\alpha\hbar k_c}{\sqrt{(\alpha\hbar k_c)^2 + m^2}}. \tag{4.36}$$

By taking the limit $k_c \to \infty$, as in section 3.5 and specifically equation (3.56), the change of the electric polarisation reads $\Delta P_x = -e\Delta\theta/(2\pi)$ in agreement with equation (3.37). Essentially, this provides an alternative derivation of the Goldstone–Wilczek formula, though for arbitrary $\Delta\theta$. One should note that while in relativistic theories taking the limit $k_c \to \infty$ is valid, in non-relativistic theories such a process generally requires particular caution. Nonetheless, the inclusion of the quadratic term regularises the integral and allows us to take such a limit, since it establishes that there is no other singularity in $E(k_x, \theta)$, apart from the one appearing for $k_x = 0$. Note that such a problem is not encountered in the case of lattice models, since Berry magnetic monopoles come in pairs, in accordance with the Nielsen–Ninomiya theorem [11]. For more details see also sections 4.2 and 4.3.3.

Given the result of equation (4.36), if $\theta(t)$ varies periodically in time, then after a single period one obtains $\Delta\theta = 2\pi$. Thus, a single unit of charge is transferred[7]. The latter quantisation stems from the fact that by taking this limit, the integration area effectively becomes a closed surface, and the Gauss law for the Berry magnetic charge applies. As a result, the charge transferred during a single period of this so-called 'Thouless pump', is proportional to the number of Berry magnetic monopoles, i.e. one in the particular case $k_x = m = 0$. In fact, as long as the closed surface encloses the existing monopoles, the result does not become modified irrespective of the exact shape of the integration surface. Such a robustness to deformations implies that the integral of the Berry curvature over a compact[8] space constitutes a topological invariant in a similar sense to the winding number discussed in section 4.1.4. In the present case this integral corresponds to the 1st Chern number [2, 5], which is defined as

$$C_1 := \int_0^{2\pi} \frac{d\theta}{2\pi} \int_{-\infty}^{+\infty} dk_x \, \Omega^-_{k_x,\theta}(k_x, \theta). \tag{4.37}$$

Straightforward calculations in the present case yield $C_1 = -1$, where we assumed $\alpha, \beta, m > 0$.

4.3.2 Relation between Chern and winding numbers

One finds a notable connection between the 1st Chern number C_1 and the winding number w_1. This can be proven via the Stokes theorem. By restricting to a periodic θ, the base manifold (k_x, θ) is compactifiable to $\mathbb{S}^1 \times \mathbb{T}^1$,[9] and we obtain

$$\begin{aligned} C_1 &\equiv \frac{1}{2\pi} \oint_{\mathbb{S}^1\times\mathbb{T}^1} d\theta dk_x \, \Omega^-_{k_x,\theta}(k_x, \theta) \\ &= \frac{1}{2\pi} \oint_{\partial(\mathbb{S}^1\times\mathbb{T}^1)} \left[dk_x A^-_{k_x}(k_x, \theta) + d\theta A^-_\theta(k_x, \theta) \right]. \end{aligned} \tag{4.38}$$

Here $\partial(\mathbb{S}^1 \times \mathbb{T}^1)$ constitutes the boundary of $\mathbb{S}^1 \times \mathbb{T}^1$. Since the boundary points become identified, the above integral should yield zero as long as $A^-_{k_x,\theta}$ are smoothly differentiable. However, direct calculation of the Berry curvature has shown the opposite. Consequently, there should be an obstruction to constructing the Berry vector potential that introduces an additional boundary that essentially is responsible for the non-zero value of the Chern number [5, 12].

This necessarily-present obstruction essentially reflects the inability to define a global gauge for the eigenvectors of equation (4.25), and thus the Berry vector potential. To carry out the integral of equation (4.38), one defines two families of eigenvectors. One for the north ($\vartheta(k_x, \theta) \leq \pi/2$), and another for the south ($\vartheta(k_x, \theta) > \pi/2$) patch, cf figure 4.5. In this manner, the vector potential becomes

[7] One has to remark that, when $\theta(t)$ is a periodic variable, the outcome for ΔP_x depends on the history that leads to the value $\Delta\theta$. That is, the system could have performed an integer number of cycles between the initial and final time-instants, t_i and t_f, respectively. As a result, the change of the electric polarisation is more precisely defined as: $\Delta P_x = -e\Delta\theta/(2\pi) \bmod \mathbb{Z}$.

[8] Or effectively compactified. See section 4.1.3.

[9] This is possible by virtue of the quadratic term $\beta(\hbar k_x)^2$, that leads to the effective compactification $\mathbb{R}^1 \mapsto \mathbb{S}^1$.

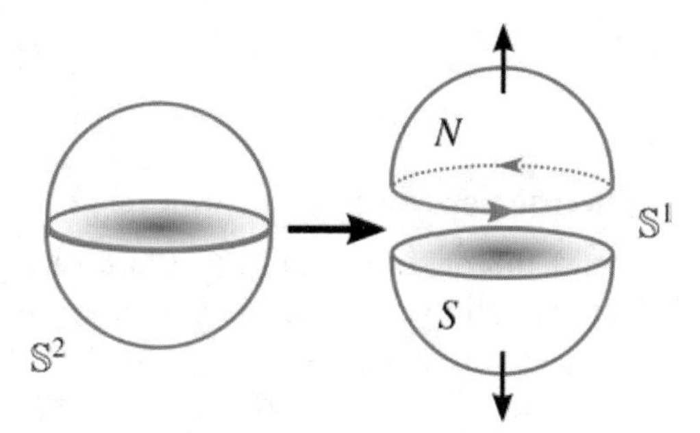

Figure 4.5. North and south patches employed to calculate the 1st Chern number using the Stokes theorem. For the patching considered in the text, the only contribution to the integral of equation (4.38) originates from value of the integrand at the equator. At the equator, the Hamiltonian of the system maps to the Hamiltonian of the unperturbed 1+1d topological insulator. As a result, the 1st Chern number coincides with the respective winding number w_1.

discontinuous for $\vartheta(k_x, \theta) = \pi/2$. Thus the initially compactified (k_x, θ)-space acquires an $\mathbb{S}^1$ boundary at the equator.

Let us proceed by considering a concrete gauge choice for the respective north- and south-patch eigenvectors, $|\phi_-^N(k_x, \theta)\rangle$ and $|\phi_-^S(k_x, \theta)\rangle$:

$$|\phi_-^N(k_x, \theta)\rangle = \begin{pmatrix} \sin\left[\frac{\vartheta(k_x, \theta)}{2}\right] e^{-i\varphi(k_x,\theta)} \\ -\cos\left[\frac{\vartheta(k_x, \theta)}{2}\right] \end{pmatrix} \quad \text{and} \quad |\phi_-^S(k_x, \theta)\rangle = e^{i\varphi(k_x,\theta)}|\phi_-^N(k_x, \theta)\rangle. \tag{4.39}$$

The two eigenvectors are related by a gauge transformation. As a result, one defines a Berry vector potential per patch $A^{-,S}(k_x, \theta)$ and $A^{-,N}(k_x, \theta)$, which according to equation (4.34) differ by $\nabla_{(k_x,\theta)}\varphi(k_x, \theta)$. Based on equation (4.38) and our choice of patching, it is only the contribution from the equator that contributes to the Chern number. At the equator $\vartheta(k_x, \theta) = \pi/2$ leading to $\theta = 0$ and $\varphi(k_x, 0) \equiv \varphi(k_x)$, with the latter angle introduced in equation (4.7) for the 1+1d topological-insulator model. Under the above conditions, we obtain:

$$C_1 = \frac{1}{2\pi}\oint_{\mathbb{S}^1} dk_x \left[A_{k_x}^{-,N}(k_x, 0) - A_{k_x}^{-,S}(k_x, 0)\right] = \frac{1}{2\pi}\int_{-\infty}^{+\infty} dk_x \frac{d\varphi(k_x)}{dk_x} \equiv w_1. \tag{4.40}$$

The above result is very important, as it shows that the Chern and winding numbers defined in $d = 2n$ and $d = 2n - 1$ spatial dimensions are equal[10], therefore revealing a connection of Hamiltonians defined in different dimensionalities. This property constitutes the basis for charting topological phases and relies on the dimensional-reduction process that we present in section 7.3. It has been employed in references

[10] Or opposite, depending on the sign in front of the term $m \sin\theta\hat{\sigma}_z$ entering the Hamiltonian of equation (4.22).

[5], [13], and others, to perform the complete topological classification of insulators and superconductors.

4.3.3 Lattice model and electric polarisation

In this paragraph we extend the previous study to the lattice model presented in equation (4.17). While for the continuum model we can perform a Fourier transform and express the polarisation operator as $\hat{P}_x = -ei\partial_{k_x}$, for lattice systems this is not possible. Instead, it is more convenient to retrieve the change of the polarisation $\Delta P_x := P_x(t_f) - P_x(t_i)$ within a time interval $\Delta t = t_f - t_i$, via integrating the current flown for the same interval, i.e. $\Delta P_x = \int_{t_i}^{t_f} dt\, J_{c,x}(t)$. King-Smith and Vanderbilt showed in [14], that, by employing Wannier functions one can retrieve the following relation for the change of polarisation (cf. equation (4.30))

$$\Delta P_x = e \int_0^{2\pi} \frac{d\theta}{2\pi} \int_{\mathrm{BZ}} dk_x\, \Omega^{-}_{k_x,\theta}(k_x, \theta). \tag{4.41}$$

The above is defined for a period of pumping effected via the time dependence of a θ parameter, as in equation (4.21). According to the properties of the Berry curvature $\Delta P_x \in \mathbb{Z}$.[11] The above formula is also derived in the Hands-on section 4.6, closely following reference [9]. By choosing the periodic gauge as in [9] and employing equation (4.38), we obtain

$$\Delta P_x = -e \int_{\mathrm{BZ}} \frac{dk_x}{2\pi} A^{-}_{k_x}(k_x, \theta)\Big|_{\theta=0}^{\theta=2\pi}. \tag{4.42}$$

We extend equation (4.17), by introducing a parameter θ, as follows:

$$\hat{H}(k_x, \theta) = \frac{\alpha\hbar}{a}\sin(k_x a)\hat{\sigma}_y + \left\{m\cos\theta - 2\beta\left(\frac{\hbar}{a}\right)^2[1 - \cos(k_x a)]\right\}\hat{\sigma}_z \\ + m\sin\theta\hat{\sigma}_x. \tag{4.43}$$

The above Hamiltonian possesses the property:

$$\hat{H}(k_x, 2\pi - \theta) = \hat{H}(k_x, -\theta) = \hat{\sigma}_z\hat{H}(-k_x, \theta)\hat{\sigma}_z \equiv e^{i\pi\hat{\sigma}_z/2}\hat{H}(-k_x, \theta)e^{-i\pi\hat{\sigma}_z/2}. \tag{4.44}$$

Based on the above, one finds from (4.42) that: $P_x(0) = -P_x(2\pi)$ and $2P_x(0) \in \mathbb{Z}$.[11] This implies that P_x essentially takes the distinct values of 0 or 1/2, since it is defined modulo an integer. The latter point is extensively discussed in [13]. For the given example, the polarisation is 1/2 only when the winding number is odd. This implies that we can alternatively classify the present system with a $\mathbb{Z}_2$ invariant $W_1 := \mathrm{Exp}(i2\pi P_x)$. Nevertheless, the $\mathbb{Z}_2$ character of the electric polarisation is independent of the presence of chiral symmetry and thus the possibility of defining a winding number. As a result, the electric polarisation can be employed to

[11] In many cases it is convenient to employ a dimensionless notation for the electric polarisation in 1d, by expressing it in units of *ea*.

topologically classify the properties of systems defined in $d = 1$ lacking chiral symmetry. Following [5], the polarisation can be identified with the so-called Chern–Simons invariant CS_1. The latter invariant can be generalised to odd spatial dimensions $d = 2n + 1$ and allows one to introduce the gauge invariant quantity

$$W_{2n+1} := \mathrm{Exp}[\mathrm{i}2\pi \mathrm{CS}_{2n+1}] \in \{-1, 1\} \tag{4.45}$$

which can be employed as a $\mathbb{Z}_2$ topological invariant in the absence of chiral symmetry. For more details see [5].

4.4 Berry phase

In the previous paragraph we came across the concepts of the Berry gauge potential and curvature. Nonetheless, the relevance of these physical quantities is not restricted to the particular calculation performed above. In fact, the Berry phase appears in different branches of physics, e.g. the Aharonov–Bohm effect [15], where the Berry flux coincides with the usual magnetic flux (see [16]). In general, the Berry vector potential plays an important role when considering the adiabatic or low-frequency response of an insulator. To clarify this statement, we consider the following Schrödinger equation

$$\hat{H}[\boldsymbol{R}(t)]|\phi(t)\rangle = \mathrm{i}\hbar|\partial_t\phi(t)\rangle \tag{4.46}$$

and the general time-dependent parameters $\boldsymbol{R}(t)$. One can then rewrite $|\phi(t)\rangle$ as

$$|\phi(t)\rangle = \sum_a c_a(t)e^{-\frac{\mathrm{i}}{\hbar}\int_{t_0}^{t} dt' E_a(t')} e^{\mathrm{i}\gamma_a(t)}|n_a[\boldsymbol{R}(t)]\rangle \tag{4.47}$$

where we introduced the so-called family of adiabatic eigenvectors of the system $|n_a(\boldsymbol{R})\rangle$, defined as

$$\hat{H}(\boldsymbol{R})|n_a(\boldsymbol{R})\rangle = E_a(\boldsymbol{R})|n_a(\boldsymbol{R})\rangle \tag{4.48}$$

and characterised by the eigenergies $E_a(\boldsymbol{R})$. Representing $|\phi(t)\rangle$ in this fashion is possible, since the adiabatic eigenstates comprise a complete basis at every time instant. The terms $\int_{t_0}^{t} dt' E_a(t')/\hbar$ and $\gamma_a(t)$ define the dynamical and Berry phases, respectively. By plugging the above expression in the Schrödinger equation one obtains

$$\begin{aligned} -\mathrm{i}\dot{c}_a(t) + c_a(t)\dot{\gamma}_a(t) &= \sum_b \mathrm{i}\langle n_a(\boldsymbol{R}) \mid \partial_t n_b(\boldsymbol{R})\rangle c_b(t) e^{\frac{\mathrm{i}}{\hbar}\int_{t_0}^{t} dt'\left[E_a(t')-E_b(t')\right]} e^{-\mathrm{i}\left[\gamma_a(t)-\gamma_b(t)\right]} \\ &\equiv \dot{\boldsymbol{R}}(t)\cdot \sum_b \mathrm{i}\langle n_a(\boldsymbol{R}) \mid \partial_{\boldsymbol{R}} n_b(\boldsymbol{R})\rangle c_b(t) e^{\frac{\mathrm{i}}{\hbar}\int_{t_0}^{t} dt'\left[E_a(t')-E_b(t')\right]} e^{-\mathrm{i}\left[\gamma_a(t)-\gamma_b(t)\right]}, \end{aligned} \tag{4.49}$$

with $\dot{\boldsymbol{R}} := d\boldsymbol{R}/dt$. The above yields an exact solution of the time-dependent problem under consideration. However, if the Fourier transform of $\boldsymbol{R}(t)$ consists of frequencies much smaller than the level splittings $E_a(\boldsymbol{R}) - E_b(\boldsymbol{R})$, then we can neglect the interlevel transitions and assume that each family of adiabatic eigenstates evolves separately. The latter implies that $c_a(t) = 1$. Under these conditions we obtain the expression for the Berry phase [8, 9]

$$\gamma_a(t_f) - \gamma_a(t_i) = \mathrm{i} \int_{\boldsymbol{R}(t_i)}^{\boldsymbol{R}(t_f)} d\boldsymbol{R} \cdot \langle n_a(\boldsymbol{R}) | \partial_{\boldsymbol{R}} n_a(\boldsymbol{R}) \rangle \equiv \int_{\boldsymbol{R}(t_i)}^{\boldsymbol{R}(t_f)} d\boldsymbol{R} \cdot \boldsymbol{A}^a(\boldsymbol{R}), \tag{4.50}$$

where we introduced the Berry gauge potential $\boldsymbol{A}^a(\boldsymbol{R}) = \mathrm{i}\langle n_a(\boldsymbol{R}) | \partial_{\boldsymbol{R}} n_a(\boldsymbol{R}) \rangle$. The expression for the Berry phase difference shows that it depends on the initial and final values $\boldsymbol{R}(t_i)$ and $\boldsymbol{R}(t_f)$. However, it becomes a gauge invariant quantity when $\boldsymbol{R}(t_i) = \boldsymbol{R}(t_f)$, which renders the path in parameter space closed. In this case, one employs the Stokes theorem and expresses the Berry phase difference as

$$\gamma_a(t_f) - \gamma_a(t_i) = \oint_C d\boldsymbol{R} \cdot \boldsymbol{A}^a(\boldsymbol{R}) = \oint_S d\boldsymbol{S} \cdot \boldsymbol{\Omega}^a(\boldsymbol{R}), \tag{4.51}$$

where we introduced the Berry curvature $\boldsymbol{\Omega}^a(\boldsymbol{R}) = \nabla_{\boldsymbol{R}} \times \boldsymbol{A}^a(\boldsymbol{R})$. Here, $C \equiv \partial S$ denotes the closed path in parameter space $\boldsymbol{R}$, and S the surface enclosed. Moreover, the vector $d\boldsymbol{S}$ points outwards of the closed surface. Following the discussion of the previous section, the Berry phase difference becomes non-zero as long as the closed surface in parameter space encloses Berry magnetic monopoles.

4.5 Hands-on: winding number in a 3+1d model

In this section we focus on the calculation of the winding number for a chiral-symmetric topological insulator in 3+1 dimensions described by the lattice model

$$\hat{H}(\boldsymbol{k}) = A \sin \boldsymbol{k} \cdot \hat{\tau}_x \hat{\boldsymbol{\sigma}} + [m - 2t(\cos k_x + \cos k_y + \cos k_z)]\hat{\tau}_z. \tag{4.52}$$

The Hamiltonian has a chiral symmetry with $\hat{\Pi} = \hat{\tau}_y$.

Task: Block off-diagonalise the Hamiltonian by employing a unitary transformation similar to the one introduced in section 4.1.5. Retrieve the eigenvectors and calculate the mapping matrix $\hat{q}(\boldsymbol{k})$. Employ equation (4.16) and integrate over the three-dimensional BZ so to compute w_3. Alternatively, identify the gap-closing $\boldsymbol{k}$ points and, after linearising the Hamiltonian in their vicinity, calculate the partial contributions, whose sum is w_3.

4.6 Hands-on: current and electric polarisation formula

In this paragraph we derive the formula for the change of the electric polarisation used in section 4.3.3. We consider a 1+1d lattice model for a topological insulator with chiral symmetry that is subjected to a generic external perturbation which induces a current $J_{c,x}(t)$, through the presence of a time-dependent parameter $\theta(t)$. As long as the perturbation allows us to consider the adiabatic picture, one can set up a time-dependent perturbation theory based on the family of adiabatic eigenstates determined by $\hat{H}(k_x, \theta)|n_a(k_x, \theta)\rangle = E_a(k_x, \theta)|n_a(k_x, \theta)\rangle$. Below, we formulate this type of perturbative approach following an analysis akin to the one presented in [9].

Task: Consider the Schrödinger equation $\mathrm{i}\hbar\partial_t|\phi_a(k_x, t)\rangle = \hat{H}[k_x, \theta(t)]|\phi_a(k_x, t)\rangle$, where $|\phi_a(k_x, t)\rangle$ define the exact eigenstates of $\hat{H}[k_x, \theta(t)]$. We continue with setting up a perturbation theory, for which the exact eigenstates coincide with the adiabatic eigenstates at zeroth order. Therefore, one rewrites

$$\begin{aligned}&-[\mathrm{i}\hbar\partial_t - E_a(k_x, \theta)]|\phi_a(k_x, t)\rangle + \hat{H}[k_x, \theta(t)]|\phi_a(k_x, t)\rangle\\&= E_a(k_x, \theta)|\phi_a(k_x, t)\rangle.\end{aligned} \tag{4.53}$$

The first term will be viewed as the perturbation on the adiabatic eigenstates $|n_a(k_x, \theta)\rangle$. At this point we expand $|\phi_a(k_x, t)\rangle$ in the following sense $|\phi_a(k_x, t)\rangle = |n_a(k_x, \theta)\rangle + \delta|\phi_a(k_x, t)\rangle + \ldots$. Show that:

$$\begin{aligned}|\phi_a(k_x, t)\rangle = |n_a(k_x, \theta)\rangle &- \mathrm{i}\hbar\dot{\theta}\textstyle\sum_{b\neq a}|n_b(k_x, \theta)\rangle\\&\times\frac{\langle n_b(k_x, \theta)|\partial_\theta n_a(k_x, \theta)\rangle}{E_a(k_x, \theta) - E_b(k_x, \theta)} + \text{h.o.t.}\,.\end{aligned} \tag{4.54}$$

At this level of approximation, the exact eigenstates are expressed as a linear combination of the adiabatic ones. Therefore we can suppress the time argument and write $|\phi_a(k_x, t)\rangle \to |\phi_a(k_x, \theta)\rangle$. By employing the definition of the adiabatic current:

$$J_{c,x}(t) = -\frac{e}{\hbar}\sum_a^{\text{occ}}\int_{\text{BZ}}\frac{dk_x}{2\pi}\langle\phi_a(k_x, \theta)|\partial_{k_x}\hat{H}(k_x, \theta)|\phi_a(k_x, \theta)\rangle, \tag{4.55}$$

show that:

$$J_{c,x}(t) = e\dot{\theta}\sum_a^{\text{occ}}\int_{\text{BZ}}\frac{dk_x}{2\pi}\,\Omega^a_{k_x,\theta}(k_x, \theta), \tag{4.56}$$

which leads to equation (4.42) for the change of the electric polarisation.

4.7 Hands-on: violation of chiral symmetry and electric polarisation

In this section we study the effects of the violation of chiral symmetry for the model of equation (4.17). For example, consider the following extension

$$\begin{aligned}\hat{H}(k_x) = \frac{\alpha\hbar}{a}\sin(k_x a)\hat{\sigma}_y &+ \left\{m - 2\beta\left(\frac{\hbar}{a}\right)^2[1 - \cos(k_x a)]\right\}\hat{\sigma}_z\\&+ \zeta\sin^3(k_x a - \delta)\hat{\sigma}_x,\end{aligned} \tag{4.57}$$

with $\delta = \{0, \pi/2\}$. The two Hamiltonians lack chiral symmetry and thus a winding number cannot be defined. Nonetheless, zero-energy bound states may still exist and their presence can be instead predicted by employing the Chern–Simons invariant CS_1, introduced in section 4.3.3.

Task 1: Investigate the gap closings of the bulk energy spectrum and infer whether the two Hamiltonians can still support topologically protected zero-energy bound states.

Task 2: Consider the further extensions

$$\hat{H}(k_x, \theta) = \frac{\alpha\hbar}{a}\sin(k_x a)\hat{\sigma}_y + \left\{m\cos\theta - 2\beta\left(\frac{\hbar}{a}\right)^2[1 - \cos(k_x a)]\right\}\hat{\sigma}_z \\ + [\zeta\sin^3(k_x a - \delta) + \sin\theta]\hat{\sigma}_x, \tag{4.58}$$

with θ a periodic parameter defined in $[0, 2\pi)$. Try to understand the results of the previous task by retrieving ΔP_x, given by equations (4.41) and (4.42).

References

[1] Nakahara M 2003 *Geometry, Topology and Physics* 2nd edn (Boca Raton, FL: Taylor and Francis)

[2] Volovik G E 2003 *The Universe in a Helium Droplet* (New York: Oxford University Press)

[3] Kaluza T 1921 *Sitzungsber. Preuss. Akad. Wiss. Berlin. Math. Phys.* **33** 966

[4] Klein O 1926 *Z. Phys.* **37** 895

[5] Ryu S, Schnyder A P, Furusaki A and Ludwig A W W 2010 *New J. Phys.* **12** 065010

[6] Atiyah M F and Singer I M 1968 *Ann. Math.* **87** 484

[7] Thouless D J 1983 *Phys. Rev.* B **27** 6083

[8] Berry M V 1984 *Proc. R. Soc. London* A **392** 45

[9] Xiao D, Chang M-C and Niu Q 2010 *Rev. Mod. Phys.* **82** 1959

[10] Jackson J D 1999 *Classical Electrodynamics* 3rd edn (New York: Wiley)

[11] Nielsen H B and Ninomiya M 1981 *Phys. Lett.* B **105** 219

[12] Bernevig B A and Hughes T L 2013 *Topological Insulators and Topological Superconductors* (Princeton, NJ: Princeton University Press)

[13] Qi X-L, Hughes T L and Zhang S-C 2008 *Phys. Rev.* B **78** 195424

[14] King-Smith R D and Vanderbilt D 1993 *Phys. Rev.* B **47** 1651(R)

[15] Aharonov Y and Bohm D 1959 *Phys. Rev.* **115** 485

[16] Resta R 2000 *J. Phys.: Condens. Matter* **12** R107

Chapter 5

Chern insulators—fundamentals

In this chapter we focus on the so-called Chern insulators, which constitute topological insulators defined in odd-dimensional spacetime, violating time-reversal and chiral symmetries. Systems belonging to this class support an anomalous Hall response. For instance, the 2 + 1d Chern insulator exhibits the quantum anomalous Hall effect, where a Hall current appears by applying an electric field without the requirement of an out-of-plane magnetic field. To introduce these systems, we first consider the 2 + 1d extension of the Jackiw–Rebbi model studied previously and make a connection to the Dirac equation and the parity anomaly. We discuss the effective action of the electromagnetic field coupled to a 2 + 1d Chern insulator, which besides the standard Maxwell term, consists of the so-called Chern–Simons term. Amongst a variety of interesting phenomena, the latter also leads to a topological Meissner effect. Further, we construct a prototypical model for a Chern insulator in 2 + 1d, which extends the 1 + 1d model of the topological insulator discussed in the previous chapter. We associate the topological aspects of this system with the Berry curvature in $\boldsymbol{k}$-space and the related 1st Chern number. In addition, we highlight the link between the Chern number and the winding number of the 1 + 1d topological insulator obtained via dimensional reduction. We extend and generalise the 2 + 1d Chern insulator model to higher dimensions. Finally, the Hands-on sections focus on the derivation of the Chern–Simons action, the second Chern number and a more general formulation of the 1st Chern number for interacting systems.

5.1 Jackiw–Rebbi model and Dirac physics in 2 + 1d

We consider the following 2 + 1d extension

$$\hat{H}(\hat{\boldsymbol{p}}, x) = \alpha(\hat{p}_x\hat{\sigma}_y - \hat{p}_y\hat{\sigma}_x) + M(x)\hat{\sigma}_z, \tag{5.1}$$

of the Hamiltonian examined in section 3.2, where we included the remaining component of the Rashba SOC introduced in section 1.2.4. We assume that the

doi:10.1088/978-1-68174-517-6ch5

system is translationally invariant along the y axis and, thus, the wavevector k_y, constitutes a good quantum number. This leads to dispersive electronic modes propagating along the y direction. The addition of the second SOC term breaks the chiral symmetry effected by $\hat{\Pi} = \hat{\sigma}_x$, which was present in the $1 + 1$d model of section 3.2. Despite the lack of chiral symmetry, this model still exhibits non-trivial topological properties, which are inherited from the $1 + 1$d Jackiw–Rebbi model.

To transparently demonstrate the above connection, we restrict ourselves to the low-energy sector of the $1 + 1$d model and solely consider the two zero-energy solutions. This approximation is valid for small wavevectors, i.e., $|\alpha\hbar k_y| \ll M > 0$. Thus, the following analysis will apply for $k_y \in [-k_c, k_c]$, with k_c an appropriate cut-off. Upon adding the second spatial dimension and the respective Rashba SOC term, these initially zero-energy solutions become dispersive according to an energy dispersion $E(k_y)$. By virtue of translational invariance, we introduce the following wavefunctions for a given k_y

$$\boldsymbol{\phi}_{0,-,k_y}(\boldsymbol{r}) = c_- e^{ik_y y} e^{-\frac{M}{\alpha\hbar}|x-x_1|} \frac{1}{\sqrt{2}} \begin{pmatrix} 1 \\ -1 \end{pmatrix} \text{ and } \boldsymbol{\phi}_{0,+,k_y}(\boldsymbol{r}) = c_+ e^{ik_y y} e^{-\frac{M}{\alpha\hbar}|x-x_2|} \frac{1}{\sqrt{2}} \begin{pmatrix} 1 \\ 1 \end{pmatrix} \tag{5.2}$$

which extend the ones obtained in equation 3.18, and are depicted in figure 5.1(a). We directly observe that the above constitute eigenvectors of the $2 + 1$d Hamiltonian

$$[\alpha(\hat{p}_x\hat{\sigma}_y - \hat{p}_y\hat{\sigma}_x) + M(x)\hat{\sigma}_z]\boldsymbol{\phi}_{0,\pm,k_y}(\boldsymbol{r}) = E_\pm(k_y)\boldsymbol{\phi}_{0,\pm,k_y}(\boldsymbol{r}), \tag{5.3}$$

only if $E_\pm(k_y) = \mp\alpha\hbar k_y$ (see figure 5.1(b)). The sign of the dispersion's slope is determined by the eigenvalue of the chiral symmetry operator labelling the zero-energy eigenvector of the $1 + 1$d model. In fact, the above modes are termed *chiral* due to the particular locking of chirality and momentum. One finds that the solution located at the left (right) edge has chirality $\sigma = -1(+1)$ and an energy dispersion $E_L(k_y) = +\alpha\hbar k_y$ ($E_R(k_y) = -\alpha\hbar k_y$). The two modes carry opposite electric currents that cancel each other out in equilibrium. However, in the presence of a constant external electric field they give rise to the so-called quantum anomalous Hall effect (QAHE) [1–7]. This phenomenon is similar to the standard integer quantum Hall

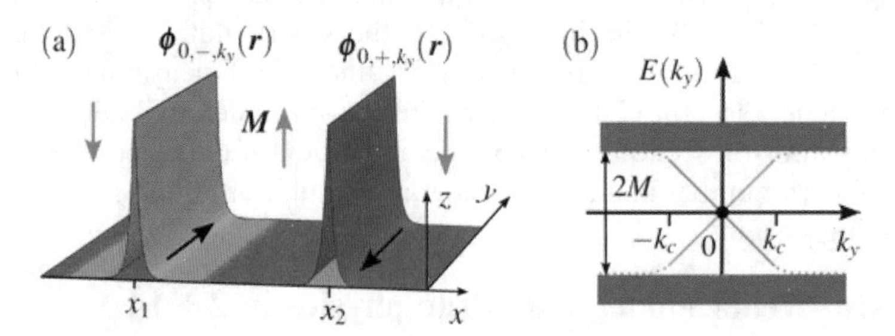

Figure 5.1. (a) Chiral edge modes appearing in the bulk energy gap of the Hamiltonian in equation (5.1), thus, $-M \leq E(k_y) \leq M$. They are trapped at the magnetic domain walls where $M(x)$ changes sign. In addition, they are characterised by opposite chirality (±) and lead to counter-propagating currents (black arrows). (b) The energy dispersion of the chiral edge modes is linear for $k_y \in [-k_c, k_c]$. Outside this window, the linear approximation breaks down and the edge modes merge with the continuum of the energy spectrum, which is schematically depicted with dark turquoise.

effect (QHE) [8, 9], albeit with a crucial difference. In contrast to the QHE, the QAHE does not require the application of an out-of-plane magnetic field B_z. The possibility of such exotic phenomena in condensed matter physics was first put forward by Semenoff [1], while Haldane [2] was the first to provide the blueprints for how to actualise the QAHE in condensed matter systems.

5.1.1 Electric charge and current responses of the chiral edge modes

In the presence of the bulk gap and $T = 0$, it is solely the chiral edge modes[1] that dictate the low-energy behaviour of the system. The same holds for the response to weak external fields. Let us now consider a strip geometry as in figure 5.1(a) and assume that the modes are sharply localised at the domain wall. This situation takes place for very small $\alpha\hbar/M$ and the edge modes can be viewed as strictly 1 + 1d systems. We minimally-couple the chiral edge modes to an electromagnetic potential (see section 2.2), which leads to the modified energy dispersions

$$\begin{aligned} E_L^{V,A_y}(k_y) &= +\,\alpha[\hbar k_y + eA_y(x_1)] - eV(x_1) \quad \text{and} \\ E_R^{V,A_y}(k_y) &= -\,\alpha[\hbar k_y + eA_y(x_2)] - eV(x_2). \end{aligned} \tag{5.4}$$

Here, we picked a gauge for the potentials which respects the translational invariance along the y axis. Specifically, we consider the constant electric and magnetic fields: $E_x \equiv [V(x_1) - V(x_2)]/L_x$ and $B_z \equiv [A_y(x_2) - A_y(x_1)]/L_x$, with $L_x = x_2 - x_1$. We now calculate the contribution of the chiral edge modes to the electric charge and current densities, at linear order with respect to the external fields. We restrict ourselves to zero temperature. After observing that only the electric (magnetic) field leads to a non-zero electric current (excess charge), we write:

$$J_{c,y}^{\text{edge}} = -\frac{e}{\hbar}\int d(\alpha\hbar k_y)\,\frac{f[\alpha\hbar k_y - eV(x_1)] - f[-\alpha\hbar k_y - eV(x_2)]}{2\pi L_x L_z} = -\sigma_H E_x, \tag{5.5}$$

$$\partial_t\rho_c^{\text{edge}} = -\,\partial_y J_{c,y}^{\text{edge}} = \sigma_H\partial_y E_x = \sigma_H\partial_t B_z \Rightarrow \rho_c^{\text{edge}} = \rho_{c,0}^{\text{edge}} + \sigma_H B_z. \tag{5.6}$$

In the above, we took into account that the Chern insulator is actually a three-dimensional system, with a small thickness L_z in the z direction. The so-called Hall response [10–13] is proportional to the Hall conductivity

$$\sigma_H = \frac{1}{L_z}\frac{e^2}{h}, \tag{5.7}$$

with e^2/h defining the unit of conductance. In contrast, a strictly two-dimensional system is characterised by the Hall conductance $G_H = e^2/h$.

---

[1] The chiral edge modes appear only if suitable magnetic solitons are introduced. See section 3.2. If these are absent, the bulk is a topologically-trivial band insulator and does not support a QAHE.

5.1.2 Chiral edge modes in the quantum Hall effect: Laughlin's argument

In the above paragraph we concluded that the Hall conductance is quantised due to the presence of an integer number of conducting edge channels. While a similar picture also holds for the case of the QHE, this was not immediately recognised at the time of its discovery. At first sight, the observation of a non-zero conductance in a two-dimensional system was puzzling, since it appeared to go against the predictions of the theory of Anderson localisation [14]. According to the latter, every disordered two-dimensional conductor should undergo a transition to an insulating phase. Nonetheless, as it was later shown (see [15]), the integer QHE constitutes an exception.

The possibility of a QHE requires the presence of delocalised modes. The latter assumption lies in the heart of Laughlin's argument [10], and the subsequent considerations by Halperin [11, 16], invoked for understanding the quantisation of the Hall conductance based on general grounds. For this purpose, Laughlin considered for his analysis a convenient experimental QHE setup, which allowed him to retrieve the desired quantisation. As shown in figure 5.2, he considered a two-dimensional gas of electrons in a ribbon geometry. The electrons feel a perpendicular magnetic field $\boldsymbol{B} = B_z\check{z}$, assumed to be generated by a vector potential in the Landau gauge, i.e. $\boldsymbol{A} = B_z x\check{y}$. In addition, the two edges are kept at a potential difference $\Delta V = -E_x L_x$. To calculate the steady-state current $J_{c,y}$, Laughlin introduced an auxiliary flux Φ, so that $J_{c,y}(\Phi) = dE(\Phi)/d\Phi$. Here $E(\Phi)$ denotes the energy of the system. In turn, the flux can be viewed as the result of a constant probe vector potential $\delta\boldsymbol{A} = \delta A_y\check{y}$, with $\Phi = -L_y\delta A_y$ and L_y the circumference (length) of the ribbon. By virtue of the homogeneity of the system along the y direction, one applies periodic boundary conditions and thus ends up with the quantised wavenumber $k_y = 2\pi s/L_y$ with $s = 0, 1, 2, \ldots$ (see section 1.1.5). Via the minimal coupling discussed in section 2.2, we find

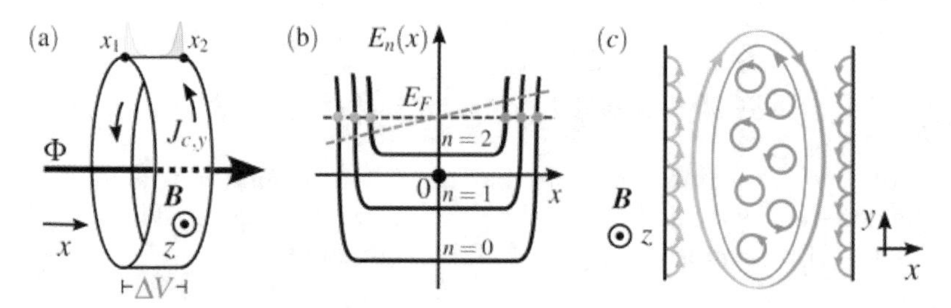

Figure 5.2. (a) Laughlin's ribbon setup, employed to explain the quantisation of the Hall conductance in the standard quantum Hall effect. As in figure 5.1, chiral edge modes also appear in the standard quantum Hall effect. (b) In the bulk, the Landau levels ($n \in \{0, 1, 2, 3, \ldots\}$) can be viewed as flat bands in k_y space and render the infinite system insulating. When boundaries are introduced, the Landau levels bend upwards. Therefore, the previously occupied levels cross the Fermi-energy level (E_F). This leads to an integer number of chiral edge modes, localised near the boundaries as shown in (a) with orange and purple. A potential drop across the boundaries $\propto\Delta V$, is equivalent to a spatially-varying Fermi-energy level. The spatial inhomogeneity yields an imbalance of an integer number of occupied states at each side. This charge deficit leads to the quantised Hall response. (c) Chiral edge modes viewed as edge electrons propagating in skipping orbits, cancelling the bulk orbital magnetisation current. See section 5.1.3.

$$k_y \mapsto k_y - \frac{e}{\hbar}\frac{\Phi}{L_y} + \frac{eB_z x}{\hbar} = \frac{2\pi}{L_y}(s - \nu_\Phi) + \frac{eB_z x}{\hbar} \tag{5.8}$$

where ν_Φ defines the number of flux quanta piercing the ribbon, where $\Phi_0 := h/e$ defines the flux quantum. From the above we infer that for an integer number of flux quanta, we recover the original energy spectrum. However, this is achieved via the so-called *spectral-flow*, i.e., the energy levels obtained in the absence of the flux become shifted in its presence.

To this end, we proceed with calculating the current. We start from the Hamiltonian of a two-dimensional electron gas coupled to the external electromagnetic fields

$$\hat{H}_s(\hat{p}_x, \Phi) = \frac{\hat{p}_x^2}{2m} + \frac{1}{2}m\omega_c^2[x + x_s(\Phi)]^2 - ex_s(\Phi)E_x + \frac{m}{2}\left(\frac{E_x}{B_z}\right)^2, \tag{5.9}$$

where we introduced the cyclotron frequency $\omega_c = e|B_z|/m$ and the flux-dependent guiding centers

$$x_s(\Phi) := (s - \nu_\Phi)\frac{L_x}{\nu_{B_z}} + \frac{eE_x}{m\omega_c^2} \quad \text{with} \quad \nu_{B_z} := B_z L_x L_y/\Phi_0. \tag{5.10}$$

Moreover, we define the magnetic length ℓ_B, through $2\pi\ell_B^2 := \Phi_0/B_z$. The above Hamiltonian is of the quantum harmonic oscillator type, thus yielding the Landau levels:

$$E_{n,s}(\Phi) = \hbar\omega_c\left(n + \frac{1}{2}\right) - ex_s(\Phi)E_x + \frac{m}{2}\left(\frac{E_x}{B_z}\right)^2. \tag{5.11}$$

We observe that the energy levels depend linearly on Φ, and for $E_x = 0$ each Landau level energy is independent of the quantum number s. This renders each Landau level multiply degenerate. The number of the degeneracy is given by $L_x L_y/2\pi\ell_B^2 = \nu_{B_z}$. We are now in a position to calculate the current $\Delta J_{c,y}$, flown while threading a single flux quantum through the ribbon. After taking into account the Landau-level degeneracy [16], we obtain: $\Delta J_{c,y} = -G_H \Delta V$ with $G_H = N_{\rm occ}\frac{e^2}{h}$. Here $N_{\rm occ}$ defines the number of occupied Landau levels.

The quantisation of the conductance can be attributed to the edge states appearing due to confinement. Near the boundaries, the energy of the Landau levels increases due to the positive confining potential. See figure 5.2(b). As a result, the Fermi level is crossed at each boundary side $N_{\rm occ}$ times. When applying ΔV, $N_{\rm occ}$ deficit edge charges appear that become transferred from one edge to the other, thus, leading to Hall response. The charge transfer and the communication between the two edges happens via delocalised modes extending throughout the sample [11, 16]. This so-called quantum Hall transition is similar to the topological transition studied in section 4.1.1 and figure 4.2, which occurs through the delocalisation of the edge states.

5.1.3 Connection to Dirac physics and parity anomaly

Similar to the 1 + 1d Jackiw–Rebbi model, where the fractional charge trapped in a mass soliton was understood in terms of the chiral anomaly and the Goldstone–

Wilczek formula, also here, the quantised Hall current and excess charge densities can be related to the so-called *parity anomaly*[2] appearing for the 2 + 1d Dirac equation [17–19]. This phenomenon can also take place for massless Dirac fermions, in which case, parity is dynamically broken by quantum corrections [19]. As a result, the so-called Chern–Simons term [17–21] appears in the action of the electromagnetic potential which couples to the Dirac electrons. To connect to the Dirac equation, we rewrite the time-dependent Schrödinger equation of the model of equation (5.1) as follows

$$[\alpha(\hat{p}_x\hat{\sigma}_y - \hat{p}_y\hat{\sigma}_x) + M\hat{\sigma}_z]\boldsymbol{\phi}(\boldsymbol{r}, t) = \mathrm{i}\hbar\partial_t\boldsymbol{\phi}(\boldsymbol{r}, t) \quad \Rightarrow \tag{5.12}$$

$$[-\mathrm{i}\alpha\hbar(\partial_x\hat{\sigma}_z\hat{\sigma}_y - \partial_y\hat{\sigma}_z\hat{\sigma}_x) + M]\boldsymbol{\phi}(\boldsymbol{r}, t) = \mathrm{i}\hbar\partial_t\hat{\sigma}_z\boldsymbol{\phi}(\boldsymbol{r}, t). \tag{5.13}$$

At this stage we introduce the $\hat{\gamma}$ matrices[3]

$$\hat{\gamma}^0 = \hat{\sigma}_z, \quad \hat{\gamma}^1 = -\mathrm{i}\hat{\sigma}_x \quad \text{and} \quad \hat{\gamma}^2 = -\mathrm{i}\hat{\sigma}_y. \tag{5.14}$$

After setting $\alpha = \hbar = 1$ for convenience, we obtain the Lagrangian density for the respective Grassmann fields (see also section 2.6)

$$L(\boldsymbol{r}, t) = \tilde{\boldsymbol{\psi}}(\boldsymbol{r}, t)\big(\mathrm{i}\hat{\gamma}^\mu\partial_\mu - M\big)\boldsymbol{\psi}(\boldsymbol{r}, t). \tag{5.15}$$

In the above, parity is conserved for $M = 0$. By integrating out the Dirac electrons we find the effective action for the electromagnetic field. Apart from the Maxwell term, it acquires the so-called Chern–Simons term

$$S_{\mathrm{CS}}^{(2+1)}(\boldsymbol{r}, t) = \frac{\theta}{4\pi}\varepsilon^{\mu\nu\lambda}A_\mu\partial_\nu A_\lambda \quad \text{with} \quad \mu, \nu, \lambda = 0, 1, 2. \tag{5.16}$$

In the above $\mathrm{sgn}(\theta) = \mathrm{sgn}(M)$. Similar to the results of figure 4.4, $|\theta| = 1/2$ for a single species of Dirac electrons, while $\theta \in \mathbb{Z}$ for an even number of species of Dirac electrons. In general, $\theta \in \mathbb{Z}$ for a Hamiltonian defined in a compact or effectively compactified space, cf section 4.1.3.

The Chern–Simons term violates parity which, according to section 1.1.1, is defined as a σ_v mirror symmetry operation: $(x, y) \mapsto (x, -y)$ or $(x, y) \mapsto (-x, y)$. As a result, it leads to a current, which for a constant θ assumes the form

$$J^\mu_{c,\mathrm{CS}} = -\frac{\delta S_{\mathrm{CS}}^{(2+1)}}{\delta A_\mu} = -\frac{\theta}{2\pi}\varepsilon^{\mu\nu\lambda}\partial_\nu A_\lambda. \tag{5.17}$$

For a strictly two-dimensional system the above induces a non-zero bulk Hall current with conductance $G_{xy} = \theta/(2\pi)$, since θ is in units of $e^2/\hbar$. Nonetheless, this current does not constitute a so-called *transport* current. It rather corresponds to the bound current generated by the electrons moving in closed circular orbits due

[2] See section 1.1.1.

[3] Note that in 2 + 1d spacetime we cannot introduce a chirality operator $\hat{\gamma}^5$. This is a general property of the Dirac equation when defined in an odd-dimensional spacetime.

to the violation of TRS effected by a non-zero M (see figure 5.2(c)). The latter current constitutes the *orbital-magnetisation*[4] current which is divergence-free, i.e. $\nabla \cdot \boldsymbol{J}_{c,\mathrm{CS}} = 0$. Via the continuity of the electric charge, reflected in equation (2.14), one finds that this current does not lead to any net electric charge transfer. However, coupling the system to a measuring apparatus, unavoidably introduces edges and thus chiral edge modes. The edge modes can be viewed as electrons moving according to the so-called *skipping orbits*, as shown in figure 5.2(c). These electrons are responsible for mediating a net Hall response since their orbits are not closed.

5.1.4 Maxwell–Chern–Simons action and topological Meissner effect

As already mentioned in the section above, and to be explicitly shown in the Hands-on section 5.6, the effective action of the electromagnetic field coupled to a 2 + 1d Chern insulator contains the Chern–Simons action (see equation (5.16)). The complete action for the electromagnetic potential reads

$$S_{\mathrm{MCS}} = \int d^3x\, L_z\left(-\frac{1}{4}F^{\mu\nu}F_{\mu\nu} + \frac{\sigma_{xy}}{4}\varepsilon_{\mu\nu\lambda}A^{\mu}F^{\nu\lambda} - J^{\mu}_{c,f}A_{\mu}\right), \tag{5.18}$$

where we reduced the Minkowski spacetime by one dimension, via integrating over the finite but very small thickness L_z, of the otherwise, two-dimensional-assumed system. Thus, $\sigma_{xy} = G_{xy}/L_z$, while the speed of light is set to $c = 1$. Here $J^{\mu}_{c,f}$ denotes the free-current density. See also section 6.1.3. In the following paragraphs we investigate the effects of the Chern–Simons term on the equations of motion and the propagation of the photon field.

Emergence of the chiral edge modes

Let us now study how the presence of the Chern–Simons term naturally implies the emergence of chiral edge modes. For this purpose we extremise the above action and find the equations of motion

$$\partial_{\nu}F^{\nu\mu} = -\,\sigma_{xy}\varepsilon^{\mu\nu\lambda}\partial_{\nu}A_{\lambda} + J^{\mu}_{c,f}. \tag{5.19}$$

In the case of two insulating homogeneous regions with different values for the Hall conductivity, the bulk free current is zero, i.e., $J^{\mu}_{c,f} = 0$. Nonetheless, boundary currents appear [5, 22]. By integrating the above equations of motion, one obtains the following boundary condition

$$\left(F^{+}_{\mu i} - F^{-}_{\mu i}\right)\check{n}^{i} = \left(\sigma^{+}_{xy} - \sigma^{-}_{xy}\right)\varepsilon_{\mu i\lambda}\check{n}^{i}A^{\lambda} + J^{\mathrm{surf}}_{c,f,\mu}. \tag{5.20}$$

Here $\check{\boldsymbol{n}}$ defines the unit vector normal to the surface and the potentials are evaluated at the same time instant. With $^{\pm}$ we represent the quantities above and below the boundary, parallel and antiparallel to $\check{\boldsymbol{n}}$. In the above we took into account the fact that the potentials are continuous across the boundary. This does not need to hold for the electromagnetic fields. However, if we are solely interested

[4] For more details see sections 6.1.3 and 6.2.2.

in the linear response of the system, we can focus on the edge current and excess charge density. Thus, we can drop the free currents and retain the CS contribution:

$$J^{\text{surf}}_{c,\text{CS},\mu} = \left(\sigma^{+}_{xy} - \sigma^{-}_{xy}\right)\varepsilon_{\mu i\lambda}\tilde{n}^{i}A^{\lambda}. \tag{5.21}$$

Let us now consider a setup similar to figure 5.1(a) and 5.2(c), in which, there exist two boundaries at x_1 and $x_2 > x_1$, with normal vector $\check{\boldsymbol{n}} = \check{\boldsymbol{x}}$ at $x = x_2$. Assume that the conductivity is σ_{xy} in the inner part, i.e. $\forall\, x \in (x_1, x_2)$, and σ'_{xy} in the outer part ($x < x_1$ and $x > x_2$). The total edge current reads

$$J^{\text{edge}}_{c,\mu} = (\sigma_{xy} - \sigma'_{xy})\varepsilon_{\mu 1\lambda}\frac{A^{\lambda}(x_1) - A^{\lambda}(x_2)}{L_x} \Rightarrow \tag{5.22}$$

$$\rho^{\text{edge}}_{c} - \rho^{\text{edge}}_{c,0} = -(\sigma_{xy} - \sigma'_{xy})B_z \quad \text{and} \quad J^{\text{edge}}_{c,y} = (\sigma_{xy} - \sigma'_{xy})E_x. \tag{5.23}$$

One observes that the number of edge modes is given by the difference of the Hall conductivity; essentially the difference between the topological invariants characterising the inner and outer parts. This constitutes a manifestation of the bulk-boundary correspondence, also encountered in section 4.1.1. In addition, if for simplicity we assume that $\sigma'_{xy} = 0$, i.e., the outer part corresponds to the vacuum or a topologically-trivial insulator, the edge current is opposite to the bulk orbital-magnetisation current. Such a behaviour can be inferred from figure 5.2(c). The edge currents need to cancel out the bulk Hall current, since the trivial insulator has a zero orbital magnetisation. In the presence of the external fields, the edge current additionally yields the experimentally-measured transport current.

Topological photon mass and Meissner effect

Let us now explore the propagation of the photon for an infinite material characterised by a non-zero bulk σ_{xy}. We can rewrite the equations of motion of equation (5.19) in terms of the electric and magnetic fields $E_{x,y}$ and B_z. We obtain:

$$\begin{aligned}
&\frac{\partial E_x}{\partial x} + \frac{\partial E_y}{\partial y} - \sigma_{xy}B_z = 0,\\
&\frac{\partial E_x}{\partial t} - \frac{\partial B_z}{\partial y} + \sigma_{xy}E_y = 0,\\
&\frac{\partial E_y}{\partial t} + \frac{\partial B_z}{\partial x} - \sigma_{xy}E_x = 0.
\end{aligned} \tag{5.24}$$

Interestingly, the first equation reads $\nabla \cdot \boldsymbol{E} = \sigma_{xy}B_z$, thus, implying $\rho^{\text{excess}}_{c} = \sigma_{xy}B_z$. Therefore, the magnetic flux leads to excess charge density, in accordance with the findings of section 5.1.1.

We proceed with analysing the consequences of the Chern–Simons term on the phonon propagation. We Fourier transform the fields and obtain a homogeneous system of coupled equations. By setting the determinant equal to zero, we obtain the dispersion for the photon which reads

$$\omega(\boldsymbol{q}) = \pm c\sqrt{\boldsymbol{q}^2 + (Z_0\sigma_{xy})^2}, \tag{5.25}$$

where we restored the speed of light and introduced the vacuum impedance $Z_0 = \sqrt{\mu_0/\varepsilon_0}$. Remarkably, we observe that the photon obtained a mass term [5]. Nonetheless, gauge invariance is preserved intact both for an infinite or a finite-sized system, by virtue of the edge modes in the latter case. In fact, this mass has a topological origin and is distinct compared to the one obtained in superconducting systems that stems from a term $\propto A_\mu A^\mu$ in the action (for instance, see [16]). In the situation considered here, the mass arises due to a coupling between the potentials and the fields.

These two distinct mechanisms for generating the photonic mass lead to different types of the so-called Meissner effect. This effect is associated with the screening of the magnetic field. The Meissner effect is considered to be a characteristic signature of conventional s-wave superconductors and it is described by the so-called penetration depth of the static magnetic field. Notably, the superconductor screens the magnetic field independently of its orientation. However, the topological Meissner effect mediated by a Chern insulator is anisotropic and appears only for out-of-plane magnetic fields [23]. The related penetration depth is given by $1/(Z_0\sigma_{xy})$.

5.2 Chern insulator in 2 + 1d

Following the same strategy as for the Jackiw–Rebbi model, we extend the 1 + 1d topological-insulator model introduced in the previous chapter to 2 + 1d, by adding the remaining Rashba SOC component. This is performed in the following two paragraphs for the continuum and lattice implementations.

5.2.1 Continuum model

Here we extend the model introduced in equation (4.1) of the previous chapter, and obtain

$$\hat{H}(\hat{\boldsymbol{p}}) = \alpha(\hat{p}_x\hat{\sigma}_y - \hat{p}_y\hat{\sigma}_x) + \left(m - \beta\hat{\boldsymbol{p}}^2\right)\hat{\sigma}_z. \tag{5.26}$$

Note that we also included a quadratic $\hat{p}_y^2$ dependence in $M(\hat{\boldsymbol{p}})$. In a similar fashion to section 4.1.1, it is straightforward to show that the present Chern insulator model harbours chiral edge modes when termination boundaries are introduced, as long as $m > 0$ for $\beta > 0$. To retrieve these modes we can confine ourselves to the low-energy sector. Following the previous argumentation, the chiral edge modes are associated with a gap closing at $\boldsymbol{k} = \boldsymbol{0}$. Therefore, for the subsequent analysis we can focus near the vicinity of small momenta and solely retain the linear in momentum SOC term, $\hat{p}_y\hat{\sigma}_x$, while at the same time we neglect the quadratic term $\hat{p}_y^2\hat{\sigma}_z$. Under these conditions, the wavefunctions for the chiral edge modes are directly obtainable and assume the form

$$\phi_{0,-,k_y}(\boldsymbol{r}) \propto e^{ik_y y} \sin[\zeta(x - x_1)]e^{-(x-x_1)/\xi} \frac{1}{\sqrt{2}} \begin{pmatrix} 1 \\ -1 \end{pmatrix}, \tag{5.27}$$

$$\phi_{0,+,k_y}(\boldsymbol{r}) \propto e^{ik_y y} \sin[\zeta(x - x_2)]e^{+(x-x_2)/\xi} \frac{1}{\sqrt{2}} \begin{pmatrix} 1 \\ 1 \end{pmatrix}, \tag{5.28}$$

with ζ and ξ defined in section 4.1.1. The corresponding energy dispersions take the form $E_{\pm}(k_y) = \mp \alpha \hbar k_y$, as also encountered earlier in the 2 + 1d Jackiw–Rebbi model of section 5.1. Once again, we find that the solution located at the left (right) edge has chirality $\sigma = -1(+1)$ and an energy dispersion $E_L(k_y) = +\alpha \hbar k_y$ ($E_R(k_y) = -\alpha \hbar k_y$), as shown in figure 5.1.

To conclude, we study the effects of a finite length L_x, on the chiral edge modes. In this case, the Hamiltonian couples the two chiral edge modes leading to the modified dispersions

$$E(k_y) = \pm\sqrt{(\alpha \hbar k_y)^2 + m^2} \quad \text{with} \quad m \propto \mathrm{Exp}(-L_x/\xi). \tag{5.29}$$

Evidently, the hybridisation of the two spatially-separated chiral edge modes gives rise to the energy spectrum of a relativistic massive particle. Therefore, the effective mass m can be safely neglected only for $L_x \to +\infty$ and small momenta $\hbar k_y$.

5.2.2 Lattice model

For completeness, we also present an example of a lattice version of the 2 + 1d Jackiw–Rebbi Hamiltonian of equation (5.1). By extending the 1 + 1d lattice model of equation (4.17), we have

$$\begin{aligned} \hat{H}(\boldsymbol{k}) = &\frac{\alpha \hbar}{a}[\sin(k_x a)\hat{\sigma}_y - \sin(k_y a)\hat{\sigma}_x] \\ &+ \left\{ m - 2\beta \left(\frac{\hbar}{a}\right)^2 [2 - \cos(k_x a) - \cos(k_y a)] \right\} \hat{\sigma}_z. \end{aligned} \tag{5.30}$$

The above Hamiltonian is defined on a two-dimensional torus $\mathbb{T}^2$.

5.3 Quantised Hall conductance and Chern number—bulk approach

Similar to the 1 + 1d topological insulators, also the Chern insulators above can be characterised by topological invariant quantities defined for the bulk. In this particular case one introduces a 1st Chern number in (k_x, k_y)-space, which plays a similar role to the winding number of section 4.1.3. As a matter of fact, the Hall conductance is proportional to the Chern number, which underlies the quantisation of the former in units of e^2/h (see also section 4.3.1). The quantisation of G_H is also expected based on the conclusions of section 5.1.1. In the following paragraphs we retrieve G_H based on the adiabatic eigenstates of the bulk model and results presented in section 4.3.

5.3.1 Bulk eigenstates

To infer G_H, we employ the adiabatic approach. For this purpose it is sufficient to retrieve the bulk eigenstates, in analogy to section 4.3. For the Fourier transformed version of the Hamiltonian in equation (5.26), and the model of equation (5.30)[5], we can write

$$\hat{H}(\boldsymbol{k}) \equiv E(\boldsymbol{k})\check{\boldsymbol{g}}(\boldsymbol{k}) \cdot \hat{\boldsymbol{\sigma}}, \tag{5.31}$$

where we introduced the $\boldsymbol{g}$-vector

$$\boldsymbol{g}(\boldsymbol{k}) = E(\boldsymbol{k})(\sin \vartheta(\boldsymbol{k})\cos \varphi(\boldsymbol{k}),\ \sin \vartheta(\boldsymbol{k})\sin \varphi(\boldsymbol{k}),\ \cos \vartheta(\boldsymbol{k})) \tag{5.32}$$

parametrised with $\varphi(\boldsymbol{k}) \in [0, 2\pi)$ and $\vartheta(\boldsymbol{k}) \in [0, \pi]$, as in equation (4.23). We also set $E(\boldsymbol{k}) = |\boldsymbol{g}(\boldsymbol{k})|$. The corresponding eigenvectors read

$$|\phi_+(\boldsymbol{k})\rangle = \begin{pmatrix} \cos\left[\dfrac{\vartheta(\boldsymbol{k})}{2}\right] \\ \sin\left[\dfrac{\vartheta(\boldsymbol{k})}{2}\right] e^{i\varphi(\boldsymbol{k})} \end{pmatrix} \text{ and } |\phi_-(\boldsymbol{k})\rangle = \begin{pmatrix} \sin\left[\dfrac{\vartheta(\boldsymbol{k})}{2}\right] e^{-i\varphi(\boldsymbol{k})} \\ -\cos\left[\dfrac{\vartheta(\boldsymbol{k})}{2}\right] \end{pmatrix}. \tag{5.33}$$

The above bulk eigenstates satisfy $\hat{H}(\boldsymbol{k})|\phi_\pm(\boldsymbol{k})\rangle = E_\pm(\boldsymbol{k})|\phi_\pm(\boldsymbol{k})\rangle$ with $E_\pm(\boldsymbol{k}) = \pm E(\boldsymbol{k})$. One notes that the phase $\varphi(\boldsymbol{k})$, becomes undefined for either $\vartheta(\boldsymbol{k}) = 0$ or $\vartheta(\boldsymbol{k}) = \pi$. A similar inability to define a gauge globally was also encountered in section 4.3. The common denominator in both cases is that time-reversal and chiral symmetries are broken. As a result, both systems are characterised by a non-zero integral of the Berry curvature and, thus, a non-zero 1st Chern number C_1.

5.3.2 Adiabatic Hall transport and Berry curvature

At this point we are in a position to revisit the dissipationless transport examined in section 5.1.1 from a bulk perspective. We employ the already obtained results of section 4.3 and the formula derived in section 4.6 for the adiabatic current. In fact, one observes a similarity between the two Hamiltonians:

$$\begin{aligned} \hat{H}(k_x, \theta) &= \alpha\hbar k_x\hat{\sigma}_y + [m \cos\theta - (\hbar k_x)^2]\hat{\sigma}_z + m \sin\theta\hat{\sigma}_x \\ \hat{H}(k_x, k_y) &= \alpha\hbar k_x\hat{\sigma}_y + [m - \beta(\hbar k_y)^2 - \beta(\hbar k_x)^2]\hat{\sigma}_z - \alpha\hbar k_y\hat{\sigma}_x. \end{aligned} \tag{5.34}$$

Evidently, the two Hamiltonians can be mapped to each other by viewing θ as a function of k_y. Therefore, one finds that a Chern insulator in $d = 2$, is equivalent to a

[5] Where $|\phi\rangle$ should be replaced by Bloch states $|u\rangle$.

Thouless pump of a topological insulator in $d = 1$. In this manner, the formula derived in equation (4.56), i.e.

$$J_{c,x}(t) = e\dot{\theta} \sum_{a}^{\text{occ}} \int_{-\infty}^{+\infty} \frac{dk_x}{2\pi} \Omega^a_{k_x,\theta}(k_x, \theta) \tag{5.35}$$

can be directly employed, after re-expressing it by taking into account that $k_y \mapsto \theta$. The wavevector k_y is here assumed to be time-dependent due to the presence of a constant electric field that yields $\hbar\dot{k}_y = -eE_y$. By defining $\Omega^a_z(\boldsymbol{k}) := \Omega^a_{k_x,k_y}(\boldsymbol{k})$, one directly finds the adiabatic Hall current

$$J_{c,x}(k_y) = -\frac{e^2}{h} E_y \sum_{a}^{\text{occ}} \int_{-\infty}^{+\infty} dk_x \, \Omega^a_z(\boldsymbol{k}). \tag{5.36}$$

After taking into account the contributions from all k_y, we find that the Hall conductance is given as

$$G_{xy} = -\frac{e^2}{h} C_1 \quad \text{with} \quad C_1 = \sum_{a}^{\text{occ}} \int \frac{d\boldsymbol{k}}{2\pi} \Omega^a_z(\boldsymbol{k}) \in \mathbb{Z}. \tag{5.37}$$

In the case under consideration $C_1 = 1$, since we have a single monopole at the point $k_x = k_y = m = 0$ of the three-dimensional parameter space (k_x, k_y, m).

Notably, the above formalism also applies to the lattice model of equation (5.30), where the $\boldsymbol{k}$-space integration now is over $\mathbb{T}^2$. However, there is an additional and more important difference. For the continuum model one pumps a single unit of charge, in principle, after infinite time since $\boldsymbol{k}$-space becomes compactified at infinity. In practice though, the contribution to the Berry curvature originates from the neighbourhood of the gap-closing point $\boldsymbol{k} = \boldsymbol{0}$, and thus there exists a cut-off wavevector scale that renders the pumping period finite. In stark contrast, the lattice model is defined in the 1st Brillouin zone, where $k_y + 2\pi n/a \equiv k_y$ with $n \in \mathbb{Z}$. Therefore, Newton's law $\hbar\dot{k}_y = -eE_y$ yields a pumping period $T_{\text{pump}} = \Phi_0/(aE_y)$. In addition, the $\boldsymbol{g}$-vector in the lattice model can perform an integer number of full twists within one period (see section 4.2). As a result, one can pump more than one units of charge per cycle.

5.3.3 Homotopy mapping and Chern number

In our first encounter of the 1st Chern number in section 4.3, we argued that it becomes quantised as it constitutes a topological invariant quantity [12, 13]. In this paragraph, we provide a geometrical understanding of this property by examining the Chern number's structure for the model under discussion. By straightforward calculations one finds that the Berry curvature for the two families of eigenvectors read

$$\Omega^{\pm}_z(\boldsymbol{k}) = \frac{\mp 1}{2} \check{\boldsymbol{g}}(\boldsymbol{k}) \cdot \left(\frac{\partial \check{\boldsymbol{g}}(\boldsymbol{k})}{\partial k_x} \times \frac{\partial \check{\boldsymbol{g}}(\boldsymbol{k})}{\partial k_y} \right). \tag{5.38}$$

For $T = 0$ only the − band is occupied and thus contributes to the Chern number. This is true as long as the chemical potential lies inside the bulk energy gap. Therefore we have

$$C_1 = \int \frac{d\boldsymbol{k}}{4\pi}\, \check{\boldsymbol{g}}(\boldsymbol{k}) \cdot \left(\frac{\partial \check{\boldsymbol{g}}(\boldsymbol{k})}{\partial k_x} \times \frac{\partial \check{\boldsymbol{g}}(\boldsymbol{k})}{\partial k_y} \right). \tag{5.39}$$

To show why this integral takes only quantised values, we have to investigate the behaviour of $\check{\boldsymbol{g}}(\boldsymbol{k})$ at the single wavevector leading to a Berry monopole $\boldsymbol{k} = \boldsymbol{0}$, and at infinity. Since the $\check{\boldsymbol{g}}$-vector reads

$$\check{\boldsymbol{g}}(\boldsymbol{k}) = \left(\frac{-\alpha\hbar k_y}{\sqrt{(\alpha\hbar\boldsymbol{k})^2 + [m - \beta(\hbar\boldsymbol{k})^2]^2}}, \frac{\alpha\hbar k_x}{\sqrt{(\alpha\hbar\boldsymbol{k})^2 + [m - \beta(\hbar\boldsymbol{k})^2]^2}}, \frac{m - \beta(\hbar\boldsymbol{k})^2}{\sqrt{(\alpha\hbar\boldsymbol{k})^2 + [m - \beta(\hbar\boldsymbol{k})^2]^2}} \right), \tag{5.40}$$

we find:

$$\check{\boldsymbol{g}}(\boldsymbol{0}) = (0, 0, \mathrm{sgn}(m)) \text{ and } \check{\boldsymbol{g}}(|\boldsymbol{k}| \to +\infty) = (0, 0, -1). \tag{5.41}$$

Note that for $m > 0$ ($m < 0$) the orientation of the $\check{\boldsymbol{g}}(\boldsymbol{k})$ at the origin and at infinity are opposite (the same), leading to a skyrmion [4, 24] (trivial) configuration. In figure 5.3(a) we depict the respective skyrmionic profile. As in section 4.1.3, since the $\check{\boldsymbol{g}}(\boldsymbol{k})$ vector is oriented along the same direction for all the points at infinity of the two-dimensional $\boldsymbol{k}$-space, all these points become essentially equivalent. In this manner, $\mathbb{R}^2$ compactifies to $\mathbb{S}^2$.[6] Furthermore, the term $\check{\boldsymbol{g}} \cdot \left(d_{k_x}\check{\boldsymbol{g}} \times d_{k_y}\check{\boldsymbol{g}} \right)$ provides the infinitesimal solid angle covered upon an infinitesimal variation of $\check{\boldsymbol{g}}(\boldsymbol{k})$, since

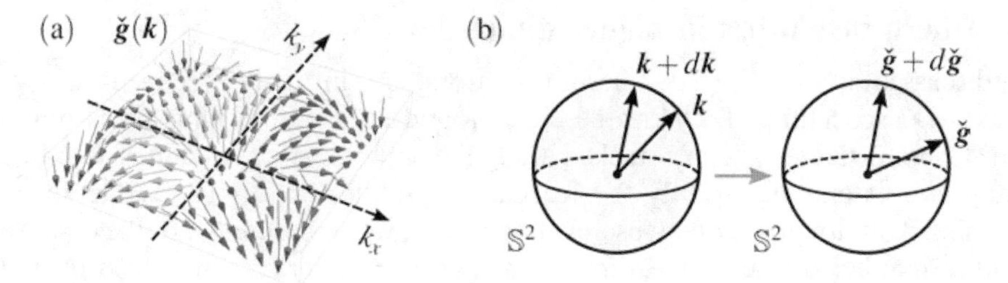

Figure 5.3. (a) Skyrmion configuration [24] of the unit vector $\check{\boldsymbol{g}}(\boldsymbol{k})$. The $\check{\boldsymbol{g}}$-vector varies smoothly in space, so that its orientation at the 'center' of $\boldsymbol{k}$-space $\boldsymbol{k} = \boldsymbol{0}$, is antiparallel to its orientation at the boundary (infinity). Note that the $\check{\boldsymbol{g}}$-vector has exactly the same orientation at infinity, thus, rendering the whole boundary equivalent to a single point. This effectively compactifies $\boldsymbol{k}$-space from $\mathbb{R}^2$ to a two-sphere $\mathbb{S}^2$. (b) Mapping of the $\mathbb{S}^2$-compactified $\boldsymbol{k}$-space, to the $\mathbb{S}^2$ target space. The homotopy mapping $\mathbb{S}^2 \mapsto \mathbb{S}^2$ is described by the 1st Chern number $C_1 \in \mathbb{Z}$. The Hall conductance is proportional to C_1 and thus becomes quantised in units of e^2/h.

[6] For the lattice version, the $\boldsymbol{k}$-space is already compact since it is defined on a Brillouin zone.

$$C_1 = \int \frac{dk}{4\pi} \sin[\vartheta(\boldsymbol{k})] \left[\frac{\partial\varphi(\boldsymbol{k})}{\partial k_y} \frac{\partial\vartheta(\boldsymbol{k})}{\partial k_x} - \frac{\partial\varphi(\boldsymbol{k})}{\partial k_x} \frac{\partial\vartheta(\boldsymbol{k})}{\partial k_y} \right] \equiv \frac{1}{4\pi} \iint d\varphi d\vartheta \sin\vartheta. \quad (5.42)$$

Due to the boundary condition at infinity, when we integrate over the entire $\boldsymbol{k}$-space the $\check{\boldsymbol{g}}$-vector will vary and cover an integer times the complete solid angle 4π, providing in the general case a total solid angle of $4\pi n$ with $n \in \mathbb{Z}$. In this manner, $C_1 \in \mathbb{Z}$ and leads to the quantisation of the Hall conductance. One may directly observe that for $m > 0$ we obtain $C_1 = 1$ which leads to a single chiral edge mode per edge of the material, as a result of the bulk-boundary correspondence. The topological invariant quantity is the integer index of the homotopy mapping $\mathbb{S}^2 \mapsto \mathbb{S}^2$ [4], which is shown in figure 5.3(b).

Finally, we note that while the Chern number is always an integer, G_{xy} is only quantised when $T = 0$ and the chemical potential lies within the bulk energy gap. This is since, in the most general case, the Hall conductance is given by the expression

$$G_{xy}(T, \mu) = -\frac{e^2}{h} \sum_a \int \frac{dk}{2\pi} \Omega_z^a(\boldsymbol{k}) f[E_a(\boldsymbol{k})], \quad (5.43)$$

where we considered the grand-canonical statistical ensemble. Notably, for a generic Hamiltonian of the form: $\hat{H}(\boldsymbol{k}) = E(\boldsymbol{k})\check{\boldsymbol{g}}(\boldsymbol{k}) \cdot \hat{\boldsymbol{\sigma}}$, we have $\Omega_z^-(\boldsymbol{k}) = -\Omega_z^+(\boldsymbol{k})$ and thus we obtain

$$G_{xy}(T, \mu) = -\frac{e^2}{h} \int \frac{dk}{2\pi} \Omega_z^-(\boldsymbol{k}) \big\{ f[-E(\boldsymbol{k})] - f[E(\boldsymbol{k})] \big\}. \quad (5.44)$$

Since the Fermi–Dirac distribution takes value in the interval [0, 1], the Hall conductance takes values in the interval $[-1, 1]$ in units of $G_{xy}(T = \mu = 0)$.

5.4 Chern insulators in higher dimensions

As discussed above, 2 + 1d Chern insulators are in reality defined in 3 + 1d Minkowski spacetime. They can be either found as ultrathin films or as multiple $d = 2$ Chern insulator systems stacked along the remaining spatial dimension. In this section we extend the strictly 2 + 1d Chern insulator to 4 + 1d. While such an extension does not appear very useful at first sight, this is certainly not the case. First of all, a number of the spatial dimensions considered here, do not need to be the physical ones. Instead, they can be synthetic [25], emerging from the periodic dependence of the Hamiltonian on a respective number of parameters. In analogy to sections 4.3 and section 5.3.2, a 4 + 1d Chern insulator can describe a Thouless pump defined for a 3 + 1d model with chiral symmetry, similar to the 2 + 1d Chern insulator corresponding to a Thouless pump of a topological 1 + 1d system (see chapter 4). Furthermore, studying 4 + 1d Chern insulators can provide key information for the properties of topological insulators in 3 + 1d with chiral symmetry. A similar connection was found for 2 + 1d Chern insulators via linking the Chern and winding numbers involved in equation (4.40).

5.4.1 Chern-insulator model in 4 + 1d

We start from the $\hat{\Gamma}$ matrices (see also section 1.2.1):

$$\hat{\Gamma}_1 = \hat{\tau}_x \hat{\sigma}_x, \quad \hat{\Gamma}_2 = \hat{\tau}_x \hat{\sigma}_y, \quad \hat{\Gamma}_3 = \hat{\tau}_x \hat{\sigma}_z, \quad \hat{\Gamma}_4 = \hat{\tau}_z \quad \text{and} \quad \hat{\Gamma}_5 = \hat{\tau}_y. \tag{5.45}$$

With the use of the respective five-dimensional vector operator $\hat{\mathbf{\Gamma}}$, we write the prototypical continuum model Hamiltonian for the 4 + 1d Chern insulator:

$$\hat{H}(\boldsymbol{k}) = \boldsymbol{g}(\boldsymbol{k}) \cdot \hat{\mathbf{\Gamma}}, \tag{5.46}$$

where, in this case, the $\boldsymbol{g}$-vector is defined as

$$\begin{aligned} &g_1(\boldsymbol{k}) = \alpha\hbar k_x, \quad g_2(\boldsymbol{k}) = \alpha\hbar k_y, \quad g_3(\boldsymbol{k}) = \alpha\hbar k_z, \\ &g_4(\boldsymbol{k}) = m - \beta(\hbar \boldsymbol{k})^2 \ \text{and} \ g_5(\boldsymbol{k}) = \alpha\hbar k_w. \end{aligned} \tag{5.47}$$

The wavevector is extended to four spatial dimensions, i.e. $\boldsymbol{k} = (k_x, k_y, k_z, k_w)$. By virtue of the special form of the Hamiltonian, and in analogy to the 1st Chern number in 2 + 1d, we introduce the 2nd Chern number [4, 26, 27]:

$$C_2 = \frac{3}{8\pi^2} \int d\boldsymbol{k}\ \varepsilon_{mnlrs} \check{g}^m(\boldsymbol{k}) \frac{\partial \check{g}^n(\boldsymbol{k})}{\partial k_x} \frac{\partial \check{g}^l(\boldsymbol{k})}{\partial k_y} \frac{\partial \check{g}^r(\boldsymbol{k})}{\partial k_z} \frac{\partial \check{g}^s(\boldsymbol{k})}{\partial k_w}. \tag{5.48}$$

We took into account that the solid angle in d spatial dimensions, subtended by $\mathbb{S}^{d-1}$, is defined as $\Omega_d = 2\pi^{d/2}/\Gamma(d/2)$ with $\Gamma(z)$ denoting the so-called 'gamma' function. Therefore, $\Omega_5 = 8\pi^2/3$. The 2nd Chern number effects the mapping $\mathbb{S}^4 \mapsto \mathbb{S}^4$ and, as a result, $C_2 \in \mathbb{Z}$. The same holds for the mapping $\mathbb{T}^4 \mapsto \mathbb{S}^4$, relevant for the lattice versions of this model.

5.4.2 Second Chern number and non-Abelian Berry gauge potentials

Similar to section 4.1.3 and also in the paragraph above, we relied on intuition and the suggestive form of the Hamiltonian to introduce a topological invariant. However, a direct algebraic construction of the topological invariant can be found through the expression of C_2 in terms of the Berry curvature. In the case of the 2 + 1d Chern insulator, C_1 was expressed as the flux of the Berry curvature arising from the U(1) Berry vector potential of the occupied band. A similar analysis can be performed for the case of the 2nd Chern number, with the difference that in the minimal implementation of a 4 + 1d Chern insulator, there exist two occupied bands. This implies that we can define a Berry vector potential which is a matrix in band space:

$$A_{ab}(\boldsymbol{k}) := \mathrm{i}\langle \phi_a(\boldsymbol{k}) | \partial_{\boldsymbol{k}} \phi_b(\boldsymbol{k}) \rangle. \tag{5.49}$$

Note that for two occupied bands, the above potential can be decomposed into its U(1) and SU(2) parts: $\hat{\boldsymbol{A}}(\boldsymbol{k}) = \boldsymbol{A}(\boldsymbol{k})\hat{1}_\zeta/2 + \overline{\boldsymbol{a}}(\boldsymbol{k}) \cdot \hat{\boldsymbol{\zeta}}/2$, with the unit ($\hat{1}_\zeta$) and Pauli ($\hat{\boldsymbol{\zeta}}$) matrices defined in the space spanned by the occupied bands.

In the general case, the corresponding non-Abelian field strength tensor is defined as

$$\hat{F}_{ij}(\boldsymbol{k}) = \partial_{k_i}\hat{A}_{k_j}(\boldsymbol{k}) - \partial_{k_j}\hat{A}_{k_i}(\boldsymbol{k}) - \mathrm{i}[\hat{A}_{k_i}(\boldsymbol{k}), \hat{A}_{k_j}(\boldsymbol{k})]. \tag{5.50}$$

For a single occupied band, the strength tensor yields the U(1) Berry curvature $\Omega^a_{k_i,k_j}(\boldsymbol{k}) = \partial_{k_i}A^a_{k_j}(\boldsymbol{k}) - \partial_{k_j}A^a_{k_i}(\boldsymbol{k})$. The 2nd Chern number can be expressed in terms of the non-Abelian field strength tensor as follows [27]

$$C_2 = \int \frac{d\boldsymbol{k}}{32\pi^2}\, \varepsilon_{mnlr}\, \mathrm{Tr}[\hat{F}_{mn}(\boldsymbol{k})\hat{F}_{lr}(\boldsymbol{k})] := \int d\boldsymbol{k}\; C_2[\hat{F}(\boldsymbol{k})], \tag{5.51}$$

with $C_2[\hat{F}(\boldsymbol{k})]$ defining the so-called 2nd Chern character [26].

5.4.3 4 + 1d Chern–Simons action and four-dimensional quantum Hall effect

In complete analogy to the 2 + 1d case discussed in section 5.1.3, we find that the action of the U(1) electromagnetic field coupled to a 4 + 1d Chern insulator contains the respective Chern–Simons term[7] that reads [26, 27]:

$$S^{(4+1)}_{\mathrm{CS}} = \frac{C_2}{96\pi^2}\int d^5x\; \varepsilon_{\mu\nu\lambda\rho\sigma}A^{\mu}F^{\nu\lambda}F^{\rho\sigma}. \tag{5.52}$$

The above equation is in appropriate units, similar to section 5.1.3, where θ was expressed in units of $e^2/\hbar$, thus, allowing the association of θ with $-C_1$. Here, the bulk Chern–Simons five-current assumes the form:

$$J^{\mu}_{c,\mathrm{CS}}(\boldsymbol{r}, t) = -\frac{\delta S^{(4+1)}_{\mathrm{CS}}}{\delta A_{\mu}} = -\frac{C_2}{32\pi^2}\varepsilon^{\mu\nu\lambda\rho\sigma}F_{\nu\lambda}F_{\rho\sigma}. \tag{5.53}$$

In fact, the current involves the $\boldsymbol{E}\cdot\boldsymbol{B}$ term, which gives rise to the $d = 4$ quantum Hall effect, with a Hall response generated for both charge and vector current. For more details see [27].

5.4.4 Generalisation to arbitrary dimensions

The above results can be generalised to $d = 2n$ ($n \in \mathbb{N}$) spatial dimensions, by defining the respective Chern number C_n. Following [27] we have:

$$C_n = \frac{1}{n!2^n(2\pi)^n}\int_{\mathbb{S}^{2n}} d^{2n}k\; \varepsilon_{mnlr\ldots stuv}\, \mathrm{Tr}[\hat{F}_{mn}\hat{F}_{lr}\cdots\hat{F}_{st}\hat{F}_{uv}], \tag{5.54}$$

with $\hat{F}$ denoting the non-Abelian field strength tensor restricted to the occupied bands subspace. Note that for lattice systems $\mathbb{S}^{2n} \mapsto \mathbb{T}^{2n}$. Furthermore, one introduces the corresponding Chern–Simons action for the electromagnetic field which reads [27]

[7] The present findings are directly extended to non-Abelian gauge fields. For $d = 2n$ spatial dimensions, the effective action of the gauge field is given by the Chern–Simons form Q_{2n+1}. For examples, see equation (5.59) and reference [4].

$$S_{\mathrm{CS}}^{(2n+1)} = \frac{(-1)^n C_n}{(n+1)!(2\pi)^n} \int d^{2n+1}x\ \varepsilon^{\mu\nu\lambda\rho\sigma\cdots\zeta\xi} A_\mu \partial_\nu A_\lambda \partial_\rho A_\sigma \cdots \partial_\zeta A_\xi. \tag{5.55}$$

One observes that the Chern–Simons term can be rewritten as

$$S_{\mathrm{CS}} \propto \int d^{2n+1}x\ \varepsilon^{\mu\nu\lambda\rho\sigma\cdots\zeta\xi} A_\mu F_{\nu\lambda} F_{\rho\sigma} \cdots F_{\zeta\xi}. \tag{5.56}$$

As mentioned earlier, the Chern–Simons action is manifestly not gauge invariant for a system with boundaries due to the presence of A_μ. The latter implies that electronic boundary modes are required to appear in order to preserve gauge invariance, similar to sections 5.1.1 and 5.1.4.

5.5 Dimensional reduction: chiral anomaly

As pointed out in section 4.3, and reflected in equation (4.40), there exists a connection between the winding number in 1 + 1d and the 1st Chern number in 2 + 1d. As a reminder, we found that

$$\begin{aligned} C_1 &= \sum_a^{\mathrm{occ}} \int_{\mathbb{S}^2} \frac{d\boldsymbol{k}}{2\pi} \Omega_z^a(\boldsymbol{k}) = \sum_a^{\mathrm{occ}} \oint_{\partial\mathbb{S}^2} \frac{d\boldsymbol{k}}{2\pi} \cdot \boldsymbol{A}^a(\boldsymbol{k}) \equiv \sum_a^{\mathrm{occ}} \int_{\mathbb{S}^1} \frac{dk_x}{2\pi} A_{k_x}^a(k_x, k_y = 0) \\ &= w_1. \end{aligned} \tag{5.57}$$

In a similar fashion, one can connect the second Chern number C_2, with the winding number w_3 introduced in section 4.1.5 and calculated in section 4.5. We have

$$C_2 = \int_{\mathbb{S}^3} \frac{d\boldsymbol{k}}{8\pi^2} \varepsilon_{mnl} \operatorname{Tr}\left[\hat{A}_{k_m}(\boldsymbol{k}) \partial_{k_n} \hat{A}_{k_l}(\boldsymbol{k}) - \mathrm{i}\frac{2}{3} \hat{A}_{k_m}(\boldsymbol{k}) \hat{A}_{k_n}(\boldsymbol{k}) \hat{A}_{k_l}(\boldsymbol{k}) \right]\Bigg|_{k_w=0} = w_3. \tag{5.58}$$

This shows that we can connect the topological invariant of different models of two successive spacetime dimensionalities, as long as they have the proper symmetry connection; i.e. the low-dimensional model in $2n - 1$ spatial dimensions should preserve chiral symmetry and the higher-dimensional one in $2n$ spatial dimensions should violate it, thus, simultaneously breaking time-reversal symmetry. In general, the connection is established by introducing the so-called *Chern–Simons form*:

$$\int_{\mathbb{S}^{2n}} d\boldsymbol{k}\ C_n[\hat{F}(\boldsymbol{k})] = \oint_{\mathbb{S}^{2n-1}} d\boldsymbol{k}\ Q_{2n-1}(\hat{A}(\boldsymbol{k}), \hat{F}(\boldsymbol{k})), \tag{5.59}$$

that also determines the form of the Chern–Simons action. From equation (5.58) Q_3 reads:

$$Q_3 = \frac{1}{8\pi^2} \varepsilon_{mnl} \operatorname{Tr}\left[\hat{A}_{k_m}(\boldsymbol{k}) \partial_{k_n} \hat{A}_{k_l}(\boldsymbol{k}) - \mathrm{i}\frac{2}{3} \hat{A}_{k_m}(\boldsymbol{k}) \hat{A}_{k_n}(\boldsymbol{k}) \hat{A}_{k_l}(\boldsymbol{k}) \right]. \tag{5.60}$$

The connection of the two types of models is further reflected in the effective action of the electromagnetic field. For Chern insulators this is given by equation (5.55), with a structure following from the Chern–Simons form. The coefficient is proportional to the ensuing Chern number.

We now extend this dimensional reduction approach employed for the topological invariant, to the effective action of the electromagnetic field. We consider a 2 + 1d Chern insulator and apply the dimensional reduction approach (see [27]). Thus, we assume that $A_y(x, y, t) = A_y + \delta A_y(x, t)$ with A_y = constant, while V and A_x are independent of y. Equation (5.16) yields

$$S_{\rm CS}^{(2+1)} = -\frac{C_1}{4\pi}\int d^3x\, \varepsilon^{\mu\nu\lambda}A_\mu\partial_\nu A_\lambda \;\Rightarrow\; S_{\rm CA}^{(1+1)} = -\frac{C_1\big(A_y\big)}{4\pi}\int d^2x\, \delta A_2\varepsilon^{2\mu\nu}F_{\mu\nu} \tag{5.61}$$

with $C_1(A_y)$ in units of conductance and satisfying $\int dA_y C_1(A_y) = C_1$. Note that we have modified the label of the action in the lower-dimensional system, i.e. $S_{\rm CA}^{(1+1)}$, following from the initials of chiral anomaly.

One observes that the respective Lagrangian density $L_{\rm CA}^{(1+1)}(x, t)$ is proportional to $\delta A_y E_x$, and based on the results of section 3.4 one finds that $\theta(x, t) \propto \delta A_y$, and $P_x(x, t) = -C_1(A_y)\delta A_y/(2\pi)$. Nevertheless, the low-dimensional model can be obtained via reduction by setting $A_y = 0$ or $A_y = \pi$ for a lattice model. This leads to the two distinct values of $P_x(x, t)$ for a bulk system, which are 0 or 1/2, as discussed in section 4.3.3. A similar analysis for a Chern insulator in $d = 2n$ spatial dimensions leads to a chiral anomaly action for a chiral symmetric topological insulator in $d = 2n - 1$ spatial dimensions, for which the effective electromagnetic action contains a term [19, 26, 27]

$$S_{\rm CA}^{(2n)} = \int d^{2n-1}x\, P_{2n-1}(x)\varepsilon_{\mu\nu\lambda\rho\ldots\tau\nu\zeta\xi}F^{\mu\nu}F^{\lambda\rho}\ldots F^{\tau\nu}F^{\zeta\xi}, \tag{5.62}$$

where we introduced a generalised polarisation field $P_d(x)$, taking values 0 and 1/2 for a bulk system. Note that P_3 is the so-called magnetoelectric polarisation. For more details see [27] and section 7.2.4.

5.6 Hands-on: Chern–Simons action

In this section we provide guidance to obtain the Chern–Simons action for the electromagnetic potential coupled to a 2 + 1d Chern insulator. For this purpose, consider the following generic Hamiltonian $\hat{H}(\boldsymbol{k}) = \boldsymbol{g}(\boldsymbol{k}) \cdot \hat{\boldsymbol{\sigma}}$, studied extensively throughout this chapter.

Task 1: Employ the results of sections 2.2 and 2.6 to show that the Matsubara action reads:

$$S_A^{(2)} = \frac{1}{2}T\sum_{\omega_s}\int\frac{d\boldsymbol{q}}{(2\pi)^2}\, A^\mu(\boldsymbol{q}, {\rm i}\omega_s)\Pi_{\mu\nu}^{\rm cc}(\boldsymbol{q}, {\rm i}\omega_s)A^\nu(-\boldsymbol{q}, -{\rm i}\omega_s) \tag{5.63}$$

where we introduced the properly-extended polarisation tensor $\Pi_{\mu\nu}^{\rm cc}$:

$$\Pi^{\rm cc}_{\mu\nu}(\boldsymbol{q}, {\rm i}\omega_s) = T\sum_{k_n}\int \frac{d\boldsymbol{k}}{(2\pi)^2} \, {\rm Tr}\Big[\hat{G}^{(0)}(\boldsymbol{k}, {\rm i}k_n)\hat{\upsilon}_\mu(-\boldsymbol{q}, -{\rm i}\omega_s)\hat{G}^{(0)}(\boldsymbol{k}+\boldsymbol{q}, {\rm i}k_n + {\rm i}\omega_s)\hat{\upsilon}_\nu(\boldsymbol{q}, {\rm i}\omega_s)\Big], \tag{5.64}$$

and the so-called matrix *vertex* functions $\hat{\upsilon}_\mu(\boldsymbol{q}, {\rm i}\omega_s) = \hat{\upsilon}_\mu(-\boldsymbol{q}, -{\rm i}\omega_s)$, with

$$\hat{\upsilon}_0(\boldsymbol{q}, {\rm i}\omega_s) = -e\hat{1}_\sigma \text{ and } \hat{\upsilon}_i(\boldsymbol{q}, {\rm i}\omega_s) = \frac{e}{\hbar}\frac{\partial \boldsymbol{g}(\boldsymbol{k})}{\partial k^i}\cdot\hat{\boldsymbol{\sigma}}. \tag{5.65}$$

Task 2: Based on the analysis of sections 2.6 and 3.5, consider the hydrodynamic limit at zero temperature and perform an expansion of the Matsubara action by keeping terms only up to linear order in $(\boldsymbol{q}, {\rm i}\omega_s)$. To facilitate the calculation, one can express the matrix Green function in terms of band projectors as in section 3.5. To perform the traces, use the identities of the Pauli matrices:

$$\mathrm{Tr}[\hat{\sigma}_n\hat{\sigma}_m] = 2\delta_{n,m}, \tag{5.66}$$

$$\mathrm{Tr}[\hat{\sigma}_n\hat{\sigma}_m\hat{\sigma}_l] = 2{\rm i}\varepsilon_{nml} \tag{5.67}$$

$$\mathrm{Tr}[\hat{\sigma}_n\hat{\sigma}_m\hat{\sigma}_l\hat{\sigma}_s] = 2(\delta_{n,m}\delta_{l,s} - \delta_{n,l}\delta_{m,s} + \delta_{n,s}\delta_{m,l}). \tag{5.68}$$

Show that the polarisation tensor reads

$$\Pi^{\rm cc}_{\mu\nu}(q) = -\frac{e^2}{h}\int\frac{d\boldsymbol{k}}{4\pi}\,\check{\boldsymbol{g}}(\boldsymbol{k})\cdot\left[\frac{\partial\check{\boldsymbol{g}}(\boldsymbol{k})}{\partial k_x}\times\frac{\partial\check{\boldsymbol{g}}(\boldsymbol{k})}{\partial k_y}\right]{\rm i}\varepsilon_{\mu\lambda\nu}q^\lambda, \tag{5.69}$$

with the three-vector $q^\mu = (\omega, \boldsymbol{q})$. After Fourier transforming to coordinate space, show that

$$S_{\rm CS} = \frac{\theta}{4\pi}\int d^3x\varepsilon^{\mu\nu\lambda}A_\mu\partial_\nu A_\lambda \text{ with } \theta = -\frac{e^2}{\hbar}C_1. \tag{5.70}$$

5.7 Hands-on: Chern number for interacting systems

In this section we derive an alternative formula for the 1st Chern number, which can be employed in the presence of interactions. As we have remarked in section 2.5, interactions can be introduced via a non-zero self-energy term in the single-particle Green function of the Hamiltonian. Below we describe how to obtain the desired expression.

Task 1: Start from the expression of the static Hall conductance:

$$G_{xy} = G_{xy}(\boldsymbol{q}=\boldsymbol{0}, \omega\to 0) = \lim_{\omega\to 0}\frac{{\rm i}\Pi^{\rm cc}_{xy}\big(\boldsymbol{0}, {\rm i}\omega_s\to\hbar\omega\big)}{\omega} = {\rm i}\hbar\left.\frac{\Pi^{\rm cc}_{xy}\big(\boldsymbol{0}, {\rm i}\omega_s\big)}{{\rm i}\omega_s}\right|_{\omega_s=0} \tag{5.71}$$

Show that the 1st Chern number is given by the formula

$$C_1 = -T\sum_{k_n}\int\frac{d\boldsymbol{k}}{2\pi}\,\mathrm{Tr}\left[\hat{G}^{(0)}(\boldsymbol{k}, \mathrm{i}k_n)\frac{\partial\hat{H}^{(0)}(\boldsymbol{k})}{\partial k_x}\frac{\partial\hat{G}^{(0)}(\boldsymbol{k}, \mathrm{i}k_n)}{\partial k_n}\frac{\partial\hat{H}^{(0)}(\boldsymbol{k})}{\partial k_y}\right] \tag{5.72}$$

where the $^{(0)}$ denotes the Hamiltonian and respective Green function of the system in the absence of the external fields.

Task 2: We are interested in the zero-temperature limit, in which the fermionic Matsubara frequencies k_n, can be represented by a continuous frequency variable k_0, and thus the respective Matsubara sum becomes an integral over k_0. Under these conditions, show that the Chern number reads

$$C_1 = -\frac{1}{8\pi^2}\int d^3k\;\varepsilon^{0ij}\,\mathrm{Tr}\left\{\hat{G}^{(0)}(k)\frac{\partial\left[\hat{G}^{(0)}(k)\right]^{-1}}{\partial k^i}\frac{\partial\hat{G}^{(0)}(k)}{\partial k^0}\frac{\partial\left[\hat{G}^{(0)}(k)\right]^{-1}}{\partial k^j}\right\} \tag{5.73}$$

with $d^3k = d\boldsymbol{k}dk_0$ and $k = (\boldsymbol{k}, k_0)$. We additionally took into account that $[\hat{G}^{(0)}(\boldsymbol{k}, \mathrm{i}k_0)]^{-1} = \mathrm{i}k_0 - \hat{H}^{(0)}(\boldsymbol{k})$. By further employing the relation $\partial\hat{G} = -\hat{G}\partial\hat{G}^{-1}\hat{G}$ we can rewrite the first Chern number as [4]

$$C_1 = -\frac{1}{24\pi^2}\int d^3k\;\varepsilon^{\mu\nu\lambda}\,\mathrm{Tr}\left(\hat{G}\frac{\partial\hat{G}^{-1}}{\partial k^\mu}\hat{G}\frac{\partial\hat{G}^{-1}}{\partial k^\nu}\hat{G}\frac{\partial\hat{G}^{-1}}{\partial k^\lambda}\right), \tag{5.74}$$

where we suppressed the $^{(0)}$ index and the k argument for simplicity. While this expression was built up via a bottom-up approach relying on the absence of interactions, it can be extended to the interacting case. Here, the Berry magnetic monopoles are defined by the poles of the single-particle Green function, where k_0 augments the parameter space. Interestingly, one notes that the above formula is identical to the w_3 winding number given by equation (4.16), after making the correpondence $\{\hat{q}, k_x, k_y, k_z\} \mapsto \{\hat{G}, k_x, k_y, k_0\}$. The latter connection provides a straightforward route to generalise the expression of the Chern number for interacting systems in higher dimensions.

5.8 Hands-on: second Chern number

In this last Hands-on section of the chapter, we focus on the calculation of the 2nd Chern number for the prototypical model of section 5.4.1.

Task: Calculate the 2nd Chern number for the model above, via the expressions of equations (5.48), (5.51) and (5.58). To facilitate the calculations it is convenient to consider a flat-band deformation of the original Hamiltonian, see [27]. By virtue of the topological invariance of the Chern number against Hamiltonian deformations that do not close the bulk energy gap, one can consider that all the occupied (empty) bands have a $\boldsymbol{k}$-independent energy dispersion: $E_{\mathrm{occ}}(\boldsymbol{k}) = -\varepsilon$ ($E_{\mathrm{empty}}(\boldsymbol{k}) = \varepsilon$). Here ε is a positive constant satisfying $\varepsilon > |\mu|$, to guarantee that the system is an insulator.

References

[1] Semenoff G W 1984 *Phys. Rev. Lett.* **53** 2449

[2] Haldane F D M 1988 *Phys. Rev. Lett.* **61** 2015

[3] Yakovenko V M 1990 *Phys. Rev. Lett.* **65** 251

[4] Volovik G E 2003 *The Universe in a Helium Droplet* (Oxford: Clarendon Press)

[5] Fradkin E 2013 *Field Theories of Condensed Matter Physics* 2nd edn (Cambridge: Cambridge University Press)

[6] Yu R, Zhang W, Zhang H-J, Zhang S-C, Dai X and Fang Z 2010 *Science* **329** 61

[7] Chang C-Z *et al* 2013 *Science* **340** 167

[8] von Klitzing K, Dorda G and Pepper M 1980 *Phys. Rev. Lett.* **45** 494

[9] Prange R and Girvin S 1990 *The Quantum Hall effect* (New York: Springer)

[10] Laughlin R B 1981 *Phys. Rev.* B **23** 5632(R)

[11] Halperin B I 1982 *Phys. Rev.* B **25** 2185

[12] Thouless D J, Kohmoto M, Nightingale M P and den Nijs M 1982 *Phys. Rev. Lett.* **49** 405

[13] Niu Q, Thouless D J and Wu Y-S 1985 *Phys. Rev.* B **31** 3372

[14] Abrahams E, Anderson P W, Licciardello D C and Ramakrishnan T V 1979 *Phys. Rev. Lett.* **42** 673

[15] Pruisken A M M 1987 *Field theory, Scaling and the Localization Problem in Quantum Hall Effect* ed R E Prange and S M Girvin (New York: Springer)

[16] Altland A and Simons B D 2010 *Condensed Matter Field Theory* 2nd edn (Cambridge: Cambridge University Press)

[17] Niemi A J and Semenoff G W 1983 *Phys. Rev. Lett.* **51** 2077

[18] Redlich A N 1984 *Phys. Rev. Lett.* **52** 18

[19] Weinberg S 1996 *The Quantum Theory of Fields* (Cambridge: Cambridge University Press)

[20] Froehlich J and Kerler T 1991 *Nucl. Phys.* B **354** 369

[21] Deser S, Jackiw R and Templeton S 1982 *Phys. Rev. Lett.* **48** 975

[22] Wen X-G 1991 *Phys. Rev.* B **43** 11025

[23] Kotetes P and Varelogiannis G 2008 *Phys. Rev.* B **78** 220509(R)

[24] Skyrme T H R 1962 *Nucl. Phys.* **31** 556

[25] Celi A, Massignan P, Ruseckas J, Goldman N, Spielman I B, Juzeliūnas G and Lewenstein M 2014 *Phys. Rev. Lett.* **112** 043001

[26] Ryu S, Schnyder A P, Furusaki A and Ludwig A W W 2010 *New J. Phys.* **12** 065010

[27] Qi X-L, Hughes T L and Zhang S-C 2008 *Phys. Rev.* B **78** 195424

Chapter 6

Chern insulators—applications

In this chapter we shift gears and focus on more applied topics related to 2+1d Chern insulators. We discuss the dynamical Hall response of such systems, as well as how to probe the Hall conductivity via the polar Kerr effect. We further examine the response of a Chern insulator to an external out-of-plane magnetic field, and consider both high- and low-field limits. In the former case we study the formation of Landau levels and investigate how their hierarchy becomes modified across the quantum phase transition, separating the topologically trivial and non-trivial phases. In the low-field scenario we concentrate on the concept of orbital magnetisation and how to derive it via a Green-function approach. The next topic on the agenda regards thermoelectric and thermal transport properties, as well as the role of orbital magnetisation in defining a transport current. Finally, the Hands-on sections address magnetic-field-induced Chern systems, and the thermoelectric response of the archetypical Chern insulator model introduced by Haldane.

6.1 Dynamical anomalous Hall response and polar Kerr effect

The goal of the following paragraphs is to study aspects of the dynamical response of Chern insulators. In particular, we focus on the polar Kerr effect, which can be employed to detect the presence of a non-zero intrinsic Hall conductivity $\sigma_{xy}(\omega)$. We first retrieve $\sigma_{xy}(\omega)$, and later on, provide a basic introduction to the Kerr effect. We introduce the Kerr angle θ_K, which constitutes the respective quantity measured in such magneto-optical experiments. We conclude with linking the Kerr angle to the dynamical Hall conductivity.

6.1.1 Dynamical anomalous Hall conductivity

In this paragraph we obtain an expression for the dynamical Hall conductivity, which holds beyond the hydrodynamic limit of section 5.6. We consider that the

doi:10.1088/978-1-68174-517-6ch6

system has a thickness L_z along the z axis, thus $\sigma_{xy}(\omega) = G_{xy}(\omega)/L_z$, as found in section 5.1.1. We start from the definition for the Hall conductance

$$G_{xy}(\omega) := G_{xy}(\boldsymbol{q} = \boldsymbol{0}, \omega) = \frac{\mathrm{i}\Pi^{cc}_{xy}(\boldsymbol{0}, \mathrm{i}\omega_s \mapsto \hbar\omega + \mathrm{i}0^+)}{\omega} \tag{6.1}$$

with the polarisation tensor given by equation (2.44). To perform the trace appearing in Π^{cc}_{xy}, we introduce the unit operator built up from the basis states, $|\phi_a(\boldsymbol{k})\rangle$. These constitute eigenstates of the Hamiltonian $\hat{H}^{(0)}(\boldsymbol{k})$ describing the system in the absence of external perturbations, i.e. $\hat{H}^{(0)}(\boldsymbol{k})|\phi_a(\boldsymbol{k})\rangle = E_a(\boldsymbol{k})|\phi_a(\boldsymbol{k})\rangle$. We employ the relation holding for $a \neq b$

$$\left\langle \phi_a(\boldsymbol{k}) \middle| \nabla_{\boldsymbol{k}} \hat{H}^{(0)}(\boldsymbol{k}) \middle| \phi_b(\boldsymbol{k}) \right\rangle = [E_b(\boldsymbol{k}) - E_a(\boldsymbol{k})] \left\langle \phi_a(\boldsymbol{k}) \middle| \nabla_{\boldsymbol{k}} \phi_b(\boldsymbol{k}) \right\rangle. \tag{6.2}$$

Applying the Matsubara-summation result of equation (2.56) and relying on the antisymmetry of the Hall conductance under $x \leftrightarrow y$, yields:

$$G_{xy}(\omega) = -\frac{e^2}{\hbar} \sum_{a,b} \int \frac{d\boldsymbol{k}}{(2\pi)^2} \frac{[E_a(\boldsymbol{k}) - E_b(\boldsymbol{k})]^2 f[E_a(\boldsymbol{k})] \mathrm{i}\varepsilon_{zij}}{[E_a(\boldsymbol{k}) - E_b(\boldsymbol{k})]^2 - (\hbar\omega + \mathrm{i}0^+)^2} \times \left\langle \partial_{k_i} \phi_a(\boldsymbol{k}) \middle| \phi_b(\boldsymbol{k}) \right\rangle \left\langle \phi_b(\boldsymbol{k}) \middle| \partial_{k_j} \phi_a(\boldsymbol{k}) \right\rangle. \tag{6.3}$$

One observes that in the static limit $\hbar\omega + \mathrm{i}0^+ \mapsto 0$, the above coincides with the static Hall conductance expressed as a sum of the integral over the Berry curvature of the occupied bands for $T = 0$, cf equation (5.37). Nevertheless, we can also express $G_{xy}(\omega)$ in a more compact manner, i.e.,

$$G_{xy}(\omega) = -\frac{e^2}{\hbar} \sum_a \int \frac{d\boldsymbol{k}}{(2\pi)^2} \mathrm{i}\varepsilon_{zij} \left\langle \partial_{k_i} \phi_a(\boldsymbol{k}) \middle| \hat{L}^{(0)}_a(\boldsymbol{k}, \omega) \middle| \partial_{k_j} \phi_a(\boldsymbol{k}) \right\rangle f[E_a(\boldsymbol{k})]. \tag{6.4}$$

with the operator

$$\hat{L}^{(0)}_a(\boldsymbol{k}, \omega) := \frac{\left[\hat{H}^{(0)}(\boldsymbol{k}) - E_a(\boldsymbol{k})\right]^2}{\left[\hat{H}^{(0)}(\boldsymbol{k}) - E_a(\boldsymbol{k})\right]^2 - (\hbar\omega + \mathrm{i}0^+)^2}. \tag{6.5}$$

For $\omega = 0$ one recovers the result of equation (5.37).

6.1.2 Polar Kerr effect

The Kerr effect [1] constitutes a very common means of probing the dielectric properties of a system. While there exist several types of Kerr effects, here we focus on the case where an external magnetic field, oriented for instance along the z axis, and a circularly-polarised electromagnetic wave propagating along the same axis, are simultaneously applied to the system. The transmission and reflection coefficients of the circularly-polarised electromagnetic wave depend on the orientation of the magnetic field, leading to the so-called phenomenon of circular-polarisation birefringence. Nevertheless, for materials with intrinsic magnetisation one expects to observe birefringence even in the absence of the magnetic field. Thus, these systems exhibit the so-called polar Kerr effect [2–5]. This effect can become non-negligible in ferromagnets with strong SOC, or as we discuss here, in systems with non-zero σ_H, e.g. Chern insulators or metals. In fact, the polar Kerr effect can be used to diagnose the topologically non-trivial properties of the bulk, without relying on the edge transport mediated by the related boundary modes. Below we present the theory of the polar Kerr effect after introducing the required basic concepts.

6.1.3 Dielectric tensor and circular-polarisation birefringence

In this section we explore the dielectric properties of a Chern medium and focus on the propagation of circularly-polarised light through it. For this purpose we start from Maxwell's equations in matter [6]:

$$\nabla \cdot \boldsymbol{D}(\boldsymbol{r}, t) = \rho_{c,f}(\boldsymbol{r}, t), \tag{6.6}$$

$$\nabla \cdot \boldsymbol{B}(\boldsymbol{r}, t) = 0, \tag{6.7}$$

$$\nabla \times \boldsymbol{E}(\boldsymbol{r}, t) = -\partial_t \boldsymbol{B}(\boldsymbol{r}, t), \tag{6.8}$$

$$\nabla \times \boldsymbol{H}(\boldsymbol{r}, t) = \boldsymbol{J}_{c,f}(\boldsymbol{r}, t) + \partial_t \boldsymbol{D}(\boldsymbol{r}, t). \tag{6.9}$$

In the above, $\rho_{c,f}(\boldsymbol{r}, t)$ and $\boldsymbol{J}_{c,f}(\boldsymbol{r}, t)$ denote the free carriers' charge and current densities. Moreover, we have introduced the fields

$$\boldsymbol{D}(\boldsymbol{r}, t) = \varepsilon_0 \boldsymbol{E}(\boldsymbol{r}, t) + \boldsymbol{P}(\boldsymbol{r}, t) \quad \text{and} \quad \boldsymbol{H}(\boldsymbol{r}, t) = \frac{1}{\mu_0}\boldsymbol{B}(\boldsymbol{r}, t) - \boldsymbol{M}(\boldsymbol{r}, t), \tag{6.10}$$

with $\boldsymbol{P}$ denoting the electric polarisation, $\boldsymbol{D}$ the dielectric displacement, $\boldsymbol{M}$ the magnetisation and $\boldsymbol{H}$ the *actual* magnetic field. Note that for non-magnetic materials $\boldsymbol{M} = \boldsymbol{0}$, and, thus, the distinction between $\boldsymbol{B}$ and $\boldsymbol{H}$ is unimportant. Nonetheless, if $\boldsymbol{M}$ is non-negligible, as for instance in ferromagnets, such a distinction is crucial and the field $\boldsymbol{B}$ is termed *magnetic induction*.

In the polar Kerr effect one considers that the Chern medium is irradiated by a circularly-polarised electromagnetic wave propagating along the z axis. In this case only the in-plane components of the electric and magnetic fields are non-zero. Moreover, the bound charge $\rho_{c,b} = -\nabla \cdot \boldsymbol{P}$, and the magnetisation contribution $\nabla \times \boldsymbol{M}$ to the total bound current $\boldsymbol{J}_{c,b} = \nabla \times \boldsymbol{M} + \partial_t \boldsymbol{P}$, are zero by virtue of the translational invariance which is assumed for the infinite (bulk) quasi two-dimensional system. This symmetry remains unbroken even in the presence of the assumed electromagnetic field. The charge density and currents of free carriers also drop out, i.e., $\rho_{c,f} = 0$ and $\boldsymbol{J}_{c,f}(\boldsymbol{r}, t) = 0$. The current essentially stems from the time dependence of the polarisation entering the dielectric displacement, and we have $\boldsymbol{J}_c(\boldsymbol{r}, t) = \partial_t \boldsymbol{P}(\boldsymbol{r}, t)$.

Under these conditions, we transfer to Fourier space and re-express the polarisation-induced current using Ohm's law, i.e., $\boldsymbol{J}_c(\boldsymbol{q}, \omega) = \bar{\boldsymbol{\sigma}}(\boldsymbol{q}, \omega)\boldsymbol{E}(\boldsymbol{q}, \omega)$. The dielectric displacement reads $\boldsymbol{D}(\boldsymbol{q}, \omega) = \varepsilon_0\bar{\boldsymbol{\varepsilon}}(\boldsymbol{q}, \omega)\boldsymbol{E}(\boldsymbol{q}, \omega)$, with:

$$\bar{\boldsymbol{\varepsilon}}(\boldsymbol{q}, \omega) = \bar{\mathbf{1}} - \frac{\bar{\boldsymbol{\sigma}}(\boldsymbol{q}, \omega)}{\mathrm{i}\omega\varepsilon_0}. \tag{6.11}$$

The above leads to the equations of motion describing the propagation of the electromagnetic field

$$\begin{pmatrix} \left(\frac{cq}{\omega}\right)^2 - \varepsilon_{xx}(q, \omega) & -\varepsilon_{xy}(q, \omega) \\ -\varepsilon_{yx}(q, \omega) & \left(\frac{cq}{\omega}\right)^2 - \varepsilon_{yy}(q, \omega) \end{pmatrix} \begin{pmatrix} E_x(q, \omega) \\ E_y(q, \omega) \end{pmatrix} = \begin{pmatrix} 0 \\ 0 \end{pmatrix}, \tag{6.12}$$

with q and $cq = \omega$ denoting the wavenumber and frequency for propagation in vacuum, since $\boldsymbol{E}(\boldsymbol{r}, t) \propto e^{\mathrm{i}(qz-\omega t)}$. For the remainder we focus on materials with a sufficiently-high degree of rotational symmetry, in order to guarantee that $\varepsilon_{xx} = \varepsilon_{yy} \equiv \varepsilon$ and $\varepsilon_{xy} = -\varepsilon_{yx} \equiv \mathrm{i}\varepsilon_H$. For more details see section 2.9. The relative longitudinal and Hall dielectric constants read

$$\varepsilon(\omega) = 1 - \frac{\mathrm{Im}[\sigma(\omega)]}{\omega\varepsilon_0} + \mathrm{i}\frac{\mathrm{Re}[\sigma(\omega)]}{\omega\varepsilon_0} \quad \text{and} \quad \varepsilon_H(\omega) = \frac{\sigma_{xy}(\omega)}{\omega\varepsilon_0}, \tag{6.13}$$

where we additionally considered the dynamical limit $\boldsymbol{q} \to \mathbf{0}$. The homogeneous system of coupled equations in equation (6.12) yields the eigensolutions $E_\pm(q, \omega)$, with

$$\boldsymbol{E}(z, t) = E_+(q, \omega)e^{\mathrm{i}(q_+z-\omega t)}\frac{\check{x} - \mathrm{i}\check{y}}{\sqrt{2}} + E_-(q, \omega)e^{\mathrm{i}(q_-z-\omega t)}\frac{\check{x} + \mathrm{i}\check{y}}{\sqrt{2}} \tag{6.14}$$

where we additionally introduced

$$\varepsilon_\pm(\omega) = \varepsilon(\omega) \pm \varepsilon_H(\omega) \quad \text{and} \quad cq_\pm = \omega\sqrt{\varepsilon_\pm(\omega)}. \tag{6.15}$$

The index + (−) defines the right (left) circularly-polarised field $E_\pm$, transforming according to the $\ell = 1$ irreducible representation of the group of rotations generated by the angular momentum operator $\hat{L}_z$, with $m_\ell = +1$ ($m_\ell = -1$).[1] Given the above, we additionally introduce the complex refractive indices $N_\pm(\omega) = \sqrt{\varepsilon_\pm(\omega)} := n_\pm + \mathrm{i}\kappa_\pm$, with $n_\pm$ and $\kappa_\pm$ defining their real and imaginary parts.

6.1.4 Kerr-angle formula

To study the polar Kerr effect it is necessary to infer the reflection and transmission coefficients for a circularly-polarised electromagnetic wave incident on the Chern

[1] A similar classification holds in the case of discrete rotations about the z axis (see also chapter 1).

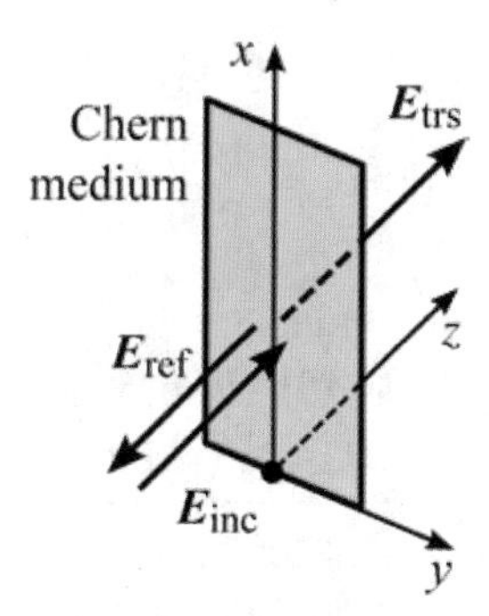

Figure 6.1. Polar Kerr effect setup. A circularly-polarised electromagnetic wave is incident on a Chern medium with a non-zero anomalous Hall conductivity. The incident beam ($\boldsymbol{E}_{\text{inc}}$) leads to a reflected ($\boldsymbol{E}_{\text{ref}}$) and a transmitted wave ($\boldsymbol{E}_{\text{trs}}$). The reflection and transmission coefficients depend on the circular polarisation of the electromagnetic wave due to the TRS breaking in the Chern system. The arising difference is experimentally quantified by measuring the so-called Kerr angle θ_K.

system (see figure 6.1). The electric field of the incident wave, propagating here along the z axis, reads

$$\boldsymbol{E}_{\text{inc}}(z,\,t) = E_{\text{inc},+}(q,\,\omega)e^{\mathrm{i}(qz-\omega t)}\frac{\check{x} - \mathrm{i}\check{y}}{\sqrt{2}} + E_{\text{inc},-}(q,\,\omega)e^{\mathrm{i}(qz-\omega t)}\frac{\check{x} + \mathrm{i}\check{y}}{\sqrt{2}}. \tag{6.16}$$

The incidence onto the Chern system leads to the following reflected and transmitted beams

$$\boldsymbol{E}_{\text{ref}}(z,\,t) = E_{\text{ref},+}(q,\,\omega)e^{-\mathrm{i}(qz+\omega t)}\frac{\check{x} - \mathrm{i}\check{y}}{\sqrt{2}} + E_{\text{ref},-}(q,\,\omega)e^{-\mathrm{i}(qz+\omega t)}\frac{\check{x} + \mathrm{i}\check{y}}{\sqrt{2}}, \tag{6.17}$$

$$\boldsymbol{E}_{\text{trs}}(z,\,t) = E_{\text{trs},+}(q,\,\omega)e^{\mathrm{i}(q_+z-\omega t)}\frac{\check{x} - \mathrm{i}\check{y}}{\sqrt{2}} + E_{\text{trs},-}(q,\,\omega)e^{\mathrm{i}(q_-z-\omega t)}\frac{\check{x} + \mathrm{i}\check{y}}{\sqrt{2}}. \tag{6.18}$$

The corresponding magnetic fields are obtained via the Maxwell equation $\nabla \times \boldsymbol{E} = -\partial_t \boldsymbol{B}$. We introduce the reflection and transmission coefficients $E_{\text{ref},\pm} = r_\pm E_{\text{inc},\pm}$ and $E_{\text{trs},\pm} = t_\pm E_{\text{inc},\pm}$, respectively. By considering the continuity of the electric and magnetic fields at the interface $z = 0$, one finds:

$$E_{\text{inc},\pm} + E_{\text{ref},\pm} = E_{\text{trs},\pm} \quad \text{and} \quad B_{\text{inc},\pm} - B_{\text{ref},\pm} = B_{\text{trs},\pm} \tag{6.19}$$

and we obtain

$$1 + r_\pm = t_\pm \quad \text{and} \quad 1 - r_\pm = t_\pm N_\pm(\omega). \tag{6.20}$$

This coupled system of equations leads to

$$r_\pm(\omega) = \frac{1 - N_\pm(\omega)}{1 + N_\pm(\omega)} \quad \text{and} \quad t_\pm(\omega) = \frac{2}{1 + N_\pm(\omega)}. \tag{6.21}$$

The Kerr angle is given by the expression[2] [1]:

$$\theta_K = \frac{\theta_- - \theta_+}{2} \quad \text{with} \quad \tan\theta_\pm = \frac{\mathrm{Im}[r_\pm(\omega)]}{\mathrm{Re}[r_\pm(\omega)]} \equiv \frac{-2\kappa_\pm}{1 - n_\pm^2 - \kappa_\pm^2}. \tag{6.22}$$

If the differences $\Delta n = n_+ - n_-$ and $\Delta\kappa = \kappa_+ - \kappa_-$, are small, we can perform an expansion about $n_\pm = n$ and $\kappa_\pm = \kappa$, with $n + i\kappa = \sqrt{\varepsilon}$. One finds [1, 3]:

$$\theta_K \approx \frac{(1 - n^2 + \kappa^2)\Delta\kappa + 2n\kappa\Delta n}{(1 - n^2 + \kappa^2)^2 + (2n\kappa)^2}, \tag{6.23}$$

with the differences given by

$$\Delta n \approx \frac{n\mathrm{Re}(\sigma_{xy}) + \kappa\mathrm{Im}(\sigma_{xy})}{(n^2 + \kappa^2)\omega\varepsilon_0} \quad \text{and} \quad \Delta\kappa \approx \frac{n\mathrm{Im}(\sigma_{xy}) - \kappa\mathrm{Re}(\sigma_{xy})}{(n^2 + \kappa^2)\omega\varepsilon_0}. \tag{6.24}$$

6.1.5 Polar Kerr effect in a 2 + 1d Chern insulator

We are now in a position to study the polar Kerr effect appearing in a 2 + 1d Chern insulator, such as the one studied in section 5.3.1. One finds two bands labelled by the index ±, for which: $\Omega_z^-(\boldsymbol{k}) = -\Omega_z^+(\boldsymbol{k})$, $E_\pm(\boldsymbol{k}) = \pm E(\boldsymbol{k})$ and $E(\boldsymbol{k}) > 0$. For positive frequencies equation (6.4) yields the Hall conductance

$$\mathrm{Re}\big[G_{xy}(\omega)\big] = -\frac{e^2}{h}\mathcal{P}\int \frac{d\boldsymbol{k}}{2\pi}\frac{[2E(\boldsymbol{k})]^2}{[2E(\boldsymbol{k})]^2 - (\hbar\omega)^2}\Omega_z^-(\boldsymbol{k})\big\{f[-E(\boldsymbol{k})] - f[E(\boldsymbol{k})]\big\}, \tag{6.25}$$

$$\mathrm{Im}\big[G_{xy}(\omega)\big] = -\frac{e^2}{h}\frac{f(-\hbar\omega/2) - f(\hbar\omega/2)}{2}\int d\boldsymbol{k}\, E(\boldsymbol{k})\Omega_z^-(\boldsymbol{k})\delta[\hbar\omega - 2E(\boldsymbol{k})] \tag{6.26}$$

where we made use of the Dirac identity discussed in section 2.7. Note that the above expression simplifies further if $\mu = 0$, since then, $f(-\varepsilon) = 1 - f(\varepsilon)$.

Given the above, one can employ equation (6.23) and calculate the Kerr angle. In an experiment, the energy of the electromagnetic waves $\hbar\omega$, can be of the order of few hundreds meV. For instance, for a Sagnac inteferometer [2] $\hbar\omega = 0.8$eV. In this regime, the dielectric properties of an electronic system are mainly dictated by the plasmonic contribution. The plasmons [8] constitute collective excitations of the electronic system appearing due to the Coulomb interaction. They appear above a threshold frequency ω_p, and when they govern the system's dielectric properties we have $\kappa \approx 0$ and

$$\varepsilon(\omega) \approx 1 - \left(\frac{\omega_p}{\omega}\right)^2 \quad \text{and} \quad 1 - n^2 \approx \left(\frac{\omega_p}{\omega}\right)^2. \tag{6.27}$$

By assuming that $\omega \gg \omega_p$ and thus $n \simeq 1$, we find the approximate expression for the Kerr angle

[2] Note, that, one can define a similar angle for the transmitted fields, related to the so-called Faraday effect [1], which can be also used to detect a non-zero anomalous Hall conductance [7].

$$\theta_K \approx \frac{1}{n(1-n^2)}\frac{\mathrm{Im}\left[\sigma_{xy}(\omega)\right]}{\omega\varepsilon_0} \approx -\frac{\alpha_s}{qL_z}\left(\frac{\omega}{\omega_p}\right)^2 \int dk\ E(\boldsymbol{k})\Omega_z^-(\boldsymbol{k})\delta[\hbar\omega - 2E(\boldsymbol{k})], \quad (6.28)$$

with α_s denoting the fine-structure constant $\alpha_s = e^2/(4\pi\varepsilon_0\hbar c)$, where we also set the chemical potential equal to zero for the Chern medium.

6.2 Chern insulators in an external magnetic field

Another important aspect of 2 + 1d Chern insulators, stemming from their non-zero anomalous Hall conductance, is their coupling to an out-of-plane magnetic field B_z. The Chern insulator response to the latter shows a diversity of interesting phenomena for a broad range of field strength. Nonetheless, here we investigate the extreme cases of low- and high-field. The response to a weak external field can be employed to infer the intrinsic orbital magnetisation present in a Chern system, which, as we show later, is an important quantity when discussing thermoelectric and thermal transport phenomena. On the other hand, in the case of a so-called quantising magnetic field the system forms Landau levels. As we show below, the Landau-level energy spectrum encodes information regarding the topological properties of the system.

6.2.1 High-field limit and the formation of Landau levels

In this paragraph we first discuss the case in which a strong out-of-plane magnetic field leads to the formation of Landau levels. We here consider the Chern insulator model introduced earlier

$$\hat{H}(\hat{\boldsymbol{p}}) = \boldsymbol{g}(\hat{\boldsymbol{p}})\cdot\hat{\boldsymbol{\sigma}} \quad \text{with} \quad \boldsymbol{g}(\hat{\boldsymbol{p}}) = \left(-\alpha\hat{p}_y,\, \alpha\hat{p}_x,\, m - \beta\hat{\boldsymbol{p}}^2\right). \quad (6.29)$$

We adopt the Landau gauge and consider the vector potential $\boldsymbol{A}(\boldsymbol{r}) = (0, B_z x, 0)$ which generates a $\boldsymbol{B} = B_z\hat{\boldsymbol{z}}$ field. The latter is introduced to the Hamiltonian via the minimal-coupling recipe of section 2.2, ie., $\hat{\boldsymbol{p}} \to \hat{\boldsymbol{\pi}} = \hat{\boldsymbol{p}} + e\boldsymbol{A}(\boldsymbol{r})$. One directly finds that the Hamiltonian satisfies

$$\hat{H}^2(\hat{\boldsymbol{\pi}}) = \alpha^2\hat{\boldsymbol{\pi}}^2 + (m - \beta\hat{\boldsymbol{\pi}}^2)^2 + 2\beta e\hbar B_z[\hat{H} - (m - \beta\hat{\boldsymbol{\pi}}^2)\hat{\sigma}_z] + e\hbar B_z\alpha^2\hat{\sigma}_z. \quad (6.30)$$

Being one step before obtaining the Landau-level eigenenergies, we note that the eigenstates of the Hamiltonian can be labelled by the eigenstates of $\hat{\sigma}_z$ ($\sigma = \pm 1$) and the eigenstates of $\hat{\boldsymbol{\pi}}^2$. Let us examine the eigenvalues of the latter. Since translational invariance is preserved along the y axis, we Fourier transform $\hat{p}_y \to \hbar k_y$, and obtain

$$\alpha^2\hat{\boldsymbol{\pi}}^2 = \frac{\hat{p}_x^2}{1/\alpha^2} + (\alpha e B_z)^2\left(x + \frac{\hbar k_y}{eB_z}\right)^2, \quad (6.31)$$

which corresponds to the Hamiltonian of a quantum-mechanical oscillator of inverse mass $2\alpha^2$ and cyclotron frequency $\omega_c = 2\alpha^2 eB_z$. Its eigenvalues are $\hbar\omega_c(n + 1/2)$ with $n = 0, 1, 2, 3, \ldots$, and they are degenerate with respect to the various so-called guiding center coordinates $\hbar k_y/eB_z$ (see also section 5.1.2). Conclusively, we obtain

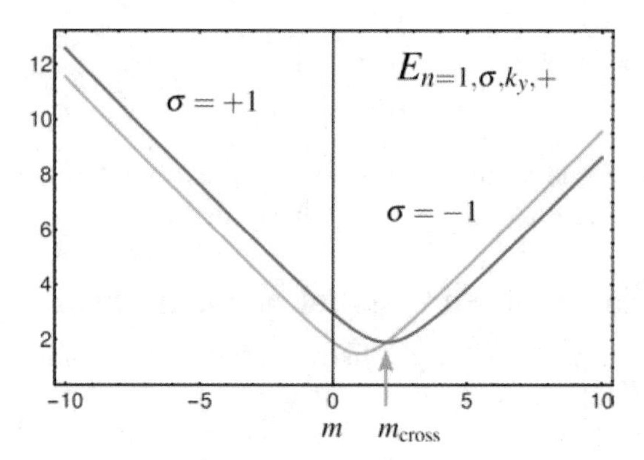

Figure 6.2. Energy spectrum of the $n = 1$ electron-like Landau levels for the Chern insulator model of equation (6.29), upon varying the parameter m. The two levels are distinguished by the quantum number $\sigma = \pm 1$. Here we have chosen $\alpha = \beta = \hbar\omega_c = 1$. Since $\beta > 0$, the system lies in the topologically non-trivial (trivial) phase for $m > 0$ ($m < 0$) for zero field. After equation (6.33), a level crossing is obtained for $m_{\text{cross}} = 2$, which is solely accessible in the non-trivial phase.

$$E_{n,\sigma,k_y,\pm} = \frac{\beta}{\alpha^2}\frac{\hbar\omega_c}{2} \pm \sqrt{\hbar\omega_c\left(n + \frac{1+\sigma}{2}\right) + \left[m - \frac{\beta}{\alpha^2}\hbar\omega_c\left(n + \frac{1+\sigma}{2}\right)\right]^2}. \tag{6.32}$$

The presence of the field leads to electron- and hole-like levels ($\pm$); for the two values of σ they exhibit a different dependence on the quantum number n. In fact, this difference in the Landau-level spectrum holds key information regarding the transition from the topologically trivial to the non-trivial phase, i.e., from $m < 0$ to $m > 0$. Specifically, we observe that the two families of electron-like[3] Landau levels $\sigma = \pm 1$, never cross in the trivial phase. In stark contrast, there can exist a single level crossing in the topologically non-trivial phase. This occurs for:

$$\beta m_{\text{cross}} = \frac{\alpha^2}{2} + \left(\frac{\beta}{\alpha}\right)^2 \hbar\omega_c\left(n + \frac{1}{2}\right). \tag{6.33}$$

As we observe, such a crossing can only exist for $\beta m > 0$, i.e. solely in the non-trivial Chern insulator phase. We show the behaviour of the electron-like $n = 1$ Landau level energies and crossings for $\alpha = \beta = \hbar\omega_c = 1$ in figure 6.2. Remarkably, Landau level crossings similar to the ones discussed here have been experimentally found in topological insulators respecting TRS [9], and were theoretically understood as signatures of the bulk topological phase transition [9, 10].

6.2.2 Theory of orbital magnetisation—a Green-function method

We continue with the antipodal limit, in which a weak out-of-plane field B_z is applied to the Chern system. While for the continuum model above it was straightforward to obtain the energy spectrum in a closed form, this appears to be formidable for crystalline systems. The reason can be traced back to the behaviour of the Hamiltonian matrix

[3] Similarly for the hole-like ones.

elements describing the coupling to the external vector potential. To clarify this statement, we consider a free electron Hamiltonian in the presence of a crystalline potential $V(\boldsymbol{r})$, properly extended to incorporate the effects of an external magnetic field:

$$\left\{\frac{[\hat{\boldsymbol{p}} + e\boldsymbol{A}(\boldsymbol{r})]^2}{2m} + V(\boldsymbol{r})\right\}\phi(\boldsymbol{r}) = E\phi(\boldsymbol{r}). \tag{6.34}$$

We consider a homogeneous magnetic field and we pick the following gauge for the vector potential:

$$\boldsymbol{A}(\boldsymbol{r}) = \frac{1}{2}\boldsymbol{B} \times \boldsymbol{r}. \tag{6.35}$$

In the weak-coupling limit we may discard the diamagnetic term. On the other hand, the paramagnetic coupling reads $\hat{H}_{\boldsymbol{B}}^{\mathrm{p}} = -\hat{\boldsymbol{m}}_{\mathrm{orb}} \cdot \boldsymbol{B}$, where we introduced the orbital magnetic moment operator

$$\hat{\boldsymbol{m}}_{\mathrm{orb}} := -\frac{e}{2}\hat{\boldsymbol{r}} \times \hat{\boldsymbol{v}} \equiv -\frac{e}{2m}\hat{\boldsymbol{L}}. \tag{6.36}$$

The presence of the position operator in the above definition renders the expansion of $\hat{\boldsymbol{m}}_{\mathrm{orb}}$, in terms of the ensuing Bloch eigenstates, a challenging task. Thus, it is equally cumbersome to calculate the orbital magnetisation [11] of a system, which corresponds to the average orbital moment.

The above obstacle appears to be very relevant for Chern systems, which possess a non-zero intrinsic orbital magnetisation linked to the non-zero Berry curvature [12–15]. At zero temperature, the spontaneous or intrinsic magnetisation of the system can be understood through the already discussed Chern–Simons term in section 5.1.3. The definition of the magnetisation reads

$$\boldsymbol{M}_{\mathrm{orb}}(\boldsymbol{r}) = -\left.\frac{\delta\Omega^{\mathrm{orb}}}{\delta\boldsymbol{B}(\boldsymbol{r})}\right|_{\mu,T}, \tag{6.37}$$

with the thermodynamic potential of the grand-canonical ensemble Ω^{orb}, obtained by considering solely the orbital coupling to the external field. We find that the orbital magnetisation generated by the Chern–Simons term of equation (5.16), is oriented out of the plane, i.e., $\boldsymbol{M}_{\mathrm{orb}} = M_z\hat{\boldsymbol{z}}$. Given the definition above, we find $M_z(\boldsymbol{r}) = -\theta V(\boldsymbol{r})/(4\pi)$. Therefore, a non-zero scalar potential induces a non-zero magnetisation when $\theta \neq 0$, i.e., when $\sigma_H \neq 0$.

While such systems have been so far extensively studied via the semiclassical formalism, see [15], we here attempt to shed light on these phenomena via a Green-function approach. This method can simplify the investigation of dynamical transport effects. Note that the authors of reference [16] followed a similar approach within a quasiclassical Green-function framework.

To this end, we continue with the calculation of the Chern system's Green function in the presence of an out-of-plane magnetic field B_z. The latter is generated by a vector potential[4] $\boldsymbol{A}(\boldsymbol{r}) = \boldsymbol{A}(\boldsymbol{q})e^{i\boldsymbol{q}\cdot\boldsymbol{r}}$. Since the orbital coupling to the electrons is

[4] Here, we consider a single $\boldsymbol{q}$ from the outset, which rescales the Fourier transforms defined in section 2.2.

non-local, it is convenient to consider here that the magnetic field is given by the zero wavevector ($\boldsymbol{q} \to \boldsymbol{0}$) limit of $\boldsymbol{A}(\boldsymbol{q})$, as done also in [14]. Thus, the field and potential are related via $B_z = \mathrm{i} \lim_{\boldsymbol{q}\to\boldsymbol{0}}[q_x A_y(\boldsymbol{q}) - q_y A_x(\boldsymbol{q})]$.

The corresponding Hamiltonian term describing the paramagnetic part of the electromagnetic coupling reads (see sections 2.2, 2.6, 3.5 and 5.6)

$$\hat{H}_A^{\mathrm{p}} = \int d\boldsymbol{k}\, \hat{\psi}^\dagger(\boldsymbol{k}+\boldsymbol{q}) \frac{e}{\hbar} \frac{\partial \hat{H}^{(0)}(\boldsymbol{k})}{\partial \boldsymbol{k}} \cdot \boldsymbol{A}(\boldsymbol{q}) \hat{\psi}(\boldsymbol{k}), \tag{6.38}$$

where the $\hat{H}^{(0)}(\boldsymbol{k})$ defines the single-particle Hamiltonian describing the Chern system in the absence of the magnetic field, with eigenvectors satisfying $\hat{H}^{(0)}(\boldsymbol{k})|\phi_a(\boldsymbol{k})\rangle = E_a(\boldsymbol{k})|\phi_a(\boldsymbol{k})\rangle$. We then retrieve the Matsubara matrix Green function through the Dyson equation

$$\begin{aligned}\hat{G}(\boldsymbol{k}, \boldsymbol{k}+\boldsymbol{q}, \mathrm{i}k_n) = {} & \hat{G}^{(0)}(\boldsymbol{k}, \mathrm{i}k_n)\delta(\boldsymbol{q}) \\ & + \frac{e}{\hbar} \int d\boldsymbol{q}'\, \hat{G}^{(0)}(\boldsymbol{k}, \mathrm{i}k_n) \frac{\partial \hat{H}^{(0)}(\boldsymbol{k})}{\partial \boldsymbol{k}} \cdot \boldsymbol{A}(-\boldsymbol{q}')\hat{G}(\boldsymbol{k}+\boldsymbol{q}', \boldsymbol{k}+\boldsymbol{q}, \mathrm{i}k_n).\end{aligned} \tag{6.39}$$

At first order with respect to the vector potential, and with an eye to the $\boldsymbol{q} \to \boldsymbol{0}$ limit, we can replace the last Green function with:

$$\begin{aligned}\hat{G}(\boldsymbol{k}+\boldsymbol{q}', \boldsymbol{k}+\boldsymbol{q}, \mathrm{i}k_n) & \approx \hat{G}^{(0)}(\boldsymbol{k}+\boldsymbol{q}, \mathrm{i}k_n)\delta(\boldsymbol{q}-\boldsymbol{q}') \\ & \approx \left[\hat{G}^{(0)}(\boldsymbol{k}, \mathrm{i}k_n) + \boldsymbol{q} \cdot \frac{\partial \hat{G}^{(0)}(\boldsymbol{k}, \mathrm{i}k_n)}{\partial \boldsymbol{k}}\right]\delta(\boldsymbol{q}-\boldsymbol{q}').\end{aligned} \tag{6.40}$$

After also employing the relation $\partial\hat{G} = -\hat{G}(\partial\hat{G}^{-1})\hat{G}$, we obtain the self-energy at linear order with respect to the magnetic field

$$\hat{\Sigma}^{(1)}(\boldsymbol{k}, \boldsymbol{k}+\boldsymbol{q}, \mathrm{i}k_n) = \frac{e}{\hbar} q_i A_j(-\boldsymbol{q}) \frac{\partial \hat{H}^{(0)}(\boldsymbol{k})}{\partial k_j} \hat{G}^{(0)}(\boldsymbol{k}, \mathrm{i}k_n) \frac{\partial \hat{H}^{(0)}(\boldsymbol{k})}{\partial k_i}. \tag{6.41}$$

We assume that only one vector-potential component is non-zero, and set $A_x(q_y) = \mathrm{i}B_z/q_y$. By employing equation (6.2), we project the self-energy onto the eigenvector $|\phi_a(\boldsymbol{k})\rangle$, which yields:

$$\begin{aligned}\Sigma_a^{(1)}(\boldsymbol{k}, \mathrm{i}k_n) & = \frac{eB_z}{2\hbar\mathrm{i}} \sum_b \frac{[E_b(\boldsymbol{k}) - E_a(\boldsymbol{k})]^2}{\mathrm{i}k_n + \mu - E_b(\boldsymbol{k})} \\ & \times \left[\langle \partial_{k_x}\phi_a(\boldsymbol{k})|\phi_b(\boldsymbol{k})\rangle \langle \phi_b(\boldsymbol{k})|\partial_{k_y}\phi_a(\boldsymbol{k})\rangle - x \leftrightarrow y\right] \\ & = \frac{eB_z}{2\hbar\mathrm{i}}\left[\langle \partial_{k_x}\phi_a(\boldsymbol{k})| \frac{[\hat{H}^{(0)}(\boldsymbol{k}) - E_a(\boldsymbol{k})]^2}{\mathrm{i}k_n + \mu - \hat{H}^{(0)}(\boldsymbol{k})} |\partial_{k_y}\phi_a(\boldsymbol{k})\rangle - x \leftrightarrow y\right]\end{aligned} \tag{6.42}$$

At this stage one can readily obtain the Matsubara Green fuction at first order in terms of the perturbation:

$$
\begin{aligned}
[G_a^{(1)}(\boldsymbol{k}, \mathrm{i}k_n)]^{-1} &= [G_a^{(0)}(\boldsymbol{k}, \mathrm{i}k_n)]^{-1} - \Sigma_a^{(1)}(\boldsymbol{k}, \mathrm{i}k_n) \\
&\approx \left[1 - \frac{e}{\hbar}\boldsymbol{\Omega}_a(\boldsymbol{k}) \cdot \boldsymbol{B}\right]\left[\mathrm{i}k_n + \mu - E_a(\boldsymbol{k})\right] \\
&+ \frac{eB_z}{2\hbar\mathrm{i}}\left[\left\langle \partial_{k_x}\phi_a(\boldsymbol{k}) \,\middle|\, \left[\hat{H}^{(0)}(\boldsymbol{k}) - \mathrm{i}k_n - \mu\right] \,\middle|\, \partial_{k_y}\phi_a(\boldsymbol{k}) \right\rangle - x \leftrightarrow y\right].
\end{aligned} \tag{6.43}
$$

One observes that for weak magnetic fields $\mathrm{i}k_n + \mu \approx E_a(\boldsymbol{k})$, i.e., we are energetically still in the vicinity of $E_a(\boldsymbol{k})$, corresponding to the eigenvector of interest. Thus, one finds the pole of the Green function

$$
\left[1 - \frac{e}{\hbar}\Omega_a(\boldsymbol{k}) \cdot \boldsymbol{B}\right]\left[\mathrm{i}k_n + \mu - E_a(\boldsymbol{k})\right] + \boldsymbol{m}_a(\boldsymbol{k}) \cdot \boldsymbol{B} \approx 0 \tag{6.44}
$$

where we introduced the orbital magnetic moment of the a-th Bloch eigenvector, defined as $\boldsymbol{m}_a(\boldsymbol{k}) = m_a(\boldsymbol{k})\check{z}$ [15]. Only its z component is non-zero, and is given by:

$$
m_a(\boldsymbol{k}) = \frac{e}{2\hbar\mathrm{i}}\varepsilon_{zij}\left\langle \partial_{k_i}\phi_a(\boldsymbol{k}) \middle| \left[\hat{H}^{(0)}(\boldsymbol{k}) - E_a(\boldsymbol{k})\right] \middle| \partial_{k_j}\phi_a(\boldsymbol{k}) \right\rangle. \tag{6.45}
$$

The above approximation yields the modified energy dispersions at linear order with respect to the field

$$
E_a^{\boldsymbol{B}}(\boldsymbol{k}) = E_a(\boldsymbol{k}) - \boldsymbol{m}_a \cdot \boldsymbol{B}. \tag{6.46}
$$

Notably, the presence of the self-energy additionally introduces an overall factor to the Green function, since along the lines of the Fermi-liquid theory [17] it can be approximately rewritten as

$$
[G_a^{(1)}(\boldsymbol{k}, \mathrm{i}k_n)]^{-1} \approx \frac{\mathrm{i}k_n + \mu - E_a^{\boldsymbol{B}}(\boldsymbol{k})}{Z_a(\boldsymbol{k})} \quad \text{with} \quad \frac{1}{Z_a(\boldsymbol{k})} = 1 - \left.\frac{\partial\Sigma_a^{(1)}(\boldsymbol{k}, \varepsilon)}{\partial\varepsilon}\right|_{\varepsilon+\mu=E_a^{\boldsymbol{B}}(\boldsymbol{k})}. \tag{6.47}
$$

Based on equation (6.43), we can readily identify the Z-factor [17], that reads:

$$
\frac{1}{Z_a(\boldsymbol{k})} \approx 1 - \frac{e}{\hbar}\boldsymbol{\Omega}_a(\boldsymbol{k}) \cdot \boldsymbol{B} \qquad \text{while} \qquad Z_a(\boldsymbol{k}) \approx 1 + \frac{e}{\hbar}\boldsymbol{\Omega}_a(\boldsymbol{k}) \cdot \boldsymbol{B}. \tag{6.48}
$$

The presence of the $Z_a(\boldsymbol{k})$ leads to a modification of the density of states [15] for a given state $|\phi_a(\boldsymbol{k})\rangle$, since it enters the expression for the spectral function (see section 2.5). Moreover, knowing the quasiparticle poles and the respective Z-factor of the Green function, readily yields the thermodynamic potential[5]

[5] One can, for instance, follow reference [18] and the derivation of the thermodynamic potential of non-interacting electrons.

$$\Omega = -T\sum_a \int dk\, Z_a(\boldsymbol{k})\ln\left(1 + e^{-\left[E_a^B(\boldsymbol{k})-\mu\right]/T}\right)$$
$$= -T\sum_a \int dk\left[1 + \frac{e}{\hbar}\boldsymbol{\Omega}_a(\boldsymbol{k})\cdot\boldsymbol{B}\right]\ln\left(1 + e^{-\left[E_a^B(\boldsymbol{k})-\mu\right]/T}\right). \tag{6.49}$$

Therefore, one obtains the zero-field orbital magnetisation[6]

$$\boldsymbol{M} = -\frac{1}{(2\pi)^2}\frac{\partial\Omega}{\partial\boldsymbol{B}}\bigg|_{\boldsymbol{B}=0}$$
$$= \sum_a \int\frac{d\boldsymbol{k}}{(2\pi)^2}\left\{\boldsymbol{m}_a(\boldsymbol{k})f[E_a(\boldsymbol{k})] + \frac{e}{\hbar}\boldsymbol{\Omega}_a(\boldsymbol{k})T\ln\left[1 + e^{-\left(E_a(\boldsymbol{k})-\mu\right)/T}\right]\right\}. \tag{6.50}$$

Note that at $T = 0$ the magnetisation reads

$$\boldsymbol{M} = \sum_a \int\frac{d\boldsymbol{k}}{(2\pi)^2}\left\{\boldsymbol{m}_a(\boldsymbol{k}) - \frac{e}{\hbar}\boldsymbol{\Omega}_a(\boldsymbol{k})[E_a(\boldsymbol{k}) - \mu]\right\}f[E_a(\boldsymbol{k})]. \tag{6.51}$$

6.3 Anomalous thermoelectric and thermal Hall transport

In the remainder of this chapter we explore aspects of bulk thermoelectric and thermal transport [15, 19–22] of Chern insulators and metals, with particular focus on their anomalous Hall response. We first obtain the expressions for α_{xy} and κ_{xy}, based on the linear response with respect to the paramagnetic coupling to the gravitational and electric fields, discussed in section 2.8. Notably, within our approach, we also account for the diathermal contributions discussed in section 2.8.2.

6.3.1 Thermoelectric conductivity tensor

We proceed with calculating the coefficient $T\alpha_{xy}$. By employing equations (2.44) and (2.78), we obtain the following energy-charge polarisation tensor[7] defined in the Matsubara-frequency domain:

$$\Pi_{xy}^{\mathrm{Ec}}(\mathrm{i}\omega_s) = -\frac{e}{\hbar^2}T\sum_{k_n}\int\frac{d\boldsymbol{k}}{(2\pi)^2 L_z}$$
$$\mathrm{Tr}\left[\hat{G}^{(0)}(\boldsymbol{k}, \mathrm{i}k_n)\hat{H}^{(0)}(\boldsymbol{k})\frac{\partial\hat{H}^{(0)}(\boldsymbol{k})}{\partial k_x}\hat{G}^{(0)}(\boldsymbol{k}, \mathrm{i}k_n + \mathrm{i}\omega_s)\frac{\partial\hat{H}^{(0)}(\boldsymbol{k})}{\partial k_y}\right]. \tag{6.52}$$

For purposes of notational convenience, we set $L_z = 1$ in the remainder. Since, here, we investigate the response to a constant electric field generated by a time-dependent

[6] Note that we have to divide by $(2\pi)^2$ ‘by hand’, due to the rescaling mentioned in footnote 4.

[7] The polarisation tensor is also related to the quantities Y_n, introduced in equation (2.98).

vector potential $\boldsymbol{E}(\omega) = \mathrm{i}\omega\boldsymbol{A}(\omega)$, it is convenient to antisymmetrise the polarisation tensor as follows

$$\Pi_{xy}^{\mathrm{Ec}}(\mathrm{i}\omega_s) \mapsto \frac{\Pi_{xy}^{\mathrm{Ec}}(\mathrm{i}\omega_s) - \Pi_{xy}^{\mathrm{Ec}}(-\mathrm{i}\omega_s)}{2}. \tag{6.53}$$

Given the above, and in analogy to the methods applied to obtain equation (6.4), we find the 'parathermal' contribution to the thermoelectric coefficient

$$\begin{aligned} T\alpha_{xy}^{\mathrm{P}}(\omega) = \frac{e}{\hbar}\sum_a \int \frac{d\boldsymbol{k}}{(2\pi)^2} \mathrm{i}\varepsilon_{zij} \\ \langle \partial_{k_i}\phi_a(\boldsymbol{k})| \left\{ E_a(\boldsymbol{k}) - \mu + \frac{\hat{H}^{(0)}(\boldsymbol{k}) - E_a(\boldsymbol{k})}{2} \right\} \hat{L}_a^{(0)}(\boldsymbol{k}, \omega)|\partial_{k_j}\phi_a(\boldsymbol{k})\rangle f[E_a(\boldsymbol{k})], \end{aligned} \tag{6.54}$$

with $\hat{L}_a^{(0)}(\boldsymbol{k}, \omega)$ introduced in equation (6.5). Note that the chemical potential appeared by subtracting the particle current in order to retrieve the heat current (see section 2.8.1). For $\hbar\omega = 0$, one finds that $\alpha_{xy}^{\mathrm{P}} := \alpha_{xy}^{\mathrm{P}}(0)$ reads

$$T\alpha_{xy}^{\mathrm{P}} = \sum_a \int \frac{d\boldsymbol{k}}{(2\pi)^2} \left\{ \frac{e}{\hbar}[E_a(\boldsymbol{k}) - \mu]\Omega_a(\boldsymbol{k}) - m_a(\boldsymbol{k}) \right\} f[E_a(\boldsymbol{k})]. \tag{6.55}$$

For the final expression of the coefficient, one is required to also add the 'diathermal' contribution to the heat current, as has been already announced in section 2.8.2. This is performed in section 6.3.3.

6.3.2 Thermal conductivity tensor

A similar procedure yields the off-diagonal thermal conductivity tensor element κ_{xy}. Here, one introduces the energy-energy polarisation tensor (see also footnote 7)

$$\begin{aligned} \Pi_{xy}^{\mathrm{EE}}(\mathrm{i}\omega_s) = \frac{1}{\hbar^2} T \sum_{k_n} \int \frac{d\boldsymbol{k}}{(2\pi)^2 L_z} \\ \mathrm{Tr}\left[\hat{G}^{(0)}(\boldsymbol{k}, \mathrm{i}k_n)\hat{H}^{(0)}(\boldsymbol{k}) \frac{\partial \hat{H}^{(0)}(\boldsymbol{k})}{\partial k_x} \hat{G}^{(0)}(\boldsymbol{k}, \mathrm{i}k_n + \mathrm{i}\omega_s)\hat{H}^{(0)}(\boldsymbol{k}) \frac{\partial \hat{H}^{(0)}(\boldsymbol{k})}{\partial k_y} \right]. \end{aligned} \tag{6.56}$$

We set $L_z = 1$ once again. To proceed, we employ equation (2.82) and the resulting equivalences: $-\nabla T \equiv T\nabla\Phi \equiv T\partial_t\boldsymbol{\Phi}$, which imply, that, the induced heat currents are proportional to the linear combinations $\bar{\boldsymbol{\kappa}}(\nabla T + T\nabla\Phi)$ and $\bar{\boldsymbol{\kappa}}(\nabla T + T\partial_t\boldsymbol{\Phi})$ respectively. Therefore, following the same procedure as in the previous paragraphs, one obtains the parathermal component:

$$\begin{aligned} T\kappa_{xy}^{\mathrm{P}}(\omega) = -\frac{1}{\hbar}\sum_a \int \frac{d\boldsymbol{k}}{(2\pi)^2} \mathrm{i}\varepsilon_{zij} \langle \partial_{k_i}\phi_a(\boldsymbol{k})|E_a(\boldsymbol{k})\hat{H}^{(0)}(\boldsymbol{k})\hat{L}_a^{(0)}(\boldsymbol{k}, \omega)|\partial_{k_j}\phi_a(\boldsymbol{k})\rangle f[E_a(\boldsymbol{k})] \\ + \frac{2\mu}{e}\left[T\alpha_{xy}^{\mathrm{P}}(\omega) - \frac{\mu}{e}\sigma_{xy}(\omega) \right] + \left(\frac{\mu}{e}\right)^2 \sigma_{xy}(\omega). \end{aligned} \tag{6.57}$$

The second line stems from obtaining the heat current using the energy current, as discussed in section 2.8.1. By putting all the contributions together, we find

$$T\kappa_{xy}^{\mathrm{p}}(\omega) = -\frac{1}{\hbar}\sum_a \int \frac{d\boldsymbol{k}}{(2\pi)^2}\, \mathrm{i}\varepsilon_{zij} \\ \langle \partial_{k_i}\phi_a(\boldsymbol{k})|[\hat{H}^{(0)}(\boldsymbol{k}) - \mu]\hat{L}_a^{(0)}(\boldsymbol{k}, \omega)|\partial_{k_j}\phi_a(\boldsymbol{k})\rangle[E_a(\boldsymbol{k}) - \mu]f[E_a(\boldsymbol{k})]. \tag{6.58}$$

For zero frequencies we have $\kappa_{xy}^{\mathrm{p}} := \kappa_{xy}^{\mathrm{p}}(0)$

$$eT\kappa_{xy}^{\mathrm{p}} = -\sum_a \int \frac{d\boldsymbol{k}}{(2\pi)^2}\left\{\frac{e}{\hbar}[E_a(\boldsymbol{k}) - \mu]\Omega_a(\boldsymbol{k}) - 2m_a(\boldsymbol{k})\right\}[E_a(\boldsymbol{k}) - \mu]f[E_a(\boldsymbol{k})]. \tag{6.59}$$

In analogy to the thermoelectric coefficient α_{xy}, we also need to include the diathermal contribution to infer the final expression of κ_{xy}. This task is performed below.

6.3.3 Diathermal contributions to the conductivities and transport current

The study of the conductivities in the above two paragraphs has been left incomplete. Here, we conclude the program of the thermal and thermoelectric transport, and obtain the final expressions for the respective coefficients in the static case. For this purpose we need to add the diathermal contributions to the already calculated parathermal currents. After equation (2.80), we have

$$J_{c,x}^{\mathrm{d}} = -\langle \hat{J}_{c,x}\hat{y}\rangle\left(-\frac{\partial\Phi}{\partial y}\right) \quad \text{and} \quad J_{E,x}^{\mathrm{d}} = -2\langle \hat{J}_{E,x}\hat{y}\rangle\left(-\frac{\partial\Phi}{\partial y}\right) - \langle \hat{J}_{c,x}\hat{y}\rangle E_y. \tag{6.60}$$

Let us first start with the electric charge current. In this case the expectation value resembles the expression of the orbital magnetic moment defined in equation (6.36). This becomes more transparent by antisymmetrising with $x \leftrightarrow y$. Therefore, one finds the final expression for the thermoelectric Hall conductivity

$$T\alpha_{xy}(0) = T\alpha_{xy}^{\mathrm{p}}(0) - \frac{e}{2}\left\langle \hat{x}\hat{v}_y - \hat{y}\hat{v}_x\right\rangle \equiv T\alpha_{xy}^{\mathrm{p}}(0) + M_z, \tag{6.61}$$

with M_z the orbital magnetisation given by equation (6.50). Thus, we conclude that the thermoelectric Hall conductance reads

$$T\alpha_{xy}(0) = \frac{e}{\hbar}\sum_a \int \frac{d\boldsymbol{k}}{(2\pi)^2}\left\{[E_a(\boldsymbol{k}) - \mu]\, f[E_a(\boldsymbol{k})] + T\ln\left[1 + e^{-(E_a(\boldsymbol{k})-\mu)/T}\right]\right\}\Omega_a(\boldsymbol{k}) \tag{6.62}$$

The above formula can be rewritten by introducing the electronic entropy $S_a(\boldsymbol{k})$, i.e.

$$\alpha_{xy}(0) = \frac{e}{\hbar}\sum_a \int \frac{d\boldsymbol{k}}{(2\pi)^2}\, S_a(\boldsymbol{k})\Omega_a(\boldsymbol{k}). \tag{6.63}$$

Having included the diathermal currents, we have managed to obtain the so-called *transport* thermoelectric current, i.e., the measurable current when we attach leads to the material. Adding the magnetisation has cancelled out the divergence-free bound current $\nabla \times \boldsymbol{M}$ (see sections 5.1.3 and 6.1.3).

In a similar fashion, the thermal conductance coefficient κ_{xy} is corrected by a term including the orbital magnetisation, and another one proportional to $\langle \hat{J}_{E,x}\hat{y} \rangle$, which plays a similar role to the orbital magnetic moment where now the electric charge is replaced by energy. Such a term can be retrieved following the same route for calculating the orbital magnetisation in section 6.2.2, but when considering the response to the gravitational field. This task goes beyond the scope of this section and the interested reader can turn to reference [21]. Following this reference, we present the final expression

$$
\begin{aligned}
eT\kappa_{xy} = -\sum_a \int \frac{d\boldsymbol{k}}{(2\pi)^2} \left\{ \frac{e}{\hbar}[E_a(\boldsymbol{k}) - \mu]\Omega_a(\boldsymbol{k}) - 2m_a(\boldsymbol{k}) \right\} \\
\times [E_a(\boldsymbol{k}) - \mu] f[E_a(\boldsymbol{k})] - 2\int_{-\infty}^{\mu} d\mu' M_z(\mu').
\end{aligned}
\tag{6.64}
$$

6.4 Hands-on: magnetic-field-induced Chern systems

The orbital coupling of a Chern material to a magnetic field can lead to additional interesting phenomena, such as, the induction of the topological phase itself [23, 24]. For this to happen, suitable interactions are required. For concreteness, let us discuss once again the prototypical model of equation (6.29). We consider that in the absence of the magnetic field $m = 0$. However, we assume the presence of an *effective* four-fermion interaction[8], which is repulsive ($g > 0$) in the channel generating a finite m, as the one below:

$$
V = \frac{g}{2} \frac{1}{(2\pi)^2} \int d\boldsymbol{k} \int d\boldsymbol{k}' \, [\hat{\psi}^\dagger(\boldsymbol{k})\hat{\sigma}_z \hat{\psi}(\boldsymbol{k})][\hat{\psi}^\dagger(\boldsymbol{k}')\hat{\sigma}_z \hat{\psi}(\boldsymbol{k}')]. \tag{6.65}
$$

Our goal is to demonstrate that, by applying an out-of-plane magnetic field B_z, the mass term m is induced in the presence of the interaction. This phenomenon is a variant of the so-called magnetic catalysis [25], in which, dynamical symmetry breaking takes place when Dirac electrons are subjected to a strong magnetic field that leads to the formation of Landau levels. In this case, an infinitesimally-weak interaction can lead to a non-zero mass. Here, we find ourselves on the diametrically opposite limit, where the field is weak, and the required interaction may be strong depending on the given electron density.

Task 1: The presence of the repulsive interaction implies that fluctuations of m about zero contribute to the free energy of the system. Perform the mean-field decoupling in the direct channel (see footnote 4 of chapter 3) by defining m as

[8] One should only view the particular interaction as an apparatus facilitating the mean-field theory and should not try to trace its microscopic origin. This, however, does not imply that such a type of interaction cannot appear in a material. It is simply out of the present scope to explore such a possibility.

$$m = g \int \frac{d\boldsymbol{k}}{(2\pi)^2} \langle \hat{\psi}^{\dagger}(\boldsymbol{k}) \hat{\sigma}_z \hat{\psi}(\boldsymbol{k}) \rangle. \tag{6.66}$$

Show that, at lowest order, the thermodynamic potential of the mean-field Hamiltonian contains the term[9]

$$\Omega(\tilde{m}) = \frac{g}{1 + g\chi_m} \frac{\tilde{m}^2}{2}, \tag{6.67}$$

where $\tilde{m}$ is the so-called conjugate field of m, defined as

$$\tilde{m} := -\frac{\partial \Omega}{\partial m} \quad \text{and} \quad \chi_m = -\frac{\partial^2 \Omega}{\partial m^2}. \tag{6.68}$$

We introduced the susceptibility to the m field χ_m. Note that m and $\tilde{m}$ are connected via the Legendre transform: $\Omega(\tilde{m}) = \Omega(m) + \tilde{m}m$. Show that

$$m = \frac{g\tilde{m}}{1 + g\chi_m}. \tag{6.69}$$

The field $\tilde{m}$ constitutes a so-called Landau order parameter [17] and, if $g < 0$, it can spontaneously become non-zero. If such a transition takes place, m also becomes non-zero. In contrast, if $g > 0$, $\tilde{m}$ can only become non-zero in the presence of the external magnetic field. In this case, the variables $\tilde{m}$ and m, play a similar role to the magnetisation and magnetic induction in magnetic systems [6]. On the other hand, if $g = 0$, one cannot induce a Chern phase via coupling to an external magnetic field, since, while $\tilde{m} \neq 0$, we find $m = 0$.
Task 2: Investigate the effects of an external field B_z, i.e., oriented along the out-of-plane axis z. Show that the total free energy for $T = 0$ becomes

$$\Omega = -\left(\frac{1}{g} + \chi_m\right)\frac{m^2}{2} - \mu \sum_a \int \frac{d\boldsymbol{k}}{(2\pi)^2} \frac{eB_z}{\hbar} \Omega_a(\boldsymbol{k}) f[E_a(\boldsymbol{k})], \tag{6.70}$$

with $\Omega_a(\boldsymbol{k})$ and $E_a(\boldsymbol{k})$ depending on m. Here, we chose to express the thermodynamic potential in terms of m, since this can be non-zero in the presence of the external field. To find the induced m, we extremise Ω, i.e.:

$$\frac{\partial \Omega}{\partial m} = 0 \quad \Rightarrow \quad m = -\frac{\partial}{\partial m}\left\{\frac{g\mu}{1 + g\chi_m} \frac{eB_z}{\hbar} \sum_a \int \frac{d\boldsymbol{k}}{(2\pi)^2} \Omega_a(\boldsymbol{k}) f[E_a(\boldsymbol{k})]\right\}. \tag{6.71}$$

One observes that if $\mu = 0$ the orbital magnetisation is zero, and the same holds for m. This is in accordance with the predictions of the Chern–Simons term and sections 5.1.3 and 6.2.2. Find m by employing the self-consistency equation (6.71).

[9] This expression holds as long as $g > 0$. For $g = 0$, we once again introduce m, which now functions as an external field. Following a similar procedure yields: $\Omega(\tilde{m}) = \tilde{m}^2/(2\chi_m)$.

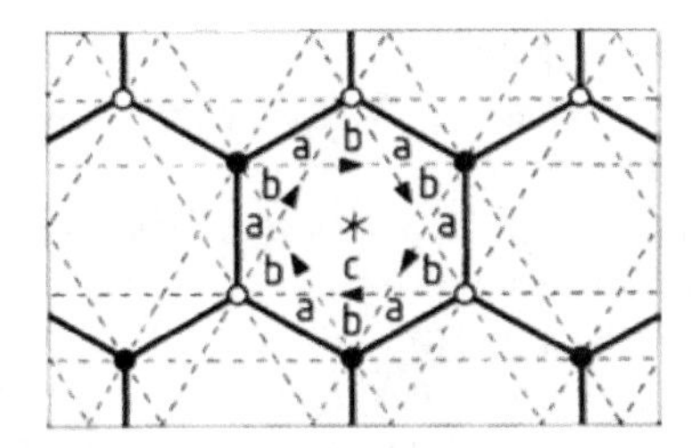

Figure 6.3. Electrons hop on nearest and next-nearest neighbours on a hexagonal lattice. There exist two sublattices, A and B, shown with open and solid circles. The unit-cell is centered on '*' and is bounded by the nearest-neighbour hexagon. The electrons feel a sublattice-dependent on-site energy. A TRS breaking inhomogeneous magnetic field is also added. Adapted from reference [26].

6.5 Hands-on: thermoelectric transport in the Haldane model

In this paragraph we study the thermoelectric response of the first model proposing the realisation of parity anomaly and the quantum anomalous Hall effect in condensed matter physics. This is the so-called Haldane model [26] and describes electrons hopping on a hexagonal lattice, as in figure 6.3. The hexagonal lattice is bipartioned into the sublattices A and B. Haldane assumed non-zero hopping between nearest and next-nearest neighbours, with respective energy scales t_1 and t_2. Following Semenoff [27], he also included an inversion-symmetry-breaking on-site energy $\pm M$, for the A, B sites. More importantly, he considered TRS breaking by adding an inhomogeneous $B_z(\boldsymbol{r})$ field, leading to zero net flux through a unit-cell. The presence of the field solely introduces a phase to the next-nearest hopping term, i.e., $t_2 \mapsto t_2 e^{i\phi}$. Remarkably, this last step is the sufficient ingredient for realising parity anomaly, which was not achievable in Semenoff's proposal.

Task: Show that the Haldane $\boldsymbol{k}$-space Hamiltonian reads, $\hat{H}(\boldsymbol{k}) = \varepsilon(\boldsymbol{k}) + \boldsymbol{g}(\boldsymbol{k}) \cdot \hat{\boldsymbol{\sigma}}$, with $\varepsilon(\boldsymbol{k}) = 2t_2 \cos\phi \sum_{s=1}^{3} \cos(\boldsymbol{k} \cdot \boldsymbol{b}_s)$ and

$$\boldsymbol{g}(\boldsymbol{k}) = \left(t_1 \sum_{s=1}^{3} \cos(\boldsymbol{k} \cdot \boldsymbol{a}_s),\ t_1 \sum_{s=1}^{3} \sin(\boldsymbol{k} \cdot \boldsymbol{a}_s),\ M - 2t_2 \sin\phi \sum_{s=1}^{3} \sin(\boldsymbol{k} \cdot \boldsymbol{b}_s) \right). \tag{6.72}$$

Here, the $\hat{\boldsymbol{\sigma}}$ Pauli matrices act on the sublattice space. The vectors $\boldsymbol{a}_{1,2,3}$ connect a point of sublattice B with its three nearest neighbours of sublattice A, while the vectors $\pm\boldsymbol{b}_{1,2,3}$ connect nearest neighbours on the same sublattice. Calculate the static thermoelectric coefficient α_{xy}.

References

[1] Bennett H S and Stern E A 1965 *Phys. Rev.* **137** A448
[2] Kapitulnik A, Xia J, Schemm E and Palevski A 2009 *New J. Phys.* **11** 055060
[3] Mineev V P 2007 *Phys. Rev. B* **76** 212501
[4] Lutchyn R M, Nagornykh P and Yakovenko V M 2008 *Phys. Rev. B* **77** 144516
[5] Kotetes P, Aperis A and Varelogiannis G 2014 *Phil. Mag.* **94** 3789
[6] Jackson J D 1999 *Classical Electrodynamics* 3rd edn (New York: Wiley)
[7] Maciejko J, Qi X-L, Drew H D and Zhang S-C 2010 *Phys. Rev. Lett.* **105** 166803

[8] Grosso G and Parravicini G P 2014 *Solid State Physics* 2nd edn (Amsterdam: Elsevier)
[9] König M, Buhmann H, Molenkamp L W, Hughes T L, Liu C-X, Qi X-L and Zhang S C 2008 *J. Phys. Soc. Jpn.* **77** 031007
[10] Bernevig B A and Hughes T L 2013 *Topological Insulators and Topological Superconductors* (Princeton, NJ: Princeton University Press)
[11] Hirst L L 1997 *Rev. Mod. Phys.* **69** 607
[12] Xiao D, Shi J and Niu Q 2005 *Phys. Rev. Lett.* **95** 137204
[13] Ceresoli D, Thonhauser T, Vanderbilt D and Resta R 2006 *Phys. Rev. B* **74** 024408
[14] Shi J, Vignale G, Xiao D and Niu Q 2007 *Phys. Rev. Lett.* **99** 197202
[15] Xiao D, Chang M-C and Niu Q 2010 *Rev. Mod. Phys.* **82** 1959
[16] Shindou R and Balents L 2006 *Phys. Rev. Lett.* **97** 216601
[17] Negele J W and Orland H 2008 *Quantum Many-Particle Systems* (Boulder, CO: Westview Press)
[18] Altland A and Simons B D 2010 *Condensed Matter Field Theory* 2nd edn (Cambridge: Cambridge University Press)
[19] Luttinger J M 1964 *Phys. Rev.* **135** A1505
[20] Jonson M and Girvin S M 1984 *Phys. Rev. B* **29** 1939
[21] Oji H and Streda P 1985 *Phys. Rev. B* **31** 7291
[22] Cooper N R, Halperin B I and Ruzin I M 1997 *Phys. Rev. B* **55** 2344
[23] Zhu J-X and Balatsky A V 2002 *Phys. Rev. B* **65** 132502
[24] Kotetes P and Varelogiannis G 2010 *Phys. Rev. Lett.* **104** 106404
[25] Shovkovy I A 2013 *Magnetic Catalysis: A Review* (Lecture Notes in Physics vol 871) (Berlin: Springer) p 13
[26] Haldane F D M 1988 *Phys. Rev. Lett.* **61** 2015
[27] Semenoff G W 1984 *Phys. Rev. Lett.* **53** 2449

Chapter 7

$\mathbb{Z}_2$ topological insulators

In the previous chapters we mainly focused on topological-insulator models in 1 + 1d (2 + 1d), which admit a $\mathbb{Z}$ classification in terms of a winding (Chern) number by virtue of the presence (violation) of chiral symmetry. If chiral symmetry becomes violated in the 1 + 1d case, a winding number cannot be defined and the resulting $\mathbb{Z}_2$ topological classification is performed via the so-called Chern–Simons invariant, associated with the electric polarisation. We further showed, that, the two classes of topological systems are related via dimensional reduction. In the present chapter we proceed with exploring a branch of topological systems distinct to the aforementioned ones. Those discussed here, are instead descendants of a Chern insulator in 4 + 1d. They constitute the TR-invariant (TRI) topological insulators. Such materials have been observed experimentally in both two and three spatial dimensions, and their characteristic feature is the presence of Kramers-degenerate boundary modes, the so-called helical surface states. We first introduce the 2 + 1d TRI topological insulators via a bottom-up approach, relying on two copies of a 2 + 1d Chern insulator. Later, we obtain a model for a 3 + 1d TRI topological insulator via effecting dimensional reduction on a 4 + 1d Chern insulator model. We obtain the surface states for both systems and highlight their properties. We additionally discuss the effects of warping on the surface states and their protection against backscattering. We introduce a $\mathbb{Z}_2$ topological invariant following the construction by Fu and Kane, and also express it in terms of the non-Abelian Berry potential. We conclude with a discussion of phenomena related to the magnetoelectric polarisation and the fractional quantum anomalous Hall effect on the topological insulator's surface. Finally, the Hands-on sections elaborate on the quasiparticle interference pattern from the surface of warped topological insulators and the concept of topological Kondo insulators.

doi:10.1088/978-1-68174-517-6ch7

7.1 $\mathbb{Z}_2$ topological insulators in 2 + 1 dimensions

The first appearance of TRI topological insulators [1, 2] was in the context of graphene [3]. Kane and Mele [4, 5] proposed that the presence of SOC can lead to a gap in graphene's bulk energy spectrum, thus, rendering this single-layered material topologically non-trivial, i.e., harbouring helical edge modes [6]. Nevertheless, the strength of the SOC was proven to be too weak to observe such a phenomenon in this system [1, 2]. Shortly after, Bernevig, Hughes and Zhang [7] (BHZ) came up with an alternative TRI topological material candidate, the HgTe/CdTe quantum wells. The theoretical predictions of the so-called BHZ model were experimentally confirmed [8], thus, unearthing the first 2 + 1d TRI topological insulator. This discovery lead to further theoretical works and material predictions [9–13], resulting in the experimental discovery of 3 + 1d TRI topological insulators [14–17]. The first materials to be identified belonged to the family of $Bi_{1-x}Sb_x$. Since then, a vast number of TRI topological insulators have been discovered [1, 2], e.g. Bi_2Se_3 and Bi_2Te_3.

7.1.1 Bottom-up construction based on Chern insulators: BHZ model

Having studied the 2 + 1d models for a Chern insulator, it is straightforward to construct a prototypical model for a TRI topological insulator in the same dimensionality, by simply doubling the degrees of freedom. In the minimal case, one starts from a Chern insulator, e.g., the one presented in equation (5.26), and further adds its time-reversed copy. The first 2 + 1d model Hamiltonian for a TRI topological insulator of this type, was exactly the one described by the BHZ model [7]:

$$\hat{H}_{\mathrm{BHZ}}(\boldsymbol{k}) = \begin{pmatrix} \hat{H}(\boldsymbol{k}) & \hat{0} \\ \hat{0} & \hat{H}^{*}(-\boldsymbol{k}) \end{pmatrix} \quad \text{with} \qquad (7.1)$$
$$\hat{H}(\boldsymbol{k}) = \varepsilon(\boldsymbol{k}) + \alpha\hbar(k_x\hat{\sigma}_x + k_y\hat{\sigma}_y) + [m - \beta(\hbar\boldsymbol{k})^2]\hat{\sigma}_z,$$

$\beta > 0$ and $\varepsilon(\boldsymbol{k}) = C - D\boldsymbol{k}^2$. In the insulating regime, $\varepsilon(\boldsymbol{k})$ takes values within the bulk gap, and, thus, it can be dropped as it does not affect the topological properties of the system. The four-state basis consists of hybridised s and p orbitals, stemming from the conduction and top valence bands, respectively. Within the common picture for the bandstructure of III–V semiconductors discussed in chapter 1, the conduction band electrons originate from the $j = 1/2$ s-orbital states, while the top valence bands from the $j = 3/2$ heavy and light holes. Nonetheless, for quantum wells, confinement pushes two out of the total of six states towards higher energies [7], thus, resulting in a low-energy subspace spanned by the four states: $|E1, m_j = 1/2\rangle$, $|H1, m_j = 3/2\rangle$, $|E1, m_j = -1/2\rangle$ and $|H1, m_j = -3/2\rangle$. By employing the $\hat{\boldsymbol{\sigma}}$ and $\hat{\boldsymbol{\tau}}$ Pauli matrices, defined in $\{E1, H1\}$ and $\{m_j, -m_j\}$ spaces, we write:

$$\hat{H}_{\mathrm{BHZ}}(\boldsymbol{k}) = \alpha\hbar\left(k_x\hat{\tau}_z\hat{\sigma}_x + k_y\hat{\sigma}_y\right) + [m - \beta(\hbar\boldsymbol{k})^2]\hat{\sigma}_z. \qquad (7.2)$$

The topological properties of the specific quantum wells depend on which one of the two types becomes dominant, the HgTe or CdTe. As shown in figure 7.1, the bandstructure of HgTe is inverted compared to the one encountered in CdTe, which

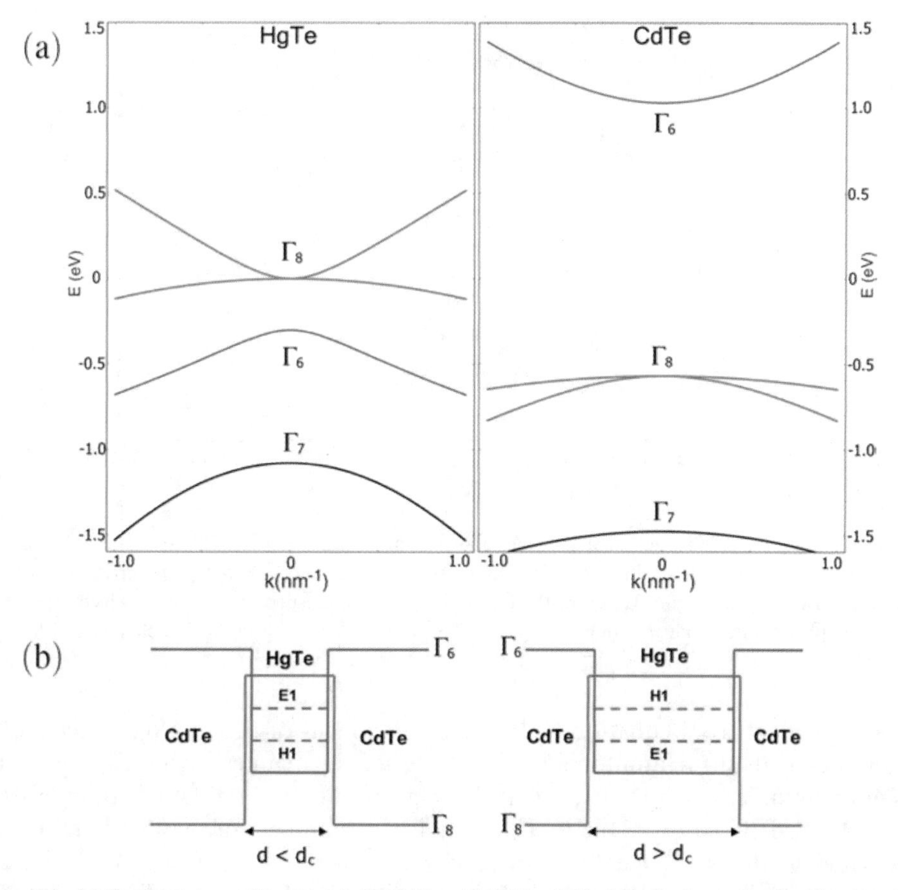

Figure 7.1. (a) Bandstructure of isolated bulk HgTe and CdTe. CdTe follows the typical bandstructure of semiconductors discussed in chapter 1, and shown in figure 1.2. In stark contrast, the bandstructure of bulk HgTe is inverted compared to the CdTe one, i.e., the Γ_6 conduction band lies below the Γ_8 top valence band. (b) Bandstructure of HgTe/CdTe quantum wells. The width of the middle region determines if the HgTe bandstructure characteristics will prevail, thus, leading to a topologically non-trivial phase. The topological phase transition occurs for a critical width d_c, for which the parameter m of equation (7.1) changes sign. Adapted from [7].

follows the usual motif previously discussed in section 1.2.1. Thus, by fabricating HgTe/CdTe quantum wells of different width d, one can impose one or the other type of bandstructure. In the Hamiltonian of equation (7.1), m is the only parameter which depends on d. Remarkably, we have already shown in section 5.2. that the parameter m controls the topological properties of each block, and the appearance of a non-zero 1st Chern number. The dependence of $m(d)$ is such, that, for $d > d_c$, one has $m > 0$. In this regime, the quantum well is characterised by the inverted bandstructure, which is typical to HgTe, and as a result, the system transits to the topologically non-trivial phase. In fact, the transition to the topologically non-trivial phase by tuning the quantum-well width, constituted a prediction of the BHZ model [7], that was experimentally verified in [8] (see figure 7.2).

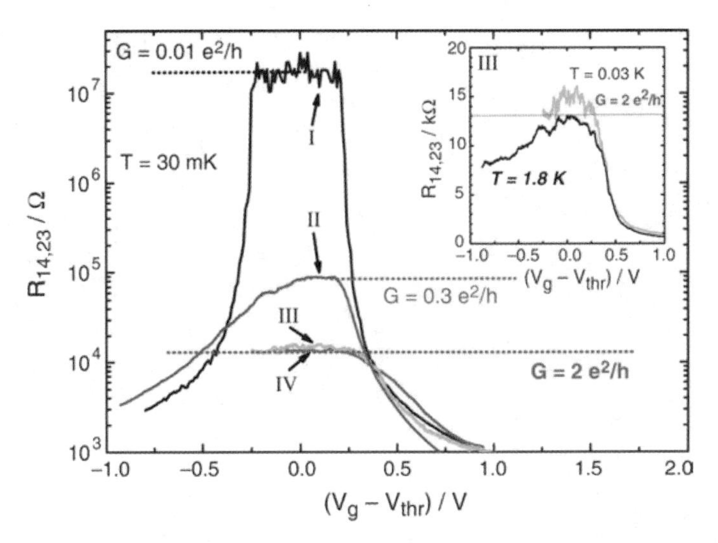

Figure 7.2. Longitudinal four-terminal resistance upon varying the gate voltage. The latter effectively controls the position of the Fermi level. In case I, $d < d_c$, and the system lies in the topologically-trivial phase, thus, exhibiting insulating behaviour. In contrast, the cases III and IV correspond to $d > d_c$, in which helical edge modes are present and conduct current of two units of conductance. The spin-filtered character of the edge modes is not probed here, and it is difficult to detect in transport measurements. Adapted from [8].

As is evident from equation (7.2), the z component of the spin operator $\hat{S}_z \propto \hat{\tau}_z$, commutes with the Hamiltonian. Therefore, its eigenvalues, labelled here by the quantum number $\tau = \pm 1$, can also label the eigenstates and the 1st Chern number of each block, denoted $C_{1,\tau}$. Due to TRS, the Chern numbers of the two blocks share the same modulus $|C_{1,\tau}| = 1$ but have opposite signs. As a result, the quantum anomalous Hall effect discussed in chapter 5 cannot take place, since it is related to the charge density and the total Chern number $\sum_\tau C_{1,\tau}$, which is, however, equal to zero here. Nonetheless, the so-called quantum spin Hall effect is accessible, since it is linked to the difference of the Chern numbers: $\sum_\tau \tau C_{1,\tau} = \pm 2$, i.e., the so-called spin Chern number. The emergence of this effect is associated with the presence of helical, instead of chiral, edge modes. For a comparison see sections 3.2 and 5.1. Given a strip geometry, cf figure 7.3(a), each Hamiltonian block yields a single branch of chiral edge modes with dispersion $E^{\tau}_{R,L}(k) = \pm\tau\alpha\hbar k$. Conclusively, at a single edge, one finds counter propagating Kramers-degenerate modes, as in figure 7.3(a, b). Since each branch is characterised by a particular spin projection, these modes are also called spin-filtered.

7.1.2 Violation of chiral symmetry and $\mathbb{Z}_2$ topological invariant

In the paragraph above we constructed a 2 + 1d TRI topological insulator model by simply putting together two decoupled Chern insulator models. Nonetheless, TRI hybridisation terms connecting the two blocks are also allowed in the general case. The goal of this section is to study the implications of the additional TRI coupling terms and identify a topological invariant to characterise the topological phase, since

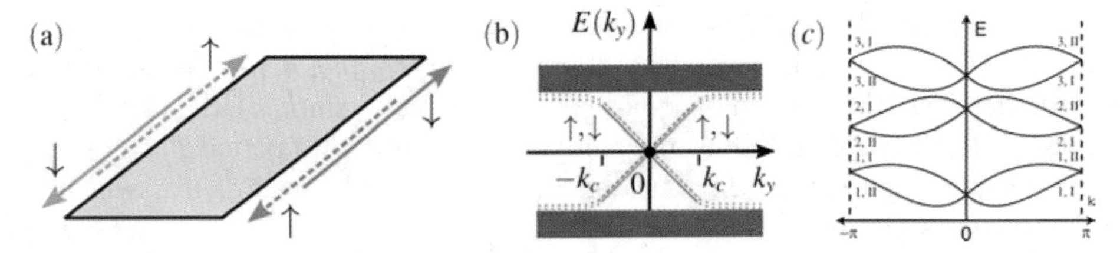

Figure 7.3. (a) Helical edge modes appearing at the boundaries separating vacuum from a 2 + 1d TRI topological insulator lying in its topologically non-trivial phase. On each edge, one finds counter-propagating modes with opposite spin projection. One can equivalently view the TRI topological insulator as two Chern insulator copies with opposite 1st Chern numbers (see equation (7.2)). In the general case, the $\mathbb{Z}_2$ character of the ensuing topological invariant, to be introduced in equation (7.17), implies that only a single Kramers pair per edge is topologically protected. (b) Energy dispersions of the helical edge modes. With orange and purple we illustrate modes located at the left and right edge, respectively. The orange and purple modes with the same slope are spin degenerate. For $|k_y| < |k_c|$, the edge mode dispersion is linear. For larger k_y the edge modes merge with the bulk modes, depicted here with dark turqoise (compare with figure 5.1). (c) Structure of the bulk energy levels for a 1 + 1d TRI topological insulator with SOC. There exist pairs of Kramers-degenerate bands, labelled by I and II, which are degenerate at the inversion-symmetric wavevectors $k_I = 0, \pi$. Figure (c) was adapted from reference [6].

the spin Chern number cannot be defined in this case. Motivated by the $k_z = 0$ version of the model Hamiltonian for TRI topological insulators[1] found in reference [18], we consider the following variant of the BHZ model:

$$\begin{aligned}\hat{H}_{d=2}(\boldsymbol{k}) &= \alpha\hbar(k_x\hat{\tau}_x\hat{\sigma}_y - k_y\hat{\tau}_x\hat{\sigma}_x) + [m - \beta(\hbar\boldsymbol{k})^2]\hat{\tau}_z \\ &+ \gamma k_x(k_x^2 - 3k_y^2)\hat{\tau}_x\hat{\sigma}_z + \zeta k_y(k_y^2 - 3k_x^2)\hat{\tau}_y.\end{aligned} \tag{7.3}$$

Since the above Hamiltonian lacks chiral symmetry, one cannot define a 1st Chern number and, thus, the spin Chern number is not employable in this general case. Nonetheless, the Hamiltonian above is manifestly TRI with $\hat{\mathcal{T}} = \mathrm{i}\hat{\sigma}_y\hat{\mathcal{K}}$. This situation is reminiscent of the one encountered in the model of the Hands-on section 4.7. There, the Hamiltonian lacked chiral symmetry and while a winding number was not accessible, one could still classify the system by employing the electric polarisation leading to a $\mathbb{Z}_2$ classification. This was made possible by virtue of the symmetry discussed in equation (4.44).

Since equation (7.3) implies that a problem of a similar nature is in order, we will try to resolve this issue by introducing a suitable kind of polarisation, the so-called $\mathbb{Z}_2$ polarisation. The presentation will follow closely the original analysis of Fu and Kane in [6]. These authors investigated the $\mathbb{Z}_2$ classification of a 2 + 1d TRI topological insulator, by alternatively studying a pumping problem of a 1 + 1d TRI topological insulator. By mapping the time variable to the second momentum, an equivalence between the two situations was made possible. A similar connection was found between the Thouless pump in 1 + 1d and the Chern insulator of 2 + 1d, as discussed in sections 4.3 and 5.3.2. To facilitate the following presentation, we therefore assume that k_y becomes a time-dependent variable due to the presence of a

[1] Such as Bi_2Se_3 and Bi_2Te_3.

constant electric field E_y. In this manner, we can instead view the Hamiltonian of equation (7.3) as a time-dependent Hamiltonian $\hat{H}_{\text{TRI}}(k_x, k_y) \to \hat{H}_{\text{TRI}}(k_x, t)$. By considering the lattice version of equation (7.3), e.g., $k_y \mapsto \sin(k_y a)$, the variable k_y is defined on the first BZ and, thus, the pumping is periodic with period $T = \Phi_0/(aE_y)$. Consequently, there exist two time instants[2], $t_1 = 0$ and $t_2 = T/2$, for which the Hamiltonian is TRI.

Under the above conditions, and following equation (4.42), one can define the electric polarisation in terms of the Berry vector potential of the occupied bands

$$P_x = \sum_a^{\text{occ}} \int_{-\pi}^{+\pi} \frac{dk_x}{2\pi} A_{k_x}^a(k_x). \tag{7.4}$$

By virtue of TRS, one can split the above term into partial contributions of Bloch wavefunctions that constitute Kramers partners. Following reference [6] closely, we define[3]

$$\left|u_a^{\text{I}}(k_x)\right\rangle = -e^{-i\chi_a(-k_x)}\hat{\mathcal{T}}\left|u_a^{\text{II}}(k_x)\right\rangle \quad \text{and} \quad \left|u_a^{\text{II}}(k_x)\right\rangle = e^{-i\chi_a(k_x)}\hat{\mathcal{T}}\left|u_a^{\text{I}}(k_x)\right\rangle, \tag{7.5}$$

via employing $\hat{\mathcal{T}}^2 = -\hat{1}$. At this stage, one defines the partial electric polarisation contributions

$$P_x^s = \sum_a^{\text{occ}'} \int_{-\pi}^{+\pi} \frac{dk_x}{2\pi} A_{k_x}^{a,s}(k_x) \equiv \sum_a^{\text{occ}'} \int_0^{\pi} \frac{dk_x}{2\pi}\left[A_{k_x}^{a,s}(k_x) + A_{-k_x}^{a,s}(-k_x)\right], \tag{7.6}$$

with $A_{k_x}^{a,s}(k_x) := i\langle u_a^s(k_x)|\partial_{k_x} u_a^s(k_x)\rangle$. The sum above is over the occupied states, for a given Kramers partner denoted with $s = \text{I, II}$. Via employing

$$i\left\langle u_a^s(k_x)\right|\hat{\mathcal{T}}^\dagger \partial_{k_x}\hat{\mathcal{T}}\left|u_a^s(k_x)\right\rangle \equiv A_{-k_x}^{a,s}(-k_x), \tag{7.7}$$

obtained by letting $\hat{\mathcal{T}}$ explicitly act on the state vector, we find that

$$A_{-k_x}^{a,\text{I}}(-k_x) = A_{k_x}^{a,\text{II}}(k_x) - \partial_{k_x}\chi_a(k_x), \tag{7.8}$$

which yields the partial electric polarisation

$$P_x^{\text{I}} = \frac{1}{2\pi} \sum_a^{\text{occ}'} \left\{ \int_0^{\pi} dk_x \sum_{s=\text{I,II}} A_{k_x}^{a,s}(k_x) - [\chi_a(\pi) - \chi_a(0)] \right\}. \tag{7.9}$$

One introduces the so-called sewing matrix $\hat{\text{W}}(k_x)$, with elements

$$\text{W}_{ab}(k_x) = \langle u_a(k_x)|\hat{\mathcal{T}}|u_b(k_x)\rangle. \tag{7.10}$$

[2] The authors of [6] considered the most general Hamiltonian and, thus, demanded the presence of two time instants $t_{1,2}$, for which, TRS is restored, without these necessarily coinciding with 0 and $T/2$.

[3] The authors of [6] follow a different definition for the action of TR on the Bloch wavefunctions compared to the one chosen here. Here we follow the definition of section 1.1.5, according to which, TR inverts $\boldsymbol{k}$.

In the above, a and b label occupied states. For the choice made in equation (7.5), the sewing matrix decomposes in a direct product of 2×2 submatrices of the form

$$\mathrm{W}_a(k_x) = \begin{pmatrix} 0 & e^{\mathrm{i}\chi_a(k_x)} \\ -e^{\mathrm{i}\chi_a(-k_x)} & 0 \end{pmatrix}, \tag{7.11}$$

in the basis of $\{\mathrm{II}, \mathrm{I}\}$. Here, a denotes a given occupied Kramers pair of states. For the inversion-symmetric[4] points of the BZ, $k_x = 0, \pi$, the sewing matrix becomes antisymmetric. Thus, it can be characterised by a Pfaffian Pf(W), for which, $[\mathrm{Pf}(\mathrm{W})]^2 = \det(\mathrm{W})$. The Pfaffian of a 2×2 antisymmetric matrix is essentially the upper right off-diagonal element, and for the two inversion-symmetric points one finds: $e^{\mathrm{i}\chi_a(0)}$ and $e^{\mathrm{i}\chi_a(\pi)}$. Therefore,

$$\frac{\mathrm{Pf}[\mathrm{W}_a(\pi)]}{\mathrm{Pf}[\mathrm{W}_a(0)]} = \mathrm{Exp}\left\{\mathrm{i}[\chi_a(\pi) - \chi_a(0)]\right\}. \tag{7.12}$$

By repeating the same procedure for P_x^{II}, one finds that $P_x^{\mathrm{II}} = P_x^{\mathrm{I}}$ modulo an integer. A similar indeterminancy was encountered in section 4.3.3 for the electric polarisation. To this end, we introduce the $\mathbb{Z}_2$ polarisation as the difference

$$P_\theta = P_x^{\mathrm{I}} - P_x^{\mathrm{II}} \tag{7.13}$$

that can be rewritten with the help of equation (7.8) as

$$\begin{aligned} P_\theta &= \frac{1}{2\pi}\sum_a^{\mathrm{occ}'}\left\{\sum_{s=\mathrm{I,II}}\int_0^\pi dk_x\left[A_{k_x}^{a,s}(k_x) - A_{-k_x}^{a,s}(-k_x)\right] + 2\mathrm{i}\ln\left\{\frac{\mathrm{Pf}[\mathrm{W}_a(\pi)]}{\mathrm{Pf}[\mathrm{W}_a(0)]}\right\}\right\} \\ &= \frac{1}{2\pi\mathrm{i}}\sum_a^{\mathrm{occ}'}\left\{\int_0^\pi dk_x\,\mathrm{Tr}\left[\mathrm{W}_a^\dagger(k_x)\partial_{k_x}\mathrm{W}_a(k_x)\right] - 2\ln\left\{\frac{\mathrm{Pf}[\mathrm{W}_a(\pi)]}{\mathrm{Pf}[\mathrm{W}_a(0)]}\right\}\right\}. \\ &= \frac{1}{2\pi\mathrm{i}}\left\{\int_0^\pi dk_x\frac{d}{dk_x}\ln\left\{\det\left[\hat{\mathrm{W}}(k_x)\right]\right\} - 2\ln\left\{\frac{\mathrm{Pf}[\hat{\mathrm{W}}(\pi)]}{\mathrm{Pf}[\hat{\mathrm{W}}(0)]}\right\}\right\}, \end{aligned} \tag{7.14}$$

where we employed the operator identity $\mathrm{Tr}\ln\hat{\mathrm{W}} \equiv \ln\det\hat{\mathrm{W}}$, since $\hat{W}^\dagger = \hat{W}^{-1}$. As a result, the first term yields only a boundary contribution at $k_x = 0, \pi$. By assuming that the branches of $\det[\hat{\mathrm{W}}]$ are chosen so that the branch at $k_x = 0$ evolves continuously along the path of integration into the branch chosen at $k_x = \pi$, we can write

$$e^{\mathrm{i}\pi P_\theta} = \frac{\sqrt{\det[\hat{\mathrm{W}}(\pi)]}}{\sqrt{\det[\hat{\mathrm{W}}(0)]}}\frac{\mathrm{Pf}[\hat{\mathrm{W}}(0)]}{\mathrm{Pf}[\hat{\mathrm{W}}(\pi)]} \quad\Rightarrow\quad (-1)^{P_\theta} = \prod_{k_I=0,\pi}\frac{\sqrt{\det[\hat{\mathrm{W}}(k_I)]}}{\mathrm{Pf}[\hat{\mathrm{W}}(k_I)]}. \tag{7.15}$$

Notably, the above result is obtained at each time instant of the adiabatic pump. As also discussed in the context of the electric polarisation, only differences of the

[4] Which satisfy: $Ik_x = -k_x \equiv k_x$. These are also occasionally termed as the TRI wavevectors or momenta.

polarisation are gauge invariant. Therefore, for the particular pump we can introduce the $\mathbb{Z}_2$ invariant as follows

$$(-1)^{P_\theta(t=T/2)-P_\theta(t=0)} = \prod_{k_I=0,\pi} \prod_{s=1,2} \frac{\sqrt{\det[\hat{\mathrm{W}}(k_I, t_s)]}}{\mathrm{Pf}[\hat{\mathrm{W}}(k_I, t_s)]}. \tag{7.16}$$

Thus, the above product contains contributions only from the TRI (k_x, t) points of the adiabatic-pump process. It is now straightforward to extend the result of the 1 + 1d pump to the case of a 2 + 1d TRI topological insulator, by simply replacing the time instants $t_{1,2}$, that yield a TRI Hamiltonian, with the inversion-symmetric k_y points, i.e., 0 and π. Consequently, the $\mathbb{Z}_2$ invariant is given by

$$\nu_{d=2} := \prod_{\boldsymbol{k}_I} \frac{\sqrt{\det[\hat{\mathrm{W}}(\boldsymbol{k}_I)]}}{\mathrm{Pf}[\hat{\mathrm{W}}(\boldsymbol{k}_I)]} := \prod_{\boldsymbol{k}_I} \delta_{\boldsymbol{k}_I}, \tag{7.17}$$

with the four inversion-symmetric wavevectors: $\boldsymbol{k}_I = \{(0, 0), (0, \pi), (\pi, 0), (\pi, \pi)\}$. The $\mathbb{Z}_2$ nature of the above invariant implies that there can only be a *single* topologically-protected Kramers pair of modes per edge. An even number of Kramers pairs of modes per edge can be always gapped out without violating TRS [1, 2].

While the above form of the invariant is the one most commonly used, the first definition introduced by Kane and Mele in [4] was in terms of the Pfaffian of the following antisymmetric matrix:

$$\mathrm{m}_{ab}(\boldsymbol{k}) := \langle u_a(-\boldsymbol{k})|\hat{\mathcal{T}}|u_b(\boldsymbol{k})\rangle. \tag{7.18}$$

In fact, one has the definition

$$\nu_{d=2} = \frac{1}{2\pi \mathrm{i}} \oint_C d\boldsymbol{k} \cdot \nabla_{\boldsymbol{k}} \ln\left\{\mathrm{Pf}\left[\hat{\mathrm{m}}(\boldsymbol{k})\right]\right\} \mathrm{mod}2, \tag{7.19}$$

where the line integral is considered for a path $\mathcal{C}$ encircling half of the BZ. Essentially, the above winding number counts the number of zeros of the respective Pfaffian in half of the BZ. If the Pfaffian above does not have only isolated zeros, then an appropriate $\mathrm{i}0^+$ has to be included in the logarithm (see [4]).

7.2 $\mathbb{Z}_2$ topological insulators in 3 + 1 dimensions

In analogy to the 2 + 1d topological insulator model above, similar topological systems are also accessible in three spatial dimensions. To study them, we start from the model Hamiltonian of [18], obtained via $\boldsymbol{k} \cdot \boldsymbol{p}$ perturbation theory. This gives us the opportunity to also investigate the properties of the topologically-protected helical surface states. Moreover, below we discuss the extension of the Fu–Kane $\mathbb{Z}_2$ topological invariant to 3 + 1d systems, with or without inversion symmetry. As we demonstrate, the latter can be also connected to the so-called Wilson loop of the non-Abelian Berry gauge potential, defined for the occupied Kramers pairs of Bloch wavefuctions.

7.2.1 Crystal structure and model Hamiltonian

The prototypical 3 + 1d TRI topological insulators have been predicted to exist [9–13] in the family of Bi-based semiconductors. The theoretical predictions were experimentally confirmed [14–17] and, thus, the first 3 + 1d topological insulators, i.e., $Bi_{1-x}Sb_x$, Bi_2Se_3 and Bi_2Te_3, were discovered. As shown in figure 7.4, the crystal structure of Bi_2Se_3 is dictated by a D_{3d}^5 ($R\bar{3}m$) point-group symmetry with a unit cell consisting of a quintuple layer. The authors of [18] exploited the invariance under the latter point group operations and TRS, to retrieve the 3 + 1d model Hamiltonian for a TRI topological insulator, which reads:

$$\begin{aligned}\hat{H}_{d=3}(\boldsymbol{k}) = \alpha(\boldsymbol{k})\hbar\left(k_x\hat{\tau}_x\hat{\sigma}_y - k_y\hat{\tau}_x\hat{\sigma}_x\right) + \left[m - \beta\hbar^2\left(k_x^2 + k_y^2\right) - \beta'\hbar^2k_z^2\right]\hat{\tau}_z \\ + \gamma k_x\left(k_x^2 - 3k_y^2\right)\hat{\tau}_x\hat{\sigma}_z + \left[\alpha'(\boldsymbol{k})\hbar k_z + \zeta k_y\left(k_y^2 - 3k_x^2\right)\right]\hat{\tau}_y,\end{aligned} \tag{7.20}$$

where only up to cubic terms in $\boldsymbol{k}$ are considered here. The coefficients $\alpha(\boldsymbol{k})$ and $\alpha'(\boldsymbol{k})$ retain contributions proportional to $\{1, k_x^2 + k_y^2\}$ and $\{1, k_z^2\}$, respectively. We omitted the diagonal kinetic energy term discussed in [18], since it is irrelevant for the topological properties of the system when it exhibits a bulk energy gap. One notes that the last two terms of the above Hamiltonian include the so-called warping terms, whose appearance can be understood via the presence of a threefold rotation

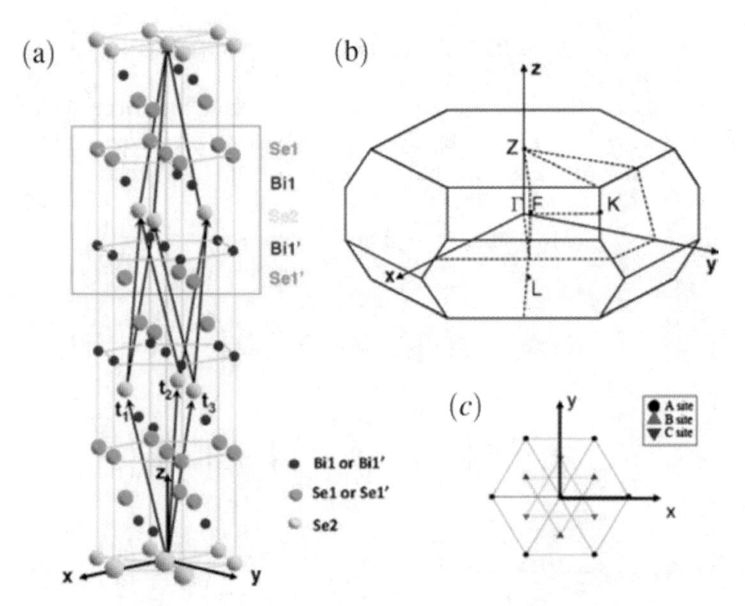

Figure 7.4. (a) Crystal structure of Bi_2Se_3, which is dictated by D_{3d}^5 point-group symmetry. The red frame encloses the unit cell, which consists of a quintuple layer. The respective lattice vectors $\boldsymbol{t}_{1,2,3}$, are defined as: $\boldsymbol{t}_1 = (\sqrt{3}a/3, 0, c/3)$, $\boldsymbol{t}_2 = (-\sqrt{3}a/6, a/2, c/3)$ and $\boldsymbol{t}_3 = (-\sqrt{3}a/6, -a/2, c/3)$. Here, a and c are the lattice constants in the xy plane and the z direction, respectively. (b) Brillouin zone of Bi_2Se_3 and high-symmetry points and $\boldsymbol{k}$-space directions. (c) In-plane coordinate lattice structure and the three possible atom locations. Adapted from reference [18].

axis, cf section 1.1.4 and table 1.5. As we discuss in section 7.2.3, the warping effects can have a significant impact on the properties of the helical surface modes. Their consequences are experimentally detectable in quasiparticle interference experiments [17, 19, 20]. Furthermore, in the presence of strong repulsive interactions, warping can be pivotal for the spontaneous development of magnetism on the topological insulator's surface [19, 21–23]. Nonetheless, in the section below, we momentarily neglect the warping effects in order to explore the properties of the conventional helical surface states.

7.2.2 Surface states for negligible warping

The situation is similar to the 2 + 1d TRI topological insulators, where only a single pair of helical states appear at the system's boundary, as dictated by the $\mathbb{Z}_2$ invariant. However, in the present case we do not obtain helical *edge* modes but rather helical *surface* modes. As we show below, the latter are described by a massless Dirac-like Hamiltonian. In order to explicitly demonstrate the existence of helical surface modes for the model above, we follow the procedure discussed in section 4.1.1 and confine the system in a box ($z \in [z_1, z_2]$) along the z axis. Along the other two spatial dimensions x and y, we assume translational invariance. Moreover, we solely retain the terms of equation (7.20) which are up to quadratic order in $\boldsymbol{k}$. Thus, $\alpha(\boldsymbol{k}) = \alpha = \text{constant}$ and $\alpha'(\boldsymbol{k}) = \alpha' = \text{constant}$. We set $k_x = k_y = 0$ and obtain the Hamiltonian:

$$\hat{H}^{(2)}_{d=3}(k_x = k_y = 0, k_z) = \alpha'\hbar k_z\hat{\tau}_y + [m - \beta'(\hbar k_z)^2]\hat{\tau}_z, \tag{7.21}$$

which is identical to the one of equation (4.2), after performing the mapping $(k_x, \alpha, \beta, \hat{\boldsymbol{\sigma}}) \leftrightarrow (k_z, \alpha', \beta', \hat{\boldsymbol{\tau}})$. The above harbours zero-energy solutions for $m\beta' > 0$. These constitute eigenstates of $\hat{\tau}_x$ and, for $m\beta' > 0$ and $\alpha' > 0$, read:

$$\boldsymbol{\phi}_{0,-,+,k_x=k_y=0}(z) \propto \sin[\zeta'(z - z_1)]e^{-(z-z_1)/\xi'}\frac{1}{\sqrt{2}}\begin{pmatrix} 1 \\ -1 \end{pmatrix} \otimes \begin{pmatrix} 1 \\ 0 \end{pmatrix}, \tag{7.22}$$

$$\boldsymbol{\phi}_{0,-,-,k_x=k_y=0}(z) \propto \sin[\zeta'(z - z_1)]e^{-(z-z_1)/\xi'}\frac{1}{\sqrt{2}}\begin{pmatrix} 1 \\ -1 \end{pmatrix} \otimes \begin{pmatrix} 0 \\ 1 \end{pmatrix}, \tag{7.23}$$

$$\boldsymbol{\phi}_{0,+,+,k_x=k_y=0}(z) \propto \sin[\zeta'(z - z_2)]e^{+(z-z_2)/\xi'}\frac{1}{\sqrt{2}}\begin{pmatrix} 1 \\ 1 \end{pmatrix} \otimes \begin{pmatrix} 1 \\ 0 \end{pmatrix}, \tag{7.24}$$

$$\boldsymbol{\phi}_{0,+,-,k_x=k_y=0}(z) \propto \sin[\zeta'(z - z_2)]e^{+(z-z_2)/\xi'}\frac{1}{\sqrt{2}}\begin{pmatrix} 1 \\ 1 \end{pmatrix} \otimes \begin{pmatrix} 0 \\ 1 \end{pmatrix}, \tag{7.25}$$

with (ζ', ξ') corresponding to (ζ, ξ) defined in section 4.1.1, after performing the mapping discussed. Note that TRS leads to Kramers pairs of zero-energy solutions localised at $z_{1,2}$, which are labelled by the eigenvalues of $\hat{\sigma}_z$, $\sigma = \pm 1$. In fact, the two solutions on a given surface have opposite spin polarisations. At this point we obtain the helical surface modes by allowing k_x and k_y to be non-zero. By projecting the

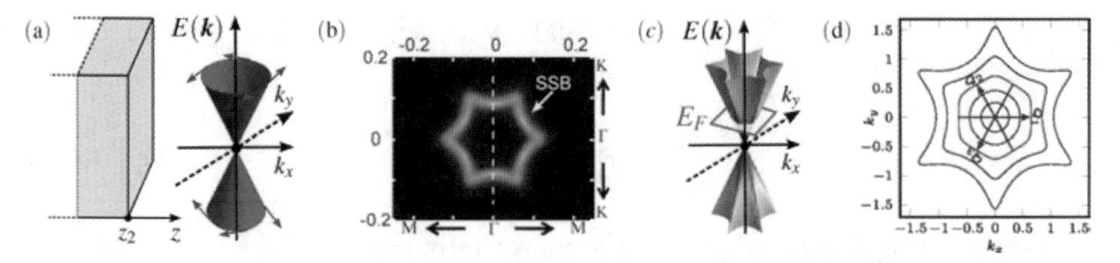

Figure 7.5. (a) Dirac-type helical modes living on the $z = z_2$ surface of a 3 + 1d TRI topological insulator. The energy dispersion follows the massless Dirac relativistic spectrum. The electron-spin configuration, shown with blue arrows, winds in the opposite sense in the upper and lower helicity bands. (b) Fermi surface of the helical boundary modes of Bi_2Te_3, experimentally inferred via angle resolved photoemission spectroscopy [16]. Here, one finds a Fermi surface with hexagonal symmetry, instead of the circular one encountered in (a). The latter difference is attributed to the so-called warping effects. (c) Energy dispersion of the helical surface modes in the presence of hexagonal warping. (d) Theoretically obtained Fermi surface upon varying the Fermi level. For the value marked in (c) with red, the Fermi surface becomes a nearly perfect hexagon. The three pairs of facets are nested with wavevectors $\boldsymbol{Q}_{1,2,3}$. In the presence of suitable interactions, nesting can drive a magnetic instability and lead to magnetic skyrmion lattice phases [19, 23]. Figure (b) and (a, c, d) have been adapted from references [16] and [23], respectively.

Hamiltonian of equation (7.20) on to the space spanned by the four wavefunctions above, we directly obtain the form of the Hamiltonian describing the surface states near $z = z_{1,2}$:

$$\hat{H}^{\tau}_{\text{surf}}(k_x, k_y) = \tau\alpha\hbar(k_x\hat{\sigma}_y - k_y\hat{\sigma}_x). \tag{7.26}$$

The above leads to the massless relativistic energy spectrum $E(\boldsymbol{k}) = \pm\alpha\hbar|\boldsymbol{k}|$. The respective eigenstates are also referred to as *helicity* eigenstates (see also figure 7.5(a)). Note that the spectrum is identical for both $\tau = \mp 1$, which is the quantum number labelling the helical modes living on the surface located at $z_{1,2}$. In the above, we neglected the Hamiltonian part which is quadratic in momenta and leads to an exponentially decaying mixing term (see equation (5.29)). The same procedure can be carried out if we assume confinement along the x and y directions, with the respective wavefuctions having a different structure than the ones above. Finally, let us remark that the Dirac cones necessarily appear in pairs, *though non-locally*, in accordance with the Nielsen–Ninomiya theorem, as discussed in section 3.1. This ensures that fractionalisation effects on the τ degree of freedom appear only locally, and not globally.

7.2.3 Consequences of warping and π Berry phase

In the above paragraph we neglected the cubic warping terms and studied the topologically-protected helical surface states appearing on the $z = z_{1,2}$ boundaries. Given the structure of the unit cell's lattice vectors[5] $\boldsymbol{t}_{1,2,3}$, the $z = z_2$ boundary corresponds to the [111][6] surface [18, 19]. The latter surface is characterised by C_{3v}

[5] See figure 7.4(a).

[6] For the Miller notation see footnote 2 of chapter 1.

point-group symmetry. Therefore, in analogy to section 1.2.4, we find that the Hamiltonian incorporating the warping effects reads:

$$\hat{H}^{\tau}_{\text{surf,warping}}(k_x, k_y) = \frac{(\hbar \boldsymbol{k})^2}{2m} - E_F + \tau\left[\alpha\hbar(k_x\hat{\sigma}_y - k_y\hat{\sigma}_x) + \gamma k_x\left(k_x^2 - 3k_y^2\right)\hat{\sigma}_z\right], \quad (7.27)$$

where we additionally included a kinetic energy term measured relative to the energy of the Fermi level E_F.

The consequences of warping effects have been already observed experimentally, see [16, 17]. In figure 7.5(b) we show the experimentally inferred [16] Fermi surface of the helical modes living on the [111] surface of Bi_2Te_3. The Fermi surface assumes a hexagonal shape by virtue of the combined C_{3v} and TR symmetries, also reflected in the form of the energy dispersion (see figure 7.5(c)). Interestingly, upon varying the chemical potential, e.g. via doping, the Fermi surface can become an almost perfect hexagon with three pairs of nested facets with wavevectors $\boldsymbol{Q}_{1,2,3}$ as depicted with red colour coding in figure 7.5(d). Fu [19] argued that, in the presence of suitable interactions, the surface states can spontaneously exhibit magnetic order. In addition, he performed a symmetry classification of the accessible magnetic phases. The authors of [21] and [22] examined several properties of the single- and triple-$\boldsymbol{Q}$ magnetic phases. It was later shown in [23] that, in the presence of an on-site Hubbard interaction and an almost perfectly hexagonal Fermi surface, the helical surface states will spontaneously violate TRS by transiting to a magnetic skyrmion lattice phase. However, a magnetic transition on the topological insulator's surface takes place for a Hubbard interaction of energy strength of few eV and, thus, appears to be a non-feasible scenario for the semiconducting materials discovered so far. Nonetheless, finding such strongly interacting topological insulator materials would open perspectives for engineering topological superconductivity, see [23, 24].

At this point, it is worth pointing out the crucial feature of the helical surface states that reflects their topological character. Certainly, this also has to constitute a topological feature, distinguishing this type of conducting states from the ones found in conventional metals. This characteristic is the presence of a π Berry phase or equivalently a Dirac point at $\boldsymbol{k} = \boldsymbol{0}$. This touching point of the upper and lower helicity bands, shown in figure 7.5(a, c), cannot be removed as long as the bulk system lies in the topologically non-trivial phase and TRS is preserved. In fact, one can quantitatively describe this topological feature in terms of the line integral of the Berry vector potential, defined for the family of the helicity eigenstates. The latter line integral is along a path enclosing the Dirac singularity. In the vicinity of the Dirac point, the warping term is negligible and only the Rashba term contributes. Thus, by parametrising the remaining spin-dependent part of the Hamiltonian in equation (7.27) as: $E(\boldsymbol{k})[\cos\varphi(\boldsymbol{k})\hat{\sigma}_x + \sin\varphi(\boldsymbol{k})\hat{\sigma}_y]$, we find that

$$\oint_C d\boldsymbol{k} \cdot \boldsymbol{A}(\boldsymbol{k}) = -\frac{1}{2}\oint_C d\varphi = \pi. \quad (7.28)$$

In fact, via the Stokes theorem, we find that also the Berry curvature should be non-zero. However, one finds that the latter is singular, i.e., defined only for $\boldsymbol{k} = 0$, thus, $\Omega_z(\boldsymbol{k}) = \pi\delta(\boldsymbol{k})$. This implies that there exists a magnetic monopole associated with the Dirac point, or more precisely a vortex with a single unit of vorticity. Similar bandstructure vortices appear in graphene, as well as in the so-called Weyl semimetals, which are defined in 3 + 1d. Such systems are further discussed in the next chapter.

The presence of the π Berry phase plays a crucial role in the topological robustness of the surface states against weak disorder. It further allows circumventing Anderson's localisation (see the discussion of section 5.1.2). As a result, the surface states exhibit the so-called weak antilocalisation [1, 2]. Such a phenomenon cannot take place in usual metals, since there, the arising Dirac points come in pairs due to the Nielsen–Ninomiya theorem (see also section 4.3.1). In the present situation, we still have pairs of Dirac points, but spatially separated. The appearance of the antilocalisation effects can be also viewed as a result of the immunity of the helical surface states against weak backscattering mediated by interactions or the addition of non-magnetic impurities. In fact, the latter property is also detectable in quasiparticle interference experiments [17]. Remarkably, warping effects can provide a loophole for actually observing an interference pattern, which is otherwise absent for a Dirac cone (see figure 7.6(a) and (b)). The modified Fermi surface, shown in figure 7.5(b)–(d), consists also of non-convex parts which give rise to an

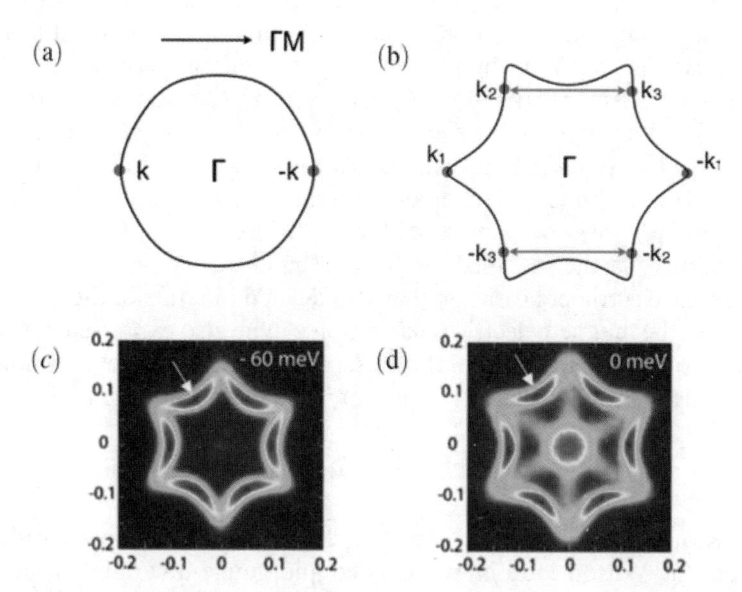

Figure 7.6. (a, b) The red arrows show the allowed scattering processes when a single impurity is deposited on the topological surface. The red dots denote the Fermi-surface points expected to contribute to the quasiparticle interference pattern for scattering along the x direction. Scattering between TR-related $\boldsymbol{k}$ points is not possible and, thus, no interference is expected for (a), at least, at lowest order. In contrast, the non-convexity of the Fermi surface in (b), allows scattering between TR-unrelated points, e.g., $\boldsymbol{k}_2$ and $\boldsymbol{k}_3$, which lead to interference. (c, d) Experimentally-inferred interference pattern for Bi_2Te_3 [17]. The figures (a, b) and (c, d) were adapted from references [19] and [17].

interference pattern. See [19, 20] and figure 7.6(c) and (d). This pattern constitutes a distinctive signature of the warped helical surface states [17]. For further details, see the Hands-on section 7.4.

7.2.4 Magnetoelectric polarisation and $\mathbb{Z}_2$ topological invariants in 3 + 1d

We may now proceed with further understanding the topological properties of such systems from a bulk perspective. For simplicity, we first omit the warping terms. In this case, the Hamiltonian of equation (7.20) reads

$$\begin{aligned}\hat{H}'_{d=3}(\boldsymbol{k}) &= \alpha(\boldsymbol{k})\hbar\left(k_x\hat{\tau}_x\hat{\sigma}_y - k_y\hat{\tau}_x\hat{\sigma}_x\right) + \alpha'(\boldsymbol{k})\hbar k_z\hat{\tau}_y \\ &+ \left[m - \beta\hbar^2\left(k_x^2 + k_y^2\right) - \beta'\hbar^2k_z^2\right]\hat{\tau}_z.\end{aligned} \tag{7.29}$$

Evidently, nelecting the warping effects leads to an emergent chiral symmetry, with operator $\hat{\Pi} = \hat{\tau}_x\hat{\sigma}_z$. Therefore, in this limit, we can block off-diagonalise the Hamiltonian and introduce a winding number w_3 in order to describe the topological properties of the system. The winding number of such a Hamiltonian has been already studied in the Hands-on section 4.5. Alternatively, the winding number and the topological properties of the present model can be understood via the 2nd Chern number of a 4 + 1d Chern insulator model obtained via dimensional extension (see section 5.5).

In analogy to the case of the 1 + 1d topological-insulator model studied in section 4.3.3, also here, apart from the winding number, one can additionally define a $\mathbb{Z}_2$ Chern–Simons invariant. In section 4.3.3 the $\mathbb{Z}_2$ invariant was identified with the electric polarisation, while in the present case, it is instead associated with the so-called magnetoelectric polarisation P_3. In fact, if the warping terms are not neglected, chiral symmetry is broken and the system can be characterised only via the magnetoelectric polarisation, i.e., the $\mathbb{Z}_2$ invariant. A similar analysis can be performed for the extended BHZ model presented in equation (7.3). Thus, the P_3 magnetoelectric polarisation constitutes the 3 + 1d extension of the $\mathbb{Z}_2$ polarisation P_θ, defined earlier via the so-called topological band theory.

Similar to the electric polarisation that was defined in terms of the Chern–Simons form in 1 + 1d, the magnetoelectric polarisation can be also expressed in terms of the Berry vector potential. However, in the present case one encounters the non-Abelian version. One defines a so-called $\mathbb{Z}_2$ invariant via the Wilson loop [25]

$$W_C = \frac{1}{2}\,\mathrm{Tr}\,\mathcal{P}_C\,\mathrm{Exp}\left[-\mathrm{i}\oint_C d\boldsymbol{k}\cdot\hat{\boldsymbol{A}}(\boldsymbol{k})\right], \tag{7.30}$$

where C denotes the path in $\boldsymbol{k}$ space and $\mathcal{P}_C$ denotes the path ordering. By construction, the Wilson loop is well defined and gauge invariant for an arbitrary closed path in $\boldsymbol{k}$-space. Thus, we can take advantage of the latter property and choose a loop that maps to itself under TR. In this manner, one can show [25] that the Wilson loop becomes equal to

$$W_C = \prod_{\boldsymbol{k}_I} \mathrm{sgn}\left\{\mathrm{Pf}[\hat{W}(\boldsymbol{k}_I)]\right\}, \tag{7.31}$$

where $\hat{W}$ corresponds to the sewing matrix of section 7.1.2. In three spatial dimensions the inversion-symmetric wavevectors are eight, and are defined by all the coordinate combinations $k_{x,y,z} = 0, \pi$. In 2 + 1d, this simply boils down to the Fu–Kane $\mathbb{Z}_2$ invariant. Therefore, we simply obtain

$$\nu_{d=3} := \prod_{\boldsymbol{k}_I} \frac{\sqrt{\det[\hat{W}(\boldsymbol{k}_I)]}}{\mathrm{Pf}[\hat{W}(\boldsymbol{k}_I)]} \equiv \prod_{\boldsymbol{k}_I} \delta_{\boldsymbol{k}_I}, \tag{7.32}$$

where the $\boldsymbol{k}_I$ are properly extended to the case of three spatial dimensions. In the case of inversion symmetry, the above expression further simplifies [11], and every $\delta(\boldsymbol{k}_I)$ can be written as

$$\delta_{\boldsymbol{k}_I} = \prod_{a}^{\mathrm{occ}'} \xi_a(\boldsymbol{k}_I), \tag{7.33}$$

where $\xi_a(\boldsymbol{k}_I)$ denotes the eigenvalue of the inversion operator when acting on the a-th occupied band. Note that we take into account only one of the two Kramers partners since they share the same inversion eigenvalue [11].

Notably, the invariant of equation (7.32) is usually termed 'strong' and is given by the product of $\delta_{\boldsymbol{k}_I}$, with contributions from all the inversion-symmetric wavevectors. However, one can define three additional, so-called 'weak' invariants, by employing the products of $\delta_{\boldsymbol{k}_I}$ involving four wavevectors lying within the same plane. We denote the latter three invariants with $\nu_{1,2,3}$, and the strong with ν_0. The former rely on the presence of translational invariance and do not constitute a genuine topological feature in three spatial dimensions. Instead, each weak invariant can be seen as a topological index characterising a stacking of layered TRI topological insulators. As a result, disorder can significantly affect the helical edge modes associated with these invariants[7]. In stark contrast, the helical edge modes related to the strong invariant ν_0 are robust when weak disorder is introduced. Conclusively, one characterises a TRI band topological insulator using the set of four $\mathbb{Z}_2$ invariants: $(\nu_0; \nu_1, \nu_2, \nu_3)$. Figure 7.7 shows examples of such topological phases.

7.3 Dimensional reduction from a 4 + 1d Chern insulator and magnetoelectric coupling

Most of the discussion so far has built upon the so-called topological band theory. In the remainder, we adopt an alternative view, relying on aspects of topological field theory. We first focus on the topological response of the electromagnetic field coupled to such a topological insulator, obtained via applying dimensional reduction on a 4 + 1d Chern insulator. In addition, we discuss the fractional quantum Hall conductance characterising the helical surface states, and conclude with the

[7] See also [26] for further aspects of weak topological insulators.

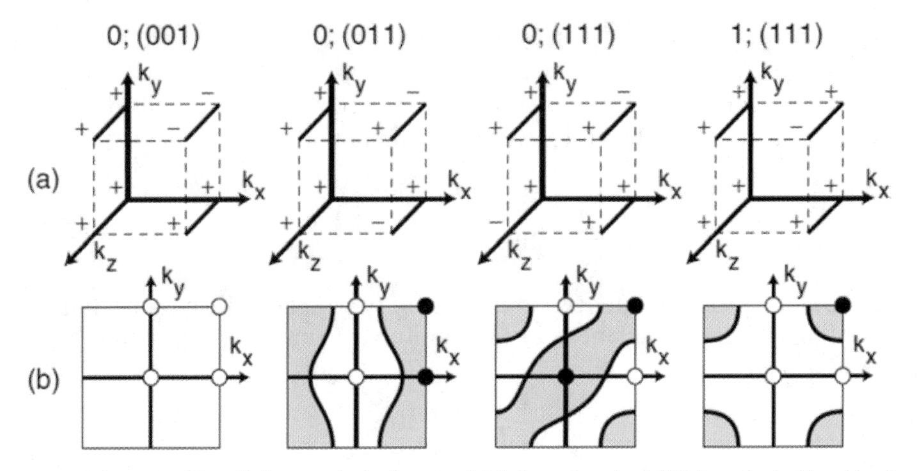

Figure 7.7. Diagrams of four distinct topologically non-trivial phases for a 3 + 1d TRI topological band insulator. In (a) the four indices denote the values (0 or 1) of the $\mathbb{Z}_2$ invariants ln $\nu_{0,1,2,3}/(\pi i)$ defined in [9], which are equivalent to (ν_0; ν_1, ν_2, ν_3). The signs ± correspond to the signs of the δ_{k_I} defined in equation (7.17), for the given topological phase. Part (b) characterises the surface (001) in each one of the phases of (a). The open and closed circles denote the ± sign of the products of the δ_{k_I} of two $\boldsymbol{k}_I$ points connected by a solid line in (a), respectively. The shaded regions and solid lines indicate possible Fermi arcs enclosing the $\boldsymbol{k}_I$ points. Adapted from reference [9].

engineering of a quantum anomalous Hall material via magnetically doping a TRI topological insulator.

7.3.1 Dimensional reduction

In section 5.5 we discussed how one obtains the effective action for the electromagnetic field coupled to a 1 + 1d topological insulator possessing chiral symmetry, via the dimensional reduction approach effected on a 2 + 1d Chern insulator. Our program followed [27] closely and the dimensional reduction was performed at the level of the effective electromagnetic action. A similar approach is pursued here. Our starting point is a 4 + 1d Chern insulator model described by the Hamiltonian:

$$\begin{aligned}\hat{H}_{d=4}(\boldsymbol{k}) = \alpha(\boldsymbol{k})\hbar\left(k_x\hat{\tau}_x\hat{\sigma}_y - k_y\hat{\tau}_x\hat{\sigma}_x\right) + \Big[m - \beta\hbar^2\left(k_x^2 + k_y^2\right) - \beta'\hbar^2k_z^2 \\ -\beta''\hbar^2k_w^2\Big]\hat{\tau}_z + \gamma g(k_w)\hat{\tau}_x\hat{\sigma}_z + [\alpha'(\boldsymbol{k})\hbar k_z + \zeta f(k_w)]\hat{\tau}_y.\end{aligned} \tag{7.34}$$

By comparing the above Hamiltonian to the one of equation (7.20), one observes that the main changes are the addition of a quadratic term $k_w^2\hat{\tau}_z$ and the replacement of the warping terms with two functions $f(k_w)$ and $g(k_w)$. These depend only on the k_w wavevector, which is associated with the added spatial dimension. Since for a Chern insulator TRS needs not to be preserved, the functions $f(k_w)$ and $g(k_w)$ are in general neither even nor odd under inversion. Nevertheless, the condition $g(k_w) = -g(-k_w)$ is sufficient to yield a non-zero 2nd Chern number. Here we assume that both functions, $f(k_w)$ and $g(k_w)$, are odd under I, thus, the Hamiltonian becomes TRI, but is still lacking chiral symmetry.

Via the Fourier transform $\boldsymbol{k} \mapsto \hat{\boldsymbol{p}}$, we can couple the electrons minimally to the U(1) electromagnetic field (see section 2.2). After integrating out the electrons, we find a Chern–Simons contribution to the effective action of the five-potential A_μ

$$S_{\mathrm{CS}}^{(4+1)} = \frac{C_2}{24\pi^2} \int d^5x \ \varepsilon^{\mu\nu\lambda\rho\sigma} A_\mu \partial_\nu A_\lambda \partial_\rho A_\sigma, \tag{7.35}$$

which is proportional to the second Chern number C_2 (see equation (5.55)). To perform the dimensional reduction we assume $f(k_w) = 0$, $g(k_w) = k_w$ and replace $\hat{p}_w + eA_w$ with a field θ, that depends only on (x, y, z). The latter dependence is also assumed for V and $A_{x,y,z}$. We then split the field θ into two parts, i.e., $\theta = \theta_0 + \delta\theta$, with $\theta_0 =$ constant. Integrating out the electrons leads to the effective action [27]

$$S_{\mathrm{eff}} = \frac{C_2(\theta_0)}{16\pi} \int d^4x \ \varepsilon^{\mu\nu\lambda\rho} \delta\theta F_{\mu\nu} F_{\lambda\rho}, \tag{7.36}$$

with $C_2(\theta_0)$ satisfying $\int_0^{2\pi} d\theta_0 C_2(\theta_0) \in \mathbb{Z}$. Similar to the case of the electric polarisation, we have

$$S_{\mathrm{eff}} = \int d^4x \ P_3(\boldsymbol{r}, t) \varepsilon^{\mu\nu\lambda\rho} F_{\mu\nu} F_{\lambda\rho}, \tag{7.37}$$

where we introduced the magnetoelectric polarisation. Notably, for a system with TRS the magnetoelectric polarisation has to be a constant, and in fact taking the values 0 or π,[8] depending on whether the insulator is topologically trivial or not.

The $\mathbb{Z}_2$ TRI topological insulators obtained above, are termed first descendants of the 4 + 1d Chern insulator. One can repeat this procedure once more, to obtain the second descendants. In order to retrieve the second descendants, which correspond to the 2 + 1d TRI topological insulators, one further replaces the canonical momentum $\hat{p}_z + eA_z$, with a field ϕ. The latter is decomposed into two parts $\phi = \phi_0 + \delta\phi$, with $\phi_0 =$ constant. We thus find [27]

$$S_{\mathrm{eff}} = C_2(\theta_0, \phi_0) \int d^3x \ \varepsilon^{\mu\nu\lambda} (\partial_\mu \delta\theta)(\partial_\nu \delta\phi) A_\lambda. \tag{7.38}$$

Therefore, the current J_c^μ is proportional to $\varepsilon^{\mu\nu\lambda}(\partial_\nu \delta\theta)(\partial_\lambda \delta\phi)$. The latter can be viewed as the $d = 2$ extension of the Goldstone–Wilczek formula in equation (3.26).

7.3.2 Magnetoelectric polarisation domain wall and quantum anomalous Hall effect

The chiral anomaly action obtained in the previous paragraph, and also discussed in section 5.5, describes the magnetoelectric coupling for a 3 + 1d TRI topological insulator. For an inhomogeneous magnetoelectric polarisation the effective action of the electromagnetic field can be rewritten as

[8] In appropriate units.

$$S_{\text{eff}} \propto \int d^4x\, (\partial_\rho P_3)\varepsilon^{\rho\mu\nu\lambda} A_\mu F_{\nu\lambda}. \tag{7.39}$$

To this end, we consider a z-dependent P_3, so that a spatial domain wall forms. The latter is specified via the boundary conditions: $P_3(z \to -\infty) = \pi$ and $P_3(z \to +\infty) = 0$. Thus, one can assume the form $P_3(z) = \pi[1 - \tanh(z/\ell)]/2$, where the lengthscale ℓ defines the intermediate region, across which, a transition from a topologically trivial to a non-trivial insulator occurs. In this region, the TRS is necessarily violated. For $\ell \to 0$ the domain wall is very steep and P_3 changes in a stepwise manner. As a result, a sharp interface is formed. According to the above expression, we obtain:

$$S_{\text{interface}} \propto \int d^3x\, \varepsilon^{\mu\nu\lambda} A_\mu F_{\nu\lambda}, \tag{7.40}$$

thus leading to the Chern–Simons action for the 2 + 1d interface. The above expression reflects the violation of TRS at the interface and the presence of massive Dirac surface states. The mass depends on the microscopic mechanism realising the domain wall. However, as a reminder to the reader, the Chern–Simons coefficient and the Hall conductance depend only on the sign of the mass. Remarkably, in the particular case $G_H = e^2/(2h)$. This is consistent with the parity anomaly for an odd number of massive Dirac fermions. For more details see section 5.1.3.

From the above, one infers that the proximity of a single surface of a three-dimensional TRI topological insulator to a ferromagnetic insulator will violate TRS and induce a mass to the Dirac surface states that, in turn, will be governed by a Chern–Simons action. The latter will lead to the corresponding anomalous Hall response. In this manner, a $d = 3$ TRI topological insulator can be employed for engineering a 2 + 1d Chern insulator and realising the QAHE.

A similar strategy was followed by the authors of [28], in order to experimentally realise the long-sought-after *integer* QAHE. For this purpose, they considered a thin 3 + 1d TRI insulator, and in particular the Cr-doped $(Bi,Sb)_2Te_3$ compound (see figure 7.8(a)). TRS was broken in this hybrid system due to the out-of-plane ferromagnetic alignment of the Cr magnetic moments. As a result, each one of the

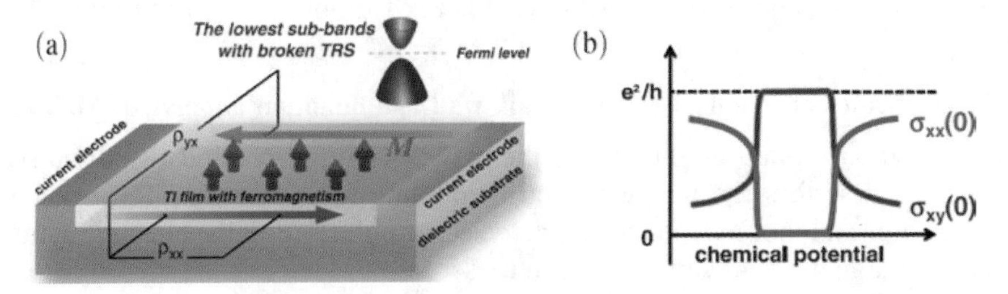

Figure 7.8. (a) Platform for engineering a quantum anomalous Hall system via magnetically doping a thin film of a TRI topological insulator. This was experimentally realised in [28] via a Cr-doped $(Bi,Sb)_2Te_3$ compound. (b) For a window of the chemical potential and in the zero magnetic-field limit, the longitudinal conductance is zero, while at the same time the Hall conductance is quantised to a single unit. Adapted from reference [28].

two surfaces of the thin 3 + 1d TRI topological insulator contributes to the Hall current with half a unit of conductance. The latter led to the observation of the integer QAHE, obtained in the zero-magnetic-field limit, with $G_H = e^2/h$ (see figure 7.8(b)).

7.4 Hands-on: quasiparticle interference on the topological surface

In this paragraph we further explore the properties of the helical modes described by equation (7.27). We focus on the profile of the related local density of states (LDOS), obtained upon adding a single impurity, see [29]. The LDOS can be probed via scanning tunneling microscopy. By Fourier transforming the spatial profile of the LDOS one can identify wavevectors, for which, one obtains a constructive or destructive interference. For energies at the Fermi level one can obtain information regarding the Fermi surface's $\boldsymbol{k}$-space structure [17, 19, 20]. The LDOS is given by the trace of the spatially-resolved spectral function defined in section 2.5: $A(\boldsymbol{r}, E) := -2\mathrm{Im}\left\{\mathrm{Tr}[\hat{G}^{\mathrm{R}}(\boldsymbol{r}, \boldsymbol{r}' \to \boldsymbol{r}, E)]\right\}$. The thermal spin-matrix Green function satisfies:

$$\begin{aligned}\hat{G}(\boldsymbol{r}, \boldsymbol{r}', ik_n) = \hat{G}^{(0)}(\boldsymbol{r} - \boldsymbol{r}', ik_n) + \int d\boldsymbol{r}_1 \int d\boldsymbol{r}_2\, \hat{G}^{(0)}(\boldsymbol{r} - \boldsymbol{r}_1, ik_n) \\ \times \hat{\mathrm{T}}(\boldsymbol{r}_1, \boldsymbol{r}_2, ik_n)\hat{G}^{(0)}(\boldsymbol{r}_2 - \boldsymbol{r}', ik_n),\end{aligned} \tag{7.41}$$

where $\hat{G}^{(0)}$ denotes the Green function in the absence of the impurity. $\hat{\mathrm{T}}(\boldsymbol{r}_1, \boldsymbol{r}_2, ik_n)$ defines the so-called T-matrix [29], given by the basis-independent expression: $\hat{\mathrm{T}} := \hat{\mathrm{U}} + \hat{\mathrm{U}}\hat{G}^{(0)}\hat{\mathrm{T}} \equiv \hat{\mathrm{U}} + \hat{\mathrm{T}}\hat{G}^{(0)}\hat{\mathrm{U}}$, with $\hat{\mathrm{U}}$ the operator corresponding to the impurity potential defined as: $\hat{H}_{\mathrm{U}} = \int d\boldsymbol{r}_1 \int d\boldsymbol{r}_2 \hat{\boldsymbol{\psi}}^{\dagger}(\boldsymbol{r}_1)\hat{\mathrm{U}}(\boldsymbol{r}_1, \boldsymbol{r}_2)\hat{\boldsymbol{\psi}}(\boldsymbol{r}_2)$.

Task 1: Consider an electric-charge-neutral impurity located at $\boldsymbol{r} = 0$, which couples to the electrons locally, i.e., $\hat{\mathrm{U}}(\boldsymbol{r}_1, \boldsymbol{r}_2) = \mathrm{U}\delta(\boldsymbol{r}_1)\delta(\boldsymbol{r}_1 - \boldsymbol{r}_2)\hat{1}_{\sigma}$. Show that the T-matrix satisfies:

$$\hat{\mathrm{T}}(\boldsymbol{k}_1, \boldsymbol{k}_2, ik_n) \equiv \hat{\mathrm{T}}(ik_n) = \frac{\mathrm{U}}{(2\pi)^2}\left[\hat{1}_{\sigma} - \mathrm{U}\int \frac{d\boldsymbol{k}}{(2\pi)^2}\, \hat{G}^{(0)}(\boldsymbol{k}, ik_n)\right]^{-1}. \tag{7.42}$$

Task 2: Based on Task 1, consider a warped Fermi surface as in figure 7.5(b). Project equation (7.41) onto the upper helicity eigenstate of equation (7.27). Calculate the LDOS for $E = E_F$. What is the contribution of wavevectors connecting TR-related $\boldsymbol{k}$ points of the Fermi surface? Explain your findings based on the topological protection of the helical surface states against weak backscattering. What changes in the case of a magnetic impurity? For the LDOS calculation see also [20].

7.5 Hands-on: topological Kondo insulator

The scope of the present section is to focus on a particular class of interacting TRI topological insulators, the so-called topological Kondo insulators (TKIs) [30]. The TKI is the result of the hybridisation between itinerant ($\hat{c}_{\alpha}(\boldsymbol{k})$) and localised ($\hat{f}_{\alpha}(\boldsymbol{k})$)

electrons. The former (latter) originate from s or d (f) atomic orbitals which form wide (narrow) bands. The TKI Hamiltonian consists of the following two parts:

$$\hat{H}_0 = \int dk \sum_{\alpha=\uparrow,\downarrow} \left\{ \varepsilon_c(\boldsymbol{k})\hat{c}_\alpha^\dagger(\boldsymbol{k})\hat{c}_\alpha(\boldsymbol{k}) + \varepsilon_f(\boldsymbol{k})\hat{f}_\alpha^\dagger(\boldsymbol{k})\hat{f}_\alpha(\boldsymbol{k}) + \sum_{\beta=\uparrow,\downarrow} \left[\hat{c}_\alpha^\dagger(\boldsymbol{k})\boldsymbol{V}(\boldsymbol{k}) \cdot \hat{\boldsymbol{\sigma}}_{\alpha\beta}\hat{f}_\beta(\boldsymbol{k}) + \text{h.c.} \right] \right\}, \tag{7.43}$$

$$\hat{V}_{\text{int}} = U_f \sum_n \hat{f}_{n,\uparrow}^\dagger \hat{f}_{n,\uparrow} \hat{f}_{n,\downarrow}^\dagger \hat{f}_{n,\downarrow}.$$

In the above, we introduced the energy dispersions: $\varepsilon_{c,f}(\boldsymbol{k}) = -t_{c,f}(\cos k_x + \cos k_y + \cos k_z)/6 - \varepsilon_{c,f}$ and the hybridisation vector function $\boldsymbol{V}(\boldsymbol{k}) = -\boldsymbol{V}(-\boldsymbol{k})$. The second part describes the repulsive Hubbard interaction of the localised electrons, with strength $U_f > 0$. The latter is expressed in terms of the direct-lattice operators $\hat{f}_{n,\alpha}$, obtained via a discrete Fourier transform from $\hat{f}_\alpha(\boldsymbol{k})$, cf section 2.2.

Task : Follow reference [30] and employ the so-called slave-boson mean-field theory. According to the latter, show that the mean-field TKI Hamiltonian is solely given by $\hat{H}_0$, though with renormalised $\boldsymbol{V}$ and t_f, i.e., $\boldsymbol{V} \mapsto \tilde{\boldsymbol{V}} = \sqrt{1 - n_f}\,\boldsymbol{V}$ and $t_f \mapsto \tilde{t}_f = (1 - n_f)t_f$, with n_f the mean-field derived occupation of the f-electrons. For the mean-field TKI Hamiltonian, calculate the $\mathbb{Z}_2$ invariants $\nu_{0,1,2,3}$, introduced in section 7.2.4.

References

[1] Hasan M Z and Kane C L 2010 *Rev. Mod. Phys.* **82** 3045
[2] Qi X-L and Zhang S-C 2011 *Rev. Mod. Phys.* **83** 1057
[3] Novoselov K S *et al* 2005 *Nature* **438** 197
[4] Kane C L and Mele E J 2005 *Phys. Rev. Lett.* **95** 146802
[5] Kane C L and Mele E J 2005 *Phys. Rev. Lett.* **95** 226801
[6] Fu L and Kane C L 2006 *Phys. Rev.* B **74** 195312
[7] Bernevig B A, Hughes T L and Zhang S-C 2006 *Science* **314** 1757
[8] König M *et al* 2007 *Science* **318** 766
[9] Fu L, Kane C L and Mele E J 2007 *Phys. Rev. Lett.* **98** 106803
[10] Moore J E and Balents L 2007 *Phys. Rev.* B **75** 121306(R)
[11] Fu L and Kane C L 2007 *Phys. Rev.* B **76** 045302
[12] Roy R 2009 *Phys. Rev.* B **79** 195322
[13] Zhang H *et al* 2009 *Nat. Phys.* **5** 438
[14] Hsieh D *et al* 2008 *Nature* **452** 970
[15] Xia Y *et al* 2009 *Nat. Phys.* **5** 398
[16] Chen Y L *et al* 2009 *Science* **325** 178
[17] Alpichshev Z *et al* 2010 *Phys. Rev. Lett.* **104** 016401
[18] Liu C-X *et al* 2010 *Phys. Rev.* B **82** 045122

[19] Fu L 2009 *Phys. Rev. Lett.* **103** 266801
[20] Lee W-C, Wu C, Arovas D P and Zhang S-C 2009 *Phys. Rev.* B **80** 245439
[21] Jiang J-H and Wu S 2011 *Phys. Rev.* B **83** 205124
[22] Baum Y and Stern A 2012 *Phys. Rev.* B **85** 121105 and Baum Y and Stern A 2012 *Phys. Rev.* B **86** 195116
[23] Mendler D, Kotetes P and Schön G 2015 *Phys. Rev.* B **91** 155405
[24] Fu L and Kane C L 2008 *Phys. Rev. Lett.* **100** 096407
[25] Ryu S, Schnyder A P, Furusaki A and Ludwig A W W 2010 *New J. Phys.* **12** 065010
[26] Ringel Z, Kraus Y E and Stern A 2012 *Phys. Rev.* B **86** 045102
[27] Qi X-L, Hughes T L and Zhang S-C 2008 *Phys. Rev.* B **78** 195424
[28] Chang C-Z *et al* 2013 *Science* **340** 167
[29] Balatsky A V, Vekhter I and Zhu J-X 2006 *Rev. Mod. Phys.* **78** 373
[30] Dzero M, Xia J, Galitski V and Coleman P 2016 *Annu. Rev. Condens. Matter Phys.* **7** 249

IOP Concise Physics

Topological Insulators

Panagiotis Kotetes

Chapter 8

Topological classification of insulators and beyond

In the present chapter we unify and generalise the results found earlier for the diverse classes of topological systems. This is achieved via the introduction of two antiunitary symmetries on top of the chiral symmetry broadly discussed so far in this book. These additional symmetries are the generalised time reversal and charge conjugation, which, together with chiral symmetry, give rise to ten symmetry classes and yield the topological classification of non-interacting insulators in arbitrary dimensions. This scheme is also suitable for classifying mean-field-derived Hamiltonians, describing superconductors with a gapped energy spectrum. Here, we introduce the corresponding topological-classification table and discuss the interconnection of the various symmetry classes and their topological invariants. In fact, as we discuss, the additional presence of unitary symmetries can modify the classification scheme above, and further allow for new types of topological phases of matter, e.g., the so-called crystalline topological insulators. Moving beyond insulating materials, we also focus on topologically non-trivial systems with a gapless energy spectrum, such as graphene and Weyl semimetals. The presentation of the topological properties of systems with both gapped and gapless bulk energy spectrum, additionally allows us to introduce the concept of twist defects. These can render the appearance of boundary states or modes possible, even in systems with a trivial strong topological invariant. We conclude with briefly mentioning important topics and directions that we do not cover in this book, such as: (i) topological superconductivity and Majorana fermions, (ii) the effects of interactions on the topological classification, (iii) the notion of symmetry-protected phases and topological order and (iv) topological phases in driven systems. Finally, the Hands-on sections focus on (i) the non-Abelian Berry curvature of semiconductors described by the Luttinger model and (ii) the so-called Floquet topological insulators.

doi:10.1088/978-1-68174-517-6ch8

8.1 Generalised antinunitary symmetries and symmetry classes

So far throughout this book, we exploited the chiral and TR symmetries to infer the topological properties of the insulators under discussion. In order to generalise this approach, it is required to: (i) extend the notion of TRS and (ii) introduce an additional antiunitary symmetry, termed charge conjugation (CC). In fact, the topological classification [1–9] to be discussed, assumes that there can be no other symmetries apart from the TR, CC and chiral symmetries. The latter, are, respectively, effected by the operators $\hat{\Theta}$, $\hat{\Xi}$ and $\hat{\Pi}$. These three symmetries can be present even for disordered systems, which renders this topological-classification scheme particularly suitable for investigating Hamiltonians described within the framework of the so-called Random Matrix Theory. For a review see [10, 11].

We append the requirements for a single-particle Hamiltonian to exhibit the two antiunitary symmetries. We have the defining relations[1]

$$\hat{\Theta}^{\dagger}\hat{H}(\hat{\boldsymbol{p}}, \hat{\boldsymbol{r}})\hat{\Theta} = +\hat{H}(\hat{\boldsymbol{p}}, \hat{\boldsymbol{r}}) \quad \Rightarrow \quad [\hat{H}(\hat{\boldsymbol{p}}, \hat{\boldsymbol{r}}), \hat{\Theta}] = \hat{0}, \tag{8.2}$$

$$\hat{\Xi}^{\dagger}\hat{H}(\hat{\boldsymbol{p}}, \hat{\boldsymbol{r}})\hat{\Xi} = -\hat{H}(\hat{\boldsymbol{p}}, \hat{\boldsymbol{r}}) \quad \Rightarrow \quad \{\hat{H}(\hat{\boldsymbol{p}}, \hat{\boldsymbol{r}}), \hat{\Xi}\} = \hat{0}. \tag{8.3}$$

Note, that, for a system with both symmetries present, one finds that also a chiral symmetry is induced and generated by $\hat{\Pi} \equiv \hat{\Xi}\hat{\Theta}$, thus, satisfying

$$\hat{\Pi}^{\dagger}\hat{H}(\hat{\boldsymbol{p}}, \hat{\boldsymbol{r}})\hat{\Pi} = -\hat{H}(\hat{\boldsymbol{p}}, \hat{\boldsymbol{r}}) \quad \Rightarrow \quad \{\hat{H}(\hat{\boldsymbol{p}}, \hat{\boldsymbol{r}}), \hat{\Pi}\} = \hat{0}. \tag{8.4}$$

Nonetheless, as discussed in the previous chapters, a Hamiltonian can possess a chiral symmetry without the necessary presence of TRS and/or CCS. By taking into account that $\hat{\Xi}^2 = \pm\hat{1}$ and $\hat{\Theta}^2 = \pm\hat{1}$, one defines the ten symmetry classes announced earlier, which we present in table 8.1.

The ten symmetry classes are divided into complex and real. The A and AIII classes span the complex ones. This distinction relies on the fact that both TRS and CCS are absent in the complex case. Consequently, the terms appearing in the single-particle Hamiltonian do not transform according to the IRs of the complex-conjugation operator $\hat{K}$. In the classification table, d denotes the number of spatial dimensions of the system and $\mathbb{Z}$ or $\mathbb{Z}_2$ corresponds to the topological invariant which can be defined for the given class and dimensionality. The latter classification only contains the so-called 'strong' invariants, i.e., the ones not residing on translational invariance or other unitary symmetries (see the discussion of section 7.2.4). To

[1] In the case of a crystalline system, the antiunitary symmetry operators have the following action on the single-particle Bloch Hamiltonian:

$$\hat{U}_{\Theta}^{\dagger}\hat{H}^{*}(-\boldsymbol{k})\hat{U}_{\Theta} = \hat{H}(\boldsymbol{k}) \quad \text{and} \quad \hat{U}_{\Xi}^{\dagger}\hat{H}^{*}(-\boldsymbol{k})\hat{U}_{\Xi} = -\hat{H}(\boldsymbol{k}). \tag{8.1}$$

In the above, we expressed the antiunitary symmetry operators as products of a unitary operator and complex conjugation, i.e., $\hat{\Theta} = \hat{U}_{\Theta}\hat{K}$ and $\hat{\Xi} = \hat{U}_{\Xi}\hat{K}$. This is analogous to writing the TR operator as $\hat{\mathcal{T}} = i\hat{\sigma}_y\hat{K}$.

Table 8.1. Tenfold symmetry and topological classification table valid for: (i) non-interacting insulators and (ii) superconductors with a fully gapped energy spectrum described at the mean-field level. The ten symmetry classes are distinguished using the outcome of $\hat{\Theta}^2$, $\hat{\Xi}^2$ and $\hat{\Pi}^2$, i.e., the squares of the operators associated with the generalised time-reversal, charge-conjugation and chiral symmetries. The symmetries are broken if the outcome is $\hat{0}$. The first two classes, A and AIII, comprise the so-called *complex* classes. The remaining correspond to the *real* classes. d denotes the number of spatial dimensions, and is restricted to the interval [1, 8] by virtue of the Bott periodicity [13]. The table only contains the strong topological invariants, while in the presence of unitary symmetries weak invariants can be additionally defined. For class A (AIII) the $\mathbb{Z}$ invariants are given by a Chern (winding) number. The $\mathbb{Z}$ topological invariants for the real classes, highlighted with blue and red, are also given by the same Chern or winding number characterising the respective complex class for the given d. Effecting dimensional reduction on the latter cases, yields the first and second $\mathbb{Z}_2$ descendants. The remaining sequence found for the real classes, highlighted here with magenta, involves systems characterised by an even $\mathbb{Z}$ invariant related to a Chern or winding number. These types of topological phases possess no parent or descendant phases. The construction of the present table relies on the classification tables of references [5, 14].

	Symmetries:			d							
Class	$\hat{\Theta}^2$	$\hat{\Xi}^2$	$\hat{\Pi}^2$	1	2	3	4	5	6	7	8
A	$\hat{0}$	$\hat{0}$	$\hat{0}$	–	$\mathbb{Z}$	–	$\mathbb{Z}$	–	$\mathbb{Z}$	–	$\mathbb{Z}$
AIII	$\hat{0}$	$\hat{0}$	$\hat{1}$	$\mathbb{Z}$	–	$\mathbb{Z}$	–	$\mathbb{Z}$	–	$\mathbb{Z}$	–
AI	$\hat{1}$	$\hat{0}$	$\hat{0}$	–	–	–	$2\mathbb{Z}$	–	$\mathbb{Z}_2$	$\mathbb{Z}_2$	$\mathbb{Z}$
BDI	$\hat{1}$	$\hat{1}$	$\hat{1}$	$\mathbb{Z}$	–	–	–	$2\mathbb{Z}$	–	$\mathbb{Z}_2$	$\mathbb{Z}_2$
D	$\hat{0}$	$\hat{1}$	$\hat{0}$	$\mathbb{Z}_2$	$\mathbb{Z}$	–	–	–	$2\mathbb{Z}$	–	$\mathbb{Z}_2$
DIII	$-\hat{1}$	$\hat{1}$	$\hat{1}$	$\mathbb{Z}_2$	$\mathbb{Z}_2$	$\mathbb{Z}$	–	–	–	$2\mathbb{Z}$	–
AII	$-\hat{1}$	$\hat{0}$	$\hat{0}$	–	$\mathbb{Z}_2$	$\mathbb{Z}_2$	$\mathbb{Z}$	–	–	–	$2\mathbb{Z}$
CII	$-\hat{1}$	$-\hat{1}$	$\hat{1}$	$2\mathbb{Z}$	–	$\mathbb{Z}_2$	$\mathbb{Z}_2$	$\mathbb{Z}$	–	–	–
C	$\hat{0}$	$-\hat{1}$	$\hat{0}$	–	$2\mathbb{Z}$	–	$\mathbb{Z}_2$	$\mathbb{Z}_2$	$\mathbb{Z}$	–	–
CI	$\hat{1}$	$-\hat{1}$	$\hat{1}$	–	–	$2\mathbb{Z}$	–	$\mathbb{Z}_2$	$\mathbb{Z}_2$	$\mathbb{Z}$	–

realise a topological phase transition, the bulk energy spectrum is required to close and reopen. However, this only holds as long as the system resides in the same symmetry class, both before and after the transition. Otherwise, a gap closing is not required [12].

In table 8.1, we restrict to $d \in [1, 8]$, since there exists a periodicity in terms of d. We find a period of 2 and 8, for the complex and real classes, respectively. This property is referred to as the Bott periodicity [13]. The classes A, AI and AII, were first explored by Wigner [1] and Dyson [2]. They focused on the symmetry classification of systems based on TRS and deepened in the properties of disordered system, thus, opening the field of Random Matrix Theory. Four additional nonstandard classes were discovered by Altland and Zirnbauer [3, 4] after introducing the notion of CCS. A catalyst for the further development of the symmetry and topological classification [5–9] was the experimental discovery of the TRI topological insulators already discussed in chapter 7.

8.2 The art of topological classification

In this paragraph we proceed with analysing table 8.1. Our approach is very closely connected to the preceding chapters. This connection allows us, for instance, to directly sort out the topological properties of the complex A and AIII classes.

8.2.1 Complex symmetry classes

The symmetry class A essentially describes all the Chern insulators. For these, one can define the respective n-th Chern number $C_n \in \mathbb{Z}$, accessible for all spacetimes with an even number of spatial dimensions. For more details see the discussion of section 5.4.4. In contrast, the systems belonging to the symmetry class AIII are dictated by chiral symmetry and, thus, one can define a winding number $w_n \in \mathbb{Z}$. As discussed in section 4.1.5, a winding number is only defined for an odd number of spatial dimensions $d - 1$, and is equal to the Chern number of the corresponding Chern insulator defined for d spatial dimensions. For further information turn to section 5.5.

8.2.2 Real symmetry classes

The topological properties of the real symmetry classes depend, to a large extend, on the features of the complex classes. As an example, consider a Chern insulator Hamiltonian in $d = 4$, belonging to the symmetry class A. As we remarked in section 7.3.1, the 2nd Chern number can be non-zero even in the presence of TRS. Since $\hat{\mathcal{T}}^2 = -\hat{1}$, its addition effects the transition A→ AII. Nevertheless, both symmetry classes are characterised by a $\mathbb{Z}$ topological invariant for $d = 4$. Therefore, the entries highlighted with red for the real classes in table 8.1 can be immediately viewed as extensions of the symmetry class A cases. A similar situation takes place for the symmetry class AIII and its related winding number, in the presence of additional symmetries. The respective entries for which we can directly extend the approach employed for AIII class systems are highlighted in table 8.1 with blue. Conclusively, one finds a regular pattern of dimensional reduction, accompanied by a symmetry class shift, according to which, the topological classification of the complex classes is directly applicable to the real ones. These cases correspond to the $\mathbb{Z}$ entries of the table which are coloured red and blue.

In addition, in section 7.3.1 we also pointed out that one can obtain the TRI topological insulators in $d = 2,3$, via dimensional reduction effected on the $d = 4$ Chern insulator. This system supports a non-zero 2nd Chern number even if the class switches to AII. Thus, applying successive dimensional reductions to a system dictated by a real symmetry class, and a $\mathbb{Z}$ topological invariant, leads to the so-called first and second $\mathbb{Z}_2$ topological insulator descendants. As a result, one retrieves a second pattern obtained via dimensional reduction, but this time without any symmetry class modification. According to this pattern, one encounters the sequence $\{\mathbb{Z}_2, \mathbb{Z}_2, \mathbb{Z}\}$, which we further analyse in the next subsection.

The last feature of the classification table remaining to be understood regards the entries with $2\mathbb{Z}$ [5]. As it appears, these particular cases are special since they are not directly derived by dimensional reduction, and at the same time they are sterile, i.e., do not lead to any direct topologically non-trivial descendants. To understand these types of cases better, let us construct an example of a symmetry class C Hamiltonian in $d = 2$. For this purpose we employ the spin-Pauli matrices $\hat{\boldsymbol{\sigma}}$ and consider the generic expression for the Hamiltonian $\hat{H}(\boldsymbol{k}) = \boldsymbol{g}(\boldsymbol{k}) \cdot \hat{\boldsymbol{\sigma}}$. For a class C Hamiltonian a CCS is present, which is effected by an operator satisfying $\hat{\Xi}^2 = -\hat{1}$. For the particular example the only possibility is $\hat{\Xi} = \mathrm{i}\hat{\sigma}_y\hat{K}$. Since $\hat{\Xi}^\dagger\hat{\boldsymbol{\sigma}}\hat{\Xi} = -\hat{\boldsymbol{\sigma}}$, one finds that the $\boldsymbol{g}$-vector has to satisfy the constraint $\boldsymbol{g}(\boldsymbol{k}) = \boldsymbol{g}(-\boldsymbol{k})$. Notably, such a Hamiltonian cannot be constructed using a single copy of Dirac-type Hamiltonians. However, one can, for instance, consider the Hamiltonian specified by the $\boldsymbol{g}$-vector:

$$\boldsymbol{g}_{\mathrm{C}}(\boldsymbol{k}) = \left(k_x^2 - k_y^2, 2k_xk_y, m^2 - \boldsymbol{k}^4\right). \tag{8.5}$$

The 1st Chern number is well-defined for the above Hamiltonian, and takes the value $C_1 = 2$, thus leading to a double winding (see also section 4.2). Note that the z component contains a quartic term, necessary for the compactification of $\boldsymbol{k}$ space, from $\mathbb{R}^2$ to $\mathbb{S}^2$. Nonetheless, similar to the bottom-up construction of section 7.1.1, here, we can also obtain a class C Hamiltonian in terms of two blocks of Dirac-type Hamiltonians. By labelling the two blocks using $\tau = \pm 1$, we consider[2]:

$$\boldsymbol{g}_{\mathrm{C}}^{\tau}(\boldsymbol{k}) = \left(\alpha\hbar k_y, -\alpha\hbar k_x, m - \beta(\hbar\boldsymbol{k})^2\right). \tag{8.6}$$

While each subblock actually belongs to the symmetry class D, there must exist a CCS operator satisfying $\hat{\Xi}^2 = -\hat{1}$, which ensures that the two blocks lead to the same partial 1st Chern number, thus, with a sum equal to $|C_1| = 2$. The role of $\hat{\Xi}$ becomes more transparent only if we add block-mixing terms, which however preserve the C symmetry class. Nevertheless, the particular block construction shows transparently that systems of this type can be viewed as consisting of an even number of Dirac–Schrödinger models, each of which, is always leading to a Chern or winding number of a single unit.

At this stage, we proceed with exploring the possibility of obtaining topologically non-trivial phases for the symmetry class C system above, but for one spatial dimension less. Thus, we apply the dimensional reduction method on the model of equation (8.5), via for instance setting $k_y = 0$,[3] and find

$$\boldsymbol{g}_{\mathrm{C}}(k_x) = \left(k_x^2, 0, m^2 - k_x^4\right). \tag{8.7}$$

For the $d = 1$ Hamiltonian above, we can define a winding number w_1, which is, however, identically zero. Essentially, the two-component $\boldsymbol{g}_C$ vector does not wind as we sweep k_x. This result constitutes an immediate consequence of the symmetry constraints related to the given class. The latter explains the absence of descendants for the $2\mathbb{Z}$ systems appearing in the topological classification of table 8.1.

[2] There exist alternative choices for the $\boldsymbol{g}_{\mathrm{C}}^{\tau}(\boldsymbol{k})$-vector which yield the same Chern number. They are obtained by multiplying any two components of the $\boldsymbol{g}_{\mathrm{C}}^{\tau}$-vector by τ, e.g., $\boldsymbol{g}_{\mathrm{C}}^{\tau}(\boldsymbol{k}) = \left(\tau\alpha\hbar k_y, -\tau\alpha\hbar k_x, m - \beta(\hbar\boldsymbol{k})^2\right)$.

[3] Note, that, alternatively setting $k_x = 0$, leads to similar results.

8.2.3 $\mathbb{Z}_2$ classification and relative Chern and winding numbers

Let us now proceed with discussing the cases classified by a $\mathbb{Z}_2$ invariant. The present analysis will generalise the discussion of sections 4.3.3, 5.5 and 7.3, where the $\mathbb{Z}_2$ topological invariants were connected to the electric and magnetoelectric polarisations $P_{1,3}$. Via dimensionally extending to $4 + 1$d, one can relate the magnetoelectric polarisation to the 2nd Chern number of the corresponding higher dimensional Chern insulator. The dimensional extension is achieved by considering an additional spatial dimension characterised by the wavenumber k_w. Alternatively, one can employ a pumping process, via introducing a time-dependent parameter θ, playing the role of k_w. For the remainder we employ the second approach, so to connect to the notation of [6], which we follow closely in this paragraph. This also allows us to relate to the dimensional-reduction approach of section 7.3.

There exist two families of TRI insulators in $3 + 1$d, obtained by setting $\theta = 0, \pi$ in the $4 + 1$d Chern insulator model. The two values distinguish two topological classes of TRI insulators, the topologically trivial and non-trivial, corresponding to $P_3 = 0$ and $P_3 = 1/2$, in appropriate units. Thus, the $\mathbb{Z}_2$ topological classification can be implemented via dimensional extension, or equivalently, via an interpolation between the two families of insulators defined for $\theta = 0$ and $\theta = \pi$. The latter crucially relies on the possibility of defining a Chern number for the interpolation. In fact, the possibility of defining a Chern number for the given symmetry class in $4 + 1$d, implies that we can always find an interpolation that respects the symmetries of the given class and at the same time preserves the gap in the bulk energy spectrum.

To implement the interpolation and define a Chern number, it is required to introduce $\theta \in [0, 2\pi)$, which leads to a closed interpolating path: $\theta = 0 \to \theta = \pi \to \theta = 2\pi \equiv 0$. Nevertheless, the presence of a discrete antiunitary symmetry sets constraints on the value of the integral of the Berry curvature for the two path segments $\theta \in [0, \pi]$ and $\theta \in [\pi, 2\pi]$. In fact, the presence of the discrete symmetry implies that $P_3(\theta) = -P_3(2\pi - \theta)$, thus leading to $|C_2| = 2P_3$. The above discussion implies that all the equivalent interpolations are characterised by Chern numbers differing by an even integer, since P_3 is defined modulo an integer (see also section 4.3.3). This motivates us to introduce a $\mathbb{Z}_2$ invariant that only depends on the Hamiltonians defined for $\theta = 0$ and $\theta = \pi$, i.e.:

$$N_3[\hat{H}(\boldsymbol{k}, \theta = 0), \hat{H}(\boldsymbol{k}, \theta = \pi)] = (-1)^{C_2(\theta)} \quad \Rightarrow \quad N_3[\hat{H}_1(\boldsymbol{k}), \hat{H}_2(\boldsymbol{k})] = (-1)^{C_2(\theta)}, \tag{8.8}$$

being independent of the interpolation employed to calculate C_2. The above is also termed a relative Chern parity, essentially performing a comparison between the two Hamiltonians $\hat{H}_{1,2}(\boldsymbol{k})$ interpolated via employing a parameter θ. Thus, if we desire to infer if a Hamiltonian is topologically trivial or not, we can compare it to a Hamiltonian whose topological class is aleady known. The simplest choice is to make use of a trivial Hamiltonian $\hat{H}_0(\boldsymbol{k})$, as a reference for defining the relative Chern parity. Under these conditions, one can define the $\mathbb{Z}_2$ topological invariant for the $d = 3$ Hamiltonian $\hat{H}(\boldsymbol{k})$ as:

$$\nu_0 := N_3[\hat{H}_0(\boldsymbol{k}), \hat{H}(\boldsymbol{k})], \tag{8.9}$$

which coincides with the strong $\mathbb{Z}_2$ invariant defined by Fu and Kane, that was presented in section 7.2.4.

A similar procedure can be followed to obtain the second $\mathbb{Z}_2$ descendants. In this case, one aims at finding a topological invariant to classify Hamiltonians defined in (k_x, k_y) space, which are additionally invariant under the antiunitary symmetry defined by the ensuing symmetry class. Similar to the 3 + 1d case, one can introduce a variable $\varphi \in [0, 2\pi)$ playing the role of k_z, in order to define an interporation between the two families of insulators, with Hamiltonians given by $\hat{H}(k_x, k_y, \varphi = 0)$ and $\hat{H}(k_x, k_y, \varphi = \pi)$. Note, that, the corresponding symmetry has to be preserved during this process. Therefore, we wish to identify a topological invariant

$$N_2\left[\hat{H}\left(k_x, k_y, \varphi = 0\right), \hat{H}\left(k_x, k_y, \varphi = \pi\right)\right] \tag{8.10}$$

with the property that is independent of the particular Hamiltonian $\hat{H}(k_x, k_y, \varphi)$, or any other $\hat{H}'(k_x, k_y, \varphi)$, employed for the interpolation.

Based on the discussion above, we can introduce the relative Chern parity for two different interpolations as

$$N_3\left[\hat{H}\left(k_x, k_y, \varphi, \theta = 0\right), \hat{H}\left(k_x, k_y, \varphi, \theta = \pi\right)\right] = (-1)^{C_2(\varphi,\theta)}, \tag{8.11}$$

via the mapping $\hat{H}(k_x, k_y, \varphi) \mapsto \hat{H}(k_x, k_y, \varphi, \theta = 0)$ and $\hat{H}'(k_x, k_y, \varphi) \mapsto \hat{H}(k_x, k_y, \varphi, \theta = \pi)$. In this manner, we have:

$$N_3\left[\hat{H}\left(k_x, k_y, \varphi\right), \hat{H}'\left(k_x, k_y, \varphi\right)\right] = (-1)^{C_2(\varphi,\theta)}. \tag{8.12}$$

However, starting from the rhs and exchanging the roles of θ and φ we have

$$(-1)^{C_2(\varphi,\theta)} \equiv N_3\left[\hat{H}\left(k_x, k_y, \varphi = 0, \theta\right), \hat{H}\left(k_x, k_y, \varphi = \pi, \theta\right)\right]. \tag{8.13}$$

Since $\hat{H}(k_x, k_y, \varphi)$ and $\hat{H}'(k_x, k_y, \varphi)$ both constitute interpolations between $\hat{H}_1(k_x, k_y)$ and $\hat{H}_2(k_x, k_y)$, corresponding to $\varphi = 0, \pi$, one obtains

$$(-1)^{C_2(\varphi,\theta)} \equiv N_3\left[\hat{H}_1\left(k_x, k_y, \theta\right), \hat{H}_2\left(k_x, k_y, \theta\right)\right]. \tag{8.14}$$

Since the Hamiltonians $\hat{H}_{1,2}(k_x, k_y)$ are independent of θ, they constitute trivial Hamiltonians with respect to the ν_0 invariant. This leads to $(-1)^{C_2(\varphi,\theta)} = 1$. As a result, we find $N_3\left[\hat{H}(k_x, k_y, \varphi), \hat{H}'(k_x, k_y, \varphi)\right] = 1$, implying that the quantity in equation (8.10) is independent of the chosen interpolation. Consequently, we can employ a Hamiltonian $\hat{H}_0(k_x, k_y)$, that, we know is topologically trivial, and define the topological invariant, N_2, for any given Hamiltonian $\hat{H}(k_x, k_y)$, in the following manner:

$$\nu_{d=2} := N_2\left[\hat{H}_0\left(k_x, k_y\right), \hat{H}\left(k_x, k_y\right)\right] = \nu_0\left[\hat{H}\left(k_x, k_y, \varphi\right)\right] \equiv (-1)^{P_\theta}, \tag{8.15}$$

with $\hat{H}(k_x, k_y, \varphi)$ interpolating $\hat{H}_0(k_x, k_y) \equiv \hat{H}(k_x, k_y, \varphi = 0)$ and $\hat{H}(k_x, k_y) \equiv \hat{H}(k_x, k_y, \varphi = \pi)$.

Example. In order to render the above analysis more transparent, let us carry out the $\mathbb{Z}_2$ classification for a Hamiltonian assuming a similar form to the one examined in the Hands-on section 4.7. We here consider

$$\hat{H}(k_x) = A \sin k_x \hat{\sigma}_y + (m - 2t \cos k_x)\hat{\sigma}_z + B \sin^3 k_x \hat{\sigma}_x. \tag{8.16}$$

The above Hamiltonian is lacking chiral symmetry but is characterised by a CCS with $\hat{\Xi} = \hat{\sigma}_x \hat{K}$. According to the topological-classification table 8.1, we find that this Hamiltonian belongs to the symmetry class D, supporting a $\mathbb{Z}_2$ topological invariant. This follows from the possibility of defining a $\mathbb{Z}$ invariant in $d = 2$. To obtain a $\mathbb{Z}_2$ classification as above, we employ an interpolation $\hat{H}(k_x, \theta)$ from a topologically trivial Hamiltonian $\hat{H}_0(k_x) \equiv \hat{H}(k_x, \theta = 0)$ and the Hamiltonian of equation (8.16) given by $\hat{H}(k_x) \equiv \hat{H}(k_x, \theta = \pi)$. The interpolating Hamiltonian has to belong to the symmetry class D in the augmented (k_x, θ) space and, thus, it is required to satisfy:

$$\hat{\sigma}_x \hat{H}^*(-k_x, -\theta)\hat{\sigma}_x = -\hat{H}(k_x, \theta). \tag{8.17}$$

By taking into account the above constraint, we here make the following choice for the interpolation Hamiltonian

$$\begin{aligned} \hat{H}(k_x, \theta) = & A \sin k_x \hat{\sigma}_y + [m - 2t \cos k_x + \Lambda(1 + \cos\theta)/2]\hat{\sigma}_z \\ & + \left(B \sin^3 k_x + B' \sin\theta\right)\hat{\sigma}_x. \end{aligned} \tag{8.18}$$

For $\theta = \pi$ we directly obtain the Hamiltonian under investigation, while for $\theta = 0$, we find:

$$\hat{H}_0(k_x) = A \sin k_x \hat{\sigma}_y + (\Lambda + m - 2t \cos k_x)\hat{\sigma}_z + B \sin^3 k_x \hat{\sigma}_x \equiv \boldsymbol{g}(k_x) \cdot \hat{\boldsymbol{\sigma}}. \tag{8.19}$$

We ensure that the above Hamiltonian is topologically trivial, by considering that $|\Lambda + m| > 2|t|$. The latter condition forbids gap closings and, thus, does not allow for any topological phase transitions. At the same time, $\hat{H}_0$ can be adiabatically connected to a trivial Hamiltonian with a $\boldsymbol{g}$-vector polarised along the z direction, by taking the limit $\Lambda \to +\infty$. It is straighforward to calculate the 1st Chern number for the interpolation Hamiltonian $\hat{H}(k_x, \theta)$, after introducing the Berry vector potential with components $A_{k_x}(k_x, \theta)$ and $A_\theta(k_x, \theta)$ of the occupied band.

To avoid the straightforward but tedious calculations, here, we instead: (i) identify the gap closing points, (ii) linearise the Hamiltonian about them, and (iii) add up the partial Chern numbers. There exist four gap closing points located at the BZ's high-symmetry points[4]: $\Gamma(0, 0)$, $X(\pi, 0)$, $Y(0, \pi)$ and $M(\pi, \pi)$. We linearise the Hamiltonian about these points and obtain:

[4] The nomenclature for the gap closing points in (k_x, θ) space, follows the one employed for the high-symmetry points in the BZ of a tetragonal crystal.

$$\hat{H}_{\Gamma}(k_x, \theta) = +Ak_x\hat{\sigma}_y + (\Lambda + m - 2t)\hat{\sigma}_z + B'\theta\hat{\sigma}_x, \tag{8.20}$$

$$\hat{H}_{\mathrm{X}}(k_x, \theta) = -Ak_x\hat{\sigma}_y + (\Lambda + m + 2t)\hat{\sigma}_z + B'\theta\hat{\sigma}_x, \tag{8.21}$$

$$\hat{H}_{\mathrm{Y}}(k_x, \theta) = +Ak_x\hat{\sigma}_y + (m - 2t)\hat{\sigma}_z - B'\theta\hat{\sigma}_x, \tag{8.22}$$

$$\hat{H}_{\mathrm{M}}(k_x, \theta) = -Ak_x\hat{\sigma}_y + (m + 2t)\hat{\sigma}_z - B'\theta\hat{\sigma}_x. \tag{8.23}$$

For each one of the above Hamiltonians, and in the vicinity of the respective expansion point, the $\boldsymbol{g}$-vector gives rise to half of a skyrmion structure in (k_x, θ) space (also refer to figure 5.3(a)). This half skyrmion configuration is termed a meron, and contributes with a fractional unit $\pm 1/2$ to the Chern number. Thus, we find the meron contributions:

$$C_1^{\Gamma} = -\frac{1}{2}\,\mathrm{sgn}\,(AB')\,\mathrm{sgn}(\Lambda + m - 2t), \tag{8.24}$$

$$C_1^{\mathrm{X}} = +\frac{1}{2}\,\mathrm{sgn}\,(AB')\,\mathrm{sgn}(\Lambda + m + 2t), \tag{8.25}$$

$$C_1^{\mathrm{Y}} = +\frac{1}{2}\,\mathrm{sgn}\,(AB')\,\mathrm{sgn}(m - 2t), \tag{8.26}$$

$$C_1^{\mathrm{M}} = -\frac{1}{2}\,\mathrm{sgn}\,(AB')\,\mathrm{sgn}(m + 2t), \tag{8.27}$$

thus resulting to the Chern number

$$\begin{aligned} C_1 &= -\,\mathrm{sgn}\,(AB')\frac{\mathrm{sgn}(2t + m) + \mathrm{sgn}(2t - m)}{2} \\ &= -\,\mathrm{sgn}\,[AB'(2t + m)]\frac{1 + \mathrm{sgn}[(2t)^2 - m^2]}{2}, \end{aligned} \tag{8.28}$$

where we took into account the condition $|\Lambda + m| > 2|t|$. Therefore, $|C_1| = 1$ for $|2t| > |m|$, which determines the parameter regime, for which, a system described by the Hamiltonian under discussion transits to the $\mathbb{Z}_2$ topologically non-trivial phase. Our results are also consistent with the ones obtained in the Hands-on section 4.7 by calculating the electric polarisation.

Notably, the contributions to the topological invariant originate from the points with $\theta = \pi$, i.e., related to gap closings at $k_x = 0, \pi$ of the Hamiltonian of interest. This was anticipated since the reference Hamiltonian for calculating this invariant was chosen to be trivial. Moreover, note that the gap closing points satisfy $Ik_x \equiv k_x$, i.e., constitute inversion-symmetric points. As a matter of fact, the gap closing points necessarily have to be inversion-symmetric due to the presence of the CCS. Interestingly, by virtue of the CCS, we can define an antisymmetric matrix [15] for the k_I points, similar to the sewing matrix introduced in section 7.1.2. The matrix under discussion, termed here $\hat{B}_{k_I}$, is obtained by employing CCS, i.e.,

$$\hat{\sigma}_x \hat{H}^*(k_I)\hat{\sigma}_x = -\hat{H}(k_I) \quad \Rightarrow \quad [\hat{\sigma}_x \hat{H}(k_I)]^{\mathrm{T}} = -\hat{\sigma}_x \hat{H}(k_I). \tag{8.29}$$

From the above, we define $\hat{B}_{k_I} := \hat{U}_\Xi \hat{H}(k_I)$, where $\hat{U}_\Xi = \hat{\sigma}_x$. Since an antisymmetric matrix can be characterised by its Pfaffian, we can construct a $\mathbb{Z}_2$ invariant as follows

$$\mathcal{M} = \operatorname{sgn} \prod_{k_I} \operatorname{Pf}[\hat{B}_{k_I}]. \tag{8.30}$$

In the present example: $\hat{B}_0 = -(m - 2t)\mathrm{i}\hat{\sigma}_y$ and $\hat{B}_\pi = -(m + 2t)\mathrm{i}\hat{\sigma}_y$, thus, leading to

$$\mathcal{M} = \operatorname{sgn}[m^2 - (2t)^2]. \tag{8.31}$$

The system enters the topologically non-trivial phase when $\mathcal{M} < 0$, which is in complete agreement with the result obtained via the interpolation and the Chern number.

8.2.4 Weak topological invariants and flat bands

In section 7.2.4 we introduced the so-called weak topological invariants, by relying on the crystalline periodicity. In fact, as we discussed in the same section and presented in figure 7.7, one can obtain topologically non-trivial phases, in which, the strong topological invariant is trivial, while, at least one of the weak invariants is not. Such phases are not contained in the classification table 8.1. However, it is straightforward to define weak invariants by considering that a number of the wavevector components constitute parameters of the Hamiltonian, thus, effectively reducing the dimensionality of spacetime. The weak invariant is then defined as a strong invariant in the reduced dimensionality.

In order to clarify the above statement, let us consider the following two-dimensional extension of the Hamiltonian in equation (8.16):

$$\hat{H}(k_x, k_y) = A \sin k_x \hat{\sigma}_y + \left(m - 2t \cos k_x - 2t \cos k_y\right)\hat{\sigma}_z + B \sin^3 k_x \hat{\sigma}_x. \tag{8.32}$$

The Hamiltonian above resides in class D, similar to its 1d cousin, and can be classified using the 1st Chern number. Direct calculation yields $C_1 = 0$. However, if k_y is viewed as a parameter, then we can define a $\mathbb{Z}_2$ topological invariant for a given one-dimensional subsystem labelled by k_y, i.e.

$$\mathcal{M}(k_y) = \operatorname{sgn}\left[\left(m - 2t \cos k_y\right)^2 - (2t)^2\right]. \tag{8.33}$$

For the k_y values leading to $|m - 2t \cos k_y| < 2|t|$, one obtains a single zero-energy bound state per edge. Therefore, there exists a finite window of states with $E(k_y) = 0$, centered at $k_y = \pm\pi/2$. Since these modes modes are non-dispersive, they give rise to the so-called flat bands, shown in figure 8.1.

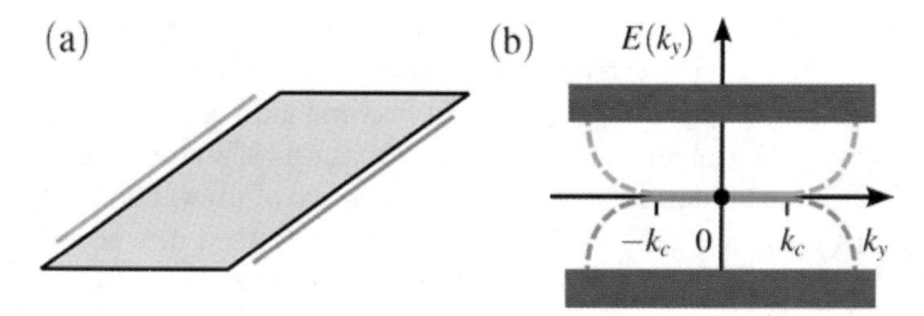

Figure 8.1. (a) Edge modes protected by a weak topological invariant by virtue of translational invariance. They are non-chiral, thus, the orange and purple lines lack a direction arrow. (b) Corresponding energy spectrum. One obtains a flat band for $k_y \in [-k_c, k_c]$. The edge mode energy branches merge with the continuum for large k_y.

Let us now investigate further properties of the emerging flat bands. The discussion will be facilitated by considering the continuum version of the above Hamiltonian, defined in the following fashion

$$\hat{H}(k_x, k_y) = \alpha\hbar k_x\hat{\sigma}_y + [m - \beta(\hbar \boldsymbol{k})^2]\hat{\sigma}_z, \tag{8.34}$$

where for convenience we dropped the cubic term in k_x. For the given model, and $m, \beta > 0$, flat bands appear at each edge for all $k_y \in [-k_c, k_c]$ with $\hbar k_c = \sqrt{m/\beta}$. Depending on the value of k_c, the flat bands can merge with the continuum for large k_y (see figure 8.1). Note, that, the addition of any term that can render the 1st Chern number non-zero, will lead to a finite slope for the flat bands, thus, rendering them chiral. This can be immediately verified by adding a term: $\propto k_y\hat{\sigma}_x$. Based on the analysis of section 5.2, we obtain the chiral spectrum[5] $E_\sigma(k_y) = -\sigma\alpha\hbar k_y$, with σ labelling the eigenstates of the chiral-symmetry operator $\hat{\sigma}_x$, dictating the Hamiltonian of equation (8.34).

One should note that the flat 'bands' are also present for a system where boundaries exist in both directions. In this case, we consider a hard-wall condition, according to which the wavefunctions at the boundaries should vanish. Thus, the wavefunctions for the flat 'bands' at the two edges read

$$\boldsymbol{\phi}_{0,-,n}(\boldsymbol{r}) \propto \sin\left(\frac{n\pi y}{L_y}\right)\sin\left[\zeta(x - x_1)\right]e^{-(x-x_1)/\xi}\frac{1}{\sqrt{2}}\begin{pmatrix}1\\-1\end{pmatrix}, \tag{8.35}$$

$$\boldsymbol{\phi}_{0,+,n}(\boldsymbol{r}) \propto \sin\left(\frac{n\pi y}{L_y}\right)\sin\left[\zeta(x - x_2)\right]e^{+(x-x_2)/\xi}\frac{1}{\sqrt{2}}\begin{pmatrix}1\\1\end{pmatrix} \tag{8.36}$$

with ζ and ξ defined in a similar fashion to the ones of sections 4.1.1 and 5.2. In the above, we introduced the length of the system along the y axis L_y, and took into account that the hard-wall condition leads to the quantisation: $k_y = n\pi/L_y$, with $n = 1, 2, 3, \ldots$.

As already mentioned, the appearance of the edge modes with a flat energy dispersion relies on the translational invariance of the system along the y axis.

[5] For sufficiently small k_y (see figure 5.1(b)).

Consequently, these edge modes are vulnerable to disorder, for instance, introduced through $m = m(y)$. In this case, the above wavefunctions labelled by the different n will hybridise and their energy spectrum will become gapped. The latter reveals the topologically-trivial character of the model of equation (8.34) in $d = 2$. Furthermore, flat bands are also particularly unstable in the presence of interactions, since they are energy degenerate. As a result, the flat bands can be gapped out, see [16], since the interactions provide another route for driving a topological phase transition without closing the bulk energy gap.

8.2.5 Topological classification with unitary symmetries

As we have mentioned in the sections above, the topological-classification table deals with systems lacking any unitary symmetries, left only with the option of being characterised by TR, CC or chiral symmetries. Nonetheless, the topological classification can be performed even when unitary symmetries are present [5]. The way to proceed in such cases is to first diagonalise the Hamiltonian in irreducible blocks via employing the irreducible representations of the group defined by the set of unitary symmetry operators present. Having completed the latter, one can then apply the usual classification methods in order to topologically categorise each one of the Hamiltonian blocks.

We now consider several examples of this type. We start with the simplest case and consider a Hamiltonian $\hat{H}$, which is characterised by a unitary symmetry O. The latter is effected by an operator $\hat{O} = e^{i\theta}\hat{U}_O$, satisfying $\hat{U}_O^2 = \hat{1}$. The discussion and results to follow are independent of the choice of θ. By virtue of the structure of $\hat{U}_O$, and along the lines of section 1.1.4, there exist two irreducible representations $|\pm\rangle$, for which $\hat{U}_O|\pm\rangle = \pm|\pm\rangle$. Therefore, we can write $\hat{H} = \hat{H}_+ \oplus \hat{H}_-$, with $|\pm\rangle$ corresponding to the irreducible representations of $\hat{U}_O$. If the Hamiltonian is lacking any other unitary and antiunitary symmetry, then each block belongs to symmetry class A. Note that a similar construction was employed in section 7.1.1, for the bottom-up construction of a 2 + 1d TRI topological insulator. A simple implementation for such a scenario, with $\hat{U}_O = \hat{\tau}_z$, can be provided using the following Hamiltonian

$$\hat{H}(\boldsymbol{k}) = \alpha\hbar k_x\hat{\sigma}_y - \alpha\hbar k_y\hat{\sigma}_x + [m - \beta(\hbar\boldsymbol{k})^2]\hat{\tau}_z\hat{\sigma}_z. \tag{8.37}$$

If, instead, there exists for instance a CCS with $\hat{\Xi}$ on top of a unitary symmetry effected by the operator $\hat{O}$, then one can define an additional CCS operator $\hat{\Xi}' = \hat{O}\hat{\Xi}$, since the latter satisfies $(\hat{\Xi}')^\dagger\hat{H}\hat{\Xi}' = -\hat{H}$. The presence of two CCS operators does not allow one to perform a symmetry and topological classification of the Hamiltonian, unless we first block diagonalise it. However, we can obtain a hint by examining $\hat{\Xi}^2$ and $(\hat{\Xi}')^2$. If the latter have the same sign then all the irreducible Hamiltonian blocks share the same symmetry class. Otherwise, the symmetry class of each block has to be inferred separately, since one of the CSS would lead to the symmetry class D and the other to C, thus, resulting in ambiguity. Let us now study a concrete example. We can again examine the properties of the

Hamiltonian presented above, but now we assume that the elements are constrained to have a CCS with $\hat{\Xi} = \hat{\sigma}_x\hat{K}$. In addition, we consider the extension

$$\hat{H}'(\boldsymbol{k}) = \alpha\hbar k_x\hat{\sigma}_y - \alpha\hbar k_y\hat{\sigma}_x + [m - \beta(\hbar\boldsymbol{k})^2]\hat{\eta}_x\hat{\tau}_z\hat{\sigma}_z + \gamma\hbar k_x\hat{\eta}_y\hat{\tau}_z\hat{\sigma}_z + \delta\hbar k_x\hat{\eta}_x\hat{\tau}_x + \zeta\hbar k_y\hat{\eta}_z\hat{\tau}_y\hat{\sigma}_z, \tag{8.38}$$

where we enlarged the Hilbert space by also considering the abstract η-space and introducing the Pauli matrices $\boldsymbol{\eta}$. The above Hamiltonian continues to have the CSS effected by $\hat{\Xi} = \hat{\sigma}_x\hat{K}$. Even more, one finds further symmetries, as for instance the CCS effected by $\hat{\Xi}' = \hat{\eta}_z\hat{\tau}_y\hat{\sigma}_x\hat{K}$ which satisfies $(\hat{\Xi}')^2 = -\hat{1}$. Thus, one finds a unitary symmetry, with $\hat{U}_O = \hat{U}_\Xi\hat{U}_{\Xi'} = \hat{\eta}_z\hat{\tau}_y$. The latter squares to $\hat{1}$. We can block diagonalise the above Hamiltonian via the unitary transformation $\hat{U} = (\hat{\eta}_z\hat{\tau}_y + \hat{\tau}_z)/\sqrt{2}$, which essentially modifies only the last two terms, i.e., $\hat{\eta}_x\hat{\tau}_x \mapsto \hat{\eta}_y$ and $\hat{\eta}_z\hat{\tau}_y\hat{\sigma}_z \mapsto \hat{\tau}_z\hat{\sigma}_z$, and leads to two blocks labelled by $\tau = \pm 1$, which enumerates the eigenvalues of $\hat{\tau}_z$. Therefore, we have

$$\hat{H}_\tau'(\boldsymbol{k}) = \alpha\hbar k_x\hat{\sigma}_y - \alpha\hbar k_y\hat{\sigma}_x + \tau[m - \beta(\hbar\boldsymbol{k})^2]\hat{\eta}_x\hat{\sigma}_z + \tau\gamma\hbar k_x\hat{\eta}_y\hat{\sigma}_z + \delta\hbar k_x\hat{\eta}_y + \tau\zeta\hbar k_y\hat{\sigma}_z. \tag{8.39}$$

By inspection, one concludes that the above Hamiltonian blocks belong to the symmetry class A. Therefore, we conclude that the presence of the unitary symmetry converted the classes C and D into A. For more details regarding such symmetry-class conversion, see [17].

8.2.6 Crystalline topological insulators

A similar symmetry-class conversion was essentially employed by Fu [18], in order to define a new class of topological systems, the so-called crystalline topological insulators. Fu relied on the combination of a Θ symmetry, with $\hat{\Theta}^2 = \hat{1}$, and the invariance under fourfold rotations generated by $\hat{C}_4$,[6] to show that the product $\hat{\Theta}' = \hat{\Theta}\hat{C}_4$ qualifies as a generalised TRS operator squaring to $-\hat{1}$. As a result, the symmetry class of the systems becomes AII, identical to the one dictating the TRI topological insulators discussed in chapter 7. Therefore, topologically-protected edge states are expected. To clarify the above, let us consider the system initially employed by Fu, consisting of a two-orbital model for spinless electrons defined in a two-atom unit cell. The considered two orbitals are assumed to belong to a two-dimensional irreducible representation of the point group.

For the remainder, we consider a D_{4h} point group symmetry and, thus, possible candidates for the required two orbitals are, for instance, the $\{p_x, p_y\}$ or $\{d_{xz}, d_{yz}\}$. For our discussion, either selection leads to the same results. To represent Hamiltonians in the orbital basis, we employ the $\hat{\boldsymbol{\tau}}$ Pauli matrices. Moreover, following Fu, we consider that the two-atom unit cell originates from a stacking of

[6] For more details see sections 1.1.1, 1.1.2, 1.1.4 and 1.3.

two different layers along the z axis, see figure 8.2. We denote the arising sublattice degrees of freedom 'a' and 'b'. These, remain unaffected under a fourfold rotation in the xy plane. We employ the $\hat{\boldsymbol{\eta}}$ Pauli matrices to represent Hamiltonians in sublattice space. Within the above framework, and by recalling equation (1.16), a C_4 rotation[7] in the combined orbital-sublattice space, is effected by the operator: $\hat{C}_4 = -\mathrm{i}\hat{\tau}_y$. From the above, it becomes apparent that if the Hamiltonian is additionally invariant under K, then the operator $\hat{\Theta}' := \mathrm{i}\hat{\tau}_y\hat{K}$, also leads to a generalised TR with $(\hat{\Theta}')^2 = -\hat{1}$. Remarkably, the latter form is reminiscent of the usual expression of the TR operator $\hat{\mathcal{T}} = \mathrm{i}\hat{\sigma}_y\hat{K}$, with the spin replaced by the orbital degree of freedom.

To verify the above, we consider the Hamiltonian proposed by Fu, which in our notation reads

$$\begin{aligned}\hat{H}(\boldsymbol{k}) = &\left[t_1\left(\cos k_x + \cos k_y\right) + 2t_2 \cos k_x \cos k_y\right]\hat{\eta}_z + t_1\left(\cos k_x - \cos k_y\right)\hat{\eta}_z\hat{\tau}_z \\ &+ 2t_2 \sin k_x \sin k_y\hat{\eta}_z\hat{\tau}_x + \left[t_1' + 2t_2'\left(\cos k_x + \cos k_y\right) + t_z' \cos k_z\right]\hat{\eta}_x \\ &- t_z' \sin k_z\hat{\eta}_y.\end{aligned} \tag{8.40}$$

It is straightforward to verify that $\hat{\Theta} = \hat{K}$, $\hat{C}_4$ and $\hat{\Theta}' = \hat{C}_4\hat{\Theta}$ constitute symmetries of the above Hamiltonian. The presence of the Θ' symmetry allows us to introduce a $\mathbb{Z}_2$ invariant residing to the Pfaffian of the inversion-symmetric wavevectors of the BZ, $\boldsymbol{k}_I$, which are simultaneously invariant under C_4.[8] These points will be referred to as $\boldsymbol{k}_{I,C_4}$ and consist of the $\Gamma(0, 0, 0)$, $\mathrm{M}(\pi, \pi, 0)$, $\mathrm{Z}(0, 0, \pi)$ and $\mathrm{A}(\pi, \pi, \pi)$ points shown in figure 8.2(b). The sewing matrix in the present case is determined via:

$$\hat{\tau}_y\hat{H}^*\left(\boldsymbol{k}_{I,C_4}\right)\hat{\tau}_y = \hat{H}\left(\boldsymbol{k}_{I,C_4}\right) \quad \Rightarrow \quad \left[\hat{\tau}_y\hat{H}\left(\boldsymbol{k}_{I,C_4}\right)\right]^{\mathrm{T}} = -\hat{\tau}_y\hat{H}\left(\boldsymbol{k}_{I,C_4}\right). \tag{8.41}$$

The above implies that we can introduce the sewing matrix for the $\boldsymbol{k}_{I,C_4}$ points as, $\hat{W}_{\boldsymbol{k}_{I,C_4}} := \mathrm{i}\hat{\tau}_y\hat{H}(\boldsymbol{k}_{I,C_4})$, which reads

$$\begin{aligned}\hat{W}_{\boldsymbol{k}_{I,C_4}} = &\left[t_1\left(\cos k_x + \cos k_y\right) + 2t_2 \cos k_x \cos k_y\right]\hat{\eta}_z\mathrm{i}\hat{\tau}_y \\ &+ \left[t_1' + 2t_2'\left(\cos k_x + \cos k_y\right) + t_z' \cos k_z\right]\hat{\eta}_x\mathrm{i}\hat{\tau}_y,\end{aligned} \tag{8.42}$$

with the above to be evaluated at $\boldsymbol{k}_{I,C_4}$. Since the matrix is antisymmetric, its Pfaffian is well defined. For the given $\boldsymbol{k}_{I,C_4}$ points, we obtain

$$\begin{aligned}\mathrm{Pf}\left[\hat{W}_{\boldsymbol{k}_{I,C_4}}\right] = &-\left[t_1\left(\cos k_x + \cos k_y\right) + 2t_2 \cos k_x \cos k_y\right]^2 \\ &- \left[t_1' + 2t_2'\left(\cos k_x + \cos k_y\right) + t_z' \cos k_z\right]^2.\end{aligned} \tag{8.43}$$

[7] Here we adopt the active view of symmetry transformations discussed in section 1.1.2.

[8] Alternatively, one can block diagonalise the Hamiltonian as in section 8.2.5 via employing the irreducible representations of $\hat{C}_4$. This can be conveniently achieved by extending the spinor to the $\{\boldsymbol{k}, C_4\boldsymbol{k}\}$ space and appropriately halving the BZ. In the resulting extended spinor formalism, one needs to only consider the contributions from the inversion-symmetric points.

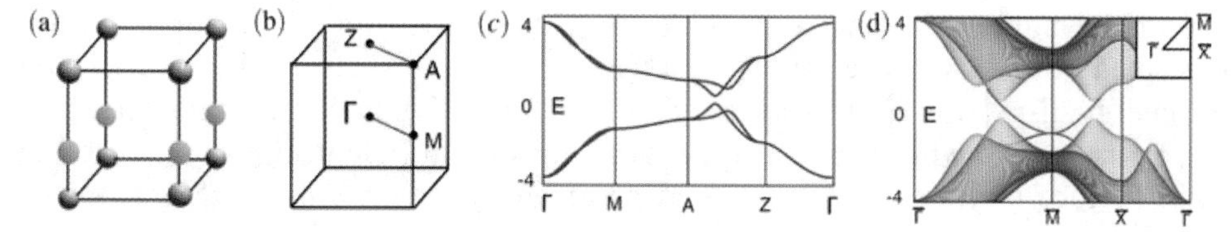

Figure 8.2. (a) Two-atom unit cell for the crystalline topological-insulator model proposed by Fu [18], which is described by equation (8.40). (b) Brillouin zone corresponding to (a). The illustrated subset of the high-symmetry points: Γ, M, A and Z, are simultaneously invariant under inversion and C_4 rotations. (c) Bandstructure for the tight-binding model of equation (8.40). (d) Topologically-protected boundary modes for a (001) surface, exhibiting a so-called quadratic band crossing, which leads to a 2π Berry phase. For the numerical details see reference [18]. Figures adapted from reference [18].

Since the $\mathbb{Z}_2$ invariant is defined as the sign of the product of the four Pfaffians and, as we observe above, each Pfaffian is negative, the invariant is always positive for the given model. In this case, the system lies in the topologically non-trivial phase, with topologically-protected modes, as shown in figure 8.2(d). The latter boundary modes appear only for surfaces that respect C_4 symmetry. For the (001) surface, which perserves the latter symmetry, one obtains a Hamiltonian for the surface states of the following form [18]:

$$\hat{H}_{(001)}(\boldsymbol{k}) = \frac{k_x^2 + k_y^2}{2m_0} + \frac{k_x^2 - k_y^2}{2m_1}\hat{\tau}_z + \frac{k_x k_y}{2m_2}\hat{\tau}_x. \tag{8.44}$$

We remark that the above form of the Hamiltonian respects C_4 symmetry, since $\{k_x^2 - k_y^2, \hat{\tau}_z\}$ and $\{k_x k_y, \hat{\tau}_x\}$ transform, respectively, according to the B_{1g} and B_{2g} irreducible representations of D_{4h}. The presence of the latter irreducible representations leads to a quadratic band touching instead of a Dirac point, which was discussed, for instance, in section 7.2.3. Interestingly, this difference leads here to a 2π Berry phase. Furthermore, the above surface states are vulnerable to perturbations that violate K and/or C_4 symmetries and, thus, spoil the topological protection.

Before concluding this section, let us mention that the first experimentally-discovered crystalline topological insulator was SnTe [19], while a number of materials have been detected after that, see [20]. Furthermore, several types of topological insulators have been proposed [21–26], which apart from rotations, rely also/or on other point-group operations [21, 23–25] or translations [22]. The latter space-group operations can be also combined with TR [24, 25].

8.3 Topological classification of gapless systems

While the primary goal of this book is the investigation of the topological properties of insulators and superconductors with a fully-gapped energy spectrum, here we discuss gapless systems, which can also support topologically non-trivial phases. For a topological classification of such systems see [27, 28]. Such a possibility has been already brought to the surface in section 7.2.3, with the study of the topologically-protected boundary modes of 3 + 1d TRI topological insulators. As we discussed, the energy dispersion of the surface states contains a Dirac point, characterised by a π Berry phase, which constitutes a quantity of a topological nature. Notably, these surface states cannot be treated as an independent bulk system, since the presence of a single bulk Dirac cone violates the Nielsen–Nimoya theorem, cf section 3.1. Luckily, Dirac cones are not only found on the surfaces of topological insulators, but they also emerge in bulk semimetals, in which, they always appear in pairs.

A system well-known to satisfy the above requirement is graphene [29], consisting of two Dirac cones per spin. Setting aside its unique material properties, graphene constitutes a prototype for semimetals in $d = 2$. Nevertheless, the topological properties of 2 + 1d semimetals are more fragile than the ones exhibited by their 3 + 1d analogs, the so-called Weyl semimetals [30–32]. In the following section we examine the topological aspects of both types of systems and point out their differences. Afterwards, we briefly discuss the general topological classification of gapless systems.

8.3.1 2 + 1d semimetals—graphene

Compared to the systems that we have studied so far in this book, graphene bears similarities to the 2 + 1d TRI topological insulators of chapter 7, with the crucial difference that it is lacking a gap in its bulk bandstructure. Graphene's bandstructure [29] coincides with that of graphite, with the only difference being, that, graphene is only a single layer thick. It crystallises in a hexagonal C_{6v} structure and in the usual case it is described by electrons hopping between nearest neighbours on the arising honeycomb lattice (see figure 8.3(a)). Notably, a similar lattice was considered in the Haldane model presented in the Hands-on section 6.5.

In the low energy limit, graphene's Hamiltonian can be effectively described by two block diagonal 2×2 Hamiltonians of the form

$$\hat{H}(\boldsymbol{k}) = \begin{pmatrix} \hat{H}_{\boldsymbol{K}}(\boldsymbol{k}) & 0 \\ 0 & \hat{H}_{\boldsymbol{K}'}(\boldsymbol{k}) \end{pmatrix} \quad \text{with} \quad \hat{H}_{\boldsymbol{K}}(\boldsymbol{k}) = \alpha\hbar\boldsymbol{k}\cdot\boldsymbol{\eta} \quad \text{and} \quad \hat{H}_{\boldsymbol{K}'}(\boldsymbol{k}) = \alpha\hbar\boldsymbol{k}\cdot\boldsymbol{\eta}^*, \tag{8.45}$$

where we have introduced the reciprocal-lattice vectors

$$\boldsymbol{K} = \left(\frac{2\pi}{3a}, \frac{2\pi}{3\sqrt{3}a}\right) \quad \text{and} \quad \boldsymbol{K}' = \left(\frac{2\pi}{3a}, -\frac{2\pi}{3\sqrt{3}a}\right). \tag{8.46}$$

In the above, $\boldsymbol{k} = (k_x, k_y)$ is measured relative to the wavevectors $\boldsymbol{K}$ and $\boldsymbol{K}'$, at which, the energy spectrum exhibits gap closings. The $\hat{\boldsymbol{\eta}}$ Pauli matrices are defined in the sublattice space arising when decomposing the honeycomb lattice into two triangular sublattices, as in figure 8.3(a). Lastly, the electron spin solely yields a

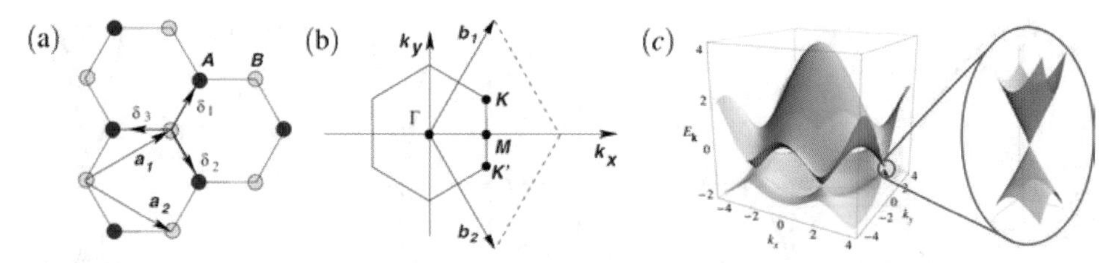

Figure 8.3. (a) Structure of the honeycomb lattice, consisting of two interpenetrating triangular sublattices A and B. The $\boldsymbol{a}_1$ and $\boldsymbol{a}_2$ denote the unit-cell lattice vectors and $\boldsymbol{\delta}_{1,2,3}$ the vectors connecting nearest neighbours. (b) 1st Brillouin zone. The Dirac cones are located at $\boldsymbol{K}$ and $\boldsymbol{K}'$. (c) Graphene's bandstructure and emergent massless relativistic Dirac spectrum. This figure was adapted from [29].

twofold degeneracy. The low energy electronic behaviour of undoped graphene can be understood by restricting to the vicinity of these two $\boldsymbol{k}$-space points. As shown in figure 8.3(b), $\boldsymbol{K}$ and $\boldsymbol{K}'$ are connected by the mirror symmetry $y \mapsto -y$, which further implies that their respective Hamiltonians are connected by TR. This is reminiscent of our bottom-up construction of 2 + 1d TRI topological insulators in section 7.1.1.

In order to explore possible non-trivial topological properties of the Hamiltonian we calculate the respective Berry vector potential. According to section 5.3.1, the eigenstates of $\hat{H}_{\boldsymbol{K}}(\boldsymbol{k})$ and $\hat{H}_{\boldsymbol{K}'}(\boldsymbol{k})$ are determined from the following expressions:

$$|\phi_{\boldsymbol{K},\pm}(\boldsymbol{k})\rangle = \frac{1}{\sqrt{2}}\begin{pmatrix} \pm 1 \\ e^{+i\varphi(\boldsymbol{k})} \end{pmatrix} \quad \text{and} \quad |\phi_{\boldsymbol{K}',\pm}(\boldsymbol{k})\rangle = \frac{1}{\sqrt{2}}\begin{pmatrix} \pm 1 \\ e^{-i\varphi(\boldsymbol{k})} \end{pmatrix}, \tag{8.47}$$

which lead to the massless relativistic energy dispersions shown in figure 8.3(c), with: $E_\pm(k) = \pm\alpha\hbar k$. Here we introduced $\tan[\varphi(\boldsymbol{k})] = k_y/k_x$ and $k = |\boldsymbol{k}|$. We readily obtain

$$\boldsymbol{A}^{\pm}_{\boldsymbol{K}}(\boldsymbol{k}) = -\frac{1}{2}\frac{\partial\varphi(\boldsymbol{k})}{\partial\boldsymbol{k}} \quad \text{and} \quad \boldsymbol{A}^{\pm}_{\boldsymbol{K}'}(\boldsymbol{k}) = +\frac{1}{2}\frac{\partial\varphi(\boldsymbol{k})}{\partial\boldsymbol{k}} = -\boldsymbol{A}^{\pm}_{\boldsymbol{K}}(\boldsymbol{k}). \tag{8.48}$$

We observe that the Berry vector potential is non-zero and independent of the band index $\pm$. We can rewrite the result for the Berry vector potential in the more illustrative form

$$\boldsymbol{A}^{\pm}_{\boldsymbol{K}}(\boldsymbol{k}) = \frac{1}{2}\left(\frac{k_y}{k^2}, -\frac{k_x}{k^2}\right) \equiv \frac{1}{2k}(\sin\varphi(\boldsymbol{k}), -\cos\varphi(\boldsymbol{k})). \tag{8.49}$$

One observes that the Berry vector potential is oriented perpendicular to the radial unit vector $\check{\boldsymbol{k}} = (\cos\varphi(\boldsymbol{k}), \sin\varphi(\boldsymbol{k}))$. Thus, it possesses a non-zero circulation (vorticity) in $\boldsymbol{k}$ space:

$$\oint_C d\boldsymbol{k} \cdot \boldsymbol{A}^{\pm}_{\boldsymbol{K}}(\boldsymbol{k}) = -\frac{1}{2}\oint_C d\varphi = n\pi \quad \text{with} \quad n \in \mathbb{Z}. \tag{8.50}$$

The above line integral of the Berry vector potential yields the vorticity of φ. In the present case, one obtains $n = 1$, i.e., a π Berry phase, similar to section 7.2.3. According to the Stokes theorem a non-zero vorticity implies a non-zero Berry

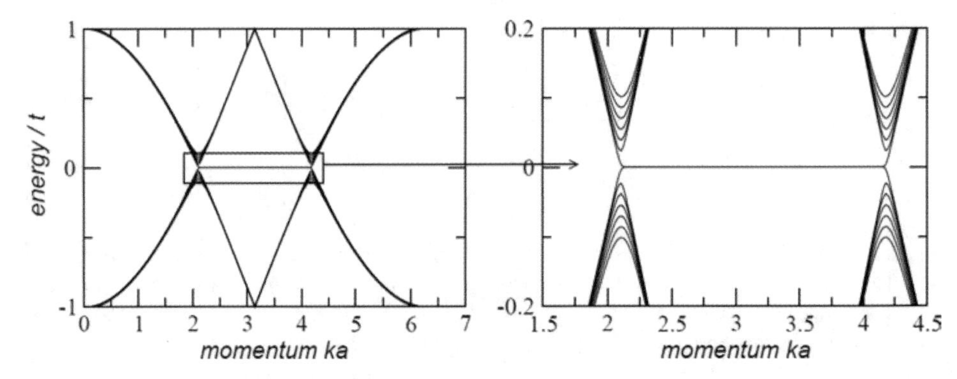

Figure 8.4. Non-dispersive edge states in a graphene zig-zag nanoribbon. The flat band modes can be viewed as the helical modes of a TRI topological insulator in the zero-energy-gap limit [35]. See discussion below equation (8.34). This figure was adapted from [29].

curvature. In the particular case, the non-zero Berry circulation stems from the $\boldsymbol{k} = 0$ wavevector, for which, we obtain a singularity in the vector potential. Consequently, the Berry curvature is singular and reads $\Omega^{\pm}_{\boldsymbol{K},z}(\boldsymbol{k}) = n\pi\delta(\boldsymbol{k})$. We remark that this result is compatible with TRS, as well as with the fact that the direct calculation of the Berry curvature yields $\boldsymbol{\Omega}^{\pm}_{\boldsymbol{K}}(\boldsymbol{k}) = \nabla_{\boldsymbol{k}} \times \boldsymbol{A}^{\pm}_{\boldsymbol{K}}(\boldsymbol{k}) = 0$, because the Berry curvature is non-zero at isolated $\boldsymbol{k}$-points which have zero measure.

From the above one notices an important difference between topological semimetals and insulators. For the former, the Berry vector potential and curvature become singular, in contrast to their gapped counterparts. Therefore, instead of integrating the Berry vector potential over the entire $\boldsymbol{k}$ space, we instead have to integrate over a surface that encloses the singular point of the bandstructure[9] [33]. The latter integral yields the so-called topological charge of the defect. The topological charges carried by the vortices located at $\boldsymbol{K}$ and $\boldsymbol{K}'$ are not protected against perturbations that open an energy gap. This can take place by violating TR or inversion symmetries [34], and more generally by introducing disorder. The latter, hybridises the electrons from the $\boldsymbol{K}$ and $\boldsymbol{K}'$ points, thus, opening an energy gap at these points simultaneously.

Based on the possibility of defining a topological invariant for graphene, i.e., the π Berry phase, one expects that also graphene could support some type of protected edge modes. Indeed, in the so-called zig-zag termination configuration of graphene nanoribbons[10], one finds non-dispersive (flat) edge modes [29], see figure 8.4, which can be viewed as the limiting case of the dispersive helical edge modes of a TRI $2+1$d topological insulator [35].

8.3.2 Weyl semimetals

The systems extending graphene-like semimetals to $d = 3$, are referred to as Weyl semimetals. For a review see [30]. A Weyl semimetal is characterised by an energy

[9] Other types of gap closings, apart from the point-like ones discussed here, are also possible. For example, line defects.

[10] There exist no edge states for the so-called armchair type of termination in graphene nanoribbons.

bandstructure exhibiting band touchings at an even number of isolated points [36], called Weyl points. Similar to the non-dispersive edge modes connecting the two Dirac cones of graphene shown in figure 8.4, Weyl semimetals harbour the so-called topological surface Fermi arcs [30], connecting two Weyl points. In the vicinity of a Weyl point $\boldsymbol{k}_0$, the Hamiltonian assumes the general form:

$$\hat{H}_{\mathrm{Weyl}}(\boldsymbol{k}) = v_F \hbar \check{\boldsymbol{k}}_0 \cdot (\boldsymbol{k} - \boldsymbol{k}_0)\hat{1} + \alpha\hbar(\boldsymbol{k} - \boldsymbol{k}_0) \cdot \hat{\boldsymbol{\sigma}}. \tag{8.51}$$

Notably, a semimetal consisting of Weyl points for which $v_F = 0$, is called a *Type-I* Weyl semimetal. This can occur if $\boldsymbol{k}_0$ is an inversion-symmetric wavevector. The first Weyl semimetal ever discovered, TaAs [37, 38], belongs to this class. However, if v_F is non-zero and $|v_F| > |\alpha|$, we obtain the so-called *Type-II* Weyl semimetals [39]. For instance, $MoTe_2$ falls into this category [40]. The presence of absence of the diagonal term defines the type of the Fermi surface emerging when the chemical potential is tuned at the energy of a Weyl point. For a Type-I Weyl point one finds that the Fermi surface consists solely of the touching point, while in the Type-II case, a Weyl point constitutes the touching point between an electron and a hole pocket that cross the Fermi level (see figure 8.5).

Similar to graphene's nodal points which are characterised by a non-zero vorticity, as discussed in section 8.3.1, every Weyl point carries a topological charge n, defined as

$$\oint_S d\boldsymbol{k} \cdot \boldsymbol{\Omega}_{\boldsymbol{k}_0}(\boldsymbol{k}) = 2\pi n \quad \text{with} \quad n \in \mathbb{Z}, \tag{8.52}$$

with S a closed surface enclosing the Weyl point and $\boldsymbol{\Omega}_{\boldsymbol{k}_0}(\boldsymbol{k})$ the respective Berry curvature. For a single Weyl point carrying a single unit of topological charge, the Berry curvature is given by:

$$\boldsymbol{\Omega}_{\boldsymbol{k}_0}(\boldsymbol{k}) = \frac{\boldsymbol{k} - \boldsymbol{k}_0}{2|\boldsymbol{k} - \boldsymbol{k}_0|^3}. \tag{8.53}$$

Notably, the 'Gauss' law of equation (8.52) can be extended to a collection of Weyl points enclosed by S, with a total topological charge n. In this case, the Berry curvature is given by the sum of the partial Berry curvatures related to each one of the Weyl points. While in the above we discussed a Weyl point of a single unit of

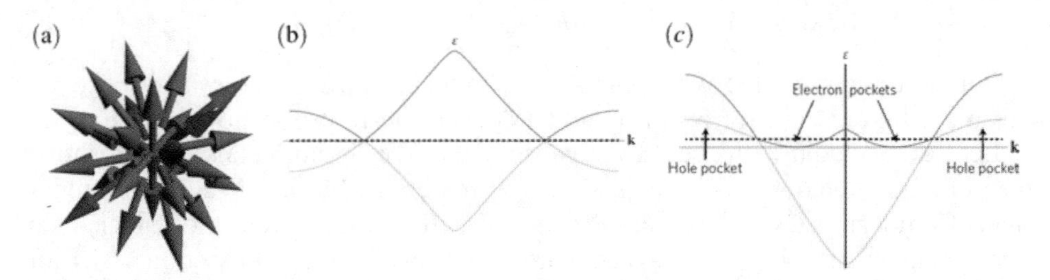

Figure 8.5. (a) Hedgehog profile of the singular Berry curvature vector stemming from a Weyl point. (b, c) Bandstructure of a Type-I (-II) Weyl semimetal. For the former (latter) only the Weyl points (and electron and/or hole pockets) cross the Fermi level when the chemical potential is tuned at the energy of the Weyl points. Figures are adapted from [30].

topological charge, there exist multi-Weyl materials [41] in which a single Weyl point carries multiple units of topological charge. These systems can be stabilised in the presence of additional crystalline symmetries. The simplest example is the double-Weyl point at $\boldsymbol{k}_0 = 0$, with an effective Hamiltonian which reads

$$\hat{H}_{\text{Double-Weyl}}(\boldsymbol{k}) \propto \left(k_x^2 - k_y^2,\, 2k_xk_y,\, k_z\right) \cdot \hat{\boldsymbol{\sigma}}. \tag{8.54}$$

Based on previous paragraphs, it is straighforward to infer that the presence of the two $\ell = 2$ harmonics $k_x^2 - k_y^2$ and $2k_xk_y$, will lead to a double winding in the xy plane, thus, also yielding a two-unit topological charge concentrated at $\boldsymbol{k}_0$.

So far, the requirements and the underlying mechanism for obtaining Weyl points have remained obscure in our discussion. An intuitive way for answering these open questions is to turn to the relativistic equation where Weyl 'points', or more precisely particles, were first encountered, i.e., the Dirac equation. In the Weyl basis, Dirac's equation in $d = 3$ reads

$$\hat{H}_{\text{Dirac}}(\boldsymbol{k}) = \begin{pmatrix} \alpha\hbar\boldsymbol{k}\cdot\hat{\boldsymbol{\sigma}} & m \\ m & -\alpha\hbar\boldsymbol{k}\cdot\hat{\boldsymbol{\sigma}} \end{pmatrix}. \tag{8.55}$$

For a zero mass m, the above Hamiltonian is decomposed into two Weyl-fermion Hamiltonians characterised by opposite topological charges and, thus, one obtains a Dirac semimetal, where four bands touch. While a Dirac semimetal is an interesting phase of matter on its own, in order to obtain two isolated Weyl subsystems from the above model, we are either required to violate inversion or TR symmetry.

A simple example of a Hamiltonian leading to a Weyl semimetal can be obtained from the above Dirac semimetal ($m = 0$) via violating TRS. After introducing the $\hat{\boldsymbol{\tau}}$ Pauli matrices for representing the 4×4 Hamiltonian, we have

$$\hat{H}(\boldsymbol{k}) = \left(\alpha\hbar k_x,\, \alpha\hbar k_y,\, \frac{\hbar^2\left(k_z^2 - k_F^2\right)}{2m}\right) \cdot \hat{\tau}_z\hat{\boldsymbol{\sigma}}, \tag{8.56}$$

where we replaced k_z with an even function of k_z, thus, leading to the violation of TRS. As a result, one obtains band touchings for the Weyl points $\boldsymbol{k}_{\text{Weyl}} = (0, 0, \pm k_F)$, about which, we can linearise the Hamiltonian above and find

$$\hat{H}_{(0,0,\pm k_F)}(\boldsymbol{k}) = \left(\alpha\hbar k_x,\, \alpha\hbar k_y,\, \pm v_F\hbar(k_z \mp k_F)\right) \cdot \hat{\tau}_z\hat{\boldsymbol{\sigma}}. \tag{8.57}$$

To further obtain isolated Weyl points, one should lift the degeneracy in τ-space. For this purpose, one can, for instance, add a constant term proportional to $\hat{\sigma}_z$.

Before concluding this section regarding the Weyl semimetals, let us shortly mention a key characteristic of this class of materials. Since a Weyl Hamiltonian stems from a massless Dirac Hamiltonian, it is useful to remind the reader of the electromagnetic response of massless Dirac electrons discussed in sections 3.4.1 and 5.5. In $d = 1$, we found that the effective action of the electromagnetic field acquires an additional term proportional to the electric field E_x, which constitutes the F^{01} component of the corresponding field strength tensor $F^{\mu\nu}$. Similarly, in $d = 3$, chiral anomaly leads to an effective action of the form $\varepsilon_{\mu\nu\lambda\rho}F^{\mu\nu}F^{\lambda\rho} \propto \boldsymbol{E} \cdot \boldsymbol{B}$. Notably, such a term was retrieved in the case of $3 + 1$d topological insulators, with the coefficient of this term being given by the magnetoelectric polarisation. In the case of Weyl semimetals, the chiral anomaly term proportional to $\boldsymbol{E} \cdot \boldsymbol{B}$ is also present, but its coefficient is generally spatially and/or temporarily dependent. One finds that the anomalous term influences the system's response as long as the chemical potential is tuned near the energy at which the Weyl points appear.

8.4 Topological classification of insulators and defects

In the paragraph above we pointed out that, apart from insulators, also gapless systems with band touchings in their energy spectrum admit a topological classification. The topological invariant is related to the respective charge of each touching point, calculated by means of a Chern or winding number defined for a surface enclosing the nodal point. Nonetheless, such kinds of defects are not specific to $\boldsymbol{k}$ space and can be also defined in coordinate $\boldsymbol{r}$ space. For instance, one can introduce a magnetic vortex to a coplanar magnetisation field, $\boldsymbol{M}(\boldsymbol{r}) = |M(\boldsymbol{r})|(0, \cos\varphi(\boldsymbol{r}), \sin\varphi(\boldsymbol{r}))$. In such a situation, $|M(\boldsymbol{r})|$ becomes zero at a given point in the two-dimensional space $\boldsymbol{r} = (x, y)$, about which, the phase field possesses a non-zero vorticity. It is important to understand the interplay of the combined presence of such coordinate space topological defects and a topologically non-trivial $\boldsymbol{k}$-space bandstructure. Since defects play the role of boundaries, it is of a particular interest to investigate the conditions, under which, such coordinate defects can trap topologically-protected excitations.

Such an exploration has been systematically performed by Teo and Kane in [8]. As they showed, the combined classification of insulators in the presence of defects can be still obtained from table 8.1 after replacing $d \mapsto d - D$. D denotes the dimensionality of the sphere $\mathbb{S}^D$, which can enclose the defect. Therefore, one obtains an augmented space in which a topological invariant has to be defined. In the whole approach it is assumed that such a defect appears due to the adiabatic change of some parameter with respect to the coordinates. This allows one to treat the spatial dependence of the Hamiltonian solely as a parametric dependence of the $\boldsymbol{k}$-space Hamiltonian. The resulting Hamiltonian is defined in the combined space $\hat{H}(\boldsymbol{k}, \boldsymbol{r})$, as long of course as the variation in coordinate space is sufficiently slow compared to a characteristic lengthscale of the bandstructure, such as the system's lattice constant or the inverse Fermi wavelength.

We now exemplify the above approach, by considering the model of equation (8.32) after the replacement $k_x \mapsto 3k_x$. The latter leads to non-dispersive threefold degenerate edge modes, protected by a weak topological invariant. We additionally consider that there exists a dislocation as in figure 8.6. The dislocation is parametrised by an adiabatic parameter s leading to a smooth displacement of the lattice sites along the y axis. Thus, the Hamiltonian now becomes $\hat{H}(\boldsymbol{k}, s)$ with $s \in [0, 1]$ so that the single-particle wavefunctions become shifted as follows $|\phi(\boldsymbol{k}, \boldsymbol{r} - s\boldsymbol{b})\rangle$ with $\boldsymbol{b}$ denoting the so-called Burgers vector of the dislocation. By direct inspection we observe that the dislocation will trap three zero-energy bound states as shown in

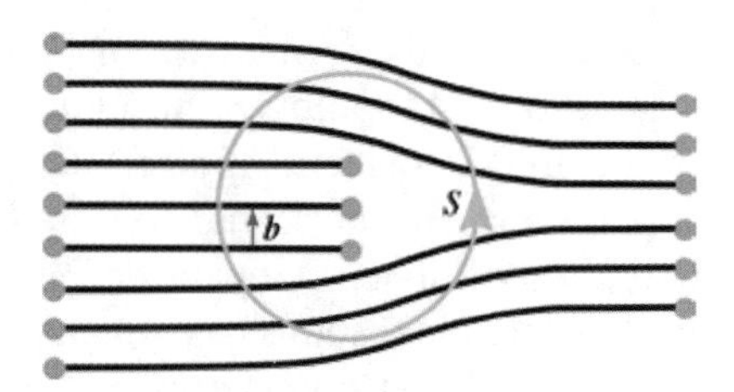

Figure 8.6. Graphical representation of the model in equation (8.32) for $k_x \mapsto 3k_x$ in the limit of decoupled chains. A dislocation, parametrised by the Burgers vector $\boldsymbol{b}$, acts as a twist defect, and traps three zero-energy bound states. The latter are protected by means of a topological invariant defined in the combined $(\boldsymbol{k}, s)$-space, with s parametrising the $\mathbb{S}^1$ sphere enclosing the defect (see also [8, 42].)

figure 8.6. This result is immediately understood if we consider that the system consists of decoupled chains. Based on table 8.1, we have that the $\boldsymbol{k}$-space Hamiltonian belongs to the BDI symmetry class, and is defined in $d = 2$, while $D = 1$. Thus $d - D = 1$ and one can define a $\mathbb{Z}$ topological invariant.

8.5 Topological superconductors and Majorana fermions

Similar to the spin-filtered boundary modes of TRI topological insulators, there exist topological systems where the charge of the electron is halved at the boundary. The corresponding charge-neutral boundary modes are termed as Majorana quasiparticles [15, 43, 44], in analogy to the homonymous charge-neutral particles introduced by Ettore Majorana [45] in high-energy physics. In fact, the boundary modes found in the models described by equations (4.1), (4.57) and (8.16), correspond to Majorana quasiparticles, if the Pauli matrices are considered to act on the so-called Nambu space. The latter describes electron (particle) and hole (antiparticle) degrees of freedom created in coordinate space by the operators $\hat{\psi}^\dagger(\boldsymbol{r})$ and $\hat{\psi}(\boldsymbol{r})$. Such a description does not only hold for effectively-spinless electrons. In the general situation one obtains the so-called Bogoliubov-de Gennes (BdG) mean-field Hamiltonian describing at the same time spin-singlet and -triplet Cooper pairing[11]:

$$\hat{H}_{\rm BdG} = \frac{1}{2}\int d\boldsymbol{r}\ \hat{\boldsymbol{\Psi}}^\dagger(\boldsymbol{r})\hat{H}_{\rm BdG}(\hat{\boldsymbol{p}}, \boldsymbol{r})\hat{\boldsymbol{\Psi}}(\boldsymbol{r}) + \text{constant}, \tag{8.58}$$

where the factor of 1/2 enforces the correct counting of the degrees of freedom, and we introduced

$$\hat{H}_{\rm BdG}(\hat{\boldsymbol{p}}, \boldsymbol{r}) = \begin{pmatrix} \hat{H}(\hat{\boldsymbol{p}}) & \hat{\Delta}(\hat{\boldsymbol{p}}, \boldsymbol{r}) \\ [\hat{\Delta}(\hat{\boldsymbol{p}}, \boldsymbol{r})]^\dagger & -\hat{H}^{\rm T}(-\hat{\boldsymbol{p}}) \end{pmatrix} \quad \text{with} \quad \hat{\boldsymbol{\Psi}}(\boldsymbol{r}) = \begin{pmatrix} \hat{\psi}(\boldsymbol{r}) \\ \hat{\psi}^\dagger(\boldsymbol{r}) \end{pmatrix}. \tag{8.59}$$

[11] The formation of Cooper pairs is reflected in the non-zero expectation values of the matrix elements: $\Delta_{ab} = \sum_{c,d} V_{abcd} \int d\boldsymbol{r} \langle \hat{\psi}_c(\boldsymbol{r}) f(\hat{\boldsymbol{p}}, \boldsymbol{r}) \hat{\psi}_d(\boldsymbol{r}) \rangle$, when the system lies in the superconducting phase. The matrix $\hat{\Delta}(\hat{\boldsymbol{p}}, \boldsymbol{r}) = \hat{\Delta} f(\hat{\boldsymbol{p}}, \boldsymbol{r})$ defines the so-called superconducting order parameter and V corresponds to the interaction driving the transition. The form of the order parameter is determined by $f(\hat{\boldsymbol{p}}, \boldsymbol{r})$ (see [46, 47]).

By employing the Nambu-space Pauli matrices $\hat{\tau}$, one finds that the BdG Hamiltonian is dictated by a CCS[12] with $\hat{\Xi} = \hat{\tau}_x \hat{K}$. Since $\hat{\Xi}^2 = \hat{1}$, the above Hamiltonian can belong to the BDI, D or DIII symmmetry classes of table 8.1.

The built-in CCS of the BdG Hamiltonian implies that *all* its quasiparticle excitations constitute Majorana excitations. To clarify this statement, we momentarily assume that the electron-spin is frozen, for instance due to the application of a strong Zeeman field. The eigenoperators of the BdG Hamiltonian $\hat{\gamma}_{E_n}$, corresponding to energy E_n, have the following general form

$$\hat{\gamma}_{E_n} = \int d\boldsymbol{r} \left[u_n(\boldsymbol{r})\hat{\psi}(\boldsymbol{r}) + v_n(\boldsymbol{r})\hat{\psi}^\dagger(\boldsymbol{r}) \right]$$
$$\text{and} \quad \hat{\gamma}_{-E_n} = \int d\boldsymbol{r} \left[v_n^*(\boldsymbol{r})\hat{\psi}(\boldsymbol{r}) + u_n^*(\boldsymbol{r})\hat{\psi}^\dagger(\boldsymbol{r}) \right] \equiv \hat{\gamma}_{E_n}^\dagger \tag{8.60}$$

with the second equation following from the CCS. The eigenoperators satisfy the anticommutation relation $\left\{ \hat{\gamma}_{E_n}, \hat{\gamma}_{E_m} \right\} = \hat{1}\delta_{n,m}$. The relation $\hat{\gamma}_{E_n}^\dagger = \hat{\gamma}_{-E_n}$ reflects the Majorana character and charge-neutrality of the excitations.

For zero energy we find $\hat{\gamma}_{E=0}^\dagger = \hat{\gamma}_{E=0}$ and $2\hat{\gamma}_{E=0}^2 = \hat{1}$. The latter implies that the operator $\hat{\gamma}_{E=0}$ does not describe ordinary fermions[13] but *anyons* instead. Anyons are found in effectively two-dimensional systems and are not described by the usual fermionic or bosonic exchange statistics. In fact, they can be exploited to perform fault-tolerant topological quantum computing [48–50]. Majoranas constitute the simplest type of non-Abelian anyons and their potential application in quantum computing has sparked an intense theoretical and experimental search for realising them in condensed matter systems. Recent experimental efforts [51–54] have revealed signatures of Majorana quasiparticles in hybrid semiconductor-superconductor devices [55, 56], while their unusual exchange-statistics property has not been demonstrated yet. In fact, when two Majorana quasiparticles are exchanged the respective operators transform as $(\hat{\gamma}_1, \hat{\gamma}_2) \mapsto (-\hat{\gamma}_2, \hat{\gamma}_1)$ [50]. The latter process is also referred to as *braiding*, and possesses a topological character since it is independent of the particle exchange path.

Braiding zero-energy Majorana quasiparticles implements the required quantum-computing operations by acting on the quantum bits, i.e., the fundamental blocks of information. In fact, braiding operations are required to take place adiabatically, otherwise decoherence is introduced to the system [57–62], which degrades the quality of information processing. Notably, zero-energy Majorana quasiparticles always come in pairs and have to be sufficiently spatially-separated. Two Majorana operators can define a standard zero-energy fermionic operator: $\hat{d}_{12} = (\hat{\gamma}_1 + i\hat{\gamma}_2)/\sqrt{2}$,

[12] Note that a CCS symmetry of this type can be also found in non-superconducting systems. However, in the latter case, the CCS can be always broken, in contrast to superconducting systems that are described at the mean-field level by the BdG formalism above. In fact, employing the BdG formalism of equation (8.58) is not necessary for describing all superconductors. Importantly, the superconductors that do not need to be described within the BdG formalism, are exactly the ones, which do not harbour Majorana quasiparticles (see also [25]).

[13] If $\hat{\gamma}_{E=0}$ would instead be an operator corresponding to standard fermions, it should satisfy $\hat{\gamma}_{E=0}^2 = \hat{0}$.

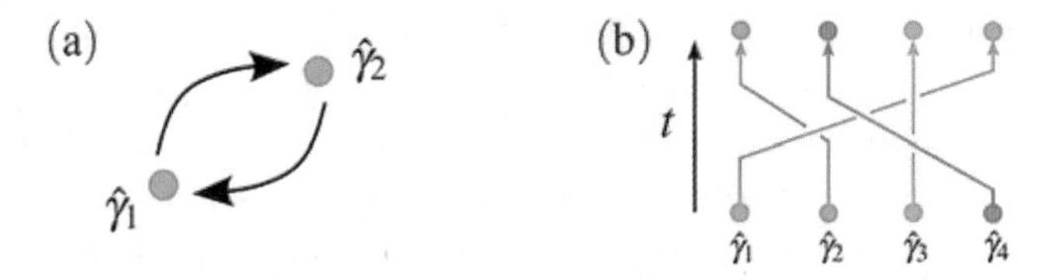

Figure 8.7. (a) Braiding of two Majorana fermions and non-Abelian exchange statistics $(\hat{\gamma}_1, \hat{\gamma}_2) \mapsto (-\hat{\gamma}_2, \hat{\gamma}_1)$. (b) Topologically-protected quantum computation via the adiabatic exchange of four Majorana fermions comprising a logical quantum bit. t denotes the time.

satisfying $\hat{d}_{12}^2 = \hat{0}$. Occupying or leaving vacant the state corresponding to $\hat{d}_{12}$ yields a twofold degenerate many-body ground state. Similarly, when $2N$ Majoranas are present, one finds a 2^N ground-state degeneracy. In order to construct a so-called logical quantum bit, four Majoranas are required.

Apart from the zero-energy Majorana quasiparticles, a topological superconductor can also host chiral or helical Majorana boundary modes. In the former (latter) case the Majorana operators satisfy $\hat{\gamma}_k^\dagger = \hat{\gamma}_{-k}$ $(\hat{\gamma}_{k,\sigma}^\dagger = \hat{\gamma}_{-k,\sigma}$ with $\sigma = \pm 1$ labelling the Kramers partners) and have the energy spectrum $E_k \propto \pm k$ $(E_{k,\sigma} \propto \pm \sigma k)$. Very recent experiments have reported the detection of the chiral Majorana edge modes [63]. However, to the best of our knowledge, the helical Majorana boundary modes appear to be elusive. For a detailed analysis of the topic, see [14, 25, 64–71].

8.6 Further topics and outlook

The main aspect that has been under-represented throughout this book, regards the role of interactions in topological systems. First of all, the presence of interactions requires the reformulation of the topological invariants in terms of Green functions [33, 72, 73], cf section 5.7. Even more importantly, the presence of interactions can modify the topological invariants characterising the symmetry classes of table 8.1. For example, taking into account the effects of interactions for the BDI systems in $d = 1$ [74], yields a $\mathbb{Z}_8$, instead of a $\mathbb{Z}$ topological invariant. The complete topological classification of interacting systems is uncharted research territory, promising to open perspectives for the discovery of new exotic phases and non-Abelian quasiparticles. The latter is crucial for crafting a universal topological quantum computer, e.g., based on Fibonacci anyons [49]. Another prominent research path concerns topological driven systems [75–79], which we discuss in section 8.8. Finally, note that throughout this work we have restricted to the so-called symmetry-protected topological systems [80–85], which have to be contrasted with systems featuring *topological order* and constitute a playground for non-Abelian anyons. The latter (former) topological systems are characterised by long (short) range correlations and a ground state degeneracy which appears without (only in) the presence of boundaries.

8.7 Hands-on: Berry magnetic monopoles in hole-like semiconductors

In this section we investigate the emergence of Berry magnetic monopoles in the bandstructure of the heavy- and light-hole bands described by the Luttinger

Hamiltonian of equation (1.27). In analogy to the discussion of section 8.3.2 concerning the Weyl points, the Γ-point constitutes a source of singular Berry curvature. Here, one can focus on each type of hole band separately, see [86]. Due to TRS and the concomitant double degeneracy of each hole band for all $\boldsymbol{k}$, the relevant Berry curvature is an SU(2) matrix defined in the degenerate subspace.

Task: Follow section 5.4.2 and calculate the singular SU(2) Berry gauge potential and curvature for the heavy- and light-hole bands. First, consider that the $\boldsymbol{g}$-vector is defined in an abstract space, and extract the monopole charge by integrating the 2nd Chern character defined in equation (5.51), over an $\mathbb{S}^4$ sphere enclosing the singular point $|\boldsymbol{g}| = 0$. Later on, project the SU(2) Berry curvature onto the three-dimensional $\boldsymbol{k}$ space.

8.8 Hands-on: Floquet topological insulator

We conclude with the discussion of periodically-driven topologically non-trivial systems, the so-called Floquet topological insulators. The single-particle Hamiltonian in this case satisfies $\hat{H}(t + T_{\text{drive}}) = \hat{H}(t)$, with T_{drive} denoting the period of the time-dependent perturbation. The situation is analogous to the case of spatially-periodic Hamiltonians, cf section 1.1.5. There, we employed the Bloch theorem to express the time-independent wavefunctions as $\boldsymbol{\phi}_{\boldsymbol{k}}(\boldsymbol{r}) = \boldsymbol{u}_{\boldsymbol{k}}(\boldsymbol{r})e^{\mathrm{i}\boldsymbol{k}\cdot\boldsymbol{r}}$. $\boldsymbol{u}_{\boldsymbol{k}}(\boldsymbol{r})$ is periodic and $\boldsymbol{k}$ is restricted to the 1st Brillouin zone. In a similar fashion, the Floquet theorem allows us to express the time-dependent wavefunctions as $\boldsymbol{\phi}_n(t) = \boldsymbol{u}_n(t)e^{-\mathrm{i}\varepsilon_n t/\hbar}$, with $\boldsymbol{u}_n(t + T_{\text{drive}}) = \boldsymbol{u}_n(t)$. The ε_n define the so-called quasienergies, in analogy to $\boldsymbol{k}$ corresponding to the quasimomentum. The quasienergies are restricted to the interval $(-\hbar\pi/T_{\text{drive}}, \hbar\pi/T_{\text{drive}}] = (-\hbar\omega_0/2, \hbar\omega_0/2]$, with ω_0 the driving frequency.

Task 1: Consider the 1 + 1d topological insulator Hamiltonian of equation (4.2), in the additional presence of an external magnetic field: $\boldsymbol{B}(t) = \boldsymbol{B}(e^{\mathrm{i}\omega_0 t} + e^{-\mathrm{i}\omega_0 t})$. Here, the magnetic field is assumed to couple to the electrons *solely* via the Zeeman term $\propto \boldsymbol{B}(t) \cdot \hat{\boldsymbol{\sigma}}$. In the spirit of section 3.3.2 and reference [77] perform the rotating wave approximation (RWA) via the following decomposition of the time-dependent wavefunction

$$\boldsymbol{\phi}(k_x, t) \approx c_+(k_x, t)e^{-\mathrm{i}\omega_0 t/2}|k_x, +\rangle + c_-(k_x, t)e^{\mathrm{i}\omega_0 t/2}|k_x, -\rangle, \tag{8.61}$$

with $|k_x, \pm\rangle$ the eigenvectors of the Hamiltonian in equation 4.2, and the coefficients $c_\pm(k_x, t)$ varying slowly in time, with frequencies much smaller than ω_0. Obtain the RWA Hamiltonian and infer the constraints on $\boldsymbol{B}$ and ω_0, so that the system transits from the topologically trivial to the non-trivial phase when the drive is present. How is the $k_x = 0$ contribution to the topological invariant modified when the driving is added? See also [77].

Task 2: Consider the model of equation (8.16), with the difference here that m is time-dependent, i.e., $m \mapsto m(t) = m\sum_n \delta(t - nT_{\text{kick}})$ and T_{kick} the period of the 'kicking'. Consider that the $\hat{\boldsymbol{\sigma}}$ Pauli matrices act in Nambu space, which was discussed in section 8.5. Explain why the kicking allows for Majorana quasiparticles with energy $\hbar\pi/T_{\text{kick}}$. In analogy to the RWA above, one can introduce an effective static Hamiltonian $\hat{H}_{\text{eff}}(k_x)$, via the evolution operator for a single period:

$$\hat{U}(k_x, T_{\text{kick}}) = \hat{\mathcal{T}}_{T_{\text{kick}}} \operatorname{Exp}\left[-\frac{\mathrm{i}}{\hbar} \int_0^{T_{\text{kick}}} d\tau \hat{H}(k_x, \tau) \right] := \operatorname{Exp}\left[-\frac{\mathrm{i}}{\hbar} \hat{H}_{\text{eff}}(k_x) T_{\text{kick}} \right], \quad (8.62)$$

with $\hat{\mathcal{T}}_{T_{\text{kick}}}$ the time-ordering operator for a kicking period. Determine $\hat{H}_{\text{eff}}(k_x)$ and calculate the respective $\mathbb{Z}_2$ invariant, $\mathcal{M}$. Strikingly, according to [76, 78], this invariant is not sufficient to describe the topological properties of a generic Floquet system, since the time evolution generally affects the possible presence of Majorana edge states.

References

[1] Wigner E P 1955 *Ann. Math.* **62** 548

[2] Dyson F J 1962 *J. Math. Phys.* **3** 1199

[3] Zirnbauer M R 1996 *J. Math. Phys.* **37** 4986

[4] Altland A and Zirnbauer M R 1997 *Phys. Rev.* B **55** 1142

[5] Schnyder A P, Ryu S, Furusaki A and Ludwig A W W 2008 *Phys. Rev.* B **78** 195125

[6] Qi X-L, Hughes T L and Zhang S-C 2008 *Phys. Rev.* B **78** 195424

[7] Kitaev A 2009 *AIP Conf. Proc.* **1134** 22

[8] Teo J C Y and Kane C L 2010 *Phys. Rev. B* **82** 115120

[9] Essin A M and Gurarie V 2011 *Phys. Rev. B* **84** 125132

[10] Akemann G, Baik J and Di Francesco P (ed) 2011 *The Oxford Handbook of Random Matrix Theory* (Oxford University Press)

[11] Beenakker C W J 1997 *Rev. Mod. Phys* **69** 731

[12] Ezawa M, Tanaka Y and Nagaosa N 2013 *Sci. Rep.* **3** 2790

[13] Bott R 1959 *Ann. Math.* **70** 313

[14] Hasan M Z and Kane C L 2010 *Rev. Mod. Phys* **82** 3045

[15] Kitaev A Y 2001 *Phys.-Usp.* **44** 131

[16] Potter A C and Lee P A 2014 *Phys. Rev. Lett.* **112** 117002

[17] Joyner C H, Müller S and Sieber M 2014 *EPL* **107** 50004

[18] Fu L 2011 *Phys. Rev. Lett.* **106** 106802

[19] Tanaka Y *et al* 2012 *Nat. Phys.* **8** 800

[20] Ando Y and Fu L 2015 *Annu. Rev. Condens. Matter Phys.* **6** 361

[21] Fang C, Gilbert M J and Bernevig B A 2012 *Phys. Rev.* B **86** 115112

[22] Slager R-J, Mesaros A, Juričić V and Zaanen J 2013 *Nat. Phys.* **9** 98

[23] Chiu C-K, Yao H and Ryu S 2013 *Phys. Rev.* B **88** 075142

[24] Shiozaki K and Sato M 2014 *Phys. Rev. B* **90** 165114

[25] Kotetes P 2013 *New J. Phys.* **15** 105027

[26] Chiu C-K, Teo J C Y, Schnyder A P and Ryu S 2016 *Rev. Mod. Phys.* **88** 035005

[27] Matsuura S, Chang P-Y, Schnyder A P and Ryu S 2013 *New J. Phys.* **15** 065001

[28] Schnyder A P and Brydon P M R 2015 *J. Phys.: Condens. Matter* **27** 243201

[29] Castro Neto A H *et al* 2009 *Rev. Mod. Phys.* **81** 109

[30] Burkov A A 2016 *Nat. Mater.* **15** 1145

[31] Murakami S 2007 *New J. Phys.* **9** 356

[32] Wan X-G, Turner A M, Vishwanath A and Savrasov S Y 2011 *Phys. Rev.* B **83** 205101
[33] Volovik G E 2003 *The Universe in a Helium Droplet* (Oxford: Clarendon Press)
[34] Xiao D, Yao W and Niu Q 2007 *Phys. Rev. Lett.* **99** 236809
[35] Imura K-I, Mao S, Yamakage A and Kuramoto Y 2011 *Nanoscale Res. Lett.* **6** 358
[36] Herring C 1937 *Phys. Rev.* **52** 365
[37] Xu S-Y *et al* 2015 *Science* **349** 613
[38] Lv B Q *et al* 2015 *Phys. Rev. X* **5** 031013
[39] Soluyanov A A *et al* 2015 *Nature* **527** 495
[40] Huang L *et al* 2016 *Nat. Mater.* **15** 1155
[41] Fang C, Gilbert M J, Dai X and Bernevig B A 2012 *Phys. Rev. Lett.* **108** 266802
[42] Barkeshli M, Jian C-M and Qi X-L 2013 *Phys. Rev.* B **87** 045130
[43] Read N and Green D 2000 *Phys. Rev. B* **61** 10267
[44] Ivanov D A 2001 *Phys. Rev. Lett.* **86** 268
[45] Majorana E 1937 *Nuovo Cimento* **14** 171
[46] Bardeen J, Cooper L N and Schrieffer J R 1957 *Phys. Rev.* **108** 1175
[47] de Gennes P G 1966 *Superconductivity of Metals and Alloys* (Advanced Book Classics) (Boulder, CO: Perseus Books)
[48] Kitaev A Y 2003 *Ann. Phys.* **303** 2
[49] Nayak C, Simon S H, Stern A, Freedman M and Das Sarma S 2008 *Rev. Mod. Phys.* **80** 1083
[50] Alicea J, Oreg Y, Refael G, von Oppen F and Fisher M P A 2011 *Nat. Phys.* **7** 412
[51] Mourik V *et al* 2012 *Science* **336** 1003
[52] Albrecht S M *et al* 2016 *Nature* **531** 206
[53] Deng M T *et al* 2016 *Science* **354** 1557
[54] Nichele F *et al* 2017 *Phys. Rev. Lett.* **119** 136803
[55] Lutchyn R M, Sau J D and Das Sarma S 2010 *Phys. Rev. Lett.* **105** 077001
[56] Oreg Y, Refael G and von Oppen F 2010 *Phys. Rev. Lett.* **105** 177002
[57] Cheng M, Galitski V and Das Sarma S 2011 *Phys. Rev. B* **84** 104529
[58] Goldstein G and Chamon C 2011 *Phys. Rev. B* **84** 205109
[59] Budich J C, Walter S and Trauzettel B 2012 *Phys. Rev. B* **85** 121405
[60] Schmidt M J, Rainis D and Loss D 2012 *Phys. Rev. B* **86** 085414
[61] Karzig T, Refael G and von Oppen F 2013 *Phys. Rev. X* **3** 041017
[62] Scheurer M S and Shnirman A 2013 *Phys. Rev. B* **88** 064515
[63] He Q-L *et al* 2017 *Science* **357** 294
[64] Qi X-L and Zhang S-C 2011 *Rev. Mod. Phys.* **83** 1057
[65] Alicea J 2012 *Rep. Prog. Phys.* **75** 076501
[66] Leijnse M and Flensberg K 2012 *Semicond. Sci. Technol.* **27** 124003
[67] Beenakker C W J 2013 *Annu. Rev. Condens. Matter Phys.* **4** 113
[68] Bernevig B A and Hughes T L 2013 *Topological Insulators and Topological Superconductors* (Princeton, NJ: Princeton University Press)
[69] Chamon C, Goerbig M O, Moessner R and Cugliandolo L F (ed) 2014 *Topological Aspects of Condensed Matter Physics* (Lecture Notes of the Les Houches Summer School Vol 103) (Oxford Scholarship Online)
[70] Elliott S R and Franz M 2015 *Rev. Mod. Phys.* **87** 137
[71] Sato M and Ando Y 2017 *Rep. Prog. Phys.* **80** 076501
[72] Wang Z, Qi X-L and Zhang S-C 2010 *Phys. Rev. Lett.* **105** 256803

[73] Gurarie V 2011 *Phys. Rev.* B **83** 085426
[74] Fidkowski L and Kitaev A 2010 *Phys. Rev.* B **81** 134509
[75] Oka T and Aoki H 2009 *Phys. Rev.* B **79** 081406
[76] Kitagawa T, Berg E, Rudner M and Demler E 2010 *Phys. Rev.* B **82** 235114
[77] Lindner N H, Refael G and Galitski V 2011 *Nat. Phys.* **7** 490
[78] Nathan F and Rudner M S 2015 *New J. Phys.* **17** 125014
[79] Potter A C, Morimoto T and Vishwanath A 2016 *Phys. Rev.* X **6** 041001
[80] Haldane F D M 1983 *Phys. Rev. Lett.* **50** 1153
[81] Wen X-G 1990 *Int. J. Mod. Phys.* B **4** 239
[82] Turner A M, Pollmann F and Berg E 2011 *Phys. Rev.* B **83** 075102
[83] Chen X, Gu Z-C, Liu Z-X and Wen X-G 2012 *Science* **338** 1604
[84] Lu Y-M and Vishwanath A 2012 *Phys. Rev. B* **86** 125119
[85] Senthil T 2015 *Annu. Rev. Cond. Matter Phys.* **6** 299
[86] Murakami S, Nagaosa N and Zhang S-C 2004 *Phys. Rev.* B **69** 235206

Index

doi:10.1088/978-1-68174-517-6ch9

◎

编辑手记

让我们先来看一个科学史上的成功案例.

巴丁、库珀、施瑞弗三人中巴丁是资深阅历广的老一辈科学家.他在大学期间学的是电子工程专业,做过地球物理方面的工作,研究生时攻读数学物理专业,后转向固体物理学专业.1947年与肖克利、布拉坦合作发明晶体管,因而获1956年诺贝尔物理学奖.20世纪50年代他的兴趣又从半导体转向超导,带领库珀、施瑞弗两个年轻人,从斯德哥尔摩领奖归来不到一年,一举拿下物理学中长期未能攻克的又一个堡垒.

库珀1951年获文学学士,1953年获理学硕士,1954年获哲学博士.他是量子场论和粒子理论方面的专家,在普林斯顿和杨振宁一起工作.老谋深算的巴丁感到,超导理论的突破需要精通量子场论的人,1955年就经杨振宁推荐把库珀这位"东方的量子技师"请来合作.库珀用不到一年时间从头熟悉超导问题后,就以"库珀对"的理论立下第一宗汗马功劳.库珀成名后始终保持对人文学科的兴趣,凭着他文学艺术的功底为人文学科的学生讲物理学,努力把人文和科学融合起来.

施瑞弗当时是巴丁的研究生.巴丁拿出十个问题供他选择,并建议他研究第十个问题——超导.施瑞弗征求另一位老师的意见,这位老师问他:"你今年多少岁?"他答:"二十多一点."这位老师说:"那浪费一两年不要紧."于是施瑞弗就选择了超导这个老大难问题.施瑞弗把解决超导体内所有电子的多体问题,形象地说成10^{23}个电子设计舞蹈动作找波函数.这个问题实在是太难了,施瑞弗一度曾想把论文题目转到铁磁性方面.此时巴丁正要出发到斯德哥尔摩去领奖,他劝施瑞弗再继续坚持一个月,终于没有功败垂成.

施瑞弗描述当时的环境说:搞理论的有各式各样的人,搞原子核物理、场论、固体理论等,常常三三两两一起讨论,什么问题都容易找到答案,有时即使旁听别人的讨论也受益匪浅.在餐厅里还可遇到搞实验的,为了互相交流,一顿饭常吃上一两个小时.看来,开放而宽松的环境、浓厚而自由的学术气氛,对发展科学是至关重要的.而专业知识狭隘、学术上划地为牢和固步自封的人很难在科学上干出什么大事.

这个案例对我国目前的科研环境具有多方面的意义,但这属于另一个层面的话题了,我们还是回到本书.

本书是一部英文版的物理学专著,中文书名或可译为《拓扑绝缘体》.

本书的作者为帕纳约蒂斯·科泰特斯(Panagiotis Kotetes),美国科学家,他在中国科学院理论物理研究所开展了五年的教学工作.2015—2018年他在哥本哈根大学尼尔斯·玻尔研究所从事博士后研究,本书的主要撰写地是哥本哈根大学.他的第一个博士后职位是在卡尔斯鲁厄理工学院,在那里他工作了五年.他在希腊雅典国立技术大学完成了他的职业课程、硕士和博士学位的研究.他的研究兴趣和活动内容涵盖了拓扑系统、非常规超导、奇异磁性和量子计算等主题.

正如作者在前言中所述:

本书旨在向读者介绍拓扑物质领域,该领域最近有了快速且重要的发展,这种情况从2016年物理学诺贝尔奖颁发给David J. Thouless,F. Duncan M. Haldane和J. Michael Kosterlitz也能反映出来.虽然我们与拓扑现象的第一次相遇可以追溯到20世纪80年代初Klaus von Klitzing对整数量子霍尔效应的探测,但该领域的系统性进展是在对时间反演不变量的拓扑绝缘体预测和随后的实验之后才发生的.这一进步引发了其他拓扑系统的发展,包括固有超导体和人造超导体、半金属、超流体等.目前,对拓扑现象的研究正在推动物理学中很大一部分的研究活动的开展.

我对这个领域的研究最初始于在雅典攻读博士学位期间,是由Georgios Varelogiannis指导的.然而,编写本书早期版本的契机来得稍晚,当时我是卡尔斯鲁厄的Gerd Schon小组的博士后研究员.在那里,Gerd让我有机会成为他在2013—2014年冬季学期教授的"凝聚态物理课程"的客座讲师.作为对这些课程的支持,我准备了一套笔记,这些笔记构成了本书的种子,这是两卷系列的第一部分,同一个出版社将在不久的将来出版一本名为《拓扑超导体》的书.本书主要是我在尼尔斯·玻尔研究所的博士后任职期间写成的,如果没有我的顾问 Karsten Flensberg 和 Brian Andersen 的大力支持和理解,我不可能完成这项任务.

本书是为想熟悉拓扑物质领域的研究生或研究者设计的,尽管预备知识的列表很长,但我试图使这本书变得简洁且自成一体.鉴于篇幅有限,我引用了适当的教科书或评论,以帮助非专业读者理解本书阐述中的不可避免的缺口.人们自然会问这本书有什么新颖之处,与现有文献相比如何.当然了,现在已经有很多我推荐的优秀材料,比如,J. K. Asbóth,L. Oroszlány 和 A. Pályi;B. A. Bernevig 和 T. Hughes;E. Fradkin;S-Q Shen;T. Stanescu;G. Volovik 等人的书,C. Chamon,M. Goerbig,R. Moessner 和 L. Cugliandolo 主编的莱苏什暑期学校讲义;A. Akhmerov 等人整理的代尔夫特理工大学拓扑在线公开课程.我在整个手稿中都引用了很优秀的评论,还有一些其他的作品,由于篇幅有限,很遗憾在此无法提及.针对这个问题的明显答案是,一本书永远无法涵盖一个研究主题,如果人们需要,替代演示文稿总是会派上用场.具体来说,在介绍该主题时,我更加重视对称性的概念,尝试将不同章节和实际操作部分紧密地联系起来,并尝试对技术方面和计算进行透彻阐释.此外,我还整理了一些很少以组合方式呈现的主题,例如酉对称、热量和热电输运、轨道磁化理论的新方法、极性克尔效应以及最后的十倍分类.事实上,为了使展示更具教学性,我尝试按照自下而上的方法安排材料,避免采用历史线性结构.

完成这本书结束了我学术研究的漫长旅程,在那里我学到了比我想象中更多的东西.我希望我可以将其中的一些知识传授给本书的读者.最后,我提前感谢任何可以教会我更多知识,并帮助我进一步改进本书内容的反馈的人.

本书的目录为:

下面,就本书做一点科普.
先介绍一下 2.10 节出现的 Boltzmann 输运方程,如下:

Ludwig Edward Boltzmann(1844 年 2 月 20 日—1906 年 9 月 5 日),奥地利物理学家、哲学家、热力学和统计物理学的奠基人之一. 作为一名物理学家,他最伟大的功绩是发展了通过原子的性质(例如:原子量,电荷量,结构等)来解释和预测物质的物理性质(例如:黏性,热传导,扩散等)的统计力学,并且从统计意义对热力学第二定律进行了阐释.

气体分子运动论的基本方程,在研究稀薄气体动力学时起着很大的作用.

气体分子运动论的基本方程,因 Boltzmann 于 1872 年首先提出而得名. 它是一个非线性积分微分方程,用于描述气体分子速度分布函数的变化. 它对研究稀薄气体动力学有重要意义.

根据质点分子或光滑球分子速度分布函数 $f(x,v,t)$ 的定义,在时刻 t,x 邻近的物理空间体积元(记为 $\mathrm{d}x=\mathrm{d}x_1\mathrm{d}x_2\mathrm{d}x_3$)内,速度在靠近 v 的速度空间元(记为 $\mathrm{d}v=\mathrm{d}v_1\mathrm{d}v_2\mathrm{d}v_3$)内的分子数目是 $f\mathrm{d}x\mathrm{d}v$. 对于单一组元且相互作用势为球对称的气体分子,如果作用在分子上的外力为 F,那么速度分布函数 f 满足下述 Boltzmann 方程

$$\frac{\partial f}{\partial t}+v\frac{\partial f}{\partial x}+\frac{F}{m}\frac{\partial f}{\partial v}=\int(f'f''_1-ff_1)gb\mathrm{d}b\mathrm{d}\varepsilon\mathrm{d}v_1=Q(f,f) \tag{1}$$

式中,m 为分子质量;g 为碰撞前速度分别为 v,v_1 的两个分子的相对速度值;b 为假设第一个分子静止时,第二个分子运动轨迹的渐近线到第一个

分子重心的垂直距离;ε 为第二个分子运动轨迹平面同通过第一个分子重心并与相对速度平行的某一固定平面之间的夹角. $f_1 = f(v_1)$, $f' = f(v')$, $f'_1 = (v'_1)$, v, v_1 是碰撞前速度为 v', v'_1 的两个分子碰撞后的速度. 式(1) 中右端是对所有可能的 b, ε, v_1 之值求积分, 称为碰撞积分, 代表由于分子相互碰撞引起的 f 的变化. 上述方程只适用于质点分子或光滑球对称分子.

对于一般刚性分子,速度分布函数有六个位置变量和六个速度变量. 因为,若分子为非球对称的刚体,除用 x 决定平动外,还需有三个决定方位的角变量. 如果分子表面不光滑,那么除速度 v 外,尚需有表征分子运动状态的三个角速度分量.

直接求解 Boltzmann 方程十分困难,平衡状态的 Maxwell 速度分布实际上是已知的仅有的精确解

$$f_0 = \left(\frac{\rho}{m}\right)\left(\frac{h}{\pi}\right)^{3/2}\exp(-hv^2) \tag{2}$$

式中, $h = l2kTm$, k 为 Boltzmann 常数; ρ, v 和 T 为平衡态下的密度、速度和温度. 为了简化,通常提出所谓模型方程,用以近似地代替 Boltzmann 方程. 最简单而常用的方程是所谓 B-G-K-W 方程,它用 $v(f_0 - f)$(称为弛豫项)代替 Boltzmann 方程右端的碰撞积分,即

$$\frac{\partial f}{\partial t} + v\frac{\partial f}{\partial x} + \frac{F}{m}\frac{\partial f}{\partial v} = v(f_0 - f) \tag{3}$$

式中, v 为碰撞频率, f_0 为由式(2) 给出的局部 Maxwell 分布. 但式中可调参量 ρ, v, T 应与由解 f 给出的局部密度、速度、温度相同. B-G-K-W 方程满足质量、动量、能量在碰撞前后守恒和其解渐近趋于 Maxwell 分布的条件. 它比较简单,且能给出与 Boltzmann 方程近似的结果,因而常被用来代替 Boltzmann 方程.

再介绍一下 3.2 节中的 $SO(10)$ 模型中的一类 Jackiw-Rebbi 零能束缚态.

复旦大学的李新洲、中国科学技术大学的汪克林、山西大学的张鉴祖三位教授早在 1985 年就讨论了 $SO(10)$ 大统一模型的球对称磁单极,以及在这个模型中的 Jackiw-Rebbi 型费米子 - 磁单极零能束缚态问题①. 结果表明, 这个理论中并不存在这样的零能束缚态,如下:

'tHooft-Polyakov 磁单极的存在,是自发对称性破缺的非 Abel 规范理

① 摘自《高能物理与核物理》,1985,9(6):653-659.

论中最令人注目的事实之一.人们对于在'tHooft-Polyakov 型磁单极外势中或 Julia-Zee 双子外势中费米子的束缚态问题表现出很大的兴趣. T. F. Walsh, P. Weisz 和 Li XinZhou 等指出,对于 $SU(5)$ 球对称基本磁单极,并不存在 Jackiw-Rebbi 型费米子 - 磁单极的零能束缚态. T. F. Walsh 等采用的是 Dirac 旋量和磁单极的吴 - 杨假设,Li XinZhou 等采用的是 Weyl 旋量和 Dokos-Tomaras 的 $SU(5)$ 磁单极.

他们考察了 $SO(10)$ 大统一理论的球对称基本磁单极情形,结果发现,在这个模型中 Jackiw-Rebbi 型费米子 - 磁单极零能束缚态也不存在.

考察一个 $SO(10)$ 规范理论,它被两个 Higgs 多重态的真空期望值自发破缺到 $SU(3)_c\times U(1)_{cm}$. 一个 Higgs 多重态 $\boldsymbol{\Phi}$ 处于 $SO(10)$ 的 $\overline{45}$ 共轭表示,另一个 Higgs 多重态 $\boldsymbol{H}$ 处于 $SO(10)$ 的 $\underline{10}$ 维表示. $\boldsymbol{\Phi}$ 和 $\boldsymbol{H}$ 的真空期望值是

$$\langle\boldsymbol{\Phi}\rangle = a\left[\begin{array}{ccccc:ccccc} & & & & & 2 & & & & \\ & & & & & & 2 & & & \\ & & & & & & & 2 & & \\ & & & & & & & & -3 & \\ & & & & & & & & & -3 \\ \hdashline -2 & & & & & & & & & \\ & -2 & & & & & & & & \\ & & -2 & & & & & & & \\ & & & 3 & & & & & & \\ & & & & 3 & & & & & \end{array}\right],\langle\boldsymbol{H}\rangle = \begin{bmatrix}0\\0\\0\\0\\b\\0\\0\\0\\0\\b\end{bmatrix} \tag{1}$$

系统的 Lagrange 密度是

$$\mathscr{L} = -\frac{1}{2}\mathrm{tr}(\boldsymbol{W}_{\mu\nu}\boldsymbol{W}^{\mu\nu}) + \mathrm{tr}(\mathscr{D}_{\mu}\boldsymbol{\Phi})^2 + | D_u\boldsymbol{H}|^2 - V(\boldsymbol{\Phi},\boldsymbol{H})$$

$$\boldsymbol{W}_{\mu\nu} = \partial_{\mu}\boldsymbol{W}_{\nu} - \partial_{\nu}\boldsymbol{W}_{\mu} + ig[\boldsymbol{W}_{\mu},\boldsymbol{W}_{\nu}]$$

$$\boldsymbol{W}_{\mu} = \frac{1}{2}\boldsymbol{W}_{\mu}^{ij}\boldsymbol{T}^{ij},\boldsymbol{W}^{ij} = -\boldsymbol{W}^{ji},i,j = 1,2,\cdots,10$$

$$\begin{gathered} D_{\mu}\boldsymbol{H} = \partial_{\mu}\boldsymbol{H} + ig\boldsymbol{W}_{\mu}\boldsymbol{H} \\ \mathscr{D}_{\mu}\boldsymbol{\Phi} = \partial_{\mu}\boldsymbol{\Phi} + ig[\boldsymbol{W}_{\mu},\boldsymbol{\Phi}] \end{gathered} \tag{2}$$

$SO(10)$ 的生成元 $\boldsymbol{T}^{ij}$ 可以写成

$$\boldsymbol{T} = \begin{pmatrix}\boldsymbol{A}+\boldsymbol{C} & \boldsymbol{B}+\boldsymbol{D}\\ \boldsymbol{B}-\boldsymbol{D} & \boldsymbol{A}-\boldsymbol{C}\end{pmatrix} \tag{3}$$

此处,$\boldsymbol{A},\boldsymbol{B},\boldsymbol{C}$ 是 5×5 的反对称矩阵,而 $\boldsymbol{D}$ 是 5×5 的对称矩阵.

Wilkinson 和 Goldhaber 曾证明了如下的定理：设大统一群 G 被破缺到一个子群 H. 令 $\boldsymbol{L}$ 是轨道角动量的生成元，$\boldsymbol{T}$ 是 G 的 $SU(2)$ 子群的生成元，而 $\boldsymbol{I}$ 是 H 的 $SU(2)$ 子群的生成元，此处，$\boldsymbol{I}$ 满足 $[\boldsymbol{I},\boldsymbol{Q}_M]=\boldsymbol{0}$. 于是，当且仅当 $\boldsymbol{Q}_M=\boldsymbol{I}_3-\boldsymbol{T}_3$ 时，弦位势可以被规范变换到在 $\boldsymbol{L}+\boldsymbol{T}$ 变换下是球对称的位势.

在磁单极核的外部，可通过 Dirac 弦位势

$$\boldsymbol{A}_D=\boldsymbol{Q}_M(1-\cos\theta)\frac{\hat{\phi}}{r\sin\theta} \tag{4}$$

定义一个电荷算符 $\boldsymbol{Q}_M$，这里 $\boldsymbol{Q}_M$ 是 $SO(10)$ 的某一表示中的矩阵. 显然，$\boldsymbol{Q}_M$ 必定是电磁荷算符 $\boldsymbol{Q}_{cm}$ 和色生成元 $\boldsymbol{Q}_c$ 的线性组合. 应用四个条件：

(1) 点磁单极的稳定性条件；

(2) Dirac 量子化条件；

(3) 层子电荷的三重性条件；

(4) Wilkinson 和 Goldhaber 定理，

容易求得最小磁荷是 $g=1/2e$，最小的无色磁荷是 $g=3/2e$.

将 $SU(2)$ 的三个生成元嵌入到 $SO(10)$ 中，我们的目的是要找到满足下列等式的一般假设

$$[\boldsymbol{L}_i+\boldsymbol{T}_i,\boldsymbol{W}_j]=ie_{ijk}\boldsymbol{W}_k,\boldsymbol{W}_0=\boldsymbol{0},[\boldsymbol{L}_i+\boldsymbol{T}_i,\boldsymbol{\Phi}]=\boldsymbol{0},(\boldsymbol{L}_i+\boldsymbol{T}_i)\boldsymbol{H}=\boldsymbol{0} \tag{5}$$

为了考察上述解的对称性，要求这些解在最大可能解下是不变的，同时要求这些解在 $SO(10)$ 的最大可能子群 S 的变换下也是不变的. 而且 S 应与球对称性 $\boldsymbol{L}+\boldsymbol{T}$ 相容. 这就是说，若 $\boldsymbol{S}_l$ 为 S 的生成元，则应有

$$[\boldsymbol{S}_l,\boldsymbol{\Phi}]=0,\boldsymbol{S}_l\boldsymbol{H}=\boldsymbol{0},[\boldsymbol{S}_l,\boldsymbol{W}_j]=\boldsymbol{0},[\boldsymbol{S}_l,\boldsymbol{L}_j+\boldsymbol{T}_j]=\boldsymbol{0} \tag{6}$$

对于 $eg=1/2$ 磁单极情形，在将 $SU(2)$ 的表示嵌入到 $SO(10)$ 的10维表示时，存在两种情形：(a) $\underline{10}\to 2(2)+6(1)$；(b) $\underline{10}\to 4(2)+2(1)$. 显然，情形(b) 的磁单极是不稳定的，并且会衰变到情形(a).

考虑 $SU(2)$ 嵌入到 $SO(10)$ 的情形 $\underline{10}\to 2(2)+6(1)$，由

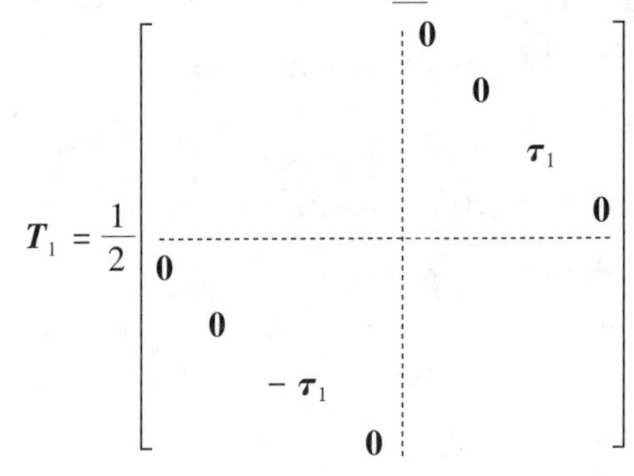

$$\boldsymbol{T}_1=\frac{1}{2}\left[\begin{array}{cccc|cccc} & & & & \boldsymbol{0} & & & \\ & & & & & \boldsymbol{0} & & \\ & & & & & & \boldsymbol{\tau}_1 & \\ & & & & & & & \boldsymbol{0} \\ \hline \boldsymbol{0} & & & & & & & \\ & \boldsymbol{0} & & & & & & \\ & & -\boldsymbol{\tau}_1 & & & & & \\ & & & \boldsymbol{0} & & & & \end{array}\right]$$

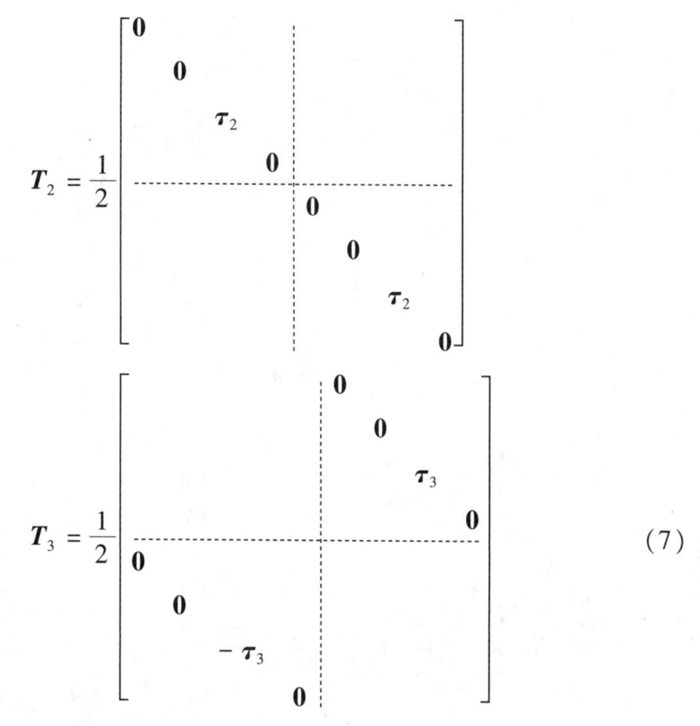

$$
\boldsymbol{T}_2=\frac{1}{2}\left[\begin{array}{cccc|cccc}
\mathbf{0} & & & & & & & \\
& \mathbf{0} & & & & & & \\
& & \boldsymbol{\tau}_2 & & & & & \\
& & & \mathbf{0} & & & & \\
\hline
& & & & \mathbf{0} & & & \\
& & & & & \mathbf{0} & & \\
& & & & & & \boldsymbol{\tau}_2 & \\
& & & & & & & \mathbf{0}
\end{array}\right]
$$

$$
\boldsymbol{T}_3=\frac{1}{2}\left[\begin{array}{cccc|cccc}
& & & & \mathbf{0} & & & \\
& & & & & \mathbf{0} & & \\
& & & & & & \boldsymbol{\tau}_3 & \\
& & & & & & & \mathbf{0} \\
\hline
\mathbf{0} & & & & & & & \\
& \mathbf{0} & & & & & & \\
& & -\boldsymbol{\tau}_3 & & & & & \\
& & & \mathbf{0} & & & &
\end{array}\right] \tag{7}
$$

给出,此处,$\boldsymbol{\tau}_a(a=1,2,3)$ 为内禀 Pauli 矩阵. 取 S 是由

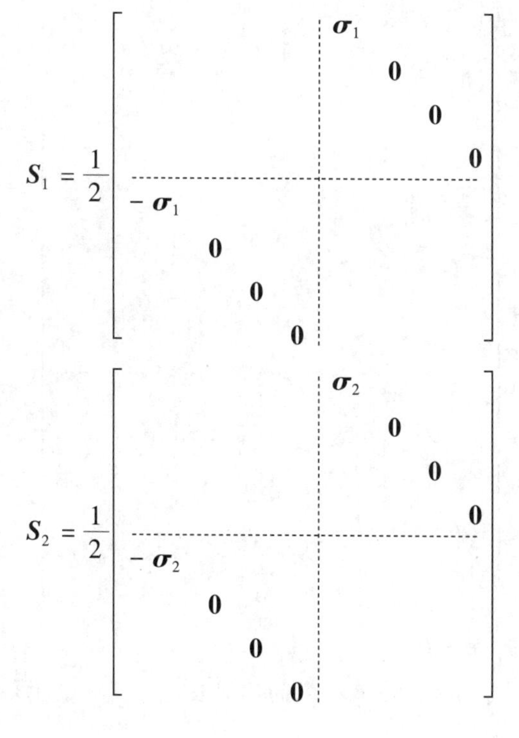

$$
\boldsymbol{S}_1=\frac{1}{2}\left[\begin{array}{cccc|cccc}
& & & & \boldsymbol{\sigma}_1 & & & \\
& & & & & \mathbf{0} & & \\
& & & & & & \mathbf{0} & \\
& & & & & & & \mathbf{0} \\
\hline
-\boldsymbol{\sigma}_1 & & & & & & & \\
& \mathbf{0} & & & & & & \\
& & \mathbf{0} & & & & & \\
& & & \mathbf{0} & & & &
\end{array}\right]
$$

$$
\boldsymbol{S}_2=\frac{1}{2}\left[\begin{array}{cccc|cccc}
& & & & \boldsymbol{\sigma}_2 & & & \\
& & & & & \mathbf{0} & & \\
& & & & & & \mathbf{0} & \\
& & & & & & & \mathbf{0} \\
\hline
-\boldsymbol{\sigma}_2 & & & & & & & \\
& \mathbf{0} & & & & & & \\
& & \mathbf{0} & & & & & \\
& & & \mathbf{0} & & & &
\end{array}\right]
$$

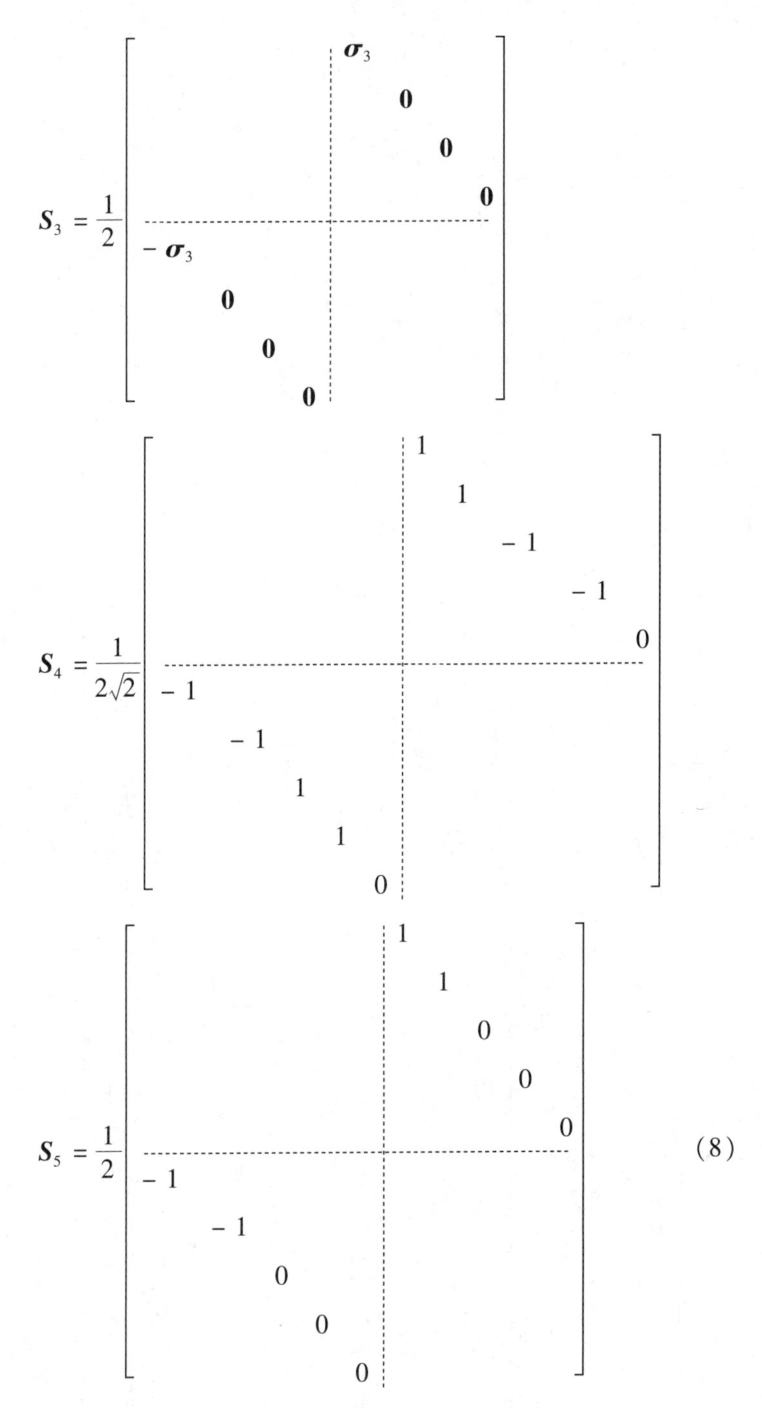

生成的 $SU(2)\times U(1)\times U(1)$.

球对称的、S 不变的并且在 $\boldsymbol{r}$ 和 $\boldsymbol{T}$ 的同时反演下是不变的场位形的普

遍形式是

$$H(\boldsymbol{r})=\frac{1}{g}\begin{bmatrix}0\\0\\0\\0\\h(r)\\0\\0\\0\\0\\h(r)\end{bmatrix}$$

$$\Phi(\boldsymbol{r})=\frac{1}{g}\left[\begin{array}{cccc|cccc}
0 & & & & \Phi_1(r) & & & \\
 & 0 & & & & \Phi_1(r) & & \\
 & & \Phi_3(r)(\hat{y}\boldsymbol{\tau}_2) & & & & [\Phi_2(r)+\Phi_3(r)(\hat{x}\boldsymbol{\tau}_1+\hat{z}\boldsymbol{\tau}_3)] & \\
 & & & 0 & & & & -2[\Phi_1(r)+\Phi_2(r)] \\
\hline
-\Phi_1(r) & & & & 0 & & & \\
 & -\Phi_1(r) & & & & 0 & & \\
 & & -[\Phi_2(r)+\Phi_3(r)(\hat{x}\boldsymbol{\tau}_1+\hat{z}\boldsymbol{\tau}_3)] & & & & \Phi_3(r)(\hat{y}\boldsymbol{\tau}_2) & \\
 & & & 2[\Phi_1(r)+\Phi_2(r)] & & & & 0
\end{array}\right]$$

$$W_i(\boldsymbol{r})=(\boldsymbol{T}\times\hat{r})_i\frac{K(r)-1}{gr} \tag{9}$$

以上,函数$K(r)\times\phi_j(r)\,(j=1,2,3)$和$h(r)$是实函数,沿$\hat{z}$方向,$\Phi$和$H$在无穷远处分别趋于它们的真空期望值$\langle\Phi\rangle$和$\langle H\rangle$. 由此,得

$$h(r)\to bg,\phi_1(r)\to 2ag,\phi_2(r)\to-\frac{1}{2}ag,\phi_3(r)\to\frac{5}{2}ag$$

$$K(r)\to\exp(-cr)\,(r\to\infty) \tag{10}$$

Lorentz 群的旋量表示(1/2,0) 和(0,1/2) 由两分量的 Weyl 旋量实现. 容易看到,若χ_L的变换如同表示(1/2,0),则$\chi_R^c=-\sigma^2\chi_L^*$的变换如同表示(0,1/2). 类似地,若$\chi_R$的变换如同表示(0,1/2),则$\chi_L^c=\sigma^2\chi_R^*$的变换如同表示(1/2,0).

在$SO(10)$大统一模型中,每一代层子和轻子填入$SO(10)$的旋量表示. 在这个理论中,用左手 Weyl 场描述费米子是方便的.

在旋量表示中,$SO(10)$的45个生成元$\boldsymbol{\Sigma}_{ij}=-\boldsymbol{\Sigma}_{ji}$可通过$32\times32$的矩阵$\boldsymbol{\Gamma}_i(i=1,2,\cdots,10)$构造出来

$$\boldsymbol{\Sigma}_{ij}=\frac{1}{2i}(\boldsymbol{\Gamma}_i\boldsymbol{\Gamma}_j-\boldsymbol{\Gamma}_j\boldsymbol{\Gamma}_i) \tag{11}$$

此处

$$\begin{aligned}
\boldsymbol{\Gamma}_1&=\boldsymbol{\sigma}_1\times\boldsymbol{\sigma}_1\times\mathbf{1}\times\mathbf{1}\times\boldsymbol{\sigma}_2\\
\boldsymbol{\Gamma}_2&=\boldsymbol{\sigma}_1\times\boldsymbol{\sigma}_2\times\mathbf{1}\times\boldsymbol{\sigma}_3\times\boldsymbol{\sigma}_2\\
\boldsymbol{\Gamma}_3&=\boldsymbol{\sigma}_1\times\boldsymbol{\sigma}_1\times\mathbf{1}\times\boldsymbol{\sigma}_2\times\boldsymbol{\sigma}_3\\
\boldsymbol{\Gamma}_4&=\boldsymbol{\sigma}_1\times\boldsymbol{\sigma}_2\times\mathbf{1}\times\boldsymbol{\sigma}_2\times\mathbf{1}\\
\boldsymbol{\Gamma}_5&=\boldsymbol{\sigma}_1\times\boldsymbol{\sigma}_1\times\mathbf{1}\times\boldsymbol{\sigma}_2\times\boldsymbol{\sigma}_1\\
\boldsymbol{\Gamma}_6&=\boldsymbol{\sigma}_1\times\boldsymbol{\sigma}_2\times\mathbf{1}\times\boldsymbol{\sigma}_1\times\boldsymbol{\sigma}_2\\
\boldsymbol{\Gamma}_7&=\boldsymbol{\sigma}_1\times\boldsymbol{\sigma}_3\times\boldsymbol{\sigma}_1\times\mathbf{1}\times\mathbf{1}\\
\boldsymbol{\Gamma}_8&=\boldsymbol{\sigma}_1\times\boldsymbol{\sigma}_3\times\boldsymbol{\sigma}_2\times\mathbf{1}\times\mathbf{1}\\
\boldsymbol{\Gamma}_9&=\boldsymbol{\sigma}_1\times\boldsymbol{\sigma}_3\times\boldsymbol{\sigma}_3\times\mathbf{1}\times\mathbf{1}\\
\boldsymbol{\Gamma}_{10}&=\boldsymbol{\sigma}_2\times\mathbf{1}\times\mathbf{1}\times\mathbf{1}\times\mathbf{1}
\end{aligned} \tag{12}$$

这些 $\boldsymbol{\Gamma}_i$ 遵守 Clifford 代数

$$\boldsymbol{\Gamma}_i\boldsymbol{\Gamma}_j+\boldsymbol{\Gamma}_j\boldsymbol{\Gamma}_i=2\delta_{ij}(i,j=1,2,\cdots,10) \tag{13}$$

在以上各式中,$\boldsymbol{\sigma}$ 是自旋 Pauli 矩阵,$\mathbf{1}$ 是 2×2 单位矩阵. 手征算符为

$$\boldsymbol{\Lambda}=-i\boldsymbol{\Gamma}_1\boldsymbol{\Gamma}_2\cdots\boldsymbol{\Gamma}_{10}=\boldsymbol{\sigma}_3\times\mathbf{1}\times\mathbf{1}\times\mathbf{1}\times\mathbf{1} \tag{14}$$

它把 32 维旋量表示 $\boldsymbol{\psi}$ 分解为 $\boldsymbol{\psi}_L$ 和其共轭部分 $\boldsymbol{\psi}_R$

$$\boldsymbol{\psi}_L=\frac{1}{2}(\mathbf{1}+\boldsymbol{\Lambda})\boldsymbol{\psi},\boldsymbol{\psi}_R=\frac{1}{2}(\mathbf{1}-\boldsymbol{\Lambda})\boldsymbol{\psi} \tag{15}$$

此处

$$\widetilde{\boldsymbol{\psi}}_L=(u_1,u_2,u_3,v_l,d_1,d_2,d_3,e^-,d_1^c,d_2^c,d_3^c,e^+,-u_1^c,-u_2^c,-u_3^c,-v_l)_L \tag{16}$$

生成元 $\boldsymbol{\Sigma}_{ij}$ 满足代数

$$[\boldsymbol{\Sigma}_{ij},\boldsymbol{\Sigma}_{kl}]=2i(\delta_{ik}\boldsymbol{\Sigma}_{jl}+\delta_{jl}\boldsymbol{\Sigma}_{ik}-\delta_{il}\boldsymbol{\Sigma}_{jk}-\delta_{jk}\boldsymbol{\Sigma}_{il}) \tag{17}$$

在所选择的基上,32×32 的矩阵 $\boldsymbol{\Sigma}_{ij}$ 分解为在对角线上的两个 16×16 的矩阵. 下面,我们将应用作用于 $\boldsymbol{\psi}_L$ 上的 16×16 的矩阵. 在这个表象中,$\boldsymbol{T}$ 也是 16×16 的矩阵

$$(T_1)_{ij}=\begin{cases}\dfrac{1}{2}, & 若(i,j)=(2,13),(8,11),(11,8),(13,2)\\ -\dfrac{1}{2}, & 若(i,j)=(1,14),(7,12),(12,7),(14,1)\\ 0, & 其他情形\end{cases}$$

$$(T_2)_{ij}=\begin{cases}\dfrac{i}{2}, & 若(i,j)=(1,14),(8,11),(12,7),(13,2)\\ -\dfrac{i}{2}, & 若(i,j)=(2,13),(7,12),(11,8),(14,1)\\ 0, & 其他情形\end{cases} \tag{18}$$

$$(T_3)_{ij}=\begin{cases}\dfrac{1}{2}, & 若(i,j)=(7,7),(8,8),(13,13),(14,14)\\ -\dfrac{1}{2}, & 若(i,j)=(1,1),(2,2),(11,11),(12,12)\\ 0, & 其他情形\end{cases}$$

Higgs 场与费米子 ψ_L^c 的 Yukawa 耦合为

$$\mathscr{L}_{\text{Yukawa}}=G\boldsymbol{\psi}_L^c H\boldsymbol{\psi}_L+h\cdot c \tag{19}$$

此处,$\boldsymbol{\psi}_L$ 是左手场,而 $\boldsymbol{\psi}_L^c$ 是右手场.

综合上面的结果,容易得出层子和轻子的运动方程为

$$\bar{\boldsymbol{\sigma}}_\mu D_\mu\boldsymbol{\psi}_R-GH(\boldsymbol{r})\boldsymbol{\psi}_L=0,\boldsymbol{\sigma}_\mu D_\mu\boldsymbol{\psi}_L+GH(\boldsymbol{r})\boldsymbol{\psi}_R=0 \tag{20}$$

以上,$\boldsymbol{\sigma}_\mu=(\mathbf{1},\boldsymbol{\sigma})$,$\bar{\boldsymbol{\sigma}}_\mu=(\mathbf{1},-\boldsymbol{\sigma})$.

Jackiw-Rebbi 型费米子 - 磁单极束缚态是将非常重的磁单极处理为外势(即略去费米子对磁单极的反作用),并且对于零能束缚态,因子 $\exp(-\mathrm{i}Et)=1$,可设费米子场与 t 无关,故(20) 化为

$$\left[\boldsymbol{\sigma}\cdot\nabla-\frac{\mathrm{i}}{2}\frac{K(r)-1}{r}\boldsymbol{\sigma}\cdot(\boldsymbol{T}\times\hat{r})\right]\boldsymbol{\psi}_{R(L)}+GH(r)\boldsymbol{\psi}_{L(R)}=0 \tag{21}$$

应用(9) 和(18),求得

$$\begin{aligned}&\left[\boldsymbol{\sigma}\cdot\nabla-\frac{\mathrm{i}}{2}\frac{K(r)-1}{r}\boldsymbol{\sigma}\cdot(\boldsymbol{\tau}\times\hat{r})\right]\begin{bmatrix}d_3\\e^+\end{bmatrix}_{R(L)}+Gh(r)\begin{bmatrix}d_3\\e^+\end{bmatrix}_{L(R)}=0\\&\left[\boldsymbol{\sigma}\cdot\nabla-\frac{\mathrm{i}}{2}\frac{K(r)-1}{r}\boldsymbol{\sigma}\cdot(\boldsymbol{\tau}\times\hat{r})\right]\begin{bmatrix}u_2^c\\u_1\end{bmatrix}_{R(L)}+Gh(r)\begin{bmatrix}u_2^c\\u_1\end{bmatrix}_{L(R)}=0\end{aligned} \tag{22}$$

若定义 2 × 2 矩阵

$$\boldsymbol{\phi}_{R,L}=(d_3,e^+)_{R,L}\boldsymbol{\tau}^2 \tag{23}$$

利用关系式 $\boldsymbol{\tau}^{\mathrm{T}}\boldsymbol{\tau}^2=-\boldsymbol{\tau}^2\boldsymbol{\tau}$,由(22) 可得 $\boldsymbol{\phi}_{R,L}$ 的矩阵方程

$$\boldsymbol{\sigma}\cdot\nabla\boldsymbol{\phi}_{R(L)}+\frac{\mathrm{i}}{2}\frac{K(r)-1}{r}\boldsymbol{\sigma}\cdot\boldsymbol{\phi}_{R(L)}(\boldsymbol{\tau}\times\hat{r})+Gh(r)\boldsymbol{\phi}_{L(R)}=0 \tag{24}$$

由此式可见,已不必再区分 $\boldsymbol{\sigma}$ 矩阵和 $\boldsymbol{\tau}$ 矩阵.

因为 2 × 2 的矩阵 $\boldsymbol{\phi}_{R,L}$ 的一般形式为

$$\boldsymbol{\phi}_L(\boldsymbol{r})=\sigma_\mu L_\mu(\boldsymbol{r}),\boldsymbol{\phi}_R(\boldsymbol{r})=\sigma_\mu R_\mu(\boldsymbol{r}) \tag{25}$$

故可将(24) 改写为

$$\begin{aligned}&\partial_j L_j+\frac{K(r)-1}{r}L_j\hat{r}_j+Gh(r)R_0=0\\&\partial_j L_0+i\varepsilon_{ikj}\partial_i L_k-\frac{K(r)-1}{r}L_0\hat{r}_j+Gh(r)R_j=0\end{aligned} \tag{26}$$

$$\partial_j R_j + \frac{K(r)-1}{r} R_j \hat{r}_j + Gh(r) L_0 = 0$$

$$\partial_j R_0 + i\varepsilon_{ikj} \partial_i R_k - \frac{K(r)-1}{r} R_0 \hat{r}_j + Gh(r) L_j = 0$$

由方程组(26),容易得出下面的等式

$$\partial_j (L_0 R_j^* + L_j R_0^*) - i\varepsilon_{ikj} \partial_i (R_k^* L_j) + Gh(r)(|R_0|^2 + R_j R_j^* + |L_0|^2 + L_j L_j^*) = 0 \tag{27}$$

将式(27)积分,得

$$\int d^3\chi h(r)(|R_0|^2 + R_j R_j^* + |L_0|^2 + L_j L_j^*) = 0 \tag{28}$$

由于稳定磁单极应使总能量极小,应用基态波函数中不存在节点的一般论述,容易看到,$h(r)$ 应不改变符号. 所以,式(28)给出 $L_\mu(\boldsymbol{r})$ 和 $R_\mu(\boldsymbol{r})$ 恒等于零.

综上所述,对于通常质量标度的费米子,我们证明了对于标准的 $SO(10)$ 大统一理论球对称基本磁单极不存在 Jackiw-Rebbi 型费米子 - 磁单极零能束缚态.

在以上讨论中 Higgs 属于$\underline{10}$ 和$\underline{45}$ 表示起关键作用. 但我们指出,考虑 $\underline{120}$ 和$\underline{126}$ Higgs 后,并不改变本文的结果.

再介绍一下 4.3 节中的 Berry 曲率,如下①:

二维材料中由 Berry 曲率诱导的新型磁学响应是近年来的新兴领域. 这些二维材料所表现出的磁学特性及量子输运与 Berry 曲率直接相关,而 Berry 曲率又与晶体的对称性、电子的轨道磁性、自旋轨道耦合以及磁电效应等息息相关. 研究这些新型磁性响应一方面有益于研究不同量子效应间的耦合作用,另一方面可探索量子效应在电子与信息器件领域的应用. 香港科技大学电子与计算机工程系的刘雨亭,香港科技大学物理系的贺文宇、刘军伟、邵启明四位研究人员在 2021 年介绍了近几年来二维材料中新型磁响应的实验研究进展,特别介绍了二硫化钼和石墨烯等材料中的谷霍尔和磁电效应、低对称性的二碲化钨等材料中的量子非线性霍尔以及转角石墨烯中的反常霍尔和量子反常霍尔效应. 他们结合二维材料的晶体结构以及电子结构,介绍了这些新奇现象的现有物理解释,回顾了相关研究的最新发展,讨论了其中尚未理解的现象,并做出展望.

① 摘自《物理学报》,2021,70(12):1-2.

材料的磁性对于传感、存储、电子及医学等领域有着极其重要的意义.磁性材料往往来自电子的自旋产生的有序磁矩结构.令人惊奇的是许多不具有自旋磁矩的二维材料,如转角石墨烯,可表现出诸如反常霍尔效应和量子反常霍尔效应等传统磁性材料所具有的性质,其原因在于这些二维材料拥有轨道磁性.与传统磁性材料所具有的自旋磁性相比,轨道磁性一般比较小且不容易在宏观尺寸被观测到.因此它的作用往往被忽略.然而,近些年的研究表明许多二维材料的新奇量子输运特性与轨道磁性息息相关.

为什么轨道磁性的研究会与二维材料的发展有着密切的联系呢?从量子力学角度,轨道磁性源于非零的Berry曲率,而非零的Berry曲率在对称性上要求材料不能同时具有时间反演和空间中心对称性.最近涌现的二维材料如单层过渡金属硫化物和放在硼氮衬底上的石墨烯都具有诸如质量型 Dirac 的谷能带结构,在这些谷能带结构中常含有非零的 Berry 曲率和谷轨道磁矩,因此运动的载流子可产生非零磁矩.以单层过渡金属硫化物为例,其二维布里渊区为六边形.在六个转角即 Dirac 点附近的电子的低能有效性质可用有质量的 Dirac 费米子来描述,而有质量的 Dirac 费米子具有非零的 Berry 曲率.即使有非零的 Berry 曲率,高晶体对称性如 C_3 等会保证总的可观测效应为零.所以要观测轨道Berry曲率诱导的磁性响应需要低对称性材料.

二维材料的晶体结构对称性点群种类丰富,而且具有很好的可调控性.二维材料的晶体结构可选择性十分广泛.磁电效应或者谷霍尔效应要求材料不具备对称中心,并且打破一些镜面对称性,使得材料的晶体结构选择限定于21个非中心对称点群中的某些低对称性点群.从低对称性的单斜晶系,比如$1T'$相的过渡金属硫族化合物、三卤化锆或钛等,到高对称性的四方晶系,比如硒化铊和砷化镉,以及六方晶系,比如石墨烯和$2H$相的过渡金属硫族化合物.并且,二维材料的对称性依赖于层厚和堆叠方式,比如多层硫化钼点群为具备对称中心的 D_{6h},而单层硫化钼点群则降为不具备对称中心的 D_{3h}.二维材料的晶体结构的对称性还可以被外界条件调控,如衬底、门电压、应力等.比如,硼氮衬底上的石墨烯破坏了 C_3 和空间反演对称性,双层石墨烯加垂直电场破坏了空间反演对称性,加面内应力破坏了镜面对称性.

研究二维材料 Berry 曲率诱导的磁性响应的重要意义在于其易与二维材料本身晶体对称性和量子效应耦合,产生有趣的物理现象.许多二维材料具有高电子迁移率和低载流子浓度的特性,对磁场的效应比较大,很容易显现量子霍尔效应,比如石墨烯和黑磷.二维材料的电子结构奇异,可以实现量子自旋霍尔效应和量子反常霍尔效应,比如二碲化钨和过渡金属五碲化物中存在量子自旋霍尔效应.另外,二维过渡金属硫化物材料具有自旋谷锁定效应,使得谷间的电子散射被抑制,导致很长的自旋能谷弛豫时间.轨道磁性与晶体低对称性和强自旋轨道耦合结合,产生了谷磁电效应和量子非线性霍尔效应.同样,宏观的轨道磁矩和量子效应的内在物理耦合使我们能够在二维材料中观测到反常霍尔效应和量子反常霍尔效应.

二维材料及异质结的磁性响应同时与自旋电子学紧密联系.自旋霍尔效应可将普通电流转换为自旋电流.它是许多自旋电子器件的应用基础,如自旋轨道力矩磁性随机存储器.自旋霍尔效应与轨道磁性具有共同的微观机制——Berry 相效应.这种紧密联系进一步说明了轨道磁性在电子与信息器件领域的应用前景.

二维材料及异质结的磁性响应是近几年兴起的研究课题,本文将探讨一系列与电子轨道磁矩或电流引起的磁学响应相关的物理现象,包括:(1) 谷霍尔和磁电效应;(2) 量子非线性霍尔效应;(3) 转角双层石墨烯中的反常霍尔效应和量子反常霍尔效应.这三种霍尔效应的物理本质均与 Berry 曲率诱导的磁响应息息相关.而为了获得这些磁响应,材料必须具有降低的晶体维度或对称性.本文将介绍这些新奇现象现有的物理解释,回顾相关研究的最新发展,讨论其中尚未理解的现象,并做出展望.

材料的磁电性对于磁场探测及信息存储有着重要的应用前景.材料的磁电性常见于多铁材料或者铁磁/铁电复合材料.然而由于蜂窝结构的二维材料中价带和导带中电子的运动被限制在能带谷中,磁矩和磁电性可以在非磁性材料中实现.这里能带谷指的是材料导带或者价带区域的极小值或极大值点.有趣的是,材料电子在不同的谷中可以有不同的状态,如谷轨道磁矩、谷锁定的自旋磁矩等,如果能够控制这些由谷标记的电子的状态,便可以将信息存储在这些谷中从而获得应用.当二维材料的空间反演对称性被打破时,石墨烯和某些过渡金属硫族化合物的 K 和 K' 能带谷附近存在富集的非零 Berry 曲率.当载流子在 K 谷中运动时,由于 Berry 曲率作用,载流子受到等效磁场作用获得横向速度,产生了谷霍尔效应.由于 Berry 曲率在 K 和 K' 谷中的符号也相反,因此 K 和 K' 谷中的载流子具有相反的横向速度.

谷霍尔效应首先在石墨烯中被提出.单层石墨烯本身的晶体结构为高对称性的蜂窝结构,具有反演对称中心,不存在能谷磁矩,以至于无法使用光学或电学激发的方法直接在单层石墨烯中观察到谷霍尔效应.相较于单层本征石墨烯,在衬底上的单层石墨烯往往和单层二硫化钼一样不具有反演对称中心,电子在其K和K'的能带谷中,具有大小相等和方向相反的能谷磁矩.如果加面内电场,由于谷霍尔效应的存在,样品边缘可以积累谷极化载流子(能谷磁矩).这种边缘能谷磁矩可被光学克尔效应探测到.如果用极化光只激发其中一类谷中的载流子,也可以观测到谷霍尔效应.Mak 等在单层二硫化钼中观察到了谷霍尔效应.他们通过照射不同手性的偏振光,可以选择性地激发单层二硫化钼K或K'能带谷中的载流子.他们的实验印证了观察到谷霍尔效应的关键在于打破材料的中心反演对称性.Gorbachev 等在石墨烯超晶格中通过非局域电学测量,观察到了谷霍尔效应.在石墨烯超晶格中,单层石墨烯与氮化硼以 A/B 方式层叠时打破了晶格.

最后介绍一下4.4 节中的 Berry 相位,如下:

吴铭群教授2013 年深入地讨论了 Berry 相位、分数电荷及手徵反常之间的关系.由 Goldstone-Wilczek 模型,当空间中有孤子时,此时的真空会因为被该外场所极化而具有分数电荷.且有趣的是,该分数电荷亦可由其他拓扑相关的物理概念——Berry 相位及手徵反常——来解释,如是观之,似乎可以将这些物理观念合而为一.然而,最近 Fujikawa 在深入研究(3+1)维系统中的pi介子衰变成2个光子的物理现象时发现,Berry相位以及手徵反常二者之间其实是具有基本差异的.且习知手徵反常是与温度不相关的,然而分数电荷却是会随温度而变的.这让他开始省思,在(1+1)维的 Goldstone-Wilczek 模型中,上述的习知将分数电荷、Berry 相位及手徵反常三者物理概念解释为相互关联可能是有问题的.为了更清楚地了解三者的异同,他研究了 Goldstone-Wilczek 模型中,该三者拓扑物理量随温度变化的关系.

首先他推广了 Schaposnik 所提出的方法来研究分数电荷以及手徵反常二者之间在有限温度时的变化关系.他发现,利用该方法可将分数电荷分解成两个项,其中一项为与温度无关的项,此项即对应手徵反常;另一项则为温度修正项.所以,由此方法,他可以清楚地看出二者的差异是来自该温度修正项.为了进一步将本方法所得到的结果与其他方法所得到的温度修正做比较,他也计算了该温度修正项.他发现该温度修正项即为熟知的大质量 Schwinger 模型.其1环的结果即与其他方法所得到的结果有漂亮的对应.

接下来,他发现上述方法也可以用于研究(2+1)维的另一拓扑量诱导 Chern-Simons 项手徵反常的关系.这可溯源到 Fosco, Rosini,

Schaposnik 等人之前的研究.他们研究了一个有外加静磁场的(2+1)维特殊系统,并发现该系统的诱导 Chern-Simons 项的温度修正亦可由手性旋转所产生的 Jacobian 计算而得.换句话说,诱导 Chern-Simons 项与手徵反常高度相关.为了进一步了解该关系,他试着探讨另外的外场.他选择研究当外场是一个随时间而变的电场时的情况.他发现,此时诱导 Chern-Simons 项的温度修正并非由 Jacobian 所推得.有趣的是他发现该温度修正亦与大质量 Schwinger 模型相关.他也计算了 1 环的温度修正,并发现该方法所得到的温度修正与其他方法所得到的结果相吻合.他认为,上述诱导 Chern-Simons 项在不同外场时的差异,是与真空极化在 $p=0$ 处的不可解析特性息息相关.

最后,他再度回到(1+1)维的 Goldstone-Wilczek 模型,并讨论其分数电荷与 Berry 相位的关系.他利用数据降维以及二次量子化的方法发现,在温度为 0 时,分数电荷可联结到相空间 Berry 相位.在有限温度时,由于温度效应,其对应关系亦不存在.

本书有一定的原创性和前沿性,这在国内出版界属稀缺品种.

古人说,作论有三不必,二不可.前人所已言,众人所易知,摘拾小事无关系处,此三不必作也.巧文刻深,以攻前贤之短,而不中要害;取新出奇,以翻昔人之案,而不切情实,此二不可作也.古人作文如此,我们工作室在选择科技图书出版时亦如此.

刘培杰

2024 年 3 月 26 日

于哈工大

国外优秀物理著作原版丛书（第一辑）

论全息度量原则

——从大学物理到黑洞热力学（英文）

On the Principle of Holographic Scaling—From College Physics to Black Hole Thermodynamics

●［美］利奥·罗德里格斯（Leo Rodriguez）
●［美］珊珊·罗德里格斯（Shanshan Rodriguez）
著

HITP
哈爾濱工業大學出版社
HARBIN INSTITUTE OF TECHNOLOGY PRESS

黑版贸审字 08－2021－044 号

图书在版编目(CIP)数据

论全息度量原则:从大学物理到黑洞热力学＝On the Principle of Holographic Scaling:From College Physics to Black Hole Thermodynamics:英文/(美)利奥·罗德里格斯(Leo Rodriguez),(美)珊珊·罗德里格斯(Shanshan Rodriguez)著.—哈尔滨:哈尔滨工业大学出版社,2024.10

(国外优秀物理著作原版丛书.第一辑)

ISBN 978－7－5767－1341－1

Ⅰ.①论…　Ⅱ.①利… ②珊…　Ⅲ.①物理学－英文
Ⅳ.①O4

中国国家版本馆 CIP 数据核字(2024)第 073687 号

LUN QUANXI DULIANG YUANZE:CONG DAXUE WULI DAO HEIDONG RELIXUE

策划编辑　刘培杰　杜莹雪
责任编辑　刘立娟　刘家琳　李　烨　张嘉芮
封面设计　孙茵艾
出版发行　哈尔滨工业大学出版社
社　　址　哈尔滨市南岗区复华四道街 10 号　邮编 150006
传　　真　0451－86414749
网　　址　http://hitpress.hit.edu.cn
印　　刷　哈尔滨博奇印刷有限公司
开　　本　787 mm×1 092 mm　1/16　印张 63.75　字数 1 195 千字
版　　次　2024 年 10 月第 1 版　2024 年 10 月第 1 次印刷
书　　号　ISBN 978－7－5767－1341－1
定　　价　378.00 元(全 6 册)

Contents

Preface

An instructor enters into a general physics class and starts the lecture by posing the question: 'What is mass?' A naively simplistic question that more often than not generates a myriad of answers along the lines of: 'All the stuff an object is made of.' The instructor's next question is then: 'Well, what is all that stuff?' to which students typically say: 'All the matter...' Eliciting the response from the instructor: 'And what is matter?' This game will continue until students realize that they do not really know what this vocabulary and seemingly familiar concept of everyday life truly is. Mass in physics is a traditionally notorious topic that can elude physics students all the way up to graduate level. However this and more such topics provide an opportunity to implement contemporary cutting-edge physics research into the undergraduate classroom. This is the premise of our book: to utilize results and concepts from holographic duality research as a pedagogical tool to make difficult concepts in physics accessible in the introductory and advanced physics curriculum.

Not typically part of an undergraduate education in physics, courses on general relativity have been on the rise and have been in continued demand by respective students. In this book, we wish to share some techniques inspired by, and derived from, modern gravitational research (holographic duality) that help bring advanced topics in gravity theory into the undergraduate classroom. This serves to satisfy the excitement and curiosity of students about topics covering black hole physics and their properties. The techniques included have been developed over several years and implemented in teaching general relativity and gravitation to undergraduates at primarily undergraduate serving institutions including Grinnell College and Assumption College; and PhD granting universities which include the University of Iowa and Worcester Polytechnic Institute.

Contemporary research can be a great resource, especially when introducing new and effective pedagogical tools into the undergraduate physics classroom. Research is where physics is pushed to its limits in the hopes of making significant discoveries about fundamental laws of nature. Thus, our overarching goal, with our own students (research and in-class) and with this book is to equip students with the tools, skills and habits in order for them to independently expand their intellectual horizons within a research model and in turn accumulate knowledge on their own.

Acknowledgements

We would like to thank our families, colleagues and friends who supported and encouraged us throughout the process of completing this work: L. Ramdas Ram-Mohan, Vincent G. J. Rodgers, Sylvester James Gates Jr., Jacob D Willig-Onwuachi, Edward Dix, Charles Cunningham, Ping Li, Xinzhuang Zhang, Leo L. Rodriguez, Himelda Krchov, James Vanderveer and Kirby Ziesmer.

LR would especially like to thank Vincent G. J. Rodgers and Sylvester James Gates Jr. who first encouraged, mentored and provided opportunity for teaching advanced theoretical physics concepts to intellectually young students within the Student Summer Theoretical Physics Research Session's program.

Additionally, LR thanks Sujeev Wickramasekara, who passed away suddenly on December 28th, 2015. Dr. Wickramasekara served as LR's post-doctoral supervisor at Grinnell College from 2011–2013. A superior mentor, teacher, collaborator and friend, Dr. Wickramasekara encouraged LR to further develop a unique teaching and scholarship paradigm and provided multiple opportunities for LR to offer pedagogically valuable undergraduate learning opportunities in topics on advanced theoretical/mathematical physics. Dr. Wickramasekara had an enlightening vision for theoretical physics scholarship and physics pedagogy. His scientific prowess, enlightening thoughts and person are greatly missed and never forgotten!

We also thank our students who braved to work on our research while learning within our theoretical physics experiential learning paradigm: Dominic Chang, Emma Machado, Majumder Swarnadeep, Zachary Chester, Bryannah Voydatch, Nathaniel Mione and Raid Suleiman.

Author biographies

Leo Rodriguez

Leo Rodriguez received his PhD in physics from the University of Iowa in 2011, after which he served as an HHMI postdoctoral research fellow at Grinnell College and held subsequent faculty positions at Grinnell College and Assumption College. Leo rejoined the physics faculty at Grinnell College in the fall of 2018. Leo's research interests focus on the thermodynamic properties of black holes, how they relate to quantum gravity and how they are encoded in dual conformal field theories. In addition to his research interests, Leo spends much time teaching physics and mathematics within the undergraduate curriculum and enjoys making difficult and esoteric topics accessible and understandable to future physicists. Outside of his professional career, Leo enjoys spending time with his family, training Judo and fishing for trout.

While in graduate school, and upon the recommendation of his all-wise PhD advisor (Professor Vincent Rodgers), Leo sought out Shanshan as a study partner for the University of Iowa's physics PhD qualifying exam. Today, they are the proud parents of two and in conjunction lead a collaborative research effort in black hole thermodynamics which includes students from Worcester Polytechnic Institute, Assumption College and Grinnell College.

Shanshan Rodriguez

Shanshan Rodriguez received her PhD in physics from the University of Iowa in 2010 in the field of theoretical space plasmas, specifically magnetic reconnection. She subsequently joined NASA's Goddard Space Flight Center in Greenbelt MD as a research engineer where she worked at the Detector System Branch till 2004, followed by a faculty position at Worcester Polytechnic Institute. Shanshan's research interests focus on numerical techniques and the implementation/optimization of computational tools in theoretical astrophysics. Shanshan joined the physics faculty at Grinnell College in 2018 and has spent much time teaching physics and developing undergraduate courses/labs in astrophysics. Outside of her work, Shanshan enjoys conducting outreach in K-12 science education and summer camps, cooking and crafting accessories.

Chapter 1

Introduction

As we continue to (theoretically) probe physics at a deeper and more fundamental level, we find that the Universe likes to store measurable quantities in a holographic manner [1]. Similar to an actual hologram, which encodes a three-dimensional image onto a two-dimensional sheet, complicated and advanced theories like string theory reveal that our description of certain gravitational theories in higher dimensions contain information about four-dimensional gauge theories [2]. Similar examples include the fact that black hole entropy scales holographically with their two-dimensional horizon areas (and not extensively with volume), or how a four-dimensional quantum gravity theory can be fully captured by a two-dimensional conformal field theory in the near vicinity of its solution space [3, 4].

The idea that the Universe is holographic in nature has been well established by these esoteric research fields (string theory, quantum gravity, black hole thermodynamics, etc) and, with it, the perceived view that these advanced topics are reserved to a select group of specialists with expertise in the respective research fields. Our goal will be to shatter this paradigm by demonstrating the holographic principle in a manner suitable for, and accessible to, undergraduate physics students. In particular we implement holography to extract complicated black hole properties, which can be mapped and dealt with in a familiar Newtonian setting. This brings modern research into the classroom, which is an important addition to the general relativity and gravitation pedagogy, since students are always eager to learn and gain exposure to topics that would normally seem far out of reach of their intellectual horizons. It also serves to continue to motivate developments towards more inclusive undergraduate physics curriculums, which would include introductory courses on relativity, gravitation and cosmology.

1.1 The four forces and the geometric Universe

As far as we can tell to low energies, the Universe seems to be completely governed by just four fundamental forces:

doi:10.1088/2053-2571/ab06bdch1

- Electromagnetic force
  - Governs the electric interaction between charged matter and radiation. This force is mostly responsible for engineering electronic devices such as cell phones, computers, and anything that runs on electricity.
- Weak nuclear force
  - The force that deals with radioactivity, beta decay and their applications including medical imaging such as positron emission tomography.
- Strong nuclear force
  - This force is responsible for binding the nuclei of atoms together. It must be stronger than the electric repulsion in order to keep protons bound together in the respective length scale of atomic nuclei. The strong force's binding energy is realized in nuclear reactors and nuclear weapons.
- Gravitational force
  - A large scale attractive force between matter and energy. Keeps us on the Earth, the Earth around the Sun, the Sun in the Milky-Way Galaxy. Gravity also acts across the general large scale structure of the Universe.

Table 1.1 summarizes a relative comparison between the four forces, scaled in comparison to the gravitational force [5, 6]. This comparison reveals a dichotomy between gravity and the other three forces. This stark contrast in force is one reason why the task of quantizing gravity is very cumbersome, to say the least. Another reason is based on the geometric nature of our Universe and how we describe each force mathematically. That is, the three forces, $E\&M$, weak and strong, are actually very similar in their mathematical formulations and described by the overarching physical theory referred to as gauge quantum field theory (quantum field theory in short) or Yang–Mills theory. Common shorthands for these quantum field theories include *QFT*, the standard model or $U(1) \times SU(2) \times SU(3)$, where the latter refers to the symmetry gauge groups of each of the three gauge forces. The most popular, commonly accepted and accurate description of gravity (especially since the recent gravitational wave detections/discoveries [7, 8]) is Einstein's theory of general relativity (GR). GR describes gravity as the metric measure of geometry of spacetime, that is, the gravitational field itself is a description of the geometry of spacetime, whose dynamical features are a result of how energy and mass is distributed within it.

To witness the similarity between the three gauge theory forces, let us delve superficially into their mathematical formulations starting with $E\&M$. Quantum

Table 1.1. The four fundamental forces.

Force	$E\&M$	Weak	Strong	Gravity
Relative strength ($\sim 10^{-15}$ m)	10^{36}	10^{25}	10^{38}	1

field theoretically we interpret the electromagnetic force between two electrons, for example, as an exchange of a force carrier particle. In the case of $E\&M$, the two electrons will communicate their repulsive desire by exchanging a photon between them. This interaction is symbolized by Feynman diagrams, where the most basic one representing this interaction is depicted in figure 1.1. Mathematically the $E\&M$ force carrier (photon) γ is describe by a $U(1)$ four-vector represented by A_μ. $U(1)$ is called the gauge group and is a shorthand for unitary ($U^\dagger U = 1$) one-by-one matrices, which has one degree of freedom, i.e. one choice to make for the entry of a one-by-one matrix. A simple example of such a matrix is $e^{-i\Lambda}$, since

$$(e^{-i\Lambda})^\dagger e^{-i\Lambda} = e^{+i\Lambda}e^{-i\Lambda} = 1.$$

The fact that $U(1)$ is the electromagnetic gauge group is not a coincidence, but relates to the number of force carrier particles, in other words the number of degrees of freedom of the gauge group counts the number of force carrier particles. Thus, since the degrees of freedom of $U(1)$ is precisely one, that implies one force carrier particle for $E\&M$ namely the photon γ. Now, since the photon has spin one, it must be mathematically described by a four-vector, where four denotes one temporal and three spacial dimensions (spacetime). The four-vector is indexed by one Greek index μ, which runs from 0, …, 3 and 0 indexes time and 1, 2, 3 index space. This gives us a pattern between geometric object and the spin of the particles they represent, namely the number indices correspond to the integer spin value of the respective particle:

$$\underbrace{\Phi}_{\text{scalar}\to\text{spin } 0}, \quad \underbrace{A_\mu}_{\text{vector}\to\text{spin } 1}, \quad \underbrace{g_{\mu\nu}}_{\text{tensor}\to\text{spin } 2} \quad \cdots$$

For $E\&M$, A_μ is also referred to as the vector potential and is comprised of $A_0 = -\varphi$ the scalar potential (voltage) and $(A_1, A_2, A_3) = \vec{A}$ the magnetic vector potential. Following this description of $E\&M$, let us turn our attention to the weak force, which has three known force carrier particles, namely the neutral spin one Z-boson, the positive spin one W^+-boson and the negative spin one W^--boson. Since they are all spin one, they should be described by vectors with one spacetime index:

$$\left\{A_\mu^Z, A_\mu^{W+}, A_\mu^{W-}\right\}$$

and since there are three, they should have a gauge group with three degrees of freedom. A good choice for such a gauge group is the special unitary two-by-two

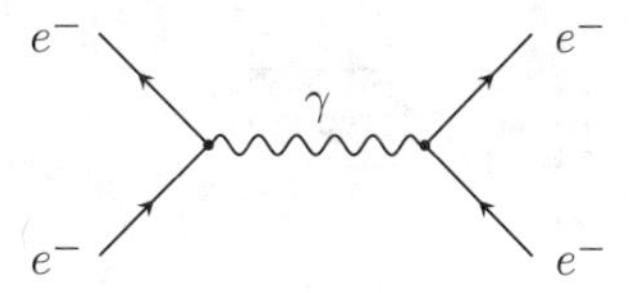

Figure 1.1. Quantum field theoretic process of two electrons repelling each other by exchanging a photon, γ. In this process the photon is referred to by any of the following: force carrier particle, gauge boson, gauge connection or virtual photon.

matrices called $SU(2)$. A two-by-two matrix has in general $2^2 = 4$ degrees of freedom, but since it is special it has unit determinant:

$$\left| \begin{pmatrix} a & b \\ c & d \end{pmatrix} \right| = ad - bc = 1, \tag{1.1}$$

and thus one of the entries a, b, c, d may be solved for and determined in terms of the other three. Thus the actual degrees of freedom of $SU(2)$ is $2^2 - 1 = 3$. In general a vector theory with an $SU(N)$ gauge group will exhibit $N^2 - 1$ number of force carrier particles. So, the strong force is known to have eight gluons as its force carrier particles, each with spin one. That means

$$N^2 - 1 = 8 \Rightarrow N = 3,$$

i.e. the strong force is a $SU(3)$ four-vector theory! To summarize (see table 1.2) the difference between the three gauge theory forces is reflected in their respective gauge groups, since they are all spin one vector theories. The combination of the three gauge theory forces comprise the standard model of particle physics without any contributions from gravity due to its comparatively feeble strength. However, there is a subtlety, in their pure mathematical formulations each vector theory is only distinguishable by means of their gauge groups. That is, each theory, purely mathematically speaking, exhibits massless gauge bosons. This is obviously known to be untrue from experimental measurements and this symmetry of indistinguishability must somehow be broken. The breaking of this symmetry is accomplished via the introduction of a new gauge boson through the Anderson–Higgs mechanism and endows the three gauge theory force particle carriers with their appropriate masses [9, 10]. This symmetry broken theory called the standard model is denoted by the Cartesian product of each gauge group:

$$\text{Standard model} \equiv U(1) \times S(2) \times S(3).$$

What about gravity? We might extrapolate from the above discussion that gravity should also be a vector theory with gauge group $SU(4)$, maybe, but where would the fun be in that …? It turns out, which we will discover in much more detail in later chapters, that gravity is a tensor theory of spin two. That is, the force carrier particle of gravity (the graviton) is described by a tensor field $g_{\mu\nu}$ with gauge group comprising the collection of general coordinate transformations in four dimensions called the diffeomorphism group. Written as a four-by-four symmetric matrix the

Table 1.2. Gauge theory.

Force	4-vector force carrier	Gauge group
$E\&M$	γ	$U(1)$
Weak	$Z, W^{\pm}$	$SU(2)$
Strong	8-gluons	$SU(3)$

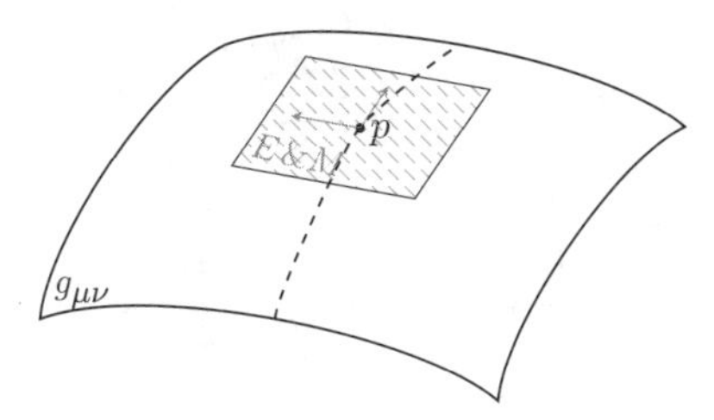

Figure 1.2. Simplistic depiction of the geometry of the Universe.

ground state graviton for a stationary gravitating mass is just the Schwarzschild back hole:

$$g_{\mu\nu} = \begin{pmatrix} -\left(1 - \frac{2GM}{r}\right) & 0 & 0 & 0 \\ 0 & \left(1 - \frac{2GM}{r}\right)^{-1} & 0 & 0 \\ 0 & 0 & r^2 & 0 \\ 0 & 0 & 0 & r^2 \sin^2\theta \end{pmatrix},$$

which is already very complicated and whose details/mysteries will be revealed in later chapters. Why gravity diverges in mathematical description in comparison to the gauge theory forces has to do with the fact that within Einstein's formulation of general relativity, gravity is synonymous with the geometry of spacetime. To try and illustrate this geometric dichotomy, let us draw a simplistic picture of the Universe in two dimensions, as done in figure 1.2, including just gravity and $E\&M$. As can be seen in figure 1.2, gravity describes the base-space (spacetime) upon which gauge theories are tangent to. In other words at some point p on the base-space we can always span a tangent plane by choosing two linearly independent tangent vectors. Thus, the space of the vector theory is fibered to gravity at point p. In technical terms we can therefore say that gauge theory is a principle vibration to gravity or spacetime. One immediate question that comes to mind in this toy description of the Universe is what if we picked a different point than p? Clearly we can fiber another $E\&M$ to some new point and then we would have two $E\&M$ theories? This question is easily resolved by placing two electrons at competing fibered $E\&M$s at differing points, as depicted in figure 1.3. From this new picture, we see that the two individual fibered $E\&M$s are connected by the photon, since the two electrons would exchange the photon in order to mediate their respective repulsion of each other. This is why another term for the photon is gauge connection and the connected collective $E\&M$s at each point of spacetime are referred to as the single principle fiber bundle.

1.2 QFT + GR

The tremendous success of the standard model is in great part due to its constituent quantum field theories of the three gauge forces: quantum electrodynamics (QED),

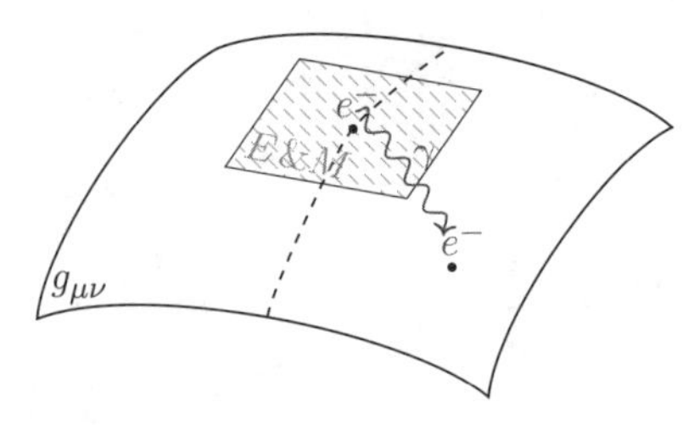

Figure 1.3. Two electrons at different points in spacetime exchange a photon to mediate the electromagnetic interaction.

electro-weak and quantum chromodynamics *QCD*. The development of these theories spanning multiple decades of hard work, developed specific quantization techniques that ignored gravity all together and rightfully so due to gravity's feebleness. This resulted in specific quantum field theoretic methods applicable to the fibrations of spacetime and not spacetime itself. In other words, another way to interpret the problem of quantum gravity is how exactly to translate modern quantization methods from the fiber bundle to the base-space itself. But what would happen if we ignored this geometric/force-strength dichotomy and applied techniques from modern quantum field theory to the gravitational field? The answer is summed up in a very important topic in QFT known as renormalization and the fact that gravity is non-renormalizable! Renormalization in a well-defined QFT deals with the fact that in perturbation theory, certain individual (or all) terms contribute infinite or divergent pieces. Using perturbation theory in QFT is similar to how a Taylor series is used to approximate a smooth function to a given order:

$$e^x \sim 1 + x + \frac{x^2}{2} + \frac{x^3}{6} + \frac{x^4}{24} + \frac{x^5}{120} + \frac{x^6}{720} + \mathcal{O}(x^7), \tag{1.2}$$

except instead of expanding a function we expand a quantum statistical partition function, labeled $Z(A_\mu)$. For example if we are considering *QED*, and instead of a series of polynomials we have a series of Feynman diagrams depicting all possible particle interactions at each order

$$\begin{aligned} Z_{E\&M}(A_\mu) &= \int \mathcal{D}A e^{iS(A)} \sim \; + \; + \; + \; + \dots \\ &= \sum_{\text{Vertex}} \{\text{All Possible Diagrams}\} \end{aligned} \tag{1.3}$$

where we have written the partition function as a path integral over all possible $U(1)$ vectors A_μ as an example for *QED*, each diagrammatic contribution is calculated using the Feynman rules for the respective QFT and we are using vertex instead of order. Comparing each diagram to the actual respective mathematical contribution in the Taylor series of the partition function leads to the Feynman rules for the specific theory. These Feynman rules are formulated in terms of the diagrammatic language and nomenclature defined in figure 1.4. These diagrams may contribute divergent terms in the series, but known renormalization procedures developed for gauge theories are very successful in isolating them from actual measurable physical results. This is not the case for gravity, which we explore next.

Let us consider a very, very small perturbation to flat (zero curvature) spacetime and look at the most simple Feynman diagrams of graviton–graviton scattering. This may initially seem like a weird concept, graviton–graviton scattering, but remember from baby-physics and Newton's law of gravity, that anything with mass or energy (also remembering special relativity) sources gravity and thus, the quantum field theoretically interacts with gravitons. So, that means that since gravitons have energy, they must source gravity and thus interact with themselves! This is unlike $E\&M$, where the photon has no electric charge and thus does not interact with itself. Such a theory, whose gauge boson(s) do not self-interact, is referred to as a linear theory. Gravity is notoriously known to be non-linear. The most simplistic (ignoring squared momentum factors of derivative squared perturbations) and toy Feynman rules for this scenario of graviton–graviton scattering, are [11] the following.

- Start with factors of momentum squared p^2, for each vertex.
- Multiply by factors of $\frac{1}{p^2}$, for each propagator.
- Integrate over d^4p for every closed loop.

Let us apply these rules to some of the first few diagrams of perturbative gravity for graviton–graviton scattering. We have

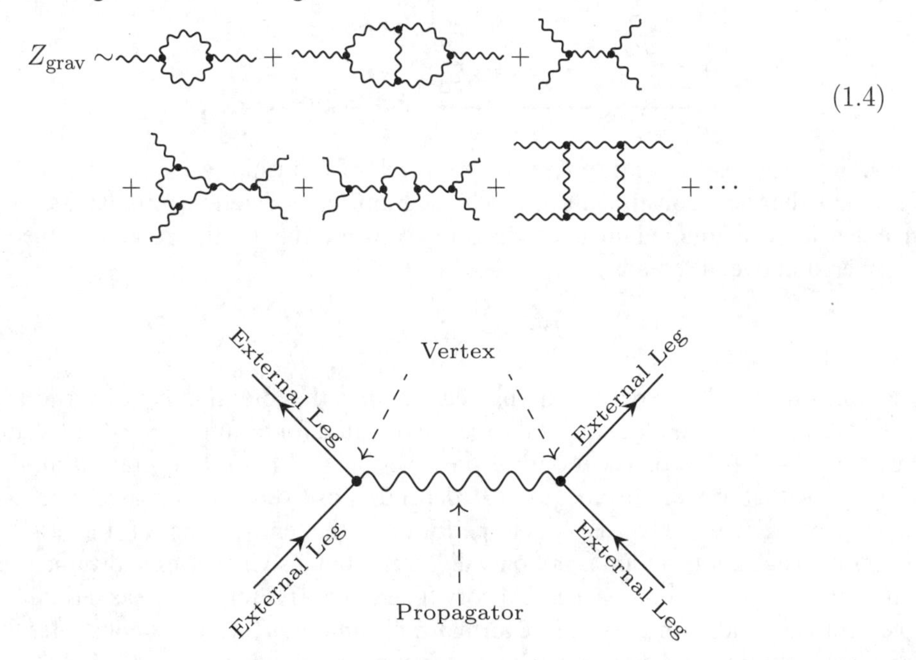

Figure 1.4. Feynman diagram language: incoming and outgoing particles are referred to as external legs, the force carrier particle (photon) is referred to as the propagator and the interaction points or order of diagram are labeled as vertex.

where we have listed the diagrams in no particular order. Next, applying the above Feynman rules, keeping in mind that we are really only interested in total powers of momentum, yields the following:

$$Z_{\text{grav}} \sim \underbrace{\qquad}_{\frac{(p^2)^2}{(p^2)^2} d^4p \sim d^4p} + \underbrace{\qquad}_{\frac{(p^2)^4}{(p^2)^5}(d^4p)(d^4p) \sim \frac{(d^4p)(d^4p)}{p^2}} + \underbrace{\qquad}_{\frac{(p^2)^2}{p^2} \sim p^2} + \underbrace{\qquad}_{\frac{(p^2)^4}{(p^2)^4} d^4p \sim d^4p} + \underbrace{\qquad}_{\frac{(p^2)^4}{(p^2)^4} d^4p \sim d^4p} + \underbrace{\qquad}_{\frac{(p^2)^4}{(p^2)^4} d^4p \sim d^4p} + \cdots \tag{1.5}$$

and if we now count total powers of momenta, which we call the degree of divergence D when evaluated over all possible energies or frequencies, we have for each diagram respectively in equation (1.5):

$$Z_{\text{grav}} \sim \underbrace{\qquad}_{D=4} + \underbrace{\qquad}_{D=6} + \underbrace{\qquad}_{D=2} + \underbrace{\qquad}_{D=4} + \underbrace{\qquad}_{D=4} + \underbrace{\qquad}_{D=4} + \cdots \tag{1.6}$$

Now, we are able to extrapolate a graphical relationship between D and the Feynman diagram constituents in each diagram. This relationship classifies the divergencies encountered in each diagram contributing to the total toy theory considered above and reads

$$D = 4 - E + n, \tag{1.7}$$

where E is the total number of external legs and n is the total number of vertices in each diagram. So, again looking at the second diagram for example, $E = 2$, $n = 4$ and thus $D = 4 - 2 + 4 = 6$. Try it out for the other ones. What is important from this analysis, is that we discovered that D depends positively or increases with the number of vertices. That is, when performing the entire sum of the Taylor expansion, each contribution is not only divergent, but also increases in divergencies with increasing vertex/order. Such a theory (like gravity) is referred to as ultraviolet divergent and renders gravity non-renormalizable and thus further demonstrates the previously mentioned dichotomy between gravity and gauge theory.

1.3 Cosmological evolution

We might be feeling a bit discouraged from the previous sections about the existence of a quantum field theory of gravity, or even question if such a theory is needed. Given gravity's feebleness compared to gauge theory, we might be set with using the standard model for all things small and GR for all things large and not worry about quantum gravity. This is fine for the most part, but there are at least two cases where the Universe provides us with a hint or even an example where quantum gravity is relevant. These two cases include black holes and the very early Universe described by the Big Bang cosmology paradigm. Starting with Hubble's observations, in which measurements were made of the first known galaxies beyond our own, Hubble was able to conclude that on the large scale all galaxies are receding away from each other and thus the Universe is expanding. He did so by taking galactic spectra and determining that they all exhibited a redshift of some degree. This redshift allowed Hubble to calculate a galaxy's respective recession velocity and comparing that to their apparent distance (determined via Cepheid variable stars) he discovered a linear relationship [5], depicted in figure 1.5. This linear relationship is called Hubble's law:

$$V = H_0 D, \tag{1.8}$$

where H_0 is the slope and is referred to as Hubble's constant. Hubble's law allows us to naively determine the age of the Universe by inverting the slope, i.e. $t_{\text{Universe}} \sim \frac{1}{H_0}$. Of course this is not an accurate calculation, since the age of the Universe depends on what type of matter dominates across the cosmos. Hubble's law only says the Universe should be expanding, but it does not tell us about the rate of this expansion. For the most part of modern astronomy, it was assumed that this rate should be mostly affected by gravity, i.e. the mutual attraction of all matter in the Universe should perhaps slow the expansion down. In order to take this into account, for determining the age of the Universe, requires some general relativistic analysis and solving the associated field equations for a cosmological metric ansatz. The result is that the age of the Universe is still on the order of one over the Hubble constant, but weight by a unit-less factor $F(M, \Lambda)$ which in turn depends on the

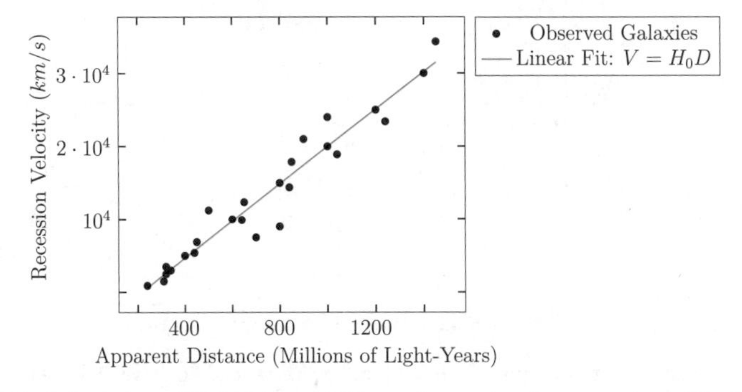

Figure 1.5. A depiction of Hubble's results from discovering the expansion of the Universe.

fractional contribution of regular gravitating matter and radiation denoted by M and of dark matter and dark energy denoted by Λ:

$$t_{\text{Universe}} = \frac{F(M, \Lambda)}{H_0}. \tag{1.9}$$

Dark energy is the term given to matter that could cause the expansion rate to increase or even to increase at an accelerated rate. Figure 1.6 depicts four different scenarios of Universes with different fractional matter contributions. To determine which one best describes our Universe we need to determine how abundant dark matter and dark energy is. Dark matter seems to be a mysterious invisible matter that only interacts gravitationally. Discovered first by Vera Rubin by studying galactic rotation curves and subsequently plotting stellar orbital velocities versus the radial distance to galactic center, as depicted in figure 1.7, Rubin noticed a trend of

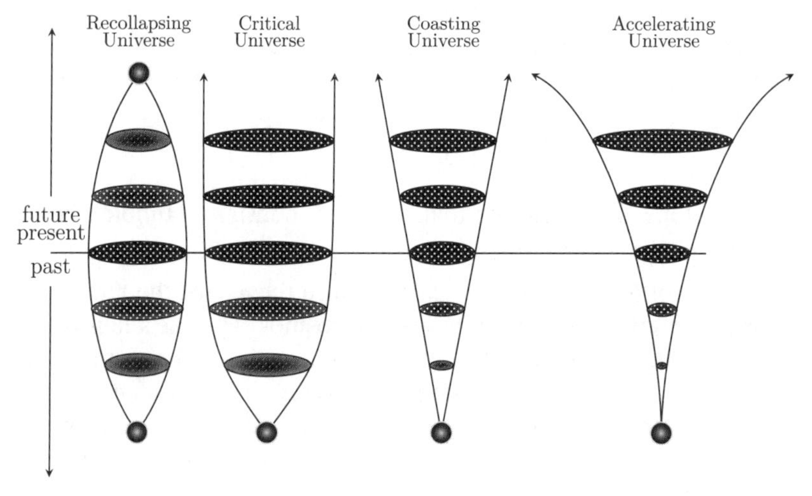

Figure 1.6. Four possible Universes are shown, depending on the relative abundance between gravitating matter versus dark energy. The Universe on the far right is what we typically refer to as a dark energy dominated Universe with an accelerated increasing expansion rate.

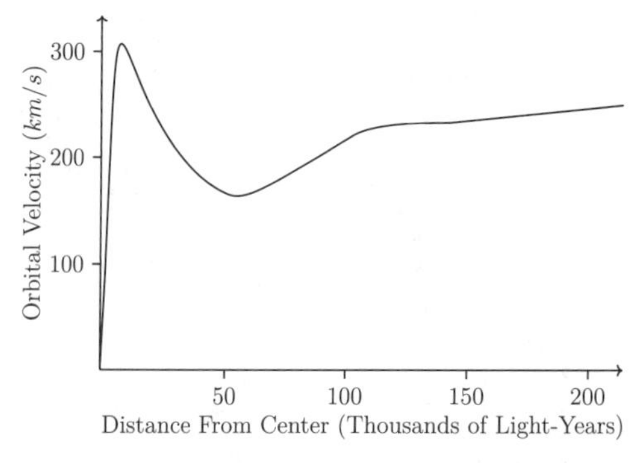

Figure 1.7. Typical galactic rotation curve. Most spiral galaxies tend to have flat rotation curves, which indicate large amounts of dark matter in their halos.

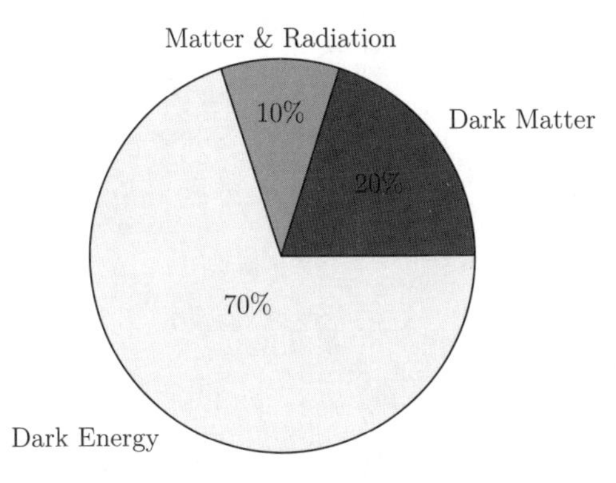

Figure 1.8. Fractional matter contribution in our Universe. It looks dark ...

flattening rotation curves for stars further and further away from their galactic centers [5, 12]. This indicated a much larger gravitating mass than optically discernible and thus coined the term: dark matter. It turns out that long term studies have revealed that our Universe seems to have a substantial amount of dark matter compared to regular matter. In fact the fractional decomposition of each type is depicted in figure 1.8 and we see that most of our Universe is comprised of dark stuff. Determining the amount of dark energy requires measuring distances to objects across the entire large scale structure of the Universe or in other words determining a measure of how the Universe scales in time. This has been done by using standard candles, something with a known brightness, and measuring its apparent brightness in order to calculate how far the standard candle must be to have faded to the measured apparent brightness. A very useful and most commonly used standard candle is a type 1a supernova. Such supernovae result when a white dwarf star in a binary orbit with a regular star accretes enough matter from its companion that it reaches a limiting balance between gravity trying to collapse the white dwarf and the repulsive electron degeneracy pressure acting against gravity. This mass limit is always the same and is given by the Chandrasekhar mass limit of around 1.4 solar masses. This mass limit specifies the total luminosity possible for such explosions, making them ideal tools to measure distances across the cosmos. The results of these studies [5, 13] are depicted in figure 1.9, which imply that our Universe is dark energy dominated with a fractional contribution given by the pie-chart in figure 1.8 and that its age is about $t_{\text{Universe}} = \frac{F(M, \Lambda)}{H_0} = 14$ billion years.

What can this discussion on cosmological evolution tell us about quantum gravity? Well, if we reverse time, then we should expect everything in the Universe to converge to one singular point, called the Big Bang, from which we assume it all originated (see figure 1.10). This Big Bang or singular point of origin of our Universe would require a unified quantum field theoretic description of all forces in order to fully capture the physics and grant us understanding. In other words, if we wish to fully understand the origin of the Universe encapsulated within our physical theories, a quantum theory of gravity is needed.

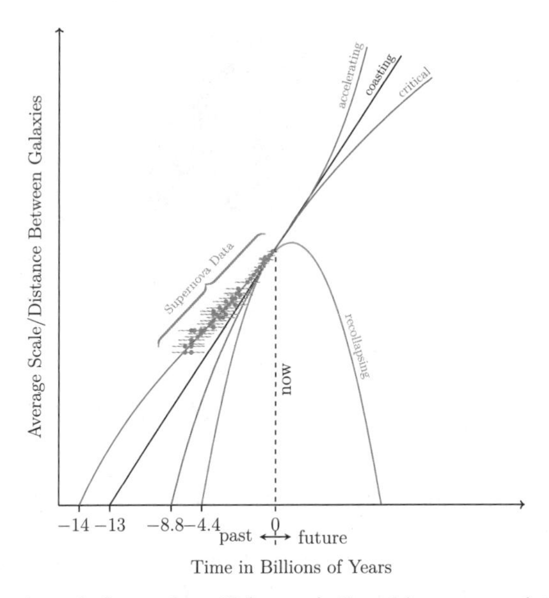

Figure 1.9. Determining the scale factor of our Universe via Type 1A supernova data. We see that the data tends to follow the accelerating inverse model and implies an age around 14 billion years.

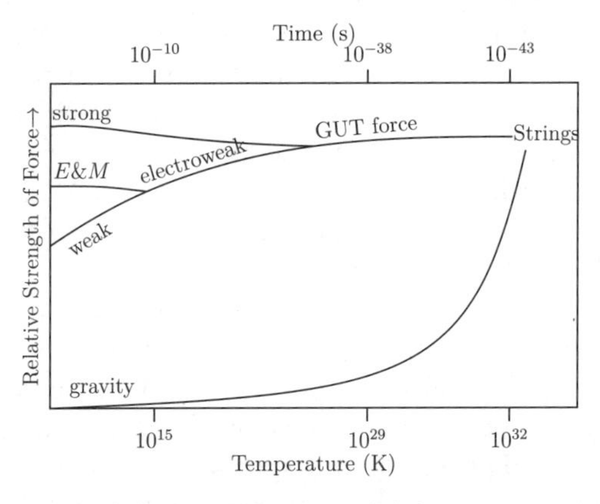

Figure 1.10. Depiction of the force hierarchy and their convergence at the Big Bang.

1.4 Black holes and holographic scaling

We will be discussing black holes, their physics and holography in almost every chapter of this monograph, so we do not wish to reveal too much here in the introduction, but for completeness of this chapter we want to mention some aspects of how black holes necessitate a quantum gravity. Considered a main-staple constituent and astrophysical object [14], black holes are another physical phenomena the Universe throws in our lap hinting at a quantum gravity. Combining large amounts of mass in extremely small sizes, black holes seem to be the perfect playground for quantum gravitational phenomena to manifest. This manifestation is best seen in what are called the laws of black hole mechanics, which closely resemble the laws of thermodynamics. As previously mentioned, we will delve into these details in later chapters, but what we can state here is that a comparison between these sets of laws implies that black holes are actually gray. That is, they exhibit thermodynamic properties and radiate at a thermal bath of temperature T_{Hawking} proportional to their surface gravity κ and exhibit entropy S proportional to their horizon area A [11]:

$$S = \frac{A}{4\hbar G} \qquad T_H = \frac{\hbar\kappa}{2\pi}, \tag{1.10}$$

where $\hbar$ is Planck's constant and G is Newton's. From the above formulae, we see two things.

- Black hole thermodynamics combines both Planck's and Newton's constants, thus they are quantum gravitational in origin!
- Black holes are holographic in nature, since their entropy scales with area and not volume, which is paradigm shifting since entropy is an extensive thermodynamic variable.

Thus, what we can conclude from the above is that black holes imply the existence of a quantum gravity and provide themselves as a tool for probing such theory.

References

[1] Maldacena J M 1999 The large N limit of superconformal field theories and supergravity *Int. J. Theor. Phys.* **38** 1113–33

[2] Horowitz G T and Polchinski J 2006 *Gauge/gravity Duality* (arXiv:gr-qc/0602037)

[3] Button B K, Rodriguez L and Wickramasekara S 2013 Near-extremal black hole thermodynamics from AdS_2/CFT_1 correspondence in the low energy limit of 4D heterotic string theory *J. High Energy Phys.* **1310** 144

[4] Rodriguez L and Yildirim T 2010 Entropy and temperature from black-hole/near-horizon-CFT duality *Class. Quant. Grav.* **27** 155003

[5] Bennett J O, Donahue M O, Schneider N and Voit M 2016 *The Cosmic Perspective* 8th edn (London: Pearson)

[6] Wolfson R 2015 *Essential University Physics* (London: Pearson)

[7] LIGO Scientific, Virgo Collaboration and Abbott B P *et al* 2016 Observation of gravitational waves from a binary black hole merger *Phys. Rev. Lett.* **116** 061102

[8] LIGO Scientific, Virgo Collaboration and Abbott B P *et al* 2018 *GWTC-1: A Gravitational-Wave Transient Catalog of Compact Binary Mergers Observed by LIGO and Virgo during the First and Second Observing Runs* (arXiv:1811.12907)

[9] ATLAS Collaboration and Aad G *et al* 2012 Observation of a new particle in the search for the Standard Model Higgs boson with the ATLAS detector at the LHC *Phys. Lett.* B **716** 1–29

[10] CMS Collaboration and Chatrchyan S *et al* 2012 Observation of a new boson at a mass of 125 GeV with the CMS experiment at the LHC *Phys. Lett.* B **716** 30–61

[11] Rodriguez L 2011 Black-hole/near-horizon-CFT duality and 4 dimensional classical spacetimes *PhD Thesis* University of Iowa https://doi.org/10.17077/etd.p4388g7i

[12] Sofue Y and Rubin V 2001 Rotation curves of spiral galaxies *Ann. Rev. Astron. Astrophys.* **39** 137–74

[13] Supernova Cosmology Project Collaboration and Kowalski M *et al* 2008 Improved cosmological constraints from new, old and combined supernova datasets *Astrophys. J.* **686** 749–78

[14] Bambi C 2018 Astrophysical black holes: a compact pedagogical review *Ann. Phys.* **530** 1700430

Chapter 2

ADM mass and holographic scaling in general physics

2.1 Vector notation and coordinates

In this section we quickly review and introduce (for completeness) some basic mathematical concepts on three-dimensional vector algebra that we do not shy away from teaching in an algebra-based introductory physics course. We note, though, that the entirety of this material is introduced gradually over a year-long sequence. This section will also serve to define conventions and specific notation used throughout the manuscript.

We employ standard Cartesian unit vectors in a right-handed coordinate system starting from the bracket/component notation for a vector $\vec{A}$ (the graph of which is depicted in figure 2.1):

$$\vec{A} = (A_x, A_y, A_z). \tag{2.1}$$

The unit vector notation follows from the fact that vectors add by components and obey the distributive property for arbitrary scaling factors, i.e.

$$\begin{aligned}\vec{A} &= (A_x, A_y, A_z) = (A_x, 0, 0) + (0, A_y, 0) + (0, 0, A_z)\\ &= A_x(1, 0, 0) + A_y(0, 1, 0) + A_z(0, 0, 1)\\ &= A_x\hat{\imath} + A_y\hat{\jmath} + A_z\hat{k},\end{aligned} \tag{2.2}$$

where we have defined the unit vectors:

$$\begin{aligned}\hat{\imath} &= (1, 0, 0)\\ \hat{\jmath} &= (0, 1, 0)\\ \hat{k} &= (0, 0, 1).\end{aligned} \tag{2.3}$$

The standard dot and cross products of vectors then follow simply by distribution and defining the individual operations tabulated below:

doi:10.1088/2053-2571/ab06bdch2

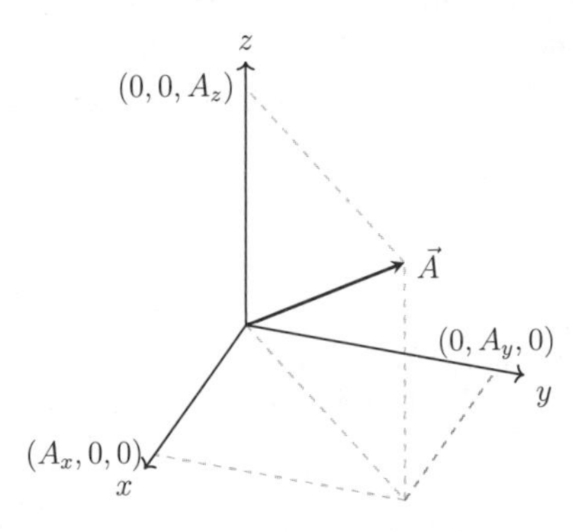

Figure 2.1. Cartesian vector plot.

$$\begin{aligned}
\hat{\imath}\cdot\hat{\imath} &= 1 & \hat{\jmath}\cdot\hat{\imath} &= 0 & \hat{k}\cdot\hat{\imath} &= 0\\
\hat{\imath}\cdot\hat{\jmath} &= 0 & \hat{\jmath}\cdot\hat{\jmath} &= 1 & \hat{k}\cdot\hat{\jmath} &= 0\\
\hat{\imath}\cdot\hat{k} &= 0 & \hat{\jmath}\cdot\hat{k} &= 0 & \hat{k}\cdot\hat{k} &= 1\\
\hat{\imath}\times\hat{\imath} &= 0 & \hat{\jmath}\times\hat{\imath} &= -\hat{k} & \hat{k}\times\hat{\imath} &= \hat{\jmath}\\
\hat{\imath}\times\hat{\jmath} &= \hat{k} & \hat{\jmath}\times\hat{\jmath} &= 0 & \hat{k}\times\hat{\jmath} &= -\hat{\imath}\\
\hat{\imath}\times\hat{k} &= -\hat{\jmath} & \hat{\jmath}\times\hat{k} &= \hat{\imath} & \hat{k}\times\hat{k} &= 0.
\end{aligned} \tag{2.4}$$

The above tables also provide ample opportunity to introduce/practice the right-hand rule and are incongruent with the usual definitions of dot and cross products in terms of the vector magnitudes and the angle between them. The dot product of a vector with itself defines the length or magnitude via the usual Pythagorean theorem:

$$\sqrt{\vec{A}\cdot\vec{A}} = |\vec{A}| = \sqrt{A_x^2 + A_y^2 + A_z^2}\,. \tag{2.5}$$

Of course this implies that the unit vectors have length one, hence the name. Next, we define the standard spherical and cylindrical coordinate maps depicted in figure 2.2(a) and (b):

$$\text{Spherical}\begin{cases} |\vec{A}| = \sqrt{A_x^2 + A_y^2 + A_z^2} = r\\ A_x = r\cos\phi\sin\theta\\ A_y = r\sin\phi\sin\theta\\ A_z = r\cos\theta \end{cases} \tag{2.6}$$

$$\text{Cylindrical}\begin{cases} r = \sqrt{A_x^2 + A_y^2}\\ A_x = r\cos\phi\\ A_y = r\sin\phi\\ A_z = z \end{cases}. \tag{2.7}$$

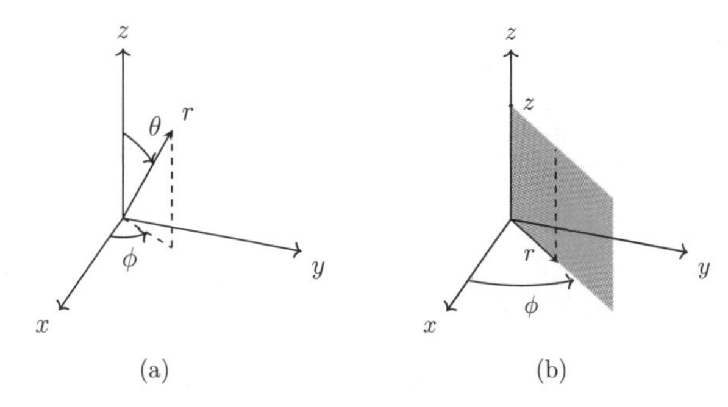

Figure 2.2. (a) Spherical coordinates. (b) Cylindrical coordinates.

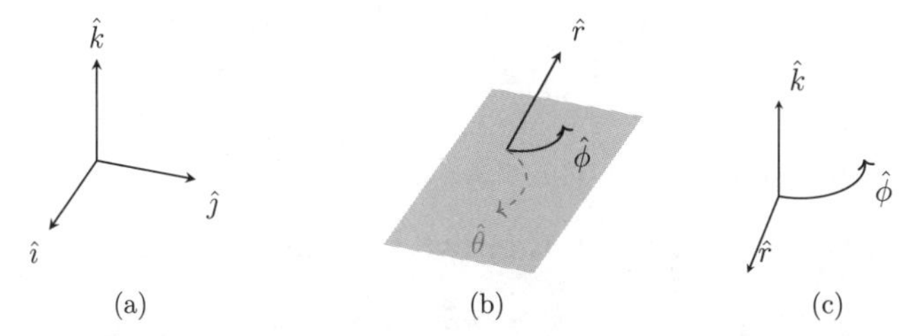

Figure 2.3. (a) Cartesian unit vectors. (b) Spherical unit vectors. (c) Cylindrical unit vectors.

Analogous to $\hat{\imath}$, $\hat{\jmath}$ and $\hat{k}$ we define the unit vectors for spherical and cylindrical coordinates:

$$\begin{aligned} &\text{Spherical } \{\hat{r},\ \hat{\theta},\ \hat{\phi}\} \\ &\text{Cylindrical } \{\hat{r},\ \hat{\phi},\ \hat{k}\}. \end{aligned} \tag{2.8}$$

At this point we do not delve into the curvelinear maps between the Cartesian unit vectors and equation (2.8), but rely on geometric visualizations/concepts to solidify the use of the new unit vectors. The hardest concept to deal with (in a general physics classroom) is the fact that some of the new unit vectors in equation (2.8) are curved as depicted in figure 2.3. Yet, we still impose similar orthogonality conditions as we did for the cartesian unit vectors, i.e.

$$\begin{array}{lll} \hat{r}\cdot\hat{r} = 1 & \hat{\theta}\cdot\hat{r} = 0 & \hat{\phi}\cdot\hat{r} = 0 \\ \hat{r}\cdot\hat{\theta} = 0 & \hat{\theta}\cdot\hat{\theta} = 1 & \hat{\phi}\cdot\hat{\theta} = 0 \\ \hat{r}\cdot\hat{\phi} = 0 & \hat{\theta}\cdot\hat{\phi} = 0 & \hat{\phi}\cdot\hat{\phi} = 1 \\ \hat{r}\times\hat{r} = 0 & \hat{\theta}\times\hat{r} = -\hat{\phi} & \hat{\phi}\times\hat{r} = \hat{\theta} \\ \hat{r}\times\hat{\theta} = \hat{\phi} & \hat{\theta}\times\hat{\theta} = 0 & \hat{\phi}\times\hat{\theta} = -\hat{r} \\ \hat{r}\times\hat{\phi} = -\hat{\theta} & \hat{\theta}\times\hat{\phi} = \hat{r} & \hat{\phi}\times\hat{\phi} = 0 \end{array} \tag{2.9}$$

and

$$
\begin{aligned}
\hat{r}\cdot\hat{r} &= 1 & \hat{\phi}\cdot\hat{r} &= 0 & \hat{k}\cdot\hat{r} &= 0\\
\hat{r}\cdot\hat{\phi} &= 0 & \hat{\phi}\cdot\hat{\phi} &= 1 & \hat{k}\cdot\hat{\phi} &= 0\\
\hat{r}\cdot\hat{k} &= 0 & \hat{\phi}\cdot\hat{k} &= 0 & \hat{k}\cdot\hat{k} &= 1\\
\hat{r}\times\hat{r} &= 0 & \hat{\phi}\times\hat{r} &= -\hat{k} & \hat{k}\times\hat{r} &= \hat{\phi}\\
\hat{r}\times\hat{\phi} &= \hat{k} & \hat{\phi}\times\hat{\phi} &= 0 & \hat{k}\times\hat{\phi} &= -\hat{r}\\
\hat{r}\times\hat{k} &= -\hat{\phi} & \hat{\phi}\times\hat{k} &= \hat{r} & \hat{k}\times\hat{k} &= 0.
\end{aligned}
\tag{2.10}
$$

Conversely, we could start with defining figure 2.3 and tabulate the results of equations (2.4), (2.9) and (2.10) via the right-hand rule, which we have found to be instructive for students.

2.2 Vectorized area

We define the area vector $\vec{A}$ in terms of two ingredients:

$$\vec{A} = A\hat{n}, \tag{2.11}$$

where A is some area and $\hat{n}$ is the unit normal vector to A. For example, take the flat rectangular sheet with sides a and b as depicted in figure 2.4. We may orient the sheet to lie in the x–y plane and thus the obvious choice of $\hat{n}$ is $\hat{k}$ and the vectorized area becomes

$$\vec{A} = ab\hat{k}. \tag{2.12}$$

Other useful general physics examples include the following.

- The circular sheet/disk

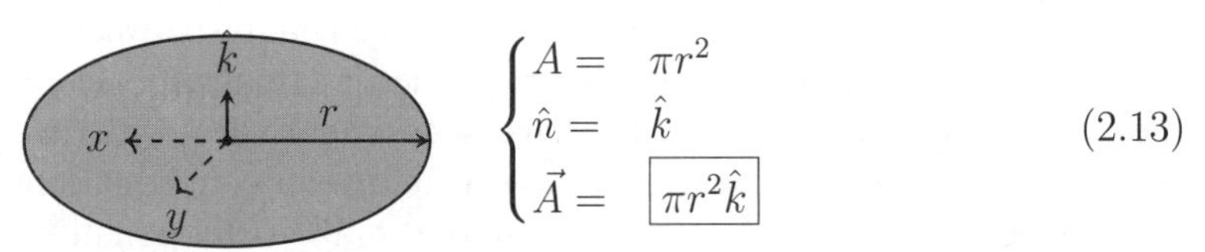

$$\begin{cases} A = & \pi r^2 \\ \hat{n} = & \hat{k} \\ \vec{A} = & \boxed{\pi r^2\hat{k}} \end{cases} \tag{2.13}$$

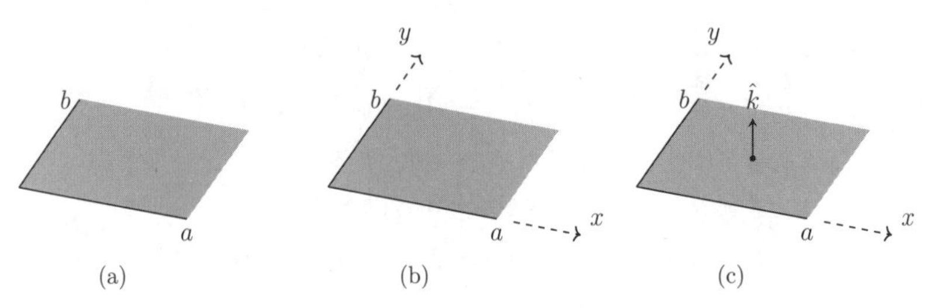

Figure 2.4. (a) Rectangular sheet with area $A = ab$. (b) Orienting the sheet in the x–y plane. (c) The obvious choice for $\hat{n} = \hat{\imath}\times\hat{\jmath} = \hat{k}$.

- The two-sphere (S^2)

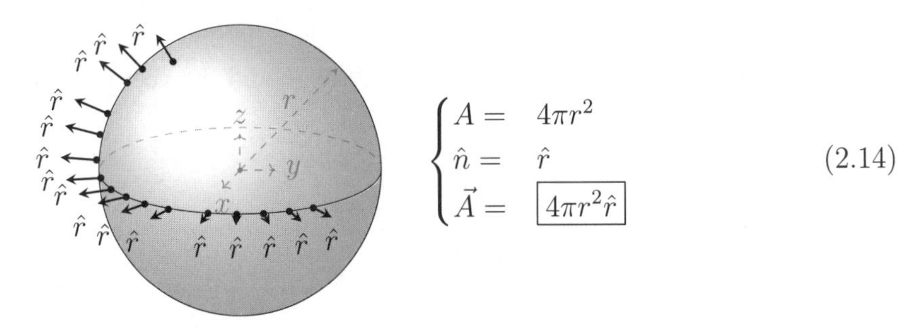

$$\begin{cases} A = & 4\pi r^2 \\ \hat{n} = & \hat{r} \\ \vec{A} = & \boxed{4\pi r^2 \hat{r}} \end{cases} \tag{2.14}$$

- The capped cylinder

  The capped cylinder is a bit tricky, since it is not differentiable at the edges and thus $\vec{A} = A\hat{n}$ only applies to each surface segment (the two circular caps and the cylindrical sheet) respectively. The issue of differentiability of the edges is beyond the general physics audience, but we can deal with it by defining a piecewise continuous area vector:

$$\begin{cases} A = & \begin{cases} A_1 = \pi r^2 & z \leq 0 \\ A_2 = 2\pi r l & 0 < z < l \\ A_3 = \pi r^2 & z \geq l \end{cases} \\ \hat{n} = & \begin{cases} \hat{n}_1 = -\hat{k} & z \leq 0 \\ \hat{n}_2 = \hat{r} & 0 < z < l \\ \hat{n}_3 = \hat{k} & z \geq l \end{cases} \\ \vec{A} = & \boxed{\begin{cases} -\pi r^2 \hat{k} & z \leq 0 \\ 2\pi r l \hat{r} & 0 < z < l \\ \pi r^2 \hat{k} & z \geq l \end{cases}} \end{cases} \tag{2.15}$$

The above examples are incongruent with the usual definition of planer area in terms of the cross product of two vectors spanning a parallelogram:

$$\begin{cases} A = & |\vec{a}||\vec{b}|\sin(\theta) \\ \vec{A} = & |\vec{a}||\vec{b}|\sin(\theta)\hat{n} \\ \hat{n} = & \frac{\vec{a}\times\vec{b}}{|\vec{a}\times\vec{b}|} = \frac{\vec{a}\times\vec{b}}{|\vec{a}||\vec{b}|\sin(\theta)} \\ \Rightarrow \vec{A} = & \vec{a}\times\vec{b} \end{cases} \tag{2.16}$$

Of course the curved two-sphere in the above examples is more complicated and requires an integral/differential extension of the parallelogram definition of area, however the result is the same, since S^2 exhibits a high degree of symmetry and isotropy. In fact almost all examples in general physics involving curved surfaces tend to have a large amount of symmetry, whose vectorized areas have a simple geometric/algebraic form conceptually understandable and derivable without knowledge of calculus. Next we can define flux, the field flow through a surface area, in the usual way via the dot product of a vector $\vec{E}$ and the vectorized area $\vec{A}$:

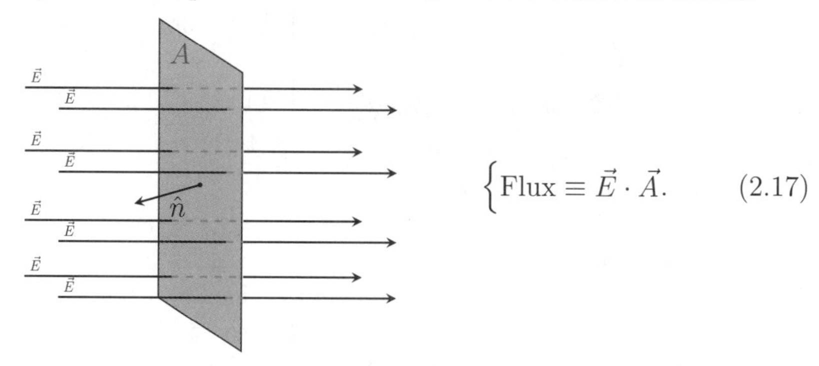

$$\left\{\text{Flux} \equiv \vec{E} \cdot \vec{A}. \right. \tag{2.17}$$

As can be seen above, the value of the flux can change by changing the size of A, the orientation of $\hat{n}$ to $\vec{E}$ or by variations in the field $\vec{E}$.

2.3 Gaussian surfaces

Gauss's law (and similarly Ampère's) is a rich subject with a plethora of (learnable) physics oozing from every pen stroke of its constituents. Even if the full pedagogical worth of it cannot be fully captured in an algebraic analysis, it still has the ability to teach rich structure of the fundamental nature of our Universe. Every fundamental force exhibits a Gauss law constraint and it, together with Ampère's law, are the only true Maxwell field equations, since the other two are consequences of the nilpotent derivative operator ($\nabla \times (\nabla \cdot\ ) \equiv 0$ and $\nabla \cdot (\nabla \times\ ) \equiv 0$). We can draw on the previous sections to define a geometric/algebraic version of Gauss's law suitable for the general physics audience:

$$\vec{E} \cdot \vec{A}_{\odot} = \frac{q_{\text{encl}}}{\epsilon_0}, \tag{2.18}$$

where $\vec{E}$ is the electric field, ϵ_0 is the electric vacuum permittivity, q_{encl} is the enclosed source charge and $\vec{A}_{\odot}$ is a Gaussian surface vector. $\vec{A}_{\odot}$ is defined analogously as the vectorized area above:

$$\vec{A}_{\odot} = A_{\odot}\hat{n}, \tag{2.19}$$

where now $A_{\odot}$ is a Gaussian surface. Without delving into any details or rigors of the topology of bounded manifolds, we will make the simplistic/heuristic definition:

- Gaussian surface: A surface that encloses or bounds a volume.

This simplistic definition helps us avoid discussing compactness and closure of topological spaces, yet illustrates the necessary idea and concept. For example a basketball (S^2) is a Gaussian surface (*GS*) since it can enclose or bound a bowling ball (3-ball):

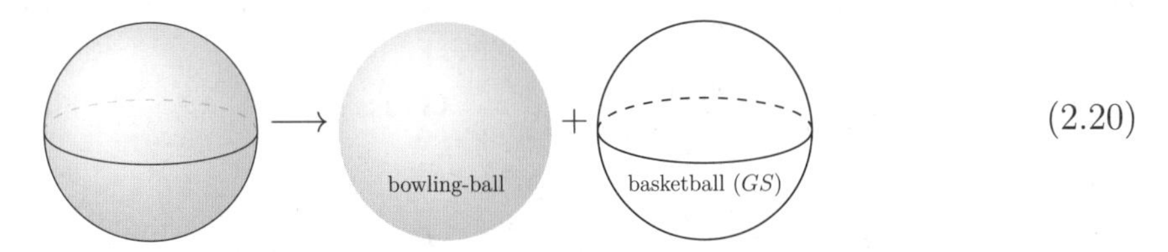

(2.20)

Referring back to equation (2.18), we see that Gauss' law relates the electric flux through a closed surface to the enclosed source charge modulo a constant, i.e. for the specific case of a point charge:

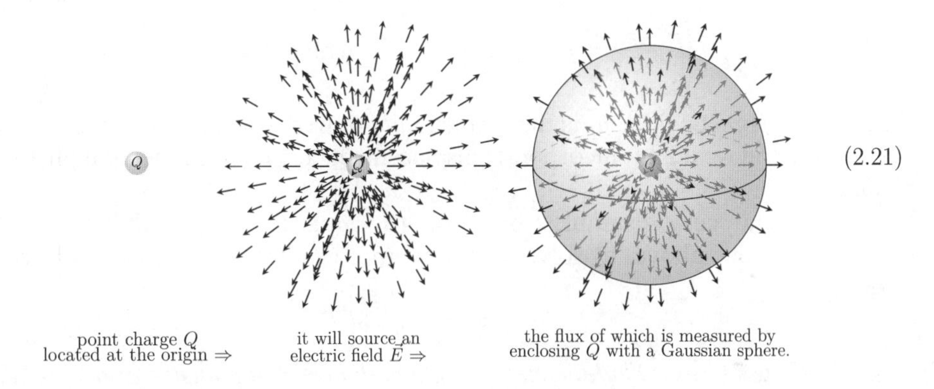

point charge Q located at the origin ⇒
it will source an electric field E ⇒
the flux of which is measured by enclosing Q with a Gaussian sphere.

(2.21)

The above *GS* definition encompasses Amperian loops (*AL*) as well by extending the conceptual idea of {surface, volume} to a more general {bulk, boundary} one. In other words, an *AL* is a one-dimensional *GS* that encloses or bounds a two-dimensional volume:

circle enclosing a disk ⟶ bulk + boundary (*AL*) (2.22)

We will denote a given *AL* by $L_\odot$ and the corresponding vector quantity $\vec{L}_\odot$. Building on the fact that position and distance vectors are already part of the standard general physics curriculum, an *AL* extends this concept to closed paths (loops). The most commonly used example, depicted in equation (2.22) above, is the circular *AL*:

$$\begin{cases} L_\odot = & 2\pi r \\ \hat{u} = & \hat{\phi} \\ \vec{L}_\odot = & L_\odot \hat{u} = \boxed{2\pi r \hat{\phi}} \end{cases}, \tag{2.23}$$

where $\hat{u}$ is the unit parallel vector to $L_\odot$ with respect to the right-hand rule. Given the above definitions, we may write down a sort of algebraic/geometric (baby) version of Maxwell's equations in terms of $\vec{A}_\odot$ and $\vec{L}_\odot$:

$$\vec{E} \cdot \vec{A}_\odot = \frac{q_{\text{encl}}}{\epsilon_0} \qquad \text{Gauss's law} \tag{2.24}$$

$$\vec{B} \cdot \vec{A}_\odot = 0 \qquad \vec{B} \text{ is Solenoidal} \tag{2.25}$$

$$\vec{E} \cdot \vec{L}_\odot = -\frac{\Delta(\vec{B} \cdot \vec{A})}{\Delta t} \qquad \text{Faraday's law} \tag{2.26}$$

$$\vec{B} \cdot \vec{L}_\odot = \mu_0 I_{\text{encl}} \qquad \text{Ampère's law,} \tag{2.27}$$

where $\vec{B}$ is the magnetic field, μ_0 is the permeability of free space, I_{encl} is the enclosed source current and we do not incorporate displacement current in the general physics curriculum.

We list a few easy examples on how to implement the baby form of Maxwell's equations above, to solve scenarios appropriate for the college physics classroom.

Example 1 (Gauss's law). *Find the electric field both inside and outside a solid 3-ball with total charge Q uniformly distributed throughout its volume.*

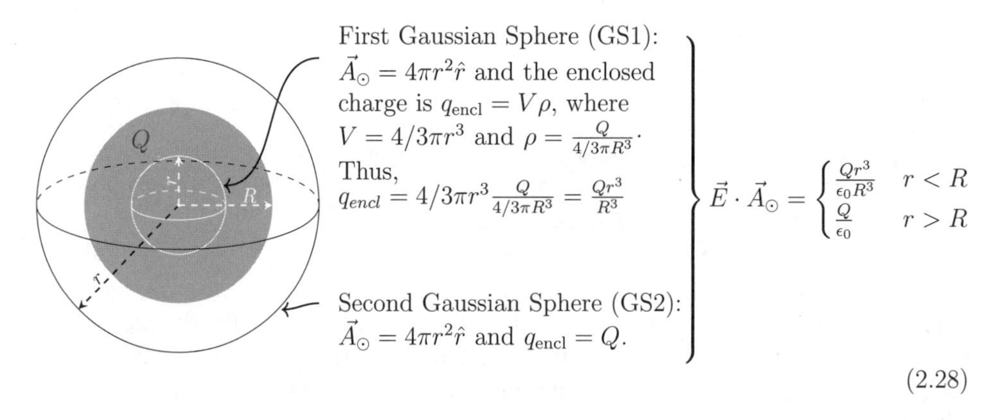

Q
r
R
r
First Gaussian Sphere (GS1):
$\vec{A}_\odot = 4\pi r^2 \hat{r}$ and the enclosed
charge is $q_{\text{encl}} = V\rho$, where
$V = 4/3\pi r^3$ and $\rho = \frac{Q}{4/3\pi R^3}$.
Thus,
$q_{encl} = 4/3\pi r^3 \frac{Q}{4/3\pi R^3} = \frac{Qr^3}{R^3}$
Second Gaussian Sphere (GS2):
$\vec{A}_\odot = 4\pi r^2 \hat{r}$ and $q_{\text{encl}} = Q$.
$\vec{E} \cdot \vec{A}_\odot = \begin{cases} \frac{Qr^3}{\epsilon_0 R^3} & r < R \\ \frac{Q}{\epsilon_0} & r > R \end{cases}$

(2.28)

From equation (2.28) we see that the electric field is a piecewise continuous function, which we solve for by making the symmetry motivated ansatz $\vec{E} = E(r)\hat{r}$, thus

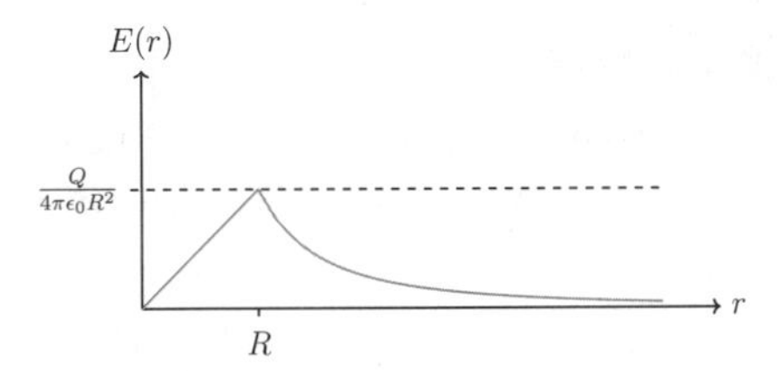

Figure 2.5. The magnitude of the electric field increases linearly within the solid 3-ball and outside, where $r > R$, it falls off according to an inverse square law.

$$\vec{E} \cdot \vec{A}_{\odot} = \frac{1}{\epsilon_0} \begin{cases} \frac{Qr^3}{R^3} & r < R \\ Q & r > R \end{cases}$$

$$E(r) 4\pi r^2 (\hat{r} \cdot \hat{r}) = \frac{1}{\epsilon_0} \begin{cases} \frac{Qr^3}{R^3} & r < R \\ Q & r > R \end{cases}$$

$$\Rightarrow E(r) = \frac{1}{\epsilon_0} \begin{cases} \frac{Qr}{4\pi R^3} & r < R \\ \frac{Q}{4\pi r^2} & r > R \end{cases} \tag{2.29}$$

$$\Rightarrow \vec{E}(r) = \boxed{\frac{\hat{r}}{\epsilon_0} \begin{cases} \frac{Qr}{4\pi R^3} & r < R \\ \frac{Q}{4\pi r^2} & r > R \end{cases}}.$$

From the above solution, we obtain the standard behavior of the magnitude of $\vec{E}$ *(depicted in figure 2.5) for the solid sphere.* $E(r)$ *increases linearly within the 3-ball and behaves like a point charge outside.*

Example 2 (Ampère's law). *Find the magnetic field a vertical distance r from the center of a current-carrying wire.*

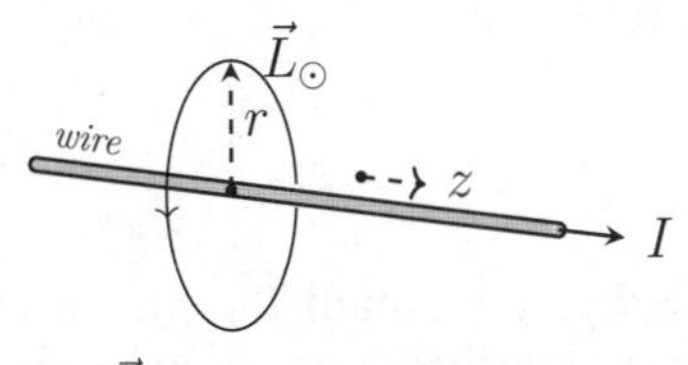

AL ($\vec{L}_{\odot}$) enclosing the current I. $\vec{B}$ should flow along $\vec{L}_{\odot} = 2\pi r\hat{\phi}$: thus, we assume $\vec{B} = B(r)\hat{\phi}$ and and solve for $B(r)$.

$$\begin{cases} \vec{B} \cdot \vec{L}_{\odot} = & \mu_0 I_{encl} \\ B(r) 2\pi r \left(\hat{\phi} \cdot \hat{\phi}\right) = & \mu_0 I \\ B(r) = & \frac{\mu_0 I}{2\pi r} \\ \Longrightarrow \vec{B} = & \boxed{\frac{\mu_0 I}{2\pi r}\hat{\phi}} \end{cases} \tag{2.30}$$

2.4 Mass in college physics

Let us start with the ideal scenario of a point particle near the Earth's surface and draw the respective free body diagram and sum the forces, neglecting air resistance, drag and friction:

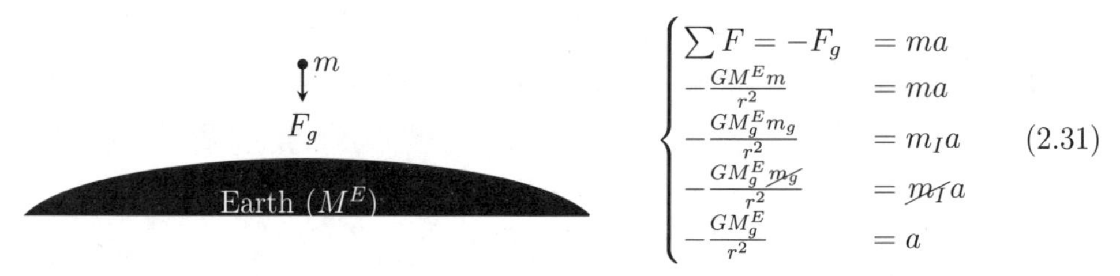

$$\begin{cases} \sum F = -F_g & = ma \\ -\frac{GM^E m}{r^2} & = ma \\ -\frac{GM_g^E m_g}{r^2} & = m_I a \\ -\frac{GM_g^E \cancel{m_g}}{r^2} & = \cancel{m_I} a \\ -\frac{GM_g^E}{r^2} & = a \end{cases} \tag{2.31}$$

In this ideal scenario, the only force that acts upon the object is the gravitational force which we included via Newton's universal law (where G is Newton's constant) on the left of the sum of forces in equation (2.31). In the third step above, we included the subscripts g and I, which stand for gravitational and inertial, and then continued with a typical general physics analysis and simplification. Yet by adding the above subscripts, this simplistic scenario presents an ample pedagogical opportunity to teach some fundamental physics/concepts about the Universe. The fact that we canceled m_g and m_I, which implies that we assumed their equivalence, usually receives no further attention even though there is no known fundamental reason to make this assumption. On the left side of the sum of forces above we have gravitational mass m_g:

- a sort of gravitational analog to electric charge, i.e. that which sources gravity.

On the right side we have the completion of Newton's second law; *the displacement in an object's velocity is proportional to the force acted upon,* where the proportionality constant is designated as the inertial mass m_I:

- an object's property to resist change in its state of motion.

m_g and m_I seem to originate from two different concepts, however they do seem to be classical equivalents supported by imperial data. The standard measurement of $g = 9.81\,\mathrm{m\,s^{-2}}$ in any freshman physics course, for example, relies on their equivalence as seen from the result above in equation (2.31):

$$a = -\frac{GM_g^E}{r^2}. \tag{2.32}$$

Substituting the standard values for G, M_E and $r = R_E$, we obtain the numerical table result $a = -9.81\,\mathrm{m\,s^{-2}}$. In fact we should only expect a divergence of m_g and m_I at the quantum gravitational scale, particularly where gravitational back reactions become significant. In other words, $m_g = m_I$ is safe for all of classical physics. However, we find it pertinent to educate our general physics audience about this equivalence and the concept of inertial versus gravitational mass and their origins.

Now, there are traditional methods for defining/dealing with the concept of mass that are usually developed and taught (in most cases) at the graduate physics level:

- Gravitational mass

  Nöther's theorem allows the identification of a conserved rank two current, based on general diffeomorphism symmetry, whose temporal component is equal to mass density [1].

  The Arnowitt, Deser and Misner (*ADM*) procedure identifies the same conserved quantity as Nöther's theorem, but from an asymptotic analysis of the initial time symmetric data of the general relativistic gravitational field [2].
- Inertial mass

  Beyond Newton's second law, the Weinberg–Salam model provides a quantum field theoretic definition of mass via the Higgs mechanism. The virtue of the absence of gravity in the standard model implies by default that any mass definition incorporated must refer to inertial properties [3].

However, keeping in mind with the premise of this section, the above list is far beyond the general undergraduate physics curriculum and we will need to find an alternative yet equally rigorous definition of mass suitable for the respective audience in this section. To do so, we will return to Gauss's law!

2.5 The gravielectric duality

Equation (2.24) provides a geometric way of defining electric charge in terms of electric flux through a *GS*, as long as the *GS* is chosen to completely enclose the respective charge. A simple choice for such a *GS* is a two-sphere with an infinite radius, however, in order to avoid immediate or initial divergences it is better to define this type of *GS* in terms of a limit, i.e. $\lim_{r\to\infty}\vec{A}_{\odot}$. This is analogous to the *ADM* procedure of general relativity. In other words, if we compute the flux of $\vec{E}$ through a Gaussian sphere S^2 with infinite radius, we are guaranteed to capture the entire charge enclosed, modulo a constant. That is, we may define the *ADM* charge:

$$q_{ADM} = \lim_{r\to\infty} \epsilon_0 \vec{E} \cdot \vec{A}_{\odot}. \tag{2.33}$$

This geometric definition of electric charge can be made pliable to mass via the gravielectric duality that can be extrapolated by inspection of Newton's and Coulomb's universal force laws:

$$\begin{cases} F_g &= -\dfrac{GMm}{r^2} \\ F_C &= \dfrac{kQq}{r^2} \end{cases}, \tag{2.34}$$

from which we obtain the map outlined in table 2.1. From this gravielectric duality we are able to obtain Gauss's law for Newtonian gravity:

Table 2.1. Gravielectric duality.

Charge	q	$\leftrightarrow$	m
Coupling	$k = (4\pi\epsilon_0)^{-1}$	$\leftrightarrow$	$-G$
Field	$\vec{E} = \vec{F_C}/Q$	$\leftrightarrow$	$\vec{g} = \vec{F_g}/M$

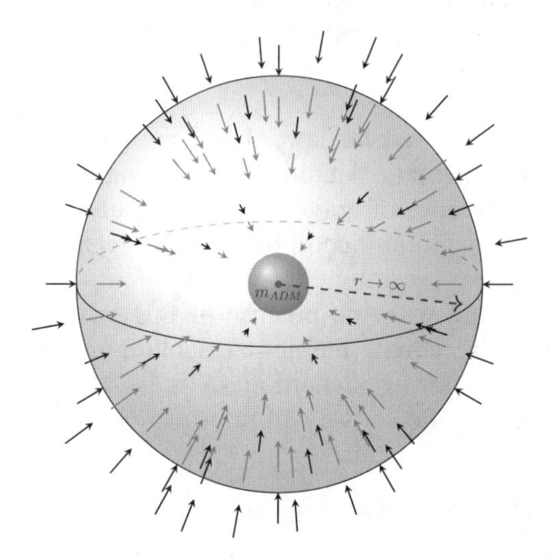

Figure 2.6. Gravitational field flux through a Gaussian sphere with radius going to infinity. The $\vec{g}$ arrows point towards the source m_{ADM} due to gravity's pure attractive nature.

$$\vec{g} \cdot \vec{A}_{\odot} = -4\pi G m_{\text{encl}} \tag{2.35}$$

and thus a way of defining a notion of *ADM* mass in general physics:

$$\boxed{m_{ADM} = -\lim_{r \to \infty} \frac{\vec{g} \cdot \vec{A}_{\odot}}{4\pi G}.} \tag{2.36}$$

This statement above (2.36) is *holographic* in nature. It states that

- a stationary gravitational *ADM* charge in four dimensions (space and time) is encoded via the flux of its field through a closed two-dimensional surface at radial infinity as depicted in figure 2.6.

Of course we are using *ADM* loosely here, since we have not shown that equation (2.36) truly is the conserved gravitational charge of general relativity and also we are restricting ourselves to pure algebraic geometric definitions with high degrees of symmetry. However, we will make our claims rigorous and justify the use of *ADM* as we progress through this book.

Table 2.2. Mapping the geometric and algebraic college physics definitions to analogous expressions in university physics.

Quantity	College		University
Vec. area	$\vec{A} = A\hat{n}$	$\rightarrow$	$\vec{A} = \int d\vec{A}$
Flux	$\vec{E} \cdot \vec{A}$	$\rightarrow$	$\int \vec{E} \cdot d\vec{A}$
GS	$\vec{A}_{\odot}$	$\rightarrow$	$\oint d\vec{A}$
AL	$\vec{L}_{\odot}$	$\rightarrow$	$\oint d\vec{L}$
Gauss's law	$\vec{g} \cdot \vec{A}_{\odot} = -4\pi G m_{\text{encl}}$	$\rightarrow$	$\oint \vec{g} \cdot d\vec{A} = -4\pi G m_{\text{encl}}$

2.6 Mass in university physics

The concepts from the previous section are easily elevated to a calculus based introductory physics course by mapping the pure geometric/algebraic college physics definitions to their calculus analogs as listed in table 2.2. From this table we are able to write a more rigorous holographic definition for m_{ADM} in the university physics setting:

$$\boxed{m_{ADM} = -\frac{\lim_{r\to\infty}}{4\pi G}\oint_{S^2} \vec{g} \cdot d\vec{A}.} \tag{2.37}$$

This definition is actually very powerful, far reaching and derived rigorously from a general relativistic analysis in section 6.7 for asymptotically flat spacetimes.

2.7 On black holes in college/university physics

In this section we highlight some avenues, following from the ideas of Michell and Laplace [4], on how to introduce very basic black hole physics concepts appropriate for the general physics classroom. This particular discussion can be easily motivated by examining Newton's gravitational force law:

$$F_g = -\frac{GMm}{r^2} \tag{2.38}$$

and asking what happens in the limit as $r \to 0$. This, even at the Newtonian level, reveals the ultraviolet (small r) divergent behavior of gravity in the general physics classroom. A similar question that follows is at what radial point is the Newtonian gravitational force so strong, but not infinite, that escape becomes a physical impossibility? In other words, at what distance from the center of a gravitating body would one need to violate energy conservation in order to escape back to infinity?

2.8 Schwarzschild black hole

The above thought questions allow us to define a simplistic model of a black hole as the two-dimensional spherical surface from which the gravitational escape velocity is equal to the speed of light. This surface is called the black hole event horizon, or just horizon, and is located at a radial distance r_H from the center of the black hole as

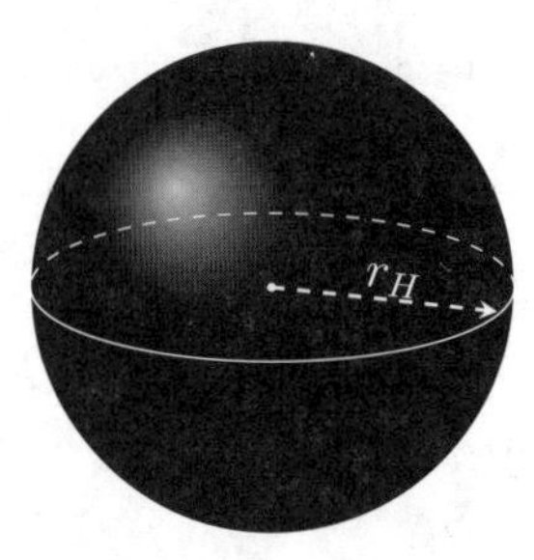

Figure 2.7. General physics model of a black hole as a spherical surface or horizon with Schwarzschild radius r_H.

depicted in figure 2.7. This defines the ground state configuration, or as its commonly known the Schwarzschild black hole, and in this specific case r_H is also referred to as the Schwarzschild radius. We can employ energy conservation (work kinetic energy theorem) for a gravitational test charge m to determine the Schwarzschild radius in terms of specific black hole parameters. If the test charge is located on the horizon initially and we wish it to escape to infinity, then we have

$$\begin{aligned} W_{\text{gravity}} &= \Delta K \\ -\int_{r_H}^{\infty}\left(-\frac{GMm}{r^2}\right)dr &= \frac{1}{2}mc^2 \\ \frac{GM}{r_H} &= \frac{1}{2}c^2, \end{aligned} \tag{2.39}$$

where W is work, K is kinetic energy, M is the black hole gravitational charge and we have employed the virial theorem $F = -dU/dr$. Solving for r_H in equation (2.39) yields

$$\boxed{r_H = \frac{2GM}{c^2}.} \tag{2.40}$$

This is the famous Schwarzschild radius result for a stationary static black hole with no auxiliary charges. Of great interest in class seems to be what the respective Schwarzschild radius of the Earth is. Meaning, to what radius would the entire Earth need to be compressed to in order for it to for a black hole? Using the above result with the table value of the Earth's mass we obtain about 9 mm and similarly for the Sun we obtain about 3 km.

2.9 Rotating/Kerr black hole

An interesting addition to the above analysis is to incorporate angular momentum. As is well known, most astrophysical objects rotate about a central axis. Again, we can incorporate this attribute in a simplistic way by taking the black hole to rotate at angular frequency $\omega = L/I \approx L/(MR^2) = a/r_H^2$, where a is the black hole angular momentum per unit mass. This is a reasonable approximation, given the constraints of college or university physics, and is close to the actual frequency obtained from a full general relativistic analysis. The horizon structure can now be obtained by equatorially augmenting (2.39) to include rotational kinetic energy (orbital at r_H only) of the test charge:

$$\begin{aligned} -\int_{r_H}^{\infty}\left(-\frac{GMm}{r^2}\right)dr &= \frac{1}{2}mv^2 + \frac{1}{2}mr_H^2\left(\frac{a}{r_H^2}\right)^2 \\ \frac{GM}{r_H} &= \frac{1}{2}v^2 + \frac{1}{2}\frac{a^2}{r_H^2} \\ \Rightarrow 0 &= v^2r_H^2 - 2GMr_H + a^2, \end{aligned} \tag{2.41}$$

where we have used v instead of c, since the total kinetic energy is now split into linear and rotational variants. From the above result we see that we will end up with two horizon radii in the case for rotating black holes, which are given by

$$\boxed{r_{\pm} = \frac{GM \pm \sqrt{(GM)^2 - (av)^2}}{v^2}.} \tag{2.42}$$

This is very close to the the actual result obtained for the horizon radii from a full general relativistic analysis, yet derived from a university physics point of view. The actual correct result replaces v with c on the right side of $r_{\pm}$ and signals a clear failure of Newtonian gravity, despite its shear pedagogical worth.We see that in the case of rotating black holes there are two horizons, an inner and an outer one as depicted in figure 2.8. The region between r_- and r_+, and the region just outside of r_+ exhibit interesting properties that are probably beyond the general physics classroom. However, the concept of an extremal black hole can be discussed/introduced, since the term inside the radical of equation (2.42) must be greater or equal to zero. Thus, there is a maximum amount of angular momentum that a black hole may be astrophysically endowed with!

We should comment on the fact that we obtained the near correct horizon structure in equation (2.42) by making an approximate/simplistic educated assumption about the black hole angular frequency ω. This has to do with the fact that a full general relativistic analysis reveals that the horizon geometry of a rotating black hole is a two-sphere or radius $\sqrt{r_+^2 + a^2/c^2}$ and not r_+ alone. Also, our simplistic assumption may be too far an extrapolation for general physics students (university or college) and may require a guiding discussion, however it is a worthwhile one in order to bring black holes into the respective classroom to an eager audience.

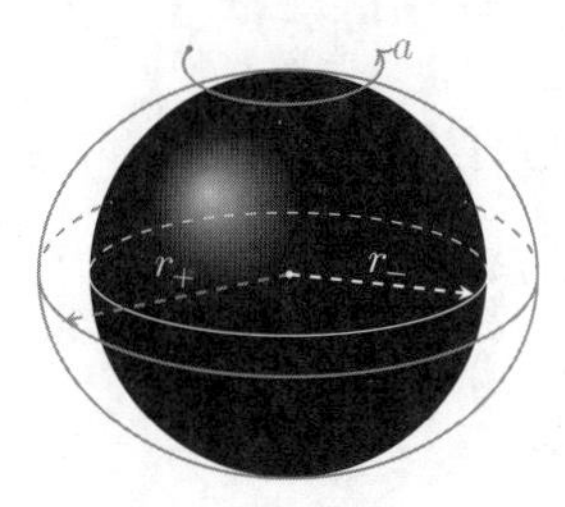

Figure 2.8. A simple model of a rotating black hole with inner and outer horizons.

2.10 Effective Newtonian black hole potentials and *ADM* mass

Looking back at equations (2.39) and (2.41) and making the substitutions $r_H \to r$ and $c \to v$, we are able to extrapolate the effective (approximate) Newtonian potentials for the Schwarzschild and Kerr black holes, by collecting all the terms involving black hole parameters on one side, i.e. mass and angular momentum:

$$\underbrace{\overbrace{\frac{GM}{r}}^{\text{Schwarzschild}}}_{-U_{\text{eff}}} = \frac{1}{2}v^2, \quad \underbrace{\overbrace{\frac{GM}{r} - \frac{1}{2}\frac{a^2}{r^2}}^{\text{Kerr}}}_{-U_{\text{eff}}} = \frac{1}{2}v^2. \tag{2.43}$$

Thus, in our chosen cases we obtain

$$U_{\text{eff}} = -\begin{cases} \dfrac{GM}{r} & \text{Schwarzschild} \\ \dfrac{2GMr - a^2}{2r^2} & \text{Kerr} \end{cases}, \tag{2.44}$$

from which we obtain the Newtonian fields:

$$\vec{g} = -\frac{\partial U_{\text{eff}}}{\partial r}\hat{r} = -\begin{cases} \dfrac{GM}{r^2}\hat{r} & \text{Schwarzschild} \\ \dfrac{GMr - a^2}{r^3}\hat{r} & \text{Kerr} \end{cases}. \tag{2.45}$$

The above results are in alignment with actual Newtonian approximations for small a compared to GM/c^2 derived from a weak field limit of general relativity. We will derive the full Newtonian fields later, but for now we can use the above result in equation (2.37) to show that the mass parameter M is the *ADM* black hole mass. A quick calculation gives

$$\begin{aligned} m_{ADM} &= -\frac{\lim_{r\to\infty}}{4\pi G}\oint_{S^2}\vec{g}\cdot d\vec{A} \\ &= \frac{\lim_{r\to\infty}}{4\pi G}\begin{cases} \displaystyle\int_0^{2\pi}\int_0^{\pi}\left(\frac{GM}{r^2}\hat{r}\right)\cdot(r^2\sin\theta d\theta d\phi\hat{r}) & \text{Schwarzschild} \\ \displaystyle\int_0^{2\pi}\int_0^{\pi}\left(\frac{GMr - a^2}{r^3}\hat{r}\right)\cdot(r^2\sin\theta d\theta d\phi\hat{r}) & \text{Kerr} \end{cases} \\ &= \frac{\lim_{r\to\infty}}{4\pi G}\begin{cases} \displaystyle\int_0^{2\pi}\int_0^{\pi} GM\sin\theta d\theta d\phi & \text{Schwarzschild} \\ \displaystyle\int_0^{2\pi}\int_0^{\pi}\frac{GMr - a^2}{r}\sin\theta d\theta d\phi & \text{Kerr} \end{cases} \\ &= \begin{cases} M & \text{Schwarzschild} \\ M & \text{Kerr} \end{cases}. \end{aligned} \tag{2.46}$$

Note, that we could have used equation (2.36) to arrive at the above result, due to the high degree of symmetry of the respective examples.

References

[1] Liu J T and Sabra W 2005 Mass in anti-de Sitter spaces *Phys. Rev.* D **72** 064021
[2] Misner C W, Thorne K S and Wheeler J A 1970 *Gravitation* (San Francisco, CA: Freeman)
[3] Pokorski S 2005 *Gauge Field Theories* (Cambridge: Cambridge University Press)
[4] Montgomery C, Orchiston W and Whittingham I 2009 Michell, Laplace and the origin of the black hole concept *J. Astronomical History and Heritage* **12** 90–6

Chapter 3

Nöther's theorem: *E*&*M* and gravity

We will now shift our focus towards the junior and senior level audience of the undergraduate physics major. Our goal will be to expand on several topics from above and in particular introduce two-dimensional holographic representations of the canonical asymptotically flat spacetimes that normally have been reserved for the advanced graduate student. Until now we have avoided any metric discussion of gravity, mainly for pedagogical reasons. However, for more advanced undergraduate students interested in learning about general relativity and its various constituent topics, it is imperative to expose and introduce the full geometrodynamical formulation of gravity. This is especially important for students who intend to pursue research and or graduate study in the respective topic.

We will make our presentation assuming a working knowledge of vector calculus (div, grad, curl, etc) [1], sophomore level modern physics, sophomore/junior level classical mechanics and junior level electricity and magnetism. Some of the more difficult new paradigms which we will introduce in this section include the concept of field variations, i.e. functional derivatives with respect to a physical field, and tensor notation/calculus. However, if students have completed an intermediate-level course on classical mechanics then the concept of a field variation is easily explained as an extension of the functional derivatives of, say, $\dot{x}^2$ in the kinetic term of the Lagrangian while applying the Euler–Lagrange equation of motion. The introduction of tensor or index notation is a bit more cumbersome and will require specific strategies of presentation and connection to what students are familiar with in order to maintain pedagogical worth and maximize comprehension.

3.1 Nöther's theorem: *E*&*M*

We mentioned Nöther's theorem briefly in our initial discussion of section 2.4:

Theorem 1 (Nöther). *Continuous symmetries of the action in a physical theory generate conserved quantities.*

doi:10.1088/2053-2571/ab06bdch3

Though, we did not expound upon it due to its obvious inherent complexity. However, this theorem is an extremely powerful pedagogical tool to help explain why gravity is geometry. In other words, the standard identification of the metric tensor as the gravitational field is usually stated without much justification or discussion in most traditional courses on general relativity (even at the graduate level). There is a rich amount of fundamental physics that is lost in such traditional lectures, which will provide an avenue for introducing to advanced undergraduate students in this section. To begin and motivate our discussion of gravity's geometrodynamical formulation via Nöther's theorem, we will first look to Maxwell's electrodynamics as a guide which will help us cut through the esoteric curtains of Nöther's theorem and thus, general relativity.

To begin, let us look at Maxwell's equation:

$$\begin{aligned} \nabla \cdot \vec{E} &= \frac{\rho}{\epsilon_0} \\ \nabla \times \vec{E} &= -\frac{\partial \vec{B}}{\partial t} \\ \nabla \cdot \vec{B} &= 0 \\ \nabla \times \vec{B} &= \mu_0 \vec{J} + \mu_0 \epsilon_0 \frac{\partial \vec{E}}{\partial t}, \end{aligned} \tag{3.1}$$

which, similar to Lagrangian mechanics, we are interested in interpreting as a set of equations of motion. However, unlike in classical mechanics, the above equations are not point-particle-like, but are comprised of fields. That means we are tasked with making analog identifications in order to map the familiar Lagrangian mechanics of point-particles to a Lagrangian field theory of Maxwell's electrodynamics. Let us back-track a bit to classical mechanics and the point-particle Lagrangian:

$$L = T - V, \tag{3.2}$$

where T is kinetic energy and $V = V(x)$ is some potential energy function. Now, we will introduce an outline that will be useful (and humorous while introducing in the classroom) in constructing physical theories that students can apply and use to determine relevant equations of motion and field content of specific physical interactions.

Outline 1 (how to physics).

1. *Construct a functional that is left invariant with respect to a specific chosen group of linear transformations.*
   - *For the above back-track of the point-particle, the functional will be the action integral:*

$$S = \int dt L. \tag{3.3}$$

- *The group of transformations that leave this functional invariant are the group of Galilean transformations. By invariant, we mean that the form of the resulting equations of motion, i.e. the Euler–Lagrange equations, will always take the same form, up to total derivative terms. Total derivative terms in the action integral/functional are dropped in congruence with Hamilton's principle of stationary action. In other words, we keep the end points of all possible paths $x(t)$ fixed and thus variations δx evaluated on the boundary vanish identically.*

2. *Identify the physical field of interest.*
   - *Again, for the above back-track of the point-particle, the physical field of interest is just the position x.*
3. *Vary the functional with respect to the identified field and set equal to zero, yielding the equations of motion.*
   - *For the point-particle we have the variation of S with respect to x. This amounts to simply taking derivatives with respect to x, implementing the chain and product rules of calculus. However, it is instructive to keep the variational notation δ, which will be analogous once we switch to actual field variations such as electromagnetic and or gravitational field variations. For the point-particle assume the standard form of kinetic energy, $T = \frac{1}{2}m\dot{x}^2$ and the action reads*

$$S = \int dt \left\{ \frac{1}{2}m\dot{x}^2 - V(x) \right\}. \tag{3.4}$$

Next we perform the variation with respect to x and set equal to zero:

$$\delta S = \int dt\delta \left\{ \frac{1}{2}m\dot{x}^2 - V(x) \right\} = 0 \tag{3.5}$$

$$= \int dt \left\{ \frac{1}{2}m2\frac{dx}{dt}\frac{d\delta x}{dt} - \frac{\delta V(x)}{\delta x}\delta x \right\}, \tag{3.6}$$

which we can rewrite using integration by parts as

$$\delta S = \int dt \left\{ -m\frac{d^2x}{dt^2}\delta x + m\frac{d}{dt}\left(\frac{dx}{dt}\delta x\right) - \frac{\delta V(x)}{\delta x}\delta x \right\}. \tag{3.7}$$

Now, since we are holding the end points of all paths $x(t)$ fixed, the total derivative term will vanish after integration and evaluating δx at the boundary, yielding

$$\delta S = \int dt \left\{ -m\frac{d^2x}{dt^2}\delta x - \frac{\delta V(x)}{\delta x}\delta x \right\} \tag{3.8}$$

$$= \int dt \left\{ -m\frac{d^2x}{dt^2} - \frac{\delta V(x)}{\delta x} \right\} \delta x = 0, \tag{3.9}$$

thus, the equation of motion is

$$-m\frac{d^2x}{dt^2} - \frac{\delta V(x)}{\delta x} = 0 \tag{3.10}$$

$$-\frac{\delta V(x)}{\delta x} = m\frac{d^2x}{dt^2} \tag{3.11}$$

$$\Rightarrow \boxed{F = ma.} \tag{3.12}$$

Next, let us apply the above outline to the electromagnetic field. We need to construct a functional that is left invariant with respect to some chosen group of linear transformations. A convenient choice are actually two groups, the groups of Lorentz and gauge (we will explain soon) transformations. Also, keep in mind that $L = T - V$ has units of energy. A good possible choice is the energy density Lorenz invariant:

$$\mathcal{L} = \frac{1}{2}\left(\epsilon_0 \vec{E}^2 - \frac{1}{\mu_0}\vec{B}^2\right) - \varphi\rho + \vec{A}\cdot\vec{J}, \tag{3.13}$$

where φ is the electromagnetic scalar potential and $\vec{A}$ is the electromagnetic vector potential, which yield $\vec{E}$ and $\vec{B}$ via the standard identifications:

$$\begin{aligned} \vec{E} &= -\nabla\varphi - \frac{\partial \vec{A}}{\partial t} \\ \vec{B} &= \nabla \times \vec{A}. \end{aligned} \tag{3.14}$$

Since $\mathcal{L}$ has units of energy per unit volume we identify it as the Lagrangian density, i.e.

$$L = \int dV \mathcal{L} \tag{3.15}$$

and thus the action reads

$$\begin{aligned} S &= \int dV\, dt\, \mathcal{L} \\ &= \int dV\, dt \left\{ \frac{1}{2}\left(\epsilon_0 \vec{E}^2 - \frac{1}{\mu_0}\vec{B}^2\right) - \varphi\rho + \vec{A}\cdot\vec{J} \right\}, \end{aligned} \tag{3.16}$$

which is commonly referred to as the Maxwell action. Next, we need to identify the field, or maybe fields. We might be tempted to choose $\vec{E}$ and $\vec{B}$, but there are some reasons that we should consider the potentials φ and $\vec{A}$ as the correct choice of electromagnetic field pair. First, both $\vec{E}$ and $\vec{B}$ are derivations of the potentials. Also, there is a more fundamental reason, related to what is called gauge symmetry (as previously mentioned). Looking back to equation (3.14) we are free to shift, or gauge transform, φ and $\vec{A}$ by an arbitrary function $\Lambda(x, t)$ via the following:

$$\begin{aligned} \varphi &\rightarrow \tilde{\varphi} = \varphi - \frac{\partial \Lambda}{\partial t} \\ \vec{A} &\rightarrow \tilde{\vec{A}} = \vec{A} + \nabla\Lambda, \end{aligned} \tag{3.17}$$

which leaves $\vec{E}$ and $\vec{B}$ unchanged:

$$\begin{cases} \tilde{\vec{E}} = -\nabla\left(\varphi - \dfrac{\partial \Lambda}{\partial t}\right) - \dfrac{\partial}{\partial t}(\vec{A} + \nabla\Lambda) \\ \tilde{\vec{B}} = \nabla \times (\vec{A} + \nabla\Lambda) \end{cases}$$
$$\begin{cases} \tilde{\vec{E}} = -\nabla\varphi + \nabla\dfrac{\partial \Lambda}{\partial t} - \dfrac{\partial}{\partial t}\vec{A} - \dfrac{\partial}{\partial t}\nabla\Lambda \\ \tilde{\vec{B}} = \nabla \times \vec{A} + \nabla \times \nabla\Lambda \end{cases}$$
$$\begin{cases} \tilde{\vec{E}} = -\nabla\varphi - \dfrac{\partial}{\partial t}\vec{A} + \left(\nabla\dfrac{\partial}{\partial t} - \dfrac{\partial}{\partial t}\nabla\right)\Lambda. \\ \tilde{\vec{B}} = \nabla \times \vec{A} + \nabla \times \nabla\Lambda \end{cases} \tag{3.18}$$
$$\begin{cases} \tilde{\vec{E}} = -\nabla\varphi - \dfrac{\partial}{\partial t}\vec{A} \\ \tilde{\vec{B}} = \nabla \times \vec{A} \end{cases}$$
$$\begin{cases} \tilde{\vec{E}} = \vec{E} \\ \tilde{\vec{B}} = \vec{B} \end{cases}$$

Now, we have just shown that $\vec{E}$ and $\vec{B}$ are unaffected by gauge transforming φ and $\vec{A}$, i.e. $\vec{E}$ and $\vec{B}$ are gauge invariant assuming that ∇ and $\dfrac{\partial}{\partial t}$ commute and ∇ is nilpotent:

$$\begin{aligned} &\text{commutative} \quad \left\{ \nabla\frac{\partial}{\partial t} - \frac{\partial}{\partial t}\nabla = 0 \right. \\ &\text{nilpotent} \quad \begin{cases} \nabla \times \nabla \equiv \text{curl(grad)} = 0 \\ \nabla \cdot \nabla \times \equiv \text{div(curl)} = 0 \end{cases} \end{aligned} \tag{3.19}$$

So, if we choose the scalar and vector potential as the electromagnetic fields, what do we do with $\vec{E}$ and $\vec{B}$? Well from their presence in the Lorentz force law:

$$\vec{F} = q(\vec{E} + \vec{v} \times \vec{B}) \tag{3.20}$$

Table 3.1. Electromagnetic field definitions.

Electromagnetic	Geometric objects
Field	φ and $\vec{A}$
Field strength	$\vec{E}$ and $\vec{B}$

we can relabel them as the electromagnetic field strengths (as cataloged in table 3.1)! Now that we have correctly relabeled our electromagnetic field content, we are ready to vary the action (3.16) with respect to our fields φ and $\vec{A}$ in order to retrieve the equations of motion (field equations). Let us start with the scalar potential.

- Variation with respect to φ:

$$\begin{aligned} \delta S &= \int dV\, dt \delta \mathcal{L} = 0 \\ &\Rightarrow \int dV\, dt \delta \left\{ \frac{1}{2} \left(\epsilon_0 \vec{E}^2 - \frac{1}{\mu_0} \vec{B}^2 \right) - \varphi \rho + \vec{A} \cdot \vec{J} \right\} = 0. \end{aligned} \tag{3.21}$$

Now, referring back to the point-particle back-track above, we will perform variations with respect to φ by treating it as a calculus variable and thus just taking derivatives of any objects in the action that involves φ by implementing the product and chain rule as necessary:

$$\begin{aligned} \int dV\, dt \delta \left\{ \frac{1}{2} \left(\epsilon_0 \vec{E}^2 - \frac{1}{\mu_0} \vec{B}^2 \right) - \varphi \rho + \vec{A} \cdot \vec{J} \right\} = 0 \\ \int dV\, dt \left\{ \frac{1}{2} \left(\epsilon_0 \delta \vec{E}^2 - \frac{1}{\mu_0} \delta \vec{B}^2 \right) - \delta(\varphi \rho) + \delta(\vec{A} \cdot \vec{J}) \right\} = 0. \end{aligned} \tag{3.22}$$

Out of the four terms varied above, only two depend on φ. In other words, we take the sources ρ and $\vec{J}$ to be independent of the scalar and vector potential and thus we have

$$\begin{aligned} \int dV\, dt \left\{ \frac{1}{2} \left(\epsilon_0 \delta \vec{E}^2 - \frac{1}{\mu_0} (0) \right) - \delta \varphi \rho + 0 \right\} = 0 \\ \int dV\, dt \left\{ \frac{1}{2} (\epsilon_0 2 \vec{E} \cdot \delta \vec{E}) - \delta \varphi \rho \right\} = 0. \end{aligned} \tag{3.23}$$

Next, let us look at the variation of the electric field with respect to the scalar potential:

$$\delta \vec{E} = \delta \left(-\nabla \varphi - \frac{\partial \vec{A}}{\partial t} \right) = -\nabla(\delta \varphi), \tag{3.24}$$

where we assume that spacial and temporal derivatives commute with respect to field variations. Substituting the above back into the variation of the action we have

$$\int dV\,dt\,\{(-\epsilon_0\vec{E}\cdot\nabla(\delta\varphi)) - \delta\varphi\rho\} = 0. \tag{3.25}$$

Again, just as in the point-particle back-track, we need to free the $\delta\varphi$ of the spacial derivative operator in order to factor it out of each term and thus revealing the equation of motion. This can again be accomplished by integration by parts and applying Hamilton's principle:

$$\int dV\,dt\,\{(\epsilon_0\nabla\cdot\vec{E})\delta\varphi - \epsilon_0\nabla\cdot(\vec{E}\delta\varphi) - \delta\varphi\rho\} = 0. \tag{3.26}$$

The above total derivative term can be integrated using Gauss's divergence theorem:

$$-\epsilon_0\int\nabla\cdot(\vec{E}\delta\varphi)dV = -\epsilon_0\oint_{\partial V}(\vec{E}\delta\varphi)\cdot d\vec{A} \tag{3.27}$$

and relates the total derivative term to field variations on the boundary ∂V, which are zero in application of Hamilton's principle. Thus we are left with

$$\begin{aligned}&\int dV\,dt\,\{(\epsilon_0\nabla\cdot\vec{E})\delta\varphi - \delta\varphi\rho\} = 0\\&\int dV\,dt\,\{(\epsilon_0\nabla\cdot\vec{E}) - \rho\}\delta\varphi = 0\\&\Rightarrow\boxed{\nabla\cdot\vec{E} = \frac{\rho}{\epsilon_0},}\end{aligned} \tag{3.28}$$

yielding Gauss's law and the first of the four Maxwell equations in (3.1).

- Variation with respect to $\vec{A}$:

$$\begin{aligned}\delta S &= \int dV\,dt\delta\mathcal{L} = 0\\&\Rightarrow\int dV\,dt\delta\left\{\frac{1}{2}\left(\epsilon_0\vec{E}^2 - \frac{1}{\mu_0}\vec{B}^2\right) - \varphi\rho + \vec{A}\cdot\vec{J}\right\} = 0\\&\int dV\,dt\left\{\frac{1}{2}\left(\epsilon_0 2\vec{E}\cdot\delta\vec{E} - \frac{2}{\mu_0}\vec{B}\cdot\delta\vec{B}\right) - (0) + \delta\vec{A}\cdot\vec{J}\right\}\\&= 0.\end{aligned} \tag{3.29}$$

As before, we need to look at the variation of the electromagnetic field strengths, which both depend on $\vec{A}$:

$$\begin{aligned}\delta\vec{E} &= \delta\left(-\nabla\varphi - \frac{\partial\vec{A}}{\partial t}\right) = -\frac{\partial\delta\vec{A}}{\partial t}\\\delta\vec{B} &= \delta(\nabla\times\vec{A}) = \nabla\times(\delta\vec{A})\end{aligned} \tag{3.30}$$

and substituting these results back into the variation of the action yields

$$\int dV\,dt \left\{ \left[\epsilon_0 \vec{E} \cdot \left(-\frac{\partial \delta \vec{A}}{\partial t} \right) - \frac{1}{\mu_0} \vec{B} \cdot (\nabla \times (\delta \vec{A})) \right] + \delta \vec{A} \cdot \vec{J} \right\} = 0. \quad (3.31)$$

As before we will employ integration by parts to remove the spacial and temporal derivative operators from $\delta \vec{A}$. The first term follows nearly identical to the point-particle back-track and where total time derivative terms are dropped in congruence with Hamilton's principle. The middle term is a bit more subtle, in that we need to employ the triple product rule vector identity (which is proven in example 3):

$$-\frac{1}{\mu_0} \vec{B} \cdot (\nabla \times (\delta \vec{A})) = -\frac{1}{\mu_0} (-\nabla \cdot (\vec{B} \times \delta \vec{A}) + (\nabla \times \vec{B}) \cdot \delta \vec{A}). \quad (3.32)$$

Similarly to equation (3.27), Gauss's divergence theorem relates the total divergence term to field variations of $\vec{A}$ evaluated on the boundary ∂V which does not contribute to variations of the action as per usual application (by now) of Hamilton's principle. Thus, equation (3.31) reduces to

$$\int dV\,dt \left\{ \left[\epsilon_0 \frac{\partial \vec{E}}{\partial t} \cdot \delta \vec{A} - \frac{1}{\mu_0} (\nabla \times \vec{B}) \cdot \delta \vec{A} \right] + \delta \vec{A} \cdot \vec{J} \right\} = 0$$

$$\int dV\,dt \left\{ \left[\epsilon_0 \frac{\partial \vec{E}}{\partial t} - \frac{1}{\mu_0} (\nabla \times \vec{B}) \right] + \vec{J} \right\} \cdot \delta \vec{A} = 0 \quad (3.33)$$

$$\Rightarrow \boxed{\nabla \times \vec{B} = \mu_0 \vec{J} + \mu_0 \epsilon_0 \frac{\partial \vec{E}}{\partial t},}$$

yielding Ampère's law and the last Maxwell equation in (3.1)!

We have now shown that the Euler–Lagrange equations of motion of the Maxwell action (3.16) is in fact the first and last Maxwell equation of (3.1). What about the other two Maxwell equations in (3.1)? These are actually nothing more than exact identities, called Bianchi identities, due to the nilpotency, (3.19), of ∇. Looking back at the electromagnetic field strength definitions in terms of the potentials (3.14):

$$\begin{aligned} \vec{E} &= -\nabla \varphi - \frac{\partial \vec{A}}{\partial t} \\ \vec{B} &= \nabla \times \vec{A}, \end{aligned} \quad (3.34)$$

we see, assuming spacial and temporal derivatives commute, that

$$\begin{cases} \nabla \times \vec{E} &= -\nabla \times \nabla\varphi - \dfrac{\partial \nabla \times \vec{A}}{\partial t} \\ \nabla \cdot \vec{B} &= \nabla \cdot \nabla \times \vec{A} \end{cases}$$

$$\begin{cases} \nabla \times \vec{E} &= -\text{curl}(\text{grad}(\varphi)) - \dfrac{\partial \nabla \times \vec{A}}{\partial t} \\ \nabla \cdot \vec{B} &= \text{div}(\text{curl}(\vec{A})) \end{cases} \tag{3.35}$$

$$\begin{cases} \nabla \times \vec{E} &= -\dfrac{\partial \vec{B}}{\partial t} \\ \nabla \cdot \vec{B} &= 0 \end{cases}$$

Thus yielding the final two Maxwell equations of (3.1).

Now, let us return to Nöther's theorem 1. We have now constructed one part of this theorem, namely the Maxwell action (3.16) for which the Maxwell equations (3.1) are the resulting Euler–Lagrange equations of motion. For the other part, let us return to the gauge transformations of (3.17), which left $\vec{E}$ and $\vec{B}$ unchanged or invariant after transformation. Our desire is that this particular group of transformations also leave the action invariant, i.e. borrowing the language of abstract algebra, we want gauge transformations to be a continuous symmetry of the action S. That is,

$$S(\varphi, \vec{A}) = S\left(\tilde{\varphi}, \tilde{\vec{A}}\right), \tag{3.36}$$

and see what the physical consequences are for requiring S to be invariant with respect to gauge transformations. Let us begin:

$$\begin{aligned} S\left(\tilde{\varphi}, \tilde{\vec{A}}\right) &= \int dV\, dt \left\{\frac{1}{2}\left(\epsilon_0 \tilde{\vec{E}}^2 - \frac{1}{\mu_0}\tilde{\vec{B}}^2\right) - \tilde{\varphi}\rho + \tilde{\vec{A}}\cdot\vec{J}\right\} \\ &= \int dV\, dt \left\{\frac{1}{2}\left(\epsilon_0 \vec{E}^2 - \frac{1}{\mu_0}\vec{B}^2\right) - \tilde{\varphi}\rho + \tilde{\vec{A}}\cdot\vec{J}\right\} \\ &= \int dV\, dt \left\{\frac{1}{2}\left(\epsilon_0 \vec{E}^2 - \frac{1}{\mu_0}\vec{B}^2\right) - \left(\varphi - \frac{\partial \Lambda}{\partial t}\right)\rho \right. \\ &\quad \left. + (\vec{A} + \nabla\Lambda)\cdot\vec{J},\right\} \\ &= \int dV\, dt \left\{\frac{1}{2}\left(\epsilon_0 \vec{E}^2 - \frac{1}{\mu_0}\vec{B}^2\right) - \varphi\rho + \vec{A}\cdot\vec{J}\right\} \\ &\quad + \int dV\, dt \left\{\frac{\partial \Lambda}{\partial t}\rho + \nabla\Lambda\cdot\vec{J}\right\} \\ &= S(\varphi, \vec{A}) + \int dV\, dt \left\{\frac{\partial \Lambda}{\partial t}\rho + \nabla\Lambda\cdot\vec{J}\right\}. \end{aligned} \tag{3.37}$$

We see that the only thing obstructing $S(\varphi, \vec{A}) = S(\tilde{\varphi}, \tilde{\vec{A}})$ is last term above, i.e. for gauge transformations to be a continuous symmetry we require

$$\int dV\, dt \left\{ \frac{\partial \Lambda}{\partial t}\rho + \nabla\Lambda \cdot \vec{J} \right\} = 0. \tag{3.38}$$

Now, we should keep in mind that the gauge transformation is a specific variation in φ and $\vec{A}$:

$$\begin{cases} \tilde{\varphi} = \varphi - \dfrac{\partial \Lambda}{\partial t} = \varphi + \delta\varphi \\ \tilde{\vec{A}} = \vec{A} + \nabla\Lambda = \vec{A} + \delta\vec{A} \end{cases}. \tag{3.39}$$

That is, in application of Hamilton's principle; equation (3.38) becomes after integration of parts and up to total derivatives:

$$\begin{gathered} -\int dV\, dt \left\{ \Lambda\frac{\partial \rho}{\partial t} + \Lambda\nabla \cdot \vec{J} \right\} = 0 \\ -\int dV\, dt \Lambda \left\{ \frac{\partial \rho}{\partial t} + \nabla \cdot \vec{J} \right\} = 0 \\ \Rightarrow \boxed{\frac{\partial \rho}{\partial t} + \nabla \cdot \vec{J} = 0.} \end{gathered} \tag{3.40}$$

In other words, the requirement that gauge transformations be a symmetry of the action yields the continuity equation:

$$\nabla \cdot \vec{J} = -\frac{\partial \rho}{\partial t}, \tag{3.41}$$

which is the statement of conservation of charge. This completes Nöther's theorem for $E\&M$, which we summarized in table 3.2.

3.2 Nöther's theorem: Newtonian gravity

We now want to apply our lessons learned from the previous section about electrodynamics ($E\&M$) to the gravitational field. Of course, as we will find out

Table 3.2. Nöther's theorem for $E\&M$.

Action:	$S(\varphi, \vec{A}) = \int dV\, dt \left\{ \frac{1}{2}\left(\epsilon_0 \vec{E}^2 - \frac{1}{\mu_0}\vec{B}^2 \right) - \varphi\rho + \vec{A} \cdot \vec{J} \right\}$
Continuous symmetry:	Gauge transformation, $\varphi \to \tilde{\varphi}$ and $\vec{A} \to \tilde{\vec{A}}$
Invariant objects:	$\vec{E} = \tilde{\vec{E}}$, $\vec{B} = \tilde{\vec{B}}$ and $S(\varphi, \vec{A}) = S(\tilde{\varphi}, \tilde{\vec{A}})$
Generated conserved quantity:	$\nabla \cdot \vec{J} = -\frac{\partial \rho}{\partial t} \Rightarrow$ charge

very shortly, the gravitational case is much more complicated than $E\&M$, but we are able to learn a great deal in analogy from the previous sections. Let us return to the gravielectric duality from table 2.1 and extend it by considering an auxiliary velocity vector field $\vec{V}$, such that the Newtonian gravitational field strength becomes

$$\vec{g} = -\nabla\Phi - \frac{\partial \vec{V}}{\partial t}, \tag{3.42}$$

similarly to the electric field strength, where Φ is the Newtonian potential in units of energy per unit mass. This addition of an auxiliary velocity vector field gives rise to a gravimagnetic (Coriolis) field strength $\vec{K}$, such that

$$\vec{K} = \nabla \times \vec{V}. \tag{3.43}$$

This gives further rise to a set of electro-gravitomagnetic field equations akin to Maxwell's equations:

$$\begin{aligned} \nabla \cdot \vec{g} &= -4\pi G\rho \\ \nabla \times \vec{g} &= -\frac{\partial \vec{V}}{\partial t} \\ \nabla \cdot \vec{K} &= 0 \\ \nabla \times \vec{K} &= \mu_g \vec{\mathcal{P}} - \frac{\mu_g}{4\pi G}\frac{\partial \vec{g}}{\partial t}, \end{aligned} \tag{3.44}$$

where $\vec{\mathcal{P}}$ is the mass–current flux density and μ_g is the Newtonian gravitational analog of the electromagnetic permeability of free space. Now, as with Maxwell's equations, the above must originate from an action principle in accord with outline 1 and summarized via Nöther's theorem as we did for $E\&M$ in table 3.2. Due to the similarity of the field equations, the analysis for Newton would follow identically as it did for Maxwell in the previous section. However, the results would need to be interpreted differently, which can easily be done by enhancing the gravielectric duality to include $\vec{K}$ as completed in table 3.3. Thus, following the enhancing gravielectric duality, we are able to construct Nöther's theorem for Newtonian gravity, which we summarize in table 3.4. We see that similar to $\vec{E}$ and $\vec{B}$, $\vec{g}$ and $\vec{K}$

Table 3.3. Gravielectric and magnetic duality.

Charge	q	$\leftrightarrow$	m
Current flux density	$\vec{J}$	$\leftrightarrow$	$\vec{\mathcal{P}}$
Coupling	$k = (4\pi\epsilon_0)^{-1}$	$\leftrightarrow$	$-G$
Permeability	μ_0	$\leftrightarrow$	μ_g
Field strength	$\vec{E}$ & $\vec{B}$	$\leftrightarrow$	$\vec{g}$ & $\vec{K}$
Field	φ & $\vec{A}$	$\leftrightarrow$	Φ & $\vec{V}$

Table 3.4. Nöther's theorem for Newtonian gravity.

Action	$S(\Phi, \vec{V}) = \int dV\, dt \left\{ \frac{1}{2}\left(-\frac{1}{4\pi G}\vec{g}^2 - \frac{1}{\mu_g}\vec{K}^2 \right) - \Phi\rho + \vec{V} \cdot \vec{\mathcal{P}} \right\}$
Continuous symmetry	Gauge transformation, $\Phi \to \tilde{\Phi}$ and $\vec{V} \to \tilde{\vec{V}}$
Invariant objects	$\vec{g} = \tilde{\vec{g}}$, $\vec{K} = \tilde{\vec{K}}$ and $S(\Phi, \vec{V}) = S(\tilde{\Phi}, \tilde{\vec{V}})$
Generated conserved quantity	$\nabla \cdot \vec{\mathcal{P}} = -\frac{\partial \rho}{\partial t} \Rightarrow$ Mass

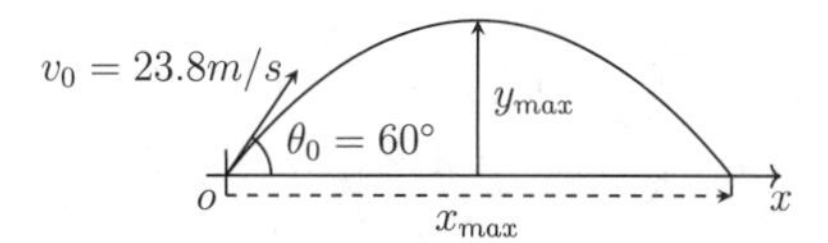

Figure 3.1. A quick x and y kinematic equation analysis yields $y_{\max} = 21.7$ m at time $t_{y_{\max}} = 2.10$ s and $x_{\max} = 50.0$ m at time $t_{\rm tot} = 4.20$ s. Now, performing the same calculation in polar coordinates $(r(t), \theta(t))$, where $r(t) = \sqrt{(x(t))^2 + (y(t))^2}$ and $\theta(t) = \tan^{-1}\frac{y(t)}{x(t)}$. We have that the maximum height is just $r(t_{y_{\max}})\sin\theta(t_{y_{\max}}) = 21.7$ m and the maximum range is $r(t_{\rm tot})\cos\theta(t_{\rm tot}) = 50.0$ m, in an agreement with our previous answers.

are too gauge invariant by the interchange of $\varphi \to \Phi$ and $\vec{A} \to \vec{V}$ in equation (3.17) and this gravitational–gauge symmetry yields mass as the conserved quantity. This is where the gravielectric and magnetic duality, within the Newtonian gravity paradigm, finds its limitations. This is due to the fact that we would like to see Nöther's theorem for gravity yield energy and momentum conservation in addition to mass conservation. Though we could lump energy and mass into the same category. This can be achieved by recognizing that gravitational–gauge symmetry is just a specific case of a coordinate transformations, which we demonstrate in appendix A, a larger class of symmetries which we call diffeomorphism symmetry. Diffeomorphism symmetry is just the statement that physics doesn't really care about how we humans describe it. For example, the projectile in figure 3.1 is indifferent as to whether we choose Cartesian or polar coordinates to describe its trajectory. If we set out to compute the maximum height $y_{\max}$ and maximum range $x_{\max}$ we better obtain the same result irrespective of choice of coordinates. As previously mentioned, it is this diffeomorphism symmetry, the fact that physics is invariant/indifferent to how we describe it, that generates energy, mass and momentum conservation. To summarize this conservation law into one equation requires the introduction of tensor or index notation, which we will do next as we augment or reconstruct Nöther's theorem for gravity given this new information. We should mention though, that the full action principle for (general relativistic) gravity will still elude us in this discussion due to its severe complexity beyond the undergraduate curriculum.

3.3 Nöther's theorem and gravity: the metric tensor

Let us restart our thought process by pondering a geometric (point set topological) object that is indifferent or invariant with respect the general diffeomorphisms, i.e. coordinate transformations. A possible candidate is the inner (dot) product or *metric* between two Cartesian vectors $\vec{A}$ and $\vec{B}$:

$$\vec{A} \cdot \vec{B} = A^x B^x + A^y B^y + A^z B^z \tag{3.45}$$

and instead of indexing over x, y and z we will switch to 1, 2, and 3, which allows us to introduce some notational simplifications:

$$\begin{aligned}\vec{A} \cdot \vec{B} &= A^1 B^1 + A^2 B^2 + A^3 B^3 \\ &= \sum_{i=1}^{3} A^i B^i \\ &= A^i B^i,\end{aligned} \tag{3.46}$$

where in the last line above we have invoked Einstein's summation conventions by dropping the summation symbol and keeping an implied summation over i since it is repeated or common to each term in A and B. Now looking back to the first line in equation (3.46), we are free to expand the right side using matrix notation as

$$\begin{aligned}A^1 B^1 + A^2 B^2 + A^3 B^3 &= (A^1 \quad A^2 \quad A^3) \begin{pmatrix} B^1 \\ B^2 \\ B^3 \end{pmatrix} \\ &= (A^1 \quad A^2 \quad A^3) \begin{pmatrix} 1 & 0 & 0 \\ 0 & 1 & 0 \\ 0 & 0 & 1 \end{pmatrix} \begin{pmatrix} B^1 \\ B^2 \\ B^3 \end{pmatrix}.\end{aligned} \tag{3.47}$$

This may naively seem like an over complication, but it reveals an interesting and fundamental structure of the physical geometry of our Universe. Using index notation and Einstein's summation convention again, the last line in equation (3.47) reads

$$(A^1 \quad A^2 \quad A^3) \begin{pmatrix} 1 & 0 & 0 \\ 0 & 1 & 0 \\ 0 & 0 & 1 \end{pmatrix} \begin{pmatrix} B^1 \\ B^2 \\ B^3 \end{pmatrix} = A^i \eta_{ij} B^j, \tag{3.48}$$

where we have introduced the Cartesian metric tensor η_{ij}, similar in this case to the Kronecker delta $\delta_i^{\,j}$, which is equal to unity for all $i = j$ and otherwise equals zero whenever $i \neq j$. Now, the dot product between two vectors should be diffeomorphic, i.e. the projected length of one vector onto another should not depend on what coordinate system we choose to measure it. That is, employing the inverse Jacobian of transformation $\frac{\partial x'^i}{\partial x^j}$ [1] to A^i and B^i to transform from coordinates $x^i = \{x^1, x^2, x^3\}$ to new coordinates $x'^i = \{x'^1, x'^2, x'^3\}$ and the Jacobian transformation $\frac{\partial x^i}{\partial x'^j}$ to transform η_{ij} we see that [2]

$$\begin{aligned}\vec{A}' \cdot \vec{B}' &= A'^i \eta'_{ij} B'^j \\ &= A^a \frac{\partial x'^i}{\partial x^a} \left(\frac{\partial x^c}{\partial x'^i} \eta_{cd} \frac{\partial x^d}{\partial x'^j} \right) B^b \frac{\partial x'^j}{\partial x^b},\end{aligned} \tag{3.49}$$

where, of course, we know from sophomore level linear algebra and unitary transformations, that we need one transformation matrix for vectors and two for matrices. Rearranging terms above, we have

$$\begin{aligned}\vec{A}' \cdot \vec{B}' &= A^a \frac{\partial x'^i}{\partial x^a} \left(\frac{\partial x^c}{\partial x'^i} \eta_{cd} \frac{\partial x^d}{\partial x'^j} \right) B^b \frac{\partial x'^j}{\partial x^b} \\ &= \left(A^a \frac{\partial x'^i}{\partial x^a} \frac{\partial x^c}{\partial x'^i} \right) (\eta_{cd}) \left(B^b \frac{\partial x'^j}{\partial x^b} \frac{\partial x^d}{\partial x'^j} \right) \\ &= \left(A^a \delta_a^c \right) (\eta_{cd}) (B^b \delta_b^d) \\ &= (A^c)(\eta_{cd})(B^d) = \vec{A} \cdot \vec{B},\end{aligned} \tag{3.50}$$

that is, since $\frac{\partial x}{\partial x'}$ and $\frac{\partial x'}{\partial x}$ are inverses to each other, $\vec{A} \cdot \vec{B}$ is the same or symmetric with respect to general diffeomorphisms or coordinate transformations. At this point let us step through a similar thought process as we did for $E\&M$ and Newtonian gravity. Looking back at equation (3.50) we performed a general diffeomorphism to the Cartesian metric, η_{ij}:

$$g_{ij} = \eta'_{ij} = \frac{\partial x'^c}{\partial x^i} \eta_{cd} \frac{\partial x'^d}{\partial x^j}, \tag{3.51}$$

where we have now introduced the general metric on the left above as g_{ij}. However, after transforming η to g we see that the dot product $\vec{A} \cdot \vec{B}$ remained unaffected. Though we should mention that we also needed to change $\vec{A}$ and $\vec{B}$ in conjunction to demonstrate diffeomorphism invariance of the dot product, but those vectors are arbitrary and we can focus our attention on the metric contribution. Again, looking back to previous sections, we identified the physical fields of each respective theory as the objects that changed under specific gauge transformations, whereas the objects that remained invariant we relabeled as field strengths. Following this thought process, we have

$$
\left.\begin{array}{ll}
\text{Electromagnetic field} & \begin{cases} \varphi \to \tilde{\varphi} = \varphi - \dfrac{\partial \Lambda}{\partial t} \\ \vec{A} \to \tilde{\vec{A}} = \vec{A} + \nabla \Lambda \end{cases} \\
\text{Newtonian gravitational field} & \begin{cases} \Phi \to \tilde{\Phi} = \Phi - \dfrac{\partial \Lambda}{\partial t} \\ \vec{V} \to \tilde{\vec{V}} = \vec{V} + \nabla \Lambda \end{cases}
\end{array}\right\} \Rightarrow \tag{3.52}
$$

$$
\text{Einstein gravitational field} \quad \left\{ \boxed{g_{ij} \to g_{ij}' = \frac{\partial x'^c}{\partial x^i} g_{cd} \frac{\partial x'^d}{\partial x^j}.} \right.
$$

In other words, we are experiencing Einstein's interpretive leap first hand by recognizing the metric g_{ij} as the (general relativistic) gravitational field! As previously mentioned, we have not addressed the true diffeomorphism invariant field strengths of gravity, which are spacetime curvature scalars. This is due to the shear difficulty and depth of the required differential geometric knowledge required to make this complete connection. However, we are able to gain a rigorous appreciation/understanding of Einstein's gravitational field (metric) within the undergraduate curriculum by restricting ourselves to two dimensions and making contact to a very familiar fact, namely the Pythagorean theorem.

Let us return to the dot product and its diffeomorphic invariance. More intuitively the length of each vector $|\vec{A}|$ and $|\vec{B}|$ projected onto each other should only depend on their relative orientation or angle between $\vec{A}$ and $\vec{B}$ given by the commonly taught form:

$$
\vec{A} \cdot \vec{B} = |\vec{A}||\vec{B}|\cos\alpha, \tag{3.53}
$$

where α is the angle between $\vec{A}$ and $\vec{B}$. This implies that the magnitude/length and therefore length squared of a given vector $|\vec{A}|^2 = \vec{A} \cdot \vec{A}$ must be diffeomorphism invariant. Let us focus on an infinitesimal length $ds^2 = d\vec{s} \cdot d\vec{s}$ in a Cartesian x–y coordinate system, as depicted below:

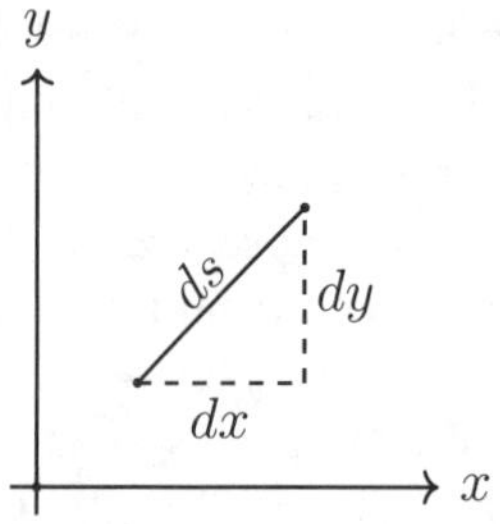

$$
\left.\right\} \begin{array}{l}
ds^2 = dx^2 + dy^2 \\
ds^2 = \underbrace{\begin{pmatrix} dx & dy \end{pmatrix}}_{dx^i} \underbrace{\begin{pmatrix} 1 & 0 \\ 0 & 1 \end{pmatrix}}_{\eta_{ij}} \underbrace{\begin{pmatrix} dx \\ dy \end{pmatrix}}_{dx^j} \\
ds^2 = \boxed{dx^i \eta_{ij} dx^j.}
\end{array} \tag{3.54}
$$

We see that by direct inspection the infinitesimal version of the Pythagorean theorem allowed us to read off the Cartesian gravitational field η_{ij}. But what about the gravitational field of a curved two-dimensional surface? Let us examine a two-sphere of constant radius r embedded in a Cartesian coordinate system, the surface of which can be described by the radial vector $\vec{R} = (r\sin\theta\cos\phi,\ r\sin\theta\sin\phi,\ r\cos\theta)$. Next, draw a small arclength ds along its surface, which we did (in a convenient way) below and construct the Pythagorean theorem for ds^2:

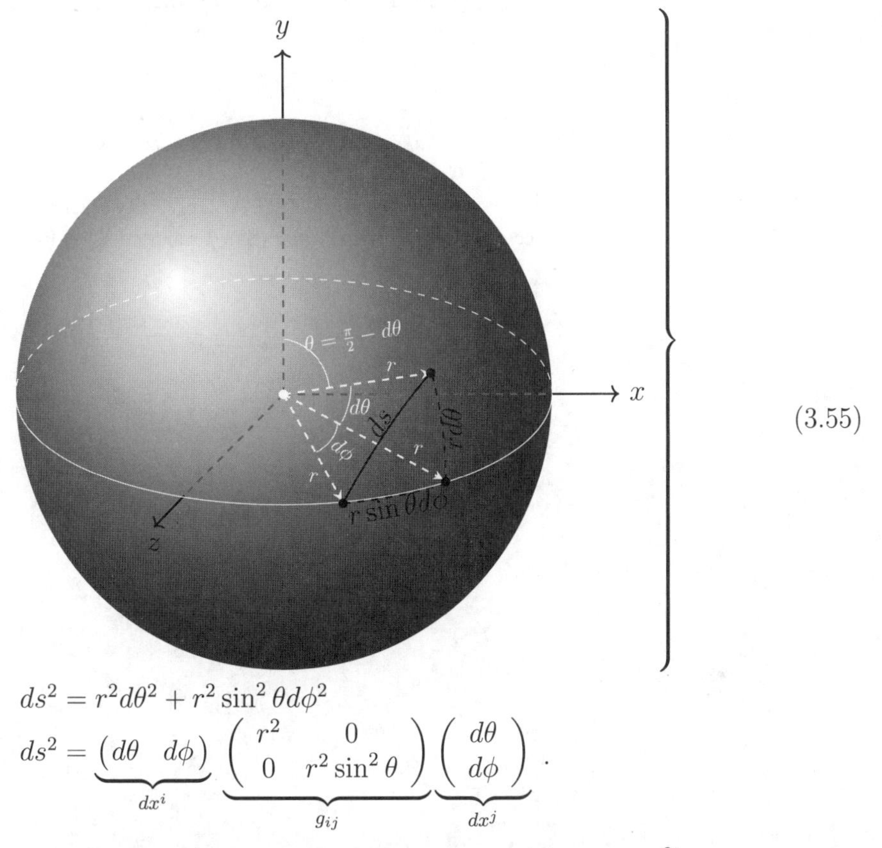

Thus, we see that the gravitational field of the two-sphere (S^2) is

$$g_{ij} = \begin{pmatrix} r^2 & 0 \\ 0 & r^2\sin^2\theta \end{pmatrix}, \tag{3.56}$$

which may seem like a non-intuitive concept, since g_{ij} above only encodes geometry. However, this is only due to the dimensionality we have chosen and the simple fact that in two dimensions there are no classical matter configurations which may source a gravitational field [3]. This stresses two conclusions:

- Einstein's revolutionary leap of interpreting gravity as the geometry of spacetime.
- Any black holes arising in two dimensions must have quantum gravitational implications or origins.

The fact that the gravitational field in equation (3.56) is purely geometric is easily seen by integrating out the square-root of its determinant, which we label $\sqrt{g}$:

$$\int_0^{\pi}\int_0^{2\pi} \sqrt{g}\, d\phi d\theta = \int_0^{\pi}\int_0^{2\pi} r^2 \sin\theta d\phi d\theta = \boxed{4\pi r^2}, \tag{3.57}$$

yielding the surface area of S^2. This is, of course, a general statement irrespective of dimensionality.

References

[1] Schey H M 2004 *Div, Grad, Curl, and All That: An Informal Text on Vector Calculus* (New York: Norton)

[2] Moore T A 2013 *A General Relativity Workbook* (Mill Valley, CA: University Science Books)

[3] Misner C W, Thorne K S and Wheeler J A 1970 *Gravitation* (San Francisco, CA: Freeman)

IOP Concise Physics

On The Principle of Holographic Scaling

From college physics to black hole thermodynamics

Leo Rodriguez and Shanshan Rodriguez

Chapter 4

Tensor calculus on manifolds

The purpose of this chapter is to provide a short introduction to the mathematical topics on differential geometry needed for a first course or study of general relativity and field theory at the undergraduate level. To demystify the geometry and topology of physical theory in a rigorous manner (in conjunction with appendix B), while maintaining pedagogical worth to a physics audience. With this in mind, mathematical theorems are addressed with rigorous proofs only if they help to solidify the concepts at hand and do not stray off into far esoteric mathematical tangents. Of course, not everything covered in this chapter is a mastery-must, but the idea is that we maintain completeness at a mathematical rigor that we believe is suitable. Readers who want to venture past this point and delve into more topics on physics are welcome to do so and return to this chapter as needed for mathematical clarification. The content of this chapter is aimed at students who are well versed in vector calculus and for a more comprehensive study of the subject we refer to the encompassing literature on this topic [1–14].

4.1 Index notation

In the previous chapter we saw how index notation arose as a way to help us isolate and read off the metric tensor and how to translate the vector dot product into matrices and then into indexed objects such as the metric, while employing Einstein's summation convention for repeated indices. Looking back to section 3.3 we introduced index notion for Cartesian vectors by looking at the dot product. Let us repeat a similar analysis, but starting with a vector written in unit vector notation:

$$\begin{aligned}\vec{A} &= A^x\hat{\imath} + A^y\hat{\jmath} + A^z\hat{k} \\ &= A^x\frac{\partial\vec{R}}{\partial x} + A^y\frac{\partial\vec{R}}{\partial y} + A^z\frac{\partial\vec{R}}{\partial z} \\ &= A^1\frac{\partial\vec{R}}{\partial x^1} + A^2\frac{\partial\vec{R}}{\partial x^2} + A^3\frac{\partial\vec{R}}{\partial x^3},\end{aligned} \tag{4.1}$$

doi:10.1088/2053-2571/ab06bdch4

where we are indexing $\{x, y, z\}$ with $\{1, 2, 3\}$, $\vec{R} = (x, y, z)$, $\frac{\partial \vec{R}}{\partial x} = \hat{\imath}$, $\frac{\partial \vec{R}}{\partial y} = \hat{\jmath}$ and $\frac{\partial \vec{R}}{\partial z} = \hat{k}$. Next, rewriting the above using an index and summation symbol, we have

$$\begin{aligned}\vec{A} &= A^1 \frac{\partial \vec{R}}{\partial x^1} + A^2 \frac{\partial \vec{R}}{\partial x^2} + A^3 \frac{\partial \vec{R}}{\partial x^3} \\ &= \sum_{i=1}^{3} A^i \frac{\partial \vec{R}}{\partial x^i} = A^i \frac{\partial \vec{R}}{\partial x^i},\end{aligned} \tag{4.2}$$

where we employed Einstein's summation convention in the last line. Now, let us reexamine the dot product between $\vec{A}$ and $\vec{B}$:

$$\begin{aligned}\vec{A} \cdot \vec{B} &= \left(A^i \frac{\partial \vec{R}}{\partial x^i}\right) \cdot \left(B^j \frac{\partial \vec{R}}{\partial x^j}\right) \\ &= \left(\frac{\partial \vec{R}}{\partial x^i}\right) \cdot \left(\frac{\partial \vec{R}}{\partial x^j}\right) A^i B^j \\ &= \eta_{ij} A^i B^j.\end{aligned} \tag{4.3}$$

Exercise 1. *Use equation* (2.4) *to show that*

$$\left(\frac{\partial \vec{R}}{\partial x^i}\right) \cdot \left(\frac{\partial \vec{R}}{\partial x^j}\right) = \eta_{ij} = \begin{pmatrix} 1 & 0 & 0 \\ 0 & 1 & 0 \\ 0 & 0 & 1 \end{pmatrix}. \tag{4.4}$$

This gives us a way of translating between vector notation and index notation for the dot product. In particular let us define the covector $A_j = \eta_{ij} A^i$, then the dot product is simply

$$\vec{A} \cdot \vec{B} = A_j B^j. \tag{4.5}$$

Obviously in Cartesian coordinates where $g_{ij} = \eta_{ij}$ there is no difference between A^i and A_j, but in subsequent scenarios on curved spaces, we will see a substantial difference and expand our notation between the two geometric objects.

To translate the cross product into index notation let us introduce the completely antisymmetric Levi-Civita symbol:

$$\epsilon_{ijk} = \begin{cases} +1 & \text{for even permutations of } (i, j, k) \\ -1 & \text{for odd permutations of } (i, j, k) \\ 0 & \text{for any } i = j, \;\; i = k, \;\; j = k \end{cases}. \tag{4.6}$$

With this to hand we may rewrite the components of $\vec{A} \times \vec{B}$ as

$$\vec{A} \times \vec{B} = \epsilon^i{}_{jk} A^i B^j \frac{\partial \vec{R}}{\partial x^j} = \epsilon^i{}_{jk} A^i B^j \partial_j \vec{R}, \tag{4.7}$$

where we introduced the shorthand notation $\partial_j = \frac{\partial}{\partial x^j}$.

Exercise 2. *Verify equation (4.7) by direct computation, using the properties of the Levi-Civita symbol, for each component of* $\vec{A} \times \vec{B}$.

Example 3. *Let us verify the triple product rule vector identity:*

$$\vec{B} \cdot (\nabla \times \vec{A}) = -\nabla \cdot (\vec{B} \times \vec{A}) + (\nabla \times \vec{B}) \cdot \vec{A}, \tag{4.8}$$

using what we have learned thus far about index notation. We have

$$\begin{aligned} \vec{B} \cdot (\nabla \times \vec{A}) &\Rightarrow B_i \epsilon_{ijk} \partial_j A_k \\ &= \epsilon_{ijk} B_i \partial_j A_k \\ &= \epsilon_{ijk} \left\{ \partial_j (B_i A_k) - \partial_j B_i A_k \right\} \\ &= -\epsilon_{jik} \partial_j (B_i A_k) + \epsilon_{jik} \partial_j B_i A_k \end{aligned} \tag{4.9}$$

$$\Rightarrow \boxed{\vec{B} \cdot (\nabla \times \vec{A}) = -\nabla \cdot (\vec{B} \times \vec{A}) + (\nabla \times \vec{B}) \cdot \vec{A}.}$$

Exercise 3. *Prove the following vector identities:*

(a) $\vec{A} \times \vec{B} \cdot \vec{C} \times \vec{D} = (\vec{A} \cdot \vec{C})(\vec{B} \cdot \vec{D}) - (\vec{A} \cdot \vec{D})(\vec{B} \cdot \vec{C})$

(b) $\nabla \times (\vec{A} \times \vec{B}) = (\nabla \cdot \vec{B})\vec{A} + (\vec{B} \cdot \nabla)\vec{A} - [(\nabla \cdot \vec{A})\vec{B} + (\vec{A} \cdot \nabla)\vec{B}]$

(c) $\nabla \times (\nabla \varphi) = 0$

(Hint: $\epsilon_{ijk} \epsilon^{i}{}_{rs} = \eta_{jr} \eta_{ks} - \eta_{js} \eta_{kr}$. *)*

4.2 Some point set topology and manifolds

Definition 1 (metric space). *A metric space is a set* M *with a distance function or metric* d: $M \times M \to \mathbb{R}$ *such that for all* $x, y \in M$:

1. $d(x, y) \geqslant 0$.
2. $d(x, y) = 0$ *if and only if* $x = y$.
3. $d(x, y) = d(y, x)$.
4. $d(x, y) + d(y, z) \geqslant d(x, z)$.

Example 4. *The most common metric for* $\mathbb{R}^n$ *is*

$$\|\vec{x} - \vec{y}\| = \sqrt{(x_1 - y_1)^2 + \cdots + (x_n - y_n)^2}. \tag{4.10}$$

From the above definition we can center an open ball B_a at $y \in M$ with radius ϵ such that $B_a(\epsilon) = \{x \in M | d(x, y) < \epsilon\}$. We are in the position to define an open set.

Definition 2. *A subset $U \subseteq M$ is said to be open if and only if for every $y \in U$ there exists an open ball $B_a(\epsilon)$ centered at y such that $\epsilon > 0$. A subset $U \subseteq M$ is said to be closed if its complement is open.*

Let $U_i \subseteq M$ be a collection of open sets, then

$$\begin{cases} \bigcup_{i=1}^{N} U_i & \text{arbitrary } N \\ \bigcap_{i=1}^{N} U_i & \text{finite } N \end{cases}$$

are open. We can now discuss what it means for M to have a topology.

Definition 3 (topological space). *A topological space is a set M with a distinguished collection of subsets, which are called the open sets. Those open sets must satisfy the following.*

- *M and the empty set are open.*
- *For all U* and: *V open, $U \cap V$ is open.*
- *The union of any collection of open sets is open.*

M is called a topological space and the open sets define the topology of M.

Another useful notion is a neighborhood (*nghbhd*) of a point $x \in M$, which is any open subset U of M such that $x \in U$. A mapping F between topological spaces M and N is said to be continuous if for all $n \in N$ and $m \in M$, where $F(m) = n$, the inverse image of the neighborhood of n is the neighborhood of m. In other words, $F^{-1}(\text{nghbhd}(n)) = \text{nghbhd}(m)$.

Definition 4 (covering). *A collection of sets U_i is said to form a covering of a subset A of a topological space M if every $m \in A$ belongs to some member of U_i, and thus $A \subseteq \bigcup_i U_i$. U_i is called an open cover if every member of the collection is open. A subset of $\{U_i\}$ which covers A is called a subcover of $\{U_i\}$.*

Definition 5 (compact). *A topological space M is said to be compact if the open cover $\{U_i\}$ contains a finite subcover.*

Theorem 2 (Heine–Borel). *A compact subset of $\mathbb{R}^n$ is closed and bounded.*

Definition 6 (homeomorphism). *Let $f: X \to Y$ for topological spaces X and Y. f is called a homeomorphism if f and f^{-1} are continuous.*

Definition 7 (countability axioms). *A topological space satisfies the* ***first countability axiom****, and is called first countable, if every point possesses a countable neighborhood* *basis. It satisfies the* ***second countability axiom****, and is called second countable, if it possesses a countable basis for the topology.*

Definition 8 (Hausdorff). *A topological space is called Hausdorff if any two distinct points have disjoint neighborhoods.*

Exercise 4. *Show that the closure of a set is a closed set.*

Definition 9 (manifold). *An n-dim (smooth-) manifold is a second countable Hausdorff space M^n together with a collection of maps called charts such that*

1. *a chart is a homeomorphism ϕ: $U \to U' \subseteq \mathbb{R}^n$ where U is open in M^n and U' is open in $\mathbb{R}^n$;*
2. *each point $x \in M$ is in the domain of some chart;*
3. *for charts ϕ: $U \to U' \subset \mathbb{R}^n$ and ψ: $V \to V' \subset \mathbb{R}^n$ the change of coordinates $\phi\psi^{-1}$: $\psi(U \cap V) \to \phi(U \cap V)$ is smooth; and*
4. *the collection of charts is maximal with properties 1, 2, and 3, see figure 4.1.*

Definition 10 (diffeomorphism). *A diffeomorphism is a map ϕ: $M \to N$ between manifolds M and N, such that ϕ is injective and ϕ and ϕ^{-1} are smooth.*

Definition 11. *A smooth parametrized curve on an n-dim manifold M is a smooth map γ: $(a, b) \to M$ for $(a, b) \subset \mathbb{R}$. The curve is said to pass through the point p at $t = t_o$ if $\gamma(t_o) = p$, for all $t_o \in (a, b)$.*

4.3 Vectors, tensors and fields

In this section we will refer to contravariant vectors as tangent vectors, or just vectors (their duals will be abbreviated as covectors instead of covariant vectors). The notion of a vector is intrinsic to most as an object tangent to some curve, surface, or embedding in three space. Yet, this notion is not covariant. In order to define vectors on manifolds it is imperative to introduce some covariant object since any given point may have overlapping charts. This is done via the directional derivative.

Definition 12 (vector). *Let M be a smooth manifold and γ: $\mathbb{R} \to M$ a smooth curve such that $\gamma(0) = p \in M$. Next let U be an open nghbhd of p and f: $U \to \mathbb{R}$ be smooth. Then the directional derivative of f along γ at p is given by*

$$D_\gamma(f) = \frac{d}{dt} f(\gamma(t))|_{t=0} \tag{4.11}$$

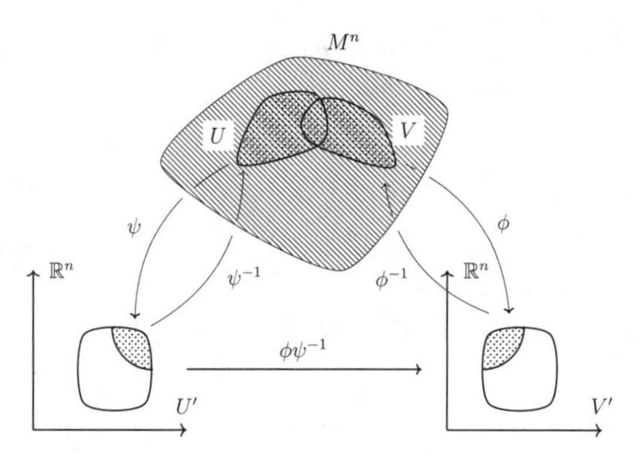

Figure 4.1. Chart mappings.

D_γ is called the tangent vector to γ at p, and for distinct γ & γ' $D_\gamma = D_{\gamma'}$ if at p $D_\gamma(f) = D_{\gamma'}(f)$.

Definition 13 (tangent space). *Let M be a smooth manifold and $p \in M$. Then $T_p(M)$ denotes the vector space of tangent vectors to M at p.*

Definition 14 (germ). *A germ of a smooth real-valued function f at p of a smooth manifold is the equivalence class of f under the relation $f_1\tilde{}f_2 \Leftrightarrow f_1(x) = f_2(x)$, $\forall x$ in some nghbhd of p.*

From definition 14 we notice that D_γ is defined on the *germ* of *f*. Calling $D = D_\gamma$, clearly

1. $D(af + bg) = aDf + bDg$; linearity.
2. $D(fg) = f(p)D(g) + g(p)D(f)$; Leibniz rule.

for all *f* and *g* smooth real-valued functions and for all $a, b \in \mathbb{R}$. Thus, *D* forms a *derivation* of the algebra of smooth real-valued functions on *M* at *p*. Next, we will extend our analysis to some coordinate chart on M^n. Let $x^1, \ldots, x^n$ be local coordinates of a chart on *M* located at *p*. Further, let $x^\mu = x^\mu(t)$ form a family of curves $\gamma^\mu(t)$. Then, for every $f: U \to \mathbb{R}$, where *U* is an open *nghbhd* of *p*, we have

$$D_\gamma(f) = \frac{d}{dt} f(\gamma^\mu)|_{t=0} \tag{4.12}$$

$$= \frac{\partial f}{\partial x^\mu}\frac{d\gamma^\mu}{dt}|_{t=0}, \tag{4.13}$$

which implies

$$D_\gamma = \frac{d\gamma^\mu}{dt}\frac{\partial}{\partial x^\mu}|_{p,t=0}, \tag{4.14}$$

thus, every first-order derivation forms a vector $\vec{X} = X^\mu \partial_\mu$. We will sometimes refer to just X^μ, the components of $\vec{X}$, as a vector which is not incorrect, but it should be noted that X^μ is dependent upon a specific choice of coordinates. Next, consider $a^\mu = (0, \dots, 0, 1, 0, \dots, 0)$ where 1 is located at the μth index. Taking the directional derivative along a^μ we get

$$D_{a^\mu} = \frac{\partial}{\partial x^\mu}|_p, \tag{4.15}$$

which implies the following theorem:

Theorem 3 (coordinate basis). $\mathrm{Span}\left(\frac{\partial}{\partial x^\mu}|_p\right) = T_p(M)$.

Clearly theorem 3 implies $(a\vec{X} + b\vec{Y})(f) = a\vec{X}(f) + b\vec{Y}(f)$.

We have now accomplished our goal of introducing vectors in a covariant fashion. In other words, for coordinates x, x' near p we have

$$X^\mu \frac{\partial}{\partial x^\mu} = X'^\mu \frac{\partial}{\partial x'^\mu}. \tag{4.16}$$

Thus, following equation (4.16) we obtain the transformation law:

$$\boxed{X^\nu = X'^\mu \frac{\partial x^\nu}{\partial x'^\mu}.} \tag{4.17}$$

Exercise 5. *Let $M \subseteq \mathbb{R}^n$ be a smooth manifold and f be a smooth real-valued function defined on a nghbhd of $p \in M^n$ and be constant on M. Show that ∇f is normal (or dual) to $T_p(M)$ at p.*

Definition 15 (cotangent space). *The dual space to $T_p(M)$ containing all covectors or* 1 – forms *denoted $T_p^*(M)$.*

We define the action of a covector on a vector by the inner product $\langle , \rangle$, i.e. for all $\eta \in T_p^*(M)$ and $\vec{X} \in T_p(M)$:

$$\langle \eta, \vec{X} \rangle = \eta_\mu X^\mu. \tag{4.18}$$

Following exercise 5 and equation (4.18) the action of a vector on a real-valued function at p on M can now be realized as an inner product. Thus, for $df \in T_p^*(M)$ and $\vec{X} \in T_p(M)$ the directional derivative reads

$$\vec{X}(f) = \langle df, \vec{X} \rangle \tag{4.19}$$

Next, take the differentials of coordinates x^μ belonging to some chart of M and acting on the tangent basis ∂_μ we get

$$\langle dx^{\mu}, \partial_{\nu}\rangle = \partial_{\nu}(dx^{\mu}) = \frac{\partial x^{\mu}}{\partial x^{\nu}} = \delta_{\nu}{}^{\mu}. \tag{4.20}$$

Thus, dx^{μ} is the dual of ∂_{μ}, which implies the following:

Theorem 4 (dual basis). Span$(dx^{\mu}) = T_p^*(M)$.
Furthermore, for any $\eta \in T_p^*(M)$, η takes the form

$$\eta = \eta_{\mu} dx^{\mu}, \tag{4.21}$$

with components η_{μ}. Holding equation (4.21) invariant on M, i.e.

$$\eta_{\mu} dx^{\mu} = \eta_{\mu}' dx'^{\mu}, \tag{4.22}$$

we deduce the transformation law for covectors, namely

$$\boxed{\eta_{\mu} = \eta_{\nu}' \frac{\partial x'^{\nu}}{\partial x^{\mu}}.} \tag{4.23}$$

We are now in a position to introduce the concept of a tensor.

Definition 16 (tensor). *A tensor of rank* (r, s) *is a multi-linear functional* A *such that*

$$A\colon \underbrace{T_p^*(M) \times \cdots \times T_p^*(M)}_{r} \times \underbrace{T_p(M) \times \cdots \times T_p(M)}_{s} \to \mathbb{R}. \tag{4.24}$$

We denote the tensor space at p *of* M *by* $T_p^{(r,s)}(M)$ *with dimension* n^{r+s}.

In other words, an (r, s) tensor, denoted $A^{\mu_1 \cdots \mu_s}_{\nu_1 \ldots \nu_r}$, is a multi-linear functional acting on the respective basis:

$$A^{\mu_1 \cdots \mu_s}_{\nu_1 \ldots \nu_r} = A(dx^{\mu_1}, \ldots, dx^{\mu_s}, \partial_{\nu_1}, \ldots, \partial_{\nu_r}) \tag{4.25}$$

Now, following equations (4.17) and (4.23) we obtain the transformation law for a mixed tensor. By treating dx^{μ} and ∂_{μ} as vector and covector respectively we obtain

$$A\left(dx^{\nu_1}, \ldots, dx^{\nu_s}, \frac{\partial}{x^{\mu_1}}, \ldots, \frac{\partial}{x^{\mu_r}}\right)$$
$$= A\left(\frac{\partial x^{\nu_1}}{\partial x'^{\alpha_1}} dx'^{\alpha_1}, \ldots, \frac{\partial x^{\nu_s}}{\partial x'^{\alpha_s}} dx'^{\alpha_s}, \frac{\partial x'^{\beta_1}}{\partial x^{\mu_1}} \partial'_{\beta_1}, \ldots, \frac{\partial x'^{\beta_r}}{\partial x^{\mu_r}} \partial'_{\beta_r}\right).$$

And since A is multi-linear we have

$$
\begin{aligned}
&A(dx^{\nu_1}, \ldots, dx^{\nu_s}, \partial_{\mu_1}, \ldots, \partial_{\mu_r}) \\
&= \left(\frac{\partial x^{\nu_1}}{\partial x'^{\alpha_1}} \cdots \frac{\partial x^{\nu_s}}{\partial x'^{\alpha_s}}\right)\left(\frac{\partial x'^{\beta_1}}{\partial x^{\mu_1}} \cdots \frac{\partial x'^{\beta_r}}{\partial x^{\mu_r}}\right) A(dx'^{\alpha_1}, \ldots, dx'^{\alpha_s}, \partial'_{\beta_1}, \ldots, \partial'_{\beta_r}) \\
&= \left(\frac{\partial x^{\nu_1}}{\partial x'^{\alpha_1}} \cdots \frac{\partial x^{\nu_s}}{\partial x'^{\alpha_s}}\right)\left(\frac{\partial x'^{\beta_1}}{\partial x^{\mu_1}} \cdots \frac{\partial x'^{\beta_r}}{\partial x^{\mu_r}}\right) A'^{\alpha_1 \ldots \alpha_s}_{\beta_1 \ldots \beta_r}.
\end{aligned}
$$

Thus, the transformation law reads

$$
\boxed{A^{\nu_1 \ldots \nu_s}_{\mu_1 \ldots \mu_r} = \left(\frac{\partial x^{\nu_1}}{\partial x'^{\alpha_1}} \cdots \frac{\partial x^{\nu_s}}{\partial x'^{\alpha_s}}\right)\left(\frac{\partial x'^{\beta_1}}{\partial x^{\mu_1}} \cdots \frac{\partial x'^{\beta_r}}{\partial x^{\mu_r}}\right) A'^{\alpha_1 \ldots \alpha_s}_{\beta_1 \ldots \beta_r}.} \tag{4.26}
$$

We should also note that the transformation law follows equally from the chain rule applied to the respective basis.

Definition 17 (tensor field). *A tensor field A on a smooth manifold M is an assignment of a tensor at each point $p \in M$. In other words, $A: M \to T_p^{(r,s)}(M)$ such that $A(p) \in T_p^{(r,s)}(M)$ at p.*

Definition 18 (tensor bundle). *The set $T^{(r,s)}(M)$ consisting of all tensor spaces over the smooth manifold M.*

4.4 The Lie derivative

Let $\vec{X}$ be a vector field on a manifold M and let $p \in M$ and $q \in M$ have coordinates x^μ and $x^\mu + dx^\mu = x^\mu + X^\mu dt$ respectively. Now, for some vector $\vec{Y}$, transforming to q we have

$$
\begin{aligned}
Y'^\alpha(q) &= \frac{\partial x'^\alpha}{x^\beta} Y^\beta(x) \\
&= (\delta^\alpha_\beta + \partial_\beta X^\alpha dt) Y^\beta \\
&= Y^\alpha(x) + Y^\beta(x) \partial_\beta X^\alpha dt
\end{aligned} \tag{4.27}
$$

and evaluating the untransformed vector $\vec{Y}$ at q gives

$$
\begin{aligned}
Y^\alpha(q) &= Y^\alpha(x + dx) \\
&= Y^\alpha(x) + \partial_\beta Y^\alpha(x) dx^\beta \\
&= Y^\alpha(x) + X^\beta \partial_\beta Y^\alpha(x).
\end{aligned} \tag{4.28}
$$

We define the Lie derivative, denoted $\mathcal{L}_X Y^\alpha$, as the measure of change of $\vec{Y}$ as it transforms infinitesimally along $\vec{X}$. In other words,

$$\begin{aligned}\mathcal{L}_X Y^\alpha &= \frac{Y^\alpha(q) - Y'^\alpha(q)}{dt} \\ &= \frac{Y^\alpha + X^\beta \partial_\beta Y^\alpha dt - Y^\alpha - Y^\beta \partial_\beta X^\alpha dt}{dt} \\ &= \frac{X^\beta \partial_\beta Y^\alpha dt - Y^\beta \partial_\beta X^\alpha dt}{dt}\end{aligned}$$

which yields

$$\mathcal{L}_X Y^\alpha = X^\beta \partial_\beta Y^\alpha - Y^\beta \partial_\beta X^\alpha \tag{4.29}$$

for the Lie derivative of $\vec{Y}$ along $\vec{X}$. The above steps can be repeated for the covariant vector η to produce

$$\mathcal{L}_X \eta_\alpha = X^\beta \partial_\beta \eta_\alpha + \eta_\beta \partial_\alpha X^\beta \tag{4.30}$$

the Lie derivative of η along $\vec{X}$. Clearly $\mathcal{L}_X$ forms a derivation, which implies the Leibniz rule for mixed and multi indexed tensors:

$$\begin{aligned}\mathcal{L}_X F^{\mu\nu} &= X^\alpha \partial_\alpha F^{\mu\nu} - F^{\alpha\nu} \partial_\alpha X^\mu - F^{\mu\alpha} \partial_\alpha X^\nu \\ \mathcal{L}_X F^\mu{}_\nu &= X^\alpha \partial_\alpha F^\mu{}_\nu - F^\alpha{}_\nu \partial_\alpha X^\mu + F^\mu{}_\alpha \partial_\nu X^\alpha \\ \mathcal{L}_X F_{\mu\nu} &= X^\alpha \partial_\alpha F_{\mu\nu} + F_{\alpha\nu} \partial_\mu X^\alpha + F^\mu{}_\alpha \partial_\nu X^\alpha\end{aligned} \tag{4.31}$$

and of course we add additional derivatives of $\vec{X}$ for every additional index on F with the appropriate sign and index structure depending on whether it is raised or lowered.

4.5 Exterior algebra

Definition 19 (p-form). *A tensor of type $(0, p)$ is called a p-form if, as a multi-linear functional, it changes sign for every interchange of any pair of its arguments:*

$$\eta(v^1, \dots, v^i, v^j, \dots, v^p) = -\eta(v^1, \dots, v^j, v^i, \dots, v^p).$$

Thus, if π is any permutation of $1, \dots, p$ then

$$\eta(v^{\pi(1)}, \dots, v^{\pi(p)}) = (-1)^\pi \eta(v^1, \dots, v^p),$$

which in components reads

$$\eta_{\mu_{\pi(1)} \cdots \mu_{\pi(p)}} = (-1)^\pi \eta_{\mu_1 \dots \mu_p}.$$

We will denote the space of all p-forms, to a manifold, as $\wedge^p T^*M$.

Definition 20 (exterior/wedge product). *Let $\alpha \in \wedge^r T^*M$ and $\beta \in \wedge^s T^*M$. We define the exterior product, $\wedge$: $(\wedge^r T^*M, \wedge^s T^*M) \to \wedge^{r+s} T^*M$ between α and β as*

$$\alpha \wedge \beta = \alpha_{\mu_1 \dots \mu_r} \beta_{\nu_1 \dots \nu_s} dx^{\mu_1} \wedge \cdots \wedge dx^{\mu_r} \wedge dx^{\nu_1} \wedge \cdots \wedge dx^{\nu_s} \tag{4.32}$$

Example 5.

- $r = s = 1$

$$(\alpha \wedge \beta)_{\mu\nu} = \frac{1}{2}(\alpha_\mu\beta_\nu - \beta_\nu\alpha_\mu)$$

- $r = 1 \;\; s = 2$

$$\begin{aligned}(\alpha \wedge \beta)_{\mu\nu\gamma} &= \frac{1}{6}(\alpha_\mu\beta_{\nu\gamma} - \alpha_\mu\beta_{\gamma\nu} + \alpha_\nu\beta_{\gamma\mu} - \alpha_\nu\beta_{\mu\gamma} + \alpha_\gamma\beta_{\mu\nu} - \alpha_\gamma\beta_{\nu\mu}) \\ &= \frac{1}{3}(\alpha_\mu\beta_{\nu\gamma} + \alpha_\nu\beta_{\gamma\mu} + \alpha_\gamma\beta_{\mu\nu}).\end{aligned}$$

Following definition 20, we obtain the following properties for the exterior product of p dual basis:

(i) $dx^{\mu_1} \wedge \cdots \wedge dx^{\mu_i} \wedge dx^{\mu_j} \wedge \cdots \wedge dx^{\mu_p} = 0$ for all $i = j$.

(ii) $\wedge_{i=1}^{p} dx^{\mu_{\pi(i)}} = (-1)^{\pi} \wedge_{i=1}^{p} dx^{\mu_i}$.

(iii) $\wedge_{i=1}^{p} dx^{\mu_i}$ is linear in dx^{μ_i}.

Following these properties, we express any p-form covariantly as

$$\eta = \eta_{\mu_1 \dots \mu_p} dx^{\mu_1} \wedge \cdots \wedge dx^{\mu_p} \tag{4.33}$$

and the space $\wedge^p T^*M$ together with the binary operation $\wedge$ form the associative algebra called the **exterior algebra**.

Exercise 6. *Show that for r and s forms* α *and* β*;* $\alpha \wedge \beta = (-1)^{rs}\beta \wedge \alpha$.

Definition 21 (interior product). *Let* $\vec{X} \in TM$ *and* $\eta \in \wedge^p\, T^*M$ *we define the interior product,* $i_X\colon \wedge^p\, T^*M \to \wedge^{p-1}\, T^*M$ *as*

$$i_X\eta \equiv X^\mu \eta_{\mu\mu_1 \dots \mu_{p-1}}$$

with the anti-derivation property

$$i_X(\eta \wedge \xi) = (i_X\eta) \wedge \xi + (-1)^r \eta \wedge (i_X\xi)$$

for η *an r-form.*

4.6 Exterior derivative

The exterior derivative maps a p-form to a $p + 1$-form, i.e.

$$d\colon \wedge^p\, T^*M \to \wedge^{p+1}\, T^*M. \tag{4.34}$$

Thus, given equation (4.34), we define the action of d on a p-form as

$$d\alpha \equiv d(\alpha_{\mu_1 \ldots \mu_p}) \wedge dx^{\mu_1} \wedge \cdots \wedge dx^{\mu_p} \tag{4.35}$$

$$= \partial_\nu \alpha_{\mu_1 \ldots \mu_p} dx^\nu \wedge dx^{\mu_1} \wedge \cdots \wedge dx^{\mu_p}, \tag{4.36}$$

which implies the component expression

$$(d\alpha)_{\mu_1 \ldots \mu_{p+1}} = \partial_{\mu_1} \wedge \alpha_{\mu_2 \ldots \mu_{p+1}}. \tag{4.37}$$

From the definition, (4.35), of the action of d on forms we obtain the following properties:

- Linearity: $d(\alpha + \beta) = d\alpha + d\beta$.
- Nilpotent: $d^2 = 0$.
- Leibniz: $d(\alpha \wedge \beta) = d\alpha \wedge \beta + (-1)^r \alpha \wedge d\beta$ ($\forall\ \alpha \in \wedge^r TM$).

Exercise 7. *Write out the components of $d\alpha$ explicitly in a Cartesian coordinate system for α a 0-form, 1-form, and 2-form.*

Theorem 5. *Let $\vec{X}$ and $\vec{Y}$ be vectors and η a p-form, then*

$$i_{[X,Y]}\eta = \mathcal{L}_X(i_Y\eta) - i_Y(\mathcal{L}_X\eta) \tag{4.38}$$

$$\mathcal{L}_X\eta = i_X d\eta + d(i_X\eta) \tag{4.39}$$

Proof. Left as an exercise. □

4.7 Orientation

In this section we will introduce the concept of an oriented manifold, but before we can do so, we must first introduce the notion of an oriented vector space. The orientability of a manifold will then follow from the orientability of its tangent bundle, which, of course, is a collection of vector spaces.

Let V be a vector space with bases $\{e_i\}$ and $\{f_i\}$ respectively such that

$$f_i = \alpha_i^{\ j} e_j, \tag{4.40}$$

where $\alpha_i^{\ j}$ is a continuous map between e_i and f_i. In other words, e_i is continuously deformed into f_i while remaining a basis at each stage. We then say that e_i and f_i are of the same orientation if $\det(\alpha_i^{\ j}) > 0$. This defines two equivalence classes of bases; those with $\det(\alpha_i^{\ j}) > 0$ and those with $\det(\alpha_i^{\ j}) < 0$ and to orient V we are left with the task of choosing one of the two orientation classes. A familiar example is the case of $V = \mathbb{R}^3$ where we choose the right-hand rule orientation to be positive.

Exercise 8. *Show that equation (4.40) with the condition* $\det(\alpha_i^{\ j}) > 0$ *defines an equivalence relation.*

Definition 22 (oriented vector space). *An oriented vector space V is a space together with an equivalence class of allowable bases. This entails choosing a basis of V to determine the orientation and those bases equivalent to the choice will be called* ***positively oriented*** *frames.*

Lemma 1. *Let ω be an n-form such that $\dim V = n$. Let $\{e_i\}$ be a basis of V, then for all $\vec{v}_i \in V$*

$$\vec{v}_i = \sum_j \alpha_i^{\ j} e_j$$

and thus

$$\omega(\vec{v}_1, \dots, \vec{v}_n) = \det(\alpha_i^{\ j})\omega(e_1, \dots, e_n). \tag{4.41}$$

Proof. From definition 19 and since ω is multi-linear we can write

$$\begin{aligned}\omega(\vec{v}_1, \dots, \vec{v}_n) &= \sum_{j_1 \cdots j_n} \alpha_1^{\ j_1} \dots \alpha_n^{\ j_n} \omega(e_1, \dots, e_n) \\ &= \sum_{\{\pi\}} (-1)^\pi \alpha_{\pi(1)}^{\ j_1} \dots \alpha_{\pi(n)}^{\ j_n} \omega(e_1, \dots, e_n)\end{aligned}$$

for π a permutation of of $j_1 \dots j_n$. Yet, by definition:

$$\sum_{\{\pi\}} (-1)^\pi \alpha_{\pi(1)}^{\ j_1} \dots \alpha_{\pi(n)}^{\ j_n} = \det(\alpha_i^{\ j}).$$

□

Corollary 1. *A non-vanishing n-form ω has the same sign on two bases if they have the same orientation. Thus, choosing an n-form $\omega \neq 0$ orients V and two distinct n-forms ω_1 and ω_2, not equal to zero, determine the same orientation if and only if $\omega_1 = \lambda\omega_2$ $\forall$ $\lambda > 0$.*

Proof. Following lemma 1 if $\lambda > 0$ then $\lambda\omega$ has the same sign on any equivalent basis, whereas the contrary holds for $\lambda < 0$. □

Next, we will apply these concepts to manifolds.

Definition 23. *Let M^n be a smooth n-dimensional manifold. If it is possible to define an n-form $\omega \in \wedge^n (T^*M)$ which is not zero at any point $p \in M^n$, then M^n is said to be oriented by the choice of ω.*

Remark 1. *A natural choice for ω is the n-volume-form:*

$$vol^n = \sqrt{g}\,dx^1 \wedge \cdots \wedge dx^n. \tag{4.42}$$

Yet, we have not discussed the nature of the object in the radical and thus, leave the justification of equation (4.42) for a later time.

Clearly, choosing an orientation via the choice of a non-zero n-form ω infers the orientation of the tangent basis $\{\partial\}$ by virtue of corollary 1. This leads to the more natural method of orienting a manifold. Let U_x be a coordinate patch of M with coordinates x. We may orient each tangent space at each point of U_x by choosing the basis $\{\partial_1, \dots, \partial_n\}$ and all its equivalents to be positively oriented. Now, if a point lies in an overlap of two patches $U_x \cap U_y$ the bases are related by $\partial_y = \partial_x \frac{\partial x}{\partial y}$ and thus, the two bases have the same orientation if the Jacobian determinant is positive, i.e. $\det\left(\frac{\partial x}{\partial y}\right) > 0$.

Definition 24 (orientable manifold). *A smooth manifold is said to be orientable if it can be covered by coordinate patches having positive Jacobian determinants in each overlap.*

4.8 Stokes' theorem

We will introduce the concept of Stokes' theorem as it relates to known theorems from vector calculus. We will not prove the specific theorem as that is a task beyond this book. Stokes' theorem relates information in the bulk of a manifold to information on its respective boundary. Therefore we need a concept of a manifold $\mathcal{M}$ with boundary $\partial\mathcal{M}$, similarly to how we defined Gaussian surfaces back in chapter 2. As an example take the closed three ball of radius R to be $\mathcal{M}$ depicted in figure 4.2. It consists of two constituents, namely the open ball of all point $x^2 + y^2 + z^2 < R^2$ and the boundary $\partial\mathcal{M} \equiv S^2$.

Theorem 6 (Stokes' theorem). *Let $\mathcal{M}$ be an oriented manifold with boundary $\partial\mathcal{M}$ and let η be a $(p-1)$-form on $\mathcal{M}$. Then*

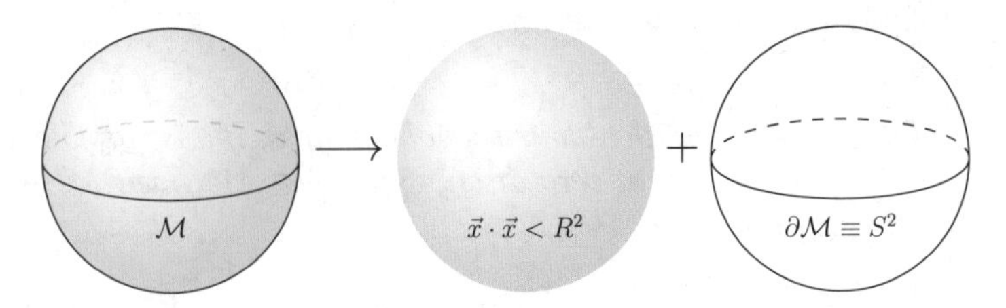

Figure 4.2. $\mathcal{M}$ depicted as the union of its bulk and boundary constituents.

$$\int_{\mathcal{M}} d\eta = \int_{\partial\mathcal{M}} \eta \tag{4.43}$$

to demystify theorem 6 let us write down what it states for $p = 1, 2, 3$ in a Cartesian coordinate system $x^1 = x$, $x^2 = y$, and $x^3 = z$.

Example 6 (p = 1). *In this case η is a zero-form or a scalar function, $\eta = \varphi(x, y, z)$. Calculating the exterior derivative:*

$$\begin{aligned} d\eta &= \partial_\mu \varphi dx^\mu \\ &= (\nabla\varphi) \cdot d\vec{R} \end{aligned} \tag{4.44}$$

where $\vec{R} = x\hat{x} + y\hat{y} + z\hat{z}$. Thus, Stokes' theorem reads

$$\int_R (\nabla\varphi) \cdot d\vec{R} = \varphi \mid_{\partial R}, \tag{4.45}$$

which we know from three-dimensional vector analysis as the fundamental theorem of vector calculus.

Example 7 (p = 2). *Now, $\eta = \varphi_\mu dx^\mu = \varphi_1 dx + \varphi_2 dy + \varphi_3 dz$ and a quick calculation reveals*

$$d\eta = (\partial_x\varphi_2 - \partial_y\varphi_1)dx \wedge dy + (\partial_z\varphi_1 - \partial_x\varphi_3)dz \wedge dx + (\partial_y\varphi_3 - \partial_z\varphi_2)dy \wedge dz \tag{4.46}$$

$$= (\nabla \times \vec{F}) \cdot d\vec{S}, \tag{4.47}$$

where the vector $\vec{F} = \varphi_1\hat{x} + \varphi_2\hat{y} + \varphi_3\hat{z}$ and the directed surface element $d\vec{S} = (dy \wedge dz)\hat{x} + (dz \wedge dx)\hat{y} + (dx \wedge dy)\hat{z}$. We can also rewrite η in terms of $\vec{F}$ as

$$\eta = \vec{F} \cdot d\vec{R}. \tag{4.48}$$

Thus, from equations (4.47) and (4.48), Stokes' theorem states

$$\int_S (\nabla \times \vec{F}) \cdot d\vec{S} = \oint_{\partial S} \vec{F} \cdot d\vec{R}, \tag{4.49}$$

which is also known as the three-dimensional Stokes' theorem.

Example 8 (p = 3). *In this case, η has the form*

$$\begin{aligned} \eta &= Qdx \wedge dy + Pdz \wedge dx + Cdy \wedge dz \\ &= \vec{F} \cdot d\vec{S} \end{aligned} \tag{4.50}$$

and

$$d\eta = (\partial_z Q + \partial_y P + \partial_x C)dx \wedge dy \wedge dz \\ = (\nabla \cdot \vec{F})dV \tag{4.51}$$

for $\vec{F} = C\hat{x} + P\hat{y} + Q\hat{z}$ *and* $dV = dx \wedge dy \wedge dz$. *Substituting equations (4.50) and (4.51) into theorem 6 we obtain*

$$\int_V (\nabla \cdot \vec{F})dV = \oint_{\partial V} \vec{F} \cdot d\vec{S}, \tag{4.52}$$

or Gauss's divergence theorem.

4.9 Riemannian mainfolds

Definition 25. *A* Riemannian metric *on a smooth manifold* $\mathcal{M}$ *assigns to each point* $p \in \mathcal{M}$ *an inner product on* $T_p(\mathcal{M})$ *such that* $\langle , \rangle$ *is*

- symmetric,
- bilinear,
- positive-definite,
- smooth.

To familiarize ourselves with the concepts above, let $\vec{A}$ and $\vec{B}$ be elements of $T_p(\mathcal{M})$, then the first three points imply the following:

$$\langle \vec{A}, \vec{B} \rangle = \langle \vec{B}, \vec{A} \rangle \tag{4.53}$$

$$\langle \alpha\vec{A}, \beta\vec{B} \rangle = \alpha\beta\langle \vec{A}, \vec{B} \rangle \qquad \forall\ \alpha \text{ and } \beta \in \mathbb{R} \tag{4.54}$$

$$\langle \vec{A}, \vec{A} \rangle = 0 \qquad \text{implies } \vec{A} = 0. \tag{4.55}$$

To realize the last point we will need the coordinate expression of a vector as defined in section 4.3. We have

$$\langle \vec{A}, \vec{B} \rangle = \left\langle A^\mu \partial_\mu, B^\nu \partial_\nu \right\rangle \tag{4.56}$$

$$= A^\mu B^\nu \left\langle \partial_\mu, \partial_\nu \right\rangle \quad \text{(bilinearity)} \tag{4.57}$$

equation (4.57) introduces a new object:

$$g_{\mu\nu} = \left\langle \partial_\mu, \partial_\nu \right\rangle \tag{4.58}$$

a symmetric (2, 0) tensor called the *metric tensor*. Next, let us exploit the tensorial properties of $g_{\mu\nu}$ to derive some useful properties. Let $\vec{A} \in T_p(\mathcal{M})$ and $\eta \in T_p^*(\mathcal{M})$, then from equation (4.18) we have

$$\langle \eta, \vec{A} \rangle = A^{\mu}\eta_{\mu} \tag{4.59}$$

and following definition 25 we get

$$A^{\mu}\eta_{\mu} = A^{\mu}\eta^{\nu}g_{\mu\nu} \tag{4.60}$$

$$\Rightarrow \eta_{\mu} = \eta^{\nu}g_{\mu\nu}. \tag{4.61}$$

From this we see that

$$g_{\mu\nu}\colon T_p(\mathcal{M}) \longrightarrow T_p^*(\mathcal{M}) \tag{4.62}$$

$$g^{\mu\nu}\colon T_p^*(\mathcal{M}) \longrightarrow T_p(\mathcal{M}). \tag{4.63}$$

In equation (4.63) we have introduced a new object, $g^{\mu\nu}$, called the *inverse metric.* To see this, let us rewrite equation (4.61) into the following:

$$\begin{aligned} \eta_{\mu} &= \eta^{\nu}g_{\mu\nu} \\ &= \eta_{\alpha}g^{\alpha\nu}g_{\mu\nu} \end{aligned}$$

thus,

$$g^{\alpha\nu}g_{\mu\nu} = \delta_{\mu}^{\alpha}. \tag{4.64}$$

Next, let $g_{\mu\nu}^{\prime}$ and $g_{\mu\nu}$ be related by a coordinate transformation, i.e.

$$g_{\mu\nu}^{\prime} = \frac{\partial x^{\alpha}}{\partial x^{\prime\mu}}\frac{\partial x^{\beta}}{\partial x^{\prime\nu}}g_{\alpha\beta}.$$

Taking the determinant of both sides, we get

$$g^{\prime} = \det\left(\frac{\partial x^{\alpha}}{\partial x^{\prime\mu}}\right)\det\left(\frac{\partial x^{\beta}}{\partial x^{\prime\nu}}\right)g,$$

where $g \equiv \det(g_{\mu\nu})$. Now, we recognize $\det\left(\frac{\partial x^{\beta}}{\partial x^{\prime\nu}}\right)$ as nothing but the Jacobian J and thus we have shown

$$\frac{g^{\prime}}{g} = J^2. \tag{4.65}$$

We have now justified the statement of remark 1, since from calculus we know that

$$d^n x = J\, d^n x^{\prime},$$

which implies

$$\sqrt{g}\, d^n x = \sqrt{g^{\prime}}\, d^n x^{\prime}. \tag{4.66}$$

Thus, we have shown that $\sqrt{g}\, d^n x$ is a natural invariant n-form.

Definition 26 (Riemannian manifold). *If a smooth manifold $\mathcal{M}$ admits a Riemannian metric $g_{\mu\nu}$ then it is called a* ***Riemannian manifold****.*

Example 9 (two-sphere with radius r). *A two-sphere is a two-dimensional smooth manifold with coordinates $x^\mu = (\phi, \theta)$ and $\mathbb{R}^3$ embedding given by the Euclidean vector*

$$\vec{R} = (r\cos\phi\sin\theta,\ r\sin\phi\sin\theta,\ r\cos\theta). \tag{4.67}$$

From equation (4.58) the metric reads

$$g_{\mu\nu} = \left\langle \partial_\mu \vec{R},\ \partial_\nu \vec{R} \right\rangle, \tag{4.68}$$

From which we obtain the individual components

$$\begin{aligned} g_{11} &= (r\cos\phi\cos\theta,\ r\sin\phi\cos\theta,\ -r\sin\theta)\cdot(r\cos\phi\cos\theta,\ r\sin\phi\cos\theta,\ -r\sin\theta) \\ &= r^2 \\ g_{12} = g_{21} &= (r\cos\phi\cos\theta,\ r\sin\phi\cos\theta,\ -r\sin\theta)\cdot(-r\sin\phi\sin\theta,\ r\cos\phi\sin\theta,\ 0) \\ &= 0 \end{aligned}$$

and

$$\begin{aligned} g_{22} &= (r\cos\phi\cos\theta,\ r\sin\phi\cos\theta,\ -r\sin\theta)\cdot(r\cos\phi\cos\theta,\ r\sin\phi\cos\theta,\ -r\sin\theta) \\ &= r^2\sin^2\theta. \end{aligned}$$

Thus, the metric of a two-sphere with radius r reads

$$g_{\mu\nu} = \begin{pmatrix} r^2 & 0 \\ 0 & r^2\sin^2\theta \end{pmatrix}. \tag{4.69}$$

We can now compute the invariant volume 2-form:

$$\begin{aligned} \boldsymbol{vol}^2 &= \sqrt{g}\ d\phi d\theta \\ &= r^2\sin\theta\ d\phi d\theta. \end{aligned} \tag{4.70}$$

4.10 Covariant differentiation

A covariant derivative is an operator which preserves symmetries. In physics the two main such symmetries are *diffeomeorphisms* and *gauge invariance.* In this section we will construct a differential operator to preserve the former, i.e. a derivative operator which transforms tensorially. We might think ∂_α would suffice as a covariant operator, and it does in the case of Cartesian coordinates, but let us see what happens upon a general coordinate transform. In other words, take ∂'_μ acting on the vector A'^ν and performing a coordinate transformation we get

$$\frac{\partial}{\partial x'^\mu} A'^\nu = \frac{\partial x^\alpha}{\partial x'^\mu}\frac{\partial}{\partial x^\alpha}\left(\frac{\partial x'^\nu}{\partial x^\beta} A^\beta\right) \tag{4.71}$$

$$= \frac{\partial x^{\alpha}}{\partial x'^{\mu}} \left(\frac{\partial^2 x'^{\nu}}{\partial x^{\alpha} \partial x^{\beta}} A^{\beta} + \frac{\partial x'^{\nu}}{\partial x^{\beta}} \partial_{\alpha} A^{\beta} \right) \tag{4.72}$$

$$= \underbrace{\frac{\partial x^{\alpha}}{\partial x'^{\mu}} \frac{\partial x'^{\nu}}{\partial x^{\beta}} \partial_{\alpha} A^{\beta}}_{\text{tensorial}} + \underbrace{\frac{\partial x^{\alpha}}{\partial x'^{\mu}} \frac{\partial^2 x'^{\nu}}{\partial x^{\alpha} \partial x^{\beta}} A^{\beta}}_{\text{non tensorial}}. \tag{4.73}$$

We see the term $\frac{\partial x^{\alpha}}{\partial x'^{\mu}} \frac{\partial^2 x'^{\nu}}{\partial x^{\alpha} \partial x^{\beta}} A^{\beta}$ in equation (4.73) destroys covariance. We can find a simple solution by introducing a new derivative operator, ∇_{α}, which is an augmentation of ∂_{α} by a non-tensor $\Gamma^{\alpha}{}_{\mu\nu}$ called a connection. The job of the connection is to restore covariance, i.e.

$$\nabla'_{\mu} A'^{\nu} = \partial'_{\mu} A'^{\nu} + \Gamma'^{\nu}{}_{\mu\gamma} A'^{\gamma} \tag{4.74}$$

$$\begin{aligned} &= \frac{\partial x^{\alpha}}{\partial x'^{\mu}} \frac{\partial x'^{\nu}}{\partial x^{\beta}} \partial_{\alpha} A^{\beta} + \frac{\partial x^{\alpha}}{\partial x'^{\mu}} \frac{\partial^2 x'^{\nu}}{\partial x^{\alpha} \partial x^{\beta}} A^{\beta} \\ &\quad + \frac{\partial x^{\alpha}}{\partial x'^{\mu}} \frac{\partial x'^{\nu}}{\partial x^{\beta}} \Gamma^{\beta}{}_{\alpha\gamma} A^{\gamma} - \frac{\partial x^{\alpha}}{\partial x'^{\mu}} \frac{\partial^2 x'^{\nu}}{\partial x^{\alpha} \partial x^{\beta}} A^{\beta}, \end{aligned} \tag{4.75}$$

which implies that

$$\nabla'_{\mu} A'^{\nu} = \frac{\partial x^{\alpha}}{\partial x'^{\mu}} \frac{\partial x'^{\nu}}{\partial x^{\beta}} \nabla_{\alpha} A^{\beta} \tag{4.76}$$

and

$$\Gamma'^{\nu}{}_{\mu\gamma} = \frac{\partial x^{\alpha}}{\partial x'^{\mu}} \frac{\partial x'^{\nu}}{\partial x^{\beta}} \frac{\partial x^{\delta}}{\partial x'^{\gamma}} \Gamma^{\beta}{}_{\alpha\delta} - \frac{\partial x^{\alpha}}{\partial x'^{\mu}} \frac{\partial^2 x'^{\nu}}{\partial x^{\alpha} \partial x^{\delta}} \frac{\partial x^{\delta}}{\partial x'^{\gamma}}. \tag{4.77}$$

Following the outline above, we obtain

$$\nabla_{\mu} A^{\nu} = \partial_{\mu} A^{\nu} + \Gamma^{\nu}{}_{\mu\gamma} A^{\gamma} \tag{4.78}$$

$$\nabla_{\mu} B_{\nu} = \partial_{\mu} B_{\nu} - \Gamma^{\gamma}{}_{\mu\nu} B_{\gamma} \tag{4.79}$$

$$\nabla_{\mu} T^{\alpha}{}_{\beta} = \partial_{\mu} T^{\alpha}{}_{\beta} + \Gamma^{\alpha}{}_{\mu\nu} T^{\nu}{}_{\beta} - \Gamma^{\nu}{}_{\mu\beta} T^{\alpha}{}_{\nu}. \tag{4.80}$$

In the literature it is most common to introduce the connection via the concept of *parallel transport*, which is the task of transporting a vector along a curve on a smooth manifold as depicted in figure 4.3. In other words, as we continuously move a vector along a curve on a manifold it will change by some amount $(\nabla_{\alpha} - \partial_{\alpha}) A^{\beta}$, which is precisely the measure of the connection since

$$(\nabla_{\alpha} - \partial_{\alpha}) A^{\beta} = \Gamma^{\beta}{}_{\alpha\mu} A^{\mu}. \tag{4.81}$$

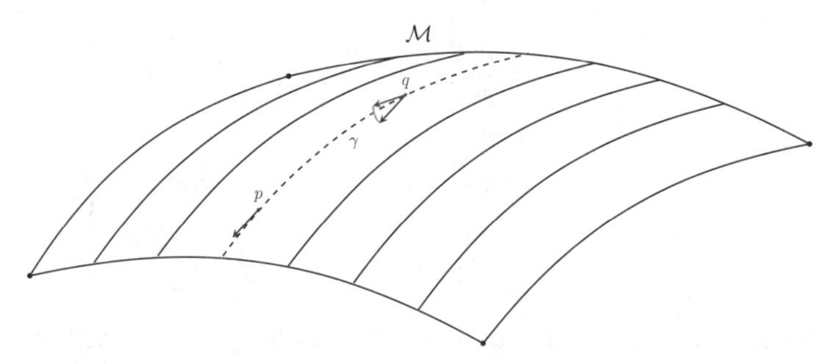

Figure 4.3. Parallel transport of a vector (blue) from point p to point q along curve γ. The green arc depicts the change in the vector as measured by $\Gamma^{\alpha}{}_{\mu\nu}$, where the red vector is the actual tangent vector at q.

From this we obtain the *parallel transport equation*:

$$\delta A^{\beta} = \Gamma^{\beta}{}_{\alpha\mu} A^{\mu}\, dx^{\alpha} \tag{4.82}$$

where δA^{β} denotes the measure of change of the components of $\vec{A}$ during its parallel transport.

Up to its transformation law, we have left the connection unspecified. We will now single out a specific connection by requiring the metric to be covariantly constant, i.e.

$$\nabla_{\alpha} g_{\mu\nu} = 0. \tag{4.83}$$

We should note that this condition only singles out a specific connection convenient for physics, since it is equivalent to the *principle of equivalence* of general relativity. Given equation (4.83) we obtain the so-called Christoffel connection:

$$\Gamma^{\alpha}{}_{\mu\nu} = \frac{1}{2} g^{\alpha\gamma}\left(\partial_{\mu} g_{\nu\gamma} + \partial_{\nu} g_{\mu\gamma} - \partial_{\gamma} g_{\mu\nu}\right). \tag{4.84}$$

Exercise 9. *Show that the non-zero Christoffel connection for equation (4.69) are*

$$\Gamma^{1}{}_{22} = -\sin\theta\cos\theta \tag{4.85}$$

$$\Gamma^{2}{}_{12} = \Gamma^{2}_{21} = \cot\theta. \tag{4.86}$$

4.11 Geodesic equation

A geodesic extremizes the path between two points on a manifold, i.e. stationary points of the functional:

$$s = \int d\lambda \left(-g_{\mu\nu}\frac{dx^{\mu}}{d\lambda}\frac{dx^{\nu}}{d\lambda}\right)^{\frac{1}{2}}. \tag{4.87}$$

Varying the above equation, we obtain

$$\delta s = -\int d\lambda \frac{1}{2}\left(-g_{\mu\nu}\frac{dx^{\mu}}{d\lambda}\frac{dx^{\nu}}{d\lambda}\right)^{-\frac{1}{2}}\delta\left(g_{\mu\nu}\frac{dx^{\mu}}{d\lambda}\frac{dx^{\nu}}{d\lambda}\right) \tag{4.88}$$

and restricting to geodesics on a Riemannian manifold we may choose $\lambda \to \tau$, where τ[1] measures proper time. We should note that we are free to choose any parameter since the functional (4.87) is reparametrization invariant. The convenience of choosing proper time fixes the term $\frac{dx^{\mu}}{d\tau}$ to be the four-velocity U^{μ} and further implies the standard normalization $U^{\mu}U_{\mu} = -1$. Given these choices, equation (4.88) simplifies to

$$\delta s = -\frac{1}{2}\int d\tau\, \delta\left(g_{\mu\nu}\frac{dx^{\mu}}{d\tau}\frac{dx^{\nu}}{d\tau}\right). \tag{4.89}$$

Let us compute equation (4.89) for the case of a two-sphere. We have

$$\delta s = -\frac{1}{2}\int d\tau\, \delta\left(\frac{d\theta}{d\tau}\frac{d\theta}{d\tau} + \sin^2\theta\frac{d\phi}{d\tau}\frac{d\phi}{d\tau}\right) \tag{4.90}$$

and holding ϕ fixed and varying with respect to θ gives

$$\begin{aligned}\delta_{\theta}s &= -\frac{1}{2}\int d\tau\, \delta\left(2\frac{d\theta}{d\tau}\frac{d\delta\theta}{d\tau} + 2\sin\theta\cos\theta\frac{d\phi}{d\tau}\frac{d\phi}{d\tau}\delta\theta\right)\\ &= -\frac{1}{2}\int d\tau\, \delta\left(-\frac{d^2\theta}{d\tau^2} + \sin\theta\cos\theta\frac{d\phi}{d\tau}\frac{d\phi}{d\tau}\right)(2\delta\theta)\end{aligned}$$

and setting $\delta s = 0$ we obtain the equation:

$$\frac{d^2\theta}{d\tau^2} - \sin\theta\cos\theta\frac{d\phi}{d\tau}\frac{d\phi}{d\tau} = 0. \tag{4.91}$$

Next, we hold θ fixed and vary with respect to ϕ yielding

$$\begin{aligned}\delta_{\phi}s &= -\frac{1}{2}\int d\tau\left(2\sin^2\theta\frac{d\phi}{d\tau}\frac{d\delta\phi}{d\tau}\right)\\ &= -\frac{1}{2}\int d\tau\left(2\sin\theta\cos\theta\frac{d\theta}{d\tau}\frac{d\phi}{d\tau} + \sin^2\theta\frac{d^2\phi}{d\tau^2}\right)(-2\delta\phi)\end{aligned}$$

and once more setting $\delta s = 0$ we obtain the equation:

$$\frac{d^2\phi}{d\tau^2} + \cot\theta\left(\frac{d\theta}{d\tau}\frac{d\phi}{d\tau} + \frac{d\phi}{d\tau}\frac{d\theta}{d\tau}\right) = 0. \tag{4.92}$$

[1] This is the standard choice made when formulating general relativity.

From exercise 9 we see that the equation of motion for the action (4.87) on a two-sphere takes the form

$$\frac{d^2\theta}{d\tau^2} + \Gamma^{\theta}{}_{\phi\phi}\frac{d\phi}{d\tau}\frac{d\phi}{d\tau} = 0 \tag{4.93}$$

$$\frac{d^2\phi}{d\tau^2} + \Gamma^{\phi}{}_{\theta\phi}\frac{d\theta}{d\tau}\frac{d\phi}{d\tau} + \Gamma^{\phi}{}_{\phi\theta}\frac{d\phi}{d\tau}\frac{d\theta}{d\tau} = 0 \tag{4.94}$$

or in general

$$\frac{d^2x^{\mu}}{d\tau^2} + \Gamma^{\mu}{}_{\alpha\beta}\frac{dx^{\alpha}}{d\tau}\frac{dx^{\beta}}{d\tau} = 0. \tag{4.95}$$

Equation (4.95) is called the *geodesic equation.*

4.12 Curvature

From previous sections we saw that vectors change as they are parallel transported along a surface, most noticeable if parallel transported along a closed path as depicted in figure 4.4. That change is measured by the connection $\Gamma^{\alpha}{}_{\mu\nu}$, which is computed in terms of the metric with equation (4.84). Now, if $g_{\mu\nu} = \eta_{\mu\nu}$, i.e. flat, then obviously from equation (4.84) all $\Gamma^{\alpha}{}_{\mu\nu} = 0$ identically. That is, vectors do not change in flat geometry. A better and more rigorous notion of this change is called curvature and in particular the failure of the covariant derivative to commute. That is, moving a vector $\vec{V}$ first along infinitesimal path $\nabla_{\vec{X}}$ and then along $\nabla_{\vec{Y}}$ should differ from moving along $\nabla_{\vec{Y}}$ first and then along $\nabla_{\vec{X}}$ only if the respective surface is

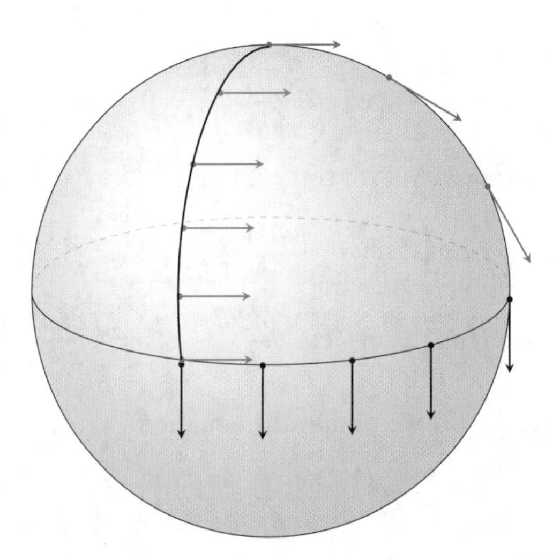

Figure 4.4. Parallel transporting of a vector on a two-sphere initially at $(0, \pi/2)$ in black, to $(\pi/2, \pi/2)$ then to the north pole (blue vector) and then back to its starting point (red vector). Upon returning we see it has rotated by an angle of $\pi/2$ due to the curved nature of the two-sphere.

curved, i.e. $\Gamma^{\alpha}{}_{\mu\nu} \neq 0$. This difference is measured by the commutator of the covariant derivative and written in components as

$$\left[\nabla_{\mu}, \nabla_{\nu}\right] V^{\rho} = R^{\rho}{}_{\sigma\mu\nu} V^{\sigma}, \tag{4.96}$$

where on the right side we have introduced a new tensor called the Riemann curvature tensor. Expanding the right side above and following the rules for adding Christoffel connections as detailed in previous sections, we obtain

$$\begin{aligned} \left[\nabla_{\mu}, \nabla_{\nu}\right] V^{\rho} = \partial_{\mu}(\partial_{\nu} V^{\rho} + \Gamma^{\rho}{}_{\nu\lambda} V^{\lambda}) - \Gamma^{\lambda}{}_{\mu\nu}(\partial_{\lambda} V^{\rho} + \Gamma^{\rho}{}_{\lambda\alpha} V^{\alpha}) \\ + \Gamma^{\rho}{}_{\mu\lambda}(\partial_{\nu} V^{\lambda} + \Gamma^{\lambda}{}_{\nu\alpha} V^{\alpha}) - \mu \leftrightarrow \nu. \end{aligned} \tag{4.97}$$

Next, assuming $\Gamma^{\rho}{}_{\mu\nu} = \Gamma^{\rho}{}_{\nu\mu}$, i.e. zero torsion, and after some lengthy but straightforward tensor analysis we obtain

$$R^{\rho}{}_{\sigma\mu\nu} = \partial_{\mu}\Gamma^{\rho}{}_{\nu\sigma} - \partial_{\nu}\Gamma^{\rho}{}_{\mu\sigma} + \Gamma^{\rho}{}_{\mu\lambda}\Gamma^{\lambda}{}_{\nu\sigma} - \Gamma^{\rho}{}_{\nu\lambda}\Gamma^{\lambda}{}_{\mu\sigma} \tag{4.98}$$

and satisfies the following:

$$\text{Symmetry relationships} \begin{cases} R_{\rho\sigma\mu\nu} = g_{\rho\alpha} R^{\alpha}{}_{\sigma\mu\nu} \\ R_{\rho\sigma\mu\nu} = - R_{\sigma\rho\mu\nu} \\ R_{\rho\sigma\mu\nu} = - R_{\rho\sigma\nu\mu} \\ R_{\rho\sigma\mu\nu} = R_{\mu\nu\rho\sigma} \end{cases} \tag{4.99}$$

$$\text{Bianchi identities} \begin{cases} \nabla_{[\alpha} R^{\rho}{}_{|\sigma|\mu\nu]} &= \nabla_{\alpha} R^{\rho}{}_{\sigma\mu\nu} + \nabla_{\mu} R^{\rho}{}_{\sigma\nu\alpha} + \nabla_{\nu} R^{\rho}{}_{\sigma\alpha\mu} = 0 \\ R^{\rho}{}_{[\sigma\mu\nu]} &= 0. \end{cases} \tag{4.100}$$

Given the full Riemann tensor we can define the contracted versions which are relevant to general relativity and lower dimensional geometries. These are defined respectively as

$$\text{Ricci curvature tensor} \left\{ R^{\rho}{}_{\mu\rho\nu} = R_{\mu\nu} \right. \tag{4.101}$$

$$\text{Ricci curvature scalar} \left\{ g^{\mu\nu} R_{\mu\nu} = R. \right. \tag{4.102}$$

Exercise 10. *Use the results from exercise 9 to show that*

$$R_{\mu\nu} = \begin{pmatrix} 1 & 0 \\ 0 & \sin^2\theta \end{pmatrix} \text{ and} \tag{4.103}$$

$$R = \frac{2}{r^2} \tag{4.104}$$

for the two-sphere.

Exercise 11. *In his development of the general theory of relativity, Einstein looked closely at equation* (4.100) *and noticed that if we contract* ρ, μ *and* α, σ *yielding*

$$g^{\alpha\sigma}\left(\nabla_\alpha R^\rho{}_{\sigma\rho\nu} + \nabla_\rho R^\rho{}_{\sigma\nu\alpha} + \nabla_\nu R^\rho{}_{\sigma\alpha\rho}\right) = 0, \tag{4.105}$$

we obtain the very interesting equation:

$$\nabla_\alpha\left[g^{\alpha\mu}\left(R_{\mu\nu} - \frac{1}{2}g_{\mu\nu}R\right)\right] = 0. \tag{4.106}$$

Verify this result.

The above exercise yields a new curvature tensor called the Einstein tensor:

$$G_{\mu\nu} = R_{\mu\nu} - \frac{1}{2}g_{\mu\nu}R. \tag{4.107}$$

This combination of Ricci curvatures is unique in that its divergence vanishes identically: $\nabla_\mu G^\mu{}_\nu = 0$. That is, as Einstein concluded, it must be proportional to a conserved current in accord with Nöther's theorem. The immediate rank two tensor known to Einstein to be conserved that came to mind was the energy momentum tensor $T_{\mu\nu}$. This led Einstein to initially guess the correct form of his gravitational field equation:

$$G_{\mu\nu} = \alpha T_{\mu\nu}, \tag{4.108}$$

where the proportionality constant was fixed by Einstein to be $\alpha = 8\pi G$ (in units where $c = 1$) after requiring that his equation reproduces Newtonian gravity in a weak field limitation.

References

[1] Nakahara M 2003 *Geometry, Topology and Physics* (London: Taylor and Francis)
[2] Nair V P 2005 *Quantum Field Theory: A Modern Perspective* (Berlin: Springer)
[3] Szekeres P 2004 *Modern Mathematical Physics* (Cambridge: Cambridge University Press)
[4] Munkres J 1999 *Topology* (Englewood Cliffs, NJ: Prentice-Hall)
[5] Janich K 1980 *Topology* (Berlin: Springer)
[6] Janich K 2001 *Vector Analysis* (Berlin: Springer)
[7] Bredon G E 1993 *Topology and Geometry* (Berlin: Springer)
[8] do Carmo M P 1992 *Riemannian Geometry* (Boston, MA: Birkhauser)
[9] Boothby W M 2003 *Differential Manifolds and Riemannian Geometry* (New York: Academic)
[10] Madsen I and Tornehave J 1997 *From Calculus to Cohomology* (Cambridge: Cambridge University Press)
[11] Frankel T 2004 *The Geometry of Physics* (Cambridge: Cambridge University Press)
[12] Strichartz R S 2000 *The Way of Analysis* (Burlington, MA: Jones and Bartlett)
[13] Poisson E 2004 *A Relativist's Toolkit* (Cambridge: Cambridge University Press)
[14] Vonk M 2005 *A Mini-course on Topological Strings* (arXiv:hep-th/0504147)

Chapter 5

Lagrangian field theory

In section 3.1 we had our first decent dose of Lagrangian field theory applied to the electromagnetic field in a traditional vector calculus formulation. The specific Lagrangian field theory was developed by following the 'How to Physics' (outline 1). That will be our starting point in this chapter, except that we need to elevate the outline to a covariant-tensorial formulation of respective physical fields. Of course, this chapter could easily span an entire book on its own and we refer to [1, 2] for further reading on the specific topic as needed. We will continue to work in units where $c = \hbar = 1$.

5.1 *E&M*

We begin by defining the single one-form four-vector potential

$$A = A_\mu dx^\mu, \tag{5.1}$$

where Greek indices will run from 0, 1, 2, 3 and index both time $\to 0$ and space $\to 1, 2, 3$. Additionally, Latin indexes i, j will only index spacial degrees of freedom. The components of the four-vector potential are given as

$$\begin{aligned} A_0 &= -\varphi \\ A_i &= (\vec{A})_i, \end{aligned} \tag{5.2}$$

i.e. the temporal component of A_μ is defined as the negative electromagnetic scalar potential and the spacial components are the components of the three dimensional vector potential. Next, taking the exterior derivative of A we obtain

$$dA = F_{\mu\nu} dx^\mu \wedge dx^\nu \tag{5.3}$$

$$\Rightarrow F_{\mu\nu} = \partial_\mu A_\nu - \partial_\nu A_\mu. \tag{5.4}$$

doi:10.1088/2053-2571/ab06bdch5

$F_{\mu\nu}$ geometrically represents the curvature two form of A_μ and is commonly referred to as the electromagnetic field strength tensor. Since $\mathcal{F} = F_{\mu\nu}dx^\mu \wedge dx^\nu$ is a curvature two form, it satisfies similar Bianchi identities as the Riemann curvature tensor given in equation (4.100). The Bianchi identities for $\mathcal{F}$ are easily obtained by virtue of the fact that $d^2 = 0$, i.e. the exterior derivative is nilpotent and thus

$$d\mathcal{F} = H_{\alpha\mu\nu}dx^\alpha \wedge dx^\mu \wedge dx^\nu = 0 \tag{5.5}$$

$$\Rightarrow H_{\alpha\mu\nu} = \partial_\alpha F_{\mu\nu} + \partial_\mu F_{\nu\alpha} + \partial_\nu F_{\alpha\mu} = 0. \tag{5.6}$$

Exercise 12. *Use the fact that* $F_{\mu\nu} = \partial_\mu A_\nu - \partial_\nu A_\mu$ *to show by direct computation that*

$$\partial_\alpha F_{\mu\nu} + \partial_\mu F_{\nu\alpha} + \partial_\nu F_{\alpha\mu} = 0.$$

Now, $F_{\mu\nu}$ has some interesting properties, which we can see by looking at specific components. Clearly $F_{\mu\nu}$ is antisymmetric:

$$F_{\mu\nu} = -F_{\nu\mu}$$

and thus has a zero diagonal. Next, let $\mu = 0$ and $\nu = i$ then equation (5.4) reads

$$\begin{aligned} F_{0i} = -F_{i0} &= \partial_0 A_i - \partial_i A_0 \\ &= \partial_0 A_i + \partial_i \varphi \\ &= -(-\nabla\varphi - \partial_t \vec{A})_i \\ &= -E_i, \end{aligned} \tag{5.7}$$

i.e. from equation (3.4) we see that the spacetime part of $F_{\mu\nu}$ are electric field components. Next, let us look at the space–space components:

$$\begin{aligned} F_{ij} &= \partial_i A_j - \partial_i A_j \\ &= \epsilon_{ij}{}^k \epsilon_k{}^{nm} \partial_n A_m \\ &= \epsilon_{ij}{}^k (\nabla \times \vec{A})_k \\ &= \epsilon_{ij}{}^k B_k, \end{aligned} \tag{5.8}$$

that is, the spacial components of $F_{\mu\nu}$ are occupied by magnetic field components and thus the electromagnetic field strength tensor takes the form

$$F_{\mu\nu} = \begin{pmatrix} 0 & -E_1 & -E_2 & -E_3 \\ E_1 & 0 & B_3 & -B_2 \\ E_2 & -B_3 & 0 & B_1 \\ E_3 & B_2 & -B_1 & 0 \end{pmatrix}. \tag{5.9}$$

Now, remembering equation (3.18) and since $F_{\mu\nu}$ only depends upon $\vec{E}$ and $\vec{B}$ it too must be gauge invariant. This is easily verified by rewriting the individual gauge transformations defined in equation (3.17) as one single one in terms of differential form notation:

$$\tilde{A} = A + d\Lambda \tag{5.10}$$

$$= \left(A_\mu + \partial_\mu \Lambda\right) dx^\mu, \tag{5.11}$$

which is ẹquivalent to equation (3.17), since $A_\mu + \partial_\mu \Lambda = (-(\varphi - \partial_0 \Lambda), A_i + \partial_i \Lambda)$. Next, we can compute $\tilde{\mathcal{F}}$:

$$\begin{aligned} \mathcal{F} &= dA \\ \Rightarrow \tilde{\mathcal{F}} &= d\tilde{A} \\ &= d(A + d\Lambda) \\ &= dA + d^2\Lambda \\ \tilde{\mathcal{F}} &= dA = \boxed{\mathcal{F}}, \end{aligned} \tag{5.12}$$

in other words, the gauge symmetry of the electromagnetic field strength tensor is a result again of the nilpotentcy of the exterior derivative *d*.

Next, we will construct the Lagrangian field theory of A_μ following outline 1. The two symmetries that we are concerned with are obviously gauge, but also general coordinate transformations, i.e. diffeomorphisms. Following from equation (4.66) we see that our action integral becomes

$$S = \int d^4x \sqrt{g}\, \mathcal{L}, \tag{5.13}$$

where $\mathcal{L}$ is the same as in equation (3.13), however we need to rewrite it in terms of our new defined tensors $F_{\mu\nu}$ and A_μ. This is easily accomplished by defining the four current $J^\mu = (\rho, J^i)$, where ρ is charge density and $J_i = (\vec{J})_i$ is the three-vector current. Then, remembering that $c = 1$:

$$\mathcal{L} = \frac{1}{2}\left(\vec{E}^2 - \vec{B}^2\right) - \varphi\rho + \vec{A} \cdot \vec{J} \tag{5.14}$$

$$= -\frac{1}{4} F_{\mu\nu} F^{\mu\nu} + J^\mu A_\mu. \tag{5.15}$$

Exercise 13. *Show that*

$$F_{\mu\nu} F^{\mu\nu} = 2\left(\vec{B}^2 - \vec{E}^2\right)$$

by direct computation. (Hint: use equations (5.7), (5.8) *and the fact that* $\epsilon_{ijk}\epsilon^{ijn} = 2\delta_k{}^n$*.)*

Thus our Maxwell action reads

$$S = -\frac{1}{4}\int d^4x\sqrt{g}\{F_{\mu\nu}F^{\mu\nu} - 4J^\mu A_\mu\}, \tag{5.16}$$

which we now vary with respect to A_μ to compute the electromagnetic field equations. To continue our analysis we vary everything in the Maxwell action that depends on A_μ:

$$\begin{aligned}\delta S &= -\frac{1}{4}\int d^4x\sqrt{g}\,\delta\{F_{\mu\nu}F^{\mu\nu} - 4J^\mu A_\mu\}\\ &= -\frac{1}{4}\int d^4x\sqrt{g}\left\{\delta\left(F_{\mu\nu}F^{\mu\nu}\right) - 4J^\mu\delta A_\mu\right\}\end{aligned} \tag{5.17}$$

and using the fact that $\delta(F_{\mu\nu}F^{\mu\nu}) = 2\delta F_{\mu\nu}F^{\mu\nu}$ we have

$$\begin{aligned}\delta S &= -\frac{1}{4}\int d^4x\sqrt{g}\{2\delta F_{\mu\nu}F^{\mu\nu} - 4J^\mu\delta A_\mu\}\\ &= -\frac{1}{4}\int d^4x\sqrt{g}\left\{2\delta\left(\nabla_\mu A_\nu - \nabla_\nu A_\mu\right)F^{\mu\nu} - 4J^\mu\delta A_\mu\right\}\\ &= -\frac{1}{4}\int d^4x\sqrt{g}\left\{2\left(F^{\mu\nu}\nabla_\mu\delta A_\nu - F^{\mu\nu}\nabla_\nu\delta A_\mu\right) - 4J^\mu\delta A_\mu\right\},\end{aligned} \tag{5.18}$$

which after using the fact that $F^{\mu\nu} = -F^{\nu\mu}$ and reindexing, reads

$$\delta S = -\frac{1}{4}\int d^4x\sqrt{g}\{4F^{\mu\nu}\nabla_\mu\delta A_\nu - 4J^\mu\delta A_\mu\}. \tag{5.19}$$

Exercise 14. *Show that* $F_{\mu\nu}$ *is concomitant for symmetric connection. That is, for*

$$\Gamma^\alpha{}_{\mu\nu} = \Gamma^\alpha{}_{\nu\mu},$$

we have that

$$F_{\mu\nu} = \nabla_\mu A_\nu - \nabla_\nu A_\mu = \partial_\mu A_\nu - \partial_\nu A_\mu.$$

Next, integrating by parts we have up to total derivatives (which we drop in accord with Hamilton's principle):

$$\delta S = \frac{1}{4}\int d^4x\sqrt{g}\{4\nabla_\mu F^{\mu\nu} + 4J^\nu\}\delta A_\nu = 0, \tag{5.20}$$

which yields the field equation:

$$\nabla_\mu F^{\mu\nu} = -J^\nu, \tag{5.21}$$

i.e. the Maxwell field equations in covariant form!

Exercise 15. *Show that the above field equation encodes the first and last of the Maxwell equation as written in vector notation in equation* (3.1). *Assume*

$$g_{\mu\nu} = \eta_{\mu\nu} = \begin{pmatrix} -1 & 0 & 0 & 0 \\ 0 & 1 & 0 & 0 \\ 0 & 0 & 1 & 0 \\ 0 & 0 & 0 & 1 \end{pmatrix},$$

where $\eta_{\mu\nu}$ *is the four-dimensional flat Minkowski metric and thus* $\nabla_\mu \to \partial_\mu$.

Finally, requiring the Maxwell action to be invariant with respect to local gauge transformations yields

$$\begin{aligned} S(A) = S(\tilde{A}) &= -\frac{1}{4}\int d^4x\sqrt{g}\left\{F_{\mu\nu}F^{\mu\nu} - 4J^\mu\tilde{A}_\mu\right\} \\ &= -\frac{1}{4}\int d^4x\sqrt{g}\left\{F_{\mu\nu}F^{\mu\nu} - 4J^\mu A_\mu - 4J^\mu\nabla_\mu\Lambda\right\} \\ &= S(A) - \int d^4x\sqrt{g}\nabla_\mu J^\mu\Lambda, \end{aligned} \tag{5.22}$$

up to total derivatives. In other words, the requirement that $S(A) = S(\tilde{A})$ yields the conserved current J^μ and covariant continuity equation:

$$\nabla_\mu J^\mu = 0. \tag{5.23}$$

5.2 Energy momentum tensor

Let us summarize Nöther's theorem for $E\&M$ in a covariant manner as obtained in the previous section. We have the field represented as the four-vector potential A_μ which generates continuous local gauge transformations $A_\mu \to A_\mu + \partial_\mu\Lambda$ leaving $F_{\mu\nu}$ and $S(A)$ invariant. The resulting generated conserved current (Nöther current) is J^μ:

$$\begin{matrix} \underbrace{A_\mu \to A_\mu + \partial_\mu\Lambda}_{\text{Gauge transformation}} & \underbrace{S(A) = S(\tilde{A})}_{\text{Gauge symmetry}} \\ \underbrace{\nabla_\mu J^\mu = 0}_{\text{Continuity equation}} & \underbrace{A_\mu}_{\text{Symmetry generator}} \end{matrix}. \tag{5.24}$$

At this point, it is common to introduce the functional generator of the Nöther current. That is, from the previous section we witnessed that the derivation of the covariant Maxwell equation $\nabla_\mu F^{\nu\mu} = J^\nu$ by variation with respect to A_μ, yielded the conserved quantity J^μ. In other words, the Nöther current J^μ is generated by the functional generator $\frac{\delta}{\delta A_\mu}$ and the two are synonymous:

$$J^\mu \equiv \frac{\delta}{\delta A_\mu}. \tag{5.25}$$

Now, from section 3.2 we learned that the gravitational field $g_{\mu\nu}$ is the generator of diffeomorphism symmetry and the resulting conserved quantities are energy, mass and momentum, which are summarized within one single Nöther current $T_{\mu\nu}$:

$$T_{\mu\nu} = \begin{pmatrix} \text{Energy/Mass density} & \text{Momentum flux density} & & \\ \text{Momentum} & & & \\ \text{Flux} & \text{Stress} & \& & \text{Strain} \\ \text{Density} & \text{Stress} & \& & \text{Strain} \end{pmatrix}. \tag{5.26}$$

and called the energy momentum tensor.

Example 10 (Perfect fluid).

$$T_{\mu\nu} = (\rho + P)U_\mu U_\nu + P\eta_{\mu\nu},$$

where P is pressure and U^μ is the four-velocity.

Example 11 ($E\&M$).

$$T_{\mu\nu} = \begin{pmatrix} \text{Energy density} & \text{Pointing vector} & & \\ \text{Pointing} & & & \\ & & \text{Pressure of} & \\ \text{Vector} & \vec{E} & \& & \vec{B} \end{pmatrix}.$$

Looking to the $E\&M$ analogy, we identify the functional generator of $T_{\mu\nu}$ with variations with respect to the inverse gravitational field, in other words:

$$T_{\mu\nu} \equiv \frac{-2}{\sqrt{-g}} \frac{\delta}{\delta g^{\mu\nu}}, \tag{5.27}$$

where the weight $\frac{-2}{\sqrt{-g}}$ is a tensor, convention restoring factor and $g \to -g$ is due to the fact that physical gravitational fields have negative (or Lorentzian signature) determinants. Now, due to the differential geometric formulation of Riemannian manifolds presented in previous sections and the fact that the electromagnetic field is defined in terms of lowered indices, the inverse gravitational field (metric) turns out to be the canonical field choice for gravity in the Lagrangian field theory of gravity.

Exercise 16.

(a) *Show that*

$$\delta\sqrt{-g} = -\frac{1}{2}\sqrt{-g}\, g_{\mu\nu}\delta g^{\mu\nu} \tag{5.28}$$

(Hint: first show that for a non-singular square matrix M, $\ln(\det M) = \operatorname{tr}(\ln M) \Rightarrow \frac{1}{\det M}\delta(\det M) = \operatorname{tr}(M^{-1}\delta M)$.*)*

(b) *Show that*

$$\delta g_{\mu\nu} = -g_{\mu\alpha}g_{\nu\beta}\delta g^{\alpha\beta}. \tag{5.29}$$

Example 12 (sourceless $E\&M$). *We will compute the energy momentum tensor for the electromagnetic field strength using the functional generator* $\frac{-2}{\sqrt{-g}}\frac{\delta}{\delta g^{\mu\nu}}$, *i.e. varying the sourceless Maxwell action:*

$$S = -\frac{1}{4}\int d^4x\sqrt{-g}\{F_{\mu\nu}F^{\mu\nu}\} \tag{5.30}$$

with respect to $g^{\mu\nu}$. *First we need to rewrite the action in terms of canonical definitions of each tensor:*

$$S = -\frac{1}{4}\int d^4x\sqrt{-g}\{g^{\alpha\mu}g^{\beta\nu}F_{\alpha\beta}F_{\mu\nu}\}, \tag{5.31}$$

which reveals the gravitational field dependence in each term. J^μ *and* A_μ *are both canonical in their definitions and thus there is no metric contribution in the last term of the Maxwell action. We now vary the above action in canonical form with respect to the inverse gravitational field:*

$$\begin{aligned}
\delta S &= -\frac{1}{4}\int d^4x\Big[\delta\sqrt{-g}\{g^{\alpha\mu}g^{\beta\nu}F_{\alpha\beta}F_{\mu\nu}\} + \sqrt{-g}\{\delta(g^{\alpha\mu}g^{\beta\nu})F_{\alpha\beta}F_{\mu\nu}\}\Big]\\
&= -\frac{1}{4}\int d^4x\Big[-\frac{1}{2}\sqrt{-g}\,g_{\lambda\rho}\delta g^{\lambda\rho}\{F_{\mu\nu}F^{\mu\nu}\}\\
&\quad ,+\sqrt{-g}\{\delta g^{\alpha\mu}g^{\beta\nu}F_{\alpha\beta}F_{\mu\nu} + g^{\alpha\mu}\delta g^{\beta\nu}F_{\alpha\beta}F_{\mu\nu}\}\Big]\\
&= -\frac{1}{4}\int d^4x\sqrt{-g}\Big[-\frac{1}{2}g_{\lambda\rho}\{F_{\mu\nu}F^{\mu\nu}\} + \{g^{\beta\nu}F_{\lambda\beta}F_{\rho\nu} + g^{\alpha\mu}F_{\alpha\lambda}F_{\mu\rho}\}\Big]\delta\\
&\quad g^{\lambda\rho}\\
&= -\frac{1}{4}\int d^4x\sqrt{-g}\Big[-\frac{1}{2}g_{\lambda\rho}\{F_{\mu\nu}F^{\mu\nu}\} + \{2g^{\beta\nu}F_{\lambda\beta}F_{\rho\nu}\}\Big]\delta g^{\lambda\rho}.
\end{aligned} \tag{5.32}$$

Thus from equation (5.27) *we have the Maxwell energy momentum tensor given by*

$$\boxed{T_{\lambda\rho} = -\frac{1}{4}g_{\lambda\rho}F_{\mu\nu}F^{\mu\nu} + g^{\beta\nu}F_{\lambda\beta}F_{\rho\nu}.} \tag{5.33}$$

Exercise 17. *The action for a scalar field in flat spacetime reads*

$$S = -\frac{1}{2}\int d^4x\left\{\partial_\mu\varphi\partial^\mu\varphi + m^2\varphi^2\right\}. \tag{5.34}$$

(a) *Rewrite equation* (5.34) *in general covariant form.*
(b) *Compute the field equation by varying equation* (5.34) *with respect to* φ.
(c) *Compute the energy momentum tenors:* $T_{\mu\nu} = -\frac{2}{\sqrt{-g}}\frac{\delta S}{\delta g^{\mu\nu}}$.

5.3 General relativity

Now we combine all we have learned so far into building a gravity theory *à la* Einstein and following outline 1.

- Symmetry:
  - As aforementioned and themed throughout the previous chapter, symmetry of general relativity is *diffeomorphism symmetry* also known as general covariance and coordinate invariance.
- From *E&M* we learned that the field equations are second order in the field, i.e. the Maxwell equations are second order in A_μ. We should expect the same for gravity in that the field equations should be second order in $g_{\mu\nu}$.
- The Lagrangian density should be second order in the metric and diffeomorphism invariant, a first choice is the Ricci scalar curvature:

$$\mathcal{L} = \frac{1}{2\kappa}R, \tag{5.35}$$

  where $\kappa = 8\pi G$ is the gravitational coupling and is fixed by requiring general relativity to reproduce Newtonian gravity in the weak field limit. G is Newton's constant, of course.

The above summarized yields the Einstein–Hilbert action for gravity:

$$\boxed{S_{\text{EH}} = \frac{1}{2\kappa}\int d^4x\sqrt{-g}\,R.} \tag{5.36}$$

Example 13 (Vacuum Einstein field equation). *Let us vary the Einstein–Hilbert action with respect to the canonical inverse gravitational field to determine the vacuum (gravity only) Einstein field equation. To begin, we will rewrite equation* (5.36) *in canonical form:*

$$\begin{aligned} S_{\text{EH}} &= \frac{1}{2\kappa}\int d^4x\sqrt{-g}\,R \\ &= \frac{1}{2\kappa}\int d^4x\sqrt{-g}\,g^{\mu\nu}R_{\mu\nu}. \end{aligned} \tag{5.37}$$

Next, we vary with respect to $g^{\mu\nu}$:

$$\begin{aligned}
\delta S_{\text{EH}} &= \frac{1}{2\kappa}\int d^4x\delta\left(\sqrt{-g}\,g^{\mu\nu}R_{\mu\nu}\right) \\
&= \frac{1}{2\kappa}\int d^4x\left(\delta\sqrt{-g}\,g^{\mu\nu}R_{\mu\nu} + \sqrt{-g}\,\delta g^{\mu\nu}R_{\mu\nu} + \sqrt{-g}\,g^{\mu\nu}\delta R_{\mu\nu}\right) \\
&= \frac{1}{2\kappa}\int d^4x\bigg(-\frac{1}{2}\sqrt{-g}\,g_{\mu\nu}\delta g^{\mu\nu}R + \sqrt{-g}\,\delta g^{\mu\nu}R_{\mu\nu} \\
&\qquad + \sqrt{-g}\,g^{\mu\nu}\delta R_{\mu\nu}\bigg) \\
&= \frac{1}{2\kappa}\int d^4x\left(\sqrt{-g}\left\{R_{\mu\nu} - \frac{1}{2}g_{\mu\nu}R\right\}\delta g^{\mu\nu} + \sqrt{-g}\,g^{\mu\nu}\delta R_{\mu\nu}\right) = 0.
\end{aligned} \tag{5.38}$$

The last term above turns out to be proportional to the set of covariant derivatives (see [3] section 4.1 for the lengthy details):

$$\delta R_{\mu\nu} = \nabla_\lambda \delta\Gamma^\lambda{}_{\mu\nu} - \nabla_\nu \delta\Gamma^\lambda{}_{\mu\lambda} \tag{5.39}$$

and thus

$$\begin{aligned}
\delta S_{\text{EH}} = \frac{1}{2\kappa}\int d^4x\bigg(&\sqrt{-g}\left\{R_{\mu\nu} - \frac{1}{2}g_{\mu\nu}R\right\}\delta g^{\mu\nu} \\
&+ \sqrt{-g}\,g^{\mu\nu}\left\{\nabla_\lambda\delta\Gamma^\lambda{}_{\mu\nu} - \nabla_\nu\delta\Gamma^\lambda{}_{\mu\lambda}\right\}\bigg) = 0,
\end{aligned} \tag{5.40}$$

which, up to total derivatives, is equivalent to

$$\begin{aligned}
\delta S_{\text{EH}} = \frac{1}{2\kappa}\int d^4x\bigg(&\sqrt{-g}\left\{R_{\mu\nu} - \frac{1}{2}g_{\mu\nu}R\right\}\delta g^{\mu\nu} \\
&+ \nabla_\lambda\left(\sqrt{-g}\,g^{\mu\nu}\right)\delta\Gamma^\lambda{}_{\mu\nu} - \nabla_\nu\left(\sqrt{-g}\,g^{\mu\nu}\right)\delta\Gamma^\lambda{}_{\mu\lambda}\bigg) = 0.
\end{aligned} \tag{5.41}$$

To simplify the last line above, we need to remember the metric compatibility condition (4.83) *and thus we are left with*

$$\delta S_{\text{EH}} = \frac{1}{2\kappa}\int d^4x\sqrt{-g}\left\{R_{\mu\nu} - \frac{1}{2}g_{\mu\nu}R\right\}\delta g^{\mu\nu} = 0, \tag{5.42}$$

and yielding the vacuum Einstein field equation of gravity:

$$R_{\mu\nu} - \frac{1}{2}g_{\mu\nu}R = 0. \tag{5.43}$$

Exercise 18. *Use the metric compatibility condition* (4.83) *to show that* $\nabla_\alpha\sqrt{-g} = 0$.

If we add matter in the form of some $\mathcal{L}_M$ to the Einstein–Hilbert action (5.36), i.e.

$$S = S_{\rm EH} + S_M = \frac{1}{2\kappa}\int d^4x\sqrt{-g}\,R + \int d^4x\sqrt{-g}\,\mathcal{L}_M. \tag{5.44}$$

and perform the functional variation with respect to the inverse metric again we obtain

$$\begin{aligned}
\delta S &= \frac{1}{2\kappa}\int d^4x\sqrt{-g}\left\{R_{\mu\nu} - \frac{1}{2}g_{\mu\nu}R\right\}\delta g^{\mu\nu} + \int d^4x\delta\left(\sqrt{-g}\,\mathcal{L}_M\right)\\
&= \frac{1}{2\kappa}\int d^4x\sqrt{-g}\left\{R_{\mu\nu} - \frac{1}{2}g_{\mu\nu}R\right\}\delta g^{\mu\nu} + \int d^4x\frac{\delta}{\delta g^{\mu\nu}}\left(\sqrt{-g}\,\mathcal{L}_M\right)\delta g^{\mu\nu}\\
&= \frac{1}{2\kappa}\int d^4x\sqrt{-g}\left\{R_{\mu\nu} - \frac{1}{2}g_{\mu\nu}R\right\}\delta g^{\mu\nu} +\\
&\quad -\frac{1}{2}\int d^4x\sqrt{-g}\left\{\frac{-2}{\sqrt{-g}}\frac{\delta}{\delta g^{\mu\nu}}\left(\sqrt{-g}\,\mathcal{L}_M\right)\right\}\delta g^{\mu\nu} = 0,
\end{aligned} \tag{5.45}$$

which after implementing equation (5.27) becomes

$$\begin{aligned}
\delta S &= \frac{1}{2\kappa}\int d^4x\sqrt{-g}\left\{R_{\mu\nu} - \frac{1}{2}g_{\mu\nu}R\right\}\delta g^{\mu\nu} - \frac{1}{2}\int d^4x\sqrt{-g}\left\{T_{\mu\nu}\right\}\delta g^{\mu\nu}\\
&= \frac{1}{2}\int d^4x\sqrt{-g}\left\{\frac{1}{\kappa}\left(R_{\mu\nu} - \frac{1}{2}g_{\mu\nu}R\right) - T_{\mu\nu}\right\}\delta g^{\mu\nu} = 0.
\end{aligned} \tag{5.46}$$

Thus, yielding the full Einstein field equation of gravity:

$$\begin{gathered}
R_{\mu\nu} - \frac{1}{2}g_{\mu\nu}R = \kappa T_{\mu\nu}\\
\boxed{R_{\mu\nu} - \frac{1}{2}g_{\mu\nu}R = 8\pi G T_{\mu\nu},}
\end{gathered} \tag{5.47}$$

again in units where $c = 1$. Looking back to Einstein's initial guess at his field equation (4.108), we see that $\alpha = \kappa$.

Remark 2. *The gravitational action* (5.36) *carries Hilbert's name (Nöther's research colleague) because it was Hilbert who first implemented the functional variation in conjunction with Hamilton's principle to arrive at the same field equation as Einstein's initial guess [4]. However, as Hilbert admitted, it was Einstein who computed the perihelion procession of Mercury, the bending of light around the Sun and even the first gravitational wave solution 100 years before their discovery [5], which rightly bestowed most of gravity's credit to Einstein. After seeing Hilbert's derivation of his field equation, Einstein was so captivated with its elegance and the implementation of* *covariant Lagrangian field theory that he adopted Hilbert's derivation in his subsequent lectures on his theory.*

References

[1] Swanson M S 2015 *Classical Field Theory and the Stress-Energy Tensor* (San Rafael, CA: Morgan and Claypool) https://iopscience.iop.org/book/978-1-6817-4121-5

[2] Pons J M 2011 Noether symmetries, energy-momentum tensors, and conformal invariance in classical field theory *J. Math. Phys.* **52** 012904

[3] Poisson E 2004 *A Relativist's Toolkit* (Cambridge: Cambridge University Press)

[4] Isaacson W 2008 *Einstein: His Life and Universe* reprint edn (New York: Simon & Schuster)

[5] LIGO Scientific, Virgo Collaboration and Abbott B P *et al* 2016 Observation of gravitational waves from a binary black hole merger *Phys. Rev. Lett.* **116** 061102

Chapter 6

Black hole thermodynamics and holographic scaling

6.1 Black holes

In the creation of the most widely accepted and successful theory of gravity, general relativity, Einstein paid close attention to encode Newton's gravity as a limiting theory at low energy scales. To see this embedding consider the Schwarzschild line element:

$$ds^2 = g_{\mu\nu}dx^\mu dx^\nu = -f(r)dt^2 + \frac{dr^2}{f(r)} + r^2 d\Omega^2, \tag{6.1}$$

where $f(r) = 1 - \frac{2GM}{r}$, $d\Omega^2 = d\theta^2 + \sin^2\theta d\varphi$ is the unit sphere line element. Equation (6.1) is a vacuum solution to the Einstein field equation

$$R_{\mu\nu} = 8\pi G\left(T_{\mu\nu} - \frac{1}{2}g_{\mu\nu}T\right), \tag{6.2}$$

where $T = g^{\mu\nu}T_{\mu\nu}$ is the trace of $T_{\mu\nu}$. In other words, the Ricci curvature tensor is flat ($R_{\mu\nu} = 0$) everywhere except at the origin, where the Schwarzschild solution exhibits a curvature singularity.

Exercise 19. *Show that equation* (5.47) *may be rewritten into the form presented above in equation* (6.2).

Physically this implies a point mass M at rest located at the origin and zero matter ($T_{\mu\nu} = 0$) elsewhere. Now, for a point mass M we know that $\Phi = -\frac{GM}{r}$ and thus we see that Newton is contained in Einstein's theory via

$$f(r) = 1 + 2\Phi \tag{6.3}$$

doi:10.1088/2053-2571/ab06bdch6

or, more generally, for spherically symmetric and asymptotically flat spacetimes:

$$\Phi = \frac{1}{2}(g_{00} - 1). \tag{6.4}$$

This identification ensures that the Einstein field equations of general relativity reduce to the Poisson equation of Newton's gravity in the point particle case to low energy, i.e. restricting our class of diffeomorphisms to the time independent Galilei group and implementing a Newton–Cartan formalism [1]. The boundary conditions implemented to solve the Einstein field equations yielding the solution (6.1) are encoded into Newton's gravity for the point particle case. To see this let us evaluate the Laplace equation for the point particle of mass M:

$$\nabla^2\Phi = -GM\nabla^2\left(\frac{1}{r}\right) = 4\pi GM\delta^3(r), \tag{6.5}$$

where we have implemented the fact that $1/r$ is proportional to the Green's function of ∇^2. This implies a mass density $\rho = M\delta^3(r)$, which is precisely the condition set forth on $T_{\mu\nu}$ for the Schwarzschild solution, in that $T_{ij} = T_{0j} = 0$ and $T_{00} \sim \rho$ at the origin and zero everywhere else. This implies an equivalence between the Poisson equation for Newton's gravity and the zero-zero component of the Einstein field equation. The Schwarzschild solution is said to describe a black hole due to several properties, but most significantly due to the fact that it exhibits coordinate and curvature singularities. The notion of coordinate singularity may be summarized as the concept of black hole (Killing) horizon, which we will discuss in detail now.

Exercise 20. *Show that the Schwarzschild solution* (6.1) *exhibits a coordinate singularity at $r = 0$ by computing the Kretschmann invariant with value*

$$R_{\alpha\mu\beta\nu}R^{\alpha\mu\beta\nu} = \frac{48G^2M^2}{r^6}.$$

In our black hole discussion, we will follow [2] and consider a black hole naively as any region in spacetime, i.e. a pair $\left(\mathcal{M},\, g_{\mu\nu}\right)$ consisting of a Riemannian manifold and metric, that cannot be mapped to conformal infinity. This includes for the most part a curvature singularity hiding behind a Killing horizon or coordinate singularity. There are more rigorous black hole definitions summarized into the isolated horizon formalism [3], however our more simplistic and more approachable definition will suffice for our purposes. As an illustration of the our above definition let us examine the Schwarzschild metric in Kruskal coordinates [4]:

$$ds^2 = -(1 - 2GM/r)dudv + r^2d\Omega^2, \tag{6.6}$$

where we employed the transformations $u = t - r^*$, $v = t + r^*$ and $r^* = \int dr \frac{1}{1 - \frac{2GM}{r}}$. Due to the spherical symmetry of the Schwarzschild solution, we can drop the

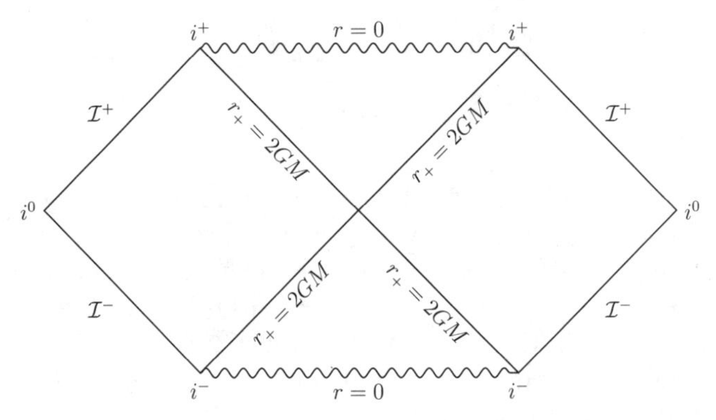

Figure 6.1. Conformal map of the Schwarzschild metric. The surface $r = 2GM$ is located at $U = V = 0$, the singularity at $r = 0$ is the line $U + V = \pm\frac{\pi}{2}$, and the points $(U, V) = (\pm\frac{\pi}{2}, \pm\frac{\pi}{2})$ are the limits of $r \to \pm\infty$.

two-sphere gravitational field contribution and focus on the r-t components. We can conformally map (6.6) to Minkowski spacetime via the transformations

$$U = \tan\frac{u}{\sqrt{2GM}} \quad \text{and} \quad V = \tan\frac{v}{\sqrt{2GM}} \tag{6.7}$$

$$T = V + U \quad \text{and} \quad R = V - U \tag{6.8}$$

modulo a conformal factor $2\cos U \cos V$. The transformations (6.7) rescale $\pm\infty$ to $\pm\frac{\pi}{2}$ and thus the entire Schwarzschild spacetime can be contained on a finite region of the U–V plane as depicted in the conformal diagram of figure 6.1 with labels defined in table 6.1. In figure 6.1 the region bounded by the lines $r = 2GM$ and $r = 0$ is the black hole and time-like curves emanating from here terminate at the horizon and never reach $\mathcal{I}^+$, thus a black hole is said to be present.

6.2 Killing vectors and horizons

Given a Riemannian spacetime metric $g_{\mu\nu}$ a vector ξ is called Killing if

$$\mathcal{L}_\xi g_{\mu\nu} = 0, \tag{6.9}$$

where $\mathcal{L}_\xi$ denotes the Lie derivative along $\vec{\xi}$. Writing equation (6.9) out in components yields the Killing equation:

$$\begin{aligned} \xi^\alpha \nabla_\alpha g_{\mu\nu} - g_{\alpha\nu}\nabla_\mu \xi^\alpha - g_{\mu\alpha}\nabla_\nu \xi^\alpha &= 0 \\ \nabla_\mu \xi_\nu + \nabla_\nu \xi_\mu &= 0 \end{aligned}. \tag{6.10}$$

In most spacetimes considered there are two such vectors, a time-like and an angular-like, $\xi^\alpha_{(t)} \equiv t^\alpha = \frac{\partial x^\alpha}{\partial t}$ and $\xi^\alpha_{(\phi)} \equiv \Omega_H \phi^\alpha = \frac{\partial x^\alpha}{\partial \phi}$. If the spacetime in question is spherically symmetric, then $\vec{\xi}_{(\phi)} = 0$ and if the spacetime is endowed with angular momentum then Ω_H is the horizon angular frequency. Combining these two gives a general Killing vector:

Table 6.1. Labels for figure 6.1.

Label	Name	Definition
$\mathcal{I}^+$	Future null ∞	$v = \infty$, u = finite
$\mathcal{I}^-$	Past null ∞	$u = -\infty$, v = finite
i^o	Spatial ∞	$r = \infty$, t = finite
i^+	Future time-like ∞	$t = \infty$, r = finite
i^-	Past time-like ∞	$t = -\infty$, r = finite

$$\xi^\alpha = a\xi^\alpha_{(t)} + b\xi^\alpha_{(\phi)} \tag{6.11}$$

$$= t^\alpha + \Omega_H \phi^\alpha, \tag{6.12}$$

where $\Omega_H = -\frac{g_{t\phi}}{g_{\phi\phi}}|_{r^+}$. A Killing vector ξ also satisfies a non-homogeneous wave equation, which is obtained by its action on the Riemann tensor:

$$-R^{\mu}{}_{\nu\alpha\beta}\xi^\nu = (\nabla_\beta \nabla_\alpha - \nabla_\alpha \nabla_\beta)\xi^\mu. \tag{6.13}$$

This can be rearranged via the Jacobi identity to yield

$$\nabla_\mu \nabla_\alpha \xi_\beta = -R_{\mu\nu\alpha\beta}\xi^\nu \tag{6.14}$$

and contracting over α and μ gives

$$\Box \xi_\beta = -R_{\beta\nu}\xi^\nu. \tag{6.15}$$

There are so-called isometries (conserved quantities) that are manifested in the inner product between the four velocity and the Killing vectors, i.e.

$$u_\alpha \xi^\alpha = u_\alpha \xi^\alpha_{(t)} + u_\alpha \xi^\alpha_{(\phi)} \tag{6.16}$$

$$= \tilde{E} + \tilde{L}, \tag{6.17}$$

where $\tilde{E}$ and $\tilde{L}$ are the conserved quantities energy and angular momentum. This is an equivalent statement as Nöther's theorem. Equation (6.9) also implies that ξ satisfies a geodesic and geodetic equation, i.e.

$$\begin{cases} (\nabla_\beta \xi^\alpha)\xi^\beta = 0 & r > r^+ \\ (\nabla_\beta \xi^\alpha)\xi^\beta = \kappa \xi^\alpha & r = r^+ \end{cases}. \tag{6.18}$$

Next, we will define the notion of a Killing horizon and its relevance in black hole physics. Let $\vec{\xi}$ be a Killing vector, then a Killing horizon is the surface defined by

$$\xi_\alpha \xi^\alpha = 0, \tag{6.19}$$

i.e. the surface on which the Killing vectors become null generators. A trivial example is the Killing vector $\vec{\xi} = x\partial_t + t\partial_x$ in Minkowski space. In this case, $\vec{\xi}$ is null on the surface $x = \pm t$, thus defining a Killing horizon. Another example is the event horizon of a black hole, which we will denote by $\xi_\mu \xi^\mu|_{r^+} = 0$. It is important to note that every event horizon defines a Killing horizon, but not every Killing horizon defines an event horizon. This is easily verified by the example in Minkowski space, which exhibits no coordinate or curvature singularities. To every Killing horizon we can associate a quantity κ called the surface gravity. It is the force required, by an observer at infinity, to hold a particle (of unit mass) stationary at the horizon. Yet, before calculating κ explicitly we need to first make some general statements about the geometry of the horizon. Combining equations (6.18), and (6.19) we conclude that the Killing vectors define a congruence of null geodesics at the horizon, which is necessarily hypersurface orthogonal. Thus we may employ Frobenius' theorem [4], which states that for null generators ξ^α on a horizon, the congruence is hypersurface orthogonal if and only if

$$(\nabla_{[\beta}\xi_\alpha)\xi_{\gamma]} = 0. \tag{6.20}$$

Expanding equation (6.20) and acting on it with $\nabla^\beta \xi^\alpha$ and making use of equations (6.10) and (6.18) yields

$$\begin{aligned}(\nabla^\beta \xi^\alpha)(\nabla_\beta \xi_\alpha)\xi_\gamma &= -2\kappa^2 \xi_\gamma \\ \Longrightarrow \kappa^2 &= -\frac{1}{2}(\nabla^\beta \xi^\alpha)(\nabla_\beta \xi_\alpha),\end{aligned} \tag{6.21}$$

which allows us to determine the surface gravity of any given spacetime ds^2 by extracting the Killing vectors thereof.

6.3 Surface element for null generators

Let Σ be a null surface then we may write the directed surface element as

$$d\Sigma_\mu = -\xi_\mu \sqrt{h}\, d^3y, \tag{6.22}$$

where h_{ab} is the metric of the hypersurface (horizon) and $y^a = (\lambda, \theta^1, \theta^2)$ are its coordinates and ξ_μ acts as the normal vector. We are implying an embedding of the hypersurface such that for some auxiliary null vector N^α[1]. The metric is given by

$$g^{\alpha\beta} = -2\xi^{[\alpha}N^{\beta]} + h^{ab}e_a^\alpha e_b^\beta, \quad \text{completeness relation} \tag{6.23}$$

where

$$e_a^\alpha = \frac{\partial x^\alpha}{\partial y^a} \tag{6.24}$$

are the pull backs from the original spacetime to the embedding. Given equation (6.24) we can rewrite equation (6.22) as

$$d\Sigma_\mu = \xi^\nu dS_{\mu\nu} d\lambda, \tag{6.25}$$

[1] $N^2 = 0$, $N_\alpha \xi^\alpha = -1$.

where

$$dS_{\mu\nu} = 2\xi_{[\mu}N_{\nu]}\sqrt{h}\,d^2\theta. \tag{6.26}$$

We can draw a direct relation between equations (6.22) and (6.26) via Stokes' theorem, which states that for any $p - 1$ form ω:

$$\int_{\partial\mathcal{M}} \omega = \int_{\mathcal{M}} d\omega. \tag{6.27}$$

A direct consequence of equation (6.27) for any antisymmetric contravariant two tensor is that

$$\int_{\Sigma} (\nabla_\beta B^{\alpha\beta}) d\Sigma_\alpha = \frac{1}{2} \oint_{\partial\Sigma} B^{\alpha\beta} dS_{\alpha\beta}. \tag{6.28}$$

Equation (6.28) has the form of a Gauss theorem and it will come in handy when calculating mass and angular momentum transfer across a horizon.

6.4 The laws of black hole mechanics

The laws of black hole mechanics were formulated by Bardeen, Carter and Hawking during the period from 1971–73 [5], though Israel presented the first rigorous proof of the third law in 1986 [6]. In this section we work in units of $G = c = \hbar = 1$ and the laws are as follows.

Law 1 (Zeroth law of black hole mechanics). *The zeroth law states that the surface gravity of a stationary black hole is constant cross the event horizon, i.e. we need to show that*

$$\boxed{\partial_\mu \kappa = 0 \text{ and } (\partial_\mu \kappa) e_a^\mu = 0,} \tag{6.29}$$

where $e^\mu{}_a$ *is the pull back from the four-dimensional spacetime to the horizon. This may be established by differentiating equation* (6.21) *and employing equation* (6.14) *to yield*

$$\xi^\mu \partial_\mu \kappa = 0, \tag{6.30}$$

which shows the first equation in (6.29). *The second equation can be determined by assuming geodesic completeness on the horizon and thus, since* $(\partial_\mu \kappa) e^\mu{}_a = 0$ *in the bifurcation two-sphere*[2] *of the spacetime, it holds across the entire horizon.*

Law 2 (First law of black hole mechanics). *Our starting point will be the Komar formulae [4]:*

[2] $(u, v) = (0, 0)$ in the conformal diagram of figure 6.1.

$$M_{\rm tot} = \frac{-1}{8\pi} \oint_S (\nabla^\alpha \xi^\beta_{(t)}) dS_{\alpha\beta}$$
$$J_{\rm tot} = \frac{1}{16\pi} \oint_S (\nabla^\alpha \xi^\beta_{(\phi)}) dS_{\alpha\beta} \tag{6.31}$$

from which we will derive the Smarr formula and expressions for the mass and angular momentum transfer across the horizon. Taking the Komar formulae and applying them across the horizon of a black hole we get

$$M_{BH} = \frac{-1}{8\pi} \oint_{\mathcal{H}} (\nabla^\alpha t^\beta) dS_{\alpha\beta}$$
$$J_{BH} = \frac{1}{16\pi} \oint_{\mathcal{H}} (\nabla^\alpha \varphi^\beta) dS_{\alpha\beta}, \tag{6.32}$$

where BH stands for black hole and $\mathcal{H}$ denotes its horizon. Next, consider

$$\begin{aligned} M_{BH} - 2\Omega_H J_{BH} &= \frac{-1}{8\pi} \oint_{\mathcal{H}} (\nabla^\alpha t^\beta) dS_{\alpha\beta} - \frac{1}{8\pi} \oint_{\mathcal{H}} (\nabla^\alpha \Omega_H) \varphi^\beta dS_{\alpha\beta} \\ &= \frac{-1}{8\pi} \oint_{\mathcal{H}} (\nabla^\alpha \xi^\beta) dS_{\alpha\beta} \\ &= \frac{-1}{4\pi} \oint_{\mathcal{H}} (\nabla^\alpha \xi^\beta) \xi_\alpha N_\beta dS \\ &= \frac{-1}{4\pi} \oint_{\mathcal{H}} \kappa \xi^\beta N_\beta dS \\ &= \frac{\kappa}{4\pi} A_{BH}, \end{aligned}$$

thus arriving at the general Smarr formula:

$$M_{BH} = \frac{\kappa}{4\pi} A_{BH} + 2\Omega_H J_{BH}. \tag{6.33}$$

Next, considering the integral $\oint_S (\nabla^\alpha \xi^\beta) dS_{\alpha\beta}$ and using equation (6.28) we have

$$\begin{aligned} \oint_S (\nabla^\alpha \xi^\beta) dS_{\alpha\beta} &= 2 \int_\Sigma \nabla_\alpha \nabla^\alpha \xi^\beta d\Sigma_\beta \\ &= 2 \int_\Sigma R^\alpha{}_\beta \xi^\beta d\Sigma_\alpha \end{aligned} \tag{6.34}$$

and invoking the Einstein field equation we have

$$\oint_S \nabla^\alpha \xi^\beta dS_{\alpha\beta} = -16\pi \int_\Sigma \left(T_{\alpha\beta} - \frac{1}{2} g_{\alpha\beta} T \right) n^\alpha \xi^\beta \sqrt{h}\, d^3y, \tag{6.35}$$

where n^α is the outward normal to the surface. Now we can now rewrite the Komar formulae as

$$M = 2 \int_\Sigma \left(T_{\alpha\beta} - \frac{1}{2} g_{\alpha\beta} T \right) n^\alpha \xi^\beta_{(t)} \sqrt{h}\, d^3y \tag{6.36}$$

and

$$J = -\int_{\Sigma}\left(T_{\alpha\beta} - \frac{1}{2}g_{\alpha\beta}T\right)n^{\alpha}\xi^{\beta}_{(\varphi)}\sqrt{h}\,d^{3}y. \tag{6.37}$$

Next, we introduce the vector currents:

$$\epsilon^{\alpha} = -T^{\alpha}{}_{\beta}\xi^{\beta}_{(t)} \tag{6.38}$$

and

$$l^{\alpha} = T^{\alpha}{}_{\beta}\xi^{\beta}_{(\varphi)}, \tag{6.39}$$

which interpret to the energy and angular momentum flux. In fact for a perfect fluid $T^{\alpha\beta} = \rho u^{\alpha}u^{\beta}$ *they take the form* $\epsilon^{\alpha} = \tilde{E}j^{\alpha}$ *and* $l^{\alpha} = \tilde{L}j^{\alpha}$*, where* j^{α} *is the current density* ρu^{α} *and* $\tilde{E}$ *and* $\tilde{L}$ *are the conserved quantities* (6.16)*. From energy conservation we know that* j^{α} *is divergenceless and thus we conclude that*

$$\oint_{\partial V}\epsilon^{\alpha}d\Sigma_{\alpha} = 0 \text{ and } \oint_{\partial V}l^{\alpha}d\Sigma_{\alpha} = 0 \tag{6.40}$$

by Stokes' theorem. Equation (6.40) *is the statement that the total energy transfer across a closed surface* ∂V *is conserved. If we take a partition* H *of this closed surface we have that the energy transfer through it is*

$$\delta M = -\int_{\mathcal{H}} T^{\alpha}{}_{\beta}\xi^{\beta}_{(t)}d\Sigma_{\alpha} \text{ and } \delta J = \int_{\mathcal{H}} T^{\alpha}{}_{\beta}\xi^{\beta}_{(\varphi)}d\Sigma_{\alpha}. \tag{6.41}$$

Using equation (6.41) *we can now derive the first law of black hole mechanics as follows. Given the linear combination* $\delta M - \Omega_H\delta J$*, we have*

$$\begin{aligned}\delta M - \Omega_H\delta J &= -\int_{\mathcal{H}} T^{\alpha}{}_{\beta}(t^{\beta} + \Omega_H\varphi^{\beta})d\Sigma_{\alpha}\\ &= \int_{\mathcal{H}} T_{\alpha\beta}\xi^{\alpha}\xi^{\beta}dSd\lambda.\end{aligned} \tag{6.42}$$

Next, we need to relate the integrand to the geometric evolution of the horizon to complete the integration over λ*. This may be done via Raychaudhuri's equation* $\frac{d\theta}{d\lambda} = \kappa\theta - 8\pi T_{\alpha\beta}\xi^{\alpha}\xi^{\beta}$ *[4], which is an evolution equation for the expansion parameter* θ *of a two-dimensional medium, i.e.* θ *is the fractional rate of change of the congruence's cross-sectional area* ($\theta = \frac{1}{dS}\frac{ddS}{d\lambda}$ *and* $\theta|_{\lambda=\pm\infty} = 0$)*. Thus, substituting for* $T_{\alpha\beta}\xi^{\alpha}\xi^{\beta}$ *yields*

$$
\begin{aligned}
\delta M - \Omega_H \delta J &= -\frac{1}{8\pi} \int d\lambda \oint_{\mathcal{H}} \left(\frac{d\theta}{d\lambda} - \kappa\theta \right) dS \\
&= \frac{\kappa}{8\pi} \int \oint_{\mathcal{H}} \theta dS d\lambda \\
&= \frac{\kappa}{8\pi} \int \oint_{\mathcal{H}} \frac{1}{dS} \frac{ddS}{d\lambda} dS d\lambda \\
&= \frac{\kappa}{8\pi} \oint_{\mathcal{H}} dS|_{\pm\infty} \\
&= \frac{\kappa}{8\pi} \delta A
\end{aligned}
\tag{6.43}
$$

and thus arriving at the first law of black hole mechanics:

$$
\boxed{\delta M = \frac{\kappa}{8\pi} \delta A + \Omega_H \delta J.}
\tag{6.44}
$$

Law 3 (Second law of black hole mechanics). *The second law was established by Hawking in 1971 [7], which states that the area of a black hole can never decrease:*

$$
\boxed{\delta A \geqslant 0.}
\tag{6.45}
$$

This follows directly from the focusing theorem [4, 7] and from the observation that the null generators' geodesics have no future endpoints in the given spacetime. The focusing theorem states that assuming the strong energy condition $R_{ab}k^a k^b \geqslant 0$, for null vector k^a, an initially negative expansion θ implies that the generators will converge in a caustic at $\theta = -\infty$. This is a contradiction to the initial observation and we conclude that $\theta \geqslant 0$.

Law 4 (Third law of black hole mechanics). *The third law follows from the weak energy condition*

$$
T_{\mu\nu} u^\mu u^\nu > 0
\tag{6.46}
$$

for an observer moving with four velocity u^α in a bounded energy momentum tensor $T_{\mu\nu}$. It states that the surface gravity of a black hole cannot be reduced to zero in a finite advanced time $v = t + r^*$. The third law may be illustrated by help of the general Vaidya spacetime [8] given by the line element:

$$
ds^2 = -f dv^2 + 2dvdr + r^2 d\Omega^2,
\tag{6.47}
$$

where $f = 1 - \frac{2m(v)}{r} + \frac{q^2(v)}{r^2}$. This metric describes a spacetime in which the mass m and charge q vary with time due to some fictitious irradiating charged null dust with

$$
T^{\mu\nu} = T^{\mu\nu}_{\text{dust}} + T^{\mu\nu}_{U(1)},
\tag{6.48}
$$

where

$$\begin{cases} T^{\mu\nu}_{\text{dust}} = \rho l^\mu l^\nu & \rho = \dfrac{1}{4\pi r^2}\dfrac{\partial}{\partial v}\left(m - \dfrac{q^2}{2r}\right) \\ T^{\mu\nu}_{U(1)} = P\,\text{diag}(-1, -1, 1, 1) & P = \dfrac{q^2}{8\pi r^4} \end{cases}. \tag{6.49}$$

For a charged spacetime the surface gravity vanishes in the case of extremality, i.e. $m(v_0) = q(v_0)$ *for some advanced time* $v_0 < \infty$. *Assume an observer is restricted to moving along the radial direction then* $T^{\mu\nu}u^\mu u^\nu = \rho\left(\frac{dv}{d\tau}\right)^2 + P$. *Now since we require equation* (6.46) *this implies* $\rho > 0$. *In particular we have the relation on the horizon* $r^+ = m\sqrt{m^2 - q^2}$*:*

$$4\pi(r^+)^3\rho(r^+) = m\dot{m} - q\dot{q} + \sqrt{m^2 - q^2}\,\dot{m} > 0, \tag{6.50}$$

where the dot means differentiation with respect to v. *The above equation implies that*

$$m(v_0)\dot{\Delta}(v_0) > 0, \tag{6.51}$$

where $\Delta = m - q$. *This means that if we assume the black hole becomes extremal in some advanced time* v_0 *then* Δ *must be decreasing, i.e. there exists* v_0 *for which*

$$\dot{\Delta}(v_0) < 0 \tag{6.52}$$

and Δ *will become zero in a finite time, but this is a contradiction to equation* (6.51). *Thus the weak energy condition prevents the black hole from becoming extremal in a finite advanced time.*

6.5 Black hole thermodynamics

The laws of black hole mechanics, laws 1–3, bear a striking resemblance to the laws of thermodynamics[3] with $\kappa\sim$ temperature, $A\sim$ entropy and $M\sim$ internal energy. This duality was first noticed by Bekenstein [9] and solidified by Hawking [10] by examining quantum processes near black holes. The exact duality is outlined in table 6.2. Classically, black holes were considered a region in spacetime from which escape was impossible. Yet, by combining a general relativistic and quantum field theoretic description of the region just outside the event horizons Hawking and his contemporaries demonstrated that black holes are not so black after all. Instead, they behave as thermodynamic objects, where the horizon area encodes information about the quantum spacetime, thus their relevance to the study of quantum gravity. Another interesting fact is that black hole entropy scales with the area instead of volume as compared to a traditional thermodynamic system. These interesting facts

[3] In the special case of an isolated system in thermal equilibrium.

Table 6.2. Black hole thermodynamic duality.

Thermodynamics	Black hole
Temperature T	$\frac{\hbar}{2\pi}\kappa$
Entropy S	$\frac{A}{4\hbar G}$
Energy E	M
Thermo. equilib.	κ = constant
First law	First law
$\delta S > 0$	$\delta A > 0$

about black hole thermodynamics are summarized in the principle of holographic scaling.

Principle 1 (Holographic scaling).

- *The entropy depends on area and not volume.*
- *The horizon area encodes information at the quantum level.*
- *Any viable quantum gravity candidate should reproduce the duality of table 6.2.*

6.6 2D holographic representations of 4D black holes

In this section we provide a brief and simple example of holography between a four-dimensional black hole and gravity in two dimensions coupled to matter fields. This type of duality seems to be universal and is manifest in the near horizon regime of most stationary spacetimes. By horizon we mean the radial distance, $r = r_+$, at which the black hole's gravitational escape velocity is equal to the speed of light. The idea of holographic duality in this context is that in the near horizon the physics of four-dimensional pure gravity exhibits an equivalent representation in terms of gravity in two dimensions coupled to a two-dimensional scalar field ψ and two-dimensional electromagnetic gauge potential $\mathcal{A} = A_\mu dx^\mu$, also known as a Kaluza–Klein reduction, i.e.

$$ds^2_{4D} \xrightarrow{r\sim r_+} ds^2_{2D},\ \ \psi,\ \ \mathcal{A}. \tag{6.53}$$

We should keep in mind that we are asserting a map between four-dimensional and two-dimensional physics only in the near vicinity of a stationary black hole. This seems an almost obvious conclusion for spherical symmetric spacetimes, where the argument can be made that due to the high degree of isotropy in angular coordinates θ and ϕ, only dynamics in (t, r) are significant. However, even for axisymmetric (rotating) spacetimes the map (6.53) still holds due to gravitational effects, which cause angular degrees of freedom to redshift away in the near horizon [11–13].

We will demonstrate the duality map (6.53) with a specific example starting with the ground state black hole, the Schwarzschild spacetime:

$$ds^2_{\text{Schwarz}} = -\left(1 - \frac{2GM}{r}\right)dt^2 + \left(1 - \frac{2GM}{r}\right)^{-1} dr^2 + r^2 d\theta^2 + r^2 \sin^2\theta d\phi^2, \quad (6.54)$$

where we have chosen units such that the speed of light is scaled to unity, i.e. $c = 1$. The above spacetime exhibits the usual coordinate singularity (Schwarzschild horizon) defined by the radial solution of

$$\left(1 - \frac{2GM}{r}\right) = 0 \Rightarrow r_+ = 2GM. \quad (6.55)$$

This singularity will be referred to as the black hole horizon or just horizon. Now, the metric (6.54) may be recast into a Kaluza–Klein form:

$$\begin{aligned} ds^2_{KK} &= -f(r)dt^2 + f(r)^{-1}dr^2 + \lambda^2 e^{-2\psi(r)}[\\ &\quad d\theta^2 + \sin^2\theta(d\phi + A(r)dt)^2] \\ &= ds^2_{2D} + \lambda^2 e^{-2\psi(r)}\left[d\theta^2 + \sin^2\theta\left(d\phi + A_\mu dx^\mu\right)^2\right], \end{aligned} \quad (6.56)$$

where we have introduced new functions or fields:

$$\begin{aligned} f(r) &= \left(1 - \frac{2GM}{r}\right) \\ \psi(r) &= \ln\frac{2GM}{r} \\ A(r) &= \text{const.} \end{aligned} \quad (6.57)$$

and λ is chosen such that $\lambda^2 e^{-2\psi(r)} = r^2$. Even though ds^2_{KK} has off-diagonal cross terms ($dtd\phi$ terms) it is actually equivalent to ds^2_{Schwarz} since $A(r)$ is constant, it represents a linear phase shift, $\phi \to \phi + \text{At}$. Since the spacetime ds^2_{Schwarz} is spherically symmetric, a linear phase shift in any angular direction will not affect the respective geometry or more importantly the respective physics.

The shear power of the above holographic duality (Kaluza–Klein reduction) becomes evident when trying to analyze quantum properties of four-dimensional black holes. Pure gravity in four dimensions is known to be highly ultraviolet divergent, making quantum gravitational investigations notoriously hard, if not impossible. However, the quantum field theory of equation (6.53) in two dimensions is well known and exhibits finite solutions. This allows the study of four-dimensional quantum gravity in terms of a two-dimensional quantum field theory of equation (6.53) in the near horizon. In the above Schwarzschild example, the Kaluza–Klein reduction gravitationally singles out just the $r - t$ part of equation (6.54), i.e.

$$\begin{aligned} ds_{2D} &= -\left(1 - \frac{2GM}{r}\right)dt^2 + \left(1 - \frac{2GM}{r}\right)^{-1} dr^2 \\ &= -f(r)dt^2 + \frac{1}{f(r)}dr^2, \end{aligned} \quad (6.58)$$

which is commonly referred to as the s-wave of equation (6.54). This terminology is rooted in the fact that the $l = m = 0$ contribution to the spherical harmonics, which are employed in the Kaluza–Klein reduction of more complicated black holes, have no angular dependence. As already mentioned, the near horizon two-dimensional s-wave black holes seem to inherit a tremendous amount of information from their four-dimensional parent spacetimes. Studies [11, 14, 15] of their respective quantum theories yield agreeing black holed thermodynamic quantities as their four-dimensional parent spacetimes. Of particular interest is implementing the near horizon two-dimensional s-wave black holes to compute four-dimensional black hole entropy and temperature:

$$\begin{cases} S = \dfrac{A_{r_+}}{4G\hbar} & \text{Entropy} \\ T = \dfrac{\hbar\kappa}{2\pi} & \text{Temperature} \end{cases}, \tag{6.59}$$

where A_{r_+} is the four-dimensional horizon area and $\kappa = -f'(r_+)/2$ is the horizon surface gravity (analogously to how g is the surface gravity at the Earth's surface). It is clear that both entropy and temperature depend on Newton's and Planck's constants and thus are quantum gravitational in nature and the recovery of equation (6.59) from a two-dimensional theory, attests to the utilitarian usefulness of the holographic principle.

We will limit our analysis in this book to four classical, static and asymptotically flat spacetimes. Two of which are spherically symmetric with line element:

$$ds^2_{\text{Schwarz}} = -f(r)dt^2 + \frac{1}{f(r)}dr^2 + r^2 d\theta^2 + r^2 \sin^2\theta d\phi^2, \tag{6.60}$$

with

$$f(r) = \begin{cases} 1 - \dfrac{2GM}{r} & \text{Schwarzschild} \\ 1 - \dfrac{2GM}{r} + \dfrac{GQ^2}{r^2} & \text{Reissner–Nordström} \end{cases} \tag{6.61}$$

where the Reissner–Nordström (RN) solution represents a Schwarzschild black hole endowed with electric charge q, such that $Q = \dfrac{q}{4\pi\epsilon_0}$. The other two are axisymmetric with line element:

$$ds^2 = -\left(\frac{\Sigma\Delta}{\Pi}\right)dt^2 + \Sigma\left(\frac{dr^2}{\Delta} + d\theta^2\right) + \frac{\sin^2\theta}{\Sigma}\Pi\left(d\phi - \frac{2GMardt}{\Pi}\right)^2, \tag{6.62}$$

where

$$\begin{aligned} &\Sigma = r^2 + a^2\cos^2\theta \\ &\Pi = (r^2 + a^2)^2 - \Delta a^2 \sin^2\theta \\ &\Delta = \begin{cases} r^2 - 2GMr + a^2 & \text{Kerr} \\ r^2 - 2GMr + a^2 + GQ^2 & \text{Kerr–Newman} \end{cases} \end{aligned} \tag{6.63}$$

Table 6.3. Black hole horizons: solving the equations for which the line element exhibits a singularity yields the radial location of the horizons.

Black hole	Singularity	Horizon
Schwarzschild	$f(r) = 0$	$r_+ = 2GM$
RN	$f(r) = 0$	$r_\pm = GM \pm \sqrt{(GM)^2 - GQ^2}$
Kerr	$\Delta(r) = 0$	$r_\pm = GM \pm \sqrt{(GM)^2 - a^2}$
KN	$\Delta(r) = 0$	$r_\pm = GM \pm \sqrt{(GM)^2 - (a^2 + GQ^2)}$

and the parameter a is a measure of angular momentum per unit mass. The Kerr solution represents a black hole with mass and angular momentum, while Kerr–Newman (KN) represents a black hole with mass, angular momentum and charge. The latter three black holes are rarely covered/discussed in a typical undergraduate general relativity setting. This is mainly due to their complicated features brought about by considering angular momentum and charge parameters in addition to mass. A first measure of how the complications come about can be seen in the location of their respective horizons, summarized in table 6.3. We see that for all black holes, except for Schwarzschild, there are inner and outer horizons and constraints on how much charge or angular momentum a black hole may possess. The special case when the radical terms in $r_\pm$ vanish correspond to the maximum amount of charge or angular momentum. In this case r_+ and r_- coincide and the respective black hole is said to be extremal. In addition to the multiple horizon complication, the off-diagonal terms in the Kerr and KN spacetimes bring about other complications, especially when trying to compute respective ADM mass or stable circular orbits. However, implementing a Kaluza–Klein reduction procedure in the near horizon to obtain two-dimensional dual black holes will greatly simplify many calculations and help bring those more complicated black holes into the classroom. Obtaining the two-dimensional dual black hole (6.58) requires a partial wave decomposition of a probe scalar field in the near horizon. Consider a single free scalar field φ in the background of a four-dimensional black hole $g_{\mu\nu}$ with action:

$$\begin{aligned} S_{\text{free}} &= \frac{1}{2}\int d^4x\sqrt{-g}\,g^{\mu\nu}\partial_\mu\varphi\partial_\nu\varphi \\ &= -\frac{1}{2}\int d^4x\,\varphi\Big[\partial_\mu\big(\sqrt{-g}\,g^{\mu\nu}\partial_\nu\big)\Big]\varphi. \end{aligned} \tag{6.64}$$

The above functional is reduced to a two-dimensional theory by expanding the four-dimensional scalar field in terms of spherical harmonics:

$$\varphi(t, r, \theta, \phi) = \sum_{lm}\varphi_{lm}(r, t)Y_l^m(\theta, \phi), \tag{6.65}$$

where φ_{lm} has the form of a complex interacting two-dimensional scalar field. Integrating out angular degrees of freedom, transforming to tortoise coordinates $dr^* = f(r)dr$ and considering the region very close to r_+, we find that the two-dimensional action is much reduced. This is due to the fact that all interaction, mixing and potential terms ($\sim l(l+1)\ldots$) are weighted by a factor of $f(r(r^*)) \sim e^{2\kappa r^*}$, which vanishes exponentially fast as $r \to r_+$. This leaves us with an infinite collection of massless charged scalar fields in the very near horizon region, with $U(1)$ gauge charge equal to the azimuthal quantum number $e = m$ and remnant functional:

$$S = -\frac{r_+^2 + a^2}{2} \int d^2x\, \varphi_{lm}^* \left[-\frac{1}{f(r)}(\partial_t - im\mathcal{A}_t)^2 + \partial_r f(r) \partial_r \right] \varphi_{lm}. \tag{6.66}$$

Thus, arriving at the two-dimensional holographic analogue fields given by

$$g^{(2)}{}_{\mu\nu} = \begin{pmatrix} -f(r) & 0 \\ 0 & \dfrac{1}{f(r)} \end{pmatrix} \tag{6.67}$$

$$\mathcal{A} = \mathcal{A}_t dt. \tag{6.68}$$

For the back holes in consideration here, the above procedure yields the following respective $f(r)$ functions [11–13]:

$$f(r) = \begin{cases} 1 - \dfrac{2GM}{r} & \text{Schwarzschild} \\ 1 - \dfrac{2GM}{r} + \dfrac{GQ^2}{r^2} & \text{RN} \\ \dfrac{\Delta}{r^2 + a^2} & \text{Kerr \& KN} \end{cases}. \tag{6.69}$$

The resulting $f(r)$ functions for the Kerr and KN spacetimes is surprisingly simple and as previously mentioned, encodes a tremendous amount of information about its four-dimensional parent that is now easily accessible to undergraduates using simple techniques which will be outlined next.

Now, in contrast to our approximate classical analysis in sections 2.8 and 2.9, we are able to compute the correct Newtonian (or weak gravity field) potentials given by the relationship [16]

$$U_{\text{eff}} = \Phi = -\frac{1}{2}(g_{00} + 1) = \frac{1}{2}(f(r) - 1). \tag{6.70}$$

Omitting the Schwarzschild case, since it is identical to the approximate analysis, we have the Newtonian potentials and field strengths for the other asymptotically flat spacetimes given by

$$\Phi = \begin{cases} -\dfrac{G(2Mr - Q^2)}{2r^2} & \text{RN} \\ -\dfrac{GMr}{r^2 + a^2} & \text{Kerr} \\ -\dfrac{G(2Mr - Q^2)}{2(r^2 + a^2)} & \text{KN} \end{cases} \tag{6.71}$$

$$\vec{g} = \begin{cases} -\dfrac{GM}{r^2}\hat{r} + \dfrac{GQ^2}{r^3}\hat{r} & \text{RN} \\ -\dfrac{GM(r^2 - a^2)}{(r^2 + a^2)^2}\hat{r} & \text{Kerr} \\ -\dfrac{G(r(Mr - Q^2) - Ma^2)}{(r^2 + a^2)^2}\hat{r} & \text{KN} \end{cases} \tag{6.72}$$

and substituting into equation (2.37) we are left with a simple exercise in vector calculus (suitable for undergraduates) to show that

$$m_{ADM} = -\frac{\lim_{r\to\infty}}{4\pi G}\oint_{\partial V}\vec{g}\cdot d\vec{A} = \begin{cases} M & \text{RN} \\ M & \text{Kerr}, \\ M & \text{KN} \end{cases} \tag{6.73}$$

yielding the gravitating mass parameter of each black hole solution respectively.

6.7 On *ADM* mass in general relativity

We will begin this section by first showing rigorously that our definition (2.37) is the same as the *ADM* mass of general relativity. We begin by considering spacetimes that exhibit asymptotically flat time symmetric initial data, i.e. $g_{ij} \approx \delta_{ij} + \mathcal{O}\left(\frac{1}{r}\right)$ which includes the Schwarzschild, RN, Kerr and KN black hole solutions. For such spacetimes, the *ADM* mass of general relativity reads [16, 17]

$$M_{ADM} = \frac{1}{16\pi G}\lim_{r\to\infty}\oint \delta^{ij}\left(\partial_i g_{jk} - \partial_k g_{ij}\right)n^k dS. \tag{6.74}$$

Here, dS is a topological two-sphere with unit normal n^k. Equation (6.74) is not a covariant statement, but is evaluated for asymptotic Euclidean coordinates. For asymptotically flat spacetimes (which include the previously mentioned ones) for example, we can obtain the asymptotically flat time symmetric initial data metric by setting $dt = 0$, Taylor expanding, and performing radial redefinitions to obtain

$$ds_3^2 = g_{ij}dx^i dx^j = g(r)(dx^2 + dy^2 + dz^2), \tag{6.75}$$

where $g(r) = 1 - 2\Phi + \mathcal{O}\left(\frac{1}{r^2}\right)$. Using this in equation (6.74), we obtain

$$\begin{aligned} &\delta^{ij}\left(\partial_i g_{jk} - \partial_k g_{ij}\right)n^k = 4\partial_i\Phi n^i \Rightarrow \\ M_{ADM} &= \frac{1}{16\pi G}\lim_{r\to\infty}\oint \delta^{ij}\left(\partial_i g_{jk} - \partial_k g_{ij}\right)n^k dS \\ &= \frac{1}{16\pi G}\lim_{r\to\infty}\oint 4\partial_i\Phi n^i dS \\ &= -\frac{1}{4\pi G}\lim_{r\to\infty}\oint \vec{g}\cdot d\vec{S}, \end{aligned} \tag{6.76}$$

validating our definition of ADM mass in equation (2.37), for the appropriate choice of a Gaussian surface.

As previously mentioned, the application of equation (6.74) to RN, Kerr and KN is a daunting task in the full theory of general relativity and without the knowledge of their s-wave $f(r)$ functions. However, by looking at differential geometric formulation of Newtonian gravity, also known as Newton–Cartan gravity, we can introduce a simpler formula for computing ADM mass suitable for introductory undergraduate courses. By introducing a universal (Galilean-affine) time $\tau = \lambda t + b$, where λ and b are constants, we are free to rewrite Newton's second law as

$$F = m\frac{d^2\vec{x}}{dt^2} = -m\nabla\Phi \Rightarrow \tag{6.77}$$

$$\frac{d^2x^i}{d\tau^2} + \eta^{ij}\frac{\partial\Phi}{\partial x^j}\left(\frac{dt}{d\tau}\right)^2 = 0. \tag{6.78}$$

Comparing equation (6.77) to the general autoparallel equation (geodesic equation):

$$\frac{d^2x^\mu}{d\tau^2} + \Gamma^\mu{}_{\alpha\beta}\frac{dx^\alpha}{d\tau}\frac{dx^\beta}{d\tau} = 0 \tag{6.79}$$

implies the non-zero Newton–Cartan connection:

$$\Gamma^i_{00} = \eta^{ij}\frac{\partial\Phi}{\partial x^j} = -g^i. \tag{6.80}$$

This non-vanishing connection and its relation to the gravitational potential implies that, just as in general relativity, matter introduces a curvature to the Newtonian spacetime and that trajectories of test particles under the influence of gravity are simply autoparallel curves in this curved spacetime. In fact using equation (4.98) we find the only non-vanishing component:

$$R_{00} = \partial_l\Gamma^l{}_{00} = \partial_l\partial^l\Phi \tag{6.81}$$

and combining this result with the Poisson equation of Newtonian gravity:

$$\nabla^2\Phi = 4\pi G\rho, \tag{6.82}$$

we obtain the relationship between matter density and the curvature of spacetime in Newton–Cartan theory, i.e. Einstein's equation:

$$\begin{aligned} -\partial_i g^i = R_{00} &= 4\pi G\rho \\ R_{00} &= 4\pi G T_{00}. \end{aligned} \tag{6.83}$$

Exercise 21. *Start with equation* (2.35) *and derive equation* (6.82).

This implies an alternate form for ADM mass in terms of $\Gamma^i{}_{00}$:

$$-\partial_i g^i = R_{00} = \partial_l\Gamma^i{}_{00} = 4\pi G\rho \tag{6.84}$$

$$\Rightarrow \boxed{m_{ADM} = \frac{\lim_{r\to\infty}}{4\pi G} \oint_{\partial V} \Gamma^{i}{}_{00}\eta_{ij} dA^{j}.} \tag{6.85}$$

Though this equivalence is made in terms of the Newton–Cartan connection, we extend it to the Christoffel connection of full general relativity, which is relatively easy to validate by considering the Schwarzschild solution, but extremely difficult for the others.

Next, we apply equation (6.85) to compute the ADM masses of the Schwarzschild, RN, Kerr and KN spacetimes. Using equation (4.84), we obtain the relevant Christoffel connections:

$$\Gamma^{r}_{tt} = \begin{cases} \dfrac{GM(r-2GM)}{r^3} & \text{Schwarzschild} \\ \dfrac{GM(r^2 + G(Q^2 - 2Mr))}{r^4} & \text{RN} \\ \dfrac{GM(a^2 + r(r-2GM))(r^2 - a^2\cos^2\theta)}{(r^2 + a^2\cos^2\theta)^3} & \text{Kerr} \\ \dfrac{GM(a^2 + r^2 + G(Q^2 - 2GMr))(r^2 - a^2\cos^2\theta)}{(r^2 + a^2\cos^2\theta)^3} & \text{KN} \end{cases} \tag{6.86}$$

and substituting into equation (6.85) yields

$$m_{ADM} = \frac{\lim_{r\to\infty}}{4\pi G} \int_0^{2\pi} \int_0^{\pi} \Gamma^{r}{}_{tt} r^2 \sin\theta d\theta d\phi \tag{6.87}$$

$$= \begin{cases} M & \text{Schwarzschild} \\ M & \text{RN} \\ M & \text{Kerr} \\ M & \text{KN} \end{cases} \tag{6.88}$$

after some lengthy but analytically solvable integration. Despite the lengthy integration, this is a much more approachable method for computing the ADM mass compared to applying equation (6.74). The computation of Christoffel connections has become standard in most undergraduate introductions to general relativity, also the respective calculations/integration can be much simplified by implementing software such as Mathematica or Maple and provide avenues for implementing computational tools into the course curriculum.

6.8 4D *ADM* mass and holographic Γ

An interesting feature of equation (6.85) is that it holds true in two dimensions for the s-wave black holes (6.58) of each respective four-dimensional black hole considered above. In other words applying either side formula below:

$$\begin{array}{ccc} 4D & \text{Near Horizon Holographic Map} & 2D \\ \frac{\lim_{r\to\infty}}{4\pi G}\oint_{\partial V}\Gamma^{i}{}_{00}\eta_{ij}dA^{j} & \xrightarrow{\hspace{8cm}} & \lim_{r\to\infty}\frac{r^2}{G}\underbrace{\Gamma^{r}{}_{tt}}_{s-\text{wave}} \end{array} \tag{6.89}$$

for a given black hole, reproduces the same gravitating ADM mass. This is one (of numerous) example where the holographic principle demonstrates its advantage (shear power). First off, just integrating the four-dimensional $\Gamma^{r}{}_{tt}$ over the two-sphere does not yield anything equal to the $\underbrace{\Gamma^{r}{}_{tt}}_{s-\text{wave}}$, which instead is derived by applying equation (4.84) to equation (6.58). Yet, the right side of equation (6.89) is much easier to evaluate compared to its left side. In fact it makes the ADM mass calculation for the four common asymptotic back holes much more accessible since the calculation of the $\underbrace{\Gamma^{r}{}_{tt}}_{s-\text{wave}}$ can be done once in terms of $f(r)$ and then evaluated for each s-wave black hole in equation (6.69) and the limit of the right side of equation (6.89) is a basic calculus exercise.

Exercise 22 (4D ADM mass from 2D near horizon holography). *Using equations* (4.84), (6.58) *and* (4.84), *show that*

$$m_{ADM}\lim_{r\to\infty}\frac{r^2}{G}\underbrace{\Gamma^{r}_{tt}}_{s-\text{wave}} = M$$

for each respective black hole.

6.9 Holographic black hole thermodynamics

In section 6.5 we learned that *black holes are actually gray*, i.e. they radiate at a thermal bath with temperature $T_H = \dfrac{\kappa}{2\pi}$ (see table 6.2 for $\hbar = 1$). Now, from principle 1 we also learned that entropic thermodynamic behavior is holographic in nature since entropy is traditionally an extensive variable and thus should depend on volume, but in the case of black holes it depends on area instead. To which, we must conclude that T_H must have holographic origins too, though not obvious, correct [15]. This should make intuitive sense, since the entire thermodynamic behavior of black holes is associated with the black hole horizon, that implies that the holographic near horizon s-wave black holes (6.58) should encapsulate the full thermodynamic behavior of their four-dimensional parent spacetimes [2, 13–15, 18]. In other words, since T_H is proportional to κ in four dimensions, we can assert that this relationship holds true in the near horizon holographic limit:

$$T_H^{(4D)} = \frac{\kappa^{(2D)}}{2\pi}, \tag{6.90}$$

where $\kappa^{(2D)}$ is the surface gravity derived from equation (6.58).

Exercise 23. *The holographic two-dimensional analogue black hole* (6.58) *exhibits one time-like Killing vector* $\vec{\xi} = \partial_t$. *Show that*

(a) $\nabla_\mu \xi_\nu = \begin{pmatrix} 0 & \dfrac{f'(r)}{2} \\ -\dfrac{f'(r)}{2} & 0 \end{pmatrix}$,

(b) *and thus, using* (6.21): $\kappa^{(2D)} = \dfrac{1}{2} f'(r_+)$.

From the above exercise, we see that the surface gravity of each holographic two-dimensional analogue black hole is completely determined by one half of the first derivative of $f(r)$ evaluated on the respective horizon. Thus, for the black holes considered previously we have

$$T_H^{(4D)} = \frac{f'(r_+)}{4\pi}, \tag{6.91}$$

for the Hawking temperature of each respective black hole. This result is a much easier and attainable one compared to computing κ in four dimensions using (6.21), especially for the Kerr and KN spacetimes.

6.10 Concluding remarks

We have provided some new pedagogical tools for introducing several complicated and advanced topics of gravity theory via results from contemporary research on holographic duality. We have focused our efforts on the four asymptotically flat spacetimes of general relativity with mass, charge and angular momentum. Our aim is that the above results may be useful to an undergraduate student population learning about relativity and gravitation. Black holes are always an exciting topic and being able to learn about *ADM* mass, horizons, surface gravity and orbits by exploiting their 2D near horizon holographic duals, thus drastically simplifying complicated calculations, is an exciting prospect for students. In particular, having the holographic $f(r)$ functions for the Kerr and KN spacetimes makes them accessible to undergraduate students. In addition our above analysis brings cutting edge research into the classroom and highlights the holographic nature of our Universe.

In equation (2.37) we introduced a simple form for computing *ADM* mass in a Newtonian setting, which relied on extracting the respective Newtonian potentials from the 2D near horizon holographic dual black holes. In equation (6.85) we formulate a simple *ADM* mass in full general relativity in terms of Γ^r_{tt}, motivated from a Newton–Cartan formulation, in particular equation (6.83). The equivalent statement in general relativity is the Einstein equation, which for the Schwarzschild solution reads simply $R_{\mu\nu} = 0$, i.e. $T_{00} = 0$ and one only recovers a non-zero *ADM* mass by considering holographic boundary terms to $T_{\mu\nu}$ [19]. This seems to be encoded into the Newton–Cartan formulation in which equation (6.85) is rooted.

References

[1] Rodriguez L, Germaine-Fuller J S and Wickramasekara S 2014 *Newton-Cartan Gravity in Noninertial Reference Frames* (arXiv:1412.8655)

[2] Rodriguez L 2011 Black-hole/near-horizon-CFT duality and 4 dimensional classical space-times *PhD Thesis* University of Iowa https://doi.org/10.17077/etd.p4388g7i

[3] Ashtekar A, Beetle C and Fairhurst S 1999 Isolated horizons: A generalization of black hole mechanics *Class. Quant. Grav.* **16** L1–7

[4] Poisson E 2005 *A Relativist's Toolkit* (Cambridge: Cambridge University Press)

[5] Bardeen J M, Carter B and Hawking S W 1973 The four laws of black hole mechanics *Commun. Math. Phys.* **31** 161–70

[6] Israel W 1986 Third law of black-hole dynamics: a formulation and proof *Phys. Rev. Lett.* **57** 397–9

[7] Hawking S W 1971 Gravitational radiation from colliding black holes *Phys. Rev. Lett.* **26** 1344–6

[8] Vaidya P C 1953 Newtonian time in general relativity *Nature* **171** 260–1

[9] Bekenstein J D 1973 Black holes and entropy *Phys. Rev.* D **7** 2333–46

[10] Hawking S W 1975 Particle creation by black holes *Commun. Math. Phys.* **43** 199–220

[11] Robinson S P and Wilczek F 2005 A relationship between Hawking radiation and gravitational anomalies *Phys. Rev. Lett.* **95** 011303

[12] Iso S, Umetsu H and Wilczek F 2006 Anomalies, Hawking radiations and regularity in rotating black holes *Phys. Rev.* D **74** 044017

[13] Rodriguez L and Rodriguez S 2017 On the near-horizon canonical quantum microstates from AdS_2/CFT_1 and conformal Weyl gravity *Universe* **3** 56

[14] Button B K, Rodriguez L and Wickramasekara S 2013 Near-extremal black hole thermodynamics from AdS_2/CFT_1 correspondence in the low energy limit of 4D heterotic string theory *J. High Energy Phys.* **1310** 144

[15] Rodriguez L and Yildirim T 2010 Entropy and temperature from black-hole/near-horizon-CFT duality *Class. Quant. Grav.* **27** 155003

[16] Misner C W, Thorne K S and Wheeler J A 1970 *Gravitation* (San Francisco, CA: Freeman)

[17] Brewin L 2007 A Simple expression for the ADM mass *Gen. Relat. Grav.* **39** 521–8

[18] Button B K, Rodriguez L, Whiting C A and Yildirim T 2011 A near horizon CFT dual for Kerr-Newman-*AdS Int. J. Mod. Phys.* A **26** 3077–90

[19] Balasubramanian V and Kraus P 1999 A stress tensor for anti-de Sitter gravity *Commun. Math. Phys.* **208** 413–28

Appendix A

Gravitational gauge transformation as a change in coordinates

Define the four vector potential $V_\mu = \{-\Phi, \vec{V}\}$ for constant $\vec{V} = (V_x, V_y, V_z)$ and the coordinate shift from $x^\mu = \left\{t, x + \frac{\Lambda}{3V_x}, y + \frac{\Lambda}{3V_y}, z + \frac{\Lambda}{3V_z}\right\}$ to $x'^\mu = \{t, x, y, z\}$, where $\Lambda = \Lambda(t, \vec{x})$. Now, performing the coordinate transformation on V_μ gives:

$$V'_\nu = V_\mu \frac{\partial x^\mu}{\partial x'^\nu} = \left\{-\left(\Phi - \frac{\partial \Lambda}{\partial t}\right), \vec{V} + \nabla\Lambda\right\}, \tag{A.1}$$

which yields one example of a gravitational–gauge transformation on V_μ.

doi:10.1088/2053-2571/ab06bdch7

Appendix B

Basic group theory

Definition 27 (group). *A group G is a set closed under a binary operation *, such that for every $g, h \in G \Rightarrow g^*h \in G$, satisfying the following (group) axioms:*

GA1(associativity) $\quad g^*(h^*k) = (g^*h)^*k, \ \forall \ g, h, k \in G.$

GA2(identity) $\quad \exists \ e \in G \ s.t. \ e^*g = g^*e = g, \ \forall \ g \in G.$

GA3(inverse) $\quad \forall \ g \in G \ \exists \ g^{-1} \in G, \ s.t. \ g^*g^{-1} = g^{-1*}g = e.$

Exercise 24. *Use the group axioms to show that*

- *e is unique.*
- *$\forall \ g \in G \ g^{-1}$ is unique.*
- $(g^*h)^{-1} = h^{-1*}g^{-1.}$

Definition 28 (subgroup). *H forms a subgroup of G, denoted $H \subseteq G$, if*

(a) $\forall \ h, k \in H \Rightarrow h^*k \in H.$

(b) $e \in H.$

(c) $\forall \ h \in H \Rightarrow h^{-1} \in H.$

Example 14. *The set $\mathbb{Z}$ or $\mathbb{R}$ with addition as the binary operation forms a group.*

Example 15. *The set $\mathbb{R}/\{0\}$ with multiplication as the binary operation forms a group.*

doi:10.1088/2053-2571/ab06bdch8

B.1 Homomorphisms and isomorphisms

Definition 29 (homomorphism). *A map ϕ between groups G and G′, denoted ϕ: $G \to G'$, is homomorphic if* $\forall\, a, b \in G\ \&\ \forall\, \phi(a) = a', \phi(b) = b' \in G'$ *the product* $ab \to a'b'$ *under ϕ, i.e.*

$$\phi(ab) = \phi(a)\phi(b)$$

Theorem 7. *For ϕ: $G \to G'$ a homomorphism the following holds:*

$$\begin{cases} \phi\colon e \to e' & \forall\ e, e' \in G, G' \\ \phi\colon g^{-1} \to g'^{-1} & \forall\ g^{-1}, g'^{-1} \in G, G' \end{cases}.$$

Proof.

- Let $g \in G$ and $g' \in G'$. Next, take

$$\begin{aligned} g' &= \phi(g) \\ &= \phi(eg) \\ &= \phi(e)\phi(g) \\ &\Rightarrow \phi(e) = e'. \end{aligned}$$

- Finally, since ϕ is homomorphic we have:

$$\begin{aligned} \phi(g^{-1})\phi(g) &= \phi(g^{-1}g) \\ &= \phi(e) \\ &= e'. \end{aligned}$$

□

Definition 30 (isomorphism). *Let ϕ: $G \to G'$ be a homomorphism, then an isomorphism between G and G′, denoted $G \cong G'$, exists if ϕ is a bijection.*

Exercise 25. *If ϕ: $G \to G'$ and ψ: $G' \to G''$ are isomorphisms, show that $\phi \circ \psi$ is an isomorphism.*

B.2 Automorphisms and normal subgroups

Definition 31 (automorphism). *An automorphism is an isomorphism onto itself, i.e.*

$$\phi: G \rightarrow G$$

for ϕ an isomorphism.

Example 16. *A trivial one is the identity map.*

An interesting class of automorphisms is the conjugation map $C_g: G \rightarrow G$, which maps $a \rightarrow gag^{-1}$ for all $g, a \in G$. We can readily construct a group of conjugations by defining the inverse

$$(C_g)^{-1} = C_{g^{-1}}$$

and the identity by

$$\mathbf{I} = C_e. \tag{B.1}$$

Definition 32 (coset). *For every subgroup H of G and $a \in G$ we define the left coset as*

$$aH.$$

Basically, we take the binary composition of every element in H by a and the resulting set is not necessarily a group.

Example 17 (vector quotient space). *Let V be a vector space and W a subspace of V. The quotient space, denoted V/W, is the set of equivalence classes $[v]$, such that for all $v_1, v_2 \in V$ $(v_1 - v_2) \in W$. In other words, $v_1 = v_2 + w$, for some $w \in W$. We can denote the space of such equivalence classes by the coset $[v] + W$.*

Any two cosets are either equal or disjoint. To see this, take cosets aH and bH and suppose there exists an element g of G such that

$$g = ah_1 = bh_2 \quad \forall\, h_1, h_2 \in H$$

then $ah = bh_2h_1^{-1}h$, which implies $aH \subseteq bH$ and $bH \subseteq aH$. Thus $aH \cap bH = \emptyset$ or $aH = bH$. The cosets gH provide a particular way of partitioning the group G. This is seen by the fact that since H is a group the element g of G is also contained in gH and thus, the cosets of H form a family of disjoint subsets covering G.

Theorem 8 (Lagrange). *If G is a finite group of order n (number of elements in G, denoted $|G|$) then the order of every subgroup H is a divisor of n.*

Proof. First we note that there exists an injection between gH and H, for if we assume there exist distinct $h_1, h_2 \in H$ such that

$$gh_1 = gh_2, \tag{B.2}$$

we quickly obtain a contradiction in our assumption. We see this when hitting (B.2) by g^{-1} from the left and thus obtaining $h_1 = h_2$. This shows not only the injection between the sets gH and H, but also says that gH contains $|H|$ many elements. Now, since the cosets partition G this implies that n must be a multiple of $|H|$. □

Definition 33. *The order m of a group element g is the number of successive binary operation of g with itself yielding e, i.e. $g^m = e$.*

Corollary 2. *The order of any group element is a divisor of the group order.*

Proof. First we note that for all g in G the set $\{g, g^2, \dots, g^m = e\}$ forms a (cyclic-) subgroup of G. Thus, by Theorem 8 the order of $\{g, g^2, \dots, g^m = e\}$ must divide $|G|$, which implies that m divides $|G|$. □

Exercise 26. *Let G be a finite group of prime order p.*

- *Show that G is trivial (the groups $\{e\}$ and G are the only subgroups).*
- *G is a cyclic group.*

B.3 Normal subgroups

Definition 34 (normal). *Let N be a subgroup of G. N is called normal if N is invariant under inner automorphisms, i.e. for all $g \in G$*

$$gNg^{-1} = N. \tag{B.3}$$

When multiplying cosets together the result is not necessarily another coset, unless the subgroup is normal. Let N be a normal subgroup of G and $g, g' \in G$. Looking at the product of the cosets gives:

$$\begin{aligned} gNg'N &= gg'(g')^{-1}Ng'N \\ &= gg'NN \\ &= (gg')N. \end{aligned}$$

Now, we are in the position of realizing a group structure by defining associativity

$$(gNg'N)g''N = (gg'g'')N \quad \forall g, g', g'' \in G,$$

the identity

$$eN = N,$$

and the inverse

$$(gN)^{-1} = g^{-1}N.$$

Thus the cosets of normal subgroups N form a group called the **factor group** denoted G/N (the set of all cosets of normal subgroups).

B.4 Group action

Definition 35 (action). *A (left) action of a group G on a set X is a homomorphism ϕ of G into the group of transformations of X,*

$$\phi: G \to \text{Transf}(X)$$

denoted $\phi(g)(x)$ or gx.

Basically, in physics an action of a group on a set entails finding an irreducible representation of the group in the set.

Example 18. *The cyclic group $\{a, a^2 = e\}$ acts on the set of reals $\mathbb{R}$ as*

$$ax = -x \quad \text{and} \quad a^2x = x,$$

for all $x \in \mathbb{R}$.

Consider the map $\rho: G \to Transf(X)$. ρ is called an anti-homomorphism if

$$\rho(gh) = \rho(h)\rho(g),$$

for all $g, h \in G$. The notion of the anti-homomorphism allows for a right action, i.e. $\rho(g)(x) = xg$. Next, let G have an action on the set X. Then, we define the orbit of Gx, for all $x \in X$, as the set of all points in X, that can be reached from x by Gx. The action of G on X is called transitive if X is the orbit of some Gx, for all $x \in X$.

Definition 36 (isotropy). *The subgroup of G, which map $x \to x$ for all $x \in X$, is called the isotropy group of G and is denoted G_x. It consists of all elements of G, such that*

$$gx = x.$$

⊙编辑手记

本书是一部英文版的物理学专著，中文书名或可译为《论全息度量原则：从大学物理到黑洞热力学》.

本书的作者有两位，一位是利奥·罗德里格斯(Leo Rodriguez)教授，他于2011年获得爱荷华大学物理学博士学位，之后在格林内尔学院担任HHMI博士后研究员，再之后在格林内尔学院和艾斯普森大学担任教职. 2018年秋天，他重新加入格林内尔学院的物理系. 他的研究兴趣集中在黑洞的热力学性质，它们与量子引力的关系，以及它们如何被编码到对偶共形场论中.

另一位为珊珊·罗德里格斯(Shanshan Rodriguez)，她于2010年在爱荷华大学获得物理学博士学位，研究方向为理论空间等离子体，特别是磁重联. 随后，她加入了美国宇航局的戈达德太空飞行中心并担任研究工程师，她在探测器系统分部工作至2004年，随后在伍斯特理工学院担任教职. 她的研究兴趣集中在数值技术和理论天体物理学中计算工具的实现或优化. 她于2018年加入格林内尔学院物理系，并花了大量时间教授物理和开发天体物理学的本科课程.

正如作者在前言中所述：

> 一位教师走进普通的物理课堂并通过提出以下问题开始讲课："什么是质量?"一个天真简单的问题通常会产生无数的答案，例如："构成物体的所有物质."教师的下一个问题是："那么，这些物质是什

么?”学生通常会说:“所有的物质……”引出教师的回答:“什么是物质?”这个游戏将一直持续到学生意识到他们并不真正知道这些词汇和日常生活中看似熟悉的概念到底是什么时才能结束.物理学中的质量历来是一个臭名昭著的话题,物理学专业的学生可能一直到研究生阶段都搞不明白这个概念.然而,本书给这个主题和更多类似的主题提供了一个机会,让读者可以在本科课堂中实现对当代前沿物理学的研究.这是我们这本书的前提:把全息对偶研究的结果和概念作为教学工具,使物理学中较难的概念在入门和高级物理课程中易于理解.

广义相对论课程通常不是物理学本科教育的一部分,但它的重要性一直在加强,并且学生对这门课程一直有需求.在本书中,我们希望分享一些受现代引力研究(全息对偶)启发和衍生的技术,这些技术有助于将引力理论的高级主题带入本科课堂中,这有助于满足学生对涵盖黑洞物理及其性质等主题的好奇心,其所包括的技术已经发展了几年,并在主要为本科生服务的机构(包括格林内尔学院和艾斯普森大学)中向本科生讲授广义相对论和万有引力;还包括可以授予博士学位的大学,比如爱荷华大学和伍斯特理工学院.

当代研究可以是一个很好的资源,尤其是在将新的、有效的教学工具引入本科物理课堂时,其研究是将物理学推向极限以期做出关于自然的基本规律的重大发现.因此,我们与我们自己的学生(研究中和课堂中的)以及本书的总体目标是为学生提供工具、技能和习惯,以便他们在研究模型中独立拓展自己的知识视野,进而积累知识.

本书的目录如下:

我们是一个数学工作室，所以在确定所要引进图书的价值判断上，一个主要指标是数学工具用得怎么样，本书真正打动笔者的是第 3 章的诺特定理和第 4 章的流形上的张量微积分. 囿于笔者的学识水平，要介绍好诺特定理有点难度，而外代数就容易一点了. 在国内现有的分析教程中，中国科学技术大学的徐森林教授的教材公认是写外代数最佳的，他的教材本工作室已经再版了. 另一本网友强烈推荐的武汉大学早年的经典分析教程已近绝版，所以这里摘录一段，待有机会再重新出版.

一、外代数

在未定义外代数之前，让我们首先介绍代数的概念.

1. 代数的概念

设 $\mathbf{K}=\mathbf{R}$ 或 $\mathbf{C}$.

定义 1.1　一个代数就是一个二元组 $(\mathscr{A},\tau)$，这里：

(1) $\mathscr{A}$ 是一个 $\mathbf{K}-$ 向量空间.

(2) $\tau:\mathscr{A}\times\mathscr{A}\to\mathscr{A},(A,B)\mapsto A\tau B$ 是一个双线性映射，它满足结合律

$$\forall A,B,C\in\mathscr{A},(A\tau B)\tau C=A\tau(B\tau C)$$

τ 称为代数积.

若任意 $A,B\in\mathscr{A},A\tau B=B\tau A$，则称 τ 是可交换的.

若存在 $e\in\mathscr{A}$ 使得任意 $A\in\mathscr{A},A\tau e=e\tau A$，则称 τ 有单位元 e.

例 1　考虑连续函数空间 $C^0([a,b],\mathbf{K})$.

(1) $C^0([a,b],\mathbf{K})$ 是一个 $\mathbf{K}-$ 向量空间.

(2) 定义映射 $\tau:C^0([a,b],\mathbf{K})\times C^0([a,b],\mathbf{K})\to C^0([a,b],\mathbf{K})$ 为

$$\forall f,g\in C^0([a,b],\mathbf{K}),f\tau g=fg$$

显然，τ 是一个双线性映射，并且满足结合律

$$\forall f,g,h\in C^0([a,b],\mathbf{K}),(f\tau g)\tau h=f\tau(g\tau h)=fgh$$

因此 $(C^0([a,b],\mathbf{K}),\tau)$ 是一个代数，这个代数积 τ 是可交换的，并且有单位元 1.

例 2　考虑 $n\times n$ 实矩阵集合 $M(n,n)$.

(1)$M(n,n)$ 是一个 $\mathbf{R}$－向量空间.

(2) 定义映射 $\tau: M(n,n)\times M(n,n)\to M(n,n)$ 为

$$\forall \boldsymbol{A},\boldsymbol{B}\in M(n,n),\boldsymbol{A}\tau\boldsymbol{B}=\boldsymbol{AB}$$

则 τ 是一个双线性映射,并且满足结合律

$$\forall \boldsymbol{A},\boldsymbol{B},\boldsymbol{C}\in M(n,n),(\boldsymbol{A}\tau\boldsymbol{B})\tau\boldsymbol{C}=\boldsymbol{A}\tau(\boldsymbol{B}\tau\boldsymbol{C})=\boldsymbol{ABC}$$

因此$(M(n,n),\tau)$是一个代数,这个代数积 τ 不是可交换的,但 τ 有单位元 $\boldsymbol{E}$,即单位矩阵.

例 3　考虑 n 个变量的系数在 $\mathbf{K}$ 中取值的多项式集合 $\mathbf{K}[X_1,X_2,\cdots,X_n]=\mathscr{P}$.

(1)$\mathbf{K}[X_1,X_2,\cdots,X_n]$ 是一个 $\mathbf{K}$－向量空间.

(2) 若我们用 $\mathscr{P}_k$ 表示变量 $X_1,X_2,\cdots,X_n$ 的 k 次齐次多项式集合,则 $\mathscr{P}_k$ 也是 $\mathbf{K}$－向量空间.

设 $\boldsymbol{\alpha}=(\alpha_1,\alpha_2,\cdots,\alpha_n)$,这里 $\alpha_i\in\{0,1,2,\cdots,k\}$,$|\boldsymbol{\alpha}|\triangleq\alpha_1+\alpha_2+\cdots+\alpha_n=k$.我们令

$$X^{\boldsymbol{\alpha}}=X_1^{\alpha_1}X_2^{\alpha_2}\cdots X_n^{\alpha_n}$$

则所有这种形式的单项式$\langle X^{\boldsymbol{\alpha}}\mid|\boldsymbol{\alpha}|=k\rangle$形成 $\mathscr{P}_k$ 的一个基.

显然对于 $\mathscr{P}$,我们有下述直和

$$\mathscr{P}=\bigoplus_{k=0}^{+\infty}\mathscr{P}_k$$

(3) 定义映射 $\tau:\mathscr{P}\times\mathscr{P}\to\mathscr{P}$ 为

$$\forall P_n,P_m\in\mathscr{P},P_n\tau P_m=P_n\cdot P_m$$

则 τ 显然是双线性映射,它满足结合律

$$\forall P_n,P_m,P_k\in\mathscr{P},(P_n\tau P_m)\tau P_k=P_n\tau(P_m\tau P_k)=P_n\cdot P_m\cdot P_k$$

因此$(\mathscr{P},\tau)$是一个代数,此代数积 τ 也是可交换的,并且有单位元 1.

有了代数的概念,现在我们可以来介绍外代数的概念.

2.外代数的概念

首先我们证明关于双线性映射的下述引理.

引理 1.1　设 X,Y 是维数分别为 n 与 m 的两个 $\mathbf{K}$－向量空间,它们的基分别为$\langle a_i\rangle$与$\langle b_j\rangle$,Z 是任一 $\mathbf{K}$－向量空间.若 $f:X\times Y\to Z$ 是任一映射,则 f 是双线性的,当且仅当存在 $n\times m$ 个 $c_{ij}\in Z$ 使得

$$\forall x=\sum_{i=1}^{n}\xi_i a_i\in X,\forall y=\sum_{j=1}^{m}\mu_j b_j\in Y$$

$$f(x,y)=\sum_{i=1}^{n}\sum_{j=1}^{m}\xi_i\mu_j c_{ij}$$

若 f 是双线性的,则 $c_{ij}=f(a_i,b_j)$.

证明 首先假设 f 是双线性映射. 于是

$$f(x,y)=f\left(\sum_{i=1}^{n}\xi_i a_i,\sum_{j=1}^{m}\mu_j b_j\right)$$

$$=\sum_{i=1}^{n}\sum_{j=1}^{m}\xi_i\mu_j f(a_i,b_j)$$

令 $c_{ij}=f(a_i,b_j)$ 即可.

反之,设存在 $n\times m$ 个 $c_{ij}\in Z$,使得任意 $x=\sum_{i=1}^{n}\xi_i a_i\in X$,任意 $y=\sum_{j=1}^{m}\mu_j b_j\in Y$,且

$$f(x,y)=\sum_{i=1}^{n}\sum_{j=1}^{m}\xi_i\mu_j c_{ij}$$

我们来证明 f 是双线性的. 为此设

$$x=\sum_{i=1}^{n}\xi_i a_i,x'=\sum_{i=1}^{n}\xi'_i a_i\in X,a_i\in \boldsymbol{X}$$

则 $x+x'=\sum_{i=1}^{n}(\xi_i+\xi'_i)a_i,\alpha x=\sum_{i=1}^{n}(\alpha\xi_i)a_i$,于是

$$f(x+x',y)=\sum_{i=1}^{n}\sum_{j=1}^{m}(\xi_i+\xi'_i)\mu_j c_{ij}$$

$$=\sum_{i=1}^{n}\sum_{j=1}^{m}\xi_i\mu_j c_{ij}+\sum_{i=1}^{n}\sum_{j=1}^{m}\xi'_i\mu_j c_{ij}$$

$$=f(x,y)+f(x',y)$$

$$f(\alpha x,y)=\sum_{i=1}^{n}\sum_{j=1}^{m}(\alpha\xi_i)\mu_j c_{ij}=\alpha\sum_{i=1}^{n}\sum_{j=1}^{m}\xi_i\mu_j c_{ij}=\alpha f(x,y)$$

因此 f 关于 x 是线性的. 同理可证 f 关于 y 也是线性的. 故 f 是双线性映射.

附注 这个引理表明,为了定义双线性映射 f,我们只需定义 $f(a_i,b_j)$ 即可. $c_{ij}=f(a_i,b_j)$ 称为双线性映射 f 关于基 $\langle a_i\rangle$ 与 $\langle b_j\rangle$ 的系数.

当 $X=Y$ 时,我们可取 $\langle a_i\rangle=\langle b_j\rangle$,于是 $c_{ij}=f(a_i,a_j)$.

此引理可推广到任一 k 重线性映射上去,具体的叙述我们留给读者.

设 $(e_1,e_2,\cdots,e_n)$ 是 $\mathbf{R}^n$ 的标准正规正交基,它的对偶基底为 $(\mathrm{d}x_1,\mathrm{d}x_2,\cdots,\mathrm{d}x_n)$. 我们令

$$G_0=\varnothing,\mathrm{d}x_{\varnothing}=1$$

$$G_k=\{\boldsymbol{i}=(i_1,i_2,\cdots,i_k)\mid 1\leqslant i_1<i_2<\cdots<i_k\leqslant n\},1\leqslant k\leqslant n$$

(1)k 重线性形式 $\mathrm{d}x_{\boldsymbol{i}}:\mathbf{R}^n\times\mathbf{R}^n\times\cdots\times\mathbf{R}^n\to\mathbf{R}(\boldsymbol{i}\in G_k)$.

设 $\mathrm{d}x_{\boldsymbol{i}}\triangleq\mathrm{d}x_{i_1}\wedge\mathrm{d}x_{i_2}\wedge\cdots\wedge\mathrm{d}x_{i_k}$. 我们定义 k 重线性形式 $\mathrm{d}x_{\boldsymbol{i}}$: $\mathbf{R}^n\times\mathbf{R}^n\times\cdots\times\mathbf{R}^n\to\mathbf{R}$ 如下

$$\forall(e_{j_1},e_{j_2},\cdots,e_{j_k})\in\mathbf{R}^n\times\mathbf{R}^n\times\cdots\times\mathbf{R}^n$$

$$\begin{aligned}\mathrm{d}x_{\boldsymbol{i}}(e_{j_1},e_{j_2},\cdots,e_{j_k})&=\mathrm{d}x_{i_1}\wedge\mathrm{d}x_{i_2}\wedge\cdots\wedge\mathrm{d}x_{i_k}(e_{j_1},e_{j_2},\cdots,e_{j_k})\\&=\begin{vmatrix}\mathrm{d}x_{i_1}(e_{j_1})&\mathrm{d}x_{i_1}(e_{j_2})&\cdots&\mathrm{d}x_{i_1}(e_{j_k})\\\mathrm{d}x_{i_2}(e_{j_1})&\mathrm{d}x_{i_2}(e_{j_2})&\cdots&\mathrm{d}x_{i_2}(e_{j_k})\\\vdots&\vdots&&\vdots\\\mathrm{d}x_{i_k}(e_{j_1})&\mathrm{d}x_{i_k}(e_{j_2})&\cdots&\mathrm{d}x_{i_k}(e_{j_k})\end{vmatrix}\\&=\det\mathrm{d}x_{i_l}(e_{j_s})\end{aligned}$$

(2) 向量空间 $\wedge_k(\mathbf{R}^n)$.

若 $k=0$,我们规定 $\wedge_0(\mathbf{R}^n)=\mathbf{R}$. 它的基为 $\mathrm{d}x_{\varnothing}=1$.

若 $k\in\{1,2,\cdots,n\}$,则由所有 k 重线性形式 $\mathrm{d}x_{\boldsymbol{i}}(\boldsymbol{i}\in G_k)$ 生成的 $\mathbf{R}$—向量空间记为 $\wedge_k(\mathbf{R}^n)$.

由 $\wedge_k(\mathbf{R}^n)$ 的定义可知, $\wedge_k(\mathbf{R}^n)$ 的维数 $\dim(\wedge_k(\mathbf{R}^n))\leqslant\mathrm{C}_n^k$.

下面我们来证明:$\dim(\wedge_k(\mathbf{R}^n))=\mathrm{C}_n^k$.

为此我们只需证明$\{\mathrm{d}x_{\boldsymbol{i}}\mid\boldsymbol{i}\in G_k\}$ 线性无关.

假设存在常数 $c_{\boldsymbol{i}}$ 使得

$$\sum_{\boldsymbol{i}\in G_k}c_{\boldsymbol{i}}\mathrm{d}x_{\boldsymbol{i}}=\sum_{\boldsymbol{i}\in G_k}c_{\boldsymbol{i}}\mathrm{d}x_{i_1}\wedge\mathrm{d}x_{i_2}\wedge\cdots\wedge\mathrm{d}x_{i_k}=0$$

则由 $\mathrm{d}x_{\boldsymbol{i}}$ 的定义,对任意$(e_{j_1},e_{j_2},\cdots,e_{j_k})\in\mathbf{R}^n\times\mathbf{R}^n\times\cdots\times\mathbf{R}^n$,我们有

$$\begin{aligned}0&=\sum_{\boldsymbol{i}\in G_k}c_{\boldsymbol{i}}\mathrm{d}x_{i_1}\wedge\mathrm{d}x_{i_2}\wedge\cdots\wedge\mathrm{d}x_{i_k}(e_{j_1},e_{j_2},\cdots,e_{j_k})\\&=\sum_{\boldsymbol{i}\in G_k}c_{\boldsymbol{i}}\det(\mathrm{d}x_{i_l}(e_{j_s}))\end{aligned}$$

由于

$$\mathrm{d}x_{i_l}(e_{j_s})=\begin{cases}1,&\text{若 }i_l=j_s\\0,&\text{若 }i_l\neq j_s\end{cases}$$

故

$$\det(\mathrm{d}x_{i_l}(e_{j_s}))=\begin{cases}1,&\text{若}(i_1,i_2,\cdots,i_k)=(j_1,j_2,\cdots,j_k)\\0,&\text{若}(i_1,i_2,\cdots,i_k)\neq(j_1,j_2,\cdots,j_k)\end{cases}$$

由此可知,若取$(j_1,j_2,\cdots,j_k)=\boldsymbol{i}$,则

$$c_{\boldsymbol{i}}=0$$

此即表明$\{\mathrm{d}x_{\boldsymbol{i}}\mid\boldsymbol{i}\in G_k\}$ 线性无关.

附注 若 $k>n$,则我们也可以类似地定义 k 重线性形式

$dx_i = dx_{i_1} \wedge dx_{i_2} \wedge \cdots \wedge dx_{i_k}$,但这时由于 $i_1, i_2, \cdots, i_k$ 中至少有两个是相同的,故

$$dx_i(e_{j_1}, e_{j_2}, \cdots, e_{j_k}) = dx_{i_1} \wedge dx_{i_2} \wedge \cdots \wedge dx_{i_k}(e_{j_1}, e_{j_2}, \cdots, e_{j_k})$$
$$= \det(dx_{i_l}(e_{j_s})) = 0$$

从而 $dx_i = 0$,因此由 dx_i 生成的向量空间 $\wedge_k(\mathbf{R}^n) = 0$.

(3) 向量空间 $\wedge(\mathbf{R}^n)$.

向量空间 $\wedge_k(\mathbf{R}^n)$ 的直和记为 $\wedge(\mathbf{R}^n)$,即

$$\wedge(\mathbf{R}^n) = \bigoplus_{k=0}^{n} \wedge_k(\mathbf{R}^n)$$

$\wedge(\mathbf{R}^n)$ 的维数为

$$\dim(\wedge(\mathbf{R}^n)) = C_n^0 + C_n^1 + \cdots + C_n^n = 2^n$$

$\wedge(\mathbf{R}^n)$ 的基为 $\{dx_{\boldsymbol{i}} \mid \boldsymbol{i} \in G_k, k = 0, 1, \cdots, n\}$,即:

$dx_{\varnothing} = 1$;

$dx_1, dx_2, \cdots, dx_n$;

$dx_i \wedge dx_j (1 \leqslant i < j \leqslant n)$;

$\vdots$

$dx_{i_1} \wedge dx_{i_2} \wedge \cdots \wedge dx_{i_k} (1 \leqslant i_1 < i_2 < \cdots < i_k \leqslant n)$;

$\vdots$

$dx_1 \wedge dx_2 \wedge \cdots \wedge dx_n$.

(4) 双线性映射 $\overline{\wedge}: \wedge(\mathbf{R}^n) \times \wedge(\mathbf{R}^n) \to \wedge(\mathbf{R}^n)$.

任意 $\boldsymbol{i} \in G_k, \boldsymbol{j} \in G_m$,若 $\boldsymbol{i}$ 与 $\boldsymbol{j}$ 不相交,则我们用 $\langle \boldsymbol{i}, \boldsymbol{j} \rangle$ 表示向量

$$(i_1, i_2, \cdots, i_k, j_1, j_2, \cdots, j_m)$$

的逆序数,而用 $\boldsymbol{i} \vee \boldsymbol{j}$ 表示对此向量重新按上升次序排列所得的 $k + m$ 维向量. 于是 $\boldsymbol{i} \vee \boldsymbol{j} \in G_{k+m}$.

现在我们定义一个双线性映射 $\overline{\wedge}: \wedge(\mathbf{R}^n) \times \wedge(\mathbf{R}^n) \to \wedge(\mathbf{R}^n)$ 为

$$\forall dx_{\boldsymbol{i}} \in \wedge_k(\mathbf{R}^n), dx_{\boldsymbol{j}} \in \wedge_m(\mathbf{R}^n)$$

$$dx_{\boldsymbol{i}} \overline{\wedge} dx_{\boldsymbol{j}} = \begin{cases} 0, & \text{若 } \boldsymbol{i}, \boldsymbol{j} \text{ 相交} \\ (-1)^{\langle \boldsymbol{i}, \boldsymbol{j} \rangle} dx_{\boldsymbol{i} \vee \boldsymbol{j}}, & \text{若 } \boldsymbol{i}, \boldsymbol{j} \text{ 不相交} \end{cases}$$

关于映射 $\overline{\wedge}$ 的性质,我们有下述定理.

定理 1.1 $\forall f, g, h \in \wedge(\mathbf{R}^n), \forall a \in \mathbf{R}$,有以下性质:

(1)$(af) \overline{\wedge} g = f \overline{\wedge} (ag) = a(f \overline{\wedge} g)$.

(2)$(f \overline{\wedge} g) \overline{\wedge} h = f \overline{\wedge} (g \overline{\wedge} h)$.

(3)$g \overline{\wedge} f = (-1)^{km} f \overline{\wedge} g$,其中 $f \in \wedge_k(\mathbf{R}^n), g \in \wedge_m(\mathbf{R}^n)$.

(4)$f \overline{\wedge} 1 = 1 \overline{\wedge} f = f$.

证明 由 $\overline{\wedge}$ 的双线性性,我们只需对

$$f = \mathrm{d}x_i, g = \mathrm{d}x_j, h = \mathrm{d}x_k$$

验证上述性质即可.

性质(1) 是显然的.

下面我们来证明性质(2),即证明

$$(\mathrm{d}x_i \overline{\wedge} \mathrm{d}x_j) \overline{\wedge} \mathrm{d}x_k = \mathrm{d}x_i \overline{\wedge} (\mathrm{d}x_j \overline{\wedge} \mathrm{d}x_k)$$

若 $\boldsymbol{i},\boldsymbol{j},\boldsymbol{k}$ 不是互不相交的,则由定义知,上式左右两边都等于0.

若 $\boldsymbol{i},\boldsymbol{j},\boldsymbol{k}$ 互不相交,则

$$\begin{aligned}(\mathrm{d}x_i \overline{\wedge} \mathrm{d}x_j) \overline{\wedge} \mathrm{d}x_k &= (-1)^{\langle i,j\rangle} \mathrm{d}x_{i\vee j} \overline{\wedge} \mathrm{d}x_k \\ &= (-1)^{\langle i,j\rangle+\langle i\vee j,k\rangle} \mathrm{d}x_{(i\vee j)\vee k} \\ &= (-1)^{\langle i,j,k\rangle} \mathrm{d}x_{i\vee j\vee k}\end{aligned}$$

$$\begin{aligned}\mathrm{d}x_i \overline{\wedge} (\mathrm{d}x_j \overline{\wedge} \mathrm{d}x_k) &= \mathrm{d}x_i \overline{\wedge} ((-1)^{\langle j,k\rangle} \mathrm{d}x_{j\vee k}) \\ &= (-1)^{\langle j,k\rangle} \mathrm{d}x_i \overline{\wedge} \mathrm{d}x_{j\vee k} \\ &= (-1)^{\langle j,k\rangle+\langle i,j\vee k\rangle} \mathrm{d}x_{i\vee (j\vee k)} \\ &= (-1)^{\langle i,j,k\rangle} \mathrm{d}x_{i\vee j\vee k}\end{aligned}$$

因此性质(2) 成立.

现在证明性质(3),即证明

$$\mathrm{d}x_j \overline{\wedge} \mathrm{d}x_i = (-1)^{|i||j|} \mathrm{d}x_i \overline{\wedge} \mathrm{d}x_j$$

这里设 $\boldsymbol{i} \in G_k, \boldsymbol{j} \in G_m$, $|\boldsymbol{i}| = k$, $|\boldsymbol{j}| = m$.

若 $\boldsymbol{i}$ 与 $\boldsymbol{j}$ 相交,则此等式显然成立.

若 $\boldsymbol{i}$ 与 $\boldsymbol{j}$ 不相交,则由于

$$\mathrm{d}x_j \overline{\wedge} \mathrm{d}x_i = (-1)^{\langle j,i\rangle} \mathrm{d}x_{i\vee j}$$

$$\mathrm{d}x_i \overline{\wedge} \mathrm{d}x_j = (-1)^{\langle i,j\rangle} \mathrm{d}x_{i\vee j}$$

故要证明的等式为

$$(-1)^{\langle j,i\rangle} \mathrm{d}x_{i\vee j} = (-1)^{|i||j|+\langle i,j\rangle} \mathrm{d}x_{i\vee j}$$

由此可知,我们只需证明

$$(|\boldsymbol{i}||\boldsymbol{j}| + \langle \boldsymbol{i},\boldsymbol{j}\rangle) \bmod 2 = \langle \boldsymbol{j},\boldsymbol{i}\rangle$$

(详细证明过程留给读者)这个等式表明:从$(\boldsymbol{j},\boldsymbol{i})$ 变换为$\boldsymbol{i} \vee \boldsymbol{j}$ 的逆序数$\langle \boldsymbol{j},\boldsymbol{i}\rangle$ 与首先从$(\boldsymbol{j},\boldsymbol{i})$ 变换为$(\boldsymbol{i},\boldsymbol{j})$ 的置换次数$|\boldsymbol{i}||\boldsymbol{j}|$ 加上从$(\boldsymbol{i},\boldsymbol{j})$ 变换为 $\boldsymbol{i} \vee \boldsymbol{j}$ 的逆序数$\langle \boldsymbol{i},\boldsymbol{j}\rangle$ 之和是模 2 同余的.

最后性质(4) 是显然的.

附注 (1) 此性质表明映射 $\overline{\wedge}$ 满足结合律(它有单位元 1,但 $\overline{\wedge}$ 不是可交换的),因此($\wedge(\mathbf{R}^n)$, $\overline{\wedge}$) 形成一个代数.

(2) 若我们取 $\boldsymbol{i} = (i_1, i_2, \cdots, i_k)$, $\boldsymbol{j} = (j_1, j_2, \cdots, j_m)$,并且

$$1 \leqslant i_1 < i_2 < \cdots < i_k < j_1 < j_2 < \cdots < j_m \leqslant n$$

则 $\boldsymbol{i} \vee \boldsymbol{j} = (i_1, i_2, \cdots, i_k, j_1, j_2, \cdots, j_m)$, $\langle \boldsymbol{i}, \boldsymbol{j} \rangle = 0$, 从而

$$\begin{aligned} \mathrm{d}x_i \overline{\wedge} \mathrm{d}x_j &= (-1)^{\langle i,j \rangle} \mathrm{d}x_{i \vee j} \\ &= \mathrm{d}x_{i \vee j} \\ &= \mathrm{d}x_{i_1} \wedge \mathrm{d}x_{i_2} \wedge \cdots \wedge \mathrm{d}x_{i_k} \wedge \mathrm{d}x_{j_1} \wedge \mathrm{d}x_{j_2} \wedge \cdots \wedge \mathrm{d}x_{j_m} \end{aligned}$$

因此今后我们就改记 $\overline{\wedge}$ 为 $\wedge$.

定义 1.2　如上定义的双线性映射 $\wedge: \wedge(\mathbf{R}^n) \times \wedge(\mathbf{R}^n) \to \wedge(\mathbf{R}^n)$ 称为 $\wedge(\mathbf{R}^n)$ 的外积, $(\wedge(\mathbf{R}^n), \wedge)$ 称为 $\mathbf{R}^n$ 上的外代数, 通常简称 $\wedge(\mathbf{R}^n)$ 为 $\mathbf{R}^n$ 上的外代数或 Grassmann 代数. $f \in \wedge_k(\mathbf{R}^n)$ 称为一个 $k-$ 形式.

由 $\wedge(\mathbf{R}^n)$ 的定义可知, $\wedge(\mathbf{R}^n)$ 中的任一元素 f 具有下述一般形式

$$f = \sum_{k=0}^{n} \sum_{i \in G_k} c_i \mathrm{d}x_i, c_i \in \mathbf{R}$$

特别地, 若 $f \in \wedge_k(\mathbf{R}^n)$, 则

$$f = \sum_{i \in G_k} c_i \mathrm{d}x_i = \sum_{i \in G_k} c_i \mathrm{d}x_{i_1} \wedge \mathrm{d}x_{i_2} \wedge \cdots \wedge \mathrm{d}x_{i_k}$$

下面我们特别来看一看具体的外代数 $\wedge(\mathbf{R}^2)$ 与 $\wedge(\mathbf{R}^3)$.

例 4　外代数 $\wedge(\mathbf{R}^2) = \bigoplus_{k=0}^{2} \wedge_k(\mathbf{R}^2)$.

$\wedge_1(\mathbf{R}^2)$: 它有基 $\mathrm{d}x, \mathrm{d}y$. $\forall f, g \in \wedge_1(\mathbf{R}^2)$, 且

$$f = \alpha \mathrm{d}x + \beta \mathrm{d}y, g = \gamma \mathrm{d}x + \delta \mathrm{d}y, \alpha, \beta, \gamma, \delta \in \mathbf{R}$$

由于

$$\mathrm{d}x \wedge \mathrm{d}x = 0, \mathrm{d}y \wedge \mathrm{d}y = 0, \mathrm{d}y \wedge \mathrm{d}x = -\mathrm{d}x \wedge \mathrm{d}y$$

故

$$f \wedge f = g \wedge g = 0$$

$$\begin{aligned} f \wedge g &= (\alpha \mathrm{d}x + \beta \mathrm{d}y) \wedge (\gamma \mathrm{d}x + \delta \mathrm{d}y) \\ &= \alpha\gamma \mathrm{d}x \wedge \mathrm{d}x + \alpha\delta \mathrm{d}x \wedge \mathrm{d}y + \beta\gamma \mathrm{d}y \wedge \mathrm{d}x + \beta\delta \mathrm{d}y \wedge \mathrm{d}y \\ &= (\alpha\delta - \beta\gamma) \mathrm{d}x \wedge \mathrm{d}y \end{aligned}$$

$\wedge_2(\mathbf{R}^2)$: 它有基 $\mathrm{d}x \wedge \mathrm{d}y$. 于是

$$\forall h \in \wedge_2(\mathbf{R}^2), h = a \mathrm{d}x \wedge \mathrm{d}y$$

因此 $f \wedge g \in \wedge_2(\mathbf{R}^2)$.

例 5　外代数 $\wedge(\mathbf{R}^3) = \bigoplus_{k=0}^{3} \wedge_k(\mathbf{R}^3)$.

$\wedge_1(\mathbf{R}^3)$: 它有基 $\mathrm{d}x, \mathrm{d}y, \mathrm{d}z$.

$\wedge_2(\mathbf{R}^3)$: 它有基 $\mathrm{d}x \wedge \mathrm{d}y, \mathrm{d}y \wedge \mathrm{d}z, \mathrm{d}z \wedge \mathrm{d}x$.

$\wedge_3(\mathbf{R}^3)$: 它有基 $\mathrm{d}x \wedge \mathrm{d}y \wedge \mathrm{d}z$.

因此,若 $f,g\in\wedge_1(\mathbf{R}^3)$,$h\in\wedge_2(\mathbf{R}^3)$,$k\in\wedge_3(\mathbf{R}^3)$,则

$$f=a\mathrm{d}x+b\mathrm{d}y+c\mathrm{d}z,\ g=\alpha\mathrm{d}x+\beta\mathrm{d}y+\gamma\mathrm{d}z$$
$$h=\lambda\mathrm{d}x\wedge\mathrm{d}y+\mu\mathrm{d}y\wedge\mathrm{d}z+\xi\mathrm{d}z\wedge\mathrm{d}x$$
$$k=\eta\mathrm{d}x\wedge\mathrm{d}y\wedge\mathrm{d}z$$

从而

$$\begin{aligned}f\wedge g&=(a\mathrm{d}x+b\mathrm{d}y+c\mathrm{d}z)\wedge(\alpha\mathrm{d}x+\beta\mathrm{d}y+\gamma\mathrm{d}z)\\&=a\alpha\mathrm{d}x\wedge\mathrm{d}x+a\beta\mathrm{d}x\wedge\mathrm{d}y+a\gamma\mathrm{d}x\wedge\mathrm{d}z+\\&\quad b\alpha\mathrm{d}y\wedge\mathrm{d}x+b\beta\mathrm{d}y\wedge\mathrm{d}y+b\gamma\mathrm{d}y\wedge\mathrm{d}z+\\&\quad c\alpha\mathrm{d}z\wedge\mathrm{d}x+c\beta\mathrm{d}z\wedge\mathrm{d}y+c\gamma\mathrm{d}z\wedge\mathrm{d}z\\&=(a\beta-b\alpha)\mathrm{d}x\wedge\mathrm{d}y+(b\gamma-c\beta)\mathrm{d}y\wedge\mathrm{d}z+\\&\quad(c\alpha-a\gamma)\mathrm{d}z\wedge\mathrm{d}x\end{aligned}$$

$$\begin{aligned}f\wedge h&=(a\mathrm{d}x+b\mathrm{d}y+c\mathrm{d}z)\wedge(\lambda\mathrm{d}x\wedge\mathrm{d}y+\mu\mathrm{d}y\wedge\mathrm{d}z+\xi\mathrm{d}z\wedge\mathrm{d}x)\\&=a\lambda\mathrm{d}x\wedge\mathrm{d}x\wedge\mathrm{d}y+a\mu\mathrm{d}x\wedge\mathrm{d}y\wedge\mathrm{d}z+a\xi\mathrm{d}x\wedge\mathrm{d}z\wedge\mathrm{d}x+\\&\quad b\lambda\mathrm{d}y\wedge\mathrm{d}x\wedge\mathrm{d}y+b\mu\mathrm{d}y\wedge\mathrm{d}y\wedge\mathrm{d}z+\\&\quad b\xi\mathrm{d}y\wedge\mathrm{d}z\wedge\mathrm{d}x+c\lambda\mathrm{d}z\wedge\mathrm{d}x\wedge\mathrm{d}y+\\&\quad c\mu\mathrm{d}z\wedge\mathrm{d}y\wedge\mathrm{d}z+c\xi\mathrm{d}z\wedge\mathrm{d}z\wedge\mathrm{d}x\\&=(a\mu+b\xi+c\lambda)\mathrm{d}x\wedge\mathrm{d}y\wedge\mathrm{d}z\end{aligned}$$

于是 $f\wedge g\in\wedge_2(\mathbf{R}^3)$,$f\wedge h\in\wedge_3(\mathbf{R}^3)$.

显然,我们有

$$f\wedge f=g\wedge g=0,h\wedge h=0,k\wedge k=0,f\wedge k=0,h\wedge k=0$$

例 6 设 $f=\mathrm{d}x_1\wedge\mathrm{d}x_2+\mathrm{d}x_3\wedge\mathrm{d}x_4\in\wedge(\mathbf{R}^4)$. 则

$$\begin{aligned}f\wedge f&=(\mathrm{d}x_1\wedge\mathrm{d}x_2+\mathrm{d}x_3\wedge\mathrm{d}x_4)\wedge(\mathrm{d}x_1\wedge\mathrm{d}x_2+\mathrm{d}x_3\wedge\mathrm{d}x_4)\\&=\mathrm{d}x_1\wedge\mathrm{d}x_2\wedge\mathrm{d}x_1\wedge\mathrm{d}x_2+\mathrm{d}x_1\wedge\mathrm{d}x_2\wedge\mathrm{d}x_3\wedge\mathrm{d}x_4+\\&\quad\mathrm{d}x_3\wedge\mathrm{d}x_4\wedge\mathrm{d}x_1\wedge\mathrm{d}x_2+\mathrm{d}x_3\wedge\mathrm{d}x_4\wedge\mathrm{d}x_3\wedge\mathrm{d}x_4\\&=\mathrm{d}x_1\wedge\mathrm{d}x_2\wedge\mathrm{d}x_3\wedge\mathrm{d}x_4+\mathrm{d}x_1\wedge\mathrm{d}x_2\wedge\mathrm{d}x_3\wedge\mathrm{d}x_4\\&=2\mathrm{d}x_1\wedge\mathrm{d}x_2\wedge\mathrm{d}x_3\wedge\mathrm{d}x_4\end{aligned}$$

3. 外代数的提升

设 $f\in\wedge(\mathbf{R}^m)$,$T\in\mathscr{L}(\mathbf{R}^n,\mathbf{R}^m)$ 是任一线性映射. 假设 T 在 $\mathbf{R}^n$ 与 $\mathbf{R}^m$ 的标准正规正交基下的矩阵为

$$\boldsymbol{M}=\begin{pmatrix}a_{11}&a_{12}&\cdots&a_{1n}\\a_{21}&a_{22}&\cdots&a_{2n}\\\vdots&\vdots&&\vdots\\a_{m1}&a_{m2}&\cdots&a_{mn}\end{pmatrix}$$

对任意 $x=(x_1,x_2,\cdots,x_n)\in\mathbf{R}^n$,若令 $T(x)=y=(y_1,y_2,\cdots,y_m)\in\mathbf{R}^m$,则

$$y_i=a_{i1}x_1+a_{i2}x_2+\cdots+a_{in}x_n, i=1,2,\cdots,m$$

从而

$$\mathrm{d}y_i=a_{i1}\mathrm{d}x_1+a_{i2}\mathrm{d}x_2+\cdots+a_{in}\mathrm{d}x_n, i=1,2,\cdots,m$$

现在设 $f\in\wedge(\mathbf{R}^m)$ 的表达式为

$$f=\sum_{k=0}^{m}\sum_{i\in G_k}c_i\mathrm{d}y_{i_1}\wedge\mathrm{d}y_{i_2}\wedge\cdots\wedge\mathrm{d}y_{i_k}$$

将 $\mathrm{d}y_{i_l}=a_{i_l1}\mathrm{d}x_1+a_{i_l2}\mathrm{d}x_2+\cdots+a_{i_ln}\mathrm{d}x_n$ 代入上式的右端得到

$$\sum_{k=0}^{m}\sum_{i\in G_k}c_i\left(\sum_{s=1}^{n}a_{i_1s}\mathrm{d}x_s\right)\wedge\left(\sum_{s=1}^{n}a_{i_2s}\mathrm{d}x_s\right)\wedge\cdots\wedge\left(\sum_{s=1}^{n}a_{i_ks}\mathrm{d}x_s\right)$$

利用外积的性质将为0的项删去,对非零项的脚标重新按上升次序排列后,我们将得到 $\wedge(\mathbf{R}^n)$ 中的一个元素.

定义 1.3 设映射 $T^*:\wedge(\mathbf{R}^m)\to\wedge(\mathbf{R}^n)$ 定义如下

$$\forall f=\sum_{k=0}^{m}\sum_{i\in G_k}c_i\mathrm{d}y_{i_1}\wedge\mathrm{d}y_{i_2}\wedge\cdots\wedge\mathrm{d}y_{i_k}$$

$$T^*f=\sum_{k=0}^{m}\sum_{i\in G_k}c_i\left(\sum_{s=1}^{n}a_{i_1s}\mathrm{d}x_s\right)\wedge\left(\sum_{s=1}^{n}a_{i_2s}\mathrm{d}x_s\right)\wedge\cdots\wedge\left(\sum_{s=1}^{n}a_{i_ks}\mathrm{d}x_s\right)$$

则映射 T^* 称为映射 T 的提升,而 T^*f 称为 f 关于 T 的提升.

例 7 设 $f=\mathrm{d}y_1\wedge\mathrm{d}y_2+2\mathrm{d}y_2\wedge\mathrm{d}y_3$. $T\in\mathscr{L}(\mathbf{R}^2,\mathbf{R}^3)$ 的矩阵为

$$\boldsymbol{M}=\begin{pmatrix}1&0\\0&-1\\2&1\end{pmatrix}$$

计算 f 关于 T 的提升 T^*f.

这时我们有

$$\mathrm{d}y_1=\mathrm{d}x_1$$
$$\mathrm{d}y_2=-\mathrm{d}x_2$$
$$\mathrm{d}y_3=2\mathrm{d}x_1+\mathrm{d}x_2$$

因此

$$\begin{aligned}T^*f&=\mathrm{d}x_1\wedge(-\mathrm{d}x_2)+2(-\mathrm{d}x_2)\wedge(2\mathrm{d}x_1+\mathrm{d}x_2)\\&=-\mathrm{d}x_1\wedge\mathrm{d}x_2-4\mathrm{d}x_2\wedge\mathrm{d}x_1-2\mathrm{d}x_2\wedge\mathrm{d}x_2\\&=3\mathrm{d}x_1\wedge\mathrm{d}x_2\end{aligned}$$

关于映射 T^* 的性质,我们有下述定理.

定理 1.2 $\forall T\in\mathscr{L}(\mathbf{R}^n,\mathbf{R}^m), S\in\mathscr{L}(\mathbf{R}^m,\mathbf{R}^p)$,则:

(1) $T^*:\wedge(\mathbf{R}^m)\to\wedge(\mathbf{R}^n)$ 是线性映射.

(2) $\forall f,g\in\wedge(\mathbf{R}^m)$, $T^*(f\wedge g)=T^*f\wedge T^*g$.

(3) $\forall f\in\wedge(\mathbf{R}^p)$, $(S\circ T)^*f=T^*(S^*f)$.

证明 (1) 由 T^* 的定义可知,T^* 是线性的.

(2) 由 T^* 的线性性可知,我们只需证明,$\forall \mathrm{d}y_i, \mathrm{d}y_j \in \wedge(\mathbf{R}^m)$,则

$$T^*(\mathrm{d}y_i \wedge \mathrm{d}y_j) = T^*(\mathrm{d}y_i) \wedge T^*(\mathrm{d}y_j)$$

这由 T^* 的定义知它是显然的.

(3) 同理由 T^* 与 S^* 的线性性,我们只需证明

$$\forall \mathrm{d}z_I \in \wedge(\mathbf{R}^p), (S \circ T)^* \mathrm{d}z_I = T^*(S^* \mathrm{d}z_I)$$

为此设 $\mathrm{d}z_I = \mathrm{d}z_{l_1} \wedge \mathrm{d}z_{l_2} \wedge \cdots \wedge \mathrm{d}z_{l_n}$.

$$\forall x = (x_1, x_2, \cdots, x_n) \in \mathbf{R}^n$$

$$T(x) = y = (y_1, y_2, \cdots, y_m) = \begin{pmatrix} a_{11} & a_{12} & \cdots & a_{1n} \\ a_{21} & a_{22} & \cdots & a_{2n} \\ \vdots & \vdots & & \vdots \\ a_{m1} & a_{m2} & \cdots & a_{mn} \end{pmatrix} \begin{pmatrix} x_1 \\ x_2 \\ \vdots \\ x_n \end{pmatrix}$$

$$S(y) = z = (z_1, z_2, \cdots, z_p) = \begin{pmatrix} b_{11} & b_{12} & \cdots & b_{1m} \\ b_{21} & b_{22} & \cdots & b_{2m} \\ \vdots & \vdots & & \vdots \\ b_{p1} & b_{p2} & \cdots & b_{pm} \end{pmatrix} \begin{pmatrix} y_1 \\ y_2 \\ \vdots \\ y_m \end{pmatrix}$$

则

$$\mathrm{d}y_j = a_{j1}\mathrm{d}x_1 + a_{j2}\mathrm{d}x_2 + \cdots + a_{jn}\mathrm{d}x_n, j = 1, 2, \cdots, m$$

$$\mathrm{d}z_s = b_{s1}\mathrm{d}y_1 + b_{s2}\mathrm{d}y_2 + \cdots + b_{sm}\mathrm{d}y_m, s = 1, 2, \cdots, p$$

于是

$$\begin{aligned} \mathrm{d}z_s &= b_{s1}\left(\sum_{i=1}^{n} a_{1i}\mathrm{d}x_i\right) + b_{s2}\left(\sum_{i=1}^{n} a_{2i}\mathrm{d}x_i\right) + \cdots + \\ &\quad b_{sm}\left(\sum_{i=1}^{n} a_{mi}\mathrm{d}x_i\right) \\ &= \left(\sum_{j=1}^{m} b_{sj}a_{j1}\right)\mathrm{d}x_1 + \left(\sum_{j=1}^{m} b_{sj}a_{j2}\right)\mathrm{d}x_2 + \cdots + \\ &\quad \left(\sum_{j=1}^{m} b_{sj}a_{jn}\right)\mathrm{d}x_n \end{aligned}$$

从而

$$\begin{aligned} (S \circ T)^* \mathrm{d}z_I &= (S \circ T)^* \mathrm{d}z_{l_1} \wedge \mathrm{d}z_{l_2} \wedge \cdots \wedge \mathrm{d}z_{l_h} \\ &= \left(\sum_{i=1}^{n}\left(\sum_{j=1}^{m} b_{l_1 j}a_{ji}\right)\mathrm{d}x_i\right) \wedge \\ &\quad \left(\sum_{i=1}^{n}\left(\sum_{j=1}^{m} b_{l_2 j}a_{ji}\right)\mathrm{d}x_i\right) \wedge \cdots \wedge \\ &\quad \left(\sum_{i=1}^{n}\left(\sum_{j=1}^{m} b_{l_h j}a_{ji}\right)\mathrm{d}x_i\right) \end{aligned}$$

$$
\begin{aligned}
&=\Big(\sum_{j=1}^{m} b_{l_1 j}\Big(\sum_{i=1}^{n} a_{ji}\,\mathrm{d}x_i\Big)\Big)\wedge \\
&\quad\Big(\sum_{j=1}^{m} b_{l_2 j}\Big(\sum_{i=1}^{n} a_{ji}\,\mathrm{d}x_i\Big)\Big)\wedge\cdots\wedge \\
&\quad\Big(\sum_{j=1}^{m} b_{l_h j}\Big(\sum_{i=1}^{n} a_{ji}\,\mathrm{d}x_i\Big)\Big) \\
&=\Big(\sum_{j=1}^{m} b_{l_1 j} T^*\,\mathrm{d}y_j\Big)\wedge \\
&\quad\Big(\sum_{j=1}^{m} b_{l_2 j} T^*\,\mathrm{d}y_j\Big)\wedge\cdots\wedge \\
&\quad\Big(\sum_{j=1}^{m} b_{l_h j} T^*\,\mathrm{d}y_j\Big) \\
&=\Big(T^*\Big(\sum_{j=1}^{m} b_{l_1 j} T^*\,\mathrm{d}y_j\Big)\Big)\wedge \\
&\quad\Big(T^*\Big(\sum_{j=1}^{m} b_{l_2 j} T^*\,\mathrm{d}y_j\Big)\Big)\wedge\cdots\wedge \\
&\quad\Big(T^*\Big(\sum_{j=1}^{m} b_{l_h j} T^*\,\mathrm{d}y_j\Big)\Big) \\
&=T^*\Big(\Big(\sum_{j=1}^{m} b_{l_1 j}\,\mathrm{d}y_j\Big)\wedge \\
&\quad\Big(\sum_{j=1}^{m} b_{l_2 j}\,\mathrm{d}y_j\Big)\wedge\cdots\wedge \\
&\quad\Big(\sum_{j=1}^{m} b_{l_h j}\,\mathrm{d}y_j\Big)\Big) \\
&=T^*(S^*\,\mathrm{d}z_I)
\end{aligned}
$$

例 8 设 $f=\mathrm{d}z_1\wedge\mathrm{d}z_2-\mathrm{d}z_3\wedge\mathrm{d}z_4\in\wedge_2(\mathbf{R}^4)$. 线性映射 $T\in\mathscr{L}(\mathbf{R}^2,\mathbf{R}^3)$ 与 $S\in\mathscr{L}(\mathbf{R}^3,\mathbf{R}^4)$ 的矩阵分别为

$$
\boldsymbol{A}=\begin{pmatrix}1&1\\0&1\\1&2\end{pmatrix},\boldsymbol{B}=\begin{pmatrix}1&0&1\\0&1&1\\1&0&-1\\1&1&0\end{pmatrix}
$$

试直接验证 $(S\circ T)^*f=T^*(S^*f)$.

$$
\boldsymbol{C}=\boldsymbol{BA}=\begin{pmatrix}2&3\\1&3\\0&-1\\1&2\end{pmatrix}
$$

是线性映射 $S\circ T$ 的矩阵. 从而

$$(\mathrm{d}z_1,\mathrm{d}z_2,\mathrm{d}z_3,\mathrm{d}z_4)^{\mathrm{T}}=\begin{pmatrix}2&3\\1&3\\0&-1\\1&2\end{pmatrix}\begin{pmatrix}\mathrm{d}x_1\\\mathrm{d}x_2\end{pmatrix}$$

$$(\mathrm{d}z_1,\mathrm{d}z_2,\mathrm{d}z_3,\mathrm{d}z_4)^{\mathrm{T}}=\begin{pmatrix}1&0&1\\0&1&1\\1&0&-1\\1&1&0\end{pmatrix}\begin{pmatrix}\mathrm{d}y_1\\\mathrm{d}y_2\\\mathrm{d}y_3\end{pmatrix}$$

$$(\mathrm{d}y_1,\mathrm{d}y_2,\mathrm{d}y_3)^{\mathrm{T}}=\begin{pmatrix}1&1\\0&1\\1&2\end{pmatrix}\begin{pmatrix}\mathrm{d}x_1\\\mathrm{d}x_2\end{pmatrix}$$

或

$$\mathrm{d}z_1=2\mathrm{d}x_1+3\mathrm{d}x_2,\mathrm{d}z_2=\mathrm{d}x_1+3\mathrm{d}x_2$$
$$\mathrm{d}z_3=-\mathrm{d}x_2,\mathrm{d}z_4=\mathrm{d}x_1+2\mathrm{d}x_2$$
$$\mathrm{d}z_1=\mathrm{d}y_1+\mathrm{d}y_3,\mathrm{d}z_2=\mathrm{d}y_2+\mathrm{d}y_3$$
$$\mathrm{d}z_3=\mathrm{d}y_1-\mathrm{d}y_3,\mathrm{d}z_4=\mathrm{d}y_1+\mathrm{d}y_2$$
$$\mathrm{d}y_1=\mathrm{d}x_1+\mathrm{d}x_2,\mathrm{d}y_2=\mathrm{d}x_2,\mathrm{d}y_3=\mathrm{d}x_1+2\mathrm{d}x_2$$

于是

$$\begin{aligned}(S\circ T)^*f&=(2\mathrm{d}x_1+3\mathrm{d}x_2)\wedge(\mathrm{d}x_1+3\mathrm{d}x_2)+\mathrm{d}x_2\wedge(\mathrm{d}x_1+2\mathrm{d}x_2)\\&=6\mathrm{d}x_1\wedge\mathrm{d}x_2+3\mathrm{d}x_2\wedge\mathrm{d}x_1+\mathrm{d}x_2\wedge\mathrm{d}x_1\\&=2\mathrm{d}x_1\wedge\mathrm{d}x_2\end{aligned}$$

$$\begin{aligned}S^*f&=(\mathrm{d}y_1+\mathrm{d}y_3)\wedge(\mathrm{d}y_2+\mathrm{d}y_3)-(\mathrm{d}y_1-\mathrm{d}y_3)\wedge(\mathrm{d}y_1+\mathrm{d}y_2)\\&=\mathrm{d}y_1\wedge\mathrm{d}y_2+\mathrm{d}y_1\wedge\mathrm{d}y_3+\mathrm{d}y_3\wedge\mathrm{d}y_2-\\&\quad\mathrm{d}y_1\wedge\mathrm{d}y_2+\mathrm{d}y_3\wedge\mathrm{d}y_1+\mathrm{d}y_3\wedge\mathrm{d}y_2\\&=-2\mathrm{d}y_2\wedge\mathrm{d}y_3\end{aligned}$$

$$\begin{aligned}T^*(S^*f)&=-2\mathrm{d}x_2\wedge(\mathrm{d}x_1+2\mathrm{d}x_2)\\&=-2\mathrm{d}x_2\wedge\mathrm{d}x_1\\&=2\mathrm{d}x_1\wedge\mathrm{d}x_2\end{aligned}$$

因此

$$(S\circ T)^*f=T^*(S^*f)=2\mathrm{d}x_1\wedge\mathrm{d}x_2$$

二、微分形式

1. 微分形式的定义

定义 2.1　设 $U\subset\mathbf{R}^n$ 是任一开集.

(1)U 上的一个微分形式就是映射 $\omega:U\to\wedge(\mathbf{R}^n)$. 于是

$$\forall x \in U, \omega(x) = \sum_{k=0}^{n} \sum_{i \in G_k} c_i(x) \mathrm{d}x_{i_1} \wedge \mathrm{d}x_{i_2} \wedge \cdots \wedge \mathrm{d}x_{i_k}$$

这里 $c_i: U \to \mathbf{R}$ 是一个依赖于 ω 的函数，称为微分形式 ω 的系数函数.

我们用 $\Omega(U)$ 表示 U 上的所有微分形式组成的集合.

(2) 设 ω 是 U 上的一个微分形式. 若 $\omega(U) \subset \wedge_k(\mathbf{R}^n)$，则我们称 ω 是一个 k— 微分形式. 于是

$$\forall x \in U, \omega(x) = \sum_{i \in G_k} c_i(x) \mathrm{d}x_{i_1} \wedge \mathrm{d}x_{i_2} \wedge \cdots \wedge \mathrm{d}x_{i_k}$$

我们用 $\Omega_k(U)$ 表示 U 上的所有 k — 微分形式组成的集合.

附注 由微分形式的定义及外代数 $\wedge(\mathbf{R}^n)$ 的性质可知，若 $k > n$，则 $\Omega_k(U) = \{0\}$.

若 $k = 0$，则 $\Omega_0(U)$ 就是所有定义在 U 上的实值函数的集合.

定义 2.2 设 ω 是 U 上的任一微分形式，我们称 ω 在 U 上是 $C^p(p \geqslant 0)$ 类的，如果 ω 的系数函数 $c_i: U \to \mathbf{R}(\boldsymbol{i} \in G_k, k = 0, 1, \cdots, n)$ 在 U 上是 C^p 类的.

例 1 如下定义的各微分形式 $\omega_i (i = 1, 2, 3)$

$$\omega_1(x) = \frac{x\mathrm{d}y - y\mathrm{d}x}{x^2 + y^2}, (x, y) \in \mathbf{R}^2 - \{(0, 0)\}$$

$$\omega_2(x) = x\mathrm{d}y \wedge \mathrm{d}z + y\mathrm{d}z \wedge \mathrm{d}x + z\mathrm{d}x \wedge \mathrm{d}y, (x, y, z) \in \mathbf{R}^3$$

$$\omega_3(x) = r\sin\theta\cos\varphi \mathrm{d}r \wedge \mathrm{d}\theta \wedge \mathrm{d}\varphi$$

$$(r, \theta, \varphi) \in \mathbf{R}_+ \times [0, 2\pi] \times [0, \pi]$$

是在各自定义域上 C^∞ 类的 1 — 微分形式、2 — 微分形式与 3 — 微分形式.

例 2 设 $f: U \to \mathbf{R}$ 是任一可微映射，由于

$$\forall x \in U, \mathrm{d}f(x) = \frac{\partial f}{\partial x_1}(x)\mathrm{d}x_1 + \frac{\partial f}{\partial x_2}(x)\mathrm{d}x_2 + \cdots + \frac{\partial f}{\partial x_n}(x)\mathrm{d}x_n$$

故 $\mathrm{d}f(x) \in \wedge_1(\mathbf{R}^n)$，从而 f 的微分映射 $\mathrm{d}f$ 是 U 上的一个 1 — 微分形式.

因此微分形式是微分映射概念的推广.

2. 外微分形式代数

首先我们在 $\Omega(U)$ 上定义加法与数乘运算如下：

$\forall \omega, \widetilde{\omega} \in \Omega(U), \forall \alpha \in \mathbf{R}$，令

$$\omega = \sum_{k=0}^{n} \sum_{i \in G_k} c_i \mathrm{d}x_i, \widetilde{\omega} = \sum_{k=0}^{n} \sum_{i \in G_k} \widetilde{c}_i \mathrm{d}x_i$$

则

$$\omega+\widetilde{\omega}\xlongequal{\triangle}\sum_{k=0}^{n}\sum_{i\in G_k}(c_i+\widetilde{c}_i)\mathrm{d}x_i$$

$$\alpha\omega\xlongequal{\triangle}\sum_{k=0}^{n}\sum_{i\in G_k}(\alpha c_i)\mathrm{d}x_i$$

于是 $\Omega(U)$ 关于这两个运算形成一个 $\mathbf{R}-$ 向量空间.

下面我们来定义一个映射 $\wedge:\Omega(U)\times\Omega(U)\rightarrow\Omega(U)$ 如下

$$\forall\,\omega=\sum_{k=0}^{n}\sum_{i\in G_k}c_i\mathrm{d}x_i,\widetilde{\omega}=\sum_{k=0}^{n}\sum_{i\in G_k}\widetilde{c}_i\mathrm{d}x_i\in\Omega(U)$$

$$\omega\wedge\widetilde{\omega}=\theta\in\Omega(U)\Leftrightarrow\forall\,x\in U$$

$$\theta(x)=\Big(\sum_{k=0}^{n}\sum_{i\in G_k}c_i(x)\mathrm{d}x_i\Big)\wedge\Big(\sum_{k=0}^{n}\sum_{i\in G_k}\widetilde{c}_i(x)\mathrm{d}x_i\Big)$$

根据外代数 $\wedge(\mathbf{R}^n)$ 上的外积的双线性性可知,上述映射 $\wedge:\Omega(U)\times\Omega(U)\rightarrow\Omega(U)$ 是双线性的,并且满足结合律

$$\forall\,\omega,\widetilde{\omega},\widetilde{\widetilde{\omega}}\in\Omega(U),(\omega\wedge\widetilde{\omega})\wedge\widetilde{\widetilde{\omega}}=\omega\wedge(\widetilde{\omega}\wedge\widetilde{\widetilde{\omega}})$$

因此$(\Omega(U),\wedge)$形成一个代数.

定义 2.3　代数$(\Omega(U),\wedge)$称为外微分形式代数,或简称 $\Omega(U)$ 为外微分形式代数,$\wedge$ 称为 $\Omega(U)$ 上的外积.

由外代数 $\wedge(\mathbf{R}^n)$ 上的外积 $\wedge$ 的性质可直接推出 $\Omega(U)$ 上的外积的下述性质.

定理 2.1　$\forall\,\omega,\widetilde{\omega}\in\Omega(U),\forall\,f:U\rightarrow\mathbf{R}$,则:

(1)$(f\omega)\wedge\widetilde{\omega}=\omega\wedge(f\widetilde{\omega})=f(\omega\wedge\widetilde{\omega})$.

(2) 若 $\omega\in\Omega_k(U),\widetilde{\omega}\in\Omega_m(U)$,则 $\omega\wedge\widetilde{\omega}=(-1)^{km}\widetilde{\omega}\wedge\omega$.

此定理的证明我们留给读者.

例 3　设 $\omega=y\mathrm{d}x-x\mathrm{d}y$,且

$$\widetilde{\omega}=x^2\mathrm{d}y\wedge\mathrm{d}z+y^2\mathrm{d}z\wedge\mathrm{d}x+z^2\mathrm{d}x\wedge\mathrm{d}y$$

则

$$\begin{aligned}\omega\wedge\widetilde{\omega}&=(y\mathrm{d}x-x\mathrm{d}y)\wedge(x^2\mathrm{d}y\wedge\mathrm{d}z+y^2\mathrm{d}z\wedge\mathrm{d}x+z^2\mathrm{d}x\wedge\mathrm{d}y)\\&=x^2y\mathrm{d}x\wedge\mathrm{d}y\wedge\mathrm{d}z+y^3\mathrm{d}x\wedge\mathrm{d}z\wedge\mathrm{d}x+\\&\quad yz^2\mathrm{d}x\wedge\mathrm{d}x\wedge\mathrm{d}y-x^3\mathrm{d}y\wedge\mathrm{d}y\wedge\mathrm{d}z-\\&\quad xy^2\mathrm{d}y\wedge\mathrm{d}z\wedge\mathrm{d}x-xz^2\mathrm{d}y\wedge\mathrm{d}x\wedge\mathrm{d}y\\&=(x^2y-xy^2)\mathrm{d}x\wedge\mathrm{d}y\wedge\mathrm{d}z\end{aligned}$$

3. 微分形式的提升

设 $U\subset\mathbf{R}^n,V\subset\mathbf{R}^m$ 是两个开集,$f:U\rightarrow V$ 是任一 C^p 类映射. 于是

$\forall x \in U, \mathrm{d}f(x) \in \mathscr{L}(\mathbf{R}^n, \mathbf{R}^m)$.

定义 2.4 设映射 $f^*: \Omega(V) \to \Omega(U)$ 定义如下

$$\forall \omega \in \Omega(V), f^*(\omega) \in \Omega(U) \Leftrightarrow$$

$$\forall x \in U, f^*(\omega)(x) = (\mathrm{d}f(x))^*(\omega) \in \wedge(\mathbf{R}^n)$$

映射 f^* 称为 f 的提升,$f^*(\omega)$ 称为微分形式 ω 关于 f 的提升.

实际求 $f^*(\omega)$ 的方法如下:

设 $\omega = \sum_{i \in G_k} c_i \mathrm{d}y_i \in \Omega_k(V)$,于是

$$\forall y \in V, \omega(y) = \sum_{i \in G_k} c_i(y) \mathrm{d}y_{i_1} \wedge \mathrm{d}y_{i_2} \wedge \cdots \wedge \mathrm{d}y_{i_k}$$

现在设 $f = (f_1, f_2, \cdots, f_m)$,则

$$\mathrm{d}y_i = \frac{\partial f_i}{\partial x_1}\mathrm{d}x_1 + \frac{\partial f_i}{\partial x_2}\mathrm{d}x_2 + \cdots + \frac{\partial f_i}{\partial x_n}\mathrm{d}x_n, i = 1, 2, \cdots, m$$

因此 ω 关于 f 的提升 $f^*(\omega)$ 为

$$\begin{aligned}\forall x \in U, f^*(\omega)(x) = & \sum_{i \in G_k} c_i(f(x))\left(\sum_{j=1}^{n} \frac{\partial f_{i_1}}{\partial x_j}(x)\mathrm{d}x_j\right) \wedge \\ & \left(\sum_{j=1}^{n} \frac{\partial f_{i_2}}{\partial x_j}(x)\mathrm{d}x_j\right) \wedge \cdots \wedge \\ & \left(\sum_{j=1}^{n} \frac{\partial f_{i_k}}{\partial x_j}(x)\mathrm{d}x_j\right)\end{aligned}$$

例 4 设 $f: U \to V$ 是任一 C^1 类映射,ω 是 V 上的任一 0-微分形式,则

$$\forall x \in U, f^*(\omega)(x) = \omega(f(x)) = (\omega \circ f)(x)$$

从而

$$f^*(\omega) = \omega \circ f$$

例 5 设 $f: U \to V$ 是任一 C^1 类映射,$g: V \to \mathbf{R}$ 是任一 C^1 类函数.于是 $\omega = \mathrm{d}g$ 是 V 上的 1-微分形式,并且

$$\mathrm{d}g = \frac{\partial g}{\partial y_1}\mathrm{d}y_1 + \frac{\partial g}{\partial y_2}\mathrm{d}y_2 + \cdots + \frac{\partial g}{\partial y_m}\mathrm{d}y_m$$

从而 ω 关于 f 的提升 $f^*(\omega)$ 为

$$\begin{aligned}\forall x \in U, f^*(\omega)(x) &= \sum_{i=1}^{m} \frac{\partial g}{\partial y_i}(f(x))\left(\sum_{j=1}^{n} \frac{\partial f_i}{\partial x_j}(x)\mathrm{d}x_j\right) \\ &= \sum_{j=1}^{n}\left(\sum_{i=1}^{m} \frac{\partial g}{\partial y_i}(f(x)) \frac{\partial f_i}{\partial x_j}(x)\right)\mathrm{d}x_j \\ &= \sum_{j=1}^{n} \frac{\partial(g \circ f)}{\partial x_j}(x)\mathrm{d}x_j \\ &= \mathrm{d}(g \circ f)(x)\end{aligned}$$

因此

$$f^*(\omega)=\mathrm{d}(g\circ f)$$

例 6 设 ω 是 $\mathbf{R}^2$ 上的 1－微分形式，定义为

$$\omega(x,y)=x\mathrm{d}y+y\mathrm{d}x,\forall(x,y)\in\mathbf{R}^2$$

$\varphi:[0,2]\to\mathbf{R}^2$ 是如下定义的 C^1 类映射

$$\forall t\in[0,2],\varphi(t)=(t,t^2)$$

则 ω 关于 φ 的提升 $\varphi^*(\omega)$ 为

$$\begin{aligned}\forall t\in[0,2],\varphi^*(\omega)(t)&=t\mathrm{d}(t^2)+t^2\mathrm{d}(t)\\&=3t^2\mathrm{d}t\end{aligned}$$

例 7 设 ω 是如下定义的 $\mathbf{R}^3$ 上的 3－微分形式

$$\omega=\mathrm{d}x\wedge\mathrm{d}y\wedge\mathrm{d}z$$

映射 $f:[0,+\infty]\times[0,2\pi]\times[0,\pi]\to\mathbf{R}$ 定义为

$$\forall(r,\theta,\varphi)\in[0,+\infty]\times[0,2\pi]\times[0,\pi]$$

$$f(r,\theta,\varphi)=(r\sin\varphi\cos\theta,r\sin\varphi\sin\theta,r\cos\varphi)$$

由于

$$\mathrm{d}x=\mathrm{d}(r\sin\varphi\cos\theta)=\sin\varphi\cos\theta\mathrm{d}r-r\sin\varphi\sin\theta\mathrm{d}\theta+r\cos\varphi\cos\theta\mathrm{d}\varphi$$

$$\mathrm{d}y=\mathrm{d}(r\sin\varphi\sin\theta)=\sin\varphi\sin\theta\mathrm{d}r+r\sin\varphi\cos\theta\mathrm{d}\theta+r\cos\varphi\sin\theta\mathrm{d}\varphi$$

$$\mathrm{d}z=\cos\varphi\mathrm{d}r-r\sin\varphi\mathrm{d}\varphi$$

故 ω 关于 f 的提升 $f^*(\omega)$ 为

$$\begin{aligned}&\forall(r,\theta,\varphi)\in[0,+\infty]\times[0,2\pi]\times[0,\pi]\\&f^*(\omega)(r,\theta,\varphi)\\=&(\sin\varphi\cos\theta\mathrm{d}r-r\sin\varphi\sin\theta\mathrm{d}\theta+r\cos\varphi\cos\theta\mathrm{d}\varphi)\wedge\\&(\sin\varphi\sin\theta\mathrm{d}r+r\sin\varphi\cos\theta\mathrm{d}\theta+r\cos\varphi\sin\theta\mathrm{d}\varphi)\wedge\\&(\cos\varphi\mathrm{d}r-r\sin\theta\mathrm{d}\varphi)\\=&(-r\sin\varphi\sin\theta)(r\cos\varphi\sin\theta)\cos\varphi\mathrm{d}\theta\wedge\mathrm{d}\varphi\wedge\mathrm{d}r+\\&(r\cos\varphi\cos\theta)(r\sin\varphi\cos\theta)\cos\varphi\mathrm{d}\varphi\wedge\mathrm{d}\theta\wedge\mathrm{d}r+\\&(\sin\varphi\cos\theta)(r\sin\varphi\cos\theta)(-r\sin\varphi)\mathrm{d}r\wedge\mathrm{d}\theta\wedge\mathrm{d}\varphi+\\&(-r\sin\varphi\sin\theta)(\sin\varphi\sin\theta)(-r\sin\varphi)\mathrm{d}\theta\wedge\mathrm{d}r\wedge\mathrm{d}\varphi\\=&(-r^2\sin\varphi\cos^2\varphi\sin^2\theta-r^2\sin\varphi\cos^2\varphi\cos^2\theta-\\&r^2\sin\varphi\sin^2\varphi\cos^2\theta-r^2\sin\varphi\sin^2\varphi\sin^2\theta)\mathrm{d}r\wedge\mathrm{d}\theta\wedge\mathrm{d}\varphi\\=&-(r^2\sin\varphi\cos^2\varphi+r^2\sin\varphi\sin^2\varphi)\mathrm{d}r\wedge\mathrm{d}\theta\wedge\mathrm{d}\varphi\\=&-r^2\sin\varphi\mathrm{d}r\wedge\mathrm{d}\theta\wedge\mathrm{d}\varphi\end{aligned}$$

关于映射 f 的提升 f^* 的性质，可用下述定理来描述.

定理 2.2 设 $f:U\to V$ 是任一 C^1 类映射，则：

(1)$f^*:\Omega(V)\to\Omega(U)$ 是线性映射.

(2)$\forall\omega,\tilde{\omega}\in\Omega(V),f^*(\omega\wedge\tilde{\omega})=f^*(\omega)\wedge f^*(\tilde{\omega})$.

(3) 若 $g:V\to W(\subset\mathbf{R}^k)$ 是另一 C^1 类映射，则 $\forall\omega\in\Omega(W)$，$(g\circ f)^*(\omega)=f^*(g^*(\omega))$.

证明 此定理可直接由提升 f^* 的定义及定理 1.2 推出.

例 8 设 $\omega\in\Omega_2(\mathbf{R}^3)$ 定义如下

$$\forall(x,y,z)\in\mathbf{R}^3,\omega(x,y)=y\mathrm{d}x\wedge\mathrm{d}z$$

映射 $f:[0,2\pi]\times[0,\pi]\to\mathbf{R}^2$ 与 $g:\mathbf{R}^2\to\mathbf{R}^3$ 定义为

$$\forall(\theta,\varphi)\in[0,2\pi]\times[0,\pi]$$

$$f(\theta,\varphi)=(\cos\theta\cos\varphi,\sin\theta\cos\varphi)$$

$$\forall(x,y)\in\mathbf{R}^2,g(x,y)=(x,y,\sqrt{1-x^2-y^2})$$

试直接验证$(g\circ f)^*(\omega)=f^*(g^*(\omega))$.

由 g 与 f 的定义知

$$\forall(\theta,\varphi)\in[0,2\pi]\times[0,\pi]$$

$$(x,y,z)=g\circ f(\theta,\varphi)=(\cos\theta\cos\varphi,\sin\theta\cos\varphi,\sin\varphi)$$

于是

$$x=\cos\theta\cos\varphi,y=\sin\theta\cos\varphi,z=\sin\varphi$$

因此

$$\mathrm{d}x=-\sin\theta\cos\varphi\mathrm{d}\theta-\cos\theta\sin\varphi\mathrm{d}\varphi$$

$$\mathrm{d}y=\cos\theta\cos\varphi\mathrm{d}\theta-\sin\theta\sin\varphi\mathrm{d}\varphi$$

$$\mathrm{d}z=\cos\varphi\mathrm{d}\varphi$$

由此计算得到

$$\begin{aligned}(g\circ f)^*(\omega)&=\sin\theta\cos\varphi(-\sin\theta\cos\varphi\mathrm{d}\theta-\\&\quad\cos\theta\sin\varphi\mathrm{d}\varphi)\wedge(\cos\varphi\mathrm{d}\varphi)\\&=\sin^2\theta\cos^3\varphi\mathrm{d}\varphi\wedge\mathrm{d}\theta\end{aligned}$$

另外，对映射 g，我们有

$$\forall(x,y)\in\mathbf{R}^2,(x,y,z)=g(x,y)=(x,y,\sqrt{1-x^2-y^2})$$

于是

$$x=x,y=y,z=\sqrt{1-x^2-y^2},\forall(x,y)\in\mathbf{R}^2$$

从而

$$\mathrm{d}z=-\frac{x}{\sqrt{1-x^2-y^2}}\mathrm{d}x-\frac{y}{\sqrt{1-x^2-y^2}}\mathrm{d}y$$

因此

$$g^*(\omega)=y\mathrm{d}x/\left(-\frac{x}{\sqrt{1-x^2-y^2}}\mathrm{d}x-\frac{y}{\sqrt{1-x^2-y^2}}\mathrm{d}y\right)$$

$$= -\frac{y^2}{\sqrt{1-x^2-y^2}}\mathrm{d}x \wedge \mathrm{d}y$$

所以

$$f^*(g^*(\omega)) = -\frac{\sin^2\theta\cos^2\varphi}{\sin\varphi}(-\sin\theta\cos\varphi\mathrm{d}\theta - \cos\theta\sin\varphi\mathrm{d}\varphi) \wedge$$
$$(\cos\theta\cos\varphi\mathrm{d}\theta - \sin\theta\sin\varphi\mathrm{d}\varphi)$$
$$= -\frac{\sin^2\theta\cos^2\varphi}{\sin\varphi}(\sin^2\theta\sin\varphi\cos\varphi\mathrm{d}\theta \wedge \mathrm{d}\varphi -$$
$$\cos^2\theta\sin\varphi\cos\varphi\mathrm{d}\varphi \wedge \mathrm{d}\theta)$$
$$= -\frac{\sin^2\theta\cos^2\varphi}{\sin\varphi}(\sin^2\theta\sin\varphi\cos\varphi +$$
$$\cos^2\theta\sin\varphi\cos\varphi)\mathrm{d}\theta \wedge \mathrm{d}\varphi$$
$$= -\sin^2\theta\cos^3\varphi\mathrm{d}\theta \wedge \mathrm{d}\varphi$$
$$= \sin^2\theta\cos^3\varphi\mathrm{d}\varphi \wedge \mathrm{d}\theta$$

因此

$$(g \circ f)^*(\omega) = f^*(g^*(\omega)) = \sin^2\theta\cos^3\varphi\mathrm{d}\varphi \wedge \mathrm{d}\theta$$

三、微分形式的外微分

从上面知道，若 $f: U(U \subset \mathbf{R}^n) \to \mathbf{R}$ 是 C^1 类函数，则 f 是 C^1 类的 0—微分形式，而 f 的微分映射 $\mathrm{d}f$ 是 U 上的 C^0 类的 1—微分形式. 下面我们将微分算子 d 推广到外微分形式代数 $\Omega(U)$ 的部分子集上.

为此我们用 $\Omega^{(p)}(U)$ 表示 $\Omega(U)$ 的所有 $C^p(p \geqslant 1)$ 类的微分形式的集合.

1. 外微分算子的定义

定义 3.1　设映射 $\mathrm{d}: \Omega^{(p)}(U) \to \Omega(U)$ 定义如下

$$\forall \omega = \sum_{k=0}^{n}\sum_{i \in G_k} c_i \mathrm{d}x_i, \mathrm{d}\omega = \sum_{k=0}^{n}\sum_{i \in G_k} \mathrm{d}c_i \wedge \mathrm{d}x_i$$

则 d 称为 $\Omega^{(p)}(U)$ 的外微分算子，$\mathrm{d}\omega$ 称为 ω 的外微分.

由这个定义可知：

(1) 若 ω 是 C^p 类的，则 $\mathrm{d}\omega$ 是 C^{p-1} 类的.

(2) 若 $\omega \in \Omega_k(U)$，即 $\omega = \sum_{i \in G_k} c_i \mathrm{d}x_i$，则

$$\mathrm{d}\omega = \sum_{i \in G_k} \mathrm{d}c_i \wedge \mathrm{d}x_i$$
$$= \sum_{i \in G_k}\left(\sum_{j=1}^{n}\frac{\partial c_i}{\partial x_j}\mathrm{d}x_j\right) \wedge \mathrm{d}x_{i_1} \wedge \mathrm{d}x_{i_2} \wedge \cdots \wedge \mathrm{d}x_{i_k}$$

$$= \sum_{i \in G_k} \sum_{j=1}^{n} \frac{\partial c_i}{\partial x_j} \mathrm{d}x_j \wedge \mathrm{d}x_{i_1} \wedge \mathrm{d}x_{i_2} \wedge \cdots \wedge \mathrm{d}x_{i_k}$$

因此 $\mathrm{d}\omega \in \Omega_{k+1}(U)$. 特别地,当 $\omega \in \Omega_n(U)$ 时,$\mathrm{d}\omega = 0$.

例 1　设 $\omega = f\mathrm{d}x + g\mathrm{d}y$ 是 $U \subset \mathbf{R}^2$ 上的一个 C^p 类的 1— 微分形式,则

$$\begin{aligned} \mathrm{d}\omega &= \mathrm{d}f \wedge \mathrm{d}x + \mathrm{d}g \wedge \mathrm{d}y \\ &= \left(\frac{\partial f}{\partial x}\mathrm{d}x + \frac{\partial f}{\partial y}\mathrm{d}y\right) \wedge \mathrm{d}x + \left(\frac{\partial g}{\partial x}\mathrm{d}x + \frac{\partial g}{\partial y}\mathrm{d}y\right) \wedge \mathrm{d}y \\ &= \left(\frac{\partial g}{\partial x} - \frac{\partial f}{\partial y}\right) \mathrm{d}x \wedge \mathrm{d}y \end{aligned}$$

例 2　设 $\omega = f\mathrm{d}x + g\mathrm{d}y + h\mathrm{d}z$ 是 $U \subset \mathbf{R}^3$ 上的一个 C^p 类的1—微分形式,则

$$\begin{aligned} \mathrm{d}\omega &= \mathrm{d}f \wedge \mathrm{d}x + \mathrm{d}g \wedge \mathrm{d}y + \mathrm{d}h \wedge \mathrm{d}z \\ &= \left(\frac{\partial f}{\partial x}\mathrm{d}x + \frac{\partial f}{\partial y}\mathrm{d}y + \frac{\partial f}{\partial z}\mathrm{d}z\right) \wedge \mathrm{d}x + \\ &\quad \left(\frac{\partial g}{\partial x}\mathrm{d}x + \frac{\partial g}{\partial y}\mathrm{d}y + \frac{\partial g}{\partial z}\mathrm{d}z\right) \wedge \mathrm{d}y + \\ &\quad \left(\frac{\partial h}{\partial x}\mathrm{d}x + \frac{\partial h}{\partial y}\mathrm{d}y + \frac{\partial h}{\partial z}\mathrm{d}z\right) \wedge \mathrm{d}z \\ &= \left(\frac{\partial g}{\partial x} - \frac{\partial f}{\partial y}\right) \mathrm{d}x \wedge \mathrm{d}y + \left(\frac{\partial h}{\partial y} - \frac{\partial g}{\partial z}\right) \mathrm{d}y \wedge \mathrm{d}z + \\ &\quad \left(\frac{\partial f}{\partial z} - \frac{\partial h}{\partial z}\right) \mathrm{d}z \wedge \mathrm{d}x \end{aligned}$$

例 3　设 $\omega = f\mathrm{d}x \wedge \mathrm{d}y + g\mathrm{d}y \wedge \mathrm{d}z + h\mathrm{d}z \wedge \mathrm{d}x$ 是 $U \subset \mathbf{R}^3$ 上的一个 C^p 类的 2— 微分形式,则

$$\begin{aligned} \mathrm{d}\omega &= \mathrm{d}f \wedge \mathrm{d}x \wedge \mathrm{d}y + \mathrm{d}g \wedge \mathrm{d}y \wedge \mathrm{d}z + \mathrm{d}h \wedge \mathrm{d}z \wedge \mathrm{d}x \\ &= \left(\frac{\partial f}{\partial x}\mathrm{d}x + \frac{\partial f}{\partial y}\mathrm{d}y + \frac{\partial f}{\partial z}\mathrm{d}z\right) \wedge \mathrm{d}x \wedge \mathrm{d}y + \\ &\quad \left(\frac{\partial g}{\partial x}\mathrm{d}x + \frac{\partial g}{\partial y}\mathrm{d}y + \frac{\partial g}{\partial z}\mathrm{d}z\right) \wedge \mathrm{d}y \wedge \mathrm{d}z + \\ &\quad \left(\frac{\partial h}{\partial x}\mathrm{d}x + \frac{\partial h}{\partial y}\mathrm{d}y + \frac{\partial h}{\partial z}\mathrm{d}z\right) \wedge \mathrm{d}z \wedge \mathrm{d}x \\ &= \left(\frac{\partial g}{\partial x} + \frac{\partial h}{\partial y} + \frac{\partial f}{\partial z}\right) \mathrm{d}x \wedge \mathrm{d}y \wedge \mathrm{d}z \end{aligned}$$

2. 外微分算子 d 的性质

我们用 $\Omega_k^{(p)}(U)$ 表示 $U(U \subset \mathbf{R}^n)$ 上的所有 C^p 类的 k—微分形式的集合.

定理 3.1　$\forall p \in \mathbf{N}, \forall k, l \in \mathbf{N} \cup \{0\}$,则:

(1) 外微分算子 $\mathrm{d}:\Omega^{(p)}(U)\to\Omega^{(p-1)}(U)$ 是线性的.

(2) 若 $\omega\in\Omega_k^{(p)}(U),\tilde{\omega}\in\Omega_l^{(p)}(U)$,则

$$\mathrm{d}(\omega\wedge\tilde{\omega})=(\mathrm{d}\omega)\wedge\tilde{\omega}+(-1)^k\omega\wedge(\mathrm{d}\tilde{\omega})$$

(3) 若 $p\geqslant 2$,则 $\mathrm{d}^2=\mathrm{d}\circ\mathrm{d}=0$.

证明 (1)d 的线性性可直接由定义推出. 我们来证明(2). 由 d 的线性性可知,我们只需对

$$\omega=c_i\mathrm{d}x_{i_1}\wedge\mathrm{d}x_{i_2}\wedge\cdots\wedge\mathrm{d}x_{i_k}$$
$$\tilde{\omega}=\tilde{c}_j\mathrm{d}x_{j_1}\wedge\mathrm{d}x_{j_2}\wedge\cdots\wedge\mathrm{d}x_{j_l}$$

进行证明即可. 根据外积的性质及定义

$$\begin{aligned}\omega\wedge\tilde{\omega}&=c_i\mathrm{d}x_i\wedge\tilde{c}_j\mathrm{d}x_j\\&=c_i\tilde{c}_j\mathrm{d}x_i\wedge\mathrm{d}x_j\\&=\begin{cases}0, & \text{若 } \boldsymbol{i} \text{ 与 } \boldsymbol{j} \text{ 相交}\\(-1)^{\langle i,j\rangle}c_i\tilde{c}_j\mathrm{d}x_{i\vee j}, & \text{若 } \boldsymbol{i} \text{ 与 } \boldsymbol{j} \text{ 不相交}\end{cases}\end{aligned}$$

于是

$$\mathrm{d}(\omega\wedge\tilde{\omega})=\begin{cases}0, & \text{若 } \boldsymbol{i} \text{ 与 } \boldsymbol{j} \text{ 相交}\\(-1)^{\langle i,j\rangle}\mathrm{d}(c_i\tilde{c}_j)\wedge\mathrm{d}x_{i\vee j}, & \text{若 } \boldsymbol{i} \text{ 与 } \boldsymbol{j} \text{ 不相交}\end{cases}$$

另外

$$\begin{aligned}&(\mathrm{d}\omega)\wedge\tilde{\omega}+(-1)^k\omega\wedge(\mathrm{d}\tilde{\omega})\\=&(\mathrm{d}c_i\wedge\mathrm{d}x_i)\wedge(\tilde{c}_j\mathrm{d}x_j)+(-1)^k(c_i\mathrm{d}x_i)\wedge(\mathrm{d}\tilde{c}_j\wedge\mathrm{d}x_j)\\=&(\mathrm{d}c_i\tilde{c}_j)\wedge\mathrm{d}x_i\wedge\mathrm{d}x_j+(c_i\mathrm{d}\tilde{c}_j)\wedge\mathrm{d}x_i\wedge\mathrm{d}x_j\\=&(\mathrm{d}c_i\tilde{c}_j+c_i\mathrm{d}\tilde{c}_j)\wedge\mathrm{d}x_i\wedge\mathrm{d}x_j\\=&\begin{cases}0, & \text{若 } \boldsymbol{i} \text{ 与 } \boldsymbol{j} \text{ 相交}\\(-1)^{\langle i,j\rangle}\mathrm{d}(c_i\tilde{c}_j)\wedge\mathrm{d}x_{i\vee j}, & \text{若 } \boldsymbol{i} \text{ 与 } \boldsymbol{j} \text{ 不相交}\end{cases}\end{aligned}$$

因此

$$\mathrm{d}(\omega\wedge\tilde{\omega})=(\mathrm{d}\omega)\wedge\tilde{\omega}+(-1)^k\omega\wedge(\mathrm{d}\tilde{\omega})$$

(3) 同理,由 d 的线性性,我们只需证明

$$\forall\,\omega=c_i\mathrm{d}x_i,\mathrm{d}^2(\omega)=\mathrm{d}\circ\mathrm{d}(\omega)=0$$

根据外微分与外积的定义,我们有

$$\begin{aligned}\mathrm{d}\omega&=\mathrm{d}c_i\wedge\mathrm{d}x_i\\&=\left(\sum_{j=1}^n\frac{\partial c_i}{\partial x_j}\mathrm{d}x_j\right)\wedge\mathrm{d}x_i\\&=\sum_{j=1}^n\left[\begin{cases}0, & \text{若 } j \text{ 与 } \boldsymbol{i} \text{ 相交}\\(-1)^{\langle j,i\rangle}\dfrac{\partial c_i}{\partial x_j}\mathrm{d}x_{j\vee i}, & \text{若 } j \text{ 与 } \boldsymbol{i} \text{ 不相交}\end{cases}\right]\end{aligned}$$

由此可知,不论 j 与 i 相交与否,下式总是成立的

$$\mathrm{d}\omega = \sum_{j=1}^{n} \frac{\partial c_i}{\partial x_j} \mathrm{d}x_j \wedge \mathrm{d}x_i$$

同理,我们又有

$$\begin{aligned} \mathrm{d}^2(\omega) &= \mathrm{d}\left(\sum_{j=1}^{n} \frac{\partial c_i}{\partial x_j} \mathrm{d}x_j \wedge \mathrm{d}x_i\right) \\ &= \sum_{j=1}^{n} \mathrm{d}\left(\frac{\partial c_i}{\partial x_j} \mathrm{d}x_j \wedge \mathrm{d}x_i\right) \\ &= \sum_{j=1}^{n} \sum_{l=1}^{n} \frac{\partial^2 c_i}{\partial x_j \partial x_l} \mathrm{d}x_l \wedge \mathrm{d}x_j \wedge \mathrm{d}x_i \end{aligned}$$

由于 ω 是 $C^p(p \geqslant 2)$ 类的,所以 c_i 在 U 上是 C^p 类的,从而 $\frac{\partial^2 c_i}{\partial x_j \partial x_l}$ 在 U 上连续.根据偏导数的 Schwarz 定理,我们有

$$\frac{\partial^2 c_i}{\partial x_j \partial x_l} = \frac{\partial^2 c_i}{\partial x_l \partial x_j}, \forall j, l = 1, 2, \cdots, n$$

因此由

$$\mathrm{d}x_j \wedge \mathrm{d}x_l = \begin{cases} 0, & 若\ l = j \\ -\mathrm{d}x_l \wedge \mathrm{d}x_j, & 若\ l \neq j \end{cases}$$

知 $\mathrm{d}^2(\omega) = 0$.

定理 3.2　设 $f: U \to V(V \subset \mathbf{R}^m)$ 是任一 C^2 类映射,$\omega \in \Omega^{(p)}(V)$.则

$$f^*(\mathrm{d}\omega) = \mathrm{d}(f^*\omega)$$

特别地,下述交换图式成立

$$\begin{array}{ccc} \Omega_k^{(p)}(V) & \xrightarrow{f^*} & \Omega_k^{(p)}(U) \\ \mathrm{d}\downarrow & & \downarrow\mathrm{d} \\ \Omega_{k+1}^{(p-1)}(V) & \xrightarrow{f^*} & \Omega_{k+1}^{(p-1)}(U) \end{array}$$

证明　由 f^* 与 d 的线性性,我们只需证明

$$\forall \omega = c_i \mathrm{d}y_i \in \Omega_k^{(p)}(V), f^*(\mathrm{d}\omega) = \mathrm{d}(f^*\omega)$$

首先我们注意到第二部分中例 4 的结论,若 $\varphi: V \to \mathbf{R}$ 是任一 C^1 类函数,则

$$f^*(\varphi) = \varphi \circ f$$

下面我们来证明

$$\mathrm{d}(f^*(\varphi)) = f^*(\mathrm{d}\varphi) \tag{$*$}$$

事实上

$$\begin{aligned} & f^*(\mathrm{d}\varphi) \\ = & f^*\left(\frac{\partial \varphi}{\partial y_1}\mathrm{d}y_1 + \frac{\partial \varphi}{\partial y_2}\mathrm{d}y_2 + \cdots + \frac{\partial \varphi}{\partial y_m}\mathrm{d}y_m\right) \end{aligned}$$

$$= \frac{\partial(\varphi \circ f)}{\partial y_1}\Big(\sum_{j=1}^{n} \frac{\partial f_1}{\partial x_j}\mathrm{d}x_j\Big) + \frac{\partial(\varphi \circ f)}{\partial y_2}\Big(\sum_{j=1}^{n} \frac{\partial f_2}{\partial x_j}\mathrm{d}x_j\Big) + \cdots +$$

$$\frac{\partial(\varphi \circ f)}{\partial y_m}\Big(\sum_{j=1}^{n} \frac{\partial f_m}{\partial x_j}\mathrm{d}x_j\Big)$$

$$= \left[\frac{\partial(\varphi \circ f)}{\partial y_1}\frac{\partial f_1}{\partial x_1} + \frac{\partial(\varphi \circ f)}{\partial y_2}\frac{\partial f_2}{\partial x_1} + \cdots + \frac{\partial(\varphi \circ f)}{\partial y_m}\frac{\partial f_m}{\partial x_1}\right]\mathrm{d}x_1 +$$

$$\left[\frac{\partial(\varphi \circ f)}{\partial y_1}\frac{\partial f_1}{\partial x_2} + \frac{\partial(\varphi \circ f)}{\partial y_2}\frac{\partial f_2}{\partial x_2} + \cdots + \frac{\partial(\varphi \circ f)}{\partial y_m}\frac{\partial f_m}{\partial x_2}\right]\mathrm{d}x_2 + \cdots +$$

$$\left[\frac{\partial(\varphi \circ f)}{\partial y_1}\frac{\partial f_1}{\partial x_n} + \frac{\partial(\varphi \circ f)}{\partial y_2}\frac{\partial f_2}{\partial x_n} + \cdots + \frac{\partial(\varphi \circ f)}{\partial y_m}\frac{\partial f_m}{\partial x_n}\right]\mathrm{d}x_n$$

$$= \frac{\partial(\varphi \circ f)}{\partial x_1}\mathrm{d}x_1 + \frac{\partial(\varphi \circ f)}{\partial x_2}\mathrm{d}x_2 + \cdots + \frac{\partial(\varphi \circ f)}{\partial x_n}\mathrm{d}x_n$$

$$= \mathrm{d}(\varphi \circ f)$$

$$= \mathrm{d}(f^* \varphi)$$

现在利用刚才所证明的式(*)得到

$$\begin{aligned} f^*(\mathrm{d}\omega) &= f^*(\mathrm{d}c_i \wedge \mathrm{d}y_i) \\ &= f^*(\mathrm{d}c_i \wedge \mathrm{d}y_{i_1} \wedge \mathrm{d}y_{i_2} \wedge \cdots \wedge \mathrm{d}y_{i_k}) \\ &= f^*(\mathrm{d}c_i) \wedge f^*(\mathrm{d}y_{i_1}) \wedge f^*(\mathrm{d}y_{i_2}) \wedge \cdots \wedge f^*(\mathrm{d}y_{i_k}) \\ &= \mathrm{d}(f^* c_i) \wedge f^*(\mathrm{d}y_{i_1}) \wedge f^*(\mathrm{d}y_{i_2}) \wedge \cdots \wedge f^*(\mathrm{d}y_{i_k}) \end{aligned}$$

另外,利用定理 3.1 的关系式 $\mathrm{d}(\omega \wedge \tilde{\omega}) = (\mathrm{d}\omega) \wedge \tilde{\omega} + (-1)^k \omega \wedge (\mathrm{d}\tilde{\omega})$ 及 $\mathrm{d}f^*(\mathrm{d}y_{i_l}) = \mathrm{d}(\mathrm{d}(f^* y_{i_l})) = 0 (l = 1, 2, \cdots, k)$,我们得到

$$\begin{aligned} \mathrm{d}(f^* \omega) &= \mathrm{d}(f^* c_i \wedge f^*(\mathrm{d}y_i)) \\ &= \mathrm{d}(f^* c_i \wedge f^*(\mathrm{d}y_{i_1}) \wedge f^*(\mathrm{d}y_{i_2}) \wedge \cdots \wedge f^*(\mathrm{d}y_{i_k})) \\ &= \mathrm{d}(f^* c_i) \wedge f^*(\mathrm{d}y_{i_1}) \wedge f^*(\mathrm{d}y_{i_2}) \wedge \cdots \wedge f^*(\mathrm{d}y_{i_k}) + \\ &\quad (f^* c_i) \wedge \mathrm{d}(f^*(\mathrm{d}y_{i_1}) \wedge f^*(\mathrm{d}y_{i_2}) \wedge \cdots \wedge f^*(\mathrm{d}y_{i_k})) \\ &= \mathrm{d}(f^* c_i) \wedge f^*(\mathrm{d}y_{i_1}) \wedge f^*(\mathrm{d}y_{i_2}) \wedge \cdots \wedge f^*(\mathrm{d}y_{i_k}) \end{aligned}$$

因此

$$f^*(\mathrm{d}\omega) = \mathrm{d}(f^* \omega)$$

例 4 设 ω 是 $V(V \subset \mathbf{R}^n)$ 上如下定义的 C^1 类的 $n-1$-微分形式

$$\omega = \sum_{i=1}^{n} (-1)^{i-1} f_i \mathrm{d}y_1 \wedge \mathrm{d}y_2 \wedge \cdots \wedge \widehat{\mathrm{d}y_i} \wedge \cdots \wedge \mathrm{d}y_n$$

这里 $f_i: V \to \mathbf{R} (i = 1, 2, \cdots, n)$ 是 n 个 C^1 类映射. 又设 $\varphi: U(U \subset \mathbf{R}^n) \to V$ 是任一 C^1 类映射. 试计算 $\varphi^*(\mathrm{d}\omega)$.

根据 φ^* 与 d 的定义,我们有

$$\varphi^*(\mathrm{d}\omega)$$

$$= \varphi^*\Big(\sum_{i=1}^{n} (-1)^{i-1} \mathrm{d}f_i \wedge \mathrm{d}y_{i_1} \wedge \mathrm{d}y_{i_2} \wedge \cdots \wedge \widehat{\mathrm{d}y_i} \wedge \cdots \wedge \mathrm{d}y_n\Big)$$

$$= \varphi^*\left(\sum_{i=1}^{n}(-1)^{i-1}\left(\sum_{j=1}^{n}\frac{\partial f_i}{\partial y_j}\mathrm{d}y_j\right)\wedge \mathrm{d}y_{i_1}\wedge \mathrm{d}y_{i_2}\wedge\cdots\wedge\widehat{\mathrm{d}y_i}\wedge\cdots\wedge\mathrm{d}y_n\right)$$

$$=\varphi^*\left(\sum_{i=1}^{n}\frac{\partial f_i}{\partial y_i}\mathrm{d}y_1\wedge\mathrm{d}y_2\wedge\cdots\wedge\mathrm{d}y_n\right)$$

$$=\sum_{i=1}^{n}\frac{\partial(f_i\circ\varphi)}{\partial y_i}\varphi^*(\mathrm{d}y_1)\wedge\varphi^*(\mathrm{d}y_2)\wedge\cdots\wedge\varphi^*(\mathrm{d}y_n)$$

$$=\sum_{i=1}^{n}\frac{\partial(f_i\circ\varphi)}{\partial y_i}\left(\sum_{j=1}^{n}\frac{\partial\varphi_1}{\partial x_j}\mathrm{d}x_j\right)\wedge\left(\sum_{j=1}^{n}\frac{\partial\varphi_2}{\partial x_j}\mathrm{d}x_j\right)\wedge\cdots\wedge\left(\sum_{j=1}^{n}\frac{\partial\varphi_n}{\partial x_j}\mathrm{d}x_j\right)$$

$$=\sum_{i=1}^{n}\frac{\partial(f_i\circ\varphi)}{\partial y_i}\frac{D(\varphi_1,\varphi_2,\cdots,\varphi_n)}{D(x_1,x_2,\cdots,x_n)}\mathrm{d}x_1\wedge\mathrm{d}x_2\wedge\cdots\wedge\mathrm{d}x_n$$

因此，$\forall x\in U$，有

$$\varphi^*(\mathrm{d}\omega)(x)=\sum_{i=1}^{n}\frac{\partial f_i}{\partial y_i}(\varphi(x))\frac{D(\varphi_1,\varphi_2,\cdots,\varphi_n)}{D(x_1,x_2,\cdots,x_n)}(x)\cdot \mathrm{d}x_1\wedge\mathrm{d}x_2\wedge\cdots\wedge\mathrm{d}x_n$$

3. 微分形式的恰当性与闭性

我们知道，若 ω 是 U 上的 C^p 类的 $k-1-$ 微分形式，则 $\mathrm{d}\omega$ 是 U 上的 C^{p-1} 类的 $k-$ 微分形式. 现在研究的是它的逆问题：给定一个 U 上的 $k-$ 微分形式 ω，是否存在 U 上的一个 $k-1-$ 微分形式 θ 使得 $\mathrm{d}\theta=\omega$？如果存在，θ 是否唯一？

首先我们给出下述定义.

定义 3.2　设 $\omega\in\Omega^{(p)}(U)$.

(1) 若 $\mathrm{d}\omega=0$，则称 ω 是闭的.

(2) 若存在 $\theta\in\Omega^{(p+1)}(U)$ 使得 $\mathrm{d}\theta=\omega$，则称 ω 是恰当的，并且 θ 称为 ω 的原微分形式.

定理 3.3　设 $\omega\in\Omega^{(p)}(U)$.

(1) 若 ω 是恰当的，则 ω 是闭的.

(2) 若 ω 是恰当的，并且 $\mathrm{d}\theta=\omega$，则 $\forall\tilde{\theta}\in\Omega^2(U)$，也有 $\mathrm{d}(\theta+\mathrm{d}\tilde{\theta})=\omega$.

证明　(1) 若 ω 是恰当的，则存在 $\theta\in\Omega^{(p+1)}(U)$ 使得 $\mathrm{d}\theta=\omega$，于是

$$\mathrm{d}\omega=\mathrm{d}(\mathrm{d}\theta)=\mathrm{d}^2\theta=0$$

即 ω 是闭的.

(2) 由于 ω 是恰当的，并且 $\mathrm{d}\theta=\omega$，故由 d 的线性性及 $\mathrm{d}^2=0$ 得到

$$\mathrm{d}(\theta+\mathrm{d}\tilde{\theta})=\mathrm{d}\theta+\mathrm{d}(\mathrm{d}\tilde{\theta})=\omega+\mathrm{d}^2\tilde{\theta}=\omega$$

由这个定理可知,若ω的原微分形式存在,则它并不是唯一的.

关于使 $d\theta=\omega$ 的θ的存在性问题,一般说来,只能有局部存在性结论,即下述定理.

定理 3.4(Poincaré **引理**) 设 $U\subset\mathbf{R}^n$ 是任一开集,$\omega\in\Omega^{(p)}(U)$ 是任一闭微分形式,则 $\forall x_0\in U$,存在x_0的一个邻域$V\subset U$ 使得$\omega|_V$ 是恰当的.

证明 由 d 的线性性,我们不妨假设 $\omega\in\Omega_k^{(p)}(U)$.

任取一点 $x_0\in U$,由于U是开集,故存在 $\delta>0$,使得 $V=B(x_0,\delta)\subset U$. $\forall x\in V$,x 可以表示成下述形式

$$x=x_0+z,z\in B(0,\delta)$$

设 $\boldsymbol{i}=(i_1,i_2,\cdots,i_k)\in G_k$. 考虑如下定义的$V$上的 $k-1$— 微分形式 β_i

$$\beta_i=\sum_{s=1}^{k}(-1)^{s-1}z_{i_s}dx_{i_1}\wedge\cdots\wedge\widehat{dx_{i_s}}\wedge\cdots\wedge dx_{i_k}$$

这里 $z_{i_s}=x_{i_s}-x_{0i_s}(s=1,2,\cdots,k)$.

直接计算得到

$$d\beta_i=kdx_{i_1}\wedge dx_{i_2}\wedge\cdots\wedge dx_{i_k}=kdx_i$$

现在我们定义映射 $I:\Omega_m^{(p)}(V)\to\Omega_{m-1}^{(p)}(V)$ 如下

$$\forall\omega=\sum_{i\in G_m}c_idx_i\in\Omega_m^{(p)}(V),\forall x\in V$$

$$I(\omega)(x)=\sum_{i\in G_m}\left(\int_0^1 t^{m-1}c_i(x_0+tz)dt\right)\beta_i$$

显然,I是线性的,并且 $I(0)=0$.

下面我们分别计算 $d(I(\omega|_V))$ 与 $I(d(\omega|_V))$.

对于 $d(I(\omega|_V))$,我们有

$$\begin{aligned}d(I(\omega\mid V))(x)=&\sum_{i\in G_k}d\left(\int_0^1 t^{k-1}c_i(x_0+tz)dt\right)\wedge\beta_i+\\&\sum_{i\in G_k}\left(\int_0^1 t^{k-1}c_i(x_0+tz)dt\right)d\beta_i\\=&\sum_{i\in G_k}\left(\sum_{j=1}^{n}\frac{\partial}{\partial x_j}\left(\int_0^1 t^{k-1}c_i(x_0+tz)dt\right)dx_j\right)\wedge\beta_i+\\&\sum_{i\in G_k}\left(\int_0^1 t^{k-1}c_i(x_0+tz)dt\right)d\beta_i\\=&\sum_{i\in G_k}\sum_{j=1}^{n}\left(\int_0^1 t^{k-1}\frac{\partial}{\partial x_j}c_i(x_0+tz)dt\right)\cdot dx_j\wedge\beta_i+\\&\sum_{i\in G_k}\left(k\int_0^1 t^{k-1}c_i(x_0+tz)dt\right)dx_i\qquad(*)\end{aligned}$$

对于 $I(\mathrm{d}(\omega|_V))$，由于 $\mathrm{d}\omega=0$，我们有

$$0=I(\mathrm{d}(\omega|_V))=I\left(\sum_{i\in G_k}\mathrm{d}c_i\wedge\mathrm{d}x_i\right)$$

$$=I\left(\sum_{i\in G_k}\sum_{j=1}^{n}\frac{\partial c_i}{\partial x_j}\mathrm{d}x_j\wedge\mathrm{d}x_i\right)$$

若 j 与 $\boldsymbol{i}$ 相交，则 $\mathrm{d}x_j\wedge\mathrm{d}x_i=0$，故我们不妨假设 $\forall j=1,2,\cdots,n$，j 与 $\boldsymbol{i}$ 不相交，于是

$$\mathrm{d}x_j\wedge\mathrm{d}x_i=(-1)^{\langle i,j\rangle}\mathrm{d}x_{j\vee i}$$

$$=(-1)^{m_j}\mathrm{d}x_{l_1}\wedge\cdots\wedge\mathrm{d}x_{l_{m_j+1}}\wedge\cdots\wedge\mathrm{d}x_{l_{k+1}}$$

$$0\leqslant m_j\leqslant k$$

这里 $\mathrm{d}x_{l_{m_j+1}}=\mathrm{d}x_j(j=1,2,\cdots,n)$. 由 I 的定义，我们有

$$0=I\left(\sum_{i\in G_k}\sum_{j=1}^{n}(-1)^{m_j}\frac{\partial c_i}{\partial x_j}\mathrm{d}x_{l_1}\wedge\mathrm{d}x_{l_{m_j+1}}\wedge\cdots\wedge\mathrm{d}x_{l_{k+1}}\right)$$

$$=\sum_{i\in G_k}\sum_{j=1}^{n}\left(\int_0^1 t^k\frac{\partial c_i}{\partial x_j}(x_0+tz)\mathrm{d}t\right)(-1)^{m_j}\beta_{\boldsymbol{l}}$$

$$\boldsymbol{l}=(l_1,\cdots,l_{m_j+1},\cdots,l_{k+1})$$

但是

$$(-1)^{m_j}\beta_{\boldsymbol{l}}=(-1)^{m_j}\sum_{s=1}^{k+1}(-1)^{s-1}z_{l_s}\mathrm{d}x_{l_1}\wedge\cdots\wedge\widehat{\mathrm{d}x_{l_s}}\wedge\cdots\wedge\mathrm{d}x_{l_{k+1}}$$

$$=(-1)^{m_j}\cdot(-1)^{m_j+1-1}z_{l_{m_j+1}}\mathrm{d}x_{l_1}\wedge\cdots\wedge\widehat{\mathrm{d}x_{l_{m_j+1}}}\wedge\cdots\wedge\mathrm{d}x_{l_{k+1}}+$$

$$(-1)^{m_j}\cdot\sum_{\substack{s=1\\ s\neq m_j+1}}^{k+1}(-1)^{s-1}z_{l_s}\mathrm{d}x_{l_1}\wedge\cdots\wedge\widehat{\mathrm{d}x_{l_s}}\wedge\cdots\wedge\mathrm{d}x_{l_{k+1}}$$

$$=z_j\mathrm{d}x_{i_1}\wedge\mathrm{d}x_{i_2}\wedge\cdots\wedge\mathrm{d}x_{i_k}-\mathrm{d}x_j\wedge$$

$$\sum_{s=1}^{k}(-1)^{s-1}z_{i_s}\mathrm{d}x_{i_1}\wedge\cdots\wedge\widehat{\mathrm{d}x_{i_s}}\wedge\cdots\wedge\mathrm{d}x_{i_k}$$

$$=z_j\mathrm{d}x_{\boldsymbol{i}}-\mathrm{d}x_j\wedge\beta_{\boldsymbol{i}}$$

因此

$$0=\sum_{i\in G_k}\sum_{j=1}^{n}\left(\int_0^1 t^k\frac{\partial c_i}{\partial x_j}(x_0+tz)\mathrm{d}t\right)(z_j\mathrm{d}x_i-\mathrm{d}x_j\wedge\beta_i)$$

$$(**)$$

将$(*)$与$(**)$两式相加得到

$$\mathrm{d}(I(\omega|_V))(x)=\sum_{i\in G_k}\sum_{j=1}^{n}\left(\int_0^1 t^k\frac{\partial c_i}{\partial x_j}(x_0+tz)\mathrm{d}t\right)\mathrm{d}x_j\wedge\beta_i+$$

$$\sum_{i\in G_k}\left(k\int_0^1 t^{k-1}c_i(x_0+tz)\mathrm{d}t\right)\mathrm{d}x_i+$$

$$\sum_{i\in G_k}\sum_{j=1}^{n}\left(\int_0^1 t^k\frac{\partial c_i}{\partial x_j}(x_0+tz)\mathrm{d}t\right)\cdot$$
$$(z_j\mathrm{d}x_i-\mathrm{d}x_j\wedge\beta_i)$$
$$=\sum_{i\in G_k}\left[\int_0^1(kt^{k-1}c_i(x_0+tz)+\right.$$
$$\left.t^k\sum_{j=1}^{n}\frac{\partial c_i}{\partial x_j}(x_0+tz)z_j)\mathrm{d}t\right]\mathrm{d}x_i$$
$$=\sum_{i\in G_k}\left[\int_0^1(kt^{k-1}c_i(x_0+tz)+\right.$$
$$\left.t^k\frac{\mathrm{d}}{\mathrm{d}t}c_i(x_0+tz))\mathrm{d}t\right]\mathrm{d}x_i$$
$$=\sum_{i\in G_k}\left[\int_0^1\frac{\mathrm{d}}{\mathrm{d}t}(t^kc_i(x_0+tz))\mathrm{d}t\right]\mathrm{d}x_i$$
$$=\sum_{i\in G_k}c_i(x_0+z)\mathrm{d}x_i$$
$$=\omega\mid_V(x).$$

由此可知，若令 $\theta=I(\omega\mid_V)$，则 $\mathrm{d}\theta=\omega\mid_V$.

附注　若 $U\subset\mathbf{R}^n$ 是一星形区域，则存在定义在整个 U 上的 C^{p+1} 类微分形式 θ 使得 $\mathrm{d}\theta=\omega$.

事实上，由于 U 是星形区域，故存在一点 $x_0\in U$ 使得 $\forall x\in U,(1-t)x_0+tx\in U(\forall t\in[0,1])$. 现在将上述定理的证明过程中的 $k-1-$ 微分形式 β_i 改为

$$\beta_i=\sum_{s=1}^{k}(-1)^{s-1}(x_{i_s}-x_{0i_s})\mathrm{d}x_{i_1}\wedge\cdots\wedge\widehat{\mathrm{d}x_{i_s}}\wedge\cdots\wedge\mathrm{d}x_{i_k}$$

而线性映射 $I:\Omega_m^{(p)}(U)\to\Omega_{m-1}^{(p)}(U)$ 的定义改为

$$\forall\omega=\sum_{i\in G_m}c_i\mathrm{d}x_i\in\Omega_m^{(p)}(U),\forall x\in U$$

$$I(\omega)(x)=\sum_{i\in G_m}\left(\int_0^1 t^{k-1}c_i((1-t)x_0+tx)\mathrm{d}t\right)\beta_i$$

然后重复上述论证过程即可证得

$$\mathrm{d}(I(\omega))=\omega$$

下面我们特别来研究 1－微分形式的恰当性.

4.1－微分形式的恰当性

定理 3.5　设 $U\subset\mathbf{R}^n$ 是一开集，并且

$$\omega=f_1\mathrm{d}x_1+f_2\mathrm{d}x_2+\cdots+f_n\mathrm{d}x_n$$

是定义在 U 上的 $C^p(p\geqslant1)$ 类的 1－微分形式. 我们考虑下述条件：

(1) 存在 C^{p-1} 类的 0－微分形式 θ 使得 $\mathrm{d}\theta=\omega$.

(2)$\mathrm{d}\omega=0$.

(3) $\dfrac{\partial f_i}{\partial x_j}=\dfrac{\partial f_j}{\partial x_i}, \forall i,j=1,2,\cdots,n$.

则有：(1)⇒(2)⇔(3). 特别地，若 U 是星形区域，则(1)⇔(2)⇔(3).

证明 (1)⇒(2) 由定理 3.3 推出.

(2)⇔(3)：由外微分 d 的定义有

$$
\begin{aligned}
\mathrm{d}\omega &= \mathrm{d}f_1 \wedge \mathrm{d}x_1+\mathrm{d}f_2 \wedge \mathrm{d}x_2+\cdots+\mathrm{d}f_n \wedge \mathrm{d}x_n \\
&= \left(\sum_{j=1}^{n} \frac{\partial f_1}{\partial x_j}\mathrm{d}x_j\right) \wedge \mathrm{d}x_1+\left(\sum_{j=1}^{n} \frac{\partial f_2}{\partial x_j}\mathrm{d}x_j\right) \wedge \mathrm{d}x_2+\cdots+ \\
&\quad \left(\sum_{j=1}^{n} \frac{\partial f_n}{\partial x_j}\mathrm{d}x_j\right) \wedge \mathrm{d}x_n \\
&= \sum_{1\leqslant i<j\leqslant n}\left(\frac{\partial f_j}{\partial x_i}-\frac{\partial f_i}{\partial x_j}\right)\mathrm{d}x_i \wedge \mathrm{d}x_j
\end{aligned}
$$

因此

$$
0=\mathrm{d}\omega \Leftrightarrow \frac{\partial f_i}{\partial x_j}-\frac{\partial f_j}{\partial x_i}=0, \forall i,j=1,2,\cdots,n
$$

当 U 是星形区域时，由定理 3.4 的附注知(2)⇒(1). 因此(1)⇔(2)⇔(3).

下面一个定理指出了 1－微分形式的任意两个原微分形式(如果存在的话)之间的关系.

定理 3.6 设 $U\subset \mathbf{R}^n$ 是任一连通开集，则

$$
\omega=f_1\mathrm{d}x_1+f_2\mathrm{d}x_2+\cdots+f_n\mathrm{d}x_n
$$

是定义在 U 上的 C^p 类的 1－微分形式. 若 Θ,θ 是 ω 的两个原微分形式，则存在常数 C 使得

$$
\Theta-\theta=C
$$

证明 因为 ω 是 U 上的 1－微分形式，所以 Θ 与 θ 是定义在 U 上的可微函数.

设 $x_0\in U$，令 $\Theta(x_0)-\theta(x_0)=C$，我们来证明

$$
\forall x\in U, \Theta(x)-\theta(x)=C
$$

为此我们注意到 U 是道路连通的，于是存在连续映射 $\varphi:[0,1]\to U$ 使得 $\varphi(0)=x_0, \varphi(1)=x$.

定义映射 $\psi:[0,1]\to \mathbf{R}$ 如下

$$
\forall t\in[0,1], \psi(t)=(\Theta-\theta)\circ\varphi(t)
$$

我们有

$$
\psi(0)=(\Theta-\theta)\circ\varphi(0)=\Theta(x_0)-\theta(x_0)=C
$$

$$
\psi(1)=(\Theta-\theta)\circ\varphi(1)=\Theta(x)-\theta(x)
$$

令

$$a = \sup\{t \in [0,1] \mid \psi(t) = C\}$$

则 $0 \leqslant a \leqslant 1$. 由 ψ 的连续性知，$\psi(a) = C$，即 $C = (\Theta - \theta) \circ \varphi(a)$.

若 $a < 1$，则由 U 的开性，存在 $\delta > 0$ 使得 $B(\varphi(a), \delta) \subset U$，$B(\varphi(a), \delta)$ 是 $\mathbf{R}^n$ 中的凸集，并且由于在 U 上，$\mathrm{d}(\Theta - \theta) = 0$，故由有限增量定理，我们有：$\forall x \in B(\varphi(a), \delta)$，则

$$\begin{aligned} &\| (\Theta - \theta)(x) - (\Theta - \theta)(\varphi(a)) \| \\ &\leqslant \sup_{z \in \varphi(a)x} \| \mathrm{d}(\Theta - \theta)(z) \| \, \| x - \varphi(a) \| \end{aligned}$$

即

$$\forall x \in B(\varphi(a), \delta), (\Theta - \theta)(x) = (\Theta - \theta)(\varphi(a)) = C$$

另外，因为 $a < 1$，所以存在 $\eta > 0$ 使得

$$[a, a + \eta] \subset [0,1], \varphi([a, a + \eta]) \subset B(\varphi(a), \delta)$$

因此

$$\forall t \in [a, a + \eta], \psi(t) = (\Theta - \theta)(\varphi(t)) = C$$

特别地，我们有 $\psi(a+\eta) = C$，这与 a 的定义矛盾，故必有 $a=1$，此即表明

$$C = \psi(1) = (\Theta - \theta)(\varphi(1)) = \Theta(x) - \theta(x)$$

推论 1.1 设 $U \subset \mathbf{R}^n$ 是任一星形开集，ω 是 U 上的任一 C^p $(p \geqslant 1)$ 类的闭 1－微分形式，则：

(1)ω 是恰当的.

(2)ω 的任意两个原微分形式相差一个常数.

证明 星形开集 U 必是连通的. 因此此推论就是定理 3.5 与定理 3.6 的直接结论.

下面我们用具体例子介绍求原微分形式的方法.

5. 求原微分形式的方法

例 5 设 $\omega \in \Omega_1(\mathbf{R}^2)$ 定义如下：$\forall (x, y) \in \mathbf{R}^2$

$$\omega(x, y) = (3x^2 + 2xy + y^2)\mathrm{d}x + (x^2 + 2xy + 3y^2)\mathrm{d}y$$

试证明 ω 的原微分形式 θ 存在，并求出 θ.

事实上，由于 $\mathbf{R}^2$ 是星形区域，并且满足

$$\frac{\partial(3x^2 + 2xy + y^2)}{\partial y} = 2x + 2y = \frac{\partial(x^2 + 2xy + 3y^2)}{\partial x}$$

故根据定理 3.5 知，1－微分形式 ω 在 $\mathbf{R}^2$ 上有原微分形式 θ 存在.

设 $\mathrm{d}\theta = \omega$. 于是我们有：$\forall (x, y) \in \mathbf{R}^2$

$$\frac{\partial \theta}{\partial x}(x, y)\mathrm{d}x + \frac{\partial \theta}{\partial y}(x, y)\mathrm{d}y = (3x^2 + 2xy + y^2)\mathrm{d}x +$$

$$(x^2+2xy+3y^2)\mathrm{d}y$$

由此得到方程组

$$\begin{cases}\dfrac{\partial\theta}{\partial x}(x,y)=3x^2+2xy+y^2\\[2ex]\dfrac{\partial\theta}{\partial y}(x,y)=x^2+2xy+3y^2\end{cases}$$

由第一个方程得到

$$\begin{aligned}\theta(x,y)&=\int(3x^2+2xy+y^2)\mathrm{d}x+\varphi(y)\\&=x^3+x^2y+xy^2+\varphi(y)\end{aligned}\qquad(*)$$

这里 $\varphi:\mathbf{R}\to\mathbf{R}$ 是任一 C^1 类的函数.

为了确定 φ,将 θ 的表达式代入上述方程组的第二个方程得到

$$x^2+2xy+\varphi'(y)=x^2+2xy+3y^2$$

因此

$$\varphi'(y)=3y^2\text{ 或 }\varphi(y)=y^3+C$$

这里 C 为任一常数. 最后将 φ 代入式(*)即得到我们所求的 ω 的原微分形式 θ 的表达式

$$\theta(x,y)=x^3+x^2y+xy^2+y^3+C,\forall(x,y)\in\mathbf{R}^2$$

例 6　设在 $U=\mathbf{R}^2-\{(0,0)\}$ 上的 1－微分形式 ω 定义为

$$\omega(x,y)=(y^2-x^2+2xy)\mathrm{d}x+(y^2-x^2-2xy)\mathrm{d}y$$
$$\forall(x,y)\in\mathbf{R}^2-\{(0,0)\}$$

试证明:

(1) 在 U 上不存在 ω 的任何原微分形式.

(2) 存在关于 $r=\sqrt{x^2+y^2}$ 的 C^1 类函数 $\varphi:\mathbf{R}_+\to\mathbf{R}$ 使得 1－微分形式 $\Omega=\varphi\omega$ 在 U 上有原微分形式存在.

事实上,如果在 U 上 ω 有原微分形式 θ 存在,即 $\mathrm{d}\theta=\omega$,那么由 ω 是 C^∞ 类的知,θ 也是 C^∞ 类的,于是 $\mathrm{d}\omega=\mathrm{d}^2\theta=0$,但是直接计算得到

$$\begin{aligned}\mathrm{d}\omega&=\mathrm{d}(y^2-x^2+2xy)\wedge\mathrm{d}x+\mathrm{d}(y^2-x^2-2xy)\wedge\mathrm{d}y\\&=[(-2x+2y)\mathrm{d}x+(2y+2x)\mathrm{d}y]\wedge\mathrm{d}x+\\&\quad[(-2x-2y)\mathrm{d}x+(2y-2x)\mathrm{d}y]\wedge\mathrm{d}y\\&=(2y+2x)\mathrm{d}y\wedge\mathrm{d}x+(-2x-2y)\mathrm{d}x\wedge\mathrm{d}y\\&=-4(x+y)\mathrm{d}x\wedge\mathrm{d}y\end{aligned}$$

因此在 U 上 $\mathrm{d}\omega\neq0$,此矛盾说明 ω 在 U 上不存在任何的原微分形式.

为了证明存在关于 $r=\sqrt{x^2+y^2}$ 的 C^1 类函数 $\varphi:\mathbf{R}_+\to\mathbf{R}$ 使得

$\Omega=\varphi\omega$ 在 U 上有原微分形式存在，我们设 $\mathrm{d}\theta=\Omega$. 由此得到

$$\frac{\partial\theta}{\partial x}(x,y)=\varphi(r)(y^2-x^2+2xy)$$

$$\frac{\partial\theta}{\partial y}(x,y)=\varphi(r)(y^2-x^2-2xy)$$

由于 φ 是 C^1 类的，故 θ 也是 C^1 类的，从而 $\frac{\partial^2\theta}{\partial x\partial y}=\frac{\partial^2\theta}{\partial y\partial x}$，此即为

$$\varphi'(r)\ \frac{y}{r}(y^2-x^2+2xy)+\varphi(r)(2y+2x)$$

$$=\varphi'(r)\ \frac{x}{r}(y^2-x^2-2xy)+\varphi(r)(-2x-2y)$$

整理后得到

$$r\varphi'(r)+4\varphi(r)=0$$

解此一阶微分方程得

$$\varphi(r)=\frac{C}{r^4}, C \text{ 为常数}$$

取 $C=1$，则

$$\begin{cases}\dfrac{\partial\theta}{\partial x}(x,y)=\dfrac{1}{r^4}(y^2-x^2+2xy)\\ \dfrac{\partial\theta}{\partial y}(x,y)=\dfrac{1}{r^4}(y^2-x^2-2xy)\end{cases}$$

由第一个方程对 x 积分得到

$$\theta(x,y)=\int\frac{1}{r^4}(y^2-x^2+2xy)\mathrm{d}x+h(y)$$

$$=\int\frac{y^2-x^2+2xy}{(x^2+y^2)^2}\mathrm{d}x+h(y)$$

$$=\frac{x-y}{x^2+y^2}+h(y)$$

这里 $h:\mathbf{R}\to\mathbf{R}$ 是 C^1 类函数.

现在将 θ 代入 $\frac{\partial\theta}{\partial y}(x,y)=\frac{1}{r^4}(y^2-x^2-2xy)$ 中以确定函数 h.

直接计算得到

$$\frac{y^2-x^2-2xy}{(x^2+y^2)^2}=\frac{y^2-x^2-2xy}{(x^2+y^2)^2}+h'(y)$$

于是

$$h'(y)=0, \forall\, y\in\mathbf{R}$$

或 $h(y)=C, \forall\, y\in\mathbf{R}$. 因此最后我们求得

$$\theta(x,y)=\frac{x-y}{x^2+y^2}+C, \forall\,(x,y)\in\mathbf{R}^2-\{(0,0)\}$$

这就是我们希望求的1－微分形式$\Omega=\varphi\omega$的一族原微分形式.这里的函数φ通常称为ω的积分因子.

诺特定理、广义相对论、黑洞热力学是人类智慧的杰出成果,值得我们赞美.

笔者曾经读到一首诗是这样总结一代人的:

我看到对平凡事物的赞美,
变成了对崇高之物的嘲笑,
最后变成对卑贱之物的偏好,
这也许就是我们的时代特征!

刘培杰
2023年7月9日
于哈工大

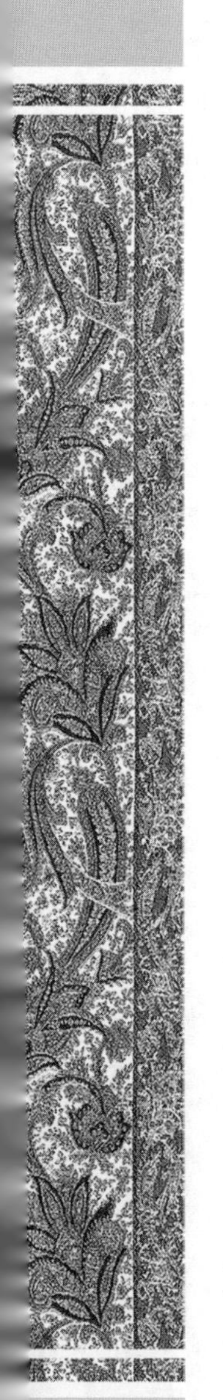

国外优秀物理著作
原版丛书（第一辑）

量化测量——无所不在的数字（英文）

Quantifying Measurement—The Tyranny of Numbers

●[英] 杰弗里·H. 威廉姆斯（Jeffrey H. Williams）著

HITP
哈爾濱工業大學出版社
HARBIN INSTITUTE OF TECHNOLOGY PRESS

黑版贸登字 08-2021-042 号

图书在版编目(CIP)数据

量化测量:无所不在的数字=Quantifying Measurement: The Tyranny of Numbers:英文/(英)杰弗里·H. 威廉姆斯(Jeffrey H. Williams)著. —哈尔滨:哈尔滨工业大学出版社,2024. 10
(国外优秀物理著作原版丛书. 第一辑)
ISBN 978-7-5767-1341-1

Ⅰ. ①量…　Ⅱ. ①杰…　Ⅲ. ①物理学-英文
Ⅳ. ①O4

中国国家版本馆 CIP 数据核字(2024)第 073688 号

LIANGHUA CELIANG: WUSUOBUZAI DE SHUZI

策划编辑　刘培杰　杜莹雪
责任编辑　刘立娟　刘家琳　李　烨　张嘉芮
封面设计　孙茵艾
出版发行　哈尔滨工业大学出版社
社　　址　哈尔滨市南岗区复华四道街 10 号　邮编 150006
传　　真　0451-86414749
网　　址　http://hitpress. hit. edu. cn
印　　刷　哈尔滨博奇印刷有限公司
开　　本　787 mm×1 092 mm　1/16　印张 63.75　字数 1 195 千字
版　　次　2024 年 10 月第 1 版　2024 年 10 月第 1 次印刷
书　　号　ISBN 978-7-5767-1341-1
定　　价　378.00 元(全 6 册)

(如因印装质量问题影响阅读,我社负责调换)

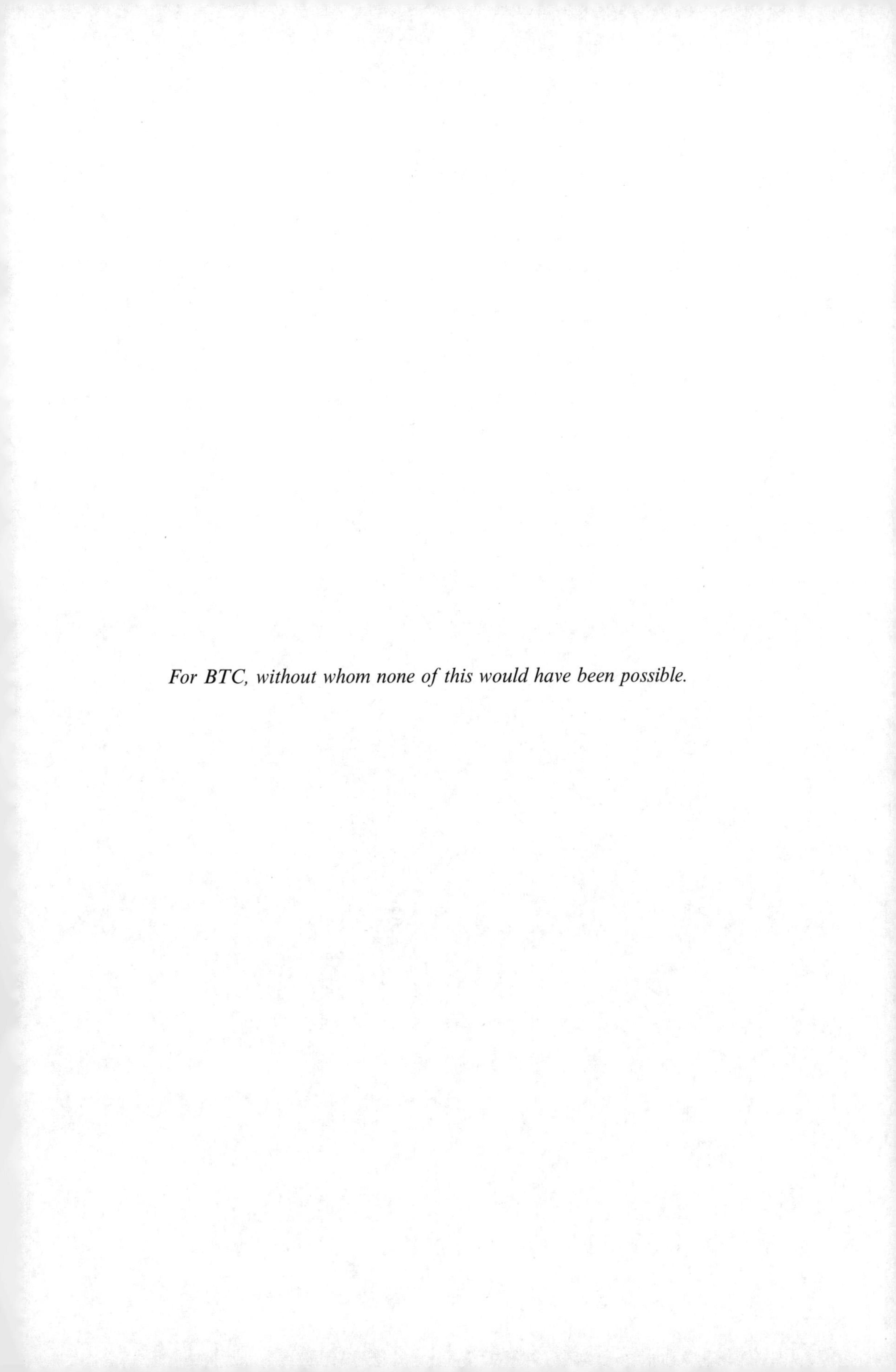

For BTC, without whom none of this would have been possible.

Contents

Introduction

Just about everyone who has ever studied science has done an experiment. Theories have their place, but they are like fashions in that they change with time, and are only of relevance until such time as someone devises an experiment to test their veracity. Then, the results of a well-designed experiment last forever; just think of Archimedes in his bath, or Galileo on the leaning Tower of Pisa or Isaac Newton and his prism in a darkened room containing only a shaft of sunlight. Experimentation is the essential motor that drives the advance of science. This dominance of the experimental result as the final arbiter in a consideration of the possible theoretical models of behaviour is as true in particle physics as it is in biochemistry and medicine. Indeed, those branches of science that are most relevant to our society involve only experiments. Our health and civilization are based on the results of experiments, and we train future scientists by getting them to undertake established experiments, and then to design their own, novel experiments. But how do you design an experiment, and what should your considerations be on deciding whether an experiment has yielded the sought after, or indeed any useful result?

This might all seem self-evident; however, it is true to say that an appreciation of the importance of the quantification of measurement is not as widespread as it should be. Too many reports of experiments leave too much unexplained or unappreciated, and the errors and levels of precision quoted often suggest that they have been estimated purely as an afterthought. The truth is that there is always an uncertainty in each and every experiment.

Imagine a set of data consisting of measurements of something plotted against time. The data is very 'noisy' with a wide-scatter of points, with the line of data points zigzagging up and down wildly, but there is perhaps a suggestion, or a hint of a general downward trend with time. If this plot represented the number of people out of work, you could be the Prime Minister seeking justification for your government's economic policies; on the other hand, if the data represented the polarization state of the cosmic microwave background radiation, you could be a cosmologist seeking evidence for gravitational waves, yet the problem is the same in both instances. That is, trying to find the needle of useful data you believe to be hidden in the haystack of obscuring, background noise.

All experiments involve trying to measure the thing of interest against a background of unwanted noise, which can have a multitude of sources. Sometimes the amount of noise is small compared to what it is that you wish to measure, and the desired result is clearly seen and the measurement readily made. However, perhaps most frequently, the level of noise is greater than the signal of interest, and various techniques must be adopted to permit the detection system to discriminate between the signal and the noise. This is the basis of experimental design, which allows one to determine the limiting sensitivity of the proposed experiment, and so to answer the questions, is this apparatus worth building to do the experiment that interests me, or is there a different design that leads to a more sensitive apparatus?

Lord Rutherford famously said, 'If your experiment needs statistics, you ought to have done a better experiment.' There is some truth in this critique of experimental design, but an analysis using statistics of your measurements can also be considered as a useful post-mortem examination; the statistical analysis may reveal why the experiment did not work. Often in science, the important thing is not so much the measurement of new facts and figures, or the design of a new experiment, but to discover new ways of thinking about those facts and measurements.

In this volume, we will examine an unexpected by-product of the metric survey of the 1790s. Much is made today, by detractors of the Metric System, of Pierre Méchain's 'error'. This was his response to discovering that his measurements were not as precise as those made by his surveying colleague. Poor Méchain was accused of fudging or manipulating his data (an unpardonable sin as far as scientists were, and are concerned) so as to make it appear that his results were more precise than they actually were. The actual reason for this variation in precision is the subject of this book, and we will see how a detailed analysis of the origins of experimental uncertainly led to the development of the science of statistics, and to the modern method of data analysis. It is not my intention to present a detailed description of the statistical and probabilistic methods used by present day scientists to analyze their measured data, as there are many standard texts and the Internet contains many of these texts as freely accessible pdf files. What I do wish to present in these pages is how we think about experiments, and the difference between a quantitative and a qualitative approach to looking at measurements.

Author biography

Jeffrey H Williams

Jeffrey Huw Williams was born in Swansea, Wales, on 13 April 1956, he gained his PhD in chemical physics from Cambridge University in 1981. His career has been in the physical sciences. First, as a research scientist in the universities of Cambridge, Oxford, Harvard and Illinois, and subsequently as a physicist at the Institute Laue-Langevin, Grenoble, one of the world's leading centres for research involving neutrons and neutron scattering.

Jeffrey Williams has published more than sixty technical papers and invited review articles in the peer-reviewed literature. However, he left research in 1992 and moved to the world of science publishing and the communication of science by becoming the European editor for the physical sciences for the AAAS's Science. Subsequently, he was the Assistant Executive Secretary of the International Union of Pure and Applied Chemistry, the agency responsible for the advancement of chemistry through international collaboration. Most recently, 2003–2008, he was the head of publications at the *Bureau international des poids et mesures* (BIPM), Sèvres. The BIPM is charged by the Metre Convention of 1875 with ensuring world-wide uniformity of measurements and their traceability to the International System of Units (SI). It was during these years at the BIPM that he became interested in, and familiar with the origin of the Metric System, its subsequent evolution into the SI, and the coming transformation into the Quantum-SI.

Since retiring, Williams has devoted himself to writing; in 2014 he published *Defining and Measuring Nature: The make of all things* in the IOP Concise Physics series. This publication outlined the coming changes to the definitions of several of the base units of the SI, and the evolution of the SI into the Quantum-SI. Last year Williams published *Order from Force: A natural history of the vacuum* in the IOP Concise Physics series. This title looks primarily at intermolecular forces, but also explores how ordered structures, whether they are galaxies or crystalline solids, arise via the application of a force.

IOP Concise Physics

Quantifying Measurement

Jeffrey H Williams

Chapter 1

The tyranny of numbers

1.1 Why we measure things

Science may be defined in many ways, with as many definitions as there are interested or dis-interested social groups. But those of us with experience of having trained and worked as scientists think of science as the quantitative study of the complex, coupled relationships that may, or may not exist between observed events. Anything that can be measured, anything that can be weighed, anything that can be numbered, anything that can be expressed mathematically—the readings on dials, the 'clicks' and signals coming from a counter or detector can all be considered as part of the enterprise of science. And you can be sure that if you do not measure something, then it will not be included in any scientific analysis, and will likely be lost or forgotten. Today, our scientific knowledge allows us to rationalize the phenomena we see around us, and to make predictions about phenomena as yet unobserved. What was once considered to be magical, is today considered to be rational. Indeed, one could say that magic exists only until it is rationalized by science, when it becomes a banal fact. This is the power of reproducible science; it allows you to be confident about your conclusions.

Conversely, there is no room in the scientific world-view for the inexact, the un-contingent, the immeasurable, the imponderable or the undefined. A process that can be repeated time after time, a system that can be reproduced and analyzed, these are the concepts that go to make up science, and not the individual, the unique, the elusive thing or phenomenon that can never occur a second time. Yet even as some measurements are so complex, or expensive that they cannot be repeated endlessly we must still have confidence in these rare events. Hence our confidence must also rest upon the theory behind the measurement, and on the methodology and practice of the experiment.

But one should also take care not to go to the other extreme and say that if we can measure something, then it must be important; that its significance or utility comes solely from the fact that it is susceptible to measurement. Whether you are part of a huge team of research physicists observing the Higgs Boson at the Large Hadron Collider at CERN, a technician in a hospital scanning for tumours in thousands of mammographs, or an investor staring at a graph on a computer screen trying to decide whether you should sell your shares today or hang on for another day or so, hoping to make more money, you are all making measurements or observations and are taking decisions, often extremely important decisions, on the basis of those measurements. Each measurement is important to some individual, and the reasons for this personal interest are many and varied.

Measurements are made everywhere, for all of us. This is perhaps best seen in manufacturing and commerce, but the intricate and invisible networks of services, suppliers and communications upon which our society is dependent also rely on metrology for their efficient and reliable operation. International time coordination; for example, involves the most precise of all routine measurements (we can define the second to better than one part in 10^{15}), permitting synchronization of computer networks for: communications, banking, satellite navigation systems that allow accurate location via the Global Positioning System, and among many applications enable aircraft to land in poor visibility and the motorist to find his way home (so precision in time measurement and in the measurement of distance are paramount to our contemporary society). Likewise, our health depends upon accurate diagnosis and the ability to deliver effective treatment based on the precise measurement of quantities of drugs, of ionizing radiation, or of viral nucleic acid in our blood; and the subsequent detailed measurements of the effect of these drugs and radiation on the infectious agents and rogue cell in our bodies.

Many physical and chemical measurements affect the quality of the world in which we live. Incorrect and/or imprecise measurements with regard to the changing environment, or of the levels of pollutants entering the biosphere can lead to the wrong decisions being taken, or to no decision at all being taken by politicians and industrialists, which can have serious consequences, costing a great deal of money and even lives. It is important therefore to have reliable and accurate measurements. It is estimated that in Europe today about six percent of our total gross national products are spent on measurements of one kind or another. Metrology has become, whether we realize it or not, an essential part of our lives. Everything from coffee to concrete, and from recreational drugs to prescribed drugs is bought according to weight, and flows of water, electricity, heat and natural gas are metered. Bathroom scales measure our weight and affect our humour and mood, and police speed traps can affect our finances. The pilot observes his altitude, course and fuel consumption; the authorities monitor the bacterial content of our food and water, and the surgeon primes his laser to cut into our bodies to the nearest fraction of a millimetre. Indeed, it is probably not too much of an exaggeration to say that today it would be difficult to think of anything without referring in some way to weights and measures.

Our world is measured and calibrated, and we are subject to the tyranny and authority of these numbers. This is the way things are, and our education system should train us for living in this quantitative world. To function in society, you need to possess an idea or a sense of how to estimate distance, time, mass and value. If this is not the case, then you will become a victim of; for example, those politicians who are prepared to manipulate statistics to further their own ends at election times.

1.2 A little history

The origin of the system of weights and measures most widely used today can be traced to two events; the creation and implementation of the decimal Metric System in France during the years immediately following the French Revolution, and the development of mass production using interchangeable parts in the industrial revolution. These two events were not, however, directly linked. The Metric System was not created in order to facilitate the mass production of engineered products, and the early development of mass production did not rely upon the new metric or decimal units of measurement. Indeed, the first nations to exploit industrial mass production were the UK and the USA, who used inches and pounds as units in their industries. The Metric System arose from an attempt to unify and bring order to the confusion generated by the multitude of units then in use in France for trade and commerce, and to embrace the grand philosophical concept of constructing a set of units that were in some way derived from nature, and unrelated to material objects or artefacts. The development of mass production, on the other hand, was related to the need to produce as many mechanical devices, or guns, or as much screw-thread as possible in the shortest time.

The Metric System came from the bloodiest period of the French Revolution. Civil war was raging in France while the *savants* were putting the finishing touches to the new Metric System, which became mandatory throughout France in April 1795. Unfortunately for the *savants* who created the Metric System, their splendid philosophical idea was not readily accepted by the ordinary people of France. And it was not until 1840 that the Metric System was finally adopted by France as the sole legal system of measurement. By that time, however, manufacturing industries in the UK and in the USA had already become completely locked into the familiar national standards of the inch and the pound. However, with time the Metric System has become the world's system of weight and measures. The origins of this modern language of science have been presented elsewhere (Williams J H 2014 *Defining and Measuring Nature: The make of all things* (San Rafael, CA: Morgan & Claypool)). Here we will use the origin of the Metric System as the starting point to investigate how scientists quantify measurements; that is, how do we determine what is or is not worth measuring, and how well or how badly has something been measured?

1.3 Surveying

In the 17th Century, *savants* in England and then in France had shown how a new system of weights and measures could be based on a single universal measurement; a measurement of length. This universal length measurement could be defined in terms of the dimensions of the Earth, rather than the length of a monarch's arm or foot, and could then be used to define the other quantities needed by technology; that is, mass, as derived from the weight of a defined volume of pure water.

At the time of the French Revolution, a science commission was set up by the French *Académie des sciences* to determine the practicalities of creating such a new universal system of weights and measures, and this commission recommended a measurement of the new standard of length, the metre, based on a detailed survey along the meridian extending from Dunkirk to Barcelona, which had already been surveyed and measured by the Abbé Nicolas-Louis de Lacaille and César-Francois Cassini in 1739. The commission calculated that if they could measure a significant piece of the meridian, the rest could be estimated. Both ends of the line to be surveyed needed to be at sea level, and as near to the middle of the Pole-to-Equator Quadrant as possible to minimize errors. The meridian chosen is about a tenth of the distance (about one thousand kilometres) from the Pole to the Equator and it runs through Dunkirk, Paris and Barcelona, so most of the distance to be surveyed lay conveniently inside France.

The *Académie des sciences* may have decided that the metre would be exactly a ten millionth of the distance between the North Pole and the Equator, but their choice also defined this distance as being precisely 10 000 000 metres. Unfortunately, an error was made in the commission's initial estimation, because the wrong value was used in correcting for our planet's oblateness. We now know that this Quadrant of the Earth is 10 000 957 metres. One should never forget, that these *savants* were not only setting out to create what they saw as a philosophically coherent system of units based on the dimensions of the Earth, but they were also imposing models and views about the character of the Earth. In 1791, a handful of mathematicians, guided by the writings of Sir Isaac Newton, imposed a definite shape and size to our planet; the Earth shrank and became precisely known.

In the afternoon of 19 June 1791, Jean-Dominique de Cassini (head of the Royal Observatory) had secured an audience with King Louis XVI for some of the members of the metric commission of the *Académie des sciences*. At six in the evening, Cassini, Adrien-Marie Legendre, Pierre François-André Méchain and Jean-Charles de Borda (the inventor of the repeating circle which was hoped would increase the level of precision possible in surveying and thus allow the determination of the metre) presented themselves at the *Palace du Tuileries*. A small group of eminent astronomers and mathematicians had come to convince the now constitutional King Louis XVI that the metre was something worth achieving.

History has not been kind to Louis XVI. He has a reputation for having been naïve and something of a simpleton, but he had hidden talents. The king was a skilled instrument (watch) maker and something of a cartographer. The king also took a close interest in the cost and necessity of the proposed survey. Turning to the head of the Royal Observatory he asked, 'How's that, Monsieur Cassini? Will you again measure the meridian your father and grandfather measured before you? Do you think you can do better than they?' Monsieur Cassini (the third generation of a dynasty of directors of the Royal Observatory) was not unused to conversing with the monarch. 'Sire, I would not flatter myself to think that I could surpass them had I not a distinct advantage. My father and grandfather's instruments could but measure to within fifteen seconds (of a minute of a degree); the instrument of Monsieur Borda here can measure to within one second;' see figure 1.1. (The details of these conversations were published in the *comptes rendue* of the *Académie des sciences*.)

King Louis XVI gave his formal approval for the new survey of the Dunkirk–Barcelona meridian. Then in the early hours of the next day, the king and his family attempted to escape from France (the 'Flight to Varennes'), but they were arrested, returned to Paris and then imprisoned. What must the king have been thinking when

Figure 1.1. A Borda repeating circle; the world's first high-precision measuring device (Exhibit in the Mathematisch-Physikalischer Salon (Zwinger), Dresden, Germany; the image is in the public domain). The first repeating circle was built by Étienne Lenoir in Paris c. 1790; the design was subsequently perfected by Jean-Charles de Borda. Details of the finesse of this instrument may be seen on the Borda circle at the Royal Maritime Museum, Greenwich, at http://collections.rmg.co.uk/collections/objects/42288.html
A surveyor using this instrument would be interested in determining the angle between a cardinal direction and an object: a hill, a church tower, etc. The telescope is trained and centred on the object of interest and the angle between the long axes of the telescope gives the angle that defines the location of the object.

he was contemplating his escape and the suppression of the revolution, while trying to comprehend the mathematics on which he was being lectured by his visitors? However, under arrest or not, Louis XVI was still the king, and from his prison cell he issued the proclamation that directed the two eminent astronomers Jean-Baptiste Delambre (1749–1822) and Pierre Francois André Méchain (1744–1804), to undertake the surveying necessary to determine the length of the metre by precisely measuring the distance from Dunkirk to Barcelona. The king also issued orders to Baron Gaspard Clair François Marie-Riche de Prony to produce new trigonometry tables, with a greater degree of precision, which would be needed to calculate the new universal measure from the surveying work of Delambre and Méchain.

The survey from Dunkirk to Barcelona was a major undertaking. Antoine Lavoisier called it 'the most important mission that any man has ever been charged with'; the measurements were designed to have been completed within a few months, yet it took the two surveyors from May 1792 to September 1798 to complete the work. The technical difficulties were compounded by more practical problems, such as civil and international wars. France was in uproar with some cities restoring governments favourable to the monarchy.

The surveying method used by Delambre and Méchain was triangulation, see figure 1.2, where they had to accurately measure the angles in each of the many hundreds of triangles into which they had subdivided the territory to be surveyed. The surveyors emphasized the decimal aspect of the new Metric System by discarding the traditional (Babylonian system where angles are sub-divided into sixtieth parts) degrees and minutes of angular measurement, and instead divided the

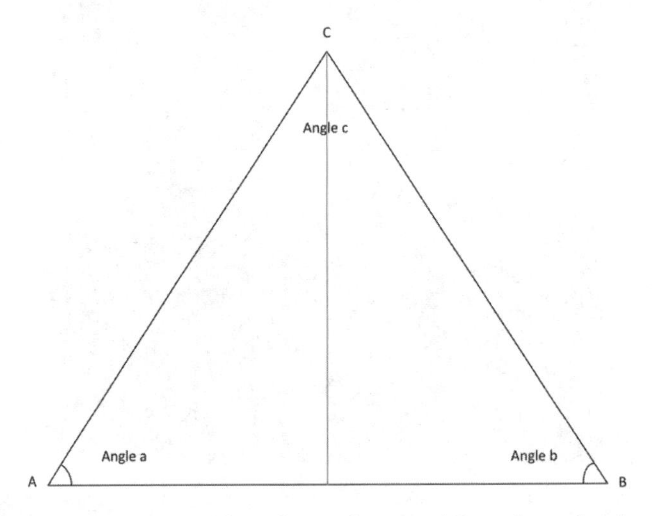

Figure 1.2. Triangulation: the known positions of two points, A and B, can be used to determine the relative position of a nearby point, C. The calculation involves the distance between the two known points, and the angles between the known points and the un-known point (as measured by the Borda circle). First, angle c is determined (we know angles a and b in triangle ABC), then AC and BC may be calculated by the Law of Sines. The altitude of the triangle or the offset of C from the horizontal line AB is found from AC.sin(a) or BC.sin(b); the offset in the opposite direction could be determined by using the cosines of a and b.

quadrant into one hundred grads that were then sub-divided into one thousand arc-minutes, that were in turn sub-subdivided decimally into arc-seconds.

At the completion of the measurements of each of the angles, in each of the triangles that made up the distance between Dunkirk and Barcelona, the surveyors had to measure as precisely as possible the length of one of the sides of one of these triangles. Then using trigonometry, they could calculate all of the distances, in all of the triangles. Consequently, two of the most important measurements were made by repeatedly laying a set of two-toise (a toise was a distance of about two yards—a fathom in English) platinum rulers mounted end to end on large wooden stabilizing blocks along two straight, flat roads.

In preparation for this final calculation of the metre, in September 1795 Méchain found a suitably flat stretch of road near Perpignan, the length of which he measured precisely with his platinum toise measuring sticks. This flat stretch of road was found from measurements made over several weeks by precisely placing the measuring sticks one in front of the other, to be a little over 6000 toise. This distance was one of the sides of one of the triangles being surveyed, and it would serve as a test for the quality of the angle measurements.

This final test of the quality of the surveying, and hence of the precision with which the metre had been determined, came in 1798 when Delambre took almost two months to measure the length of one of the sides of one of the surveyed triangles in the countryside near Paris. Again, the precise measurement of length was made with a pair of two-toise platinum measuring sticks. After this measurement of distance was completed in September 1798, the two surveyors met to compare their two length measurements. From the measured distance in the countryside near Paris, 6075.90 toise, they calculated a theoretical value for the length of the side of the triangle near Perpignan, which they calculated to be 6006.198 toise. Méchain then revealed that he had actually measured this distance to be 6006.27 toise. As the surveyors commented, 'the difference is negligible.'

But even after almost seven years of field measurements made under the most arduous of conditions, the astronomers' carefully calculated metre turned out to be no more precise than the preliminary estimate of the *Académie des sciences*, which the French authorities, to the annoyance of many *savants* (including the two surveyors), had promoted in the interim as a provisional unit. Sadly, because of the political situation in France and the physical and financial hardship of the surveying process, the two surveyors and their teams took a lot longer to complete their task than had originally been expected. As time went on, the politicians in Paris realized that to satisfy the demands of the people for a new system of weights and measures they could not wait for the precise measurements of Delambre and Méchain, so they adopted a provisional metre based upon the survey of the same meridian carried out by César-Francois Cassini, and made and distributed wooden and steel 'provisional metre-sticks'.

Today, we think nothing of making measurements of distance over extended sections of the globe. We have a network of satellites in orbit around the Earth, and this investment has made global positioning something that you have in your car when you are going out for dinner. But what of the measurements of those two late-18th Century surveyors, Delambre and Méchain, who set out into a war zone with the very latest of surveying equipment and measured the distance from Dunkirk to Barcelona with the greatest of precision? The angle measuring repeating circles, which had been designed by Jean-Charles de Borda and used by the two surveyors to measure the meridian (see figure 1.1) were an enormous improvement over previous surveying instruments. In themselves, the circles were very stable by virtue of being massive and large (made of dense metals); however, the problem was that they had usually to be mounted well above ground level to make a measurement. And consequently, the instability of the wooden platforms that the precision repeating circles were mounted upon were the main source of the error in the measurement; particularly, when the weather was stormy and windy.

To gain an idea of the precision of the metric survey, consider some data measured by Delambre near Dunkirk. Here, he needed to establish the latitude of his observations, and to do this he made detailed measurements of the transits of various well-known stars. Delambre made dozens of observations, which gave him a latitude of 51° 2′16.66″; a value which changed by a fraction of an arc-second when he removed what he perceived to be the least-reliable measurement. Let us therefore assume a precision of 0.1 arc-second; to relate this estimate of measurement uncertainty to a distance, consider the circumference of the perfectly spherical Earth as being 40 000 kilometres. By dividing this circumference by 360, we determine the distance equivalent to measurement of a single degree, then by dividing by sixty and then again by sixty we determine the distance equivalent to a single arc-second; which gives an uncertainty in their measurements of about 3 m in the measurement of distance.

That is, these late-18th Century surveyors were capable of determining the location of an object on the surface of the Earth to a precision of about fourteen feet; provided they had sufficient time. But this remains an amazing feat when one considers that a commercial GPS system is only accurate to about three-times this level of precision; and the GPS represents an enormous investment, in terms of money and time in space-age technology.

The final results of the survey not only confirmed the value of the provisional metre established in 1793, but also produced something that was not anticipated; genuine new science, which we will explore in the coming chapters. Delambre and Méchain had not set out to do basic research, but merely to improve the precision of something that had previously been measured. But what they discovered was that the Earth was even less spherical than Sir Isaac Newton had calculated. According to surveying data determined by French *savants* in the mid-18th Century in Peru and France, the eccentricity of the Earth was roughly 1/300; that is, the Earth's radius to the Poles was 1/300 or 0.3% shorter than a radius to the Equator. Delambre and Méchain had found that this eccentricity (from Dunkirk to Barcelona) was about 1/150, or twice as great as had previously been thought.

The difference between the theory of Newton and the measurements of the surveyors had revealed that the Earth's surface was a patchwork of coupled segments—it was not uniform. As Méchain commented to a colleague, 'Our observations show that the Earth's curve is nearly circular from Dunkirk to Paris, more elliptical from Paris to Evaux, even more elliptical from Evaux to Carcassonne, then returns to the prior ellipticity from Carcassonne to Barcelona, So why did He who moulded our globe with his hands not take more care … ?'; a splendid question, which modern geology can now answer.

As in the scientific advances of the late-19th Century, when scientists attempted to measure something familiar merely with greater precision, they invariably discovered new science; so with the metric survey. In the late-19th Century, increased precision in measurement led to the creation of quantum mechanics and relativity, in the late-18th Century, increased precision of measurement allowed the creation of 19th Century science.

1.4 Other surveys

The European Enlightenment had come up with the idea of constructing a new decimal metrology based on a single measurement of length. Such ideas, however, have a long history, and it is to Ancient China that we must turn for the first consistent use of decimal weights and measures; particularly, in the decrees of the first emperor, Chin Shih Huang Ti in 221 BCE.

Given the size of China, it is perhaps not surprising that an early effort was also made in fixing terrestrial length measurements in terms of astronomical measurements or observations. It was an early idea of Chinese *savants*, going back before the time of Confucius (551–479 BCE), that the shadow-length of a standard height (an 8 foot gnomon), at the summer solstice increased by 1 inch for every thousand *li* (a length measurement equivalent to 1500 *chi* or Chinese feet) north of the Earth's 'centre', and decreased by the same proportion as one went south. This rule of thumb remained current until the Han Dynasty (205 BCE–220), when detailed surveying of the expanding Chinese Empire showed it to be incorrect. But it was not until the Tang Dynasty (618–907) that a systematic effort was made to determine a range of latitudes. This extensive Tang survey had the objective of correlating the lengths of terrestrial and celestial measures by finding the number of *li* that corresponded to 1° of polar altitude; that is, terrestrial latitude; thereby fixing the length of the *li* in terms of the Earth's circumference. This Chinese meridian survey takes its place in history between the lines of Eratosthenes (c 200 BCE), and those of the astronomers of the Caliph, al-Ma'mūm (c 827), but more than a thousand years before the metric survey.

The majority of these Chinese surveying measurements were undertaken between 723 and 726 by the Astronomer-Royal, Nankung Yüeh, and his assistant, I-Hsing, a Buddhist monk. The survey was carried out at eleven sites along a meridian running from the Great Wall in the north to Indo-China in the south; a distance of 7973 *li* or about 2500 km. The main result of this field work was that the difference in shadow-length was found to be close to 4 in for each 1000 *li* north and south, and that the terrestrial distance corresponding to 1° of polar altitude was calculated to be 351 *li* and 80 *bu* (the *bu* was a measure of between 5–6 *chi*). The imperial surveyors had achieved their goal of defining a terrestrial unit of length, intended for use throughout the empire, in terms of the dimensions of 'Heaven and Earth'; that is, 1/351 of a degree.

This survey is today practically unknown, yet it represents an outstanding achievement given the spaciousness and amplitude of its plan and organization, and one of the earliest uses of advanced mathematics which was needed to compute the final result. These results were known in 18th Century Europe, as they were commented upon by Leonard Euler and later by Pierre Simon de Laplace. While the metric survey obtained a routine precision of about 1 part in 10^6 in distance, the much earlier Chinese survey could boast only of a precision of 1 part in 10^3. The Tang value of the *li* gives a modern equivalence of 323 m, but the earlier standard Han *li* is very different at 416 m.

Further reading

For further information about all aspects of the history of the Metric System mentioned in this chapter, please see Williams J H 2014 *Defining and Measuring Nature: The make of all things* (San Rafael, CA: Morgan & Claypool).

In addition, there are many useful entries in the online encyclopaedia, Wikipedia, about the topics touched upon in this chapter. And the best place to find further details about science in historical China is Needham J 1962 *Science and Civilization in China* (Cambridge: Cambridge University Press); the survey mentioned in this chapter is taken from volume 4 (part 1), pages 42–55

Quantifying Measurement

Jeffrey H Williams

Chapter 2

The error in all things

2.1 Introduction

In 1792, two internationally renowned French mathematicians and astronomers set out from Paris; one headed north, the other south to measure the distance between Dunkirk and Barcelona with as much precision as was afforded by their technology. These two *géomètres* were seeking to lay the groundwork for a new universal system of measurement, by creating a new standard unit of length to be derived from nature herself. Forget the distance from the tip of the king's nose to the tip of his thumb, the *aune* and the ell, the confusing patchwork of local measures in *ancien régime* France, which only served to institutionalize fraud. The new standard of measurement, as befitted the spirit of the age of universal rights and ideals, would be based on something *universal*: the size of the Earth. The result of these endeavours would be a system of weights and measures appropriate for all people; irrespective of where they lived, or under what type of government.

Making long-distance measurements of great precision and accuracy is never easy. Engineers and geodesists today would use the Global Positioning System or laser range-finding theodolites to do this sort of work. Best practice in the late-18th Century involved fitting together a number of different kinds of observations, each demanding work in the field, or on top of a mountain with large, heavy instruments of brass and glass (see figure 1.1). Astronomical observations fixed the terrestrial coordinates of the endpoints of the line to be measured, but these points then had to be squared with a set of observations made along the line itself: first, the angles of triangles sighted from hilltop to hilltop along the entire distance of interest; second, the actual paced-off length of one side of one of those triangles, a length which could then be projected through the whole chain of triangles using trigonometry. It was an immensely complex business, demanding patience, fortitude, good physical health, fine eye-sight and a ready hand for kilometres of long arithmetic calculations.

Sadly, the French Revolution and the wars triggered by that event did not facilitate such an enterprise. As Méchain and Delambre set about their task, the

doi:10.1088/978-1-6817-4433-9ch2

countries around France declared war on France, which was a particular problem for Méchain's surveying team working in Spain. Indeed, the saga of the surveying process is one of the great epics of science. The surveying team would arrive in a small town and present their ormolu commission papers to surly peasant mobs led by sly ambitious local politicians, only to discover that the government who had issued these papers has since dissolved, and the politicians who had signed the papers with a great flourish had already been executed as counter-revolutionaries.

Not only did the surveyors have to deal with local peasants who had not the slightest interest in what they were doing or who were of the opinion that these educated gentlemen from Paris were themselves counter-revolutionary agents, but the measurements were not easy to make. Often it was necessary to construct a platform around the spire of a local church or on top of a hill so as to be able to see the next hill, or castle, or mountain in the chain of surveying points. Then, making the line of sight measurements of the angles between one point and another point became difficult because of the weather; storms, snow, mist and rain all contributed to the difficulty of the enterprise. In addition, the French currency became worthless through hyper-inflation and so the surveyors had to beg and steal food, they fell ill and Méchain almost died in a fall from an observation platform. All of these hardships contributed to the quality of the measurements they were making.

I am sure that any of my readers who have spent a longish period of time trying to make a measurement of something that is not easy to measure, will agree with me that there were good days in the laboratory, and then there were bad days in the laboratory. That is, on some days of measurements, you knew everything was working well, and that you were gathering 'good' data, but on another day something will not have been working so well and the data will not have seemed so consistent, and you will know that these newer data are less reliable. But how (and why) do you distinguish between the, perhaps, subjective good data and the less-good data? How do you sort and characterize your data? And if you have been making measurements over seven years, you will have a lot of data to sort. This was the other great problem for the surveyors; how to qualify and then quantify their measurements? No experimenters had previously gathered such a quantity of internationally important data, and European and American *savants* were waiting for these results. Delambre and Méchain were the first to undertake such an extended exercise; these two astronomer mathematicians were the first true scientists, the first to make high-precision measurements of anything.

Both surveyors survived the seven-years of field work, dodging war zones and the guillotine. They brought back data that enabled an international committee of mathematicians and geographers (assembled from those countries with which Frances was not at war—there were not many) to arrive at a value for the quarter meridian, and from it to derive the length of what would be called the *mètre*. A length that was for almost two centuries defined by a flat bar of platinum-iridium alloy, and was the basis of the metric system of weights and measures. Today, however, this length is defined by the speed of light (299 792 458 metres per second) and the definition of the second. Yet when the data brought back to Paris by the two surveyors was analyzed, the shape of the Earth was discovered to be a lot less regular than had been imagined. No elegant geometrical curve could accommodate the survey's data. Several centuries of scientific debate and theorizing about the form of the globe (was it a sphere, an oblate ellipsoid or something egg-shaped ... and if so, why?) was ending in a tautology: the shape of the Earth was the shape of the Earth. A unique, irregular, lop-sided thing. The seven-years of field work transformed the world by giving to science a new morphology for the globe; that irregular spheroid today called the geoid.

2.2 Méchain's 'error' in greater detail and least-squares

Every so often, much is made of a supposed error in the data measured by Méchain; an error that was, supposedly, kept a secret. However, the truth is that this secret is what the French call *le secret de polichinelle*; that is, not a secret at all, but something everyone knows. No one was seeking to hide anything.

Was there a hidden error that corrupted the pure, philosophical metre? Is that metre-stick wrong because of some French double-dealing? Of course, not. Admittedly, your metre is wrong, in that the distance from the North Pole to the Equator via Paris and Barcelona is more than 10 000 000 metres. And yes, there was some double-dealing in the observations used to calculate the metre's length; Méchain, seeking to fix the southern endpoint of his line, ran into some discrepancies in the results he calculated from his observations, and rather than report them directly, he fussed and fudged for years while trying to clean up the data. Anxiety over the whole affair, pushed Méchain to the edge of madness, and hounded him to his death in 1804.

But Méchain's fussing and fudging had only a minuscule effect on the length of the metre itself, a distortion completely lost in the mathematics that were required once it became clear that the Earth has an irregular form. So, what then is one ten-millionth of a quarter meridian? This, of course, depends on where you are and on what you want to measure.

How then, could Méchain's 'error', which came to light after his death, possibly have any influence or be of any importance? In *The Measure of All Things*, Ken Alder comments that the world was changed by the 'error itself'. As an experienced astronomer who is credited with the discovery of many stars and comets, Méchain could not bring himself to accept that he had made an error, but in fact, unbeknownst to him, in the complex set of interconnected measurements he had accumulated over seven years, among the many thousands of individual observations, which constituted his data he had caught a first glimpse of the phenomenon of error and of uncertainty. There was no mistake that he could find and simply correct. As he spent years fretting over his data, and losing his physical and mental health in the process, he was merely chasing a spectre. And the ghost he kept almost seeing as he stared at the columns of numbers was nothing less than the limits of his ability, or the ability of any measurement scientist to get the 'right answer' to the problem they are investigating.

After Méchain's death, when his colleagues re-worked his thousands of calculations and observations, they did not find an error but the concept of measurement uncertainty, and, in the early years of the 19th Century, they created a theory to deal with it, a metric of metrics. Science itself was in this way transformed. The plight of Pierre Méchain will make sense to any scientist familiar with the standard laboratory tools of error analysis, those mathematical techniques that enable researchers to separate precision from accuracy, and to assess what is attainable in a specific measurement. Méchain, however, had no such tools to hand.

Science would not be the same after the metric survey; left behind was the world of *savants* like Méchain and his co-surveyor for whom the pursuit of enlightenment was a simple confrontation with the demon of error personifying a lack of enlightenment and rationality. For them, error was like sin, and to resist temptation they bound their spirits to the rectitude of their measurements. But the metric survey transcended any previous collection of observations. A new form of science arose, and with it the new man, the scientist. That is, someone more functionary than wise sage or *savant*, a professional who calibrated himself along with his instruments. Gone were the high-priests of Reason, who had been the souls of their measuring devices. Modern science was born, along with the idea of an inherent unavoidable uncertainty in measurement.

2.3 The metric survey

The geodetic survey done in Peru in the early-18th Century by a group of French scientists serves as a good example of how such surveys were conducted at that time. The survey consisted of two teams, who perform the same measurements and calculations, and then compared their results. In Peru, the French were trying to measure the length of a degree of arc near the equator and, using a French length unit, the toise which is 6 Paris feet or 6.39 English feet, one team got a result for a degree of longitude of 56 749 toise and the other team's result was 56 768 toise.[1]

Pierre Méchain was a meticulous astronomer, but given to obsessive attention to detail. His own measurements did not agree exactly nor meet his exacting standards, though they are, in retrospect, some of his best. He ended up fudging results, changing figures to make himself look better or no worse than his co-surveyor, but in reality he was trying to cover up errors that were, to him, intolerable. When Delambre received Méchain's raw data, after Méchain had died of yellow fever while trying to correct his observations by surveying further south than Barcelona, the data was not in bound notebooks but on scraps of paper with erasures, lack of dates, etc (see figure 2.1). Delambre carried forward his colleague's cover-up, cleaning up the data to make it publishable. He did this because he found that Méchain's final results were correct, erring mostly in the size of variations among his observations. The erasures and corrections did not affect the final result, but made Méchain's work look as if it had been better performed than it actually had been. Various instrumental problems, such as excessive wear and lack of calibration, were also observed.

Twenty-five years after Méchain's death, a young astronomer named Jean-Nicolas Nicollet (1786–1843) resolved the enigma of the 'error'. The problem was

[1] This is a difference of 19 toise, or 0.0335%. That's about 120 (English) feet difference over about 68.73 miles; again a remarkably precise result given the period and the locality.

Figure 2.1. Méchain's data with notes by Delambre. Between 1806 and 1810, Delambre reconstructed Méchain's logbook by pasting into a bound register the loose sheets on which Méchain had recorded his data. Delambre organized the sheets into chronological order, retraced Méchain's pencilled data in ink, and indicated the provenance of each document (that is, Delambre followed what is still sound practice). On this particular page Delambre has pasted Méchain's observations from Barcelona for 15 December 1793. In the margin Delambre notes: '*Here are some changes that Méchain has made to the angle measurements for which it is difficult to imagine a legitimate rationale*'. He goes on to explain that Méchain's calculations on this page leave no doubt whatsoever that the corrections are not legitimate, but serve only to make the data appear more precise than they actually are. (From the *Archives de l'Observatoire de Paris*).

that Méchain and his contemporaries did not make a principled distinction between precision (that is, the internal consistency of results) and accuracy (that is, the degree to which those results approached the 'right answer'—whatever that might be).

The definitions for precision and accuracy given above are the key. Look at the Peru survey, where there was a difference of 19 toise between two results, this is a situation that occurs regularly in sequences of measurements. The second time you measure something the result may not agree with the first time it was measured, or other subsequent measurements. What is causing the problem? If you are surveying in the Andes, could it have been the weather? It was warmer on day two than day one, so maybe the metal tape measure or ruler had expanded with the heat, or was the measurement more difficult to make on the second day, because you were not feeling well, or there had been an avalanche and the flat surface you needed to measure was not as flat as before. These are all sources of error, and they would perturb your measurement. Maybe on day one you did the measurement, and on day two you had a colleague repeat the measurement. There can be individual human variations in the way that measurements are set up, calibrated and used. So how can anything be measured accurately and with precision?

The method that has been developed in the last 200 years to analyze or to quantify such perturbations of a measurement is called statistical analysis, and includes techniques and concepts for dealing with multiple measurements. It is often the case that a detailed analysis of sets of measurements can require a detailed statistical analysis, which can be a lot less interesting than doing the measurements themselves, and this was probably what caused Lord Ernest Rutherford to make his famous comment, 'If you need statistics to analyse your experimental data, you should have done a better experiment'. But in modern science, it is not possible to avoid statistics, and it is a requirement for good practice (and it is demanded by the editors of journals, who decide when error is too great to bear).

It was subsequently discovered that in no instance had Méchain's alterations distorted the final result by more than two arc-seconds, meaning that his adjustments were minor compared to the uncertainties caused by the observer's inability to correct entirely for the refraction of light in the Earth's atmosphere. He had edited his results, not to alter the final outcome, but to keep up appearances as an astronomer. Delambre wrote, 'Undoubtedly Méchain was wrong not to publish these observations as he found them, and to modify them in such a way as to make them appear more precise and consistent than they were. But he always chose his final values in such a way as to ensure that the average was not altered, so there was no real harm in his action, except for the fact that another observer who published unadulterated numbers would be judged less capable and careful'. This is almost a definition of art, rather than the application of scientific principles. What was needed was a sound set of rules upon which an experimentalist could base his choice of which of his measurements they felt to be the least, or the most reliable.

On the subject of Méchain's 'error', Delambre preferred to blame the Earth, rather than his colleague. As he pointed out, the meridian project had confirmed that the shape of the Earth was irregular, and that not all meridians were equal. Delambre hypothesized that Méchain's readings had been distorted by local irregularities in the Earth's crust or by nearby mountains. This concern was not new, as Isaac Newton had tried to estimate the gravitational pull of mountains. But all of this was speculation as 18th Century technology did not permit measurements sufficiently precise to determine the effect of mountains on the value of the local acceleration due to gravity. Today, however, the science of geodesy consists principally of mapping these gravitational effects. Physicists and engineers estimate the pull of mountains on rockets and missiles. Some of the maps that chart the contours of the geoid are classified as military secrets.

Pierre Méchain and his contemporaries did not make a principled distinction between precision and accuracy. The two are not the same; precise results may appear 'reliable' in the sense that they give very nearly the same answer when measured again and again; yet they may lack validity in that they deviate consistently from the 'right answer'. Of course, in practice, distinguishing between the two can be extremely difficult as the 'right answer' is unknown.

Repetition of measurements using the Borda circle was designed to improve precision by reducing those errors that stemmed from the imperfect senses of the observer or the imperfect construction of the instrument's gauge (the sort of errors we would today characterize as falling into a random distribution). The Borda circle, however, was still subject to errors caused by the design of the instrument; the sort of errors we would today characterize as those constant (or systematic) errors which make results inaccurate, whatever their level of precision. Constant errors generally go undetected, as long as they stay constant. And in an intuitive way, Méchain and Delambre like their contemporaries understood this, hence their vigilance about maintaining a consistent setup for their apparatus from one series of observations to the next. What they failed to appreciate, however, was that the same repetition that enhanced precision might reduce accuracy. For instance, constant manipulation of the circle might wear down the instrument's central axis and, over time, cause the circle to tilt ever so slightly from the perpendicular. It was this unanticipated drift in the constant error, Nicollet suggested, that was the source of Méchain's discrepancies. Without a concept of error to help him identify the source of this contradiction, Méchain was confounded.

Oddly enough, Nicollet noted, it was Méchain's own obsessiveness which made it possible to confirm the cause of the discrepancy, and to correct for it. One needed to compensate for any change in the instrument's verticality by balancing the data for stars which passed north of the zenith (the highest point of the midnight sky) against those which passed south of it. Because Méchain had measured so many extra stars, such an operation was possible from the recorded data.

To calculate the latitude, Méchain had first calculated the average latitude implied by each star he measured, and then averaged all the averages, giving equal weight to each. Nothing could be simpler (or more naïve). Nicollet, by contrast, first analyzed the data for the stars Méchain had measured that passed north of the zenith, taking the average of the average latitude implied by each. Then Nicollet separately did the same thing for the stars Méchain had measured to the south of the zenith. Clustered in this manner, the results seemed to lack precision; at Barcelona the average latitude implied by the north-going stars differed from the average latitude implied by the south-going stars by 1.5 arc-seconds. At Fontana de Oro they differed by 4.2 arc-seconds. But when the northern average and the southern average were themselves combined at each location, they suggested a remarkable accuracy; the combined latitude for the Fontana de Oro agreed with the combined latitude from Barcelona to within 0.25 arc-seconds. Nicollet had demonstrated that there was no discrepancy, and that Méchain's reported value for Barcelona was within 0.4 arc-seconds of the answer indicated by his data; when properly analysed after a post-mortem of Méchain's surveying instrument.

2.4 Least-squares

Hindsight and time are ideal for solving problems that were originally thought to be mysteries. But the confusions about the results of the metric survey did start *savants* thinking about the nature of the 'right answer', and about the possible existence of the 'correct value' of a natural phenomenon. It may be more of a theological question to ask if perfection exists in nature awaiting discovery, but Delambre was neither a believer nor an atheist. He was a sceptic, for whom perfect knowledge lay beyond man's grasp; so why should anyone expect him to produce a perfect metre?

In coming to terms with the imperfections and limits of the experimenter's art, Delambre had a powerful new tool at his disposal, one which he and Méchain had inadvertently inspired, but which he alone had lived to see. Previously, *savants* had sought to fit imperfect data to a perfect curve. Astronomers had agreed that the Earth was an oblate ellipsoid, but they had been unable to agree on its degree of eccentricity, which now seemed, moreover, to vary from place to place (the same would also have been the case for the Chinese survey of the 8th Century mentioned in chapter 1). Assume that the data had been gathered by fallible (but exacting) investigators using fallible (but ingenious) instruments on a (possibly) lumpy irregular Earth, and then ask yourself: what was the best curve through the data, and how much did the data deviate from that curve? This was the question asked by the mathematician Adrien-Marie Legendre (1752–1833).

A contemporary of Laplace and Delambre, Legendre was elected to the *Académie des sciences* at the age of thirty. In 1788 he showed the geodesists on the Paris–Greenwich surveying expedition how to correct for the curvature of their triangles. Appointed with Cassini and Méchain to the metric survey, he withdrew in favour of Delambre. Later he was one of the *savants* who calculated the length of the metre for the international commission who received the data of the surveyors. He was as baffled by the result (an unexpected eccentricity of 1/150) as were the other members of the commission. Legendre's answer, the method of least-squares, has since become a standard tool of statistical analysis. It was also among the most important breakthroughs in mathematical science; not because it produced new knowledge of nature, but because it produced a new way of conceptualizing and quantifying error.

For centuries *savants* had felt entitled to use their intuition and experience to publish their single 'best' observation as the measure of a phenomenon. During the course of the 18th Century, *savants* had increasingly come to believe that the arithmetic mean of their measurements offered the most balanced view of their results. Yet many *savants* continued to feel, like Méchain, that any measurement that strayed too far from the mean ought to count for less than those near to it, and hence could be discarded.

Adrien-Marie Legendre suggested a practical solution; that the best curve would be the one that minimized the square of the value of the difference of each data point from that theoretical curve. This was a general rule, and it was also a tractable calculation that could be widely understood. Legendre's least-squares method played off the intuition that the best result should strike a balance among divergent data, much as the centre of gravity defines the balance point of an object. As he noted, the least-squares method also justified choosing the arithmetic mean in the simplest of cases. It gave *savants* a workable method for weighting and sorting data.

In 1805, just as Delambre was completing the first volume of his biography of the metric survey, Legendre tried out his method on what was now the world's most famous data set, the one he had puzzled over ever since Delambre and Méchain had handed it to the international commission in 1798. Legendre assumed that the Earth's meridian traced out an ellipse; he then used the least-squares rule to find the eccentricity that would minimize the square of each latitude's deviation from that curve as it connected the data points. In this way, he observed that the deviations of the various latitudes from that optimal curve remained sufficiently large to be ascribed to the figure of the Earth and not to the data; it was the Earth that was difficult to understand, not the data.

Four years after Legendre's paper, the great mathematician Karl Friedrich Gauss (1777–1855) claimed that he had been using the least-squares rule, which he called 'my method', for nearly a decade. As often happens, this simultaneous discovery was no coincidence. Indeed, Gauss was working on the same data set, the data from the metric survey, which had been published in Germany in 1799. They were also reading the same mathematicians, especially Laplace. However, as so often happens, this simultaneous discovery prompted a bitter dispute over priority. In this instance, there seems little doubt that the two men arrived at the method independently, although it was Legendre who published first. But there is no doubt that it was Gauss who pointed out the method's important meaning.

Gauss had predicted the position of an asteroid that has been lost to observation while passing behind the Sun. The asteroid in question, Ceres, had been observed 24 times early in 1801 by Giuseppe Piazzi who published the positions. Gauss used the observations to calculate where and when the asteroid would re-appear. He was correct to within half a degree. Among the mathematical tools he used in this truly impressive calculation was the least-squares method, to match observations with a model developed from the observations assuming that the experimental or observational errors followed a normal distribution. He took the observations of others, and used statistics and probability theory to construct a theoretical model, which he used to predict where Ceres would be some time into the future; that is, to define the path and the velocity of the asteroid on the path. Gauss got it right, thereby demonstrating his own genius and the power of his technique, as nothing proves the value of new mathematical models like predictions that are validated by subsequent experiment. Thus was modern physical science born.

Adrien-Marie Legendre presented his method of least-squares as workable and plausible. Gauss justified it by showing that it gave the most probable value in those situations where the errors were distributed along a 'bell curve' (known today as a normal or Gaussian distribution), see figure 2.2. This probability-based approach enabled Laplace to show, in 1810–11, that the least-squares method had the following advantages: it best reduced the error as the number of observations increased; it indicated how to distinguish between random errors (which define precision) and constant errors (which define accuracy); and it suggested how likely it was that the chosen curve was the best curve. In their search for an illusory perfection, *savants* had learned not only how to distinguish between different kinds of error, but also that error could be approached with quantitative confidence. The years between 1805 and 1811 saw the rise of a new scientific theory; not a theory of nature, but a theory of error. It was this theory that would allow Nicollet to redeem

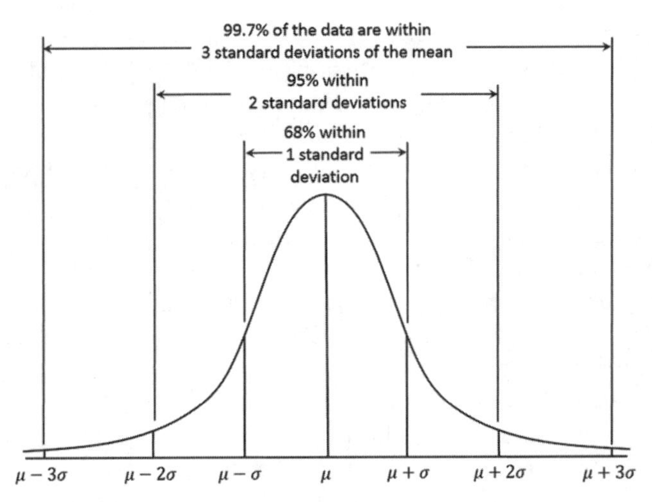

Figure 2.2. Curve representing a Gaussian or Normal 'Bell' distribution, where the mean of the distribution is μ and the standard deviation is σ. For such a normal distribution, the values less than one standard deviation away from the mean account for 68.27% of the set; while two standard deviations from the mean account for 95.45%; and three standard deviations account for 99.73% (this is the integrated area under the curve).

Méchain's reputation by distinguishing between those errors that were random and those that were systematic.

During the course of the next century science learned to manage uncertainty, to look at measurements as having an intrinsic uncertainty. The fields of statistics and probability that would one day emerge from the insights of Legendre, Laplace, and Gauss would transform the physical sciences, inspire the biological sciences, and give birth to the social sciences. In the process, *savants* became scientists.

Men like Delambre, Laplace, and Legendre began as *savants*, but they were changed by the struggle to quantify the uncertainty that always exists in studying nature. They would be some of the first non-philosophers and non-theologians to ask: how confident are we that we know what we think we know? They sought to rid themselves of value judgments about nature, and to perform measurements with detachment. Pierre Méchain lived and died a *savant*. Measurement mattered to him as much as it did to the *ancien régime* peasants, bakers, and ordinary families. Whether it was the location of a star or the weight of a loaf of bread, measurement expressed value. Measuring something was a moral act. For the *savant*, the pattern of the heavens revealed an aspect of an overarching structure. To measure the shape of the Earth or the position of a star was to fix its place in that overall pattern of things, just as the weight of a loaf of bread sustained the just price for bread and helped to maintain a social order.

In 1792, the extreme radical Jean-Paul Marat had been the first person to tag *savants* with the name *scientifiques* when he referred sneeringly to the academicians' self-serving project to measure the Earth in order to create uniform weights and measures. But the name stuck, even though we have continually to justify our existence.

2.5 Statistical methods

One important concept in statistics is the mean of a data set: commonly referred to as an average. For the measurements made by the Peruvian survey mentioned above, there were two measurements of a distance, one from each surveying team These two distances, x_1 and x_2 would have then been added together and divided by the number of measurements to derive the mean. But more information is provided by indicating the range of the two measurements. The two measurements in question were 56 749 toise and 56 768 toise; the average being 56 758.5 toise. But these uncertainties in the measurement could also be written as 56 749 ± 9.5 toise and 56 768 ± 9.5 toise; the amount that may be added or subtracted from the measured value indicates the range of values that could be expected if the measurement were repeated again. But such an approach to data analysis was not able to help Méchain in analysing his data.

Subsequent to his re-evaluation of Méchain's data, Nicollet used the same kinds of statistical methods as had Méchain to deal with the discrepancies in Méchain's observations, and was able to show that these discrepancies were not errors, but variations within the tolerable for a precise and accurate understanding of the material needed to establish the correct length of the metre. Statistics are not measurements, but a formulation that permits the building of models from measurements. There are many statistical methods available for the analysis of data, which allow scientists to identify and quantify the unavoidable variance or variation of measurements. If these methods are used correctly, others may dispute some parts of the questions being addressed, but not how the conclusions were reached through the statistical analysis. If one starts with an assumption, proceeds to gather data about that assumption, develops an hypothesis, then performs a statistical analysis on the data to see if the initial assumption is sound, any questions that may be asked are about the initial assumption, and not about the statistical procedure.

The method of least-squares is today a standard technique in regression analysis for the approximate solution of overdetermined systems; that is, to sets of equations in which there are more equations than there are unknowns. The name signifies that the overall solution minimizes the sum of the squares of the errors made in the results of every single equation or data point. The best fit in the least-squares sense minimizes the sum of squared residuals; a residual being the difference between an observed value and the fitted value provided by a model.

Least-squares problems fall into two categories: linear or ordinary least-squares and non-linear least-squares, depending on whether or not the residuals are linear in all unknowns. The linear least-squares problem occurs in statistical regression analysis; it has a closed-form solution. The non-linear problem is usually solved by iterative refinement; at each iteration the system is approximated by a linear one, and thus the core calculation is similar in both cases.

The method of least-squares grew out of the fields of astronomy and geodesy as astronomers and mathematicians sought to provide solutions to the challenges of navigation, as the accurate description of the behaviour of celestial bodies was the key to enabling ships to navigate in the open sea. The method was the culmination of several advances that took place during the course of the 18th Century. Firstly by *savants* attempting to combine numerous observations taken under the same conditions, as opposed to simply trying ones best to observe and record a single observation as accurately as possible. The approach was known as the method of averages, and was notably used by the German astronomer Tobias Mayer (1723–62) while studying the librations of the Moon in 1750, and by Pierre Simon de Laplace in his work in explaining the differences in motion of Jupiter and Saturn in 1788.

Then *savants* tried to combine observations made under different conditions. The method came to be known as the method of least absolute deviation. It was derived by the polymath, poet and priest Roger Joseph Boscovich (1711–87) in his work on the shape of the Earth in 1757, and by Pierre Simon de Laplace working on the same problem in 1799. And finally, there was the concept of probability and the development of criteria that could be evaluated numerically to determine when the solution with the minimum error has been achieved. Laplace tried to specify a mathematical form of the probability density for errors and define a method of estimation that minimized the error of estimation. For this purpose, he used a symmetric two-sided exponential distribution we now call the Laplace distribution to model the error distribution, and used the sum of absolute deviation as the error of estimation.

As we have seen, the first clear and concise presentation of what today we call the method of least-squares was made by Legendre in 1805. The technique can be described as an algebraic procedure for fitting linear equations to data, and Legendre demonstrated the new method by analyzing the same data that Laplace had examined, the data accumulated by Delambre and Méchain to define the shape of the Earth. The value of Legendre's method of least-squares was immediately recognized by leading astronomers and geodesists of the time.

After reading Gauss's successful mathematical modelling of the orbit of Ceres, Laplace was able to use the same methods to prove the central limit theorem, and used it to give a large sample justification for the method of least-squares analysis and the normal distribution. In 1822, Gauss was able to state that the least-squares approach to regression analysis is optimal in the sense that in a linear model where the errors have a mean of zero; that is, are uncorrelated and have equal variances, the best linear unbiased estimator of the coefficients is the least-squares estimator; this is the Gauss–Markov theorem.

The problem consists in adjusting the parameters of a model function to best fit a data set (that is, data fitting; which as we will see later is very much the basis of modern physical science). A simple data set consists of n points or data pairs (x_i, y_i), $i = 1, \ldots, n$, where x_i is an independent variable and y_i is a dependent variable whose value is found by observation. The model function has the form $f(x, \boldsymbol{\beta})$, where m adjustable parameters are held in the vector $\boldsymbol{\beta}$. The goal is to find the parameter values for the model which best fits the data. The least-squares method finds its optimum when the sum, S, of squared residuals

$$S = \sum_{i=1}^{n} r_i^2$$

is a minimum. A residual being defined as the difference between the actual value of the dependent variable and the value predicted by the model. Each data point has one residual, r_i.

$$r_i = y_i - f(x_i, \boldsymbol{\beta}).$$

Both the sum and the mean of the residuals are equal to zero.

An example of a model would be a straight line in two dimensions. Denoting the y-intercept as β_0 and the slope as β_1, the model function may be written as $f(x, \boldsymbol{\beta}) = \beta_0 + \beta_1 x$.

A data point may consist of more than one independent variable; for example, fitting a plane to a set of measurements, the plane is a function of two independent variables. In the most general case, there may be one or more independent variables and one or more dependent variables at each data point.

The minimum of the sum of squares is found by setting the gradient to zero. Since the model contains m parameters, there are m gradient equations:

$$\frac{\partial S}{\partial \beta_j} = 2\sum_i r_i \frac{\partial r_i}{\partial \beta_j} = 0, \quad j = 1, \ldots, m,$$

and as $r_i = y_i - f(x_i, \boldsymbol{\beta})$, the gradient equations become

$$-2\sum_i r_i \frac{\partial f(x_i, \boldsymbol{\beta})}{\partial \beta_j} = 0, \quad j = 1, \ldots, m.$$

The gradient equations are the central element of all least-squares problems; each particular problem requires particular expressions for the model and its partial derivatives.

A fitting model (also called a regression model) is linear when the model comprises a linear combination of the parameters; that is,

$$f(x, \beta) = \sum_{j=1}^{m} \beta_j \phi_j(x),$$

where the function φ_j is a function of x.

Letting

$$X_{ij} = \frac{\partial f(x_i, \boldsymbol{\beta})}{\partial \beta_j} = \phi_j(x_i),$$

we can see that here the least-square estimate (or estimator, in the context of a random sample), $\boldsymbol{\beta}$ is given by

$$\underline{\boldsymbol{\beta}} = \left(X^T X\right)^{-1} X^T \mathbf{y}.$$

There is no closed-form solution to a non-linear least squares problem. Instead, readily available numerical algorithms are used to find the value of the parameters $\boldsymbol{\beta}$ that minimizes the function of interest. Most algorithms involve choosing initial values for the various parameters. Then, the parameters are refined iteratively; that is, the values are obtained by successive approximations:

$$\beta_j^{k+1} = \beta_j^k + \Delta\beta_j,$$

where k is a number defining the order of the iteration, and the vector of increments $\Delta\beta_j$ is termed the shift vector. With some commonly used algorithms, at each stage of iteration the model may be linearized by approximation to a first-order Taylor series expansion about β^k:

$$\begin{aligned} f(x_i, \boldsymbol{\beta}) &= f^k(x_i, \boldsymbol{\beta}) + \sum_j \frac{\partial f(x_i, \boldsymbol{\beta})}{\partial \beta_j}\left(\beta_j - \beta_j^k\right) \\ &= f^k(x_i, \boldsymbol{\beta}) + \sum_j J_{ij}\Delta\beta_j. \end{aligned}$$

The Jacobian $\mathbf{J}$ is a function of constants, the independent variable and the parameters, so it changes from one iteration to the next. The residuals are given by

$$r_i = y_i - f^k(x_i, \boldsymbol{\beta}) - \sum_{k=1}^{m} J_{ik}\Delta\beta_k = \Delta y_i - \sum_{j=1}^{m} J_{ij}\Delta\beta_j.$$

To minimize the sum of squares of r_i, the gradient equation is set to zero and solved for $\Delta\beta_j$:

$$-2\sum_{i=1}^{n} J_{ij}\left(\Delta y_i - \sum_{k=1}^{m} J_{ik}\Delta\beta_k\right) = 0,$$

which, on rearrangement, become m simultaneous linear equations:

$$\sum_{i=1}^{n}\sum_{k=1}^{m} J_{ij}J_{ik}\Delta\beta_k = \sum_{i=1}^{n} J_{ij}\Delta y_i \qquad (j = 1, \ldots, m).$$

These may be written in matrix notation as

$$\left(\mathbf{J}^{\mathrm{T}}\mathbf{J}\right)\Delta\boldsymbol{\beta} = \mathbf{J}^{\mathrm{T}}\Delta\boldsymbol{y},$$

which are the defining equations of the Gauss–Newton algorithm.

Differences between linear and non-linear least-squares: The model function, f, in linear least-squares (LLSQ) is a linear combination of parameters of the form $f = X_{i1}\beta_1 + X_{i2}\beta_2 + \cdots$ The model may represent a straight line, a parabola or any other linear combination of functions. In non-linear least-squares (NLLSQ) the parameters appear as functions such as β^2 or $e^{\beta x}$. If the derivatives $\partial f/\partial\beta_j$ are either constant or depend only on the values of the independent variable, the model is linear in the parameters. Otherwise the model is non-linear. Some of the differences between LLSQ and NLLSQ are given below, but fuller details may be found in any standard textbook.

- Algorithms for finding the solution to a NLLSQ problem require initial values for the parameters, LLSQ does not.

- Like LLSQ, solution algorithms for NLLSQ often require that the Jacobian be calculated. Analytical expressions for the partial derivatives can be complicated. If analytical expressions are impossible to obtain, either the partial derivatives must be calculated by numerical approximation or an estimate must be made of the Jacobian.
- In NLLSQ non-convergence (that is, failure of the algorithm to find a minimum) is a common phenomenon whereas the LLSQ is globally concave so non-convergence is not an issue.
- NLLSQ is usually an iterative process. The iterative process has to be terminated when some defined convergence criterion has been satisfied. LLSQ solutions can be computed using direct methods, although problems with large numbers of parameters are usually solved by iterative methods.
- In LLSQ the solution is unique, but in NLLSQ there may be multiple minima in the sum of squares.

These differences should be considered whenever a non-linear least squares problem is being investigated.

The method of least-squares is often used to generate estimators and other statistics in regression analysis. Consider a simple harmonic spring obeying Hooke's law, where the extension of a spring y is proportional to the applied force, F. This is the basis of the analysis of molecular vibrations in infrared spectroscopy, and of the dynamics of engineered structures. We can write

$$y = f(F, k) = kF,$$

which constitutes the model, where F is the independent variable. To estimate the force constant, k, a series of n measurements with different forces will generate a data set (F_i, y_i), $i = 1, \ldots, n$, where y_i is a measured spring extension. Each experimental observation will contain some error. If we denote this error ε, we may specify an empirical model for our observations,

$$y_i = kF_i + \varepsilon_i.$$

There are many methods that could be used to estimate the unknown parameter k. Noting that the n equations in the m variables in our data comprise an over-determined system with one unknown and n equations, we may choose to estimate k using least-squares. The sum of squares to be minimized is

$$S = \sum_{i=1}^{n} (y_i - kF_i)^2.$$

The least-squares estimate of the force constant, k, is given by

$$\hat{k} = \frac{\sum_i F_i y_i}{\sum_i F_i^2}.$$

Here it is assumed that application of the force causes the spring to expand and, having derived the force constant by least-squares fitting, the extension can be predicted from Hooke's law. This is how a great many, much more complex problems are investigated today (see chapter 9).

In such an analysis, the investigator specifies an empirical model; for example, a straight line model which is used to test if there is a linear relationship between dependent and independent variables. If a linear relationship is found to exist, the variables are said to be correlated. However, correlation does not prove causation, as both variables may be correlated with other, hidden, variables, or the dependent variable may 'reverse cause' the independent variables, or the variables may be otherwise spuriously correlated. For example, suppose a correlation is 'found' between the number of deaths by drowning and the volume of soft-drinks sold at a beach. Yet, both the number of people going swimming and the volume of drinks sold increase with the rising temperature, and presumably the number of deaths by drowning is correlated with the number of people going swimming. Perhaps it is an increase in the number of swimmers that causes both variables to increase.

In order to make statistical tests on a set of measurements it is necessary to make assumptions about the nature of the experimental errors. A common (but not necessary) assumption is that the errors can be described by a normal distribution. This assumption is supported by the central limit theorem in many cases. However, even if the errors are not normally distributed, a central limit-like theorem will often, nonetheless imply that the parameter estimates will be approximately normally distributed as long as the sample is reasonably large. So given that the error mean is independent of the independent variables, the distribution of the error term is not an important issue in regression analysis.

In a least-squares calculation with unit weights, or in linear regression, the variance on the jth parameter, denoted $\mathrm{var}(\hat{\beta}_j)$, is usually estimated with

$$\mathrm{var}\left(\hat{\beta}_j\right) = \sigma^2\left([X^{\mathrm{T}}X]^{-1}\right)_{jj} \approx \frac{S}{n-m}\left([X^{\mathrm{T}}X]^{-1}\right)_{jj},$$

where the true residual variance σ^2 is replaced by an estimate based on the minimized value of the sum of squares objective function S. The denominator, $n - m$, is the statistical degree of freedom.

Confidence limits can be found if the probability distribution of the parameters is known, or if an asymptotic approximation can be made, or assumed. Likewise statistical tests on the residuals can be made if the probability distribution of the residuals is known, or assumed. The probability distribution of any linear combination of the dependent variables can be derived if the probability distribution of experimental errors is known, or assumed. Inference is particularly straightforward if the errors are assumed to follow a normal distribution, which implies that the parameter estimates and residuals will also be normally distributed conditional on the values of the independent variables.

Further reading

Details of the mathematics of the least-squares method, and how it may be applied to a range of problems may be found on the Internet. The online encyclopaedia, Wikipedia, has several excellent entries for the statistical methods used in data analysis; particularly, the least-squares method. For specific applications of the methodology to individual problems, the best place to start would be to Google the specific application.

With regard to the details of the epic surveying adventure of Delambre and Méchain to define the length of the metre, a reader could do no better than read *The Measure of All Things* by Ken Adler, published by Abacus, 2004. This is a readable, non-scientific but detailed account of the metric survey. Even though the volume is long at 370 pages, with an additional 200 pages of notes and references, it is informative and easy to read, with some thought-provoking observations on the difficulties of undertaking detailed scientific work in two countries that are actually at war.

Chapter 3

A language for measurement

3.1 Introduction

In previous chapters, we saw how Pierre Méchain and his contemporaries did not, in their measurements, make a distinction between precision; that is, the consistency of results, and accuracy; that is, the degree to which these results approached the 'right answer.' The 'right answer' being the great unknown. The modern nomenclature of measurement science seeks to avoid the possible ambiguities that may arise through the use of terms such as accuracy and the 'right answer'.

Having to use a new, but established nomenclature to describe something with which you feel you are familiar, and which you describe adequately in your own manner, can often seem like being forced to use a new system of weights and measures, or even a new language. It is irritating and confusing, but strictures being imposed on how you describe science and technology are there for the best of reasons. That is, that we may be confident that we are all talking about the same thing, and in the same manner.

A measurement tells you about a property of something you are investigating, giving it a number and a unit; for example, x grams cm^{-3} or y kelvin. Measurements are made using an instrument of some kind; rulers, stop-clocks, chemical balances, thermometers, or even the Large Hadron Collider at CERN are all measuring instruments.

3.2 The quality of measurements

Evaluating the quality of a measurement is an essential step on the route to drawing sensible conclusions from that measurement, and, consequently, scientists have evolved a special vocabulary to help them think clearly about their data. Key terms that describe the quality of measurements are:

- validity;
- accuracy;

doi:10.1088/978-1-6817-4433-9ch3

- precision (repeatability or reproducibility);
- measurement uncertainty.

3.2.1 Validity

A measurement is valid if it measures what it is supposed to be measuring. If a factor or a condition in a measurement is uncontrolled, the measurements may not be valid; for example, if you are investigating the heating effect (a measure of power, P) of an electric current, I, flowing through a wire with an electrical resistance, R, by increasing the current (that is, you are exploring the equation $P = I^2R$), the resistance of the wire may change as it is heated by the flowing current. This heating effect may well invalidate your measurements, as R is a function of temperature, and so your results (at a level of precision determined by the magnitude of the temperature dependence of R) could be skewed.

3.2.2 Accuracy

This describes how closely a measurement approaches the 'true value' of a physical quantity. The 'true' value of a measurement is the value that would be obtained by a perfect measurement; that is, in a perfect or ideal world. As the true value is not known, accuracy can only ever be a qualitative term. Many measured quantities have a range of values rather than one 'true' value. For example, a collection of electrical resistors all marked 1 kΩ will have a range of values because they are mass produced, but the mean value of a collection of such resistors should be close to 1 kΩ. The variation enables you to identify: a mean, a range and the distribution of values across the range.

3.2.3 Precision

The closeness of agreement between replicated measurements on the same or similar objects under specified conditions. We could also say that this is the extent to which a measurement replicated under the same conditions gives a consistent result.

3.2.4 Measurement uncertainty

The uncertainty of a measurement is the doubt that exists about its value. For any measurement, even the most carefully undertaken measurements, there is always a margin of uncertainty. (This will be the central topic of much of this volume.) The uncertainty about a measurement has two aspects:

- the width of the margin, or interval. This is the range of values within which one expects the true value to be found. (Note: this is not necessarily the range of values one might obtain when taking measurements of that property, which may include outliers.)
- confidence level; that is, how sure the experimenter is that the true value lies within that margin or interval.

Uncertainty in measurements can be reduced trivially by using an instrument that has a scale with smaller divisions. For example, if you use a ruler with a centimetre scale then the uncertainty in a measured length is likely to be, of order, plus and minus one centimetre. A ruler with a millimetre scale would reduce the uncertainty in length to, of order, plus and minus one millimetre. This was the idea behind the use of a decimal metric scale, as opposed to the traditional Babylonian angle divisions for the surveying instruments used in the metric survey (the Borda circles), see section 1.3.

3.3 Measurement errors

It is important not to confuse the terms 'error' and 'uncertainty'. Error refers to the difference between a measured value and the true value (which we do not know) of the physical quantity being measured. Whenever possible we try to correct for any known errors; for example, by applying corrections from calibrations. But any error whose value we do not know is a source of uncertainty. Measurement errors can arise from two sources: a random component, where repeating the measurement gives an unpredictably different result; and a systematic component, where the same influence affects the result for each of the repeated measurements.

Every time a measurement is made under what seem to be the same conditions, random effects can still influence the measured value. A series of measurements therefore produces a scatter of values about a mean value. The influence of variable factors may change with each measurement, changing the mean value. Increasing the number of observations generally reduces the uncertainty in the mean value. Systematic errors (measurements that are either consistently too large, or consistently too small) can result from:

- poor technique (for example, carelessness with parallax corrections when reading a scale);
- zero error of an instrument (for example, a balance that has been improperly set for zero mass with nothing on the weighing pan);
- poor calibration of an instrument (for example, every volt of signal being measured is either too large or too small).

Correcting for systematic errors will improve accuracy; as Jean-Nicolas Nicollet did when re-analyzing the data collected by Pierre Méchain. But sometimes you can only find a systematic error by measuring the same value by a different method.

Let us now look further at accuracy and precision. Precision is a description of random errors; a measure of statistical variability. Accuracy has two definitions: most commonly, it is a description of systematic errors, a measure of statistical bias; but alternatively, the International Organization for Standardization (ISO) defines accuracy as describing both types of observational error mentioned above (preferring the term trueness for the common definition of accuracy). Schematically, this may be represented as:[1]

[1] See Wikipedia's entry for accuracy and precision for further details.

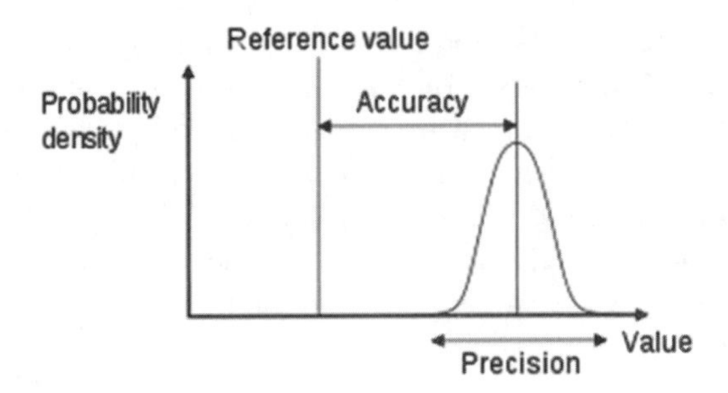

Accuracy is the proximity of a measurement result to the true or reference value; precision is the repeatability or reproducibility of the measurement; that is, the range of measurement results.

In science, the accuracy of a measurement system is the degree of closeness of measurements of a quantity to that quantity's true value. The precision of a measurement system, related to reproducibility and repeatability, is the degree to which repeated measurements under unchanged conditions show the same value. Although the two words precision and accuracy can be synonymous in a colloquial sense, they are deliberately contrasted in the context of measurement science.

A measurement system can be accurate but not precise, precise but not accurate; a measurement system can be neither or both. For example, if an experiment contains a systematic error (a poor calibration), then increasing the sample size generally increases precision but does not improve accuracy. The result would be a consistent, yet inaccurate set of results from the flawed experiment; Méchain's problem in his analysis of the data from the metric survey. Eliminating the systematic error improves accuracy but does not change precision. In addition to accuracy and precision, measurements may also have a measurement resolution, which is the smallest change in the underlying physical quantity that produces a response in the measurement.

A measurement system is considered valid if it is both accurate and precise. Related terms include bias (non-random or directed effects caused by a factor or factors unrelated to the independent variable) and error (random variability). This terminology is also applied to indirect measurements; that is, values obtained by a computational procedure from observed data. In numerical analysis, accuracy is also the nearness of a calculation to the true value; while precision is the resolution of the representation, typically defined by the number of decimal places after the decimal separator.

Mathematical or statistical literature prefers to use the terms bias and variability instead of accuracy and precision, which is the preferred vocabulary of the laboratory scientist: bias is the amount of inaccuracy, and variability is the amount of imprecision. In industrial instrumentation, accuracy is usually the measurement tolerance, or transmission of the instrument and defines the limits of the errors made when the instrument is used in normal operating conditions.

Ideally a measurement device is both accurate and precise, with measurements all close to and tightly clustered around the true value; certainly this should be the case if the experiment has been designed for a particular measurement (as we will see in chapter 8). The accuracy and precision of a measurement process is usually established by repeatedly measuring some traceable reference standard. Such standards are ultimately defined in the International System of Units and maintained by the BIPM (*Bureau internationale des poids et mesures*), near Paris.

This also applies when measurements are repeated and averaged. In this case, the term standard error is applied as: the precision of the average is equal to the known standard deviation of the process divided by the square root of the number of measurements involved in the averaging process. Further, the central limit theorem tells us that the probability distribution of the averaged measurements will be closer to a normal distribution than that of individual measurements.

Precision is sometimes stratified into:

- repeatability: the variation arising when all efforts are made to keep conditions constant by using the same instrument and operator, and repeating measurements during a short time period; and
- reproducibility: the variation arising using the same measurement process among different instruments and operators, and over longer time periods.

A shift in the meaning of accuracy and precision appeared with the publication of the ISO 5725 series of standards in 1994, which is also reflected in the 2008 edition of the International Vocabulary of Metrology (VIM) published by the BIPM; particularly, items 2.13 and 2.14. According to ISO 5725-1, the general term accuracy is used to describe the closeness of a measurement to the true value. When the term is applied to sets of measurements of the same quantity, it involves a component of random error and a component of systematic error. In this case trueness is the closeness of the mean of a set of measurement results to the actual (true) value and precision is the closeness of agreement among a set of results.

ISO 5725-1 and VIM also avoid the use of the term bias, previously specified in BS 5497-1, because it has different connotations outside the fields of science and engineering, as in medicine and law.

Consider the following explanation of measurement uncertainty taken from the chemical literature. The nomenclature is different from that often seen in physics, but this explanation does introduce some useful ideas and figure 3.1 is particularly informative.

Measurement is the process of experimentally obtaining the value of a quantity (that is, a value with an associated unit and quantity, as in x moles per litre). The quantity that is measured is termed the measurand. In chemistry the measurand is usually the content (that is, concentration) of some chemical entity in some solvent. The chemical entity under study is termed the analyte. In principle, the aim of a measurement is to obtain the true value of the measurand. Every effort can be made to optimize the measurement procedure in such a way that the measured value is as close as possible to the true value. However, the measurement results will be just an estimate of the true value, as the actual true value will (almost) always remain unknown. Therefore, we cannot know exactly how near our measured value is to the true value—our estimate always has some associated uncertainty.

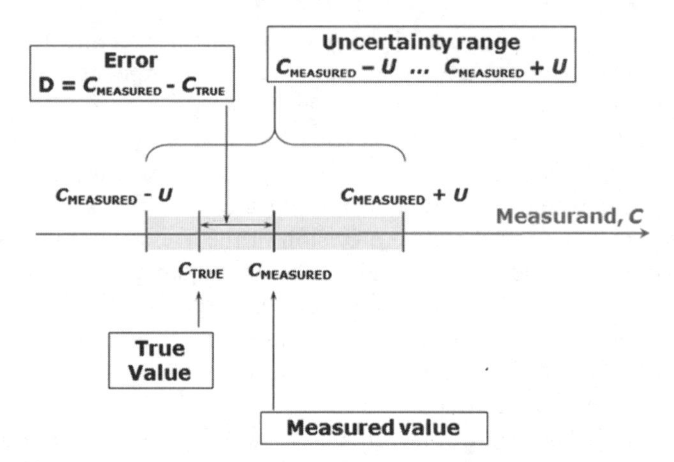

Figure 3.1. Interrelations between the concepts involved in defining the quality of a measurement: the true value (which is hardly known in practice, and certainly not in research), the measured value, the error (the difference between the measured value and the true value; of which the latter can never be known with full certainty, so this quantity is something of an abstraction) and the measurement uncertainty (which may be computed).

The difference between the measured value and the true value is called the error. Errors can be positive or negative. As we saw earlier, error can be regarded as being composed of two parts, random error and systematic error. Like the true value of the measurand, the true value of the error is also not known (hence the need for the probabilistic methodology in the study of measurement uncertainty.

The quality of the measurement result, its accuracy, is characterized by the measurement uncertainty, which defines an interval around the measured value C_{MEASURED}, where the true value C_{TRUE} lies with some estimated probability. The measurement uncertainty U itself is the half-width of that interval and is always non-negative. This is illustrated in figure 3.1, which should be compared with the schematic on page 3-4.

Measurement uncertainty is always associated with some probability, but it is not possible to define an uncertainty interval in such a way that the true value of the measurand lies within it with 100% probability. The measurement uncertainty expressed above, is in some contexts also known as the absolute measurement uncertainty. This means that the measurement uncertainty is expressed in the same units as the measurand. It is sometimes more useful to express measurement uncertainty as relative measurement uncertainty, U_{rel}, which is the ratio of the absolute uncertainty U_{abs} and the measured value y,

$$U_{\mathrm{rel}} = U_{\mathrm{abs}}/y.$$

Relative uncertainty is a dimensionless or unit-less quantity, which is sometimes also expressed as a 'per cent'.

Measurement uncertainty is different from error in that it does not express a difference between two values, and it does not have a sign. Therefore, it cannot be used for correcting the measurement result, and cannot be regarded as an estimate of the error. Instead measurement uncertainty can be thought of as an estimate of what is the highest probable absolute difference between the measured value and the true value. However, both the true value and error (random and systematic) are abstract concepts. Their exact values cannot be determined. But these concepts are nevertheless useful, because their estimates can be determined and hence compared between; for example, different experimenters or different laboratories. As said above, our measured value can only ever be an estimate of the true value.

Further reading

Measurement uncertainty is a vast subject, and if you Google it you will be swamped by responses. You will also discover that there is no single international language or vocabulary of terms and definitions; however, at www.bipm.org/en/publications/guides/vim.html you can find all 108 pages of the international vocabulary of metrology (the VIM), including useful diagrams to explain the terms and concepts. With regard to the points outlined in the last part of this chapter; particularly, with reference to figure 3.1, the ten minute video at www.youtube.com/watch?v=BogGbA0hC3k is well worth watching.

Chapter 4

What is it that we measure, and what does it tell us?

Since we are assured that the all wise Creator has observed the most exact proportions, of number, weight and measure in the make of all things, the most likely way therefore to get any insight into the nature of those parts of the creation, which come within our observation, must in all reason be to number, weight and measure.

Stephen Hales, 1677–1761

Before we look in detail at how measurements are quantified, let us consider briefly some very different types of measurements. In this chapter, we will first consider two standard measurements; that is, where the experimenter is measuring or monitoring one characteristic property of a system while some other property of that system is varied. Then we will look at two non-standard measurements, which although not made in what you might consider to be a scientific laboratory are, in fact, more important (for different reasons) than measurements made in a classic laboratory setting.

4.1 A classic laboratory experiment

In figure 4.1 we see the birefringence induced in a high-pressure sample of carbon dioxide by a large, constant applied electric-field gradient, as a function of the temperature (room temperature to 123 °C) of the gas sample. This data is taken from the author's PhD thesis, and represents several weeks of measurements in the summer of 1979. In chapter 8, we will look in detail at this experiment when we consider the design of an experiment. But here, I wish to draw the reader's attention to the classic nature of this set of measurements; it is what most people envisage as

doi:10.1088/978-1-6817-4433-9ch4

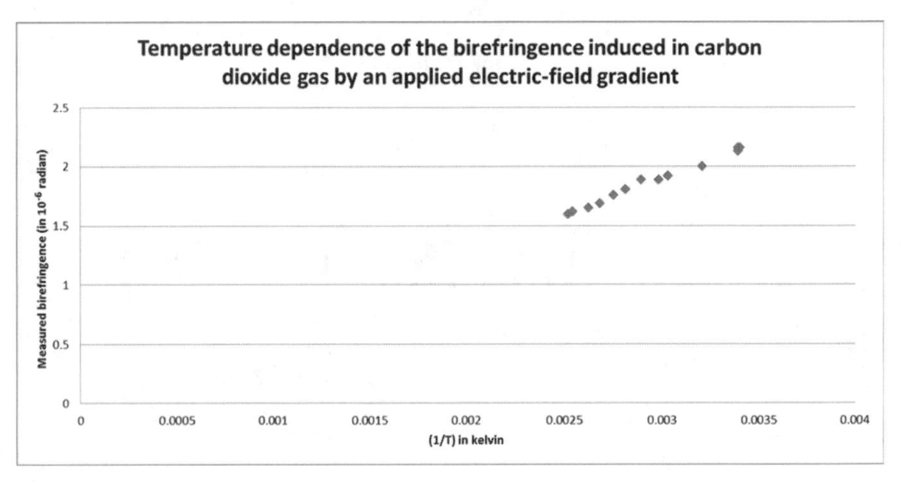

Figure 4.1. A classic laboratory measurement: the temperature dependence of the birefringence induced in a high-pressure sample of carbon dioxide by an applied electric-field gradient. The measured birefringence is plotted against the reciprocal of the absolute temperature to demonstrate the presence of both a temperature dependent term and a temperature independent term in the observed measurement. The data is taken from the author's PhD thesis, and is published in Battaglia M *et al* 1981 *Mol. Phys.* **43** 1015.

an experiment. When the apparatus was working well, a measurement of the induced birefringence took about an hour. However, it took rather longer than this to change the sample temperature as the high-pressure gas cell was made of two concentric cylinders of stainless steel, each 1 m in length, and these tubes were heated by circulating externally heated anti-freeze through the space between the cylinders.

The applied electric-field gradient orients the molecules, and the medium becomes birefringent; so linearly polarized light propagating through the oriented medium becomes elliptically polarized. The induced retardation is, of order, 10^{-6} radian for available laboratory electric-field gradients and pressures of several atmospheres of carbon dioxide. There are two terms that contribute to the measured effect; one of which is temperature-dependent and another which is independent of temperature. Undertaking measurements of the induced birefringence at a variety of temperatures therefore allows one to measure both these contributing factors.

From the data in figure 4.1, one is able to extract both the temperature-dependent term contributing to the measurement; that is, the slope of the line joining the measured points, and the temperature-independent term contributing to the measurement; that is, the intercept on the vertical axis of the line joining the data points. There is noise or scatter on the data displayed in this figure, and this noise will be the major component to determining the uncertainly in the determination of the slope of the line, and will certainly dominate the determination of the intercept as the extrapolation is over such a long distance (the lever arm principle, as can be seen in figure 4.1), and as you go to higher temperatures, the signal decreases and so the signal/noise ratio is not moving in your favour. In actual fact, the uncertainty in the slope of the line gives a quadrupole moment for CO_2 of $-15.3 \pm 0.7 \times 10^{-40}$ cm^2; an uncertainty of 4.5%. Given this level of uncertainty, and the range of the extrapolation, it is not possible to say anything more than that the temperature independent term (determined from the intercept of the data in figure 4.1) contributes less than 4.5% to the measured birefringence at room temperature.

This data is typical of a classic laboratory experiment, where there is some measured parameter, which is a function of temperature, $f(T)$ and is plotted against sample temperature. There are two unknown quantities (the temperature-independent term and the temperature-dependent term), and so we require, at least, two equations to determine both unknowns. The experiment is relatively straightforward and can be repeated to establish reproducibility, and to generate sufficient data to obtain decent statistics on the final derived value; the measurements are made within the known precision of the apparatus as we will discuss in chapter 8.

Let us now consider a straightforward measurement, which has huge significance but cannot be repeated endlessly.

4.2 Precision measurements made infrequently

Figure 4.2 displays data representing the temporal evolution of the masses of the six official copies of the International Prototype of the Kilogram (IPK), with respect to the IPK (which in the data displayed in figure 4.2 is labelled K), and which by definition cannot change its mass. The vertical scale is in micrograms (that is, 10^{-6} g or 10^{-9} kg). For two of the copies or *témoins* of the IPK (N43 and N47), however, there are only two real measurements, as all prototypes were assumed to have the same mass in 1889, and N43 and N47 were assumed to have no deviations from perfection in 1947 when they were first measured. For the remaining data sets (the other mass standards) there are only three data points on the figure as perfection was assumed back in 1889. This assumption of perfection in mass in 1889 and in 1947 is of course nonsense. Each of these mass standards would have had a unique mass at that time—it simply was not measured.

This graph appears simple (six objects, most of which are slowly increasing in mass with time), but this simplicity arises because of a lack of data. In 1889, only the

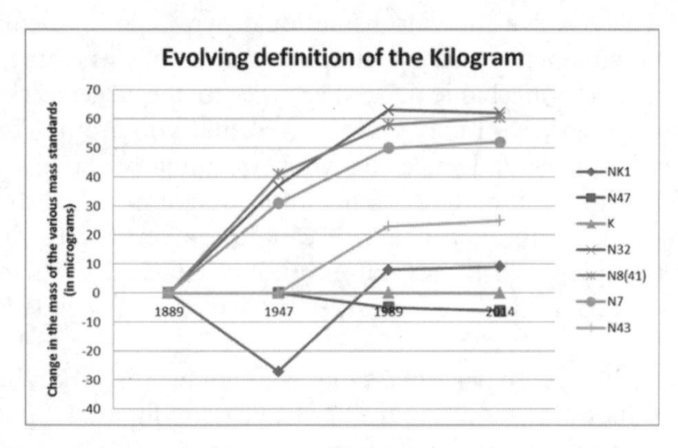

Figure 4.2. The evolution of the masses of the six official copies of the International Prototype of the Kilogram, with respect to the International Prototype of the Kilogram (which in the data displayed here is labelled K). The data is taken from the website of the BIPM (www.bipm.org/en/bipm/mass/ipk/).

IPK would have possessed a mass of precisely 1 kg, because that mass was defined by international law as being precisely 1 kg. By comparison, all the other mass standards would have had a unique, measurable mass, different from 1 kg. Yet for all the sparseness of the data presented in this graph, this image carries a huge responsibility. The measurements are straightforward, one is measuring the force (F) generated by placing a mass (m) in the Earth's gravitational field (g); that is, $F = mg$, Newton's celebrated equation.

The present hierarchy of measurement standards for mass closely resembles a religious dogma. At the highest point is the omnipotent object which is, as far as the world is concerned, indivisible and in which one must believe and have faith, but without ever touching or even seeing it. The perfect mass of this object, precisely 1 kg, is fixed by international diplomatic treaty, the Metre Convention of 1875, and not by mere experiment (and the Metre Convention made no mention of the uncertainty of this definition).

This near-sacred object is the IPK. This memento from the early-days of the Metric System is one of science's most valuable, but also one of its most derided objects; it is kept in a secure vault near Paris surrounded by six identical copies or *témoins*. Thus protected, the IPK reigns supreme over the world's measurements of mass. Every hill of beans, every human, every milligram of medication and recreational drug; in short, the great globe itself and even the smallest of sub-atomic particles, that can be weighed, must be gauged and compared against the mass of that small, glittering, dense object made of an incorruptible alloy, the IPK. And if you believe what you read in the popular press from time to time, this object is mysteriously changing its mass.

These platinum-iridium prototypes are used every so often to monitor how the stability of these mass artefacts is being maintained, and to provide calibration certificates to the rest of the metric world so that we may all see that 'the kilogram' is in its vault and that all is right with the metric world. The good news of the continued stability and perfection of the kilogram then passes down from these closely-guarded objects to their copies in the various Member States of the Metre Convention, and then down the various national pyramids of scientific endeavour and precision measurement to, for example, the humble weighing scale in your bathroom or the electo-mechanical balance next to the vegetables in your local supermarket. In this way, you may be reassured that you are not fooling yourself after a holiday of over-indulgence; as your weighing scale has certificates of calibration linking it to the perfect essence of a kilogram locked away in France. And it is through this unbroken chain of calibration certificates that uncertainty enters mass metrology; the IPK has no uncertainty attached to its mass, but those weighing scales in your local supermarket do have an associated measurement uncertainty.

The IPK will, of course, have a mass that is different from 1 kg, but it cannot be measured. The values of mass along the vertical axis in figure 4.2 give the range of values within which one would be likely to discover the true mass of the IPK, if it could be measured; and will be found, when the kilogram has been redefined by a non-artefact based definition.

So, how much is a kilogram? Well, as it turns out nobody can say for sure; at least, not in a way that won't change ever so slightly over time. And that's not so good for a fundamental standard that the world depends upon to define mass. Of course, such statements in the popular press beg the question, which is never addressed; getting lighter compared to what? The inference is that the kilogram is not a sound unit. That it is getting lighter compared to a more stable mass. But how can this be determined? If there were a more stable mass it would be the IPK, and the present IPK would become just another mass standard. At present, given the artefact-basis of mass metrology, some object has to occupy that solitary position at the apex of the pyramid of mass measurements. The possible instability in the physical characteristics of a metrological artefact is the essential problem with all artefact-based systems of weights and measures, which is why they are re-calibrated every quarter-century or so.

Since it was placed in service in 1889, the IPK has been used during three measurement campaigns or 'periodic verifications of national prototypes of the kilogram'. The most recent such verification was carried out in 2014. Over more than a century, the masses of the official copies are seen to be increasing, with respect to the mass of the IPK (which by definition cannot change its mass and so appears as a horizontal line in figure 4.2 labelled K); mass standard NK1 has lost and gained mass, and standard N47 is losing mass. Interestingly, it is relatively easy to understand how a mass standard can acquire additional mass—we live in polluted cities. But it is not at all straightforward to come up with an explanation of how a platinum–iridium alloy can lose mass with time; although part of that problem disappears if we assume that the mass of mass standard NK1 in 1889 was in fact, well below the mass of the IPK, and similarly for the mass of N47 in 1947.

By definition, the mass of the IPK cannot vary, but given that the masses of its copies are fluctuating, then the mass of the IPK must also be changing, because the IPK and its copies are all stored, and have been stored in the same manner, they are made of the same material and were all made in the same manner. And given that the masses of the copies are mostly increasing, then as they are weighted against the IPK, the mass of the IPK must be decreasing. For the scientists who rely on the continued stability of the base unit of mass of the SI for precise measurements, this inconstant metric mass is a nuisance. Our ability to precisely measure an electric current or a quantity of gas flowing through a pipeline is dependent upon the precision with which the unit of mass is known, and any instability in the precision with which we are able to define the base unit of mass perturbs such calculations; calculations which are worth hundreds of billions of Euros every year.

The problem of the IPK losing mass (it is not) arises because we have an artefact-based system of mass metrology based on seven mass standards that are all becoming increasingly contaminated at different rates. It is a combination of the absorption of contaminants, insufficient data (mass measurements) and the principle of conservation of mass that is creating the confusion.

The SI is the pivot from which hang all measurements, no matter what the area of investigation or the location of the measurement. Whether it is the accuracy of your bathroom scales, the amount of electricity you have consumed in the last month, or the reliability of a petrol pump, there is an unbroken chain of calibration certificates that leads back to realizations of the seven base units of the SI. Science has moved a long way since the IPK became the basis of mass metrology, and this object is becoming increasingly anachronistic; as can be seen in the carefully measured data in figure 4.2. The surest way of stabilizing mass metrology, and to remove our dependence on the drifting values of the masses of a collection or artefacts is to do away with the IPK altogether. Greater precision would also be brought to the SI if the unit of mass were defined by a constant of nature rather than an object; no matter how carefully conserved. The choice of the physics community is to redefine the unit of mass in terms of Planck's constant, and the choice of the chemistry community is to redefine the kilogram via Avogadro's number (for details about the redefinition of the SI, please see Willilams J H 2014 *Defining and Measuring Nature; The make of all things* (San Rafael, CA: Morgan & Claypool).

Let us now consider data that represent many hundreds of measurements made on blood samples from a single patient.

4.3 An overabundance of uncertain data

The data given in figure 4.3 represent routine laboratory measurements made on the blood of someone living with human immunodeficiency virus (HIV). This data set is unique; that is, a set of the same measurement data over the same period from another person living with HIV would look very different. The data in figure 4.3 represent the life of an individual, and no other individual would generate an identical set of results. This is the problem when designing clinical trials to test the efficacy of new medications, which we shall look at in chapter 7. We are all distinct, whereas all carbon dioxide molecules behave in the same way when stimulated in the same manner (see figure 4.1), and all mass standards become coated in pollutants (figure 4.2). But no two humans respond in exactly the same way to the same amount of a single drug; this is termed genetic variability.

The data in figure 4.3(*a*) covers a period of about fifteen years. Each data point represents the measurements made by an automated blood sampler in the laboratory of a large hospital, which takes blood that a nurse has taken from a patient's arm, dilutes it and scatters laser light from the diluted sample; the intensity of scattered light is proportional to the number of CD4 cells in the blood (in one microlitre of blood). The automated apparatus is quoted as having a 10% error, which arises principally in the automated dilution procedure. Each blood sample required a new syringe, and these mass-produced syringes have a 10% uncertainty in their volume (±5%). This uncertainty propagates through to the final value of the number of CD4 cells in one microlitre of the patient's blood.[1]

Obviously, the data in figure 4.3(*a*) is hugely important to the patient involved, and to his or her clinician who is responsible for their successful treatment. Indeed, it is from such data that decisions are taken about whether or not a particular

[1] For your interest, the lower the CD4 concentration, the more at risk of illness and death is the patient; and levels below 200 (on the scale in this figure) are considered to be dangerously low.

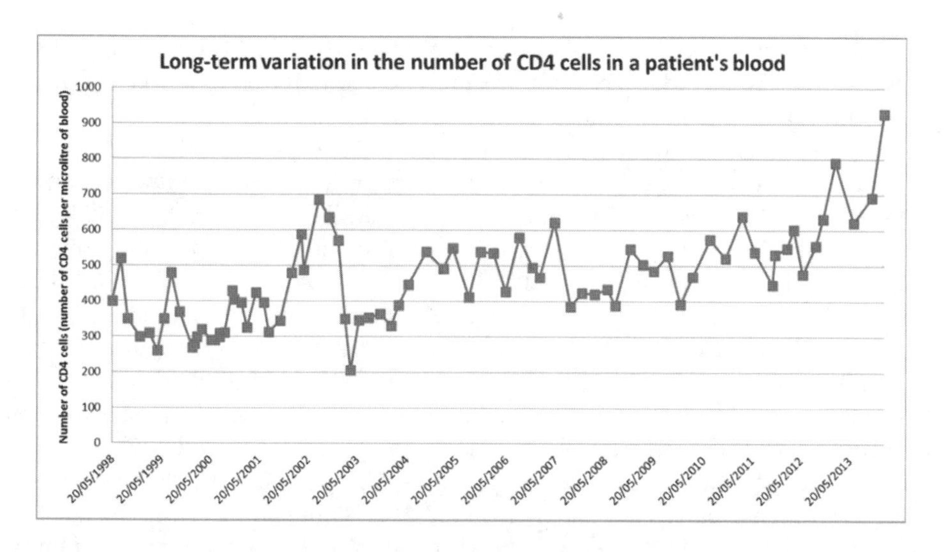

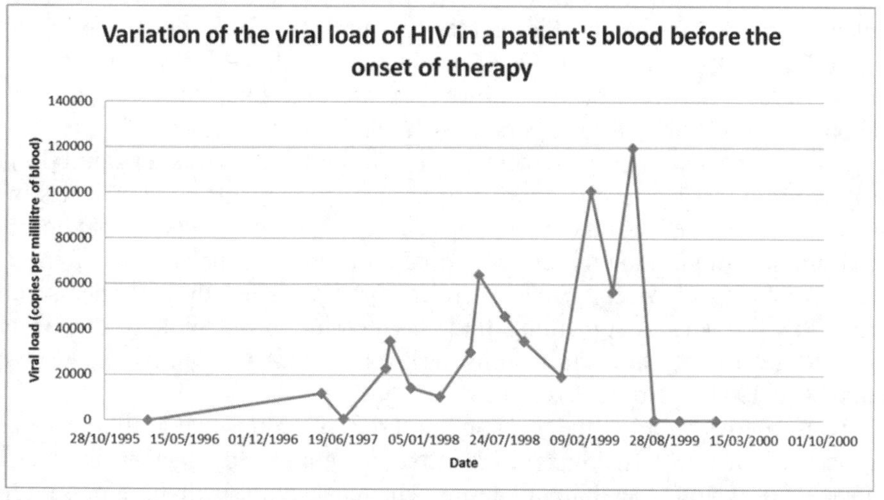

Figure 4.3. (*a*) A more complex set of measurements: the evolution over many years of the number of CD4 cells (an essential component of the human immune system) in a patient living with human immune deficiency virus (HIV). The units are number of CD4 cells per microlitre of blood, and the uncertainty of the measurement is ±5%. (*b*) An even more uncertain (±50%) set of measurements, but which are of great importance to the patient: the evolution of the amount of HIV virus (the viral load) in the patient's blood before the onset of treatment. The units are copies of HIV nucleic acid per millilitre of blood.

treatment or combination of anti-HIV drugs is failing for this patient (the clinician is not trained to consider measurement uncertainties, but he or she can see if the data is trending upwards or downwards with time). But knowing that there is a 10% error on the measured value is very important as there is no point in changing the therapy for a patient who has successive CD4 counts of 510, 460 and 470 (per microlitre of blood), as these values are all within the overlapping error limits. And indeed, we can see from the data in figure 4.3(*a*) how the values of successive measurement can see-saw quite dramatically over six or four month intervals (by as much as ±200 cells per microlitre or up to ±30% of the total number of cells). But we can also see periods in figure 4.3(*a*) when the data was pretty consistent over a period of a year or more, with almost identical values being measured.

And what of the dramatic fluctuations seen in this data set? Why does the data fluctuate so wildly at some times, yet is relatively constant at other times, given that the uncertainty on each measurement is 10%? Well, we know that CD4 levels fluctuate diurnally in all of us; that is, the levels are higher in the morning (perhaps by up to 20%) than in the evening, and that women have higher levels of CD4 cells than men (of the same age) and that pregnant women have very high levels of CD4 cells. Depending upon the sex of the patient who supplied the anonymous data in figure 4.3(*a*), all these factors will operate to generate the fluctuations seen in the data. The diurnal fluctuations will be superimposed on the patient's measurements like a background noise in a classic laboratory measurement. It will always be there as we cannot demand that the patient always give blood at exactly the same time of the day over many years. The other reason the data is fluctuating is that the CD4 cell is the basis of the body's defence against invading micro-organisms and internal problems such as cancer, and the greater the number of infections, the lower the levels of CD4 cells. Figure 4.3(*a*) is also the record of how the patient was feeling over the time period displayed; how they were feeling both physically and psychologically, as stress also lowers CD4 levels.

Figure 4.3(*b*) displays measurements of the viral load (the virus is again HIV) in a patient's blood just before the onset of treatment to control the HIV levels in the patient; this is another routine blood measurement carried out in all large hospitals. A measurement of viral load, is an indirect measurement of the amount of virus in a quantity of blood, and relies on a chemical process called the polymerase chain reaction (PCR). This is something that has revolutionized biology and forensic science in the last 25 years, and won its discoverer, the American biochemist, Kary Mullis (born 1944) a Nobel Prize in 1993.

PCR is useful because the genetic material of every living thing possesses sequences of chemical building blocks that are combined together in a unique manner. PCR exploits the ability of certain chemical catalysts that are present in all living organisms to make exact copies of genetic material. It copies the process which happens when the genetic material is being transferred from one generation of an organism to the next (cell division). Sometimes referred to as 'molecular photocopying', PCR can characterize, analyze, and synthesize any specific piece of genetic material. It works even on complicated mixtures, finding, identifying, and duplicating a particular bit of genetic material from samples of blood, hair, or tissue, or from microbes, animals, or plants, which can be many thousands of years old.

Although the chemical composition of the genetic material of HIV is now well known; even in a patient who has advanced HIV disease, the amount of virus in the blood is too low to be measured directly. So PCR is used to amplify the amount of genetic material to such a degree that it becomes measurable and quantifiable.

To begin the PCR process, an automated blood analyzer will sample a known quantity of a patient's blood, which contains some HIV and hence some HIV genetic material, and mix it with an appropriate label. This label or tag is designed so that it will only combine with the genetic material of HIV. Such a tagged or labelled piece of HIV genetic material is termed the 'template'. It is this template or labelled part of the genetic material of the virus, which is copied. Because the building blocks of the genetic material only combine together following specific rules, they will only combine together in the presence of the template to make copies of the original labelled genetic material. In this way, the original minute quantity of genetic material in the blood can be amplified to make measurable quantities.

There are three basic steps in PCR. First, the genetic material must be removed from the virus. The second step is the labelling process where the specific tag or label is attached to HIV's genetic material. The third stage is chemical amplification. The result of each stage of the amplification process is that the number of original pieces of labelled genetic material doubles. So starting from one piece of genetic material, one can make, 2, 4, 8, 16, 32, …, pieces at each stage of the chemical amplification. Each cycle of amplification takes only a few minutes, and repeating the process for just a couple of hours can generate millions of copies of a specific genetic material. As in all laboratory measurements, technical limitations apply to PCR. The most important is contamination of a patient's sample with some unwanted genetic material that could also be amplified and generate numerous copies of irrelevant genetic material. The result of such a contaminated amplification will often simply be useless, but sometimes it can lead to wrong conclusions.

Because of the potential for contamination in the PCR process, the uncertainty in the measurement of the viral load is high. The amplification process is at the heart of the PCR, and so the final measurement is given in 'copies' (copies per millilitre of blood). Thus clinicians speak of a viral load of 100 000 copies; see figure 4.3(*b*). If the viral load is undetectable, this means that there is so little HIV present in your blood that even with amplification, the PCR reaction cannot produce a measurable quantity. Such a result does not mean that there is no HIV present, only that the treatment is working and the amount of virus in the patient's blood is below the precision of the PCR equipment.

The uncertainty on the final measurement can be as high as ±50%. Thus when a viral load is measured at 100 000 copies, the actual value could be as high as 150 000 or as low as 50 000. Thus when two successive measurements are compared, for example, 200 000 and 150 000, is there a difference between these two numbers? A statistician would tell you that it is difficult to see any 'clear daylight' between these two measurements with such overlapping uncertainties. Of course, if one viral load is 200 000 and the next is 20 000, then there is clearly a difference and the anti-HIV therapy is working. This is what is seen in figure 4.3(*b*), where the viral load measurements increase erratically to the point where treatment is initiated, and then the values fall very quickly.

Not only has the PCR process made it possible to measure the amount of a virus in blood, it is now an essential tool in forensic science—the genetic fingerprint. PCR is routinely used to determine the paternity of a child, and has been used to follow the family relations of ancient Egyptian mummies. Even if the Pharaoh has been dead three thousand years, there is still enough genetic material present in the embalmed body to establish if he or she is related to the mummy in the tomb next door. Yet, as mentioned above, there are significant uncertainties generated in the PCR process, and this uncertainty becomes greater, the poorer the quality of the nucleic acid being amplified. Thus, in forensic science applied to criminal cases, it is arguments about measurement uncertainty that limits the use of PCR technology; are we sufficiently sure or certain (as determined by our laboratory measurements) that we can deprive someone of their liberty (or worse) for many years?

In these clinical measurements, we are a long way from the classic laboratory data seen in figure 4.1 or figure 4.2, yet such unique, personal medical data represent the largest set of measurements made today, and the number of such automated blood-based measurements made each year is increasing by, of order, 10%.

The data given in figure 4.3 may be personal, but such data is used (in an anonymous form) for research. Could one, for example, use such data to explore the 'placebo' effect? The medication is taken daily, so the quantity of active ingredient in the patient's body is constant, so why does the data see-saw in the manner seen in the figure? Could it be the in the summer months, the patient is feeling better than in the winter months and this has an effect upon his or her immune system? The question that has to be asked is how to extract data from such a set of measurements. What does it all mean? Does it mean anything at all … except of course to the one patient who gave all those blood samples and which form a record of his or her life? One has to be careful, however, about how you look at such complex data. When viewed in one way, it may tell you something, but when viewed in another sense, the meaning might change completely (see figure 4.4).

Now for something completely different; measurements and data sets that are studied in detail by those who run our society.

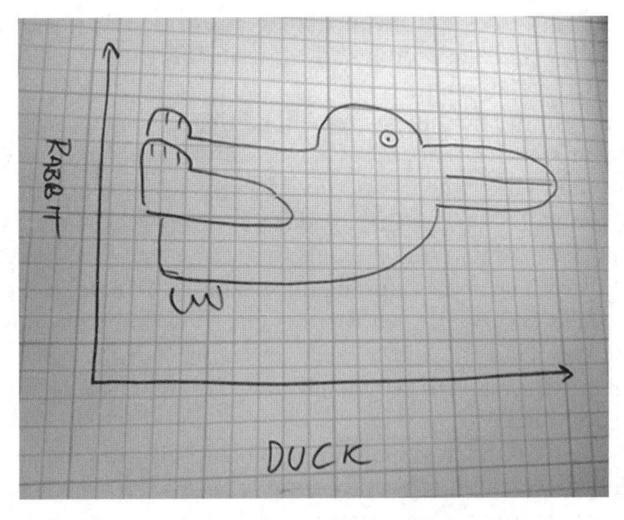

Figure 4.4. An image to remind us that a single set of data points can have a number of interpretations, depending upon how one looks at it.

4.4 What makes the world go around?

If you have a personal pension, you have an interest in knowing something about how the Stock Market functions. All pension funds are in some way linked to the Stock Market; either to 'blue chip' shares that usually fluctuate in value slowly and by small amplitudes, or to riskier stocks whose value can be all over the place.

Since the 1980s, the Stock Market has become, for the vast majority, the only method of growing an investment, apart from investing in property. The data in figure 4.5 is derived from the average value of a company on a daily basis, for the hundred largest companies in the UK (the FTSE 100). Figure 4.5(*a*) shows how this

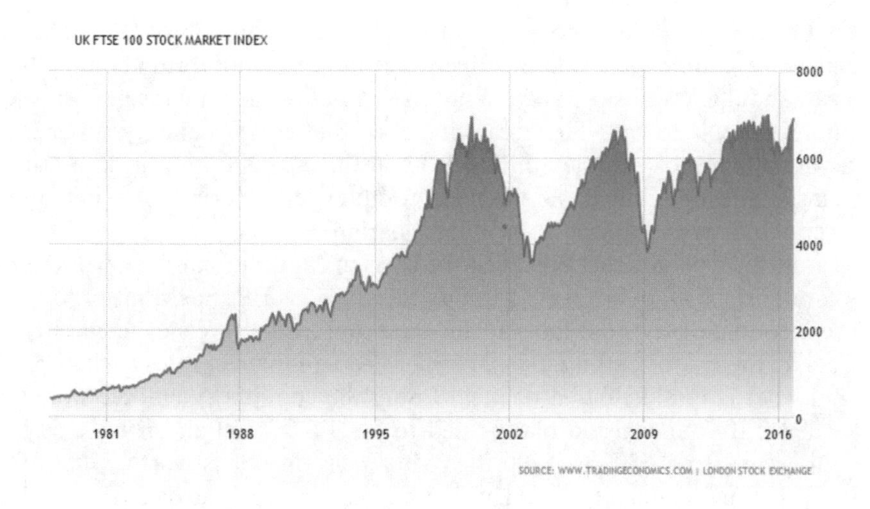

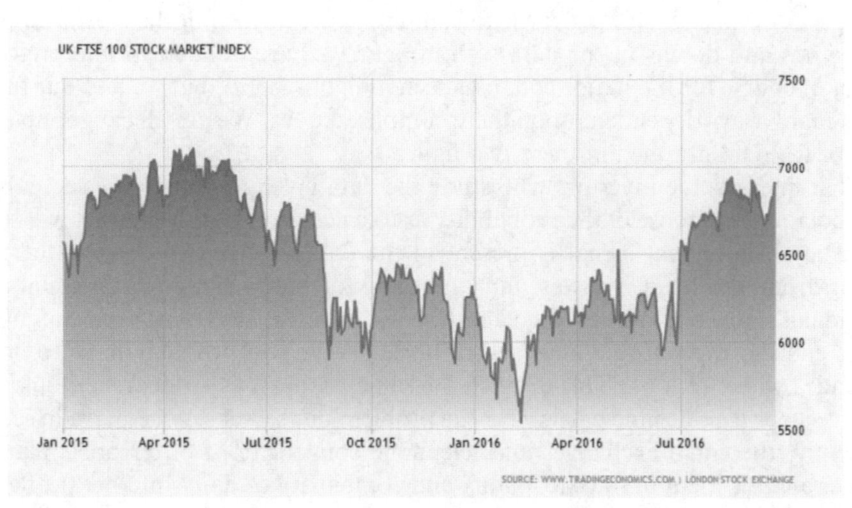

Figure 4.5. A very complex set of data (or observations, but they can still be considered as measurements): the evolution of the FSTE100 share average: (*a*) over the last 35 years, and (*b*) January 2015–August 2016. What is contributing to these fluctuations? What on earth would be the uncertainty on these observations, and how would you estimate it? Data taken from www.tradingeconomics.com.

share average has evolved to August 2016, and figure 4.5(*b*) shows a more detailed section of part (a), covering the period January 2015 to the end of August 2016.

Looking at the data set in figure 4.5(*a*), which are of huge significance and importance to the national and the international economy and which affect the lives of all of us, one can clearly see periods of spectacular growth, but also periods of spectacular loss. Up until 2001, there was a near exponential growth in the value of the FTSE100 (and the wealth of the individuals who owned those hundred companies. Then after September 2001, the market crashed taking a year to reach its nadir. It then recovered over about seven years to its 2001 value before crashing as a consequence of the sub-prime banking crisis in the USA. It then rose in a complex set of curves to the value it has today.

Interestingly, the rate of recovery; that is, the rate of growth of the FTSE 100, once underway, can be seen to be relatively constant, varying from about 400 units per year to about 600 units per year. What these rates of increase mean is, however, much more difficult to say. The great engine of the world economy is dependent upon a vast number of complex things; from an investor's optimism to a criminal trader's insider dealing, but all we see in the complex set of curves given in figure 4.5 is the overall average of all the contributing parameters.

Who uses this type of data? Well, it is the basis of capitalism and the world is now wholly capitalist; so all politicians scrutinise this data. Our Prime Minister (or rather his or her economic advisors) look at this data and at the levels of unemployment and calculate the risk of cutting income tax or keeping interest rates at their present level, and what is the likelihood of them losing the coming general election. On a personal level, if your pension plan is tied to the FSTE 100 and it falls, as it did earlier this year, from 6500 to 5700 then your pension pot is worth 5700/6500 of what it was before the markets fell. This is why many inspect the daily rises and falls of the Stock Market, waiting for it to stabilize so as to cash in their pension pots.

The ups and downs of the data seen in figure 4.5 are difficult to interpret fully, except of course for the major collapses seen in figure 4.5(*a*), but we and our futures are in many ways dependent upon these complex curves. We are all caught up in the coils of the data graphed in these two figures.

What interests the investors who study the data given in figure 4.5 is actually an extrapolation or a sense of the probability as to whether or not the market will rise or fall. They believe that signal is present in the data, but buried deep in the noise arising from the fears, worries and ambitions of many hundreds of millions of individuals. How does one even go about seeking for this electronic needle of interest in the enormous electronic haystack of noise? Well, you probably have to have a good idea where to look. This is much the same problem as a researcher looking to observe a new sub-atomic particle, or the problem our secret services have in seeking to identify the email exchanges and telephone conversations of terrorists planning murder and mayhem in the cacophony and complexity of daily Internet traffic. It is not easy, but progress is being made.

As mentioned earlier, measurements are made in myriad forms for myriad reasons. Some measurements are what one might consider to be classic scientific experiments; a unique piece of fundamental research where the result is of interest to a handful of scientists. And some are made routinely for medical reasons, where there is no fundamental law of physics to be decided or measured, but it can be a matter of life and death for an individual whose body is the subject of the measurement. Then there are the measurements and data associated with society, how the world works. Measurements of complex parameters, which have the values that are observed for enormously complex, inter-connected reasons that economists are not too good at divining, but which affect and effect us all. Figure 4.5 represents the complex interaction of multi-dimensional parameters, which with time keep providing new paradoxes, or more precisely, new data since we cannot know the 'right answer' to any question or the 'right value' of any phenomenon.

IOP Concise Physics

Quantifying Measurement

Jeffrey H Williams

Chapter 5

Measurement uncertainty

Errors using inadequate data are much better than those using no data at all.
Charles Babbage, 1792–1871

5.1 Uncertainty

All experimentation is (or should be) a voyage of discovery, or at least an investigation of something about which the investigator is uncertain and wishes to be more certain. The subject of uncertainty is vast. It involves imperfect and/or unknown information. In other words, it is a term used in subtly different ways in a number of very different fields, from insurance and philosophy to physics and information science. It applies to predictions of future events, to physical measurements that have already been made, and to the unknown.

Even among specialists in the quantitative fields, the definitions of uncertainty and risk and their measurement are varied, and this variety becomes greater when we consider these terms as used by non-specialists:

- Uncertainty: The lack of certainty. A state of having limited knowledge where it is impossible to exactly describe the existing state, a future outcome, or more than one possible outcome.
- Measurement of uncertainty: A set of possible states or outcomes where probabilities are assigned to each possible state or outcome; this includes the application of a probability density function to continuous variables.
- Risk: A state of uncertainty where some possible outcomes have an undesired effect or significant loss.
- Measurement of risk: A set of measured uncertainties where some possible outcomes are losses and the magnitudes of those losses.

doi:10.1088/978-1-6817-4433-9ch5

Of course, these definitions create as many questions as they answer. Perhaps the best recent example of the confusion arising from the term uncertainty is due to Donald Rumsfeld (United States Secretary of Defense, 2001–06) when he was commenting in a news briefing (12 February 2002) about the lack of evidence linking the government of Iraq with the supply of weapons of mass destruction to terrorist groups.

Secretary Rumsfeld stated:

Reports that say that something hasn't happened are always interesting to me, because as we know, there are known knowns; there are things we know we know. We also know there are known unknowns; that is to say we know there are some things we do not know. But there are also unknown unknowns—the ones we don't know we don't know. And if one looks throughout the history of our country and other free countries, it is the latter category that tends to be the difficult one.

Many thoughtful people are still trying to comprehend what was behind this gnomic statement. But then, you can never be certain about uncertainty.

Quantitative uses of the terms uncertainty and risk are fairly consistent through fields such as probability theory, actuarial science, and information theory. Vagueness and ambiguity are sometimes described as second-order uncertainty, where there is uncertainty even about the definitions of uncertain states or outcomes; as happens when we look at outcomes in the social sciences. The difference here is that this uncertainty is about subjective definitions and concepts, rather than an objective fact of nature.

The state of uncertainty could also be purely a consequence of a lack of knowledge, but of facts that could be observed. That is, there may be uncertainty about whether a new experimental apparatus will work, but this uncertainty can be removed with further analysis and experimentation. At the sub-atomic level, however, uncertainty may be a fundamental and unavoidable property of the Universe. In quantum mechanics, the Heisenberg Uncertainty Principle puts limits on how much an observer can ever know about the position and velocity of a quantum particle. This is the way nature is, at this level.

Uncertainty in science is something guaranteed to generate controversy, and has led to science being ridiculed in the popular press. This is due, in part, to the diversity of the public audience, and the tendency for scientists to communicate ideas poorly and ineffectively. In addition, there are often many scientific voices attempting, with varying degrees of success to provide meaningful input on a single topic; particularly, on emotive topics such as climate change. In extreme cases, depending on how an issue is reported in the public domain, discrepancies between outcomes of multiple scientific studies due to methodological differences, and their subsequent statistical analysis, are sometimes interpreted by the public as representing a lack of consensus in a situation where a consensus does in fact exist.

Indeterminacy of measurement can be said to apply to situations in which not all the parameters of the system and their interactions are fully characterized (the validity of a measurement), whereas ignorance refers to situations in which it is not known what there is to know. Such unknowns, indeterminacies and lack of information that exist in science are often transformed by the popular press into uncertainty when reported to the public in order to make the science more manageable or to tell a better 'story', since scientific indeterminacy and ignorance are difficult enough concepts for scientists to convey even among themselves. So uncertainty is often interpreted by the public as ignorance. This is why, when scientists are seeking to communicate subjects that are poorly understood, yet are of considerable interest to the public, for example, mobile phone radiation and public health, they should take particular care when specifying uncertainties associated with them, and the precision of measurements.

5.2 Uncertainty in measurements

As we saw earlier, measurement uncertainty is a non-negative parameter that describes the dispersion of the values attributed to a measured quantity. All measurements are subject to uncertainty, and a measurement result is complete only when it is accompanied by a statement of the associated uncertainty; this uncertainty has a probabilistic basis and reflects incomplete knowledge of the quantity value.

As mentioned previously, the measurement uncertainty is often taken as the standard deviation of a known probability distribution over the possible values that could be attributed to a measured quantity. Relative uncertainty is the measurement uncertainty relative to the magnitude of a particular single choice for the value for the measured quantity, when this choice is non-zero. This particular single choice is usually called the measured value, which may be optimal in some well-defined sense; for example, a mean, median, or mode. Thus, the relative measurement uncertainty is the measurement uncertainty divided by the absolute value of the measured value, when the measured value is not zero.

The purpose of measurement is to provide information about a quantity of interest, which is called the measurand. For example, the measurand might be the value of the molecular quadrupole moment of a molecule, the force generated by a new type of jet engine, or the level of 'bad' cholesterol in your blood. No measurement is exact, and no measurement is free from uncertainty. When a quantity is measured, the outcome depends on the measuring system, the measurement procedure, the skill of the operator, the environment, and a host of other factors. Even if the quantity were to be measured several times, in the same way and in the same circumstances, a different measured value would in general be obtained each time, assuming the measuring system has sufficient resolution or precision to distinguish between the values.

The dispersion of the measured values would relate to how well the measurement is performed. Their average would provide an estimate of the true value of the quantity that generally would be more reliable than a single (albeit well done) measured value. The dispersion and the number of measured values would provide information relating to the average value as an estimate of the true value. However, the true value will almost always be unknown.

The measuring system may provide measured values that are not spread about the true value, but about some value offset from it. If the instrument measuring the data has been miscalibrated, it will work well, but the results derived from the device will all be offset from the true value by the faulty calibration. This was one of the problems faced by one of the two surveyors who derived the value of the metre in the 1790s; the Borda Circle used by Pierre Méchain was worn, and this gave a systematic distortion of his measurements.

Measurement uncertainty has important economic consequences for calibration and measurement activities. In calibration reports generated by national metrological laboratories, the magnitude of the uncertainty associated with the calibration is often taken as an indication of the quality of the laboratory making the calibrations; with smaller values of uncertainty being indicative of higher standing and of higher cost. The *Guide to the Expression of Uncertainty in Measurement* (GUM) is the definitive document on this subject. The GUM has been adopted by organizations responsible for the international standardization of science and technology, and we will look at it in detail in the next chapter.

The above discussion concerns the direct measurement of a quantity. A weighing scale converts a measured extension of a spring into an estimate of the measurand; for example, the weight of a person on the scale. The particular relationship between extension and weight is determined by the calibration of the weighing scale. A measurement model then converts such a quantity value into the corresponding value of the measurand.

There are many types of measurement, and therefore many measurement models. A simple measurement model (for example, for a weighing scale, where the mass is proportional to the extension of the spring) might be sufficient for everyday use. Alternatively, a more sophisticated model of a weighing device, involving additional effects such as air buoyancy (the object being weighed in air will have displaced a volume of air and so by Archimedes' Principle, its weight will change) is capable of providing more appropriate results for industrial or scientific purposes; for example, greater precision as in the measurement of the masses of the mass standards seen in figure 4.2. In general there are often several different quantities, temperature, humidity and displacement that contribute to the definition of the measurand, and these need to be measured to define the measurand.

As well as unanalyzed (raw) data representing measured values, there is another form of data that is frequently needed in a measurement model. Such data relate to quantities representing physical constants, such as the speed of light, c, or the charge on the electron, e, each of which is also known imperfectly, but with greater precision. The items required by a measurement model to define a measurand are known as input quantities in a measurement model. The model is often referred to as a functional relationship, and the output quantity in a measurement model is the measurand of interest.

In the physical sciences, the uncertainty of a measurement, when explicitly stated, is given by a range of values likely to enclose the true value. This may be denoted graphically by error bars, or by the following notations: measured value ± the uncertainty.

Parentheses are the usual means of presenting the ± notation; for example, in stating a length one rarely encounters a fraction ($5\frac{1}{2}$metre), it would be written 5.5 m or 5.50 m, which by convention would mean that the length being investigated was precise to one tenth of a metre or one hundredth of a metre, respectively. This precision is symmetrically distributed about the last written digit; so, we could also write 5.5 ± 0.05 and 5.50 ± 0.005 or 5.50(5) and 5.500(5). If the precision is within two tenths, the uncertainty is ± one tenth, and needs to be stated explicitly. The numbers in parenthesis apply to the numeral left of themselves, and are not part of that number, but part of a notation of uncertainty, and apply to the least significant digit or figure. For instance, 1.007 94(7) represents 1.007 94 ± 0.000 07, while 1.007 94(72) represents 1.007 94 ± 0.000 72. This notation is the basis of scientific reporting.

A reading of 8000 m, with trailing zeroes and no decimal point, is ambiguous; the trailing zeroes may or may not be intended as significant figures. To avoid this ambiguity, the number could be represented as 8.0×10^3 m indicating that the first zero is significant, while 8.000×10^3 m indicates that all three zeroes are significant. However, these rules are not always followed in reporting results. Hence the need for international standardization in nomenclature and uncertainty evaluations; if you are being treated for some chronic illness, requiring regular blood tests, will an emergency blood test you have to make while on holiday be meaningful?

Often, the uncertainty of a measurement is found by repeating the measurement often enough to generate an estimate of the standard deviation of the values. Then, any single value has an uncertainty equal to the standard deviation of the ensemble. However, if the values are averaged, then the mean measurement value has a much smaller uncertainty, equal to the standard error of the mean, which is the standard deviation divided by the square root of the number of measurements. However, such a simplistic statistical analysis of measured results neglects systematic errors.

In probability theory, the normal or Gaussian distribution (as seen in figure 2.2) is a continuous probability distribution used to describe the results of measurements (from the physical size of students in a classroom to the distribution of galaxies in some part of the Universe). Normal distributions are important in statistics, and are often used to represent real-valued random variables whose distributions are not known. The normal distribution is useful because of the central limit theorem, proved by Pierre Simon de Laplace, which in its most general form, including finite variance, states that averages of random variables independently drawn from independent distributions converge to the normal distribution; that is, become normally distributed when the number of random variables is sufficiently large. Physical quantities that are expected to be the sum of many independent processes (such as measurement errors) often have distributions that are nearly normal. Moreover, many results and methods (such as propagation of uncertainty and least-squares parameter fitting) can be derived analytically in explicit form, when the relevant variables are normally distributed.

The normal distribution is sometimes termed a bell-shaped curve (see figure 2.2). However, many other distributions are also bell-shaped (such as the Cauchy, Student's *t*, and logistic distributions).

The probability density of the normal distribution is:

$$f(x|\mu, \sigma^2) = \left(1/\sqrt{2\sigma^2\pi}\right)\exp -(x-\mu)^2/2\sigma^2,$$

where μ is mean or expectation of the distribution (and also its median and mode), σ is standard deviation, and σ^2 is variance.

From the properties of such probability distributions, we are able to make statements about measurement uncertainty when the uncertainty represents the standard error of the measurement; for example, from figure 2.2 when the true value of the measured quantity falls within the stated uncertainty range 68.3% of the time. Such a probability distribution has the advantage that the quoted standard errors are easily converted to 68.3% (one sigma or one standard deviation distributed about the mean), 95.4% (two sigma or twice the standard deviation distributed about the mean), or 99.7% (three sigma or three times the standard deviation distributed about the mean) confidence intervals.

In this context, uncertainty depends on both the accuracy and precision of the measuring instrument. The lower the accuracy and precision of the instrument, the larger the uncertainty of the measurement.

Formally, the output quantity, denoted by Y, about which information is required, is often related to input quantities, denoted by $X_1, \ldots, X_N$, about which information is available, via a measurement model in the form of,

$$Y = f(X_1, \ldots, X_N)$$

where f is known as the measurement function. A general expression for a measurement model is,

$$h(Y, X_1, \ldots, X_N) = 0.$$

It is taken that a procedure exists for calculating Y given $X_1, \ldots, X_N$, and that Y is uniquely defined by this equation.

The true values of the input quantities $X_1, \ldots, X_N$, are unknown. In the GUM approach to measurement uncertainties $X_1, \ldots, X_N$, are characterized by probability distributions and treated mathematically as random variables. These distributions describe the respective probabilities of their true values lying in different intervals, to which they are assigned based on available knowledge concerning $X_1, \ldots, X_N$. Sometimes, some or all of $X_1, \ldots, X_N$ are interrelated and the relevant distributions apply to these quantities taken together.

Consider estimates $x_1, \ldots, x_N$, respectively, of the input quantities $X_1, \ldots, X_N$, obtained from a variety of sources. The probability distributions characterizing $X_1, \ldots, X_N$ are chosen such that the estimates $x_1, \ldots, x_N$, respectively, are the expectations of $X_1, \ldots, X_N$. Moreover, for the ith input quantity, consider a so-called standard uncertainty, given the symbol $u(x_i)$, defined as the standard deviation of the input quantity X_i. This standard uncertainty is said to be associated with the (corresponding) estimate x_i.

The use of available knowledge to establish a probability distribution to characterize each quantity of interest applies to the X_i and also to Y. In the latter case, the characterizing probability distribution for Y is determined by the measurement model together with the probability distributions for the X_i. The determination of the probability distribution for Y from this information is known as the propagation of distributions. Figure 5.1 depicts a measurement model $Y = X_1 + X_2$, in the case

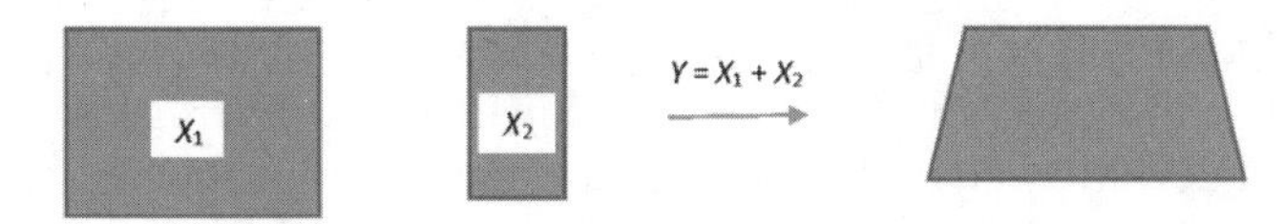

Figure 5.1. Figure depicting a measurement model, $Y = X_1 + X_2$ in the case where both inputs X_1 and X_2 are each characterized by a (different) rectangular, or uniform, probability distribution. In this case, the output function has a symmetric trapezoidal probability distribution.

where X_1 and X_2 are each characterized by a (different) rectangular, or uniform, probability distribution Y has a symmetric trapezoidal probability distribution (figure 5.1).

Once the input quantities $X_1, \ldots, X_N$ have been characterized by appropriate probability distributions, and the measurement model has been developed, the probability distribution for the measurand Y has been fully specified. In particular, the expectation of Y is used as the estimate of Y, and the standard deviation of Y as the standard uncertainty associated with this estimate.

Often an interval containing Y with a specified probability is required. Such an interval, a coverage interval, can be deduced from the probability distribution for Y. The specified probability is known as the coverage probability. For a given coverage probability, there is more than one coverage interval. The symmetric coverage interval is an interval for which the probabilities of a value to the left and the right of the interval are equal. The shortest coverage interval is an interval for which the length is shortest over all coverage intervals having the same coverage probability.

Prior knowledge about the true value of the output quantity Y can also be incorporated into the model; for example, a bathroom scale, the fact that the person's mass is positive, and that it is the mass of a person, rather than that of a motor car that is being measured. Both these conditions constitute prior knowledge about the possible values of the measurand. Such additional information can be used to provide a probability distribution for Y that can give a smaller standard deviation for Y and hence a smaller standard uncertainty associated with the estimate of Y.

The uncertainty of the result of a measurement generally consists of several components. These components are regarded as random variables and may be grouped into two categories according to the method used to estimate their numerical values.

5.3 Type A and Type B uncertainty

Knowledge about an input quantity X_i may be inferred from repeated measured values (this is termed a Type A evaluation of uncertainty: where components of the uncertainty are evaluated by statistical methods), or scientific judgement or other information concerning the possible values of the quantity (this is termed a Type B evaluation of uncertainty: where components of the uncertainty are evaluated by other means; for example, by analyzing a probability distribution).

In Type A evaluations of measurement uncertainty, the assumption is made that the distribution best describing an input quantity X, given repeated independently measured values of this quantity, is a Gaussian distribution. X then has an expectation or expected value (see below) equal to the average measured value and a standard deviation equal to the standard deviation of the average. When the uncertainty is evaluated from a small number of measured values, the corresponding distribution can be taken as a t-distribution.

For a Type B evaluation of uncertainty, it is often the case that the only available information is that X lies in a specified interval $[a,b]$. In such a case, knowledge of the quantity can be characterized by a rectangular probability distribution with limits a and b. If additional information were available, a probability distribution consistent with that additional information would be used.

Sensitivity coefficients $c_1, \ldots, c_N$ describe how the estimate y of Y would be influenced by small changes in the estimates $x_1, \ldots, x_N$ of the input quantities $X_1, \ldots, X_N$. For the measurement model $Y = f(X_1, \ldots, X_N)$, the sensitivity coefficient c_i equals the partial derivative (first order) of f with respect to X_i evaluated at $X_1 = x_1$, $X_2 = x_2$, etc. For a linear measurement model such as,

$$Y = c_1X_1 + \cdots + c_NX_N,$$

with $X_1,\ldots, X_N$ independent, a change in x_i equal to $u(x_i)$ would give a change $c_i\, u(x_i)$ in y. This statement would generally be approximate for measurement models $Y = f(X_1,\ldots, X_N)$. The relative magnitudes of the terms $|c_i|\, u(x_i)$ are useful in assessing the respective contributions from the input quantities to the standard uncertainty $u(y)$ associated with y.

The standard uncertainty $u(y)$ associated with the estimate y of the output quantity Y is not given by the sum of the $|c_i|\, u(x_i)$, but by these terms combined in quadrature, namely by

$$u^2(y) = c_1^2u^2(x_1) + \cdots + c_N^2\, u^2(x_N),$$

an expression that is generally approximate for measurement models $Y = f(X_1,\ldots, X_N)$. This is termed the law of propagation of uncertainty. When the input quantities X_i contain dependences, the above formula is augmented by terms containing covariances, which may increase or decrease $u(y)$.

By propagating the values of the variance (the squared deviation of a random variable from its mean), of the various components of the total uncertainty through a function relating the components to the measurement result, the combined measurement uncertainty is given as the square root of the resulting variance. The simplest form is the standard deviation of a repeated measurement; that is, a measurement of one unknown by an apparatus where all other parameters that influence the measured unknown are fully defined.

5.4 Propagation of uncertainty

To apply probabilistic methods to estimating errors and uncertainty in laboratory experiments, one must be able to look at the contributions of each and every element of the experiment to the overall uncertainty in the final result. This is the concept of the propagation of uncertainty.

Most commonly, the uncertainty on a quantity is expressed in terms of the standard deviation, σ, and the positive square root of variance, σ^2. The value of a quantity and its error are then expressed as an interval $x \pm u$. If the statistical probability distribution of the variable is known, or can be assumed, it is possible to derive confidence limits to describe the region within which the true value of the variable may be found. For example, the 68.3% confidence limits for a one-dimensional variable belonging to a normal distribution are ± one standard deviation from the true value; that is, there is approximately a 68.3% probability that the true value lies in the region $x \pm \sigma$.

Consider an experiment; we are studying a simple electric circuit, and knowing the electrical resistance (R) of the circuit and its measurement uncertainty, and the current flowing through the circuit (I) together with its measurement uncertainty, we wish to measure the voltage (V) using Ohm's law ($V = IR$). Knowing the uncertainties (standard deviations) in I and R, what is the uncertainty in V? In other words, given a function relationship between several measured variables; that is, $Q = f(x, y, z)$, what is the uncertainty in Q if the uncertainties in x, y and z are known?

To answer this question, we must assume that when we talk about the uncertainties in a measured variable that the value of that measured variable represents the mean of a Gaussian or normal distribution, and that the uncertainty in this variable is the standard deviation (σ) of that Gaussian or normal distribution.

To calculate the variance in Q as a function of the variances in; for example, x and y, we use the following (for further details see, for example, the entry for propagation of errors in the online encyclopaedia Wikipedia):

$$Q_Q^2 = \sigma_x^2\left(\frac{\partial Q}{\partial x}\right)^2 + \sigma_y^2\left(\frac{\partial Q}{\partial y}\right)^2 + 2\sigma_{xy}\left(\frac{\partial Q}{\partial x}\right)\left(\frac{\partial Q}{\partial y}\right).$$

If the variables, x and y are uncorrelated, that is, $\sigma_{xy} = 0$ then the last term in the above equation vanishes. However, if the variables are correlated, as they would be in the above example of using Ohm's law to determine the voltage in a circuit knowing the resistance and the current, then the last term in the above equation does not vanish and it must be evaluated.

In chapter 8, we will encounter an experiment to measure the molecular quadrupole moment of gaseous molecules. One of the major sources of uncertainty with the measurement is the total amount of light arriving at the detector (one is searching for a signal oscillating at a frequency ω against a large noisy (but un-modulated) background. The total light level at the detector arises from imperfections in the polarizing and the analyzing polarizers and in strain in the windows of optical components that leads to depolarization of the initially linearly-polarized light beam. In this experiment, the uncertainty in the measurement of the signal at frequency ω and the static background would be uncorrelated; whereas the uncertainty in the measurement of the signal at ω and in the uncertainty of the measurement of the pressure of gas in the measurement cell would be correlated.

The average or mean of several measurements, each with the same uncertainty, σ, is given by:

$$\mu = (x_1 + x_2 + \cdots x_n)/n,$$

and the uncertainties may be written as:

$$\sigma_\mu^2 = \sigma_{x1}^2\left(\frac{\partial\mu}{\partial x1}\right)^2 + \sigma_{x2}^2\left(\frac{\partial\mu}{\partial x2}\right)^2 + \cdots \sigma_{xn}^2\left(\frac{\partial\mu}{\partial xn}\right)^2 = \sigma^2\left(\frac{1}{n}\right)^2 + \sigma^2\left(\frac{1}{n}\right)^2 + \cdots \sigma^2\left(\frac{1}{n}\right)^2,$$

which may be written as

$$\sigma_\mu = \sigma/\sqrt{n}.$$

This is the well-known formula for the error in the mean. It tells us that the precision in the final measurement only increases with the square root of the number of measurements. The standard deviation, σ, is related to the probability density function or pdf (usually a normal distribution) from which the values are taken, and σ does not get smaller as we combine measurements.

5.5 Uncertainty evaluation

The main steps involved in the estimation of measurement uncertainty constitute formulation and calculation; the latter stage consisting of error propagation. The formulation stage constitutes:

1. defining the output quantity Y (the measurand),
2. identifying the input quantities upon which Y depends,
3. developing a measurement model relating Y to the input quantities, and
4. on the basis of available knowledge, assigning probability distributions, Gaussian, rectangular, etc, to the input quantities (or a joint probability distribution to those input quantities that are not independent).

The calculation stage consists of propagating the probability distributions for the input quantities through the measurement model to obtain the probability distribution for the output quantity Y, and then using this distribution to obtain:

1. the expectation of Y, taken as an estimate y of Y,
2. the standard deviation of Y, taken as the standard uncertainty $u(y)$ associated with y, and
3. a coverage interval containing Y with a specified coverage probability.

The propagation stage of uncertainty evaluation is also known as the propagation of distributions, various approaches for which are available, including:

- the GUM uncertainty framework, constituting the application of the law of propagation of uncertainty, and the characterization of the output quantity Y by a Gaussian or a t-distribution,
- analytic methods, in which mathematical analysis is used to derive an algebraic form for the probability distribution for Y, and

- a Monte Carlo method, in which an approximation to the distribution function for Y is established numerically by making random draws from the probability distributions for the input quantities, and evaluating the model at the resulting values.

For any particular problem in uncertainty evaluation, any one of these three approaches may be used, however, the first approach is generally approximate, while the second approach is exact, and the third approach provides a solution with a numerical accuracy that can be controlled.

5.6 Probability

As we saw earlier, the great contribution of Gauss to the concept of measurement uncertainty was to introduce the idea of a distribution of possible outcomes of a measurement process; that is, the observed or measured value would be distributed about the true value. In this, Gauss was extending the ideas of the French mathematician, and one of the creators of the Metric System, Pierre Simon de Laplace, who was the first to combine Newtonian mechanics and probability theory (in his *Théorie Analytique des Probabilités* of 1812).

Probability is the branch of mathematics concerned with the analysis of random variables, stochastic processes, and random phenomena; that is, modelling of non-deterministic events or measured quantities that may either be single occurrences or evolve over time in an apparently random fashion. It is not possible to predict precisely the results of random events, however, if a sequence of individual events, such as coin flipping or the rolling of dice, is influenced by other factors, such as friction, it will exhibit certain patterns, which can be studied and predicted. Two representative mathematical results describing such patterns are the law of large numbers and the central limit theorem.

As a mathematical foundation for statistics, probability theory is essential to many activities that involve quantitative analysis of large sets of data. Methods of probability theory also apply to descriptions of complex systems given only partial knowledge of their state, as in statistical mechanics. The great discovery of 20th Century physics was the probabilistic nature of physical phenomena at atomic scales, described in quantum mechanics.

The mathematical theory of probability has its roots in the world of gaming and gambling; attempts to analyze games of chance by Gerolamo Cardano in the 16th Century, and by Pierre de Fermat and Blaise Pascal in the 17th Century (the 'problem of points'). In 1657, Christiaan Huygens published a book on the modelling of games of chance (*De ratiociniis in ludo aleae*), and in 1812 Pierre Simon de Laplace published the first definitive text on probability. Initially, probability theory considered discrete events, and its methods were mainly combinatorial. Eventually, analytical considerations required the incorporation of continuous variables into the theory.

Consider an experiment that can produce a number of outcomes. The set of all outcomes is called the sample space of the experiment. The power set of the sample space is formed by considering all possible results. For example, rolling an honest die produces one of six possible results. One collection of possible results corresponds to obtaining an odd number. Thus, the subset (1,3,5) is an element of the sample space of die rolls. These collections are called events. In this case, (1,3,5) is the event that the die falls on an odd number. If the results that actually occur fall in a given event, that event is said to have occurred.

Probability is a way of assigning an event a value between zero and one, with the requirement that the event be made up of all possible results; in our example, the event (1,2,3,4,5,6) be assigned a value of one. To qualify as a probability distribution, the assignment of values must satisfy the requirement that if you look at a collection of mutually exclusive events (events that contain no common results; for example, the events (1,6), (3), and (2,4) would be mutually exclusive for our present purpose), the probability that one of the events will occur is given by the sum of the probabilities of the individual events.

The probability that any one of the events (1,6), (3), or (2,4) will occur is 5/6. This is the same as saying that the probability of event (1,2,3,4,6) is 5/6. This event encompasses the possibility of any number except five being rolled. The mutually exclusive event (5) has a probability of 1/6, and the event (1,2,3,4,5,6) has a probability of 1; that is, absolute certainty.

Discrete probability theory deals with events that occur in countable sample spaces; for example, throwing dice, experiments with decks of cards, random walk, the tossing of coins, and studies of radioactive decay. Initially, the probability of an event occurring was defined as the number of cases favourable for that event divided by the number of total possible outcomes in an equally probable sample space. For example, if the event of interest is the occurrence of an even number when a die is rolled, the probability is given by 3/6, since three faces out of the six have even numbers and each face has the same probability of appearing.

The modern definition of a probability starts with a finite or countable set, the sample space, which relates to the set of all possible outcomes in a classical sense, denoted by Ω. It is then assumed that for each element $x \in \Omega$, an intrinsic probability value $f(x)$ is attached, which satisfies the following properties:

- $f(x) \in [0,1]$ for all $x \in \Omega$, and
- $\sum_{x \in \Omega} f(x) = 1$.

That is, the probability function $f(x)$ lies between zero and one for every value of x in the sample space Ω, and the sum of $f(x)$ over all values x in the sample space Ω is equal to 1. An event is defined as any subset E of the sample space Ω. The probability of the event E is defined as

$$P(E) = \sum_{x \in \Omega} f(x).$$

So, the probability of the entire sample space is 1, and the probability of the null event is 0. Continuous probability theory deals with events that occur in a continuous sample space. If the outcome space of a random variable X is the set of real numbers ($\mathcal{R}$) or a subset thereof, then a function called the cumulative distribution function F exists, defined by $F(x) = P(X \leqslant x)$; that is, $F(x)$ gives the probability that X will be less than or equal to x.

As F is continuous, its derivative exists and integrating the derivative gives us the cumulative distribution function; then the random variable X is said to have a probability density function (pdf), or more simply a density, $f(x) = \mathrm{d}F(x)/\mathrm{d}x$.

For a set $E \subseteq \mathcal{R}$, the probability of the random variable X being in E is

$$P(X \in E) = \int_{x \in E} \mathrm{d}F(x),$$

this can be written as

$$P(X \in E) = \int_{x \in E} f(x)\,\mathrm{d}x.$$

Whereas the pdf exists only for continuous random variables, the cumulative distribution function exists for all random variables (including discrete random variables) that take values in $\mathcal{R}$.

Intuition tells us that if a fair coin is tossed many times, then roughly half of the time it will turn up heads, and the other half it will turn up tails. Furthermore, the more often the coin is tossed, the more likely it should be that the ratio of the number of heads to the number of tails will approach unity. Probability theory provides a formal version of this intuitive idea, known as the law of large numbers. The law of large numbers states that the sample average

$$X_n = (1/n)\sum_{k=1}^{n} X_k$$

of a sequence of independent and identically distributed random variables X_k converges towards their common expectation, provided that the expectation of $|X_k|$ is finite.

The central limit theorem explains the ubiquitous occurrence of the normal distribution in nature. The theorem states that the average of many independent and identically distributed random variables with finite variance tends towards a normal distribution irrespective of the distribution followed by the original random variables. Formally, let X_1, X_2,... be independent random variables with mean μ and variance $\sigma^2 > 0$. Then the sequence of random variables

$$Z_n = \sum_{i=1}^{n}(X_i - \mu)/\sigma\sqrt{n}$$

converges in a distribution of a standard normal random variable.

5.7 Expected value

In probability theory, the expected value of a random variable is, intuitively, the long-run average value of repetitions of an experiment; for example, the expected value in rolling a six-sided die is 3.5. More precisely, the law of large numbers states that the arithmetic mean of the values converges to the expected value as the number of repetitions becomes large.

The expected value of a discrete random variable is the probability-weighted average of all possible values. In other words, each possible value the random variable may assume is multiplied by its probability of occurring, and the resulting products are summed to produce the expected value. The same principle applies to a continuous random variable, except that an integral of the variable with respect to its probability density replaces the sum. For random variables, the long-tails of the distribution that represent such events prevent the summation from converging.

The expected value is a key aspect of how one characterizes a probability distribution. By contrast, the variance is a measure of dispersion of the possible values of the random variable around the expected value. The variance is the expected value of the squared deviation of the variable's value from the variable's expected value.

The expected value plays an important role in a variety of contexts: in regression analysis, one desires a formula or model to represent the observed data that gives a good estimate of the parameter(s) generating the effect of interest. The formula will give different estimates using different sets of data, so the estimate it gives is itself a random variable. A formula is considered good in this context if it is an unbiased estimator; that is, if the expected value of the estimate (the average value it would give over an arbitrarily large number of separate samples) can be shown to equal the true value of the desired parameter.

If the probability distribution of X admits a probability density function $f(x)$, then the expected value may be calculated as

$$E[X] = \int_{-\infty}^{\infty} x f(x) \, \mathrm{d}x.$$

It is possible to construct an expected value equal to the probability of an event by taking the expectation of an indicator function that is unity, if the event has occurred, and zero otherwise. This relationship can be used to translate properties of expected values into properties of probabilities; for example, using the law of large numbers to justify estimating probabilities by frequencies.

The expected values of the powers of X are called the moments of X; the moments about the mean of X are expected values of powers of $X - E[X]$. The moments of some random variables can be used to specify their distributions, via appropriate moment generating functions.

To estimate the expected value of a random variable, one could repeatedly measure the variable and compute the arithmetic mean of the results. If the expected value exists, this procedure estimates the true expected value in an unbiased manner and has the advantage of minimizing the sum of the squares of the residuals (the sum of the squared differences between the observations and the estimate). The law of large numbers demonstrates that, as the size of the sample gets larger, the variance of this estimate gets smaller.

This property is often exploited in a wide variety of applications, including general problems of statistical estimation and machine learning, to estimate (probabilistically) quantities of interest via; for example, Monte Carlo methods, since most quantities of interest can be written in terms of an expectation; for example, $P(X \in A) = E[I_A(X)]$ where $I_A(X)$ is the indicator function for set A; that is, $X \in A \to I_A(X) = 1$, and $X \notin A \to I_A(X) = 0$.

In classical mechanics, the centre-of-mass is an analogous concept to expectation (as are the electrical moments of molecules, which characterize the charge distribution that is the molecule); for example, suppose X is a discrete random variable with values x_i and corresponding probabilities p_i. Now consider a weightless rod on which are placed weights, at locations x_i along the rod and having masses p_i (whose sum is one). The point at which the rod balances is $E[X]$; the mass of a probability distribution is balanced at the expected value.

Expected values can also be used to compute the variance, by means of the computational expression for the variance,

$$\text{Variance}\,(X) = E[X^2] - \left(E[X]\right)^2.$$

An important application of the expectation value is in quantum mechanics. The expectation value of a quantum mechanical operator A, operating on a quantum state defined by state vector $|\psi\rangle$ is written as $\langle A\rangle = \langle\psi\,|A|\,\psi\rangle$ where $\langle\psi\,| = |\,\psi\rangle^*$. The uncertainty in A can be calculated using the formula $(\Delta A)^2 = \langle A^2\rangle - \langle A\rangle^2$.

Further reading

As stated earlier, the subject of measurement uncertainty is vast, but there are many standard textbooks. Some of these texts (or at least parts of them) are available on the Internet, Google them. In addition, there are many useful articles in the online encyclopaedia Wikipedia; for example, see the entries for measurement uncertainty, probability theory, and expected values. There are also freely available documents on measurement uncertainty on the websites the BIPM (www.bipm.org) listed under the heading publications.

IOP Concise Physics

Quantifying Measurement

Jeffrey H Williams

Chapter 6

Guide to the Expression of Uncertainty in Measurement (the GUM)

6.1 Introduction

As we have seen, a measurement result is complete only when accompanied by a quantitative statement of its uncertainty. This requirement for a statement of uncertainty enables one to decide if the result is adequate for its intended purpose, and to determine if it is consistent with similar results. This qualification to a measurement may seem excessive, but is necessitated by the increasing dependence of the global economy upon technical measurements.

In the economic boom that followed the Second World War, it became apparent that the levels of precision seen in the various national industries (national measurement capacities), and the uncertainties associated with those levels of precision, were a fundamental factor in international trade and commerce. If a nation wished to increase its share of global trade it had to demonstrate quantitatively that it had the capacity to make the things needed by other nations, but at a unit cost and level of precision that was advantageously competitive. This was also the requirement that forced the UK and the USA to finally abandon their own customary weights and measures, and fully embrace the Metric System.

Over the years, many different approaches to evaluating and expressing the uncertainty of the results of measurements have been developed and used by different groups of scientists and different professions. But because of the difficulty in achieving international consensus on the expression of uncertainty in measurement, in 1977 the International Committee for Weights and Measures (CIPM, *Comité international des poids et mesures*), the world's highest authority in the field of measurement science, asked the BIPM (*Bureau international des poids et mesures*), an international agency created under the Metre Convention of 1875, to address this problem of finding an appropriate means of expressing uncertainty in measurement. This project was undertaken in collaboration with other international

doi:10.1088/978-1-6817-4433-9ch6

agencies and the various national measurement laboratories, and it led to the development of a recommendation or statement as to what constituted measurement uncertainty for consideration by the international physical-science community.

The uncertainty in the result of a measurement generally consists of several components, which may be grouped into two categories according to the manner in which their numerical value is estimated. As we saw earlier, this may be stated as:

- Type A uncertainty: those components which are evaluated by statistical methods.
- Type B uncertainty: those components which are evaluated by other means.

However, it is often difficult to find a simple correspondence between the classification into categories A or B. Any detailed report of uncertainty should consist of a complete list of all the components contributing to the uncertainty, specifying for each the method used to obtain its numerical value.

The components in category A are characterized by the estimated variances σ_i^2 (or the estimated standard deviation σ_i) and the number of degrees of freedom ν_i. Where appropriate the covariances should be given, which requires a statement of how the measurand is coupled to other properties of the system under study. The components in category B should be characterised by quantities u_j^2, which may be considered approximations to the corresponding variances, the existence of which is assumed. The quantities u_j^2 may be manipulated as if they were variances, and the quantities u_j treated as if they were standard deviations. Where appropriate, the covariances should be treated in a similar way.

The combined uncertainty should be characterized by the numerical value obtained by applying the usual method for the combination of variances (as outlined previously in the Propagation of uncertainty). The combined uncertainty and its components should be expressed in the form of standard deviations.

The above recommendations or suggestions are a brief outline rather than a detailed prescription. Subsequently, the CIPM worked with a wide-range of organizations with responsibility for the international coordination of technology, to develop a guidebook to the expression of uncertainty in measurements. This project developed into the *Guide to the Expression of Uncertainty in Measurement* (GUM), which is a summary of the methods of evaluating and expressing uncertainty in measurement, which has been widely adopted by the industries of many nations, by national metrology laboratories and many international organizations (see section 6.7 for more details).

6.2 Basic definitions

Measurement equation: the case of interest is where the quantity Y being measured, the measurand, is not measured directly but is determined from N other quantities $X_1, X_2, \ldots, X_N$ through a functional relation f, the measurement equation:

$$Y = f(X_1, X_2, \ldots, X_N). \tag{6.1}$$

Included among the quantities X_i are corrections (or correction factors), as well as quantities that take into account other sources of variability, such as different experimenters, different instruments, other samples, other laboratories, and different times at which observations are made. Thus, the function $f(X_1, X_2, \ldots, X_N)$ should express not simply a physical law, but a measurement process and, in particular, it should contain all the quantities that can contribute a significant uncertainty to the final measurement result.

An estimate of the measurand or output quantity Y, denoted by y, is obtained from equation (6.1) using input estimates $x_1, x_2, \ldots, x_N$ for the values of the N input quantities $X_1, X_2, \ldots, X_N$. Thus, the output estimate y, which is the result of the measurement, is given by

$$y = f(x_1, x_2, \ldots, x_N). \tag{6.2}$$

For example, if a voltage or potential difference V is applied to the terminals of a temperature-dependent resistor that has a resistance R_0 at the defined temperature t_0 and a linear temperature coefficient of resistance b, the power P (the measurand) dissipated by the resistor at the temperature t depends on V, R_0, b, and t according to:

$$P = f(V, R_0, b, t) = V^2/R_0\left[1 + b(t - t_0)\right]. \tag{6.3}$$

The uncertainty of the measurement result y arises from the uncertainties $u(x_i)$ (or u_i for simplicity) of the input estimates x_i that enter equation (6.2). Thus, in the example of equation (6.3), the uncertainty of the estimated value of the power P comes from the uncertainties of the estimated values of the potential difference V, resistance R_0, temperature coefficient of resistance b, and temperature t. In general, components of uncertainty may be categorized according to the method used to evaluate these quantities. With a Type A evaluation involving statistical analysis of series of observations, and a Type B evaluation involving evaluation by means other than statistical analysis of series of observations.

Each component of uncertainty, irrespective of how it has been evaluated, is represented by an estimated standard deviation, also termed standard uncertainty often with symbol u_i, and equal to the positive square root of the estimated variance. Thus, an uncertainty component obtained by a Type A evaluation is represented by a statistically estimated standard deviation σ_i, equal to the positive square root of the statistically estimated variance σ_i^2, and the associated number of degrees of freedom ν_i. For such a component, the standard uncertainty is $u_i = \sigma_i$. In a similar manner, an uncertainty component obtained by a Type B evaluation is represented by a quantity u_j, which may be considered an approximation to the corresponding standard deviation; it is equal to the positive square root of u_j^2, which may be considered an approximation to the corresponding variance and which is obtained from an assumed probability distribution based on all available information. Since the quantity u_j^2 is treated like a variance and u_j like a standard deviation, for such a component the standard uncertainty is simply u_j.

6.3 Evaluating uncertainty components

A Type A evaluation of standard uncertainty may be based on any valid statistical method for treating data; for example, calculating the standard deviation of the mean of a series of independent observations; using the method of least-squares to fit a curve to data in order to estimate the parameters of the curve and their standard deviations; and carrying out an analysis of variance (ANOVA)[1] in order to identify and quantify random effects in certain kinds of measurements.

In the case where an input quantity X takes random values from a finite data set $x_1, x_2, \ldots, x_N$, with each value having the same probability, the standard deviation and the mean (μ) are:

$$\sigma = \sqrt{\frac{1}{N}\left[(x_1 - \mu)^2 + (x_2 - \mu)^2 + \cdots + (x_N - \mu)^2\right]}, \quad \text{where}$$
$$\mu = \frac{1}{N}(x_1 + \cdots + x_N), \tag{6.4}$$

or,

$$\sigma = \sqrt{\frac{1}{N}\sum_{i=1}^{N}(x_i - \mu)^2}, \quad \text{where} \quad \mu = \frac{1}{N}\sum_{i=1}^{N}x_i. \tag{6.5}$$

If, instead of having equal probabilities, the values have different probabilities, let x_1 have probability p_1, x_2 have probability p_2, …, x_N have probability p_N. In this case, the standard deviation will be

$$\sigma = \sqrt{\sum_{i=1}^{N}p_i(x_i - \mu)^2}, \quad \text{where} \quad \mu = \sum_{i=1}^{N}p_i x_i. \tag{6.6}$$

If it had been possible, this would have been the appropriate place for Pierre Méchain and his surveying colleague to have carefully considered which of their data they would have preferred to keep in their final analysis of the metric survey, and which data they would have chosen to exclude from the final calculation. Analysis of Type B uncertainty allows one to put a subjective analysis of the quality of measurements onto a firmer quantitative basis. It turns an art into a science. A Type B evaluation of standard uncertainty should be based on scientific judgment using all of the relevant information available; for example,

[1] Analysis of variance (ANOVA) is a set of statistical models used to analyze the differences among group means (such as variation among and between groups), developed by the British statistician and evolutionary biologist Ronald Fisher (1890–1962). In an analysis by ANOVA, the observed variance in a particular variable is split into components attributable to different sources of variation. In its simplest form, ANOVA provides a statistical test of whether or not the means of several groups are equal, and therefore generalizes the t-test to more than two groups. Such an analysis is useful for comparing three or more means (of groups) for statistical significance. It is conceptually similar to multiple two-sample t-tests, but is more conservative (results in less type I hypothesis testing errors) and is therefore suited to a wide range of practical problems; see chapter 7 on clinical trials.

- previous measurement data,
- experience with, or general knowledge of, the behaviour and property of relevant materials and instruments,
- manufacturer's specifications,
- data provided in calibration and other reports, and
- uncertainties assigned to reference data taken from handbooks.

Below are some examples of Type B evaluations in different situations, depending on the available information and the assumptions of the experimenter. Generally, the uncertainty is derived either from an outside source, or calculated from an assumed distribution.

6.4 Uncertainty derived from some assumed distributions[2]

Normal distribution: if the quantity of interest is modelled by a normal or Gaussian probability distribution, there are no finite limits that will contain 100% of its possible values. However, plus and minus three standard deviations about the mean of a normal distribution corresponds to 99.73% confidence limits. Thus, if the lower and upper limits, a_- and a_+, respectively, of a normally distributed quantity with mean $(a_+ + a_-)/2$ are considered to contain almost all of the possible values of the quantity; that is, 99.73% of them, then u_j is approximately $a/3$, where $a = (a_+ - a_-)/2$ is the half-width of the interval.

Uniform (rectangular) distribution: Estimate lower and upper limits a_- and a_+, respectively, for the value of the input quantity such that the probability that the value lies in the interval a_- to a_+ is 100%. Provided that there is no contradictory information, treat the quantity as if it is equally probable for its value to lie anywhere within the interval a_- to a_+; that is, model it by a uniform (that is, rectangular) probability distribution. The best estimate of the value of the quantity is then $(a_+ + a_-)/2$ with $u_j = a/\sqrt{3}$, where $a = (a_+ - a_-)/2$ is the half-width of the interval.

Triangular distribution: The rectangular distribution is a reasonable default model in the absence of any information about the distribution of the quantity of interest. But if it is known that values of the quantity in question near the centre of the limits are more likely than values close to the limits, a normal distribution or, for simplicity, a triangular distribution, may be a better and simpler model. For a triangular distribution, estimate lower and upper limits a_- and a_+ for the value of the input quantity in question, such that the probability that the value lies in the interval a_- to a_+ is 100%. Provided that there is no contradictory information, model the quantity by a triangular probability distribution. The best estimate of the value of the quantity is then $(a_+ + a_-)/2$ with $u_j = a/\sqrt{6}$, where $a = (a_+ - a_-)/2$ is the half-width of the interval.

---

[2] For further details see http://physics.nist.gov/cuu/Uncertainty/typeb.html.

6.5 Combining uncertainty components

The combined standard uncertainty of the measurement result y, designated by $u_c(y)$ and taken to represent the estimated standard deviation of the result, is the positive square root of the estimated variance $u_c^2(y)$ obtained from:

$$u_c^2(y) = \sum_{i=1}^{N}\left(\frac{\partial f}{\partial x_i}\right)^2 u^2(x_i) + 2\sum_{i=1}^{N-1}\sum_{j=i+1}^{N}\frac{\partial f}{\partial x_i}\frac{\partial f}{\partial x_j}u(x_i, x_j) \tag{6.7}$$

Equation (6.7) is based on a first-order Taylor series approximation of the measurement equation $Y = f(X_1, X_2, \ldots, X_N)$ given in equation (6.1), and is referred to as the law of propagation of uncertainty (see previous chapter). The partial derivatives of f with respect to X_i (sometimes referred to as sensitivity coefficients) are equal to the partial derivatives of f with respect to X_i evaluated at $X_i = x_i$; and $u(x_i)$ is the standard uncertainty associated with the input estimate x_i; and $u(x_i, x_j)$ is the estimated covariance associated with x_i and x_j. Equation (6.7) is often reduced to a simple form; for example, if the input estimates x_i of the input quantities X_i can be assumed to be uncorrelated, then the second term vanishes. Further, if the input estimates are uncorrelated and the measurement equation is one of the following two forms, then equation (6.7) becomes simpler still.

1 Assuming a measurement equation to consist of a sum of quantities X_i multiplied by constants a_i; that is, $Y = a_1X_1 + a_2X_2 + \cdots + a_NX_N$, and with measurement result: $y = a_1x_1 + a_2x_2 + \cdots + a_Nx_N$, the combined standard uncertainty can be written as: $u_c^2(y) = a_1^2u^2(x_1) + a_2^2u^2(x_2) + \cdots a_N^2u^2(x_N)$.

2 If the measurement equation is a product of quantities X_i, raised to powers a, b, … p, multiplied by a constant A; that is, $Y = AX_1^{\,a}\, X_2^{\,b} \ldots X_N^{\,p}$, and with measurement result: $y = Ax_1^{\,a}\, x_2^{\,b} \ldots x_N^{\,p}$, the combined standard uncertainty can be written as: $u_{c,r}^2(y) = a^2u_r^2(x_1) + b^2u_r^2(x_2) + \cdots + p^2u_r^2(x_N)$.

Here $u_r(x_i)$ is the relative standard uncertainty of x_i and is defined by $u_r(x_i) = u(x_i)/|x_i|$, where $|x_i|$ is the absolute value of x_i and x_i is not equal to zero; and $u_{c,r}(y)$ is the relative combined standard uncertainty of y and is defined by $u_{c,r}(y) = u_c(y)/|y|$, where $|y|$ is the absolute value of y and y is not equal to zero.

If the probability distribution characterized by the measurement result y and its combined standard uncertainty $u_c(y)$ is approximately normal, and $u_c(y)$ is a reliable estimate of the standard deviation of y, then the interval $y + u_c(y)$ to $y - u_c(y)$ is expected to encompass approximately 68% of the distribution of values that could reasonably be attributed to the value of the quantity Y for which y is an estimate. This implies that we can state that with a level of confidence of 68%, Y is greater than or equal to $y - u_c(y)$, and is less than or equal to $y + u_c(y)$, or $Y = y \pm u_c(y)$.

6.6 Expanded uncertainty and coverage factor

Although the combined standard uncertainty u_c is used to express the uncertainty of many measurement results, for some commercial, industrial, and regulatory applications (e.g. when health and safety are concerned), what is often required is a measure of uncertainty that defines an interval about the measurement result y, within which the value of the measurand Y is believed to lie with a defined certainty. The measure of uncertainty intended to meet this requirement is termed expanded uncertainty, symbol U, and is obtained by multiplying $u_c(y)$ by a coverage factor, k. Thus $U = ku_c(y)$ and it is confidently believed that Y is greater than or equal to $y - U$, and is less than or equal to $y + U$, or $Y = y \pm U$.

In general, the value of the coverage factor k is chosen on the basis of the desired level of confidence to be associated with the interval defined by $U = ku_c$. Typically, k is in the range two to three. When the normal distribution applies and u_c is a reliable estimate of the standard deviation of y, $U = 2\,u_c$ (i.e. $k = 2$) defines an interval having a level of confidence of approximately 95%, and $U = 3\,u_c$ (i.e. $k = 3$) defines an interval having a level of confidence greater than 99%.

The GUM is an entire way of looking at, and treating measurement uncertainty, and represents over thirty years of research and development, and of international collaboration. The principles and concepts underlying the GUM include the following:

- A measurement model relating functionally one or more output quantities, about which we are seeking information, to input quantities, about which we possess information.
- Modelling our knowledge about the measurement of a quantity in terms of a probability distribution.
- Estimate the expectation and standard deviation (standard uncertainty) of a quantity characterized by a probability distribution.
- The use of new information to update an input probability density function: Bayes' Theorem.
- Assignment of a probability density function to a quantity using the Principle of Maximum (information) Entropy.
- Determination of the distribution for an output quantity (or the joint distribution for more than one output quantity) using the propagation of distributions.

Figure 6.1 summarizes the principles of the GUM methodology. The model equation provides the basis for the propagation of the probability density functions for the input quantities, and in the case of using Gaussian uncertainty propagation, for the propagation of their expectation values and associated uncertainties. Consequently, it is the modelling of the measurement process that is the key element of modern uncertainty evaluation, irrespective of the method used for calculating the uncertainty.

The GUM is the 'gold standard' of estimating uncertainty in measurements; particularly, measurements made at an international level and which have the backing of law; for example, defining and maintaining the world standard for mass using the data displayed in figure 4.2 in section 4.2 or to define the precision and uncertainty of international atomic time (UTC or *Temps universel coordonné*). An

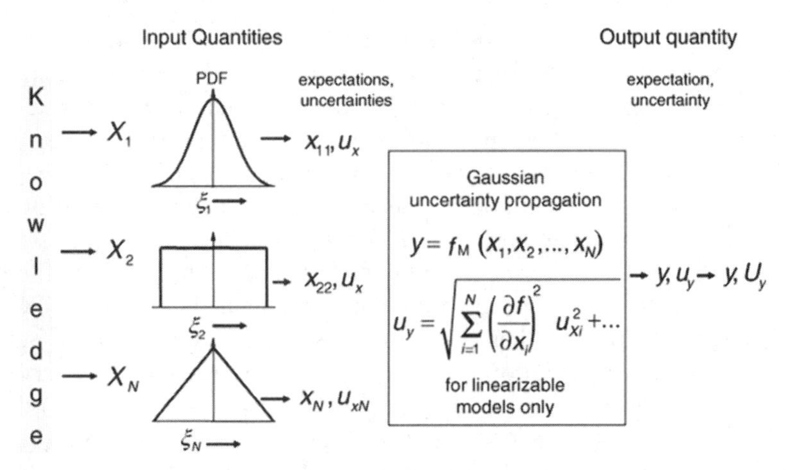

Figure 6.1. Illustration of the GUM procedure for the expression of measurement uncertainty. Symbols: Y is the measurand, $X_1 \ldots X_N$ are the input quantities, $y = E[Y]$ is the expectation of the probability density function for the measurand (taken as a best estimate), u_y is the standard uncertainty associated with y, $x_1 \ldots x_N$ are the expectations for the probability density functions for the input quantities, $u_{x1} \ldots u_{xN}$ are the standard uncertainties associated with $x_1 \ldots x_N$, U_y is the expanded measurement uncertainty. (Image taken from Sommer K D and Siebert B R L 2006 *Metrologia* **43** S200–S210; thanks to the BIPM for permission to reproduce this image.)

ordinary research student working away in a small university laboratory is not seeking such an exhaustive formulation for his or her estimation of uncertainty, but the general principles of the GUM are good practice and worth following. After all, every scientist is in some way a metrologist, we all make measurements and are trained to look at the world with an experimenter's eye; unless, of course, you prefer to do theory.

Further reading

There are many freely available Internet sources where documentation about the GUM and the application of the GUM to specific problems may be found; for example, the BIPM website www.bipm.org/en/publications/guides/gum.html provides access to a range of documents relating to the GUM (including the main document running to 134 pages including references), and the evolution of the GUM; there is also the NIST (National Institute of Standards and Technology) website at http://www.nist.gov where one may use the search facility to find documents relating to measurement uncertainty and to the GUM. In addition, at www.bipm.org/en/publications/guides/vim.html, you can also find the international vocabulary of metrology; and international and US perspectives on measurement uncertainty (http://www.physics.nist.gov/cuu/Uncertainty/international1.html) is also well worth investigating.

In addition, there are many entries on the online encyclopaedia, Wikipedia that are well worth reading; for example, the entry on standard deviation.

If the reader is interested in the GUM, I would also like to draw their attention to a special issue of the journal *Metrologia* (which is published by IOP Publishing on behalf of the BIPM) on Statistical and Probabilistic Methods for Metrology (volume 43, issue number **4**, 2006). This special issue contains a great many technical articles on the applications of the GUM; in particular, K D Sommer and B R L Siebert 2006 Systematic approach to the modelling of measurements for uncertainty evaluation *Metrologia* **43** S200.

Chapter 7

Clinical trials

Statistical thinking will one day be as necessary for efficient citizenship as the ability to read and write.

H G Wells, 1866–1946

7.1 Introduction

Clinical research includes investigating the efficacy and long-term benefit of a proposed treatment, assessing the relative benefits of competing therapies, and establishing optimal treatment combinations. Such research attempts to answer questions such as: should a man with prostate cancer undergo surgery or receive radiation, or should the clinicians merely wait watchfully? Then there are the questions about new drugs; for example, is the incidence of serious adverse effects among patients receiving a new pain-relieving therapy greater than the incidence of serious adverse effects in patients receiving the standard therapy, and how does a new anti-HIV molecule compare with those presently in use when used in combination with drugs A and B?

Before the widespread use of experimental trials, clinicians attempted to answer such questions by generalizing from their experiences with individual patients to the population at large. Clinical judgement and reasoning were applied to reports of interesting cases in the pursuit of new, best medical practice. The concept of variability among individuals and its sources may have been noted, but were not addressed formally.

In the 20th Century, the applications of statistics evolved rapidly and it began to be applied to clinical research. However, the statistical approach used in clinical research is different from that used in physics; in medicine, one is more interested in the theoretical science or formal study of the inferential process, especially the planning and analysis of experiments, surveys, and observational studies. Statistical

doi:10.1088/978-1-6817-4433-9ch7

methods provide formal accounting for variability in patients' responses to treatment; that is, statistics allow the clinicians to put on a firm quantitative basis the wide-range of responses seen in a collection of patients, when they are all given the same treatment. The use of statistics allows the clinical researcher to form reasonable and accurate inferences from collected information, and to make sound decisions in the presence of uncertainty.

Clinical and statistical reasoning are both crucial to progress in medicine. Clinical researchers must generalize from the few to the many, and combine empirical evidence with theory. In both medical and statistical sciences, empirical knowledge is generated from observations and data, and as we have seen, there is no data without uncertainty (and as we saw in chapter 4, there is generally a large uncertainty attached to automated measurements made in a hospital laboratory). Medical theory is based upon established biology and hypotheses. To establish a hypothesis requires both a theoretical basis in biology and statistical support for the hypothesis, based on the observed data and a theoretical statistical model.

An experiment may be thought of as a series of observations made under conditions controlled by a scientist. A clinical trial is an experiment testing medical treatments on human subjects. The clinical investigator controls factors that contribute to variability and bias such as the selection of subjects, application of the treatment, evaluation of outcome, and methods of analysis. The distinction of a clinical trial from other types of medical studies is the experimental nature of the trial, and the use of human trial subjects. The term clinical trial is preferred over clinical experiment, because the latter may suggest disrespect for the value of human life. But whatever it is called, it is a study that can last a long time.

The reason clinical trials are long is that clinicians are interested in the long-term (that is, over at least five years) benefits or otherwise of; for example, a new drug, together with the advent and severity of any associated side-effects. The trial may also be very large; that is, there may be over 2000 patients enrolled in the trial as one needs to investigate a significant, representative sample of humanity. Whereas all hydrogen atoms or all carbon dioxide molecules will behave in exactly the same way when stimulated in the same manner, the same is not true of humans. No two humans are alike in their biological response to medication, and men behave differently from women. So the most important question asked by those designing a clinical trial is about sample size. That is, how many individuals are needed to mask the genetic variability and individual uniqueness that has arisen through millions of years of evolution? This individual uniqueness can be considered to be the noise and uncertainty against which the clinicians are seeking to observe the effect they are after. No two people respond in the same way to the same size dose of drug X; so how many trial participants are required to average out natural genetic variability. After this, the question will be the percentage of men and women and the age groups to be studied; this is where the statistics is required, in trying to identify meaningful results in what could be a long, expensive trial/experiment.

Design is therefore the process or structure that isolates the factors of interest. Although the researcher designs a trial to control variability due to factors other than the treatment of interest, there is inherently larger variability in research involving humans than in a controlled laboratory situation. In many ways the design of a study is more important than the analysis. A badly designed trial can never be retrieved, whereas a poorly analysed one can usually be re-analyzed. Consideration of design is also important because the design of a study will govern how the data are to be analysed.

Most medical studies consider an input, which may be a medical intervention or exposure to a potentially toxic new drug, and an output, which is some measure of health that the intervention is supposed to affect. The data that will be analysed will resemble the data given in figure 4.3 in section 4.3 but for many individuals. The simplest way to categorize studies is with reference to the time sequence in which the input and output are studied. The most powerful studies are prospective studies where the paradigm is the randomized controlled trial. Here, subjects with a disease are randomized to one of two (or more) treatments, one of which may be a control treatment. The importance of randomization is that, in the long run, the clinicians know that treatment groups will be balanced in known and unknown prognostic factors. It is also important that the treatments are concurrent—that the active and control treatments occur in the same period of time.

One of the major threats to the validity of a clinical trial, and a major source of uncertainty, is compliance. Patients are likely to drop out of trials if the treatment is unpleasant, and often fail to take medication as prescribed. It is usual to adopt a pragmatic approach and analyse by intention to treat; that is, analyse the study by the treatment that the subject was assigned to, not the one they actually took.

7.2 Sample size

As mentioned earlier, this is the most important question in the design of a trial. So one of the most frequently asked questions put to a statistician about the design of the trial is the number of patients to be included. It is an important question, because if a study is too small it will not be able to answer the question posed, and would be a waste of effort. It could also be deemed unethical because patients may be put at risk with no apparent benefit. However, studies should not be too large because resources would be wasted if fewer patients would have sufficed. The sample size depends on four critical quantities: the type I and type II error rates α and β (see below), the variability of the data σ^2, and the effect size, d.[1] In a clinical trial, the effect size is the amount by which we would expect the two treatments to differ, or is the difference that would be clinically worthwhile.

Usually α and β are fixed at 5% and 20% (or 10%), respectively. A simple formula for a two group, parallel trial with a continuous outcome is that the required sample size per group is given by $n = 16\,\sigma^2/\mathrm{d}^2$ for each of the two sides, with α of 5% and β of 20%, respectively. For example, in a trial to reduce blood pressure, if a clinically worthwhile effect for diastolic blood pressure is a reduction of 5 mm Hg and

[1] The type I and type II error rates mentioned here have little to do with the Type A and Type B uncertainties we encountered earlier in this volume when looking at the analysis of measurement uncertainty. There is still a gulf between the terminology, vocabulary, procedures and definitions used by clinical statisticians and metrologists. These terms are also used in a more general way by social scientists to refer to flaws in reasoning.

the observed effect between subjects (the standard deviation) is 10 mm Hg; that is, 5 ± 5 mm Hg, and one can immediately see the difference between statistics in clinical trials and in physics experiments. This is a difference which comes essentially from the genetic variability of the subjects in the clinical trials. In such a trial, we would require $n = 16 \times 100/25 = 64$ patients per group to try and mask these natural variations, and lead to a useful result. The sample size goes up as the square of the standard deviation of the data (the variance) and goes down inversely as the square of the effect size. Doubling the effect size (looking for a bigger effect) reduces the sample size by a factor four, it is much easier to detect large effects. In practice, the sample size is often fixed by other criteria, such as finance or resources, and the formula is used to determine a realistic effect size.

7.3 Statistical hypothesis testing

As mentioned in the above footnote, the basis of the statistical approach to clinical research is different from the approach adopted in the GUM. The clinical researcher is interested in testing hypotheses and in analyzing the results of their trials to see if the hypothesis has been upheld or not. And so there is little connection between the Type A and Type B uncertainties of the GUM, and the type I and type II error of statistical hypothesis testing. Hypothesis testing does of course take place in physics and chemistry, but usually when a researcher is trying to observe or measure a phenomenon for the first time. The vocabularies used by clinical researchers and metrologists are different, yet they are both based on the same scientific principles, thereby demonstrating the ongoing need for rationalization of definitions and nomenclature.

In statistics, a null hypothesis is a statement that one seeks to nullify with evidence to the contrary. Most commonly it is a statement that the phenomenon being studied produces no effect or makes no difference; for example, 'this diet has no effect on people's weight'. Usually, an experimenter frames a null hypothesis with the intent of rejecting it; that is, intending to run an experiment which produces data that shows that the phenomenon under study does make a difference. In some cases there is a specific alternative hypothesis that is opposed to the null hypothesis, in other cases the alternative hypothesis is not explicitly stated, or is simply 'the null hypothesis is false'. In either event, this is a binary judgment, but the interpretation differs, and is a matter of dispute between the various schools of statistics.

A type I error, or error of the first kind is the incorrect rejection of a true null hypothesis. Usually a type I error leads one to conclude that a supposed effect or relationship exists when in fact it doesn't. Examples of type I errors include a test that shows a patient to have a disease when in fact the patient does not have the disease, a fire alarm going off indicating a fire when in fact there is no fire, or an experiment indicating that a medical treatment should cure a disease when in fact it does not.

A type II error, or error of the second kind is the failure to reject a false null hypothesis. Examples of type II errors would be a blood test failing to detect the disease it was designed to detect in a patient who has the disease; a fire breaking out and the fire alarm not ringing; or a clinical trial of a treatment failing to show that the treatment works when in fact it does.

In terms of false positives and false negatives, a positive result corresponds to rejecting the null hypothesis, while a negative result corresponds to failing to reject the null hypothesis. In these terms, a type I error is a false positive, and a type II error is a false negative.

When comparing two means of the results of two sets of data, and concluding that the means were different when in reality they were not different would be a type I error; concluding the means were not different when in reality they were different would be a type II error (see comments in section 4.3 about identifying real differences between successive routine blood measurements.

All statistical hypothesis tests have a probability of making type I and type II errors. For example, all blood tests for a disease will falsely detect the disease in some proportion of people who do not have the disease, and will fail to detect the disease in some proportion of people who do have it. There may be a perfectly good biological reason for this; perhaps, a set of rare genes produces a marker that triggers or nullifies the test in the same way that the vector of the disease would. This is a delicate situation requiring large groups in clinical trials to be tested fully. A test's probability of making a type I error is denoted by α. A test's probability of making a type II error is denoted by β. These error rates are traded off against each other; for any given sample set, the effort to reduce one type of error generally results in increasing the other type of error. For a given test, the only way to reduce both error rates is to increase the sample size, and this may not be feasible.

If we wished to test the hypothesis: a patient's symptoms improve after treatment A more rapidly than after a placebo treatment. The null hypothesis would be: a patient's symptoms after treatment A are indistinguishable from a placebo. The null hypothesis can be tested relatively straightforwardly, and if it is found to be the case that a patient's symptoms after treatment A are indistinguishable from a placebo (this is not as obvious as it may sound; remember the strange and unexplained 'placebo effect'), then one will need to investigate what exactly went on in the trial.

This kind of hypothesis error may happen due to staff failing to keep patients unaware of which treatment they're receiving, to uncontrolled variables, to difficulty tracking subjective symptoms like pain or for many other reasons. A type I error would falsely indicate that treatment A is substantially more effective than the placebo, whereas in a type II error the mistake is believing that treatment A has no effect. Statistical tests always involve a trade-off between:

- the acceptable level of false positives, and
- the acceptable level of false negatives.

The notions of false positives and false negatives is not limited to clinical research, but has a wide currency in the world of computers and security applications. Security and breeches of security are important considerations in keeping computer data safe, while maintaining access to that data for appropriate users. We need to consider:

- avoiding the type I errors (or false negatives) that classify authorized users as intruders, and

- avoiding the type II errors (or false positives) that classify imposters as authorized users.

False positives are routinely found every day in airport security screening, which are essentially visual inspection systems. The installed security alarms are intended to prevent weapons being brought onto aircraft, yet they are often set to such high sensitivity that the alarm goes off many times a day for minor items, such as keys, belt buckles, loose change, mobile phones, and even metal tacks stuck in the soles of shoes.

The ratio of false positives (identifying an innocent traveller as a terrorist) to true positives (detecting a would-be terrorist) is therefore high; and because almost every alarm is a false positive, the positive predictive value of these screening tests is low.

The relative cost of false results determines the likelihood that test creators allow these events to occur. As the cost of a false negative in this scenario is extremely high (not detecting a weapon or a bomb being brought onto a plane could result in hundreds of deaths) whilst the cost of a false positive is relatively low (further inspection) the most appropriate test is one with a low statistical specificity but high statistical sensitivity (one that allows a high rate of false positives in return for minimal false negatives).

Although it might be imagined that there is a close similarity between the theory of measurement uncertainty in the physical sciences and the statistical interpretation of clinical trials in medicine, the truth is that these two hugely important fields of research are far apart. They are far apart in nomenclature, and in the magnitude of the uncertainties being considered. And this difference in the magnitude of the uncertainties is the result of the inescapable uniqueness of individuals (their genetic variability), as opposed to the identical nature of all; for example, carbon dioxide molecules. In figure 4.1 in section 4.1, we see the behaviour of about 10^{24} molecules of carbon dioxide all doing the same thing when subjected to an applied electric-field gradient, whereas in figure 4.3, we see the data generated by a single person trying to survive.

It is the inevitable, large uncertainty in the biological behaviour of a group of individuals (the genetic variability, which is the result of millions of years of evolution) that does not permit a GUM or even a GUM-like approach to quantifying uncertainty in clinical trials. So, we have to use statistical inference, with its larger uncertainties, when we come to dealing with the behaviour (even the biochemical and biological responses) of groups of people. Consider the data in figure 4.3(*b*), where we have an uncertainty on each measurement of $\pm 50\%$; then imagine a statistical analysis of the results of a clinical trial involving many thousands of people being similarly treated. The uncertainties would become truly large, and perhaps meaninglessly large, as far as a physicist would be concerned, but such patients must still be treated and well treated. In physics, there is generally greater precision and smaller uncertainty on a measurement, and so sophisticated models of expressing that uncertainty may be adopted.

IOP Concise Physics

Quantifying Measurement

Jeffrey H Williams

Chapter 8

Direct measurements: quadrupole moments and stray light levels

To call in the statistician after the experiment is done may be no more than asking him to perform a post-mortem examination: he may be able to say what the experiment died of.

Ronald Aylmer Fisher, 1890–1962

8.1 Introduction

As we know, not all atoms are alike. Atoms such as fluorine and chlorine have a propensity, when covalently bonded with other atoms for attracting an additional electron into their structure and becoming slightly negatively-charged. This attractive pull for electrons is termed the atom's electronegativity (a heuristic quantity developed in chemistry by the American chemist and double Nobel Laureate, Linus Pauling, 1901–1994). Fluorine has the highest electronegativity and hydrogen has a low electronegativity; that is, a fluorine atom when covalently bonded to another atom will pull an electron from the other atom towards itself, but hydrogen will be less able to pull an electron to itself, but will allow its single electron to be pulled closer to the more electronegative atom to which it may be bonded.

What this electronegativity means is that if a hydrogen atom and a fluorine atom combine chemically to form hydrogen fluoride (HF), although the new molecule will not have an overall electrical positive- or negative-charge, there will be a slight separation of the charge-density within the molecule. Chemists talk about the fluorine end of the molecule being slightly more negatively charged than the hydrogen end of the molecule, which will be slightly more positively charged ($\delta+$ H-F $\delta-$). However, the overall electrical charge will be zero, so the molecule will be uncharged. A symmetric molecule such as hydrogen (H_2) has no such disproportionate or unbalanced electronegativity, because of the symmetry.

The consequence of such charge asymmetries in molecules is that there is a slight net attraction between any two such molecules. If one molecule has a slightly negatively-charged end and a slightly positively-charged end, then when two such molecules encounter each other there will be a slight electrostatic attraction with the slightly negative-end of one molecule seeking (for reasons of attractive electrostatic interaction) to be closer to the slightly positive-end of the other molecule. This type of weak electrostatic interaction is manifest in more energy being required to separate individual molecules from bulk quantities of those molecules; that is, the boiling temperature of HF is higher that the boiling temperatures of H_2 and of F_2.

If the charge distribution in a molecule is not perfectly spherical, as it is in an atom of helium, there will always be a small electrical moment in that molecule. These electrical moments are related to the shape or symmetry of the molecule, and as the molecule becomes increasingly symmetric; that is, on going from a linear shape to a tetrahedron and on to an octahedron and then to a sphere, the electrical moments become increasingly smaller and smaller; becoming zero in a sphere, the perfect Platonic body.

It was the Dutch-American physicist and Nobel Laureate, Peter Debye (1884–1966), who was the first to explain the nature and origin of the small electric dipole moments seen in molecules; that is, a separation of charge onto two centres in the molecule. Debye quantified the polarization in molecules and introduced the idea of a charge separation, defining in the process a new unit of measurement; one unit of electrical charge separated by one unit of length (that is, electric charge by distance). In addition, he developed a means of measuring the absolute value of molecular dipole moments, which is still in use today. The absolute value of the dipole moment is its size (or magnitude) and its sign (greater than zero or less than zero); if you know the sign of the dipole moment, you can determine which end is positively-charged and which end is negatively-charged.

The unit of distance Debye chose to work with when creating his unit for the dipole moment was the angstrom (a unit of length named after the Swedish physicist Anders Ångström; 1 Å equals 10^{-10} metre), because it was ideally suited to describing the size of the molecule; the bond between the hydrogen atom and the fluorine atom in hydrogen fluoride is about one angstrom in length. Debye defined a dipole moment as being one unit of electric charge separated by one angstrom; this convenient measure means that molecules have dipole moments of about one or two such units. This unit proved to be so useful that it was named, by Debye's colleagues, a Debye. One Debye (abbreviated as 1 D) is one unit of charge separated by one angstrom; in HF, the dipole moment is 1.86 D with the negative end on the fluorine atom.

In the HF molecule, the centres of gravity of the distribution of negative-charge and the centre of gravity of positive-charge do not coincide, as they would in a molecule of hydrogen (H_2). In such dipolar molecules, there are lines of force going from the positive-end of the molecule to the negative-end (this is a convention).

In the International System of Units (SI), a molecular dipole moment is defined in coulomb metres or C m (SI units of electric charge, the coulomb, and distance, the metre), and typical values are 3.33×10^{-30} C m, and so the SI value of the dipole moment of HF would be 6.19×10^{-30} C m. It is easier to tabulate, remember and discuss values of molecular properties which are, of order, one to five rather than something multiplied by 10^{-30}, which is the reason why there is still no single universal system of units in use by scientists; scientists tend to use whichever system of units is convenient for them.

In carbon dioxide, CO_2, which is linear with the carbon atom symmetrically placed between two more electronegative oxygen atoms, the two bond dipole moments are opposed (pointing in opposite directions) and the net dipole moment of the molecule is zero. However, the molecule is better thought of as being oval in shape and not infinitely narrow (that is, thought of as being 3-dimensional rather than 1-dimensional). In this way, we see that the CO_2 molecule is truly like a fat cigar with a finite width, and so even if the net dipole moment is zero, there is still an asymmetry in the charge distribution—there is a lot of charge-density on the long axis and some charge-density perpendicular to the long axis. This geometry defines an electric quadrupole moment. That is, the charge asymmetry in the molecule may be thought of as residing on four centres; at each end and at the middle, but off axis. So here we have a charge distributed over an area; that is, charge by distance by distance or in the SI system, coulomb metres × metres or C m^2, with values typically 3.33×10^{-40} C m^2.

The first direct method for measuring the absolute value (sign and magnitude) for a molecular quadrupole moment was devised in 1959 by A David Buckingham (born 1930) [1]. The technique suggested by Buckingham proved to be successful and Peter Debye suggested in, *Chemical and Engineering News*, the news magazine of the American Chemical Society [2] that the unit of the quadrupole moment be called the Buckingham (that is, one unit of charge distributed over one square angstrom be called one Buckingham), so the quadrupole moment of CO_2 would be about 4.5 Buckinghams or 15×10^{-40} C m^2 in the SI.[1]

After the quadrupole moment, the next highest electrical moment of a molecule is the octupole moment, which would be one unit of electric charge spread throughout a volume or one unit of electric charge in a cubic angstrom or about 3.33×10^{-50} C m^3 in the SI. (We are now starting to become seriously small.) Molecules possessing an octupole moment, arising from the asymmetry in the distribution of electronic charge, are tetrahedral in shape; for example, methane or carbon tetrachloride. The next highest electrical moment would the hexadecapole moment; such a molecule would be an octahedron, as in sulphur hexafluoride. These higher moments are very small, compared with the dipole moment of a small molecule such as water; they represent ever-smaller departures from a symmetric distribution of charge.

[1] There is an amusing corollary to this story of the naming of units. The author was a research student of Professor Buckingham, and one day, long-ago we were discussing this story of how the unit of the quadrupole moment came to be called the Buckingham (my thesis was essentially about the measurement of this molecular property), and my supervisor said to me with a smile, 'if you work out a way of measuring the octupole moment, then I will suggest that the unit be called the Williams'. To date, there is still no method for the determination of the sign and magnitude of the octupole moment.

8.2 Measuring the quadrupole moments of molecules

The experiment for measuring the molecular quadrupole moment of a molecule such as CO_2 or benzene is due to Buckingham [1]. Essentially, a large well-defined electric-field gradient is applied to a sample of the gas or vapour, of known number density, and a polarized light (laser) beam is passed through the medium in the presence of the applied electric-field gradient. If the laser light is initially plane-polarized, it will become elliptically-polarized on passage through the birefringent medium arising from the molecules oriented by the applied electric-field gradient.

This technique of measurement arises from the 19th Century observations of the Kerr effect in Glasgow in the 1870s. Under the influence of James Clerk Maxwell, John Kerr (who was a theology student) was seeking to observe the effect of an applied electric field on a beam of light. To undertake this experiment he needed to pass a beam of polarized light between the charged electrodes that generated the electric field. Of course, what he was also doing was observing the effect of the applied field on the polarization of the light beam as it passed through the medium (a dense, optically-pure lead glass) placed between the electrodes. What Kerr observed was that the initially linearly-polarized beam of light became elliptically polarized on passing through a medium subject to the applied electric field. That is, the electric field caused the medium under study to behave as a birefringent or uniaxial crystal such as calcite.

In the Kerr effect, there is no rotation of the plane of polarization of the light upon propagation through the material in the applied electric field, which would be dichroism, instead an additional plane of polarization is generated perpendicular to that of the incident light. That is, the material behaves as a birefringent or uniaxial crystal under the influence of the electric field; there are now two routes for the light to propagate through the glass.

The Kerr effect is a means of imposing order on a collection of molecules (just as the magnetic field of a permanent magnet will impose order on a sample of iron filings). The applied electric field imposes an ordered, crystal-like structure on an initially structure-less sample. In a liquid there is no permanent structural order, the molecules are tumbling over each other while keeping a constant separation, and the higher the temperature, the greater the tumbling motion and the degree of disorder. If an electric field is applied to a liquid, which possesses a large Kerr effect, the molecules become ordered by the interaction of the applied electric field and the electrical properties (in particular, the electric dipole moment) of the molecules; the molecules will line-up following the applied electric force-field. However, as this ordering effect is working against the effect of temperature, which is seeking to randomize any order, the Kerr effect is temperature-dependent; the lower the temperature, the bigger the observed Kerr effect. We may write the measured effect as being proportional to $(\mu E/k_B T)$, where E is the applied electric field, T is the sample temperature, μ is the molecule's dipole moment and k_B is the Boltzmann constant; both (μE) and $(k_B T)$ have units of energy. The Kerr effect is a means of generating complexity or order out of the chaos of the thermally driven motion in a sample of randomly oriented molecules (a gas, a liquid or a disordered glass).

The Kerr effect is an example of induced birefringence, the generation of a material which is capable of supporting two routes for the propagation of light beams. The best example in nature of such birefringence is calcite. Whereas in the Kerr effect, the applied electric field generates a partial, temperature-dependent ordering of the molecules, in calcite the ions in the crystal structure are perfectly regimented or ordered within the lattice. In particular, it is the optically anisotropic (planar) carbonate anions (CO_3^{2-}) that give to the crystal a birefringence of about 3° (that is, the difference between the parallel and perpendicular components of the refractive index are separated by 3°, or $n_{\text{parallel}} - n_{\text{perpendicular}} \sim 0.1$ radian), and allows the crystal to demonstrate a large birefringence.

If one views a pencil line through a crystal of calcite, the line appears twice because light can take two paths through the crystal (parallel or perpendicular to the oriented flat carbonate anions in the solid structure). The natural birefringence of the calcite crystal is huge and demonstrates the degree of internal structure and order of the crystal. The Kerr effect induced by application of a laboratory electric field to a sample of gas is much more modest. Whereas the natural ordering of carbonate anions in calcite generates a birefringence of 3°, the largest laboratory electric fields generate a birefringence or imposed order in gas samples of about 10^{-5} degrees. And that imposed order exists only as long as the electric field is applied; the crystal on the other hand can be chipped from a mountain and is naturally ordered.

In experiments involving the Kerr effect, the degree of ordering induced in the sample of material by the applied electric field is measured, and this allows the determination of the magnitude of the molecular dipole moment. Molecules that have a large electric dipole moment (for example, nitrobenzene and liquid crystals) have very large, easily measureable Kerr effects. In these experiments, if the measurement is made at a number of temperatures one can separate the temperature-dependent term that contribute to the measurement from the temperature-independent term. The temperature-independent term arises because the applied field distorts the molecules as well as orienting them.

Indeed, in experiments such as the Kerr effect or the quadrupole measurement, which we will consider below, one is ordering the molecules of interest by application of a uniform electric field or by application of a field gradient, respectively, and then one probes the degree of inducted orientation. This is possible because the molecules under study are anisotropically polarizable; that is, there is a difference in the degree to which the molecule is susceptible to an optical electric field (the laser beam) applied parallel or perpendicular to the main rotational axis of the molecule.

8.3 Experimental details

We will now look at the details of this experiment; in particular, at those elements that allow us to estimate the limits of the precision of the experiment for measuring electric quadrupole moments. The measurement of molecular quadrupole moments with the apparatus described here is typical of a direct measurement; that is, an experimenter measures an output parameter, and is able to calculate from this single quantity the property of interest. To extract the information of interest, there is no need for the measured data to be fitted to a computer model.

Although the experiment to measure quadrupole moments is relatively straightforward, in practical terms it is in fact quite complex. It can be thought of as several ‘black boxes’, each of which is an experiment in its own right, and each of these individual experiments must be working optimally, and they must all be working coherently to allow a measurement of a molecular quadrupole moment. The experimenter induces an effect by applying an electric-field gradient to a sample of gas, and by applying an out-of-phase birefringence signal from a calibrated source (a nulling Kerr effect cell) in the optical train after the quadrupole cell; thus measuring directly the induced effect, which is the product of the applied electric-field gradient (a known quantity) and the molecular quadrupole moment (the unknown quantity). The output signal is monitored and after applying known constants, you have a direct measurement of the property of interest; that is, there is no lengthy computer-based data fitting. All this is a matter of care and patience on the part of the experimenter, and a different experimenter might be more careful or be less patient, characteristics that will contribute to the Type B measurement uncertainty (see section 5.3).

The experimental set-up is shown schematically in figure 8.1. The light from a He–Ne laser (red light at 632.8 nm) is polarized by passage through a Glan–Thompson calcite polarizer. This linearly-polarized light then propagates the length of the 1 m quadrupole cell containing the gas or vapour subject to a stable electric-field gradient (this is generated by two wires at the same high electrical potential on either side of the laser beam inside the Earthed quadrupole cell. At the centre of the laser beam, the electric field arising from the charged wires is zero, but the field gradient is a maximum.) As this linearly-polarized light propagates through the molecules oriented by the applied electric-field gradient, it becomes elliptically-polarized, as in the Kerr effect (see images (i) and (ii) in part (b) of figure 8.1). After the quadrupole cell, there is a quarter-wave plate ($\lambda/4$), which is used to facilitate measurement, and a calibrated Kerr cell that generates an out-of-phase Kerr effect in a calibrated (liquid CS_2) Kerr cell, which is used to null the ellipticity generated in the quadrupole cell. After this Kerr cell, there is the analyzing Glan–Thompson calcite polarizer and the photomultiplier detector.

With the laboratory z axis along the light path down the axis of the quadrupole cell, the x axis is chosen such that relative to it the azimuth of the linearly-polarized laser beam emerging from the first polarizer is precisely $\pi/4$. The azimuth of this linear polarization is chosen as the reference from which the azimuths of the other optical elements are measured. The quadrupole cell is then rotated about its axis until the plane of its two wires coincides with the yz plane.

The electric field of the light beam entering the quadrupole cell may be resolved into components ε_x and ε_y, which experience refractive indices n_x and n_y, respectively, as the beam propagates through the metre-long cell. If the path-length of the beam in the applied field gradient is l, the two field components will emerge

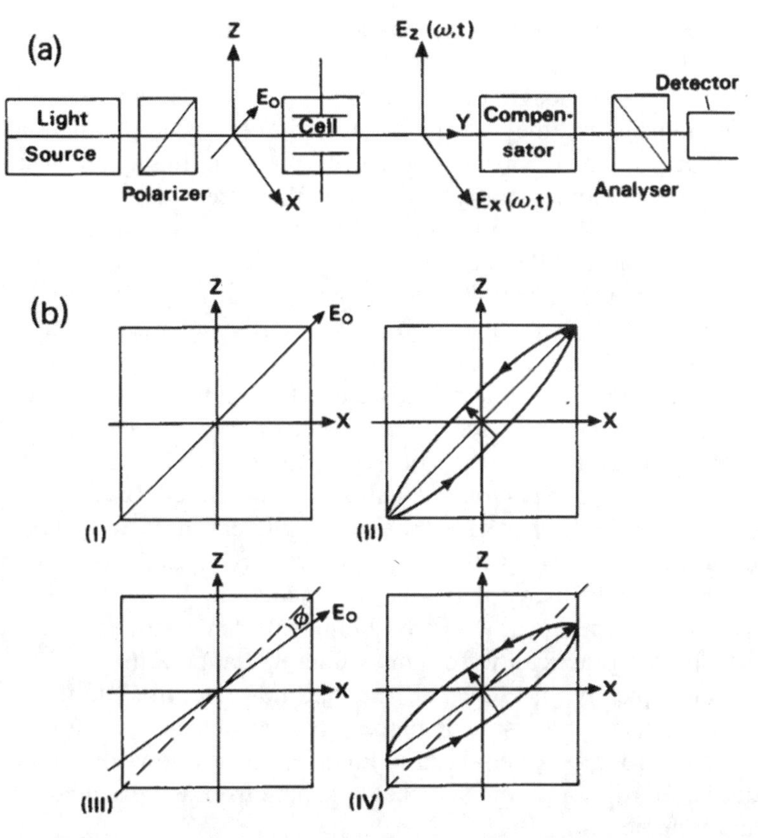

Figure 8.1. The basic components of an apparatus to measure an induced birefringence are shown in the upper part of this image. The birefringence is induced in a fluid in the cell, by application of an electric field, an electric-field gradient or by an applied magnetic field. The light source is a laser, the polarizer and analyzer are high-quality Glan–Thomson prisms. The lower part of this figure shows the polarization states in such electro-optic experiments: (i) linearly-polarized light propagating away from the observer with its electric vector at 45° or 135° to the x and z axes; (ii) elliptically-polarized light generated by the positive value of $(n_{\text{parallel}} - n_{\text{perpendicular}})$, the induced birefringence; (iii) positive rotation, of magnitude φ, of the plane of polarization produced by a positive value of $(k_{\text{parallel}} - k_{\text{perpendicular}})$, an induced dichroism; (iv) elliptically-polarized light, with its major axis rotated, produced by positive values of both $(n_{\text{parallel}} - n_{\text{perpendicular}})$ and $(k_{\text{parallel}} - k_{\text{perpendicular}})$. Images taken from Williams J H 1993 Aspects of the optical Kerr effect and Cotton–Mouton effect of solutions *Advances in Chemical Physics: Modern Nonlinear Optics* Part 2 vol 85 ed M Evans and S Kielich (Hoboken, NJ: Wiley).

from the quadrupole cell with a relative phase-difference δ given by (see images (i) and (ii) in part (*b*) of figure 8.1):

$$\delta = 2\pi\, l(n_x - n_y)/\lambda. \tag{8.1}$$

The light beam exiting the cell containing the oriented quadrupolar molecules is elliptically-polarized. But after passage through the quarter-wave plate, with its fast axis set at an azimuth of $\pi/4$, the beam emerges in a state of linear polarization with an azimuth of $\pi/4 + \delta/2$. The consequence of the combination of the quadrupole cell and the quarter-wave plate is to rotate the plane of polarization of the laser beam by δ/2 (as shown in image (iii) in part (*b*) of figure 8.1).

Through choice of the length of the quadrupole cell, the magnitude of the applied electric-field gradient, and the pressure of gas in the cell, the apparatus was designed to measure induced retardations, δ, of order, 10^{-7} radian (this is the precision of the system, and the level of uncertainty arising from the system many be seen by looking at the data given in figure 4.1 in section 4.1). This means that we are seeking to measure a field-induced difference in the refractive index of the gas, $(n_x - n_y)$, of order, 10^{-16} (see equation (8.1)). By using a blue laser instead of the red laser, we see from equation (8.1) that we could increase the sensitivity of this measurement, as there would be more waves per centimetre.

There are very few phenomena in nature that can be measured to such a precision (the definitions of the second and the metre are the only others that come to mind). However, the magnitude of δ, the retardation induced by the applied electric-field gradient, means that it cannot be measured by extinction at the analyzing polarizer (but one can see that its size is linear in the length of the cell; so the longer the cell, the bigger will be the measured effect. Likewise, the shorter the wavelength of the light used for the measurements, the bigger the effect. So, the best design for this experiment would be a long path-length for the quadrupole cell (size limited by the practicalities of controlling its temperature uniformly) and a blue laser; δ is typically, of order, 10^{-6} radian, however, modulating the applied electric-field gradient and using a phase-sensitive detector to discriminate the signal of interest from the general (un-modulated) noise allows one to measure the small induced retardation in the laser beam arising from the oriented molecules in the 1 m quadrupole cell.

For a sample of molecules such as carbon dioxide, the measured induced birefringence, $n_x - n_y$ is related to the properties of the individual molecules through the expression (first derived by Buckingham)

$$n_x - n_y = \{NE'/15\varepsilon_0\}\{(15\,\mathrm{B}/2) + (\Theta\Delta\alpha/k_\mathrm{B}T)\} \tag{8.2}$$

where N is the number density of molecules, E' the applied electric-field gradient, T the sample temperature and k_B is the Boltzmann constant, and ε_0 is the permittivity of free space (the degree to which an electric field may propagate). The first term in the second set of curly brackets gives the temperature-independent distortion of the molecule by the applied electric-field gradient, and the temperature-dependent term, $\Theta\Delta\alpha$, consists of the molecular quadrupole moment Θ and the anisotropic part of the polarizability of the molecules, $\Delta\alpha$. If the molecules have an axis of rotation C_n where $n \geqslant 3$, then $\Delta\alpha$ can be written as $\alpha_\mathrm{parallel} - \alpha_\mathrm{perpendicular}$; that is, the difference of the molecular polarizability parallel and perpendicular to the main rotational axis of the molecule.

We see from equation (8.1) and equation (8.2) how a series of measurements of the inducted birefringence, as a function of sample temperature will allow the separation and measurement of the two terms in the second set of curly brackets (this is the data displayed in figure 4.1 in section 4.1). However, to determine the sign and magnitude of Θ, one must know the sign and magnitude of $\Delta\alpha$; but the anisotropy of the molecule's polarizability is available from light scattering experiments (measuring the depolarization ratio). However, care must be taken as the errors in the measurement of $\Delta\alpha$ will carry through to the determination of Θ.

What has been outlined above is an ideal experiment, the perfect experiment. But experiments are performed in the real world, and are always imperfect. For example, the polarizers are almost crossed in the experiment so as to limit the total light level at the detector; one is looking for a small signal against a large background of noise, so any effect that increases stray light levels at the detector is to be avoided. On this point, we must consider the optical properties of the windows of the quadrupole cell and the Kerr cell to the induced birefringence. The strain birefringence induced by the window will not be modulated at the frequency of the applied field gradient in the quadrupole well, which is also the frequency of modulation of the nulling Kerr effect, but it will allow extra light to reach the detector, and so contribute to the final signal to noise problem of the measurement.

So the question becomes, is there a means of investigating in detail the contribution of the various optical components in the train of the experiment to the final measurement? That is, how do you quantify the contributions of the various optical elements to the measurement so as to maximize the measurement capacity of the experiment, by reducing the measurement uncertainty (that is, to increase the precision of the apparatus).

To determine the influence of each component on the polarization state of the light beam we can use Jones calculus. Piazza *et al* [3] have shown that the Jones matrices for certain optical components may be expressed as linear combinations of the unit and Pauli matrices

$$\mathbf{I} = \begin{pmatrix} 1 & 0 \\ 0 & 1 \end{pmatrix}, \qquad \mathbf{i} = \begin{pmatrix} i & 0 \\ 0 & -i \end{pmatrix}, \qquad \mathbf{j} = \begin{pmatrix} 0 & 1 \\ -1 & 0 \end{pmatrix}, \qquad \mathbf{k} = \begin{pmatrix} 0 & i \\ i & 0 \end{pmatrix}, \tag{8.3}$$

which combine in the following ways:

$$\mathbf{i}^2 = \mathbf{j}^2 = \mathbf{k}^2 = -1$$

$$\mathbf{ij} = \mathbf{k}, \qquad \mathbf{jk} = \mathbf{i}, \qquad \mathbf{ki} = \mathbf{j}. \tag{8.4}$$

Then the Jones matrix for a linear retarder (phase retarders introduce a phase shift between the vertical and horizontal component of the field and thus change the polarization of the beam) of retardance ρ and azimuth ϕ is given by

$$\mathbf{J}(\rho, \phi) = \cos(\rho/2)\mathbf{I} + \sin(\rho/2)\cos 2\phi\mathbf{i} + \sin(\rho/2)\sin 2\phi\mathbf{k}, \tag{8.5}$$

and that for an optical rotator having a rotation ψ is given by

$$\mathbf{R}(\psi) = \cos\psi\mathbf{I} + \sin\psi\mathbf{j}. \tag{8.6}$$

The Jones matrix for a polarizer of azimuth σ has the form

$$\mathbf{P}(\sigma) = \begin{pmatrix} \frac{1}{2}(1 + \cos 2\sigma) & \cos\sigma \sin\sigma \\ \cos\sigma \sin\sigma & \frac{1}{2}(1 - \cos 2\sigma) \end{pmatrix} \tag{8.7}$$

but which cannot be expressed in terms of the **I, i, j** and **k** matrices. The normalized Jones vector for a linearly polarized light beam of azimuth η is

$$\nu(\eta) = \begin{pmatrix} \cos\eta \\ \sin\eta \end{pmatrix}. \tag{8.8}$$

Suppose that the polarization azimuth of the light entering the quadrupole cell is $\pi/4$, and that the analyzing polarizer is offset from its exact crossed position by a small angle α to simulate addition light reaching the detector against which one is attempting to make a measurement. Then the Jones vector for the light leaving the analyzer is given by

$$\nu = \mathbf{P}\left(-\frac{\pi}{4} + \alpha\right)\mathbf{M}_n\mathbf{M}_{n-1}\ldots\mathbf{M}_1\,\nu_0, \tag{8.9}$$

where $\mathbf{M}_1$, $\mathbf{M}_2$... $\mathbf{M}_n$ are the Jones matrices for the retarders and the rotators in the optical train. Since the **I, i, j, k** matrices form a closed group under matrix multiplication, it follows that the the product $\mathbf{M}_n\mathbf{M}_{n-1}\ldots\mathbf{M}_1$ may also be expressed as a linear combination of these matrices,

$$\mathbf{M}_n\mathbf{M}_{n-1}\ldots\mathbf{M}_1 = a\mathbf{I} + b\mathbf{i} + c\mathbf{j} + \mathrm{d}\mathbf{k}. \tag{8.10}$$

If I_0 is the intensity of the light beam after passing through the first polarizer, then the intensity of the light beam at the detecting photomultiplier tube is given by

$$I = \nu^*\,\nu I_0,$$

which together with equations (8.7)–(8.10) yields

$$(I/I_0) = b^2 + c^2 + 2\alpha(ac + bd) + \alpha^2\left(a^2 - b^2 - c^2 + d^2\right),$$

in which we have retained terms to order α^2. Thus, to answer our question about how to maximize the measurement signal, we now have to find the various coefficients a, b, c and d for a given optical train.

The optical train of this experiment to measure molecular quadrupole moments is given in figure 8.1, we will now consider the addition of two additional optical components, the two windows of the quadrupole cell (these windows are typically of very dense glass and about 5 mm in thickness) and how they contribute to the measurement.

Consider the entry window of the quadrupole cell as a linear retarder of small retardance β_1 and arbitrary azimuth θ_1:

$$\mathbf{S}_1\left(\beta_1, \theta_1\right) = \cos\left(\beta_1/2\right)\mathbf{I} + \sin\left(\beta_1/2\right)\cos 2\theta_1\,\mathbf{i} + \sin\left(\beta_1/2\right)\sin 2\theta_1\,\mathbf{k}.$$

A similar expression could be written for the exit window of the quadrupole cell. Ideally, the quadrupole cell behaves as a linear retarder with retardance δ and zero azimuth. However, the cell was rotated into its experimental position, so one could allow for a small residual error in the azimuth, γ. The Jones matrix for the quadrupole cell is then:

$$\mathbf{J}q(\delta, \gamma) = \cos(\delta/2)\,\mathbf{I} + \sin(\delta/2)\cos 2\gamma\,\mathbf{i} + \sin(\delta/2)\sin 2\gamma\,\mathbf{k}.$$

So, the possible contributions of the extra, but essential optical elements in the optical train to the final measurement may be isolated and examined.

In this way, the Jones matrices for every component in the optical train may be combined using equation (8.10), and an expression for the final intensity of light reaching the detector be generated. In practice this is a long and tedious calculation, because of the many unknown quantities that appear. Making only small angle approximations and retaining terms to a certain order does simplify the algebra, but there remain many unknowns.

$$(I/I_0) = \delta\left\{C_1 - C_2 + \varepsilon\left[\mp\cos\phi + 2S_2(1 + \sin\phi)\right] + \alpha(\pm\cos\phi - 2S_2\sin\phi)\right\}$$
$$+\,\theta\left[-2(C_1S_1 + C_2S_2 + 2C_1S_2)\sin\phi \pm 2(C_1 + C_2)\cos\phi - 2\varepsilon(1 + \sin\phi) + 2\alpha\right].$$

Here

$$S_n = (\beta_n/2)\sin 2\theta_n, \quad \text{and} \quad C_n = (\beta_n/2)\cos 2\theta_n,$$

and ε and ϕ arises from the manner in which one models the quarter-wave plate. The value of θ equal to $\delta/2$ is termed θ_{null}, as it is this rotation which, for ideal optical components in perfect orientation, is required to null the birefringence generated in the quadrupole cell. This induced retardation may be determined by writing a computer programme to sample the signal coming from the phase-sensitive detector as a voltage is applied to the nulling, calibrated Kerr cell.[2] From the above equations it follows that, for a small value of ϕ, these lines will intersect at a value of θ, θ_{int}, given by

$$\theta_{\text{int}} = -(1/2)\delta(\pm 1 - 2S_2/1 + \phi).$$

Such an analysis allows one to determine the magnitude of the contribution of the windows to the measured effect, and if a systematic study of the contributions of all of the windows were made, at differing orientations relative to the direction of polarization of the incident light beam (the windows might for example possess an intrinsic strain, which could induce an ellipticity in linearly-polarized light), one may seek to minimize the contribution of the widow to the measured effect. With such a calculus, one could also investigate the most appropriate composition of the windows (quartz, hard glass or soft glass, etc) to maximize the precision of the apparatus. However, pressure resistant windows are essential.[3]

---

[2] Graham *et al* [4] have performed the most detailed analysis of the contributions of the various optical elements in this experiment to the overall noise of the experiment. For example, they found θ_{null} from the intersection of two graphs of I/I_0 versus θ corresponding to two different offsets of the quarter-wave plate.

[3] In the late-1970s when I was using such an apparatus, there were no laptop computers, and the data coming from the lock-in detector went to a buffer and was then punched onto paper tape, which was then read into the university mainframe for subsequent analysis.

Both the quadrupole moment determination and the measurement of the Kerr effect (and the optical Kerr effect that we will encounter in chapter 9) are experiments that are undertaken between crossed polarizers. In such situations, the induced retardation δ arising from: the electric field in the Kerr effect, the electric-field gradient in the quadrupole effect, the applied optical electric field in the optical Kerr effect, or an applied magnetic field as in the Cotton–Mouton effect, is determined by placing the birefringent element between crossed polarizers (see figure 8.1) with its birefringent axis oriented at 45° to the polarization direction. The intensity of the light transmitted by the analyzing polarizer is, for small δ

$$I = I_0 \sin^2(\delta/2) \sim I_0(\delta/4).$$

where I_0 is the incident light intensity; typically, δ is of order 10^{-6} radian and, because of the presence of stray birefringence of order 10^{-4} radian, which may be quantified and identified using the Jones calculus described above, a heterodyne method (modulation of the applied field and phase-sensitive detection of the output) is usually adopted for measuring the small inducted retardation. If both a small modulated birefringence $\delta(\omega)\sin\omega t$, and a large static (dc) birefringence δ_0 are placed between the crossed polarizers, then at the detector

$$I = I_0\left[\sin^2(\delta_0/2) + (1/2)\sin\delta_0\delta(\omega)\sin\omega t + \text{order}\left(\delta(\omega)\right)^2\right].$$

A phase-sensitive detector measures the component of I oscillating at the frequency ω. This term may be made larger by increasing the magnitude of δ_0. However, the noise level against which you are seeking to measure $\delta(\omega)$ will also become larger as it will be determined by the total light intensity passing the analyzing polarizer, and hence will be strongly dependent on δ_0^2. As mentioned above, in practice, δ_0 is introduced by a linear retarder and may be, of order, 0.1 radian.

At the detector (a red-sensitive photomultiplier tube), the average number of photons counted in a time interval Δt is given by $N = IQ\Delta t$, where Q is the quantum efficiency of the detector. The number of photons detected in this time interval will be randomly distributed about a mean with a standard deviation $\Delta N = \sqrt{N}$. These statistical fluctuations are called shot noise and determine the sensitivity of the detection system for a particular measurement. For a typical induced birefringence apparatus, $\delta(\omega) \ll \delta_0$, and the signal to noise ratio is:

$$\frac{S}{N} = (1/2)QI_0\Delta t\sin\delta_0\delta(\omega)/\sin(\delta_0/2)\ \sqrt{QI_0\Delta t} = \sqrt{QI_0\Delta t}\cos(\delta_0/2)\delta(\omega).$$

Hence, δ_0 should be chosen large in relation to the stray birefringence but small compared to 1 radian; the signal to noise ratio is then independent of δ_0. For example, a He–Ne laser emitting about 10^{16} photons per second at 632.8 nm, assuming $Q = 0.1$ and $\Delta t = 1$ second, we have $\delta(\omega) = 3 \times 10^{-8}$ radian as the induced retardation giving a signal/noise of unity. But we see from equation (8.1) that this sensitivity could be increased by using a longer quadrupole cell and by using a blue laser (which would increase the sensitivity in the ratio 632.8 nm/458 nm).

We have looked in some detail at how one would make a measurement of the molecular quadrupole moment of a molecule. This is what I term a direct measurement. Here one constructs an apparatus, it can be quite a complex apparatus, but one measures only one thing. In the quadrupole moment experiment, one is only interested in the magnitude and sign of the signal coming from the phase-sensitive detector, which is analyzing the signal coming from the photomultiplier tube, which is measuring the amount of light passing the analyzing polarizer. There is only one measurement of interest. As mentioned, the experiment can be repeated at a number of sample temperatures to separate the temperature-dependent and the temperature-independent terms that contribute to the induced birefringence, and which both represent information about the electronic structure of the molecule under investigation (see figure 4.1 in section 4.1). But even if a series of temperature measurements are made, each one will only be a measurement of the one single observable, the voltage coming from the phase-sensitive detector.

The observer measures one number, the retardation arising from the applied electric-field gradient, and then multiplies it by a number of other parameters (see equation (8.2)), the sample temperature, the number density of molecules in the quadrupole cell, the polarizability anisotropy of the molecule, the value of the electric-field gradient arising from the potential applied to the two wires inside the Earthed metal cylinder, all of which will possess an intrinsic measurement uncertainty. Then he or she has a value for the quadrupole moment of the molecule under study. It is a classic experiment; you measure one observable to determine one unknown. We have seen that Jones calculus may be used to analyze the contributions of the various optical components to the noise in the experiment, and how one is thus able to modify or discriminate against such sources of unwanted noise. After that, one attempts to make each component in the apparatus as reliable and precise as possible, and one then makes measurements, seeking reliability in a series of measurements, which may then be subject to some statistical analysis.

This final statistical analysis need not be complex and laborious, because analysis (as outlined above) has shown that the apparatus is not designed to readily measure induced retardations, of order, 10^{-8} radian or smaller. One may make such measurements, but you know from your analysis of the apparatus that such measurements will be subject to large amounts of random noise, and so will never be very reproducible and will, consequently, be very uncertain, and so do not warrant a great investment of time.

8.4 How many measurements do you need?

Before one gets into the details of how to analyze the data one has measured, it is as well to look briefly at how many measurements are needed.

8.4.1 Over-determined system

In mathematics, a system of equations is considered overdetermined if there are more equations than there are unknowns. An over-determined system is almost always inconsistent (that is, it has no solution) when constructed with random coefficients. However, an over-determined system will have solutions under some circumstances; for example, if some equation occurs several times in the system, or if some equations are linear combinations of the other equations. For the experimenter, this set of abstract equations becomes the set of equations that define what it is that one is measuring (one of the unknown coefficients) and the known experimental conditions and parameters (the known coefficients).

Each unknown in the set of equations can be thought of as an available degree of freedom. Each equation introduced into the ensemble can be viewed as a constraint that restricts one degree of freedom. However, for every variable giving a degree of freedom, there exists a corresponding constraint. The over-determined case occurs when the system has been over-constrained; that is, when the equations outnumber the unknowns. In contrast, the under-determined case occurs when the system has been under-constrained; that is, when the number of equations is smaller than the number of unknowns.

Consider the system of three equations and two unknowns (x_1 and x_2), which is overdetermined: $2x_1 + x_2 = -1$, $-3x_1 + x_2 = -2$, $-x_1 + x_2 = 1$. There is one solution for each pair of linear equations: for the first and second equations (0.2, −1.4), for the first and third (−2/3, 1/3), and for the second and third (1.5, 2.5). However there is no solution that satisfies all three simultaneously. If one were to graph these equations, one would observe configurations that are inconsistent, because no point is on all three graphed lines.

The only cases where the over-determined system does in fact have a solution are: one equation is linearly dependent on the others and there is an intersection, the three lines intersect at the same point, and when the three lines are superimposed (an infinity of intersections). These exceptions can occur only when the over-determined system contains enough linearly dependent equations that the number of independent equations does not exceed the number of unknowns. Here, linear dependence means that some equations can be obtained from linearly combining other equations. For example, $y = x + 1$ and $2y = 2x + 2$ are linearly dependent equations because of the constant of multiplication. There is no, one measurement that will satisfy the three equations and yield all the unknowns. A series of experiments is required. In the quadrupole moment experiment, one could write down a set of equations governing the various 'black box' components of the overall experiment, but these equations would to a large degree be dependent upon each other.

8.4.2 Under-determined system

In mathematics, a system of linear equations or a system of polynomial equations is considered under-determined if there are fewer equations than unknowns. Each unknown can be seen as an available degree of freedom. Each equation introduced into the system can be viewed as a constraint that restricts one degree of freedom. Therefore, the critical case (between over-determined and under-determined) occurs when the number of equations and the number of free variables are equal. For every variable giving a degree of freedom, there exists a corresponding constraint removing a degree of freedom. The under-determined case, by contrast, occurs when the system has been under-constrained; that is, when the unknowns outnumber the equations.

An under-determined linear system has either no solution or infinitely many solutions; for example, $x + y + z = 1$ and $x + y + z = 0$ is an under-determined system without any solution; any system of equations having no solution is said to be inconsistent. On the other hand, the system $x + y + z = 1$ and $x + y + 2z = 3$ is consistent and has many solutions. All of these solutions can be characterized by first subtracting the first equation from the second, to show that all solutions obey $z = 2$; using this in either equation shows that any value of y is possible, with $x = -1-y$.

There are algorithms available to decide whether an under-determined system has solutions, and if it has any, to express all solutions as linear functions of the variables; the simplest would be Gaussian elimination.

But back to the experimental laboratory. It is not possible to solve a problem where the number of unknowns exceeds the number of equations describing the system of interest. If you have two unknowns, which are to be determined experimentally, but you have only one equation describing the observable phenomenon, you cannot measure the two unknowns independently. The best you can do is determine a ratio of these two unknowns. If then, this ratio is used as a known quantity in some other experiment, then one needs to be careful. Ratio measurements can only be used to attribute absolute values to, for example, unknown measurement standards if, at least, one absolute reference value is known at the starting point of the ratio chain. In addition, measurement results cannot be more accurate that the standard used in the measurement process.

References

[1] Buckingham A D 1959 *J. Chem. Phys.* **30** 1580

[2] Debye P 1963 *quoted in Chem. Eng. News* **41** 40–3

[3] Piazza R, Degiorgio V and Bellini T 1986 *Opt. Commun.* **58** 400

[4] Graham C, Pierrus J and Raab R E 1989 *Mol. Phys.* **67** 939

Chapter 9

Indirect measurement: the optical Kerr effect

9.1 Introduction

In the previous chapter we looked at what I term a direct measurement; that is, one where you build an apparatus to measure one observable. This observable is then multiplied by constants and fixed or defined parameters, and you derive the quantity of interest. The error analysis of such experiments is quite straightforward as one can analyze the possible sources of noise to the one observable of interest and determine the ultimate sensitivity of the apparatus. But what happens when there is no straightforward observable to the experiment; for example, when one is making an indirect measurement of something?

By indirect measurement, I am referring to an experiment where the measured data must first be computer fitted to some theoretical model before any discussion of the measured result can be entertained; that is, experimental data fitting. Today, this is how a great many of the most significant discoveries in physics are made. It is not a scientist working in a laboratory with an apparatus he or she has built. Today, discoveries such as the Higgs boson or the observation of gravitational waves involve huge teams of scientists measuring and recording vast amounts of data, but where only a tiny part of that data contains the desired phenomenon. This may well be a phenomenon, which to be identified and quantified within the measured data requires a complex process of data fitting and data manipulation. For example, how was it determined that quarks are the ultimate basis of matter? The atom was known philosophically to the Ancient World, but was not identified until the late-19th Century, and not observed directly until the second-half of the last Century. The electron was identified in Cambridge at the end of the 19th Century by the group of Joseph John Thomson; the proton was discovered by Ernest Rutherford and his team at the end of World War I, and the neutron was discovered by James Chadwick in 1932. But what of the constituents of the proton and the neutron; how were they 'observed'?

doi:10.1088/978-1-6817-4433-9ch9

Well, the truth is that the constituent particles of the proton and the neutron (quarks) have never been 'observed' in the same way that the electron or the proton were originally observed and studied. Quarks have been studied by fitting the data that is accumulated by smashing protons together at high energy (the proton is electrically charged and so it may be accelerated electro-magnetically). Such experiments are performed at laboratories like CERN, the European organization for nuclear research, in Switzerland, and they generate vast quantities of data.

Approximately 600 million times per second, sub-atomic particles collide within the Large Hadron Collider (LHC) at CERN. Each collision generates particles that decay in complex ways into other sub-atomic particles via well-established routes that may be analyzed using quantum mechanics and the Standard Model of particle physics. Circuits record the passage of each sub-atomic particle through a detector as a series of electronic signals, and send the data to the CERN computers for digital reconstruction. The digitized summary is recorded as a 'collision event'. Then physicists must sift through the 30 petabytes (30×10^{15} bytes) or so of data produced annually to determine if the collisions have (literally) thrown out any interesting physics or new particles.

When a piece of interesting physics or a suspected new particle is identified, an army of scientists home in on that particular event and analyze all the relevant computer records. This is not a direct experiment that was designed and undertaken, it is an event that has occurred, it just happened in the collider; it was one event in 6×10^8 events per second taking place in the LHC. The intense scrutiny of the physicists might, of course, come to nothing; one was merely seeing a well-known phenomenon from an odd angle (see figure 4.4), but sometimes new physics is observed.

For the 'observation' of the quark, protons were collided at higher and higher energies so as to probe their innermost constituents. The scattering event for the collision of two protons was observed, as was the expulsion of all the various particles in the two protons; all moving outwards from the scattering centre. Physicists start with a model of what should be happening according to their current best theory, and then look for evidence of that theorized event. With the quark, it was theorized that each proton is formed of three quarks. So when two protons collide at sufficiently high energy, the scattering event should contain evidence of the interaction of nine quarks. This was what was observed in the vast amount of accumulated data. Fitting the observed data to a model of nine interacting quarks explained what was being measured in the detectors of the atom smashers. And on going to even higher energies, no new physics or unexplained events were seen in the fitted data originating from the scattering event, so it could be assumed that the model of a proton consisting of three quarks, which are indivisible, is correct. In no case was there a direct 'observation' of a quark. There was an indirect observation; one starts with a model of what you expect and you look for evidence of that model. If no evidence is found, then the model must be modified or abandoned.

This was how quarks were established as the basis of matter, and how the Higgs boson was observed. Indeed, fitting an observed event or phenomenon to a model was how gravitational waves were observed. There was no direct measurement of a Higgs boson or a gravitational wave, but the observation of a scattering event at CERN that could only be modelled (and thus explained) by assuming that the particle involved was a Higgs boson with some of the expected properties of the Higgs boson. And the signal (GW150914) seen at both LIGO gravitational wave detectors could only be explained by a model that assumed that it matched the predictions, from general relativity, for a gravitational wave emanating from the inward spiralling and merging of a pair of black holes of around 36 and 29 solar masses, and the subsequent 'ring-down' of the resulting single black hole.

So, in such indirect experiments, you begin the experiment having a complex model of what may be happening (or of what is about to happen). Of course, the person who designed the quadrupole experiment discussed in the previous chapter also had in mind a theory of what happens when linearly-polarized light passes through a birefringent medium, and this model would have been subject to evolution should the experiment have yielded unexpected, but reproducible results. That is the nature of experimentation; a theory is useful, but the results of a good experiment will last forever.

In the indirect experiment, the model is all important. It is not possible to do a 'quick and dirty' measurement to check that there is something there to be measured, subsequently, with greater care and precision. In some way, indirect experiments are unique experiments—they take a long time to do. We have been looking for gravitational waves and the Higgs boson for half a century, and I am sure that now these things have been observed, there will be more observations in the near future. But these experiments are not susceptible to the type of statistical analysis that may be made on the measurements derived from a direct experiment, such as the determination of the quadrupole moment of a molecule. My own PhD thesis was based on measurements of molecular quadrupole moments. It required a couple of years to build that apparatus, and many hundreds of measurements were then made on more than a dozen gases and vapours. During this time, a trivial statistical analysis established the precision of my measurements. But how do you establish error bars for the observation of a Higgs boson or of a gravitational wave?

9.2 The optical Kerr effect

As an example of an indirect experiment (that is, a measurement where one does not make a direct determination of the desired quantity, but must analyse the measured data via a process of data fitting), we will consider the optical Kerr effect (OKE) generated by pulsed laser systems.

As we saw earlier, the Kerr effect was traditionally used to make measurements of static molecular properties; in particular, the bulk electric susceptibility of a fluid. The laboratory electric field that is applied to the fluid to generate the birefringence is a static field. This applied field orients the molecules, and when linearly-polarized light passes through these oriented molecules it becomes elliptically-polarized. Measurement of this inducted ellipticity allows investigation of the electrical properties (electric dipole moment and the anisotropic part of the molecular polarizability) of the molecules under investigation. The applied electric field may be modulated to assist in improving the signal to noise of the measurement (as described above), but this frequency of oscillation is so low (a few hundreds of cycles per second) that the molecules are able to follow the reversing field.

In the OKE, however, the applied electric field that induced the birefringence in the fluid oscillates at optical frequencies.[1] Consider a pulsed laser with a pulse width of about 30 nanoseconds with a peak power (W) of 40 MW cm^{-2}. We may calculate the peak electric field of the light as $E = \sqrt{W/c\varepsilon_0}$ (where c is the speed of light and ε_0 is the permittivity of free space) to be 1.2×10^7 V m^{-1}, which is larger than any available laboratory applied, static electric field. The associated magnetic field, which would be a source of high-frequency magnetic birefringence is a factor of c smaller, and so is negligible. Given the sensitivity of the various experiments designed to measure induced retardations of linearly-polarized light beams by oriented molecules (as show above), it is even possible to determine the OKE induced by a continuous laser of modest power. However, in the experiments to be discussed here, the laser inducing the orientation is a pulsed laser with a pulse width of 200 fs and a peak power for orienting the molecules of interest of 10^9 W, giving an electric field, of order, 10^7 V m^{-1} in a weakly focused (beam waist, of order, 150 μm) beam.[2]

The systems investigated below are aqueous solutions of simple ions. That is, systems that it is only possible to study with the OKE. If you take an aqueous solution of sodium chloride and apply a static or dc electric field, you will not orient the ions, but you will cause the ions to move in the solution, and you will have electrolysis. Pure water is also slightly conducting, but a static Kerr effect constant has been measured for pure water at 632.8 nm of $2.96 \pm 0.08 \times 10^{-14}$ V^{-2} m. If, however, the fluid is subjected to an applied electric field oscillating at 10^{15} s^{-1} (the optical field of a pulsed laser) the dipolar contribution to the induced birefringence is lost because the molecules cannot follow the rapidly oscillating electric field, and the OKE of pure water has been found to be 2.48×10^{-16} V^{-2} m (see [2]). What is clear is the loss of the dipolar contribution to the induced birefringence. In the static measurement, the molecular dipole moment dominates, but at optical frequencies, the measurement probes only the distortion of the anisotropic part of the molecular polarizability by the applied optical field.

A schematic representation of the apparatus used for the OKE measurements that will be discussed below is given in figure 9.1. The experiments to be discussed were carried out in the European Laboratory for Nonlinear Spectroscopy, Florence, Italy. The 80 ps pulses of a 12 W mode-locked Nd:YAG laser are pulse-compressed and frequency doubled to 4 ps at 532 nm. This beam was stabilized and used to pump a dye-laser, the output of which, 250 mW at 600 nm with a 350 fs width, is again compressed to give 120 fs pulses.

[1] This orienting electric field, is the field associated with the light beam.

[2] The optical Kerr effect was predicted by Buckingham in 1956 [1], well before the advent of lasers. On being asked by the author how he (ADB) had envisaged making a measurement of this phenomenon in 1956, the response given was that the light from a WWII searchlight could have been focused down into a suitably polarizable fluid.

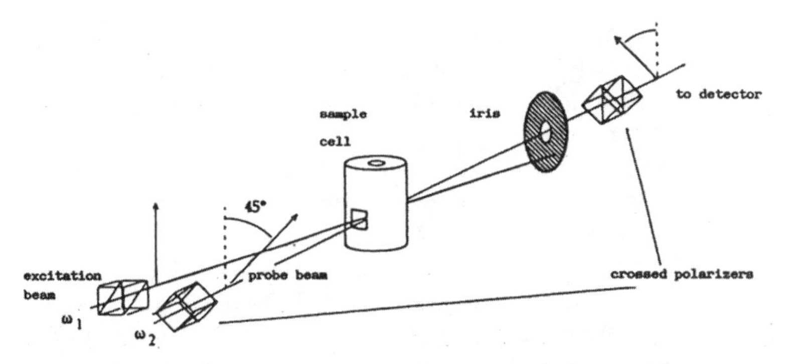

Figure 9.1. Schematic representation of a typical femtosecond optical Kerr effect laser system (the TROKE experimental set-up at the European Laboratory for Nonlinear Spectroscopy, Florence, Italy). The two incoming laser beams are at frequencies, ω_1 and ω_2; the former, the excitation beam, is the more intense and induces a birefringence in the medium under investigation, which is then probed by the weaker probe beam. The polarization of the probe beam is at 45° to that of the excitation beam and is crossed with respect to the analyzer or second polarizer. The iris defines the experimental geometry, and serves as a beam stop. Image taken from [2].

In the OKE experiments, the more powerful orienting laser pulse, which is linearly-polarized, generates an anisotropy in the refractive index of the fluid under investigation (as the applied laboratory electric field does in the static Kerr effect measurements discussed above). This induced optical anisotropy of the fluid generated by the intense optical field is then probed by a weaker, variably delayed, probe pulse that is also linearly-polarized, but with the plane of polarization at 45° to the direction of polarization of the intense orienting pulse. The birefringence induced by the applied electric field oscillating at optical frequencies is then determined by observing the modulation in the intensity of light passing the analyzing polarizer. This apparatus could readily measure an induced retardation, of order, 1×10^{-7} radian, which corresponds to a difference in the parallel and perpendicular components of the refractive index of ($n_{\text{parallel}} - n_{\text{perpendicular}}$) of about 10^{-13} at 600 nm for a 1 cm path length. The quadrupole moment experiment described in the last chapter has greater precision, in that we could measure an induced retardation, of order, 3×10^{-8} radian, but our path-length was 100 cm.

In simple organic liquids (for example, liquid benzene), investigated by the OKE before these measurements on aqueous solutions, it was seen that five contributions to the complex dynamics of the molecules in the liquid state could be identified:

i. an instantaneous, purely electronic response of the polarizability (the distribution of electrons) of the molecule, arising from the molecule's hyperpolarizability;
ii. an intermolecular vibrational contribution originating from the motion of the target molecule moving in the cage of nearest neighbour molecules; a motion decaying with the lifetime of the nearest-neighbour cage $\leqslant$200 fs;
iii. an intramolecular vibrational contribution, decaying with the dephasing time of the molecular vibrations, typically a few hundreds of femtoseconds;

iv. a term arising from molecular interactions; here relaxation is governed by the local structure or relative orientations of the molecules in the liquid state, and which decays over a few hundreds of femtoseconds;
v. a rotational contribution, the timescale of which is determined by the macroscopic properties of the fluid (sample temperature, density, viscosity, the size of the molecules, and which could range from a picosecond to tens of nanoseconds.

Thus in terms of fitting any measured data for subsequent analysis, the standard model adopted for the interpretation of the OKE in simple organic liquids involved five parameters each with a weighting coefficient; that is, there are of order 10 unknowns. The system that we chose to investigate did not consist of organic liquids, but aqueous solutions of simple salts, which we considered as being less complex in their dynamics by virtue of the stronger (ionic) intermolecular forces involved.

The OKE measures the change in refractive index of the aqueous solution due to the interaction of the ions present in the medium (and the medium itself) with the optical electric field. The time-resolved OKE signal $S(\tau)$ at a delay time τ, is given by the double integral

$$S(\tau) = \int_{-\infty}^{\infty} I_{\text{probe}}(t-\tau)\sin^2\left[\int_{-\infty}^{t} R(t-t')I_{\text{pump}}(t')\mathrm{d}t'\right]\mathrm{d}t \tag{9.1}$$

where τ is the delay between the orienting pulse of the laser and the probing pulse of the laser, $R(t)$ is the response function for the system and I_{probe} and I_{pump} are the intensities of the probe and orienting pump beams, respectively. The second integral in equation (9.1) represents the phase difference between the parallel and perpendicular components of the probe beam, induced by the pump beam (the desired observable). Assuming that this phase shift is small we may write:

$$S(\tau) = \int_{-\infty}^{\infty} I_{\text{probe}}(t-\tau)\left[\int_{-\infty}^{t} R(t-t')I_{\text{pump}}(t')\mathrm{d}t'\right]^2\mathrm{d}t. \tag{9.2}$$

The time profiles for the laser pulses were seen to be well fitted by use of a bi-exponential function with a fast rising and slow decaying form:

$$\begin{aligned}&\exp(\alpha_k t) \quad \text{for} \quad t<0\\ &\exp(-\gamma_k\alpha_k t) \quad \text{for} \quad t>0\end{aligned}$$

where $\gamma > 1$ is an asymmetry parameter and $k = 1$ for the pump beam and $k = 2$ for the probe beam, respectively. The typical model for the response function $R_s(t)$ is the sum of exponential, nuclear, or molecular decays plus an instantaneous or electronic response; that is,

$$\begin{aligned}R_s(t) &= R_{\text{electronic}}(t) + R_{\text{molecular}}(t)\\ &= n_2{}^{e}\delta(t) + \sum(n_2{}^{i}/\tau_i)\exp(-t/\tau_i)\end{aligned} \tag{9.3}$$

where τ_i is the relaxation time of the ith decay component and $\delta(t)$ is the delta-function. The other terms are as follows: n_2^e represents the instantaneous electronic component to the observed induced birefringence (component (i) from the list above), and n_2^i are the various nuclear or molecular contributions to the observed relaxation process (vibrational and rotational).

We see immediately the difference between a direct experiment, such as the measurement of a molecular quadrupole moment and the more complex indirect experiment such as a determination of the OKE of a liquid. In the direct experiment one constructs an apparatus to measure one observable; all other experimental conditions are specified in advance. The errors in the measurement of the induced retardation propagate through to the derived value of the quadrupole moment. Provided the errors in the determination of the experimental conditions are lower than the errors involved in the determination of the induced retardation, then the value of the retardation and the quadrupole moment will have much the same measurement error. If, however, there is a much greater uncertainty in; for example, the determination of the number density of molecules in the quadrupole cell, then this error will define the final uncertainty in the quadrupole moment.

In the OKE experiment of ionic solutions, however, one measures a huge quantity of data, but this huge amount of data is required because you have many more unknowns to be simultaneously determined (perhaps as many as five). The laser generates pulses of light, of order, 100 fm in length; there are a lot of these pulses in an experimental run. In both the OKE experiment and the direct quadrupole measuring experiment, although one is a static measurement and the other a dynamic measurement, we are able to measure tiny fluctuations of the refractive index of a fluid, but they are only able to tell us something about these weak effects because there are many molecules contributing to the measurement. In addition, we are measuring (repeatedly measuring in the OKE) for a long time. About 10^{24} molecules contributing to the determination of the molecular quadrupole moment, and in the OKE experiment we have about 10^{22} ions contributing to the measurement. Also, in both cases we are able to measure for a long period; in the quadrupole moment experiment we are using a continuous laser beam and so we can integrate the signal from the phase-sensitive detector by varying the time-constant of the circuitry, and in the pulsed laser OKE experiment, we can count for as many pulses as we think necessary.

Both direct and indirect measurements have their place in science, but the latter only really became possible with the advent of computers.

In figure 9.2(a) we see the OKE as determined in pure water and in three aqueous solutions of sodium chloride. As the delay time of the probing pulse can be varied with respect to the powerful pumping pulse, we are able to observe the relaxation of the oriented molecules and ions. That is, the pump laser-pulse orients the molecules via the near instantaneous electronic interaction of the molecules (water and solvated ions in these figures) with the electric field of the laser-pulse. Then we vary the delay time and observe the OKE signal decay as the molecules relax or lose their induced orientation. This relaxation process can then be used to investigate intermolecular interactions via a suitable model which must be fitted to the measured data.

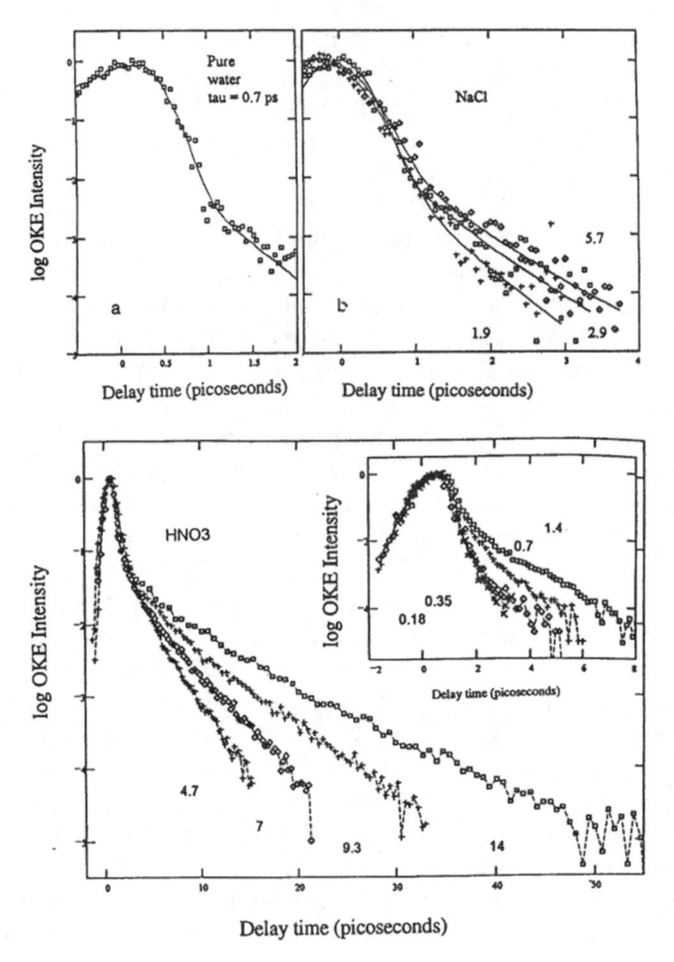

Figure 9.2. Typical relaxation data from femtosecond time-resolved optical Kerr effect measurements. First there is the intense electronic response of the sample to the applied optical electric field, then a slow fall of the signal, which contains information about the dynamics of the molecules and ions in the sample (an aqueous solution at room temperature). (*a*) The response of a sample of pure water and of three aqueous solutions of sodium chloride. (*b*) Relaxation data for eight solutions of nitric acid of varying concentration. As the nitrate anion is anisotropically polarizable, the signal (and hence signal to noise) from the nitrates is larger than that from the chlorides. The data is analyzed by assuming that the laser pulses (excitation and probe pulse) are δ-functions with infinitely sharp rise time; this is not actually the case and, consequently, there are some data before the time zero of the arrival of the probe pulse; this serves to demonstrate the uncertainty in the model. In this figure, the intensity axis has a log scale, and the time axis is in picoseconds (10^{-12} second). The data is taken from [2].

In pure water, the peak at a delay time of zero seconds is the electronic orientation of the molecules (the dipole moments of which cannot contribute). This is the distortion of the anisotropic part of the molecules' polarizability by the applied optical field. Then with time, this orientation decays away with a decay rate of 0.7 ps (for a sample of water at 18 °C). Similarly in the sodium chloride solutions, where in principle there should be no contribution from the sodium cations or the chloride anions because they are spherically symmetric. But one can clearly see the difference between the OKE of water and of the solutions, and this difference varies with solute concentration. The spread in the data points gives an idea of the measurement uncertainty; particularly in the solutions of sodium chloride, which should only generate a small OKE as there are no anisotropically polarizable ions present. However, we see in figure 9.2(*b*), the OKE generated in nitric acid of various concentrations. Here there is an anisotropically polarizable ion present (the planar NO_3^-) thereby generating a bigger signal than the solutions of isotropic ions. There is thus more signal, a better signal to noise and a lower measurement uncertainty. Figure 9.3 contains data for solutions of sodium nitrate (*a*), which generate a large OKE signal, and different concentrations of HCl (*b*), which generates a smaller OKE signal than that generated by the nitrate anions.

What is also apparent in these figures is the huge amount of data one is able to collect and subject to analysis. In the quadrupole moment experiment, one spent all day measuring one quantity half a dozen times, and then performing a trivial statistical analysis of the six measurements to determine the quadrupole moment. In the OKE experiment, in a couple of hours one can generate enough data to keep a student busy (with his or her computer fitting software) for days. What this greater quantity of data means is that one can seek to identify more than one unknown from the measured data—provided one has a suitable model.

We may write the Kerr constant for a multi-component system as

$$K_{\text{solution}} = \sum_i x_i K^{(i)} + \sum_{ij} x_i x_j K^{(ij)} + \cdots, \tag{9.4}$$

where x_i are the mole fractions of the components with $K^{(i)}$ being the corresponding Kerr constant. The first sum corresponds to the additivity of the signals arising from the various components present, and the second summation, and subsequent summations permit deviations from additivity. In the ionic solutions studied here, we can assume that at low concentrations all the sodium and chloride ions, or sodium and nitrate ions are completely solvated; that is, there is an octahedron of six water molecules around each Na^+ and Cl^- ion, and two water molecules above and below the planar nitrate ions. However, when the solutions become concentrated (6 moles per litre is very concentrated), there are barely enough water molecules to generate these solvation sheaths around all the ions. We have therefore a picture (a model) of long-range, polarizable structures established at low solute concentrations, which are disrupted by strong ionic forces produced by the addition of extra solute.

Combining the equations given above, our time-resolved OKE measurements were analyzed by means of the following equation (the model used to fit the data):

$$\begin{aligned} R_s(t) = {} & x_{\text{water}}\{n_2^{e}(\text{water})\delta(t) + (n_2^{o}(i)/\tau(i)) \exp-(t/\tau(i))\} \\ & + x_{\text{solute}}\{n_2^{e}(\text{solute})\delta(t) + (n_2^{o}(ii)/\tau(ii)) \exp-(t/\tau(ii)) \\ & + (n_2^{o}(iii)/\tau(iii)) \exp-(t/\tau(iii))\} \end{aligned} \tag{9.5}$$

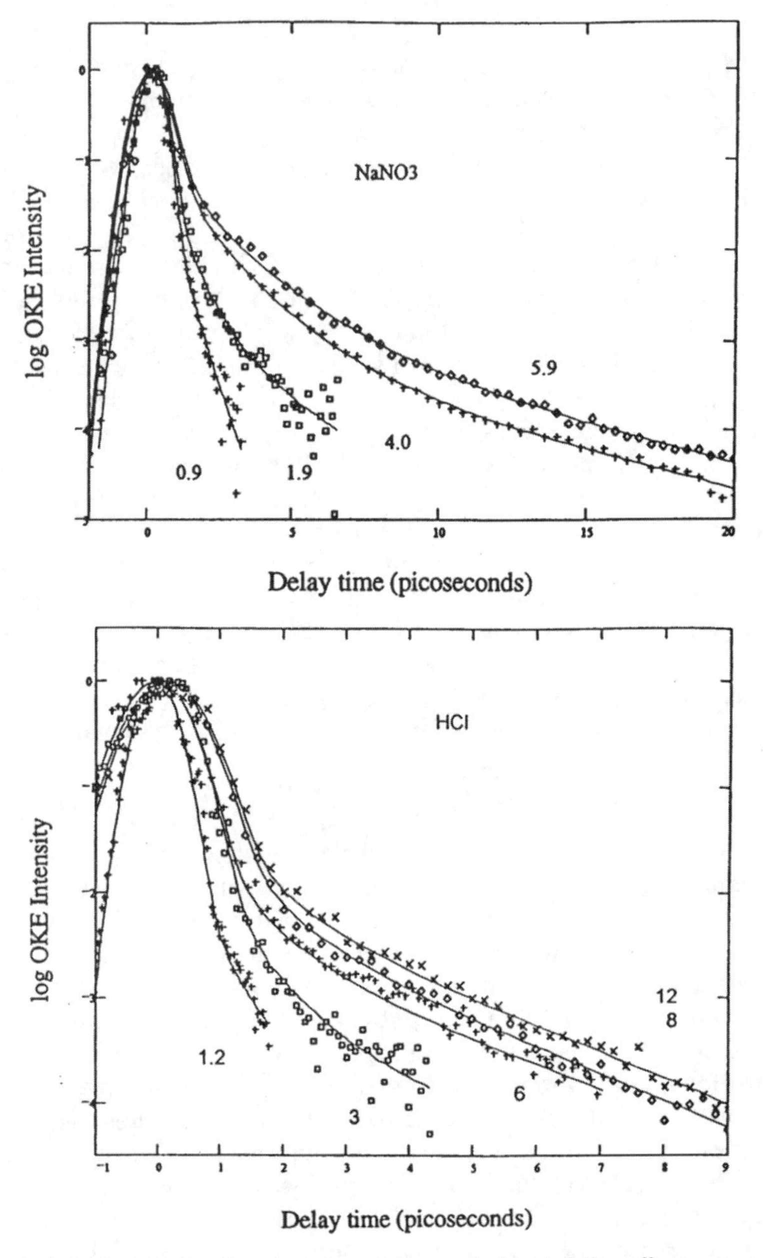

Figure 9.3. Typical relaxation data from femtosecond time-resolved optical Kerr effect measurements. First there is the intense electronic response of the sample to the applied optical field, then a slow fall of the signal, which contains information about the dynamics of the molecules and ions in the sample. (*a*) The response of four aqueous solutions of sodium nitrate. (*b*) Relaxation data for five solutions of hydrochloric acid of varying concentration. As the nitrate anion is anisotropically polarizable, the signal (and hence signal to noise) from the nitrates is larger than that from the chlorides. The data is analyzed by assuming that the laser pulses (excitation and probe pulse) are δ-functions with infinitely sharp rise time; this is not actually the case and consequently, there are some data before the time zero of the arrival of the probe pulse; this serves to demonstrate the uncertainty in the model. In this figure, the intensity axis has a log scale, and the time axis is in picoseconds (10^{-12} s). The data is taken from [2].

where n_2^e(water) and n_2^e(solute) are the instantaneous electronic contributions of the water and solute, respectively, to the measured OKE signal. In the second term on the rhs of (9.5), $n_2^o(i)$ refers to the *i*th component of the slow or molecular (non-electronic) response to the measured OKE signal, which is decaying with a time constant $\tau(i)$. Thus, there are purely electronic responses and molecular responses for both solvent and solute; in fact there are two nuclear or molecular contributions from the solute, $n_2^o(ii)$ and $n_2^o(iii)$, with time constants $\tau(ii)$ and $\tau(iii)$, respectively. Thus the mass of data represented by these indirect measurements will be fitted to an equation (9.5), which has five unknowns, and which will be determined simultaneously by the fitting process. To achieve anything from such an exercise, we require, at least five, equations that relate the values of interest to the experiment. As it happens, because we are able to vary the time between the pumping and probing pulse, we can generate the required expressions, and the various unknowns may be found by fitting the data to the model function; all the fitting being done by least-squares methodology. Figures 9.2 and 9.3 show the best fits results for the various solutions investigated.

In figures 9.2 and 9.3, we see the measured relaxation of water molecules and solvated ions after their polarizability has been ordered by the pulse or pump laser. As it happens, the dynamics of water molecules in liquid water is one of the biggest areas of research in chemical physics (it is a field of endeavour that has applications in modern medicine and in global warming), and there are many theories as to how the water molecules interact in the condensed phase (see [3]). However, the signal to noise of the data represented in figure 9.2(*a*) did not allow us to differentiate between these various mechanisms for the relaxation of water molecules in liquid water, Consequently, we left our measured relaxation constant as $\tau(i) = 700 \pm 50$ fs, and made no comment as to why the property has this value; it has an error of 50 in 700 or 7%, which is to be expected as the effect being measured is weak, even though many thousands of measurements contributed to the final result.

In the case of sodium chloride solution and hydrochloric acid, it was found that the data could be well represented by invoking only one parameter to describe the molecular relaxation, with a relaxation time constant to represent the contribution of the solute ions to the OKE; component (*ii*), $\tau(ii) \sim 1.5$ ps. However, with the nitrates ($NaNO_3$ and HNO_3), the nitrate anion is planar and geometrically anisotropic and, consequently, the observed OKE was larger than for the chlorides, and an additional molecular contribution (*iii*) also characterized by an exponential decay was required to fully represent the data. The time constant, $\tau(iii)$, was found to be a function of the sample concentration.

It is the much larger signal arising from the polarization of the nitrate anions in the nitrate systems that gave us the opportunity to fit the data to three parameters as opposed the data derived from the chloride systems; in the chlorides there was a smaller OKE and the data was, consequently, noisier and so fitting with one parameter or with three parameters made no difference to the residuals (we will encounter the problem of trying to over-fit data in the next chapter).

It is the quality of the data, the signal to noise of the experiment that allows one to fit multi-variate models to the data from a single experiment. As it happened, in the nitrates there was the instantaneous electronic contribution (clearly seen in the data) and two molecular relaxation components, (*ii*) with time constant $\tau(ii)$ of about 2.2 ps and (*iii*) with relaxation constant $\tau(iii)$. The first of these two contributions was found to be largely independent of solute concentration, while the second term was found to be strongly dependent on solute concentration, and was interpreted as the onset of a glassy-like state in the solution where there is not enough water to shield the bare ions from each other.

It is not my intention to give a detailed explanation of the type of information that may be derived from OKE experiments on aqueous ionic solutions. But merely to point out that such experiments are possible and they are typical of an enormous class of indirect measurements (for example, most particle physics experiments, all neutron scattering experiments and all crystal structure determinations involve indirect measurements), where what you are interested in comes from fitting a specific model to a large body of measured data. But even in these experiments, if the signal generated is not great, as in the OKE of pure water and in solutions of HCl and NaCl, then the signal to noise may not be sufficient to allow one to discriminate between different models, each with several unknowns, for what may be happening in the fluid, or to isolate the contributions of the various components present in the solution to the overall measured signal. In the latter case, one would need to undertake additional experiments using additional samples and varying other parameters; for example, the sample temperature.

There is always a trade-off between how many parameters you put in the theoretical model you have constructed of what it is that you think is going on in your experiment, and the quality of the measured data. If the signal to noise is poor (figure 9.2(*a*)) do not try and fit the data with a complex model; the solutions you derive from the fitting procedure (and your computer will generate solutions!) will be unstable, and will themselves be subject to large uncertainties. As the signal to noise improves (as for example in figure 9.2(*b*)) you may attempt to quantify ever more subtler effects in the data (that may or may not be there).

References

[1] Buckingham A D 1956 *Proc. Phys. Soc.* **B69** 344
[2] Santa I, Foggi P, Righini R and Williams J H 1994 *J. Phys. Chem.* **98** 7692
[3] Williams J H 2015 *Order from Force* (San Rafael, CA: Morgan and Claypool)

Chapter 10

Data fitting and elephants

With four parameters I can fit an elephant and with five I can make him wriggle his trunk.

Attributed to John von Neumann by Enrico Fermi

10.1 Introduction

As a young research professor in 1953, the eminent British-born theoretical physicist and mathematician, Freeman Dyson (born 1923) made a trip to Chicago to visit his mentor, Enrico Fermi. Dyson and his group had been calculating meson–proton scattering cross-sections using a theory (that is, a model) of the strong nuclear force known as pseudoscalar meson theory, which had been published by Fermi. Dyson was impressed and excited by the apparent agreement between his theoretical computations and the available experimental data measured by Fermi in Chicago. This agreement had emboldened Dyson to visit Fermi to present his results. Sadly for Dyson, Fermi on viewing the results was unimpressed. Dyson naturally asked why Fermi was not impressed by the apparent agreement between theory and experiment. Fermi responded by asking a question, 'To reach your calculated results, you had to introduce arbitrary cut-off procedures that are not based on solid physics or on solid mathematics … How many arbitrary parameters did you use for your calculations?' Dyson's response was four, to which Fermi famously commented, 'I remember my friend Johnny von Neumann (1903–57) used to say that with four parameters I can fit an elephant and with five I can make him wriggle his trunk'. The interview between Dyson and Fermi was over [1].

By saying this, Enrico Fermi was not insulting his enthusiastic young guest, but he was pointing out a fundamental truth of data analysis. That is, with enough arbitrary parameters, it is possible to find a curve that goes through just about any set of data points. This statement is perhaps not true for a set of truly random numbers, but apparently it is true for a set of data that when graphed would

doi:10.1088/978-1-6817-4433-9ch10

resemble the outline of an elephant, with a big fat body, distinct legs, a tail and a trunk. Then, of course, if you find an equation containing several adjustable parameters that links all the data points in your complex, exotic data set, you are naturally led to the assumption that you now understand the physics behind the force (s) that is(are) generating your odd looking set of data. Of course, there is no natural phenomenon that could generate such an outlandish dataset, the outline of an elephant, but Fermi, and presumably, von Neumann, were using this weird and wonderful idea to instruct their listeners in the dangers of putting all your trust in data fitting algorithms. When interpreting data; particularly, the results of theoretical calculations made to interpret experimental results (particularly, experimental results with appreciable uncertainties), one should never lose sight of the underlying physics.

10.2 Regression analysis

Regression analysis is the most widely used statistical tool for the investigation of relationships between variables. Usually, an investigator is seeking to learn if there is a causal effect of one variable upon another: for example, what is the relationship between income and education, or is there a temperature-independent term that contributes to the measurement of the quadrupole moment via electric-field gradient induced birefringence?

To explore such questions, the investigator assembles data on the underlying variables of interest and employs regression to estimate the quantitative effect of the causal variables upon the variable that they (may or may not) influence. The investigator also assesses the statistical significance of the estimated relationships; that is, the degree of confidence that the true relationship is close to the estimated relationship.

The techniques of regression analysis were first used by the 19th Century British biologist and statistician, and cousin to Charles Darwin, Francis Galton (1822–1911). Galton was a pioneer in the application of statistical methods to measurements in many areas of biology; specifically, in analyzing data on relative sizes of parents and their offspring in studies of plants and animals. He famously observed the following: a larger-than-average parent tends to produce a larger-than-average child, but the child is likely to be less large than the parent in terms of its relative position within its own generation. If the parent's size is x standard deviations from the mean within its own generation, then one could predict that the child's size will be $r{\cdot}x$ standard deviations from the mean within the set of children of those parents, where r is a number less than 1 (the correlation between the size of the parent and the size of the child). Such observations may be found for physical measurements throughout biology, and in humans for most measurements of cognitive and physical ability; and it led to the horrors of eugenics in the 20th Century. For Galton, regression had only this biological meaning, but his work was later extended to a more general statistical context. For the statisticians, the joint distribution of the response and explanatory variables is assumed to be Gaussian.

Francis Galton termed this phenomenon a 'regression towards mediocrity', which today is expressed as a 'regression to the mean' and it is an inescapable fact of life.

Your children can be ‘expected’ to be less exceptional (for better or worse) than you are. Your score on a final exam in a course can be ‘expected’ to be less good (or bad) than your score in the mid-term exam, relative to the rest of the class. The key word here is ‘expected’. This does not mean it’s certain that regression to the mean will occur, but that’s the way things tend to go in terms of the probability.

In data analysis, errors and residuals are two closely related and often confused measures of the deviation of an observed value of an element of a data set from its ‘theoretical value’. The error of an observed value is the deviation of the observed value from the (unobservable) true value of a quantity of interest (for example, a population mean), and the residual of an observed value is the difference between the observed value and the estimated value of the quantity of interest (for example, a sample mean). The distinction is particularly important in regression analysis.

Suppose there is a series of observations from a univariate distribution and we wish to estimate the mean of that distribution. In this case, the errors are the deviations of the observations from the population mean, while the residuals are the deviations of the observations from the sample mean.

A statistical error is the amount by which an observation differs from its expected value, the latter being based on the whole population from which the sample was chosen randomly. For example, if the mean weight of a group of individuals is 80 kg, and a randomly chosen individual has a weight of 85 kg, then the ‘error’ is 5 kg; however, if the randomly chosen individual has a weight of 78 kg then the ‘error’ is now −2 kg. The problem is that the expected value, being the mean of the entire population, is typically unobservable, and hence the statistical error cannot really be observed.

A residual (or fitting deviation), on the other hand, is an observable estimate of the unobservable statistical error. Consider the previous example of weights and suppose we have a random sample of *n* people. The sample mean could serve as a good estimator of the population mean. Then we have the difference between the weights of each person in the sample and the unobservable population mean is a statistical error, whereas the difference between the weight of each person in the sample and the observable sample mean is a residual.

The sum of the residuals within a random sample is necessarily zero, and thus the residuals are necessarily not independent. The statistical errors on the other hand are independent, and their sum within the random sample is almost certainly not zero. In regression analysis, the distinction between errors and residuals is subtle and important, and leads to the concept of studentized residuals. Given an unobservable function that relates the independent variable to the dependent variable; for example, a straight line, the deviations of the dependent variable observations from this function are the unobservable errors. If one runs a regression on some data, then the deviations of the dependent variable observations from the fitted function are the residuals.

In the statistical modelling of experimental or observational data, regression analysis is the means of estimating the relationships among the variables involved in the measurements. It has evolved into a standard procedure that is today often encountered as a ‘black box’ technology often already programmed into the computers that record experimental data. The technique includes many different ways of modelling and analyzing several variables, when the focus is on the relationship between a dependent variable and one or more independent variables (or predictors). More specifically, regression analysis helps one understand how the typical value of the dependent variable (or criterion variable) changes when any one of the independent variables is varied, while the other independent variables are held fixed. Most commonly, regression analysis estimates the conditional expectation of the dependent variable given the independent variables; that is, the average value of the dependent variable when the independent variables are fixed. In all cases, the estimation target is a function of the independent variables called the regression function. In regression analysis, it is also of interest to characterize the variation of the dependent variable around the regression function which can be described by a probability distribution.

Let us consider an example: the influence of education upon income, for which we take a large sample of individuals of similar age and ask them how long they spent in full-time education and what is their present income. Clearly, when graphed this data will generate a plot with an enormous range of uncertainty or scatter. But the question that has to be answered (and is often asked): is there a relationship between education level and income? The graph will suggest that higher values of education tend to result in higher incomes, but the relationship is not perfect; and it would be clear that knowledge of education level does not suffice for an accurate prediction of income. To apply regression analysis to this particular problem requires that we first hypothesize that earnings for each individual are determined by education, and by a collection of other factors (race, locality, sex, profession, etc.) that we term contributing noise.

Then, we write a hypothesized model relationship or equation between education (E) and income (I) as; for example,

$$I = \alpha + \beta E + \varepsilon,$$

where α is a constant amount (what one earns with zero education), β is the coefficient relating how an additional year of education influences income (assumed to be positive), and ε is the noise term representing other factors that influence earnings (factors that are unobservable, or at least unobserved). The variable I is the dependent or endogenous variable and E is termed the independent, explanatory, or exogenous variable. The parameters α and β are not observed, and regression analysis is used to produce an estimate of their value, based upon the information contained in the data set.

What we have hypothesised is that there is a straight line or linear relationship between E and I. Thus somewhere in the cloud of data points on our graph we expect to find a line defined by the equation $I = \alpha + \beta E$. The task of estimating α and β is equivalent to the task of estimating where this line is located on the axes? The answer depends in part upon what we think about the nature of the noise term ε. If we have reason to believe that ε is zero, then the line to be fitted to the scattered data points will lie somewhere in the cloud of data points. If ε has a large value (positive or negative), then the fitted line will lie above or below the cloud of points.

In figure 4.1, we see a set of measured data points that were plotted in a manner so as to answer the question; is there a temperature-independent contribution to the measured quadrupole moment as determined by electric-field gradient induced birefringence? That is, what is the best line through these data points, and is there a finite value of the intercept on the x-axis?

Regression analysis is used to search for the best line through a set of data points, thereby providing values for the coefficients in a model equation. This is the line that reflects the estimated error for each data point as the vertical distance between the value of, for example, I along the estimated line $I = \alpha + \beta E$ (generated by putting the actual values of E into our hypothesised equation or model) and the true value of I for the same observation; that is the difference between theoretical values of the variable and actual measurements of the variable.

With each possible line that might be superimposed upon the data set, a different set of estimated errors will result. Regression analysis then chooses among all the possible fitted lines by selecting the one for which the sum of the squares of the estimated errors is a minimum. This is termed the minimum sum of squared errors (minimum SSE) criterion. The intercept of the line chosen by this criterion provides the value of α, and its slope provides the value of β.

Why should we choose our line using the minimum SSE criterion? We can readily imagine other criteria that could be used; for example, minimizing the sum of errors in absolute value. One virtue of the SSE criterion is that it is straightforward to use. When one expresses the sum of squared errors to find the values of α and β that minimize it, one obtains expressions for α and β that are easy to evaluate using only the observed values of E and I in the data set. Continuing with our example, imagine that we have data on E and I for a number of individuals indexed by j. The actual value of I for the jth individual is Ij, and its estimated value for any line with intercept α and slope β will be $\alpha + \beta E_j$. The estimated error is thus $I_j - \alpha - \beta E_j$. The sum of squared errors is then $\sum_j (I_j - \alpha - \beta E_j)^2$. Minimizing this sum with respect to α requires that its derivative with respect to α be set to zero, or $-2\sum_j (I_j - \alpha - \beta E_j) = 0$. Minimizing with respect to β requires $-2\sum_j E_i(I_j - \alpha - \beta E_j) = 0$. We now have two equations with two unknowns that can be solved for α and β. But computational convenience is not the only virtue of the minimum SSE criterion; it also has some attractive statistical properties under plausible assumptions about the noise term, ε.

Many techniques for carrying out regression analysis have been developed. Familiar methods such as linear regression and ordinary least-squares regression are parametric, in that the regression function is defined in terms of a finite number of unknown parameters that are estimated from the available data. Nonparametric regression refers to techniques that allow the regression function to lie in a specified set of functions, which may be multi-dimensional. The earliest form of regression was the method of least-squares, which was published by Legendre in 1805 and by Gauss in 1809, and which was famously used to re-interpret the data obtained by Delambre and Méchain during the metric survey of the 1790s (see chapter 2).

The actual form of the regression analysis to be used depends on the form of the data generating process, and how it relates to the regression approach being used. Since the true form of the data-generating process is generally not known (it is well characterized in a direct experiment, but is less well-defined in an indirect experiment), regression analysis often depends to some extent on making assumptions about this process. These assumptions are sometimes testable if a sufficient quantity of data is available.

Regression models generally involve the following variables:

- The unknown parameters, denoted as $\boldsymbol{\beta}$, which may represent a scalar or a vector.
- The independent variables, $\mathbf{X}$.
- The dependent variable, Y.

In various fields of application, different terminologies are used in place of dependent and independent variables. A regression model relates Y to a function of $\mathbf{X}$ and $\boldsymbol{\beta}$.

$$Y \approx f(\mathbf{X}, \boldsymbol{\beta}), \text{ cf. } I = \alpha + \beta E + \varepsilon.$$

The approximation is usually formalized as $E(Y \mid \mathbf{X}) = f(\mathbf{X}, \boldsymbol{\beta})$. For regression analysis, the form of the function f must be specified. Sometimes the form of this function is based on information about the relationship between Y and $\mathbf{X}$ that does not rely on the data. If no such information is available, a convenient form for f is chosen.

Assume that the vector of unknown parameters $\boldsymbol{\beta}$ is of length k. In order to perform a regression analysis one must provide information about the dependent variable Y:

1. If N data points of the form $(Y, \mathbf{X})$ are observed, where $N < k$, most classical approaches to regression analysis cannot be performed: since the system of equations defining the regression model is underdetermined; there is not enough data to recover $\boldsymbol{\beta}$.
2. If $N = k$ data points are observed, and the function f is linear, the equations $Y = f(\mathbf{X}, \boldsymbol{\beta})$ can be solved exactly rather than approximately. This situation reduces to solving a set of N equations with N unknowns (the elements of $\boldsymbol{\beta}$), which has a unique solution as long as the $\mathbf{X}$ are linearly independent. If f is nonlinear, a solution may not exist, or many solutions may exist.
3. The most common situation is where $N > k$ data points are observed. In this case, there is enough information in the data to estimate a unique value for $\boldsymbol{\beta}$ that best fits the data in some sense, and the regression model when applied to the data can be viewed as an overdetermined system in $\boldsymbol{\beta}$. This is the situation with the optical Kerr effect discussed in the previous chapter.

In case 3, the regression analysis provides the tools for:

- Finding a solution for unknown parameters $\boldsymbol{\beta}$ that will, for example, minimize the distance between the measured and predicted values of the dependent variable Y (this is the method of least-squares).
- Under certain statistical assumptions, the regression analysis uses the surplus of information to provide statistical information about the unknown parameters $\boldsymbol{\beta}$ and predicted values of the dependent variable Y.

Consider a regression model which has three unknown parameters, β_0, β_1, and β_2. Suppose an experimenter performs several measurements all at the same value of independent variable vector **X** (which contains the independent variables X_1, X_2, and X_3). In this case, regression analysis fails to give a unique set of estimated values for the three unknown parameters; the experimenter did not provide enough information. The best one can do is to estimate the average value and the standard deviation of the dependent variable Y. Similarly, measuring at two different values of **X** would give enough data for a regression with two unknowns, but not for three or more unknowns.

If the experimenter had performed measurements at three different values of the independent variable vector **X**, then regression analysis would provide a unique set of estimates for the three unknown parameters in $\boldsymbol{\beta}$. In the case of general linear regression, this statement is equivalent to the requirement that the matrix $\mathbf{X}^\mathrm{T}\mathbf{X}$ be invertible.

When the number of measurements, N, is larger than the number of unknown parameters, k, and the measurement errors ε_i are normally distributed, then the excess of information contained in $(N - k)$ measurements is used to make statistical predictions about the unknown parameters. This excess of information is referred to as the degrees of freedom of the regression.

Classical assumptions for regression analysis include:

- The sample is representative of the population for inference prediction.
- The error is a random variable with a mean of zero, conditional on the explanatory variables.
- The independent variables are measured with no error. If this is not so, modelling may be done instead using errors-in-variables model techniques.
- The independent variables (predictors) are linearly independent; that is, it is not possible to express any predictor as a linear combination of the others.
- The errors are uncorrelated; that is, the variance–covariance matrix of the errors is diagonal and each non-zero element is the variance of the error (see the sections on propagation of uncertainty in chapters 5 and 6).
- The variance of the error is constant across observations. If not, weighted least-squares or other methods might instead be used.

These are sufficient conditions for the least-squares estimator to be useful; in particular, these assumptions imply that the parameter estimates will be unbiased, consistent, and efficient in the class of linear unbiased estimators. It is important to note, however, that actual data rarely satisfies the assumptions; that is, the method may be used even though the assumptions are not valid.

In linear regression, the model specification is that the dependent variable, y_i is a linear combination of the parameters (but need not be linear in the independent variables). For example, in simple linear regression for modelling n data points there is one independent variable: x_i, and two parameters, β_0 and β_1, we define a straight line (as we did above for education and income):

$$y_i = \beta_0 + \beta_1 x_i + \varepsilon_i \quad \text{for} \quad i = 1, \ldots, n.$$

In multiple linear regression, there are several independent variables or functions of independent variables. Adding a term in x_i^2 to the preceding regression generates a parabola,

$$y_i = \beta_0 + \beta_1 x_i + \beta_2 x_i^2 + \varepsilon_i \quad \text{for} \quad i = 1, \ldots, n.$$

This is still termed linear regression, although the expression is quadratic in the independent variable x_i. In both cases, ε_i is an error term and the subscript i indexes a particular observation. For the straight line case; given a random sample from the population, we can write the population parameters and obtain the sample linear regression model:

$$\underline{y}_i = \underline{\beta}_0 + \underline{\beta}_1 x_i.$$

The residual, $e_i = y_i - \underline{y}_i$, is the difference between the value of the dependent variable predicted by the model, $\underline{y}_i$, and the true value of the dependent variable, y_i. One method of estimation is ordinary least-squares, and as we saw above this provides estimates that minimize the sum of squared residuals, SSE, or:

$$SSE = \sum_i e_i^2.$$

Minimization of this function generates a set of simultaneous linear equations in the parameters of interest, which are solved to yield the parameter estimators, $\underline{\beta}_0$, $\underline{\beta}_1$.

In the case of simple regression, the formulas for the least-squares estimates are:

$$\underline{\beta}_1 = \sum (x_i - x_{\text{mean}})(y_i - y_{\text{mean}}) \Big/ \sum (x_i - x_{\text{mean}})^2 \text{ and } \underline{\beta}_0 = y_{\text{mean}} - \underline{\beta}_1 x_{\text{mean}}$$

where x_{mean} is the mean (average) of the x values and y_{mean} is the mean of the y values.

Under the assumption that the population error term has a constant variance, the estimate of that variance is given by:

$$\underline{\sigma}_\varepsilon^2 = \frac{SSE}{n-2}.$$

This is called the mean square error (MSE) of the regression. The denominator is the sample size reduced by the number of model parameters estimated from the same data, $(n-p)$ for p regressors or $(n-p-1)$ if an intercept is used. In this case, $p = 1$ so the denominator is $n-2$.

The standard errors of the parameter estimates are given by

$$\underline{\sigma}_{\beta 0} = \sigma_\varepsilon \sqrt{(1/n) + \left\{x_{\text{mean}}^2 \Big/ \sum (x_i - x_{\text{mean}})^2\right\}}$$

$$\underline{\sigma}_{\beta 1} = \sigma_\varepsilon \sqrt{1 \Big/ \sum (x_i - x_{\text{mean}})^2}.$$

Under the further assumption that the population error term is normally distributed, one can use these estimated standard errors to create confidence intervals and conduct hypothesis tests about the population parameters.

Once a regression model has been constructed, it is necessary to confirm the 'goodness of fit' of the model, and the statistical significance of the estimated parameters. Commonly used checks of goodness of fit include the *R*-squared, analyses of the pattern of residuals and hypothesis testing. Statistical significance can be checked by an *F*-test of the overall fit, followed by *t*-tests of individual parameters.

Interpretations of these diagnostic tests are based on the model being assumed. Although examination of the residuals can also be used to invalidate a model, the results of a *t*-test or F-test are sometimes more difficult to interpret if the model's assumptions are violated. For example, even if the error term does not follow a normal distribution, in small samples the estimated parameters will not follow normal distributions and complicate inference. With relatively large samples, however, the central limit theorem can be invoked such that hypothesis testing may proceed using asymptotic approximations.

Regression models usually predict a value of the Y variable, given known values of the $\mathbf{X}$ variables. Prediction within the range of values in the data set used for model-fitting is known as interpolation. Prediction outside this range of the data is termed extrapolation. Performing extrapolation relies strongly on the assumptions upon which the regression analysis is based. The further the extrapolation goes outside the data, the more room there is for the model to fail due to differences between the assumptions and the sample data, or the true values. When performing extrapolation, one should accompany the estimated value of the dependent variable with a prediction interval that represents the uncertainty. Such intervals tend to increase in size rapidly as the values of the independent variable(s) move further outside the range covered by the observed data.

However, this does not cover the full set of modelling errors that may be made: in particular, the assumption of a particular form for the relation between Y and $\mathbf{X}$. A properly conducted regression analysis will include an assessment of how well the assumed form is matched by the observed data, but it can only do so within the range of values of the independent variables available. This means that any extrapolation is reliant on the assumptions being made about the structural form of the regression relationship. It is best to use all available knowledge in constructing a regression model.

But whatever the caveats, regression analysis is widely used for prediction and forecasting; not only for estimating the values of variables outside of the area where related measurements have been made, but there is a substantial overlap with the design of predictive texting (there are also many economists who seek to apply such techniques to the data shown in figure 4.5 with a view to making short-term extrapolations). However this can lead to false relationships, so caution is advisable; for example, correlation does not imply causation (for example, which came first, the economic recession or increasing levels of unemployment).

10.3 Over-fitting data

As we have seen, if we determine n distinct values of x, and the corresponding values of y, it may be possible to find a curve that goes precisely through all n points (x,y) when graphed. This can be done by setting up a system of equations and solving them simultaneously. But this is not what regression methods are designed to do.

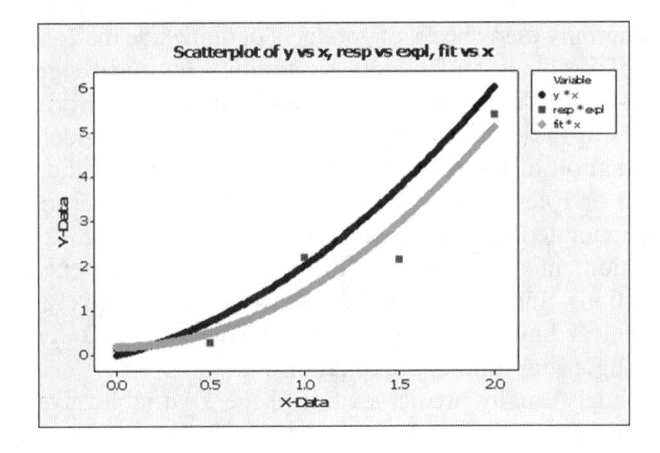

Most regression methods (for example, least-squares) estimate conditional means of the response variable given the explanatory variables. They are not expected necessarily to go through all the data points. Indeed, the residuals between the curve calculated from your model and the measured data is the means of expressing the uncertainty of the measurements, and there is always uncertainty. For example, with one explanatory variable X and response variable Y, if we fix a value x of X, we have a conditional distribution of Y given $X = x$; for example, the conditional distribution of the incomes of individuals who have had five years of education. This conditional distribution has an expected value (the population mean), which we will denote $E(Y|X = x)$; that is, the mean-income of people with five years of schooling, x. This is the conditional mean of Y given $X = x$. It depends on x, as $E(Y|X = x)$ is a function of x.

In least-squares regression, one of the model assumptions is that the conditional mean function has a specified form. Then we use the data to find a function of x which approximates the function $E(Y|X = x)$. This is different from finding a curve that goes through all the data points. Consider the following example to illustrate this point.[1] We will use simulated data: five points sampled from a joint distribution where the conditional mean $E(Y|X = x)$ is known to be x^2, and where each conditional distribution $Y|(X = x)$ is normal with standard deviation 1. Least-squares regression is used to estimate the conditional means by a quadratic curve $y = a + bx + cx^2$; that is, least-squares regression, with $E(Y|X = x) = \alpha + \beta x + \gamma x^2$ as one of the model assumptions, to obtain estimates a, b, and c of α, β, and γ, respectively, based on the data.

The above graphic shows:

- the five data points in red (one at the left is mostly hidden by the green curve)
- curve $y = x^2$ of conditional means (black)
- graph of the calculated regression equation (in green).

[1] Taken with permission and thanks from the teaching material of Professor Martha K Smith, University of Texas; ma.utexas.edu/users/mks/statmistakes/.

Note that the points sampled from the distribution do not lie on the curve expressing the means (black). Notice again, that the green curve is not exactly the same as the black curve, but is close to it. In this example, the sampled points are mostly below the curve representing the means. Since the regression curve (green) was calculated using just the five sampled points (red), the red points are more evenly distributed above and below the green curve than they are in relation to the curve representing the means (black).

Remember that in the real world, we would not know the conditional mean function (black curve); and in most problems, would not even know in advance whether it is linear, quadratic, or something else. Thus, part of the problem of finding an appropriate regression curve is trying to discover what kind of function it should be. Continuing with the example, if we (perhaps naively) try to obtain a 'good fit' to the data by trying a quartic (fourth degree) regression curve; that is, using a model assumption of the form $E(Y|X=x) = \alpha + \beta_1 x + \beta_2 x^2 + \beta_3 x^3 + \beta_4 x^4$, we obtain the following graphic.

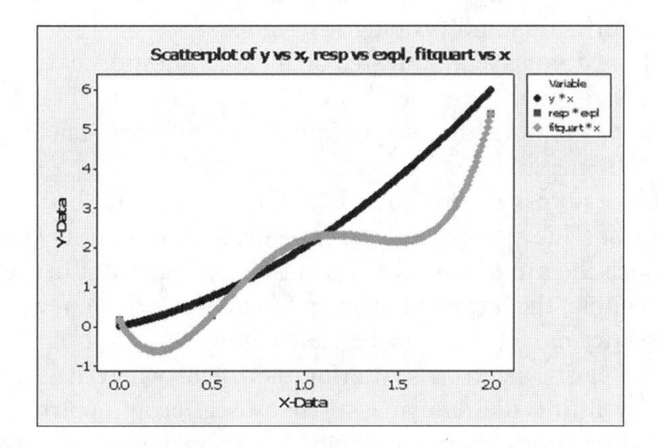

Here we barely see any of the red points, because they are all on the calculated regression curve (green). We have found a regression curve that fits all the data, but it is not a good regression curve, because what we are really trying to estimate by regression is the black curve (curve of conditional means). And for that, we have done a poor job; we have made the mistake of overfitting. We are well on the way to fitting an elephant, so to say. If we had instead tried to fit a cubic (third degree) regression curve (that is, using a model assumption of the form $E(Y|X=x) = \alpha + \beta_1 x + \beta_2 x^2 + \beta_3 x^3$), we would get something more wiggly than the quadratic fit and less wiggly than the quartic fit. However, it would still be overfitting, since (by construction) the correct model assumption for these data would be a quadratic function.

10.4 Avoiding over-fitting

As with most problems in statistics and data analysis, there are no hard and fast rules that guarantee success; it all depends upon the problem under investigation and the quality and quantity of the data. However, there are some guidelines, which apply to many types of statistical models. Of premier importance is the quality and quantity of your data. If you are gathering data (especially through an experiment), be sure to optimise the design and set-up of the apparatus before beginning to take data. There are no generally agreed methods for relating the number of observations needed to be made in an experiment to the number of independent variables in your model for that experiment, which will allow you to decide *a priori* whether your model is any good or will lead to useful results. This is also the problem with the design of clinical trials, as we saw in chapter 7. In their textbook, P I Good and J W Hardin [2] offer the following conjecture or rule of thumb on an appropriate sample size: 'If m points are required to determine a univariate regression line with sufficient precision, then it will take at least m^n observations and perhaps $n!m^n$ observations to appropriately characterize and evaluate a regression model with n variables'. That dataset could become very large, very quickly. Happily for the OKE measurements described in the previous chapter, it is relatively straightforward to acquire large quantities of data by varying the experimental conditions; for example, the delay time between the orienting and probing laser pulses.

Having mentioned von Neumann's elephant, the question naturally arises as to whether or not this is at all possible. That is, is it possible to model and define the silhouette of an elephant with four parameters, and then make the simulation wriggle its trunk by the addition of a fifth parameter?

Earlier, we came across a crestfallen Freeman Dyson who had had a few of the facts of a physicist's life pointed out to him by Enrico Fermi. I must confess to having heard much the same thing when working as a physicist in the Institut Laue-Langevin in Grenoble, the largest European facility for research involving neutrons and neutron scattering. It has to be said, however, that neutron scattering, particularly quasi-elastic neutron scattering, is a field where it is only possible to make progress by fitting the data measured by scattering neutrons from samples containing hydrogen to a range of complex (and sometimes opaque) models. One morning, while discussing with a visiting scientist the scattering data that had accumulated overnight, the question arose as to how many parameters should we use in attempting to fit the data to the model we thought appropriate. No sooner had we started talking, than someone in the group quoted the comment about the elephant. Apparently, these critics of data fitting always quote von Neumann's aphorism with a smile, even though they themselves have probably never done any data fitting more sophisticated than eye-balling a straight line through a cloud of points. But the story of the elephant will not go away, which probably means that there is some truth in it. So how easily can you fit the shape of an elephant?

James Wei was the first to demonstrate that something like the shape of an elephant could be generated by computer fitting. In 1975 [3], Wei took the outline of an elephant defined by 36 points (see figure 10.1) and used least-squares Fourier series fits of the form $x(t) = \alpha_0 + \sum_i \alpha_i \sin(it\pi/36)$ and $y(t) = \beta_0 + \sum_i \beta_i \sin(it\pi/36)$ for $i = 1,\ldots N$ to try and represent the defined outline of the elephant. Figure 10.1 shows the fits Wei obtained for $N = 5$, 10, 20 and 30; he stopped at a thirty parameter fit, which has a certain similarity to the 36-point image of an elephant which Wei had

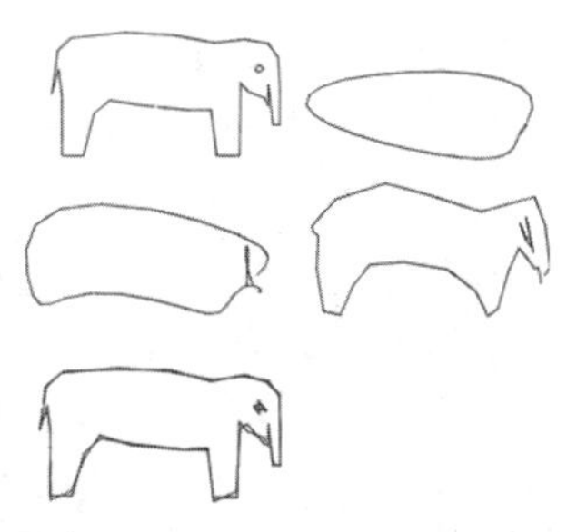

Figure 10.1. Fitting the outline of an elephant. These images are taken from the fitting of James Wei in 1975 [3]. One can clearly see the evolution of the outline of the elephant; from an egg to the embryo and then to the elephant, but Wei was using a lot more than four parameters in his fit—and there is no wiggling trunk. (*a*) James Wei's starting point, an elephant defined by 34 points (not a simulation, but a drawing). (*b*) A least-squares fit with five terms—more of an egg than an elephant. (*c*) A least-squares fit with 10 terms. (*d*) A least-squares fit with 20 terms—an embryo of an elephant? (*e*) A least-squares fit with 30 terms—now something like an elephant.

drawn to be his ideal, but as we can see, it is not too good, and is using significantly more than the five parameters mentioned by von Neumann.[2]

This work by Wei is humorous and insightful. The standard image against which Wei was seeking to generate an image was itself pretty crude. Perhaps the target image should have been a digitized photograph, but again this is only an approximation to an elephant, and we can clearly see that the generated fits do not really merit any further investigation. This straightforward exercise should, however, encourage thinking about the nature of reality, about what constitutes a true model and the 'right answer', and about the merits of 'approximate models'. Perhaps there is no such thing as the 'perfect fit', in the same way that there is no such thing as the 'right answer' or a measurement free of uncertainty.

A more recent attempt to parameterize an elephant also makes use of Fourier series. The outline of the beast is best described by a set of points $(x(t), y(t))$, where t is a parameter that can be described as the time elapsed while moving along the path of the animal's outline [4]. If motion along the animal's contour is uniform, t defines the length of an arc or a curve of the animal's body. Expanding $x(t)$ and $y(t)$ in Fourier series gives

$$x(t) = \sum_k A_k^x \cos(kt) + B_k^x \sin(kt)$$
$$y(t) = \sum_k A_k^y \cos(kt) + B_k^y \sin(kt)$$

where A_k^x, B_k^x, A_k^y and B_k^y are the coefficients of the expansion; k defining the term in the expansion and x and y labelling the expansion of the x and y function, respectively. Using such Fourier expansions it is possible to analyze shapes by

[2] Interestingly, the least-square fits of James Wei given in figure 10.1 suggest the gestation of the elephant, from an egg to the animal. Perhaps this is a new take on the now largely discredited hypothesis that in developing from embryo to adult, animals go through stages resembling successive stages in the evolution of their remote ancestors; that is, the Theory of Recapitulation ('*ontogeny recapitulates phylogeny*') of the German biologist Ernst Haeckl.

tracing the boundary and calculating the coefficients in the expansion. By truncating the expansion, the derived shape is smoothed. The question is how many truncated parameters are required to give the derived curve the outline of an elephant? From Fourier analysis, we know that $k = 0$ corresponds to the centre-of-mass of the perimeter, $k = 1$ corresponds to the best fit ellipse. The higher components trace out elliptical corrections analogous to Ptolemy's epicycles (see en.wikipedia.org/wiki/Deferent_and_epicycle). Mayer *et al* were able to generate the image seen in figure 10.2, using the expansion coefficients given in their paper.

Here the authors have only used the required four parameters, but the final result is a comical cartoon of an elephant. The real part of the fifth parameter is the 'wriggle coefficient' which determines the x-value where the trunk is attached to the body. The imaginary part of this parameter is used to make the cartoon more 'life-like' by giving it an eye.

In another approach, the programmers formed a second elephant by reflection of the distal portion of the proboscis of the first elephant, parallel to the x-axis. The wiggling was then implemented by linear interpolation between the primary and secondary elephants, using a sinusoidally varying mixing coefficient. The virtual

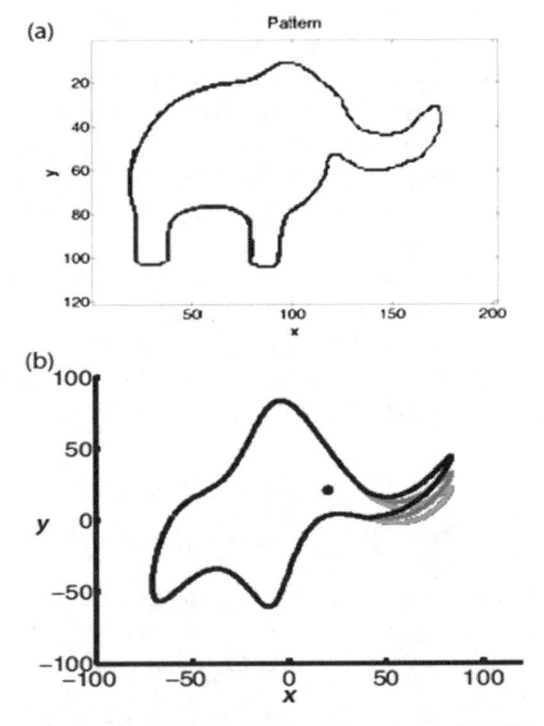

Figure 10.2. These images are taken from [4]. In these images, the real part of the fifth parameter is the 'wriggle coefficient' which determines the x-value where the trunk is attached to the body. The imaginary part of this parameter is used to make the cartoon more 'life-like' by giving it an eye. All the images in this figure and figure 10.1 are cartoon-like, but are they elephant-like?

elephant may be observed here: http://theoval.cmp.uea.ac.uk/~gcc/projects/elephant/ And interactive graphics at http://demonstrations.wolfram.com/FittingAnElephant/ and http://kourentzes.com/forecasting/2016/02/08/how-to-fit-an-elephant/ allows one to observe directly the data fitting to this strange and instructive shape.

Success in the analysis of real data, and the resulting inferences often depend crucially on the choice of a best model. Data analysis in those fields where researchers have a liking for multi-variate model fitting (I am sure that there are economists who are quite happy to construct hugely complex models to try and fit the type of data displayed in figure 4.5 as the potential rewards are too great to ignore) should be based on a strict application of Occam's razor, *Essentia non sunt multiplicanda praetor necessitatem* (entities are not to be multiplied beyond necessity); that is, represented by an accurate approximation of the data in hand. This fitting should not be thought of as a search for the 'true model'. Modelling and model selection can all too easily become an art, rather than an aid to science. And when fitting large data sets, one needs to be aware that it is possible to fall into a minimum in the fitted surface (as defined by the value of your residuals), perhaps even a deep minimum, but which is not the true minimum.

Enrico Fermi gave Freeman Dyson a tutorial on the dangers of over-fitting; something which all of us who have undertaken complex indirect experiments have been guilty of at one time or another. It is a cautionary tale in physics, warning us not to lose sight of the physics (and the concept of measurement uncertainty) when programming and fitting data. As it happens, Dyson took Fermi's point and changed his area of research. As Dyson commented over half a century later, '*I am eternally grateful to him, for destroying our illusions and telling us the bitter truth*' [1]. But now we do know that it is possible, with only four complex parameters to generate the vague outline of an elephant (with a contemporary aesthetic), and with a fifth complex parameter one can control both the position of the trunk and the position of an eye.

Further reading

There are some very useful videos on the Wikipedia entry for curve fitting, that demonstrate in real time what a scientist experiences when he or she is trying to fit his measured data to some existing model that they have programmed: https://en.wikipedia.org/wiki/Curve_fitting).

There are many standard university texts on the do's and don'ts of data fitting, and how to avoid the elephant traps of over-fitting data; for example [2, 5] (these texts, or parts of them are also available as free-access pdf files). The online encyclopaedia, Wikipedia is also a good source for further information (see entries on statistical fitting, regression analysis, curve fitting, goodness of fit, and errors and residuals). I can also recommend the following site at the University of Texas: www.ma.utexas.edu/users/mks/statmistakes/Types.html

References

[1] Dyson F 2005 *Nature* **427** 297

[2] Good P I and Hardin J W 2006 *Common Errors in Statistics (And how to avoid Them)* (New York: Wiley)

[3] Wei J 1975 *Chemtech.* February 128–9

[4] Mayer J, Khairy K and Howard J 2010 *Am. J. Phys.* **78** 648–649

[5] Ryan T 2009 *Modern Regression Methods* (New York: Wiley)

◎

编辑手记

著名数学家、物理学家 F. J. Dyson 曾说：

> 不能一劳永逸地定义数学在物理科学中的位置，数学和科学的相互关系就像科学本身的纹理那样丰富和多样.

本书是一部英文版的物理学教科书，中文书名或可译为《量化测量：无所不在的数字》.

本书的作者为杰弗里·H. 威廉姆斯（Jeffrey H. Williams），英国人，1956 年 4 月 13 日出生于威尔士斯旺西，1981 年在剑桥大学获得化学物理学博士学位. 他的职业生涯是在自然科学领域. 首先，他是剑桥大学、牛津大学、哈佛大学和伊利诺伊州大学的研究科学家，后来又在格勒诺布尔的劳厄-兰格万研究所担任物理学家，该研究所是世界领先的中子和中子散射研究中心之一. 他发表了 60 多篇技术论文，并在同行评议的文献中受邀发表多篇评论文章. 1992 年他成为美国艺术和科学研究院的《科学》杂志的欧洲编辑，进入了科学出版和科学传播领域. 随后，他担任"国际纯粹与应用化学联合会"的助理执行秘书，该机构负责通过国际合作促进化学发展.

正如作者在前言中所指出：

> 几乎每个学过科学的人都做过实验. 理论有它们的位置，但它们就像时尚一样，随着时间的推移而变化，并且只有在有人设计实验来测试它们的真实性之前才具有相关性. 然后，精心设计的实验的结果将永

远持续下去,想想洗澡的Archimedes,或者比萨斜塔上的Galilei,或者在一个只有一缕阳光的黑暗房间里的Newton和他的棱柱.实验是推动科学进步的重要动力.在考虑可能的行为理论模型时,作为最终仲裁者的实验结果占主导地位,这在粒子物理学、生物化学和医学中都是真实的.的确,那些与我们的社会最相关的科学分支只涉及实验.我们的健康和文明都基于实验结果,我们培养未来的科学家,方法是让他们进行既定的实验,然后他们可以设计自己的新颖的实验.不过你如何设计一个实验,应该考虑什么来决定一个实验是否产生了想要的结果,或者确实产生了任何有用的结果?

这一切似乎都是不言而喻的,然而,确实可以说,对测量量化重要性的认识并没有像应有的那样普遍.太多的实验报告留下了太多的未被解释的或是未被重视的内容,并且其引用的误差和精确度水平通常表明它们纯粹是事后才估计出来的.事实是每一个实验都存在不确定性.

想象一下,有一组数据是由某物的测量值按时间绘制而成的.数据非常“嘈杂”,点很分散,数据点的线呈上下曲折的走势,或者是随着时间的推移呈总体下降的趋势.如果此图代表失业人数,你可能是一位总理,为你的政府经济政策寻求正当理由;另外,如果这些数据表示宇宙微波背景辐射的偏振状态,你可能是一名宇宙学家,正在寻找引力波的证据,但在这两种情况下问题都是一样的.也就是说,试图找到你认为隐藏在模糊的背景噪声中的有用数据.

所有实验都涉及尝试在可能有多种来源的不需要的噪声背景下测量感兴趣的事物.有时,与你想要测量的噪声相比,实际的噪声量很少,并且可以清楚地看到所需的结果,还可以轻松地进行测量.然而,也许最常见的是,噪声水平大于你感兴趣的信号,你必须采用各种技术来允许检测系统区分信号和噪声.这是实验设计的基础,它允许人们确定所提议实验的极限灵敏度,然后回答问题,这个装置是否值得为我感兴趣的实验而建造,或者是否有一种不同的设计能导致更灵敏的仪器出现?

Lord Rutherford有句名言:“如果你的实验需要统计数据,那么你应该做一个更好的实验.”这种对实验设计的批评是有道理的,但使用测量统计数据进行分析也可以被认为是一个有用的事后检查;统计分析可能会揭示实验失败的原因.通常在科学中,重要的不是新事实和数据的测量,或者新实验的设计,而是对思考这些事实和测量的新方法的发现.

在本书中,将研究18世纪90年代度量测量的一个意想不到的副产品.今天,度量系统的批评者对 Pierre Méchain 的"误差"进行了大量的讨论.可怜的 Pierre Méchain 被指控伪造或操纵数据(就科学家而言,这是一个被大家普遍关注且不可原谅的罪过),以使他的结果看起来比实际更精确.这种准确性变化的实际原因是本书的主题,我们将看到对实验不确定性起源的详细分析是如何导致统计科学的发展以及现代数据分析方法的发展的.描述当今科学家用来分析其测量数据的统计和概率方法不是本书的目的,因为有许多标准文本来讨论这些内容,并且互联网包含许多这种文本(可免费访问的 PDF 文件).我希望展示的是我们如何看待实验,以及观察测量的定量和定性方法之间的区别.

我们工作室版权部主任李丹女士为了方便国内读者快速了解本书的基本内容,特翻译了本书的目录,如下:

本书既可以当作物理教材使用(因为国内目前许多高校都开始使用英文原版教材),亦可当作科普读物.西班牙作家伊莲内·瓦列霍在《书籍秘史》中写道:

> 在乱世中寻获图书宛如在深渊的边缘努力保持平衡.拥有图书相当于努力捡起宇宙散落的碎片,拼成有意义的图案;相当于面对混乱,搭建出和谐的建筑物;相当于聚沙成塔;相当于筑一道堤坝,抵挡时间的海啸.

刘培杰

2024年4月23日

于哈工大

国外优秀物理著作
原版丛书（第一辑）

Comets in the 21st Century—A Personal Guide to Experiencing the Next Great Comet

21世纪的彗星

——体验下一颗伟大彗星的个人指南

（英文）

●［美］丹尼尔·C. 布易士（Daniel C.Boice）
●［美］托马斯·霍基（Thomas Hockey） 著

哈爾濱工業大學出版社
HARBIN INSTITUTE OF TECHNOLOGY PRESS

黑版贸登字 08-2021-026 号

图书在版编目(CIP)数据

21世纪的彗星：体验下一颗伟大彗星的个人指南 = Comets in the 21st Century：A Personal Guide to Experiencing the Next Great Comet：英文/（美）丹尼尔·C. 布易士（Daniel C. Boice），（美）托马斯·霍基（Thomas Hockey）著. —哈尔滨：哈尔滨工业大学出版社，2024. 10
（国外优秀物理著作原版丛书. 第一辑）
ISBN 978 - 7 - 5767 - 1341 - 1

Ⅰ. ①2… Ⅱ. ①丹… ②托… Ⅲ. ①彗星-普及读物-英文 Ⅳ. ①P185. 81 - 49

中国国家版本馆 CIP 数据核字(2024)第 073690 号

21 SHIJI DE HUIXING：TIYAN XIAYIKE WEIDA HUIXING DE GEREN ZHINAN

策划编辑　刘培杰　杜莹雪
责任编辑　刘立娟　刘家琳　李　烨　张嘉芮
封面设计　孙茵艾
出版发行　哈尔滨工业大学出版社
社　　址　哈尔滨市南岗区复华四道街 10 号　邮编 150006
传　　真　0451 - 86414749
网　　址　http://hitpress. hit. edu. cn
印　　刷　哈尔滨博奇印刷有限公司
开　　本　787 mm×1 092 mm　1/16　印张 63. 75　字数 1 195 千字
版　　次　2024 年 10 月第 1 版　2024 年 10 月第 1 次印刷
书　　号　ISBN 978 - 7 - 5767 - 1341 - 1
定　　价　378. 00 元(全 6 册)

Contents

Preface

I knew I wanted to study the heavens, ever since peering at Comet West seeming to hover above my home. When in the 1990s it became known that another Great Comet soon would appear in the sky, I wrote *The Comet Hale–Bopp Book: Guide to an Awe-inspiring Visitor from Deep Space* (1996 (Shrewsbury, MA: ATL)). The comet did not disappoint; I recall lecturing via bull horn to a thousand people standing in front of the Adler Planetarium, seeing the comet despite the bright lights of downtown Chicago.

The goal of that book was to introduce at the most basic level this cosmic sight to the millions who would watch it. While about Comet Hale–Bopp specifically, the work contained a lot of general information on comets that stood in its own right. (Some of this material appears in the present volume.) Of course, it is by now outdated—in rode co-author Dr Daniel Boice, whom I had known for nearly forty years and who, during that time, devoted himself to the investigation of comets, to provide the leap in comet science and observation that has occurred in more recent years. Thanks for his support of this project also go to Dr William Sheehan, astronomy author extraordinaire.

This present volume was created literally around the world: Boice writing from Thailand and I working in the United States. We hope our combined effort pleases the reader, regardless of whether they have had the opportunity to see a Great Comet themselves—or have yet to.—T H, Cedar Falls, December 2018

I was born at the dawn of the space age and was inspired by NASA's Apollo missions, which eventually landed 12 people on the Moon. Afterwards, my interest in all things astronomical grew, and I eventually bought my first telescope, a used 60 mm refractor, while attending high school in San Francisco, CA. I quickly graduated to an 8 inch Dobsonian reflector that I built myself under the guidance of the legendary John Dobson and joined the amateur ranks by enrolling in the historic San Francisco Amateur Astronomers. I have been very fortunate to follow my dream to become a professional astronomer and am deeply indebted to NASA and the National Science Foundation who have supported my career and to the opportunities provided by my employer of 26 years, Southwest Research Institute (San Antonio, TX). Ultimately, I must thank the American people for this funding and their trust that their money would be well spent!

Although primarily a researcher, my passion for teaching still burned. I was able to share my knowledge and enthusiasm of astronomy with students at University of Texas at San Antonio, San Antonio Community College, and Trinity University for over 20 years. Teachers learn a lot from their students. During this time, I was fortunate to supervise graduate students, our next generation of astronomers, and participate in many foreign collaborations.

I met my co-author Dr Thomas Hockey in graduate school at New Mexico State University. I graduated three years earlier and left for Los Alamos National Laboratory in 1985 where I was converted to comet science by Comet Halley and my mentor, Dr Walter Huebner. T H joined the faculty at the University of Northern Iowa to pursue his joy of teaching and historical astronomy. Our paths diverged for several decades, even though we are both members of the American Astronomical Society (AAS) and the International Astronomical Union (IAU). I took notice in 2017 when he won the prestigious Osterbrock Book Prize from the AAS Historical Astronomy Division for his four-volume reference work, *Biographical Encyclopedia of Astronomers.* Having the chance to co-author a book with him was a no-brainer.

Special thanks go to Dr Walter Huebner, retired Institute Scientist at Southwest Research Institute, and Professor Amaury de Almeida, Chair of the Astronomy Department at the University of São Paulo, Brazil, for reviewing our manuscript. Their insightful comments have significantly improved this book. Naturally, we are responsible for any remaining errors. I share my co-author's desire that you, the reader, will enjoy our work.—D B, Puan Phu, Thailand, December 2018

Frontispiece

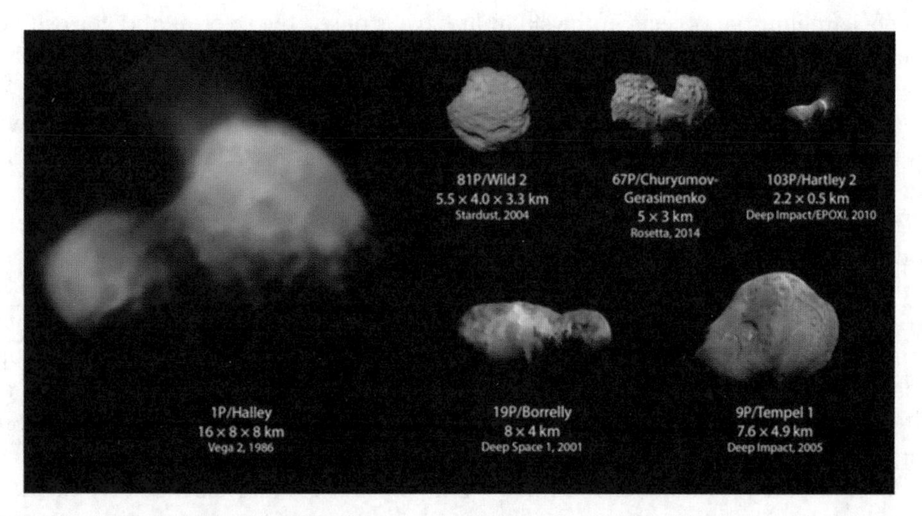

Gallery of comet nuclei visited by space probes (to scale) with year of encounter. Courtesy of The Planetary Society.

Foreword

Comets are unusual objects in the sky that have aroused the interest and curiosity of many who have seen one. What are these objects, where did they come from, what causes their strange appearance, and what do we know about them? The authors of this book are much better known than these comets and much younger; they will reveal to us many of a comet's visual phenomena in familiar terms from everyday life without getting into deep scientific explanations.

The purpose of this book is to bring comets into the living rooms of general households, to familiarize politicians with these fascinating objects when they ponder funding for comet research, to teach children and young students, and to provide teaching tools about these very unusual objects in our skies. The presentation is very comprehensive in its description of orbits around the Sun, the development of the coma (escaping atmosphere) from a comet's nucleus and source of all activities, various types of comet tails, trailing as well as leading as a comet orbits our Sun, ancient beliefs and explanations of these phenomena, and the most recent discovery of the first interstellar comet. An interstellar comet is particularly exciting, because it has the potential to reveal data about a neighboring star without going there. For example, the stellar nebula from which the star formed may have been larger or smaller than the solar nebula (thus a different density), have a different radiation field and have a somewhat different composition. This would lead to different chemical reactions and thereby affect the conditions for the origins of life when compared to the conditions in our Solar System.

Daniel C Boice is a well-known astronomer, specializing in comet science, a discipline where professionals and amateur observers work hand-in-hand to study the mysteries of these celestial bodies. Thomas A Hockey is a professor of astronomy, specializing in the history of that field, and an award-winning author. We spoke recently about the allure of comets and the impression they make on people. Many astronomers, both professional and amateur, can name a comet that influenced their interest in the sky at an early age. For Boice, it was Comet West, which he saw light up the morning sky in 1976. For Hockey, it was Comet Bennett. Bennett was so shiny that he could even see it through his dining room window, in March 1970, the month of his eleventh birthday. 'Dan', Thomas suggested, 'sky lovers don't need to tell one another their age, only the name of their comet'. For me, it was the Great Comet Ikeya–Seki, the spectacular sungrazer of 1965.

Walter F Huebner, December 2018
Division of Space Science and Engineering
Southwest Research Institute
San Antonio, TX

Author biographies

Daniel C Boice

Daniel Boice is the principal astronomer at Scientific Studies and Consulting in San Antonio, TX. Prior to his present position, he spent 26 years in the Space Science and Engineering Division at Southwest Research Institute, TX, where he performed cometary research sponsored by NASA and the National Science Foundation. Concurrently, he held a joint appointment to the Department of Physics and Astronomy faculty at the University of Texas at San Antonio, where he taught undergraduate and graduate courses for 20 years. After receiving his BS in Physics at Brigham Young University in 1975, he obtained a PhD in astronomy at New Mexico State University in 1985. While a postdoctoral fellow in the Theoretical Division at Los Alamos National Laboratory, Dr Boice developed a computer model of cometary comas that has been successfully used to interpret spacecraft data and ground-based observations of many comets. He was a member of the science team for NASA's Deep Space 1 Mission to Comet P/Borrelly. His professional activities include over 75 peer-reviewed research papers, several hundred conference reports, and serving as Past Chairs of the Physical Studies of Comets Working Group (International Astronomical Union) and Space Related Studies of Small Bodies of the Solar System (Committee on Space Research (COSPAR)). In 2000, he became a Fellow of the Royal Astronomical Society. He has spent several years abroad teaching and working with colleagues in Germany, Japan, France, and Brazil. Dr Boice is continuing his comet research to assimilate information from ground-based observations and *in situ* spacecraft measurements to develop a better global understanding of comets. When not engaged in all things comets, Daniel loves collecting books and rock 'n' roll music, board gaming, and rice farming with his family in northern Thailand.

Thomas Hockey

Thomas Hockey is a professor of astronomy in the Department of Earth and Environmental Sciences at the University of Northern Iowa (UNI), where his research interests include studies of the history of planetary astronomy in light of the modern search for planets orbiting other stars, as well as the astronomy of solar eclipses and comets, and the archeoastronomy of pre-historic peoples. After receiving his BS in Planetary Science at the Massachusetts Institute of Technology in 1980, he ventured west to obtain an interdisciplinary PhD (Astronomy, History, and Philosophy) at New Mexico State University in 1988, specializing in History of Astronomy and Science Education. Professor Hockey enjoys teaching, having taught undergraduate and graduate courses at UNI for three decades. His professional activities include many peer-reviewed research papers and several books. In 2017, he received the

Osterbrock Book Prize from the Historical Astronomy Division (HAD) of the American Astronomical Society (AAS) for his four-volume tome, *Biographical Encyclopedia of Astronomers*. He has also served as the Editor-in-Chief of the *Astronomy Education Review* and Past Chair of the HAD/AAS.

Chapter 1

Introduction

'The comet is coming!' The words are alliterative and compelling. We do not know the name of that comet yet. We do not know when it will arrive. However, eventually, another Great Comet will grace our skies—one that dazzles the naked eye and may even be seen in broad daylight. This means that you can have a one-on-one relationship with the comet, without the assistance of astronomers, television commentators, or anybody. Like others before it, the next Great Comet will be every person's comet.

Typically, a brilliant comet makes an appearance in Earth's sky every decade or so. Except for those who saw the difficult-to-observe Comet McNaught in 2007, the last Great Comet (figure 1.1), a generation grew up without really knowing what a comet is all about. We are overdue! This book is written to enhance your anticipated comet experience, or just to inform those who wish to know a little more about these celestial visitors.

Why should you make a point to experience the next Great Comet? A bright comet is an astronomical event. It is a 'one-time offer' from the heavens. Moreover, unlike eclipses or conjunctions (the alignment of planets in the sky), it is unique and usually cannot always be predicted. A comet is an awe-inspiring presentation of nature, on par with an erupting volcano, a mass migration, or a cyclone. Moreover, it is not nearly as dangerous to watch!

People have been looking up at comets for a long time, longer than we have recorded history. Comets are part of our culture; they are incorporated into our traditions and beliefs. Observations of comets are universal. They are a communal human experience. As you look up at it, perhaps from your back porch, you will be sharing it with persons around the planet, standing outside huts, mansions, pueblos, skyscrapers, yurts, condos, igloos, marinas, trailers, villas, and pup tents (figure 1.2).

A comet does not flash across the sky, as some individuals who confuse comets with meteors believe. You do not have to set your alarm clock and go to bed with

doi:10.1088/2053-2571/ab027ech1

Figure 1.1. Comet McNaught in 2007. NASA image.

Figure 1.2. Comet Ikeya-Seki in 1965. Photo taken at NASA's Jet Propulsion Laboratory (JPL) Table Mountain Observatory, reproduced with permission from James W Young, photographer.

your sneakers on to catch it. Depending upon how bright it gets, most people will be able to view a Grand Comet for weeks or months.

Furthermore, unlike the constancy of the full Moon or constellations of stars, a comet will change in appearance, perhaps night-to-night or even hour-to-hour.

The authors spoke recently about the allure of comets and the impression they make on people. Many astronomers, both professional and amateur, can name a comet that influenced their interest in the sky at an early age. For D B, it was Comet West, which he saw light up the morning sky in 1976. For T H, it was Comet Bennett (figure 1.3). Bennett was so shiny that he could even see it through his dining room

Figure 1.3. Comet Bennett in 1970. Copyright 1970 Fred Espenak, www.Astropixels.com.

window, in March 1970, the month of his eleventh birthday. 'Dan', T H suggested, 'sky lovers don't need to tell one another their age, only their comet.'

During the 1970s fad of 'disaster movies', T H watched his home town of Phoenix, AZ, 'destroyed' by a fluke comet in a television version of the genre. (With all that empty desert surrounding Phoenix, it was a remarkable shot!) Those cardboard models of familiar landmarks flying about looked silly, but foreshadowed evidence of real cometary cataclysms discovered in the 1990s.

In 1986, we joined astronomers around the world pointing out Comet Halley to the public (figure 1.4). That faint smear (no longer a Great Comet), seen through the telescope, never failed to disappoint. The fact that it was 4:00 A.M., and wintertime, did not help matters, either.

Would a bright comet ever come? Actually, there had been such a comet in 1976, Comet West (figure 1.5). However, West came less than two years after the 'hype' accompanying the fizzled Comet Kohoutek, the so-called 'comet of the century'. Many did not wish to be fooled twice and remained skeptical. Thus, most folks missed a lovely comet. (Again, early morning prominence kept the audience low.)

It almost seemed as if the comets were impishly conspiring to mislead us. Then, suddenly, it was here! Comet Hale–Bopp was discovered in 1995. One year before Hale–Bopp, Comet Hyakutake danced overhead. Hyakutake was discovered less than two months before its peak brightness in the Earth's sky in 1996. It came and went as Comet Hale–Bopp crept inexorably nearer for a 1997 performance (figure 1.6). In 2007, Comet McNaught became one of the brightest comets of the last century. Unfortunately, it remained in that portion of the Celestial Sphere only seen from far south. Thus, most of the world's population had no chance of getting a good look at it.

Figure 1.4. Comet Halley in 1986. NASA image.

Figure 1.5. Comet West in 1976. J Linder/European Southern Observatory (ESO).

Lest we give you the impression that comets in general are very rare events, the above discussion centered on Great Comets, ones that can be seen with the naked eye, some in broad daylight. They outshine Venus, the third brightest object in the heavens (following the Sun and Moon). There have been only nine such comets (McNaught included) in the past three and a half centuries (on average one every 38 years). In 2013, Comet ISON was expected to be on this list (predicted to be almost as bright as the full Moon due to its expected proximity to the Earth), but it fragmented during its close approach to our home star. During the past century, there have been 14 brilliant, showpiece comets that could be seen by the naked eye at night or with a small pair of binoculars (on average 1.4 comets per decade). Notable

Figure 1.6. Comet Hale–Bopp in 1997. NASA image. Copyright 1970 Fred Espenak, www.Astropixels.com.

naked-eye comets in the past few years include Comet PanSTARRS (C/2011 L4), the three Lovejoy comets (C/2011 W3, C/2013 R1, and C/2014 Q2), and Comet 46P/ Wirtanen[1]. Dozens of garden-variety comets are discovered each year, some bright enough to be seen with a small telescope, albeit most are out of reach for amateurs. Every night, a few comets are in the night sky to be studied by the pros. We may be awaiting the next Big One, but many comets can be enjoyed every few years or so.

You probably know that comets, like all other Solar System bodies, orbit our Sun so they periodically return to our skies when they get close to the Earth. Almost all observed comets follow elliptical orbits that penetrate well inside the orbit of Mars, some even cross the orbit of Mercury. In practice, you rarely see comets beyond Jupiter since they are very small and dim. Why this happens is explained in chapter 3.

What if comets were sentient beings or if aliens installed monitoring equipment on them; imagine the record of the Earth they would capture. Comet Halley would have an impressive photo album of the Earth taken at 76 year intervals, dating at least back from the ancient Chinese to the present. Major milestones would include:

- 240 BCE—Chinese 'broom star' seen during the Warring States Period, first historical record.
- 87 BCE—Babylonians record it in their province of the Parthian Empire.
- 12 BCE—'Hung like a sword over Rome before the death of Agrippa'—historian Dion Cassius.
- 451—Defeat of Attila the Hun by the Visigoth/Roman alliance at the Battle of Châlons.
- 684—Seen by Europeans and later published in the *Nuremberg Chronicle.*
- 1066—Depicted in the Bayeux Tapestry, Battle of Hastings.

[1] This comet nomenclature will become clear in chapter 5.

- 1301—Seen as the Star of Bethlehem in Giotto di Bondone's fresco *The Adoration of the Magi.*
- 1607—Observed by the famous German astronomer Johannes Kepler and published in his book, *De Cometis Libelli Tres*, in 1619.
- 1682—Sighted by Edmund Halley at Islington, 'Hence I dare venture to foretell, that it will return again in the year 1758'.
- 1910—Jack Johnson knocks out James Jeffries to retain his World Heavyweight Championship title, Boy Scouts founded, comet hysteria, and the first photograph of Halley.
- 1986—Challenger Space Shuttle disaster, Chernobyl catastrophe, first laptop computer from IBM (weighing 12 lbs), and international armada of space probes to Halley.

Maybe a few selfies would be thrown in, too. What will be seen in 2061 during its next return?

Is a mere comet worth all the fuss? For example, does it merit this book? You be the judge. For a few days in 1997, Comet Hale–Bopp was all the rage with TV meteorologists. (We guess weather and astronomy both involve looking up.) If you were one of those who wished that they would just get on with telling you whether it was going to rain, comets probably are not for you. On the other hand, if Comets Hale–Bopp or McNaught whetted your appetite for these celestial specters, this is the place to be. Here is what we know, so far.

Chapter 2

Roller-coaster comets

One does not need to know about pyrotechnics to appreciate a good fireworks show. Likewise, one does not need to know a lot about comets to appreciate the show they put on. However, for those of you for whom understanding enhances seeing, this chapter will give you a bit of background information on comets. If it piques your interest, you will want to read the rest of the chapters.

How do we define 'comet?' You may be surprised that astronomers have no official definition of this word. (You can find one in language dictionaries, but these are not meant for the specialist.) We accept the millennial-old traditions of our ancient relatives who noted that some 'stars' were fuzzy and had the appearance of 'hair' streaming from their 'heads'. They also moved through the heavens relative to the surrounding stars. The Greeks used the word 'κομήτης', meaning long-haired. So when astronomers discover a small Solar System body that shows signs of actively producing a surrounding cloud, called the *coma*, and/or *tail(s)*, it is classified as a comet. This simplistic definition now leads to confusion since we have several objects that share the distinction of being both an *asteroid*[1] and a comet, i.e. they were originally discovered as asteroids but later showed signs of activity or vice versa (e.g. asteroid 2060 Chiron and Comet 95P/Chiron are the same object). Some small bodies were recently discovered in the Main Asteroid Belt that displayed *dust* tails and so were named 'Main Belt comets'. Later several were shown to be activated by collisions with other bodies, not water ice *sublimation*[2] as is the case for comets, so are actually 'active asteroids', but still retain a comet designation. In fact, there may be a continuum of bodies between the inert asteroids and the active comets. Now that we have gathered details on thousands of these bodies using high-powered

[1] Asteroids are small Solar System bodies primarily made up of rock and metal. Most are found between the orbits of Mars and Jupiter, a region called the Main Asteroid Belt. More on these space rocks can be found in chapter 3.

[2] Sublimation is the chemical term to describe the change of a substance from solid to gas and vice versa without passing through the liquid phase.

doi:10.1088/2053-2571/ab027ech2

telescopes and spacecraft, this observationally motivated definition needs a makeover, much like the definition of 'planet' was revised in 2006. Comets should be classified by their physical and chemical properties, time and place of formation in the Solar System, activation mechanism, and orbital properties. Hopefully, astronomers will remedy this rather embarrassing situation soon! If, after reading this book, you would like to propose a definition, please let us know.

2.1 The paths of comets

Comets are peculiar beasts. They seem to appear in the sky suddenly, unexpectedly. They move rapidly along paths unlike any other object in the sky. They do not look like anything else in the sky, with their fuzzy heads and long tails, and their appearance may change nightly. Then, after weeks or months, they go away, some coming closer to the Sun than any other Solar System object.

It was this mysterious behavior of comets that gave them their reputation. Comets often appear as swords or daggers in the sky. In days of old, comets were considered omens of things to come; unpredictable, they were thought to be celestial prophecies of the future. Most often, the events they 'predicted' were interpreted to be bad news: a famine, a plague, the death of a monarch (figure 2.1). As there always was something bad happening somewhere—alas, there always is—the 'predictions' were right much of the time!

Then, late in the seventeenth century, Englishman Edmond Halley (1656–1743) suggested that comets follow all the rules of motion that other celestial objects do (figure 2.2). If that were true, it took some of the mystique out of comets and the fear of comets out of the minds of people.

Until he tackled comets, Edmond Halley seemed destined by history to play Watson to the great scientist Isaac Newton's (1642–1727) Sherlock Holmes (figure 2.3). The two men were contemporaries. It was Halley who encouraged his eccentric and reclusive friend to publish Newton's world-changing theories of motion and

Figure 2.1. The Bayeux Tapestry illustrates the conquest of England by the Normans in 1066. Note the comet looming over losing King Harold's head.

Figure 2.2. Edmund Halley. NASA image.

gravity. These theories demonstrated that the Universe ran according to the same scientific laws of motion that, here on the Earth, caused the apocryphal apple to fall on Newton's head.

Halley embraced gravity wholeheartedly. Newton had shown that all objects in the Solar System[3] must revolve in orbits about the Sun, under the influence of the gravitational force between them and the Sun. What about comets? Comets did not seem to be a periodic phenomenon as regular as the orbit of a planet. Before Halley, each comet seen in the sky was considered a 'one shot deal'; it appeared and disappeared, never to be seen again. Even the shifting of a comet from the evening to the morning skies (or vice versa), due to its passage around the Sun, was thought to be two different comets.

But Halley noticed, in the historical records, stories of a similarly appearing comet in the years 1531, 1607, and 1682. That was once every (about) 76 years! Maybe it was the *same* comet, thought Halley, coming near the Earth where we can see it, with that frequency.

[3] Although moons revolve about their host planet, they still orbit the Sun.

Figure 2.3. Isaac Newton. Reproduced from Bolton S K 1889 *Famous Men of Science* (New York: Crowell).

Halley used Newton's laws, and observations of the comet, to plot an orbit for it around the Sun. He then went out on a limb and predicted its return in the year 1758. Halley was dead by then, but his comet was very much alive and turned up—more-or-less on schedule—Christmas Day 1758, recognized by Nicolas-Louis de Lacaille (1713–62), a French astronomer, and the comet was named in Halley's honor in 1759.

It has been doing so ever since. Halley's Comet continues to appear in our sky approximately every 76 years. In the century just past, it put on a 'double feature', showing up in 1910 and again in 1986 (figure 2.4).

Now, we have just explained a popular misconception about Halley's Comet. Edmond Halley did not *discover* his comet. Further searches of historical records show that it was noticed as far back as the year 240 BCE. (Chinese and European sky-watchers recorded its every apparition since.) It is simply that everybody before Halley thought that it was a different comet each time. Halley did something more important than find a comet. He was the first to realize that comets orbit the Sun like planets, have periods of revolution like planets, and, in short, behave a lot like planets. Comets come and go, not as harbingers of death and obliteration, but on a regular schedule, like a celestial clock.

Figure 2.4. Comet Halley in 1910. From *Encylopaedia Britannica* (1911).

By the way, a personal quirk of T H: 'Halley' is pronounced like 'alley'. Bill Haley and his Comets ('Haley' as in 'Hale') were a popular music group in the 1950s. They might have done a lot for rock-and-roll but had nothing to do with comets! There is even some evidence to suggest that Edmond himself said 'Halley' like 'hallway' without the 'w'.

There is a good reason why it is easy to attribute the discovery of Comet Halley to Edmond Halley. Since his time, it has become the custom to name a comet after the first person, or persons, to set eyes upon it. (Alan Hale and Thomas Bopp spied their comet within minutes of each other.) The modern rule says that up to three people, anywhere in the world, can independently 'discover' a comet and still get their names attached to it as part of a hyphenated string[4]; hence, we English speakers have mouthfuls like Comet Honda–Mrkos–Pajdušaková.

Once the comet is confirmed, the name becomes official. No doubt as you read this, there are astronomers somewhere on the planet—mostly amateurs—scanning the skies with telescopes or binoculars. They wait for that first glimpse of a faint smudge against the blackness of sky, a new comet and their slice of immortality (figure 2.5). (A dozen or more are discovered each year.)

Why do comets need to be 'discovered' at all? Why can we not see comets all the time? After all, planets almost always can be seen in the sky, some with the naked eye, others by using a telescope. (The only exception is when they appear too near the Sun.) But planets travel around the Sun on paths that are very nearly circular. Their distance from the central Sun does not vary much. Because our planet Earth is comparatively close to the Sun, this means that the distance between us and another planet does not change greatly.

This is not true for comets. Comets travel in elongated elliptical paths. Their orbits are said to be eccentric. The ellipse is not centered on the Sun. Instead, the Sun occupies a point closer to one end of the ellipse. This point is one of two foci belonging to the ellipse. The other focus is empty.

---

[4] More on comet discovery and naming in chapter 5.

Figure 2.5. Don Machholz, discoverer of eleven comets. Courtesy of Michele Machholz. The eye patch is an observer's tool to prevent eye fatigue, not indicative of a missing eye or a pirate!

In its orbit, an object's closest point to the Sun is called its *perihelion*; its farthest point is its *aphelion* (figure 2.6). The difference between perihelion and aphelion is small for most planets, including the Earth. For a comet, it can be immense, perhaps by a factor of a million or more.

(An extreme example is that of the sungrazing comets, which, as their name implies, either skim by the outer layers of the Sun or—in rarer cases—crash into it (figure 2.7). The Great Christmas Comet of 2011, another comet discovered by an amateur, Comet Lovejoy actually passed through the outermost atmosphere of the Sun, its *corona*[5], and survived at least for a few days, then seems to have fragmented and disappeared!)

Obviously, it is hard to see something when it is far away. Also, comets mainly shine by reflected sunlight. Not only is sunlight less bright far from the Sun, but any light that is reflected by a distant comet is dimmed as it travels the distance between the comet and our eyes. Importantly, comets change their appearance as they approach and recede from the Sun. Their surface temperatures cool to the point where water ice sublimation, the source of the comet's activity, ceases and the comet's brightness diminishes greatly. Beyond this point in the Main Asteroid Belt, the coma and tails gradually disappear and the tiny solid body, the *nucleus*, only a few kilometers in size, is all that remains. Occasionally we see unpredictable outbursts from the nucleus that may brighten and form a new coma and tail. Comet Halley displayed such an outburst at a distance past the orbit of Saturn, as did Comet Holmes in 2007 when it was mid-way between Mars and Jupiter and brightened by about half a million times in 42 h, making it clearly visible to the naked eye. More details of this behavior are given in chapter 3.

[5] The corona is the hot but tenuous gas of electrically charged particles surrounding the Sun and extending millions of kilometers into space. It is best seen during a total solar eclipse.

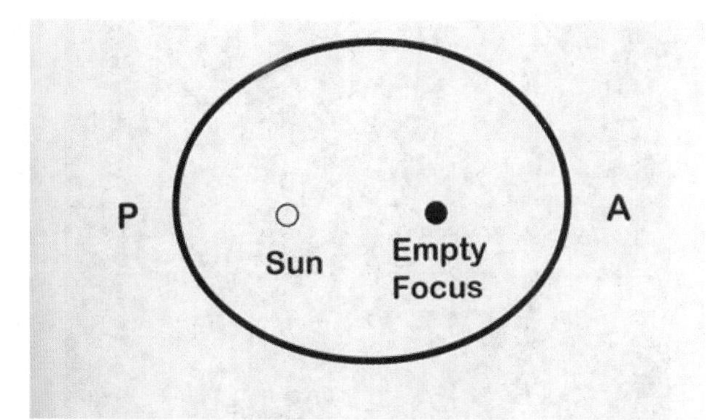

Figure 2.6. An elliptical orbit. P stands for perihelion, A stands for aphelion.

Figure 2.7. The brightest part of the Sun has been blotted out to reveal a nearby comet and the Sun's corona. NASA image.

2.2 Two kinds of comets

There are actually two types of comets based on their orbits: *short-period comets* and *long-period comets*[6]. (The *period* refers to the time it takes to orbit the Sun once.) Short-period comets may travel as far away from the Sun as Neptune, the most distant planet. Yet they also may get closer to the Sun than the Earth. As we said before, we cannot normally see comets as small nuclei when they are far away[7]. They only become visible to us when they are in our 'neighborhood', near the Sun.

[6] The recent discovery of the first interstellar comet, Oumuamua, necessitates a third class, one that is not in orbit around the Sun and mentioned below.

[7] There are a few comets that we can track throughout their entire orbits, a testament to the incredible power of modern telescopes.

Figure 2.8. Periodic Comet Encke. Copyright F H Hemmerich.

Because objects travel much faster in their orbits when they are close to the Sun, comets spend proportionately more time far away and very little time close by. (Comets may travel as fast as one-hundred kilometers per second at perihelion, but only one-hundred meters per second at aphelion.) This is why comets seem to 'appear' in our sky quickly and just as quickly 'disappear'. Short-period comets are defined—arbitrarily—as comets with periods of less than two hundred years. Other than their stretched-out orbits, short-period comets behave essentially like the major planets. They travel around the Sun within about 30° of the same plane as the Earth—this plane is called the *ecliptic*—and that of other planets. Thus, they may cross the orbits of the Earth[8] and the other planets. Short-period comets also (mostly) travel in the same direction as the major planets: counter-clockwise as viewed from the direction we call north. (Comet Halley is a rare clockwise one.)

Several distinctions are made within this group. Comets with periods of less than twenty years are known as Jupiter-family comets as their aphelia generally lie near Jupiter's orbit with low inclinations and clearly have been influenced by the giant planet's gravity. The comet with the shortest known period is Comet Encke: 3.3 years (figure 2.8). Its orbit does not reach Jupiter and is the prototype of the Encke-type comets, having orbits within Jupiter's and decoupled from it. Comets having periods greater than twenty years and less than two hundred are known as Halley-type comets. Their orbits can be highly inclined to the ecliptic and can even be in retrograde (e.g. clockwise Halley as noted above). Only one-tenth of well-studied comets are short-period.

If short-period comets travel in elongated paths, long-period comets travel in *extremely* elongated paths (figure 2.9). Their orbits are said to be highly eccentric. These comets plummet almost directly toward the Sun and execute hairpin turns at their closest approach before hurtling away. While they may get nearer to the Sun

[8] These near-Earth comets may present a collisional hazard to us and are the source of *meteor showers* as discussed in chapters 3 and 4.

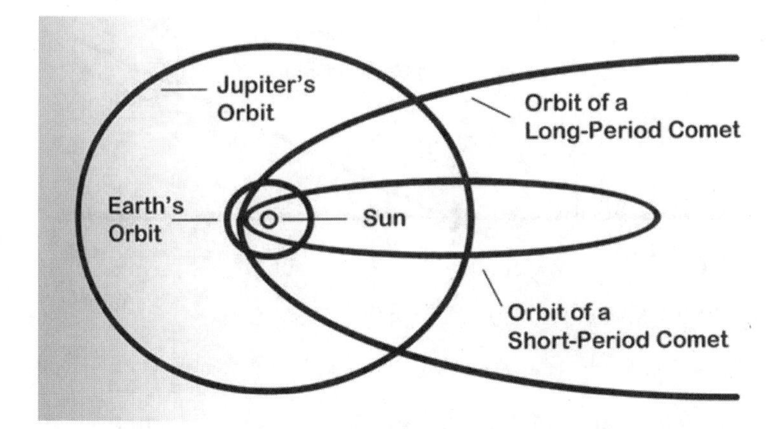

Figure 2.9. The orbits of a short-period comet and a long-period comet compared.

than any planet, at their most distant point, long-period comets are much, much farther away than Neptune. They often have periods of hundreds of thousands or even millions of years. It takes a long time to see a long-period comet twice. There are likely lots more long-period comets awaiting discovery than short-period comets. The reason is, simply, because the last time a particular long-period comet visited the inner Solar System and appeared in our sky, we were too busy keeping our cave fires lit and fending off saber-toothed cats to notice it!

Thousands of comets have been observed[9], but of the 3550 cataloged comets[10], 2723 (77%) are long-period comets. Furthermore, of the short-period comets, only 375 out of 827 have been observed more than once, a requirement to earn a number (e.g. 19P/Borrelly)[11]. Of the short-period comets, 675 are identified as members of the Jupiter-family, 99 are Halley-type, and 53 belong to the Encke-family. Great Comets like Hale–Bopp, with a period of around 2500 years, are long-period comets. On average, a dozen or so long-period comets are discovered each year (but most of these are seen only through telescopes and are quite faint). The general appearance of long-period and short-period comets in the sky is similar; differences in other properties, such as, composition, nucleus size, and shape, have not been convincingly established.

How far away do long-period comets get? When measuring lengths in the Solar System, miles or kilometers just do not 'cut it' anymore. The distances involved are too vast. Describing the expanses between the planets in kilometers makes about as much sense as quoting the distance from Chicago to Bangkok in inches! Such a number would contain a long string of zeros. A new 'yardstick' is called for.

Astronomers measure distances in the Solar System in terms of the Earth's average distance from the Sun. This distance (149 597 870.700 km exactly) is called

[9] Including more than 3000 sungrazing comets spied by satellites in the vicinity of a fixed Earth–Sun position whose primary function is to observe the Sun.

[10] Comets for which an orbit has been determined. For the current number, see the JPL Small-Body Database: https://ssd.jpl.nasa.gov/sbdb.cgi#top (Accessed 24 December 2018).

[11] More on comet designation and naming in chapter 5.

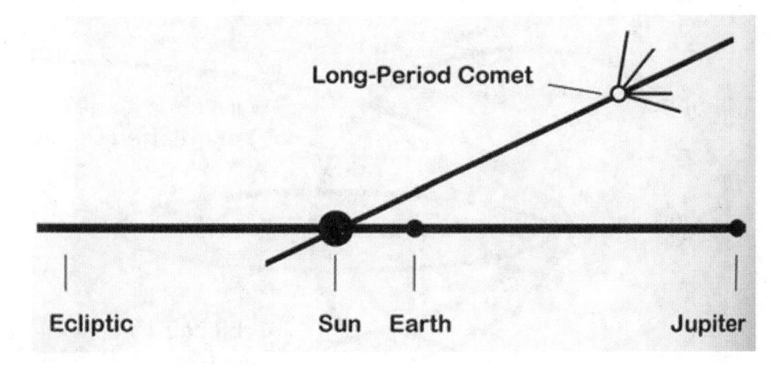

Figure 2.10. The orbit of a long-period comet may be highly inclined to the ecliptic.

one *astronomical unit* (au). It is a convenient and well-known length. Is it not sort of arbitrary to invent the astronomical unit? Yes. However, all distance units are really arbitrary—after all, somebody had to decide whose foot was exactly 'one foot' long!

Using this system, the planets conveniently range in average orbital distance from 0.4 au (Mercury) to 30.1 au (Neptune). Yet long-period comets stray so far from the Sun that even the number of astronomical units that they travel becomes huge. Long-period comets reach distances from the Sun of 50 000 au or more. Far from the limits of what many conventionally think of as the Solar System (actually the Planetary System), these objects travel up to one-fifth the distance to the next nearest star! If the Planetary System, Sun to Neptune, were the size of a baseball sitting on home plate, comets would travel beyond the outfield bleachers.

It is difficult to measure the shape of a comet's orbit based only on the small segment we can observe near the Sun. Some comet paths look like they never repeat, so-called parabolic or hyperbolic orbits, unbound to the Sun. These comets might not return at all—a cometary 'home run.' Indeed, the first interstellar comet was found in 2017, called Comet 1I/2017 U1 Oumuamua[12]. This interloper from beyond our Solar System confirmed our idea that comets may be expelled from planetary systems. A careful measurement of its motion revealed forces other than gravity arising from outgassing[13]. This raises the intriguing possibility of sampling material from other star systems without having to go there! Long-period comets do not necessarily stay in the ecliptic plane (figure 2.10). They can appear coming from any direction. Their orbits are randomly clockwise or counter-clockwise. (Hale–Bopp is a counter-clockwise comet, but it travels nearly perpendicular to the ecliptic.) Furthermore, they may be making their first visit to the inner Solar System with a fresh, frosty surface leading to greater activity and brightness. With few rules governing the orientation of their orbits, long-period comets are the 'bad boys' of the Solar System, challenging astronomers whose job it is to trace their courses. Now that we know these two types of cometary orbits, we will learn in chapter 3 that they

[12] 1I/ is the first designation of an interstellar comet, a comet that originated in another star system. 'Oumuamua' is Hawaiian for scout or first distant messenger.

[13] Called non-gravitational forces, they arise from gas and dust jets emitted from the nucleus that impart a reaction force on it. See chapter 3 for more details.

indicate two different cometary reservoirs in our Solar System. But first let us describe how orbits work. Onward!

2.3 Celestial clockwork

The key to almost every aspect of a comet is its orbit about the Sun. Its orbit defines what kind of comet it is, how long it will last, and how it will appear to us and behave throughout its lifetime. Comets execute the sort of perfect figure in space of which Olympic skaters can only dream. A comet's orbit is truly a thing of beauty. Nevertheless, for some, it can be the most difficult aspect of comets to understand.

Indeed, the physics of orbits gives movement and life to the whole Solar System. For the remainder of this chapter, we will explain how orbits work. Orbital mechanics is mathematically elegant. In this book, however, we will restrict ourselves to a qualitative description of orbits, using comets as our principal example. We promise to refrain from using a single equation! Most basically, we hope to answer the question: 'How do those things stay up there, anyway?'

First, let us acknowledge two of astronomy's giants who started the scientific revolution that set in motion the work of those discussed below. In 1543, the Polish astronomer, Nicolaus Copernicus (1473–1543), proposed the radical idea (at the time) that the Sun was at the center of the Universe (our Solar System as we know it today) and that the Earth and planets orbit it. This heliocentric (Sun-centered) system broke from the millennially held geocentric (Earth-centered) belief of the Greeks that the Earth was at the center, causing great discord with the ruling religious establishment[14]. We have known about the heliocentric Solar System since kindergarten, but remember that it took the genius of Copernicus to overturn the notion held by all of the great intellects that preceded him. (By the way, that map of the Solar System from kindergarten is still under construction as we continue to learn more about our celestial neighborhood.) Following Copernicus, was the great mathematician/astronomer, Johannes Kepler (1571–1630). Through his arduous analysis of observations of Mars, he was able to describe all of planetary motion with three simple laws. Kepler's laws of planetary motion can be summarized as follows: (1) planetary orbits are elliptical with the Sun at one of the two foci (law of ellipses), (2) the speed of planets along their orbit constantly changes, being fastest at perihelion and slowest at aphelion (law of equal areas), and (3) the orbital period is related to the planet's average distance from the Sun, meaning that the farther from the Sun, the slower the motion of the planet in its orbit (harmonic law). This achievement earned Kepler the title, 'Lawgiver of the Heavens'. Kepler did not know why his laws described planetary motion, only that they fit the data well. In fact, we know today that Kepler's laws apply to any bodies in orbital motion, be it satellites orbiting planets, extra-solar planets orbiting their host stars, binary stars, galaxies orbiting each other, etc. Newton acknowledged these great minds when he stated, 'If I have seen further, it is by standing upon the shoulders of Giants.'

[14] Aristarchus (310 BCE–230 BCE), a Greek, had formulated a heliocentric model some eighteen centuries earlier but this idea was rejected by his contemporaries.

To understand planetary and cometary orbits, we must begin by investigating two ideas. First, we need to look at motion itself. How does a thing—anything—move? Second, we need to look at a force called gravity.

2.4 A Universe in motion

In our busy modern age, motion seems natural. Is it? Aristotle (384–322 BCE), the Greek thinker who established the tradition of western science, believed not. He was not very impressed with motion. He felt that rest was the natural state of all things. Set an object in motion, and it eventually will come to rest, he observed.

Galileo Galilei (1564–1642), the arrogant and obstinate Renaissance genius, thought differently. He questioned whether either motion or rest were a 'preferred' state of objects. He did not think that there was anything more 'natural' about rest than about motion, or vice versa. This was embodied in his concept of inertia: the resistance of an object to any change in its position and state of motion. By eliminating preferred states, Galileo inaugurated the modern study of motion called dynamics. What a rebellious act to disagree with the great Aristotle whose idea held for almost two millennia.

Newton picked up where Galileo left off. (Quite literally, he was born within a year of Galileo's passing!) He adopted Galileo's concept of inertia and stated that an object at rest remains at rest and that an object in motion remains in motion—traveling in a straight line, at a constant speed. This is Newton's first law of motion: the law of inertia. The only way to put an object at rest into motion, stop a moving object, or change the speed or direction of an object already in motion is to exert a force, a push or a pull, on it. According to Newton, the change in motion produced by a force, occurs in the direction of the force and is proportional to its strength. This is Newton's second law of motion: the force law.

This makes sense. When you pick up and roll a bowling ball, it travels forward in the direction you threw it, not over your shoulder or someplace else. Furthermore, the harder you throw it, the faster the bowling ball is going to go.

What do we mean when we write 'change in motion'? The speed and direction of a moving object together are called its velocity. A change in velocity is an acceleration. So a force produces an acceleration, a change in motion.

We usually think of an acceleration as an increase in speed. It is. It can be just as easily a decrease in speed (commonly called a deceleration), however. We are less likely to think of a change in direction as an acceleration. However, a change in direction is just as much a change in velocity as is a change in speed. A given force can produce a change in speed, a change in direction, or both—depending on how the force is applied. We have all experienced this in our lives when we round a curve in a vehicle moving at a constant speed. We have to lean into the curve to oppose the force due to this change-of-direction acceleration, lest we be thrown to the outside of the curve. (This is called centripetal force.) Fortunately, our seat belt and door prevent this from happening.

Newton was on a roll and kept going. Acceleration is proportional to the force applied, declared Newton. It also is proportional to a property of every object, its mass. Mass is simply a measure of the amount of matter within an object. Given a certain force, if you try to change the motion of a large mass, you will get less acceleration than you would with a smaller mass. Even pushing very hard, it takes much effort to get a stalled car rolling. Yet, what happens when you use the same amount of force on a skateboard? Whoosh! A very rapid change in velocity results.

Do not confuse mass with weight. Weight is a force due to gravity's pull on mass. It is proportional to mass, but change gravity and you change your weight, not mass. A great recipe for weight loss is simply traveling to the Moon where your weight would decrease to 1/6 your Earth weight due to the lower gravity of the Moon. Of course, the object of weight loss is really mass loss.

Whew. That is a lot of Newton. Nonetheless, it works in our macroscopic world. These last few paragraphs describe every motion that you and we have ever seen. That is really complete, really powerful stuff. NASA uses Newton's laws[15] today to plot trajectories of spacecraft through our Solar System. Newton showed that motion is not capricious, that it follows relatively straightforward and describable rules. Newton's laws never fail us. You can bank on them because they cannot be broken. (At least this is true in the macroscopic world; it is altogether a different matter in the microscopic quantum realm, but you will need another book to learn about that!) They are also vital to understanding the orbits of comets.

Feel free to read over the last page or two again before going on. (We will wait!)

2.5 The gravity of the situation

Gravity is both simple and complex. In a sense, we all know what gravity is. Everyone who has ever fallen on his or her face has had an intimate relationship with gravity. Nevertheless, one can take an entire graduate physics course called Einstein's general theory of relativity without grappling with all the nuances of gravity.

For our purposes, gravity is simply a force, but a special kind of force. First, it is always a pull and never a push. Second, unlike most other forces, it can act at a distance, any distance. In other words, it is universal. This simple idea was extremely profound at the time. Newton had connected what we experience in our terrestrial world with what we see in the celestial one. (In the earlier example of acceleration, you had to place your hands on the rear bumper of the car to get it moving; gravity can accelerate an object without any direct physical contact between the objects.)

Gravity is produced by any object that has mass. As nearly everything that we know of (physically) has mass, this includes the whole Universe of objects! Between each pair of these objects there is the attractive force of gravity—the more mass in the pair of objects, the stronger the force. Think of that. Right now you (an object with mass) and a distant comet (another object with mass) are attracted to each other. Do you feel it? No, of course not. That is because gravity grows weaker the greater the distance between objects. It becomes infinitesimally weak at infinite distance, but never zero; we all are bound together in this Universe. What a profound notion. This could be the basis of world peace!

[15] There is a third law, the reaction law, that will be discussed shortly.

To experience the force of gravity, one need only leap up. The velocity that you give yourself with your feet is quickly decelerated away. There is a moment of 'hang time,' and then you are accelerated down toward the massive object underneath you, the Earth. You do most of the moving because the Earth is much more massive than you, but the Earth has moved, too, in the opposite direction. This is Newton's third law of motion: for every action, there is an equal and opposite reaction[16]. Technically, the gravity produced by the mass of the Earth, since it is spherical, acts as if it is emanating from the center-of-mass of the Earth, a point at its middle, some 6400 km below you. Before you get there, though, another force, that exerted by the hard floor, stops you.

Now back to our original question: with all those bodies revolving around in the heavens—planets, satellites, comets, and asteroids—why does not everything fall down?

For centuries, people wondered what might propel the planets and comets in their paths. Some speculated that it might be a magnetic force, magnetism being a popular phenomenon to dabble with in the physics of the seventeenth century.

Then Newton described gravity and explained that the same force that works in day-to-day life on the Earth also affects the planets. That is, they move in a way that is totally consistent with a force proportional to the masses involved and inversely proportional to their distances from one another[17]. (That sounds a lot like gravity.) Newton's immortality rests in his realization that the force that runs the Solar System lies—literally—beneath our feet. Armed with his powerful ideas of motion and gravity, Newton focused on understanding the why of Kepler's laws of planetary motion. He was able to show that Kepler's laws derived from his basic laws of motion and gravity. He had to invent calculus to do so, but that is beside the point. In fact, when he examined Kepler's third law, he added the total mass of the orbiting bodies into the relationship between orbital period and average distance from the Sun. This important revision gives us the primary method of estimating the total mass of orbiting celestial bodies, a kind of cosmic balance.

Wait a minute. Has not our study of gravity backfired? Our intent was to tell why objects in the sky stay there, why they do not 'fall down'. We have invented a force that seems to require that they do just that—fall toward each other! There is an attractive force between the Earth and its Moon, between the Sun and a comet, and between all other objects as well: gravity. Read on.

2.6 Looping comets

'Why do the comets not fall?' we asked. They are falling—all the time! They have to be. Recall that in the absence of a force, an object is either at rest or moving at a constant speed in a straight line. That goes for any object, anywhere in the Universe.

[16] In 1686, Newton published his three laws of motion in the *Principia Mathematica Philosophiae Naturalis*, considered by many to be the greatest scientific tome in history. If you get a chance to see an early edition, we highly recommend it.

[17] More specifically, gravitational force is inversely proportional to the square of the distance between attracting objects.

With no force acting upon it, a comet would either be static or traveling at a constant rate toward or away from us. If the latter were the case, the comet would either crash into something or disappear forever. No real comet behaves in this manner. Its speed and direction of travel are perpetually changing. (There are no 'straight' comets; their paths always curve, even if only a little.) No comet is stationary, of course.

If gravity is the force acting on the comet—and it must be—should not every comet collide with the Sun? This would be the case, if the comet started at rest. In other words, if you placed a new comet at rest with respect to (and some distance away from) the Sun, imparted no other force on it, and let it go, the comet would fall into the Sun. No question.

No, the comet can neither be unaffected by gravity nor lacking in original motion. Otherwise it would not behave as we observe comets to do.

What if the comet is not only attracted by the gravity between it and the Sun, but also has some initial velocity? If the direction of this velocity is toward or away from the Sun, it will not make any difference, but what if it is not?

We say that the direction toward or away from a center is radial. (Spokes of a bicycle wheel are aligned radially.) The direction perpendicular to this is transverse (figure 2.11). A comet that starts with a transverse velocity will be acted upon by the radial gravitational force of the Sun, perpendicular to the comet's transverse direction of travel. Its direction of travel will then change. (The comet is accelerated in the direction of the Sun by the force, but its transverse motion will be unchanged.) The path of the comet will be a curve.

The classic example is balls on a string (part of boleadoras in certain parts of the world). If you hold the balls motionless by the middle of the string, they dangle to the floor. If you give them a radial velocity (throw them), they move in a straight line (for a moment, at least). However, if you hold onto the string and give the balls a transverse velocity (perpendicular to the string, by twirling it), they travel in a curved path. The tension in the now-taut string takes on the role of gravity, in this analogy, by pulling at the balls radially in toward your hand. Yet the balls do not move inward. They stay at a constant radius from your hand (the half-length of the string). The balls continually accelerate, but never hit anything. In fact, they travel in a particular curved path that repeats itself over and over: a circle. They are constantly

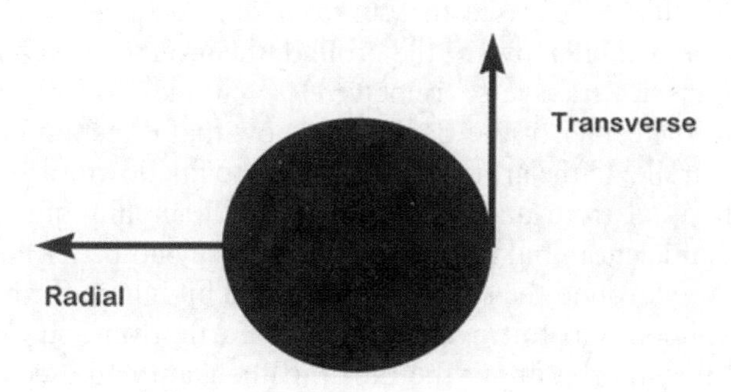

Figure 2.11. Radial and transverse motion.

'falling' to the center, but this fall is countered by their transverse motion. You could say that the balls are orbiting your hand.

To be sure, Newton's laws still apply. Just cut the string while twirling it and look at what happens. The balls take off. Invariably, they travel away in the transverse direction. We need both Newton's laws of motion and Newton's universal gravity to make an orbit.

An object moving in a curved path really is moving in two directions simultaneously: the radial direction and the transverse direction. Consider a giant pitching a baseball. Once he releases the ball, it flies at a constant speed parallel to the ground. There is no other force acting upon the ball in the transverse (horizontal) direction. (Ignore the frictional force with the air for the moment, i.e. no curve balls.) In the radial (vertical) direction, it is another story. Gravity pulls the ball to the ground. The ball accelerates radially until it hits the dirt. (It falls downward.) The two motions exist independently of each other. The ball would have been 'happy' to remain traveling at the same transverse speed indefinitely, had it not dug its chin into the ball field and come to a halt. (Except for a little deceleration caused by friction with the air, a fastball is moving at the same speed over the plate that it was leaving the pitcher's mound.)

What if the giant wants to throw the ball farther? He could give it a radial velocity during his pitch. This time, as the ball leaves his hand, it travels upward. Gravity must now decelerate this upward velocity to zero before accelerating the ball downward—this time from a good height—to *terra firma*. The ball takes much longer to strike the ground. Unfortunately, the ball makes no transverse progress across the field and falls at the pitcher's feet.

The giant's best bet is to throw the ball at an angle to the ground so that it has both transverse- and radial-velocity components. The radial component 'buys time', while the transverse velocity gains yardage.

Still, eventually, no matter what force the giant imparts on the ball, it will always 'bite the dust'. Of course, this assumes that the ground is an infinite flat plane and the ball does not achieve escape velocity, discussed below.

It surprises no one anymore to learn that the Earth is round; photographs of its globe, beamed down from space, have seen to that. But even the ancient Greeks knew about a round Earth, approximately the size we know today, by carefully measuring shadows at different locales throughout the year. Suppose our giant is really tall, so tall that he can see the curve of the Earth. He hurls the ball hard enough that, as the ball falls toward the ground, the ground curves away beneath it. The ball maintains its transverse velocity. It continues to be pulled downward (meaning toward the center of the Earth), but now that direction has changed. The ball still travels parallel to the ground. It continues to fall downward, and the ground continues to curve out from under it before the ball actually hits. If the giant has given the ball a sufficient initial transverse velocity, he had better look out! The ball will 'fall' all the way around the spherical Earth and hit him in the back of the head. If he ducks, it will continue to travel around the Earth, over and over (ignoring air friction). The ball is in orbit about the Earth. If the giant had used a slightly higher transverse velocity, the resulting orbit would have been elliptical instead of circular, and even more velocity would result in the ball permanently leaving the vicinity of the Earth (escape velocity) in a parabolic or even hyperbolic orbit. What goes up does not necessarily come down.

(When we watch the flame and clouds of smoke pouring forth from the engines of a rocket on its launching pad, all that thrust serves only one purpose: it is to impart on the rocket a velocity great enough so that the spaceship atop it will achieve orbit or beyond. This is another example of Newton's third law of motion. It is not the push of the rocket's thrust on the launch pad nor against the atmosphere that lifts the rocket upward.)

As comet nuclei coalesced in the *proto-solar nebula*, they acquired velocities that were the sum of the velocities of the material that formed them. Within this material, there was often a surplus of motion in the nebula's direction of rotation. This direction was transverse to the direction of the newly formed Sun, at the center of the nebula.

Notice that an orbit requires a precise combination of gravity and transverse velocity. Is it a coincidence that so many planets, comets, and other bodies achieved this necessary combination? Not really. We are seeing just the result of billions of years of Solar System evolution. Only those objects with the right velocities survived so that we can see them today. How many comets did not, and ended up falling into the Sun or were expelled? Probably quite a lot.

2.7 Orbits of your own

To help picture it, you can create a model of your own orbiting comet. The problem with doing so usually is that we cannot turn off the Earth's gravity to make the model! Everything falls to the floor. However, we can use the Earth's gravity to simulate the pull of the Sun on a comet using simple kitchenware. Here is how.

Get hold of a large, hemispherical-shaped salad bowl. (Glass or aluminum is best.) Mark the inside center of the bowl with a pen or sticker. This point will represent the Sun. The only additional thing you need is a marble to 'stand in' for the comet. Put both the bowl and the marble on the floor.

If you 'shoot' the marble across the floor, it will roll away from the bowl, in a straight line, at a constant speed. In other words, it will obey Newton's laws to the letter—just like any other moving object on the Earth or in space. The marble represents a comet (or anything else moving) not affected by the gravity of any other mass.

The marble in our demonstration eventually will slow and even stop. However, this is because of friction with the floor (an external force that negatively accelerates the marble). In space, there would be no such friction. You would lose your marble for good. (Colleagues have suggested from time to time that T H has lost all of his, already.)

In reality, it is impossible to avoid the gravity produced by other objects in the Universe. The salad bowl represents such an object: the Sun. Put the marble on the inside rim of the bowl. It 'falls' toward the central 'Sun.' It will do this, no matter where on the rim you place it. The salad bowl mimics the gravity of the Sun in that it causes nearby objects to be attracted radially toward a central point. The marble now simulates what would happen if a comet were placed, motionless, in the vicinity of the Sun. The real marble will overshoot the representation of the Sun, certainly. (There is really nothing in the middle of the salad bowl.) The real comet would hit, and be swallowed up by, the real Sun.

Let us try our experiment once more. This time, hold the marble on the inside rim of the bowl, and give it a little shot with your finger, in a direction more-or-less parallel to the rim. Because of the spherical shape of the bowl, the marble will 'orbit' the center of the bowl: the marble/comet will make many trips around the center/Sun (without going through it) until friction, again, decelerates it, and the marble comes to rest at the bottom of the bowl. Without friction, the marble would continue looping around forever.

This all may take a little practice. You will quickly see that the orbital speed you give the marble must be just right: if you give it too hard a 'flick,' the marble will leave the neighborhood of the Sun altogether (it will 'jump' out of the bowl). Too little force causes the marble to curve inward right away. It will not make even one orbit.

A physicist could rightly protest that the shape of the bowl does not exactly correspond to the way the gravitational force increases as you approach a massive body[18]. It still looks about right, though.

Notice that most random orbits are elliptical: the marble moves a little closer, then a little farther away from the center … a little closer, then a little farther away, a little closer … You get the point. The marble has a perihelion and aphelion with respect to the focus of its path. Slightly disparate initial trajectories will result in differently shaped ellipses. Let us leave our make-believe Universe of salad bowl stars and cometary marbles and return to the authentic Solar System.

A complication of the real Universe is this: we talk about a comet 'orbiting the Sun', but the Sun orbits the comet, too. Just as the comet moves under the grasp of the Sun's gravity, the Sun also is required to move under the influence of the comet's. Yet just as a gerbil on a seesaw with a hippopotamus does most of the moving, the comet does most of the work in its dance with the Sun. The Sun wobbles ever so imperceptibly as the comet whirls about it. Planets and comets orbit the Sun, and satellites like the Moon orbit planets, all for the same reason. Newton soon realized that planetary orbits are never circular nor elliptical, as stated by Kepler, due to these ever-changing gravitational tugs of all of the other planets, satellites, etc. As genius as Newton was, he could not accept that cometary motion could be described by periodic orbits; he preferred parabolic shapes so that the comet would never return. As we noted above, this notion would be changed by his friend, Edmund

[18] The correct shape is like an inverted cone that tapers to its apex. You may have seen one at a museum for donations. Start a coin rolling around its surface, and it will orbit the apex until exiting through a hole at the bottom.

Halley, who predicted that comets followed elliptical orbits, like planets, that periodically brought them back into our night skies.

Though comet orbits can be extremely eccentric, they are physically no different from circular orbits. There are subtle non-gravitational forces acting on the comet, too, mainly due to the gas and dust *jets* emitted from the nucleus that perturb its elliptical orbit. These will be discussed in more detail in chapter 3. As we watch a periodic comet move through our sky, remember that it is constantly falling toward (or away from) the Sun. It just will never quite make it. An orbit is the ultimate example of 'falling—with style'.

Chapter 3

What comets are all about

A quiz:

What is a comet?

A. A Lincoln–Mercury automobile.
B. A vintage aircraft.
C. A bathroom cleanser.
D. A brand of beans.
E. An ancient astronomical visitor of strange beauty.
F. All of the above.

Here we discuss the anatomy and make-up of comets. We also will talk about their origin stories. By the end of the chapter, we are ready to take a virtual journey to a comet (and return). Buckle up, we are off to see a comet! By the way, the answer to the quiz is ‘F’, but we will focus on selection ‘E’.

3.1 Parts of a comet

No two comets are alike, somewhat like people. If we could view them in a line-up, they would share certain common features but they would all appear as individuals. These features constantly change as comets approach and recede from the Sun. Their appearance as seen from the Earth adds perceived differences as well (such as *antitails*[1], faint coma features, etc). Comets continue to surprise and amaze us, but we can draw some general conclusions about the parts of a comet, starting with the nucleus.

There is another reason why it is difficult to observe a comet when it is far away, besides the fact that its light is dim: far from the Sun, a comet is small. While planets have diameters measured in thousands or tens of thousands of kilometers, comet

[1] An antitail refers to a tail that appears to point towards the Sun but is actually an illusion caused by observing the comet from a peculiar perspective, as discussed below.

doi:10.1088/2053-2571/ab027ech3

nuclei are at most city sized, not planet sized. They are too small to appear to us as anything more than a point of light. Estimates of their sizes come from various observations, including measurements from space probes made of a few comets that have come near the Earth. The most common method to estimate size is from the brightness (reflected sunlight) of the bare nucleus. Assuming the percentage of light that is reflected from the surface (albedo, see below), one can obtain the effective cross-sectional area, hence the size of an equivalent spherical object. Modeling Hubble Space Telescope images at far heliocentric distances can also extract the nucleus size. The close-approaching Comet Hyakutake's nucleus, for instance, was found to be merely 1–3 km across (figure 3.1). Based on these techniques, it seems likely that few comet nuclei are more than ten kilometers in diameter (although Comet Hale–Bopp at about 50–60 km is an important exception). The subtle distinction between the phenomenon we see in our night skies with its coma and resplendent tails ('comet') and the small solid body that gives rise to these magnificent features ('comet nucleus') is important and one that we adhere to throughout this book.

Like everything else in the Universe, comet nuclei also spin, giving them daytime and night-time. They rotate relatively rapidly, from a few hours to a few days. Some show complex rotational states where they wobble and tumble as they spin (like Halley). And their spin rate can speed up or slow down due to jet activity. Comet

Figure 3.1. Comet Hyakutake in 1996. Taken by Rick Scott and Joe Orman, near Florence Junction, Arizona. Courtesy of Night of the Comet.

Figure 3.2. The Giotto space probe. NASA image.

Tuttle–Giacobini–Kresák holds the record, slowing its spin from 20 h to 46–60 h in a two month span during 2017.

Only a few comet nuclei have had their close-up picture taken, starting with Comet Halley. A fleet of space probes met up with Halley on its 1986 return. The Soviet Union's Vega 1 and 2 space probes, from 30 000 km away, took the first pictures of a comet nucleus, and found Halley to be oddly peanut shaped. The European Space Agency's (ESA) Giotto probe imaged the nucleus from only 600 km away[2] (figures 3.2 and 3.3). More recently (2001), the National Aeronautics and Space Administration's (NASA) Deep Space 1 encountered Comet Borrelly (at a distance of 2200 km) and sent back even better nucleus images (figure 3.4). Since then, three additional spacecraft have had close encounters with comets[3]: Stardust (NASA)—Comet Wild 2 in 2004 and Comet Tempel 1 in 2011; Deep Impact (NASA)—Comet Tempel 1 in 2005 and Comet Hartley 2 in 2010; and Rosetta (ESA)—Comet Churyumov–Gerasimenko in 2014. (Stardust and Deep Impact did double duty as NASA 'recycled' these spacecraft before their fuel was expended.)

Back on the Earth, we would probably miss most comets, if it were not for the spectacle they stage as they approach the Sun. Within about three astronomical units of our star, a bright cloud surrounds the comet nucleus. This cloud is called the coma or comet head (when including the nucleus). It can extend out to hundreds of thousands of kilometers in all directions. 'Coma' comes from the Latin word for hair, and, indeed, the coma gives the comet a diffuse, 'hairy' appearance. Now it is easier to see the comet. Not only is it closer, it is a bigger target. Comets are discovered after they have produced a coma.

Very close to the Sun, the comet grows its most distinctive and most impressive feature, its tails. Tails come in three varieties based on their composition: *ions*[4] (type I), *dust* (type II), and neutral atoms (type III). The dust tail is often a bright, curved

[2] The Halley fleet also included two Japanese space probes, Suisei and Sakegake, and the repurposed NASA ISEE-3 satellite, recommissioned as the International Cometary Explorer (ICE), having first encountered Comet Giacobini–Zinner in 1985.

[3] ESA's Giotto spacecraft also visited Comet Griggs–Skjellerup in 1992 but did not provide images since its camera was damaged during the Halley flyby.

[4] Ions are electrically charged atoms and molecules.

Figure 3.3. Close-up view of the Comet Halley nucleus by Giotto. ESA image.

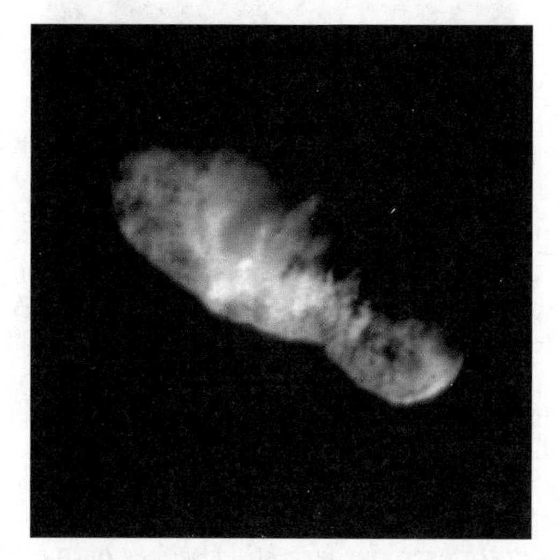

Figure 3.4. Comet Borrelly. NASA image.

fan-shaped appendage to the coma with a yellowish tinge. The ion tail is more straight, bluish in hue, and points in the opposite direction from the Sun (figure 3.5). Sometimes it narrows, bends, and then detaches from the comet entirely, only to grow back after some time like a lizard tail. Both may become over an astronomical unit long, giving them, for a short time, the distinction of being the largest structures in the Solar System[5]. Type III tails are only seen using special filters that isolate the light from specific atoms such as sodium. They are the straightest of all of the tails and also point anti-sunward. Some comets display all three types of tails, some only one and, in 2016, a comet was seen that had a well-developed coma but no tail, a Manx comet (named after the tailless cat).

[5] The Ulysses spacecraft detected ions from the ion tail of Comet Hyakutake more than 3.8 au downstream.

Figure 3.5. A comet exhibiting both ionized gas and dust tails. NASA image.

Notice how long the comet tail is compared to the coma! Pretend that the coma is the size of a standard audio cassette tape. Now (in your imagination) cut the tape. Unwind it by pulling on the end, dangling the cassette behind. You will have to pull off one hundred meters of tape (more than half an hour's worth) to represent a long comet tail.

After a comet rounds the Sun, it heads back out into the remote recesses of the Solar System. Its tail fades, and then its coma disappears (figure 3.6). The comet's adornment was only very temporary. The comet now returns to its normal drab and dormant state in the outer Solar System. It is just an inconspicuous nucleus again, too small and too faint for even our largest telescopes on the Earth to see.

If you could follow a comet as it approaches the Sun, it would appear as it is most commonly illustrated: a fuzzy coma with a graceful tail or tails trailing it. If you continue to watch the comet after it passes around the Sun, however, it will look a bit strange. The tail will lead the comet. The ion tail is like a wind sock that always points in the direction of the wind (figure 3.7). (With comets, sometimes, the tail really does wag the dog!)

Space is not a fluid; it is virtually empty. Our mental picture of the comet tail, simply following in the wake of the comet, breaks down. So, what is going on?

The key thing you might have noted, as you followed the comet on its course, was that the tail stayed on the opposite side of the nucleus from the Sun, regardless of the comet's direction of travel. This is apparent from the Earth, as well. If a comet appears in the early evening sky, its tail will point eastward. If it appears in the early morning sky, its tail will point westward. This fact led astronomers to speculate that it is being blown back by something emanating from the Sun.

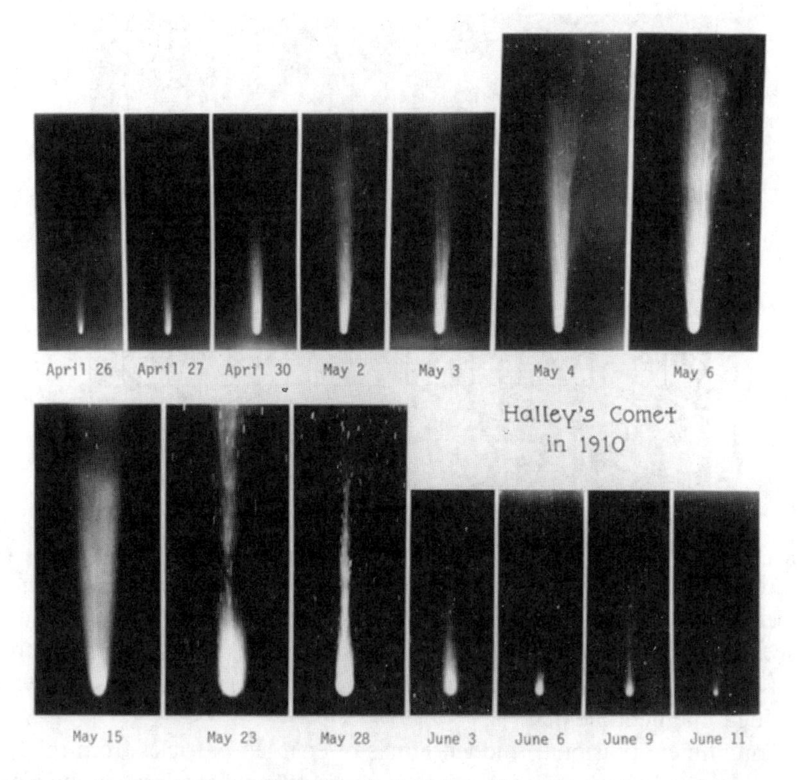

Figure 3.6. A sequence of images showing Comet Halley's ion tail lengthen and then shorten over time, as it approaches and then recedes from the Sun. 1910 apparition. NASA image.

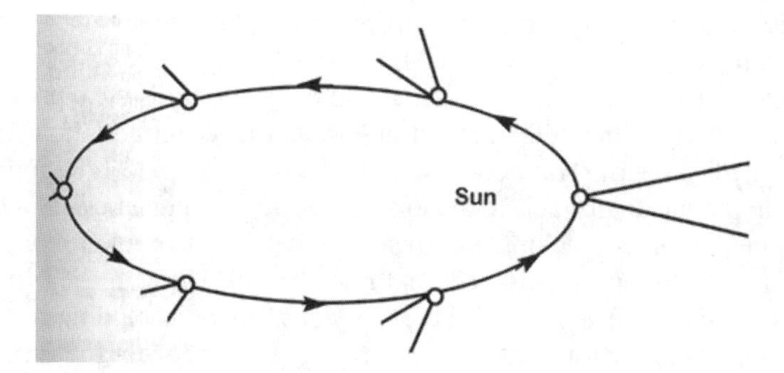

Figure 3.7. The orientation of a comet's tail with respect to the Sun.

The Sun is, of course, emitting light in all directions. We do not normally think of light as exerting a force on anything. (We do not feel a push when the car behind us turns on its headlights!) Still, light does exert a tiny force on every surface that it shines upon, though one much too small to be felt. It is this solar radiation pressure due to sunlight that pushes the dust and neutral atoms away from the Sun, forming the type II and III tails.

Figure 3.8. The aurora borealis. NASA image.

The pressure of sunlight alone is insufficient to account for the ion tail, which really does look like something being blown away from the comet by a wind. It is! In addition to light, escaping from the Sun at all times is a rarefied, but continual, stream of charged electrical particles (electrons and ions) called the solar wind, with an embedded magnetic field.

For a long time, occasional, sudden bursts of charged particles from the Sun were known to produce the aurora, or 'northern' and 'southern lights', when they interacted with the Earth's atmosphere (figure 3.8). Strong outbursts interfere with the Earth's radio communications and electrical power systems, although these are very infrequent events. The only hint that the solar wind is continuous came from comets. In 1951, the German astrophysicist Ludwig Biermann (1907–86) predicted the 'corpuscular' nature of the solar wind from comet tail observations. In the Space Age, space probes with instruments capable of detecting particle radiation and magnetic fields have proved what the comet tails have been telling us for generations, the existence of the solar wind. In 1957, the Swedish astrophysicist, Hannes Alfvén (1908–95), explained how ion tails form. Cometary ions present an obstacle to the solar wind and the magnetic fields embedded in it drape around the comet, stacking up in front and bending around the comet in the tail direction. The ions follow this draped field into the tail. He won the Nobel Prize in Physics in 1970 for his essential work in the study of *plasmas*[6], including comet ion tails.

Bends and knots in the ion tails of comets, long observed from the Earth, now can be explained, too. These phenomena signal that there has been a change in the solar wind or that the comet has moved to a different position with respect to the Sun, where the solar wind speed and orientation differ.

The ion tail points out behind the comet under the influence of the solar wind. This wind blows, on average, at 400 km s^{-1} (1 440 000 km h^{-1})! In comparison, the highest recorded wind speed ever measured in a hurricane is 300 km h^{-1}.

[6] Plasma, often called the fourth state of matter, is a gas consisting of electrons and ions.

Figure 3.9. Comet Lulin exhibits an antitail. From: http://www.jacknewton.com/Comet%20Lulin.jpg. Used with permission from Jack Newton.

The liberated dust takes a more complicated route. A grain of dust is much bigger (thousands of times; its size can be measured in microns[7]—imagine something the size of smoke particles) and more massive (billions of times) than a molecule of gas. Each dust grain is influenced by the solar wind (comet dust often departs with a net electrical charge), the pressure of sunlight, and the gravity of the Sun and planets. The resulting forces cause each dust grain to follow a slightly different orbit around the Sun than the comet does. Thus, away from the coma, the comet's dust tail (traveling more slowly than the ion tail) separates from the ion tail and has a curved, but more uniform appearance.

Occasionally, the geometry of our point of view here on the Earth is just right so that we can see the tip of the dust tail appearing seemingly on the sunward side of the coma. It looks like another, 'backwards' tail on the wrong side of the comet. However, this antitail is just a projection effect. In reality, it is behind the comet and farther from the Sun than the coma (figure 3.9).

3.2 What is a comet made of?

What is the comet made of that blows away in the solar wind? Into the twentieth century, comets were pictured as flying, frozen sand piles in space. However, why should a bunch of rocky particles exhibit a coma and tail?

A new model for comets was called for. According to the one developed by Harvard University astronomer Fred Whipple (1906–2004) in 1950, comet nuclei are nothing more than 'dirty snowballs'. They are bodies made of an amalgamation of about 50–50 percent *ice* and dust. (They are now thought to have more refractories[8], dust; 'frozen mudball' might be a better analogy.) The dirty snowball model of a

[7] 1 micron (micrometer) = one-millionth of a meter (0.000 001 m).

[8] *Refractory* material is a mineral that retains its form and properties at high temperatures, being quite suitable for making furnace bricks, as an example.

comet has now replaced the 'sandbag' model in the minds of astronomers. Moreover, recent evidence shows that nuclei are made up of many independent bodies (snowballs) loosely held together by gravity called 'rubble piles'. This is consistent with the low material strength exhibited by comets, resulting in the splitting of many nuclei (e.g. Shoemaker–Levy 9, to be discussed later). Rock and ice are common ingredients in the Solar System. In the inner Solar System, rock and metal are the principal building materials out of which planets are made. Our Earth is a fine example.

In the cold outer reaches of the Solar System, we find worlds very different from the Earth. These worlds have rock in them, but they also contain organic compounds and frozen gases (ices). The latter are made of *volatile* materials, under normal pressures, chemicals that turn to gas unless kept very cold. The planets Jupiter and Saturn, for instance, are mostly hydrogen and helium, not rock. Other worlds have layers of ice. The larger satellites that orbit the outer planets are made mostly of ice and rock. However, rock is denser than ice, so gravity has caused most of the rock to sink to the centers of these bodies, whereas the ice has floated to the outside. Hence, the surfaces of these worlds are very icy. So, it is not unusual to find ice in the Solar System.

The only difference with comets is that they are not very massive and, therefore, do not produce much gravity. Nothing sinks or floats. The dust and ice do not separate from one another. They remain a homogeneous mix. In other words, there is as much ice and dust in the outer layers of the comet nucleus as in its center—just like in a dirty snowball. (However, it must be conceded that nobody has ever tunneled into a comet nucleus to verify this.)

The snowball analogy is good in another way as well. Again, because comets have little mass, they do not produce enough gravity to compress themselves into a dense ball. Comets may be oddly shaped and very loosely held together. In fact, they often fragment or dissolve completely!

Density is defined as the mass of an object divided by the volume it takes up. The density of a spherical 'ice cube' should be about 0.9 g cm^{-3}. The density of an average comet nucleus is only half this (0.5 g cm^{-3}—the density of pine wood). Clearly, there is a lot of empty space in a comet nucleus. The material is fluffy.

We can now explain the appearance of the comet's coma and tail. As the comet nucleus nears the Sun, it warms up. It takes little heat to 'vaporize'[9] the ice near the surface of the nucleus. The ice does not melt and then boil. In the low pressure of space, the ice instead sublimates directly from solid to gas. (Solid carbon dioxide—'dry ice'—will do this at the temperatures and pressures normally encountered on the Earth (figure 3.10).)

To recap, as gas escapes the comet nucleus due to sublimation, it forms the coma cloud, mostly as neutral atoms and molecules. These are mixed with ions and electrons as the gas is ionized under the influence of solar radiation at a rate depending on the specific atomic and molecular properties. Dust freed by entrainment in this outgassing also escapes the low gravity of the comet nucleus.

[9] At many places in this book, we loosely use vaporize and sublimate interchangeably, but, please remember, we always mean sublimate!

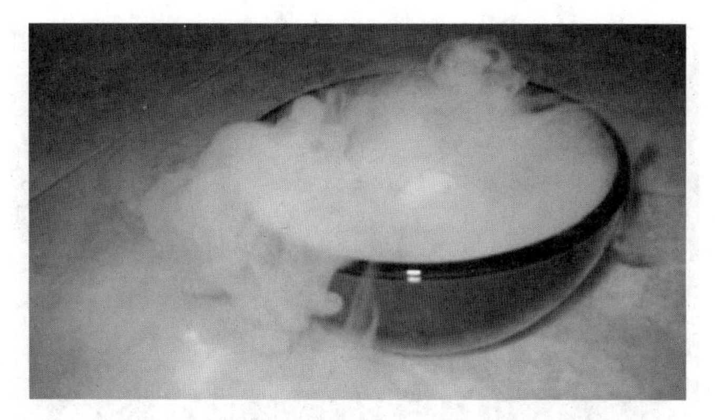

Figure 3.10. Sublimating carbon dioxide—'dry' ice.

Eventually, as the comet travels still closer to the Sun, around the orbit of Mars, the solar wind sweeps back the electrically charged ions of gas, and the pressure of sunlight pushes the dust particles away from the Sun. The comet tails are now formed as we described previously.

While dust is pretty self-explanatory, the term 'ice' needs some elaboration. Chemists call any solid made out of a volatile substance an ice. Much of the comet nucleus is in fact water ice, but other ices exist in the 'deep freeze' of the outer Solar System. Comets contain carbon-dioxide ice, carbon-monoxide ice, ammonia ice, methane ice, formaldehyde ice, and ices of other more complex chemicals, including a tincture of complicated organic molecules such as glycine, the simplest amino acid. (Given the preceding list, imagine what a piece of comet would smell like if thawed out on the Earth!) The exotic ices may bond loosely with the water ice as a hydrate or be trapped within a matrix of water ice called a clathrate.

Once turned to gas, many of these volatile chemicals glow in the presence of blue and ultraviolet light—the kind of light that makes a 'black-light' poster glow. In addition to visible light, the Sun shines in ultraviolet light. (Thankfully, our atmosphere shields the Earth from most of this potentially harmful radiation.) In space, the comet is bathed in this high-energy light and begins to glow, too. The process is called fluorescence. Photons (increments of light) of a wavelength we cannot see are absorbed by a coma molecule, and their energy is re-emitted as one or more photons of visible light. This makes the comet coma and ion tail even easier to see.

Most of the light we see from the coma actually comes from diatomic carbon (C_2) and cyanide (CN) molecules. Although they are very minor constituents of the coma, they fluoresce very effectively in sunlight, making them shine brightly. The parents of these molecules (i.e. the ices in the nucleus) are thought to be hydrocarbons (ethane C_2H_6 and acetylene C_2H_2) and small organic dust called CHON particles (CHON = carbon–hydrogen–oxygen–nitrogen-bearing molecules), respectively. An ion tail rich in carbon-monoxide ions (CO^+) can shine blue in absorbed and re-emitted sunlight.

Solar energy produces many interesting chemical reactions in the coma, transforming a dozen or so parent molecules (those contained in the nucleus ices) into a

Figure 3.11. The Moon's surface is largely dark, basaltic rock. However, a comet nucleus has even a lower albedo. (Not to scale: the comet should be about 800× smaller!) Courtesy of Robert Vanderbei, Princeton University.

plethora of thousands of siblings: reactive radicals, ions, and neutral atoms and molecules, which form from secondary gas-phase reactions. These reactions produce energy, heating and accelerating the coma gas. For instance, water vapor is broken up by sunlight into hydrogen and oxygen, forming an even larger part of the comet called the hydrogen envelope, a sphere roughly centered on the nucleus stretching more than a million kilometers in size (larger than the Sun). Hydrogen cyanide (HCN) is thought to be a molecule native to the nucleus (an ice), perhaps some in the form of a polymer associated with the dust. It is broken into hydrogen (H) and cyanide (CN), all of which can end up in the comet's tail.

This caused quite a stir in 1910, when it was announced that the Earth would actually pass through the tail of Halley's Comet. The Earth was about to be immersed in poison! Hucksters had a field day selling antidotes and gas masks to the public, but, of course, comet gas is so thin that it could never affect life on the Earth. Indeed, the Earth survived.

We have only scratched the surface of the chemistry that occurs in the coma. It is a chemist's delight to be able to study reactions that occur in the extremes of space, where vacuums and temperatures cannot be achieved in the laboratory. The study of comets is truly multidisciplinary.

3.3 More on the comet nucleus

That is it for the coma and tails. What does the comet nucleus itself look like? Astronomers refer to the ability of a surface to reflect light as a number called the *albedo*[10].

A shiny surface, like a mirror, has an albedo close to 1.0; a dark surface, e.g. black velvet, has an albedo close to 0.0. With all the ice present, you might think of a comet nucleus as a shiny, high-albedo object. It need not be (figure 3.11).

[10] Albedo is the fraction of light reflected from a surface, ranging from 0 (total absorption) to 1 (total reflection).

If you live in a climate that experiences snow in the winter, you know how beautiful a street with newly fallen snow on it can look. On the other hand, now think of what that snow looks like a few days later! It becomes dark and unattractive. Is the street that dirty? If you take some of this dark, dirty snow home and melt it, you will find that there is very little dirt and soot in it. It is mostly pure water. It only takes a tiny quantity of dark material, mixed in with ice, to cover the faces of all the ice crystals and to drastically reduce the ice's ability to reflect light.

Most of the surfaces of the icy satellites in the outer Solar System are, in fact, very dark, though they are largely water ice (figure 3.12). Of the comets for which we have close-up space probe pictures, one of the surprises in these images is that the nucleus is almost black (typical albedo = 0.04)! It is darker than asphalt—blacker than practically anything else in the Solar System. Apparently, the outer layer of the comet has been preferentially depleted of high-albedo ice by repeated heating during many passages by the Sun. What is left is a crust of low-albedo dust, etc, pieces too big to be swept easily into the comet's coma and dust tail. Plus, repeated exposure to the Sun's ultraviolet rays has caused chemical reactions on the surface of the comet that combine volatile molecules into an organic sludge with an albedo less than that of coal.

Solar wind and cosmic radiation also can cause radiation damage to the surface material, especially on long-period comets, forming a radiation crust. Some darkness is caused by the fluffy nature of this surface, where light enters and cannot find its way out, like velvet!

For all the splendor of a comet in the sky, this frozen mix of dust and ice at the heart of a comet does not sound very exciting. Some might even call it ugly. Yet to planetary scientists who study the origin of the Solar System, it is more desirable than gold.

Figure 3.12. One of planet Jupiter's large icy satellites, Callisto. NASA image.

The problem with studying how the Solar System began is that it is difficult to get hold of the original stuff out of which it was made. It is believed that the proto-Solar System started out as a cloud of gas and dust that condensed and accreted into small bodies called planetesimals. These planetesimals collided into one another to assemble the planets. The planetesimal material still exists, but it has been changed so in the process of planetary formation and evolution that it is unrecognizable. The planetesimals themselves are gone now. Almost. Comets are thought to be 'original equipment' in the Solar System, leftover planetesimals that never participated in the construction of planets. In the cold locker of the outer Solar System, they remain just as they were more than 4.6 billion years ago. This primitive matter, not significantly altered by any astronomical or geological process, is a time capsule of the chemistry at the birth of the Solar System. Planetary scientists are like Solar System archeologists, digging back in time to its earliest days. They would love to get their hands on a piece of pristine comet.

In fact, they have. The Stardust space probe of 2004 flew through the coma of Comet Wild 2, within 237 km of the nucleus (figures 3.13 and 3.14). This was a short-period comet that recently had a gravitational interaction with Jupiter, changing its orbital period from 43 to about 6 years. Aboard Stardust was a tennis racket-shaped collector with an ultralight-weight, silica-based material named aerogel, in which comet dust particles became imbedded (figure 3.15). The collector was then returned in 2006 to the Earth (via parachute) for study. Everyone can participate in the analysis of the captured dust in a citizen science project at the website http://stardustathome.ssl.berkeley.edu.

3.4 Comet scene investigation (CSI) using chemical fingerprints

Time out. How is it that we think we know what comets might be made out of? The shape and motion of comets are evident from descriptions and photographs made from the Earth, but composition is another matter.

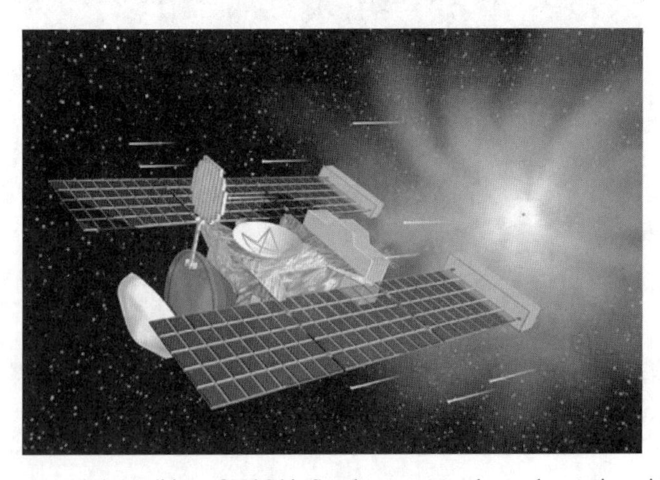

Figure 3.13. An artist's rendition of NASA's Stardust space probe rendezvousing with a comet.

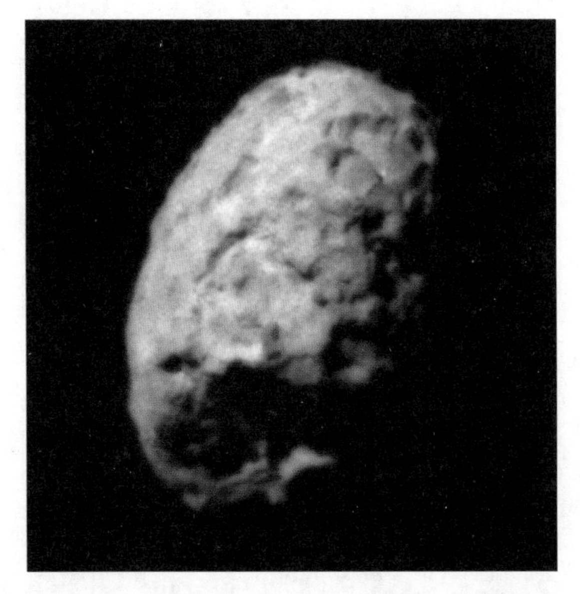

Figure 3.14. Comet Wild 2 as imaged by Stardust. NASA image.

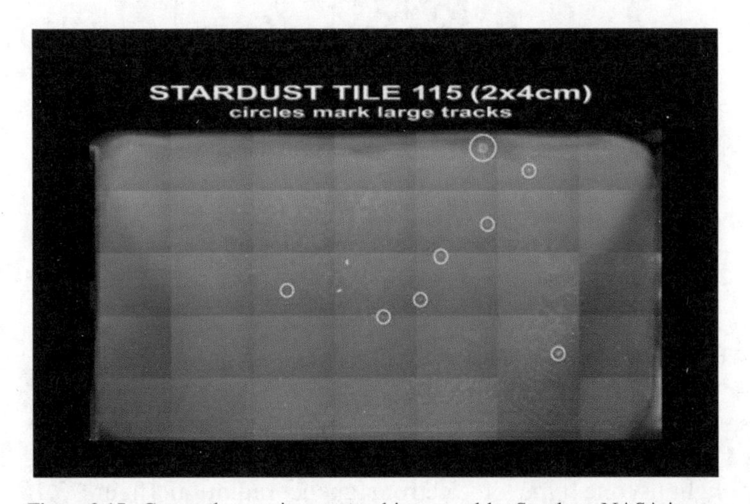

Figure 3.15. Comet dust grains captured in aerogel by Stardust. NASA image.

Astronomers are always at a disadvantage compared to their colleagues in other sciences. Chemists, geologists, or biologists can examine their subject matter up close, in a laboratory, conveniently displayed in test tubes and beakers. The subject of astronomy, the Universe beyond the Earth, is by definition detached. Except for meteorites (see below), small interstellar dust particles (IDPs) collected by high-flying aircraft and Earth-orbiting craft, and a few hundred kilograms of Moon rocks retrieved by astronauts, we cannot hold pieces of the rest of the Universe in our hand. All our information comes to us, indirectly, in the form of light and other kinds of radiation.

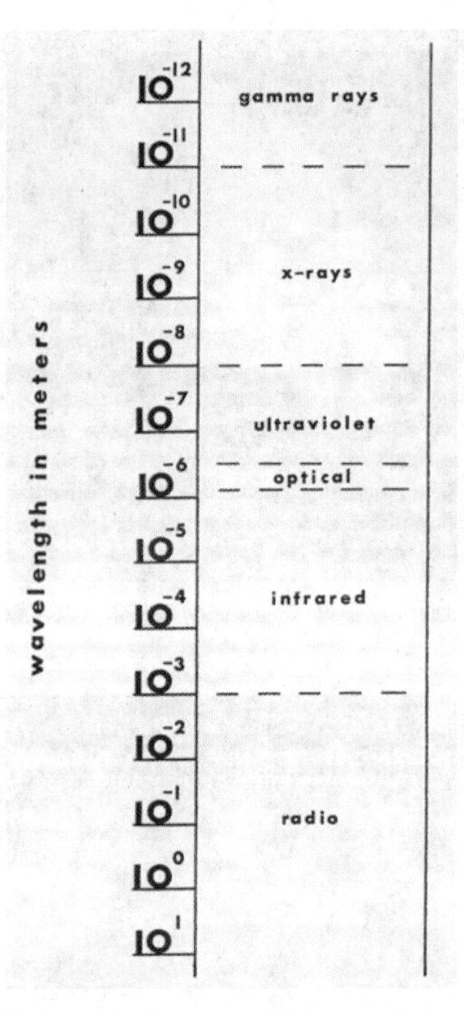

Figure 3.16. The electromagnetic spectrum. We can see only the optical portion.

Because they rely on light to such a great extent, astronomers must be intimately acquainted with the information imbedded in it. Here, we will hit the highlights. Formally, light is one type of electromagnetic radiation (figure 3.16). For our purposes, electromagnetic radiation is a wave that can travel through empty space without a medium to convey it. It does so at a constant speed (almost 300 000 km s^{-1}), the cosmic speed limit[11]. All electromagnetic radiation is alike except for its wavelength, defined as the physical distance between succeeding crests of the wave (or succeeding troughs). For instance, radio is a form of electromagnetic radiation. You and we think of radio as something very different from light. (We can see light, we cannot see radio.) Yet the only real difference is that radio has wavelengths

[11] According to Einstein's special theory of relativity, nothing in our Universe can go faster than the speed of light.

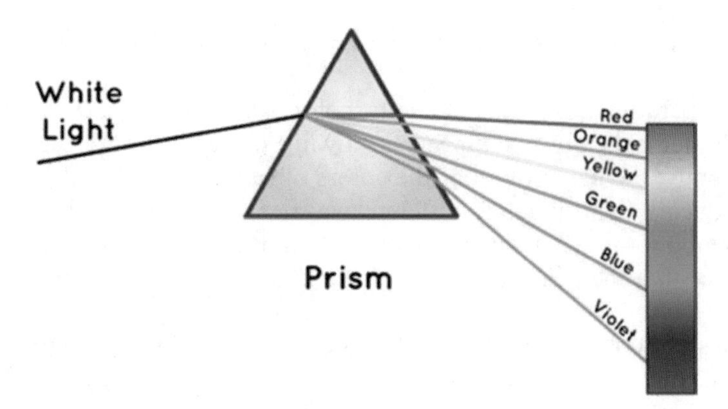

Figure 3.17. White light being dispersed into its component colors. NASA image. Illustration by: Simon Tuckett.

measured in millimeters or meters, while visible light has wavelengths measured in ten-millionths of a meter. (The wave crests are very close together.)

All visible light is not created equally. Sir Isaac Newton proved this when he shone white light through a prism. A prism separates white light into a spectrum of colors (figure 3.17). Each color has a slightly different wavelength. Another way of stating this is that wavelength is simply the quantity (number) scientists assign to the quality of color; so when you hear color, think wavelength and vice versa. There are really a tremendous number of colors, each distinguishable by wavelength, but the human eye and brain tend to group them into six 'bins' we call red, orange, yellow, green, blue, and violet. Newton described what he saw as a continuum of light but ended up using seven 'bins', including indigo between blue and violet. This conformed with the notion that seven was a 'magic number', like the seven notes of a western major musical scale.

The colors do not stop with either red (the longest wavelengths we can see) or violet (the shortest wavelengths we can see). Even longer than red is infrared (IR). Our eyes are not constructed to detect it, but our bodies are. Infrared is radiated heat. (Vipers (rattlesnakes), pythons, and boas can sense infrared from 'pit' organs on their heads—one reason you are at a distinct disadvantage when in a dark room with a rattlesnake!)

Even shorter than violet is ultraviolet (UV). The invisible ultraviolet rays of the Sun are notorious for causing sunburn and skin cancer. In a less deadly application, when you receive a re-admission hand stamp at the amusement park gate, it is made visible by an ultraviolet lamp. Honey bees can find nectar on flowers by sensing UV patterns unseen by human eyes; however, they do not enjoy beautiful sunsets since they cannot see red!

Any dense object that is heated will produce electromagnetic radiation. If it becomes hot enough, it will glow in visible light. The coil of an electric stove top will emit infrared light when set on 'low'. (It looks dark, but you can feel the radiated heat with your hand.) If the stove is turned up, the coil will glow red and then orange. It is still producing infrared, and red light for that matter, but more energy is being emitted in those wavelengths we call orange than any other, so our eyes see the coil as orange.

Every temperature has its characteristic color. The shorter the wavelength of that color, the hotter is the object. Our Sun is glowing at the temperature associated with a 5500 °C body: the blend of all of the colors of sunlight is yellow. Remember, the Sun is producing light at many different wavelengths—a prism and sunlight will create a 'rainbow'—but actually more of the light is green than any other color. Is it a coincidence that the peak sensitivity of the human eye is in the green part of the spectrum?

The Sun's light shines on all the other bodies in the Solar System. There is no other significant source of light in the vicinity. We can easily see the planets, satellites, and comets because they reflect sunlight.

Not all the Sun's light reaches, and is reflected by, the comet. Thin, cool gas in the Sun's outer atmosphere absorbs specific wavelengths of light. Moreover, the atoms and molecules that make up this gas will absorb light of only these wavelengths. This absorption leaves dark lines (the absence of color) in the spectrum of sunlight (figure 3.18). We see these lines, too, revealed in the spectra of comets. The pattern is unique to the gas. It is a spectral 'fingerprint,' identifying the gas that the light has encountered.

Physicists and chemists in the laboratory have examined the pattern of absorption lines produced by all sorts of gases and have cataloged these patterns. They do this using an instrument called a spectroscope. A spectroscope is just a prism, or other device, which spreads light out into a spectrum (a process called dispersion) so that we can see what wavelengths are there and what wavelengths are missing. This branch of science is called spectroscopy.

A common misconception is that sunlight is composed of a continuous spectrum, all of the colors of the rainbow. Not so. As we have just learned, it contains up to 20 000 absorption lines where colors have been subtracted due to the cool gases in the outer layers of the Sun. In fact, the element helium (from ήλιος, pronounced helios, the Greek word for Sun) was first discovered in 1868 by an astronomer analyzing the solar spectrum during a total eclipse, before it was found on Earth. This spectroscopy is powerful stuff!

When a thin gas is itself heated, it emits rather than absorbs light. However, it emits light only at its 'fingerprint' wavelengths. The result is a spectrum of bright emission lines alone, rather than a continuous spectrum with dark absorption lines

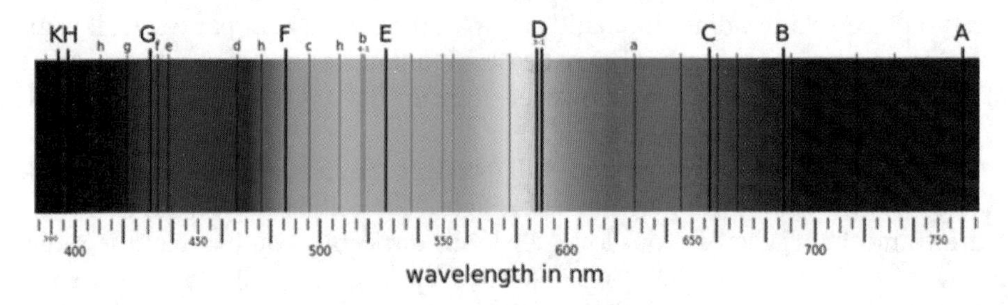

Figure 3.18. An absorption spectrum.

running through it. Mixtures of gases result in a superposition of 'fingerprints' that can be disentangled by carefully removing layer upon layer of each gas's spectrum (figure 3.19).

Using these tools, comet detectives gather clues to understanding these bodies. When an astronomer causes comet light to pass through a spectroscope, in addition to the solar absorption lines, a pattern of emission lines is seen. These lines are produced by the comet. By comparing these patterns to the patterns of known gases, the gases present in the comet can be unmasked.

Spectral emission lines are only emitted by thin gases. It is lucky for those studying comets that there is so much gas associated with comets. While spectroscopy applies only to the coma and tail of the comet, we assume that the coma and tail gases come from the materials in the nucleus. That is, the nucleus must be made of chemicals that can generate those we see in the coma and tail. Therefore, we have a clue about the composition of the nucleus, too.

Spectroscopy does not work so well for comet dust. (Dust grains emit at infrared wavelengths.) Still, comet dust scatters sunlight just as surely as a dusty car window does. By studying the component of sunlight that is scattered and comparing it to the way dust in a laboratory scatters light, we can conclude, at least, what is the size of the particles involved, if not the particles' exact chemical make-up.

In 2005 the Deep Impact space probe hurled a 370 kg copper projectile, almost 60% of the spacecraft's total mass, into low-density Comet Tempel 1 (figure 3.20). (By way of comparison, Tempel 1 is a typical, short-period comet with a size of about six kilometers.) The comet and projectile collided at a speed of ten kilometers per second (or 37 000 km h^{-1}) delivering the equivalent energy of almost 5 tons of TNT (figure 3.21).

The explosive result broke through the comet crust, created a new 150 m-diameter crater, and hurled some 10 000 tons of fresh comet material into space, causing the comet to shine six times brighter. There it was studied by Deep Impact, the Hubble Space Telescope (HST), the Spitzer Ultraviolet Telescope, and by hundreds of Earth-based telescopes in a coordinated campaign that enlisted amateurs world-wide to collect important data about Tempel 1 prior to, during, and after the collision. In addition to taking pictures, spectroscopes revealed the presence of water, silicates, clay, and other (somewhat surprising) crystal minerals. Meanwhile, Deep Impact went on to visit Comet Hartley 2 as part of a mission extension called Extrasolar Planet Observation and Deep Impact Extended Investigation (EPOXI) that also searched for exoplanets (figure 3.22). NASA is always looking to recycle, reuse, and record eco-friendly spacecraft.

Figure 3.19. Top: an absorption spectrum. Bottom: the emission spectrum of the same element.

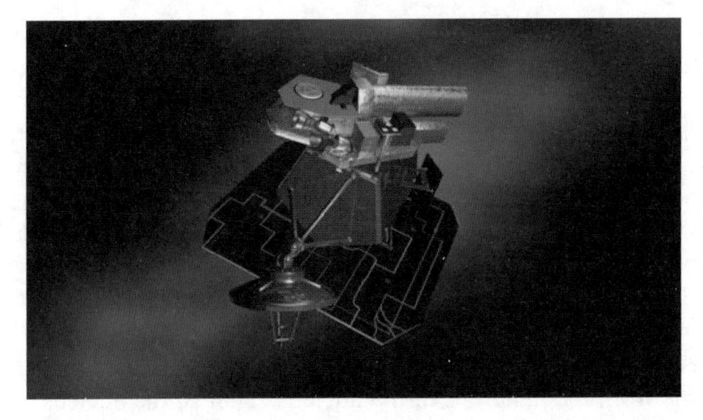

Figure 3.20. The Deep Impact space probe. NASA/JPL-Caltech image.

Figure 3.21. The explosive collision between the Deep Impact projectile and Comet Tempel 1. NASA image.

3.5 The demise of comets

Once vaporized gas and dust leave the comet nucleus, they are gone for good. Therefore, each passage of a comet near the Sun depletes the nucleus since it has no means to replenish these substances along its orbit. Every perihelion, meters of a comet's outer crust may be eroded away. The coma and tail that make a comet such a magnificent sight in our sky ultimately mean the demise of the comet. It has been

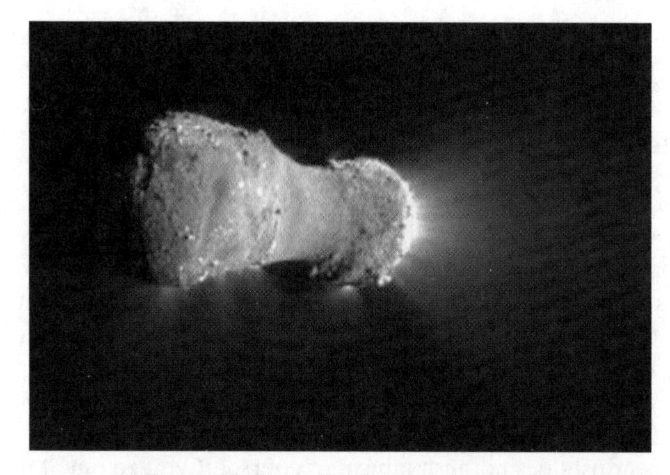

Figure 3.22. Comet Hartley 2. NASA image.

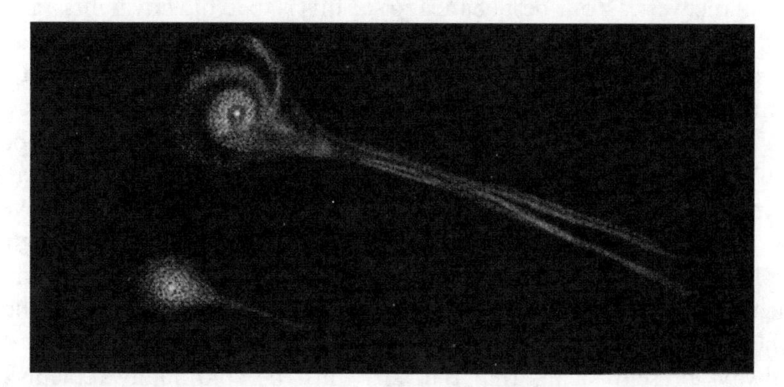

Figure 3.23. Drawing of Comet Biela. The comet split in two after its 1826 apparition.

estimated that a comet can only make one hundred to a thousand such near-Sun trips, on average, before there is nothing but grit left. The comet nucleus will expend its available volatiles before this, and with no glowing coma or tail to mark its presence, will be lost from view. Alternatively, some dust particles may fall back onto the nucleus forming a thick, insolating mantle that prevents the Sun's heat from reaching lower layers where volatiles may still exist. These dormant comets may be reactivated by a collision that breaks through the mantle, exposing volatiles in the interior. Short-period comets last, at most, half a million years.

This is not just theory. Comet Biela was discovered and charted so that its return time could be predicted with great precision (figure 3.23). Still, at the appointed time, no comet appeared.

Comets also have been observed to break up, often near the Sun. Gravity becomes weaker with distance. A comet venturing very near the Sun will experience a noticeably stronger gravitational pull on its 'front' than on its 'back'. This difference in attractive force, called a tidal force, is sometimes strong enough to tear the comet asunder. Maybe rapid sublimation at certain places on the comet nucleus has left structurally weak spots, such as exposed cavities and cracks, along which the nucleus is pulled apart. If the nucleus is a 'rubble pile', an aggregate of smaller bodies in the hundred meter range, tidal forces may overcome the self-gravity that holds them together. The result: two or more smaller comets, traveling in more-or-less the same path. Recent examples of this were Comet ISON in 2002, which broke up very near the Sun, and Comet Schwassmann–Wachmann 3 in 2006. (The latter's fragments went on to fragment themselves.)

Long-period comets tend to put on better performances in the sky than short-period comets. It is they that have the potential to become a Great Comet. The reason is that they have made far fewer trips to the purlieu of the Sun, in the course of their lives, than have their short-period cousins. While the short-period comets have had their volatile materials significantly depleted, long-period comets still have much fresh ice to produce a bright coma and long tail.

All that comet dust does not just vanish, of course. If you go out at night, just long enough after sunset for the sky to become completely dark, you may see an extremely faint glow in the west. (Your best chance to see this is to avoid city lights and choose a time when the bright Moon is not in your sky.) This glow may look like a tapering pyramid reaching up into the sky from where the Sun set. The pyramid does not stand straight up, however. It follows a path across the sky coincident with the ecliptic. You are seeing the *zodiacal light* (figure 3.24). It is called that because it appears in the zodiac, the set of twelve star constellations along the ecliptic. You can see the same thing in the East, in the early hours before sunrise. Moonless evenings in February, March, and April are the best times to see the zodiacal light in the Northern Hemisphere. (Change evenings to mornings for the Southern Hemisphere.)

Zodiacal light is produced by sunlight reflecting off dust particles in the ecliptic plane. These particles will eventually fall into the Sun. The fact that the zodiacal light is always present means that the dust must be continually replenished. The most obvious source is comets.

Figure 3.24. The zodiacal light. A Fitzsimmons/European Southern Observatory (ESO).

What happens when a comet dies? It may break into pieces, like the last sliver of a bar of soap left on the shower floor. Ironically, at this point the comet may brighten as more area of fresh ice is exposed. Perhaps, it may go out more quietly, simply fading from view. What is left then?

Besides comets, there are small rocky and metallic bodies in the Solar System called asteroids (figure 3.25). Most asteroids are thought to be leftover planetesimals (or collisional fragments thereof) created in the inner, hotter part of the proto-Solar System cloud. Most known asteroids now reside in the Main Asteroid Belt, between the orbits of Mars and Jupiter. They formed when the temperature cooled enough to allow high-melting-point refractory elements to solidify. (Examples of refractory materials include silicon, calcium, titanium, aluminum, iron, and compounds that contain these elements.) Most asteroids are mere points of light in the sky, even through a telescope. They do not exhibit a coma or tail.

If a comet has one or more large pieces in it, made of rock or metal, these pieces will go on, even after the volatile components of the comet have been exhausted. We may never see them. They may continue to follow in the comet's orbit long after the comet itself ceases to be. In short, this devolatilized or dormant cometary body may become one or more asteroids. Some asteroids on comet-like orbits are thought to be dormant comet nuclei.

However, the leftover pieces of rock may be much smaller and more numerous. The corpse of the comet may travel through space as a dissociated rubble pile. This swarm cannot be seen in visible light, but it is there, filling the orbit of what was once the comet's. We may encounter this debris if the Earth's orbit happens to cross the path of the former comet.

Meteoroids are small bits of extraterrestrial rock and metal that can strike the Earth, much smaller than asteroids (less than about three meters in size). When they do so, they are traveling at terrific speeds. During their short flight through our atmosphere, on the way to the ground, they compress the air in front of them incredibly quickly, heating it to a high temperature that is transferred to the meteor

Figure 3.25. Ida, a typical asteroid, about 60 km in the long dimension. NASA image.

causing it to glow and possibly leaving a trail of hot vapor in its wake. We can see their fiery deaths as quick, often momentary, streaks of light in the night sky. We call them 'shooting stars' or 'falling stars', but because they have absolutely nothing to do with stars, a better name is *meteor*. Many meteors are the size of grains of sand and completely burn up in the atmosphere. Larger ones may survive their fall, and can be recovered as meteorites.

Meteors appear in the sky randomly all the time. Human beings spend little time looking up at the sky; therefore, when one does happen to glance up and catch sight of a meteor, it is considered a sign of good luck, something to 'wish upon'. Yet, several times each year, the number of meteors in a night increases, in some cases significantly (figure 3.26).

Such a meteor shower is misnamed. The frequency of meteors during these days or weeks is usually not as great as the word 'shower' implies: it is one every one-to-fifteen minutes, on average. (Maybe 'meteor sprinkle' would be better!) Still, the effect is real. At one time, meteors were believed to be a strange weather phenomenon and to have nothing to do with bodies in space. Then it was noticed that shower meteors all seem to be coming from the same point in the sky. Each streak diverges from this point, called the shower radiant.

The effect is the same if one watches a line of cars, side by side, traveling toward you on the highway. At the horizon, the cars all seem to occupy a single point, but as they get closer, they appear to move outward from this point. (When this happens, it would be a good idea to stop standing in the middle of the road! Run away!) Also see figure 3.27.

The fact that shower meteors have a radiant suggests that they are produced by objects approaching us from far away. The Earth travels through the same place in its orbit at the same time each year. The fact that meteor showers are annual suggests that there are meteoroids present at this location always. Putting what we know about meteor showers and comets together, we can guess that the meteoroids that produce these showers are the spent remains of comets, a trail of comet dust so

Figure 3.26. Time-lapse image of a meteor shower. Thomas W Earle.

Figure 3.27. Both the telephone poles and the canal appear to diverge from a common, distant radiant.

to speak, expanding from the comet's path, and now intercepting the Earth. Indeed, the paths of shower meteoroid swarms have been calculated and matched up with the orbits of known comets. Meteor showers are some comets' last 'hurrah'.

When an especially dense part of the meteor swarm (the comet *trail*) is intersected by the Earth, the frequency may rise to as many as one thousand meteors per hour—a true meteor storm! Major meteor showers like this occurred in the Novembers of 1966, 1996, 2001, and 2002; they are called the Leonids after the constellation in which their radiant appears. They are due to Comet Tempel–Tuttle that had recently rounded the Sun and replenished its debris trail. The Leonids also contain many fireballs, very bright meteors that exceed the brilliance of Venus. What happens if the comet is crossing our orbit when we arrive? A spectacular collision, of course, but more on that in chapter 4.

Do not confuse the meteoroids that produce shower meteors with meteorites found on the Earth. These larger objects from space survive their hot passage through the atmosphere to land nearly intact. However, these falls are random; they do not coincide necessarily with any annual meteor shower. High-flying aircraft have been successful in capturing bits of meteor-producing grains. Analysis of the material shows it to be chemically different from meteorites. This well-traveled dust is nearly our only sample of 'comet stuff'.

Because we are always losing comets, there must be a source of a fresh supply. Where?

3.6 The origin of comets

We know what becomes of a comet that spends too much time near the Sun. But where do comets come from in the first place? What is their origin? They must have formed outside the snow line, the distance from the Sun at which ice remains frozen, currently at the distance of Jupiter's orbit.

According to Dutch astronomer Jan Oort's (1900–92) theory from the 1950s, the answer lies in the average distances of long-period-comet aphelia. That almost all go out to 50 000 au in random directions is more than a coincidence. Oort envisioned a vast shell of dirty snowballs surrounding the Solar System at that distance up to 150 000 au (more than half the distance to the nearest star and at the limit of the Sun's gravitational boundary). This shell is a reservoir of potential comet nuclei—more than a trillion of them. Today we recognize a doughnut-shaped Inner Oort Cloud (or Hills Cloud), extending from about 2000 to 20 000 au, surrounded by the Outer Oort Cloud, the spherical shell envisioned by Jan Oort (figure 3.28). By far, most of these bodies stay in the *Oort Cloud* (figure 3.29). They slowly move around the Sun, always at this vast distance, in nearly circular orbits. They are as cold as space gets: about 3 K (−270 °C). Here we make an important distinction between the Solar System and the Planetary System. Oort showed that the influence of the Sun (due to its strong gravitational pull) extends to at least 50 000 au, far beyond the realm of planets (about 30 au). So when you hear someone (like NASA) claim that a certain spacecraft is at the edge of the Solar System or has left it, they actually mean that it has left the Planetary System (at 30 au) or at best the influence of the solar wind (about 120 au). Please remind them that it has quite a ways to go before leaving the Solar System!

Figure 3.28. Astronomer Jan Oort. Copyright the Leiden Observatory, reproduced with permission.

Occasionally, however, something disturbs these loosely bound objects in the Oort Cloud. (Such a thing would have to come within 100 000 au of the Sun to produce a noticeable effect.) It may be the gravity of a traveling nearby star, an event that occurs once every 3–10 million years, or that of a wandering interstellar cloud of gas or even galactic tides[12]. (The star Gliese 720 actually will pass *through* the Oort Cloud—in a million years.)

Our Sun, along with hundreds of billions of other stars, slowly revolves around the center of a huge disk of stars called the Milky Way Galaxy. Every time the Sun passes through a dense part of the disk, the disk's gravity may nudge the Oort Cloud. Any of these mechanisms could supply us with enough comets for 10 000 years.

Regardless, nuclei are bumped out of the cloud. Some escape the Solar System entirely. It even may be possible for such a comet to travel from the influence of one star to another. (Remember, Comet Oumuamua that we discussed earlier in chapter 2? Some argue that Comet Hyakutake is a captured interstellar comet.) More likely, however, the nucleus plunges inward. The result is a long-period comet. These objects did not start out at such great distances from the Sun. They probably formed out of the denser gas and dust present in the proto-Solar System cloud, at about the distance of Neptune (30 au) and in the ecliptic plane. The modern theory of the origin of our Solar System, called the Nice model[13], posits that the giant planets originally formed much closer to the Sun and then migrated outward, unleashing a flood of comets throughout the Solar System as part of the Late Heavy Bombardment era (about four billion years ago) and formed the Oort Cloud among other Solar System features. Some have suggested that five giant planets were originally formed, and one was ejected in a case of planetary musical chairs. The Grand Tack Hypothesis suggests that Jupiter originally migrated inwards, before interacting with Saturn and reversing its course to its present location, solving some of the limitations of the Nice model. With all of this exciting work, there remains so much that we do not know about the history of our Solar System, but comets certainly provide many important clues.

Some meteor dust contains hydrated minerals, those that incorporate water. These are common on the Earth; an example is olivine. This dust must have formed nearer the Sun and then somehow migrated out to the realm of the comets. That comets contained both 'fire and ice' materials was a major finding of the Stardust Mission.

As was mentioned earlier, many of the objects so formed (planetesimals) collided with and joined the outer major planets as they grew. The left-overs were subjected to the gravitational pull of the now-massive planets and were banished to the distance of the Oort Cloud or farther.

There also remain potential comet nuclei between Neptune and about 50 au out (with the densest part lying between 40 and 48 au). This is the *Kuiper Belt* (figure 3.29), named after a pioneer planetary scientist of the 1940s and 1950s, Gerard Kuiper (1905–73). (Not the fancy belt worn by D B as one student noted on his exam!)

---

[12] Galactic tides are tidal forces exerted by the Milky Way Galaxy on our Solar System. Tidal forces are explained in chapter 4.

[13] It is a nice model but the name actually comes from the French city, Nice, where it was originally developed.

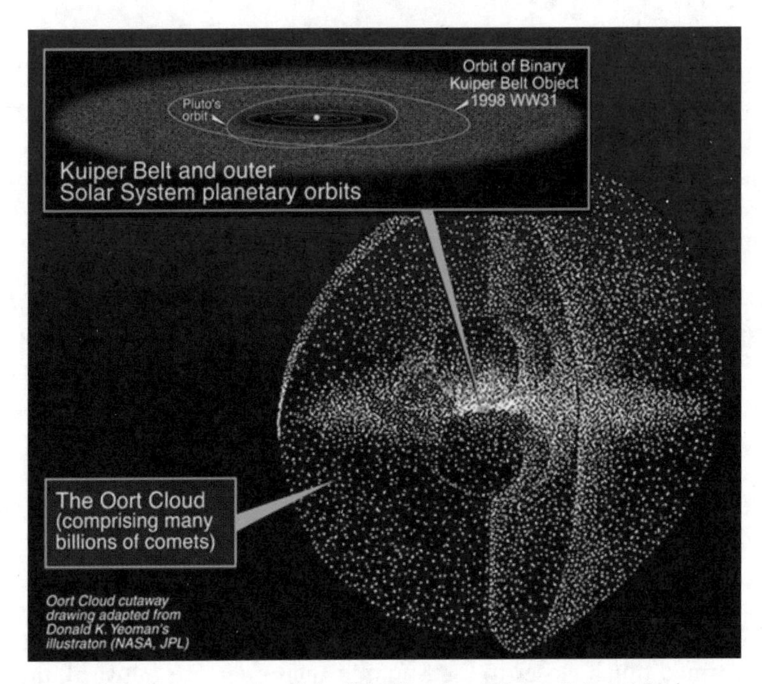

Figure 3.29. Representation of the Oort Cloud and Kuiper Belt. NASA image.

He thought of it in 1951 after asking a simple question: why should the Solar System end at Pluto (then a planet)? He could find no reason why it should. Originally considered a hypothetical hiding place for would-be comets, beginning in 1992, large Earth-based telescopes finally imaged larger members of the Kuiper Belt. The total count now is more than one thousand. Because we expect large nuclei to be rare compared to smaller ones, the Kuiper Belt, like the Oort Cloud, must be a repository for a great number of comet bodies, maybe more than 100 000 with a diameter greater than 100 km. (It is hard to tell the size of these objects, most point-like in appearance from the Earth, because we do not know their albedos.) Indeed, telescopic survey images (e.g. by Hubble, figure 3.30) suggest that hundreds of thousands more Kuiper Belt Objects (KBOs) await to be detected by telescope, while hundreds of millions more remain invisible. Moreover, it is improbable that the space between the Kuiper Belt and Outer Oort Cloud is devoid of objects, hence the Inner Oort Cloud that smoothly joins these two. The mass of the Kuiper Belt may exceed that of the main asteroid belt by hundreds of times (but still only 1/10 of an Earth mass). KBOs have satellites: one of them has three, and Pluto[14] has five. The New Horizons space probe, having flown by Pluto, will attempt to visit another KBO, 2014 MU_{69} (nicknamed Ultima Thule), on January 1, 2019 (figure 3.31). (The first results from New Horizons' successful encounter with Ultima Thule can be found in the Postscript.) During the elaborate Kabuki dance of the Giant Planets

[14] Pluto is a Dwarf Planet but is also considered to be a KBO, in a special class, called Plutinos; these are in synch with Neptune's orbit.

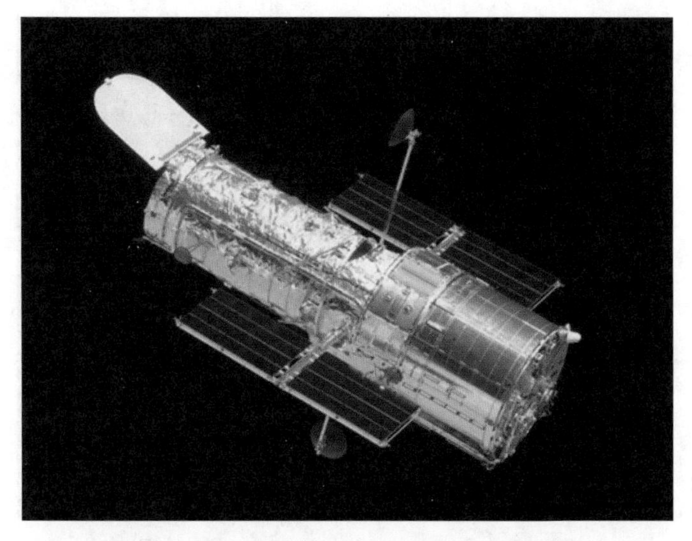

Figure 3.30. The Hubble Space Telescope. NASA image.

Figure 3.31. Launch of the New Horizons space probe in 2006. It flew by Pluto in 2015 and is now headed to a KBO, ignominiously named (486958) 2014 MU_{69} (nicknamed Ultima Thule). NASA image.

predicted by the Nice model, Uranus and Neptune shepherded the Kuiper Disk to its present location and may have even traded places. Also formed was the Scattered Disk, overlapping the Kuiper Belt and extending beyond 100 au, containing objects with very elliptical orbits and higher inclinations. The Scattered Disk is now thought to be the principal source of short-period comets.

There is a population of small bodies with characteristics of both asteroids and comets between the orbits of Neptune and Jupiter called *Centaurs*. Interactions with the outer planets make their orbits inherently unstable. Some are thought to be short-period comet nuclei from the Kuiper Belt or Scattered Disk making their way progressively into the inner Solar System. 95P/Chiron was the second Centaur discovered in 1977, joining a list now totaling over 500 objects.

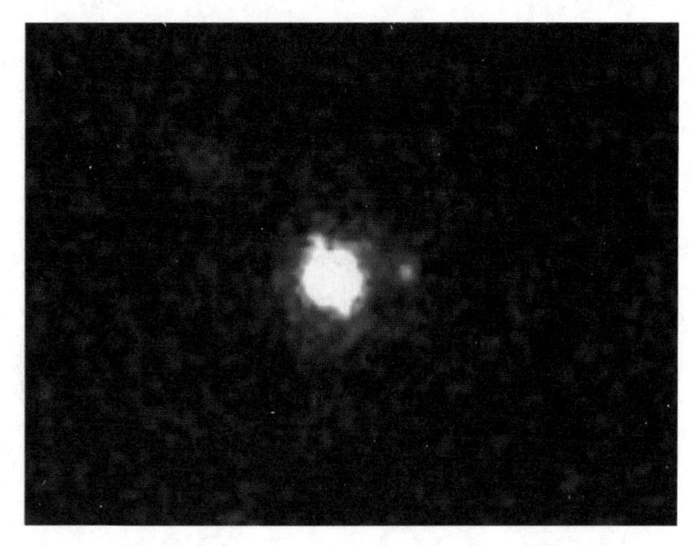

Figure 3.32. Distant KBO Eris is slightly smaller than Pluto but is more massive. Note its satellite Dysnomia on the right. Image courtesy of W M Keck Observatory.

Another place in the Solar System where comet nuclei may be sequestered are the *Trojan asteroids*. These small bodies are located at the orbit of Jupiter, in two groups about 60° ahead (Greek node) and 60° behind (Trojan node) its orbit, trapped at stable points. All are classified asteroids but many may be comet nuclei that were captured by Jupiter in the early history of the Solar System. To date, 7040 Trojans have been discovered. In 2021, NASA will launch the Lucy Mission to visit six Trojans between both groups, arriving in 2027.

In 2006, a Great Debate was held at the International Astronomical Union meeting in Prague about the definition of a planet. What precipitated the debate? It was the discovery of Eris in 2005 (figure 3.32), initially thought to be larger than Pluto (figure 3.33) and at twice its distance from the Sun. If Pluto was a planet, then surely Eris was also. Experts in the field suggested that there could be hundreds of other Plutos in the Kuiper Belt. (We now know of an additional two[15], Haumea and Makemake, joined by the largest asteroid, Ceres, in the Main Asteroid Belt.) The issue was settled by vote and 76 years after its discovery by United States astronomer Clyde Tombaugh (1906–97), Pluto lost its status as a planet, being reclassified into a new category of 'dwarf planet'. Do not let the adjective fool you; Pluto is not simply a small planet but an entirely new creature. For one thing, Pluto is small for a planet (smaller than our Moon!) and only one out of three made substantially out of ice (in addition to ice giants Uranus and Neptune; Pluto is thought to be 70% rocky, 30% water ice but without the deep hydrogen–helium atmosphere). It does have sufficient mass to pull itself into a round shape, like a planet. It also has the most noncircular orbit of them all—it crosses the orbit of Neptune. It is more inclined to the ecliptic

[15] There are four dwarf planet candidates in the Kuiper Belt as we go to press: Sedna, Quaoar, Orcus, and 2014 UZ224.

Figure 3.33. Pluto as seen by the New Horizons space probe. NASA/Johns Hopkins University Applied Physics Laboratory/Southwest Research Institute image.

than other planets. Finally, it inhabits a crowded region of the Solar System, unlike planets that have largely cleared the region of their orbits. All these things sound more like 'comet' than 'planet'. We now recognize that Pluto has been reclassified as the first of the KBOs. Pluto is not the smallest planet.

May it be the largest comet? While Pluto does sublimate a thin atmosphere when near perihelion, it never gets close enough to the Sun to form long ion and dust tails. Moreover, Pluto is too massive to allow gas and dust to freely escape its gravitational pull as is characteristic of a comet. So, we must add to our definition of a comet: a small body that cannot bind the sublimating gas and attendant dust. Pluto also has the uncometary characteristic of a large natural satellite, Charon, and four smaller ones (Hydra, Nix, Styx, and Kerberos). It is unlikely that school children will be required to unlearn this quixotically named dwarf planet any time soon.

Rather than mourn the demise of Pluto, we should celebrate our new found knowledge. This is a prime example of the scientific method at work, causing us to constantly review and revise previously held notions as new discoveries are made. In any event whether you agree with Pluto's reclassification or not, Clyde Tombaugh summed it up best: it doesn't matter what you call Pluto, it remains an important and fascinating body!

Occasionally, a long-period comet, on its trip into the Planetary System, will come near one of the major planets. Comets have so little mass that they are easily

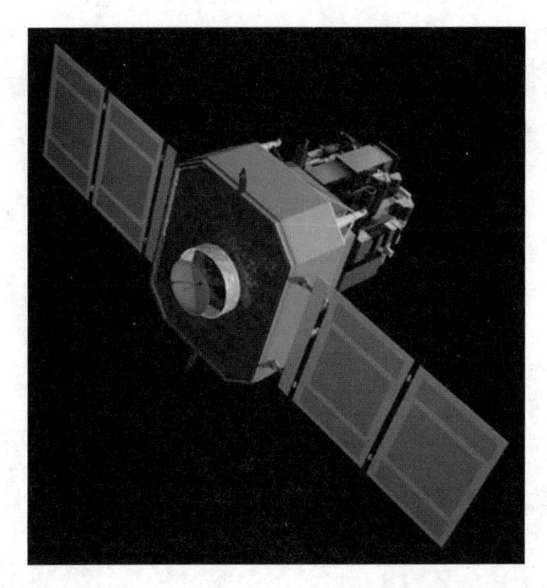

Figure 3.34. ESA's Solar and Heliospheric Observatory (SOHO). NASA drawing.

diverted from their courses by the gravity of these worlds. Oort had to account for these course changes first before he could see that the original orbits of long-period comets share a common aphelion in the Oort Cloud. Often the affected comet will be drawn in toward the Sun or discharged from the Solar System entirely.

The Solar and Heliospheric Observatory (SOHO) in space watches the environment near the Sun (figure 3.34); it has discovered well over three thousand sungrazer comets—comets that get up to seventy times closer to the Sun than the planet Mercury. Bright Comet ISON came within two solar radii of the Sun in 2013, before disintegrating (figure 3.35). SOHO's images also show dozens of comets plummeting *into* the hot Sun each year. Sometimes these suicidal comets come in groups. Were they once parts of a single, larger comet? Current thought points to a parent Great Comet, possibly seen by Aristotle in 371 BCE, for the origin of many fragments.

An example of a gravity altered orbit is that of the recent Comet McNaught, which had an initial orbital period of 6 500 000 years, but because of gravitational interactions with the other planets on its recent pass, its orbit was changed to a hyperbolic one that will eventually relax to a period of about 90 000 years, becoming a member of the Inner Oort Cloud.

Sometimes, however, the comet's orbit will be made more circular by these interactions. Its period will shorten, and it will visit the realm of the planets more frequently. There, it will have more opportunities to be affected by the planets' gravity. Eventually its orbit will be 'beaten down' near the plane of the ecliptic. The result is a short-period comet. Halley's Comet is thought to have evolved this way.

Short-period comets may also come directly from the Scattered Disk. Here, an occasional nudge from Neptune, or gravitational interactions or collisions between the KBOs themselves, change their orbits.

Figure 3.35. Comet ISON. In a comet image, the camera follows the moving comet thereby making it appear as if it is the backgrounds stars that are moving. TRAPPIST/E Jehin/ESO.

We can even explain why most short-period comets orbit the Sun in a counterclockwise direction: a comet orbiting in the same direction as the motion of the planets will spend more time near a planet than will a comet whirling by the 'wrong way'. Thus, counterclockwise comets will have more time to be diverted gravitationally by the planets than will clockwise comets.

Massive planet Jupiter is particularly effective at influencing comets. Most short-period comets have aphelia out to about the orbit of Jupiter (5.2 au). Clearly, Jupiter has modified the orbits of these comets, the so-called Jupiter-family comets.

On the other hand, there is a very strange body that goes nowhere near Jupiter. The Centaur Chiron—not to be confused with Charon, Pluto's satellite—spends most of its time between the orbits of Saturn and Uranus (average orbital radius of 14 au). It originally was classified as an asteroid, but asteroids do not usually venture this far out. Chiron seems to be an icy body. The line of demarcation between these two categories of bodies becomes fuzzier. At perihelion in 1996, Chiron formed a coma! With an orbit much more circular than normal, is chameleon-like Chiron the ultimate short-period comet? If so, it is a big one; Chiron is two-hundred-kilometers across and nearly at the boundary of having a gravitationally bound atmosphere!

3.7 A visit to a comet

A Great Comet is one that we can see easily without optical aids (figure 3.36). They are as bright as planets; some can be seen in the daytime. They are invariably long-

Figure 3.36. An example of a Great Comet from the past.

Figure 3.37. Comet Holmes' outburst in 2007, visible in the daytime!

period comets because these comets have not developed a significant mantle from too many passages around the Sun.

Obviously, most comets are dim objects, seen only through a telescope. However, comet brightening is unpredictable. In 2007, Comet Holmes was a run-of-the-mill comet, that is, until it became half a million times brighter in only 42 h (figure 3.37). What happened? Did a meteoroid impact excavate fresh material and fling it into the coma and tail? Was a pocket of fresh material suddenly exposed, with the same result? We just do not understand comet brightening and dimming well, making it difficult to predict what will become a Great Comet and what will not. Even well-known Comet Halley became brighter on its last *outward* path from the Sun at 14 au. We may see what changed when it revisits us in 2061.

But let us now not restrict ourselves to telescopic views from the Earth. Let us take an imaginary spacecraft voyage to a newly discovered comet. As we approach the comet, we aim toward the bright center of its head. This is not necessarily the nucleus. It is just the densest part of the coma. No one has ever gotten a good view of the nucleus of a comet from the Earth. When near us, the nucleus is always hidden by the coma. We have only space probe images of the nuclei of certain comets to guide us (see the frontispiece). They may be atypical, but this is an imaginary encounter, so let us continue.

Closer, we enter a cloud of hydrogen gas surrounding the comet, ten million kilometers in diameter. This cloud is invisible to our eyes. We detect it with ultraviolet sensing instruments. It is our first warning that the comet looms ahead.

Now we begin to be bombarded by cometary dust particles in the outer coma. Being hit by a dust grain does not sound very violent. But dust particles traveling at the speed of a comet may be moving orders-of-magnitude faster than the bullet of the highest-power rifle. These granules could easily pass right through us at such speeds. However, we have matched speed with the comet and are creeping up on it at a leisurely rate. Shielding built into our spacecraft further protects us.

As we enter the dusty, thickest part of the coma, the nucleus can be made out. We are used to seeing round planets and satellites. However, our comet is oddly out-of-round. It does not contain enough mass for its own gravity to have squeezed it into a space-minimizing ball.

Ours is a dense comet nucleus. (Comet densities are less than that of ice, varying between 0.1 to 0.6 g cm^{-3}; we measure the mass of our comet by how its gravity affects the course of our spaceship.) From our point of view, accustomed to the comparatively mild topography of the Earth, there may be absurdly high mountains jutting out of the comet nucleus, or impossibly deep fissures cutting into it. There are craters produced by still smaller bodies smacking into it during its long journey around the Sun. The internal strength of our compressed comet material is strong enough to support such features.

We might have ended up at a low-density comet. Here, the nucleus may not even be one solid piece; it may consist of separate pieces barely in contact with one another. Or the whole thing might be the consistency of a fluffy new fallen snow, one with porous cavities within it (porosity up to 75%)—in other words, a body that scarcely holds itself together.

Our comet nucleus rotates underneath us, with a period measured in days. As it does so, each place on the surface experiences day-time and night-time just as places on the rotating Earth do each day. On the daylight side, jets of gas and dust erupt into the comet's sky, at speeds of a thousand kilometers per hour, and feed the coma. At night they shut down. These jets come from isolated 'active spots' on the nucleus's surface, where freshly exposed ice, from a slope exposed by a landslide or small porous holes in the comet's crust, is being rapidly sublimated away by the Sun's heat. With so little gravity, this material easily escapes into the coma.

Wait a minute. Is something wrong here? The day side of the comet nucleus is facing the Sun; the night side is facing the comet's tail. No, it is true that the jets point toward the Sun, but quickly the gaseous material expands to fill the vacuum of space and flows in all directions. Ions produced by photoionization of cometary neutrals by UV sunlight are caught up in the solar wind, blowing far faster than the

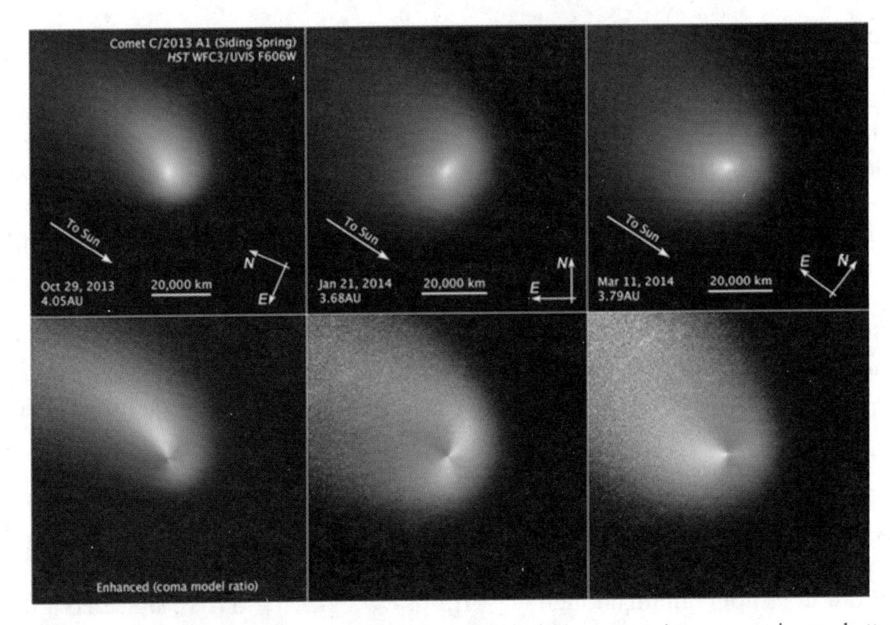

Figure 3.38. Active, rotating jets, on a comet's nucleus, in this time sequence (top—comet image; bottom—coma dust model). NASA/ESA/J-Y Li (Planetary Science Institute) image.

comet itself moves, and are turned around to blow back in the direction of the ion tail, away from the Sun.

The true path of the dust entering the coma is a spiral. The rotation of the nucleus acts like a spinning lawn sprinkler. This spiral structure sometimes can be seen within the coma of a comet from the Earth (figure 3.38). Such was the case with Comet Hale–Bopp.

Nuclear jets help explain a long-standing mystery about comets. The laws of celestial orbits are incredibly precise. It is possible to predict the position of an orbiting planet millennia in advance. This is not true of comets, however. Comets are notorious for not returning to perihelion on schedule. Sometimes they are a little late, sometimes they are a little early. Even when the perturbing effects caused by the gravity of nearby planets are considered, a comet orbit seems to change with time.

The jets are the solution to the mystery. According to Newton, for every action there is an opposite reaction. (This is the classic sailor-steps-off-boat-onto-dock/boat-moves-away-from-dock/sailor-falls-into-water effect.) As material flies off the nucleus in a jet, the comet itself is pushed a little in the opposite direction. We call this a non-gravitational force.

The gas and dust have little mass but lots of momentum. A bullet does not weigh a lot, but, if traveling fast enough, can knock over an elephant. This is the same principle upon which a rocket engine works. Hot, expanding gas, moving at high velocity, comes out of the bottom of the rocket, and the rocket moves upward. That is one explanation of how comets change their orbits: comets are rocket propelled!

In reality, a jet of gas pointing toward the Sun will not greatly affect the comet's orbit. The comet must be accelerated or decelerated in the direction in which it is traveling. However, the jets do not point directly toward the Sun. The nucleus' rotation swings the jets around to point a little forward or backward, depending on the sense of the comet's rotation compared to the comet's direction of travel. If the jet acts to speed up the nucleus a bit, the comet's orbit will increase in size, and the comet will be tardy for perihelion. If it happens to retard the nucleus, the comet's orbit will become smaller, and the comet will arrive at its appointment with perihelion ahead of schedule. We notice that the northern pole of our comet is in

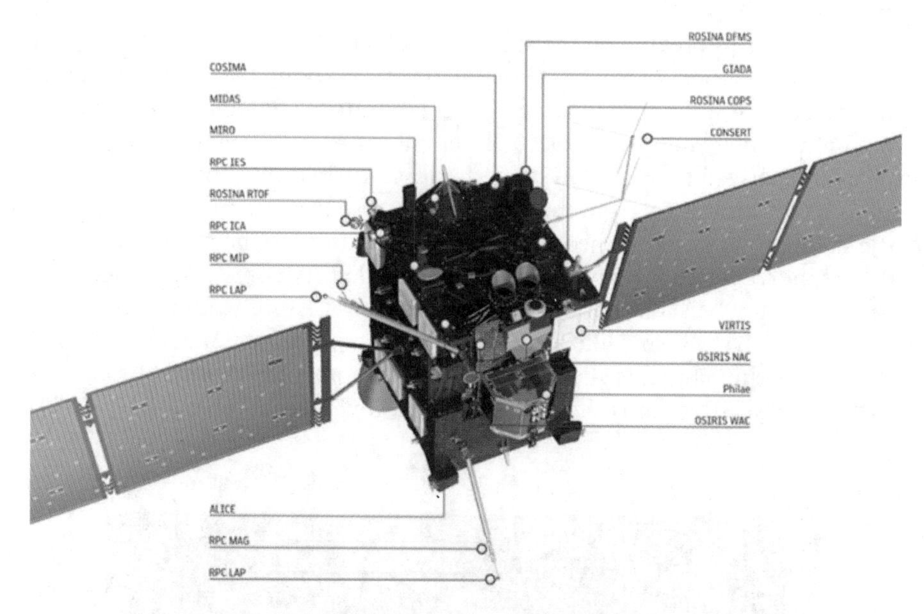

Figure 3.39. The Rosetta space probe. ESA image.

Figure 3.40. Comet Churyumov–Gerasimenko imaged by the Rosetta spacecraft. ESA/NASA image.

constant daylight. It is summer in the north. Tilt of the comet's rotation with respect to its orbit (called the obliquity) causes seasonal effects. So as the comet rounds the Sun, new, fresh surfaces are illuminated causing changes in the gas and dust production.

We are now so close to the comet nucleus that we could reach out and touch it. Should we step out of our spacecraft and plant a flag, claiming its territory as our own? Well, there are international treaties that prevent that sort of thing, but we will not even try.

There is no air to breathe on the surface of a comet, and we would be exposed to the cold and radiation of space. (For all the fuss about the comet coma, at its thickest its density still approximates a good laboratory vacuum; that of the ion tail is even thinner: a few hundred molecules per cubic meter.) Even if we properly attire ourselves in spacesuits, it is likely that the force of planting the flag is enough to propel us backwards away from the comet and our spacecraft, never to return. (With so little gravity, we would be nearly weightless standing on the nucleus of a comet.)

What is more, we might have arrived at a low-density comet. The consistency of the comet ice may be such that our meaningless flag gesture would merely result in pieces of comet flying apart, like a sneeze over talcum powder. We may also be able to achieve superhuman feats such as breaking the comet apart with our bare hands since the material strength is thought to be that low.

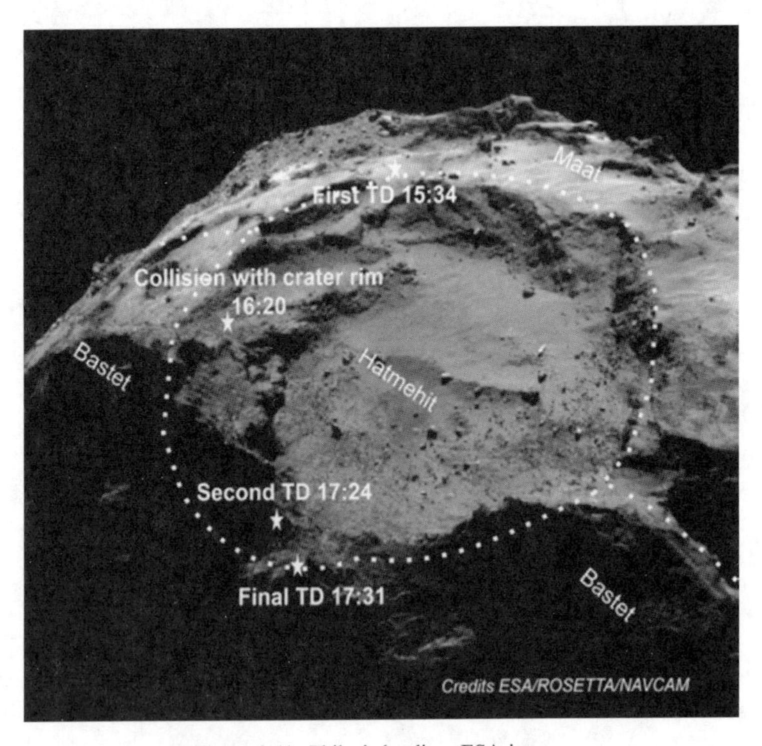

Figure 3.41. Philae's landing. ESA image.

Figure 3.42. A view from Philae. ESA image.

No, standing on a comet might be a tricky proposition. In 2014, ESA's Rosetta space probe (figure 3.39) flew along with Comet Churyumov–Gerasimenko, after zipping past Mars and a couple of asteroids. It found a dumbbell-shaped comet nucleus, perhaps the merger of two separate comets (figure 3.40).

Accompanying Rosetta was a lander named Philae. The idea was for Philae not so much to land as to grab onto the comet nucleus. For this it was equipped with harpoons and screws. However, Philae's landing failed when it could not stick to the comet properly (figure 3.41). It apparently hit a relatively hard surface just beneath the crust. Instead of coming to rest, Philae bounced across the comet terrain, and its solar energy collectors probably were left in shade (figure 3.42). Its batteries ran out shortly thereafter. Two instruments did return valuable results about the surface material, detecting 16 organic compounds, four for the first time.

At its very nucleus, we are too close to appreciate the magnificence of the whole comet. We cannot see the forest for the trees. The Earth provides the near perfect perspective from which to enjoy a comet. It is neither too close nor too far away. Moreover, we do not need an imaginary spacecraft to get us there!

Now, it is time for us to return to the Earth but not before scooping up a handful of comet material for analysis on the Earth. As we depart, we take a moment to look back at the night side of the comet. There, above the horizon and out of the shadow of the nucleus itself, we see hovering in the sky a short column of light. It is the tail of the comet, foreshortened because it is pointing almost directly away from us. Only at its top do we see it begin to split into two wisps, one faintly blue (the glowing ion tail), the other yellow-whitish (the dust tail reflecting the color of sunlight). And not a minute too soon, as it appears that a sizable part of the nucleus is splitting off and heading in our direction … Beam me up, Scotty!

Chapter 4

Comet crashes

Let us begin by stating that none of the comets we have mentioned will strike the Earth. That is worth repeating. They and all other known comets are not on track to hit the Earth—or any other planet, for that matter. Recently, the 2013 Comet Pan-STARRS came within about nine times the distance to the Moon—not very close (figure 4.1). While there are many natural threats to our lives and property, currently a comet is not among them. However, in the long run, we know that it is not a question of if a comet (or asteroid) will hit the Earth, but when it will occur. Comets have collided with the Earth in the past and will do so again at some future dates. Will we have the technology to first find them and then mitigate the threat? Read on!

We can say this with a certainty that we reserve for few other things because of the utter predictability of gravitational orbits. While comet orbits do change, they usually do not do so significantly within a single orbit, and no comet's present orbit takes it anywhere near us.

Be that as it may, a slight apprehension is understandable. Beginning at an early age, we were all admonished of the dangers presented by heavy, flying objects. The sky is falling! Do you remember the story of Chicken Little (Henny Penny)? A comet is such an object, and it is traveling very, very fast. Moreover, in our collective subconscious we may carry with us the vestiges of that primordial fear of comets as agents of mischief.

All this is not completely irrational. Some comets do end their lives prematurely. We can see the artifacts of their early deaths all over the Solar System. If a comet and a solid planet (or satellite) try to be in the same place at the same time, a collision must occur. The modern Solar System is huge and mostly empty space; but, over long intervals of time, such collisions on very rare occasions do happen. The low-mass comet is the big loser. It is destroyed in the collision, and an impact crater forms on the planetary surface. It is these craters that we see covering all the ancient exposed surfaces in the Solar System. We need only look at the round features all over our moon to see that impacts are an important geological feature on other

doi:10.1088/2053-2571/ab027ech4

Figure 4.1. Comet Pan-STARRS. The name comes from Panoramic Survey Telescope and Rapid Response System, a pair of telescopes located on the island of Maui, Hawaii, designed specifically to look for moving celestial objects such as comets. Courtesy of Ignacio Diaz Bobillo, www.pampaskies.com.

worlds. Many of these craters are caused by asteroids, not comets. When either body strikes, it is going at such a high speed that a violent explosion results. Little visible evidence is left indicating the nature of the impacting body. However, the population of asteroids decreases in the outer Solar System. Most of the craters we see on the icy satellites of Jupiter, Saturn, Uranus, and Neptune surely must be the result of comets (figure 4.2).

Have you ever been to the beach and thrown a stone into the wet sand? If so, you have noticed that the stone is buried in the sand (one end may be above surface), a small depression ('crater') is formed, and some sand is blasted out of the hole. Imagine what would happen if you threw the stone at an angle, say 45°, to the beach. The resulting 'crater' would be oval in shape with most of the excavated sand splashing ahead of the stone. Since impactors on solid planets and their satellites arrive at various angles to the surface, we would expect craters to have a variety of shapes, from circular to highly elongated ovals, if our beach analogy is correct. This is not what we observe. Impact craters are all nearly circular. What went wrong? The energy of the beach impacts was much, much lower than those in the Solar System. The hypervelocity speeds of asteroids and comets range from ten to more than one hundred kilometers per second, resulting in energies greater than nuclear explosions (in some cases many times greater than civilization's entire stock of nuclear bombs).

Figure 4.2. Mimas, a heavily cratered satellite of Saturn. Had the object that created the huge crater on the right, named Herschel, been any more massive, Mimas might have been destroyed in the collision. NASA image.

A better analogy would be throwing sticks of dynamite at the wet sandy beach. Caveat: do not try this at home!

We generally divide crater formation into three stages: (1) contact with and compression of the surface resulting in the vaporization of the impactor and much surface material (less than a second), (2) excavation and ejection of material forming crater rims and rays of material streaking radially away from the crater and a mushroom cloud growing above the impact site if an atmosphere is present (minutes to hours), and (3) modification of the nascent crater by crater wall slumping and slippage, possible crater floor rebound forming a central peak, and others depending on the size of the crater (hours to days). A general rule of thumb is that the crater size is about ten times larger than the size of the original impactor.

The outer layers of Jupiter, Saturn, Uranus, and Neptune themselves are made mostly of gas. They have no solid surface on which to form as lasting a record of comet collision as a crater. Still, these large planets must certainly be targets, as well.

In July 1994, astronomers watched a comet named D/Shoemaker–Levy 9 [SL-9] crash into Jupiter. It was the ninth comet discovered by the team of Eugene Shoemaker, Carolyn Shoemaker, and David Levy. (The D/ preceding the comet's name indicates that it is deceased.) Before this, Jupiter's prodigious gravity broke SL-9 into 21 large pieces during a close approach and temporarily captured these pieces as 'moons' on a collisional trajectory (figure 4.3). As the fragments struck, monstrous explosions were recorded by Earth telescopes and robotic space probes (figure 4.4). They left impressive, if transitory, dark spots in the atmosphere of

Figure 4.3. Comet Shoemaker–Levy 9. NASA/ESA/H Weaver and E Smith (Space Telescope Science Institute) image.

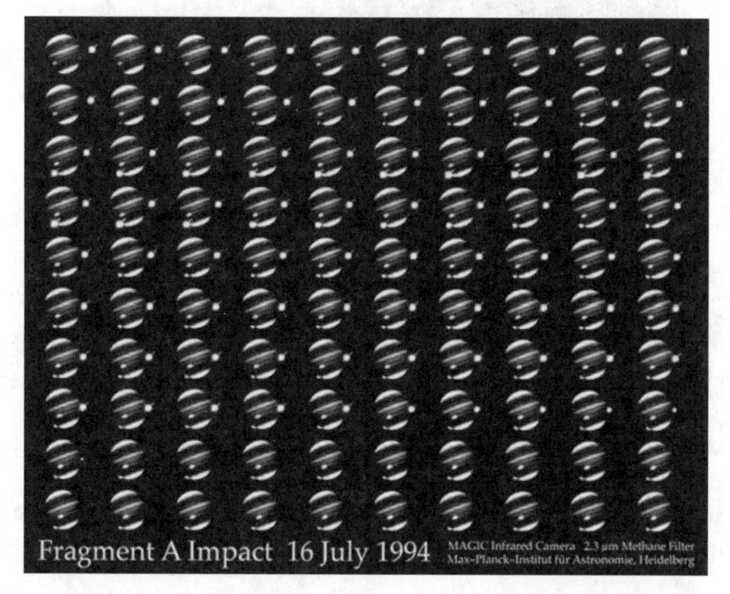

Figure 4.4. A time sequence of Shoemaker–Levy 9 fragment 'A' explosively colliding with the planet Jupiter. Movement is due to the rotation of the planet. NASA image.

Jupiter (figure 4.5). These were easily visible from the Earth, even through a small telescope.

A spectacular comet collision happening in our lifetimes was incredibly serendipitous. It was the first time the astronomers who theorize about impact processes could actually see a collision first hand—'up close' but not 'too close'. Predictions varied wildly from a 'big bang' to a 'whimper'. We learned a lot from this event and have a much better understanding of cometary impacts. (They are 'big bangs!') Since SL-9, observers (primarily amateurs) have witnessed other—presumably—comet impacts on Jupiter[1]. This reinforces the notion of Jupiter as a 'cosmic vacuum cleaner' providing increased protection for the Earth from possible collisions with

[1] Similar 'flashes' have been seen on the Moon by the Moon Impacts Detection and Analysis System (MIDAS) at https://www.uhu.es/josem.madiedo/obs/e_midas_intro.html. Many are probably impacts from comet particles since they occur during meteor showers.

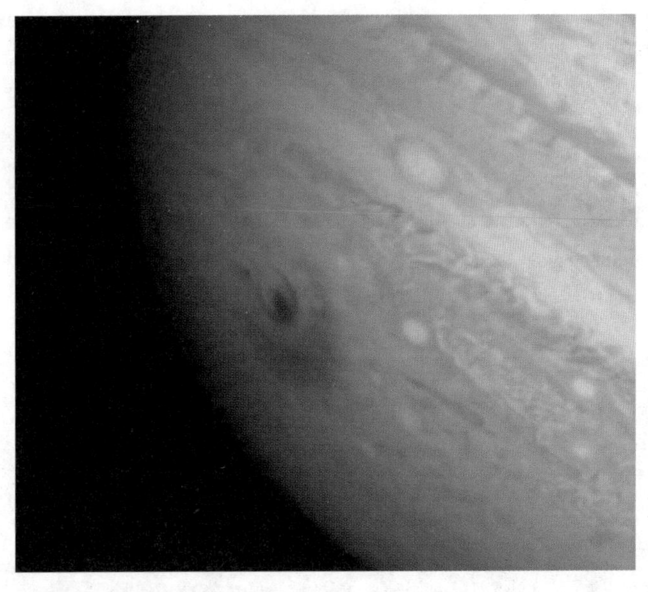

Figure 4.5. The temporary spots produced by Shoemaker–Levy 9 fragments 'D' (left) and 'G' (right) striking Jupiter in 1994. This was the first time we witnessed a collision between two Solar System objects. NASA/ESA image.

these small bodies. In 2014, Comet C/2013 A1 (Siding Spring) had a close encounter with Mars, coming within 140 497 km. Although the nucleus did not hit Mars, the planet 'flew' through the inner coma, resulting in a strong meteor shower, tons of vaporized dust in the atmosphere, and additional ions added to Mars' ionosphere. These effects were recorded first hand by the international suite of spacecraft now deployed at Mars that were unaffected by the passage.

4.1 What if?

What about collisions with our planet, the Earth? Have we been spared? A several-kilometer-wide comet really colliding with the Earth would bring unimaginable ruin to our home world. We have never seen it happen. Nothing like this has occurred in historical times. Yet what is unlikely on short time scales may be inevitable over much longer time scales.

Imagine that it is a nice warm Saturday. You have the grass mowed just so. You look forward to spending the rest of the afternoon in your hammock, watching The Game on TV.

Suddenly, an errant comet tears into the atmosphere above your head, traveling 100 000 kilometers per hour. This massive an object, about 10 km across, is barely slowed at all by the Earth's atmosphere in the second or two it takes the comet to reach the ground. It undergoes an airburst at an altitude of 90 km, fragmenting into several pieces. When the blast wave hits the surface, it does so with a force equal to more than 7500 times that of the nuclear arsenal of the human race, generating an earthquake of 10th magnitude on the Richter scale. A fireball more than

200 kilometers across is seen, two hundred times brighter than the Sun, at a distance of 500 km from the impact site. The people, trees, houses, and very rock of your city are vaporized, and a cloud of debris expands outward faster than the speed of sound. This cloud heats the local atmosphere to a temperature that ignites forest fires for more than one thousand kilometers. A crater more than one hundred kilometers across and more than a kilometer deep is carved out. Nitrogen and oxygen, the gases out of which our atmosphere is made, are chemically united to form a smog of nitrous and nitric oxides. Pulverized rock and dust are thrown upward into space. Most of it falls back, however, to deposit the equivalent of a centimeters-thick layer of dust over the entire atmosphere of the Earth, thereby blocking out sunlight. The artificial night lasts for months resulting in 'impact winter', a precipitous drop of temperature that freezes the surface. The smog is spreading meanwhile, of course. By now these noxious chemicals have combined with atmospheric water to form nitric acid. The result is a global acid rain, the strength of automobile battery fluid.

The good news? The Earth is barely disturbed and loses negligible mass. There is no noticeable shift in Earth's orbit nor its axial tilt nor rotation. And an event of this magnitude only occurs every 90 million years on the average.

Still, you thought when you awoke this morning that the only thing you had to worry about was whether the mower would start! Obviously, you should have been listening to the news since a comet of this size would have been spotted months prior to impact.

Is this story simply the rattling of a pessimist? Unfortunately, no. It is based on computer simulations performed at various research institutes and probably accurately describes what could happen if a large body encountered the Earth[2]. The catastrophe would be similar to the much-feared nuclear winter after an all-out war.

We already know that one hundred tons of interplanetary material arrives weekly on the Earth. This is in the form of harmless meteoroids and wafting dust. Large bodies are much rarer in the Solar System than small bodies. But one-in-a-million odds are no comfort if this happens to be the one-in-a-millionth year.

Scientists take this threat seriously enough to propose that a more systematic search for Earth orbit-crossing asteroids and comets be made, so that we have advanced notice of their imminent arrival, especially those large enough to inflict significant local damage (size greater than 140 m—some much greater) and that make extremely close approaches to Earth (within 19.5 lunar distances) in the next one hundred years, called potentially hazardous objects (PHOs). The United States Congress listened and, in the 1990s, mandated that by 2008, NASA find at least 90% of the celestial objects crossing the Earth's orbit larger than one kilometer (causing a global cataclysm) and funded several telescopic surveys to accomplish that goal. Since then, the search has continued for smaller, more-distant objects (still larger than 140 meters and capable of inflicting a large-scale local catastrophe). To date, about 95% of the larger objects (about 1000 including 157 PHOs) and more than

[2] Online impact simulators can be found at https://impact.ese.ic.ac.uk/ImpactEarth/index.html and http://simulator.down2earth.eu/planet.html?lang=en-US. Go ahead, play with them!

17 000 smaller ones (including almost 2000 PHOs) have been found and are undergoing characterization. It is important to know the physical properties of these objects if we hope to mitigate their hazards. Know your enemy! These lists are dominated by asteroids (since we are in the inner Solar System), but more than one hundred near-Earth, short-period comets are known.

After detection, do we have the technology to eliminate the threat by either destroying or deflecting it from a collision with the Earth? Researchers have devised a toolbox of mitigation strategies based on the matrix of warning times versus object sizes; including standoff, surface, and subsurface explosions (conventional and nuclear); hypervelocity impactors; gravitational tractors; mass drivers; laser ablation; ion beams; focused solar energy; and a host of others. Working with the Department of Defense, efforts are underway to implement the most promising strategies under the direction of NASA's Planetary Defense Coordination Office. NASA has funded the Double Asteroid Redirection Test (DART) to be launched in 2021 as a proof of concept for the kinetic impact technique to deflect asteroid (65803) Didymos in 2022.

With enough forewarning we might be able to use a nuclear-tipped missile to divert the trajectory of the incoming projectile. (This is the only good use we can think of for such a missile.) We would have to be careful, however. If we simply broke the comet into many pieces, the result could be a rain of comet chunks upon the Earth, rather than a single impact. The devastation could be made greater! Of course, we could assemble Bruce Willis and his team at the Cape … Oh drat, we do not have any more space shuttles …

Would a seek-and-destroy plan for comets be worthwhile? Is this 'know thy enemy' philosophy a responsible use of funds in otherwise fiscally tight times? We do spend money and time to avert other natural calamities: for instance, warning of and protection against tornadoes. Of course, every year we hear about a few people killed by tornadoes. The peril seems more immediate. Yet what if a deadly comet or asteroid strikes only once per million years-but kills a few million people? The odds are the same. (The odds of comet impact are calculated best by estimating how often comets have already struck the Earth in its remote past; see below.) This is clearly an international issue so an effort enlisting as many international partners is being undertaken. We are all sailing in this planetary boat together!

Perhaps we should continue to search for and study comets on such an undertaking's own scientific merit, too. Comets are intrinsically fascinating and potentially tell us a lot about our Solar System's past and possibly the origins of life on the Earth. The long history of observing comets and interpreting them, as part of our cultural heritage, tells us something about ourselves.

4.2 The comets come to Earth

All of the above sounds a little bit hypothetical. Have real comets struck the Earth in the past? The answer is that they must have. The Earth is fundamentally no different from any other planetary body with a surface. If comets and other objects have crashed down upon worlds like the Moon, there is no reason to think that the Earth should be spared. In fact, the Earth's stronger gravity should do a better job of attracting wandering comets.

Why then is the Earth's not a cratered and gouged surface like that of the Moon? Ah, 'fundamentally' no different from the Moon? Yes, but in the details there are certainly dissimilarities between our world and its satellite. These details make all the difference when it comes to searching for a record of impacts on the Earth.

First, the Earth possesses an atmosphere. Weathering—wind, water, and other forms of erosion (including life)—will eventually obliterate an impact crater on the Earth, whereas it will remain unspoiled on the dry, airless Moon.

Second, the Earth is a geologically active place. We experience earthquakes, volcanoes, and the inexorable shift in location of land masses called continental drift. This latter force of nature, in particular, changes the appearance of our planet over the eons. Pieces of the Earth's crust are always being destroyed while others are being created. (Most of this activity takes place, out of view, under the Earth's deep oceans.)

In contrast, the Moon is a geologically dead body. A globe of the Earth fabricated by imaginary navigators one hundred million years ago will do us no good today. The continents and ocean basins have changed. However, the Moon looks today much as it did a billion years ago.

Geological activity on the Earth destroys comet and asteroid impact craters. If they form at some fairly constant rate, the number that will exist on the Earth at any one time is limited. On the Moon, the craters keep piling up. With no efficient mechanism to get rid of them, the Moon just becomes more and more pockmarked.

So it is reasonable that today the craters on the Earth should be the most recent (in terms of geologic time) and, therefore, there should be relatively few. However, 'few' is not the same as 'none'. Where are they?

Well into the past century, no impact craters were recognized on the planet closest to us—the Earth. Of course, no one was looking for them! While comet impact, as a means of creating landforms on the Earth, was recognized as a theoretical (but inconsequential) possibility by geologists, it was a somewhat distasteful subject. The catastrophic nature of an impact—one moment there is no crater, the next moment there is—went against the grain of the predominant geological principle, uniformitarianism.

This principle states that changes in the appearance of the Earth are slow, approximately constant, and predictable. Uniformitarianism was a hard won concept. It was claimed on the intellectual battlefield of the nineteenth century, when political and religious forces maintained dogmatically that the Earth was created pretty much 'as is'—suddenly and with finality, in the recent geological past. This older idea was catastrophism. To many geologists, even in the 1970s, comet impacts ranked right up there with archaic deluges—irrelevant to the overall history of the Earth. With no evidence to the contrary, who could blame them?

The only impact crater recognized by the geological community at this time was the Barringer Crater in Arizona, USA (figure 4.6). It is particularly fresh (50 000 years old) and not very big. (You can visit it, right off of Interstate 40.) Compare it to the much larger lunar crater Tycho (figure 4.7).

The problem is that we are too close to the subject. Literally. It is quite easy to see the craters on the Moon from our vantage point—the ultimate bird's eye view from

Figure 4.6. The Barringer impact crater in Arizona, USA. It is almost 1.2 km in diameter and about 170 meters deep. United States Geological Survey image.

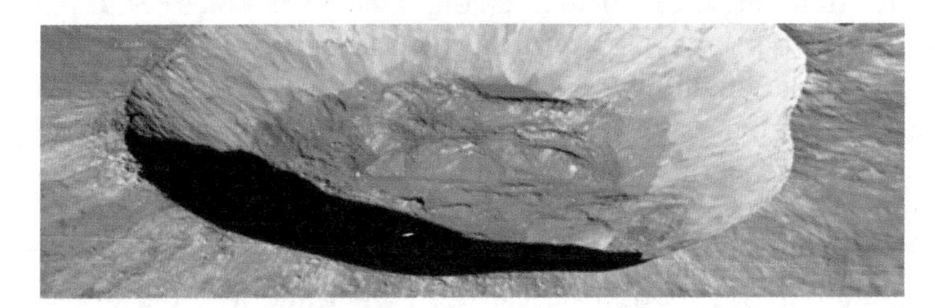

Figure 4.7. Compare the Barringer Crater to 85 km-wide crater Tycho on the Moon. NASA image.

the Earth. We do not have that perspective of our own world. Your authors have stood in the middle of a *bona fide* terrestrial impact crater. It was formed by the collision of a comet or asteroid with the Earth, fifteen million years ago, and is located in that place on the Earth we now call Germany. Yet if it had not been pointed out to us, we might never have recognized it. The Ries Crater looks like any other Bavarian valley (a little flatter than usual, maybe), surrounded by a 'coincidental' ring of hills. Since it formed, rivers have run, forests have grown, castles and villages have been built, and wars and plagues have been won and lost, all within its basin. It looks nothing like the impact craters on the Moon—today.

The scale of large impact craters can cause them to escape our notice. It was only when humans achieved the same 'overhead view' of their home planet as they had of the Moon that a number of round features began to attract notice. This became possible at the beginning of the Space Age, when artificial satellites were first launched to look down on and photograph the Earth. The intent was not to search for impact craters, but they found them anyway (figure 4.8).

Admittedly, there are other ways to produce roundish features on the Earth: volcanic calderas, salt domes, sink holes, etc. It remains necessary to inspect the sites of possible craters for the 'ground truth' of their impact nature—rocks shocked by sudden, immense blows. By this method, even if the crater has been so eroded that it no longer resembles the circular, raised, bowl-shaped feature we associate with a crater, it can be identified.

(The geology of impact craters got a boost in the cold-war era from the study of craters produced by nuclear detonations. These craters were recognized to be very similar to those produced by impacts, and the effect on the surrounding rock is much

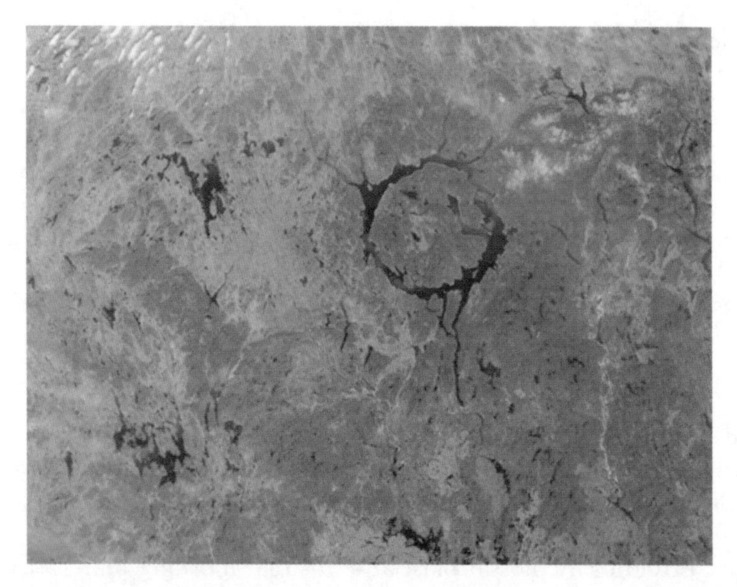

Figure 4.8. Ring-shaped Lake Manicouagan in Canada is actually an impact crater formed over 200 million years ago. It is one hundred kilometers in diameter. NASA/Goddard Space Flight Center/LaRC/JPL/Multi-angle Imaging SpectroRadiometer Team image.

the same. This idea was championed by the United States geologist, Eugene Shoemaker (1928–97) who first showed that Barringer Crater was formed by an extraterrestrial impact.)

Today, satellites and aircraft aloft with cameras, and geologists on foot with their rock hammers, have verified 190 impact craters on the Earth[3]. (Again, it is not possible to say how many were produced by comets, and how many by asteroids, inasmuch as the parent body disintegrates in the explosive impact.) The largest is the Vredefort crater in South Africa at more 300 km across and dated to be about 2 billion years old. The number keeps growing every year. Indeed, a map of the impact sites on the Earth shows a curious density distribution matching that of the human population of the Earth. Do comets aim for people? No, more likely there are impact craters in the sparsely inhabited areas of Earth, too. These places simply have not been explored as well. That is just the land. Ocean covers 7/10 of Earth's surface. There surely must be many more impact craters waiting to be discovered, once the sea bottom has been explored.

While uniformitarianism, the slow unyielding rise of mountains and gradual erosion of valleys and silting of oceans, is the primary mechanism for change on the Earth, the geology of today recognizes that Earth's history has been punctuated by sudden, brief episodes of catastrophism: cometary and asteroid impacts[4]. These events are extremely rare. When they do happen, though, they create land features in a few minutes that would ordinarily take millions of years to excavate by uniformitarian forces.

[3] Earth Impact Database maintained by the Planetary and Space Science Centre, University of New Brunswick, Canada (http://passc.net/AboutUs/index.html).

[4] Volcanoes and earthquakes also cause sudden changes.

The craters come in all sizes: from tens of meters to more than 100 km across. The smallest impact crater (13.5 m in diameter) resulted from a three-meter body that crashed in Peru in 2007. Villagers witnessed the fall and accompanying fireball. Tremors shattered windows, racked buildings, and threw a man from his bicycle.

Other than the very recent small craters, they generally date from two billion years to 'merely' tens of thousands of years ago. Their rims have been knocked down. Their floors have been filled with sand, water, or jungles. Nevertheless, they are there. The Earth cannot escape the rain of comets. The craters are their monuments.

4.3 The killer comet

If this reads like so much ancient history, consider the Tunguska Event. The name is intentionally vague. On 30 June 1908, something happened in a largely uninhabited region of Siberia, something that has become one of the great puzzles of planetary science.

The story is told of a colossal blast at 07:14 that morning that knocked people 60 km away to the ground and caused a herd of reindeer to vanish entirely. The pressure wave in the atmosphere was dutifully recorded by weather instruments across Europe. For many evenings, strollers as far away as London commented on an unusually bright sky.

The economy and politics of Russia, as well as the remoteness of the site, necessitated a nearly twenty-year delay before the first serious investigation of the Tunguska Event (figure 4.9). In 1927, a Russian scientist named Leonid Kulik (1883–1942) fought his way through dense woods, swamps, and vicious swarms of mosquitoes to visit the spot. There, he found evidence of a great forest fire and trees tipped over like bowling pins, but radially, in every direction away from the site, flattening some 2000 km^2, about the size of New York City. Yet there was no tell-tale crater.

Figure 4.9. The Tunguska site.

The Tunguska Event sounds like a twentieth-century comet impact. The glowing European skies seen as noctilucent clouds easily can be accounted for by scattered light from the sudden influx of icy particulate material from a vaporized comet. (Astronomer Fred Whipple suggested in 1930 that the glow in the sky was caused by the dust of disintegrating Comet Encke.) Where is the crater, however?

Maybe there does not have to be a crater. A small comet plummeting toward the Earth's surface could reach a point where the atmosphere is thick enough so that the comet is impeded sufficiently to blow up before hitting the ground. The result would be much the same as an above-ground nuclear test—without the radioactivity. While there would be great destruction, there would be no new crater.

It seems that a comet did hit within our great-grandparents' lifetimes. That is recent! It was not a major impact. Even so, had the Tunguska Event happened over an inhabited place, such as a city, the death and damage would have been horrifying. Luckily, it did not. There was not even a well-placed eyewitness. One wonders how many other Tunguska-like events have drawn no notice?

Something much smaller (an about 20 m near-Earth asteroid) exploded over Chelyabinsk, Siberia, in 2013. Recorded by many car dashcams as brighter than the Sun, it underwent an airburst due to its shallow angle of entry into the atmosphere. The shock wave from the explosion (equivalent to several hundred kilotons of TNT) broke windows all over the city. What is the deal with Siberia? Answer: it is geographically large[5]!

While a Tunguska Event can kill locally, a large comet impact could radically affect all life on the Earth. There is evidence that it has done just that.

Lately, popular books and media have renewed our interest in dinosaurs. (It seems that dinosaurs, of course, have been a favorite subject of eight-year-olds since time immemorial, including your trusted authors.) Moreover, the great mystery associated with these great creatures is not why they lived, but why they died. The dinosaurs ruled the Earth (or the top of its food chain, anyway) for 140 million years. (Compare this performance to *Homo sapiens*' measly half-million years so far.) Then they vanished 66 million years ago, leaving only their fossilized remains. Suddenly. Forever. Why?

Among paleontologists who study fossils, there are as many theories for the extinction of the dinosaurs as there are Godzilla movies. Maybe it was disease. Maybe it was supervolcanoes. Maybe it was normal climatic change to which the dinosaurs could not adapt. Maybe they did not go away at all, but instead evolved into something else. (Birds?)

How about this? It was the unique father/son team of scientists Walter (1940–) and Luis Alvarez (1911–88), a Nobel-prize winner in physics, who first brought attention to an unusual rock layer in the Earth. This layer is thin, but it is everywhere

[5] NASA has released a map of bolides (commonly referred to as fireballs) due to small asteroids impacting our atmosphere. It shows a random distribution around the globe: https://www.jpl.nasa.gov/news/news.php?feature=4380. You can follow fireball sightings at https://www.datastro.eu/explore/dataset/nasa-fireball-and-bolide-reports/table/?sort=peak_brightness_date_time_ut.

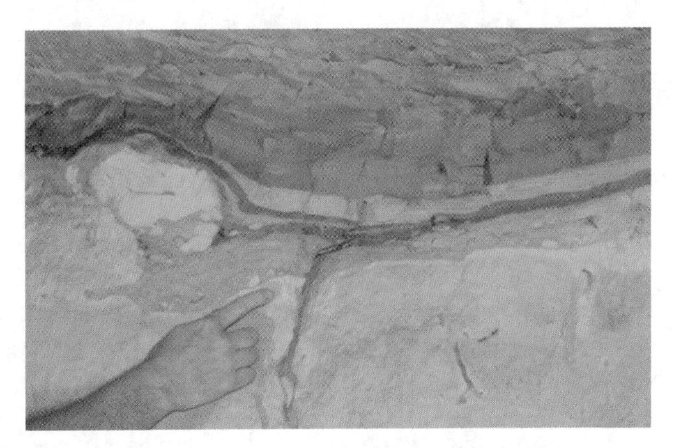

Figure 4.10. The K–Pg (Cretaceous–Paleogene, formerly known as K–T) boundary, marking the end of the dinosaurs' reign on the Earth. Courtesy of Mark A Wilson, the College of Wooster.

and contains high amounts of the element iridium—unusual for the crust of the Earth[6], yes, but not for comets and asteroids. The Alvarezes proposed that this iridium-enriched debris was laid down globally when a gigantic impactor struck the Earth. The depth of the impact layer points to a date 66 million years ago (figure 4.10).

66 million years ago? That is a familiar number. A comet impact that deposited remains of itself over the entire world is fully capable of precipitating a global disaster even greater than the one imagined at the beginning of this chapter. Before settling, the dust would have choked out sunlight for the world's plants and animals. The ecological system would have collapsed, and hardest hit would have been the mighty dinosaurs.

It is a frightening, and at the same time, fascinating scenario—dinosaurs populate the whole Earth and then are wiped out after millions of generations, on some random Tuesday afternoon. It is a lesson for us in the conceit of species.

However, there are loopholes in the theory. The fossil record does not tell us whether the dinosaurs disappeared in a given year or a given hundred thousand years. We cannot establish if they all died out at once or over a very long interval of time. So we cannot know for sure that the impact was the cause of the dinosaurs' demise; but the punch did land, and it surely did them no good. The impact may have been one more straw in a sequence of events that ultimately did in these behemoths.

Still, if a comet had something to do with the extinction of the dinosaurs, we should thank this lucky 'fuzzy star'. It wiped the ecological slate clean. The removal of the dinosaurs allowed a new type of creature to ascend—the mammals. (The mammals, which were smaller and could burrow and hibernate, were better suited to survive the impact-induced global winter.) That is, it opened the way for us.

[6] The rare-Earth element, iridium, is thought to be more abundant in the Earth's interior, where heavy elements sank when the early Earth was in a molten state (called differentiation).

If all this is so, the comet (or, to be fair, maybe asteroid) impact of 66 million years ago was an important event in (future) human history. It would be interesting to know the site of this roll-of-the-cosmic dice that 'came up sevens' for us (at the expense of the dinosaurs). The search was on for the crater.

For some years, the only known crater that dated from approximately 66 million years ago lay beneath T H's present home state of Iowa. (Much more recent glaciers bulldozed the crater flat; only well drilling accidentally revealed the presence of an impact feature beneath the unremarkable corn fields of Calhoun County.) However, that crater is rather too small and old to be the 'killer', and the density of the impact fallback points more to a larger crater in the Caribbean. The Iowa crater could at best be the result of a small piece of the comet breaking off and producing a precursor crater.

Then oil geologists in Mexico realized that there was an enormous ring-like structure in the Yucatán. It had eluded attention before because only part of it was on land. The rest extended into the Gulf of Mexico. The 180 km Chicxulub (CHEEK-Shoo-loob) impact crater has been dated to 66 million years ago. It may be the 'smoking gun' of the most famous comet impact of all time.

There have been other episodes of mass extinction in the Earth's history, besides the one of 66 million years ago. Each of these was followed by a flourish of evolutionary activity, replenishing the ecosystems with species. Natural calamities are, in fact, 'natural'.

Some scientists have proposed that there is a 26 million year periodicity to mass extinctions. If this were true, it would be tempting to look for a common periodic cause for them. If the Oort Cloud was nudged regularly with this frequency, sending a myriad of new Hale–Bopps (a particularly big comet) each time into the inner Solar System, the odds of cometary collisions with the Earth would increase dramatically every 26 million years. What could cause a periodic disruption of the Oort Cloud? Periodicity suggests some sort of orbit. Suppose the Sun has a faint companion star (as many other stars do). If such a companion were in an eccentric orbit, it might be far away most of the time, but periodically (every 26 million years?) would swoop through the Oort Cloud and initiate a storm of comets. The postulated companion star even has been given a name, Nemesis.

It is a neat chain of reasoning: a variation in the number of species seen in the fossil record leads to speculation about comet impacts and, ultimately, an unseen companion star to the Sun. The problem is that the 26 million year periodicity is by no means statistically proven. Furthermore, a careful search for Nemesis by powerful infrared telescopes has turned up nothing (and it should have been found). Regardless, while the hypothesis may be completely wrong, it is a fine example of the cross fertilization of scientific ideas between different disciplines (paleontology, biology, astronomy) that has marked the 21st century.

Long ago, comets and their tails were seen as cosmic swords hanging over the Earth. It was thought that the comet sword would inevitably bring harm to people and societies. Our modern theory of comets shows that there was a hint of truth in that medieval view. Still, are comets always to be considered harbingers of annihilation?

As we suggested earlier, rapid change on the Earth (some might say destructive change) is not always bad. Sudden, altered conditions may accelerate evolution and lead to the advancement of life and increase in the variety of species on the Earth, known as punctuated evolution. In other words, a totally safe Earth might be a very dull Earth, evolutionarily speaking.

Also controversial is the idea that comets had a role in providing the environment on the Earth capable of sustaining life in the first place. During the Earth's formation, it was much too hot for water to exist since it would have been vaporized into space. Why then do we have an abundant supply of water on the Earth's surface? Astronomers have suggested that comets coming to the Earth may be responsible for much of the Earth's supply of life-giving water[7]. Like we import much of our goods from China today, the Earth may have imported its water from the outer Solar System. This water (in frozen form) would be delivered by comets that struck the ground billions of years ago. To further investigate this idea, we must look for subtle differences in ocean water and comet water. Not all water is equal! One tell-tale indicator is the relative amount of water's hydrogen (H) atoms to its *isotope*[8], deuterium (D), called the D/H ratio. The first measurements of cometary D/H were quite different from that of Earth's oceans but these came from long-period comets. Results from two short-period comets (much more likely to deliver water to Earth) were similar to ocean water, but the recent Rosetta Mission found a higher value in the short-period Comet 67P/Churyumov–Gerasimenko. Since our statistics are small (only a dozen comets), the origin of Earth's water is still an open question.

A radical suggestion is that the very organic molecules that would (one day) unite to create life on this planet were not indigenous to the Earth; they were transported here by comets! We know that comets do contain moderately complex organic molecules. (The Stardust and Rosetta space probes have recently found the simplest amino acid, glycine, in comets.) These are the kinds of molecules found in living organisms. It is unclear, however, how many of these delicate molecules could withstand the passage through the Earth's atmosphere intact. Alternatively, they could be forged by chemistry in the fiery aftermath of the collision.

These conjectured connections between comets and life on the Earth are far from conclusive. Yet they provide a plausible argument for thinking of comets, not only as the destroyers of worlds, but also as—just possibly—sowers of life.

The scientific fact that Comet Lovejoy (just to pick one or any other comet we know about, see figure 4.11) will not endanger the Earth did not, of course, stop various supermarket periodicals from proclaiming that it would. When any new natural phenomenon is discovered, it is the habit of tabloid journalists to search for some way in which it might imperil the average citizen. (Of course, their goal is to increase sales, not accuracy.) The conjured carnage resulting from a comet falling

[7] An alternative idea is that water was sequestered in subsurface hydrated minerals that release water vapor when brought to the surface by volcanic activity.

[8] Isotopes are variants of elements with different numbers of neutrons in their nuclei but generally have the same chemical properties. Add a neutron to a hydrogen atom and you have deuterium, a heavier atom. Make water with deuterium and you have 'heavy water'.

Figure 4.11. Comet Lovejoy rounds the Sun. NASA image.

from the sky is too good to pass up. One vivid example could be found on any checkout lane during the week of 9 January 1996. There, perusers of the *Sun* read in a front page headline of a 'giant comet of century hurtling for head-on crash with Earth!' Sure enough, on inside pages, after reading that a 'werewolf girl is on the loose in the Amazon jungle' and that 'mutant killer rats threaten the US', we came to the article about the comet 'threatening the very existence of humanity'. (This Earth shattering news was modestly placed on page 16, after the article about 'wine on a stick'.)

The prediction was attributed to Johan Oftebrau of Oslo, Norway, a 'top astronomer'. 'Of course, I pray my calculations are wrong', Oftebrau is quoted in the *Sun*, 'but I've been a scientist for a long time.' Such are the claims of pseudoscientists (figure 4.12).

Everyone is entitled to make a prediction. The world of ideas is open to anyone, and the *Sun* has every right to publish its unique perspective on comets. But it is up to us to use our critical thinking skills to keep us from being fooled! As the Nobel Prize-winning physicist Richard Feynman (1918–88) said, 'The first principle is that you must not fool yourself and you are the easiest person to fool.'

The trouble comes when we are asked to select between competing predictions. How to choose? One way is to ask which prediction is the most specific (and therefore most easily could be demonstrated to be wrong, sometime in the future). The orbits calculated for comets are the product of scores of independent observations and calculations. These calculations predict, to many decimal points, where a comet will be at any given time. The methods and results of the scientists who compute these orbits are available for public inspection and critique. Most importantly, observations continue. These observations objectively verify the specific predictions of the calculated orbit. (Another way of saying this is that no one has succeeded in disproving the orbit.)

The same cannot be said for the predictions that have foretold calamitous results from comets. (No doubt there were alarms sounded in Australia at the time of

WEEKLY WORLD NEWS

RUSSIA JOINS U.S. IN FRANTIC BID TO BLOW IT OUT OF THE SKY!

HUBBLE FINDS DOOMSDAY COMET

. . . AND IT'S SPEEDING TOWARD EARTH!

HOW YOU CAN PREPARE FOR THE 20TH CENTURY'S WORST DISASTER

Figure 4.12. Another dubious headline.

Comet McNaught, as well.) In fact, the doomsayers' theories fail the test. All this sounds funny, until we read of a cult in California that committed mass suicide in response to Comet Hale–Bopp's approach.

Lastly, some have asked: if a comet did represent a threat to humankind, would the astronomical community tell us? Such a question is, perhaps, inescapable in a time when conspiracy theories are all the rage. Still, the answer must be a qualified 'yes'. On the one hand, not only would these individuals (many of whom have spent their lives studying comets) want to be the first to sound the claxon, it may be impossible to shut them up. As we have seen, planetary scientists readily accept the fact that comets do hit the Earth and, moreover, that such events are inevitable. But scientists are fallible, too, so they may 'jump the gun' with incomplete data or erroneous analysis. Fortunately, cooler minds have prevailed about this serious responsibility so we now have many checks and balances before an announcement to the press occurs. We do not want to cry 'the sky is falling'. Too many false positives would numb the public to the real next big one. There is an established protocol for announcing such a discovery to make sure that the orbit and risk are reliably assessed. If verified, the discovery goes up the chain of command through NASA to the President and the Federal Emergency Management Agency (FEMA). The Office for Outer Space Affairs (UNOOSA) of the United Nations also is notified. (UNOOSA coordinates with the International Asteroid Warning Network

(IAWN) to disseminate information concerning an extraterrestrial impact.) The final word resides with the White House. It is a distinct possibility that they may know there is one headed our way that cannot be avoided but will not let the public know due to the ensuing mass panic. D B always asked his students what they would do if the world was going to end next week. Would they still come to class? (A resounding 'No!') Would the fabric of society unravel (riots, lawlessness, etc)? Would many people want to finish their bucket lists quickly regardless of the law? There may well be a comet out there with 'our name on it'. But astronomers worldwide likely have not found it yet.

Astronomers have pondered the best way to assess the risk of impact and communicate it to the public. In 1999, they introduced the Torino Impact Hazard scale. It consists of a color-coded number between 0 (we are in the clear) and 10 ('Houston, we've got a problem,' the end is near)[9]. Currently, all Near-Earth Objects (NEOs) are at Torino scale 0[10], but we have had some causes for alarm. In 2004, the 370 m sized asteroid (99942) Apophis was discovered and initial orbit calculations gave it a 2.7% chance that it would hit the Earth on 13 April 2029. Further observations showed only a close encounter on this date but raised the possibility that the Earth's gravity could bend Apophis' orbit and result in an impact seven years later on 13 April 2036. Did we mention that both of these dates fall on Friday the 13th and that the asteroid was named for an Egyptian deity of destruction? (This would be much scarier than the movie franchise.) Fortunately, we can now rule out an impact with Apophis, but it still holds the record for the highest Torino scale rating of 4.

[9] The Torino Impact Hazard Scale: 0 (white, no hazard), 1 (green, no unusual hazard), 2–4 (yellow, merits attention by astronomers), 5–7 (orange, threatening with close encounter), 8–10 (red, collision is certain).
[10] NASA's Sentry System tracks the risks of newly found NEOs and can be found at https://cneos.jpl.nasa.gov/sentry/.

Chapter 5

Observing comets

Here we offer some practical advice for observing comets. It applies to any faint astronomical object or constellation as well. We cannot describe to you exactly how a comet will appear in our sky. Both you and we may be glimpsing something last seen by the ancient Egyptians! Despite that, it is possible to make a good guess based on the appearance of previous comets.

5.1 Eye on a comet

The comet will first show up as a fuzzy star to the naked eye. (Those with telescopes or binoculars will catch sight of it earliest, obviously.) Stars always appear point-like; they have no shape. As you look among the stars for the comet, search for one that seems diffuse and has some apparent size.

'It looks like a cotton ball' is the description one onlooker gave us of Hyakutake, the Great Comet of 1996. Indeed, the fuzzy patch that is the comet coma will be more-or-less round. It likely will appear brightest at a central condensation and will become radially fainter. This was the appearance of the Christmas Comet Wirtanen that D B spied using binoculars in December 2018, with a distinct greenish hue.

Because of this, the coma will not have a well-defined edge. Its size will depend on the darkness of the sky. The phase of the Moon can play a large role in sky brightness. It is best to observe during a new (or nearly new) moon or after moonset/before moonrise. Comet Hyakutake had an apparently large coma because it was so close—yet it was only about the size of the full moon to the naked eye. There are other astronomical objects with similar appearance to the naked eye: do not confuse a comet with the Milky Way, Andromeda Galaxy, or Large and Small Magellanic Clouds (if you live in the Southern Hemisphere).

How can you distinguish a comet from a cloud? A small, isolated cloud is uncommon. Still, if there are distant clouds in the sky while you are observing, you will readily see why the ancients were confused about whether comets belong to the Earth's atmosphere or beyond. The proof that you are sighting a comet will come

doi:10.1088/2053-2571/ab027ech5

with its lack of immediate movement on the Celestial Sphere. Clouds likely will move with respect to the background stars within twenty minutes or so. The comet's motion during this time will be negligible.

Now that you have found the comet, does it appear to have any structure? A trick astronomers use to bring out faint detail in an object is averted vision. The retina of the human eye is less sensitive to light at its center. By glancing slightly away from an object but keeping your attention to the center, you can use the more sensitive outer retina to make it look brighter.

Can you see a tail? Remember to look for it in the direction opposite the Sun. The tail may be many times longer than the diameter of the coma. It will appear brightest near the coma and become fainter farther away. How far away from the coma can you see the tail? How wide is it? Is there one tail or are there several? Are there breaks in the tail?

What is the color of the tail and coma, and does it vary? Do not be too disappointed if you do not see color in a comet. Individuals' color perception varies markedly (and you will need a bright comet). If you cannot see color in the stars, you are unlikely to see it in the comet.

Of course, if you are using binoculars or a telescope, the comet will appear brighter, larger, and more colorful. (A certain amount of light is needed to trigger your color vision; this is why your bedroom looks gray when you wake up in the middle of the night.) You also will be able to distinguish more detail. Look for structure or asymmetry in the coma. The nucleus of the comet remains hidden, but look for jets of material, originating at the nucleus, heading out into the coma.

There is a 'downside' to telescopes or binoculars, too. With a telescope or binoculars, you may not see an entire Grand Comet at once! This, we think, diminishes the effect. Therefore, no matter how sophisticated the optical aids you use are, be sure to spend some time peering at the comet with your eyes alone. The darker the site, the better. In addition, there is a certain magic about seeing the unfamiliar comet in a sky set against the silhouettes of familiar trees, mountains, or buildings. As you do so, remember that, except for a little air, there is nothing between you and the titanic comet (figure 5.1).

The following applies in particular to telescope and binocular observers: be patient. The quality of your view of a comet will change minute-to-minute and night-to-night. This is because you are looking through the Earth's blanket of atmosphere. The air is constantly in motion. Even if there is no wind at ground level, there are air currents high above you. As cells of different air density pass over you, the direction of the light passing through them is bent this way and that. This phenomenon causes the stars to vary in brightness and is called *scintillation*. Yes, stars do twinkle, but only to those of us watching them from the Earth's surface. Astronauts (e.g. aboard the International Space Station) see stars as steady points of light.

Astronomers use a not-very-technical-sounding term to describe this effect. They call it *seeing*. On some nights, the atmosphere between you and the comet changes more quickly and more drastically than on others. Thus, astronomers refer to nights of ‘bad seeing’ and ‘good seeing’. On nights of bad seeing, stars dance around in the telescope, or appear to go out of focus for an instant.

(Think of a fish tank. It is a lot easier to see objects through it clearly when the water is still than when someone is stirring up the water.)

Seeing does not affect the view of an object with apparent size (such as a comet or planet) as much as it does the stars. The effects of bad seeing on different points on the comet tend to average out. (The comet will not twinkle.) Still, if the seeing is bad when you are observing the comet, your view will be degraded. It is worth waiting for a moment when the seeing settles down. In this moment the comet will suddenly appear much clearer. You may see more detail in this one instant than in minutes of previous viewing.

The transparency of the air also can vary from night to night. By ‘transparency’, we do not mean whether it is cloudy or not. Even on a technically clear night, there can be a thin layer of material in the atmosphere, material that makes celestial objects appear dimmer or with less contrast. Your best bet, most assuredly, is to observe for as long and on as many nights as possible, to increase your chances of a very good view.

When you first detect the comet, note its location with respect to several nearby stars, preferably located in a triangle centered on the comet. Later in the evening, see if you can discern its motion in relation to these stars. As you watch the comet from night to night, this movement will become more evident. It is the true movement of the comet through the heavens.

Whatever its real motion, a comet’s tail gives it a continuous sense of motion in the sky. It looks like a dynamic entity. Thus, although you know that comet tails always point away from the Sun, a comet’s real direction of travel (after perihelion) may surprise you.

During a night, or over a set of nights, look for changes in the shape of the comet itself. It will get brighter as it approaches, and fainter as it recedes, of course. However, it may alter physically as well. New outbursts from the hidden nucleus can change the appearance of a comet quickly. Remember that some comets even have split into fragments before our eyes!

For anyone artistically inclined, attempting to draw the comet is a pleasant pastime. Make sure you note the date, time, location, seeing, etc. Include a few reference stars as context. Until the advent of astronomical photography a little more than one hundred years ago, drawing was the way astronomers recorded what they saw. An added benefit of drawing is that it requires a concentration that may ultimately cause you to see more characteristics of the comet.

(Astronomical photography of the comet is outside the scope of this book. Producing a photograph of a faint comet that matches its visual appearance is a challenge; a camera with manual exposure and aperture settings and a steady support are essential. Still, if you are a camera buff, it is worth a try! If you have a smart phone and a telescope, try aligning the camera with the eyepiece and snapping a photo. It takes some practice, but you will get a quick and simple image of the comet.)

Lastly, astronomical observing, like anything else, takes practice. If your initial view of the comet is disappointing, try again. Be an active observer. (You may want

Figure 5.1. A comet painting. Courtesy of Richard Baum.

to develop your astronomical viewing skills ahead of time by looking at other celestial objects.) For instance, fix at first on just one thing. Study the brightness of the comet and how that brightness varies. Next, concentrate on the shape and structure of the comet. Spend another session trying to decide color.

And so forth. You may be surprised that you can see more the second (third, fourth, fifth) time you see the comet than you did the first time. This might be so, even if the optimum view of the comet has passed. It is actually possible to train the eye to see more.

5.2 Where to go?

We cannot stress enough the importance of choosing a dark location from which to inspect a comet. By 'dark' we mean one far from artificial illumination of any kind. Those who were 'let down' by, for instance, a bright comet like Hyakutake usually were those who did not make the effort to find such a site.

These days, finding the ideal viewing location is not as easy to do as it once was. Even though Comet Halley was not as favorably placed in our sky in 1986 as it was during its previous 1910 apparition, expectations ran high. After all, our grandparents told us what a sight it was and not to miss it. However, our grandparents had an advantage over us that had nothing to do with the comet.

All over the world, skies were darker in 1910. Outdoor lighting in this age was in its infancy. We were much more a rural civilization. It was easier to find a dark place back then; it might even have been your front yard.

Today, we are losing the dark of night. For most urban dwellers, even the stars have all but disappeared. They have not gone anywhere—the fault lies in street lamps, spotlights, glowing fast-food signs, etc. Astronomers call this problem *light pollution*.

Now, we do not have anything against light. Light so that we can find our way at night is good. It gives us a degree of security and keeps us from running into each other! We have no wish to return to the days when we were virtually prisoners in our homes, once the Sun had set.

Everything is fine as long as artificial illumination is directed where it is intended: down onto the ground. However, too much of this lighting is directed carelessly upward. This includes lights intentionally beamed up for advertising purposes and, more commonly, improperly aimed street and yard lights. This escaped light does not just disappear as it shines upward. It scatters in the Earth's atmosphere and makes the sky glow, thereby obscuring faint astronomical objects.

Photographs of the Earth at night, taken from space, show this problem clearly. North America and Europe particularly are ablaze with light (figure 5.2). The location of every city and village can be pinpointed. The paths of interstate highways can be traced just by the glow of roadside rest stops!

This light is not doing anyone any good. It is totally wasted. Need we add that it takes energy to illuminate the stratosphere superfluously? Energy conservation begins above our heads! More pragmatically, all this light that never reaches the ground represents wasted money. In the case of public lighting, it is most often at the taxpayers' expense.

Do not get us wrong. We understand the need for security lighting to make our public spaces safe. Notwithstanding, more light is not synonymous with better light.

Figure 5.2. World-wide light pollution.

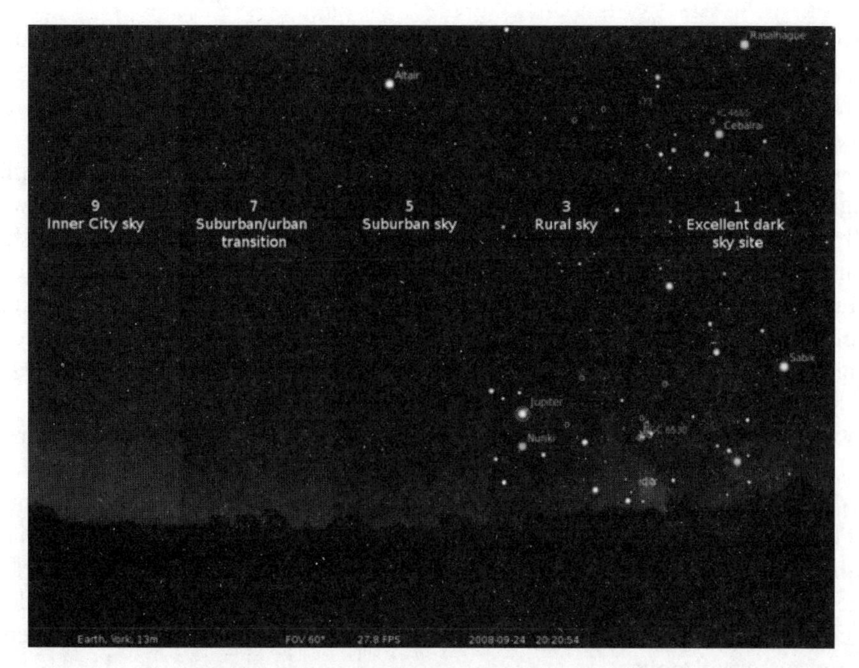

Figure 5.3. A light pollution scale. Raj Chanian.

Efficient illumination involves putting enough light where it will be useful, and no light where it will not.

The irony of light pollution is that, unlike many other forms of pollution, the problem is so easy to cure. Often, that cure is as simple as placing a reflector over the lamp to bounce light back down instead of beaming it up. Meanwhile, though, light pollution continues to rob us of a birthright no less legitimate than clean air and water—the right to enjoy a truly dark and beautiful night sky (figure 5.3).

As we dismount our soap box, let us address the subject of how to deal with light pollution: you can fight or flee. To fight light pollution in the 'big picture', you can encourage governments and businesses to install responsible illumination. More locally, you can make sure your own house is in order by inspecting lighting on your property. Are the tops of bulbs exposed? Are your lights on unnecessarily in an hour when no one is about? Do any lights shine into a neighbor's yard? (She or he may be trying to observe a comet!)

Alternatively, try this experiment. Stand under an outdoor lamp. Can you see the bulb from another lamp? If so, the lamps do not have enough shielding. (The light from the second lamp should not be reaching you; it is not required where you stand because you are, after all, standing beneath a lamp![1])

In this way, you can reduce the light pollution coming from your home. For comet watching, you can even turn everything off. You can do a great deal to

[1] More light pollution experiments can be found at https://www.globeatnight.org/dsr/. We urge you to join the growing ranks of advocates to preserve our dark skies.

improve your view of the sky by avoiding just a few nearby lights. Ultimately, however, you have little control over more-distant but more numerous lights, strung around your city or town.

For most people, their best bet for a very dark sky will be to flee—out of town, or away from any source of nocturnal illumination. Luckily, there still are such places, though citizens of major cities may have a fairly long trip. Regardless, it will be worth the effort. You will be amazed at how many stars you can distinguish at a non-light-polluted site. (An added benefit of this detour is that it may get you away from urban smog, which adversely affects the transparency of the sky.) A convenient test of your site is the Milky Way: if you can make out this band of light, you are seeing the sky as the ancients did—after they extinguished their camp fires.

If you cannot avoid light completely, find a site that has a relatively dark, unobstructed horizon in the direction of the comet. Once there, do not introduce your own light pollution! After you are settled, turn your headlights or flashlights off.

If you require a little illumination (for instance, to consult a sky chart), try putting a red filter over your flashlight. Red light does not disturb your night vision as much as other colors do. (This is the same strategy used in lighting airplane cockpits.) While you can buy special red LCD flashlights, a piece of red cellophane will do. We also have succeeded at making our own 'red light' by painting the bulb of a penlight with red fingernail polish.

Clearly, comet viewing takes a bit of planning ahead of time. This is time well spent. Why look at a cool comet from a crummy site? It is like watching the Super Bowl from the highest row of bleachers. Remember, a comet is almost certainly a 'one shot' opportunity. Few of us will be available to inspect it on its next apparition, say in the year 4018!

5.3 When to look?

A word about time: while its flowing tail gives the impression of something flying through the heavens, in reality, a comet seems to move slowly against the stellar background. This is for the same reason that a distant airliner gradually works its way through our field of vision, though we know that it is cruising at hundreds of kilometers an hour. It is very far away. Comets travel much faster than aircraft, but they are also much, much farther away. You may not notice much motion in the comet during a given night, though you will see its position changing from night to night. This is still fast for a celestial object. Only the Moon normally appears to move through the stars at a rate measurable to the naked eye in days, rather than weeks.

The final arbiter of your view during these nights of the comet will be the weather. There is little that can be done about clouds, besides crossing one's fingers. Even so, the comet will be visible long enough that all but the most dreary locales are unlikely to be clouded out altogether. (There will be comet enthusiasts aboard ships who will navigate their route to avoid overcast skies!)

Assuming the night is clear, when should you look? The simple answer is: look whenever the comet is above the horizon! Remember, however, that twilight must be ended totally (or not yet begun) to have the darkest achievable sky. Still, since comets are brightest and most well-developed when closest to the Sun in their orbit, some of the most spectacular views appear in the pre-dawn hours or evening twilight skies.

If you cannot avoid light pollution entirely, there is a certain practical advantage to observing comets after midnight. (This assumes, of course, that one is above the horizon at this time.) The reason is that more outdoor lighting is turned off as one heads into the early hours of the morning, and your sky will be a little bit darker.

All else being equal (weather, seeing conditions, atmospheric transparency, phase of the Moon, etc), the optimum time to observe a comet on a given night is when it is highest in the sky. By 'highest,' we mean farthest from the horizon. This is true even at a totally dark site. Why? It is the atmosphere again. The atmosphere always dims our view of a comet to some extent. (It would be considered a complete nuisance to astronomers if it were not for the fact that we need the atmosphere to breathe!) Astronauts, cosmonauts, and tychonauts aboard their space stations will have the ultimate view of the comet.

The Earth's atmosphere is a layer of gas surrounding the solid globe. It is, effectively, only about one hundred kilometers thick. Thus, it is very thin compared to the size of the Earth. From our point of view, then, although our geography books have taught us that the world is 'round', we can picture ourselves as standing at the bottom of a flat 'dish' of air. The 'dish' is wider and longer than it is tall. We want to stare upward through as short a column of obscuring air as possible. This path is directly overhead.

Astronomers call the width of the atmosphere they are looking through the *air mass*. Observing a celestial object at any angle with respect to the zenith (the point directly overhead) means that your air mass increases as a celestial object's angular distance from the zenith increases (figure 5.4).

If you are peering through a greater air mass, your view is dimmed, and effects of bad seeing are amplified. As you gaze closer to the horizon, the view deteriorates quickly. Try looking at the stars overhead and compare them to stars near the

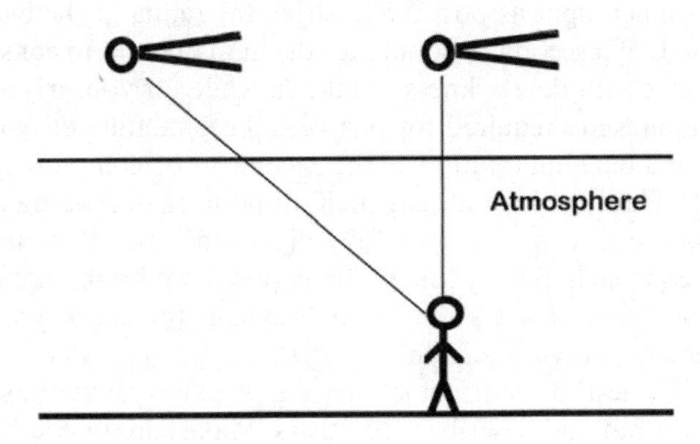

Figure 5.4. We must look at greater air mass to see the imaginary comet on the left than we do to see the imaginary comet on the right.

horizon. The ones at the horizon twinkle much more. Indeed, it is feasible to see only the brightest astronomical objects (the Sun, the Moon, and Venus) at the horizon. Even an extremely bright comet will not be visible immediately upon 'comet rise' nor just before 'comet set'.

Few people will be at a latitude such that a comet looms directly overhead. Still, the view is potentially best when the comet is halfway between rising and setting, on a line running North–South across the sky through your zenith. This line (or semicircle, really) is called your *celestial meridian*. The comet reaches its maximum apparent distance from the horizon when it crosses the meridian.

This often will occur during the day, of course. Nevertheless, if the comet is visible in the western sky, you know that its zenith angle is increasing and that the air mass is growing. Conversely, if the comet is in the eastern sky, this angle is decreasing and the air mass is shrinking.

There is one other factor that will affect your view of a comet, to which we have alluded to a couple of times before. It is the Moon. The Moon is a fascinating astronomical object to study in its own right. However, for comet watching, moonlight is worse than a bright street light. The Moon is a natural source of light pollution. It goes through its cycle of phases (new, waxing crescent, first quarter, waxing gibbous, full, waning gibbous, third quarter, waning crescent) once each month. You cannot do anything about this other than plan your comet observing session with the Moon in mind. This is easy around the time of the new moon. The Moon is not visible in the sky at all then. The full moon, however, is a disaster as far as viewing faint comets is concerned. It is up all night and very bright. The first and third quarter moons, those times when we see half of the Moon's disk illuminated, are not quite as bright as the full moon and can be avoided. The first quarter moon sets around midnight; the third quarter moon does not rise until that time. (Many calendars denote the phase of the Moon with a small symbol near the date.)

In conclusion, it may be that exactly when you observe a comet is not as important as the time you spend with it. If you have just come outdoors from a brightly illuminated building or vehicle, the iris of your eye has shrunk to a small size. Not much light is entering through your pupil. To discern faint objects, it is necessary for as much light as possible to strike the retina in the back of your eye, within the thirtieth of a second or so it takes the human brain to construct an image. When you first step into the darkness, it takes a while for your iris to open. In fact, up to twenty minutes are required for people to become fully dark adapted.

Not waiting for dark adaptation is the second most common cause of 'comet disappointment'. Plan to spend at least half an hour or so looking at the comet to make sure that your eyes become fully light-sensitive. Remember that dark adaptation is reset each time your eye is exposed to bright light, so resist the temptation to use your smart phone as a flashlight (or check your social media account) during an observing session.

Here is the 'final test' of your comet viewing: we know something best when we attempt to point it out and describe it to others. Make sure to compare your comet experiences with someone else. That other person may be a member of your family, a neighbor, a new friend, or someone halfway around the globe on the Internet. Remember, the world is small to a comet.

5.4 Expectations

As you can tell by now, astronomers toss around words like 'spectacular' and 'bright'. These words are meaningful in the context of astronomical events. Today, however, movies and television have 'cranked up' the threshold of what people perceive as spectacular. Frankly, we need many sensory stimuli to be excited about an event these days. (You cannot merely have a car crash in today's movies; it has to be a whole lot of cars-or at least a bus ... in 3D.) The media bombard us with light and color. This is quite a bit for any natural phenomenon to have to compete with. However, it is the very actuality that comets are completely natural events in our Universe, putting on their display with no help from—and with indifference to—humankind, that makes them remarkable.

Astronomers inadvertently raise expectations, too. We like to publish our 'scrapbook' pictures of the heavens. Amazing 'shots' of distant astronomical objects can be found in books, magazines, and online-in color, and with extraordinary detail. (Some of these pictures appear in this book.) However, such images were built up with sensitive detectors, over long exposures, using huge telescopes. Sometimes they are computer-enhanced.

Your view of a new comet almost certainly will be subtle. Remember, however, that what you are looking at is real. It is not recorded and edited for your consumption as is so much of what we experience today. You are watching the Universe 'live'.

IOP Concise Physics

Comets in the 21st Century

A personal guide to experiencing the next great comet!

Daniel C Boice and Thomas Hockey

Chapter 6

Hunting comets

'Hung by the heavens with black, yield day to night!
Comets, importing change of time and states,
Brandish your crystal tresses in the sky…'
Henry V, I.i, William Shakespeare

6.1 Who discovers comets?

Until very recently, most new comets were discovered by amateur astronomers. Now, the term 'amateur astronomer' may suggest to you something taken lightly, something done naively, or worse. Certainly, a phrase like 'amateur biologist' conjures up images of Doctor Frankenstein. 'Amateur nuclear physicist' sounds even scarier—the word 'crackpot' comes to mind as a synonym!

Yet, unlike many other sciences, astronomy holds the word 'amateur' in high regard. Amateur astronomers often are very serious and well educated about their avocation. Moreover, in these tight times, when professional jobs are exceptionally hard to come by, the classical distinction between amateur and professional—whether you are being paid or not—is beside the point. Few professional astronomers would argue that there are 'amateurs' who know their way around their telescopes better than some of their 'pro' colleagues[1].

Amateur astronomers occupy an important niche in the discipline. Reductionist sciences such as physics or chemistry concentrate on a few underlying principles. (These principles are both beautiful and profound, but exploring their subtlety takes years of graduate study and, for the experimentalists, outrageously expensive

[1] Nowadays, many rightly refer to 'amateurs' as 'citizen scientists'.

doi:10.1088/2053-2571/ab027ech6

equipment.) Astronomy is messier! We have to deal with the real Universe, which is inhabited by all sorts of strange and differing celestial bodies.

Contrary to popular misconception, most professional astronomers do not spend their nights patrolling the entire night sky on alert for new comets. Nothing could be further from the truth since doing so would take too long! With uncommon exceptions it is inefficient for these few professionals to spend precious hours of large telescope time (not to mention the expense) with only the hope of making some as-yet-unimagined discovery. It is too risky for their careers. Professional astronomers, for the most part, study known bodies, with the hope of bettering their understanding of these objects. Large telescopes are focused on a tiny patch of sky as part of a specific scientific investigation (rarely involving comets), leaving the rest of the sky open for discovery.

This is where amateurs come in. Compared to the professionals, amateurs have more time (because there are so many more of them). They have more telescopes (because they use smaller, cheaper instruments). Their telescopes have greater fields-of-view than the enormous ones of professionals—all the better for surveying the sky. Perhaps most importantly, amateurs have less to lose if they fail (because their after-work hours are their own)!

Amateurs are well suited for frontline search-and-discovery. While small telescopes and modest instrumentation restrict them to brighter objects, it is exactly these suddenly brighter objects for which the sky continually must be monitored: new comets, asteroids, and stars that vary abruptly in luminosity.

Professional astronomers are indebted to amateurs for mounting this celestial posse. These efforts take a long time to pay off. The greatest tool of the amateur astronomer is perseverance.

A typical comet hunter might begin scanning the sky shortly after sunset. He or she will be on the lookout for a faint 'wisp' against the near-black sky. Hours before moonrise particularly are coveted. Do not bother telephoning serious comet seekers the night of the new moon—they will not answer!

There are 'fuzzy' objects in the sky, other than comets, of course. Distant star clusters, nebulae, and galaxies sometime mimic comets. The early French comet-seeker Charles Messier (1730–1817) catalogued the fixed locations of a little over one hundred of these objects. He did so precisely to avoid the nuisance of repeatedly *confusing* them with comets. Our amateur, however, is very familiar with the sky and recognizes these old friends. She or he is looking for something that does *not* belong there.

As Earth revolves around the Sun, the Sun appears to move through the Celestial Sphere during the year. That portion of the Celestial Sphere that is in the sky after sunset and before sunrise slowly shifts. Each night, a thin new swath of dark sky emerges from twilight. It is here that our comet hunter searches, in the predawn hours, for comets that have become bright while the Sun prevented us from observing them.

Back and forth the observer scans with a telescope or binoculars, occasionally consulting a star chart. The process is meticulous and systematic; some may even call it boring. Notwithstanding, it works. Some amateurs have discovered multiple comets (or rediscovered ones that were lost).

It is not always a lonely enterprise. There are frequent star parties. Maybe not as festive as their name implies (but just as fun), 'star parties' are gatherings of amateur astronomers and their telescopes, held outdoors under a (hopefully) clear, dark sky. At a star party, one can 'mingle' from one telescope to the next, stopping to look at whatever that particular 'scope happens to be aimed at. A star party is a veritable buffet of astronomical delights.

Once a comet is discovered, a call goes out for confirmation observations. A comet cannot be attributed to its discoverers until it is impartially observed.

The International Astronomical Union's Central Bureau for Astronomical Telegrams (IAU-CBAT) is located in Cambridge, MA, at the Smithsonian Astrophysical Observatory. CBAT is the *Guinness Book of World Records* for astronomy. By world-wide consensus, all claims to astronomical discovery rest on the official distribution of the *IAU Circular*[2], announcing a discovery and when, where, and by whom it was made. CBAT is non-profit and funded through subscriptions to the *Circulars*. It operated under the auspices of the IAU until 2015 when the IAU was reorganized.

(The name 'CBAT' is anachronistic today. Astronomical information streaming in and out of CBAT now is transmitted largely by electronic mail. Telegrams have gone the way of the Pony Express.)

Reports of a new comet cause CBAT to implement a system for initially designating comets[3]. Pretend we discovered a comet. It is written down by the year (2019) and the half month of that year in which it was discovered ('H' for latter April) and 1 (for the first comet discovered in this two-week period), 2 (for the second comet discovered in this two-week period), 3 … etc. Put it all together and our comet is the alphanumeric 2019 H1. (The letter I is skipped insofar as so many confuse it with the Roman numeral for '1'.) Later, a C/ is attached if it turns out to be a long-period comet, a P/ is attached if it is a short-period one, and the family names of up to three discoverers are added: Comet C/2019 H1 (Hockey–Boice). But we do not kid ourselves. If one of us had been clouded out, stuck with a flat tire, or laid up in bed with the flu, there is no doubt that others would have found the comet in ensuing nights. And the comet would have a different name. Comets do not 'sneak up on us'; there are too many astronomical sentries on alert.

6.2 Where is our comet?

We all know how big a skyscraper ought to be, so we can judge our distance from it by sight. Astronomical bodies vary in size dramatically, and human beings had no first-hand experience with them before the Space Age. Anything that was permanently out of arm's reach (like the Moon) could be huge, but then again might just as well be the size of a pizza. How would our ancestors know?

Tycho Brahe knew. Tycho (1546–1601) was a Danish aristocrat who operated the first modern astronomical observatory in Europe worthy of the name on a sparsely

[2] http://www.cbat.eps.harvard.edu/services/IAUC.html.

[3] We describe the current system adopted by the IAU in 1994, see https://www.iau.org/public/themes/naming/#comets. If you see other designations, they are from an older, outdated system (e.g. Comet 1969i (Bennett)).

inhabited, royally owned island. It was probably just as well that it was lightly populated. Tycho seems to have mistreated pretty much anybody he came into contact with whom he did not consider his equal—and that was just about everybody. A notable exception was a young Polish assistant named Johannes Kepler who, as detailed in section 2.3, translated Tycho's magnificent data tables into a simple means of describing planetary orbits.

Tycho (usually remembered by his 'toy-sounding' first name[4]) did not know how to play well with others, but he did know comets. He 'discovered' one when he was a young man. This was at a time when there were no official records of comet discoveries. It was a naked-eye comet[5], and many people spotted it independently. Still, Tycho studied the comet, and it eventually bore his name: Tycho's Comet of 1577.

We claim that Tycho was the original comet scientist. He was the first to prove that comets are celestial bodies and not simply some sort of temporary meteorological phenomenon, contrary to the belief that went back to the ancient Greeks. Because of this, he was the earliest scientist to consider comets astronomical objects and not apparitions of doom. This is how he did it.

First, a demonstration. Hold your thumb out in front of your face. Compare it to distant objects, such as furniture across the room, or trees and houses if you are outside. (Your thumb will appear very large in comparison!) Now, blink slowly, by opening just one eye, and then just the other. (You will feel silly doing this, but nobody is watching you, and it is for a good cause.) The background objects appear no different. However, your thumb looks as if it moves back and forth. It is not really going anywhere, of course, attached to your body as it is. This apparent motion is the result of seeing your thumb alternately from two places (your left eye and your right eye). These two vantage points are separated by a distance, the distance between your eyes (about ten centimeters).

Now stretch your arm out so that your thumb is farther away from you. Repeat the blinking. What happens to the apparent shift in location of your thumb? It gets smaller. Believe it or not, behind this funny exercise you will find the basis for finding the distance to comets, as well as to planets and stars.

The farther away an object, the less *parallax* it will exhibit. Parallax is the angle through which an object seems to move when viewed from two disparate places, as reckoned against some more-distant background. If you know the distance between the two measuring stations (a distance called the baseline), you need no more than to measure the parallax angle in order to calculate the distance to the object.

The calculation requires only simple trigonometry. We will admit that if 'trig' is a word you tried to avoid in high school, the word 'simple' may not seem an appropriate adjective with which to describe it now! We will not do the calculation

[4] Every language seems to pronounce Tycho's name differently. In Danish, it is pronounced 'tee-Koh Brah,' in Latin it is 'tee-Koh Bra-Hay,' and in German it is pronounced 'two-Show Brah.'

[5] Remember that Brahe's observatory did not contain a single telescope; it was not invented yet! His observatory consisted of a large quadrant, sextants, and other instruments to make precise measurements of the positions of astronomical objects.

here. Nevertheless, in the grand scheme of mathematics, it is a reasonably straightforward application of the properties of right triangles.

This method is used routinely by surveyors: If you have seen one person standing along the highway (in an orange vest holding a pole) and another some distance away (squinting at him or her through a small telescope), measuring parallax is what they are doing. We still find it amazing that we live in a geometrically well-behaved Universe in which we can measure the distance to objects without ever touching them with the end of a tape measure.

The surveyor does not wink back and forth, of course. You noticed when you did the parallax demonstration that objects much farther than your thumb did not appear to change location perceptibly. This was because the baseline distance (inside your head) was too small. The parallax angle can be exaggerated by increasing the baseline distance. If you quickly dodge back and forth across the room, you can make more-distant objects appear to shift their position, while the farthest objects still keep their relative place. (The surveyor picks up her or his telescope and moves it to another station along a pre-measured baseline.)

Objects in the sky are farther still. As you might guess, the baseline distance has to be great to see any astronomical parallax at all (figure 6.1).

This brings us back to Tycho (figure 6.2). Tycho measured the position of the 1577 comet very precisely against the background of extremely distant stars. He then compared his determination to those made by astronomers elsewhere in Europe. (He foreshadowed the international observing campaigns that are common today.) There was no difference! A cloud-like thing anywhere between the Earth and the Moon (where the celestial domain traditionally was considered to begin) would have exhibited a parallax. Because he could find no parallax shift, Tycho could not really measure the distance to the comet. However, he could state with certainty that the distance must be on the order of that of the planets (which also did not exhibit a parallax with the angle-measuring techniques of his day). Therefore, Tycho reasoned, comet distances must be far away.

Tycho was the first to suggest that comets should be thought of like planets and treated like planets. He conceptually pointed the way to applying Newton's and Halley's science of orbits, to comets.

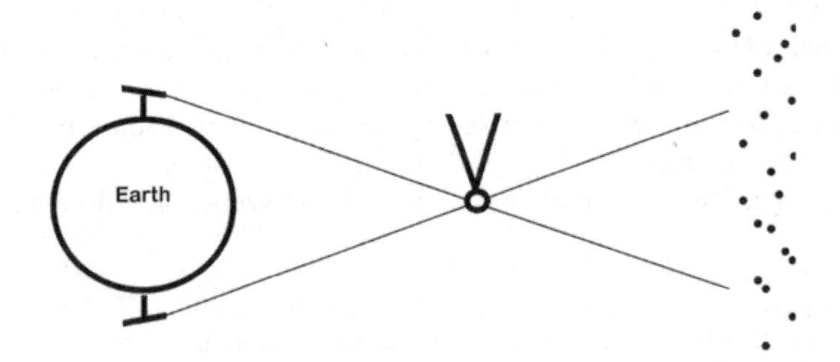

Figure 6.1. A modern measurement of parallax for a nearby comet. The angle is highly exaggerated.

Figure 6.2. Tycho Brahe at his observatory. A close inspection reveals a peculiar nose. In fact, it is prosthetic, made of brass (or gold or silver, depending on the story). Tycho lost his nose in a duel. Obviously, he was a much better astronomer than swordsman.

Tycho's comet observations had a philosophical implication, too. According to the natural science theories of the ancient Greeks (still going strong after more than fifteen-hundred years), the heavens were perfect, orderly, and unchanging. This made the realm above far removed from happenings below on the Earth, where things are often ugly, chaotic, and unpredictable! Comets, however, seem to show up and then disappear haphazardly. Maybe things in the celestial canopy were not so immutable and unlike the Earth after all. Maybe there was a commonality between the heavens and Earth—we are all part of one Universe.

When observers all over the world send CBAT measured positions of a comet, they are repeating an experiment with great historical significance. Since Tycho's time, however, some improvements have been made in the parallax technique.

First, the baseline can be longer. The distance between, say, New Mexico and Japan is a significant fraction of the diameter of the Earth! And if we are willing to wait six months, the Earth's revolution about the Sun will move us by a baseline distance of 2 au.

Second, we now can measure very small parallax angles, those that would have been imperceptible to Tycho. The reason for this latter technology is that we have a device to aid the naked eye, the astronomical telescope—a tool invented just some years after Tycho's death!

With a telescope, it is possible to measure angles in seconds of arc (or smaller). An arc second is one-sixtieth of one arc minute: 1/3600 of one degree!

Comet parallaxes now are measured routinely. Still, at the time of its discovery, no parallax was observed initially for Comet Hale–Bopp! Hale–Bopp must have been very far away. In fact, Hale–Bopp was discovered while it was beyond the orbit of the planet Jupiter (5.2 au average orbital distance). No comet had ever before been found by amateur astronomers this far from the Sun.

Once you know where a comet is in space at different times, it becomes possible to calculate its orbit. Because elliptical orbits are geometrically simple, you might think this an easy task. After all, we already know that the figure of the orbit is an ellipse, right? It is not quite that simple. One has to learn both the size of the orbit (perihelion or aphelion distance) and its shape (the ellipse's *eccentricity*). We cannot stop there. It is also necessary to compute the orientation of the ellipse. Which way does the imaginary line through the foci of the ellipse go? Because this line can point in any direction on the Celestial Sphere, two numbers are required to describe it uniquely. The ellipse still can rotate an infinite number of ways about the axis that is this line. This rotation must be specified, too.

(We have not even mentioned distortions from the true ellipse, caused by the gravity of Solar System bodies other than the Sun. Nor have we mentioned non-gravitational forces, due to the comet 'jets' exerting forces that can be counter to or in concert with the usual elliptical motion. This results in the slowing or speeding up of the comet during its orbit.)

Calculating the orbit of a comet begins to sound a lot like that nightmare algebra problem you tried to solve in school—you know, the one with trains leaving all sorts of stations at all sorts of different times and speeds! With this bad memory, you might think that the comet would have to be observed many places in its orbit in order to forecast with certainty where it will be in the future. In reality, by pulling a couple of mathematical tricks, you can get by with just a few observations.

It helps if these observations are spaced out in time. In the case of Comet Hale–Bopp, it was possible to obtain observations *back* in time. After learning of its discovery, Robert McNaught (who would later have his own name-sake Great Comet of 2007) rummaged through some photographs of the sky taken at the Anglo-Australian Observatory two years earlier. These photographs were made for purposes that had nothing to do with the still-unknown Comet Hale–Bopp. Serendipitously, Hale–Bopp happened to be in the frame of one of the pictures! Nobody would have recognized the inconspicuous comet in the image if they were not looking for it.

Once an orbit is determined, one knows where the comet will be at any time, and it is time to figure out how close it will get to the Sun and Earth, distances that will help estimation of its peak brightness. Now that the comet's period is known, one can establish whether it is a C/ comet or a P/ comet. Remember that long-period comets usually put on a flashier show.

6.3 How bright a light?

When discovered, Comet Hale–Bopp was more than a thousand times brighter than Comet Halley at an equal distance. If the lights of an oncoming truck seem bright while still far away, it is reasonable to conclude that they will be very bright when the truck passes you. However, what if, when you glanced at the distant truck, the driver had happened to 'flick' on and off its high beams? Your estimate of the lights' future brightness will be biased. The lights might not be all that bright when the truck passes you, running on its normal headlights exclusively.

Comets can 'flick.' They undergo episodic outbursts as fresh material from the nucleus is inserted into their comas. This material shines brightly until it dissipates. For a while, the comet is much brighter than it normally would be at a given distance from the Sun.

The most notorious comet of the 1970s was Comet Kohoutek (figure 6.3). Like Hale–Bopp, Comet Kohoutek was found far from the Sun. This led to speculation that it would be extremely bright in our sky when it neared the Sun. Unfortunately, Kohoutek happened to be eyed during an outburst. Soon after its discovery, the comet faded. It never achieved remarkable brightness in our sky.

Comet brightness is predicted by way of a mathematical function. It takes into account distance from the Sun and Earth, the albedo of the comet, and the behavior of comets in the past. Scientists call this method an empirical equation; others might call it an 'educated guess'. Obviously, only the first item on the list is well known.

The behavior of past comets is a real problem because no comet is really canonical. When, and the rate at which, comets 'turn on' (produce comas) depends on how an individual comet is made. Different ices, under different conditions,

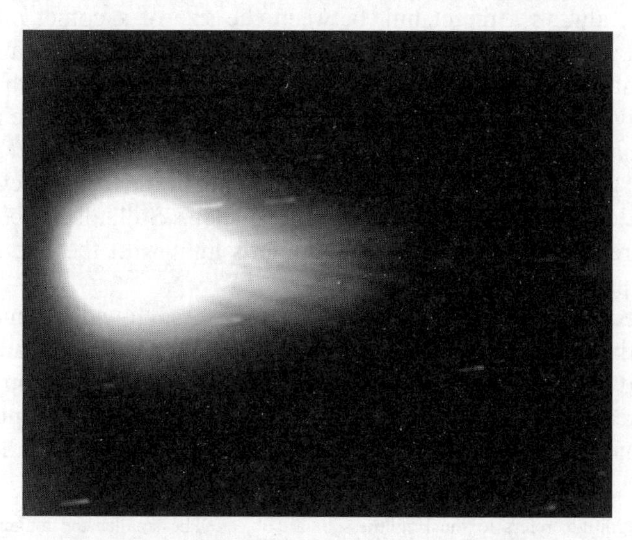

Figure 6.3. Comet Kohoutek in 1973. NASA image.

outgas at different distances from the Sun—and each fluoresces uniquely! Dust plays a major role in coma brightness too, of course, but you must convert ice first to liberate reflective dust. Dust comas are exclusively seen at large heliocentric distances, detection of gas is only seen closer to the Sun. This can lead to misidentification of new objects as comets since only dust is seen and no gas. Our current thinking is that a significant amount of ice (gas) needs to be present for an object to be a comet. Some 'Main Belt Comets', displaying dust tails, have later been found to be the result of asteroid collisions, not comets at all!

The outburst mechanisms are still unknown but the leading candidate is gas-pressure build-up in interior voids that eventually overcomes the mechanical strength of the overlaying surface material, producing an eruption. Another is the penetration of the solar heat wave deep inside the comet, converting amorphous water ice to crystalline water ice, an exothermic process that releases a large amount of energy. A third mechanism involves cometary avalanches (landslides) that expose fresh, volatile-rich surfaces, as recently observed in Comet 67P/Churyumov–Gerasimenko, the target of the Rosetta Mission.

We now understand that 'first timers' (comets not known to have passed this way before) are particularly quick to brighten at far distances and then fade as they near the Sun (e.g. Kohoutek in 1973 and Austin in 1990). They are thought to have accumulated a surface frost of very volatile ices such as carbon dioxide (CO_2) and carbon monoxide (CO) during their cold storage in the Oort Cloud that sublimates quickly, resulting in very rapid brightening. A middle-aged comet is a nice find; it has passed the Sun before, but not so often that it is nearly worn out (like Comet Halley). Quite the contrary.

When a comet brightens (forms a coma) too distant for water ice to have sublimated, it must be the result of the sublimation of ices that do so at a lower temperature. We expect these ices to be less plentiful than water ice. Thus, the comet should be 'saving itself' for the inner Solar System (and us).

There is not always a direct link between the size of a comet nucleus and the comet's brightness. Only a small part of the nucleus's surface area is actively producing coma gas and dust at any one time. A small nucleus with a proportionately large active area could produce as impressive a coma as a large nucleus with a proportionately smaller active area, such as Comet Hartley 2 with an active fraction exceeding 100%, the excess water production coming in the form of icy dust particles that release water as they travel away from the nucleus. Still, all else being equal, it is natural to correlate the *magnitude*[6] of a comet's light with the size of the nucleus itself.

Only a few comets have been visited by space probes. How do we measure the size of something that is a mere point as seen from the Earth? If you can see the coma cloud, it should be possible. The Hubble images of Comet Hale–Bopp reveal a coma cloud that increased in brightness as one radially approached the center. The center is where the hidden nucleus is presumed to be. If the nucleus really was a mere

[6] Astronomical magnitude refers to the brightness of an object. The smaller the magnitude number, the brighter the object.

mathematical point, the coma brightness would decrease smoothly as a function of radius. But sometimes it does not. There may be a slight light 'bump' at a radius near the center. By assuming that this represents the light of the nucleus itself, the coma light can be subtracted out by computer techniques to reveal a rough view of the nucleus. It is not possible to discern shape, but some appraisal of size can be made.

Rotation is a fundamental property of all Solar System bodies. It is one of the first things that gives the object itself a unique character (as opposed to properties of its orbit). A uniform comet nucleus is like a white cue ball. You can spin it fast. You can spin it slowly. It does not matter. With no marks or number on the ball, it is difficult to prove whether the sphere is turning at all. This goes for a billiard ball on the table beside you! Imagine the difficulty in deciding the rotation period of a featureless, far-off comet nucleus. Again, we are talking about all those comets that have not been visited *in situ*.

A 'jet', marking the location of a sublimation active region on the nucleus, turns the comet into an eight ball. Now there is a reference mark for timing the rotation period. Every time you see the numeral '8' rotate in front of you, one rotation period has elapsed. Simple counting tells you whether the ball is spinning fast or barely turning. Similarly, charting the angular motion of a fixed 'jet' on the comet nucleus will yield the nuclear rotation rate.

Timing the rotation of a billiard ball is more difficult, however, if the numeral happens to be at the very top or bottom of the ball. Now, even if the ball is turning very rapidly, the numeral does not seem to move much because it is so near the rotation axis of the ball. So it goes with the comet nucleus.

Still, the 'jets' reveal where the pole of the comet is. Unless the comet's axis is perpendicular or parallel to our line of sight, the 'jets' are more foreshortened on one side of the comet, compared to the other. This means that they tilt alternately a little toward and away from us.

A pinwheel will always look completely symmetrical if you face its axis—a direction perpendicular to the stick. However, its blades, too, will be foreshortened on one side if you look at the pinwheel from any other direction.

To complicate matters, no single rotation period may fit the observations of all of the 'jets'. Does a region of activity get up and walk around the surface of the comet nucleus? Not likely! The comet might not be spinning in a single mode. It might be wobbling, as well as rotating. If that were not bad enough, the nucleus may skip eruptions. That is, the nuclear vent may not produce a 'jet' each time it rotates into sunlight. Fickle is the comet.

6.4 Eureka! I've found a comet!

Do you want to discover a comet? If so, this section is for you. In it, we give you professional tips to aid you in your attempt. Your reward? It is having a celestial body officially bearing your name in the history books. Amateurs have played an important role in comet discovery and continue to do so as detailed in section 6.1. In recent years, smaller, automated telescopes have been dedicated to survey the skies for unexpected visitors to Earth's neighborhood, looking for Near-Earth Objects

(NEOs). In 1992, the United States Congress directed NASA to be on the lookout for small Solar System bodies (mainly asteroids) that might be on a collision course with the Earth, definitely bad news for us. (Think dinosaur extinction some 66 million years ago.) This is where amateurs (you!) come into play. But you better act fast, since the current professional surveys and those on the planning books[7] are set to dominate the discovery scene in the near future.

Do you live in a major urban area without dark skies or just want to avoid those pesky mosquitoes? Then discover a comet from the comfort of your indoors armchair. Several dedicated amateurs have done just that (and continue to do so) using the live online feed from the SOHO satellite that monitors our Sun. Occasionally, a previously unknown small sungrazing comet will enter the field-of view of the SOHO/LASCO instrument[8]. If you are the first to spot it (and report it), you have discovered a comet. Unfortunately, the rules of this game normally do not allow you to attach your name to it. Alternatively, the comet bears the name of the SOHO spacecraft. But hey, you just bagged a comet! These armchair comet hunters are very competitive, so your discovery will take a lot of patience and persistence like most accomplishments in life (unless you are just plain lucky).

Do you feel lucky? Look through online archives of night sky images for evidence of a comet that others have missed. These images are processed without regard for hunting moving objects. Such large repositories are managed by the Catalina Sky Survey, the Faulkes Telescope Project, Planetary Resources, and the Zooniverse[9]. There are automated telescope networks[10] with serious telescopes that offer online access for a small fee. It does not take any special technical knowledge to do this. After some practice, you are off and running. Discovering a comet can also enrich your pocketbook. The Edgar Wilson Award[11], overseen by the CBAT, annually awards up to $20 000 with a plaque for the discovery of comets by amateurs. Comet discovery can be complex and confusing for the novice, so we try to carry you up the learning curve as quickly as possible! We have provided additional resources in the appendices that give further details and instructions for the serious observer with modest resources.

6.5 Starting your comet quest

i. *Select an observing site*. The motto of the Scouts is ‘Be Prepared’ and this certainly applies to comet hunting. As the great French scientist Louis Pasteur said: ‘Chance favors the prepared mind.’ As a first step, you want to find a location with dark skies. As we discussed in section 5.2, light

[7] The Large Synoptic Survey Telescope (LSST), scheduled for ‘first light’ in 2019, is expected to discover 10 000 comets in its first year of operation. Every night the LSST will collect 30 terabytes of data.

[8] The Large Angle and Spectrometric COronagraph (LASCO) instrument is a set of three coronagraphs aboard SOHO that image the Sun’s corona from 1.1 to 32 solar radii. A coronagraph is a telescope designed to produce an artificial solar eclipse, blocking the brightness of the Sun’s disk, to reveal the extremely faint light surrounding it, called the corona.

[9] https://www.zooniverse.org/projects/mschwamb/comet-hunters.

[10] For example, Slooh (https://www.slooh.com) and iTelescope (https://www.itelescope.net).

[11] http://www.cbat.eps.harvard.edu/special/EdgarWilson.html.

pollution, the combined effects of night lights on urban skies, is a major problem in modern times for those of us wanting to make a personal connection to the night sky. The ancients had more access to dark night skies and were more connected to them, not only for personal satisfaction but because their existence depended in part on nightly observations to mark the time, seasons, location, etc. (Today, we use 'high tech' devices such as clocks or smart phones, calendars, and GPS systems.)

If you live in an urban area, then you will need to find a dark site outside the city limits for your observations. How can you judge the extent of light pollution in your area? If you are in the Northern Hemisphere and can find Polaris (the North Star), a member of the constellation Ursa Minor (the Little Bear or Dipper), then follow this procedure—remember, it is important to let your eye adapt to the dark (about twenty minutes) and to use minimal lighting afterwards (a red-light flashlight or one with a red filter). Do not use your smart phone as a flashlight as it will destroy your night vision and you will need to wait another twenty minutes or so to re-adapt. Now look for the two end stars of the dipper, Kochab and Pherkad. If you cannot see these stars, you are not far enough away from the city! Next, try to fill in fainter stars between Polaris and Kochab that outline the dipper. If you can make out all seven stars in the dipper, you have skies sufficiently dark to start your comet search (limiting to 5th magnitude[12]). (Ideally, you want 6th magnitude.) Keep in mind this technique works for the northern part of the sky but darkness can vary with the cardinal points. For example, you may have driven north of the city so that the southern skies at your observing location may still suffer from light pollution. You can usually judge the other parts of the sky once you have made the dipper estimation. Keep in mind that there can always be clouds blocking the stars that cannot be seen directly at night. And then there is the phase of moon.

ii. *Become familiar with the night sky*. Learn the constellations. They are your guides to the night sky. If you have an interest in Greek/Roman mythology, you will have the added pleasure of connecting the myths to the sky[13]. There are many books for doing so (e.g. *The Stars* by H A Rey[14] or the classic *Norton's Star Atlas*, now in its 20th edition), as well as online materials (e.g. www.SkyMaps.com); smart phone apps (e.g. *Google Sky Map* for Android and *Star Walk* for Apple iOS devices); and planetarium software (e.g. *Stellarium*, etc) for your home computer. After some basic knowledge, more parts of the sky can be learned by 'star hopping', going from a known bright star to an unknown nearby bright star following the star chart. Up to now,

[12] We are referring to the astronomical magnitude system as was described earlier. More details on finding the limiting magnitude and reporting it to a world-wide campaign can be found at https://www.globeatnight.org/5-steps.php.

[13] Did you enjoy the movie, *Clash of the Titans* (either version)? All of the major characters from that legend are memorialized in the autumn evening skies.

[14] Yes, that is the same Rey who wrote the *Curious George* books!

we have only used our naked eyes for observing. This is sufficient for making a discovery since several important comet discoveries in modern times have been made with the unaided eye. But binoculars and telescopes will enhance your chances and increase your enjoyment of the night skies.

A good challenge is the Messier marathon—attempting to view all 110 unique Messier objects in a single night. In the Northern Hemisphere, it is best done in mid-March to early April, around the time of a new moon. To prepare for the marathon, spot a few Messier objects every week so that you know what to look for. You will need at least a six-inch-aperture telescope to complete the list. (The faintest is M95, a barred spiral galaxy, at 11.4 magnitude.) Otherwise, 100 mm binoculars under optimum sky conditions can find all but M95. Using your naked eyes, you will be lucky to see 30 objects.

iii. *Choose an instrument.* Naked eyes, binoculars, or telescope, that is the question. The answer is largely determined by your budget. The larger the instrument, the brighter and more detailed the image will be. To begin, resist the temptation to go all in—do not buy too much since you may decide later that it is not for you! Start small and develop your interest. Remember that your instrument serves a general purpose to enjoy the splendor of the night skies, not only comet spotting. If you do choose a telescope, go for as large an aperture[15] as your budget will allow to see the faintest possible objects (highest limiting magnitude). This means obtaining a reflecting telescope (reflector), preferably with a low *f*/ratio (<5) and accompanying large field-of-view. It goes without saying that optical quality is essential, so go with a reputable manufacturer. Typical 'scopes in this class are called Schmidt–Cassegrain or rich-field telescopes. (5–6 inch apertures are good starters.) Objects will appear brighter. In addition, great views of the Milky Way, galaxies, nebulae, etc are afforded. Use a low magnification eyepiece initially for a larger field-of view, then switch to higher magnification eyepieces as practical. Both authors attended graduate school with Alan Hale, the co-discoverer of Comet Hale–Bopp (*the* Great Comet of the 20th century), who used his 16 inch reflector for the discovery. (Incidentally, the amateur astronomer Thomas Bopp used a 17 inch reflector.)

A sturdy mounting is a necessity. A simple 'cannon' mount, the most common known as a Dobsonian, can be operated manually to follow the diurnal motion of the sky. More sophisticated ones, called equatorial mountings, align one axis of motion with the celestial pole. A small motor geared appropriately will move the telescope about the other axis at the appropriate rate, called sidereal tracking. If you plan on becoming an astrophotographer, this arrangement with a CCD camera is essential for long-time exposures. Of course, nowadays quick photos can be snapped using the camera in your smart phone. Simply align with the eyepiece and

[15] This is the diameter of the mirror (reflecting telescope) or lens (refracting telescope).

shoot away. Post the best ones to your favorite social media account and watch for the reactions! An adaptor to keep the phone still can be added for a small cost.

Any pair of binoculars will give great sky views. Of course, going for larger aperture and lower magnification are best for celestial scanning (e.g. 50–80 mm, 7–25× magnification) and have wide fields of view. There are specialized binoculars for comet hunting (100 mm and larger), but these are for the serious amateurs and usually exceed the beginner's budget.

iv. *Practice with known comets.* Where can you find comets currently in our night skies? Due to Newton, we have great knowledge of celestial motion, including that of comets (thanks to his contemporary, Halley). Pick up a copy of *Sky and Telescope* or *Astronomy* magazines and look for their night sky sections. There are several online resources detailing the positions of comets currently in our night skies, too. Try these websites: https://in-the-sky.org/newsindex.php?feed=comets&year=2018&month=9&day=5&town=1690313, https://www.skyandtelescope.com/observing/celestial-objects-to-watch/comets/, https://theskylive.com/comets, and https://www.cfa.harvard.edu/skyreport. (More can be found using your favorite web browser.) Most have finder charts to help identify the comet among the stars. You can make your own finder charts—just make sure that they are up-to-date since the comet is in constant motion[16]. The orbit of a new comet may be slightly inaccurate due to lack of observations, so it pays to scan around its predicted location. Remember, non-gravitational forces also may cause deviations from the predictions.

Next, check the magnitude to see if it can be seen in your instrument. Comet magnitudes overestimate their visibility since their light is spread out, resulting in low surface brightness. Therefore a 4th magnitude comet may appear as a 6th magnitude object. Use averted vision for the faint ones! Our naked eyes are limited to 6.5 magnitude when fully dark adapted under optimum skies. Comets fainter than about 8th magnitude will only be seen in large binoculars (>50 mm), and modest telescopes (>6 inch) are needed to detect 11th magnitude comets with great sky conditions. If several comets can be seen that night, start with the brightest and work your way to down to the faintest. Recall our discussion at the beginning of chapter 5 about the comet's appearance, usually a bright central point with a surrounding diffuse, roundish cloud (coma) and perhaps the hint of a tail. Make your own estimation of the comet's brightness by comparing it to nearby stars. With practice, you can report your measurements to the International Comet Quarterly where they are published and you are credited. Estimating comet magnitudes is a very important activity for amateurs. These visual magnitude estimates (either by eye, binoculars, or telescopes) are used by the pros to augment their data since amateurs report with greater frequency on

[16] You can delve deeper into locating the comet using its *ephemeris*, a table with celestial coordinates (right ascension, declination), expected magnitude, and other information for a given night. Check out NASA's Horizons System for more details: https://ssd.jpl.nasa.gov/horizons.cgi.

a near-nightly basis. World-wide campaigns have been organized in support of spacecraft missions to comets and the authorship of scientific publications that resulted was shared with the amateurs[17].

After a few hours, look for motion of the comet relative to the background stars. Make sure you keep records of your observations (with sketches or photos) so you can compare the comet night after night and to others.

v. *Happy hunting*! Now you are ready to scan the skies for interlopers (those not seen on the star maps and not known to us). Start in the twilight, just before sunrise, as this is the part of the sky that is newly revealed after being blocked by the Sun for many months. Do your search systematically by scanning adjacent swarths of the sky (and with low magnification if using a telescope). When you do find the comet, you will experience the thrill of discovery at that moment and the satisfaction of successfully finding it by yourself. In the meantime, stop to 'smell the roses', to wonder and appreciate the beauty of the objects that are part of the celestial zoo. Clyde Tombaugh (1906–97), while searching for Pluto and other planetary candidates, discovered a comet (274P/Tombaugh–Tenagra), many asteroids, hundreds of variable stars, and a few star clusters and galaxies. Good luck. But, again, persistence and determination are most important.

You can also become a member of a comet hunting team. Check with your local astronomy club. It can be a great resource for those beginning their nightly adventures and have a variety of telescopes for you to test drive at a star party before deciding on your own. You also may find similar-minded people online with a few simple searches (e.g. Zooniverse) and in groups like the Planetary Society[18], the world's largest and most influential non-profit space organization, co-founded by astronomer Carl Sagan (1934–96) in 1980.

vi. *How to report a discovery*. If you do find an object you suspect to be a new comet, you will need to report it to be recognized as the discoverer. Time is of the essence; keep in mind that hundreds of other comet hunters world-wide also are scanning the skies! However, you must strike a balance between relative certainties—make certain that your discovery is real (confirming motion, appearance, etc). Reporting too soon will cause unnecessary effort if the sighting does not pan out and may hurt your credibility. Report too late, and you may lose the discovery. A brief message can be sent that you suspect a discovery to keep your foot in the door with a more detailed message to follow if confirmed later that night or the next. Whom do you contact? The Harvard-Smithsonian Minor Planets Center (MPC) is the official office that represents the International Astronomical Union (IAU) in the discovery of comets, asteroids, and other transient

[17] Astrobiologist Karen Meech (University of Hawaii) organized such a campaign in 2005 to coordinate observations of Comet Tempel 1, the target of NASA's Deep Impact Mission.

[18] http://www.planetary.org.

objects in the night sky (novas, supernovas, variable stars, etc). The MPC manages the Central Bureau for Astronomical Telegrams (CBAT) to do so. Send your report to the CBAT via email at cbatiau@eps.harvard.edu. (Telegrams are no longer accepted.) You also may use their web interface at http://www.cbat.eps.harvard.edu/DiscoveryForm.html. Your report should include the following information, or it may be rejected: your name and contact information; date and time of observation; method (naked eye, binoculars, etc); details of instrument used, if any; observational site; and the comet's position, motion, and appearance[19]. If one of the first three to report before it is officially announced, you get naming rights (in chronological order of the report), e.g. Comet (your name)–Boice–Hockey!

6.6 Epilogue

This completes our introduction to comets. We hope that you share our enthusiasm and passion for these fascinating objects that grace our skies, either as a fun pastime or for serious observation. This book is just the iceberg tip of our knowledge about comets. For some, it may be a springboard to dive into more advanced comet texts (ones that may even have equations). (See appendices F and G for suggestions.) We might even have planted the seeds for someone to become a professional comet astronomer. We do not know when the next Great Comet will appear, but now you are prepared to discover it and participate in the thrill of comet observations with a multitude of others around the world who share your interests. Who knows? Maybe you will write a book chronicling your many comet discoveries in the years to come. Good luck and clear skies!

[19] More details on reporting comet discoveries can be found at http://www.cbat.eps.harvard.edu/CometDiscovery.html.

Chapter 7

Postscript

On 1 January 2019, the New Horizons spacecraft flew by the first small body encountered in the Kuiper Belt, MU_{69} (nicknamed Ultima Thule), at a distance of 3500 km (a veritable 'cosmic hole in one'), returning many scientific treasures. It captured the first images of a primordial comet nucleus (figure 7.1), one that has never had a near encounter with our fiery star and has been in cold storage in the Kuiper Belt since the dawn of the Solar System. Preserved in this pristine state, it provides clues about our early history, some 4.6 billion years ago. This sensational mission revealed a two-lobed body joined by a small 'neck', a contact binary, like another contact binary, Comet 67P/Churyumov–Gerasimenko, much eroded and transformed by successive close encounters with the Sun. This shape likely resulted from a slow collision where the two separate bodies stuck or a collision that fragmented a larger body and two large fragments came back together. Its size was measured to be about 34 km in the longest dimension, slightly larger than the average comet nucleus (but certainly smaller than Hale–Bopp), with a spin period of 16 h or so. It shows surface properties consistent with comet nuclei (very dark with some albedo variations and a brighter ring at the neck, mottled and ruddy terrain) and similar features (hills, slopes, plateaus, and craters) but with less relief due to its unprocessed nature. Its surface is slightly reddish from irradiation of ices during its long storage stint, similar to other Kuiper Belt Objects. Its detailed composition has yet to be determined but ices of water, methanol, and organic molecules have been identified spectroscopically on its surface, mixed with rocks. At its distance of 43.4 au from the Sun, surface temperatures are frigid, about 40 K, but hypervolatile ices of carbon monoxide, molecular nitrogen, molecular oxygen, and methane have sublimated from its surface. Data will continue to be returned until August 2020 so stay tuned for more revelations about this spectacular body and perhaps a cousin, since NASA may decide to send New Horizons to another Kuiper Belt Object!

doi:10.1088/2053-2571/ab027ech7

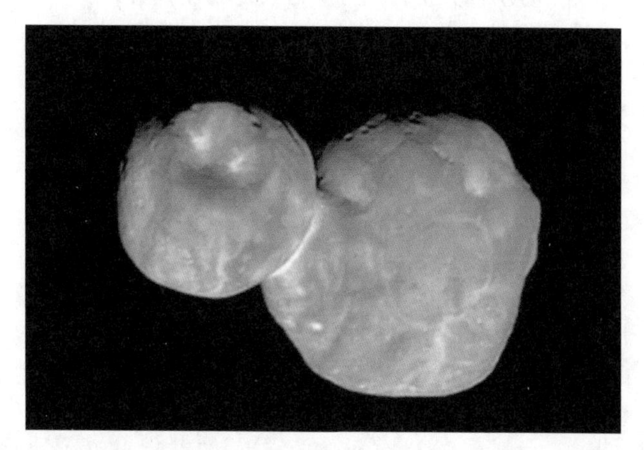

Figure 7.1. 'Cosmic snowman with a necklace', the primitive comet nucleus, MU_{69} (Ultima Thule). Its shape is that of a contact binary, two roughly round but flattened bodies joined by a neck, with a size of 34 km along the long axis. The 'necklace' is thought to contain fine grains that tumbled there by gravity or from the initial lobe merger. (NASA/John Hopkins Applied Physics Laboratory, Southwest Research Institute.)

Appendix A

Lesson suggestions for teachers

A.1 Comet orbits and tails

To illustrate the orbit of a comet about the Sun, use a table fan, Styrofoam ball, and a lightweight ribbon (such as the tape inside an old audio cassette). Pin various lengths of cut ribbon to the same point on the ball. The table fan represents the Sun. Its breeze plays the role of the solar wind. (The real solar wind should blow in all directions, of course.) The ball represents a comet nucleus surrounded by its coma. Place it at a distance from the 'Sun'. No tail (the ribbon strips) blows back away from the Sun. The ribbon hangs toward the floor and for now should be ignored. Carry the comet/ball almost directly toward the fan. Such is the appearance of a high eccentricity orbit. As the ball approaches the fan, strips of ribbon will begin to blow back away from the *faux* Sun. In other words, a comet tail grows longer the closer the comet is to the Sun. Upon reaching the fan, whip the ball quickly behind it and retrace its path, now away from the fan. The tail gradually shortens and then disappears as it retreats from the Sun, just as a real comet behaves. However, there is a difference: because the ribbon must always be blown away from the Sun, the tail now leads the comet instead of trailing it (the antitail) as it did on its inward passage. This, too, mimics the behavior of a real comet tail.

A.2 Recipe for a comet

To illustrate the physical nature of a pristine comet nucleus, use a large, glass bowl, crushed ice, some dirt, and a big spoon. Pour crushed ice into the bowl and then a layer of dirt, in roughly a 3:1 ratio. The dirt represents the refractory (dust) portion of a comet. Now comes the important part: stir the mix rapidly with the spoon. Comets, unlike planets, are not differentiated (layered). A comet is a homogeneous compilation of various ices and a refractory component. The 'comet' mix quickly

doi:10.1088/2053-2571/ab027ech8

and almost magically changes from bright, white ice to dark as each ice crystal becomes coated with a thin layer of 'dust.' It takes little dust to darken a comet drastically. Unintuitively, bodies made principally out of ice can have very low albedos. (Optionally, you can add some ground coal and dry ice to the mix. Now the 'comet' becomes darker and sublimates!)

A.3 Light pollution

Join the growing ranks of advocates to preserve our dark skies by estimating the sky darkness (limiting magnitude) at your designated observing site or at the students' homes (as homework). This can be done on any clear night; however, global campaigns are organized twice a year with instructions on submitting your results to the web. Students will gain familiarity with constellations during this exercise. This and more light pollution experiments can be found at: https://www.globeat-night.org/dsr/. See chapter 5 and section 6.5 for more details.

A.4 Learn constellations

This exercise builds on the previous one as more constellations of the night sky can be filled in. Monthly sky maps can be downloaded at: http://www.skymaps.com/. Since constellations are seasonal, concentrate on the prominent ones visible in the current evening skies. Make sketches or photos and label the bright stars and any planets that might be visiting. Zodiacal constellations are best to start with along with popular ones (e.g. Orion, Ursa Major, Ursa Minor, etc). Optionally, students can research the myth associated with each and other important facts. Section 6.5 has more details for your lesson plan.

A.5 Observations of the Moon

Observe the Moon's phase for an entire month with the naked eye using sketches or photographs from a smart phone camera. Note how the phase progressively changes and the motion of the Moon in the night sky relative to the constellations. Have the students estimate the position of the Moon in the sky (e.g. NE about 30° above the horizon). This will help you verify that the student actually made the observation rather than copying from the Internet. The students will soon notice that the Moon rises later each day so records must be taken accordingly. Learn the proper terminology to describe the phase (e.g. first quarter, not half moon; waxing gibbous vs waning gibbous, etc). Observations can be organized on a common calendar, making sure to record the time of observation also. (Optionally, if a student has knowledge of star charts, the phase and position can be recorded with respect to the stars.) Remember, some observations can be made during the day!

A.6 Observe a planet(s)

After consulting the website https://theskylive.com/planets (or another appropriate reference), chose a planet or planets that can easily be seen in the evening or morning skies with the naked eye (i.e. Mercury, Venus, Mars, Jupiter, or Saturn). Have the students make a sketch (or snap a photo) of the planet with respect to a few surrounding bright stars and the horizon (if appropriate), noting date, time, place, and general sky condition of the observation. (If exercise A.3 has been assigned previously, they can estimate the limiting magnitude. If A.4, they can identify the surrounding constellations.) Repeat this observation at later dates to note the motion of the planet relative to the stars. (Mercury and Venus—every few days; Mars—every week; Jupiter and Saturn—every month.) Observing during retrograde motion will heighten this experience.

A.7 Observe a comet

Exercises A.3–A.6 have prepared the students for their comet adventure. Consult the Internet for a bright comet (magnitude 4 or lower) that can be seen with the naked eye or with binoculars during the academic year (e.g. https://theskylive.com/comets). Make sketches noting the comet's position and brightness relative to the surrounding stars. A camera can be used if available for more accurate records. Hold the camera very still using a tripod or leaning against a steady object. Repeat the observation every night for a week or so, noting the comet's motion through the sky. Consult chapters 5 and 6, in particular sections 5.3 and 6.5, in making your lesson plan.

A.8 Observe a meteor shower

After selecting a prominent meteor shower from appendix B, depending on your academic calendar, consult the Internet for details of the shower (location in sky, best dates and times for observing, phase of the Moon, etc). This activity is best done with a small team to completely cover the sky. Assign each team member with a portion of the sky. Count how many meteors are seen every 10 min for one hour in the evening and, optionally, repeat for an hour in the early morning. Estimate the limiting magnitude (exercise A.3) every 10 min as well. This is important to determine the sky darkness (the darker the sky, the more meteors seen) and any changes in sky conditions that may occur during the one hour observation. This activity is best done lying on the ground or using lawn chairs, just don't fall asleep!

Appendix B

Approximate dates for some famous meteor showers

The shower can begin a few days before this date and last a few days after. Meteor showers are named after the constellation or nearest star from which the shower radiant appears. The IAU recognizes 112 established meteor showers.

Name	Peak night	Hourly rates	Parent object
Quadrantids	January 3	25	2003 EH_1[a]
Lyrids	April 21	10	C/1861 G1 (Thatcher)
Eta Aquariids	May 4	10–60	1P/Halley
Perseids	August 11	50–75	109P/Swift–Tuttle
Orionids	October 21	15–20	1P/Halley
Leonids	November 16	15[b]	55P/Tempel–Tuttle
Geminids	December 13	60–75	3200 Phaethon[a]

[a]Asteroid.

[b]Meteor storms (greater than 1000 meteors/hour) were seen in 1999, 2001, and 2002, but are not expected again until 2099.

doi:10.1088/2053-2571/ab027ech9

Appendix C

Typical comet visibility rates

In a typical year:

- Around a dozen comets reach perihelion.
- Tens of comets are visible to sky-watchers on the Earth.
- Two or three of these comets will be visible to amateur astronomers with small telescopes.
- One comet discernible to the naked eye will grace the sky.

doi:10.1088/2053-2571/ab027ech10

Appendix D

The brightest comets since 1935

Peak magnitude[a]	Name
(−10)	C/1965 S1 (Ikeya–Seki)
(−5.5)	C/2006 P1 (McNaught)
−3.0	C/1975 V1 (West)
(−3)	C/1947 X1 (Southern comet)
(−1)	C/1948 V1 (Eclipse comet)
(−1)	C/2011 S3 (Lovejoy)
−0.8	C/1995 O1 (Hale–Bopp)
(−0.5)	C/1956 R1 (Arend–Roland)
(−0.5)	C/2002 V1 (NEAT) (figure D1)
0.0	C/1996 B2 (Hyakutake)
0.0	C/1969 Y1 (Bennett)
(0)	C/1973 E1 (Kohoutek)
(0)	C/1962 C1 (Seki–Lines)
0.5	C/1998 J1 (SOHO)
1.0	C/1957 P1 (Mrkos)
(1.0)	C/2011 L4 (PANSTARRS)
(1)	C/1970 K1 (White–Ortiz–Bolelli)
1.7	C/1983 H1 (IRAS–Araki–Alcock)
(2)	C/1941 B2 (de Kock–Paraskevopoulos)
(2.2)	C/2002 T7 (LINEAR)
2.4	1P/1982 U1 (Halley)
(2.4)	17P/Holmes [October 2007]
2.5	C/2000 WM_1 (LINEAR)

(*Continued*)

2.7	C/1964 N1 (Ikeya)
2.8	C/2001 Q4 (NEAT)
2.8	C/1989 W1 (Aarseth–Brewington)
2.8	C/1963 A1 (Ikeya)
2.9	153P/2002 C1 (Ikeya–Zhang)
3.0	C/2001 A2 (LINEAR)
3.3	C/1936 K1 (Peltier)
(3.3)	C/2004 F4 (Bradfield)
3.5	C/2004 Q2 (Machholz)
3.5	C/1942 X1 (Whipple–Fedtke–Tevzadze)
3.5	C/1940 R2 (Cunningham)
3.5	C/1939 H1 (Jurlof–Achmarof–Hassel)
3.5	C/1959 Y1 (Burnham)
3.5	C/1969 T1 (Tago–Sato–Kosaka)
3.5	C/1980 Y1 (Bradfield)
(3.5)	C/1961 O1 (Wilson–Hubbard)
(3.5)	C/1955 L1 (Mrkos)
3.6	C/1990 K1 (Levy)
3.7	C/1975 N1 (Kobayashi–Berger–Milon)
3.9	C/1974 C1 (Bradfield)
3.9	C/1937 N1 (Finsler)

Source: *The International Comet Quarterly*. Published by the Earth and Planetary Sciences Department at Harvard University. Updated 17 November 2013. Accessed 25 May 2016.
[a] On the astronomers' magnitude scale, a smaller (more negative) number represents a brighter comet. Peak magnitudes in parentheses are uncertain.

Figure D1. Comet NEAT in 2003. NASA image.

Comets in the 21st Century
A personal guide to experiencing the next great comet!
Daniel C Boice and Thomas Hockey

Appendix E

Notable comets in history

Year	Comet	Comments	Year	Comet	Comments
1066	Halley	Portent of William the Conqueror	1106	X/1106 C1	Widely visible in day—Europe and Orient
1145	Halley	Well documented by Chinese	1264	C/1264 N1	Poured out smoke like a furnace
1378	Halley	Recorded by Chinese, Koreans, and Japanese	1402	C/1402 D1	Comet visible in broad daylight
1456	Halley	Comet was excommunicated by the Pope!	1531	Halley	Orbit computed by Halley
1556	Heller	Great fiery unusual-looking star	1577	Tycho's Comet	Observed by Tycho Brahe; tail 80° long
1607	Halley	Orbit computed by Halley	1618	C/1618 W1	Tail 104°
1661	Hevelius	6° tail and multiple nucleus structures	1680	Kirch	Maximum tail arc of 90°
1682	Halley	Epoch of Edmund Halley's observations	1689	C/1689 X1	Discovered at sea, tail 68°
1729	Sarabat	Large perihelion distance	1744	De Cheseaux	Remarkable appearance with six tails
1759	Great Comet	Passed 0.07 au from the Earth	1769	Messier	Tail length exceeded 90°
1811	Flaugergues	Unprecedented 17 month visibility	1823	Great Comet	Large sunward antitail

(*Continued*)

doi:10.1088/2053-2571/ab027ech12

1835	Boguslawski	Star-like nucleus and a broad tail	1843	Great Comet	Sungrazing comet
1858	Donati	Most beautiful comet on record	1861	Tebbutt	Daytime 'auroral glow' reported
1874	Coggia	Unusual jet features	1880	Great Southern Comet	Orbit resembles comet of 1843
1881	Great Comet	Only comet spectrum observed before 1907	1882	Great Comet	Orbit resembles comet of 1880
1887	Great Southern Comet	Orbit resembles comet of 1843	1901	Great Southern Comet	Brightness rivals that of the star Sirius

Courtesy of Karen Meech and Gary Kronk's *Cometography,* volumes 1 and 2.

Appendix F

Books on historical comets

Heidarzadeh T 2008 *A History of Physical Theories of Comets, from Aristotle to Whipple* (New York: Springer)

Metz J 1985 *Halley's Comet, 1910: Fire in the Sky* (Saint Louis, MI: Singing Bone)

Schechner S J 1997 *Comets, Popular Culture, and the Birth of Cosmology* (Princeton, NJ: Princeton University Press)

Seargent D 2009 *The Greatest Comets in History: Broom Stars and Celestial Scimitars* (New York: Springer)

Stephenson F R and Walker C B F (ed) 1985 *Halley's Comet in History* (London: British Museum)

Van Nouhuys T 1998 *The Age of Two-faced Janus: The Comets of 1577 and 1618 and the Decline of the Aristotelian World View in the Netherlands* (Leiden: Brill)

Yeomans D K 1991 *Comets: A Chronological History of Observation, Science, Myth, and Folklore* (New York: Wiley)

doi:10.1088/2053-2571/ab027ech13

Appendix G

Intermediate and advanced books on comets

Brandt J C and Chapman R D 2004 *Introduction to Comets* 2nd edn (Cambridge: Cambridge University Press)

Crovisier J and Encrenaz T 2000 *Comet Science: The Study of Remnants from the Birth of the Solar System* (Cambridge: Cambridge University Press)

Festou M C, Keller H U and Weaver H A (ed) 2004 *Comets II* (Tucson, AZ: University of Arizona Press)

Krishna Swamy K S 2010 *Physics of Comets* 3rd edn (Singapore: World Scientific)

Kronk G 1999–2017 *Cometography* vols 1–6 (Cambridge: Cambridge University Press)

Meierhenrich U 2015 *Comets and their Origin: The Tool to Decipher a Comet* (New York: Wiley)

Sagan C and Druyan A 1997 *Comet* (New York: Ballantine)

doi:10.1088/2053-2571/ab027ech14

Glossary

(The first occurrence of each entry in the text has been italicized.)

Air mass	The 'width' (or path length) of the atmosphere through which an observation is made.
Albedo	The fraction of light reflected from a surface, ranging from 0 (total absorption) to 1 (total reflection).
Antitail	A dust tail that appears to point towards the Sun but is actually an illusion caused by observing the comet from a particular perspective from the Earth.
Aphelion	The farthest point from the Sun in an orbit.
Asteroid	A small Solar System body primarily made up of rock and metals. Recent work shows that many also contain water in some form or other, making their distinction from comets less clear. Most are found between the orbits of Mars and Jupiter, a region called the Main Asteroid Belt (MAB).
Astronomical unit (au)	The average distance between the Sun and Earth during its orbit, being 149 597 870.700 km exactly.
Celestial meridian	The imaginary line running from an observer's north point on the horizon, through the point directly overhead (zenith), to the south point. It divides the hemispherical sky into eastern and western parts. Objects in the east rise during the night, achieve their greatest distance above the horizon as they cross the meridian, and set in the west.
Centaurs	Small bodies with characteristics of both asteroids and comets between the orbits of Neptune and Jupiter (e.g. 95P/Chiron), now totaling over 500 objects. Some are thought to be comet nuclei from the Kuiper Belt or Scattered Disk making their way progressively into the inner Solar System.
Coma	A constantly escaping cloud of gas (including ions) and dust surrounding an active comet, its 'atmosphere'.
Corona	A hot but tenuous gas of electrically charged particles surrounding the Sun and extending millions of kilometers into space. It is best seen during a total solar eclipse.
Dust	Small grains (less than about 1 μm) composed of refractory materials (silicates, carbonaceous substances, carbon–hydrogen–oxygen–nitrogen (CHON) material, rocks, metals), some with ices.
Eccentricity	The degree of 'flattening' of an ellipse (or orbit), ranging from 0 (circular) to near 1 (highly elliptical).

doi:10.1088/2053-2571/ab027ech15

Ecliptic	The plane of the Earth's orbit in the Solar System.
Ephemeris	A table of celestial coordinates (right ascension, declination) giving the sky positions of celestial bodies. Comet ephemerides can be found online at NASA's Horizons System, https://ssd.jpl.nasa.gov/horizons.cgi.
Ice	Any solid made out of a volatile substance, e.g. water, carbon dioxide, etc.
Ion	An electrically charged particle caused by adding or removing electrons from a neutral atom or molecule.
Isotope	A variant of an element with a different number of neutrons in its nucleus but generally having the same chemical properties (e.g. adding a neutron to a hydrogen atom yields its heavier isotope, deuterium).
Jet	A stream of sublimating gas with dust emanating from a specific surface area (active region) usually on the day-time side of the nucleus. Jets exert non-gravitational forces on the nucleus that changes its orbit and spin state.
Kuiper Belt (KB)	A doughnut-shaped disk of small bodies, including three dwarf planets, at 30 to about 50 au from the Sun, first postulated by Gerard Kuiper in 1951. The first Kuiper Belt Object (KBO), other than Pluto, was found in 1992. Now more than one thousand KBOs are known. It overlaps the Scattered Disk which extends out to at least 100 au and is the source of most short-period comets with very elliptical orbits and higher inclinations.
Light pollution	The combined lighting that illuminates the night sky, mainly in urban areas, including street lamps, spotlights, glowing fast-food signs, automobile lights, house lighting, the phase of the Moon, etc, making it difficult to see stars and other astronomical phenomena.
Long-period comet (LPC)	Comets with orbital periods exceeding 200 years, up to about a million years, at random directions and angles to the ecliptic, forming a roughly spherical cloud surrounding the Sun (the Oort Cloud).
Magnitude	In astronomy, the system used to measure the brightness of an object, the smaller (more negative) the number, the brighter the object. A difference of five magnitudes is a hundred-fold change in brightness. Examples: the Sun is −26.7 in magnitude, the full Moon is −12.6, Venus at its brightest is −4.4, Sirius (the brightest star) is −1.6, the faintest object seen by the unaided eye under the darkest skies is 6.5, the limit of the Hubble Space Telescope is 31.5, and the limit of the James Webb Space Telescope (Hubble's replacement to be launched in 2021) is 34.
Meteor	Small pieces (less than about three meters in size) of extraterrestrial rock and metal that enter the Earth's atmosphere traveling at very high speeds, and are heated to a high temperature, causing them to glow and possibly leaving a trail of hot vapor in their wake. We can see these streaks of light in the night sky, which many call 'shooting stars'. Many meteors are the size of grains of sand and completely burn up in the atmosphere. Before atmospheric entry, they are called meteoroids and if they survive to hit the ground, we call them meteorites.
Meteor shower	An increase in meteor activity over several consecutive nights each year, with meteors that appear to radiate from a common point in the night sky. The frequency of meteors can vary from a dozen (a sprinkle) to more than one thousand per hour (a storm). The source of these meteors is the dusty debris trail of comets whose orbits intersect that of the Earth.

Nucleus	The tiny solid part of a comet, measured in tens of kilometers down to a few hundreds of meters, with a typical size being about 10 km. It is composed of ices and refractories, being very dark and very porous (fluffy). It is the source of all activity that gives rise to the coma and tails.
Oort Cloud (OC)	The outer region of the Solar System containing long-period comet nuclei, first postulated by Jan Oort in 1950. It consists of the doughnut-shaped Inner Cloud (Hills Cloud) from about 2000 to 20 000 au that smoothly joins with the Kuiper Belt and the spherical Outer Cloud from about 20 000 up to 150 000 au (the limit of the Sun's gravitational boundary). Several thousand long-period comets are known to have originated from the Oort Cloud. They have extremely elliptical orbits with random orientations to the ecliptic plane.
Parallax	The angle through which an object seems to move when viewed from two disparate places, as reckoned against some, more-distant background and is inversely proportional to its distance. It is the basis for finding distances to comets, asteroids, planets, and stars.
Perihelion	The closest point to the Sun in an orbit.
Period	The amount of time to complete one orbit.
Plasma	A gas consisting of electrons and ions, often called the fourth state of matter since it behaves very differently to an ordinary (neutral) gas.
Proto-solar nebula	The interstellar gas and dust cloud out of which the Sun and its Solar System formed some 4.6 billion year ago.
Refractory	Minerals that retain their form and properties at high temperatures, such as rocks and minerals.
Scintillation	The variation in brightness of a point-like astronomical object, such as a star, when seen through Earth's atmosphere, i.e. twinkling or flickering. Extended objects, such as planets and comets, do not flicker appreciably.
Seeing	A measure of the 'blurring' of astronomical objects due to the atmosphere, principally due to scintillation, light pollution, and cloud cover.
Short-period comet (SPC)	Comets with orbital periods less than 200 years. They travel near the ecliptic plane of the Solar System and generally in the same direction as the planets, most originating from a doughnut-shaped region surrounding the Sun (the Kuiper Belt and Scattered Disk). They include Jupiter-family comets (JFC) with periods less than 20 years, Encke-type comets (ETC) with the shortest known periods whose orbits do not reach Jupiter, and Halley-type comets (HTC) with periods between 20 and 200 years. About 10% of well-studied, short-period comets (observed more than once) fall into this latter category.
Sublimation	The chemical term to describe the change of a substance from solid to gas and vice versa without passing through the liquid phase.
Tails	Long appendages made up of ions (type I, atomic and molecular ions affected by the solar wind), dust particles (type II, affected by sunlight), and gas (Type III, neutral sodium atoms affected by sunlight) that stream away from the comet's head in the inner Solar System, generally in the direction away from the Sun. Comets may exhibit one or more tails simultaneously. A Manx comet was seen in 2016 that had no tail.

Trojan asteroids	Small bodies at the orbit of Jupiter, in two groups about 60° ahead (Greek node) and 60° behind (Trojan node) its orbit, trapped at stable points. They are classified as asteroids but many may be comet nuclei that were captured by Jupiter in the early history of the Solar System. To date, 7040 Trojans have been discovered.
Volatile	Under normal pressures, chemical substances that turn to gas unless kept very cold, e.g. carbon dioxide, methane, molecular oxygen, molecular nitrogen, etc. Water ice is also considered to be a volatile.
Zodiacal light	Sunlight reflected from comet dust in the ecliptic plane, most easily seen in the western sky after sunset as an elongated cone extending up from the horizon.

◎

编辑手记

本书是一部关于天文学的英文原版著作.

在我国的古书中,最早使用“天文”一词似乎是在《易经》中.在《周易·贲卦·彖传》中有:“观乎天文,以察时变”.在《易经·系辞传》里面也有记载:“仰以观于天文,俯以察于地理;是故知幽明之故”.在《淮南子》中有《天文训》一篇,在《汉书》中有《天文志》,而在《艺文志》中有天文部分.那么在中文中天文一词究竟是什么意思呢?《淮南子·天文训》称:“文者象也”.

根据这种解释,天文就是天象,或天空的现象.

天空所发生的现象,可以分为两大类.一类是关于日月星辰的现象,即星象;另一类是地球大气层内所发生的现象,即气象.从我国历史来讲,天文学实际是研究星象和气象的两门知识.希腊文字的天文学语根和气象学语根不同.天文学的希腊语是“αστρου,γδμος”.据其语根是研究天体的科学,亦即星象学,而我国自古以来,均用天文学而不用星象学这个名称.

本书的中文书名或可译为《21世纪的彗星:体验下一颗伟大彗星的个人指南》.

本书的作者有两位:

一位为丹尼尔·C.布易士(Daniel C. Boice).他是德克萨斯州圣安东尼奥市科学研究与咨询公司的首席天文学家,他的专业活动包括75篇同行评审的研究论文和数百篇会议报告.2000年,他成为英国皇家天文学会的会员.

另一位为托马斯·霍基(Thomas Hockey).他是北爱荷华大学天文学教授.他之前写过七本书,包括《伽利略的星球》和《我们如何看天空》.他还是获奖书籍《天文学家传记百科全书》的主编.

正如作者在前言中所述:

自从观察到了似乎在我家上方的威斯特彗星以来,我就知道我想研究天空.20 世纪 90 年代,当人们知道另一颗大彗星很快就会出现在天空中时,我写了《海尔-波普彗星:来自深空的令人敬畏的访客指南》(1996 年(马萨诸塞州什鲁斯伯里:ATL)).我记得我通过扩音器向站在阿德勒天文馆前的一千人讲课时,尽管芝加哥市中心灯火通明,但我还是看到了彗星.

《海尔-波普彗星:来自深空的令人敬畏的访客指南》的目标是在最基本的层面上向数百万想看它的人介绍这种宇宙景象.虽然这本书主要介绍的是关于海尔-波普彗星的内容,但是在这本书中也包含了许多关于彗星本身的基础信息(其中一些材料会出现在本卷中).当然,这些材料现在有些陈旧了.本书的共同作者丹尼尔·C.布易士博士,我认识他将近四十年,在我认识他的这段时间里,他致力于对彗星的研究,为彗星科学和观测提供了近年来发生的新情况.感谢他对这个项目的支持,还要感谢杰出的天文学作家威廉·希恩(William Sheehan)博士.

这本书是在不同的地点创作的:布易士在泰国写作,美国工作.我们希望我们的共同努力可以取悦读者,无论他们是否有机会亲眼看到大彗星.

——托马斯·霍基,锡达福尔斯,2018 年 12 月

我出生在太空时代的黎明时期,受到美国宇航局阿波罗任务的启发,最终让 12 人登上了月球.后来,我对天文的所有事物都产生了兴趣,最终我在加利福尼亚州旧金山上高中时买了我的第一台望远镜,一台二手的 60 毫米折射式天文望远镜.我很快就获得了一个 8 英寸(1 英寸等于2.54 厘米)的多布森反射式望远镜,这是我在大名鼎鼎的约翰·多布森(John Dobson)的指导下自己制造的,并通过注册历史悠久的"旧金山业余天文学家"加入了业余行列.我很幸运能够实现成为一名专业的天文学家的梦想,并且非常感谢 NASA 以及美国国家科学基金会支持我的事业,感谢我 26 年的雇主西南研究所(德克萨斯州圣安东尼奥市)提供的机会.最后,我还要感谢美国人民的资助,感谢他们相信他们的钱会花得很好!

虽然我只是一名研究人员,但我对教学有很大的热情.二十多年来,我能够与德克萨斯大学圣安东尼奥分校、圣安东尼奥社区学院和三一大学的学生分享我的天文学知识和热忱.老师们从学生那里学到了很多东西.在此期间,我有幸指导研究生——我们的下一代天文学家,并参与了许多对外合作.

我在新墨西哥州立大学的研究生院遇到了我的合著者托马斯·霍基

博士. 我提前三年毕业, 并于 1985 年前往洛斯阿拉莫斯国家实验室, 在那里我被哈雷彗星和我的导师沃尔特 · 许布纳(Walter Huebner)博士影响去研究彗星科学. 托马斯 · 霍基成为了北爱荷华大学的教职员工, 从而可以追求他对教学和历史天文学的兴趣. 尽管我们都是美国天文学会(AAS)和国际天文学联合会(IAU)的成员, 但我们的道路已经偏离了几十年. 我在 2017 年注意到他的四卷本参考著作《天文学家传记百科全书》, 此书获得了 AAS 著名的奥斯特布罗克历史天文学图书奖. 有机会与他合著一本书是非常荣幸的事情.

特别感谢西南研究所退休的科学家沃尔特 · 许布纳博士和巴西圣保罗大学天文学系主任阿毛里 · 德 · 阿尔梅达(Amaury de Almeida)教授审阅了我们的手稿. 他们富有洞察力的评论大大改进了本书. 当然, 我们对任何剩余的错误负责. 我和我的合著者一样希望读者会喜欢我们的作品.

——丹尼尔 · C. 布易士, 泰国普安富, 2018 年 12 月

本书的中文目录翻译如下:

从目录中可见,本书是一部专门讲彗星的书,而在中国古代彗星和妖星是画等号的.根据古天文史专家陈美东教授考证①:

除了日、月、五星、恒星、银河等天体,古人对于在天空中突然的、随机的、短期出现的天体也给予了充分的注意,对于它们的本质、出现的规律等均有所讨论,并出现试图对它们做某种分类的尝试.在众多这类天体中,古人对于彗星有较明确的定义和名称.战国时期的石申就有如下的论述:

凡彗星有四名,一名孛星,二名拂星,三名扫星,四名彗星,其状不同,为殃如一.②

对此,大约成书于汉代的《黄帝占》有更明确的阐释:

彗星出见,本类星,长可二三尺,其炎如气黄白,孛孛然,名孛星.

彗星出见,本类星,长可一丈,其炎头散下垂,状如毛拂,名曰拂星.

彗星出见,可二丈至三丈,形如竹木、枝条,名曰扫星.三丈以上至十丈名曰彗星.彗扫同形,长短有差,殃灾如一.③

由此可知,彗星的长短是区别孛星等四种名称的最主要标准,此外还

---

① 摘自《中国古代天文学思想》,陈美东著,中国科学技术出版社,2007.

② 《开元占经》卷八八.

③ 《开元占经》卷八八.

有形状、颜色等判据. 石申还指出:

彗星出东南,其本类星,末类彗,长可二三尺至一丈,名曰天枪.

彗星出东北,其本类星,末类彗,长可四五尺至一丈,名曰天搀.

彗星出西北,其本类星,末类彗,长可四五尺至一丈,名曰天棓.

彗星出西南,其本类星,末类彗,长可二三丈,名曰扫星.

彗星出中央,正在人上,其本类星,末类彗,长可五六尺至一丈,名曰天戈.①

这里对于彗星的基本特征的描述是:有一个较明亮的圆形的头部("本类星"),并有长短不一的、扫帚状的尾部("末类彗"). 石申主要依据它们出现的方位,及其长短,分别赋予不同的命名. 但石申的命名仅为一家之言. 据《史记 · 天官书》载,甘德则另有说法:

其失次舍以下,进而东北,三月生天棓,长四丈,末兑. 进而东南,三月生彗星,长二丈,类彗. 退而西北,三月生天搀,长四丈,末兑. 退而西南,三月生天枪,长数丈,两头兑.

同是天棓、天搀、天枪,其出现的方位、长短、形态,均与石申所论迥异. 不过,甘德据以命名的思想则与石申并无二致. 后世各家多各是其说,所以对于彗星命名的状况一直十分混乱. 上引甘德所说"其失次舍以下"云云是指木星而言,认为天棓等的出现,决定于木星失行所处的方位,如木星本应在甲处,但却见于甲处的东北,则三个月以后则生天棓,等等. 这是彗星生成与出现同木星失行密切相关的论说. 此说在《汉书 · 天文志》中也有大同小异的记述.

巫咸曰:岁星行东行南六十日不还,以初去日六月彗星出;东北六十日不还,以初去日六月棓星出,西北六十日不还,以初去日六月搀星出;西南六十日不还,以初去日六月枪星出.②

甘德和巫咸二家说大同小异,只是认为岁差失行后出现彗星的时间不同而已. 而这种思想的产生也许还要早些,"齐伯曰:填星出入,以舍斗五十日不下,彗星出."③

关于彗星的生成,石申指出:

扫星者,逆气之所致也.④

这一说法是他对彗星致灾观点的自然延伸.

---

① 《开元占经》卷八八.

② 《开元占经》卷八八.

③ 《开元占经》卷八八.

④ 《开元占经》卷八八.

关于以下数种妖星的生成,石申认为:

女帛:东北有星长三丈而出,木水气交.

盗星:东南有星长三丈而出,木火气交.

积陵:西南有星长三丈而出,金水气交.

瑞星:四隅有星,大而赤,察之中黄数动,长可四丈,此土之气.①

石申认为这些星是五行中的金木水火等气两两互交而合成,或由土之气生成.

众多说法中应属石申学派的《黄帝占》所说稍异,如它认为女帛"五星气含之变,出东北,木水气合也"②,等等.即把石申的五行气含改变成了五星气合.

在长沙马王堆三号汉墓出土的帛书中,有一幅十分珍贵的关于彗星的画图③,内含29种形态和各异的彗星图像,它们应是楚人汇集的对彗星长期观测的成果.图像中彗尾有宽有窄,有长有短,有曲有直,彗尾的条数有多有少;彗星画成一个圆圈或圆点,有的圆圈的中心又有一个圆圈或小圆点,这可能表明当时人们已经注意到彗头又可分为彗发和彗核两个部分,而且也有不同的类型.这些图像可以与上述石申等人的记述相呼应,表明战国时期人们关于彗星的形态描述和进行分类的尝试.

西汉早期刘安等人的《淮南子·天文训》中有"鲸鱼死而彗星出"的说法.两汉之交颇有此说的赞同者,如《春秋纬·考异邮》就重申之④,而《春秋纬·演孔图》更推而广之,以为"海精死,彗星出"⑤.对于这类说法,有人解释说:"鲸鱼阴物,生于水,今出而死,是时有兵相杀之祥也,故天应之以妖彗也."⑥原来这是基于物类相感的思想而提出的彗星生成说.

西汉董仲舒以为"孛者,乃非常恶气之所生也"⑦.而刘向《说苑·辨物》曰:"搀、枪、彗、孛、旬始、枉矢、蚩尤之旗,皆五星盈缩之所生也."这些则是对石申、甘德相关论说的继承与发展.

西汉晚期的京房在《风角书·集星章》中,论及包括彗星在内的35种妖星的生成,以为它们分别由岁星、荧惑、填星、太白、辰星所生,每一行星各生7种.如:"天枪、天根、天荆、真若、天椟、天楼、天垣,皆岁星所生也.

① 《开元占经》卷八六.

② 《开元占经》卷八六.

③ 席泽宗:《马王堆汉墓帛书中的彗星图》《文物》,1978年,第2期.

④ 《开元占经》卷八八.

⑤ 《开元占经》卷八八.

⑥ 《太平御览》卷八七五.

⑦ 《文献通考》卷二八一《象纬四》.

见以甲寅,其星咸有两青方在其旁”等.它们“皆见于月旁,互有五色方云,以五寅日见”.京房又指出:“三十五星,即五行气所生,皆出于月左右方气之中,各以其所生星将出不出日数期候之.”①既言为五星所生,又言“即五行气所生”,似彼此矛盾.其实不然,察其意,京房是认为五星是地上的五行气所生(这与两汉之际流行的说法相符合),35种妖星由五星的散气所生,也可以说就是五行气所生.这实际上是地上的五行气间接生成说.这里京房还认为35种妖星的生成同月亮有关,必生于离月亮不远处,而且与月旁的五色云气互生,出现的日期则与五星晨见(或夕见)之时相吻合.至于为什么月亮周围有五色云气之类出现,《黄帝占》则有所说明:“五星将欲为彗之变,先见其气,后见其彗”②,即以为这是生成彗星过程中气的先期作用造成的.

大约成书于西汉晚期的《河图》认为,各类妖星(包括彗星)是由五星之精流散而成,如“岁星之精,流为天棓、天枪、天猾、天叩、国皇、反登、黄彗”等,每一行星之精流散生成6至10种妖星.其中苍彗、赤彗、黄彗、白彗和黑彗则分别由岁星、荧惑、填星、太白和辰星之精流散而成③.这显然是妖星的五星气散成说.

可是《河图》又认为苍彗“少阳之精”,赤彗“太阳之精”,又曰“火精”,黄彗“土精”,白彗“少阴之精”,黑彗“太阴之精”④,这却是五行生成说.两汉之际《春秋纬》也依彗星的颜色分苍彗、黑彗、白彗等,并以为它们分别由“木精”“水精”“金精”⑤等生成,同《河图》说相同.“木精”即指“少阳之精”,“水精”即指“太阴之精”,“金精”即指“少阴之精”,等等.

《孝经纬·雌雄图·三十五妖星占》对京房35妖星生成说做了修正.如“天枪星,在箕宿中,出月左方,日在甲寅,岁星将出而不出,其与日合八十日,其未出八日,必有灾云苍赤黑色之物厌日之光,青色之星,有两青方在其旁出而生天枪之星,长数丈”⑥,等等.这给彗星的生成与出现规定了更为严格的条件,除京房所说之外,还要加上必须在二十八宿中的某一特定的宿次,妖星的生成还与太阳有关,它们必须与太阳相合若干日,其出现之前若干日还有灾云遮掩日光等现象.

① 《晋书·天文志·中》.
② 《开元占经》卷八八.
③ 《晋书·天文志·中》.
④ 《开元占经》卷八八.
⑤ 《开元占经》卷八八.
⑥ 《开元占经》卷八七.

这时,关于妖星的生成说真是多种多样,人们各抒己见,做出种种猜测.《春秋纬·运斗枢》主张若干妖星是由北斗七星的散精生成的,如"璇星散为五残","玑星散为昭明"①,等等.而《春秋纬·合诚图》则认为一些妖星是由苍彗、黑彗、黄彗、赤彗、白彗之气流散而成的,如"苍彗散为五残","赤彗分为昭明"②,等等.此即彗星派生若干妖星之说.如同彗星等的命名一样,若干妖星的生成说也莫衷一是.就以这里提及的昭明为例,其生成说还有巫咸曰:"金之精";《河图》曰:"荧惑之精";《黄帝占》曰:"金之气"③,等等.

截至两汉之际,包括彗星在内的妖星生成说至少有九种:鲸鱼死彗星出说、逆气生成说、五星气散成说、五行生成说、五行合成说、五星气合成说、五星失行生成说、北斗七星生成说、彗星生成妖星说,等等,其中又以五星气散成说最为流行.宋代《中兴天文志》对此有很精辟的概括:

凡妖星,五行之乖戾气也,五行掩合陵犯,怒逆错乱、流散杂变之所生也.④

曹魏孟康在五星失行生成说与五星气散成说之间搭起了一座桥梁,把两者有机统一起来:"五星有变,则其精散为妖星也."⑤即以为五星失行是导致五星气散的原因.五星失行而生妖星的论说,可能与人们确曾观测到五星留、逆行或失次之后不久,彗星或其他发光体出现的现象,把这种偶合推广而为必然,遂以立论的.

时至今日,关于彗星的起源,天文界的看法也莫衷一是,乃是有待进一步研究的课题.中国古代的占主导地位的彗星生成说是五星气散成说,这同近现代的喷发说有某种共同之处.喷发说认为,彗星是由木星等行星或卫星上火山喷发的一些物质形成的.此外,彗星运行轨道等确实受到行星摄动力的巨大影响.所以,古人关于彗星与五星相关的推测是有合理之处的.

在两汉之交以后,关于彗星、妖星的生成几乎没有新论说出现,最为主要的进展则莫过于东汉张衡和唐代李淳风等人的有关论述.

张衡在《灵宪》中指出:

方星巡镇,必因常度,苟或盈缩,不逾于次,故有列司作使,曰老子四

---

① 《开元占经》卷八五.

② 《开元占经》卷八五.

③ 《开元占经》卷八五.

④ 《文献通考》卷二八一《象纬四》.

⑤ 《汉书·天文志》注引.

星,周伯、王逢絮、茀各一,错乎五纬之间,其见无期,其行无度,实妖星之所.①

方星即指五星而言,这里亦主五星失行而生妖星之说.张衡还认为妖星"其见无期,其行无度",这是对妖星特点的很好的概括,更重要的是,他指出这些妖星的运行轨迹是穿行于五星之间,用现今的观点看,可以说这是认为妖星是属于太阳系内的天体,这自然是一种很有意义的科学推测.

唐代李淳风在《晋书·天文志·中》指出:

孛星,彗之属也.偏指曰彗.芒气四出曰孛.孛者,孛孛然,非常恶气之所生.

这是关于区分彗孛的简明判定.恶气所生云云,乃取石申、董仲舒之说.李淳风关于彗星的最重要论述还在于:

彗星无光,傅日而为光,故夕见则东指,晨见则西指,在日南北,皆随日光而指.顿挫其芒,或长或短,光芒所及则为灾.

他提出了彗星本身并不发光,借日光的照射而发光的理论.他还指出彗尾的取向,总是背向太阳所在位置的现象,认为这一现象与他的理论是互为因果的.如果彗星自身发光,那么彗星的取向应是随机分布的,并不随太阳位置的变化而变化,而上述现象则证明彗体的发光是受到太阳的明显制约,其光源只能来自于太阳.现在我们知道,彗星确实是反射太阳光而发光的,而彗尾沿太阳相反的方向延伸,是因组成彗头的气体和微小尘埃,受到太阳风和太阳辐射压力作用的结果.所以,李淳风关于彗星发光原因和彗尾指向的论述都是符合现代科学的.

中华人民共和国成立后,我国的大学天文学教育主要是学习苏联,早期的教材由我国著名天文学家戴文赛先生引入中国,其中彗星和流星是一起讲的②.

1. 彗星的一般特征

我们所熟悉的星空景象往往被不平常形状的星体的出现所扰乱:在天穹上看见了很大的星,带着长长发光的尾巴.这是彗星,也就是所谓"发"星③.彗星大部分是出乎意外地出现的(甚至连天文学家也事先不知

① 《续汉书·天文志·上》刘昭注引.

② 摘自《普通天文学教程》(下册),И. Φ. ПОЛАК 著,戴文赛、石延汉等译,商务印书馆,1953.

③ 译者注:中国俗名叫"扫帚星".

道),它迅速地变化着形状和大小,在星空上所绘的轨迹往往和行星的运动完全不相同,经过数周或数月以后,便消失不见.

因为彗星的外形与运动和别的星体是如此的不同,所以古代的人们以为它们不是天上的现象,而是地球大气中的现象.一直到16世纪第谷才证明彗星离我们比月球还远,因此彗星是在宇宙空间中运动着的.再过了一百年,牛顿又证明彗星的运动和行星的运动一样,也依照万有引力定律.

彗星决不是罕见的事.近年来每年我们都可以发现几颗彗星,有时甚至在10颗以上.它们极大部分是属于望远镜彗星之列,也就是说,用肉眼是看不见的.

每一个在某一年所发现的彗星,首先用发现者的名字称之,然后加上年代数目,又依照发现次序加上一个拉丁字母.等到算出彗星轨道以后,便根据彗星通过近日点的先后次序改用一个罗马数字来代替拉丁字母.例如普尔科夫天文学家诺伊明所发现七个彗星中的第二个是他在1916年所发现的,又是那一年的第一个彗星,所以就先叫作诺伊明彗星1916a,后来才得到了最后的称呼:诺伊明彗星2(1916Ⅱ),这是因为在1916年所有彗星之中,它不是第一个而是第二个通过近日点的.

2. 彗星的运动

彗星是从太阳系遥远的区域跑来的,当它们在远处的时候,甚至用最大的望远镜也看不见它们,然后它们就消失在那边.因此每一个彗星我们只能够在其轨道上极小的一部分看到它,只是当它靠近太阳和地球的时候.往往可以近似地确定轨道,发现在这一部分里彗星沿着抛物线而运动,也就是说,沿着开放的无限曲线而运动(图1),假如彗星真的沿抛物线运动,它便永不再回到太阳附近来.因此,在这种情形下彗星便不能算作我们太阳系的经常成员,仅仅是偶然撞进了太阳的引力范围,向太阳“落下”,一度绕太阳画出一个抛物线,便离开太阳,以后可能再为某一个其他的“太阳”所吸引.

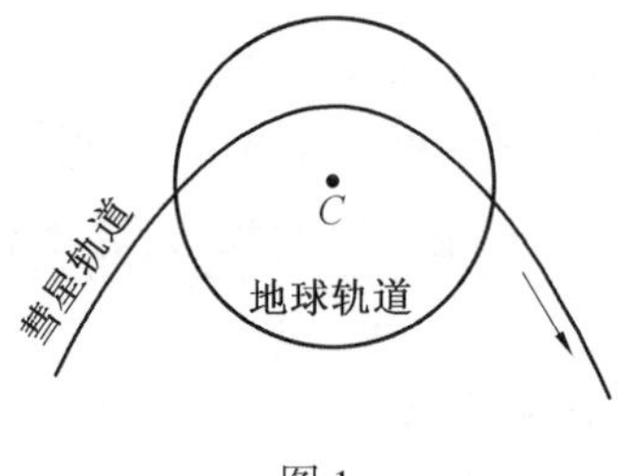

图1

但是近年来已经证明彗星的真正轨道并非抛物线，而是拉得非常长的椭圆(图2)，许多彗星要花费数万甚至数十万年才沿椭圆转一整圈，而彗星 1914 V 甚至要用 2.4×10^7 年. 就彗星在能够被我们看得见的数月之内所经过的那一段短弧来讲，这种椭圆便无法和抛物线区别开来. 彗星沿着这样的椭圆运动还是严格地遵守开普勒定律，在近日点附近的速度可能达到每秒数百千米，而在远日点附近便非常小.

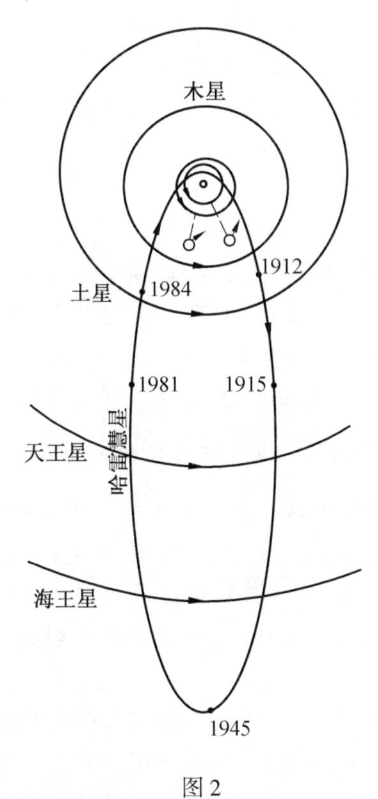

图2

因为彗星的椭圆轨道的真正形状和偏心率常是不知道的，所以无法根据抛物线运动公式来计算大部分彗星的位置，因为从数学方面来讲，这要比用椭圆运动计算简单得多. 因为抛物线的偏心率等于一，所以对于彗星轨道只要知道五个要素就够了(椭圆需要六个)，而轨道的大小不由半长径 a 决定(抛物线的半长径等于无穷大)，而用近日点距离 q 来决定(就是彗星和太阳的最短距离). 这样，彗星轨道由下列要素决定：i,Ω,π,q,t_0. 轨道和抛物线相差不太多的彗星称抛物线彗星.

3. 彗星轨道的变化

行星，特别是巨大的木星，它们的引力可以对彗星的轨道引起各种摄动. 现在讨论这种摄动中最主要的两种情况.

(甲)行星以自己的引力加速了彗星的运动. 由彗星运动公式知,速度增加以后,椭圆的长径和旋转时间也应当加大——椭圆会变得更加扁长. 如果这时候速度超过了抛物线速度,那么轨道便变成双曲线,彗星便永远脱离我们太阳系. 这种现象被观测过.

(乙)行星延缓了彗星的运动. 这时候彗星椭圆轨道的偏心率减小,长径和旋转时间也都减小. 偶尔有几次,彗星会走到距离行星很近的地方,那时摄动会变得很强,以致彗星的旋转时间缩短到只有几年. 顺着这种和抛物线相差很远的小椭圆而运动着的彗星叫作周期彗星. 彗星轨道变化的例子参看第 12 部分(戊).

4. 周期彗星

周期彗星在数量上约占所有已知彗星中的十分之一略多一些. 其中只有那些已经观测过回到太阳系来不止一次的彗星的运动才被精确地记录下来. 现在我们已知的这种彗星有 37 个. 很亮的哈雷彗星(是由与牛顿同时代的英国天文学家哈雷证明出它具有周期性的)具有这种类型的彗星中最长的周期,就是 76 年,而周期最短的只有 $3\frac{1}{3}$年,就是恩克-巴格隆特彗星. 这种彗星的极大部分具有 5 年到 7 年的旋转时间,而且它们轨道的远日点都位于木星轨道附近. 它们形成木星的"彗星族".

大部分周期彗星的轨道平面和黄道构成不太大的角度(大致和小行星的轨道差不多). 彗星运动的方向一般是顺时针转动的,和行星运动的方向相同. 哈雷彗星是少数例外之一——其运行是逆向的.

彗星椭圆的偏心率普通大于 0.5,所以彗星轨道远较行星轨道扁(参看图 2). 但是也有若干彗星轨道和小行星的轨道全无区别,也曾经找到具有和大行星同样轨道的彗星. 其中有一个运行于木星和土星轨道之间,轨道几近正圆,和行星一样,每年冲时就可以看到(施瓦斯曼-瓦赫曼彗星 1925 Ⅱ). 另一个彗星(奥特姆彗星)运行于火星及木星轨道间的几近正圆的轨道上,除它具有雾状包层以外,和小行星没有什么区别. 除哈雷彗星之外,所有其他的周期彗星都是很暗弱的.

5. 彗星的形状

所有彗星的主要部分是几近圆形的发光而透明的东西,中部的亮度常是强一些,逐渐向边缘减弱. 往往这个"中央密集部分"像是在雾中所看到的星一样,它叫作彗核,而包围核的雾状物叫做彗发. 核和发合成彗头. 望远镜中的彗星往往既没有核,甚至也看不见明显的中央密集部分.

彗头的云雾状物质向一个方向延伸,而成为尾——这是发光的带,距离头部越远,常是越弱越宽,像烟柱一般看不见什么明确的终界.彗尾位于彗星轨道平面上,而且总是朝着和太阳相反的方向,所以当彗星在离开太阳时,是以尾巴在前而运动着.有时候彗星会有几个尾,像扇子一样地张开于彗星的轨道平面上.

彗尾看得见的长度是很不一致的.望远镜中的彗星的尾巴往往或者完全没有,或者很淡弱;而亮的彗星的尾有时候伸长到90°以上(由地平直到天顶).

6.彗星的形成

彗尾并非总是彗星的附属物,它只是当彗星接近太阳(在近日点附近)的短时间内在太阳光线作用下形成的.

当彗星离开太阳较远的时候,它的形状是一个朦胧的圆点,在这个时候,它大概只是因反射太阳光而发亮.许多彗星在它能被看见的期间内,一直保持着这种形状,可是有一些彗星当它们走近太阳的时候,便开始逐渐变化.核和整个头部变得更明亮,开始从核"蒸发出"发光的气体物质,比彗发的其他部分亮.这种气流的大部分集中在核的向着太阳的那一面,向太阳延伸,像射线或者"包层"一样,它是自己发光的.

但是这些"蒸发物"并不停留在彗星的头部,在从太阳而来的斥力的作用下,它们被向远处推开,推到和太阳相反的方向,这样便形成彗尾.因此它的构成物质不断在变换,正像汽船烟囱上的烟柱一样,我们今天所看到的彗尾并不全是构成昨天彗尾的质点所构成的,一部分已经是新的质点了.

假如用照相的方法,便可以直接地看出彗尾质点的这种运动.在很亮的彗星的照片上可以看出它们的尾部具有很复杂的结构,常是断断续续的,由个别发亮的密集部分或"雾状物"所构成.假如把前后相隔数小时所摄取的一系列照片加以比较,便可以注意到这种物质团的位置在变化,甚至可以计算出它们的速度.运动的方向总是由彗星的头部沿着尾部前进:它好像是加速的,离开头部越远,速度越大.雾状物离开彗核的速度可能达到100千米每秒.彗星越走近太阳,彗尾的形成过程便进行得越猛烈,常在它通过近日点之后不久,彗尾达到最大的长度,以后逐渐减短而至全部不见,彗星便又和出现时一样,成为一个雾状的点.

7.彗星的大小和质量

彗星是太阳系中最大的物体,甚至那些最暗的完全没有尾部的望远镜中的彗星也要比地球大好多倍.有几颗彗星(其中的一颗是1811年有名的彗星)的头部比太阳还大.彗尾则长达数千万甚至数万万千米.有时候它的长度甚至超过地球轨道的直径.彗尾的宽度和厚度也达数百万千米.这样,彗星的体积是非常庞大的.

更奇特的是彗星的质量微小之至.直到现在为止,当彗星行近某一个行星的时候,没有一次能够看到彗星对于行星的运动引起甚至极微小的摄动.甚至发现过几次彗星穿过了木星的卫星之间,但是对于卫星运动的规律性却丝毫未加以破坏.因此彗星质量的测定是非常困难的,大多数彗星的质量直到如今还是未知的.

虽然如此困难,沃隆佐夫-维里亚米诺夫教授还是发明了计算大彗星的核的大小、质量和构造的巧妙方法.沃隆佐夫-维里亚米诺夫用自己的方法来研究哈雷彗星,证明这个彗星(最大的彗星中的一颗)的核的直径为60 km左右,而且是由个别大小在100 m左右的石块所构成的.

哈雷彗星的核的质量才等于不大的小行星的质量.

沃隆佐夫-维里亚米诺夫教授的研究证明彗星的质量仅及地球质量数百万分之一(也可能只有十万万分之一).

因此,它们的密度是难以想像地微小,远小于我们的大气密度.

8. 彗星的光谱

在差不多每一个亮的彗星的光谱里都可以发现连续光谱,在有利的条件下甚至可以在连续光谱上面看到夫朗和斐暗线.这表示在彗星中含有反射太阳光的固体质点.

在彗核的光谱中,也常在彗尾的光谱中,和连续光谱在一起,还可以看到自己发光的气体的明线或带.在彗星的头部曾经发现过碳 C_2,氰 CN 和若干别种碳的化合物.在彗尾中发现过电离的一氧化碳 CO^+ 和电离的氮分子 N_2^+.

这些气体发光的原因还没有得到说明,很可能这是"发光"现象,落在气体上的太阳光线"激发"气体原子,迫使它们发出光波.也很可能是从太阳飞出来的带电质点或者电子撞击了彗星气体因而使其发光,和我们的大气中的极光现象的原因类似.有时有若干彗星,离太阳还很远,会忽然增加亮度达数十倍(例如施瓦斯曼-瓦赫曼彗星),这种亮度爆发可能用上述的第二种原因来说明.

很早便已注意到彗星的光谱和陨星的光谱相同.假如把一块陨星的碎片放在真空管中,又通以电流,则真空管便开始发光,这时候所生的光谱和彗星光谱相似.

9. 彗星的本质

彗星是比较小的流星体的集合，它们占有很大的空间，因此物体间的间隔远较物体自身的大小为大. 这可以由彗星惊人的透明度而获得证明. 不仅透过尾部，甚至透过头部也可以看得见恒星，亮度一点都没有减弱. 在1910年哈雷彗星的头部正好通过了地球和太阳的中间. 当"彗星通过太阳圆面"之际，太阳表面的状况完全和寻常一样，甚至用最大的望远镜也不能看出彗星的痕迹. 假如在彗核中有直径达数千米的不透明物体，那么它们便会在太阳圆面上显示为一个黑点. 这证实了沃隆佐夫-维里亚米诺夫关于组成彗星头部的流星体的大小的结论(第7部分)，并且使我们推想望远镜中彗星的核可能是由小石块和砂粒所构成的. 彗尾则由气体和尘粒所构成，而其密度很小，可能每一立方千米的空间里包含不到十个尘粒.

在它旋转的大部分期间内，这样一个流星物质团是暗而冷的，因此是看不见的固体块和质点的集合. 当它行近太阳的时候，它最初只是因为太阳照明的增加而变得可以看见. 然后因为被太阳光加热的结果，包含于流星体空隙中的气体便从彗核跑出来而开始发光. 被彗核所抛掷出来的气体物质主要集结在彗核上向着太阳的那一边，在那里形成有些复杂的迅速运动着的大气. 太阳斥力把这种大气的质点赶离太阳，这样便出现了彗尾. 在斥力影响之下，从彗核也有小的尘粒和气体一起飞出来，这种尘粒在彗核的成分中的量是很多的. 所以彗尾是由气体及尘粒所构成的.

因此，彗尾的形成总是伴随着物质的损失，所以最后便会引起彗星的瓦解. 彗星发展尾的次数越多，瓦解便来得越早. 正因为这个原因，所以所有的短周期彗星回转到太阳来已不下数千次，现在都很微弱，差不多完全没有尾. 除彗尾的形成以外，在彗星中还发生另一种瓦解过程，就是形成彗核的流星体散开到空间去(第12部分).

10. 不同类型的彗尾

彗尾的力学理论主要是由著名的俄国天体物理学家布烈基兴的工作而创造出来的，他从1890年到1895年领导普尔科夫天文台. 这个理论的内容如下：从彗核不断有微小的质点以一定的速度被抛掷出来，它们在失去了和彗星的联系以后，运动便开始和彗核的运动不一致. 因为对这些质点有从太阳而来的两种力量在作用着：(1)牛顿的太阳引力；(2)只有对于微小质点才有显著影响的太阳斥力(第11部分). 而彗核则像别的天体一样，只在第一种力量作用之下运动着，因为第二种力量在它上面的作用是微不足道的.假若知道太阳斥力的大小和由彗核抛掷出来的质点的初始速度,那么便不难算出前几天以来被抛掷出来的质点在某一瞬间所应达到的位置,因此便可以求出该瞬间彗星的形状(图3).在图3上用A,B,C,…各字母代表彗星绕太阳(S)旋转时的不同位置.当彗星位于点A,B,C,…时,由彗核所抛掷出来的质点各自沿着轨道Aa,Bb,Cc,…而运动,当彗星位于点K之时,在该瞬间质点的位置是a,b,c,…,对观测者来说这些点便形成一条弯曲的彗尾.也可能解出相反的问题:根据观测到的彗尾的形状去计算斥力的大小.

布烈基兴证明在不同的彗星中斥力可以有不同的数值,但并不具有一切可能的数值.他研究了很多彗星,认为它们可能并成三种主要的种类(或类型)(图4).

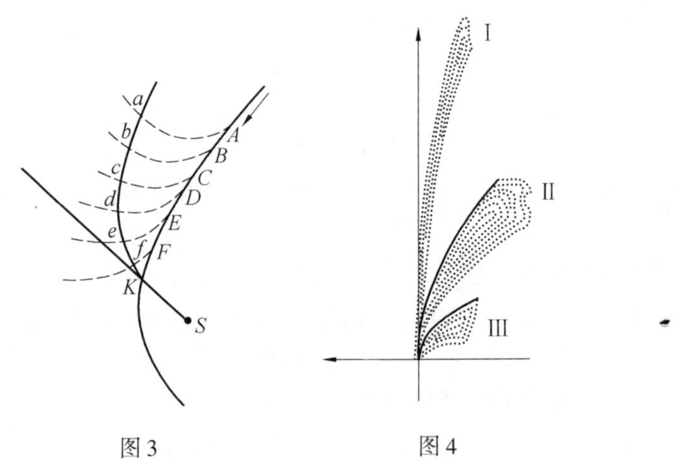

图3　　　　图4

对第一类型的彗尾,太阳的斥力超过太阳的引力约20到200倍;对第二类型,斥力等于引力的0.5到2.5倍;而对于第三类,则是从0.1到0.3倍.质点从彗核喷出来的速度对于各类型也各不相同:第一类型彗尾是每秒数千米,而对于其余两类,常不超过1 km/s.斥力越小,彗尾的曲率越大.第一类型的彗尾几乎是直线,方向很接近于彗星的向径延长的方向.第二类型的彗尾朝着和彗尾运动相反的方向向后弯曲.而在第三类型里,这个弯曲度更大.

斥力的大小和引力不同,不仅和尾部质点的质量有关,而且也和它们的物理及化学的本质有关.第一类型的彗尾是由较轻的物质所构成的(根据近代的材料,它们是由一氧化碳和氮的稀薄气体所构成的),而第二类型和第三类型,则是比较重的东西(宇宙尘).这样便可以说明,例如在同一个彗星上同时出现几条属于不同类型的尾.

在一些彗星上也看到有一种面向太阳的短扇形的尾.这种彗尾称为

反常彗尾. 显然地,它们是由更大的质点所构成的,太阳斥力对这些大质点的作用是很微弱的.

在近年来的彗星照片上发现了在若干第一类型的彗尾中,斥力比引力大数千倍. 在其他方面,布烈基被的理论能够很好地说明一切观测到的现象. 近年来,彗尾的理论由苏联科学院通信院士奥尔洛夫(С. В. Орлов)发展得更完善. 奥洛夫因其对彗星的研究工作而获得斯大林奖金.

奥尔洛夫根据力学和物理学的最新成就完成了彗星形状的现代理论,他考虑到彗星的光度测量和分光研究的结果,而创造了彗尾的物理理论. 他指出彗星头部的物质不仅为太阳所排斥,而且也为彗核所排斥. 这样便使得我们有可能近似地确定核的大小和质量;对于质量所获得的数值是地球质量的 10^{-15} 到 10^{-10}. 他精确地讨论了不同类型的彗尾的形成过程,因而得出了这样的结论:第一类型的彗尾是由气体组成,而第二、第三类型的彗尾是由尘粒组成的.

在形成第一类型彗尾的时候,某种气体分子是以连续气流的形式,在不同的时间由彗核飞出来的,但是在同一个斥力的作用下运动着,所以所有这些质点都位于等力线上,也就是说位于"相同的力"的曲线上. 因为从彗核里有不同化学性质的气体飞出来,而各种气体的分子应该位于它们自己的等力线上,所以第一类型的彗尾应该就是这种等力线的线束.

假如从彗核一下就飞出来了许多种大小不同的固体质点(尘粒),也就是说发生了爆发,那么这种同时被抛掷出来的质点群在继续运动之际应该对观测者而言是一条延长的曲线,位于曲线一端的应该是受到最大排斥加速度的质点,而位于另一端的是受到最小加速度的质点. 这种曲线叫作等时线,也就是说"相同的时间"(由彗核飞出来)的曲线. 第二类型和第三类型的彗尾是由一系列这种等时线所形成的. 它们是由尘粒组成的,而且反射太阳光谱.

11. 斥力的本质

只对第二和第三类型的彗尾已经弄清楚什么是斥力的本质——这是太阳光线的斥力,或者所谓光压. 像著名的俄罗斯物理学家列别捷夫(П. Н. Лебедев)的实验所证明的一样,光线可以在被它所照射着的物体上产生压力. 对于普通大小的物体这个压力是微不足道的,但是对于非常小的物体(尘粒、气体分子),光压力超过太阳引力好多倍,所以特别当质点跑近太阳的时候,它将被太阳所排斥,而不是被它所吸引.

事实上,假如在离太阳同一距离处有密度等于一的大小各异的一些小球,那么球表面越大,光压力也越大. 因此假如其他的条件相同,那么光

压力和球半径的平方成正比例

$$D=c_1r^2$$

太阳引力却和小球的质量成正比例，在密度不变的条件下，引力便和体积也就是说和半径的立方成正比

$$F=c_2r^3$$

斥力和引力的比率可用下式表示

$$\frac{D}{F}=\frac{c_1}{c_2}\frac{1}{r}$$

常数$\frac{c_1}{c_2}$是很小的，所以通常斥力和引力比较起来是微不足道的.但是正像上面的公式所表示的一样，假如球半径很小，斥力甚至可能超过引力.假如质点的直径只有一厘米的数十万分之一（尘粒）的话，就会发生这样的情况.

对于第一类型的彗尾，所观测到的斥力远超过理论值.毫无疑问，在这种彗尾中光压也起着重要作用，但是不能够只用光压力来解释第一类型彗尾的形状.显然，正如罗蒙诺索夫早就指出的，在这里和电有关的力量也起着作用.

12. 几个显著的彗星

（甲）哈雷彗星.这是第一个被预言其回转来的彗星.当这颗彗星在1682年出现的时候，哈雷注意到它的轨道和在1607年及1531年所观测到的彗星的轨道相类似.这使他想到所有三颗彗星只不过是同一颗彗星的三次独立的出现，它以75年左右的周期沿着很扁长的椭圆而旋转，因此他决心预测它将于1758年再度回来.假如要更精密地指出它回来的时间，必须计算在75年中的行星摄动，直到哈雷去世后这项工作才被完成.由于木星和土星摄动的结果，彗星应该延迟一年多才回来.和计算完全吻合，它在1759年再度回来，以后在1835年又回来一次.最后一次且是最显著的一次，它于1910年再度回来，在5月19日彗星通过太阳和地球中间，和地球的距离只有两千四百万千米左右，但是因为它的尾部比这个距离长得多，而且彗尾总在和太阳相反的方向，所以地球穿过彗尾.由于彗星和地球相距很近，彗尾的角长度竟超过140°.从古代的史书中可以找出2 000年中哈雷彗星回来的年代.在这2 000年中它的亮度即使减少了一些，也减少得不多，而周期则因摄动而变化于$74\frac{1}{2}$到$79\frac{1}{2}$年之间.这颗彗星下一次将于1986年回到太阳附近来（图2）.

（乙）恩克——巴格隆彗星.这是从18世纪末就已经知道的微弱的周

期彗星,它的运动特别有趣.它具有最短的周期——只有 $3\frac{1}{3}$ 年.最初证明它的周期性的恩克发现在它的运动中有加速度:就是说,每次旋转比前一次短差不多3小时.这样,它的轨道的大小逐渐减少,假如这种加速度一直继续下去,彗星最后一定会落到太阳里去.后来俄罗斯天文学家巴克隆证明彗星的加速度也在变化:现在加速度比20世纪中叶小得多,但是这种减少不是连续的,而是突然的跳跃的——在最近50年内减小了三次或者四次.加速度和它的变化都还是未解之谜,可能的原因是彗星和陨星群相碰撞.直到现在还没有找到别的彗星,其运动具有类似的不规则性.

(丙)彗星1882Ⅱ及其家族.这个彗星在它被发现以后不久便通过了太阳圆面(和1910年哈雷彗星的情况相类似).这时候它(和哈雷彗星不同)是如此地明亮,以致在白天在太阳近傍也可以用肉眼看见它,用望远镜观测它进入太阳圆面,就完全像在观测月掩星一样.它消失在太阳圆面上,它的周期在800年左右.已经被指出,在同一个轨道上还有几颗(计有7颗)很亮的彗星在运行着,它们在最近两世纪之内分别出现于不同的年代.它们在近日点的时候都异常接近太阳,当1843年彗星通过太阳表面的时候,和太阳表面相距只有130 000千米(也就是说0.1个太阳直径),所以它离开太阳表面比许多日珥的顶部还要近,这颗彗星也可以在白天看见.

显然这颗彗星“家族”的所有成员都是由一颗大彗星组成的,这颗彗星在某一个时候分裂为许多部分.这个分裂甚至在我们眼前继续进行:彗星1882Ⅱ的核最初是圆的,后来变扁,最后分成四个独立的核.当我们还能看见这些彗星的时候,它们之间的距离不断地增加,因此每一个彗核是沿着自己的轨道运动着的.假如这颗彗星一直延续到27世纪,那么当它再度回到太阳来的时候,将变成四颗彗星,一颗跟着一颗出现,彼此相隔数十年.

(丁)比拉彗星.比拉周期彗星是彗星分裂的更显著的例子.这颗微弱的彗星在1846年于数日之内分裂成两颗差不多同样形状的小彗星:它们之间的距离不断增加,当我们还能看见这些彗星的时候,它们之间的距离达到了差不多等于地球和月球间的距离.在1852年两颗彗星都再度回来,但是在这时候它们之间的距离差不多增加了10倍.从那一次起,我们从未再看到这些彗星,然而照理它们每隔 $6\frac{1}{2}$ 年便应该回来一次,所以我们认为它们已经消失了.在1872年彗星应当很接近地球,然而我们没有发现它.可是在同一年11月27日的夜里,在许多地方都看到了真正的“流星雨”,后来证明它们都是沿着比拉彗星的轨道飞行的.这颗彗星已经完全碎裂,而只剩下一个“流星群”使我们记起它,这个流星群直到现在还每年在十一月底出现,不过一年比一年来得微弱了.

(戊)列克塞尔彗星.在1770年出现了一颗彗星,肉眼可以看见,一直被观测了几个月.最初根据观测确定了它的轨道是抛物线形,因为在那个时候除了哈雷彗星,还没有看到别的具有椭圆轨道的彗星.俄国院士列克塞尔的更精确的计算证明,事实上这颗彗星是在周期只有$5\frac{1}{2}$年的椭圆轨道上运动着的,在1781年它应该会重新到达极有利于观测的情况,但是这个预言却没有实现,而一般说起来已不能够再看到这颗彗星.只是到列克塞尔去世后,拉普拉斯和勒威耶才证明这颗彗星以前是沿着几近于抛物线的轨道而运动着的,但是到了1767年,它走到很接近木星的位置,以致木星的引力把彗星的轨道改变成比较小的椭圆,而这正是列克塞尔所完全正确地计算的.过了12年(木星的周期),彗星再度走近了木星,这一次走得更近,以致穿过了木星的卫星.这时候由木星所引起的加速度的方向和第一次相遇时相反,加速度大大地增加了彗星所走的椭圆轨道的长径,以致彗星不能被地上的观测者所看到.

(己)紫金山-阿特拉斯彗星①.据2024年10月13日消息,号称2024年最值得期待的彗星——紫金山-阿特拉斯彗星经过近地点,距离地球只有0.472天文单位,约为4 390万千米.北京天文馆专家介绍,从现在起一直到26日,将是观测这颗彗星的最佳时间段.这颗彗星由中国天文台名字命名——紫金山-阿特拉斯彗星(国际编号C/2023 A3),于2023年1月9日被中国科学院紫金山天文台近地天体望远镜首次观测到,是一颗带有黄色彗尾的彗星.

据了解,紫金山-阿特拉斯彗星还被称为6万多年后才能再见的彗星.但天文专家认为,这只是人们的一个美好愿望,更大概率是这颗彗星将一去不复返了.按照德国天文学家开普勒提出的行星运动规律,所有行星绕太阳的轨道都是椭圆,太阳在椭圆的一个焦点上.按照这个规律,理论计算上,紫金山-阿特拉斯彗星将在六七万年后再回到现在这个位置,但后来的科学研究发现,这其实也属于一个特例,对行星或周期性彗星来说,它是适用的,对太阳系其他天体来讲,它们的运动轨迹都是属于圆锥曲线,就是椭圆、抛物线和双曲线.不过,就紫金山-阿特拉斯彗星目前已知的运动轨迹,这个椭圆轨道过于扁长.

---

① 摘编自“科普世界”.

13. 彗星和其他宇宙体的碰撞

因为彗星的数目非常多,它们的轨道又是各种各样的,而且又不断地在变化着,所以彗星和别的星体相碰撞,理论上是可能的,不过实际上这种碰撞应当是很少发生的,因为即使很大的彗星,其大小如果和行星际的距离相比仍是微小得很.特别地讲,彗星和地球的碰撞也是可能的,但是这种碰撞不会带给地球任何损害.假如地球和彗星尘群相碰撞,那么不过出现异常显著的"流星雨"而已.假如地球和彗核相碰撞,那么会在地球表面上落下许多陨星,这些陨星假如质量比较大,说不定会引起一些不幸的事故,但是谈不到什么大灾难.在地球久远的地质年代中,这种碰撞可能曾经发生过.此外有人假定1947年2月12日落在远东的锡霍特·阿林的陨星曾经属于某一颗彗星的核,也可能曾经是一个不大的小行星.可是它和地球的碰撞除破坏了一些地方的森林以外,并没有任何其他的后果.

至于地球和彗尾相遇的问题,因为彗尾极端稀薄,所以这种相遇发生时一定谁也不会觉察到.

彗星掉到太阳里去也可能发生,但是这也不会生出大灾难.假如下落的速度超过600 km/s,自然会发出大量的热,但是因为太阳表面是极度稀薄的,所以在彗星的动能转变成热能以前,彗星便已深深地穿入太阳的内部了,并且彗星下落之际所放出的热量和太阳所辐射出来的能量相比是微不足道的,所以太阳表面的温度丝毫也不会上升.假如彗核是由小质点构成的,那么由于它们太小,有一些小质点会在尚未落进太阳以前便被太阳光线的斥力所抛掷开(第11部分),又有一些小质点因为太阳的灼热也会变成了气体.

14. 彗星的起源问题

所有被观测到的彗星现象证明这些星体都在相当迅速的毁灭中.毁灭是循着许多不同途径而发生的.

(1)彗星碎裂:由一颗大彗星裂成若干小彗星(比拉彗星,彗星1882 Ⅱ等).

(2)彗星扩散:构成彗核的流星体逐渐分布到更大的体积里或者完全飞离彗星.

(3)彗星蒸发:每次当彗星走近太阳的时候便形成了彗尾,组成彗尾的蒸气和气体再也不会回到彗星里去.很久以来便用这一事实来说明为什么所有的短周期彗星的亮度都很弱,并且无疑地其中有许多尚在不断减弱之中.

如果假定从前这种减弱进行得和现在一样快,那么便可获得一个结论,就是在数世纪以前,有若干微弱的周期彗星应该容易被肉眼所看见.然而在古代天文学家和编年史者的记录里却没有谈到任何一个这样的彗星.因此有人提出这样一个假设,就是彗星的形成远较行星之后,而它们的形成很可能目前还在进行.

因为许多彗星轨道的远日点位于木星轨道附近,所以拉格朗日还提出了一个假设,说这些彗星是从这颗巨行星内部喷发出来的东西.最近伏谢赫斯维亚茨基(С. К. Всехсвятский)支持拉格朗日的假设.有利于这个假设的若干论证之一是:彗星的化学成分和木星大气的成分相近.不利于这个假设的最重要的反驳是:为了能够形成彗星,被抛掷出来的物质,应该从非常浓密的木星大气以大于 60 km/s 的速度飞出来才行,不然它或者掉回到行星里去,或者只形成了卫星.但是很难想像能够引起如此大的速度的力量.

当代奥尔洛夫的陨星假设显得更近乎事实.根据这个假设,彗星是由两个天体,比方说,小行星和大陨星相碰撞而形成的.这样碰撞之后,直径数千米的行星应该会飞散成小碎片,这些碎片便会在太阳系中沿着一切可能的轨道旋转.设想碎片群被许多尘粒所伴随沿着这种轨道飞行,然后走近了太阳,被太阳光所加热.温度上升的结果使碎片开始喷射出气体,陨星群便被气体和尘粒的包层所包围,这样便形成了彗星.

15. 流星

行星际空间,至少在地球轨道附近,含有大量的小而暗的物体与尘粒,它们在太阳引力作用下在各种可能的轨道上运转.它们的科学名称是流星体.地球在运行之际总是和它们相遇,只是在这种相遇的瞬间,我们才晓得了流星体的存在:它们以巨大的速度从很空的空间飞进地球的大气,由于空气的阻力,差不多立刻就变得灼热而粉碎,或者像普通所说的“烧掉”了.这种短时间的光亮的爆发就是流星的现象.

假如从相距数千米的两个地点观测同一颗流星,那么便很容易看出它在恒星间的视差位移,也就容易算出它距离地面的高度.最后知道所有的现象完成在很高的大气层里:流星远在 100 km 以上的高度便发光,通常消失在高度 70 ~ 80 km 地方.它的看得见的路程的真正长度常在 100 km 以上.从这些观测也可以计算出流星在地球大气中的速度(地心视速度).由于空气的阻力,这个速度应该比流星体在没有空气的空间中相对于地球的运动速度(地心的宇宙速度)小得多.

在没有月光的晴朗之夜,每一小时平均可以看见四颗到五颗流星.但是这个数目随一日内的时刻和季节而变化(周日变化和周年变化):在子夜以后,流星出现次数要比子夜以前差不多多两倍,秋天流星比春天多.这个现象以前被认为是证明流星是地球本身的现象,但是斯基阿巴莱利于1967年却指出这恰好证明它们的宇宙(地球以外)来源.事实上,假如地球是不动的,那么流星体从所有各方向落下的数目应该差不多是相同的.但是因为地球以将近30 km/s的速度绕太阳旋转,所以在地球上不同的地点应该会落下不同数目的流星:地球向前走,在前面的半个球面上落下比后半个球面上更多的流星.由于地球在自转,地面上各点在早晨六点钟左右出现在地球的前面.在这时流星的数目应该最大.这样便可以说明流星的周日变化.

在某一时刻地球的轨道运动所指向的那点称为奔赴点,该点位于黄道上,在太阳西面90°.大部分的流星应该出现于天空上这一点的附近,奔赴点越高出地平,所看到的流星便越多.这样便可以说明周年变化,奔赴点在秋天位于夏至点附近,因此比起春天来(就北半球而言)要在地平上高得多.

16. 流星群

每天夜里都可以看到的向着所有方向运动的流星称为偶发流星.周期流星便和它们不同,每年只在一定的日子里出现而且沿着一定的方向运动.就是说,假如我们在恒星图上记下这些流星的视轨迹而且向后面延长,那么所有的轨迹会大致聚于一点.

这些好像从同一点飞出来的流星构成所谓流星群,流星从那里朝各方向飞出去的那一点称为流星群的辐射点.相对于恒星而言这一点有一定的位置,和恒星一样东升西落.

现在已知的流星群有几百个.它们就以辐射点所在的星座的名字来命名,例如依照英仙、狮子、天龙这些星座的名字命名为英仙流星群、狮子流星群、天龙流星群等.

就流星群出现的情况来讲,它们之间相差是很大的.例如英仙流星群每年所看到的流星数目大概差不多(对于任何一个地点的观测者,每小时可看到40到50颗流星),而且最大数目的流星群是在8月11日和12日看到的.狮子流星群一般来讲数目是不多的,但是也有几年它曾在一小时之内出现数万颗甚至数十万颗,变成真正的流星雨(1799年、1833年、1866年).

这些流星雨是最壮丽的天象之一.根据一些观测者的描写"天空发亮如强大的闪电","星散布开来,像大雪暴时的雪球一般",或者"假如把所有同时看见的流星逐一计数的话,恐怕要花费几小时”.而且整个现象是完全不声不响地进行着的,没有一颗流星掉落到地面上来.

偶发流星在空间中相互间毫无关联地飞翔,反之,周期流星全体构成一群.当地球每一次和这种群相遇的时候,立刻便可以看到许多流星.该群的所有流星都是按照平行轨道飞行的,可是为了透视的关系,这些轨道看起来像是由同一点(流星群的辐射点)向四面八方扩散出去似的.

因为每一场流星雨总是在一年内的同一个日子里看到的,这表示地球每年在它自己轨道的同一点上和流星群相遇.每年类似的相遇可以这样来解释:流星群在椭圆轨道上绕太阳运行,这个轨道和地球轨道相交于一点,而且流星群分布于整个轨道之上,因此流星群具有一个连续椭圆环的形状.地球每一年所遇到的应该是环状流星群的新的部分,环的厚度是很大的,因此地球可能要经过数天才通过这一流星群.

假如流星体是或多或少均匀地分布在整个轨道之上,那么地球每一年会遇到大致同等数量的流星体,英仙流星群就是这样(图5,地球和英仙流星群的轨道).

反之,狮子流星群像云一样地密集在自己轨道上的某一部分,地球要每隔33到34年才和它相遇一次,于是在那个时候才出现特别多的“流星雨”.显然地,流星群也以同一周期绕太阳旋转一次.

密集部分沿着轨道延伸至整个椭圆的$\frac{1}{10}$乃至$\frac{1}{12}$左右,所以流星群通过和地球轨道的交点需要一年以上的时间.因此可以一连两年甚至三年看到很多的流星雨(最近一次是在1866年至1868年).在流星群的内部,甚至在它最密集的部分,个别流星体之间的距离是很大的,平均在边长

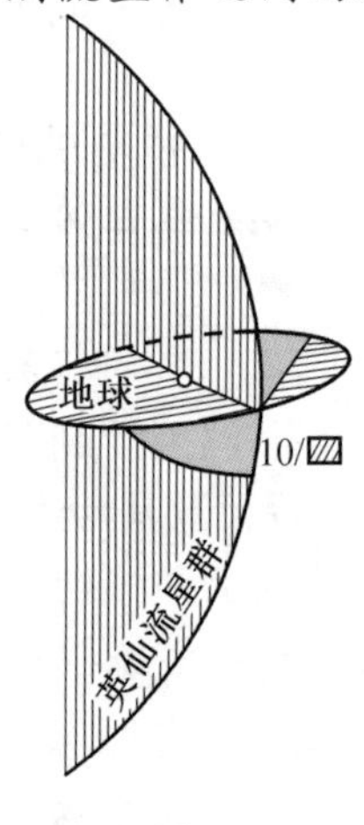

图5

150 km的立方体内才有一个流星体. 但是因为密集部分的直径将近200 000 km(假如连周围比较稀薄部分一起计算在内将达数百万千米), 所以流星体的总数量是很庞大的.

17. 流星轨道的测定

假如不要求很高的精密度,这是一个比较容易的问题. 它在一定时刻和地球相遇,这种情况便给出流星在空间中的精确位置,因为地球相对于太阳的位置是已经知道的. 地球相对于太阳的速度也是已经知道的. 假如把这个速度和流星体相对于地球的速度(已做空气阻力作用的改正)几何地相加起来,便可以求得相对于太阳的速度. 这样的计算证明就大部分的流星而言,这个日心速度或宇宙速度在它和地球相遇的点上将近42 km/s,也就是说等于离太阳那么远的抛物线速度 $v_0:v_0=v\sqrt{2}$,这里 v 是"圆"速度,和地球的速度很接近. 这样流星体在该点的速度可以认为是已知的. 辐射点的位置又告诉我们流星体的速度的方向. 知道了在某一点上速度的大小和方向,便可以用作图或计算来决定抛物线轨道. 当然,流星体真正的轨道不是抛物线而是椭圆或者双曲线. 但是要精密地计算这些轨道的要素,只有在很难得的情况下方才可能,例如,或者已经知道沿着椭圆轨道旋转的时间(例如天龙流星群或者狮子流星群),或者已经能够精密地计算当流星体和地球相遇时刻的相对于太阳的宇宙速度(第14部分).

对于一些看见它落下来的陨星,当我们进行了这一类的计算之后,好像这样几十个陨星体的轨道都是双曲线轨道,因为它们的宇宙速度都大于42 km/s. 从这里便导出一个结论:陨星是从恒星际空间飞进太阳系来的. 但是后来更精密的计算却推翻了这一点,速度好像是小于42 km/s. 现在认为所有流星体都是属于太阳系的.

18. 流星和彗星的关系

在1867年证明了八月的流星群,就是英仙流星群,运行在一个周期彗星(1862Ⅲ)的轨道上,周期约120年. 在同一年又被指出狮子流星群在另一个周期等于 $33\frac{1}{4}$ 年的彗星(1866Ⅰ)的轨道上运行,而11月流星雨重现的周期也大致是 $33\frac{1}{4}$ 年. 这种关系证明:这些流星群是由彗星所形成的,也就是说,彗星是不断在瓦解中的. 形成彗星的流星体(第9部分)慢慢地分布到整个轨道之上,最后便形成了椭圆的流星环. 因此可以说,英仙流星群比狮子流星群老;狮子流星群还远没有延伸到整个轨道上,还只占有轨道的$\frac{1}{12}$左右.

流星和彗星有密切关系的第三个例子便是比拉彗星和仙女流星群,这个流星群在1872年和1885年11月27日引起了壮丽的"流星雨"(第12部分).近年来又发现了若干类似的情况,例如贾可比尼·秦诺彗星和天龙流星群的关系.显然并不是所有的流星群都由彗星形成,有一系列的流星群和彗星没关系.

彗星分裂是起因于太阳对于彗星上离太阳比较近的部分和比较远的部分的引力作用的差别,换句话说,就是起因于太阳引潮力的作用.假如彗星相当大而且经过近日点时离太阳很近,那么最近的质点和最远的质点比较起来便走在前面,相差很多,于是彗核便沿着它的轨道伸长了(在彗星1882Ⅱ的场合曾观测到这个现象,第12部分),每一次再回来时同样的情况又出现.在这里行星的引力,特别是木星和土星的引力,也起着作用.

布烈基兴指出另一个和上述无关的流星群的起源,就是反常的彗尾(第10部分),这是固体质点从彗星的喷发,当它们从彗核飞出来的时候,像观测所证明的一样,运动得像一个稍微扩散的束.这可以说明这样的情况,就是当流星群和地球相遇的时候,它并不总是很严格地沿着平行轨道而运动的.

读天文学的书可以使你暂时从满脑子金钱、地位、名利的状态中解脱出来.仰望星空,思绪飞扬,而且Andy Warhol(1928—1987)曾说过:

一旦你不再想要某个东西,你就会得到它.

刘培杰

2024年10月13日

于哈工大

国外优秀物理著作
原版丛书（第一辑）

激光及其在玻色—爱因斯坦凝聚态观测中的应用（英文）

Lasers and Their Application to the Observation of Bose-Einstein Condensates

[加]理查德·A. 邓拉普（Richard A. Dunlap） 著

HITP
哈爾濱工業大學出版社
HARBIN INSTITUTE OF TECHNOLOGY PRESS

黑版贸登字 08－2021－034 号

图书在版编目(CIP)数据

激光及其在玻色－爱因斯坦凝聚态观测中的应用＝Lasers and Their Application to the Observation of Bose-Einstein Condensates：英文/（加）理查德·A.邓拉普（Richard A. Dunlap）著.—哈尔滨：哈尔滨工业大学出版社，2024.10
（国外优秀物理著作原版丛书.第一辑）
ISBN 978－7－5767－1341－1

Ⅰ.①激… Ⅱ.①理… Ⅲ.①激光－应用－玻色凝聚－凝聚态－观测－英文 Ⅳ.①O469

中国国家版本馆 CIP 数据核字(2024)第 073588 号

JIGUANG JI QI ZAI BOSE－AIYINSITAN NINGJUTAI GUANCE ZHONG DE YINGYONG

策划编辑　刘培杰　杜莹雪
责任编辑　刘立娟　刘家琳　李　烨　张嘉芮
封面设计　孙茵艾
出版发行　哈尔滨工业大学出版社
社　　址　哈尔滨市南岗区复华四道街 10 号　邮编 150006
传　　真　0451－86414749
网　　址　http://hitpress.hit.edu.cn
印　　刷　哈尔滨博奇印刷有限公司
开　　本　787 mm×1 092 mm　1/16　印张 63.75　字数 1 195 千字
版　　次　2024 年 10 月第 1 版　2024 年 10 月第 1 次印刷
书　　号　ISBN 978－7－5767－1341－1
定　　价　378.00 元（全 6 册）

（如因印装质量问题影响阅读，我社负责调换）

Contents

Preface

Lasers, which were developed in the early 1960s, have not only played an important role in the advancement of scientific knowledge in a wide variety of fields, but have found commercial applications in a multitude of devices that have become important in our daily lives. These applications include CD/DVD players, laser printers and fiber optic communication devices. While these devices depend largely on the monochromaticity and coherence of the light which lasers produce, other well-known applications, such as laser machining and laser fusion depend on the intensity of laser light. The first part of the present book overviews the physics of lasers and describes some of the more common types of lasers and their applications.

Part II of the book looks at the phenomenon of Bose–Einstein condensation. These condensates represent a state of matter that exists in some dilute gases at very low temperature. This state was first predicted in the 1920s by Satyendra Nath Bose and Albert Einstein. Bose–Einstein condensates were first observed experimentally in 1995 by Eric Cornell and Carl Wieman at the University of Colorado and shortly thereafter by Wolfgang Ketterle at the Massachusetts Institute of Technology. The experimental techniques used to create a Bose–Einstein condensate provide an interesting and somewhat unconventional application of lasers; that is, the cooling and confinement of a dilute gas at very low temperature.

Acknowledgements

I am grateful to Nicki Dennis for her support and encouragement in her role as Commissioning Editor and Karen Donnison for all her work as Permissions Editor. I am also grateful to Chris Benson at IOP Publishing for his careful and efficient editing of the manuscript, and Melanie Carlson and Brent Beckley at Morgan & Claypool for handling the production of the printed version of the book.

Author biography

Richard A Dunlap

Richard A Dunlap received a BS in Physics from Worcester Polytechnic Institute in 1974, an A.M. in Physics from Dartmouth College in 1976 and a PhD in Physics from Clark University in 1981. Since receiving his PhD, he has been on the Faculty at Dalhousie University. He was appointed Faculty of Science Killam Research Professor in Physics from 2001 to 2006 and served as Director of the Dalhousie University Institute for Research in Materials from 2009 to 2015. He currently holds an appointment as Research Professor in the Department of Physics and Atmospheric Science. Professor Dunlap has published more than 300 refereed research papers and his research interests have included, magnetic materials, amorphous alloys, critical phenomena, hydrogen storage, quasicrystals, superconductivity and materials for advanced batteries. Much of his work involves the application of nuclear spectroscopic techniques to the investigation of solid-state properties. His previous books include; *Experimental Physics: Modern Methods* (Oxford 1988), *The Golden Ratio and Fibonacci Numbers* (World Scientific 1997), *An Introduction to the Physics of Nuclei and Particles* (Brooks/Cole 2004), *Sustainable Energy* (Cengage, 1st edn 2015, 2nd edn 2019), *Novel Microstructures for Solids* (Morgan & Claypool 2018), *Particle Physics* (Morgan & Claypool 2018) and *The Mössbauer Effect* (Morgan & Claypool 2019).

Part I

Lasers

IOP Concise Physics

Lasers and Their Application to the Observation of Bose–Einstein Condensates

Richard A Dunlap

Chapter 1

The basic physics of lasers

1.1 Introduction

The term laser is an acronym for '**l**ight **a**mplification by **s**timulated **e**mission of **r**adiation'. The first laser was constructed in 1960 by Thomas Maiman on the basis of theoretical work by Charles Townes and Arthur Schawlow. The operation of the laser follows from the same basic principles as its predecessor, the maser (**m**icrowave **a**mplification by **s**timulated **e**mission of **r**adiation). The name laser was originally used to describe devices that worked in the visible portion of the electromagnetic spectrum. However, the term 'light' in its name is now used in a broader sense to indicate any frequency of electromagnetic radiation. Hence, devices which work in the non-visible part of the spectrum are sometimes called 'x-ray lasers', 'ultraviolet lasers', 'infrared lasers', etc as appropriate, although the term maser is still used for devices that operate in the microwave region. The present book deals primarily with optical lasers.

Lasers are distinguished from other sources of light by the fact that the light that they emit is coherent, that is the electromagnetic fields associated with the various photons in the beam are in phase. More precisely, they have a long coherence length, meaning that the beam remains coherent for a large spatial distance. The fact that the light from a laser is coherent means that it can be very intense and directional, and it also requires that the light be polarized and monochromatic. While 'monochromatic' would imply a single frequency or wavelength, it is necessary to consider the extent to which light needs to be monochromatic in order to satisfy the coherence condition.

We begin by considering the ways in which visible light can be produced and how the production method affects the degree of monochromaticity.

doi:10.1088/2053-2571/ab2f2fch1

1.2 Optical spectra

Optical photons are most commonly emitted by materials as a result of electronic processes in the material and can be classified in terms of those which produce a broad distribution of wavelengths and those which are, more-or-less, monochromatic. Analogous to the situation which produces x-rays by bremsstrahlung, thermal energy results in the emission of photons with a broad distribution of wavelengths, or a continuum spectrum. This is black body radiation and the wavelength of the maximum in the energy spectrum, in nm, is given by Wein's displacement law as

$$\lambda_{\max} = \frac{b}{T} \tag{1.1}$$

where T is the temperature in Kelvin and b is the displacement constant 2.8978×10^6 K nm. An incandescent lamp produces light that is a good approximation of black body radiation as illustrated in figure 1.1 where the maximum in the spectrum at about 660 nm corresponds to a temperature of approximately 4400 K. As the visible portion of the electromagnetic spectrum covers the range of wavelengths from about 390 to 700 nm, the figure shows that the incandescent lamp emits most of its radiation in the infrared portion of the spectrum.

Analogous to the discrete line spectra of x-ray sources which are the result of specific electronic transitions, well defined spectral lines in the optical region can also result from electronic transitions. Because optical photons are of much lower energy than x-ray photons, the electronic transitions involve outer shell electrons with small binding energies, rather than inner shell electrons, as is the case for x-ray production. The common fluorescent lamp produces spectra which contain well defined peaks, as illustrated in figure 1.1.

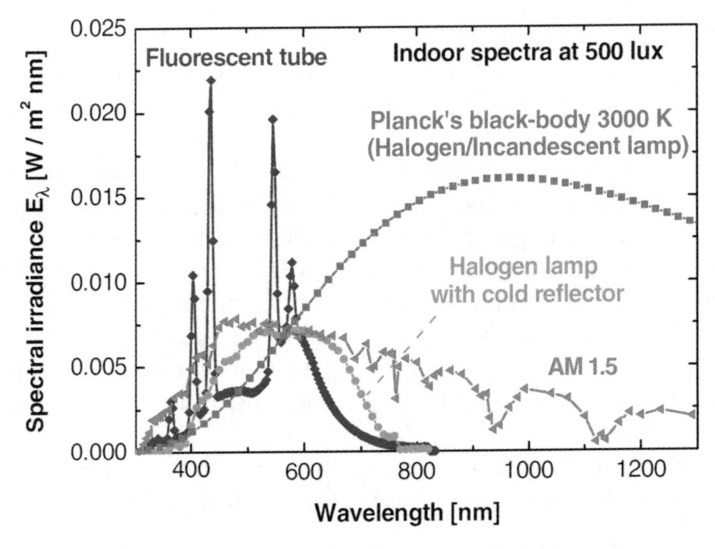

Figure 1.1. Spectra of incandescent, halogen and fluorescent lamps. AM 1.5 is the spectrum of simulated natural daylight. (Figure 2 reprinted from Virtuani *et al* 2006. Copyright (2006), with permission from Elsevier.)

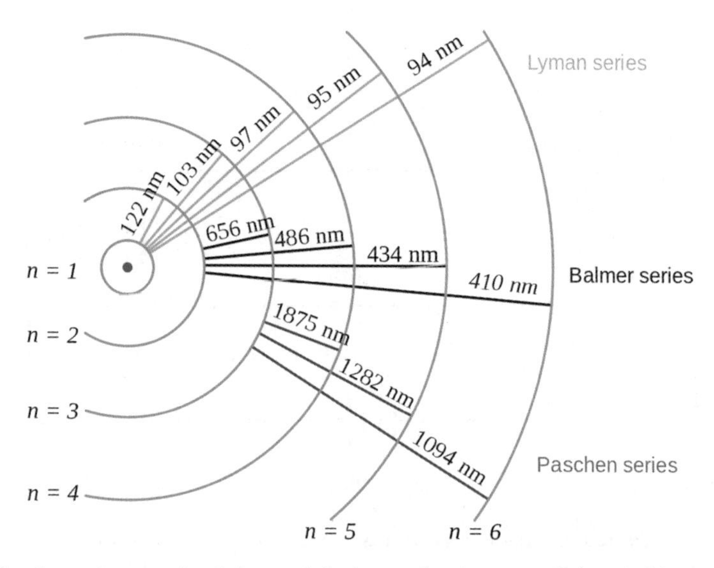

Figure 1.2. The electronic energy levels in atomic hydrogen showing some of the transitions that give rise to photon emission. The names of the various series of transitions are indicated. Szdori / wikimedia commons / CC-BY-SA 3.0 / https://creativecommons.org/licenses/by-sa/3.0/deed.en https://commons.wikimedia.org/wiki/File:Hydrogen_transitions.svg.

A quantitative example of optical line spectra follows from the calculated energy levels of a hydrogen atom, as illustrated in figure 1.2. The transitions are identified by the energy level of the final (lowest) level involved in the transition and are named after researchers who studied optical spectroscopy in the early 20th century. The energy, E, of the photon that is emitted is given as

$$E = \frac{m_e e^4}{32\pi^2 \varepsilon_0^2 \hbar^2}\left[\frac{1}{{n_2}^2} - \frac{1}{{n_1}^2}\right] \tag{1.2}$$

where n_2 is the quantum number for the initial (higher energy) state and n_1 is the quantum number for the final (lower energy) state. The wavelength of the emitted photons is related to the energy of the transition by

$$\lambda = \frac{hc}{E} \tag{1.3}$$

When the energy is in eV and the wavelength is in nm, then the constant $h.c. = 1.24 \times 10^3$ eV nm. The range of visible wavelengths corresponds to a range of energies from 3.18 to 1.77 eV. The Lyman series involves transitions that end at the ground state ($n_1 = 1$) and produce photons in the far ultraviolet. It is some of the Balmer series transitions (which end at $n_1 = 2$) that have energies corresponding to optical photons.

A typical discharge tube for producing emission spectra from a gas is illustrated in figure 1.3. A high voltage discharge excites electrons from their ground state to

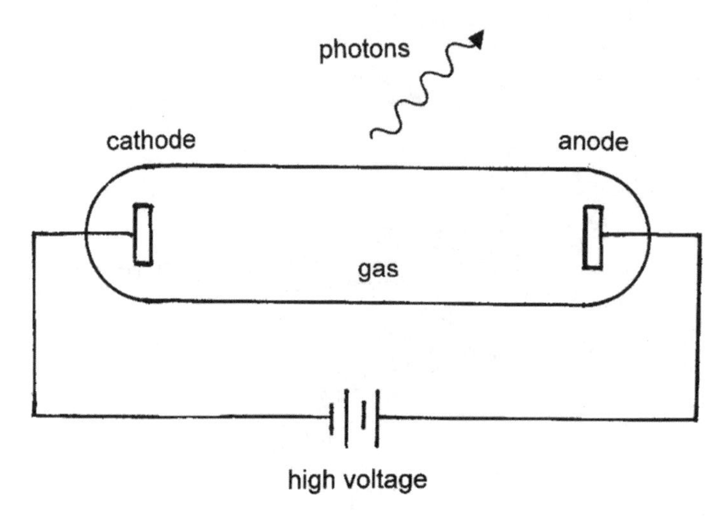

Figure 1.3. Typical gas discharge tube for producing line spectra.

excited levels. These excited levels are unstable and will spontaneously decay back to a lower energy excited state or the ground state, thereby emitting energy in the form of photons.

In cases where an atom is part of a molecule, then the details of the electronic energy levels become more complex. The energy of the system is not only a result of the energy of the electrons in their quantized energy levels but also results from the rotational and vibrational energy associated with the molecule. These rotational and vibrational levels are quantized, where the rotational levels are assigned a quantum number J and the vibrational levels are assigned a quantum number v. Figure 1.4 shows possible transitions involving rotational and vibrational energy levels. Transitions that involve a change in vibrational state, but do not involve a change in rotational state (i.e. $\Delta J = 0$), are referred to as 'Q' transitions. Transitions that involve a change of vibrational state as well as a decrease by one unit of rotational energy (i.e. $\Delta J = -1$) are referred to as 'P' transitions, while those which involve an increase in rotational energy (i.e. $\Delta J = +1$) are called 'R' transitions. These are identified in figure 1.4.

Figure 1.5 shows the measured spectrum for N_2O and illustrates the P-branch, Q-branch and R-branch of the spectrum. It is common to label the horizontal axis of such spectra with the wavenumber (usually in cm^{-1}). As the figure shows, the spectral peaks occur for wavenumbers around 600 cm^{-1}. The wavenumber in cm^{-1} may be expressed in units of energy using the relation that a photon of wavenumber 1 cm^{-1} has the energy of a photon with a wavelength of 1 cm. This relationship is given in equation (1.3) when the constant is expressed in appropriate units $h.c. = 1.24 \times 10^{-4}$ eV cm. Thus, a wavenumber of 1 cm^{-1} corresponds to an energy of 1.24×10^{-4} eV and a wavenumber of 600 cm^{-1} corresponds to an energy of $(1.24 \times 10^{-4}$ eV cm$) \times (600$ cm$^{-1}) = 0.074$ eV. The wavelength in cm is merely the

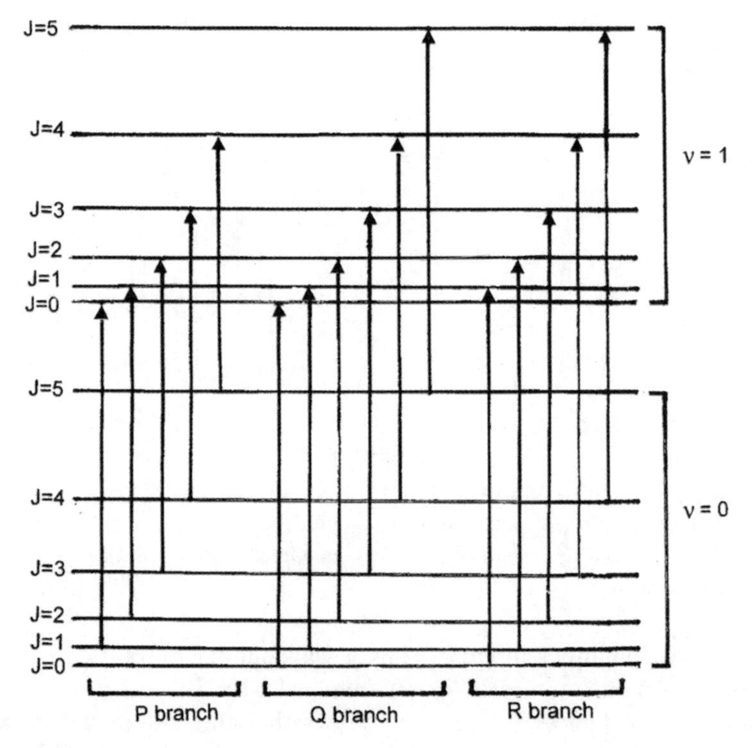

Figure 1.4. Transitions involving vibrational states (designated with the quantum number v) and rotational states (designated with the quantum number J).

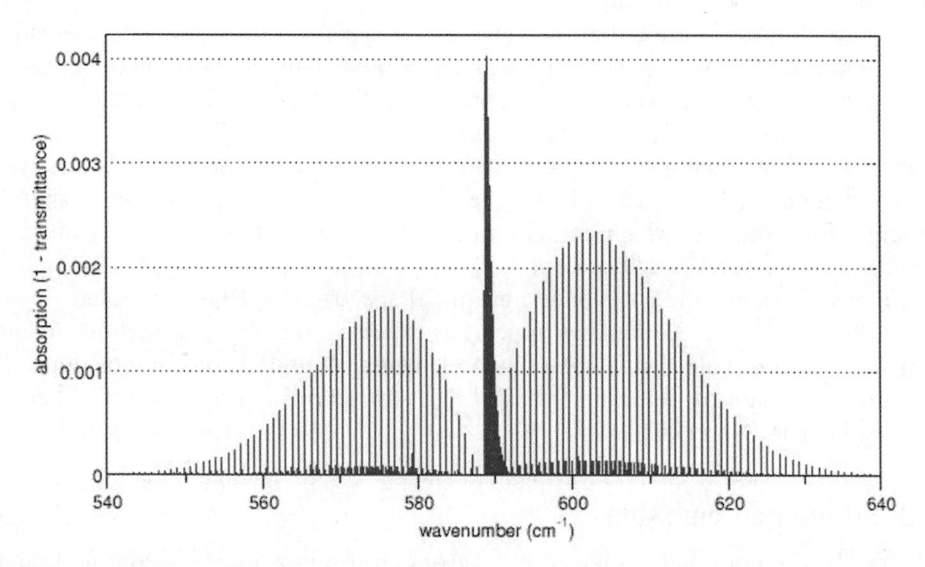

Figure 1.5. Rotational/vibrational spectrum of N_2O. Petergans / wikimedia commons / CC-BY-SA 3.0 / https://creativecommons.org/licenses/by-sa/3.0/deed.en https://en.wikipedia.org/wiki/Rotational%E2%80%93vibrational_spectroscopy#/media/File:Nu2_nitrous_oxide.png.

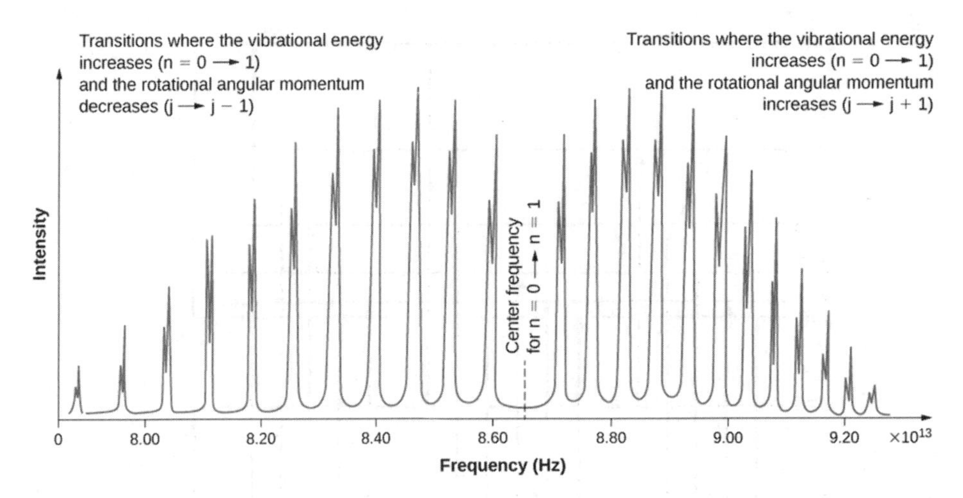

Figure 1.6. Rotational/vibrational spectrum of HCl. CC-BY-SA 4.0. Physics Libretexts.org https://phys.libretexts.org/TextBooks_and_TextMaps/University_Physics/Book%3A_University_Physics_(OpenStax)/Map%3A_University_Physics_III_-_Optics_and_Modern_Physics_(OpenStax)/9%3A_Condensed_Matter_Physics/9.2%3A_Molecular_Spectra.

inverse of the wavenumber in cm^{-1}, thus a wavenumber of 600 cm^{-1} will give a wavelength of $(600\ cm^{-1})^{-1} = 1.67 \times 10^{-3}$ cm $= 1.67 \times 10^{4}$ nm.

A comparison of figures 1.4 and 1.5 shows that the P-branch has the lowest energy and, therefore, appears on the left side of the spectrum. The Q-branch gives rise to the peak at the center of the spectrum. The R-branch corresponds to the right-hand portion of the spectrum.

The rotational/vibrational spectrum of N_2O spectrum illustrates an interesting feature that is often seen in such spectra. The P- and R-branches show the presence of a weak spectral component superimposed on the principal spectrum. Nitrogen consists of two naturally occurring isotopes, ^{14}N (about 99.7%) and ^{15}N (about 0.3%). These two isotopes have different masses and give rise to molecules with different moments of inertia. Thus, the splitting of the energy levels is slightly different for molecules which contain atoms of different nitrogen isotopes and this results in two superimposed spectra.

Figure 1.6 shows the rotational/vibrational spectrum of HCl. It is clear in this figure that the P- and R-branches appear, as expected, but the Q-branch is missing. For a linear molecule consisting of two elements, A and B, of composition AB, quantum mechanical selection rules forbid transitions with $\Delta J = 0$, thereby eliminating the Q-branch.

1.3 Stimulated emission

The previous section has shown how electronic transitions in atoms and molecules that involve discrete quantized energy levels can give rise to well defined spectral wavelengths. While this feature will satisfy, more-or-less, the requirement that a light source produces photons that are monochromatic in order to be classified as a laser, we still need to consider the nature of the electronic transition and the question of polarization. Here we begin with a consideration of the difference between spontaneous emission processes and stimulated emission processes.

Figure 1.7 shows a diagram of a spontaneous emission process. In figure 1.7(a), energy (in this case in the form of a photon) is absorbed by an atom, giving rise to the excitation of one of its electrons. As this excited state is not stable it will, after some time, decay back to the ground state as in figure 1.7(b) re-emitting a photon with energy equal to the difference in the energies of the excited and ground states.

Figure 1.8 illustrates a stimulated emission process. As in the spontaneous emission process, an atom absorbs energy, it is excited, and this excited state decays spontaneously, re-emitting a photon. This re-emitted photon encounters another atom of the same type which is in the excited state and which has not yet decayed spontaneously. The incident photon causes the excited atom to decay before it would have decayed spontaneously. As the two atoms involved are of the same type, they have the same energy levels, and the incident photon and the photon that results from the stimulated emission, will have the same energy (within certain limits, as discussed further below). There are two additional features that are of importance in this process; the two photons are not only of the same energy, but they are in phase and have the same polarization vector. Thus, they satisfy the description of radiation from a laser.

The characteristics of the radiation as described above are that it is monochromatic (to the degree discussed below) and it is intense, as the components of the beam are all in phase and interfere constructively with each other. If this process continues from a collection of the same type of atoms, then the beam will continue to increase in intensity. Light, in which the components of the beam remain in phase with one another over an extended distance, is referred to as coherent.

Two practical problems need to be resolved in order to create the situation as described above. Firstly, we need to ensure that there are a sufficient number of atoms in the excited state (rather than in the ground state) for the stimulated

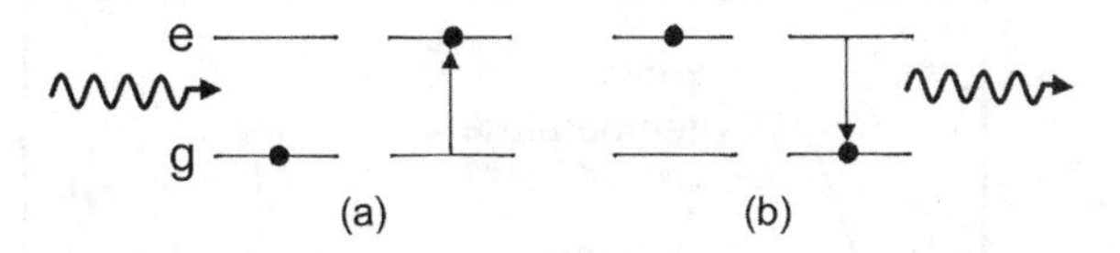

Figure 1.7. Spontaneous emission process; (a) excitation of an atom into an excited state and (b) spontaneous decay back to the ground state accompanied by photon emission.

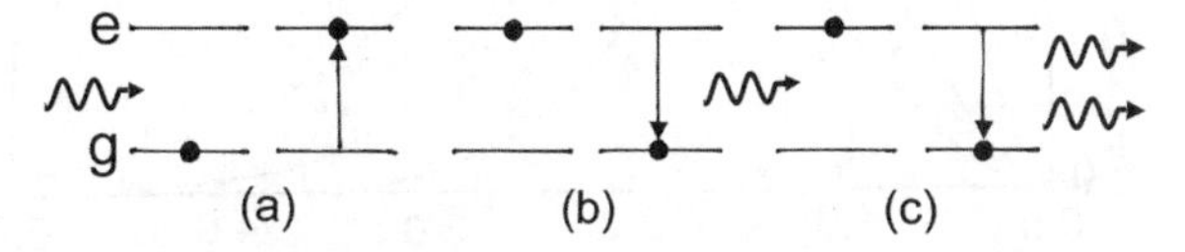

Figure 1.8. Stimulated emission process: (a) absorption of a photon and excitation of electron into excited state, (b) spontaneous decay and emission of photon from unstable excited state and (c) stimulated emission of photon from another atom in an excited state by photon from spontaneous decay in panel (b).

emission to occur frequently, and we need to ensure that there is sufficient opportunity for the beam to gain strength by successive stimulated emissions. The solutions to these two problems are the creation of a population inversion and the creation of an appropriate resonant cavity, respectively. These two requirements are discussed below.

1.4 Creating a population inversion

Consider a simple two-level quantum system as illustrated in figure 1.9(a). At zero temperature all atoms will be in their ground state. However, at any temperature, T, above absolute zero, thermal effects need to be considered. The distribution of particle velocities as a function of temperature is given by the Maxwell–Boltzmann distribution as illustrated in figure 1.10. This distribution has the functional form

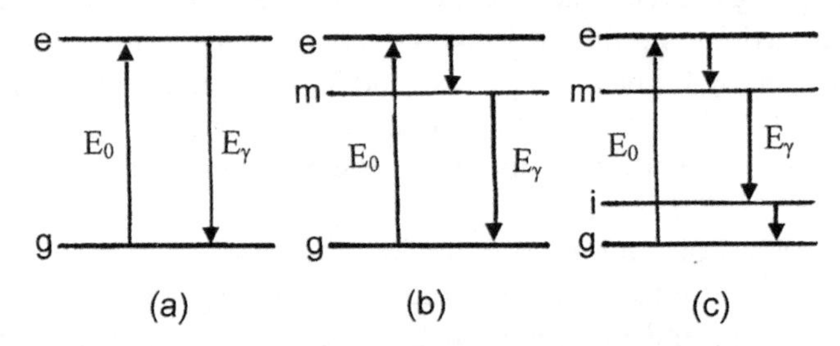

Figure 1.9. (a) Two-level system showing energy absorption, E_0, and emission, E_γ, between the ground state (g) and the excited state (e). (b) Three-level system showing energy absorption between the ground state (g) and the excited state (e), the spontaneous emission between the excited state (e) and a metastable state (m) and stimulated emission between the metastable state (m) and the ground state (g). (c) Four-level system showing absorption between the ground state (g) and excited state (e), the spontaneous emission between the excited state (e) and the metastable state (m), stimulated emission between the metastable state (m) and the intermediate state (i) and spontaneous emission between the intermediate state (i) and the ground state (g).

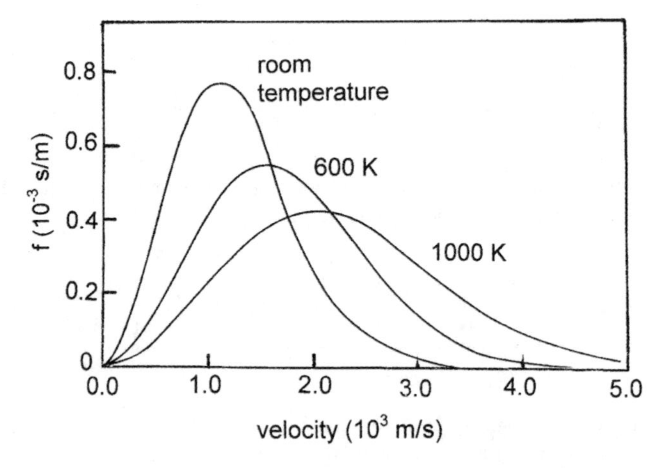

Figure 1.10. Maxwell–Boltzmann distributions at different temperatures.

$$f(v) = \left(\frac{m}{2\pi k_B T}\right)^{3/2} 4\pi v^2 \exp\left(-\frac{mv^2}{2k_B T}\right) \tag{1.4}$$

where $f(v)$ is the fraction of atoms with velocity v and k_B is the Boltzmann constant. The Maxwell–Boltzmann function may also be expressed as a distribution in energy using a change of variables and the relation $E = mv^2/2$, as

$$f(E) = 2\sqrt{\frac{E}{\pi}}\left(\frac{1}{k_B T}\right)^{3/2} \exp\left(-\frac{E}{k_B T}\right) \tag{1.5}$$

For a system consisting of discrete microstates with energies E_i, then the number of atoms in a state with energy E_i follows from the Maxwell–Boltzmann distribution as

$$\frac{N_i}{N} = \frac{\exp\left(-\frac{E_i}{k_B T}\right)}{\sum_j \exp\left(-\frac{E_j}{k_B T}\right)} \tag{1.6}$$

where N is the total number of atoms in the system. A plot of equation (1.6) at room temperature is shown in figure 1.11. It is clear that the state population decreases as a function of increasing energy. This is the normal equilibrium situation. For a population inversion we require that the population of a particular energy level is greater than that for the level below it. For the case in figure 1.9(a) atoms in the ground state can be excited into the higher energy state by inputting an energy greater than the excited state energy, E_0. This can be done by, e.g., photon irradiation or an electric discharge. Excited state atoms will spontaneously decay back to the ground state with a characteristic time (the mean life of the excited state) releasing energy in the form of a photon of energy $E_\gamma = E_0$. If a sufficient number of

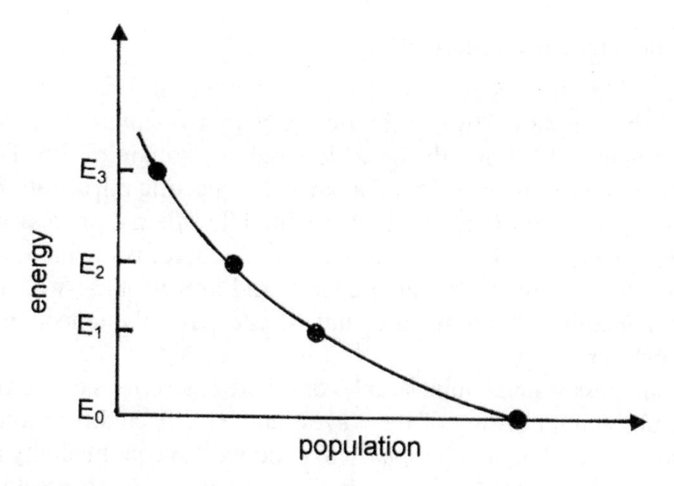

Figure 1.11. Population as a function of energy for the Maxwell–Boltzmann distribution near room temperature.

atoms are excited from the ground state in less than the mean life of the excited state, a population inversion results and stimulated emission can occur. However, this population inversion cannot be maintained as stimulated emissions will decrease the population of the excited state and increase the population of the ground state until the population inversion is eliminated.

The simplest system which is compatible with the requirements for the design of an operational laser is the three-level system, as shown in figure 1.11(b). The short-lived (e.g., $\sim 10^{-8}$ s) excited state decays spontaneously to a metastable state at a lower energy which in turn decays by stimulated emission to the ground state. The metastable state is relatively long-lived (e.g., 10^{-5}–10^{-4} s). Thus, ground state atoms are pumped up to the excited state which quickly decays to the metastable state, where atoms collect to create a population inversion with the ground state. Three-level lasers most commonly operate in a pulsed mode; a short discharge from a flash tube, for example, pumping the system up to the excited state from the ground state followed by the stimulated emission from the metastable state. It should be noted that the energy (per photon) required to excite the ground state atoms up into the excited state is greater than the energy (per photon) produced by the stimulated decay. This means, for example, that green photons might be needed to excite the transition to the excited state while red photons might be emitted during the stimulated decay process.

Lasers that operate continuously are more commonly based on a four-level system, as shown in figure 1.9(c). In this case, atoms that are pumped up to the excited state decay quickly by spontaneous emission to the relatively long-lived metastable state. Atoms collect in this metastable state and then decay by stimulated emission to a short-lived intermediate state, which quickly decays back to the ground state. Since the intermediate state decays quickly, its population remains small. The population inversion can, therefore, be maintained between the metastable state and the intermediate state and the excited state can continuously be populated by excitation of ground state atoms.

1.5 Laser modes and coherence

The second consideration in laser design is the creation of a resonant cavity so that the beam can be amplified by continuous constructive interference of the beam components. Figure 1.12 shows the basic design of a resonant cavity. The cavity has a fully reflecting mirror on one end and a partially reflecting mirror on the other end. Photons that are emitted as a result of a stimulated decay process are reflected, back and forth in the cavity, causing more stimulated decays, which produce more photons, all of which are of the same energy and are in phase with one another. Some of the coherent radiation leaks out of the partially reflecting mirror and constitutes the laser beam.

The above process sounds quite simple, but there is a serious concern; how do we maintain the phase relationship of the waves that are reflected from the ends of the cavity? In order to do this, it is necessary to produce a resonant cavity that is a half integer number of wavelengths in length. That is, we want to create a condition

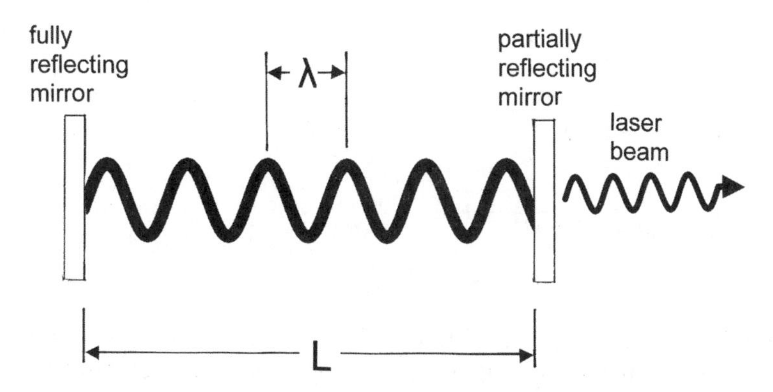

Figure 1.12. Conditions for creating a resonant laser cavity which is a half integer number of wavelengths long.

where the waves inside the cavity are standing waves. Figure 1.12 illustrates this point. For this condition to be met we require that

$$n\frac{\lambda}{2} = L \tag{1.7}$$

where n is an integer, λ is the wavelength of the radiation and L is the length of the cavity. Actually, figure 1.12 is not really to scale, as a typical wavelength would be in the range of a few hundred nanometers and a typical cavity would have a length in the centimeter or many-centimeter range. Using $L = 30$ cm and $\lambda = 500$ nm, gives $(3 \times 10^{-1}\ \text{m})/(2.5 \times 10^{-7}\ \text{m}) = 1.2 \times 10^{6}$ half wavelengths in the cavity. If we think about how we might actually create this situation, we realize that it is a very difficult task. The light, which has a wavelength of the order of a few hundred nanometers, bounces back and forth in the cavity many times. Any error in satisfying relationship (1.7) will be compounded with each reflection. Therefore, we might think that we need to know how accurately equation (1.7) has to be satisfied in order to make a functioning laser. Fortunately, we never actually have to concern ourselves with this problem, and that brings us back to a question we have never really answered; how monochromatic is monochromatic?

The decay of a quantum mechanical state is a statistical process characterized by a characteristic mean decay time, τ. As the exact time of the decay is uncertain, the energy of the state will be uncertain with a distribution that is related to the mean time by the Heisenberg uncertainty principle,

$$\Delta E \Delta t = \Delta E \tau \geqslant \frac{\hbar}{2} \tag{1.8}$$

or

$$\Delta E \geqslant \frac{\hbar}{2\tau} \tag{1.9}$$

This form of the Heisenberg uncertainty principle follows directly from the more common form involving position, x, and momentum, p;

$$\Delta x \Delta p \geqslant \frac{\hbar}{2} \tag{1.10}$$

by the relationships for position, velocity, v, and time, t;

$$x = vt \tag{1.11}$$

and momentum and energy, E,

$$p = \sqrt{2mE} \tag{1.12}$$

where m is mass. Taking the differential of equations (1.11) and (1.12) and substituting into equation (1.10) gives the form of the uncertainty relation in equation (1.9).

Using $\tau = 10^{-5}$ s for a metastable state mean lifetime in equation (1.9), we find $\Delta E \geqslant 3 \times 10^{-11}$ eV, out of a photon energy of about 2.5 eV (for $\lambda = 500$ nm). This value is actually a substantial underestimate of the actual width of the energy distribution as thermal effects substantially broaden the distribution. In a gas, this broadening results from Doppler effects due to the thermal motion. In a solid, phonon interactions broaden the energy distribution. In many materials, an energy width of around 10^{-5} eV is fairly typical. Since

$$E = \frac{hc}{\lambda} \tag{1.13}$$

then it is straightforward to show that

$$\Delta\lambda = -hc\frac{\Delta E}{E^2} \tag{1.14}$$

From the above example, we find $\Delta\lambda \sim 2 \times 10^{-3}$ nm. This is still small compared to the wavelength of 500 nm, so the photons are monoenergetic to about 1 part in (500 nm)/(2×10^{-3} nm) = 2.5×10^5.

It is now necessary to consider the resonant modes of the cavity. We need to ask: if the cavity contains n half wavelengths, then how much does the wavelength have to change to fit $n + 1$ half wavelengths in the cavity? From equation (1.7) we solve for λ:

$$\lambda = \frac{2L}{n} \tag{1.15}$$

and differentiating gives

$$\Delta\lambda = -\frac{2L}{n^2}\Delta n = -\frac{\lambda^2}{2L}\Delta n \tag{1.16}$$

So, for $\Delta n = 1$, then $\Delta\lambda = 4 \times 10^{-4}$ nm. Comparing this with the typical width of the wavelength distribution from the stimulated emission shows that there are about five cavity modes within this distribution. This relationship is illustrated in figure 1.13.

The above analysis shows that, in a typical situation, it is not necessary to adjust the length of the resonant cavity to match the wavelength of the lasing transition. No

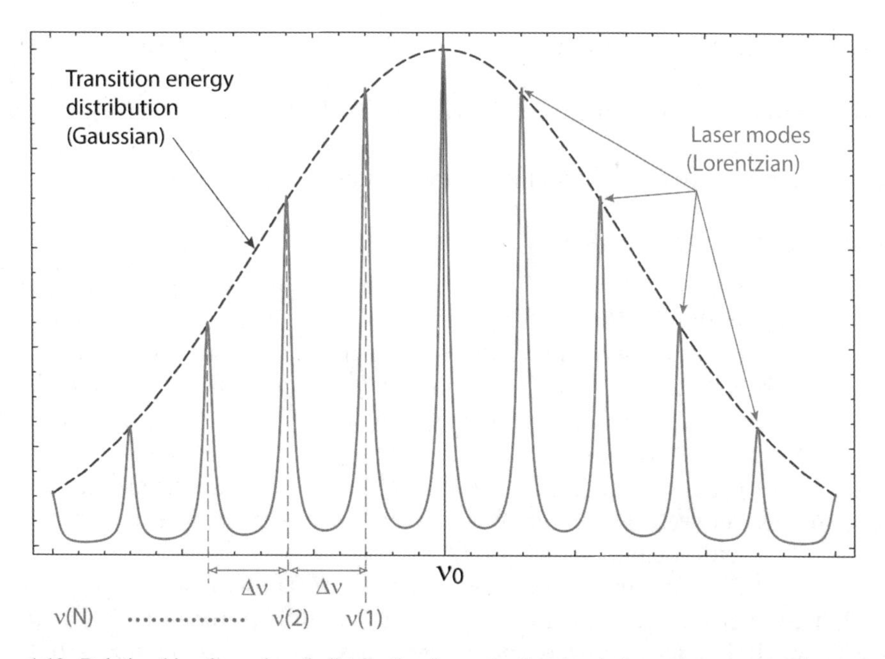

Figure 1.13. Relationship of wavelength distribution from stimulated emission and resonant cavity modes in a typical laser. Dr Wolfgang Geithner / wikimedia commons / CC BY SA 3.0 https://creativecommons.org/licenses/by-sa/3.0/deed.en.

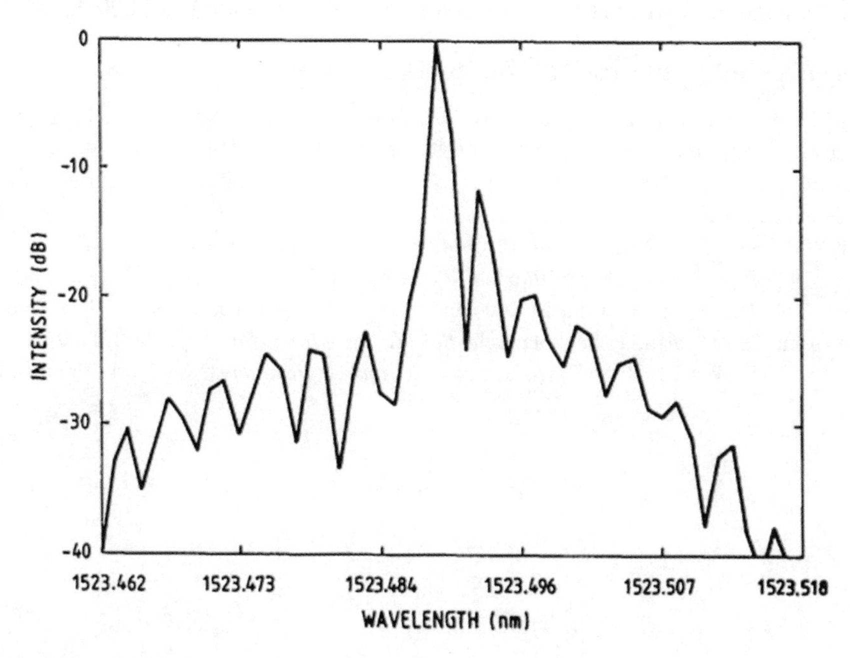

Figure 1.14. Wavelength spectrum of the lasing transition at 1523 nm (infrared) from a He–Ne laser showing the presence of cavity modes. Reprinted with permission from Junttila and Stahlberg 1990 © The Optical Society.

matter what the length of the cavity is, it will contain several resonant modes. A measured wavelength spectrum of a laser transition is illustrated in figure 1.14. Thus, the monochromatic light produced by a laser typically consists of a number of closely spaced modes within a Gaussian-like envelope. In the next two chapters we describe the principle of operation of several common types of lasers.

1.6 Problems

1.1. A laser with a lasing cavity of 0.2 m in length produces light with a wavelength of 550 nm. If the lifetime of the lasing transition is very long, then the Heisenberg linewidth of the energy distribution of the laser state will be very narrow and the resonant cavity modes will not exist. Calculate the maximum lifetime of the lasing state that is consistent with the construction of a laser.

1.2. (a) Calculate the spacing in energy (in eV) and in wavelength (in nm) between the resonant cavity modes of a laser operating at 600 nm with a cavity length of 1.0 m.
   (b) Repeat part (a) for a cavity length of 0.001 m.

1.3. Consider a hypothetical laser consisting of 10^{20} atoms of lasing material.
   (a) If the system has three energy levels with energies of −10.0 eV, −9.8 eV and −9.4 eV, find the equilibrium populations of the three levels according to Maxwell–Boltzmann statistics at room temperature.
   (b) Repeat part (a) if the lasing material is heated to 5000 K.

References and suggestions for further reading

Csele M 2004 *Fundamentals of Light Sources and Lasers* (Hoboken, NJ: Wiley)

Hecht J 1992 *The Laser Guidebook* 2nd edn (New York: McGraw Hill)

Junttila M L and Stahlberg B 1990 Laser wavelength measurement with a Fourier transform wavemeter *Appl. Opt.* **29** 3510–16

Silfvast W T 1996 *Laser Fundamentals* (Cambridge: Cambridge University Press)

Svelto O 1998 *Principles of Lasers* 4th edn (New York: Springer)

Virtuani A, Lotter E and Powalla M 2006 Influence of the light source on the low-irradiance performance of Cu(In,Ga)Se_2 solar cells *Sol. Energy Mater. Sol. Cells* **90** 2141–9

Wilson J and Hawkes J F B 1987 *Lasers: Principles and Applications* (New York: Prentice Hall)

Richard A Dunlap

Chapter 2

Types of lasers I: conventional lasers

2.1 Introduction

The earliest lasers used lasing atoms contained in a solid material. The most notable of these is the ruby laser. Later, traditional lasing transitions in gaseous and liquid media were utilized for the construction of lasers. The present chapter reviews the most important of these lasers.

2.2 Solid state lasers

Solid state lasers contain atoms that undergo a lasing transition that are contained within a solid matrix. The earliest lasers that were developed were solid state lasers; specifically, the ruby laser. Ruby is aluminum oxide (Al_2O_3), which is itself colorless, with Cr^{3+} ion impurities, which produce the characteristic red color. It is the Cr^{3+} ion which has energy levels that produce the laser radiation. Figure 2.1 shows the energy levels of the Cr^{3+} ion. There are two short-lived states that readily absorb photons at around 400 nm and 550 nm (corresponding to blue and green light, respectively). These short-lived states decay to a metastable state with a lifetime of about 3 ms. This metastable state decays back to the ground state and emits photons at 694 nm (red). In this sequence, Cr^{3+} ions are pumped from the ground state to the two short-lived states, which then populate the metastable state, creating a population inversion between the ground state and the metastable state. This population inversion results in the stimulated emission back to the ground state. Since the stimulated emission re-populates the ground state, the population inversion cannot be continuously maintained and the ruby laser must be operated in a pulsed mode.

Solid state lasers are typically pumped using photons from a xenon flash tube. These flash tubes are intense and have a substantial portion of their spectrum in the proper wavelength range as illustrated in figure 2.2. The figure shows that there are

doi:10.1088/2053-2571/ab2f2fch2

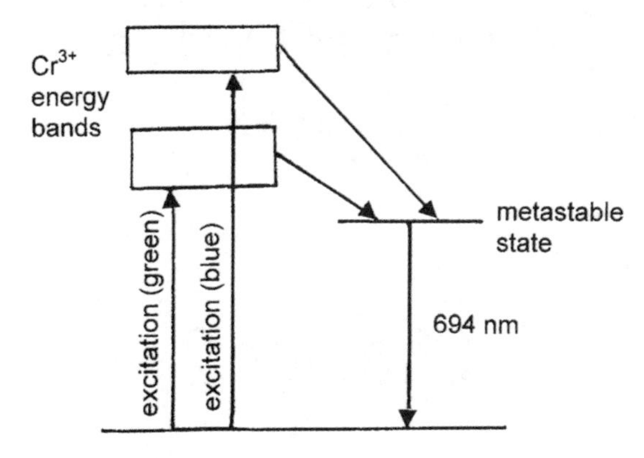

Figure 2.1. Simplified energy level diagram of the Cr^{3+} ion showing the stimulated emission at a wavelength of 694 nm.

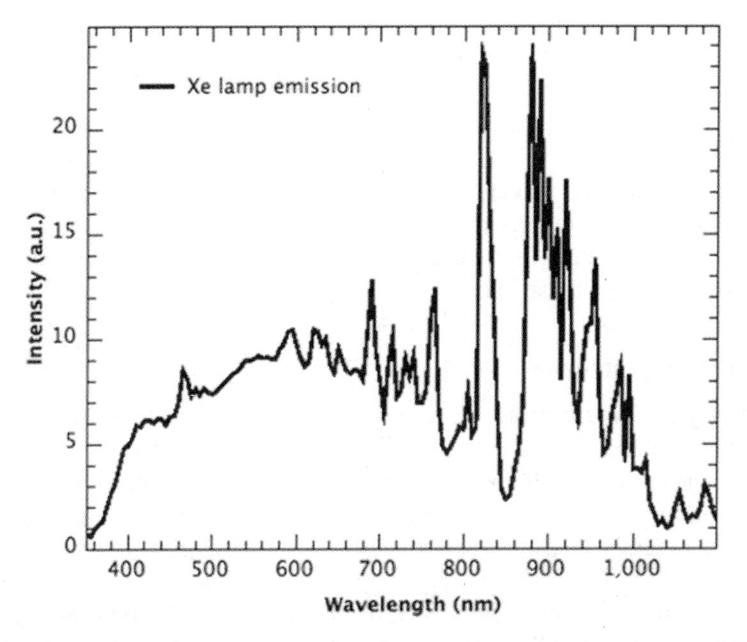

Figure 2.2. Spectral output of a xenon arc lamp. Reproduced from Presciutti *et al* 2014. CC-BY-SA.4.0.

some discrete peaks in the long wavelength infrared portion of the spectrum. In the near infrared, visible and near ultraviolet portions of the spectrum (from about 800 nm down to about 300 nm) there is a broad, mostly featureless continuum. The spectrum is quite intense in the 400–550 nm region required for pumping the ruby laser.

One of the most common solid state lasers is the Nd^{3+}:YAG laser. Nd^{3+} ions are responsible for the lasing transitions. These are incorporated into a suitable transparent matrix. YAG (yttrium aluminum garnet, $Y_3Al_5O_{12}$) is commonly used, but other

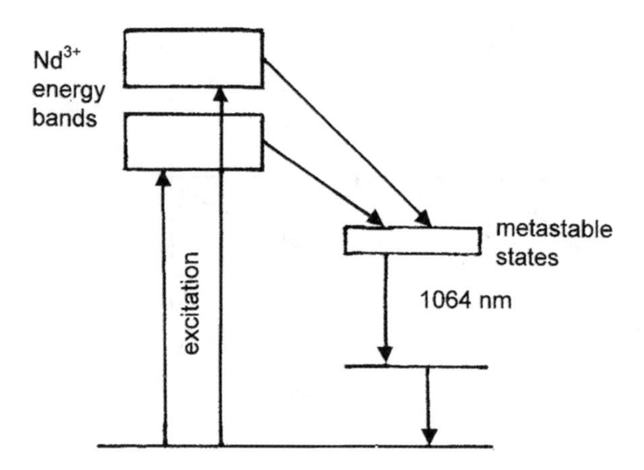

Figure 2.3. Energy levels of Nd^{3+} in YAG.

materials such as $LiYF_4$, YVO_4 and glass have also been used. The energy level diagram for the Nd^{3+} ion is shown in figure 2.3. This ion is a four-level system and can, in principle, be used either in the pulsed or the continuous mode of operation. Photons from a flash lamp in the 730–800 nm range (near infrared) are used to pump the Nd^{3+} ions from the ground state up to several excited states, as shown in the figure. These states decay spontaneously to the metastable state which decays by stimulated emission to the short-lived intermediate state. The metastable and intermediate states have some splitting, so there are several closely spaced lasing transitions at around 1064 nm.

Nd:YAG lasers have found a number of applications as discussed further below because quite powerful lasers of this type can be constructed fairly economically. The wavelength produced by Nd:YAG lasers (and all Nd^{3+} based lasers), however, is in the infrared. For some applications this is not convenient, as photons in the visible portion of the spectrum are more appropriate. One solution to deal with this difficulty is second harmonic generation, as discussed below.

2.3 Second harmonic generation

The photons produced by a Nd:YAG laser at 1064 nm have an energy of 1.17 eV. These may be converted into optical photons by second harmonic generation or frequency doubling. This process is easily explained from a quantum mechanical point of view. Two photons of 1.17 eV (wavelength of 1064 nm) can be combined to produce one photon at 2.34 eV (wavelength of 532 nm). The number of photons is reduced by a factor of two, but the energy per photon is doubled.

This process can be viewed from a more classical standpoint by looking at the effects of the electric field associated with the electromagnetic wave as it passes through a dielectric material. In a dielectric material, the positive and negative charge distributions are not coincident, and this gives rise to an electric dipole moment. Under normal conditions, this results in a polarization, P, which is proportional to the electric field, E:

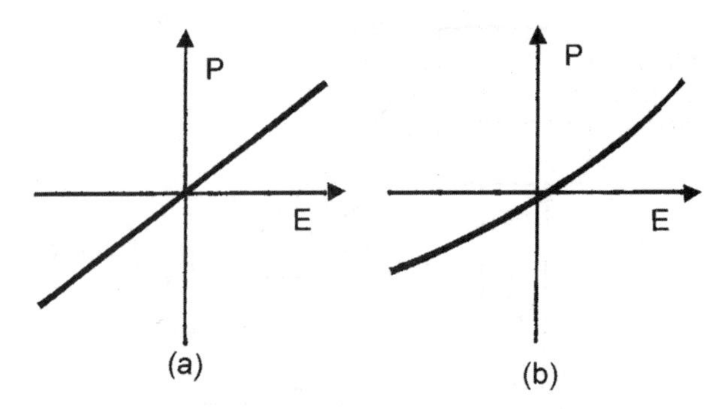

Figure 2.4. Relationship between the electric field associated with a beam of light and the induced polarization for (a) a linear medium and (b) a nonlinear medium.

$$P = \chi \varepsilon_0 E \tag{2.1}$$

where χ is the relative electric susceptibility of the material and ε_0 is the permittivity of free space. In this context, the polarization is the vector sum of the induced dipole moments in the material and is not the same as the polarization of the light beam which represents the orientation of the electric field vectors. The relationship in equation (2.1) shows a linear dependence between P and E, as illustrated in figure 2.4(a). If the electric field associated with the light oscillates sinusoidally then the induced polarization will also oscillate sinusoidally (at the same frequency) and will generate a sinusoidally oscillating electric field that will contribute to the propagating electric field associated with the light beam.

In some materials, and particularly for large electric fields, the relationship between the electric field and the polarization is nonlinear, as shown in figure 2.4(b). The nonlinear behavior of the polarization can be expressed as a Taylor expansion in powers of the electric field as

$$P = \chi_1 \varepsilon_0 E + \chi_2 \varepsilon_0 E^2 + \chi_3 \varepsilon_0 E^3 + \cdots \tag{2.2}$$

This means that if the electric field is sinusoidally varying then the induced polarization will not be a pure sine wave. Thus, the electric field associated with the polarization will also not be purely sinusoidal. This non-sinusoidal electric field can be expanded in a Fourier series of multiples of the fundamental frequency, f_0, of the electric field. Figure 2.5(a) shows the Fourier expansion of a sinusoidal electric field with a single Fourier component at the fundamental frequency. Figure 2.5(b) shows the Fourier expansion of a non-sinusoidal electric field.

The harmonics of the fundamental frequency represent components of the beam that have twice, three times, etc the frequency (and energy) of the incident beam. The harmonic at twice the fundamental frequency is generally the most intense harmonic and corresponds to second harmonic generation. Some materials are quite efficient at generating harmonics because of a significant nonlinear component given by the nonlinear terms for the polarization in equation (2.2).

The efficiency of second harmonic generation may be viewed in a quantum mechanical sense as the fraction of photons that are combined to form photons of

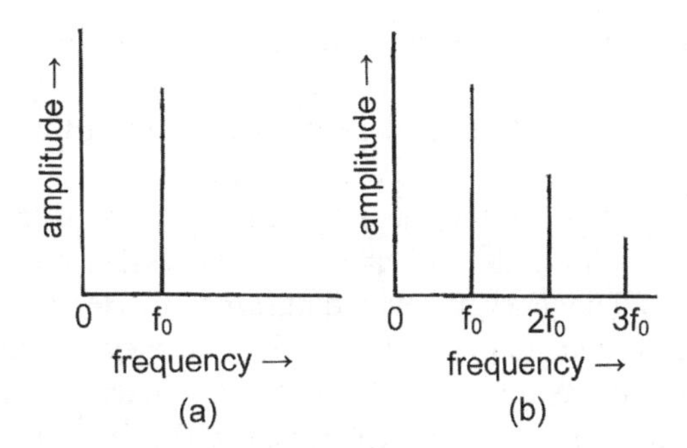

Figure 2.5. Fourier expansion of (a) a linear polarization and (b) a nonlinear polarization.

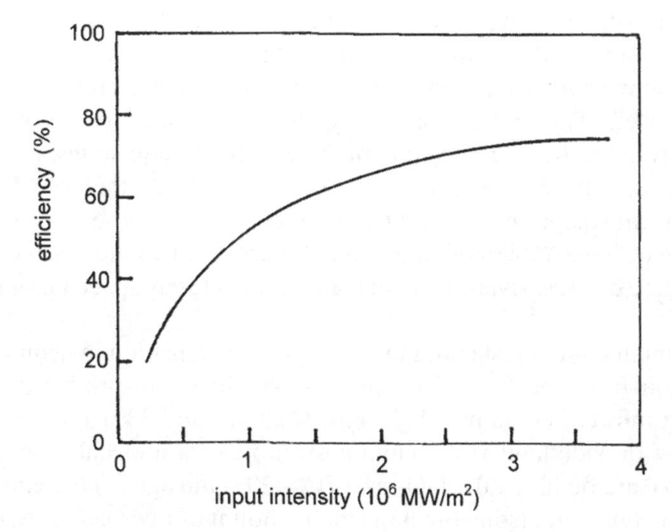

Figure 2.6. Second harmonic generation efficiency for three layers of magnesium oxide:lithium tantalate crystals for 1064 nm Nd:YAG laser radiation.

twice the energy. In the classical sense the second harmonic generation efficiency may be viewed as the ratio of the $2f_0$ beam component to the initial beam intensity. An example of second harmonic generation efficiency for the conversion of 1064 nm near infrared radiation to 532 nm green light is shown in figure 2.6. Up to some point, the polarization becomes progressively more nonlinear as a function of electric field and the second harmonic generation efficiency increases.

2.4 Gas lasers

Gas lasers use lasing atoms which are in the gaseous form, often in combination with other gas atoms or molecules. The gas may be either neutral atoms, ions or molecules. Examples of each of these are discussed in the present section. Gas lasers are analogous to gas discharge tubes as shown in figure 1.3 except they are designed to emit light by stimulated, rather than spontaneous, emission. Excitation may be induced with electric fields, as in figure 1.3, or by the application of radio frequency radiation.

The helium–neon (He–Ne) laser is the most common of the gas lasers in current use. The lasing cavity contains a mixture of about 10 parts He to 1 part Ne. Although it is actually transitions in the neon atoms that produce the laser radiation, the large amount of helium that is present is necessary in order to create the population inversion. The energy level diagrams of He and Ne are shown in figure 2.7. The energy associated with the electric discharge preferentially excites the helium atoms (because there are more of them) to two excited states, as shown in the figure. These two helium energy levels are at nearly the same energy as two levels in neon. Through atomic collisions, energy is transferred from excited helium atoms (which fall back to their ground state) to ground state neon atoms (which are pumped up into their excited states). These are the metastable laser states in the neon. The reason for not exciting neon atoms directly through an electric discharge, is that there are short-lived states (the laser intermediate states) that would be populated as well. The transfer of energy from the helium to the neon creates a population inversion between several combinations of neon states, i.e. between the 5s and 4p states, between the 5s and 3p states and between the 4s and 3p states, as illustrated in the diagram. Stimulated emission can now occur between these combinations of metastable and short-lived intermediate states. Since there is fine splitting of some of these levels, there are sometimes closely spaced lasing transitions that result.

The most intense and most commonly used lasing transition in neon is the 633 nm (red) transition from the 5s to 3p state. Weaker lines are produced in the green (543 nm) and infrared (1118 nm, 1152 nm, 1523 nm and 3391 nm).

The carbon dioxide laser is a common example of a molecular laser. This laser contains a mixture of 10%–20% CO_2 and 10%–20% nitrogen. The remainder of the gas is mostly helium with (sometimes) a small amount of hydrogen or xenon. It is the CO_2 and the nitrogen that are responsible for creating the population inversion and

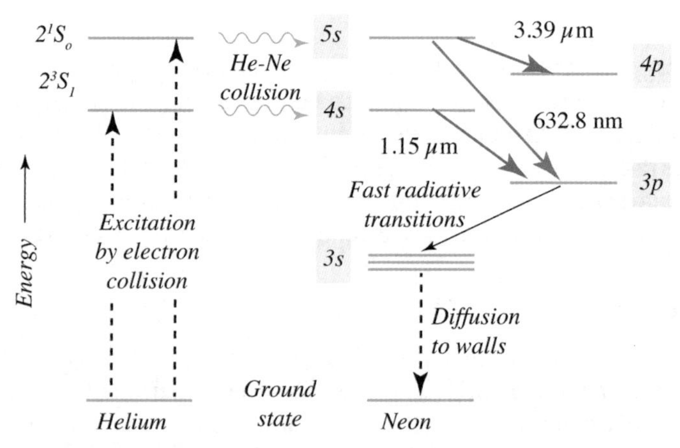

Figure 2.7. Energy levels in the He–Ne laser. ZuPanda / Wikimedia Commons / CC-BY-SA 4.0 https://creativecommons.org/licenses/by-sa/4.0/deed.en.

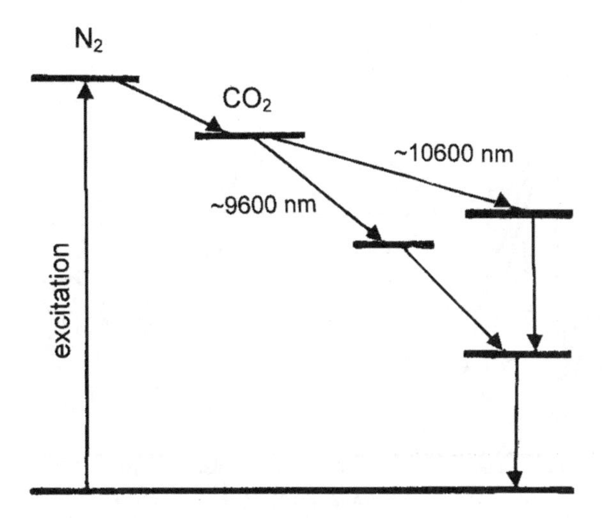

Figure 2.8. Energy levels in a CO_2 laser showing the role of nitrogen and the laser transitions at 10 600 nm and 9600 nm.

the stimulated emission. The energy diagram is shown in figure 2.8. The lasing transitions are in the CO_2 states, but the nitrogen is necessary to create the population inversion, much in the same way that helium is needed in the He–Ne laser. Energy from an electric discharge excites nitrogen molecules from their ground state to a higher energy level. This energy is transferred to CO_2 molecules through collisions and populates the metastable state. This metastable state decays by stimulated emission to one of several short-lived intermediate states. The helium is needed to help carry away excess energy from the nitrogen molecules after they have interacted with the CO_2.

2.5 Dye lasers

Dye lasers are analogous to molecular sources of optical spectra. They are referred to as dye lasers because the earliest lasers of this type utilized transitions in commercial dyes to produce laser radiation. Dyes and other organic molecules used in this type of laser have very complex electronic energy levels giving rise to a very large number of possible transitions that can emit stimulated radiation.

Dye lasers generally consist of the 'dye' molecules dissolved in a liquid organic solvent. The pumping radiation can come from a broad-band flash lamp or a traditional gas or solid state laser. Typical dye lasers contain the lasing medium in a transparent vial and this design allows for the dye to be changed to produce different wavelength outputs. A given dye typically has an almost continuous band of laser wavelengths over a range of a few tens of nm. Although the light is not 'monochromatic' in the sense that light from a solid state or gas laser would be, it is intense and coherent and can be tuned to a specific wavelength of interest. Spectra of some commercially available dyes are shown in figure 2.9, and a simple arrangement for tuning the dye laser to a specific wavelength is shown in figure 2.10.

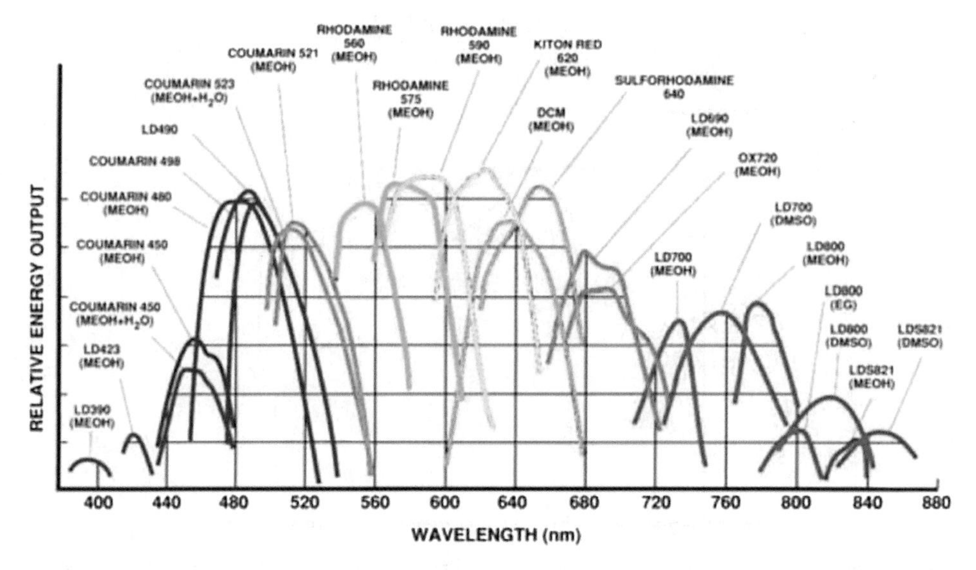

Figure 2.9. Spectral output of a dye laser using different dyes as indicated in the legend. (Figure 3 from Shankarling and Jarag 2010 reprinted by permission from Springer Verlag. Copyright Indian Academy of Sciences 2010.)

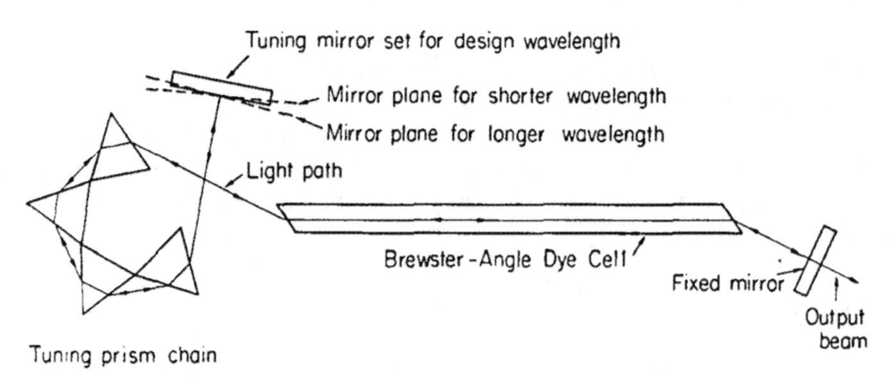

Figure 2.10. A simple method of tuning a dye laser to a specific wavelength. The tuning mirror is rotated so that only the desired wavelength is reflected back through the resonant cavity. Reprinted with permission from Strome and Webb 1971. Copyright The Optical Society.

References and suggestions for further reading

Presciutti A, Asdrubali F, Marrocchi A, Broggi A, Pizzoli G and Damiani A 2014 Sun simulators: development of an innovative low cost film filter *Sustainability* **6** 6830–46

Shankarling G S and Jarag K J 2010 Laser dyes *Resonance* **15** 804–18

Silfvast W T 1996 *Laser Fundamentals* (Cambridge: Cambridge University Press)

Strome F C and Webb J P 1971 Flashtube-pumped dye laser with multiple-prism tuning *Appl. Opt.* **10** 1348–53

Svelto O 1998 *Principles of Lasers* 4th edn (New York: Springer)

Wilson J and Hawkes J F B 1987 *Lasers: Principles and Applications* (New York: Prentice Hall)

IOP Concise Physics

Lasers and Their Application to the Observation of Bose–Einstein Condensates

Richard A Dunlap

Chapter 3

Types of lasers II: semiconducting lasers

3.1 Introduction

While traditional lasers using solid, liquid or gaseous media that contain atoms, ions or molecules that exhibit lasing transitions have been important for a number of very significant scientific and industrial applications, as discussed below in chapter 4, it was the development of semiconducting lasers, as described in the present chapter that made the use of lasers widespread in applications related to consumer electronics and communication. The basic physics of semiconducting materials and the method of producing coherent stimulated radiation from a semiconducting junction are overviewed in the present chapter.

3.2 Semiconductor physics

Semiconducting lasers produce coherent laser radiation as a result of transitions between the conduction band and the valence band of semiconducting materials. To begin our overview of semiconducting lasers we first consider the effects of impurities on the electrical properties of semiconductors. We then consider the properties of semiconducting junctions.

Silicon is a good example of a semiconducting material that will allow us to overview the way in which semiconducting devices function. Semiconducting lasers are made from more complex combinations of semiconducting materials, which will be considered in detail in section 3.4.

Silicon has a diamond structure and as a result each silicon atom is tetrahedrally bonded to its four nearest neighbors. The electronic configuration of silicon, which has 14 electrons, is $1s^2 2s^2 2p^6 3s^2 3p^2$. The band gap in silicon is 1.12 eV, making silicon a semiconductor rather than an insulator, with thermally excited electrons and holes contributing to the conductivity at room temperature. We begin with a consideration of the properties of silicon at low temperature where these thermal

doi:10.1088/2053-2571/ab2f2fch3

effects are not significant and look at the ways in which impurities can affect the electrical properties.

Consider first a piece of silicon with some phosphorus impurities. Phosphorus has 15 electrons and the electronic configuration $1s^2 2s^2 2p^6 3s^2 3p^3$. Fourteen of these phosphorus electrons will take the place of the 14 silicon electrons and will bond with the four nearest neighbors. The 15th electron is not needed for bonding and will be free to move about in the lattice. Phosphorus is referred to as a donor impurity as it donates its extra electron to the conduction band. A two-dimensional illustration of this is shown in figure 3.1(a).

If instead we add aluminum, which has 13 electrons and the configuration $1s^2 2s^2 2p^6 3s^2 3p^1$, to silicon, then each aluminum atom has one too few electrons to make the four necessary tetrahedral bonds to its nearest neighbors. This situation is illustrated in figure 3.1(b) where one bonding electron is missing and this missing bond behaves like a positively charged hole. This hole can propagate through the material by exchanging places with valence electrons from other bonds. Aluminum is referred to as an acceptor impurity as it provides a missing bond which can accept electrons from other atoms. In general, semiconductors with impurities that form electron or hole states are referred to as doped semiconductors.

We can view the situation for impurities in silicon on the basis of the dispersion relation which illustrates the band structure. A very simplified picture is shown in figure 3.2. The lowest energy band in the figure represents the two lowest energy 3p electrons and is referred to as the valence band. The highest energy band in the figure represents the next two highest 3p electrons and is referred to as the conduction band. The filled 1s–3s bands are not shown in the figure as they do not directly contribute to the electrical properties of the material.

In the case of phosphorus impurities, the silicon electrons and 14 of the phosphorus electrons are just sufficient to fill the bands up to the valence band and the extra electron from the phosphorus impurity atom must go into a state in the conduction band. This is seen in figure 3.1(a) where the number of electrons in the conduction band is exactly equal to the number of phosphorus impurities in the

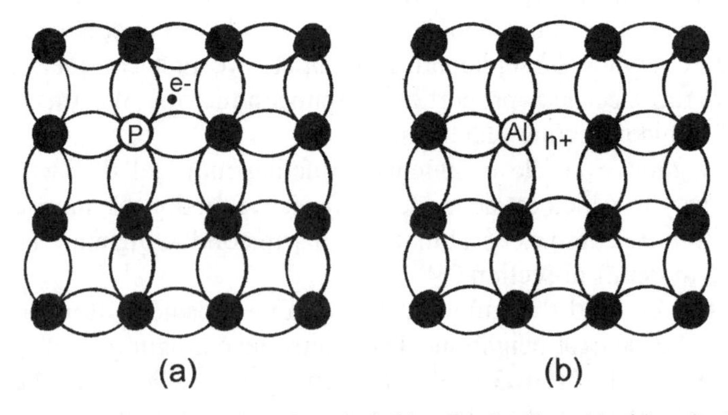

Figure 3.1. Two-dimensional representation of tetrahedral bonding in silicon (a) with a phosphorus impurity and (b) with an aluminum impurity. (Adapted from Dunlap R A 2018 with permission of Morgan & Claypool Publishers.)

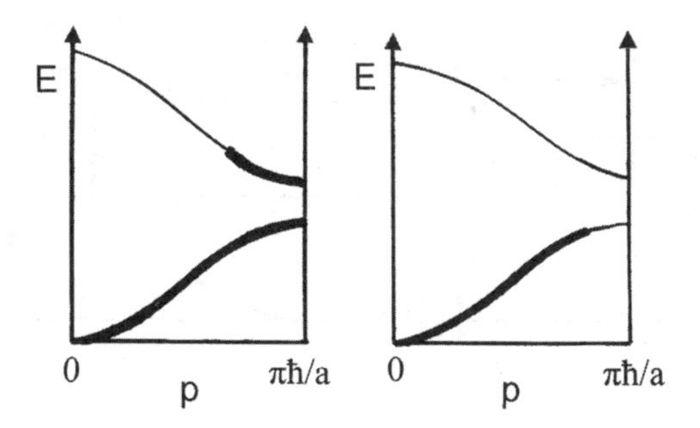

Figure 3.2. Dispersion relation showing the band structure and occupied states for (a) silicon with phosphorus impurities and (b) silicon with aluminum impurities.

sample. In the case of aluminum impurities in silicon, the band structure shown in figure 3.1(b) shows that there is one missing electron (or one hole) in the valence band for every aluminum impurity.

Thus, the introduction of impurities with the appropriate electronic configuration provides a means of creating electron states in the conduction band or hole states in the valence band of a semiconducting material. As these electrons and holes are free to move throughout the material then they can contribute to the electrical conductivity of the material by responding to any applied electric fields. This is analogous to the situation for diamond with boron impurities. In the case of silicon with phosphorus impurities, the charge is carried by free electrons (which have a negative charge). This type of material is referred to as an n-type semiconductor as it has negative charge carriers. In the case of silicon with aluminum impurities, the charge carriers are holes (which have an effective positive charge) and this type of material is referred to as a p-type semiconductor. Phosphorus and aluminum impurities in silicon do not necessarily represent the best n-type and p-type semiconducting materials, respectively, from a practical standpoint, but they are a simple example of how the valence of impurity atoms can influence electrical properties of the host material.

Since semiconducting materials are most commonly used at around room temperature, it is important to consider thermal effects. Thermal energy can cause electrons from the valence band to be excited up into the conduction band, leaving hole states behind them in the valence band. Thus, in a piece of n-type doped silicon there are a lot of free electron carriers that come from the donor impurities. There are also a smaller number of electron carriers that come from thermal excitations and an equal number of hole carriers that are formed. In the n-type semiconductor the electrons are called the majority carriers and the holes are called the minority carriers. In a p-type semiconductor the role of electrons and holes is interchanged.

The behavior of electron and hole charge carriers in a semiconductor can be viewed somewhat more quantitatively by looking at some basic physics. We begin with a free electron at rest in a piece of semiconducting material as illustrated in figure 3.3. If we close the switch in the diagram and supply a potential difference, V,

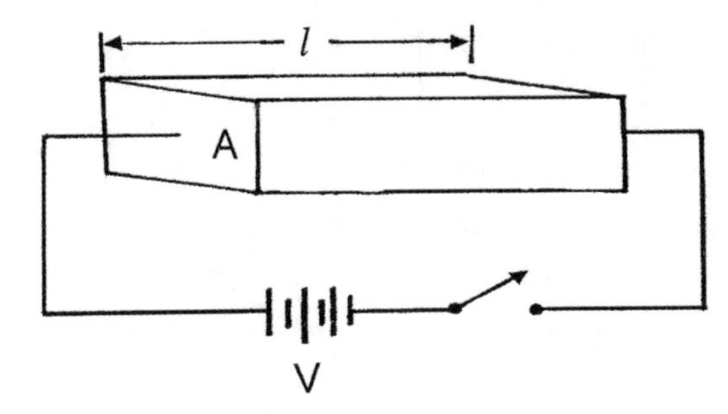

Figure 3.3. Experiment for measuring the conductivity of a sample of material with cross-sectional area A and length l.

between the two ends of the material then there will be an electric field, E, present, where

$$E = \frac{V}{l} \tag{3.1}$$

This will exert a Lorentz force, F, on the negatively charged electrons,

$$F = -eE = -\frac{eV}{l} \tag{3.2}$$

Newton's law gives the acceleration of the electron as

$$a = \frac{F}{m_e} = -\frac{eV}{m_e l} \tag{3.3}$$

where m_e is the electron mass. A time t after the switch is closed, the electron velocity will be

$$v = at = -\frac{eVt}{m_e l} \tag{3.4}$$

The minus sign merely means that the electrons will flow in the direction opposite to the electric field because of their negative charge. The electric current flowing through the sample is given as the charge transported per unit time. If n_e is the free electron density in the material, then the current can be expressed as

$$I = -en_e vA = \frac{e^2 n_e VAt}{m_e l} \tag{3.5}$$

where A is the cross-sectional area of the sample, as shown in the figure. Using Ohm's law $V = IR$, we can rewrite equation (3.5) as

$$V = \left(\frac{m_e l}{e^2 n_e At}\right) I \tag{3.6}$$

or

$$R = \frac{m_e l}{e^2 n_e A t} \tag{3.7}$$

Since the resistivity is $\rho = AR/l$ and the conductivity is $\sigma = 1/\rho$, then the conductivity may be written as

$$\sigma = \frac{e^2 n_e}{m_e} t \tag{3.8}$$

Some of this makes sense, that is, the conductivity is proportional to the charge on the carriers and their density and it is inversely proportional to their mass. However, it does not make sense that the conductivity will increase linearly with time after the switch is closed. Because the free electrons in the material are not really free, but they interact with other charges and phonons, the electrons travel for some time (the mean free time, τ_e) before interacting substantially, at which time they have to start accelerating all over again. Therefore, the conductivity reaches some equilibrium value which is written as

$$\sigma_e = \frac{e^2 n_e}{m_e} \tau_e = e n_e \mu_e \tag{3.9}$$

where $\mu_e = e\tau_e/m_e$ is defined as the mobility of the electrons.

We can repeat this derivation for the hole carriers realizing that their number density will be n_h and their mobility may be defined in terms of their own mean free time, τ_h, and effective mass, m_h, so that $\mu_h = e\tau_h/m_h$ and the conductivity from the holes will be

$$\sigma_h = e n_h \mu_h \tag{3.10}$$

We note that the sign of the charge on the holes is positive and the current flows in the opposite direction to that of the electrons. However, the Lorentz force will also be in the opposite direction. This means that the electrons and holes travel in opposite directions but contribute to a current in the same direction. This means that we can write the total conductivity as

$$\sigma = \sigma_e + \sigma_h = e n_e \mu_e + e n_h \mu_h \tag{3.11}$$

While the above discussion describes the behavior of n-type and p-type semiconducting materials, most semiconducting devices, including semiconductor lasers, utilize junctions between these two types of materials. The next section begins with a discussion of what happens when n-type and p-type semiconductors are placed in electrical contact.

3.3 Semiconducting junctions

In order to understand what happens when semiconducting materials of different types form junctions, it is first important to know where all of the charges in an n-type and in a p-type semiconductor are located and how they behave. In a piece of n-type material, we have neutral host atoms (e.g., silicon) that are combined with

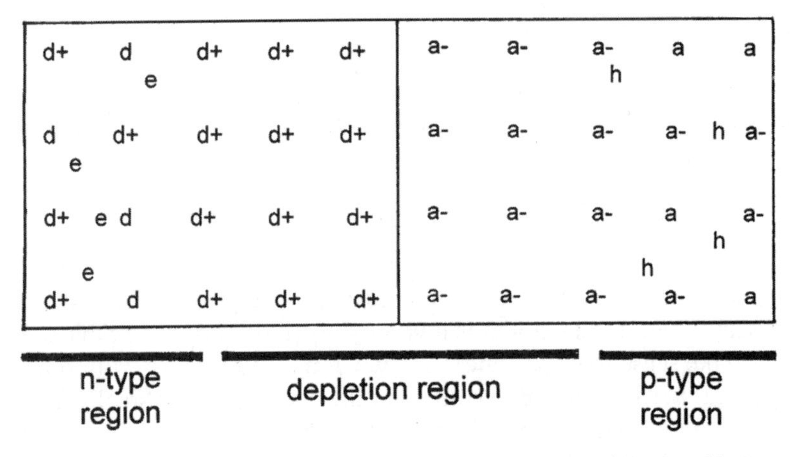

Figure 3.4. The p–n semiconducting junction or diode. (From Dunlap R A 2019 *Renewable Energy Vol. 1 – Requirements and Sources* (San Rafael, CA: Morgan & Claypool). Reproduced with permission of Roger W Watt.)

neutral donor atoms (e.g., phosphorus). Thus, the doped semiconductor must be neutral overall. There are negatively charged electrons that are freed from donor atoms. There are an equal number of donor atoms which have lost an electron and have become positive ions. There are negatively charged electrons that are thermally excited and there are an equal number of positively charged holes that are created in this process. Thus, the electrical neutrality is maintained, but the behavior of the positive and negative charges is different. There are the majority negatively charged mobile electron carriers that are the result of both donor electrons from the impurities and thermally excited electrons. There are the minority positively charged mobile holes that are the result of thermal excitations. Finally, there are the positively charged impurity ions which are part of the lattice and are not free to move. In a piece of p-type material the charges are just the opposite.

Another feature of a semiconducting material that is important to realize is that semiconducting materials are ohmic, that is, they obey Ohm's law. It is only when there are junctions formed from different types of semiconducting materials that non-ohmic behavior results. The simplest junction is a single piece of n-type material joined to a single piece of p-type material as shown in figure 3.4. This is the basic design of a diode. In the diagram 'd' represents the fixed donor impurities, 'a' represents the fixed acceptor impurities, 'e' represents the mobile negatively charged electrons and 'h' represents the mobile positively charged holes. We begin with a discussion of the behavior of the majority carriers, that is, the electrons in the n-type material and the holes in the p-type material.

There are two aspects of the majority carriers in the system that need to be considered; (1) the concentration of the majority carriers on each side of the junction and (2) the spatial distribution of the carriers. With respect to the first point, the majority electrons in the n-material will see a much smaller electron concentration in the p-material on the other side of the junction. This concentration gradient across the junction provides a driving force for electron diffusion from the high concentration in the n-material to the low concentration in the p-material to try to evenly distribute the electrons. Similarly, the majority hole carriers in the p-material will see a lower hole concentration in the n-material on the other side of the junction and this will provide a driving force for hole diffusion from the p-material to the n-material. As soon as this diffusion begins then the n-material will lose negatively charged electrons and gain positively charged holes, while the p-material will lose positively charged holes and gain negatively charged electrons. Thus, there will be an excess of positive charge on the n-side of the junction and an excess of negative charge on the p-side of the junction. This gives rise to the formation of an electric field pointing from left to right across the junction in the figure. This field will oppose the diffusion of the electrons and holes. An equilibrium situation will be set up where the resulting electric field is just sufficient to compensate for the driving force for diffusion that results from concentration gradients.

The spatial distribution of majority carriers can be viewed in the following way. The negatively charged electrons in the n-region see the negative charge of the fixed acceptor impurities on the other side of the junction and are repelled. As a result, they are redistributed in the n-region away from the junction as shown in the figure. Similarly, the holes in the p-region see the positive charge of the fixed donor impurities in the n-region and are also repelled away from the junction as shown. If we look at the majority carrier distribution as shown in figure 3.4, we readily see that there are lots of charge carriers in the region away from the junction and virtually no charge carriers in the region just on either side of the junction. This region that is depleted of charge carriers is called the depletion region or depletion layer. If we were able to measure the electrical conductivity in the various regions of the system, we would find that the regions far from the junction would have a fairly high conductivity because of the high density of free carriers. The depletion region, on the other hand, would have a very low conductivity as a result of the lack of mobile charge carriers in the region. As a result, the electric field, and hence the change in the electric potential, will occur primarily across the depletion region.

Finally, we need to look briefly at the behavior of the minority carriers. These are the electron and hole carriers that are created by thermal excitations. If a minority hole carrier in the n-region drifts (because of thermal motion) into the depletion region it will experience a force due to the presence of the electric field in this region. The electric field will push the hole across the junction to the p-region where it will become a majority carrier. Similarly, a minority electron on the p-side of the junction which drifts into the depletion region will be carried by the electric field across the junction to the n-region where it will become a majority carrier. These minority currents across the junction will unbalance the majority charge carrier configuration. As a result, majority charge carriers will flow across the junction to compensate for this imbalance. Since no net current flows in or out of the diode, then all of these majority and minority currents must cancel.

We can view the equality of majority and minority currents quantitatively as follows. The details of the two majority and two minority currents are shown in table 3.1. The total current, I_{total}, must be equal to zero;

Table 3.1. Definitions of the majority and minority currents across the diode junction. A positive current is defined as a positive flow of charge from left to right (i.e., n-type to p-type).

Current	Type	Carrier	Direction	Sign
I_1	Minority	Electrons	p → n	(+)
I_2	Majority	Electrons	n → p	(−)
I_3	Majority	Holes	p → n	(−)
I_4	Minority	Holes	n → p	(+)

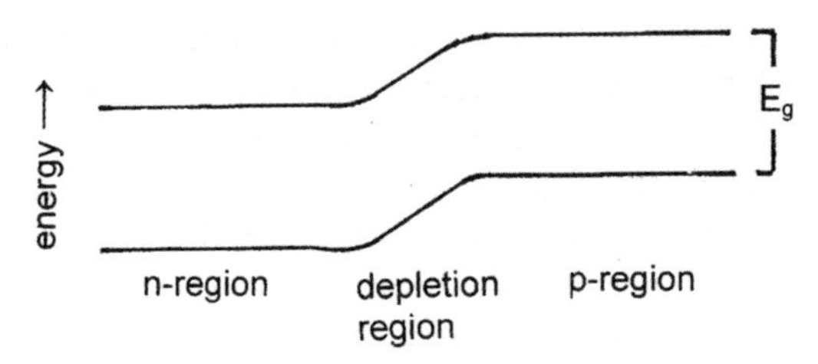

Figure 3.5. Energy bands in a diode.

$$I_{\text{total}} = I_1 + I_2 + I_3 + I_4 = 0 \tag{3.12}$$

We also know that the electron currents must cancel each other, and the hole currents must cancel each other so that

$$I_1 = -I_2 \tag{3.13}$$

and

$$I_3 = -I_4 \tag{3.14}$$

A convenient way of picturing the p–n junction is shown in figure 3.5, where energy is plotted on the vertical axis and spatial position is plotted on the horizontal axis. The figure shows the locations of the valence and conduction bands and indicates the filled and vacant states in each. A very important aspect of the energy levels that is seen in this diagram is that the levels in the p-region are shifted upward (to higher energy) from those in the n-region. This shift is caused by the electric field that is set up across the junction by the redistribution of mobile charge carriers.

If a diode is connected to an external voltage source, then the interesting features of a semiconducting junction become apparent. As the two sides of the diode are different in terms of the polarity of the charge carriers, then there are two different ways in which the external voltage can be applied to the device. These two ways are shown in figure 3.6. The symbol for the diode is a triangle with a bar. The triangle points from the p-side to the n-side of the diode, as shown in the figure. The conditions shown in the figure are referred to as the forward bias case and the reverse bias case, if the positive voltage is applied to the p-side of the diode or the n-side of the diode, respectively.

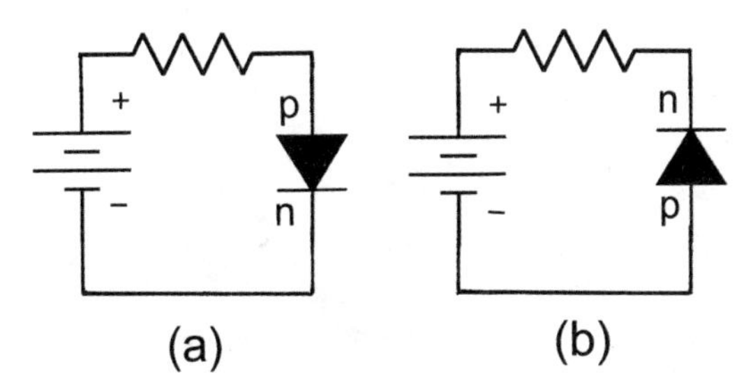

Figure 3.6. The (a) forward and (b) reverse biased diode. The symbol for the diode is a triangle with a bar, where the triangle points from the p-side to the n-side of the junction.

The application of an external voltage to the diode produces an electric field across the device. This electric field either adds to the internal field or subtracts from it. Recall that the internal field in the device is a result of excess positive charge on the n-side and excess negative charge on the p-side. In the forward bias case we see that the applied field opposes the internal field in the diode and in the reverse bias condition the external field adds to the internal field. If we refer to figure 3.5, we see that the change in the energy levels across that junction that results from the charge redistribution that creates the internal field, inhibits majority charge carrier movement across the junction. In the forward bias case, the external voltage reduces this change in energy levels and makes it easier for majority carriers to flow across the junction. In the reverse bias case, just the opposite happens, and it is more difficult for the majority carriers to flow. The minority currents are not affected by the external field as these result from thermal motion. So, from table 3.1, we see that the external field will affect currents I_2 and I_3. It can be shown that in the presence of an applied voltage, V, the majority currents for the forward (+) and reverse (−) bias conditions become

$$I_2^{\pm} = I_2 \exp\left(\pm\frac{eV}{k_B T}\right) \tag{3.15}$$

and

$$I_3^{\pm} = I_3 \exp\left(\pm\frac{eV}{k_B T}\right) \tag{3.16}$$

respectively. Combining these two equations with equations (3.12), (3.13) and (3.14), gives

$$I_{\text{total}} = (I_1 + I_4) \cdot \left[1 - \exp\left(\pm\frac{eV}{k_B T}\right)\right] \tag{3.17}$$

This equation is called the rectifier equation or diode equation and gives the total current flowing through the diode as a function of the applied voltage. Figure 3.7 shows the voltage dependence of the individual currents and the total current through the diode.

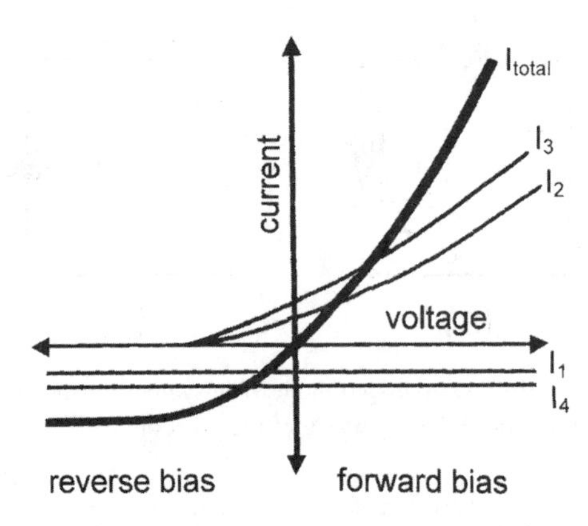

Figure 3.7. Total current and individual currents flowing through a diode as a function of bias voltage.

3.4 LEDs and semiconductor lasers

The simplest method of producing light from a semiconducting junction is by forward biasing (as illustrated in figure 3.6(a)). In this case, electrons are injected from the power supply into the n-side of the diode and holes are injected into the p-side of the diode. The forward bias increases the majority carriers across the junction and increases the forward current as in figure 3.7. When the electrons from the n-side and the holes from the p-side enter the depletion region they can recombine, where, effectively, an electron from the conduction band falls across the energy gap to fill a hole in the valence band. The change in electron energy can be liberated in the form of a photon, thereby emitting light from the depletion region.

The energy levels in the semiconductor valence and conduction bands are quantized but are very close together. The electrons in the conduction band can be in a variety of different levels and the holes in the valence band with which they recombine can also be in a variety of levels. However, it is most likely that electrons near the conduction band edge will recombine with holes near the valence band edge, meaning that the energy that produces the photon will be close to the energy of the band gap.

Figure 3.8 shows the spectrum of a red light emitting diode (LED). While the spectrum is not broad band, the range of wavelengths that are emitted is much larger than for, e.g., a gas discharge tube. This is because of the nature of the transitions between the conduction and valence bands as described above.

The wavelength of the light emitted by an LED can be varied by choosing a semiconducting material with an energy gap that is commensurate with the desired wavelength. Table 3.2 lists common semiconducting materials that are used for LEDs of different colors. Many of the materials, e.g., GaP, are referred to as 3–5 semiconductors. The semiconductors described in section 3.2 were valence 4 semiconductors (like Si). Equal proportions of a valence 3 material (e.g., Ga) and a valence 5 material (e.g., P) will act as a valence 4 semiconductor. The equivalent of a doped material can be created by adjusting the composition so that materials with more of the valence 3 component will act as a p-type material and materials with

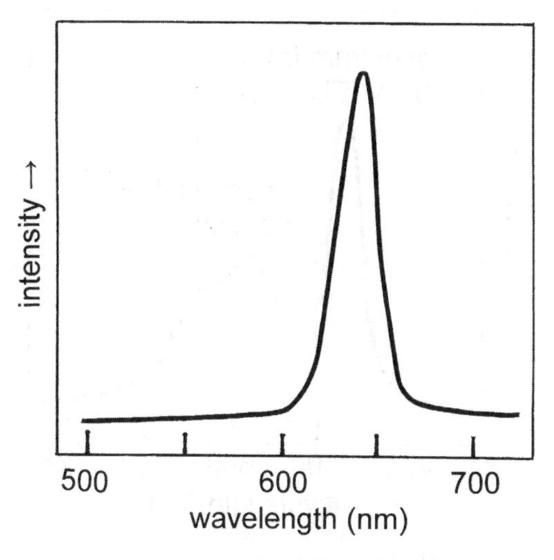

Figure 3.8. Spectrum of a red LED.

Table 3.2. Some common LED materials used to produce light of different colors.

Color	Wavelengths (nm)	Semiconducting materials
Infrared	$\lambda > 760$	GaAs GaAlAs
Red	$610 < \lambda < 760$	GaAlAs GaP GaAlInP
Orange	$590 < \lambda < 610$	GaAsP GaP GaAlInP
Yellow	$570 < \lambda < 590$	GaAsP GaP GaAlInP
Green	$500 < \lambda < 570$	GaP GaAlP GaAlInP
Blue	$450 < \lambda < 500$	GaInN ZnSe
Violet	$400 < \lambda < 450$	GaInN
Ultraviolet	$\lambda < 400$	AlN BN C (diamond) GaAlN GaAlInN GaInN

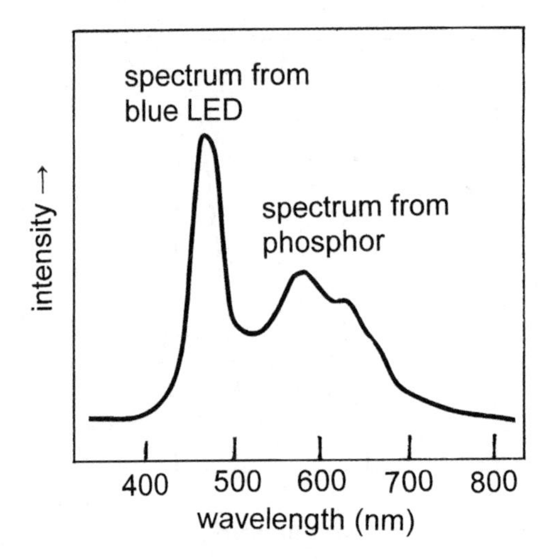

Figure 3.9. Optical spectrum of a while LED created by a blue LED incident on a broad band phosphor.

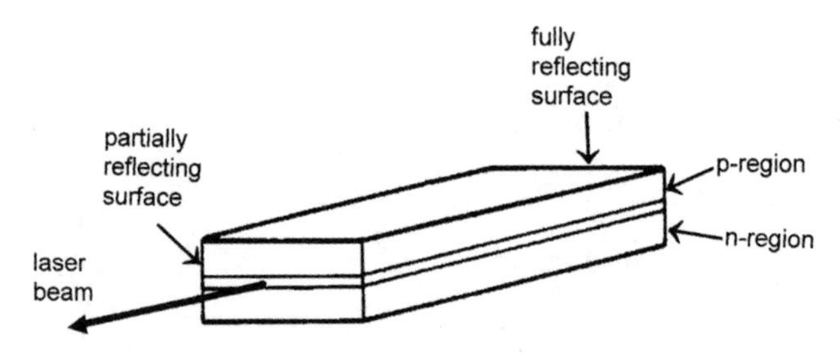

Figure 3.10. Design of a laser diode.

more of the valence 5 component will act as an n-type material. These materials will allow for the construction of p–n junctions. Adjusting the composition will also result in variations of the energy gap and hence the wavelength of the LED light.

For many applications (e.g., household lighting) white light is preferred rather than light of a single color. White LEDs can be created by two methods. A straightforward method is to combine LEDs of different colors (i.e., red, green and blue) so that the eye perceives the light as white. However, a simpler technique is to use light from a short wavelength (blue) LED to irradiate a phosphoric material, which then re-irradiates photons in a broad band over the visible region as shown in figure 3.9.

While LEDs radiate light that is more-or-less a single color, they are not as monochromatic as a laser and the light which they produce is not coherent. In order to produce the equivalent of laser radiation from an LED it is necessary to construct a resonant cavity, so that the light is amplified, and to create something equivalent to a population inversion, so that the radiation is stimulated rather than spontaneous.

A simple design for a resonant cavity in a semiconducting device is illustrated in figure 3.10. Partially and fully reflecting surfaces are made on the ends of a semiconductor so that light produced in the active junction region can be amplified. In order to improve the efficiency of the laser diode, more complex junction geometries are generally utilized which involve several layers of semiconducting materials with differing impurity levels. Stimulated emission results when there are sufficient majority carriers injected into the depletion region. Figure 3.11 shows the light output as a function of forward bias current for a typical LED and a typical laser diode. It is seen that above a threshold current the laser diode begins to produce intense stimulated emission and becomes much brighter than the LED.

The spectrum of a typical laser diode is illustrated in figure 3.12. This spectrum may be compared with the spectrum for a He–Ne laser in figure 1.14. Although the general features in these two spectra are similar, there is an important and substantial difference; the range of wavelengths is much greater for the laser diode than for the He–Ne laser. There are two important factors that are responsible for this behavior. Firstly, the energy levels of the electrons and holes, as mentioned above, are not single, well defined, levels but have a broader range of transition energies than for the atomic energy levels in a gas. Secondly, the resonant cavity is typically a fraction of a mm in length rather than tens of centimeters. The first feature means that the overall distribution of wavelengths (i.e., the envelope in figure 1.13) is much greater for the semiconducting laser than for the atomic laser. The second feature means that the spacing of the resonant modes is much greater for the semiconducting laser. This is seen by the $1/L$ dependence in equation (1.16).

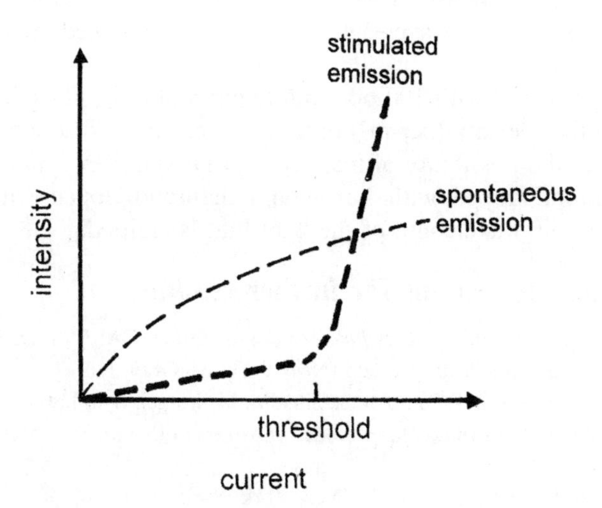

Figure 3.11. Energy output of an LED and a laser diode as a function of forward bias current.

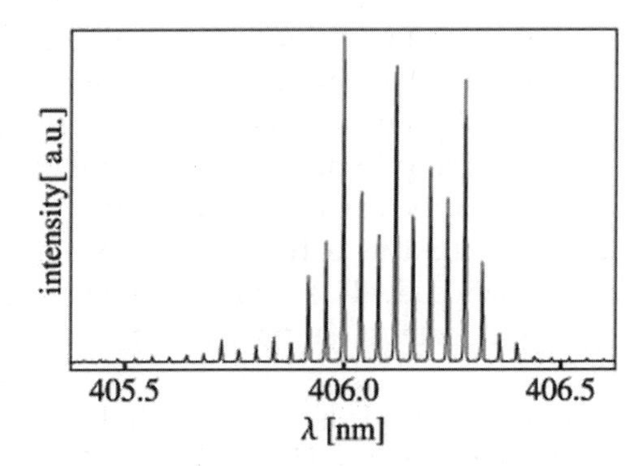

Figure 3.12. Spectrum of an (Al,In)GaN laser diode. Reprinted with permission from Meyer *et al* 2008, copyright The Optical Society.

3.5 Problems

3.1. (a) Calculate the energy (in SI) of a typical photon emitted by a red semiconducting laser.
(b) A typical red laser pointer has an output of 3 mW. Calculate the number of lasing transitions per second for such a laser.
(c) If the exit diameter of the laser pointer is 1 mm and if the optics in the human eye focuses the beam to 0.1 of its initial diameter, what is the energy incident on the retina per unit area for an exposure of 1 s? Find information about the damage threshold for red light on the human retina and compare with this result.

3.2. $Ga_{1-x}In_xAs$ is used to manufacture an LED. This material has an energy gap that varies linearly from 1.43 eV (for $x = 0$) to 0.39 eV (for $x = 1$). Derive an expression for the wavelength of the light produced by this LED as a function of x.

3.3. Silicon is doped with 100 ppm (atomic fraction) of phosphorus. Calculate the number density (per m^3) of majority electrons in the material.

3.4. Consider the possibility of making a LED from a p–n junction consisting of diamond doped with boron and diamond doped with phosphorus. Estimate the wavelength of the light that is emitted.

References and suggestions for further reading

Dunlap R A 2018 *Novel Microstructures for Solids* (San Rafael, CA: Morgan & Claypool)

Hecht J 1992 *The Laser Guidebook* 2nd edn (New York: McGraw Hill)

Meyer T, Braun H, Schwarz U T, Tautz S, Schillgalies M, Lutgen S and Strauss U 2008 Spectral dynamics of 405 nm (Al,In)GaN laser diodes grown on GaN and SiC substrate *Opt. Express* **16** 6833–45

Svelto O 1998 *Principles of Lasers* 4th edn (New York: Springer)

Wilson J and Hawkes J F B 1987 *Lasers: Principles and Applications* (New York: Prentice Hall)

Chapter 4

Laser applications

4.1 Introduction

There is an enormous number of applications for lasers that are based on one or more of the fundamental properties of laser light; intensity, monochromaticity and coherence. The details of the requirements of the application will determine the most suitable type of laser. This chapter reviews only a few of the numerous commercial, industrial and scientific applications for lasers. Part II of this book presents a detailed description of the application of lasers to the observation of an important fundamental phenomenon in statistical physics, the Bose–Einstein condensate.

4.2 Communications

A simple fiber optic cable consists of a glass core surrounded by a cladding made from glass with a different index of refraction. Light propagates along the fiber in the core by total internal reflection from the cladding which has a different index of refraction. More complex designs involve graded index fibers, where the light is gradually bent back to the center of the fiber rather than by an abrupt reflection at an interface.

Fiber optic communication systems use a modulated LED or laser diode as a source of electromagnetic radiation to transmit signals. Figure 4.1 shows the basic design of such a system. An electrical signal is used to modulate the output of the light source. The resulting optical signal is coupled to a fiber optic cable for transmission. Relay stations or regenerators can be used to extend the range or bandwidth of the transmission as discussed below. The receiver consists of a photodiode to convert the modulated light signal back into an electrical signal. The photodiode operates in a manner that is the reverse of a light emitting diode. That is, photons which are incident on the depletion region of the diode excite electrons from the valence band to the conduction band and the electron–hole

doi:10.1088/2053-2571/ab2f2fch4

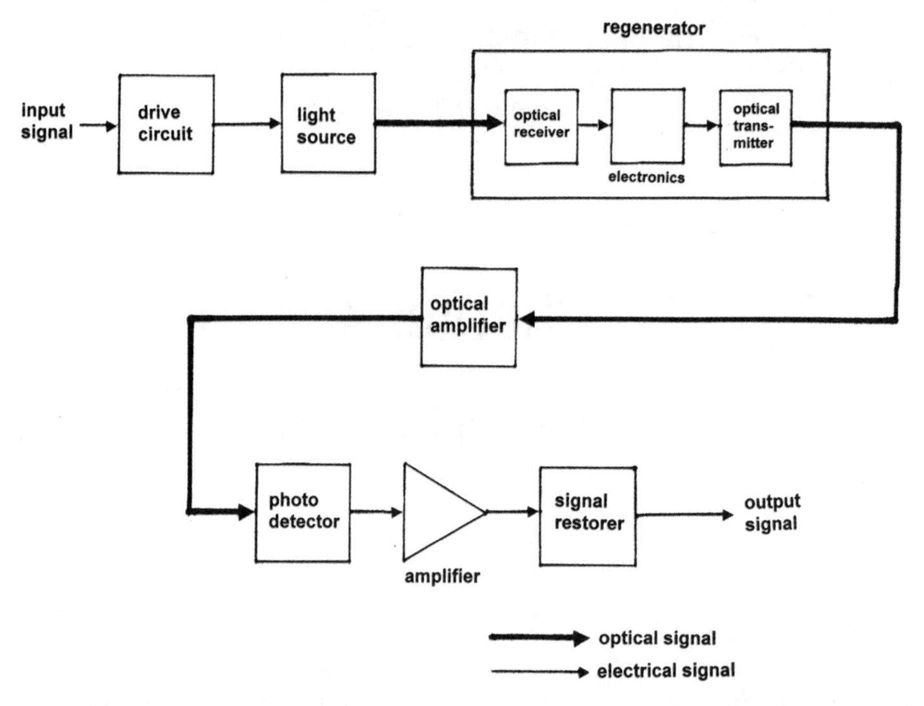

Figure 4.1. Schematic of general design of a fiber optic communication system.

pairs that are created constitute a current through the diode, which can then be measured.

Fiber optic communications have some advantages over conventional electrical conductors. These include;

- small size and low weight
- no electromagnetic interference
- high security
- low attenuation
- high bandwidth.

The lack of electromagnetic interference comes from the fact that the signal is carried by a light beam, rather than as an electric current in a wire. Since there is no electromagnetic field produced around the fiber optic cable, it is not possible to detect the signal externally, as in the case of an electrical transmission, leading to increased security of data transmission.

The limits to the distance between transmitter and receiver and to the frequency of the signal are the result of two factors; attenuation and dispersion. Attenuation comes from light absorption within the cable while dispersion comes from the properties of the glass itself and the characteristics of the light. It is typically the latter factor which limits the signal carrying ability of the fiber.

To analyze the signal transmission in an optical fiber we begin by expressing the speed of light in a medium with an index of refraction n as

$$v_p = \frac{c}{n} \tag{4.1}$$

where c is the speed of light in vacuum and v_p is the phase velocity of the wave in the medium. It is the group velocity of the wave, v_g, at which information can be transmitted. This velocity is

$$v_g = c\left(n - \lambda\frac{dn}{d\lambda}\right)^{-1} \tag{4.2}$$

where λ is the wavelength of the light source. Since the light output from an LED or laser diode is not truly monochromatic but has some spectral bandwidth, it is important to understand the relationship between the index of refraction and the wavelength. Figure 4.2 shows the index of refraction as a function of wavelength for some common optical glasses. Since the index of refraction is a function of wavelength, the group velocity of the signal is also a function of wavelength. This property means that if we send a pulse of light through an optical fiber, then the shorter wavelength component of the light will travel slower than the longer wavelength component of the light and the pulse will get stretched out spatially and temporally. The farther the pulse travels in the fiber, then the more it will be stretched out. Therefore, if we modulate a light beam in order to transmit a signal over a fiber optic cable, then the maximum frequency of modulation will be limited by the length of the cable, the dispersion of the glass fiber and the spectral width of the light source.

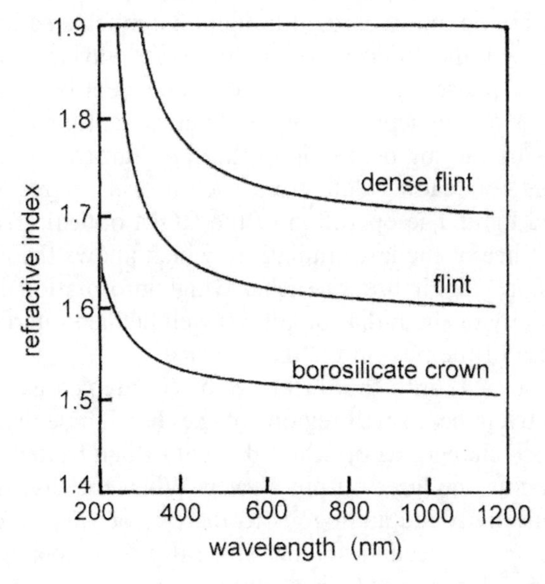

Figure 4.2. Dispersion relations for some common types of glass. The range of visable wavelengths is from about 390 nm to about 700 nm.

From a practical standpoint, the length of the cable is determined by the spacing of regenerators, as shown in figure 4.1. This distance might be a kilometer or so. On the basis of figure 4.2, it is practical to use light that is of as long a wavelength as possible, as the dispersion, $dn/d\lambda$, decreases with increasing wavelength. Perhaps the most important consideration, though, is the spectrum of the light source. If an LED is used, then figure 3.8 shows that the light will have a spectral width of about 35 nm. This spectral width translates into a maximum data transmission rate of about 500 MHz. If we use a laser diode with a spectrum as shown in figure 3.12, then the spectral width is reduced to about 2 nm and the corresponding maximum data transmission rate will be increased to about 10 GHz. This simple example clearly illustrates the advantages of the greater monochromaticity of laser radiation for data communication applications.

4.3 Optical data discs

Optical data discs are in common use in the form of CDs, DVDs and Blue-Ray discs (for example) for the storage of information for computers and audio and video devices. There are many aspects to the operation and use of optical data discs, including, the method of data encryption, the computer interface, the rotation mechanism and the head positioning mechanism. Here we will concentrate on the optical system and the application of laser diodes for writing and reading data on the disc.

There are three different designs of optical media used for data storage; read only media (ROM), recordable media and rewritable media. ROM are produced by a manufacturing process that creates pits on the surface of the disc. These pits are arranged in spiral grooves on the surface of the disc, much like the grooves on the surface of a vinyl phonograph record, although on a much smaller scale. The depth of the pits is typically of the order of about 20% of the wavelength of the laser light. The reflected beam is phase shifted relative to the incident beam and interferes in a manner that depends on the depth of the pit. Thus, data is encrypted on the disc by manufacturing pits of varying depth along the spiral groove on the disc. A simple optical arrangement for reading data from such a disk utilizes a photodiode that detects the reflected light. The operation of the ROM optical drive depends on the intensity and coherence of the laser radiation, which allows the light to be focused down onto a very small spot in order to achieve high information density on the disc. The monochromaticity of the radiation allows well defined interference effects that depend on the depth of the pits on the disc surface.

Information is 'burned' onto the surface of a recordable disc using light from a laser diode to selectively heat small regions of the disc. These discs are coated with an organic dye which changes its optical reflectivity when heated. The reflectivity is controlled by changing the laser output power. When the disc is read, regions of higher or lower reflectivity are analogous to the pits of varying depth on a ROM disc. Higher write speeds need faster heating rates and, therefore, require more powerful laser diodes. Lasers used for writing to recordable media might have an output of 200 mW, compared to lasers of about 5 mW used for reading ROM discs.

While the changes to the reflectivity of the organic dye introduced by heating the surface of a recordable disc are permanent, rewritable discs use a different mechanism that is reversible. The disc is coated with a crystalline alloy. Heating this alloy with the laser allows for the alloy to be melted and re-solidified in an amorphous phase. Since the different phases have different reflectivities, information can be stored by controlling phase formation through the laser output. Since the alloy can be re-melted by reheating, new data can overwrite old data.

4.4 Printers

The general design of a laser printer is illustrated in figure 4.3. The printer produces an image on a sheet of paper using a seven-step process, as follows:

1. raster image processing
2. charging
3. exposing
4. developing
5. transferring
6. fusing
7. cleaning.

Each of these steps is described in some detail below.

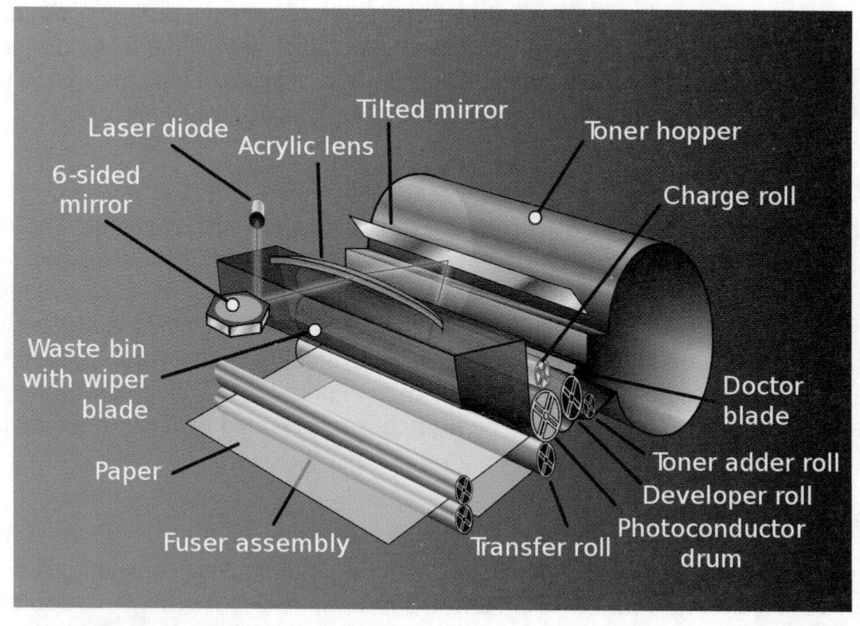

Figure 4.3. General design of a laser printer. KDS4444 / Wikimedia Commons / CC-BY-SA 4.0 https://creativecommons.org/licenses/by-sa/4.0/deed.en.

4.4.1 Raster image processing

A computer converts the object to be printed into a bitmap where each horizontal line or raster line of the image is represented by a series of bits. The computer sends each raster line to the laser printer in succession without pauses (as occur in the case of ink jet printers).

4.4.2 Charging

The image to be printed is transferred to a cylindrical printer drum which is made of a photosensitive material. This material is made of layers of semiconducting material, generally organic semiconductors, hence the usual name organic photo conductor (OPC). Before the image is transferred to the printer drum, a negative charge (electrons) is deposited on the surface of the drum. This charge is deposited from a wire (called the corona wire) which is at a high negative potential and is placed near the surface of the drum. As the drum rotates, electrons collect on the surface and form a uniform layer of charge on the surface of the drum. As the OPC is relatively non-conducting at this point the charge remains on the surface.

4.4.3 Exposing

The image is transferred to the drum by exposing selected portions of the drum to a laser beam. The laser is modulated by the information stored in the raster file and is scanned across the drum as the drum rotates. Areas where the image is intended to be dark are exposed by the laser and the laser is turned off while scanning regions which will be white in the final print. When the laser strikes the surface of the photosensitive drum, the light is absorbed by the photosensitive surface of the drum. This creates charge carriers which increase the conductivity of the drum and allow the charge on the drum to be carried away in these regions. Thus, the image to be printed becomes a pattern of non-charged dots on the surface of the drum.

4.4.4 Developing

At this time, the surface of the drum is exposed to the toner particles. The toner consists of fine particles of a coloring agent (e.g. carbon black for a black and white printer) mixed with powdered plastic. These toner particles are given a negative charge and are preferentially deposited on the regions of the drum which have been exposed by the laser and have lost their negative charge. The toner particles are repelled from the regions which have not been exposed and which still retain their negative charge. Thus, the image to be printed is represented by dark toner particles on the surface of the drum.

4.4.5 Transferring

A sheet of paper is then placed in contact with the drum and as the drum rotates the paper is moved along with the drum and the toner particles are transferred from the drum to the paper. Some printers use a positively charged electrode on the other side of the paper to help pull the negatively charged toner particles off the drum.

4.4.6 Fusing

The paper containing the image is then passed between two rollers and heated. The combination of heat and pressure melts the plastic component of the toner particles and fuses the image onto the paper.

4.4.7 Cleaning

The final step of the printing process is to remove residual toner and charge from the drum so that it will be ready to print the next page. The remaining charge is removed by exposure to a discharge lamp and the unused toned is scraped off by a soft plastic blade.

4.5 Industrial applications

The industrial use of lasers most commonly takes advantage of their high intensity and their ability to be focused onto a very small area to heat or to melt material. Principal applications include cutting, welding, selective heat treatment and additive manufacturing. For most applications, CO_2 gas lasers or Nd-based (e.g. Nd:YAG) solid state lasers are used, as these produce high output and are reasonably cost effective. For cutting applications the laser beam is focused onto a small spot on the material and a gas jet (as illustrated in figure 4.4) is used to carry away excess material. Computer-controlled movement of the laser allows for the cutting of complex patterns.

For welding applications, the ability to use CO_2 or Nd lasers in either the continuous mode or the pulsed mode allows for versatility in positioning the welds at desired locations. The high heating rates that can be achieved in laser welding are

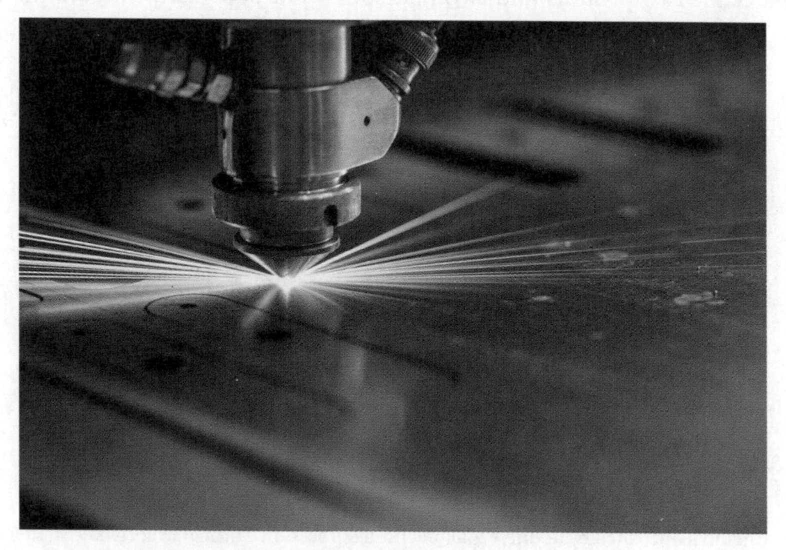

Figure 4.4. Laser cutting of metal sheet. Guryanov Andrey / Shutterstock.com https://www.shutterstock.com/image-photo/cutting-metal-sparks-fly-laser-318330188?src=X0FfSFdT53guNPwpJvi4Cg-1-14.

beneficial for welding dissimilar metals with largely different thermal conductivities (e.g. copper and stainless steel) which is difficult with conventional welding techniques.

Traditional manufacturing methods for the production of metal components include machining (that is removal of material to achieve the desired shape) and casting of molten material or powder sintering in a mold. These methods have geometric limitations, which prohibit the production of certain shapes, e.g. a single piece hollow sphere. The production of complex shapes can also be time consuming and expensive. Additive manufacturing is an alternative to these methods and follows along the lines of 3D printing. In additive manufacturing, a thin layer of powder is deposited on a build table and a laser is rastered across the surface by a computer-controlled optics system to form the cross section of the object. The build platform is then lowered, and a new layer of powder is applied, and laser melted. This approach is an effective and economical method of producing complex components or for producing individual components for prototypes. Figure 4.5 shows a component manufactured using this technique that would be difficult or impossible (and certainly expensive) to produce by conventional methods.

4.6 Inertial confinement fusion

Inertial confinement fusion (often referred to as laser fusion) is one of the approaches to producing energy by fusing light nuclei together to make a heavier nucleus. Energy is liberated because of differences in the binding energy before and after the fusion. Fusion is the method by which energy is produced in the Sun (and all other stars). Most of the energy produced by the Sun comes from the fusion of four hydrogen nuclei into one helium nucleus. This process is not straightforward and proceeds in several steps. The nucleus of the ^{1}H atom is a proton and the nucleus of the ^{4}He atom is two protons and two neutrons. (The superscript before the element

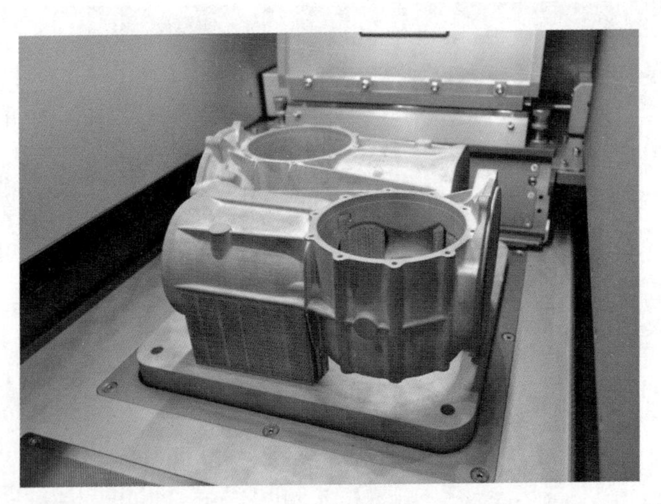

Figure 4.5. Component created by additive manufacturing processes. MarinaGrigorivna / Shutterstock.com https://www.shutterstock.com/image-photo/model-supports-created-laser-sintering-machine-766905421?src=Av3MAYQn6ra3fm2WFSlSRQ-1-62.

name gives the total number of nucleons, i.e. protons plus neutrons, in the nucleus.) So, the process in the Sun not only binds nuclei together but also converts two protons into two neutrons in the process. The most common method of energy production in the Sun begins by the fusing together of two protons:

$$\mathrm{p} + \mathrm{p} \rightarrow \mathrm{d} + \mathrm{e}^+ + \nu_\mathrm{e} \tag{4.3}$$

where d is a deuteron (that is the nucleus of a deuterium or ^{2}H atom), e^+ is the positron or antielectron and ν_e is the electron neutrino. The process in equation (4.3) is referred to as the p–p process and represents the simultaneous binding together of two protons and the conversion of one of the protons into a neutron, a positron and a neutrino. This latter process requires the weak interaction and, as a result, proceeds very slowly.

The energy associated with the p–p process can be calculated as the difference in the mass of the left-hand side and the right-hand side, where this difference in mass, Δm, is converted into energy, E, according to Einstein's mass–energy equivalence relation

$$E = \Delta mc^2 \tag{4.4}$$

where c is the speed of light. For the p–p process the energy is, therefore,

$$E = [m_d + m_e - m_p]c^2 \tag{4.5}$$

where the positron has the same mass as the electron and the electron neutrino is assumed to be massless. Using known values for the particle masses, gives the energy of the p–p process as 0.42 MeV. Normally the positron would then annihilate with an electron in the environment giving rise to (normally) two gamma rays;

$$\mathrm{e}^+ + \mathrm{e}^- \rightarrow 2\gamma \tag{4.6}$$

and yielding an additional $2m_ec^2 = 1.022$ MeV of energy, for a total of 1.44 MeV. There are a number of possible fusion processes involving deuterons, but because the Sun is still largely ^{1}H, the most likely processes will be the fusion of a deuteron and a proton;

$$\mathrm{d} + \mathrm{p} \rightarrow {}^3\mathrm{He} + \gamma \tag{4.7}$$

which releases 5.49 MeV of energy. It is important in calculating the energy for such processes, to properly account for all electron masses. The most likely route from here is the eventual fusion of two ^{3}He nuclei;

$${}^3\mathrm{He} + {}^3\mathrm{He} \rightarrow {}^4\mathrm{He} + 2{}^1\mathrm{H} + \gamma \tag{4.8}$$

which yields 12.86 MeV of energy.

A complete fusion cycle, therefore, requires the process in equation (4.3) to occur twice, the process in equation (4.7) to occur twice and the process in equation (4.8) to occur once. The input on the left-hand sides of the equations will be six protons and the final output will be a ^{4}He nucleus, two protons, two positrons and two neutrinos, leading to the net reaction

$$4\mathrm{p} \rightarrow {}^4\mathrm{He} + 2\mathrm{e}^+ + 2\nu_\mathrm{e} \tag{4.9}$$

The total energy, including the electron–positron annihilations will be $2 \times 1.44 + 2 \times 5.49 + 12.86 = 26.72$ MeV. The magnitude of this number can be appreciated if the energy is converted to Joules per kg of hydrogen; 6.5×10^{14} J kg^{-1}. This may be compared to the energy produced by a chemical reaction by burning oil; 3.8×10^{7} J kg^{-1}, and clearly indicates the interest in developing this source of energy as substantial amounts of energy can be produced from an inexpensive fuel (e.g. protons from hydrogen atoms in water).

One might think that if a reactor could be constructed that reproduced the conditions in the center of the Sun then this would represent a viable source of energy. However, this is not true. The sun will require somewhere around 20 billion years to fuse all its hydrogen into helium (and subsequently into heavier elements). If we put fuel into a reactor, then we obviously do not want to have to wait tens of billions of years to get the energy out of that fuel. Another way of looking at this is to calculate the power produced per unit volume of the Sun. The total power output of the Sun is 3.8×10^{26} W. Although the Sun's diameter is 1.392×10^{6} km, most of the energy is produced near the center, where the temperature and pressure are the highest. Figure 4.6 shows the power produced per m^3 as a function of distance from the center of the Sun. This figure shows that the reactions at the center of the Sun produce as much power per cubic meter as a few medium sized light bulbs. The only reason that the Sun produces so much power is because it is so large. The problem with implementing the reactions that occur in the Sun for a power plant on earth is that these reactions require the weak interaction and therefore, proceed very slowly.

A reasonable approach to constructing an operational fusion reactor would be to bypass equation (4.3) and start with deuterons. The most obvious reaction would be

$$d + d \rightarrow {}^{4}He + \gamma \tag{4.10}$$

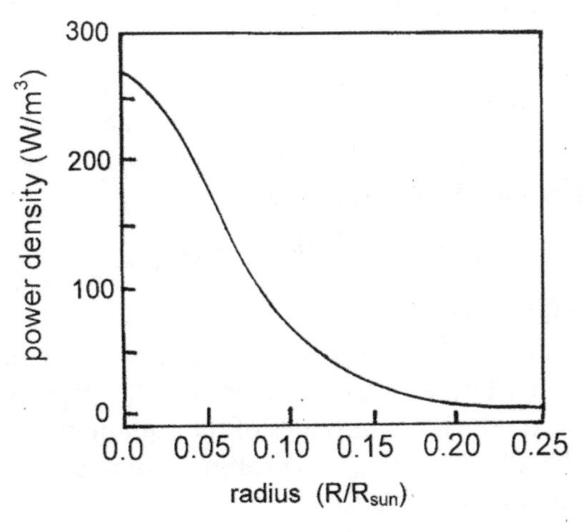

Figure 4.6. Energy production (W m^{-3}) as a function of distance from the center of the Sun as calculated from the standard solar model. Data from: http://www.sns.ias.edu/~jnb/SNdata/Export/BP2004/bp2004stdmodel.dat.

which releases 23.85 MeV of energy. Because of the large amount of energy released in this single reaction, the ^{4}He nucleus that is formed is unstable and immediately leads to one of the following reactions

$$d + d \rightarrow {}^3\text{He} + n \tag{4.11}$$

or

$$d + d \rightarrow {}^3\text{H} + p \tag{4.12}$$

The nucleus of the right-hand side of equation (4.12) is referred to as a triton (t) and the ^{3}H nucleus is tritium. These reactions yield 3.27 MeV and 4.03 MeV, respectively, and may be followed by the reaction in equation (4.8) or by

$$d + t \rightarrow {}^4\text{He} + n \tag{4.13}$$

which yields 17.59 MeV of energy. In all cases, the fusion of deuterium leads to ^{4}He and 23.85 MeV of energy. It is important in this calculation to properly account for the deuteron binding energy on the right-hand side of the equation. Although, deuterium accounts for only 0.015% of the naturally occurring hydrogen on earth, there is enough deuterium in only 0.01% of the Earth's seawater to provide all of our energy for about 4 million years at our current rate of use.

Thus, the creation of a d–d fusion reactor would provide a virtually inexhaustible supply of energy for society. The problem is to create conditions amenable to d–d fusion. The first difficulty we need to overcome is the coulombic repulsion between the positively charged nuclei. Figure 4.7 shows the coulombic potential between two nuclei as a function of distance between their centers. When the nuclei are far apart, they are repelled by the coulombic interaction. If the nuclei approach one another

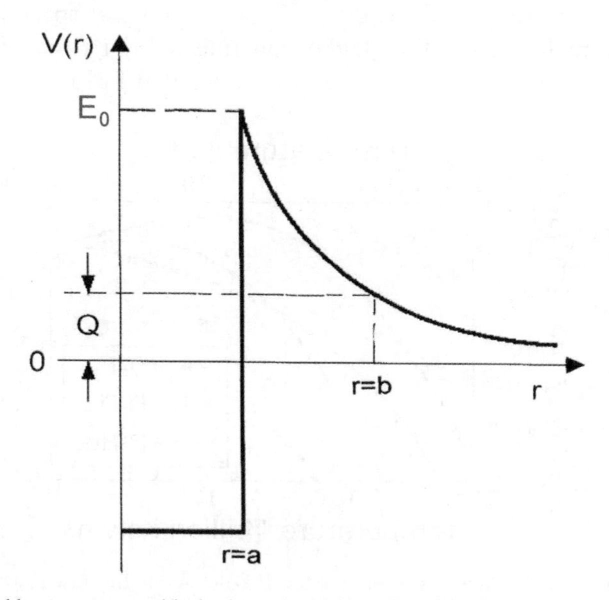

Figure 4.7. Potential between two positively charged nuclei as a function of the distance between their centers.

with some kinetic energy, Q in the figure, then they will scatter as a result of the coulombic interaction when they are a distance b (in the figure) apart. If the nuclei actually came into contact at a distance a (in the figure), then the attractive strong interaction between the nuclei would take over and they would fuse together. The energy required to do this, E_0 in the figure, is greater than that which can be achieved by any practical means. The philosophy of constructing a fusion reactor is to create an environment in which the nuclei are as dense as possible and with as great an energy as possible, so that a reasonable number of nuclei will tunnel through the coulomb barrier quantum mechanically and fuse together to produce energy. Since the temperatures associated with such energies are far above the melting points of any physical materials, the design of a fusion reactor requires a suitable mechanism for confining the collection of nuclei. At these temperatures, all atoms would be fully ionized, and the material would be a plasma consisting of positively charged nuclei and negatively charged electrons. One approach to confining the plasma is by means of magnetic fields which allow for the control of the motion of the charged particles. Another approach makes use of lasers in a technique referred to as inertial confinement. Before we continue with a description of inertial confinement, let us look at the actual conditions that are necessary to achieve a fusion reaction.

The probability of fusion (i.e. the probability for nuclei to tunnel through the coulomb barrier) increases with increasing temperature (i.e. nuclei kinetic energy). Figure 4.8 shows the reactivity for d–d, d–t and d–He^3 fusion as a function of temperature. The reactivity is defined as the average of the product of the fusion cross section, σ, and the relative velocity of the two nuclei, v, that is, $\langle \sigma v \rangle$. The fusion rate, f, is given in terms of the reactivity as,

$$f = n_1 n_2 \langle \sigma v \rangle \tag{4.14}$$

where n_1 and n_2 are the number densities of the two nuclear species.

It is clear from figure 4.8 that the fusion rate for d–t fusion is much higher (at any given temperature) than the fusion rate for d–d fusion. Thus, much lower

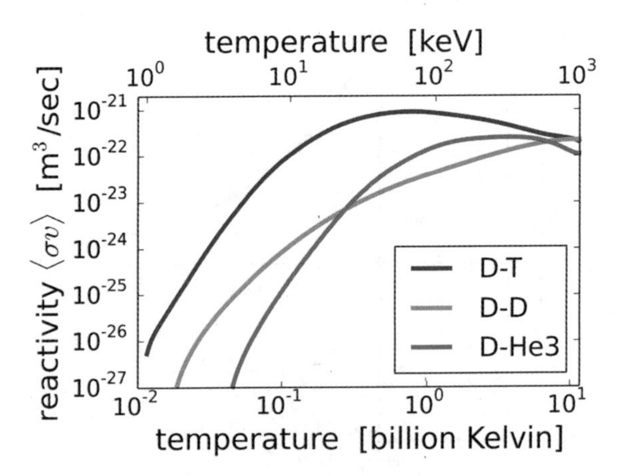

Figure 4.8. Reactivity as a function of temperature for d–d, d–t and d–He^3 fusion. Dstrozzi / Wikimedia Commons / CC-BY-SA 2.5 https://creativecommons.org/licenses/by/2.5/deed.en.

temperatures, by a factor of up to 10 or even more, are needed to achieve the same fusion rate for d–d fusion compared to d–t fusion. It is for this reason that virtually all fusion research aimed at constructing a viable power reactor deals with d–t fusion.

Once appropriate conditions are achieved, it is anticipated that nuclei may fuse by tunneling through the coulomb barrier at a sufficient rate to yield a net power output. The longer the time that the appropriate conditions are maintained, then the greater the probability that fusion will actually occur. The relevant parameter that determines the fusion rate is the Lawson parameter. This parameter is defined as the product of the number density of particles and the confinement time, $n\tau$. For d–d fusion the Lawson criterion must satisfy the condition $n\tau > 5 \times 10^{21}$ s m^{-3} in order to produce a net energy output. For d–t fusion the condition is $n\tau > 2 \times 10^{20}$ s m^{-3}. Again, this emphasizes the fact that d–d fusion will be more difficult to achieve than d–t fusion.

The objective of laser fusion experiments is to use a high-power laser to heat a very small pellet containing a mixture of deuterium and tritium to a very high temperature. In the process, the fuel is also compressed to a very high density and contained by inertial forces, hence the terminology 'inertial confinement fusion'. Inertial confinement fusion systems may be either 'direct drive' or 'indirect drive'. In the former case, the laser radiation is incident directly on a millimeter sized pellet of fuel. In the latter case, the fuel pellet is contained inside a centimeter sized cylinder called a hohlraum. The hohlraum is typically made of gold and when the laser radiation strikes the hohlraum, the gold is rapidly heated to a very high temperature and emits x-rays. The x-ray come from two processes; firstly, black body radiation which, because of the high temperatures involved, is in the x-ray region of the electromagnetic spectrum and, secondly, because of induced electronic transitions. Figure 4.9 shows a typical x-ray spectrum from a hohlraum heated by laser irradiation and illustrates the broad-band black body radiation and the series of sharp characteristic x-rays from the discrete electronic transitions.

When the laser or x-ray radiation is incident on the fuel, the fuel is heated and compressed in a series of steps, as shown in figure 4.10. The incident energy is absorbed by the fuel pellet, which is heated from the outside. This causes the outer

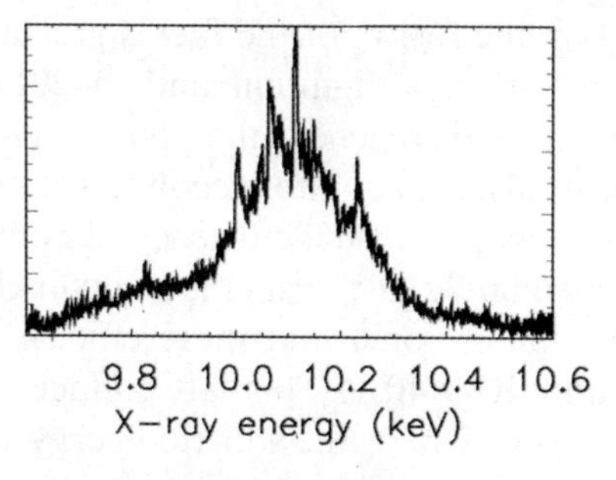

Figure 4.9. Spectrum of x-rays produced by heating a gold hohlraum with a laser. From Schneider *et al* 2008. Reproduced with permission of Canadian Science Publishing.

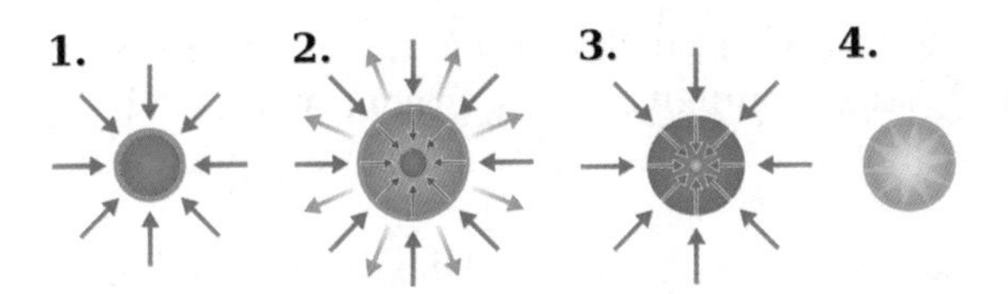

Figure 4.10. Sequence of events in the irradiation of a d–t fuel pellet; (1) laser or laser-produced x-rays incident on fuel pellet, (2) absorption of x-ray radiation and ablation of outer portion of pellet causing an inward force on the inner portion of the pellet, (3) compaction of inner portion of the pellet and (4) creation of thermonuclear fusion. https://commons.wikimedia.org/wiki/File:Inertial_confinement_fusion.svg.

portion of the pellet to expand and through inertial forces, the inner portion of the pellet is compressed. Temperatures in the range of tens of millions kelvin and densities up to 10^5 kg m^{-3} can be achieved by this approach.

The most successful laser fusion experiments have been at the National Ignition Facility located at the Lawrence Livermore National Laboratory in California. This facility utilizes a Nd-glass laser with pumping radiation provided by 7680 xenon flash lamps and the lasing transition is at 1053 nm (in the infrared). The output of this laser is more than 5×10^{14} W. This may be compared with the average total global power consumption of about 1.5×10^{13} W. The laser is pulsed with a pulse duration of about 4 ns giving the energy per pulse as $E = Pt = (5 \times 10^{14}\ \text{W}) \times (4 \times 10^{-9}\ \text{s}) = 2$ MJ. Although, this is about the same as the energy utilized by an automobile while traveling half a kilometer, it is concentrated on a target weighing a few milligrams.

The absorption of laser radiation by the fuel at 1053 nm is relatively poor but is much greater at shorter wavelengths. Harmonic generation, as described above, is useful for producing photons of shorter wavelength and greater energy. In the case of the National Ignition Facility, a combination of two potassium dihydrogen phosphate crystals are used to generate radiation at 351 nm. This wavelength corresponds to three times the fundamental frequency and is in the ultraviolet region where power absorption efficiency is near 100%.

A viable inertial confinement fusion reaction must meet several criteria. Firstly, it must produce more energy than it consumes. The basic requirement in this respect is that the energy produced by the fusion of the fuel is greater than the incident laser energy. This condition is known as ignition and in 2013 the National Ignition Facility announced that it had reached this point. However, an analysis of commercial viability must also take into account the efficiency of converting electrical energy into laser energy and the efficiency of converting heat from fusion back into electricity. Unfortunately, the former is a very inefficient process, typically about 1%, although the latter is somewhat more efficient, typically, limited by a Carnot efficiency of around 30%–40%. This latter efficiency can be improved by using direct energy conversion where the kinetic energy of the plasma's charged particles is converted directly into a voltage. Economic considerations concerning fuel production and infrastructure cost and operation are important before fusion energy can be considered a viable option for other traditional and renewable energy sources.

The above examples are a few of the very large number of applications that lasers, have found. Other notable applications include medical diagnostics and treatment, where (for example) the tunability of dye lasers allows for the optimization of the interaction of the laser radiation with particular tissues. Laser pointers and bar code readers utilize laser diodes and are in common use worldwide. Holography, which was originally only seen in scientific research laboratories, has found widespread use as a security device and now appears on the banknotes of numerous countries and on credit cards.

References and suggestions for further reading

Dunlap R A 2004 *An Introduction to the Physics of Nuclei and Particles* (Belmont, CA: Brooks/Cole)

Krane K S 1988 *Introductory Nuclear Physics* (New York: Wiley)

Schneider M B *et al* 2008 An overview of EBIT data needed for experiments on laser-produced plasmas *Can. J. Phys.* **86** 259–66

Wilson J and Hawkes J F B 1987 *Lasers: Principles and Applications* (New York: Prentice Hall)

Part II

Bose–Einstein condensates

IOP Concise Physics

Lasers and Their Application to the Observation of Bose–Einstein Condensates

Richard A Dunlap

Chapter 5

Fermions and bosons

5.1 Introduction

In the first part of this book we have seen some of the properties of particles in interacting systems. These properties include the classical behavior of particles in a gas as described by the Maxwell–Boltzmann distribution in figure 1.10 and quantum mechanical behavior of electrons in a solid. We have also seen that the behavior of systems of particles is greatly influenced by temperature. While the majority of materials used for commercial applications are used at around room temperature, many systems have behavior at low temperatures that is of considerable scientific interest. Bose–Einstein condensation is one of the phenomena which occurs in some systems at very low temperatures and is the subject of the second part of this book. It also represents an interesting and highly unusual application of lasers. We begin with an overview of the two different types of particles that exhibit quantum mechanical behavior, fermions and bosons, and their properties as a function of temperature.

5.2 Fermions, bosons and the Pauli principle

Electrons must obey the Pauli exclusion principle and this requirement determines their ability to occupy various quantum levels in the system. More precisely, the Pauli principle requires that each electron in an interacting system must have a unique set of good quantum numbers. The electrons in an atom occupy quantum mechanical energy levels designated by a principal quantum number n. In a multielectron atom, the electron–electron interactions split the levels defined by n into sub-levels designated by the angular momentum quantum number l, where $l < n$. A level described by the orbital angular momentum number l is degenerate in the z-component of the orbital angular momentum with a degeneracy of $2l + 1$. That is, there are $2l + 1$ possible orientations of the orbital angular momentum vector as shown by the example for $l = 2$ in figure 5.1.

doi:10.1088/2053-2571/ab2f2fch5

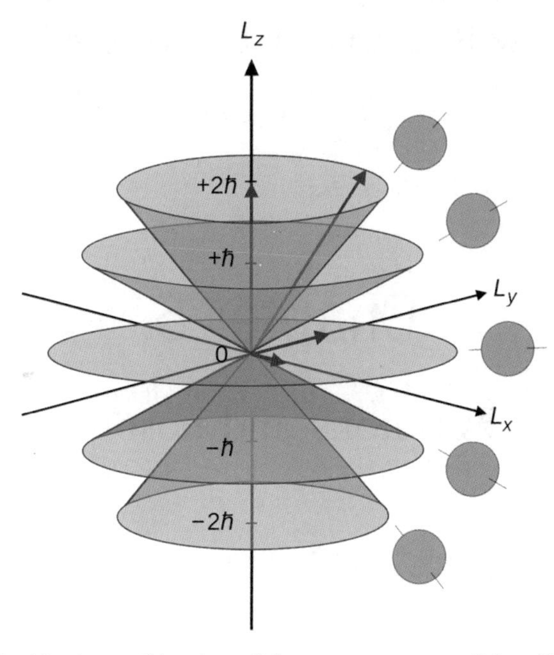

Figure 5.1. Cones defined by the possible values of the z-component, m_l, of the orbital angular momentum quantum number, l, for l = 2. https://commons.wikimedia.org/wiki/File:Vector_model_of_orbital_angular_momentum.svg.

An electron distinguished by the set of quantum numbers, n, l, and m_l, is also described by a spin quantum number s, where the intrinsic spin of the electron is $s = 1/2$. The spin has two possible orientations defined by its z-component m_s, where $m_s = -1/2$ or $m_s = +1/2$. This gives a total degeneracy of an l level as $2(2l + 1)$. Thus, a unique set of good quantum numbers n, l, m_l and m_s, defines each electron in an atom and determines how electrons will occupy the available quantum energy levels.

We also saw in chapter 3 how electrons can be thermally excited into unoccupied higher energy levels. This behavior can be explained quantitatively by the Fermi–Dirac function;

$$f(E) = \frac{2}{\exp\left[\frac{E-\mu}{k_{\mathrm{B}}T}\right] + 1} \tag{5.1}$$

as illustrated in figure 5.2. Here μ is the chemical potential (which reduces to the Fermi energy at zero temperature, see figure 5.2) and k_{B} is the Boltzmann constant. The factor of 2 in the numerator comes from the spin degeneracy.

The behavior of electrons follows the description given above because electrons are fermions. A fermion is any particle or system of particles that has a half integer spin, i.e. 1/2 in the case of the electron. Other fermions include the proton, the neutron and many atoms, as discussed below. For a particle with a spin of s, the spin angular momentum is given in SI units as $s\hbar$, where $\hbar$ is the Planck constant. It is

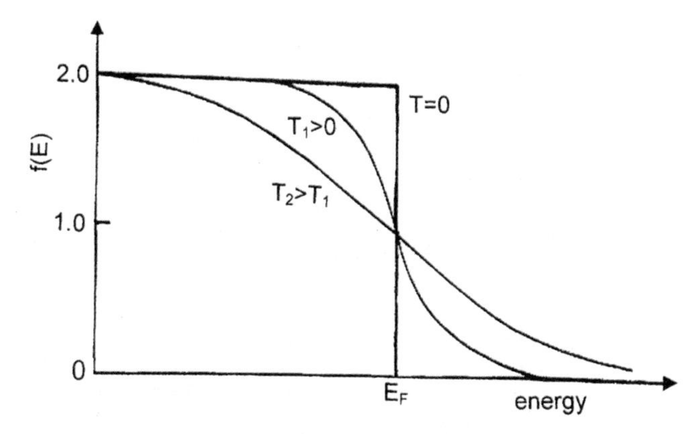

Figure 5.2. Fermi–Dirac distribution at different temperatures. The Fermi energy, E_F, is the energy separating the occupied states and the unoccupied states at $T = 0$.

interesting to note that s is dimensionless, so the Planck constant has units of angular momentum, Js.

We do not have a real fundamental understanding of why particles with half integer spin must obey the Pauli principle and follow Fermi–Dirac statistics. However, it is known that particles, or systems of particles with integer spin do not behave this way. These particles, known as bosons, after the Indian physicist Satyendra Nath Bose, are exempt from the Pauli principle and are described by Bose–Einstein statistics. In 1924 Bose suggested this model to describe the behavior of photons and it was generalized for all integer spin particles by Einstein the following year. Therefore, Bose–Einstein statistics describes the behavior of photons (spin 1) as well as all other particles or systems of particles with integer spin. The Bose–Einstein distribution has the temperature dependence

$$f(E) = \frac{g}{\exp\left[\frac{E-\mu}{k_B T}\right] - 1} \tag{5.2}$$

where g is the state degeneracy. Figure 5.3 shows the Bose–Einstein distribution at a finite temperature. A comparison of figures 5.2 and 5.3 shows a fundamental difference between Fermi–Dirac statistics and Bose–Einstein statistics that arises because of the difference in the way that the Pauli principle relates to fermions and bosons.

As the temperature for a system of bosons approaches absolute zero the form of the Bose–Einstein distribution shows that all particles will occupy the same ground state at $E = 0$. This is quite distinct from the ground state for a system of fermions, where figure 5.3 shows that the distribution is a step function which goes to zero at the Fermi energy. It is this behavior of bosons in their ground state that gives rise to the very unusual phenomenon of Bose–Einstein condensation. A fundamental concept that forms the basis of understanding why the ground state of a system of bosons behaves as it does, is the inability to distinguish particles in a quantum system, as discussed in the next section.

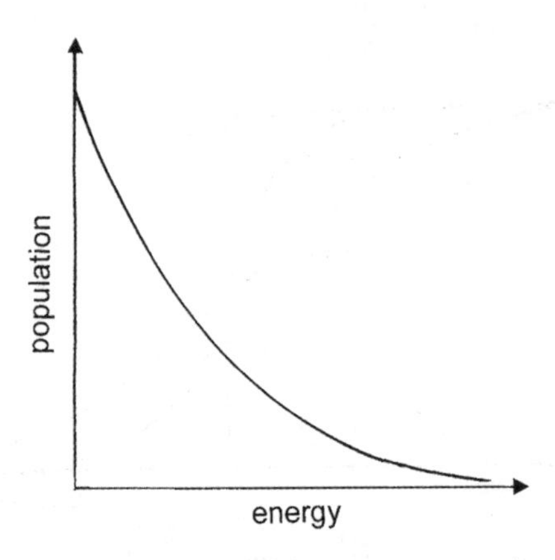

Figure 5.3. Bose–Einstein distribution at finite temperature as given by equation (5.2).

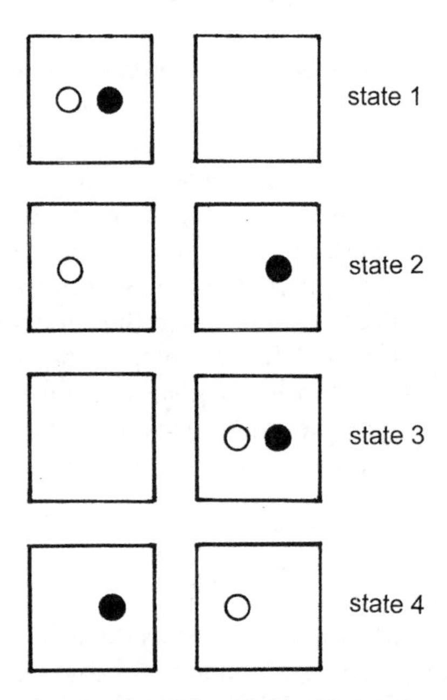

Figure 5.4. State configurations for two distinguishable objects in two states with equal energy.

5.3 Distinguishable and indistinguishable particles and quantum states

Consider two macroscopic objects that are readily identifiable in a system consisting of two states of equal energy. Figure 5.4 shows a simple example of this situation, where a white ball and a black ball can exist in one of two boxes. It is clear that there are four possible configurations for the system, as illustrated in the figure, each with equal probability. If we replace the objects in the example with identical objects, e.g. two visually identical balls, there are still four possible states each with equal probability. This is because the objects may be identical, but they are not indistinguishable. If we observe the system closely, we can keep track of which ball is in which box and we can observe changes of state by observing which ball moves where.

If we now consider a macroscopic system in which there are two states and N distinguishable objects, then we can calculate that the number of distinct configurations for the system will be 2^N, since each of the objects will have equal probability of being in either of the two states and that the probability of each configuration is the same. Since there are more configurations which have approximately equal numbers of objects in each box then it is most likely that the objects will be more-or-less equally divided between the two boxes.

If we consider the same problem for identical particles that must obey quantum mechanics, then we will find that the results are quite different because such objects are indistinguishable. Indistinguishability is a feature that results from the Heisenberg uncertainty principle. It is not possible to follow the trajectory of a particle, so if a particle changes state it is not possible to know which particle it is. In the case of two particles and two states, there are only three (rather than four) possible states; one particle in each state, two particles in one state or two particles in the other state. It is not important which particle is where, only how many particles are in each state. If we consider a system of N particles and two states, then the number of configurations is merely $N + 1$, each with equal probability, and the only relevant factor is how many particles are in each state and not which particles are where. Thus, the number of particles in one state is 0, 1, 2, 3, …, N, with the remainder in the other state.

Let us now look at a situation where the energy of the two states is not quite the same and the system is described by a ground state at zero energy and an excited state at energy, E. For distinguishable (i.e. classical) particles with $E = 0$, the particles will be equally divided between the two states. As E is increased the Maxwell–Boltzmann distribution shows that the occupancy of the excited state at a temperature T, decreases with increasing E as $\exp(-E/k_B T)$.

In the case of indistinguishable, particles there is no statistical tendency for the particles to be equally distributed between the two states when $E = 0$. Thus, when an energy difference exists between the two states, there is a greater probability (than in the case for distinguishable particles) for particles to transition into the lower energy state. It is this fundamental property of indistinguishability in quantum mechanics, combined with the fact that bosons do not have to obey the Pauli principle, that gives rise to the formation of a Bose–Einstein condensate where virtually all of the particles of a system at low temperature occupy the ground state.

5.4 What is a boson and what is not a boson?

In order to study the behavior of a system of bosons it is necessary to collect a number of bosons of the same type together and to confine these particles in some way that will enable us to investigate their properties. This is difficult with elementary particles that are bosons, e.g. photons, mesons, etc, so we need to look for some type of boson that is stable and can be collected together and confined. This means we need to look at atoms which are bosons. That is, we need atoms that have an integer spin. Although, in principle, it is simple to determine if a particular atom is a boson, it is important to understand fully how the spin of an atom arises.

The spin of an atom is the result of the spin of the nucleus and the spin associated with the atomic electrons. We begin with the nucleus. The nucleus is comprised of neutrons and protons, both of which are spin 1/2 fermions. The total spin of the nucleus, $\boldsymbol{J}_{\text{nucl}}$, is the vector sum of the spins of all the neutrons, $\boldsymbol{J}_{\text{n}}$ and all the protons, $\boldsymbol{J}_{\text{p}}$;

$$\boldsymbol{J}_{\text{nucl}} = \boldsymbol{J}_{\text{n}} + \boldsymbol{J}_{\text{p}} \tag{5.3}$$

The spin of the neutrons is the vector sum of the spins of the individual neutrons, $\boldsymbol{j}_{\text{ni}}$;

$$\boldsymbol{J}_{\text{n}} = \sum_{i}^{N} \boldsymbol{j}_{\text{ni}} \tag{5.4}$$

and similarly, for the protons;

$$\boldsymbol{J}_{\text{p}} = \sum_{i}^{Z} \boldsymbol{j}_{\text{pi}} \tag{5.5}$$

The total spin of a neutron or proton is the vector sum of its orbital angular momentum, $\boldsymbol{l}$, and its intrinsic spin, $\boldsymbol{s}$;

$$\boldsymbol{j} = \boldsymbol{l} + \boldsymbol{s} \tag{5.6}$$

For neutrons and protons in the nucleus, there is a strong coupling between the intrinsic spin and the orbital angular momentum (i.e. the spin–orbit coupling) leading to the relationship

$$j = l \pm 1/2 \tag{5.7}$$

where 1/2 is the intrinsic spin of the neutron and proton. Since l is an integer (i.e. 0 for an s-state, 1 for a p-state, etc) then for each neutron and each proton the total spin, j, from equation (5.7) will always be a half integer. Although the vector relationships of the spins of the individual neutrons and protons is not necessarily easy to determine, it is clear from the above discussion, that a bound system of neutrons and protons where the total number of particles is even, will have an integer spin, while a system where the total number of particles is odd, will have a half integer spin.

We now need to add the spin associated with the atomic electrons to the spin associated with nucleus. Analogous to the discussion above, the total spin of the atomic electrons will be the vector sum of the orbital angular momenta and intrinsic spins of all the electrons. Again, while it is not straightforward to understand the vector relationships of all these spins, it follows from the discussion above, that a system which consists of an even number of electrons will have a net spin which is an integer and a system which consists of an odd number of electrons will have a net spin which is a half integer.

It should be noted that the spin quantum numbers represent a quantized angular momentum in units of $\hbar$, which has units in the SI system of Js. Thus, although we may view electrons as quite different from neutrons and protons, in terms of mass and spatial distribution in the atom, it is appropriate to merely add (vectorially) the spin of the nucleus and the spin of the electrons to get the total spin of the atom. Thus again, while the vector relationship of the electron spin and the nuclear spin may not be known, we come to the simple conclusion that if the total number of electrons, neutrons and protons in an atom is even, then the atom will be a boson. Conversely, if the total number of electrons, neutrons and protons is odd, then the atom will be a fermion.

We now have a simple prescription for determining if a given atom is a boson or a fermion, and this will enable us to choose atoms that are bosons for a Bose–Einstein condensate experiment. Since the number of electrons, in a neutral atom, must be equal to the number of protons, then it is obvious that the number of electrons plus the number of protons must be an even number. This gives us the simple criterion for determining if an atom is a boson: all neutral atoms with an even number of neutrons will be bosons. This means that different isotopes of the same element, which differ in the number of neutrons, will be bosons when the neutron number is even and fermions when the neutron number is odd.

We can now look at some specific examples. A ^{4}He atom which consists of two neutrons, two protons and two electrons is one of the simplest atoms which is a boson. While a Bose–Einstein condensate has not been observed in ^{4}He, it does exhibit another interesting property, superfluidity, which is related to the fact that the atoms are bosons. The first true Bose–Einstein condensate was observed in 1995 by Eric Cornell and Carl Wieman for dilute ^{87}Rb atoms. A few months later Wolfgang Ketterle observed a Bose–Einstein condensate for ^{23}Na. In more recent years, Bose–Einstein condensates have been observed for dilute gases of a number of other atoms, including ^{7}Li, ^{23}Na, ^{39}K, ^{40}Ca, ^{41}K, ^{52}Cr, ^{84}Sr, ^{87}Rb, ^{86}Sr, ^{87}Rb, ^{88}Sr, ^{133}Cs, ^{164}Dy and ^{168}Er and ^{174}Yb. While roughly half of all nuclear species have an even number of neutrons in their nucleus, it is clear from the list of isotopes in which this phenomenon has been observed, that the vast majority come from specific portions of the periodic table, specifically group 1 (alkali metals), group 2 (alkaline earth metals) and the lanthanides. The reasons for this have to do with the interactions between atoms as will be discussed in further detail below.

5.5 Bose–Einstein condensation

The unusual behavior of a Bose–Einstein condensate will occur for a system of bosons which is at a temperature that is low enough that virtually all of the particles are in their ground state. It is in this state that, under the right conditions, the collection of bosons will exhibit large scale collective quantum mechanical behavior. This situation occurs when the collection of bosons is dilute enough that there is minimal interaction between individual particles and the de Broglie wavelength is large enough that the wave functions overlap. Alkali metals which have integer spin, are some of the most suitable atoms to utilize for the creation of a Bose–Einstein condensate. Alkali metals have filled electronic shells plus one additional loosely bound s-state electron. In a dilute gas there is a weak repulsive interaction between the atoms because of this outer electron and this enables the atoms to exist in their bosonic ground state with a minimum of physical interaction.

An approximate value for the temperature at which Bose–Einstein condensation occurs can be found by the following simple derivation. The de Broglie wavelength of a particle with linear momentum, p, is

$$\lambda = h/p \tag{5.8}$$

where h is the Planck constant. For non-relativistic particles of mass m, the energy is given by

$$E = \frac{p^2}{2m} \tag{5.9}$$

So

$$\lambda = \frac{h}{\sqrt{2mE}} \tag{5.10}$$

The thermal energy of a particle in three dimensions at a temperature, T, is

$$E = \frac{3}{2}k_{\mathrm{B}}T \tag{5.11}$$

Combining the above equations gives

$$\lambda = \frac{h}{\sqrt{3mk_{\mathrm{B}}T}} \tag{5.12}$$

For the situation as described above, the Bose–Einstein condensation temperature, T_{BE}, occurs when the spatial extent of the de Broglie wave becomes greater than the average distance between the atoms. For a gas with a number density of atoms n, in atoms per unit volume, then the average distance between atoms will be $n^{-1/3}$, so that the Bose–Einstein condition is met in terms of equation (5.12) when

$$\frac{1}{n^{1/3}} \leqslant \frac{h}{\sqrt{3mk_{\mathrm{B}}T_{\mathrm{BE}}}} \tag{5.13}$$

Solving the expression for temperature gives

$$T_{\mathrm{BE}} \leqslant \frac{h^2 n^{2/3}}{3mk_{\mathrm{B}}} \tag{5.14}$$

A more detailed calculation actually gives a value for the Bose–Einstein temperature which is about a factor of three higher than that in equation (5.14), but the above simple analysis provides some physical insight into the meaning for this behavior. In order to ensure that the atoms in the condensate do not interact with one another, they cannot be too close together. This puts some practical limit on the density, n, and subsequently imposes limits on how high T_{BE} can be. Typical experiments require temperatures in the nano-Kelvin range and very sophisticated experimental techniques are required to achieve temperatures this low. No single approach is sufficient to cool an experiment from room temperature down to the required nano-Kelvin range. Thus, several different cooling techniques must be used in conjunction. In the next chapter, some of the techniques that can be used to cool an experiment to very low temperature are reviewed.

5.6 Problems

5.1. Which of the following nuclides have ground state atoms which are bosons; ^{6}Li, ^{13}C, ^{27}Al, ^{57}Fe, ^{191}Ir and ^{238}U?

5.2. Consider a system of eight particles with two energy levels.
   (a) Calculate the number of microstates corresponding to each value of the total energy
      (i) if the particles are distinguishable
      (ii) if the particles are indistinguishable.
   (b) Determine the most probable particle configuration
      (i) if the particles are distinguishable
      (ii) if the particles are indistinguishable.

5.3. Tin has the largest number of stable isotopes of any element. Find information about the stable tin isotopes and determine which ones are bosons.

References and suggestions for further reading

Cornell E A and Wieman C E 1998 The Bose–Einstein condensate *Sci. Am.* 40–5
Kittel C 1969 *Thermal Physics* (Hoboken, NJ: Wiley)
Pitaevskii L P and Stringari S 2003 *Bose–Einstein Condensation* (Oxford: Clarendon)
Reif F 1965 *Fundamentals of Statistical and Thermal Physics* (New York: McGraw-Hill)

Richard A Dunlap

Chapter 6

Cooling techniques

6.1 Introduction

It is clear from the discussion in the previous chapter that the observation of a Bose–Einstein condensate requires a collection of appropriate bosons to be cooled to an appropriately low temperature. Achieving temperatures in the range of nano-Kelvin, as indicated by equation (5.14), requires the use of sophisticated cooling techniques. Suitable cooling is obtained in stages, beginning with cooling by liquid helium. This allows for cooling to a temperature of about 1.2 K by controlling the vapor pressure of the helium. The technology required to get from 1.2 K down to the nano-Kelvin range needed to form a Bose–Einstein condensate utilizes a combination of approaches including laser cooling and using lasers to spatially confine the condensate. Low temperature cooling methods are reviewed in the present chapter.

6.2 Cooling techniques: the dilution refrigerator

The simplest approach to cooling an object is to make use of the properties of the thermodynamic phase diagram. The element which has the lowest boiling point is helium. Naturally occurring helium consists of two isotopes, ^{4}He, which has a nucleus consisting of two neutrons and two protons and is 99.999 86% naturally abundant and ^{3}He, which has a nucleus consisting of one neutron and two protons and is 0.000 14% naturally abundant. Since ^{4}He is by far the most common, we might want to make use of its properties. The phase diagram of ^{4}He is illustrated in figure 6.1. At a pressure of one atmosphere, liquid helium has a temperature of 4.2 K. We can lower the temperature of liquid helium by lowering its vapor pressure with a vacuum pump. As the vapor pressure is lowered the liquid will remain in thermodynamic equilibrium with the vapor and will follow the liquid–vapor transition line, thereby lowering the temperature of the liquid. The details of the temperature–vapor pressure relationship, as illustrated in figure 6.2, show that this

doi:10.1088/2053-2571/ab2f2fch6

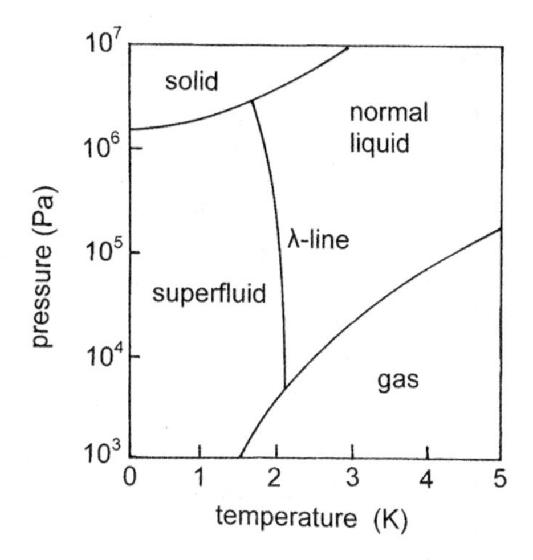

Figure 6.1. The phase diagram of ^{4}He.

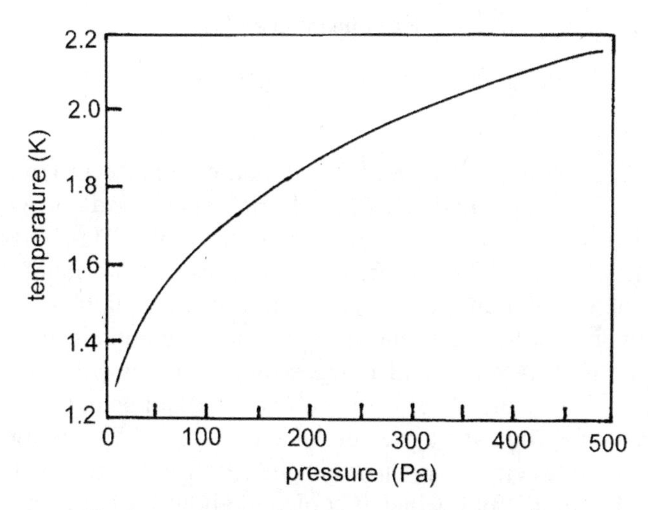

Figure 6.2. Temperature pressure relationship for ^{4}He.

method can, in principle, lower the temperature to about 1.2 K. Lowering the vapor pressure of pure ^{3}He allows the temperature to be reduced to about 0.6 K; a bit lower than for ^{4}He, but still far from the requirements for the Bose–Einstein condensate.

The general technique to achieve temperature below about 1 K is by the use of a dilution refrigerator. The dilution refrigerator utilizes the behavior of a mixture of ^{3}He and ^{4}He to transfer heat. The operation of the dilution refrigerator follows from the phase diagram shown in figure 6.3. From about 2.1 K down to 0.87 K, a mixture of ^{3}He and ^{4}He will exist in one of two phases as determined by the temperature and the composition of the mixture. At a given temperature, a mixture of ^{3}He and ^{4}He will consist of either ^{3}He in superfluid ^{4}He (for low ^{3}He concentrations) or a mixture of normal ^{3}He and ^{4}He (for higher ^{3}He concentrations). Below the tri-critical point

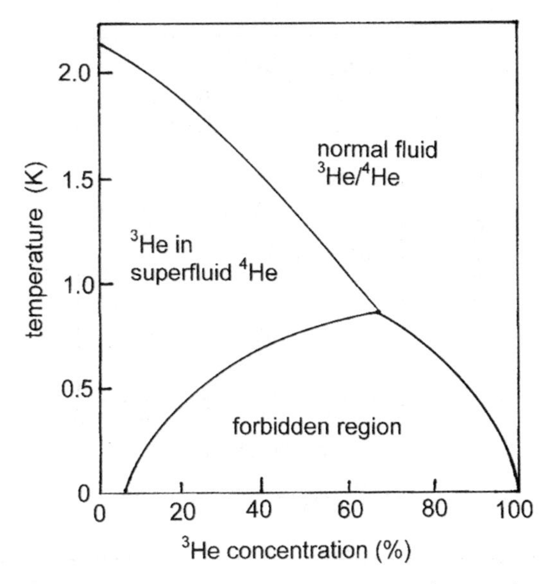

Figure 6.3. The phase diagram of a mixture of ^{3}He and ^{4}He.

at about 0.87 K, a mixture of ^{3}He and ^{4}He will phase separate into two components, one with dilute ^{3}He in superfluid ^{4}He (the ^{4}He-rich phase) and one which is normal ^{3}He in a ^{3}He-rich ^{3}He/^{4}He phase. At low temperatures, the ^{4}He-rich phase contains about 6.6% ^{3}He while the ^{3}He-rich phase approaches pure normal ^{3}He.

The general design of a dilution refrigerator is shown in figure 6.4. The working fluid is ^{3}He (analogous to ammonia in conventional commercial refrigerators or tetrafluoroethane used in household refrigerators) which is circulated through the system by pumps. The principle of operation is described below.

^{4}He at room temperature and a pressure of a few tens of kPa is pumped by the 1 K pot pump into the refrigerator, as shown in the detail in figure 6.5. The ^{4}He enters into a bath of ^{4}He (the 1 K pot) which is cooled to about 1.2 K by lowering its vapor pressure as described above. This bath condenses ^{3}He gas which enters the condenser (which is inside the 1 K pot) converting it to liquid and removing its latent heat of vaporization. The ^{3}He then passes through the primary impedance (i.e. a constricted region where its pressure is reduced). The ^{3}He is cooled by thermally coupling it through a heat exchanger to the still (operation to be described below). The ^{3}He then enters into the mixing chamber where it is combined with the proper amount of ^{4}He. This ^{3}He/^{4}He mixture phase separates, as it is below the tri-critical temperature, into the ^{4}He-rich phase containing about 6.6% ^{3}He and a nearly pure ^{3}He phase. The ^{3}He-rich phase has a lower density than the ^{4}He-rich phase and therefore floats on top of it in the mixing chamber. A tube inserted into the mixing chamber connects the region of ^{4}He-rich fluid to the still above it as shown in the illustration. It is essential that the amount and concentration of ^{3}He in the system is correct so that the interface between the two phases in the mixing chamber and the liquid–gas interface in the still are in the proper locations as shown in the figure.

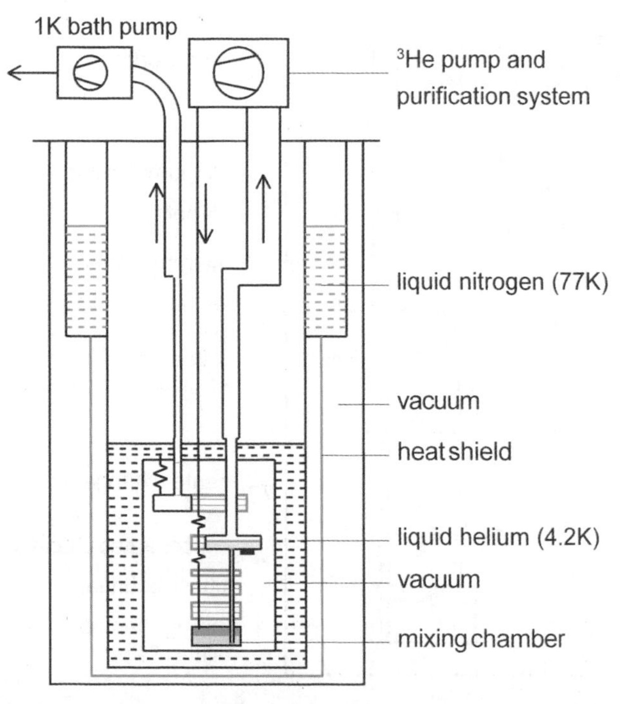

Figure 6.4. Schematic diagram of a ^{3}He–^{4}He dilution refrigerator. Adwaele / Wikimedia Commons / CC-BY-SA 3.0 https://creativecommons.org/licenses/by-sa/3.0/deed.en.

Cooling to a few mK takes place in the mixing chamber. The ^{3}He concentration gradient across the interface causes ^{3}He in the ^{3}He-rich phase on the top to mix with the ^{3}He dilute phase. As the ^{4}He in the lower part of the mixing chamber does not contribute to the ^{3}He vapor pressure, then the dilute ^{3}He in the lower part of the chamber acts like a gas. Thus, the mixing of ^{3}He from the concentrated ^{3}He phased on the top with the dilute ^{3}He on the bottom is equivalent to ^{3}He from the liquid phase evaporating, and thereby absorbing the latent heat of vaporization. In this way the mixing chamber is cooled. In the process, the evaporation of the ^{3}He forces the ^{4}He-rich phase up the tube into the still. Since the concentration of ^{3}He in the ^{4}He-rich phase is finite even as the temperature approaches absolute zero, this method of cooling has, at least in principle, no lower limit, although in actuality a few mK is the practical limit. A pump lowers the vapor pressure in the still, as shown in figure 6.5. Since ^{3}He has a lower boiling point than ^{4}He, it preferentially evaporates in the still. The ^{3}He vapor is now retuned to the condenser and the refrigeration cycle repeats. The evaporation of the ^{3}He in the still absorbs heat and is used to cool the ^{3}He as it travels through the impedance stage.

In the next section, another common method of cooling well below the temperature of a pumped helium bath, that is, adiabatic demagnetization, is discussed.

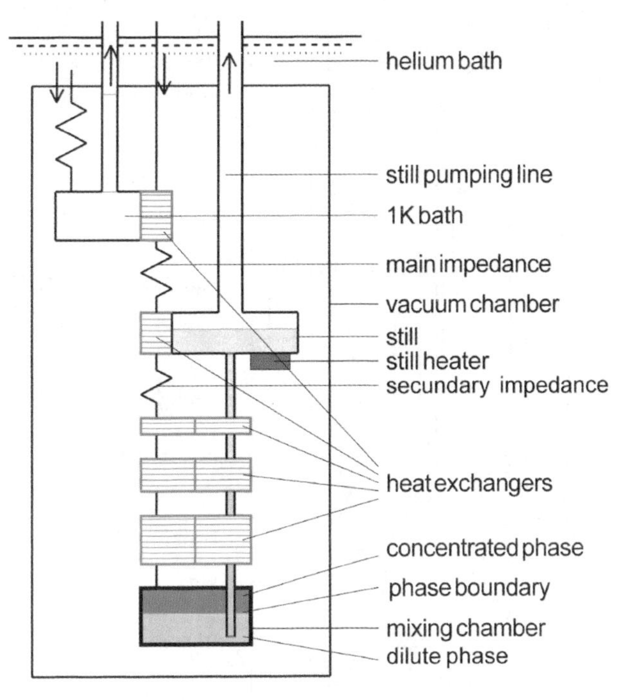

Figure 6.5. Detail of the low temperature portion of the dilution refrigerator in figure 6.4. Adwaele / Wikimedia Commons / CC-BY-SA 3.0 https://creativecommons.org/licenses/by-sa/3.0/deed.en.

6.3 Cooling techniques: adiabatic demagnetization

One of the common methods for lowering the temperature to the range of a few mK is adiabatic demagnetization. Along the lines of the dilution refrigerator, this method makes use of the entropy difference associated between ordered and disordered phases. In the case of adiabatic demagnetization, the ordered and disordered phases are related to the ordering of magnetic moments in a paramagnetic solid. We begin with a brief overview of some basic magnetic properties of atoms.

The magnetic moment associated with an atom arises primarily as a result of the spins associated with the atomic electrons. This magnetic moment, $\boldsymbol{\mu}$, may be expressed in terms of the spin, orbital and total angular momentum of the electrons, S, L and J, respectively as

$$\boldsymbol{\mu} = -g\mu_{\mathrm{B}}\boldsymbol{J} \tag{6.1}$$

where μ_{B} is the Bohr magneton;

$$\mu_{\mathrm{B}} = \frac{e\hbar}{2m_{\mathrm{e}}} \tag{6.2}$$

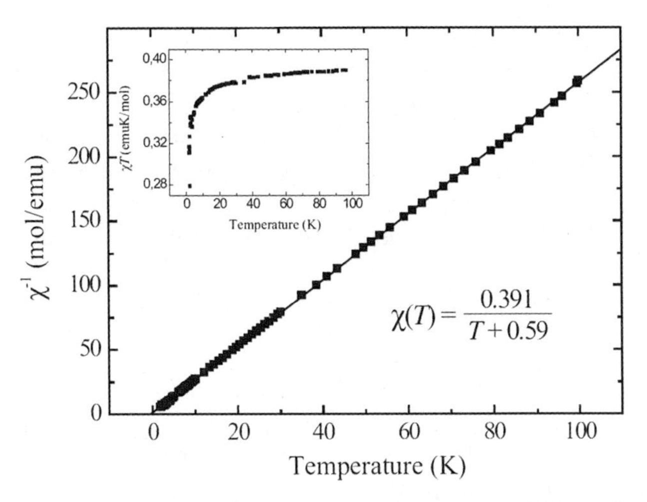

Figure 6.6. Inverse magnetic susceptibility of the copper complex diaqua-(pyridine-2,6-dicarboxylato) copper (II). Reprinted from Pérez *et al* 2017. Copyright 2017, with permission from Elsevier.

and the g-factor is defined to be

$$g = 1 + \frac{J(J+1) + S(S+1) - L(L+1)}{2J(J+1)} \tag{6.3}$$

The orientation of the magnetic moments in a material is determined by the interaction between the moments, their interaction with any external magnetic fields and the temperature. The net magnetization of a sample (per unit volume) is defined as the vector sum of the individual magnetic moments as

$$\boldsymbol{M} = \frac{1}{V}\sum \boldsymbol{\mu}_i \tag{6.4}$$

where V is the sample volume. In a paramagnetic material, which is of relevance to magnetic cooling, the interactions between the individual magnetic moments is weak and the net alignment of the magnetic moments, and hence the magnetization is determined by the temperature and the external magnetic field, $\boldsymbol{H}$. Thus, we may write

$$\boldsymbol{M} = \chi_V(T)\boldsymbol{H} \tag{6.5}$$

where $\chi_V(T)$ is the temperature dependent magnetic susceptibility per unit volume. For an ideal paramagnet the susceptibility is described by the Curie law as

$$\chi_V(T) = \frac{C}{T} \tag{6.6}$$

The validity of the Curie law is most easily observed by plotting the inverse susceptibility as a function of temperature. This is shown for a paramagnetic material that obeys the Curie law in figure 6.6.

The alignment of the magnetic moments in a paramagnetic material can be used to extract heat from the material. This process, sometimes called adiabatic demagnetization, is more generally referred to as the magnetocaloric effect, as it describes the behavior of materials in which the application of a magnetic field can result in a change in temperature. The change in temperature of a material, referred to as the refrigerant, as it is analogous to the fluid refrigerant used in Carnot cycle refrigerators, is related to the magnetization induced by an applied field H, by

$$\Delta T = -\int_{0_1}^{H} \left[\frac{T}{C(T, H)} \frac{\partial M(T, H)}{\partial T} \right]_H dH \tag{6.7}$$

where $C(T, H)$ is the heat capacity of the refrigerant as a function of temperature and applied magnetic field and the derivative is evaluated at constant field. The change in temperature of the refrigerant may be maximized by optimizing the following experimental parameters:

- using a large applied field
- using a refrigerant with a small heat capacity and
- using a refrigerant that has a large change in magnetization as a function of temperature for a given applied magnetic field.

Figure 6.7 shows the basic design of an adiabatic demagnetization refrigerator. The paramagnetic refrigerant sometimes called a 'salt pill' is shown in the diagram. The operation of the refrigerator is described by the flow diagram in figure 6.8. There are four basic steps in the principle of cooling by adiabatic demagnetization as follows:

Adiabatic magnetization: A non-magnetized paramagnetic material, the refrigerant, at temperature T is thermally isolated from its surroundings. A magnetic field, H, is then applied to induce a magnetization in the refrigerant. Since the magnetic state of the refrigerant has gone from disordered to ordered, its entropy has decreased. Since the refrigerant has been thermally isolated from its surroundings, the laws of thermodynamics require that its total energy remains unchanged. Hence, as its entropy decreases, its temperature increases to $T + \Delta T$.

Isomagnetic enthalpic transfer: The additional heat, Q, associated with the increase in temperature can be removed from the refrigerant by thermally coupling it to a heat sink by introducing a gas (i.e. helium) in the region around the refrigerant. During this process the magnetic field is maintained so that the magnetic moments remain aligned.

Adiabatic demagnetization: The heat transfer gas is removed, thus thermally isolating the refrigerant once more, and the magnetic field is removed slowly, allowing the thermal energy to randomize the orientations of the magnetic moments. This process increases the entropy of the refrigerant and thereby decreases its temperature.

Isomagnetic entropic transfer: The refrigerant, now colder than it was at the beginning of the process, is thermally coupled to the experiment to be cooled, thereby removing heat and lowering its temperature. During this time the magnetic

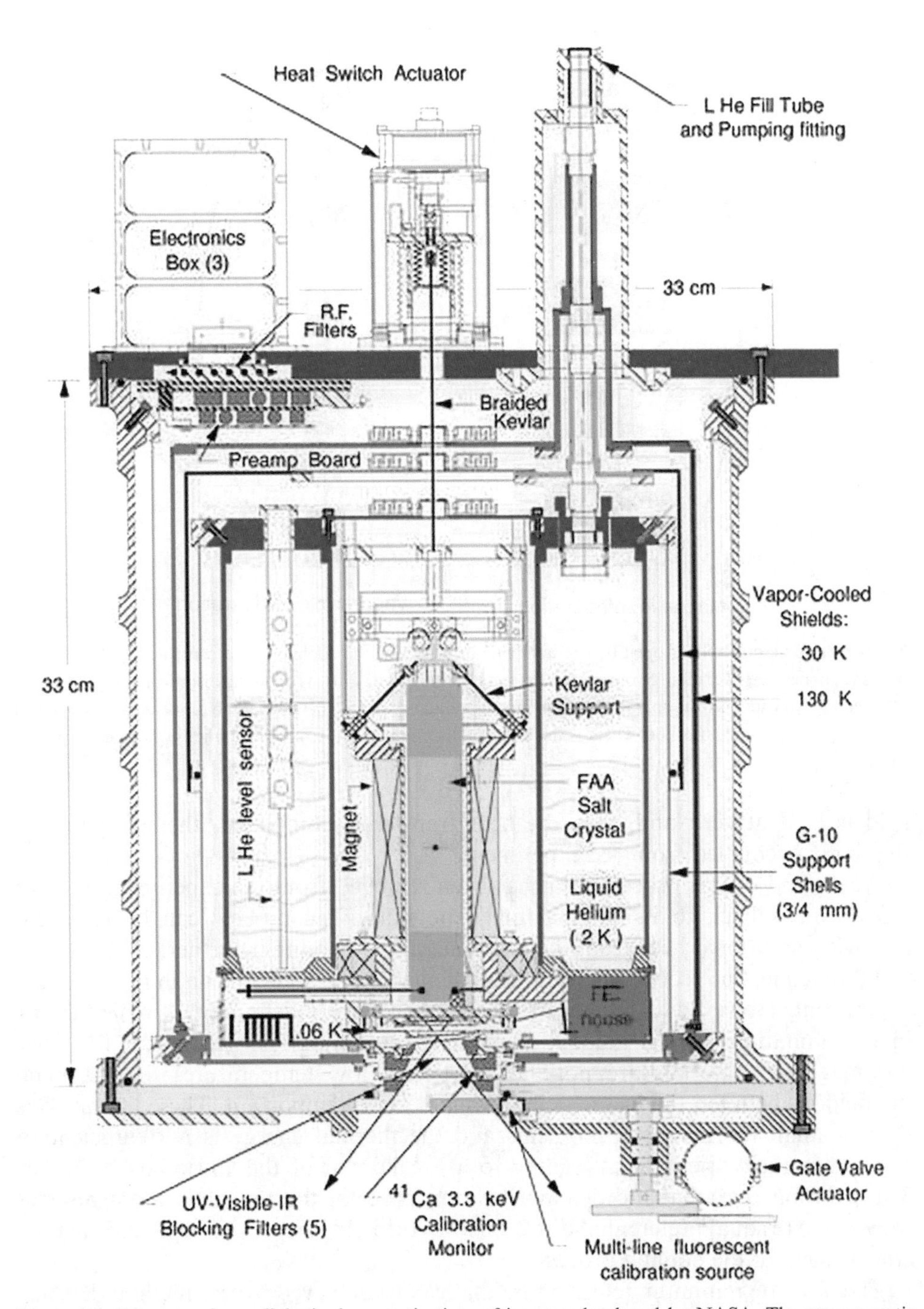

Figure 6.7. Diagram of an adiabatic demagnetization refrigerator developed by NASA. The paramagnetic material (salt crystal) is seen as the light blue material surrounded by the magnet. Courtesy of NASA https://phonon.gsfc.nasa.gov/rocket/payload/payload.html.

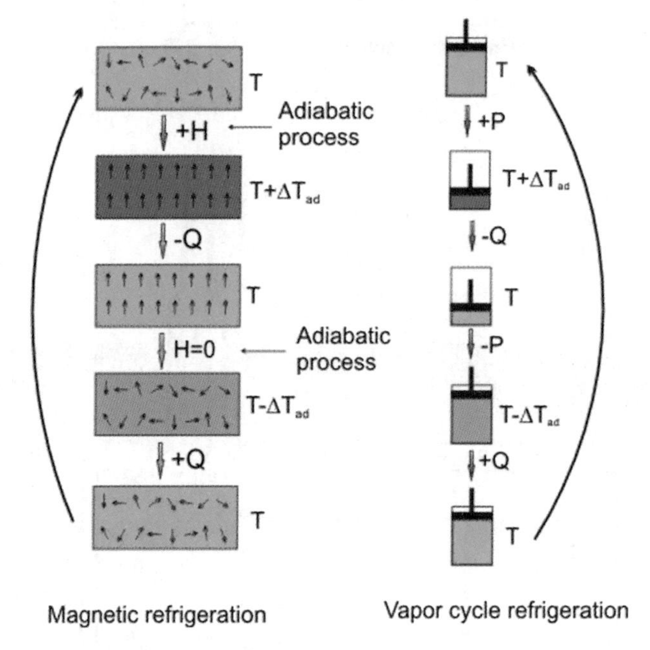

Figure 6.8. Basic principle of operation of adiabatic demagnetization cooling and a comparison with conventional vapor cycle refrigeration. From the top the states of the system are (1) beginning disordered magnetic state, (2) after adiabatic magnetization, (3) after isomagnetic enthalpic transfer, (4) after adiabatic demagnetization and (5) isomagnetic entropic transfer (which is the same as the original state). https://commons.wikimedia.org/wiki/File:MCE.gif.

field is held at zero and once the heat transfer is completed, the refrigerant is thermally decoupled from the experiment.

At the end of this process, heat has been removed from the experiment and the refrigerant is back to its original magnetic state. The process can be repeated, removing additional heat from the experiment and cooling it further.

In principle, this process can be repeated indefinitely, and will continue to cool the experiment. However, in practice this does not work indefinitely. A major factor in the limitations of the adiabatic demagnetization refrigerator, is the fact that paramagnetic materials are not ideal. At very low temperatures the magnetic interactions between the magnetic moments becomes important. These interactions tend to align the magnetic moments and the thermal energy is not sufficient to randomize their orientations leading to a breakdown of the Curie law behavior. Thus, during the adiabatic demagnetization process, the magnetic moments still show some residual preferential direction after the field is removed, thus reducing the effectiveness of the cooling process.

The ultimate minimum temperature that can be achieved by this method depends on the ability of the refrigerant to avoid being subject to coupling between the magnetic moments of the atoms. $PrNi_5$ is one of the best-known refrigerants for this purpose and allows for cooling to temperatures around 2 mK.

One way of cooling to lower temperatures that can be achieved with adiabatic demagnetization is to make use of adiabatic demagnetization associated with the very weak magnetic moments of the neutron and/or proton. If the atom has an electronic configuration where all the electron spins cancel, then there is no atomic magnetic moment and the nuclear magnetic moment may be important. Such materials are referred to as diamagnetic materials. One example of a diamagnetic material is copper. A free copper atom has the electronic configuration $[Ar]3d^{10}4s^{1}$. In copper metal the unpaired 4s electron exists in the conduction band and does not contribute to the localized atomic moment. This leads to the localized electronic configuration $[Ar]3d^{10}$, where all the electrons are in filled shells. Natural copper consists of about 69% ^{63}Cu (with 29 protons and 34 neutrons) and 31% ^{65}Cu (with 29 protons and 36 neutrons). Therefore, all copper nuclei have one unpaired proton leading to the formation of a net nuclear magnetic moment. This moment is, by analogy to equation (6.2), about three orders of magnitude smaller than a typical atomic magnetic moment because of the much greater proton mass compared with the electron mass. Thus, the magnetic coupling between the nuclear moments in diamagnets is much smaller than the coupling between atomic moments in paramagnets. This has both advantages and disadvantages. The advantage is, obviously, that nuclear adiabatic demagnetization can achieve much lower temperatures (i.e. μK) before the magnetic coupling affects the effectiveness of the method. The disadvantage is that, because of the very weak moments associated with the nucleus, much larger applied magnetic fields are necessary in order to align the nuclear moments.

While a combination of some of the techniques described above will allow an experiment to be cooled to a very low temperature, it cannot readily achieve the temperature necessary to create a Bose–Einstein condensate. All the above methods are similar in one, self-defeating, respect; they cool by transferring heat to a heat sink through a thermal link. The problem with this method is that heat can also flow into the sample from the outside through the thermal link. In order to reach the temperatures that are necessary for the formation of a Bose–Einstein condensate, we need to take a different approach, and this requires looking at temperature in a different way.

6.4 Laser cooling

It is clear that the atoms in a material are in motion as a result of thermal energy associated with them. We, therefore, commonly think of reducing this thermal motion by lowering the temperature. However, we could also think of lowering the temperature of a material by reducing the thermal motion of the atoms. This is the way we should look at temperature in order to understand how lasers can be used for cooling. In chapter 4 we looked at applications of laser light by virtue of its different unique properties; monochromaticity, coherence and intensity. It was obvious that the intensity associated with (at least some) lasers allows them to be used for heating. However, it is the monochromaticity of laser light that allows it to be used for cooling. In order to use a laser for cooling it is essential to be able to tune the wavelength (or

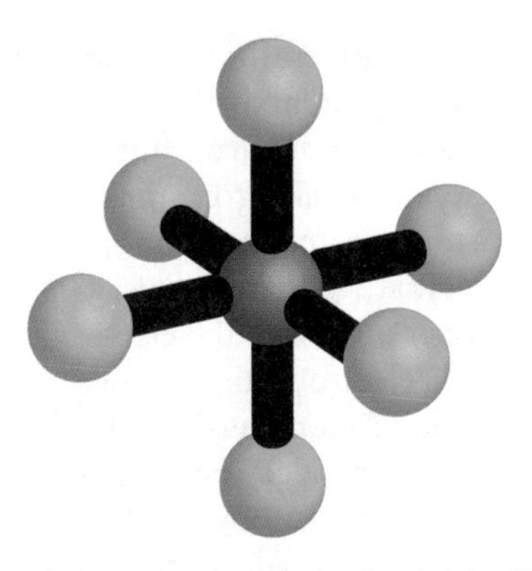

Figure 6.9. Octahedral oxygen coordination around the Ti ion in the Ti substituted sapphire structure. magnetix / Shutterstock.com https://www.shutterstock.com/image-illustration/influence-lone-electron-pairs-on-octahedral-467129483?src=-HTWncdkfvLz4AnKmKi8kA-1-3.

energy) of the laser to the correct value. We begin with a brief discussion of tunable lasers.

In chapter 2 we saw that dye lasers were tunable over a wide range of wavelengths. In the early years of laser cooling experiments, dye lasers were the most appropriate tunable lasers that were available. In more recent years, other types of lasers have been developed which can be tuned over a wide range of wavelengths. These include gas, solid state and semiconducting lasers.

Semiconducting lasers, or laser diodes, have a fairly broad distribution of wavelengths in their output, as described in chapter 3. It is also possible to adjust the output wavelength of these devices by changing their temperature, as the energy gap is a function of temperature. The output can be tuned to a very narrow wavelength using a grating along the lines of the scheme for tuning a dye laser, as shown in figure 2.10.

Probably the most useful laser that is currently available that can be tuned over a wide range of wavelengths is the titanium–sapphire laser. In fact, the titanium–sapphire laser has the largest tuning range of any laser currently known (660–1180 nm). This is sometimes referred to as a Ti:sapphire or Ti:Al_2O_3 laser, as it consists of a solid state lasing medium made of sapphire (Al_2O_3) with Ti impurities. This is analogous to the ruby laser where the lasing medium is sapphire with Cr impurities. The lasing transition in the Ti:Al_2O_3 laser occurs between electronic states of the Ti^{3+} ion. The electronic configuration of the Ti^{3+} ion is [Ar] $3d^1$. The d-state is five-fold degenerate in m_l ($m_l = -2, -1, 0, +1\ +2$). When the Ti^{3+} ion is contained in the host Al_2O_3 matrix, the charge distribution from the neighboring atoms lifts this degeneracy. The local crystallographic environment for the Ti ion is an octahedron of six oxygen as illustrated in figure 6.9. This geometry can be compared with the spatial orientation of the five-fold degenerate m_l levels for $l = 2$ as shown in figure 5.1. It can be seen that two of the m_l values (−2, +2) correspond to directions which represent bonds between the Ti and O ions along the z-axis.

The remaining three m_l values of −1, 0, +1 do not correspond to bond directions between the Ti and O ions. The directions for $m_l = -2, +2$ represent a higher energy configuration and these form an excited doublet state, while the other three values of $m_l = (-1, 0, +1)$ correspond to a lower energy triplet state. This energy level splitting of the degenerate 3d level is shown in figure 6.10. The mean energy difference between the triplet ground state and the doublet excited state corresponds to a wavelength of about 500 nm; in the green portion of the optical spectrum.

The energy levels of the Ti^{3+} ions are further complicated by the fact that the exact energies of the degenerate levels are a function of the position of the Ti ion inside the octahedral oxygen cage. Displacing the Ti ion slightly lowers the energy of the excited doublet (the so called Jahn–Teller effect) but not the ground state triplet. Thermal vibrations of the Ti^{3+} ion in the octahedral cage produce changes in the Ti^{3+} energy levels and these changes create phonons. The coupling of these phonons to the optical transitions smears the energy of the photons that are produced, giving rise to a very broad distribution of wavelengths. Because of this behavior, the Ti:Al_2O_3 laser is sometimes referred to as a vibronic laser.

In order to populate the excited state and create the necessary population inversion, the Ti:Al_2O_3 laser is optically pumped using higher energy (i.e. shorter wavelength) radiation that corresponds to the un-distorted spacing of the energy levels of about 500 nm. The most common method of optical pumping for the Ti:Al_2O_3 laser is by the use of an Argon gas laser, which has a mean output wavelength of just over 500 nm. This situation is illustrated in figure 6.11. The general design of a Ti:Al_2O_3 laser is illustrated in figure 6.12. The wavelength of the Ti:Al_2O_3 laser is tuned by adjusting the frequency of the master oscillator.

The ability of a laser to cool a collection of gas atoms, relies on the interaction between the photons of the laser beam and the atoms and our ability to control this interaction. This interaction may be described in terms of the properties of the photon. The photon obviously carries energy, but it also has momentum; $p = E/c = h/\lambda$, where E is the energy and λ is the wavelength. It is the conservation of momentum that describes the interaction. Let us look at a simple example. A laser beam is incident on a collection of gas molecules at a particular temperature. The gas molecules, by virtue of their temperature, are moving in random directions. We consider a simple one-dimensional view of this problem where the atoms are moving either parallel or anti-parallel to the direction of the laser beam. Firstly, consider a photon moving in the positive direction along the x-axis which encounters an atom moving in the negative direction as shown in figure 6.13(a). When the photon interacts with the atom the

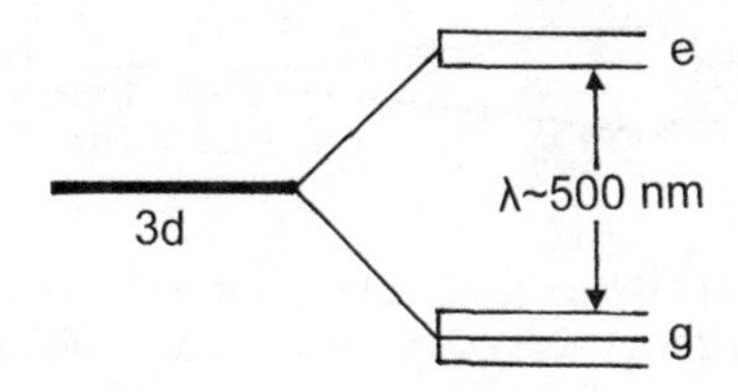

Figure 6.10. Energy level diagram of the Ti^{3+} ion showing the 3d ground state on the left and the splitting into an excited state doublet and a ground state triplet on the right.

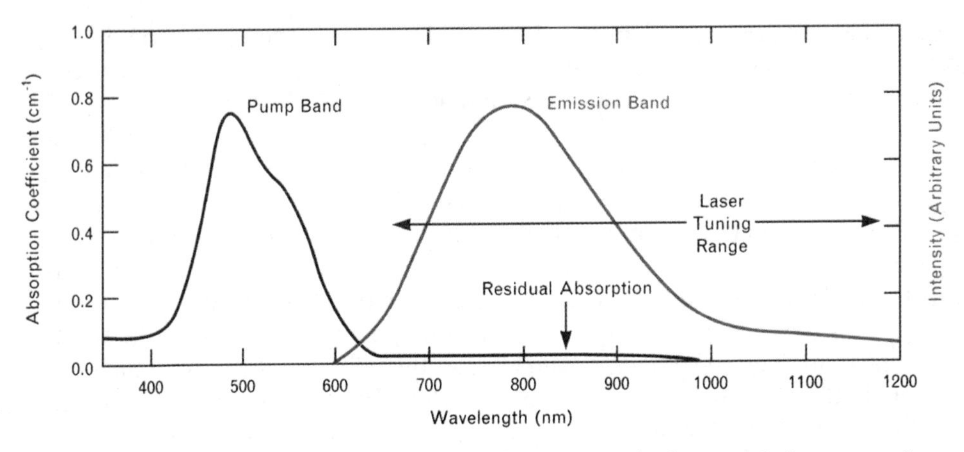

Figure 6.11. Optical spectrum of the pumping radiation from an argon ion laser and the laser output from a Ti:Al_2O_3 laser. From Wall and Sanchez 1990, reprinted with permission of MIT Lincoln Laboratory, Lexington, MA, USA.

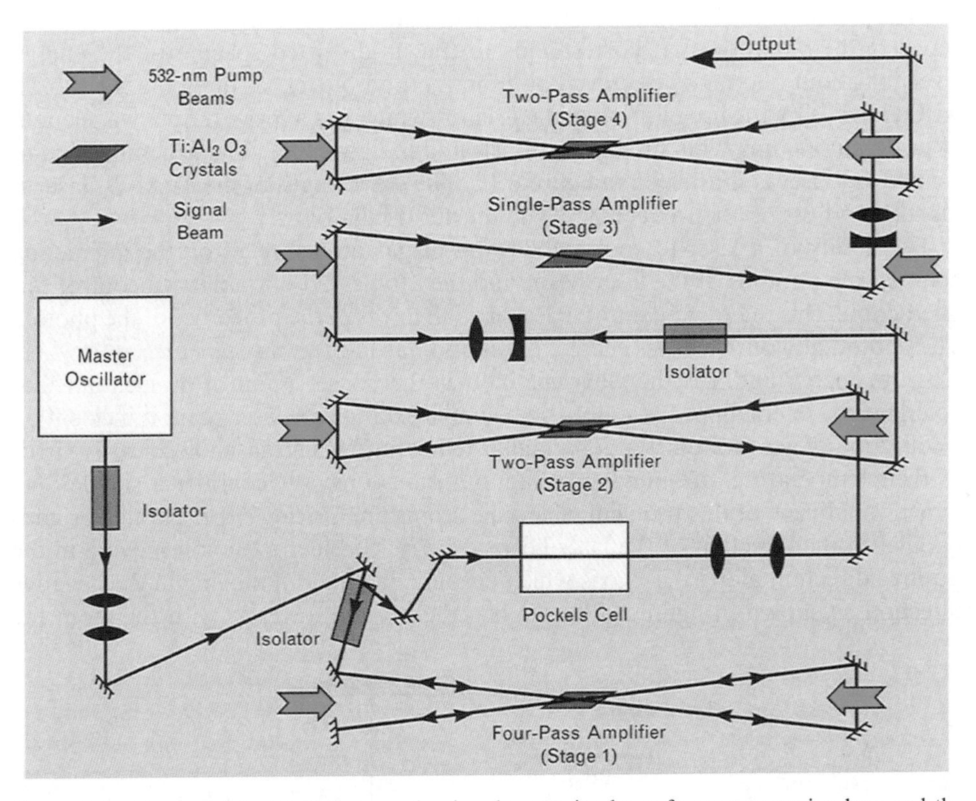

Figure 6.12. Diagram of the Ti:Al_2O_3 laser showing the pumping beam from an argon ion laser and the birefringent tuner. From Wall and Sanchez 1990, reprinted with permission of MIT Lincoln Laboratory, Lexington, MA, USA.

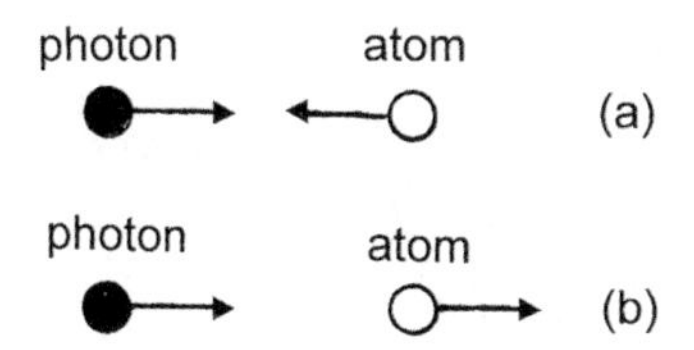

Figure 6.13. Interaction between a photon and an atom where (a) the momentum of the photon decreases the velocity of the atom and (b) the momentum of the photon increases the velocity of the atom.

momenta of the two particles will subtract leading to a reduction in the momentum (and hence the velocity) of the atom. In the opposite case, shown in figure 6.13(b), the momentum of the photon will add to the momentum of the atom, leading to an increase in the atom's velocity. A macroscopic manifestation of this basic physical behavior is referred to as radiation pressure. Although this simple example shows how we can affect the velocity of atoms using a laser, controlling the temperature of the gas is not so simple. This is because in the gas, the atoms are moving at random and the action of the laser beam will be to increase and decrease the velocity of atoms at random. To understand how we can avoid this problem, we need to look in more detail at the interaction between the photon and the atom.

Many of the Bose–Einstein condensate experiments have been performed on alkali elements. These elements have a fairly simple electronic structure with filled shells and one additional s-state electron; e.g. Na is [Ne]3s, K is [Ar]4s, Rb is [Kr]5s and Cs is [Xe]6s. This one extra s-electron has a fairly simple excited state structure where, typically the first excited state corresponds to the outer *n*s electron being excited into the *n*p orbital. The s-electron can be excited from its ground state to the first excited state by irradiation with photons of just the right energy. This energy typically corresponds to photons in the visible region of the spectrum and a tunable laser, as described above, is ideal for this purpose as the photon energy can be adjusted to match the atom's excited state energy. When the Heisenberg linewidth of the excited state energy overlaps with the width of the wavelength envelope of the laser output, then the laser radiation will be readily absorbed and will excite the atom into its excited state. This absorption will slow the atom when it is approaching the photon at the time the absorption of the photon occurs, as the momenta will partially cancel out. On the other hand, the absorption will speed the atom when it is moving away from the photon, as the momenta will add. The way in which the atoms will preferentially be slowed involves detuning the laser energy so that it is not centered on the energy of the atomic transition. Figure 6.14 illustrates this point.

Typically, the Heisenberg width of the atomic transition will be several times the width of the energy envelope of the output from the laser. The laser energy will be detuned (to lower energy than the peak of the atomic transition energy) so that the center of the laser energy envelope sits about half way down the low energy side of the atomic transition energy distribution. The cooling effect as described below is maximized when the peak in the laser energy occurs at the point of maximum slope in the curve representing the energy distribution of the atomic transition. The additional factor that we have not yet considered is the Doppler effect. The Doppler effect gives a shift in energy, ΔE;

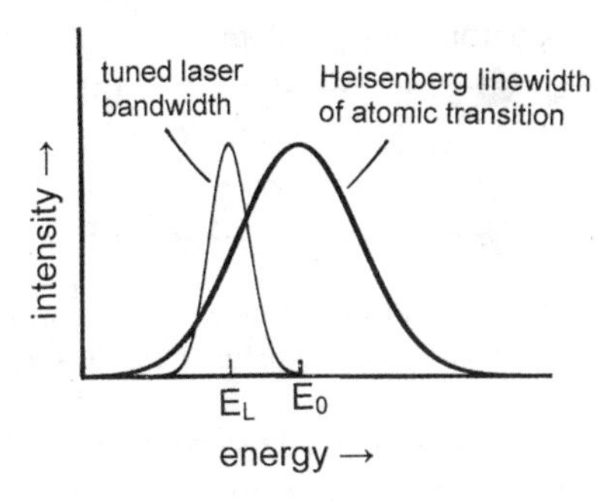

Figure 6.14. Detuning of laser energy to produce a suitable condition for Doppler cooling. E_L is the energy of the maximum in the laser output and E_0 is the center of the energy distribution associated with the atomic transition.

$$\Delta E = E_0 \frac{v}{c} \tag{6.8}$$

when the source of photons is moving with a velocity v relative to the observer. When the atom is moving towards the source of photons (i.e. the laser), then the photon energy will be Doppler shifted to a higher value; that is, towards the peak in the energy distribution of the atomic transition. This will increase the probability that absorption of the photon will occur and the probability that the atom's velocity will be affected, i.e. in this case, slowed. When the atom is moving away from the source of photons, then the photon energy will be Doppler shifted to a lower value, that is, away from the peak in the energy distribution of the atomic transition, and, therefore, the probability of absorption, and a corresponding increase in the velocity of the atom, will be decreased. Thus, although the velocity of the atom can still be increased or decreased through its interaction with the photon, it is more likely that this will occur when the atom is moving towards the laser. That means that more atoms will have their velocity decreased than increased, leading to an overall decrease in the average atomic velocity and, therefore, the temperature of the gas. In fact, the faster the atoms are moving the greater the shift of the photon energy towards the peak in the transition energy so the greater the cooling effect. This technique is sometimes known as Doppler cooling.

If two lasers are used, pointing at the collection of gas atoms from opposite directions, the atoms will be confined in the center between the lasers. This is because one laser will preferentially slow atoms moving in the negative direction but will have relatively little effect on those moving in the positive direction, while the other laser will slow atoms moving in the positive direction and will have little effect on those moving in the negative direction. If three sets of lasers pointing in opposite directions are place along the three orthogonal axes, as shown in figure 6.15, then the atoms will be confined spatially in three dimensions. This technique is referred to as laser trapping and because the force slowing the atoms is proportional to their velocity it behaves rather like viscous damping in a fluid. For this reason, Doppler cooling systems are sometimes referred to as optical molasses.

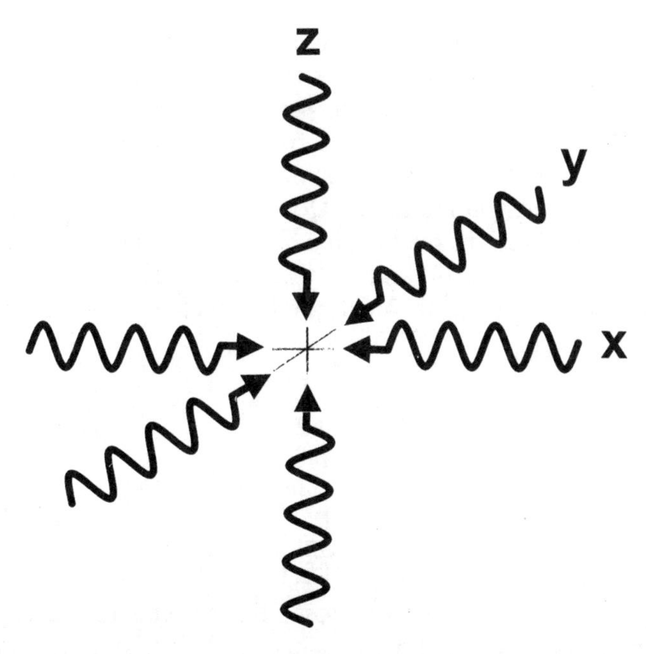

Figure 6.15. Confining atoms in three dimensions by laser trapping and cooling.

Two final questions need to be considered at this point: What is the status of the atom after the cooling has occurred and what happens to the kinetic energy that the atom has lost or the thermal energy that the gas has lost?

The atom is left in an unstable excited state after absorbing a photon. This unstable excited state will spontaneously decay back to the ground state and, in doing so, will emit a photon. Conservation of energy and momentum during this process requires that the atom recoil in the opposite direction to the direction of the emitted photon. Since, the direction of the re-emitted photon is random (that is, it is unrelated to the direction of the initial photon that was absorbed), then the recoil will be in a random direction. The isotropic re-emission means that sometimes the recoil will speed up the atom, which was just slowed, and sometimes it will further slow the atom. This effect averages to zero and, at least to a point, does not affect the overall temperature. We will see below that this will ultimately be an important factor.

In general, the photon which was absorbed was at a lower energy than the peak in the energy distribution of the atomic transition. The re-emitted photon will, on the average, be at an energy that is equal to the peak energy in this distribution. Because of the random motion of the atoms, this re-emitted energy may sometimes be red shifted and sometimes blue shifted by the Doppler effect relative to the peak energy but will, on the average be equal to the peak energy. So, the re-emitted photon will be blue shifted (that is, at a higher energy) relative to the absorbed photon. The kinetic energy loss experienced by the atom in the process of slowing down from its interaction with the absorbed photon, is exactly, on the average, radiated away in the form of the excess energy associated with the re-emitted photon.

While it may seem that there would be no limit to the temperature that can be achieved by Doppler cooling, this is not the case. As suggested above, the re-emitted photons are important in this respect as they represent the way in which the kinetic energy of the atoms is removed from the system. The ability to cool the gas by slowing the atoms is the result of shifting the laser energy towards the peak in the atomic energy distribution by the Doppler effect introduced by the thermal motion of the atoms. Thus, the relationship between the energy width of the atomic state, Γ (as shown in figure 6.14) and the thermal energy of the atoms determines the low temperature limit, T_D, to which the Doppler cooling method is effective. That is

$$k_B T_D = \frac{\Gamma}{2} \tag{6.9}$$

or

$$T_D = \frac{\Gamma}{2k_B} \tag{6.10}$$

For typical systems that are used for Bose–Einstein condensate experiments the Doppler cooling limit is around 100 μK. This is still far from the requirements for the Bose–Einstein condensate, but it turns out that laser cooling can be substantially more effective than equation (6.10) would suggest. The phenomenon behind this is Sisyphus cooling.

6.5 Sisyphus cooling

The ground state of free alkali atoms, as are typically used for Bose–Einstein condensate experiments, have a magnetic dipole moment, μ, associated with them as a result of their unpaired s-electron. This atomic magnetic moment is anti-parallel to the electron's spin (because of the electron's negative charge) and can be influenced by the application of an electric or magnetic field. The electromagnetic field associated with the laser beam lifts the degeneracy of the $s = 1/2$ ground state into $m_s = +1/2$ and $m_s = -1/2$ states with a splitting that is related to the sign of m_s and the relationship between the spin direction and the polarization vector of the light. Two light beams of the same frequency which are in phase or π radians out of phase have a net linear polarization, while two beams that are $\pi/2$ or $3\pi/2$ radians out of phase have net circular polarization. Two counterpropagating laser beams with opposite polarizations will give rise to a spatially varying polarization that will oscillate between left and right circular polarization. The spatial variations of the polarization state of the laser light give rise to a spatially varying splitting of the $m_s = +1/2$ and $-1/2$ state energies, as shown in figure 6.16.

Because the laser is detuned to the red, then the absorption of the photon, which slows the atom, always occurs from the higher sublevel of the ground state to the excited state of the atom and the re-emission of the photon during the decay back to the ground state is at higher energy (as previously explained) and leaves the atom in the lower sublevel of the ground state. As an example, consider an atom traveling

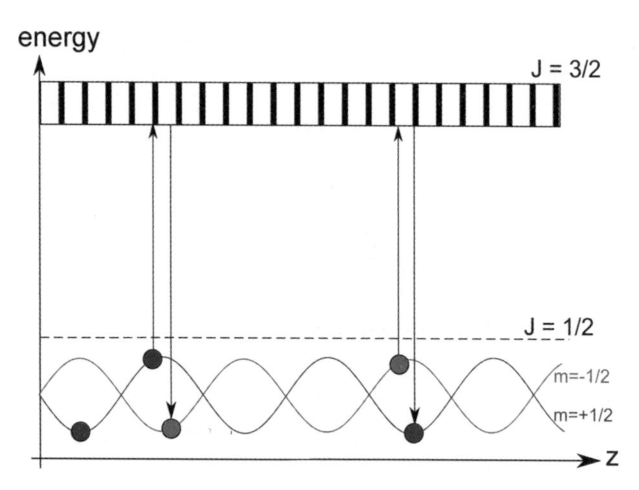

Figure 6.16. The principle of Sisyphus cooling as described in the text. https://commons.wikimedia.org/wiki/File:Sisyphus.svg.

(say) from left to right in figure 6.16. If the atom is in the $m_s = -1/2$ state, then it will preferentially absorb a photon and induce a transition into the excited state when the $-1/2$ sublevel has an energy near its maximum. When the excited state re-emits a photon, it will have a greater energy than the absorbed photon, as described above and will end up in the $m_s = +1/2$ sublevel of the ground state. These are the transitions represented by the leftmost set of arrows in the figure.

At some time later, the atom, which is now in the $m_s = +1/2$ sublevel, will be at the maximum energy and can again absorb a photon and will then decay back into the $m_s = -1/2$ state. This process is shown by the arrows on the righthand side of figure 6.16. This process will continue as the atom propagates to the right in the figure. Each time the atom climbs the energy curve of the ground state sublevel, it converts kinetic energy into potential energy. During each cycle, this potential energy is radiated away by the increased energy of the re-emitted photon. Thus, the atom is continuously climbing up a potential hill and the process is referred to as Sisyphus cooling after the character in Greek mythology who was condemned to repeatedly roll a huge boulder up a steep hill, only to see it roll back down to the bottom every time he got to the top.

Sisyphus cooling will continue to reduce the atom's thermal energy to the point where this energy becomes comparable to the energy difference between the m_s levels. As the splitting is a function of the intensity of the laser light, then decreasing the intensity of the laser will lower the ultimate temperature that can be obtained by this method. This is only possible up to a point, which brings us back to the question of the recoil energy of the atom. When the splitting becomes less than the thermal energy associated with an individual atomic recoil, then Sisyphus cooling can no longer dissipate the energy that the atom acquires during the absorption/re-emission process. We can look at this quantitatively in the following way. The conservation of linear momentum equates the recoil momentum of the atom, p_R, to the momentum of the emitted photon

$$p_R = E_0/c \tag{6.11}$$

The non-relativistic recoil energy is, therefore,

$$E_R = \frac{p_R^2}{2m} = \frac{E_0^2}{2mc^2} \tag{6.12}$$

where m is the atomic mass. Equating this energy to temperature, leads to a theoretical lower limit for Sisyphus cooling of

$$T_s = \frac{E_0^2}{2k_B mc^2} \tag{6.13}$$

For a typical alkali atom this simple calculation gives a lower limit of about 2 μK. In practice, the lower limit that can be achieved experimentally is about an order of magnitude higher than this but is still a factor of 5 or so lower than expected on the basis of simple Doppler cooling. However, we still have a considerable way to go in reducing the temperature in order to achieve a Bose–Einstein condensate.

6.6 Magneto-optic traps

Before we discuss the final stage of cooling which will allow for the formation of a Bose–Einstein condensate, it is important to look more closely at the way in which the collection of bosons (that is the dilute gas atoms) is confined within this very cold region of space. The discussion above concerning laser cooling showed how the velocity of the atoms, and hence the temperature, was lowered. This approach confined the atoms in velocity space to a point near the origin, i.e. $v_x = v_y = v_z = 0$. This confinement in velocity space, does not necessarily confine the atoms to a specific location in real space. This is particularly important in the z-direction as the atoms will be subject to the gravitational force. The atoms may be spatially confined by the use of magnetic fields. This magnetic confinement in space, combined with the velocity confinement by the Doppler cooling, is referred to as a magneto-optic trap.

As discussed above, the magnetic moment associated with a free alkali atom couples to magnetic fields and can, therefore, be influenced, not only by the electromagnetic field associated with a laser beam, but also by an externally applied magnetic field. The magnetic moment can align parallel or anti-parallel to the magnetic field and this will decrease or increase, respectively, the atom's energy. The energy difference between the two spin alignments will be

$$\Delta E = 2\mu H \tag{6.14}$$

where the lowest energy will be for a parallel alignment of the magnetic moment and the applied field. The use of anti-Helmholtz coils, as discussed below, is ideal for creating a region of very low magnetic field where the magnetic splitting is minimized, and atoms can be trapped.

Figure 6.17. Helmholtz coil geometry. Fouad A. Saad / Shutterstock.com https://www.shutterstock.com/image-vector/helmholtz-coils-286714868?src=mfz0gneeOg58AkkaK6WPJg-1-0.

Helmholtz coils are in common use as a means of producing a fairly large volume of space with a fairly uniform magnetic field. The coils consist of two thin solenoids, each with a radius R, which are separated by a distance R, as illustrated in figure 6.17. In the normal Helmholtz configuration, the currents in the two solenoids travel in the same direction and the two coils produce magnetic fields which point in the same direction and add together at the center of the solenoids. This is the location where the field is the largest and the most uniform.

Anti-Helmholtz coils use the same geometry as Helmholtz coils but the direction of the current in one of the solenoids is reversed. This produces a region of zero magnetic field at the center, with a magnetic field of increasing magnitude in all directions away from the center. The magnetic field produced by a single solenoid (or a bar magnet) is a dipole field, like the magnetic dipole moment produced by an electron's orbital angular momentum. The field produced by the anti-Helmholtz coils is referred to as a quadrupole field and the zero-field region at the center forms the magnetic quadrupole trap used for the Bose–Einstein condensate experiment. Typically, in a Bose–Einstein condensate experiment the anti-Helmholtz coils are oriented with axis of the coils along the z-direction (i.e. gravitational field direction). A magneto-optic trap, as shown in figure 6.18(a), may be created by combining the laser cooling trap (as shown in figure 6.15) to confine the atoms to near-zero velocity in three dimensions with the magnetic trap (created by anti-Helmholtz coils) to confine the atoms spatially. The importance of spin orientation will be discussed in the next section.

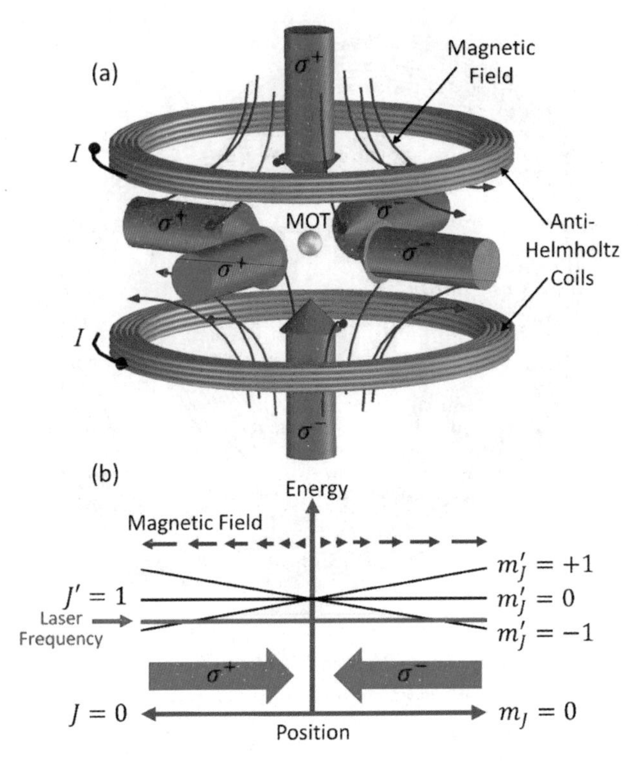

Figure 6.18. Spin-flip forced evaporative cooling from a magneto-optic trap. (a) Magneto-optic trap formed by a combination of three mutually orthogonal laser beams and a magnetic field gradient produced by anti-Helmholtz coils. (b) Energy level diagram of magnetically split energy levels around the magneto-optic trap. Reprinted from McClelland *et al* 2016, with the permission of AIP Publishing.

6.7 Forced evaporative cooling

A combination of the cooling methods as described above will lower the temperature to a few μK, still an order of magnitude or more too high for the creation of a Bose–Einstein condensate. The final step to cooling the bosons to a sufficiently low temperature uses a very simple and well know technique, that of evaporative cooling. For a classical system, the velocity distribution is given by the Maxwell–Boltzmann distribution, as illustrated in figure 1.10. It is the particles with the greatest velocity in the high velocity tail of the distribution that are lost to evaporation. These high energy particles carry away a disproportionately large fraction of the energy because the energy per particle is proportional to the square of the velocity. Thus, the evaporation of a relatively small fraction of the particles in a system can give rise to a substantial reduction in temperature. It is essential in this cooling process, that once the high velocity particles have been lost due to evaporation, the remaining particles re-equilibrate by distributing their energies to create a new, proper Maxwell–Boltzmann distribution for the new temperature. Let's look at how this approach can be used effectively at very low temperatures.

Since the laser cooling technique limits the ultimate temperature that can be achieved, it is necessary to turn off the lasers temporarily to allow the system to cool to a lower temperature. The gas atoms are now just confined within the magnetic trap. In this region of very low magnetic field created by the anti-Helmholtz coils, the magnetic splitting given by equation (6.14) is minimized and increases spatially in all directions from the center of the trap, as shown in figure 6.18(b). The lower sublevel in the figure corresponds to a parallel alignment of the atom's magnetic moment and the applied magnetic field, while the upper sublevel corresponds to an anti-parallel alignment of the moment and the applied field. It is clear from the figure that those atoms with moments that align anti-parallel to the field are in a stable energy minimum and are trapped while those with a parallel alignment are not. Transitions between the lower and upper states can be induced by the application of a radio frequency (rf) magnetic field with an appropriate angular frequency, ω, such that

$$\hbar\omega = 2\mu H \tag{6.15}$$

By controlling the duration of rf pulses, the atoms can be configured in either desired state. The atoms with anti-parallel moments and the lowest velocities sit near the minimum in the upper sublevel in the middle of the magnetic trap. Those atoms with anti-parallel moments and higher velocities climb up the potential and reside in the outer regions of the magnetic trap. By applying a tuned rf pulse of the proper frequency and duration, these higher energy atoms near the edges of the trap can undergo a spin-flip and fall down into the lower parallel moment sublevel. Once there, it is energetically favorable for them to fall off the edge of the energy curve and evaporate from the trap, carrying energy away with them. Since it is the higher velocity atoms from the anti-parallel level that were removed, the atoms that remain in the trap re-equilibrate to a lower temperature. This process can be repeated, lowering the temperature at each step at the expense of losing a small fraction of the atoms in the trap. This final step of cooling by forced evaporation can lower the temperature to the nK range necessary for the formation of a Bose–Einstein condensate.

6.8 Problems

6.1. The ^{23}Na excited state lifetime is 16 ns. Estimate the minimum temperature that can be achieved Doppler cooling using this nuclide.

6.2. A ^{87}Rb atom with a thermal velocity of 50 m s^{-1} absorbs a photon with a wavelength of 780 nm traveling in an anti-parallel direction. Calculate the change in velocity of the atom.

6.3. Calculate the magnetic moment associated with 3+ ions of Mn, Fe and Co.

References and suggestions for further reading

McClelland J J, Steele A V, Knuffman B, Twedt K A, Schwarzkopf A and Wilson T M 2016 Bright focused ion beam sources based on laser-cooled atoms *Appl. Phys. Rev.* **3** 011302

Metcalf H and van der Straten P 1999 *Laser Cooling and Trapping* (New York: Springer)

Pérez A L, Neuman N I, Baggio R, Ramos C A, Dalosto S D, Rizzi A C and Brondino C D 2017 Exchange interaction between $S = 1/2$ centers bridged by multiple noncovalent interactions: contribution of the individual chemical pathways to the magnetic coupling *Polyhedron* **123** 404–10

Phillips W D and Cohen-Tannoudji C 1990 New mechanisms for laser cooling *Phys. Today* 33–40

Richardson R C and Smith E N (ed) 1988 *Experimental Techniques in Condensed Matter Physics at Low Temperatures* (Reading, MA: Addison-Wesley)

Wall K F and Sanchez A 1990 Titanium sapphire lasers *Linc. Lab. J.* **3** 447–62

White G K and Meeson P J 2002 *Experimental Techniques in Low Temperature Physics* 4th edn (Oxford: Oxford University Press)

Richard A Dunlap

Chapter 7

The Bose–Einstein condensate

7.1 Introduction

After the process of cooling and trapping bosons as described above is completed, a collection of typically a few thousand to a few million atoms at a temperature below 100 nK is trapped in a volume which is less than a mm on a side. Once such a condition has been achieved, how do we know if the bosons have, in fact, condensed into a Bose–Einstein condensate? The present chapter reviews the methods for confirming the presence of a Bose–Einstein condensate.

7.2 Creating and identifying a Bose–Einstein condensate

The most straightforward method of confirming that the collection of bosons has formed a Bose–Einstein condensate is to show that its velocity distribution is characteristic of such a condensate and not of a classical Maxwell–Boltzmann distributed gas. Measuring the velocity distribution for the atoms is relatively straightforward. The magnetic field producing the trap is turned off, allowing the atoms to move freely. There are two things that happen to the movement of the atoms. Firstly, the atoms are subject to the gravitational force and will start to fall. Secondly, since the atoms are no longer trapped by the magnetic field, their spatial distribution will start to spread. These changes in the atomic motion can be observed using a laser. The laser is tuned to the resonant frequency of the atomic excited state. There is, therefore, a high probability of resonant photon absorption by the atoms and the accompanying population of the excited state. There is, of course, re-emission of these absorbed photons, but this is isotropic, as discussed above. As a result of this absorption, there is a large decrease in the intensity of the laser beam in the forward direction. The amount of light that is transmitted through the gas is inversely proportional to the density of the gas. The light that is transmitted through the gas can be imaged using a CCD (charge coupled device), which is the sensor used

doi:10.1088/2053-2571/ab2f2fch7

in a digital camera. Changes in the spatial distribution of atoms can be observed by following changes in the image recorded by the CCD, and these can be related to the distribution of velocities in the two-dimensions orthogonal to the propagation direction of the laser.

Figure 7.1 shows the results of two-dimensional velocity distribution measurements for a gas of ^{87}Rb atoms undergoing Bose–Einstein condensation. At the left, the temperature is above the Bose–Einstein condensation temperature for the gas. The velocity distribution is characterized by the broad Gaussian shaped Maxwell–Boltzmann distribution characteristic of a classical gas. In the center image, the gas has been cooled below the condensation temperature and a very sharp peak, due to the atoms that have condensed into a Bose–Einstein condensate, appears in the center of the velocity distribution. In the image on the right, the temperature has been lowered by removing the higher velocity atoms using forced evaporative cooling and the resulting collection of atoms has formed a nearly pure Bose–Einstein condensate.

Figure 7.2 shows another example of a Bose–Einstein condensate. In this case the bosons are $^{40}K_2$ molecules. A ^{40}K atom has 21 neutrons and 19 protons in the nucleus and 19 atomic electrons. Thus, a neutral ^{40}K atom has a total of 59 fermions and is, therefore, a fermion itself. The $^{40}K_2$ molecule consists of two ^{40}K atoms and is therefore a boson. The results shown in the figure illustrate the formation of a Bose–Einstein condensate of $^{40}K_2$ molecules.

Another interesting and important feature that is seen in figures 7.1 and 7.2 relates to the shape of cross section of the peak in a plane parallel to the velocity axes. It is observed that the Maxwell–Boltzmann distribution at high temperature has a cross section which is circular. This is because the Maxwell–Boltzmann velocity distribution is isotropic in three-dimensional space. It is seen, however, that the

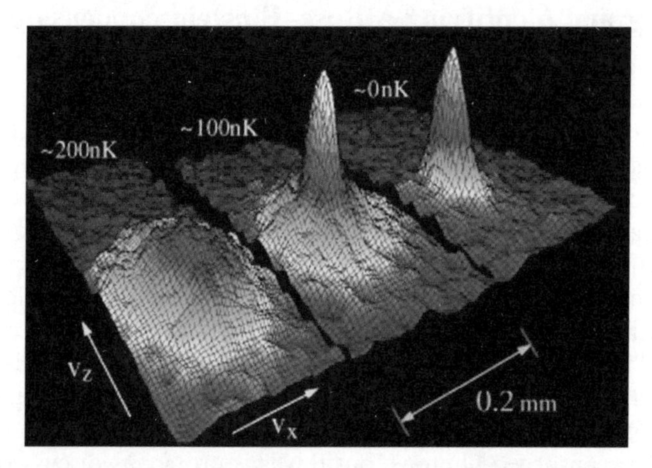

Figure 7.1. Formation of Bose–Einstein condensate as a function of temperature for a collection of ^{87}Rb atoms. Left: before the formation of the Bose–Einstein condensate showing a Maxwell–Boltzmann velocity distribution. Center: just after the formation of the condensate showing the sharp Bose–Einstein peak in the center. Right: nearly pure Bose–Einstein condensate. https://commons.wikimedia.org/wiki/File:Bose_Einstein_condensate.png.

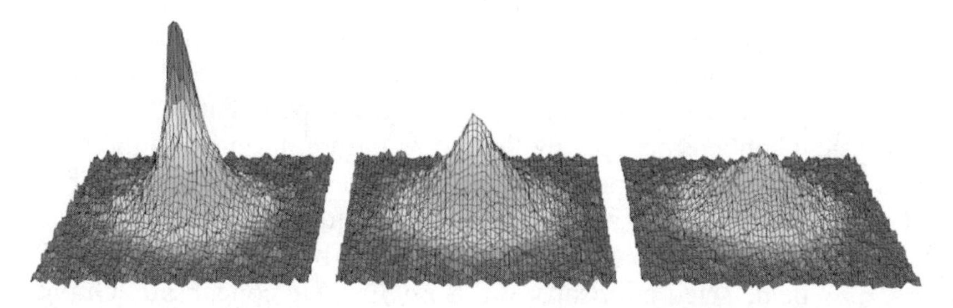

Figure 7.2. Formation of a Bose–Einstein condensate for bosonic $^{40}K_2$ molecules, from right to left. (Figure 3 reprinted with permission from Regal *et al* 2004. Copyright 2004 by the American Physical Society.)

sharp peak that arises from the Bose–Einstein condensate has a cross section that is elliptical, not circular. As can be noted, the velocities that are shown in the graphs are the velocities in the x and z directions (not x and y) and this is the reason for the elliptical shape. The velocity distribution of the Bose–Einstein condensate is not isotropic. Rather it assumes a shape that is characteristic of the spatial geometry of the magnetic field trap. This is because it is, in fact, the magnetic field distribution, not the position in real space, that defines the Bose–Einstein condensate. Based on the orientation of the magnetic coils, the magnetic field will have axial symmetry in the x–y plane but will be different in the z-direction. As a consequence of the three-dimensional characteristics of the magnetic field lines, the x and z dependences of the velocity will be different, leading to the anisotropy, as seen in the figure. In fact, some experiments intentionally alter the anisotropy of the magnetic field profile in the z-direction in order to accentuate the anisotropy of the Bose–Einstein velocity distribution as a convenient means of distinguishing between the isotropic Maxwell–Boltzmann distribution and the anisotropic Bose–Einstein distribution.

7.3 Why is it useful?

The Bose–Einstein condensate does not have obvious commercial applications, as is the case for the laser as described in chapter 4. However, it has been important in advancing the fundamental understanding of quantum systems, wave coherence and interference effects. Not only does the study of Bose–Einstein condensates validate the theoretical framework of the statistics of particles and the properties of de Broglie or matter waves, but it provides a convenient platform for investigating many properties of condensed systems. A 'lattice' can be constructed from the interference patterns of multiple laser beams. Since the atoms in a Bose–Einstein condensate can be manipulated by means of laser beams, this periodic optical lattice influences the atoms of a Bose–Einstein condensate in a way that is analogous to the way in which the crystalline lattice influences electrons in a normal solid. The optical lattice, however, can be manipulated much more easily and in ways that are not possible for a crystalline lattice. For example, the lattice periodicity, interactions and dimensionality can be controlled by the experimental parameters related to the properties of laser beams and the Bose–Einstein condensate. This ability allows for the convenient testing of solid-state models and the investigation of phenomena that are not easily accessible by conventional solid-state experiments.

Bose–Einstein condensates are, by their very nature, fragile and the use of their properties in practical devices poses substantial challenges. However, their ability to respond to very small perturbations may make them ideal for certain uses. It has been proposed that gravitational waves, as predicted by Einstein's theory of general relativity, may be detected by means of a Bose–Einstein condensate. Gravitational waves will generate phonons in a Bose–Einstein condensate and the influence of these phonons on the properties of the condensate can be observed. This would mean that interference effects in the matter waves in the condensate could be used to produce a table-top size experiment that could replace the kilometer-long optical interferometers currently used to observe gravitational waves.

The very stable wave function associated with a Bose–Einstein condensate could form the basis for a highly accurate time standard that could improve upon current technologies that utilize the frequency of electronic transitions in atoms as a basis for measuring time. Other, potentially more commercially viable applications include matter holograms, which use the interference of matter waves in the same way that optical holograms use light, quantum computers, which use the quantum state of atoms in a condensate to store information, and nanolithography, where atomic scale features on electronic devices could be created with precision-controlled condensates. All of these may someday be possible but will require substantial technical advances to become realities.

7.4 Problems

7.1. (a) Estimate the Bose–Einstein condensation temperature for a collection of ^{87}Rb atoms with a density of 10^{17} m^{-3}.
(b) Discuss the result of part (a) in the context of the results shown in the chapter.

7.2. A collection of ^{87}Rb atoms is used to create a Bose–Einstein condensate.
(a) Explain why ^{87}Rb is a boson.
(b) The gas is cooled to a temperature of about 100 nK as shown in figure 7.1. At this point additional atoms evaporate from the condensate, but its volume remains the same. Assume that the temperature is linearly proportional to the particle density. Discuss what happens to the velocity distribution.

References and suggestions for further reading

Cornell E A and Wieman C E 1998 The Bose–Einstein condensate *Sci. Am.* **278** 40–5

Davis K B, Mewes M O, Andrews M R, van Druten N J, Durfee D S, Kurn D M and Ketterle W 1995 Bose–Einstein condensation in a gas of sodium atoms *Phys. Rev. Lett.* **75** 3969–73

Griffin A, Snoke D W and Stringari S (ed) 1995 *Bose–Einstein Condensation* (Cambridge: Cambridge University Press)

Junttila M L and Stahlberg B 1990 Laser wavelength measurement with a Fourier transform wavemeter *Appl. Opt.* **29** 3510–16

Pethick C J and Smith H 2001 *Bose–Einstein Condensation in Dilute Gases* (Cambridge: Cambridge University Press)

Pitaevskii L P and Stringari S 2003 *Bose–Einstein Condensation* (Oxford: Clarendon)

Regal C A, Greiner M and Jin D S 2004 Observation of resonance condensation of fermionic atom pairs *Phys. Rev. Lett.* **92** 040403

⊙编辑手记

在联合国教科文组织的出版物(UNESCO Publications)中曾指出：

> 在所有的时代里，物理科学和技术都是数学灵感的先声，也是检验数学实践效果的标准. 数学用户的总数，包括用数学很多的用户和偏少的用户，总是比纯数学用户的数目大得多.

本书是一部国外优秀的物理教程，为英文版.

本书的中文书名或可译为《激光及其在玻色—爱因斯坦凝聚态观测中的应用》. 本书的作者为理查德·A. 邓拉普(Richard A. Dunlap)，加拿大人，达尔豪斯大学物理和大气科学系的研究教授，在伍斯特理工学院获得物理学学士学位，在达特茅斯学院获得物理学硕士学位，在克拉克大学获得物理学博士学位.

正如本书作者在前言中指出：

> 激光是在20世纪60年代初开发出来的，它不仅在很多领域中对科学知识的进步发挥了重要作用，而且在许多设备中产生了商业应用，这些设备在我们的日常生活中已经变得很重要. 这些应用包括CD/DVD播放器、激光打印机和光纤通信设备. 虽然这些设备在很大程度上取决于激

光产生的光的单色性和相干性，但其他众所周知的应用，如激光加工和激光融合，则取决于激光的强度. 本书的第一部分概述了激光的物理特性，并描述了一些更常见的激光类型及其应用；第二部分着眼于玻色—爱因斯坦凝聚态. 这些凝聚态代表了在极低的温度下存在于某些稀薄气体中的物质状态. 这种状态是在 20 世纪 20 年代由萨特延德拉·纳特·玻色(Satyendra Nath Bose)和阿尔伯特·爱因斯坦(Albert Einstein)首次预测出来的. 1995 年，科罗拉多大学的埃里克·康奈尔(Eric Cornell)和卡尔·威曼(Carl Wieman)首次通过实验观察到玻色—爱因斯坦凝聚态，此后不久，麻省理工学院的沃尔夫冈·克特勒(Wolfgang Ketterle)也观察到了该凝聚态. 用于产生玻色—爱因斯坦凝聚态的实验技术提供了一种有趣且稍微有点非常规的激光应用，即在非常低的温度下对稀薄气体进行冷却和约束.

本书的版权编辑李丹女士为方便读者阅读，特翻译了本书的目录如下：

华中师范大学的汪德新教授在其所编著的《理论物理学导论(第三卷)》中的“量子体系的状态”这一章中全面介绍了这一理论.

量子力学本质上是在科学实验的基础上建立起来的.一百多年前的微观物理现象显示微观粒子的基本特征是它的波粒二重性.下文在介绍

这些著名实验的基础上,讨论怎样用波函数描述量子体系的状态,并通过对实验事实的分析提出态叠加原理,接着提出波函数随时间变化的规律——薛定谔(Schrödinger)方程,最后通过一维无限深势阱和线性谐振子两个实例介绍定态薛定谔方程的解法.

1　微观物理现象　物理观念的飞跃

19 世纪末叶,物理学的各个分支已经建立起系统的理论:经典力学从牛顿三大定律发展为分析力学;电磁学和光学发展成为麦克斯韦(Maxwell)理论;热学在建立了以热力学定律为基础的宏观理论的同时,玻尔兹曼(Boltzmann)和吉布斯(Gibbs)又建立了称为统计物理学的微观理论.在经典物理学的辉煌成就面前,有的科学家认为物理学已大功告成了,正如绝对温标的创始人开尔文(Kelvin)在 1889 年新年贺词中所说的"19 世纪已将物理大厦全部建成,今后物理学家的任务就是修饰、完美这所大厦了."

但是,经典物理学却一直无法解释 19 世纪末到 20 世纪初的若干物理实验.当时遇到的困难主要是黑体辐射、光电效应、原子光谱的线状结构等这样一些后来被称为"微观物理现象"的问题.

本节介绍经典物理学遇到的困难,以及如何解决这些困难并促使物理观念的飞跃.

1.1　黑体辐射　普朗克的能量子假设

1859 年,基尔霍夫(Kirchhoff) 发现了热辐射定律.该定律指出:在给定温度下,物体的能谱发射率 $\varepsilon_\nu(T)$ 与它的吸收系数 $a_\nu(T)$ 之比是与物体的物理性质无关的普适函数

$$\frac{\varepsilon_\nu(T)}{a_\nu(T)} = f(\nu, T)$$

式中能谱发射率 $\varepsilon_\nu(T)$ 表示物体在给定温度下,单位时间内从单位面积辐射出单位频率间隔的电磁波能量.上式表明,一旦求得 $f(\nu, T)$,便可由物体的吸收系数 $a_\nu(T)$ 求得该物体的能谱发射率 $\varepsilon_\nu(T)$.因此,寻找 $f(\nu, T)$ 的函数形式具有重要意义.

由于照射到黑体上的辐射能被黑体完全吸收,即黑体的吸收系数 $a_\nu(T)=1$,这样,普适函数 $f(\nu, T)$ 就是黑体的能谱发射率 $\varepsilon_\nu(T)$,而能谱发射率又与该温度下平衡辐射的能量密度(单位体积内,单位频率间隔的辐射能量)ρ_ν 成正比.因此,寻找普适函数 $f(\nu, T)$ 的问题就转化为寻找黑

体辐射在$\nu \to \nu + d\nu$范围内的辐射能量密度$\rho_\nu d\nu$了.

1. 黑体辐射的实验定律

自然界没有绝对黑体.煤烟的吸收系数也只有95%.但是,一个比较深的空腔壁上的小孔可以作为黑体的近似.当光线通过小孔射入空腔以后,光线将在空腔的内表面多次来回反射而被吸收,几乎不可能通过小孔射出.这样,小孔的吸收系数就可以看作1,从小孔发出的辐射就可以看作黑体辐射了.

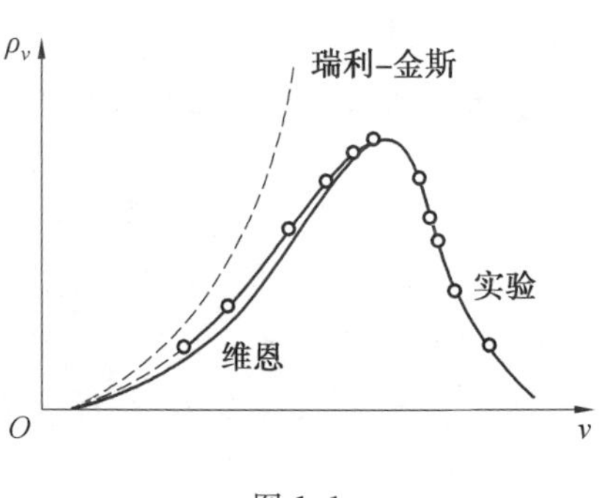

图 1.1

在实验室温度下,任何物体都会发出热辐射.实验表明:在频率$\nu \to \nu + d\nu$之间的辐射能量密度$\rho_\nu d\nu$只与空腔的温度T有关,而与空腔的形状及其组成物质无关,在$T = 1\ 500$ K时,实验曲线如图1.1.

2. 维恩公式与瑞利—金斯公式

为了解释这个实验结果,维恩(Wien)在1894年利用电动力学和热力学推出了一个公式(称为维恩公式)

$$\rho_\nu d\nu = c_1 \nu^3 e^{-c_2 \nu / T} d\nu \qquad (1.1.1)$$

式中c_1, c_2为常数.除低频部分外,维恩公式与实验曲线符合得很好,如图1.1.

1900年,瑞利(Rayleigh)和金斯(Jeans)利用统计物理学和电动力学推出另一个公式(称为瑞利—金斯公式)

$$\rho_\nu d\nu = (8\pi\nu^2 / c^3) k_B T d\nu \qquad (1.1.2)$$

式中k_B为玻尔兹曼常量,c为真空中的光速.

导出瑞利—金斯公式只有两个步骤:

(1) 电动力学的谐振腔理论已证明:在$\nu \to \nu + d\nu$之间,谐振腔单位体积内可以存在的振动模式数为

$$(8\pi\nu^2 / c^3) d\nu \qquad (1.1.3)$$

例如,在三边长分别为L_1, L_2, L_3的理想导体谐振腔中,电磁场的边界条件要求波矢的各分量满足

$$k_x = \frac{m\pi}{L_1}, k_y = \frac{n\pi}{L_2}, k_z = \frac{p\pi}{L_3} \quad (m, n, p = 0, 1, \cdots)$$

这样的一组$\{m, n, p\}$数,对应着一个可能的波矢$\boldsymbol{k}$,对应着两个可能的振动模式(因为有两个独立的偏振方向),据此可以证明式(1.1.3).

(2) 利用玻尔兹曼能量分布律计算振子(相当于一个振动模式)的平

均能量.

玻尔兹曼能量分布律指出,振子的能量处于 $\varepsilon \to \varepsilon + d\varepsilon$ 之间的概率是

$$P(\varepsilon) = \frac{e^{-\varepsilon/k_B T}}{k_B T} \tag{1.1.4}$$

由此得振子的平均能量为

$$\varepsilon = \frac{\int_0^\infty \varepsilon P(\varepsilon) d\varepsilon}{\int_0^\infty P(\varepsilon) d\varepsilon} = \frac{\int_0^\infty \varepsilon e^{-\varepsilon/k_B T} d\varepsilon}{\int_0^\infty e^{-\varepsilon/k_B T} d\varepsilon} = k_B T \tag{1.1.5}$$

这就是能量均分定理(在热平衡状态下,原子的每个振动自由度占有相同的平均能量 $k_B T$).

综合这两点,在 $\nu \to \nu + d\nu$ 之间的辐射能量密度 $\rho_\nu d\nu$ 等于 $\nu \to \nu + d\nu$ 之间单位体积内的振动模式数乘以每一振动模式的平均能量 $\bar{\varepsilon}$,即

$$\rho_\nu d\nu = \left(\frac{8\pi\nu^2}{c^3}\right) d\nu \cdot \bar{\varepsilon} = \left(\frac{8\pi\nu^2}{c^3}\right) k_B T d\nu$$

这就是瑞利—金斯公式. 遗憾的是,瑞利—金斯公式在低频部分与实验曲线符合得较好,而高频部分不符合,如图 1.1. 实际上,利用这个公式计算总辐射能密度导致

$$\int_0^\infty \rho_\nu d\nu = \int_0^\infty \left(\frac{8\pi\nu^2}{c^3}\right) k_B T d\nu \to \infty \tag{1.1.6}$$

这与事实不符,当时称之为"紫外光的灾难". 上式之所以发散,源于积分上限,与紫外光相对应.

3. 普朗克公式　能量子假设

为了解释黑体辐射现象,1900 年普朗克(Planck) 提出另一个公式(称为普朗克公式)

$$\rho_\nu d\nu = \frac{8\pi}{c^3} \frac{h\nu^3}{e^{h\nu/k_B T} - 1} d\nu \tag{1.1.7}$$

式中 c 为真空中的光速,h 为普适常量. 根据式(1.1.7) 得到的理论计算结果与实验曲线惊人地一致. 普朗克公式的建立仍然可以采取上述的两个步骤(在历史上,推导方法比较复杂),差异仅在于每一振子的平均能量的取值. 如果假定振子的平均能量为

$$\bar{\varepsilon} = \begin{cases} k_B T, 当\ \nu\ 比较小 \\ 0, 当\ \nu \to \infty \end{cases} \tag{1.1.8}$$

则"紫外光的灾难"即可避免. 这意味着,振子的平均能量必然与频率有关.

那么,$\bar{\varepsilon}(\nu)$ 应该具有怎样的函数形式呢? 如果不把能量作为连续变

量，而只取离散值$\varepsilon_n = n\varepsilon\,(n=0,1,\cdots)$；同时考虑到$\bar{\varepsilon}(\nu)$依赖于频率，只要假设$\varepsilon = h\nu$，并将式(1.1.5)的积分改为叠加，则频率为$\nu$的振子的平均能量应为

$$\bar{\varepsilon} = \frac{\sum\limits_{n=0}^{\infty} \varepsilon_n P(\varepsilon_n)}{\sum\limits_{n=0}^{\infty} P(\varepsilon_n)} = \frac{\sum\limits_{n=0}^{\infty} nh\nu \mathrm{e}^{-nh\nu/k_{\mathrm{B}}T}}{\sum\limits_{n=0}^{\infty} \mathrm{e}^{-nh\nu/k_{\mathrm{B}}T}} = \frac{h\nu}{\mathrm{e}^{h\nu/k_{\mathrm{B}}T} - 1} \tag{1.1.9}$$

在ν很小时应用指数函数展开式的头两项，在$\nu \to \infty$时利用洛必达法则，即可证明式(1.1.9)与式(1.1.8)一致.

这样，在$\nu \to \nu + \mathrm{d}\nu$之间的辐射能量密度为

$$\rho_\nu \mathrm{d}\nu = \frac{8\pi\nu^2}{c^3}\mathrm{d}\nu \cdot \bar{\varepsilon} = \frac{8\pi}{c^3}\ \frac{h\nu^3}{\mathrm{e}^{h\nu/k_{\mathrm{B}}T} - 1}\mathrm{d}\nu \tag{1.1.10}$$

这就是普朗克公式. 至于常数h，当时通过理论与实验数据相符而确定为$h = 6.385 \times 10^{-34}\ \mathrm{J \cdot s}$，称为普朗克常量.

普朗克成功的关键是认识到振子的能量应取离散值. 他把黑体看作一组连续振动的谐振子，振子的能量值只能取最小能量单位ε的整数倍. 于是，黑体与辐射场交换能量也只能以ε为单位进行. 由此可见，黑体吸收或发射电磁辐射能量的方式是不连续的，只能“量子”式地进行. 每个“能量子”的能量为

$$\varepsilon = h\nu = \hbar\omega \quad \left(\hbar = \frac{h}{2\pi}\right) \tag{1.1.11}$$

普朗克的“能量子假设”是与经典物理的基本观念根本对立的. 因为经典振子的能量正比于振幅的平方，而振幅可以连续变化，所以振子的能量也就可以连续变化. 利用式(1.1.5)的积分，可得能量均分定理. 普朗克抛弃了能量均分定理而代之以能量子假设，这是对经典物理学的革命性突破，导致量子论的创立.

遗憾的是，普朗克在推出这个公式以后的十多年里，还一直想把量子概念纳入经典理论的框架内. 尽管他采用了许多新的技巧，但是都没有成功. 后来，由于他的量子假设在其他问题上取得了巨大的成功，他才确信这个假设是正确的.

1.1.2 光电效应　爱因斯坦的光量子假设

1. 实验事实与经典理论的困难

1888年，赫兹(Hertz)在实验中发现，用紫外线照射火花隙的阴极时，放电现象较易发生. 直到1897年汤姆孙(Thomson)发现电子后，1899年勒纳德(Lenard)用实验证明：这是由于紫外光照射金属表面时有电子

逸出所造成的，这种现象称为光电效应，逸出的电子称为光电子.

(1) 当入射光的频率 $\nu \geqslant \nu_0$ 时，才有光电子产生(ν_0 称为截止频率，与金属的性质有关)；即使光强较弱，光电子也能在 10^{-9} s 内逸出.

(2) 光电子的最大动能与入射光的频率 ν 有关，而与入射光强无关.

(3) 光电子的密度与入射光强成正比.

经典理论无法解释上述结果. 按照经典电动力学，在入射光(电磁场)的作用下，电子做强迫振动，不断积聚能量. 只要光照的时间足够长，电子必能从金属表面逸出，不应受 $\nu \geqslant \nu_0$ 的限制. 而且，当入射光强较弱时，积聚能量需要较长的时间，电子不可能在 10^{-9}s 内逸出. 光电子的最大动能应决定于入射光的振幅而不是频率.

2. 光量子概念的提出与普朗克—爱因斯坦关系式

1905 年，爱因斯坦(Einstein) 为了解释光电效应，在普朗克能量子假设的基础上，提出了光量子的概念. 他认为，不仅黑体与辐射场的能量交换是量子化的，而且辐射场本身就是由光量子(即光子)组成的，每个光子均以光速 c 运动. 频率为 ν 的光波，其光子的能量和动量分别为

$$E = h\nu = \hbar\omega \tag{1.1.12}$$

$$p = \frac{E}{c} = \frac{\hbar\omega}{c} = \hbar k \tag{1.1.13}$$

式中 $E = pc$ 来自相对论的质能关系式 $E^2 = \mu_0^2 c^4 + p^2 c^2$(光子的静质量 $\mu_0 = 0$). 设 $\boldsymbol{n}$ 为光子运动方向上的单位矢量，引入波矢 $\boldsymbol{k} = k\boldsymbol{n}$，则式(1.1.13) 可写成矢量形式

$$\boldsymbol{p} = p\boldsymbol{n} = \hbar k\boldsymbol{n} = \hbar\boldsymbol{k} \tag{1.1.14}$$

公式(1.1.12) 和(1.1.14) 称为普朗克—爱因斯坦关系式.

3. 爱因斯坦方程及其对光电效应的解释

现在，利用普朗克—爱因斯坦关系式计算光电子逸出金属后的最大动能. 当金属受到频率为 ν 的光照射时，金属中的束缚电子即可吸收光子而获得能量 $h\nu$(两个或多个光子同时被一个电子吸收的概率极小). 若光子的能量 $h\nu$ 大于电子挣脱金属束缚所需要的能量 W_0(称为脱出功)，则光电子逸出金属后的最大动能为

$$\frac{1}{2}\mu_e v^2 = h\nu - W_0 = h(\nu - \nu_0) \tag{1.1.15}$$

式中 μ_e 是光电子质量，$\nu_0 = \dfrac{W_0}{h}$ 就是截止频率. 上式称为爱因斯坦方程.

利用爱因斯坦方程很容易解释光电效应的实验结果：

(1) 当 $\nu < \nu_0$ 时，无论光强有多强，电子都不能逸出金属表面；而当

$\nu > \nu_0$ 时，无论光强多微弱，光电效应也能出现.

(2) 爱因斯坦方程表明，光电子的最大动能与入射光的频率 ν 有关，而与入射光强无关.

(3) 光电子的密度应正比于入射光子的密度，即正比于入射光强.

爱因斯坦的光量子假设成功地解释了光电效应，但是开始并没有得到所有科学家的赞同. 密立根(Millikan) 为了否定爱因斯坦方程，整整花了九年时间设计了更加精密的装置，竟于 1914 年用实验完全证实了这个方程. 密立根由这个实验测得的 h 值 $6.57\times10^{-34}\mathrm{J\cdot s}$，与普朗克的 h 值 $(6.385\times10^{-34}\mathrm{J\cdot s})$ 非常接近，测量值来自不同实验，这为当时确立量子论的地位具有十分重要的意义.

1.1.3　康普顿效应　光的波粒二重性

1. 实验事实

1923 年，康普顿(Compton) 用 X 射线入射到碳或石墨的靶上，发现散射后 X 射线的波长随散射角的增加而增大，这个现象称为康普顿效应.

按照经典电动力学，电磁波被散射后仅改变其传播方向而不改变频率，故经典理论无法解释康普顿效应.

2. 光量子说的解释

康普顿按照光量子假设，把散射过程看作光子与自由电子的碰撞过程，利用能量守恒和动量守恒定律，得到与实验一致的结果，从而进一步证实了光具有粒子性.

设碰撞前电子静止，取 x 轴沿光子入射的方向；碰撞后，光子沿 θ 方向射出，电子的速度为 v，反冲角为 θ'，如图 1.2.

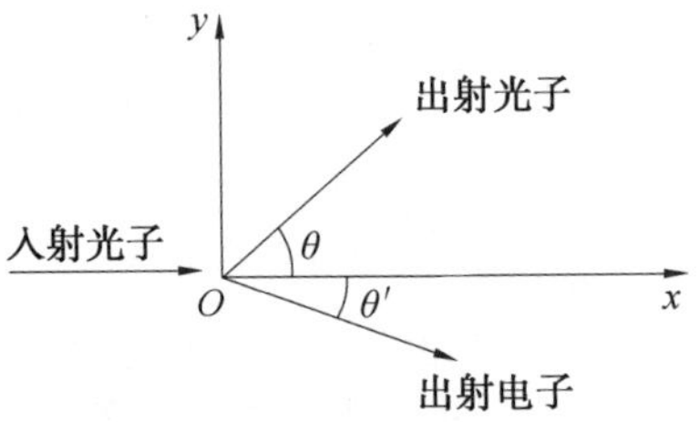

由能量守恒定律，可得

$$\hbar\omega+\mu_0c^2=\hbar\omega'+\frac{\mu_0c^2}{\sqrt{1-(v/c)^2}} \tag{1.1.16}$$

图 1.2

动量守恒定律沿 x 轴方向和 y 轴方向的投影分别为

$$\frac{\hbar\omega}{c}+0=\frac{\hbar\omega'}{c}\cos\theta+\frac{\mu_0v}{\sqrt{1-(v/c)^2}}\cos\theta' \tag{1.1.17}$$

$$0=\frac{\hbar\omega'}{c}\sin\theta-\frac{\mu_0v}{\sqrt{1-(v/c)^2}}\sin\theta' \tag{1.1.18}$$

由这三个方程消去 v 及 θ'，可得

$$\omega-\omega'=\frac{\hbar\omega\omega'}{\mu_0 c^2}(1-\cos\theta)=\frac{2\hbar\omega\omega'}{\mu_0 c^2}\sin^2\frac{\theta}{2}$$

$$\Delta\lambda=\lambda'-\lambda=\frac{2\pi c}{\omega\omega'}(\omega-\omega')=\frac{4\pi\hbar}{\mu_0 c}\sin^2\frac{\theta}{2} \tag{1.1.19}$$

由式(1.1.19)可见,散射波的波长λ'随θ的增加而增大,与实验结果完全符合.此外,当$\theta=\pi$时,$\Delta\lambda$取最大值$\Delta\lambda=4\pi\hbar/\mu_0 c=4.86\times10^{-10}\text{cm}$,它与入射光的波长无关.对于实际测量来说,$\Delta\lambda/\lambda$越大越容易测量,所以康普顿效应均采用波长比可见光短得多的X射线.

在高能物理和天体物理学中常常还会遇到所谓反康普顿效应,它是指低能光子与高能电子碰撞后得到高能光子,即光子的频率变高而波长变短的效应,它是宇宙X射线的重要来源.

1927年,康普顿因发现X射线被带电粒子散射而获得诺贝尔物理学奖.

3.光的波粒二重性

人们在19世纪已从光的干涉、衍射和偏振现象中认识到光的波动性,在20世纪初又从光电效应和康普顿效应中认识到光的粒子性.也就是说,光子具有确定的能量$\hbar\omega$和确定的动量$\hbar\boldsymbol{k}$,它像实物粒子一样作为一个整体运动,与物质相互作用时遵守粒子间相互作用的基本定律(能量守恒定律和动量守恒定律).人们通常把光具有波动和粒子的双重性质称为光的波粒二重性,这显示人们对光的本质的认识已进入一个崭新的阶段.

从数学形式来看,普朗克—爱因斯坦关系式指出光子的能量和动量分别为

$$E=\hbar\omega,\quad \boldsymbol{p}=\hbar\boldsymbol{k} \tag{1.1.20}$$

公式的左侧是反映粒子性的量$E,\boldsymbol{p}$,公式的右侧是反映波动性的量$\omega,\boldsymbol{k}$,深刻地反映出光的波动性和粒子性的有机联系.此外,从普朗克—爱因斯坦关系式还可看出,物质吸收一个光子就得到能量$\hbar\omega$,吸收两个光子就得到能量$2\hbar\omega$,能量的变化是不连续的.由此可见,能量和动量的量子化是通过$\hbar$表现出来的.换句话说,当$\hbar\to0$(指在该问题中$\hbar$可以忽略不计)时,能量的变化就由不连续过渡到连续,物理过程的量子效应也随之消失而过渡到它的经典极限.

1.1.4 原子的线状光谱 玻尔的量子论

1.原子光谱的经验公式

1885年,巴尔末(Balmer)总结了大量实验事实,发现氢原子可见光

的光谱线具有如下规律

$$\nu = cR_H\left(\frac{1}{2^2} - \frac{1}{n^2}\right) \quad (n = 3,4,5,\cdots) \tag{1.1.21}$$

1889 年,里德伯(Rydberg)在此基础上把氢原子的所有谱线归结为方程

$$\nu = cR_H\left(\frac{1}{n^2} - \frac{1}{n'^2}\right) \quad (n = 1,2,\cdots;n' = n+1,n+2,\cdots) \tag{1.1.22}$$

式中 $R_H = 109\ 677.58\ \text{cm}^{-1}$,称为里德伯常量,公式(1.1.22)称为里德伯方程.对于每一个给定的 n 值,n' 可取一切大于 n 的整数值,由此得到一组谱线,构成一个谱线系.

1908 年,里兹(Ritz)提出的组合规则指出,每一种原子都有它特有的光谱项 $T(n)$,原子发出的光谱线的频率为光速 c 乘两光谱项之差

$$\nu = c[T(n) - T(n')] \quad (n = 1,2,\cdots;n' = n+1,n+2,\cdots) \tag{1.1.23}$$

2.经典物理学面临的两个困难

(1) 经典物理学认为原子光谱应为连续谱.根据卢瑟福(Rutherford)的原子有核模型,原子中的电子环绕原子核的运动是一种加速运动,而做加速运动的电子因不断辐射能量而使其轨道半径 r 越来越小,电子绕核运动的频率 $\nu = v/2\pi r$ 也随之连续变化.按照经典电动力学,电子辐射的电磁波频率等于电子运动的频率.因此,电子辐射的电磁波频率也是连续变化的,原子光谱就不可能是线光谱.

(2) 经典物理学无法解释原子的稳定性.根据上面的讨论,随着电子不断将自己的能量辐射出去,它的动能不断减少,最后必将落到原子核中,导致原子坍缩.如果原子的初始半径为 $a \sim 10^{-10}\ \text{m}$,电子经过 $10^{-12}\ \text{s}$ 必将落到原子核上,但事实上原子是稳定的.

3.玻尔提出的量子论

1912 年,年仅 27 岁的丹麦物理学家玻尔来到卢瑟福实验室,开始研究这个问题.他将卢瑟福原子模型与普朗克—爱因斯坦的光量子理论结合起来,扬弃了经典电动力学的若干基本概念,于 1913 年提出了原子的量子论.这个理论包含了两个重要概念:

(1) 定态.原子只能稳定地处于一系列能量具有离散值的状态,称为定态.在定态中,电子沿特定的轨道运动,电子既不辐射也不吸收能量.

玻尔当时只考虑了圆形轨道,为了确定这些特定的轨道,他提出了量子化条件,即电子的轨道角动量只能是 $\hbar$ 的整数倍

$$L=\mu_e R v=n\hbar \quad (n=1,2,3,\cdots) \tag{1.1.24}$$

式中的μ_e为电子的质量，R为圆轨道的半径，v为电子的速度，n称为量子数.

1916年，索末菲(Sommerfeld)将玻尔的量子化条件推广为玻尔一索末菲量子化条件

$$\oint p\,\mathrm{d}q=nh \quad (n=1,2,3,\cdots) \tag{1.1.25}$$

式中q是广义坐标，p是相应的广义动量，积分号表示对一个运动周期的积分.

(2) 量子跃迁. 电子从一个可能的轨道过渡到另一可能的轨道是以跃迁的方式突然完成的. 当电子从能量为E_n的定态跃迁到能量为E'_n的定态时，发射或吸收的光子的频率由频率条件给出

$$h\nu=E'_n-E_n \tag{1.1.26}$$

玻尔频率条件并不是凭空提出来的. 它继承了普朗克一爱因斯坦的光量子论，并建立在以实验为依据的里兹组合规则的基础上. 如果用h乘以组合规则(1.1.26)的两边，则有

$$h\nu=hc[T(n)-T(n')]$$

上式的左边是原子辐射出来的光子能量，根据能量守恒定律，右边就应该是原子辐射前后能量之差E'_n-E_n. 由此可见，每一个光谱项$T(n)$乘以$-hc$正好等于原子某一个定态能量E_n.

4. 由玻尔理论求得氢原子的能级

设电子沿半径为a的圆轨道运动，由式(1.1.24)可得

$$v=n\hbar/\mu_e a \tag{1.1.27}$$

又因核对电子的库仑力就是电子做圆周运动的向心力，即

$$\mu_e v^2/a=e_s^2/a^2 \tag{1.1.28}$$

式中$e_s^2=e^2/4\pi\varepsilon_0$. 将上两式联立，消去$v$，即有

$$a=n^2\hbar^2/\mu_e e_s^2 \tag{1.1.29}$$

取$n=1$，得第一玻尔半径

$$a_0=\hbar^2/\mu_e e_s^2 \tag{1.1.30}$$

当电子处于第n个圆形轨道时，电子的能量为

$$E=T+U=\frac{1}{2}\mu_e v^2-\frac{e_s^2}{a}=\frac{e_s^2}{2a}-\frac{e_s^2}{a}=-\frac{\mu_e e_s^4}{2n^2\hbar^2} \tag{1.1.31}$$

由频率条件即得电子发射或吸收的光子频率为

$$\nu=\frac{E'_n-E_n}{h}=\frac{\mu_e e_s^4}{4\pi\hbar^3}\left(\frac{1}{n^2}-\frac{1}{n'^2}\right) \tag{1.1.32}$$

这就是里德伯方程.与式(1.1.22)相比较得里德伯常量

$$R_H = \frac{\mu_e e_s^4}{4\pi c h^3} \tag{1.1.33}$$

考虑到氢原子的运动是二体问题,用折合质量 $\mu=\mu_e\mu_p/(\mu_e+\mu_p)$ 代替上式中的 μ_e,与实验值符合得很好.

5. 玻尔理论的历史地位

玻尔理论取得了巨大的成功.玻尔提出的定态概念和量子化条件,不仅解决了原子的稳定性问题,而且可以求出氢原子中电子的轨道半径和能级.1914年,弗兰克(Franck)—赫兹(Hertz)实验证实了原子具有离散的能级.玻尔提出的量子跃迁概念和频率条件,解决了原子光谱的线状结构问题.它不仅从理论上给出了氢原子可见光谱中的巴耳末(Balmer)线系公式($n=2$)和红外光谱的帕邢(Paschen)线系公式($n=3$),而且预言了紫外光谱的莱曼(Lyman)线系的存在.这个预言在1914年即被莱曼的观测所证实.

玻尔理论还成功地解决了经典比热理论的困难.例如,经典理论不能解释为什么原子中的束缚电子对比热的贡献为零.玻尔理论指出原子的能级是离散的,束缚电子的基态距离第一激发态能级远大于常温下电子做无规则运动的能量.这样,电子只能处于基态,处于基态的电子的平均能量就是基态能量(与温度无关),因而对比热没有贡献.

玻尔理论不仅在量子物理学的发展过程中起过重要的作用,而且它以基本假设形式提出的若干物理思想已成为量子物理学的基石.

当然,玻尔理论也遇到了无法克服的困难.它只能解释氢原子及碱金属原子的光谱,而不能解释含有两个或两个以上价电子的原子的光谱;它只能给出氢原子光谱线的频率,而不能计算谱线的强度及这种跃迁发生的速率,更不能指出哪些跃迁能观察到以及哪些跃迁不能观察到;它只能讨论束缚态而不能讨论散射态,等等.究其原因,玻尔理论是经典力学加上与之不相容的量子化条件的产物,它并没有成为一个完整的理论体系.

玻尔理论的困难,推动着人们去寻找新的理论.

1.1.5 实物粒子的波动性　德布罗意假设

通过对光电效应和康普顿效应的研究,人们认识到光具有波粒二重性.那么,实物粒子是否也具有波动性呢?

1. 对物理学方法论的反思

1924年,当时还是研究生的青年物理学家德布罗意(De Broglie)在向巴黎大学理学院提交的博士论文中提出:“在光学上,比起波动的研究方法来，是否过于忽略了粒子的方法；在物质理论上，是否发生了相反的错误呢？是不是我们把粒子的图像想得太多，而过于忽略了波的图像？”接着，他进一步提出如下的假设：光的波粒二重性同样适用于实物粒子，正像光子有一个与它相关联的光波描述它的运动一样，实物粒子（如电子）也有一个与它相关联的物质波描述它的运动．对于具有一定能量和动量的粒子，其运动可以用一定频率和波长的物质波来描述，它们之间的联系为

$$E=\hbar\omega, \quad \boldsymbol{p}=\hbar\boldsymbol{k} \tag{1.1.34}$$

这两式与光子的普朗克－爱因斯坦关系式在形式上完全相同，称为德布罗意关系式．描述实物粒子运动的物质波称为德布罗意波，相应的波长 $\lambda=2\pi/k=h/p$ 称为德布罗意波长．

德布罗意假设实质上指出自然界存在着一种总体的对称性：光子与实物粒子都具有波粒二重性，德布罗意关系式适用于一切微观粒子．唯一的区别是，实物粒子的静止质量不为零，实物粒子的能量 $E=\sqrt{p^2c^2+\mu_0^2c^4}\neq pc$，因而式（1.1.34）中的两个式子是相互独立的，不能像光子那样，由前式直接推出后式．

历史上，德布罗意提出的式（1.1.34）还受到狭义相对论的启发：在相对论中，四维动量 $p_\mu=\left(\boldsymbol{p},\frac{\mathrm{i}}{c}E\right)$ 与四维波矢 $k_\mu=\left(\boldsymbol{k},\frac{\mathrm{i}}{c}\omega\right)$ 的数学形式相同，自然会考虑 $\boldsymbol{p}$ 与 $\boldsymbol{k}$，以及 E 与 ω 应该成比例，如果假设比例系数为 $\hbar$，则 $\hbar\boldsymbol{k}$ 与 $\hbar\omega$ 分别有动量和能量的量纲．

特别是，相位 φ 是洛伦兹不变量，德布罗意关系式 $E=\hbar\omega$ 和 $\boldsymbol{p}=\hbar\boldsymbol{k}$ 保证了

$$\varphi=\boldsymbol{k}\cdot\boldsymbol{x}-\omega t=\frac{1}{\hbar}(\boldsymbol{p}\cdot\boldsymbol{x}-Et)=\frac{1}{\hbar}p_\mu x_\mu \tag{1.1.35}$$

没有自由指标，是洛伦兹不变量．

2. 戴维逊（Davisson）－革末（Germer）实验

当然，德布罗意假设的正确性，必须得到实验事实的证实．这个实验是戴维逊和革末在 1925 年偶然做出来的，当时他们并不知道德布罗意假设．那时他们正在贝尔（Bell）电话实验室研究电子在镍表面的散射，使用的是镍多晶体．在实验过程中由于真空系统不慎漏气，镍表面发生了氧化，为了使氧化镍还原为镍，他们先将镍置于氢中加热，之后又在真空中加热，无意中使镍变成单晶结构，结果使散射的特征发生强烈的改变：原来预料随着散射角的增加，散射电子束的强度会单调下降．但实验表明，在某些角度产生了很强的电子束．后来，他们改进了实验装置，进一步发

现,电子束强度的极大值满足X射线在晶格上反射的布拉格(Bragg)关系式(反射波相长干涉条件)

$$2d\sin\theta=n\lambda \quad (n=1,2,\cdots) \tag{1.1.36}$$

式中 d 为晶格间距(如图1.3),用X光测得 $d=0.91$ Å,当布拉格角 $\theta=65^\circ$ 时得到第一级极大($n=1$).由式(1.1.36)算得

$\lambda=2d\sin\theta=2\times0.91\cdot\sin 65^\circ=1.65$ Å

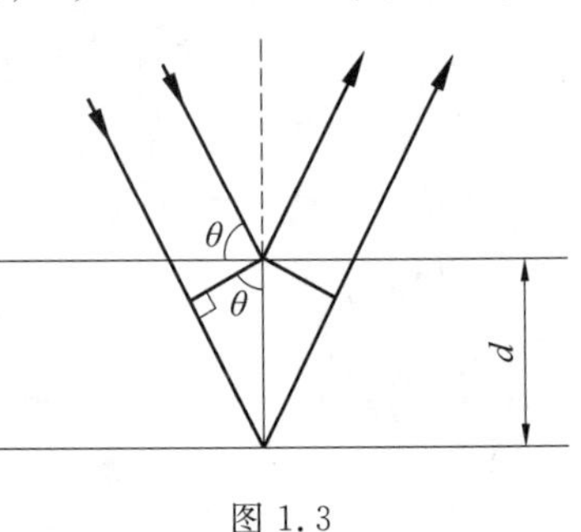

图1.3

实验中使用54 V的电势差加速电子,在 $v\ll c$ 的情况下,电子的能量 $E=p^2/2\mu$,再将 $E=eV$ 代入,求得电子的德布罗意波长为

$$\lambda=\frac{h}{p}=\frac{h}{\sqrt{2\mu E}}=\frac{h}{\sqrt{2\mu eV}}=1.67\ \text{Å}$$

与实验结果基本一致,这就证实了德布罗意假设.

计算表明,电子的波长非常短,这是实物粒子的波动性长期没有被发现的原因.光的波动性最初是从光的干涉现象发现的,因为干涉条纹的间距与光波的波长成正比,干涉条纹的间距太小就很难发现.

3.汤姆孙实验

1927年,G.P.汤姆孙利用穿透能力很强的高能电子束通过金属薄膜观察到轮廓分明的衍射现象——明暗相间的同心圆环,再次证实了德布罗意关系式.其后几十年来的大量实验事实无一例外地证明:波动性是一切微观粒子都具有的性质.

例1.1.1 试证明:粒子德布罗意波的相速 v_p 与粒子运动速度 v 满足

$$v_p v=c^2$$

解 利用德布罗意关系式 $E=\hbar\omega$,$p=\hbar k$(取 $\boldsymbol{p}=\hbar\boldsymbol{k}$ 的模),粒子的总能 $E=\mu c^2$,动量 $p=\mu v$(取 $\boldsymbol{p}=\mu\boldsymbol{v}$ 的模),可得相速

$$v_p=\frac{\lambda}{T}=\frac{\omega}{k}=\frac{E}{p}=\frac{\mu c^2}{\mu v}=\frac{c^2}{v}$$

即

$$v_p v=c^2 \tag{1.1.37}$$

例1.1.2 试证明:粒子德布罗意波包的群速度 $v_g=\dfrac{\mathrm{d}x_e}{\mathrm{d}t}=\left.\dfrac{\mathrm{d}\omega(k)}{\mathrm{d}k}\right|_{k=k_0}$(波包中心 x_e 的运动速度)等于粒子运动速度 v.

证 波包是由中心频率为 ω_0(对应的中心波数为 k_0),频移为 $\pm\Delta\omega$

的许多单色波叠加而成的. 由于波数 k 的分布有一定的范围,严格地说,粒子没有确定的动量和速度. 通常以中心波数 k_0 对应的动量 $\hbar k_0$ 来表示粒子的动量 $p=\mu v$,即

$$\mu v=\hbar k_0 \tag{1.1.38}$$

将 $\hbar\omega=\frac{p^2}{2\mu}=\frac{\hbar^2k^2}{2\mu}$ 代入 $v_g=\left.\frac{d\omega(k)}{dk}\right|_{k=k_0}$,并利用式(1.1.38),便有

$$v_g=\left.\frac{d}{dk}\frac{\hbar k^2}{2\mu}\right|_{k=k_0}=\frac{\hbar k_0}{\mu}=v \tag{1.1.39}$$

1.2 波函数的统计解释

波函数统计解释的提出可以看成是人们的认识从"初等量子论"步入"量子力学"的开端. 通常教材对量子力学的介绍有两种形式:一种是基本上按照量子力学发展过程来叙述,这种叙述自然表现出从经典物理概念到量子力学概念过渡的痕迹;另一种是采用公理化体系,完全采用现代的量子力学语言. 通常的"高等量子力学"教程都采用后者,本书作为量子力学的初等教材还是采用前者. 这样的叙述,一是有利于读者观念的转化,二是便于显示前人探索和创新的历程,特别是在当时的知识背景下,他们提出这些基本假设的想法和方式是怎样的.

本节首先引入自由粒子的波函数,随后通过对电子双缝衍射实验的讨论提出波函数的统计解释,最后给出波函数的归一化条件.

1.2.1 自由粒子的波函数

德布罗意指出,实物粒子的运动由一个与之相关联的物质波描述,物质波的角频率与波矢由德布罗意关系给出

$$\omega=E/\hbar,\quad \boldsymbol{k}=\boldsymbol{p}/\hbar \tag{1.2.1}$$

对于具有确定的能量和动量的自由粒子,描述粒子的波就是有确定的角频率和波矢的平面波

$$\psi(\boldsymbol{x},t)=Ae^{i(\boldsymbol{k}\cdot\boldsymbol{x}-\omega t)}=Ae^{\frac{i}{\hbar}(\boldsymbol{p}\cdot\boldsymbol{x}-Et)} \tag{1.2.2}$$

式中的 A 与坐标无关,$\psi(\boldsymbol{x},t)$ 称为自由粒子的波函数①.

那么,波函数到底具有什么物理意义呢?

---

① 当微观粒子处于场 $U(\boldsymbol{x},t)$ 中时,它的能量和动量就不会具有确定值,这时粒子的波函数自然比较复杂了(见 1.5 节).

1.2.2 波函数的统计解释

历史上,人们对“物质波”是有一个认识过程的.开始的时候,有人认为描述电子的波就是大量电子在空间形成像声波一样的疏密波;也有人认为粒子是波长不同的无限多平面波叠加而成的“波包”.但是,这两种看法都与实验事实相矛盾.其中一个很生动的例子就是电子双缝衍射实验,它的实验装置如图 1.4 所示.实验步骤如下:

(1) 电子枪发射强电子束.这时荧光屏马上显示出双缝衍射花样,这是电子的波动性的表现.这花样是不是由于每个电子打到荧光屏上弥散成带而形成的呢?不是的,再看实验的第二个步骤.

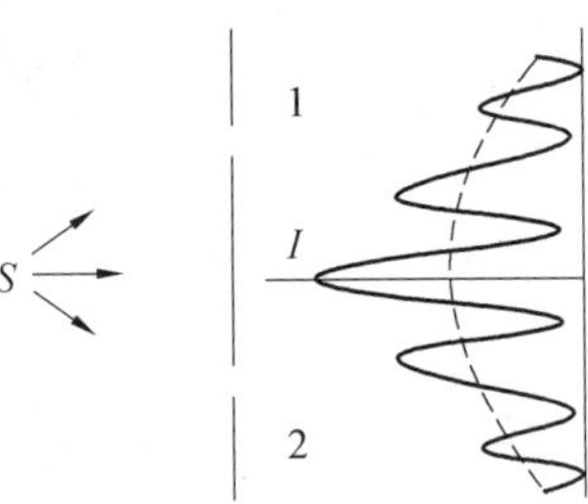

图 1.4

(2) 电子枪发射弱电子束.电子束微弱到几乎是一个一个电子射向双缝,这时发现荧光屏出现一点点的闪光.这表明电子是作为完整的颗粒一个一个地到达荧光屏的,这是电子的粒子性的表现.

在微弱电子束的情况下,电子似乎是杂乱无章地落在荧光屏上的:当一个电子落在荧光屏某处之后,下一个电子将落在什么地方是无法预言的.那么,当电子表现出粒子性的时候,是否就不表现出波动性呢?电子的运动是否没有规律可循呢?不是的,再看实验的第三个步骤.

(3) 将荧光屏换成照相底板,对微弱电子束的衍射作长时间的曝光.实验发现,得到的衍射花样与强电子束的衍射花样完全相同.

这说明在两种情况下,在出现亮条纹的地方,即波强度(振幅的模方)大的地方,到达的电子数目比较多;而在比较暗的地方,即波强度较小的地方,到达的电子数目比较少.在使用弱电子束的时候,一个电子到达亮条纹处的概率比较大,而到达比较暗的地方的概率比较小.

据此,玻恩(Born)在 1924 年提出了波函数的统计解释:在 t 时刻在 $\boldsymbol{x}$ 处体积元 $\mathrm{d}\tau$ 内找到微观粒子的概率与 $|\psi(\boldsymbol{x},t)|^2\mathrm{d}\tau$ 成正比.

这表示,描写粒子的波是概率波,波的强弱反映在空间某处发现粒子的概率的大小.由此可见,在 t 时刻在 $\boldsymbol{x}$ 点发现粒子的概率密度为

$$w(\boldsymbol{x},t)=|\psi(\boldsymbol{x},t)|^2 \tag{1.2.3}$$

因此,波函数又称为概率幅.

玻恩对波函数的统计解释深化了人们对微观粒子波粒二重性的认识:

(1) 在电子双缝衍射的实验中,电子总是作为一个整体落在荧光屏上

并呈现一点闪光，显示了电子的粒子性. 但是，它又没有经典粒子那样完全确切的轨道，我们不知道它到底穿过哪个缝以及是怎样到达荧光屏的，可见微观粒子的粒子性有别于经典的观念.

(2) 电子衍射花样显示了电子的波动性，它反映了电子运动的统计规律性. 所谓“电子的波动性”是体现在每一个电子上的，正因为每一个电子都以相同的概率分布到达荧光屏，因此整个电子束的衍射花样也按这一概率分布呈现在荧光屏上. 但是，微观粒子的波函数不是可观察的物理量. 因此，电子的波动性不像水波或声波那样反映某个物理量(如位移、压强)在空间各点随时间做周期性的变化，可见微观粒子的波动性也有别于经典的观念.

总之，波函数的统计解释把微观粒子的粒子性和波动性有机地统一起来. 其后的无数实验事实证明，统计性是微观物理现象的本质特征.

1.2.3 波函数的归一化条件

波函数的统计解释显示，$\psi(\boldsymbol{x},t)$ 存在着不确定性. 若用常数 C 乘以 $\psi(\boldsymbol{x},t)$ 得到的相对概率分布 $|C\psi(\boldsymbol{x},t)|^2$ 与 $|\psi(\boldsymbol{x},t)|^2$ 是相同的，则 $C\psi(\boldsymbol{x},t)$ 与 $\psi(\boldsymbol{x},t)$ 描述粒子的同一状态. 为消除波函数的这种不确定性，就要提出波函数的归一化条件.

考虑到非相对论量子力学只研究低能范围的问题，实物粒子不会产生或湮灭. 这样，对于一个粒子而言，在整个空间发现它的概率为 1，即

$$\int_{V_\infty} |\psi(\boldsymbol{x},t)|^2 \mathrm{d}\tau = 1 \tag{1.2.4}$$

这称为波函数的归一化条件.

如果某个波函数 $\varphi(\boldsymbol{x},t)$ 不满足归一化条件，如

$$\int_{V_\infty} |\varphi(\boldsymbol{x},t)|^2 \mathrm{d}\tau = C^2$$

则可令 $\psi(\boldsymbol{x},t)=\dfrac{1}{C}\varphi(\boldsymbol{x},t)$，由于 $\psi(\boldsymbol{x},t)$ 与 $\varphi(\boldsymbol{x},t)$ 只差一个常数因子，它们描述粒子的同一状态，故可采用 $\psi(\boldsymbol{x},t)$ 描述该粒子. 这个数学手续称为波函数的归一化，常数 $\dfrac{1}{C}$ 称为归一化常数.

但是，即使波函数 $\psi(\boldsymbol{x},t)$ 满足归一化条件，波函数还是不唯一的. 当 α 为实数时，$|\mathrm{e}^{\mathrm{i}\alpha}|^2=1$，波函数 $\mathrm{e}^{\mathrm{i}\alpha}\psi(\boldsymbol{x},t)$ 也满足归一化条件，即波函数只能准确到一个相位因子.

还有些波函数不是平方可积函数，如平面波 $\psi(\boldsymbol{x},t)=A\mathrm{e}^{\frac{\mathrm{i}}{\hbar}(\boldsymbol{p}\cdot\boldsymbol{x}-Et)}$，它的模方对全空间的积分为

$$\int_{V_\infty} |Ae^{\frac{i}{\hbar}(\boldsymbol{p}\cdot\boldsymbol{x}-Et)}|^2 d\tau = |A|^2 \int_{V_\infty} d\tau \to \infty$$

因此不能采用上面的方法进行归一化，而应采用箱归一化或归一化为 δ 函数的方法.

例 1.2.1　已知电子的波函数为 $\psi(\boldsymbol{x}) = Ne^{-\frac{R}{a_0}}$，试求：

(1) 归一化常数 N.

(2) 在球壳 $R \to R + dR$ 内找到电子的概率 $w(R)dR$.

(3) 电子的位置径向概率密度于何处取最大值.

解　(1) 由波函数归一化条件及 $\int_0^\infty x^n e^{-ax} dx = \frac{n!}{a^{n+1}} (a > 0)$，可得

$$1 = \int_{-\infty}^{\infty} |\psi(\boldsymbol{x})|^2 d\tau = N^2 4\pi \int_0^\infty e^{-\frac{2R}{a_0}} R^2 dR = N^2 4\pi \frac{2!}{(2/a_0)^3} = N^2 \pi a_0^3$$

由此得归一化常数 $N = (\pi a_0^3)^{-\frac{1}{2}}$.

(2) 在球壳 $R \to R + dR$ 内找到电子的概率为

$$w(R)dR = \left(\int_0^\pi \int_0^{2\pi} |\psi(\boldsymbol{x})|^2 d\Omega\right) R^2 dR = \left(\int_0^\pi \int_0^{2\pi} \frac{1}{\pi a_0^3} e^{-\frac{2R}{a_0}} \sin\theta d\theta d\phi\right) \cdot R^2 dR$$

$$= \frac{4}{a_0^3} e^{-\frac{2R}{a_0}} R^2 dR$$

(3) 由 $w(R)$ 取极大值条件

$$0 = \frac{dw(R)}{dR} = \frac{4}{a_0^3} \frac{d}{dR}\left(R^2 e^{-\frac{2R}{a_0}}\right) = \frac{4}{a_0^3}\left(-\frac{2}{a_0}R^2 + 2R\right) e^{-\frac{2R}{a_0}}$$

可得 $R = a_0$.

上面的讨论表明，由微观粒子的波函数可以获得关于粒子位置概率的信息；后面的讨论表明，由它还可以获得关于粒子的其他所有信息(如关于粒子其他力学量取值的概率，等等). 这就是说，微观粒子的状态可以用波函数 $\psi(\boldsymbol{x},t)$ 完全描述. 因此，波函数又常称为状态函数，简称态函数.

此外，还可以将波函数的统计解释推广到多粒子体系. 设由 N 个粒子组成的体系的波函数为 $\psi(\boldsymbol{x}_1,\boldsymbol{x}_2,\cdots,\boldsymbol{x}_N,t)$，则 $|\psi(\boldsymbol{x}_1,\boldsymbol{x}_2,\cdots,\boldsymbol{x}_N,t)|^2 d\tau_1 d\tau_2 \cdots d\tau_N$ 表示在 t 时刻发现第 1 个粒子在 $\boldsymbol{x}_1$ 处的 $d\tau_1$ 中，并且发现第 2 个粒子在 $\boldsymbol{x}_2$ 处的 $d\tau_2$ 中…… 发现第 N 个粒子在 $\boldsymbol{x}_N$ 处的 $d\tau_N$ 中的概率.

显然，多粒子体系波函数的归一化条件为

$$\int_{V_\infty} |\psi(\boldsymbol{x}_1,\boldsymbol{x}_2,\cdots,\boldsymbol{x}_N,t)|^2 d\tau_1 d\tau_2 \cdots d\tau_N = 1 \tag{1.2.5}$$

这是在 $3N$ 维空间的积分.

1.3 态叠加原理

本节从电子双缝衍射实验出发提出态叠加原理，接着利用态叠加原理讨论电子在晶体表面的衍射现象，从而引入动量表象(即以粒子动量为自变量)的波函数.

1.3.1 电子双缝衍射实验

在介绍波函数的统计解释时，已经介绍过电子双缝衍射实验. 不同的是，本节将通过比较“只打开缝 1”“只打开缝 2”“同时打开两缝”这三种情况下的衍射花样，看看遵守叠加规则的到底是强度还是态函数.

(1) 打开缝 1，关闭缝 2. 电子只能通过缝 1 到达屏，用 ψ_1 描述电子通过缝 1 后的状态. 屏上出现的衍射花样由 $I_1 \sim |\psi_1|^2$ 决定，如图 1.5 中曲线 I_1 所示.

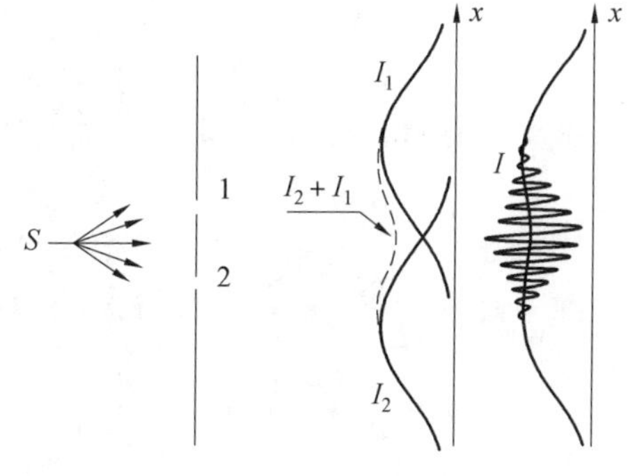

图 1.5

(2) 打开缝 2，关闭缝 1. 电子只能通过缝 2 到达屏，用 ψ_2 描述电子通过缝 2 后的状态. 屏上出现的衍射花样由 $I_2 \sim |\psi_2|^2$ 决定，如图 1.5 中曲线 I_2 所示.

(3) 同时打开缝 1 和缝 2. 用 ψ 描述电子通过双缝后的状态. 这时屏上出现的花样由 $I \sim |\psi|^2$ 决定.

但到底是 $|\psi|^2 = |\psi_1|^2 + |\psi_2|^2$，还是 $|\psi|^2 = |\psi_1 + \psi_2|^2$ 呢? 实验证明是后者，即

$$|\psi|^2 = |\psi_1 + \psi_2|^2 = |\psi_1|^2 + |\psi_2|^2 + \psi_1^* \psi_2 + \psi_1 \psi_2^* \quad (1.3.1)$$

干涉项 $\psi_1 \psi_2^* + \psi_1^* \psi_2$ 使 I_1 与 I_2 均不为零的地方出现暗纹($I=0$)，而在另一些地方出现 $I > I_1 + I_2$ 的亮纹. 由 $|\psi|^2 = |\psi_1 + \psi_2|^2$ 可推断

$$\psi = \psi_1 + \psi_2 \quad (1.3.2)$$

电子双缝衍射实验表明，概率不遵守叠加的规则，而概率幅(态函数)遵守叠加的规则.

公式(1.3.2)的物理意义是明显的. 因为两个缝都打开时(因而电子处于 ψ 态)，电子既可能处于 ψ_1 态，也可能处于 ψ_2 态；或者反过来说，若 ψ_1 和 ψ_2 是电子的可能状态，则 ψ_1 和 ψ_2 的线性叠加态 $\psi = \psi_1 + \psi_2$ 也是电子的可能状态.

由于相干叠加，使叠加态出现了原来两个态没有的新性质：例如，在原先 I_1 与 I_2 均单调变化的位置出现了 I 的极大值（亮纹），而在原先 I_1 与 I_2 均不为零的位置出现了 I 的极小值（暗纹）.

可以预料，当两个缝的几何参数或者两缝相对电子枪的位置不完全对称的情况下，电子通过双缝后的状态 ψ 可写成

$$\psi = c_1\psi_1 + c_2\psi_2 \tag{1.3.3}$$

式中的 c_1 和 c_2 是复常数. 这时屏上的强度分布为

$$I \sim |\psi|^2 = |c_1|^2|\psi_1|^2 + |c_2|^2|\psi_2|^2 + c_1^* c_2 \psi_1^* \psi_2 + c_1 c_2^* \psi_1 \psi_2^* \tag{1.3.4}$$

式中 $c_1^* c_2 \psi_1^* \psi_2 + c_1 c_2^* \psi_1 \psi_2^*$ 是 ψ_1 与 ψ_2 之间的干涉项，这里说的“干涉”（相干性叠加）是同一个电子的两个态 ψ_1 与 ψ_2 之间的干涉，而不是两个电子之间的干涉. 微弱电子束（微弱到一个一个电子入射）的双缝衍射实验结果就是一个很好的证明由于实验技术的限制，这个实验在 20 世纪 20 年代只能进行所谓的“假想实验”，直到 1989 年由殿村（Tononmura）等人使用配置了电子双棱镜的电子显微镜和灵敏电子探测系统才得以实现，这套仪器保证总有一个电子处于狭缝与屏幕之间，并能观察到干涉条纹的累积过程. 人们在这些实验的启发下，提出了态叠加原理.

1.3.2 态叠加原理

若 $\psi_1, \psi_2, \cdots, \psi_n$ 是体系的可能状态，则它们的线性叠加态

$$\psi = c_1\psi_1 + c_2\psi_2 + \cdots + c_n\psi_n \tag{1.3.5}$$

也是体系的可能状态. 当体系处于 ψ 态时，发现体系处于 ψ_k 态的概率是 $|c_k|^2$，式中 $k = 1, 2, \cdots, n$，并且

$$\sum_{k=1}^{n} |c_k|^2 = 1 \tag{1.3.6}$$

态叠加原理是量子力学的一个基本假设，它的正确性也应通过实验来证实.

1.3.3 电子在晶体表面的衍射　动量表象的波函数 $c(\boldsymbol{p}, t)$

电子在晶体表面的衍射实验如图 1.6 所示. 设电子沿垂直方向射到单晶表面，出射后将以各种不同的动量 $\boldsymbol{p}$ 运动. 由于电子离开晶体后就不再受晶体的作用，可以把电子看作自由粒子，并用波函数

$$\psi_{\boldsymbol{p}}(\boldsymbol{x}, t) = \frac{1}{(2\pi\hbar)^{3/2}} e^{\frac{i}{\hbar}(\boldsymbol{p}\cdot\boldsymbol{x} - Et)} \tag{1.3.7}$$

描述动量为 $\boldsymbol{p}$ 的电子的运动. 取 $A = (2\pi\hbar)^{-3/2}$ 的理由见后文.

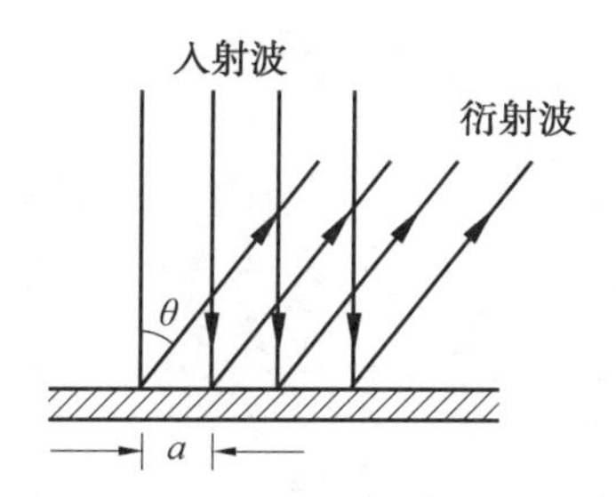

图 1.6

电子从晶体表面出射后，既可能处于态 $\psi_{\boldsymbol{p}'}(\boldsymbol{x},t)$，也可能处于态 $\psi_{\boldsymbol{p}''}(\boldsymbol{x},t)$，$\psi_{\boldsymbol{p}'''}(\boldsymbol{x},t)$，…. 按照态叠加原理，电子将处于这些态的线性叠加态

$$\psi(\boldsymbol{x},t)=\sum_{\boldsymbol{p}}c(\boldsymbol{p})\psi_{\boldsymbol{p}}(\boldsymbol{x},t) \tag{1.3.8}$$

考虑到电子的动量 $\boldsymbol{p}$ 可以连续变化，求和应改为积分

$$\begin{aligned}\psi(\boldsymbol{x},t)&=\int_{-\infty}^{\infty}c(\boldsymbol{p})\psi_{\boldsymbol{p}}(\boldsymbol{x},t)\mathrm{d}^3\boldsymbol{p}=\frac{1}{(2\pi\hbar)^{3/2}}\int_{-\infty}^{\infty}c(\boldsymbol{p})\mathrm{e}^{-\frac{\mathrm{i}}{\hbar}Et}\mathrm{e}^{\frac{\mathrm{i}}{\hbar}\boldsymbol{p}\cdot\boldsymbol{x}}\mathrm{d}^3\boldsymbol{p}\\&=\frac{1}{(2\pi\hbar)^{3/2}}\int_{-\infty}^{\infty}c(\boldsymbol{p},t)\mathrm{e}^{\frac{\mathrm{i}}{\hbar}\boldsymbol{p}\cdot\boldsymbol{x}}\mathrm{d}^3\boldsymbol{p}\end{aligned} \tag{1.3.9}$$

上式已采用记号

$$c(\boldsymbol{p},t)=c(\boldsymbol{p})\mathrm{e}^{-\frac{\mathrm{i}}{\hbar}Et} \tag{1.3.10}$$

式(1.3.9) 表明，$\psi(\boldsymbol{x},t)$ 是 $c(\boldsymbol{p},t)$ 的傅里叶(Fourier) 逆变换.

$\psi(\boldsymbol{x},t)$ 的傅里叶变换为

$$c(\boldsymbol{p},t)=\frac{1}{(2\pi\hbar)^{3/2}}\int_{-\infty}^{\infty}\psi(\boldsymbol{x},t)\mathrm{e}^{-\frac{\mathrm{i}}{\hbar}\boldsymbol{p}\cdot\boldsymbol{x}}\mathrm{d}^3\boldsymbol{x} \tag{1.3.11}$$

由此可见，$\psi(\boldsymbol{x},t)$ 与 $c(\boldsymbol{p},t)$ 是一一对应的. 既然 $\psi(\boldsymbol{x},t)$ 是完全地描述粒子状态的波函数，因此 $c(\boldsymbol{p},t)$ 也可以用来完全地描写粒子的状态. 两者的区别仅在于：$\psi(\boldsymbol{x},t)$ 是以坐标为自变量，称为坐标表象的波函数；$c(\boldsymbol{p},t)$ 是以动量为自变量，称为动量表象的波函数.

既然 $|\psi(\boldsymbol{x},t)|^2\mathrm{d}^3\boldsymbol{x}=|\psi(x,y,z,t)|^2\mathrm{d}x\mathrm{d}y\mathrm{d}z$ 是时刻 t 粒子的坐标在 $x\to x+\mathrm{d}x$，$y\to y+\mathrm{d}y$，$z\to z+\mathrm{d}z$ 范围内取值的概率；类似地，$|c(\boldsymbol{p},t)|^2\mathrm{d}^3\boldsymbol{p}=|c(p_x,p_y,p_z,t)|^2\mathrm{d}p_x\mathrm{d}p_y\mathrm{d}p_z$ 就是时刻 t 粒子的动量在 $p_x\to p_x+\mathrm{d}p_x$，$p_y\to p_y+\mathrm{d}p_y$，$p_z\to p_z+\mathrm{d}p_z$ 范围内取值的概率.

在一维的情形下，$\psi(x,t)$ 与 $c(p,t)$ 的关系为

$$\psi(x,t)=\frac{1}{(2\pi\hbar)^{1/2}}\int_{-\infty}^{\infty}c(p,t)\mathrm{e}^{\frac{\mathrm{i}}{\hbar}px}\mathrm{d}p \tag{1.3.12}$$

$$c(p,t)=\frac{1}{(2\pi\hbar)^{1/2}}\int_{-\infty}^{\infty}\psi(x,t)\mathrm{e}^{-\frac{\mathrm{i}}{\hbar}px}\mathrm{d}x \tag{1.3.13}$$

容易发现,如果在上两式中将 x 与 p,ψ 与 c 互换的同时把 i 换为 $-$i,则两式互换.

如果仅考虑两表象间某一时刻粒子波函数的关系,在式(1.3.9)和式(1.3.11)中取 $t=0$,便有

$$\psi(\boldsymbol{x})=\frac{1}{(2\pi\hbar)^{3/2}}\int_{-\infty}^{\infty}c(\boldsymbol{p})\mathrm{e}^{\frac{\mathrm{i}}{\hbar}\boldsymbol{p}\cdot\boldsymbol{x}}\mathrm{d}^3\boldsymbol{p} \tag{1.3.14}$$

$$c(\boldsymbol{p})=\frac{1}{(2\pi\hbar)^{3/2}}\int_{-\infty}^{\infty}\psi(\boldsymbol{x})\mathrm{e}^{-\frac{\mathrm{i}}{\hbar}\boldsymbol{p}\cdot\boldsymbol{x}}\mathrm{d}^3\boldsymbol{x} \tag{1.3.15}$$

在一维的情形下,$\psi(x)$ 与 $c(p)$ 的关系为(在这种情况下,已将 p_x 简写为 p)

$$\psi(x)=\frac{1}{(2\pi\hbar)^{1/2}}\int_{-\infty}^{\infty}c(p)\mathrm{e}^{\frac{\mathrm{i}}{\hbar}px}\mathrm{d}p \tag{1.3.16}$$

$$c(p)=\frac{1}{(2\pi\hbar)^{1/2}}\int_{-\infty}^{\infty}\psi(x)\mathrm{e}^{-\frac{\mathrm{i}}{\hbar}px}\mathrm{d}x \tag{1.3.17}$$

例 1.3.1 已知 $t=0$ 时粒子的状态为

$$\psi(x)=\begin{cases}\sqrt{\dfrac{2}{a}}\sin\dfrac{\pi x}{a},0\leqslant x\leqslant a\\ 0,x<0,x>a\end{cases}$$

试求粒子动量的概率密度 $|c(p)|^2$.

解 动量表象的波函数为 $c(p)=\frac{1}{(2\pi\hbar)^{1/2}}\int_{-\infty}^{\infty}\psi(x)\mathrm{e}^{-\frac{\mathrm{i}}{\hbar}px}\mathrm{d}x$,据此可求得粒子动量的概率幅

$$c(p)=\frac{1}{(2\pi\hbar)^{1/2}}\int_0^a\sqrt{\frac{2}{a}}\sin\frac{\pi x}{a}\mathrm{e}^{-\frac{\mathrm{i}}{\hbar}px}\mathrm{d}x=\frac{1}{(\pi\hbar a)^{1/2}}\ \frac{-\hbar^2\pi a}{(a^2p^2-\hbar^2\pi^2)}(\mathrm{e}^{-\frac{\mathrm{i}}{\hbar}pa}+1)$$

处于 $\psi(x)$ 态的粒子,其动量取值在 p 到 $p+\mathrm{d}p$ 范围内的概率为

$$\begin{aligned}|c(p)|^2&=\left[\frac{-\hbar^{3/2}(\pi a)^{1/2}}{a^2p^2-\hbar^2\pi^2}\right]^2(\mathrm{e}^{-\frac{\mathrm{i}}{\hbar}pa}+1)(\mathrm{e}^{\frac{\mathrm{i}}{\hbar}pa}+1)\\&=\frac{2\hbar^3\pi a}{(a^2p^2-\hbar^2\pi^2)^2}\left(1+\cos\frac{pa}{\hbar}\right)\end{aligned}$$

1.4 薛定谔方程 概率守恒定律

本节研究量子力学的动力学问题,即要寻找量子体系的状态随时间变化的规律.这个规律就是著名的薛定谔方程.薛定谔方程不是从基本原理推导出来的,而是作为一个基本假设提出来的.读者将会发现,本节寻找薛定谔方程的叙述是从猜测到假设,而不是推导.

本节首先研究在非相对论条件下方程应满足哪些条件,随后考察自

由粒子的波函数遵守怎样的方程，进而假设这个方程能推广到普遍的情形，这样便得到薛定谔方程. 无数实验事实证明，由薛定谔方程得出的结果与实验事实相符，因此上述假设成立. 接着利用分离变量法求解势场与时间无关情况下的薛定谔方程，得到定态薛定谔方程及定态解的形式. 最后，由波函数的统计解释和薛定谔方程出发，导出概率守恒定律.

1.4.1 波函数随时间变化的规律 —— 薛定谔方程

在经典力学中，力学体系运动状态随时间变化的规律是牛顿方程，如果知道力学体系的初始条件，利用牛顿方程即可求出体系在任一时刻的运动状态.

在量子力学中，量子体系的运动状态由波函数 $\psi(\boldsymbol{x},t)$ 描述. 那么，波函数随时间变化的规律是怎样的呢?

1. 在非相对论条件下，薛定谔方程应满足哪些条件?

(1) 当粒子的速度 $v \ll c$ 时，质量为 μ 的粒子的总能为 $E=\frac{\boldsymbol{p}^2}{2\mu}+U$.

(2) 波函数在任何时刻都遵守态叠加原理，因此方程应为线性齐次方程，以保证方程特解的线性组合仍然是方程的解.

(3) 方程既然要反映 $\psi(\boldsymbol{x},t)$ 随时间变化的规律，就必然包含 $\frac{\partial\psi(\boldsymbol{x},t)}{\partial t}$；但方程不应包含 $\frac{\partial^2\psi(\boldsymbol{x},t)}{\partial t^2}$，否则需要利用两个初始条件 $\psi(\boldsymbol{x},0)$ 及 $\left.\frac{\partial\psi(\boldsymbol{x},t)}{\partial t}\right|_{t=0}$ 才能确定 $\psi(\boldsymbol{x},t)$，这就意味着体系的初始状态不能由波函数 $\psi(\boldsymbol{x},0)$ 完全描述，违反了波函数完全地描述体系运动状态的基本假设.

2. 寻找最简单的自由粒子的波函数 $\psi(\boldsymbol{x},t)=A\mathrm{e}^{\frac{\mathrm{i}}{\hbar}(\boldsymbol{p}\cdot\boldsymbol{x}-Et)}$ 遵守的方程

由于条件(3) 的要求，我们来计算 $\psi(\boldsymbol{x},t)$ 对时间和空间坐标的偏导数(留给读者作为练习)，易得

$$\mathrm{i}\hbar\frac{\partial\psi(\boldsymbol{x},t)}{\partial t}=E\psi(\boldsymbol{x},t) \tag{1.4.1}$$

$$-\hbar^2\nabla^2\psi(\boldsymbol{x},t)=\boldsymbol{p}^2\psi(\boldsymbol{x},t) \tag{1.4.2}$$

式中 $\nabla^2=\frac{\partial^2}{\partial x^2}+\frac{\partial^2}{\partial y^2}+\frac{\partial^2}{\partial z^2}$ 是拉普拉斯算符.

当 $v\ll c$ 时，自由粒子的总能就是它的动能，即 $E=\frac{\boldsymbol{p}^2}{2\mu}$，乘以 $\psi(\boldsymbol{x},t)$ 便有

$$E\psi(\boldsymbol{x},t)=\frac{\boldsymbol{p}^2}{2\mu}\psi(\boldsymbol{x},t)$$

用 2μ 除式(1.4.2) 后,将(1.4.1),(1.4.2) 两式代入上式,即得

$$\mathrm{i}\hbar\frac{\partial\psi(\boldsymbol{x},t)}{\partial t}=-\frac{\hbar^2}{2\mu}\nabla^2\psi(\boldsymbol{x},t) \tag{1.4.3}$$

这就是自由粒子的波函数随时间变化的规律,即自由粒子的薛定谔方程.

通过把经典物理量替换为算符

$$E\rightarrow\mathrm{i}\hbar\frac{\partial}{\partial t},\boldsymbol{p}\rightarrow-\mathrm{i}\hbar\nabla$$

从而将经典物理量与体系波函数 $\psi(\boldsymbol{x},t)$ 的乘积

$$E\psi(\boldsymbol{x},t)=\frac{\boldsymbol{p}^2}{2\mu}\psi(\boldsymbol{x},t)$$

替换为量子算符对体系波函数 $\psi(\boldsymbol{x},t)$ 的作用

$$\mathrm{i}\hbar\frac{\partial\psi(\boldsymbol{x},t)}{\partial t}=-\frac{\hbar^2}{2\mu}\nabla^2\psi(\boldsymbol{x},t)$$

这种建立自由粒子的薛定谔方程的方法称为“一次量子化”方法[①].

3. 在势场中运动的粒子,它的波函数遵守同样的方程吗?

对于在势场中运动的粒子,它的总能为

$$E=\frac{\boldsymbol{p}^2}{2\mu}+U(\boldsymbol{x}) \tag{1.4.4}$$

同理可得

$$E\psi(\boldsymbol{x},t)=\frac{\boldsymbol{p}^2}{2\mu}\psi(\boldsymbol{x},t)+U(\boldsymbol{x})\psi(\boldsymbol{x},t) \tag{1.4.5a}$$

但是,在势场 $U(\boldsymbol{x})$ 中运动的粒子的波函数是否仍满足式(1.4.1) 和式(1.4.2) 呢? $U(\boldsymbol{x})$ 是否需要像 E 和 $\frac{\boldsymbol{p}^2}{2\mu}$ 那样换成另一个运算符号呢?其实是不知道的. 如果进一步假设可以这样做,即得

$$\mathrm{i}\hbar\frac{\partial}{\partial t}\psi(\boldsymbol{x},t)=\left[-\frac{\hbar^2}{2\mu}\nabla^2+U(\boldsymbol{x})\right]\psi(\boldsymbol{x},t) \tag{1.4.5b}$$

这就是薛定谔方程. 它完全符合上面提出的三个条件. 这个方程是薛定谔在 1926 年提出来的.

运算符号 $\left[-\frac{\hbar^2}{2\mu}\nabla^2+U(\boldsymbol{x})\right]$ 是经典力学的哈密顿函数 $H=\frac{\boldsymbol{p}^2}{2\mu}+U(\boldsymbol{x})$ 按照式(1.4.3) 的替换得来的,所以将上述运算符号称为哈密顿算符,并用 $\hat{H}$ 标记,即

$$\hat{H}=-\frac{\hbar^2}{2\mu}\nabla^2+U(\boldsymbol{x}) \tag{1.4.6}$$

---

① “二次量子化”的问题,请参看:苏汝铿. 量子力学. 2 版. 北京:高等教育出版社,2002:361-367.

4. 多粒子体系的薛定谔方程

现在，进一步将薛定谔方程推广到由 N 个粒子组成的多粒子体系. 设体系的波函数为 $\psi(\boldsymbol{x}_1,\boldsymbol{x}_2,\cdots,\boldsymbol{x}_N,t)$，体系的哈密顿量为

$$H=\sum_{i=1}^{N}\frac{p_i^2}{2\mu_i}+U(\boldsymbol{x}_1,\boldsymbol{x}_2,\cdots,\boldsymbol{x}_N) \tag{1.4.7}$$

式中 μ_i 及 p_i 为第 i 个粒子的质量和动量，$U(\boldsymbol{x}_1,\boldsymbol{x}_2,\cdots,\boldsymbol{x}_N)$ 是 N 个粒子在外场的势能及粒子间的相互作用能之和. 按照式(1.4.3) 给出的规则，体系的哈密顿算符为

$$\hat{H}=-\sum_{i=1}^{N}\frac{\hbar^2}{2\mu_i}\nabla_i^2+U(\boldsymbol{x}_1,\boldsymbol{x}_2,\cdots,\boldsymbol{x}_N) \tag{1.4.8}$$

由此得多粒子体系的薛定谔方程为

$$\mathrm{i}\hbar\frac{\partial\psi(\boldsymbol{x}_1,\boldsymbol{x}_2,\cdots,\boldsymbol{x}_N,t)}{\partial t}=\hat{H}\psi(\boldsymbol{x}_1,\boldsymbol{x}_2,\cdots,\boldsymbol{x}_N,t) \tag{1.4.9}$$

5. 关于薛定谔方程的讨论

(1) 薛定谔方程是作为一个基本假设提出来的，它的正确性已被非相对论量子力学在各方面的实验所证实. 当然，所谓“正确性”是相对的，在 $v\sim c$ 的情况下，它已被克莱因－戈尔登方程和狄拉克方程所代替.

(2) 薛定谔方程在非相对论量子力学中的地位与牛顿方程在经典力学中的地位相仿. 只要给出粒子在初始时刻的波函数，由方程即可求得粒子在以后任一时刻的波函数.

(3) 薛定谔方程只含对时间的一阶导数，为何可以描述波动过程？在经典物理中，波动方程 $u_{tt}-a^2u_{xx}=0$ 有周期性的解，而热传导方程 $u_t-a^2u_{xx}=0$ 则描述不可逆过程，没有周期性的解. 实际上，$u=\cos(kx-\omega t)$ 或 $u=\sin(kx-\omega t)$ 均不满足热传导方程，因为它们的一阶导数使相位增加 $\pi/2$

$$(\cos\theta)'=-\sin\theta=\cos(\theta+\pi/2),(\sin\theta)'=\cos\theta=\sin(\theta+\pi/2) \tag{1.4.10}$$

这样 u_t 使相位增加 $\pi/2$，u_{xx} 使相位增加 π，可见周期函数不可能满足热传导方程. 薛定谔方程虽然只含对时间的一阶导数，但在 $\frac{\partial\psi}{\partial t}$ 前面出现 $\mathrm{i}=\mathrm{e}^{\mathrm{i}\frac{\pi}{2}}$，正好使两者相位一致，因而有周期性的解. 而且，薛定谔方程中 i 因子的出现，使得波函数一般是 $\boldsymbol{x}$ 和 t 的复函数.

1.4.2 定态　定态薛定谔方程

本节讨论势场 $U(\boldsymbol{x})$ 不显含 t 的情况下薛定谔方程的解，势场显含 t

的情况在后文讨论.

1. 用分离变量法求解薛定谔方程

令

$$\psi(\boldsymbol{x},t)=\psi(\boldsymbol{x})f(t) \tag{1.4.11}$$

代入式(1.4.5),并用 $\psi(\boldsymbol{x})f(t)$ 去除,可得

$$\frac{\mathrm{i}\hbar}{f(t)}\frac{\mathrm{d}f(t)}{\mathrm{d}t}=\frac{1}{\psi(\boldsymbol{x})}\left[-\frac{\hbar^2}{2\mu}\nabla^2+U(\boldsymbol{x})\right]\psi(\boldsymbol{x}) \tag{1.4.12}$$

上式只包含 $\boldsymbol{x}$ 与 t 这两个独立变量,但等式的左边与 $\boldsymbol{x}$ 无关,等式的右边与 t 无关,可见上式两边只能等于同一常数,并记作 E(它的物理意义暂时不清楚). 由此分离出两个方程

$$\mathrm{i}\hbar\frac{\mathrm{d}f(t)}{\mathrm{d}t}=Ef(t) \tag{1.4.13}$$

$$\left[-\frac{\hbar^2}{2\mu}\nabla^2+U(\boldsymbol{x})\right]\psi(\boldsymbol{x})=E\psi(\boldsymbol{x}) \tag{1.4.14}$$

第一个方程可改写为 $\frac{\mathrm{d}f(t)}{f(t)}=-\frac{\mathrm{i}}{\hbar}E\mathrm{d}t$,积分后可得($C$ 为积分常数)

$$f(t)=C\mathrm{e}^{-\frac{\mathrm{i}}{\hbar}Et} \tag{1.4.15}$$

第二个方程要给出 $U(\boldsymbol{x})$ 才能解出,解法见下节. 将式(1.4.15)代入式(1.4.11),并将 C 归入 $\psi(\boldsymbol{x})$,即得薛定谔方程的特解

$$\psi(\boldsymbol{x},t)=\psi(\boldsymbol{x})\mathrm{e}^{-\frac{\mathrm{i}}{\hbar}Et} \tag{1.4.16}$$

当指数函数 $\mathrm{e}^{-\frac{\mathrm{i}}{\hbar}Et}$ 的指数为虚数时,它描述一个简谐振动,将它与 $\mathrm{e}^{-\mathrm{i}\omega t}$ 相比较,即得 $E=\hbar\omega$. 由德布罗意关系式可知上述常数 E 正是量子体系的能量.

既然 E 为常数,可见 $\psi(\boldsymbol{x},t)=\psi(\boldsymbol{x})\mathrm{e}^{-\frac{\mathrm{i}}{\hbar}Et}$ 是描述体系能量具有确定值的状态波函数. 我们把能量具有确定值的状态称为定态,描述定态的波函数(1.4.16)称为定态波函数,定态波函数遵守的方程(1.4.14)称为定态薛定谔方程.

2. 哈密顿算符的本征值方程

若体系的能量可以取 $E_1,E_2,\cdots,E_n$ 等值,将 $\hat{H}=-\frac{\hbar^2}{2\mu}\nabla^2+U(\boldsymbol{x})$ 代入式(1.4.14),则定态薛定谔方程可以写成下述形式

$$\hat{H}\psi_n(\boldsymbol{x})=E_n\psi_n(\boldsymbol{x})\quad(n=1,2,\cdots) \tag{1.4.17}$$

一个算符 $\hat{F}$ 作用于函数 u 得到常数乘该函数的方程称为算符 $\hat{F}$ 的本征值方程

$$\hat{F}u=fu \tag{1.4.18}$$

式中 f 称为算符 $\hat{F}$ 的本征值,u 称为算符 $\hat{F}$ 的本征值为 f 的本征函数.

显然,定态薛定谔方程就是算符 $\hat{H}$ 的本征值方程,E_n 就是算符 $\hat{H}$ 的本征值,波函数 $\psi_n(\boldsymbol{x})$ 称为 $\hat{H}$ 的本征值为 E_n 的本征函数,$\psi_n(\boldsymbol{x})$ 所描述的状态称为能量本征态.也就是说,当体系处于能量本征态 $\psi_n(\boldsymbol{x})$ 时,体系的能量具有确定值 E_n.

3. 薛定谔方程的特解和通解

将 $E=E_n$,$\psi(\boldsymbol{x})=\psi_n(\boldsymbol{x})$ 代入式(1.4.16),即得薛定谔方程的特解

$$\psi_n(\boldsymbol{x},t)=\psi_n(\boldsymbol{x})\mathrm{e}^{-\frac{\mathrm{i}}{\hbar}E_n t}\quad(n=1,2,\cdots)\tag{1.4.19}$$

由于薛定谔方程是线性齐次方程,它的通解就是这些特解的线性组合,即

$$\psi(\boldsymbol{x},t)=\sum_n c_n\psi_n(\boldsymbol{x},t)=\sum_n c_n\psi_n(\boldsymbol{x})\mathrm{e}^{-\frac{\mathrm{i}}{\hbar}E_n t}\tag{1.4.20}$$

4. 叠加系数模方 $|c_n|^2$ 的物理意义

由态叠加原理可知,处于 $\psi(\boldsymbol{x},t)$ 态的体系,其能量取 E_n 值的概率,即发现体系处于 $\psi_n(\boldsymbol{x},t)$ 态的概率为 $|c_n|^2$(假设 $\sum\limits_n |c_n|^2=1$).

既然体系处于 $\psi(\boldsymbol{x},t)$ 态时,体系的能量可以取各种不同的数值,故波函数(1.4.20) 不是定态波函数.

例 1.4.1　设粒子的两本征函数分别为

$$\psi_1(x)=c_1\mathrm{e}^{-\frac{1}{2}ax^2},\quad \psi_2(x)=c_2(x^2+b)\mathrm{e}^{-\frac{1}{2}ax^2}$$

求粒子这两状态的能级间隔.

解　由于 ψ_1 与 ψ_2 均满足定态薛定谔方程,故有

$$\psi''_1+\frac{2\mu}{\hbar^2}(E_1-U)\psi_1=0,\quad \psi''_2+\frac{2\mu}{\hbar^2}(E_2-U)\psi_2=0$$

以 ψ_2 乘前式减去 ψ_1 乘后式,即有

$$(E_2-E_1)\psi_1\psi_2=\frac{\hbar^2}{2\mu}(\psi_2\psi''_1-\psi_1\psi''_2)$$

上式对任意 x 值均成立.为方便起见,取 $x=0$ 的值进行计算,易见

$$\psi_1(0)=c_1,\quad \psi''_1(0)=-c_1a,\quad \psi_2(0)=c_2b,\quad \psi''_2(0)=2c_2-c_2ab$$

代入

$$(E_2-E_1)\psi_1(0)\psi_2(0)=\frac{\hbar^2}{2\mu}[\psi_2(0)\psi''_1(0)-\psi_1(0)\psi''_2(0)]$$

即得粒子这两状态的能级间隔

$$E_2-E_1=\frac{1}{c_1c_2b}\frac{\hbar^2}{2\mu}[-c_2bc_1a-(2c_1c_2-c_1c_2ab)]=-\frac{\hbar^2}{\mu b}$$

1.4.3 概率守恒定律

现在讨论当一个粒子在势场 $U(\boldsymbol{x})$ 中运动时,在空间各点发现它的概率密度如何随时间变化.

1. 概率守恒定律的微分形式

设粒子的波函数为 $\psi(\boldsymbol{x},t)$,t 时刻在 $\boldsymbol{x}$ 处发现粒子的概率密度为

$$w(\boldsymbol{x},t)=\psi^{*}(\boldsymbol{x},t)\psi(\boldsymbol{x},t) \tag{1.4.21}$$

概率密度随时间的变化率为

$$\frac{\partial w}{\partial t}=\frac{\partial}{\partial t}(\psi^{*}\psi)=\psi^{*}\frac{\partial\psi}{\partial t}+\frac{\partial\psi^{*}}{\partial t}\psi \tag{1.4.22}$$

为了计算它,用 $\mathrm{i}\hbar$ 除薛定谔方程两边,并取其复共轭,可得

$$\frac{\partial\psi}{\partial t}=\frac{\mathrm{i}\hbar}{2\mu}\nabla^{2}\psi+\frac{1}{\mathrm{i}\hbar}U(\boldsymbol{x})\psi,\quad \frac{\partial\psi^{*}}{\partial t}=-\frac{\mathrm{i}\hbar}{2\mu}\nabla^{2}\psi^{*}-\frac{1}{\mathrm{i}\hbar}U(\boldsymbol{x})\psi^{*} \tag{1.4.23}$$

代入式(1.4.21),随后分别加减 $\nabla\psi\cdot\nabla\psi^{*}$,再利用 $\nabla\cdot(\varphi\boldsymbol{f})=\varphi\nabla\cdot\boldsymbol{f}+\nabla\varphi\cdot\boldsymbol{f}$,即有

$$\begin{aligned}\frac{\partial w}{\partial t}&=\frac{\mathrm{i}\hbar}{2\mu}(\psi^{*}\nabla^{2}\psi-\psi\nabla^{2}\psi^{*})\\&=\frac{\mathrm{i}\hbar}{2\mu}(\psi^{*}\nabla^{2}\psi+\nabla\psi\cdot\nabla\psi^{*}-\psi\nabla^{2}\psi^{*}-\nabla\psi\cdot\nabla\psi^{*})\\&=\frac{\mathrm{i}\hbar}{2\mu}\nabla\cdot(\psi^{*}\nabla\psi-\psi\nabla\psi^{*})=-\nabla\cdot\boldsymbol{J}\end{aligned} \tag{1.4.24}$$

式中

$$\boldsymbol{J}=\frac{\mathrm{i}\hbar}{2\mu}(\psi\nabla\psi^{*}-\psi^{*}\nabla\psi) \tag{1.4.25}$$

不难发现,式(1.4.24)与经典电动力学中的电荷守恒定律

$$\frac{\partial\rho_f}{\partial t}+\nabla\cdot\boldsymbol{J}_f=0 \tag{1.4.26}$$

具有相同的形式,式(1.4.26)中的 ρ_f 是电荷密度,$\boldsymbol{J}_f$ 为电流密度矢量.既然式(1.4.24)中的 w 是概率密度,通过与式(1.4.26)对比可知 $\boldsymbol{J}$ 即为概率流密度矢量.

概率为什么会流动呢?在外场的作用下,在 $\boldsymbol{x}$ 处发现粒子的概率密度 $w(\boldsymbol{x},t)=|\psi(\boldsymbol{x},t)|^{2}$ 就有可能随时间变化,有些地方的概率密度增加了,有些地方的概率密度减小了,这表明概率在流动.

$\boldsymbol{J}$ 的物理意义也可以通过与 ρ_f 的类比得到:它的大小是单位时间内流过垂直于 $\boldsymbol{J}$ 的单位面积的概率,它的方向就是该点概率流动的方向.式(1.4.24)给出了概率守恒定律的微分形式.

2. 概率守恒定律的积分形式

将式(1.4.24)对空间 V 作体积分,可得

$$\int_V \frac{\partial w}{\partial t}\mathrm{d}\tau + \int_V \nabla\cdot \boldsymbol{J}\mathrm{d}\tau = 0$$

第一项交换微分和积分的次序,第二项利用高斯定理把体积分变成面积分,即有

$$\frac{\mathrm{d}}{\mathrm{d}t}\int_V w\,\mathrm{d}\tau = -\int_V \nabla\cdot \boldsymbol{J}\mathrm{d}\tau = -\oint_S \boldsymbol{J}\cdot \mathrm{d}\boldsymbol{\sigma} = -\oint_S J_n\mathrm{d}\sigma \quad (1.4.27)$$

上式指出,单位时间内,在 V 内发现粒子的概率的增量等于单位时间内流入 V 内的概率(负号表示流入),式(1.4.27)是概率守恒定律的积分形式.

在式(1.4.27)中令 $V\to\infty$,由于任何可实现的波函数应满足平方可积的条件($\int_{V_\infty}|\psi|^2\mathrm{d}\tau$ 有限),因此当 $R\to\infty$ 时,$\psi\to 0$ 应比 $R^{-\frac{3}{2}}$ 快,使得 J_n 沿无限大封闭面的积分为零[①],即

$$\frac{\mathrm{d}}{\mathrm{d}t}\int_{V_\infty}|\psi|^2\mathrm{d}\tau = \frac{\mathrm{d}}{\mathrm{d}t}\int_{V_\infty} w\,\mathrm{d}\tau = -\oint_{S_\infty} J_n\mathrm{d}\sigma = 0$$

上式说明,在整个空间发现粒子的概率(这是必然事件,概率为1)与时间无关.若波函数已归一化,则归一化条件 $\int_{V_\infty}|\psi|^2\mathrm{d}\tau = 1$ 不会随时间变化.

例 1.4.2 已知自由粒子的 $\psi(\boldsymbol{x},t) = A\mathrm{e}^{\frac{\mathrm{i}}{\hbar}(\boldsymbol{p}\cdot\boldsymbol{x}-Et)}$,求它的概率密度 $w(\boldsymbol{x},t)$ 和概率流密度矢量 $\boldsymbol{J}(\boldsymbol{x},t)$.

解 $w(\boldsymbol{x},t) = |\psi(\boldsymbol{x},t)|^2 = |A\mathrm{e}^{\frac{\mathrm{i}}{\hbar}(\boldsymbol{p}\cdot\boldsymbol{x}-Et)}|^2 = |A|^2$,这说明在空间各处发现自由粒子的概率相同.

$$\boldsymbol{J}(\boldsymbol{x},t) = \frac{\mathrm{i}\hbar}{2\mu}(\psi\nabla\psi^* - \psi^*\nabla\psi) = \frac{\mathrm{i}\hbar}{2\mu}\psi\nabla\psi^* + \mathrm{c.c.}$$

式中 c.c. 是表示前面式子的复共轭.利用

$$\nabla\psi^* = \sum_{k=1}^{3}\boldsymbol{e}_k\frac{\partial}{\partial x_k}A\mathrm{e}^{-\frac{\mathrm{i}}{\hbar}(\boldsymbol{p}\cdot\boldsymbol{x}-Et)} = \sum_{k=1}^{3}\boldsymbol{e}_k\left(-\frac{\mathrm{i}}{\hbar}p_k\right)\psi^* = -\frac{\mathrm{i}}{\hbar}\boldsymbol{p}\psi^*$$

代回前式,并利用 $w=|A|^2$ 及 $\boldsymbol{p}=\mu\boldsymbol{v}$,可得

$$\boldsymbol{J}(\boldsymbol{x},t) = \frac{\mathrm{i}\hbar}{2\mu}\psi\left(-\frac{\mathrm{i}}{\hbar}\boldsymbol{p}\psi^*\right) + \mathrm{c.c.} = w\frac{\boldsymbol{p}}{2\mu} + \mathrm{c.c.} = w\boldsymbol{v}$$

上式在数学形式上也与经典电动力学中电流密度矢量 $\boldsymbol{J}_f$、电荷密度 ρ_f 及

---

① 取 S_∞ 为球面,当 $R\to\infty$ 时,$\psi\to 0$ 应比 $R^{-\frac{3}{2}}$ 快,故 $J_n = \boldsymbol{J}\cdot\boldsymbol{e}_R = \frac{\mathrm{i}\hbar}{2\mu}\left(\psi\frac{\partial\psi^*}{\partial R} - \psi^*\frac{\partial\psi}{\partial R}\right)$ 趋于零应比 $R^{-\frac{3}{2}}R^{-\frac{5}{2}} = R^{-4}$ 快,而 S_∞ 仅与 R^2 成正比,易见 $\oint_{S_\infty} J_n\mathrm{d}\sigma = 0$.

电荷速度 $\boldsymbol{v}$ 的关系 $\boldsymbol{J}_f=\rho_f\boldsymbol{v}$ 相似.

3. 量子力学中的电荷守恒定律

设粒子的电荷为 e,定义 t 时刻在 $\boldsymbol{x}$ 处的电荷密度为

$$w_e(\boldsymbol{x},t)=ew(\boldsymbol{x},t)=e\mid\psi(\boldsymbol{x},t)\mid^2 \tag{1.4.28}$$

也就是说,尽管带电粒子是一个小颗粒,但是 t 时刻它在整个空间范围内都可能有电荷密度,因为 t 时刻在整个空间都有可能发现它,只是在不同位置发现它的概率大小不同而已.

类似地,定义 t 时刻在 $\boldsymbol{x}$ 处的电流密度矢量为

$$\boldsymbol{J}_e(\boldsymbol{x},t)=e\boldsymbol{J}(\boldsymbol{x},t)=\frac{\mathrm{i}e\hbar}{2\mu}(\psi\nabla\psi^*-\psi^*\nabla\psi) \tag{1.4.29}$$

用 e 乘概率守恒定律,即有

$$\frac{\partial w_e}{\partial t}+\nabla\cdot\boldsymbol{J}_e=0 \tag{1.4.30}$$

及

$$\frac{\mathrm{d}}{\mathrm{d}t}\int_V w_e\mathrm{d}\tau=-\oint_S\boldsymbol{J}_e\cdot\mathrm{d}\boldsymbol{\sigma} \tag{1.4.31}$$

两者分别是量子力学中电荷守恒定律的微分形式与积分形式.

1.5 定态薛定谔方程的解法
一维无限深势阱与线性谐振子

本节介绍定态薛定谔方程的解法.首先讨论波函数的标准条件,然后利用(1) 定态薛定谔方程;(2) 波函数的归一化条件;(3) 波函数的标准条件,求解在一维无限深势阱中运动的粒子与线性谐振子的能级和波函数.通过这两个实例得到的物理图像,将有助于对其后两章的理解.

最后,介绍反映一维定态普遍性质的"一维束缚定态的无简并定理".

1.5.1 波函数的标准条件

波函数的统计解释已经指出,归一化的波函数是概率波的振幅,那么,在数学上 $\psi(\boldsymbol{x},t)$ 应满足哪些条件呢?

1. 单值性

单值性是指 $\mid\psi(\boldsymbol{x},t)\mid$ 应是 $\boldsymbol{x},t$ 的单值函数.因为 $\mid\psi(\boldsymbol{x},t)\mid^2$ 是 t 时刻在 $\boldsymbol{x}$ 处发现粒子的概率密度,物理上要求它是唯一的,即要求 $\mid\psi(\boldsymbol{x},t)\mid$ 为单值函数,但不要求 $\psi(\boldsymbol{x},t)$ 是单值函数.

2. 有限性

有限性指在有限的空间范围内发现粒子的概率有限

$$\int_{V_0} |\psi(\boldsymbol{x},t)|^2 \mathrm{d}\tau = \text{有限值}$$

注意,它允许 $\psi(\boldsymbol{x},t)$ 存在孤立奇点 $\boldsymbol{b}$,若取 V_ε 表示以 $\boldsymbol{b}$ 为心,$\varepsilon=|\boldsymbol{R}-\boldsymbol{b}|$ 为半径的小球的体积,则要求当 $\varepsilon\to 0$ 时,$|\psi(\boldsymbol{x},t)|^2 V_\varepsilon \sim |\psi(\boldsymbol{x},t)|^2\varepsilon^3\to 0$. 这相当于要求当 $\boldsymbol{R}\to\boldsymbol{b}$ 时,只要 $|\psi(\boldsymbol{x},t)|\to\infty$ 比 $\varepsilon^{-\frac{3}{2}}\to\infty$ 慢就可以了,并不要求 $|\psi(\boldsymbol{x},t)|$ 在空间任一点有限. 这就是 $\psi(\boldsymbol{x},t)$ 有限性的含义.

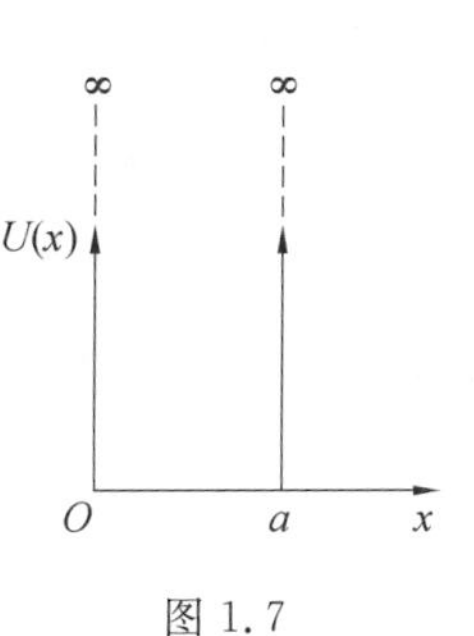

图 1.7

3. 连续性

定态薛定谔方程包含 $\psi(\boldsymbol{x},t)$ 对坐标的二阶导数,故要求 $\psi(\boldsymbol{x},t)$ 及其对坐标的一阶导数连续. 当势能曲线的跃度有限时,上述性质仍然成立,但在跃度为 ∞ 处(如图 1.7),则 $\psi'(x)$ 不连续.

1.5.2 一维无限深势阱

设质量为 μ 的粒子在下述势场中运动(如图 1.8)

$$U(x)=\begin{cases}0, 0<x<a \\ \infty, x\leqslant 0, x\geqslant a\end{cases} \tag{1.5.1}$$

无限深势阱是一个理想情况. 我们先研究电子在装有两对板极和栅极的真空管中的运动,如图 1.8(a). 两个栅极接地,两个板极的电势低于栅极. 电子在 $(0,a)$ 区域运动时不受电场的作用,当电子越过栅极进入 $(-\varepsilon,0)$ 区域或 $(a,a+\varepsilon)$ 区域时,则受到让它返回 $(0,a)$ 区域的斥力. 实验表明,当电子的能量 E 小于势能 U_0 时[如图 1.8(b)],还有可能在 $(-\varepsilon,0)$ 及 $(a,a+\varepsilon)$ 区域发现电子;而当板极与栅极之间的电势差不断增加时,在上述两区域发现电子的可能性越来越小. 如果让 $U_0\to\infty$,则过渡到图 1.7 这样的理想情况.

现在利用波函数标准条件与归一化条件求解在上述势场的定态薛定谔方程. 这类问题的求解步骤可归纳如下.

1. 写出分区的定态薛定谔方程 $\hat{H}\psi=E\psi$

$$-\frac{\hbar^2}{2\mu}\frac{\mathrm{d}^2\psi}{\mathrm{d}x^2}=E\psi \quad (0<x<a) \tag{1.5.2}$$

$$-\frac{\hbar^2}{2\mu}\frac{\mathrm{d}^2\psi}{\mathrm{d}x^2}+U_0\psi=E\psi \quad (x\leqslant 0, x\geqslant a) \tag{1.5.3}$$

在式(1.5.3)中,由于 $U_0\to\infty$,只有 $\psi=0$ 才能满足. 它的物理意义是,当势壁无限高时,不可能在势阱之外发现能量有限的粒子,故阱外波函数为零

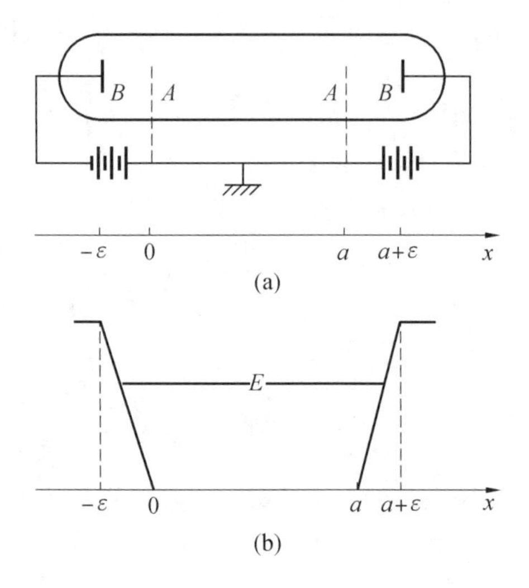

图 1.8

$$\psi(x)=0 \quad (x \leqslant 0, x \geqslant a) \tag{1.5.4}$$

2. 引入参数简化方程,得到含待定系数的解

令

$$k=\frac{\sqrt{2\mu E}}{\hbar} \tag{1.5.5}$$

由于 k 与 E 有关,只要求得 k,便可由 k 得 E.

阱内定态薛定谔方程为

$$\psi''(x)+k^2\psi(x)=0 \tag{1.5.6}$$

由此得 $0<x<a$ 区域的解

$$\psi(x)=A\sin(kx+\delta) \tag{1.5.7}$$

式中 A 和 δ 为待定常数.

3. 由波函数标准条件确定参数 k

阱外的波函数 $\psi(x)=0$,由波函数的连续性条件可得 $0=\psi(0)=A\sin\delta$,即

$$\delta=0 \tag{1.5.8}$$

类似地,由 $0=\psi(a)=A\sin ka$ 得 $ka=n\pi, n=1,2,\cdots$,即

$$k=\frac{n\pi}{a} \quad (n=1,2,\cdots) \tag{1.5.9}$$

将上两式代入式(1.5.7),得

$$\psi_n(x)=A\sin\frac{n\pi x}{a} \quad (n=1,2,\cdots) \tag{1.5.10}$$

当 n 取负值时,得到的解与 n 取正值的解线性相关,故应舍去.

当 k 取零值时，方程(1.5.6) 变为 $\psi''(x)=0$，其解为

$$\psi_0(x)=Bx+C$$

由 $0=\psi_0(0)=C$ 得 $\psi(x)=Bx$；再由 $0=\psi_0(a)=Ba$ 得 $B=0$. 将 B 和 C 代入上式得 $\psi_0(x)=0$，是平庸解，也应舍去.

4. 由波函数的归一化条件确定归一化常数 A

由

$$1=\int_{-\infty}^{\infty}|\psi_n(x)|^2\mathrm{d}x=\int_0^a|A|^2\sin^2\frac{n\pi x}{a}\mathrm{d}x=\frac{a}{2}|A|^2$$

取 A 为实数，即得 $A=\sqrt{2/a}$，由此得

$$\psi_n(x)=\begin{cases}\sqrt{\dfrac{2}{a}}\sin\dfrac{n\pi x}{a},0<x<a\\0,x\leqslant 0,x\geqslant a\end{cases}\tag{1.5.11}$$

5. 由参数 k 得粒子的能量 E

将式(1.5.9) 代入式(1.5.5) 即得

$$E=E_n=\frac{\hbar^2\pi^2n^2}{2\mu a^2}\quad(n=1,2,\cdots)\tag{1.5.12}$$

6. 解的物理意义

(1) 束缚态与离散能级.

由式(1.5.11) 可见，粒子被束缚在$(0,a)$ 的区域内，不可能到达无穷远处.

粒子被束缚在有限的空间区域(即 $\psi(\infty)=0$) 的状态称为束缚态；反之，粒子可到达无限远处的状态称为非束缚态.

由 E_n 可见，只有当粒子的能量取式中的离散值时，波函数才满足连续性条件 $\psi(a)=0$；也就是说，在无限深势阱中粒子的能量是量子化的.

为什么束缚态的能谱必然为离散谱呢？从上面的推导可看出，存在三个待定常数和参数(A,δ和k) 以及三个确定它们的方程(两个连续性条件和一个归一化条件). 因此，必然对参数 k 产生限制，也就是对 $E=\hbar^2k^2/2\mu$ 取值的限制，使得 E 只能取离散值，即能谱是离散谱. 但对于非束缚态，待定的常数和参数比确定它们的方程要多，这样对 k(因而对 E) 不构成限制，使得 E 具有连续谱.

(2) 基态的能级 $E_1=\dfrac{\hbar^2\pi^2}{2\mu a^2}\neq 0$，这是与经典粒子的一个本质区别，是微观粒子波动性的表现(没有“能量为零”的波). 在经典物理中，粒子的能量可以等于零，这意味着粒子静止，即坐标有确定值和动量为零. 而粒子的坐标和动量同时有确定值正是量子力学中的不确定性关系所不允许

的.

(3) 激发态的能级 E_n 与 n^2 成正比,能级分布不均匀.能级间隔 $\Delta E_n = E_{n+1} - E_n = \frac{\hbar^2\pi^2}{2\mu a^2}(2n+1)$;当 $n \to \infty$ 时,能级的相对间隔 $\frac{\Delta E_n}{E_n} = \frac{2n+1}{n^2}$ 趋于零.这说明,当量子数 n 很大时,能级可以看作是连续的,量子效应消失而过渡到经典情况.

(4) 在区间 $(0,a)$ 中 $\psi_n(x) = \sqrt{\frac{2}{a}}\sin\frac{n\pi x}{a}$.这表明,阱内有 $(n-1)$ 个点满足 $\psi_n(x)=0$,称为 $\psi_n(x)$ 的节点,如图 1.9 所示.

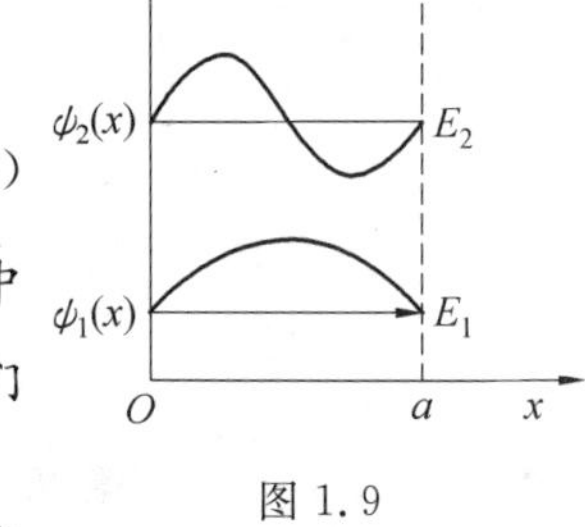

图 1.9

(5) 由于薛定谔方程是线性齐次方程,它的解的线性组合

$$\psi(x,t) = \sum_{n=1}^{\infty} c_n \psi_n(x) e^{-\frac{i}{\hbar}E_n t} \tag{1.5.13}$$

仍然是薛定谔方程的解,因此,在一维无限深势阱中粒子的态不一定是定态 $\psi_n(x)e^{-\frac{i}{\hbar}E_n t}$,也可以是它们的线性叠加态(1.5.13).

由叠加原理可知,当粒子处于 $\psi(x,t)$ 时,测量它处于 $\psi_n(x)e^{-\frac{i}{\hbar}E_n t}$ 态的概率是 $|c_n|^2$.

1.5.3 线性谐振子

设质量为 μ 的粒子在势场

$$U(x) = \frac{1}{2}kx^2 = \frac{1}{2}\mu\omega^2 x^2 \tag{1.5.14}$$

中运动,求定态薛定谔方程的解.

在经典力学中,粒子受到弹性力 $F=-kx$ 作用时,其势能为

$$U(x) = -\int_0^x F(x')\mathrm{d}x' = \frac{1}{2}kx^2$$

粒子将做简谐振动.因此,在量子力学中,也把在势场 $U=\frac{1}{2}kx^2$ 中运动的微观粒子称为线性谐振子.由式(1.5.14)可见,其势能曲线是抛物线.

讨论线性谐振子具有非常重要的意义.

(1) 许多物理体系的势能曲线可以近似看作抛物线,如双原子分子的势能曲线在稳定平衡点 a 附近的势能曲线就可以看作抛物线,如图 1.10.

(2) 复杂的振动可以分解成相互独立的谐振动,如带电粒子在均匀磁

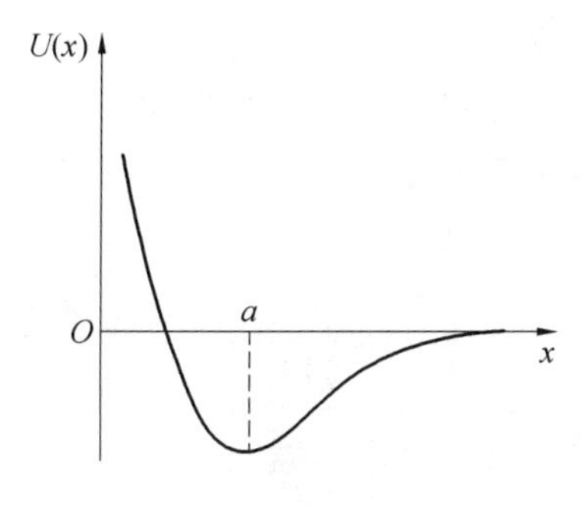

图 1.10

场中的圆周运动可分解为在相互垂直的两个方向上的谐振动，电磁场的振动可以分解为各种频率的谐振动的叠加，等等.

(3) 处理线性谐振子的方法也很典型，本节将在坐标表象进行讨论，后文将在粒子数表象进行讨论. 以后我们还会发现，也可以将谐振子量子化的方法用于电磁场量子化.

线性谐振子的哈密顿量是

$$H=\frac{p^2}{2\mu}+\frac{1}{2}\mu\omega^2x^2 \tag{1.5.15}$$

按照 1.4 节给出的替换 $p\to-\mathrm{i}\hbar\frac{\mathrm{d}}{\mathrm{d}x}$，即得线性谐振子的哈密顿算符

$$\hat{H}=-\frac{\hbar^2}{2\mu}\frac{\mathrm{d}^2}{\mathrm{d}x^2}+\frac{1}{2}\mu\omega^2x^2 \tag{1.5.16}$$

定态薛定谔方程为

$$\left(-\frac{\hbar^2}{2\mu}\frac{\mathrm{d}^2}{\mathrm{d}x^2}+\frac{1}{2}\mu\omega^2x^2\right)\psi(x)=E\psi(x) \tag{1.5.17}$$

现按下述步骤求解定态薛定谔方程.

1. 引入参数简化方程

作变换

$$\xi=\alpha x,\quad \alpha=\sqrt{\frac{\mu\omega}{\hbar}} \tag{1.5.18}$$

利用复合函数的微商公式

$$\frac{\mathrm{d}\psi}{\mathrm{d}x}=\frac{\mathrm{d}\psi}{\mathrm{d}\xi}\frac{\mathrm{d}\xi}{\mathrm{d}x}=\alpha\frac{\mathrm{d}\psi}{\mathrm{d}\xi},\quad \frac{\mathrm{d}^2\psi}{\mathrm{d}x^2}=\alpha^2\frac{\mathrm{d}^2\psi}{\mathrm{d}\xi^2}$$

引入参数

$$\lambda=\frac{2E}{\hbar\omega} \tag{1.5.19}$$

方程(1.5.17)即可简化为[①]

$$\psi''(\xi)+(\lambda-\xi^2)\psi(\xi)=0 \tag{1.5.20}$$

方程(1.5.20)是变系数二阶常微分方程,如果直接用幂级数解法求解,将 $\psi(\xi)=\sum_k c_k\xi^k$ 代入上述方程,则得到

$$(k+2)(k+1)c_{k+2}+\lambda c_k-c_{k-2}=0 \quad (k\geqslant 2)$$

因为三个待定系数纠缠在一起,由上式不能找到系数递推公式,所以不能直接用幂级数解法求解.常用的方法是先求方程在 $\xi\to\pm\infty$ 时的渐近解,然后再求方程在区间 $(-\infty,+\infty)$ 的解.这种方法具有普遍的意义.

2.求方程 $\psi''(\xi)+(\lambda-\xi^2)\psi(\xi)=0$ 在 $\xi\to\pm\infty$ 的渐近解

当 $\xi\to\pm\infty$ 时,方程简化为

$$\psi''(\xi)-\xi^2\psi(\xi)=0$$

将尝试解 $\psi(\xi)=\mathrm{e}^{q\xi^2}$ 代入上述方程,并略去数量级较小的项,即可求得 $q=\pm\frac{1}{2}$.因波函数的有限性条件要求 $\psi(\xi)$ 在 $\xi\to\pm\infty$ 时有限,故 $q=\frac{1}{2}$ 应舍去,即得渐近解

$$\psi(\xi)=\mathrm{e}^{-\frac{\xi^2}{2}} \tag{1.5.21}$$

3.由波函数标准条件确定方程 $\psi''(\xi)+(\lambda-\xi^2)\psi(\xi)=0$ 在区间 $(-\infty,+\infty)$ 的解

(1)将波函数看作渐近解 $\mathrm{e}^{-\frac{\xi^2}{2}}$ 与待求函数 $H(\xi)$ 之积,把求 $\psi(\xi)$ 的问题转化为求 $H(\xi)$ 的问题.令

$$\psi(\xi)=H(\xi)\mathrm{e}^{-\frac{\xi^2}{2}} \tag{1.5.22}$$

使 $\psi(\xi)$ 在无穷远处的行为符合渐近解的要求.将上式代入 $\psi''(\xi)+(\lambda-\xi^2)\psi(\xi)=0$,即得 $H(\xi)$ 应遵守的方程.

利用莱布尼兹公式

$$(uv)''=u''v+2u'v'+uv''$$

得

$$H''(\xi)-2\xi H'(\xi)+(\lambda-1)H(\xi)=0 \tag{1.5.23}$$

这个方程称为厄米方程.

(2)用幂级数解法求解厄米方程得 $H(\xi)$.

因 $\xi=0$ 是方程的常点,方程的解可表示为泰勒级数

① 严格而言,应采用记号 $\psi(x)=\psi[x(\xi)]=\varphi(\xi)$,但通常为简单起见,不加区别地记作 ψ 与 $\psi(\xi)$.

$$H(\xi)=\sum_{\nu=0}^{\infty}a_{\nu}\xi^{\nu} \tag{1.5.24}$$

将级数代入方程，可得

$$\sum_{\nu=0}^{\infty}a_{\nu}\nu(\nu-1)\xi^{\nu-2}-2\xi\sum_{\nu=0}^{\infty}a_{\nu}\nu\xi^{\nu-1}+(\lambda-1)\sum_{\nu=0}^{\infty}a_{\nu}\xi^{\nu}=0$$

由 ξ 的同次幂项系数之和为零，即得系数递推公式

$$a_{\nu+2}=\frac{2\nu+1-\lambda}{(\nu+2)(\nu+1)}a_{\nu} \tag{1.5.25}$$

但是，将这个级数代入 $\psi(\xi)=H(\xi)\mathrm{e}^{-\frac{\xi^2}{2}}$，当 $\xi\to\pm\infty$ 时，能否保证 $\psi(\xi)$ 有限呢？

(3) 波函数的有限性要求级数 $H(\xi)=\sum_{\nu=0}^{\infty}a_{\nu}\xi^{\nu}$ 中断为多项式.

由于级数在无穷远处的行为取决于级数相邻两项系数之比在 $\nu\to\infty$ 时的极限，由式(1.5.25)得

$$\lim_{\nu\to\infty}\frac{a_{\nu+2}}{a_{\nu}}=\lim_{\nu\to\infty}\frac{2\nu+1-\lambda}{(\nu+2)(\nu+1)}=\frac{2}{\nu}$$

它与级数 $\mathrm{e}^{\xi^2}=\sum_{k=0}^{\infty}\frac{\xi^{2k}}{k!}=\sum_{\nu=0}^{\infty}\frac{\xi^{\nu}}{(\nu/2)!}$ 的行为相同，后者相邻两项系数之比在 $\nu\to\infty$ 时的极限亦为

$$\lim_{\nu\to\infty}\frac{1/\left(\frac{\nu+2}{2}\right)!}{1/\left(\frac{\nu}{2}\right)!}=\lim_{\nu\to\infty}\frac{\left(\frac{\nu}{2}\right)!}{\left(\frac{\nu}{2}\right)!\left(\frac{\nu+2}{2}\right)}=\frac{2}{\nu}$$

这说明，当 ξ 很大时，$H(\xi)$ 的行为与 e^{ξ^2} 相同.

如果级数不中断为多项式，则不能保证 $\psi(\xi)=H(\xi)\mathrm{e}^{-\frac{\xi^2}{2}}\sim\mathrm{e}^{\xi^2}\mathrm{e}^{-\frac{\xi^2}{2}}=\mathrm{e}^{\frac{\xi^2}{2}}$ 在 $\xi\to\pm\infty$ 时有限.

由系数递推公式(1.5.25)可见，当取

$$\lambda=2n+1\quad(n=0,1,2,\cdots) \tag{1.5.26}$$

时，系数 $a_{n+2},a_{n+4},\cdots$ 均为零，这时 $H(\xi)$ 成为 n 次多项式. 若规定最高次幂的系数 $a_n=2^n$，利用系数递推公式即可求得

$$H_n(\xi)=\sum_{s=0}^{\left[\frac{n}{2}\right]}\frac{(-1)^s n!}{s!\,(n-2s)!}(2\xi)^{n-2s} \tag{1.5.27}$$

式中[ ]是取整数的符号

$$\left[\frac{n}{2}\right]=\begin{cases}n/2,\text{当 } n \text{ 为偶数}\\(n-1)/2,\text{当 } n \text{ 为奇数}\end{cases} \tag{1.5.28}$$

$H_n(\xi)$ 称为厄米多项式,它的两个线性无关的解可以这样得到:

① 令 $a_0 \neq 0, a_1 = 0$ 可得只含 ξ 的偶次幂的多项式.

② 令 $a_0 = 0, a_1 \neq 0$ 可得只含 ξ 的奇次幂的多项式.

(4) 厄米多项式的微分表达式为

$$H_n(\xi) = (-1)^n e^{\xi^2} \frac{d^n}{d\xi^n} e^{-\xi^2} \tag{1.5.29}$$

将式(1.5.29)代入式(1.5.22)即得线性谐振子的波函数

$$\psi_n(\xi) = N_n e^{-\frac{\xi^2}{2}} H_n(\xi) = N_n (-1)^n e^{\frac{\xi^2}{2}} \frac{d^n}{d\xi^n} e^{-\xi^2} \tag{1.5.30}$$

式中 N_n 为归一化常数

$$N_n = \sqrt{\frac{\alpha}{2^n n!\ \sqrt{\pi}}} \tag{1.5.31}$$

(5) 线性谐振子前面的几个波函数为

$$\psi_0(x) = \sqrt{\frac{\alpha}{\sqrt{\pi}}} e^{-\frac{1}{2}\alpha^2 x^2}, \psi_1(x) = \sqrt{\frac{2\alpha}{\sqrt{\pi}}} \alpha x e^{-\frac{1}{2}\alpha^2 x^2}$$

$$\psi_2(x) = \sqrt{\frac{\alpha}{2\sqrt{\pi}}} (2\alpha^2 x^2 - 1) e^{-\frac{1}{2}\alpha^2 x^2}, \psi_3(x) = \sqrt{\frac{\alpha}{3\sqrt{\pi}}} (2\alpha^3 x^3 - 3\alpha x) e^{-\frac{1}{2}\alpha^2 x^2} \tag{1.5.32}$$

4. 由参数 λ 得粒子能量 E

将 $\lambda = 2n + 1 (n = 0, 1, 2, \cdots)$ 与 $\lambda = 2E/\hbar\omega$ 联立,得

$$E_n = \left(n + \frac{1}{2}\right) \hbar\omega \quad (n = 0, 1, 2, \cdots) \tag{1.5.33}$$

这样,我们便从量子力学的基本假设(薛定谔方程)出发,自然地导出了普朗克的能量子假设,即振子的能量取离散值 $n\hbar\omega$. 至于 $\hbar\omega/2$ 项,旧量子论更是无法得到的.

5. 解的物理意义

(1) 谐振子的能量取离散值 $E_n = \left(n + \frac{1}{2}\right) \hbar\omega$.

原因是当 $x \to \infty$ 时线性谐振子的势能为

$$\lim_{x \to \infty} U(x) = \lim_{x \to \infty} \frac{1}{2} \mu\omega^2 x^2 = \infty$$

即粒子不能运动到无限远处,可见谐振子处于束缚态. 这正如前面所指出的那样:束缚态的能谱必为离散谱.

(2) 谐振子相邻能级的间隔 $\Delta E = E_{n+1} - E_n = \hbar\omega$ 是均匀分布的.

(3) 谐振子的基态能量 $E_0 = \frac{1}{2} \hbar\omega \neq 0$,这也是一个量子的效应,得到

实验证实.当原子发生自发辐射从高能态跃迁到低能态时,实际上是电磁场的真空态(无光子的电磁场的零振动态)与电子相互作用的结果.

(4) 线性谐振子的能级是无简并的.在本节末我们将给出一个普遍的证明.

(5) 谐振子波函数的宇称为$(-1)^n$.

由式(1.5.30)容易证明,$\psi_n(-x)=(-1)^n\psi_n(x)$,可见波函数$\psi_n(x)$的奇偶性由$n$决定,通常称谐振子波函数$\psi_n(x)$的宇称为$(-1)^n$.

(6) 与经典谐振子的比较.在经典力学里,粒子在Δx范围内出现的概率正比于粒子通过Δx所需要的时间$\Delta t=\dfrac{\Delta x}{v}$.

若粒子的$x=a\sin\omega t$,则$v=\dfrac{\mathrm{d}x}{\mathrm{d}t}=\omega\sqrt{a^2-x^2}$,故粒子在$\Delta x$出现的概率

$$W\sim\frac{\Delta x}{v}=\frac{\Delta x}{\omega\sqrt{a^2-x^2}}$$

如图1.11(b)中的虚线.粒子在原点($x=0$)的速度最大,出现的概率最小.

量子谐振子的空间位置概率分布如图1.11(a)所示,有三个明显的特点:

① 在原点发现粒子的概率要么极大(n为偶数),要么为零(n为奇数).

② 可以在经典禁区(粒子的总能小于势能的区域)发现粒子,这是一种势垒穿透效应.在图1.11(a)中,势能曲线(虚抛物线)以外的区域$|\psi|^2\neq 0$.例如,在空间发现处于基态的谐振子的概率密度分布为

$$w_0=|\psi_0(x)|^2=\frac{\alpha}{\sqrt{\pi}}\mathrm{e}^{-\alpha^2x^2}$$

对基态而言,谐振子总能等于势能的条件是$\dfrac{1}{2}\hbar\omega=\dfrac{1}{2}\mu\omega^2x^2$,解之得$x=\sqrt{\dfrac{\hbar}{\mu\omega}}=\dfrac{1}{\alpha}$.按照经典力学,粒子只能在$|x|<\dfrac{1}{\alpha}$(即$|\xi|=|\alpha x|<1$)的区域内运动.图1.11(a)表明,在$|x|>\dfrac{1}{\alpha}$即$|\xi|>1$的区域,发现粒子的概率不为零,且可算出

$$\frac{\int_1^\infty \mathrm{e}^{-\xi^2}\mathrm{d}\xi}{\int_0^\infty \mathrm{e}^{-\xi^2}\mathrm{d}\xi}=15.7\%$$

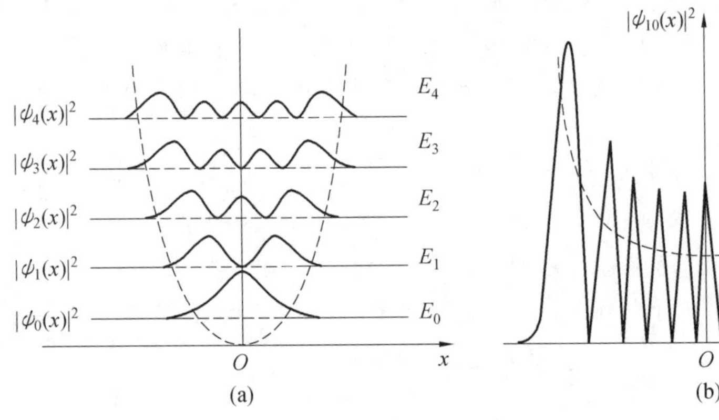

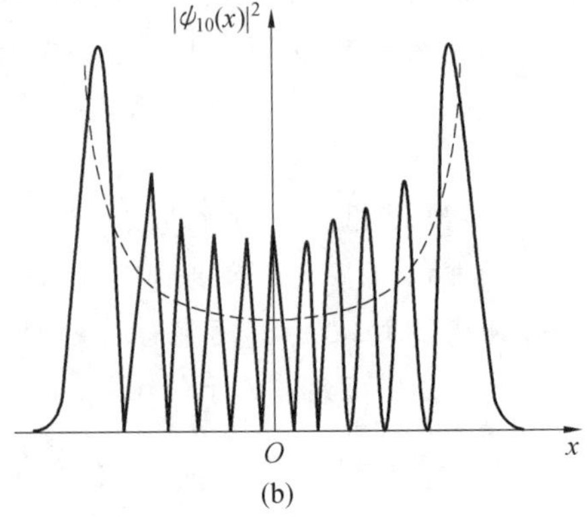

图 1.11

③ 当量子数 n 越大时,其概率密度分布与经典概率密度分布越接近,图1.11(b) 中的实线是 $n=10$ 时谐振子的位置概率密度分布,它已与经典的位置概率密度分布很接近.

1.5.4　一维束缚定态无简并定理

定理　若 $U(x)$ 在 x 有限处无奇点,则一维的束缚态无简并.

证　设 ψ_1,ψ_2 是任意两个能量为 E 的一维束缚态,若能证明 $\psi_1 = C\psi_2$(C 为常数,ψ_1 与 ψ_2 只差一个常数因子),这就证明 ψ_1 与 ψ_2 描述同一束缚态,即无简并.

薛定谔方程为 $\frac{\mathrm{d}^2\psi(x)}{\mathrm{d}x^2}+\frac{2\mu}{\hbar^2}[E-U(x)]\psi(x)=0$. 既然 $U(x)$ 无奇点,则

$$\frac{\psi''_1(x)}{\psi_1(x)}=-\frac{2\mu}{\hbar^2}[E-U(x)]=\frac{\psi''_2(x)}{\psi_2(x)}$$

取有限值. 由上式得

$$0=\psi''_1\psi_2-\psi_1\psi''_2=(\psi_2\psi'_1-\psi_1\psi'_2)'$$

积分得

$$\psi'_1\psi_2-\psi_1\psi'_2=\text{常数}$$

若 ψ_1 与 ψ_2 均为束缚态,则 $\psi_1(\infty)=0$,$\psi_2(\infty)=0$,代入上式得知常数为零. 故

$$\psi'_1\psi_2-\psi_1\psi'_2=0$$

易见 $\frac{\psi'_1}{\psi_1}-\frac{\psi'_2}{\psi_2}=0$,即 $\left(\ln\frac{\psi_1}{\psi_2}\right)'=0$,积分得

$$\psi_1=C\psi_2$$

两者描述同一束缚态,即无简并.

定理指出,一维线性谐振子的能级是无简并的.这个结论在粒子数表象和一维方势阱中要用到.

像激光与量子力学这类“高大上”的素材,国人还是很热衷的,但究竟有多少人能真正懂就是另外一个问题了.在最近一期《人物》杂志中有一篇是写许倬云的,文中称“和许倬云的谈话,有时候会陷入一种困境.他写的是‘大历史’,谈的也是‘大问题’.我们谈及中国文化的未来,他先从量子力学里的纠缠现象讲起,讲到雅利安人驯服了马匹,开始有了掳掠文化,再讲到周人的天命文化,讲到孔子的‘忠’与‘恕’,在几千年的尺度里,他比较东西方文明的差异,试图让我理解东西文明系统中的复杂脉络,在纷乱的线索中抓住核心.”

当然,许老先生还有其他令人震惊的论断,比如他曾说:“中国历史上大禹治水是真有其事,公元前2019年的那次大洪水,是喜马拉雅山底下一个冰川堰塞湖崩了!”

如此看来,科普永远在路上!

刘培杰

2024年5月13日

于哈工大

刘培杰物理工作室
已出版(即将出版)图书目录

序号	书　　名	出版时间	定　价
1	物理学中的几何方法	2017－06	88.00
2	量子力学原理.上	2016－01	38.00
3	时标动力学方程的指数型二分性与周期解	2016－04	48.00
4	重刚体绕不动点运动方程的积分法	2016－05	68.00
5	水轮机水力稳定性	2016－05	48.00
6	Lévy 噪音驱动的传染病模型的动力学行为	2016－05	48.00
7	铣加工动力学系统稳定性研究的数学方法	2016－11	28.00
8	粒子图像测速仪实用指南:第二版	2017－08	78.00
9	锥形波入射粗糙表面反散射问题理论与算法	2018－03	68.00
10	混沌动力学:分形、平铺、代换	2019－09	48.00
11	从开普勒到阿诺德——三体问题的历史	2014－05	298.00
12	数学物理大百科全书.第 1 卷(英文)	2016－01	418.00
13	数学物理大百科全书.第 2 卷(英文)	2016－01	408.00
14	数学物理大百科全书.第 3 卷(英文)	2016－01	396.00
15	数学物理大百科全书.第 4 卷(英文)	2016－01	408.00
16	数学物理大百科全书.第 5 卷(英文)	2016－01	368.00
17	量子机器学习中数据挖掘的量子计算方法(英文)	2016－01	98.00
18	量子物理的非常规方法(英文)	2016－01	118.00
19	运输过程的统一非局部理论:广义波尔兹曼物理动力学,第 2 版(英文)	2016－01	198.00
20	量子力学与经典力学之间的联系在原子、分子及电动力学系统建模中的应用(英文)	2016－01	58.00
21	动力系统与统计力学(英文)	2018－09	118.00
22	表示论与动力系统(英文)	2018－09	118.00
23	工程师与科学家微分方程用书:第 4 版(英文)	2019－07	58.00
24	工程师与科学家统计学:第 4 版(英文)	2019－06	58.00
25	通往天文学的途径:第 5 版(英文)	2019－05	58.00
26	量子世界中的蝴蝶:最迷人的量子分形故事(英文)	2020－06	118.00
27	走进量子力学(英文)	2020－06	118.00
28	计算物理学概论(英文)	2020－06	48.00
29	物质,空间和时间的理论:量子理论(英文)	2020－10	48.00
30	物质,空间和时间的理论:经典理论(英文)	2020－10	48.00
31	量子场理论:解释世界的神秘背景(英文)	2020－07	38.00
32	计算物理学概论(英文)	2020－06	48.00
33	行星状星云(英文)	2020－10	38.00

刘培杰物理工作室
已出版(即将出版)图书目录

序号	书　　名	出版时间	定　价
34	基本宇宙学:从亚里士多德的宇宙到大爆炸(英文)	2020—08	58.00
35	数学磁流体力学(英文)	2020—07	58.00
36	高考物理解题金典(第2版)	2019—05	68.00
37	高考物理压轴题全解	2017—04	48.00
38	高中物理经典问题25讲	2017—05	28.00
39	高中物理教学讲义	2018—01	48.00
40	1000个国外中学物理好题	2012—04	48.00
41	数学解题中的物理方法	2011—06	28.00
42	力学在几何中的一些应用	2013—01	38.00
43	物理奥林匹克竞赛大题典——力学卷	2014—11	48.00
44	物理奥林匹克竞赛大题典——热学卷	2014—04	28.00
45	物理奥林匹克竞赛大题典——电磁学卷	2015—07	48.00
46	物理奥林匹克竞赛大题典——光学与近代物理卷	2014—06	28.00
47	电磁理论(英文)	2020—08	48.00
48	连续介质力学中的非线性问题(英文)	2020—09	78.00
49	力学若干基本问题的发展概论(英文)	2020—11	48.00
50	狭义相对论与广义相对论:时空与引力导论(英文)	2021—07	88.00
51	束流物理学和粒子加速器的实践介绍:第2版(英文)	2021—07	88.00
52	凝聚态物理中的拓扑和微分几何简介(英文)	2021—05	88.00
53	广义相对论:黑洞、引力波和宇宙学介绍(英文)	2021—06	68.00
54	现代分析电磁均质化(英文)	2021—06	68.00
55	为科学家提供的基本流体动力学(英文)	2021—06	88.00
56	视觉天文学:理解夜空的指南(英文)	2021—06	68.00
57	物理学中的计算方法(英文)	2021—06	68.00
58	单星的结构与演化:导论(英文)	2021—06	108.00
59	超越居里:1903年至1963年物理界四位女性及其著名发现(英文)	2021—06	68.00
60	范德瓦尔斯流体热力学的进展(英文)	2021—06	68.00
61	先进的托卡马克稳定性理论(英文)	2021—06	88.00
62	经典场论导论:基本相互作用的过程(英文)	2021—07	88.00
63	光致电离量子动力学方法原理(英文)	2021—07	108.00
64	经典域论和应力:能量张量(英文)	2021—05	88.00
65	非线性太赫兹光谱的概念与应用(英文)	2021—06	68.00
66	电磁学中的无穷空间并矢格林函数(英文)	2021—06	88.00
67	物理科学基础数学.第1卷,齐次边值问题、傅里叶方法和特殊函数(英文)	2021—07	108.00
68	离散量子力学(英文)	2021—07	68.00

刘培杰物理工作室
已出版(即将出版)图书目录

序号	书　　名	出版时间	定　价
69	核磁共振的物理学和数学(英文)	2021—07	108.00
70	分子水平的静电学(英文)	2021—08	68.00
71	非线性波:理论、计算机模拟、实验(英文)	2021—06	108.00
72	石墨烯光学:经典问题的电解解决方案(英文)	2021—06	68.00
73	超材料多元宇宙(英文)	2021—07	68.00
74	银河系外的天体物理学(英文)	2021—07	68.00
75	原子物理学(英文)	2021—07	68.00
76	将光打结:将拓扑学应用于光学(英文)	2021—07	68.00
77	电磁学:问题与解法(英文)	2021—07	88.00
78	海浪的原理:介绍量子力学的技巧与应用(英文)	2021—07	108.00
79	杰弗里·英格拉姆·泰勒科学论文集:第1卷.固体力学(英文)	2021—05	78.00
80	杰弗里·英格拉姆·泰勒科学论文集:第2卷.气象学、海洋学和湍流(英文)	2021—05	68.00
81	杰弗里·英格拉姆·泰勒科学论文集:第3卷.空气动力学以及落弹数和爆炸的力学(英文)	2021—05	68.00
82	杰弗里·英格拉姆·泰勒科学论文集:第4卷.有关流体力学(英文)	2021—05	58.00
83	多孔介质中的流体:输运与相变(英文)	2021—07	68.00
84	洛伦兹群的物理学(英文)	2021—08	68.00
85	物理导论的数学方法和解决方法手册(英文)	2021—08	68.00
86	非线性波数学物理学入门(英文)	2021—08	88.00
87	波:基本原理和动力学(英文)	2021—07	68.00
88	光电子量子计量学.第1卷,基础(英文)	2021—07	88.00
89	光电子量子计量学.第2卷,应用与进展(英文)	2021—07	68.00
90	复杂流的格子玻尔兹曼建模的工程应用(英文)	2021—08	68.00
91	电偶极矩挑战(英文)	2021—08	108.00
92	电动力学:问题与解法(英文)	2021—09	68.00
93	自由电子激光的经典理论(英文)	2021—08	68.00
94	曼哈顿计划——核武器物理学简介(英文)	2021—09	68.00
95	粒子物理学(英文)	2021—09	68.00
96	引力场中的量子信息(英文)	2021—09	128.00
97	器件物理学的基本经典力学(英文)	2021—09	68.00
98	等离子体物理及其空间应用导论.第1卷,基本原理和初步过程(英文)	2021—09	68.00
99	伽利略理论力学:连续力学基础(英文)	2021—10	48.00
100	高中物理答疑解惑65篇	2021—11	48.00

刘培杰物理工作室
已出版(即将出版)图书目录

序号	书　名	出版时间	定　价
101	纽结与物理学:第二版(英文)	2022－09	118.00
102	磁约束聚变等离子体物理:理想 MHD 理论(英文)	2023－03	68.00
103	相对论量子场论.第1卷,典范形式体系(英文)	2023－03	38.00
104	相对论量子场论.第2卷,路径积分形式(英文)	2023－06	38.00
105	相对论量子场论.第3卷,量子场论的应用(英文)	2023－06	38.00
104	涌现的物理学(英文)	2023－05	58.00
105	量子化旋涡:一本拓扑激发手册(英文)	2023－04	68.00
106	非线性动力学:实践的介绍性调查(英文)	2023－05	68.00
107	静电加速器:一个多功能工具(英文)	2023－06	58.00
108	相对论多体理论与统计力学(英文)	2023－06	58.00
109	经典力学.第1卷,工具与向量(英文)	2023－04	38.00
110	经典力学.第2卷,运动学和匀加速运动(英文)	2023－04	58.00
111	经典力学.第3卷,牛顿定律和匀速圆周运动(英文)	2023－04	58.00
112	经典力学.第4卷,万有引力定律(英文)	2023－04	38.00
113	经典力学.第5卷,守恒定律与旋转运动(英文)	2023－04	38.00
114	对称问题:纳维尔－斯托克斯问题(英文)	2023－04	38.00
115	摄影的物理和艺术.第1卷,几何与光的本质(英文)	2023－04	78.00
116	摄影的物理和艺术.第2卷,能量与色彩(英文)	2023－04	78.00
117	摄影的物理和艺术.第3卷,探测器与数码的意义(英文)	2023－04	78.00
118	具有连续变量的量子信息形式主义概论	2024－10	378.00(全6册)
119	拓扑绝缘体	2024－10	378.00(全6册)
120	论全息度量原则:从大学物理到黑洞热力学	2024－10	378.00(全6册)
121	量化测量:无所不在的数字	2024－10	378.00(全6册)
122	21世纪的彗星:体验下一颗伟大彗星的个人指南	2024－10	378.00(全6册)
123	激光及其在玻色－爱因斯坦凝聚态观测中的应用	2024－10	378.00(全6册)

联系地址:哈尔滨市南岗区复华四道街10号　哈尔滨工业大学出版社刘培杰物理工作室
网　　址:http://lpj.hit.edu.cn/
邮　　编:150006
联系电话:0451－86281378　　13904613167
E-mail:lpj1378@163.com